DISCARDED

WILDFOWL IN GREAT BRITAIN

WILDFOWL
IN GREAT BRITAIN

SECOND EDITION

Illustrated by Sir Peter Scott

MYRFYN OWEN

G.L. ATKINSON-WILLES

D.G. SALMON

with contributions by
G.V.T. Matthews and M.A. Ogilvie

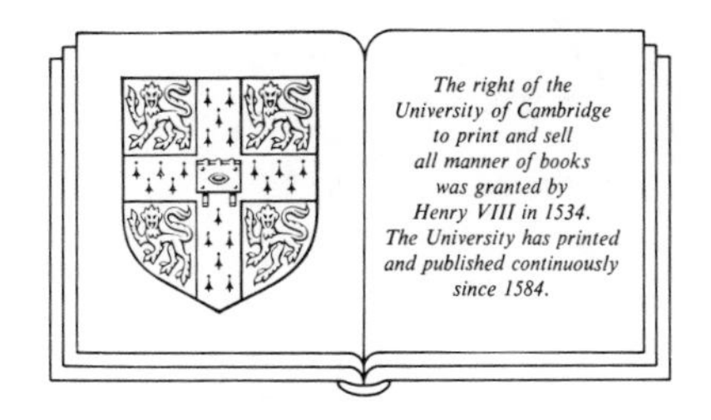
The right of the
University of Cambridge
to print and sell
all manner of books
was granted by
Henry VIII in 1534.
The University has printed
and published continuously
since 1584.

CAMBRIDGE UNIVERSITY PRESS
Cambridge
London New York New Rochelle
Melbourne Sydney

Published by the Press Syndicate of the University of Cambridge
The Pitt Building, Trumpington Street, Cambridge CB2 1RP
32 East 57th Street, New York, NY 10022, USA
10 Stamford Road, Oakleigh, Melbourne 3166, Australia

First published by Her Majesty's Stationery Office 1963
Second edition published by Cambridge University Press 1986

**Reproduced, printed and bound in Great Britain by
Hazell Watson & Viney Limited,
Member of the BPCC Group,
Aylesbury, Bucks**

British Library cataloguing in publication data

Wildfowl in Great Britain – 2nd ed.

1. Water-birds – Great Britain
I. Owen, Myrfyn II. Atkinson-Willes, G.L.
III. Salmon, D.G. IV. Matthews, G.V.T.
V. Ogilvie, M.A.
598.4'1'0941 QL690.G7

Library of Congress cataloging in publication data

Owen, Myrfyn
Wildfowl in Great Britain.

Bibliography: p.
Includes index.
1. Anatidae. 2. Waterfowl – Great Britain.
3. Birds, Protection of – Great Britain. 4. Birds –
Great Britain. I. Atkinson-Willes, G.L. (George L.)
II. Salmon, D.G. III. Title.
QL696.A520927 1986 598.4'10941 86-2281

ISBN 0 521 30986 7

Contents

P.S.1984

Location maps

Fig	Page	Area

Foreword

WILLIAM WILKINSON, *Chairman of the Nature Conservancy Council*

As one who, in his time, has been a wildfowl counter, both in Britain for the Wildfowl Trust and abroad for the International Waterfowl Research Bureau, it is a particular pleasure to welcome the completion of this Domesday Book of the wildfowl and their wetland habitats. Since its predecessor was published in 1963, the range and bulk of data have greatly increased, thanks to the efforts of several thousand dedicated volunteer workers. The Wildfowl Trust, supported by the Nature Conservancy Council, has played the central role in collecting and analysing the data, but their sister organisations, the British Trust for Ornithology and the Royal Society for the Protection of Birds, have been more than supportive, as have been the Naturalists' Trusts and Ornithological Societies throughout the country. Indeed this is a very British achievement, basically one of free enterprise but underpinned by a, relatively modest, funding from Government sources.

The vulnerability of wetlands to modern technological developments was recognised by the drawing up in 1971 at Ramsar, in Iran, of the Convention on Wetlands of International Importance Especially as Waterfowl Habitat. With 38 countries now adhering to the convention, and more than 300 wetlands covering some 19 million hectares afforded special protection, the international effort has clearly had considerable success. In the United Kingdom 19 wetlands of international importance have been designated for the Ramsar List and our aim is to extend that status to cover the 132 which satisfy the agreed criteria. The data presented in this book are of crucial importance in this task. Having so many hard facts neatly and clearly available is also of enormous value to the NCC's Regional Officers, and their non-governmental counterparts, in combating threats to valuable areas. The level of awareness of local authorities has been increased by the assemblage of data of this kind, enabling them to build essential environmental safeguards into their planning procedures. The importance of maintaining regular and extensive coverage of Britain, to this end, cannot be overestimated.

The international dimension of wildfowl conservation is well illustrated in this book, particularly by the clear maps of ringing recoveries and migration routes. In their migrations, wildfowl pass over many national borders and use chains of wetlands throughout their range. More than any other factor, they emphasise the need for international action in conserving wetlands, and the benefits of such conservation accrue particularly to the fortunate countries endowed with these habitats. By their very presence, the spectacular flocks of wildfowl provide one of the strongest arguments to convince the layman, planner and politician that wetlands are not wastelands, but a very precious though vulnerable part of our national heritage.

Many wildfowl have shown adaptability to a changing scene in the countryside and have turned to feeding on agricultural land when their natural feeding grounds have been destroyed or rendered untenable. They thus are in the forefront of what I regard as the most important conservation issue of the present time, the need to work out rational and harmonious policies for the rural estate which recognise the legitimate rights of the various interests, be they agriculture or wildlife, tourism or wilderness.

Throughout known history, wildfowl have been a quarry for the hunter, and many a long-time conservationist has begun as a wildfowler. In a crowded island, and with modern weapons and increased mobility, the hunter must accept restrictions in time, through open seasons, and in space, through refuges, if he is to continue as an acceptable predator in the ecosystem. The planning of an adequate network of wildfowl refuges was one of the initial stimuli for collecting together the data in this volume and its predecessor. The wildfowlers, and in particular their coordinating organisation, the British Association for Shooting and Conservation, have lent considerable support to the research and monitoring work of the Wildfowl Trust.

Wildfowl thus impinge on many aspects of modern nature conservation and it is fitting that their story should be set out in such adequate detail. The name of Slimbridge has become synonymous with conservation since Sir Peter Scott founded the Wildfowl Trust there in 1946. This volume will surely be reckoned as one of the finest fruits of the team that he has inspired over the years. I would wish to commend them on their painstaking labours and their ability to put over an enormous amount of data in a way that makes it readily accessible and understandable.

Acknowledgements

Although it is organised nationally from Slimbridge, the Wildfowl Counts scheme is heavily dependent on Regional and County Organisers, who are responsible for the coverage of waters in their own areas. Their contribution cannot be underestimated and we record our special gratitude to them. In Scotland the counts are organised under the auspices of the Scottish Ornithologists' Club, formerly through a central organiser – successively Miss E.V. Baxter and Miss L.J. Rintoul, Miss E.A. Garden and Miss V.M. Thom – latterly through a network of Regional Organisers, as in England and Wales. Many thousands of individual counters have, over the twenty years, contributed to the enormous quantity of data on which this book is based. They have given freely of their time and travelled, at their own expense, often considerable distances to cover their areas. The quality and completeness of the data, especially for the most important resorts, is a tribute to their efforts. We hope that this volume will emphasise the value of this work, and will provide a spur to its continuation and expansion.

The ready cooperation of the British Trust for Ornithology, through the provision of data from their Birds of Estuaries Enquiry and ringing scheme, has proved immensely valuable.

We are grateful to the Wildfowl Trust for the opportunity of taking part in this work and in particular to its Honorary Director, Sir Peter Scott, who has shown continued interest and has kindly provided the line drawings which help to lighten the text. Our efforts have been guided by the Trust's Deputy Director, Professor G.V.T. Matthews, and we are grateful to him for his encouragement and support throughout. We received much valuable guidance from our Scientific Advisory Committee, and in particular from D.R. Langslow of the Nature Conservancy Council and C.J. Cadbury and G.J. Thomas of the Royal Society for the Protection of Birds.

The survey of wildfowl and their habitats in Part II is a substantial work, which would not have been achieved without the help of our colleagues G.V.T. Matthews and M.A. Ogilvie, who wrote a considerable proportion of this section. We are also extremely grateful to M. Smart and to G.M. Williams and A. Henderson, who also took part in the writing of Part II. The contributions of the additional writers are detailed on p.26.

Our other colleagues at the Wildfowl Trust have been a constant source of advice and assistance. In particular Joyce Portlock, over a period of three years, typed the manuscript and edited it several times on a word processor. We are also grateful to Angela Wenger and Pauline Jackson for additional secretarial help.

The Nature Conservancy Council have, since 1954, financed the central organisation of the counts and wildfowl ringing activities through generous grants and contracts. They have also covered some of the costs of producing this volume.

Finally, we are especially grateful to the following individuals (mainly wildfowl counters and Regional Organisers), who have provided additional data and/or made valuable comments on early drafts:

D. Andrew
S. Angus
T.P. Appleton
I. Armstrong
D.N. Arnold
G.A. Arnold
M.A. Arnold
K.M. Atkinson
L. Batten
M. Bell
T.H. Bell
R. Billings
P.M.M. Bircham
R. Bone
C.J. Booth
H. Boyd
A.T. Bramhall
A.W. Brown
L.M. Brown
J. Brucker
N.J. Bucknell
N.J. Buxton
C.J. Cadbury
B. Campbell
C.R.G. Campbell
L.H. Campbell
N. Campbell
G. Catley
J. Clark
P.E. Clement
M.R. Coates
D. Codd
P.M. Collett
A.H. Cook
W.A. Cook
J. Cudworth

Acknowledgements

W.A.J. Cunningham
A. Currie
W.F. Curtis
R.H. Davies
S. Davies
P. Davis
A. Deadman
A.R. Dean
M.E. Dennis
R.H. Dennis
T. Dixon
A. Dobbs
A. Doulton
H. Dow
B. Draper
A. Duncan
G.M. Dunnett
N.J. Elkins
P. Ellis
S. Evans
P. Ferns
R. Findon
J. Fitzpatrick
D. Forshaw
A. Fox
R. Foyster
I.S. Francis
M. George
H.B. Ginn
R. Gomes
A. Goodall
F.C. Gribble
J. Harradine
G.M. Harrison
R.J. Haycock
C.G. Headlam
A. Henderson
D. Henshilwood
D. Herringshaw
M. Heubeck
J. Howard
G. Howells
M.J. Hudson
J. Humphrey
J. Hunt
M. Jones
F.D. Kelsey
R. King
P. Kinnear
R. Knight
P.J. Knights
M.V. Labern
A. Langford
D. Lea
R. Leach
R. Leavett
R. Lovegrove
B. Martin
J. Martin
F. Mawby
A. Mayo
H. Milne
S. Moon
D. Moore
P. Morley
A.J. Morris
G.P. Mudge
R.D. Murray
P. Naylor
S. Newton
M.J. Nugent
D.A. O'Connor
N. Odin
P.J. Oliver
A.J. O'Neil
J. Partridge
B.S. Pashby
C. Prentice
D. Price
C.E. Ranson
E.I.S. Rees
I. Rees
P. Reynolds
M.G. Richardson
P.W. Richardson
D.H.V. Roberts
S. Roddis
J.C. Rolls
W. Russell
A.K.M. St Joseph
C.J.D. Shackles
D. Smallshire
S.R. South
C.J. Spray
J. Stevenson
D.A. Stroud
P. Stuttard
S. Taylor
P.J. Tilbrook
P. Toynton
C.R. Tubbs
A. Venables
A.F.G. Walker
T. Wall
C. Walley
A.C. Warne
G. Waterhouse
R. Wilcox
R. Williams
J. Wilson
C. Wright
G.M. Wright

Myrfyn Owen
G.L. Atkinson-Willes
D.G. Salmon
The Wildfowl Trust
Slimbridge
Gloucester

Conventions and abbreviations

A number of terms are used conventionally throughout the text, and are defined as follows:

Season means the period between mid-September and mid-March, unless otherwise defined. Only the year in which the season began is normally quoted, e.g. 1970 means September 1970 to March 1971, and 1970-1975 means 1970-71 to 1975-76 inclusive.

1-200 means 100-200; 2-3,000 means 2,000-3,000; 15-20,000 means 15,000-20,000, and so on. Numbers below 100 are written in full.

100 Mallard and Teal means 100 Mallard and 100 Teal; a total of 100 Mallard and Teal means that the two species together total 100.

References in the text are given as bracketed numbers; the full list appears on p.551. For example, in paragraph one of Introduction to the Survey "(279)" refers to Hutchinson 1979. In the tables and figures references are given in conventional form (e.g. Ogilvie 1981).

Maximum Counts also appear in brackets, e.g. "Warwick Castle Park Lake with 99 (375) Mallard" on p.55. Here "99" denotes the regular count of Mallard at the site, "(375)" the maximum recorded there (see also p.27).

BASC
British Association for Shooting and Conservation
BTO
British Trust for Ornithology
IWRB
International Waterfowl Research Bureau
LNR
Local Nature Reserve
NCC
Nature Conservancy Council
NNR
National Nature Reserve

RSPB
Royal Society for the Protection of Birds
SOC
Scottish Ornithologists' Club
SSSI
Site of Special Scientific Interest
SWT
Scottish Wildlife Trust
WAGBI
Wildfowlers' Association of Great Britain and Ireland

Location maps:
Letters and numbers on the border refer to the National Grid 100km squares and divisions thereof.

Part I

Introduction

Introduction to the survey

Two decades have passed since the publication, in 1963, of the First Edition of *Wildfowl in Great Britain* (edited by G.L. Atkinson-Willes, published by HMSO) – the first comprehensive survey into wildfowl habitats, stocks and prospects in Britain. Because of sparsity and irregularity of cover it proved impossible to include Ireland, either in the previous volume or in this work, but a separate study was carried out there recently (279).

The aim of the original survey was to provide a basis for conservation planning following the 1954 Protection of Birds Act and the formation by the then Nature Conservancy of the Wildfowl Conservation Committee to advise on wildfowl conservation and exploitation. Prior to this there had been much debate on the status of wildfowl and the effects of shooting, with conflicts arising largely from the lack of objectively gathered information on numbers and distribution. *Wildfowl in Great Britain* summarised the information collected during 14 years of Wildfowl Counts and provided a basis for future planning.

The volume more than adequately fulfilled its objectives and it continued until recently to provide basic data for cases of both national and local conservation. The last 20 years have, however, seen major changes not only in the habitat and conservation of wildfowl but also, partly as a result of this, in the status of most wildfowl species wintering in Britain. Following numerous requests from individuals and organisations for a new review the Nature Conservancy Council (NCC) and the Wildfowl Trust decided in 1979 that a complete reassessment should be undertaken. The volume of data meant that analysis was impossible except by computer and over the following two years all counts since 1960-61 were typed in and stored on a microcomputer at Slimbridge.

The broad outline of the survey follows closely that of 1963 but developments in techniques and in the extent of data available gave rise to changes of emphasis. Part I gives a summary of the techniques used to collect information relevant to the understanding of wildfowl populations, distribution and conservation. A much wider range of data is available to aid the present assessment than was the case in the early 1960s.

Part II, being a treatment of the country region by region, still forms the core of the survey. Although this section is longer than that in the First Edition because of the increased quantity of data, it is necessarily condensed. Summary tables are given wherever possible, providing not only the most recent data but also those from earlier years for comparison. Although the Wildfowl Counts scheme has provided the bulk of the information, the use of data from local bird reports and from special projects, published and unpublished, has been extensive. The growth in interest in active ornithology, fostered and initiated by such organisations as the British Trust for Ornithology (BTO) and the Royal Society for the Protection of Birds (RSPB) as well as the Wildfowl Trust, has been largely responsible for the increased volume of such data.

As was realised when the previous survey was carried out, it is impossible to consider the wildfowl of Great Britain in isolation. Most do not breed here and our stock is only part of the western Palearctic or north-west European population of the species. The species accounts in Part III include an assessment of the species in the range as a whole, based on data from the International Wildfowl Counts and other surveys organised through the International Waterfowl Research Bureau (IWRB). The seasonal movements of the birds have been analysed by the use of ringing recoveries as well as the monthly pattern of Wildfowl Counts. For those species breeding in Britain figures given by the Atlas survey of the BTO (542) have been combined with autumn and winter count data to give an up-to-date estimate of the breeding population.

Part IV discusses the conservation of wildfowl in a changing environment. Many of the changes are a direct result of man's activities and not all are deleterious. The increase in the area of inland waters, particularly as gravel pits and reservoirs, has had a major effect on many species, and the efforts of conservation and other organisations to maximise the value of these habitats by suitable management has played a large part. The success of conservation efforts has, in the case of geese, led to an increase in the conflict between farmers and the birds but the problems are now better understood and constant attempts are being made to alleviate them. Threats to sea ducks from oil pollution have increased following developments in the North Sea and the increase in tanker traffic. The loss of habitat through industrial and other development, particularly on estuaries, gives constant cause for concern.

As well as providing a better picture of the wildfowl situation in England, Scotland and Wales than has hitherto been possible, this survey also points out gaps in our knowledge. Some concern particular parts of the country such as north-west Scotland, notoriously difficult to cover, others the distribution of birds at certain times of year, particularly during the moult. What we need increasingly are ways of forecasting the effect of habitat loss or changes in its quality on wildfowl numbers and distribution. Since such predictions are always complex because of the involvement of many factors outside the area of interest and because subtle social and behavioural mechanisms may be involved, they must be based on thorough and relatively long-term studies. Similarly, our understanding of population dynamics is very important in decision making in conservation, and since many wildfowl are relatively long-lived these studies take time. However, applied decisions must be based on a sound knowledge of bird biology and progress in this field must be maintained. Developments over the last twenty years have clearly illustrated the importance of detailed monitoring internationally, nationally and on a local scale. While this must be carried on and continually improved, we must also concentrate on detailed biological studies which enable us to interpret these masses of data in the interests of the birds. The present survey assesses the progress made in the last two decades and will, we hope, provide a stepping-stone for the next two.

Wildfowl Counts

History and organisation

The Wildfowl Count network was set up in 1947 by the International Wildfowl Inquiry Committee and a Central Organiser was appointed. The aim was to cover as many waters as possible once in each winter month, September – March. The network was made up of volunteer observers and coverage was inevitably incomplete, but special efforts were made to obtain regular counts from the most important sites. The Wildfowl Trust took responsibility for the counts in 1954 and the Organiser (G.L. Atkinson-Willes) moved to Slimbridge. By this time the number of waters counted had risen to more than 500 and about 700 volunteer counters were involved. By the early 1960s more than 2,000 waters had been covered at some time although the average number of counts received was 5-600 in each month.

An International Wildfowl Count scheme was set up by the International Waterfowl Research Bureau at Slimbridge in the mid 1960s, aiming to cover the whole of the western Palearctic range in January each year, and in some years in November or March. When these counts began, with the first full survey in 1967, there was a major impetus to increase the coverage in Britain even further and Regional Organisers recruited more observers. The usual number of counts made in January (the month with most complete coverage) rose to 1,100-1,300 and in other months to 700-900. By 1982, more than 4,000 waters had been covered at some time since 1960 and in the early 1980s more than 1,100 individual counters were involved. The original network consisted almost entirely of amateur volunteers but, as many of the most important areas became reserves, there was an increasing proportion of professional input. Nevertheless, the vast majority of counters are still amateurs carrying out the counts in their own time and travelling at their own expense. Even the Regional Organisers, who are responsible for a major part of the organisation, are still almost all amateurs.

A further boost to the count coverage occurred in 1969, when the Birds of Estuaries Enquiry was launched by the British Trust for Ornithology (485). The Enquiry, though concentrating mainly on waders, includes all birds on estuaries by means of monthly counts organised in a similar way to the Wildfowl Counts. Many areas previously unrecorded for wildfowl have been covered. The wildfowl data from the Enquiry have been made available at all times to the Wildfowl Trust (which co-sponsored the scheme for some years), and have been used extensively in the present work.

The advent of the International Wildfowl Counts made the British figures, which represented coverage of only part of most populations and flyways, much more meaningful. Not only was it now possible to assess the importance of Britain as a wintering area for the different species but individual sites could be assessed in an international context (see p.527). The coverage of international counts in north-west Europe, the Mediterranean and North Africa has been remarkably good, with 13,380 sites being counted in January during the ten-year period 1967-1976 (24). Efforts have been concentrated on January counts, but surveys have also been conducted in November and March to assess changes in distribution at those times of year. These additional counts should be extremely valuable for assessing changes from year to year according to winter severity.

The monthly British counts received at Slimbridge used to be transcribed from the original forms into ledgers. Hand analysis of the whole of the data was too laborious so the assessment of trends from year to year was made using a sample of some 200 so-called "priority sites". Results from these were returned during the winter and trends assessed by comparison with numbers counted in a master year. Although most major concentrations were included in priority counts, the index proved to be a rather variable estimator of trends. Since the priority scheme was also rather costly, it came to an end in 1979-80.

Between 1978 and 1980 all counts from 1960-61 to 1979-80 were entered into a computer at Slimbridge so that comprehensive analyses could be carried out for this survey. Seasonal counts were also entered as they were received from 1979-80 onwards and an analysis of each winter was made within a few months of the end of the season. Each counter now receives promptly a copy of an annual report (524). These reports also include wader counts organised by the BTO under the Birds of Estuaries Enquiry.

Accuracy and representativeness of the counts

Wildfowl counters are experienced ornithologists, many of whom have been active as counters or involved in other bird surveys for many years. With the exception of a few sites, covered by organised teams of counters, bird concentrations are not large, seldom more than a thousand individuals of a single species. A test of accuracy involving 117 observers of varying experience estimating numbers in goose flocks showed that substantial errors were made even by experienced counters. Since there was no bias, however, errors tended to cancel each other out and the mean of all counts was within 10% of the true figure (332). These observers were asked to make their estimates within 30 seconds and would not be able to make repeat counts as under most field conditions. Another test on counts of waders (484), showed that there was consistent bias, with most observers overestimating numbers when these were small and underestimating large flocks. In another comparison of counts with photographs, underestimating error even with very large wader flocks averaged only 5% for experienced observers (229).

These tests would tend to exaggerate the errors under most conditions since birds are difficult to count on photographs in 30 seconds—the time allowed. Many counters make several counts, from different viewpoints or over a period of two or three hours. We can say with some confidence, therefore, that the Wildfowl Counts provide us with a reasonably accurate picture of wildfowl numbers and distribution. Where there are errors, numbers are likely to have been underestimated.

It is impossible using volunteers to make very frequent counts and a test using almost daily counts for three years at Durleigh Reservoir showed that numbers of wildfowl varied considerably over a monthly period and the counts made on the count dates deviated by up to 50% from the average for the whole month (332). The deviation varied with the species, being high for mobile species such as Pochard (51%) and low for the more sedentary ones such as Tufted Duck (25%) and Mute Swan (13%). However, over a long series of counts fluctuations cancelled each other out and averages gave a good indication of the importance of the site. Most of the fluctuations were short term so that for most species fortnightly counts were no less variable than monthly ones. Even with weekly counts the numbers of Mallard and Tufted Ducks remained variable. Much of the variation was removed when adjacent waters were combined, indicating that local movements were partly responsible.

Monthly counts are too infrequent to give detailed information on site use by wildfowl, but very many more counts would be necessary to achieve this. The tests of accuracy that have been made indicate that when averages over a series are used the counts give a reliable and representative picture, especially when coverage of numerous adjacent sites is good.

Count coverage

Because of their locations or habits some species pose severe problems for regular counting and attempts are made to cover these in ways other than by Wildfowl Counts. Geese are usually away from their estuarine or lake roost sites during the day and are frequently missed. Only for Brent Geese do the counts give a realistic picture. Although virtually all grey goose roosts are covered, no more than half the Greylag and Pinkfoot populations are ever included in the count totals. Grey geese are covered by an autumn census of roosts and feeding grounds, and in 1982 and 1983 a March count was also made. Because these geese are concentrated in autumn these censuses give an accurate total for the post-breeding population. Wildfowl Count data are used to indicate distribution at other times of year. There has been no regular census of Greenland Whitefronts, but a general picture of the situation in the 1970s has been drawn from all available count data (516). In 1983 coordinated counts were made in Ireland and in Scotland. European Whitefronts occur on very few sites and information is obtained from local observers.

Barnacle Geese are counted in autumn in the Solway and on Islay but the remainder of the Greenland population is censused only occasionally from the air. The first aerial survey of this species in Scotland was carried out in 1957 (79) and the population has been censused every four or five years since. Only about half the Canada Geese in Britain are counted in winter but there have been two complete summer censuses, organised by the Wildfowl Trust, in 1968 and 1976. Mute Swans are similarly censused in summer, the latest count being in 1978. Bewick's Swans are well covered by the counts but the scattered

Whooper Swans are not. A special census of Whoopers in 1980 yielded just under twice as many as were included in Wildfowl Counts at the same time.

Sea ducks pose special problems in that not often are they easily counted from the shore. Regular aerial surveys are made in north-east Scotland by Aberdeen University and these results are included in the counts. Special boat surveys have also been carried out in connection with oil developments in the Moray Firth (361), and in the Northern Isles. It is very likely, however, that there are many flocks, especially off the west coast, which have never been found.

Regular counts in summer have not been attempted but a special survey of breeding ducks was carried out in the late 1960s (621). The Atlas survey of the BTO (542) was much more comprehensive but not quantitative. However, a combination of Atlas data with autumn counts of the most important breeding ducks does provide an estimate of population which is better than that achieved previously. A survey was carried out in 1980 on inland waters, but this did not provide data on national totals. However, a comparison of matched waters with the 1960s indicated increasing trends in most species (598).

Coverage of moulting concentrations is patchy; many groups of moulting sea ducks are probably unknown and only recently have summer flocks of Shelducks been found moulting in some of our large estuaries (see p.394).

As indicated above, winter coverage in terms of sites counted has continually improved since the network was first established. Ideally, each site should be counted seven times in each season, but this level is achieved for rather a small proportion of haunts, though usually the more important ones. Theoretically, the number of counts expected would be 140 for each site over a 20-year period. This is not the true figure for all sites since some were created during the period and others were lost and could not be counted. However, the proportion of the potential counts achieved gives an idea of the minimum quality of the coverage.

Of the 513,000 potential counts on the 3,663 sites between 1960-61 and 1979-80, 107,000 were actually made (21%). This seems an extremely low level of coverage but it is caused by very many sites in remote areas being counted only once or twice. In the early years counters were encouraged to explore their local waters for important places, and in many cases sites were visited once or twice and found to have no wildfowl, so coverage ceased. There is a substantial difference between months, with the coverage in the best month, January, being 27.2% as opposed to 16.8% in the worst, September. There are also regional differences, with January coverage in south-east England reaching 33% as opposed to 14% in the Scottish Highlands and Hebrides. The corresponding figures for September were only 23 and 8%.

Despite the apparent inadequacy of coverage, most large concentrations of birds are in well-counted areas, so the proportion of the total of each species on all sites which are counted each year is high—70-80% for most species (see p.342). This is because many ducks are very gregarious, concentrating on the larger sites which are well covered. Data are, however, not complete enough adequately to describe many of the less important areas in Part II.

Wildfowl ringing

Knowledge of the origins and movements of wildfowl is essential if they are to be effectively managed or conserved. The Wildfowl Inquiry Committee was initially responsible for coordinating and stimulating efforts (591). By 1937 the ringing scheme had been transferred to the management of the BTO and all subsequent rings bore the address of the British Museum (Natural History).

The Wildfowl Trust was making a substantial contribution to the ringing effort soon after its establishment in 1946, and in 1954 responsibility for wildfowl ringing was transferred from the Wildfowl Inquiry Committee to the Trust. Table 1 gives the number of wildfowl ringed by the Trust and in Britain as a whole, to the end of 1981. Apart from the Mute Swan, the subject of a large number of local population studies, the Trust has been responsible for the vast majority of geese and swans ringed. Expeditions to the Arctic breeding grounds of Barnacle and Pink-footed Geese have added considerably to the total ringed by the Wildfowl Trust. The majority of ringed ducks have been marked at the Trust's trapping stations, but a few species have been the subject of special studies elsewhere. For example, most of the Eiders and many of the Shelducks have been marked in connection with population studies on the Ythan Estuary, Aberdeen, and the Goosanders have largely been caught by a ringing group in Northumberland (337).

The majority of marked birds have been ringed in the last two decades, largely as a result of new trapping stations coming into operation and new methods of catching yielding greater numbers of certain species. This is particularly true of inland diving ducks and swans (see below).

Table 1 also shows the recovery rates of ringed wildfowl. These are extremely high in comparison with rates for other birds, because many wildfowl are quarry species and most shot birds are retrieved and the rings reported. The rates shown in the table are minimal since many ringed birds are still alive, but the figures do allow comparison between species. Although protected, the sedentary and conspicuous Mute Swan not surprisingly yields the highest recovery rate. The low recovery rate of the migratory Bewick's Swans might be expected since the winter mortality in England is only 10% of the annual mortality rate for the species (445). A much higher proportion of Whooper Swans should be recovered since they migrate shorter distances and nest in accessible places. However, as nearly all their ringing has been very recent the recovery rate is rather low.

Recovery rates for geese are generally high; those for the grey geese are total recoveries since very few have been ringed in the last 15 years. The low rates for Barnacle and Brent Geese reflect their long migrations into remote areas and the high proportion of their mortality outside Britain—about 60% of the annual mortality of Barnacle Geese (442). Both species are protected and it is unlikely that rings are returned from birds illegally shot even when this happens in Britain.

A relatively high proportion of ringed ducks are generally recovered; the overall recovery rate (i.e. when sufficent time has elapsed for all ringed ducks to have died) is around 20% for most species. The rates in the table are depressed for Gadwall, Pochard and Tufted Duck because much of the ringing of these species has been very recent. Despite its conspicuousness the protected Shelduck is much less likely to be reported than are quarry species.

Information from ringing recoveries has been used in this volume to compile maps which illustrate the origins of species wintering in Britain. These are presented in the species sections in Part III. This chapter presents historical information on the major ringing stations, and describes the catching and marking techniques used and the changes in emphasis and scope of the ringing effort over the years.

Catching methods and success

Duck ringing

The majority of ducks ringed in Britain have been caught in duck decoys, originally constructed to

Table 1. *Total numbers of ducks, geese and swans ringed by the Wildfowl Trust since establishment in 1946, the total ever ringed in Britain and the number of recoveries to the end of 1981. The Wildfowl Trust totals include geese caught and ringed by Trust expeditions overseas and a small number of geese and swans which have been marked with plastic rings only.*

	Wildfowl Trust	British total	Recoveries	Recovery rate (%)*
Mute Swan	2842	35597	11961	33.6
Bewick's Swan	1201	1201	88	7.3
Whooper Swan	148	158	14	8.9
Pink-footed Goose	21716+	11844	3489	29.5
Bean Goose	0	1	1	—
White-fronted Goose#	581	615	199	32.4
Greylag Goose	1786	2171	517	23.8
Canada Goose	1321	28858	5971	20.7
Barnacle Goose	3548+	1781	192	10.8
Brent Goose$	1217[3]	249	35	14.1
Shelduck	446	5163	501	9.7
Mandarin Duck	0	33	3	—
Wigeon	4895	7571	1300	17.2
Gadwall	1875	2141	261	12.2
Teal	51199	66040	11808	17.9
Mallard	99192	123857	20905	16.9
Pintail	4323	5182	778	15.0
Garganey	393	417	68	16.3
Blue-winged Teal	1	1	1	—
Shoveler	1810	2043	358	17.5
Red-crested Pochard	8	26	6	—
Pochard	3403	3881	336	8.7
Tufted Duck	9725	12633	1516	12.0
Ring-necked Duck	2	2	1	—
Scaup	68	152	33	21.7
Eider	65	14808	1855	12.5
Long-tailed Duck	0	16	3	—
Common Scoter	1	41	5	—
Velvet Scoter	0	4	0	—
Goldeneye	24	99	11	—
Smew	9	9	0	—
Goosander	4	673	103	15.3
Red-breasted Merganser	8	74	8	—
Ruddy Duck	13	18	3	—
Total swans	4191	36956	12063	
Total geese	30169	45519	10404	
Total ducks	177464	244884	39863	
Grand total	211824	327359	51330	

* Recovery rate is calculated on recoveries received to date (British total – excluding foreign-ringed birds). For species where ringing is continuing eventual recovery rate will be higher (see text).
\+ Many caught overseas and marked with foreign rings.
\# *A.a.albifrons* plus 3 *A.a.flavirostris*.
$ *B.b.bernicla* plus 1 *B.b.hrota*.
[3] Most marked with plastic rings only.

catch ducks for the market. The idea was first developed in the Netherlands and the word decoy is derived from the Dutch words "eende" (duck) and "kooi" (trap). The decoy consists of a small secluded pond, set in woodland, from which radiate four to eight "pipes"—curved extensions of the pond which are covered with netting hung over semi-circular hoops. The opening of the pipe may be 5-8m wide and the hoops 2-5m tall, tapering away from the pond to a small catching-up net, 15-20m away from the opening and out of sight of the pond (Fig 1). A series of overlapping reed screens about 2m high run along the outside of the curve and shorter screens, known as "dog leaps", link the taller ones. The screens act in the manner of a sunblind so that the decoyman is visible to ducks in the pipe but is hidden from the pond, which remains undisturbed by catching operations.

Decoys are established close to feeding grounds of ducks, which retreat to the pond to roost in safety. In many decoys tame ducks are kept in order to lure the wild birds onto the pond. The success of the catching operation depends on the habit of ducks of "mobbing" predators. When a land predator approaches a pond all the ducks face it and follow its movement, though staying at a safe distance. The decoyman trains a dog, usually a brown medium-sized breed resembling a fox and traditionally known as "Piper", to run between the screens, jumping over the dog leaps and showing itself to the ducks. As the dog moves along the pipe it leads the ducks away from the pond. The hidden decoyman watches through peep-holes in the screens and when the ducks are sufficently far in he shows himself between the screens, causing the birds to fly or swim down the net tunnel and into the catching-up net (Fig 1). Ducks take off into the wind and catching is most successful when the wind blows down the pipe towards the pond. Each decoy has several pipes so that ducks can be caught under all wind conditions.

All mammalian predators elicit the mobbing response and success has been achieved using cats and ferrets and even stuffed foxes or stoats held on long poles and moved to simulate the live animal. Ducks can sometimes be lured far enough down the pipe merely by providing food, usually waste grain, weed seeds or potatoes.

The first decoys were built in the Netherlands in the late 16th century and as late as 1956 about a

Fig. 1. Plan view of Borough Fen Decoy, Cambridgeshire, and a diagram of the working of a single pipe. The "Piper" dog is shown jumping over one of the dog leaps while the decoyman watches the progress of the ducks up the pipe through a peep-hole in one of the reed screens. (Drawing by J.B. Blossom.)

hundred remained in commercial operation, catching about 300,000 ducks annually. More than 200 were built in Britain, following the Dutch model, but most went out of use before this century. Some were taken over for ringing during the 1940s and 1950s and the last commercial decoy, at Nacton, Suffolk, was converted to a ringing station in 1967 (see below).

Because of the considerable work, upkeep and time involved in operation, most decoys used for ringing have been operated professionally, or by a decoyman who was employed at least partly for the purpose. As their effectiveness has declined some decoys have ceased operation because the costs could not be justified.

Ducks are also caught in baited traps, mainly based on the design developed at Abberton Reservoir, Essex (604). These traps were either 4m × 4m × 2m or 2m × 2m × 1.3m, and were designed to be moveable with reasonable ease by one or two people. The trap consists of a cage of wire-netting around a wooden frame, with one or more funnels for the ducks to enter and a door for the ringer to retrieve his catch. The funnels make entrance easy but once inside the ducks are unable to escape (Fig 2).

The trap is placed on the margin of a pond or lake so that the funnels are in the water. Where the water level is constant the trap can be fixed but where the level is variable, as at Abberton, the trap has to be moved frequently. Bait in the form of corn or small seeds is regularly placed in the trap, with a little sprinkled outside to attract the birds to the vicinity. The traps are visited daily and the birds collected for ringing.

Incubating female ducks can be caught on the nest by cautious approach and the use of a hand net. Broods of Goosanders have been caught in Northumberland by erecting a mist net or wader net across a stream or river and driving the broods into it (337). Cannon and rocket nets can be used with success for duck catching, particularly at loafing sites where the birds are densely packed. Substantial numbers of Eiders, Shelducks and Wigeon have been caught in cannon nets. Spring-loaded clap nets were developed by the Wildfowl Trust to catch loafing ducks but have been little used. Pneumatically powered ("phutt") nets, fired in a manner similar to cannon nets, have been tried but have not been used successfully for ducks.

More than 164,000 of the 177,466 ducks (92%) ringed by the Trust have been caught at only 5 stations, and the annual catches at these are given in Table 2. There follows a brief historical account of these and other stations used for duck catching.

Slimbridge, Gloucestershire
Most of the catching at Slimbridge has been carried out in the Berkeley New Decoy, completed in 1843 (534). The decoy has 4 pipes, one at each corner of a small (0.35ha) pool, in plan view having the shape of a "skate's egg". It was used commercially until 1929, but although partly reconditioned in 1937 was little used

Fig. 2. The Wainwright duck trap used originally at Abberton and later at other ringing stations. This example has one funnel, but larger traps with multiple entrances are also in common use. (Drawing by J.B. Blossom.)

until the arrival of the Wildfowl Trust in 1946. Over the next three years the decoy was completely refurbished and catching for ringing began in earnest. A dog was used in the early years, but later ducks were lured into the pipes by feeding only.

The majority of the ducks caught have always been Mallard, with a few Teal and the occasional Pintail and Shoveler. Catches in the 19th century had exceeded 1,000 in three seasons but the largest numbers were caught in 1961 and 1962 (Table 2). Subsequently, especially as other catching stations were increasing in importance, catching intensity decreased. Ducks also visited the pool in smaller numbers following the creation of a large number of well-fed pools in the surrounding enclosures of the Wildfowl Trust. Consequently the catch dwindled to nothing in the mid 1970s. The decoy was again reconditioned in the late 1970s and catching restarted, the total reaching 900 in 1981, including 250 Teal.

In 1976 a trap designed for catching Bewick's Swans (see below) began contributing to the duck catch and accounted for most of the species other than Mallard and Teal. Substantial numbers of Pintail, Shoveler, Gadwall, Pochard and Tufted Ducks are now caught annually in this trap. Slimbridge is now the second most important catching station for ducks in Britain.

Abberton Reservoir, Essex

This 500ha reservoir was completed in 1940 and very soon became an important roost for ducks. Much of the perimeter is steep and concrete lined, but about a quarter is natural and suitable for duck traps. In 1949 Major General C. B. Wainwright designed traps (Fig 2) and established a small ringing station at Abberton. The traps were so successful that within a few years

Table 2. *Number of ducks ringed at five major ringing stations run by the Wildfowl Trust, 1946-1982.*

	Abberton Reservoir	Borough Fen Decoy	Deeping Lake	Nacton Decoy	Slimbridge
1946-60	24810	12369	0	0	8934
Year's mean*	1908	1031	0	0	596
1961	3239	1909	0	0	2368
1962	2818	1969	0	0	2018
1963	2211	1650	0	0	842
1964	2084	2886	13	0	600
1965	2082	1278	74	0	574
1966	1167	2550	257	0	462
1967	2527	3150	533	1639	994
1968	1978	2152	383	1798	329
1969	2623	1817	622	1905	751
1970	2583	1393	336	1235	449
1971	2247	1278	379	1193	248
1972	2536	766	621	1107	228
1973	2261	1267	900	1123	31
1974	1901	1159	912	834	30
1975	2008	1505	1361	1233	0
1976	1672	439	791	732	618 +
1977	2300	489	1252	625	550
1978	2093	220	676	493	588
1979	2766	78	664	435	1032
1980	1780	115	546	679	1284
1981	2508	221	432	539	1949
1982	2097	20	822	61	1682
Total	74291	40680	11574	15631	26561

* Mean for years of operation Abberton 1948-1960
Borough Fen 1949-1960
Slimbridge 1946-1960

+ Decoy refurbished and catching started in the large cage trap designed for catching swans. Numbers in subsequent years include those caught in the swan trap.

more than a thousand ducks were being caught annually, with the bulk made up of Teal. The Wildfowl Trust became involved with the ringing in 1954 and the station continues to be the most effective of the catching stations.

In the first 15 years of catching, Teal accounted for 70% of the birds caught (605), but this proportion has declined drastically to less than 20%. Mallard now predominate in the catch but increasing numbers of Tufted Ducks and Pochard have been caught in recent years. The large size of the reservoir and its proximity to the Blackwater Estuary means that a large number of species are recorded and a wide variety of ducks are caught, including most of the British-ringed Garganey and Smew, and most of the Shelduck and Scaup ringed in England. Twelve Ruddy Ducks were caught in 1981 following the rapid spread of this species into south-east England.

Abberton is managed as a reserve by the Essex Water Company, which also contributes to the running of the ringing station there. The Ringing Officer, outposted from the Wildfowl Trust, also helps to warden the bird sanctuary for the Water Company—an arrangement which has proved convenient and successful for both sides. In 1968 the reservoir was declared a statutory Bird Sanctuary and it has recently been nominated as one of the British wetlands of international importance for wildfowl under the Ramsar Convention (p.538).

Borough Fen Decoy, Cambridgeshire

This is the most ancient decoy in Britain, being in operation in 1640; records of catches at Borough Fen are available for as far back as 1776. The decoy, set in a secluded wood in the vast East Anglian fenland, has eight pipes radiating from a 1ha pond. The catching performance in the past cannot be assessed in detail since species were not separated and ducks other than Mallard were counted only as "half ducks". The maximum recorded was 450 dozen and 8 "ducks" in 1804-05 (a minimum of 5,408 but probably 7,000-8,000 birds). Numbers caught have always fluctuated widely, with no apparent trend from the earliest records to the 1950s (129).

Set in an area with abundant food, especially in autumn when most Mallard feed on stubble grain, but little standing water, Borough Fen attracted up to a thousand ducks to roost. These were lured into the pipes, at least one of which was usable in any wind direction, by dogging. Mallard have always formed the bulk of the catch with important numbers of Teal also ringed. Substantial numbers of Wigeon, Pintail and Shoveler were caught formerly, but very few have appeared since the late 1940s.

In the 1970s the catch at Borough Fen fell sharply as the number of ducks roosting on the pond declined. It also became difficult to keep the water level high because of the lowering of the water table in the surrounding area. There ceased to be a full-time decoyman in 1978 and the decoy continued to be operated from the neighbouring waterfowl gardens at Peakirk. The site was designated by the Department of Environment as an Ancient Monument in 1976 and, by arrangement with the owners, continues to be maintained by the Trust and used for ringing on a small scale.

Deeping Lake, Lincolnshire

This is a flooded gravel pit a few miles from Borough Fen Decoy. Catching began at Deeping in the 1950s, when the ringing totals were included with those of Borough Fen. Ducks are caught in large fixed cage traps on an island in the lake, which holds substantial numbers of diving ducks. Catching was taken over by the Trust's decoyman at Borough Fen in 1964 and from then on activities became more intensive. About three-quarters of the catch consists of Mallard and the majority of the remainder are Tufted Ducks with a few Teal and Pochard.

Nacton Decoy, Suffolk

This decoy, known formerly as the Orwell Park Decoy, was acquired on a 21-year lease by the Wildfowl Trust in 1968 (ringing having commenced there, and the killing of caught birds having ceased, a year earlier). The last decoy in Britain to be used commercially, it was built around 1830 and, unusually, has two working ponds. Three pools were originally dug as mill reservoirs and the central one, of about 1ha, was adapted to form the main decoy pond in the traditional "skate's egg" style. A fifth pipe, designed for catching Teal, leads from a 0.4ha pool in a secluded part of the wood.

The bag records, kept by the Orwell Park Estate since 1895, are the most detailed and accurate of any British decoy (334). Well over a quarter of a million ducks have been caught at Nacton since 1895 and 15,631 ringed (Table 2). Only once during the years of commercial use did the take fail to reach 1,000, although the high levels of the 1920s, when the annual total averaged 6,300 birds, were never repeated. The total of 9,303 for 1925-26 is the highest authenticated total for any British decoy. The ducks are decoyed with a dog and since 1967-68, when the Wildfowl Trust took over catching operations, the ducks have been ringed and released. The catch continued to decline, however, with the last six seasons yielding 400-600 ducks annually. This caused the Trust to give up the lease in 1982.

Nacton Decoy had always caught a high proportion of Wigeon and Teal and substantial numbers of Pintail were trapped from 1940 onwards. Prior to 1967 only 910 Pintail had been ringed in Britain but Nacton surpassed this in 3 seasons and a total of 2,666 (51% of the British total) had been ringed up to 1981. The decline of Nacton Decoy was also probably due to the increase in areas of undisturbed waters in its neighbourhood.

Other ringing stations

The decoy at Abbotsbury, Dorset, caught 200 ducks in the mid 1960s and, following a period of inactivity, resumed operation in 1976 and has since caught some 1,400 ducks, of which 74% were Teal and 11% Pintail.

Dersingham Decoy, Norfolk, was first built in 1818 but was dismantled in 1870 owing to problems with maintaining the water level and a decline in the value of the catch (307). Cage traps were built there in 1963 and a pipe reconstructed in 1965. The catch increased to 1,479 in 1967, and a total of 4,942 ducks had been caught by the time catching ceased in 1970. About 1,400 Teal were caught, but the bulk of the catch was of Mallard.

A five-year study of ducks breeding at Loch Leven, Kinross, began in 1966 and this included catching incubating females using a hand net as well as cage-trapping. A total of 1,919 ducks were caught, about half of which were Mallard and most of the remainder Tufted Ducks, with a few Gadwall, Wigeon and Shoveler.

The cage traps in a gravel pit at Blunham, Bedfordshire, have made a substantial contribution to the number of diving ducks ringed since they began operation in 1979. In the three years to 1981 the catch of 2,621 included 1,222 Pochard, 1,231 Tufted Ducks and 87 Gadwall. The 1979 total also included a Ring-necked Duck, only the second of this species to be ringed in Britain.

Over the last 20 years the duck trapping effort has gradually shifted from decoys to cage traps, as Fig 3 shows. The proportion of the total caught in decoys has declined from about two-thirds in the late 1960s to a quarter recently, in spite of the fact that as many decoys are still operating as at any time during the period. The main reason for the decline in decoys is thought to be the increasing number of water-bodies available in the lowlands of England and the lessening of disturbance in many places as a result of conservation activities. It has also been suggested that the change from killing to ringing ducks in decoys may cause wariness of the pipes to be transmitted by caught birds (334), and this may well be a factor.

Overall numbers caught annually have remained remarkably constant at between 4,000 and 7,000 birds except for three good years in the late 1960s when Nacton became operational and Dersingham Decoy caught large numbers. The decline through the 1970s was halted when catching began at Blunham in 1979. The increasing use of cage traps has the advantage that they can be sited in many places and that a much wider diversity of ducks, especially diving ducks, can be caught in them.

Swan ringing

The Mute Swan, being easy to attract to bait and catch in small numbers, has been extensively ringed. Most have been caught one by one using long-handled hooks which are used to grip the bird's neck. Larger flocks have been caught by herding them into confined places or by rounding up flightless birds. The

Fig. 3. Total Wildfowl Trust duck catch and the percentage of that catch made in decoys (the remainder in cage traps), 1961-1981.

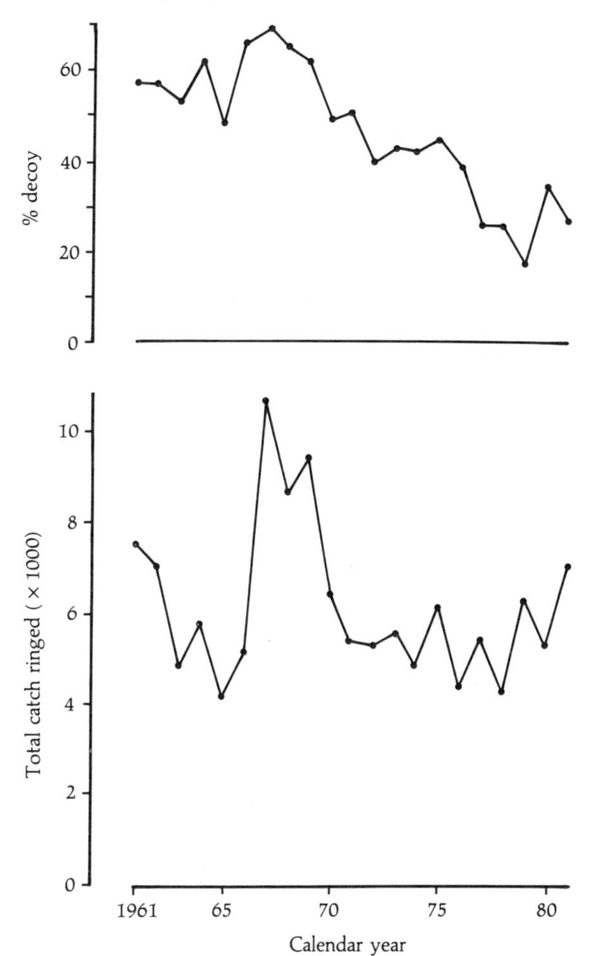

largest round-up of swans in Britain was on the Chesil Fleet in Dorset in 1980 when 831 flightless birds were rounded up by using canoes and rowing boats (468). Another round-up in 1982 yielded 642 swans. Mutes have also been caught in large numbers in the English Midlands, at Montrose Basin, and in the Outer Hebrides. In recent years many studies of local movements and population dynamics, notably in the Midlands, the Oxford area, at Abbotsbury on the Chesil Fleet, and in the Outer Hebrides, have involved intensive marking. It is probable that about a quarter of all British Mute Swans are carrying rings.

The first Bewick's Swan was ringed in Britain in 1961, but relatively few were ringed until the mid 1960s; only a handful of Whooper Swans had been caught before 1980. The much larger catches since then have resulted almost entirely from the building of specially designed traps at three of the Wildfowl Trust's centres. At Slimbridge, Bewick's Swans have been visiting a pool within the enclosures for two decades and have become used to the proximity of buildings and other tall structures. In the early years over a hundred were caught following collisions with trees, and an attempt was made in 1967 to catch birds with flight nets as they left the lake. In 1969 an arrangement similar to a decoy pipe was built in a corner of the pool; over the next few years this was modified to form a large cage trap with a 4m-wide entrance (181). The trap is heavily fed and the swans are caught when a gate is swung to close off the entrance. They are driven up the pipe, together with large numbers of ducks which are usually also caught, to a specially designed holding and catching area. Over a hundred swans can be caught at once but the catch is usually limited to reduce the processing time. Over 1,200 Bewick's Swans have been caught in this trap and many of these more than once.

Following the success of the Slimbridge trap, others were built at Caerlaverock and at Welney. Both were operated successfully in 1980, the one at Caerlaverock catching mainly Whooper Swans, that at Welney Bewick's Swans. The largest catch at Caelaverock, in February 1982, consisted of 78 Whooper, 37 Bewick's and 27 Mutes, and at Welney a record 53 Bewick's and 13 Whooper Swans were taken in the same month. The layout of these traps differs from that at Slimbridge in that swans follow the feeder into a net-covered channel which has doors at both ends. These are closed simultaneously once the birds have settled to feed in the channel.

Goose ringing

Two principal methods are used for catching geese—nets propelled over feeding flocks, and summer round-ups on the moulting areas while the birds are flightless. Early attempts at catching geese with clap nets, used successfully in the Netherlands, largely failed in Britain. Sir Peter Scott devised a rocket-propelled net in 1948, later developed by the Wildfowl Trust in cooperation with the British Army. The aim is to propel the net over geese on the ground and the system consists of two 55 x 18m nets, of small mesh to prevent the birds becoming entangled. The net is reinforced with thicker cords and around the margins flaps are sewn in to form pockets to prevent birds walking out. The nets are also divided by pockets into two sections which prevent large numbers of birds accumulating in one spot.

The rockets are made from modified 25-pounder shells. They are loaded with fluted cordite sticks and placed on ramps set in holes in the ground at 45° such that the top of the rocket is at ground level. Six rockets are placed behind each net. The charge is fired electrically, and as the rockets move up the ramps they pick up 1m-long wire traces which are attached to the leading edge of the net. The other edge is attached to anchored rubber cords which prevent the net taking off or being torn by the power of the rockets. An asymmetric weight at the base of the rockets flattens the trajectory and ensures that the time between firing and the landing of the rockets is extremely short – around one and a half seconds.

The whole "set" is meticulously camouflaged with material from the catching field. The fuel igniters are connected by an electrical cable to a firing box in a hide, up to 400m away but giving a good view of the "catching area" (the area potentially to be covered by the nets) which is marked on the ground with unobtrusive markers such as turves or sticks. The two nets are laid either end to end, firing in the same direction, or parallel and some 45m apart, firing towards each other. The face-to-face method is more effective at catching the birds but is more easily detectable by them. The parallel set is usually used on stubbles where camouflage is easy, and the end-to-end set on grass.

Nets are laid at night in a field used by geese the previous day and the birds are lured into or near the catching area by the judicious placing of stuffed decoys. If the geese land densely in the area the catch can be large—nearly 500 Pinkfeet have been caught at a single firing. Catches of more dispersed feeders such as Whitefronts are smaller but concentration can be induced by the use of bait or by laying string along the ground to "funnel" the geese towards the catching area. Nearly all the Pinkfeet, Greylags, Barnacles and White-fronted Geese ringed in Britain have been caught by rocket nets, as have a few hundred Brents.

Cannon nets operate on the same principle as rocket nets except that the projectiles are fired from mortars. The power of these projectiles is much less than that of rockets and the net is narrower, only 10-15m across. Cannon nets have been used with success for catching Brent Geese. A recent development is the so-called "phutt" net, which is a cannon net fired pneumatically. The catching area is small but the system has the advantage that it is easily transportable and almost silent so that it can be used in situations sensitive to disturbance. Also, no explosives are involved so transporting the equipment across national boundaries is not a problem. The system is still being developed but Barnacle Geese have been caught by the Wildfowl Trust in very small numbers on islands in Norway.

The technique of rounding up flightless geese was probably first used by Eskimos to catch Snow Geese for food. Birds are driven by a number of men into a prepared catching area – a C-shaped pen with "wings" of netting funnelling the birds into it. In some cases geese can be driven onto a pool and held there by part of the team while one or two members construct the pen. The most successful expedition to ring British geese was in 1953, when no fewer than 12,310 Pinkfeet were rounded up in Thjosarver, Iceland, the majority in only 10 days of intensive catching (537). Recently visits to Barnacle breeding grounds in Spitsbergen yielded 1,519 birds in 1977 (455) and 1,213 in 1981.

Of the two species breeding in Britain the Canada Goose has been rounded up in very large numbers and many local populations are caught annually in connection with long-term population studies. Greylags are more difficult to catch since they tend to scatter when driven but some small catches were made in south-west Scotland and in the Outer Hebrides in the late 1960s and a few feral birds have been caught in southern England.

Most goose populations wintering in Britain have now been marked extensively and their movements and migration routes are well established. Marking of grey geese has now ceased but catching of Canada Geese continues to the extent that at least a quarter of the British stock are carrying rings. A quarter of the Spitsbergen Barnacles wintering in the Solway are also marked.

Marking methods

Early leg rings were made of aluminium, having the address of the body to be informed as well as a unique serial number or group of letters. Rings were closed over the leg by hand and fastened with an overlapping clip. Because of problems with wear these have now been replaced by alloy rings which are durable enough to remain on the bird for its lifetime, except in extreme conditions. Alloys including aluminium, manganese, nickel and stainless steel have been used. Young wildfowl cannot be ringed until they are well grown and small metal tags are used to mark ducklings. These have a wire loop which is passed through the patagium—the flap of skin which links the two inner joints of the wing. Reports from these two methods depend on the birds being handled again, either as dead birds or as recaptures.

More conspicuous marks have since been developed, which allow information on ringed birds to be collected without the need to examine them in the hand. Plastic patagial tags, neckties and coloured plastic chicken rings had been used to differentiate populations or to follow groups but a breakthrough came in the late 1960s when engraved rings were designed at the Wildfowl Trust (402). These are made of laminated plastic ("darvic") of two contrasting colours, the upper one a thin layer removed by the engraving. The rings are made as large as the bird will carry and blanks are cut out of a large plastic sheet. Unique number or letter codes are engraved on the rings which are then softened by heating and moulded into shape. Specially modified pliers are used to open the rings to put them on the leg. The elasticity of the material and a squeeze with ringing pliers ensures the shape is regained. The edges of the band overlap by 10 to 15mm and this overlap can be bonded to prevent re-opening. Bonding is essential for larger rings for swans and large geese but smaller ones stay in place without gluing.

These plastic rings were designed for use on swans but have now been used on goose species as small as the Brent. Ring codes can be read, using a telescope, from a distance of up to 200m in good light, and use of these rings has enabled observers to make multiple resightings, essential for many kinds of study. The proportion of birds resighted may exceed 90% annually, even after a long migration; a much higher return rate than from recoveries of metal-ringed birds. Rings are not suitable for birds which spend a large proportion of their time in water or often move into stubbles or long grass. Neck-collars have been designed along the same lines as the rings and these have been used on Mute Swans in the Hebrides and a few on Canada Geese in southern England. The use of neck-collars is common in North America, but adverse public reaction, chiefly to their unsightliness, has restricted their use in Britain. For some species there may also be problems in that the presence of the collar might influence the behaviour of the bird.

Plumage dyes have been used to mark geese and

swans but, since there is a limited number of fast dyes available, this can only be used for groups rather than individuals. The most common use of dyes is to attract attention to birds marked in other ways so that observers can concentrate on trying to read the leg ring or neck-collar.

Undoubtedly the most modern and sophisticated marking method is radio-tagging, whereby the bird is fitted with a small radio transmitter attached with a harness or glued onto the feathers. The observer moves around the study area with a receiver which picks up transmitted signals and locates the bird. Radios have been used successfully to monitor the movements of duck broods by tagging the female, in a study of duck production and survival carried out by the Game Conservancy (262). Attempts to mark ducks and swans to follow winter movements have, as yet, been unsuccessful.

Information from catching and marking

Most ringing schemes were begun to determine the distribution of birds at different times of year, relying on the recovery of a ringed bird and the reporting of the recovery to the ringing centre. The number of birds caught for ringing is small in comparison with population totals so recovery of the bird alive (recapture) is rare and most analyses rely on reports of birds recovered dead. Information on mortality rates can also be deduced from the timing and rate of recovery. Forms of visible marking do not rely on dead recoveries to provide data and the use of these has increased in recent years.

Other information can be gathered when the bird is in the hand, not only its age and sex but also its body condition and health. The remainder of this section outlines the kind of information that is collected from caught birds and gives examples of how this information provides us with a better understanding of wildfowl movements and general biology—knowledge which is vital to the compilation of Parts III and IV of this book.

Ringing recoveries

Before we can assume that any map of recoveries gives a true picture of the movements of the British population, we must realise that bias can be introduced both in the ringing and in the recovery distributions (464). As described above, most British duck ringing has been concentrated at a very few centres and most of these are in southern England. Any recovery map would therefore be biased if the behaviour of southern English birds were different from those further north. This is almost certain to be true for species which arrive here both from Iceland and from continental Europe and also for those that move out from Britain in winter. Thus 71% of winter recoveries of Wigeon ringed in Iceland were in the British Isles, two-fifths of these in Ireland, two-fifths in Scotland and only a fifth in England and Wales (160). Since most Wigeon are ringed in southern England it is not surprising that Iceland is under-represented in the recovery map (Fig 94). Tufted Ducks from Loch Leven winter in Ireland whereas those from England and Wales are either sedentary or move to France and the Netherlands. Since so many were ringed at Loch Leven in the late 1960s and early 1970s the map of this species is more representative of its true distribution.

When examining important quantitative problems, such as determining the proportion of immigrants in the winter population, such sampling problems are difficult to overcome. An analysis of Mallard recoveries (75) showed that a higher proportion of birds ringed at Abberton were migrants than those marked at Borough Fen and Slimbridge. These last two ringing stations gave estimates of about 30% for the proportion of immigrants in Britain, which were reasonably consistent with estimates made by other means.

Another major problem is bias in the probability of recovery. Most recoveries of dead birds are the result of shooting and there is a high probability that such recoveries are reported. The main problem is that shooting seasons vary from country to country and the probability of recovery is largely determined by this. On the other hand, birds dying from other causes are more likely to be found in well-populated areas of western Europe than in the vast wildernesses of Scandinavia and the USSR. As an example Fig 4 shows the distribution of recoveries in May and June of Teal ringed in the Netherlands. The distribution of those shot is contrasted with those found dead. It is clear that using shot birds the breeding area seems to be well to the east, where shooting was allowed, whereas most birds are found dead in the more populated areas (many of these could of course have been shot illegally and not retrieved). Similarly, the centre of gravity of recoveries in July tends to be to the south and west of those later in the autumn although this is contrary to the general flow. This is because July shooting was allowed only in countries in south-west Europe and as soon as the shooting season opens further north the centre of gravity moves there (465).

The usual way of avoiding these problems in comparisons between species and birds of different age and sex is to restrict the ringing location to one centre (416) and/or the finding method either to shot or found dead (76, 464, 465). The recovery maps in Part III are, however, intended to show the origins of ducks wintering in Britain. Thus all recoveries have been included, but to appreciate the anomalies and omissions the maps must be used in conjunction with the text.

Fig. 4. The recovery pattern of Dutch-ringed Teal in May and June: a) shot birds only and b) those found dead. Redrawn from Perdeck and Classon (1980).

The use of ringing recoveries to examine the causes of death and to calculate mortality rates is similarly fraught with problems, of a slightly different kind from those outlined above. Thus a maximum of 11% of reported deaths of adult Pinkfeet were due to natural causes but, because shot birds were much more likely to be reported than those which died of natural causes, this was thought to be a considerable underestimate (62). Similarly, 75% of recoveries of the protected Spitsbergen Barnacle Goose population were of birds said to be shot, but the real level of shooting mortality was probably below 50% (443).

One of the major problems of using ringing recoveries to estimate mortality rates is that the reporting rate (the likelihood of a dead bird being reported) varies both in time and space. This is because of different shooting traditions, "weariness" in continually reporting ringed birds shot, or deliberate withholding of information by shooters. An analysis of these problems is beyond the scope of the present description but various models have been devised to minimise their effect (62, 90). Where differences between adult and juvenile or between sexes are being examined, reporting rate problems can be considered negligible.

Ringing recoveries have also provided a great deal of information on the principles of migration and orientation (333).

In only a few cases is it possible to use recapture data in a meaningful way, but ringing of Pink-footed Geese was so intensive in Britain in the 1950s that a large number of recaptures was obtained. These were used, together with recoveries, to look at the seasonal movements within Britain and to investigate the loyalty of birds to their wintering area (61). The results showed differences between birds of different ages and between different areas depending on whether they were autumn staging posts or terminal wintering areas to which the geese were more loyal. Recaptures of female ducks at nest sites have been used to study the loyalty of females to their nesting sites from year to year (380).

Live resightings

For many years Bewick's Swans have been identified individually at Slimbridge by their bill markings, but this technique is not possible on many species and it does require very experienced observers. Mute and Bewick's Swans were the first wildfowl to be marked with "darvic" plastic rings, a process which enabled detailed individual life histories to be drawn up and made population studies more effective. The Mute Swan population at Abbotsbury, Dorset, has almost all its individuals marked and details of movements, pairing and breeding success are recorded annually. The marking of Bewick's Swans, in conjunction with the use of plumage dyes, enabled observations of Slimbridge wintering birds to be made at sites both in Britain and in continental Europe (181). A small number of Whooper Swans collar-marked in Iceland have produced a number of interesting reports, with some being seen as far south as the Ouse Washes, Norfolk (83).

Several populations of Canada Geese have been marked with darvic rings and their local movements monitored in detail. A study in Yorkshire has been in progress for many years and now involves extensive individual marking. A study in Nottinghamshire is aimed at assessing the impact of the geese on agricultural crops, for which an indication of the mobility of various sections of the population is essential. A long-term population dynamics study on Solway Barnacle Geese relies heavily on information collected on the 2,000 or so marked geese. The performance of the rings has been such that annual loss rate (assessed on recaptured geese also ringed with metal bands) is as low as 0.14%, at least for the first five years. Since the resighting rate of ringed individuals averages 94% annually, accurate estimates of mortality can be made merely by monitoring the disappearance of ringed geese from the population (443).

A study set up by the enquiry into the third London Airport at Foulness in the 1970s was aimed at monitoring the movements of Brent Geese throughout their range in England. More than 1,200 geese were marked with darvic leg rings and many thousands of sightings were made, in staging areas in the Netherlands and Germany in spring, as well as on the wintering grounds. These showed that Foulness was an important autumn gathering area for geese that wintered in other parts of eastern England, and, surprisingly, the English wintering birds overflew Dutch staging areas (used by Brent wintering in France) in spring and stayed in the German part of the Wadden Sea (519). Such information, vital to conservation policy for the species, could not have been gathered without the use of readily visible marks.

Population levels, age and sex ratios

Since wildfowl catching is done opportunistically, it might be expected that the number of birds caught would be related to the number available, i.e. the population level, provided the catching effort remained constant. There are indications that this is the case with certain species; at least large fluctuations in population show up in the ringing totals (211). The performance of decoys in recent years has declined despite an increase in most species, so the overall totals caught are not related to population size. If we compare the number of ducks caught at Abberton Reservoir with the average number counted there in the 7 winter counts in the years since 1970 there is a good correlation for Mallard ($r=0.844$) but not for Teal ($r=0.061$). We must conclude, therefore, that it is dangerous to use numbers caught to indicate trends in duck numbers even on a very local scale.

There is no evidence of sex bias in any catching technique although there are several possible factors which might induce such a bias. Sex ratios of caught birds can thus be used fairly reliably to indicate the sex ratio of the population from which the sample is taken. Since there is no evidence of large-scale differential movements of the sexes in Mallard in winter, it is not surprising that the sex ratio of the catch at Borough Fen Decoy in the late 1950s and 1960s was very close to 1:1 (415). There is known to be a marked differential movement in Pochard, with males staying further to the north and females overflying them to winter in milder regions (526). This shows up well in catches in that the sex ratio of 1,016 Pochard caught at Slimbridge during the winters 1977-78 to 1981-82 was almost 3 males : 1 female (73% of the birds caught were males).

Distinguishing between adults and immatures of most duck species in the field is very difficult or impossible and there are no other very reliable ways of assessing the breeding success variation from year to year—very important in determining population changes. From 1959-60 to 1970-71 the proportion of juvenile birds in the catch at Borough Fen Decoy in July, August and September was compared with the size of the autumn population of Mallard on local waters but no correlation was obtained (416). The proportion of young in the catch was extremely high (99% in one year), reflecting the greater tendency for young birds to be trapped by decoying. There is no evidence of better results from other ringing stations and all forms of catching provide too biased a sample to use as any indication of the age ratio of the population as a whole.

Measurements and estimates of condition

Since ringers have been catching birds they have also been weighing and measuring them, and much of the information on structural size and body weight of wildfowl comes from measurements of trapped birds. Apart from telling us how large birds are, measurements provide an indication of the variability of a population and may help to delimit subpopulations within a species as well as show size differences between birds of different age and sex.

Whereas ducks and geese are fully grown

(though not necessarily at adult weight) when they arrive in the wintering grounds in autumn, young Bewick's Swans are still smaller than adults (182). Young ducks achieve adult weight in their first autumn (446), and geese do so in their second, but swans in their second winter are still substantially below adult weight. The rate of maturation is, of course, related to the age at which the different groups begin to breed. There is also a relationship to the extent of parental care (532). Ducks are independent of their parents soon after fledging; geese remain in families through the first winter whereas swan families often remain associated in the second or even third winters.

Attempts to distinguish between British and immigrant Mallard in the catch at Borough Fen failed to produce conclusive evidence that these were different in size (450). One surprising finding was, however, that juveniles reared in different seasons differed significantly in size, probably related to the earliness of the season. Early hatched young have longer to grow to fledging than later ones, which have to accelerate development at the expense of eventual body size.

Analyses of body weights give an insight into the changes in condition brought about by changes in feeding conditions or by severe weather. The normal pattern of weight change in wildfowl is that body weight increases considerably soon after arrival on the wintering grounds, when days are long and food in the form of seeds or grain is plentiful. The birds reach peak weight in November or December and begin to lose it again in the cold weather of mid-winter (44, 182, 446). Weight gain begins again in spring as days lengthen in the build-up to the breeding season. The establishment of the normal pattern of weight change is important in monitoring the effect of severe weather and particularly in governing the timing of a shooting or catching ban. The necessary research into the "danger level" has not, however, been carried out for most species and restrictions must be based on other criteria (see p.530).

Moult

Though most people catching birds for ringing routinely record the progress of moult this has been a neglected field of wildfowl research until recently. The length of the flightless period, such a vulnerable time, is not accurately known for most species, though several early estimates had suggested 24-26 days for Mallard (69). Recent studies have produced accurate estimates of 25 days for Barnacle Geese (452) and 32-34 days for Mallard (449). Detailed studies of flightless ducks caught at Abberton continue and these are primarily aimed at investigating the weight variations during the moult and annual variations in moult timing in both sexes in relation to the timing of breeding and its success.

Injury and disease

With the use of fluoroscopic and X-ray techniques it is possible to determine the proportion of wild birds carrying lead shot in their tissues, which is related to the amount of shooting to which the population is exposed (shooting pressure). Early work on Pink-footed and Greylag Geese (168) indicated that 42 and 37% of adults birds, respectively, were carrying shot. In the Bewick's Swan, a species protected for many years throughout its range, 34% of those X-rayed alive at Slimbridge were carrying shot—a proportion disturbingly close to that in quarry species (184). Similarly, 20% of adult Barnacle Geese, protected on the Solway since 1954, were found to be carrying shot in 1975 and 1976. It is also possible to determine whether birds contain ingested lead shot in their gizzards but checks at Slimbridge have shown that the results are not completely reliable.

Caught birds have also been used to monitor background disease levels by blood sampling or by examining throats or excreta. The studies have included diseases affecting humans or animals, for which wildfowl are potential carriers, as well as those affecting the birds themselves.

Since the First Edition was published wildfowl ringing has become more intensive and increased its emphasis on special studies rather than on opportunistic catching and ringing. The migrations of most species are well established but the sea ducks are little ringed and their movements are not understood. It seems unlikely that this gap will be filled in the foreseeable future but the increasing amount of data accumulating from the numerous population studies, especially on geese and swans, should lead to better insights into population dynamics in the next few years.

Information from shot birds

Traditionally, wildfowlers have contributed much to our knowledge of wildfowl numbers, distribution and habits, although early information was not as rigidly quantitative as we might have liked. The early writings and reminiscences of shooters, such as Colonel Hawker (254), John Millais (344) and others (285), often provide the only accounts of the abundance of wildfowl against which present numbers and distribution can be judged. Later, wildfowlers began to contribute scientific data on habits and foods which were valuable in the understanding of the requirements of the birds (107, 110). When biologists began studying ecological aspects of wildfowl in detail in the late 1950s and 1960s, wildfowlers became part of the collecting mechanism; information or samples supplied by them still provide a valuable source of scientific information.

Duck production surveys

Whereas geese and swans can be aged by observation in the field and valuable information on production and population dynamics so gained, ducks are only reliably aged in the hand. We have seen that the age ratios of caught ducks give unreliable estimates of the proportion of young in the population (p.18) and considerable efforts have been expended in attempting to collect such information from shot birds. A preliminary key to the sex and age identification of wildfowl from wings only was published in 1964 by the United States Fish and Wildlife Service (117) and this provided a stimulus to a pilot study in Britain in 1965 (246). Wildfowlers were asked to send in a single wing from ducks they shot and these were examined by experts who classified them by species, sex and age (adult or first winter). Initially the vast majority of wings were from Kent, but the survey was expanded through the Wildfowlers' Association (WAGBI), now the British Association for Shooting and Conservation (BASC), to cover the whole country. By the mid 1970s the wings received regularly exceeded 4,000 annually, largely made up of Mallard, Teal and Wigeon. The number of wings collected is shown in Table 3.

The main object of the survey is to assess the success of the breeding season by the percentage of young birds in the bag. Ageing and sexing criteria are well enough detailed to make this possible for most species (80). Because young birds are more vulnerable to shooting than adults there is an age bias in the bag and it is important to use the data only to make comparisons rather than to indicate absolute production levels. There is also a regional and seasonal bias in sampling. If the method is to give any useful results it ought to do so for Mallard. Most Mallard in Britain are shot in the autumn and their age composition ought to relate to the total numbers of Mallard in Britain in autumn – i.e. the production from our own breeding stock. On a national scale the result is not encouraging—the Mallard index for September is not significantly correlated with the percentage of young Mallard in the bag. That might be expected if the production of young were density-dependent – i.e. there was good production in years when there were fewer breeding adults, resulting in a similar autumn population. In that case a negative correlation between March numbers and breeding success would be expected, but this relationship was similarly not significant. Neither was there a relationship between the January index and breeding success, although this would not be expected since the proportion of immigrants in the sample varies with weather abroad rather than with breeding success (80).

Looking at the data on a more local level there is a significant correlation between the proportion of juveniles in the bag in Kent and the September index for south-east England (Spearman rank correlation coefficient $rs = 0.717$ $P < 0.02$). The index and percentage young follow each other fairly closely (Fig 5) at least until 1974. Good breeding seasons subsequently failed to cause an increase in the September index, but, since this was a period of low March numbers, that may not be surprising. The lack of correlation using the national sample thus appears to be due to sampling inadequacy rather than a real failure of the method. Indeed, substantial regional differences in the composition of the bag have been demonstrated (80). The national production index thus gives a poor indication of the number of ducks in autumn and winter, but, since there is a high correlation between breeding success in different areas (i.e. a successful season is so everywhere although absolute production may differ), it does give a good indication of the quality of the breeding season. Because volunteers contribute the wings, it is unlikely that the size or the representativeness of the sample can be much increased.

As Fig 5 shows, the age ratio has fluctuated but there is no general trend over the period, and this is true for the national sample both for Mallard and for other species of ducks (80). There is close concordance between the production of the three common dabbling ducks, which is not surprising since many of our wintering birds originate from the same general area in western Europe and the USSR.

The wing samples can also be used to determine the sex ratio of birds in the bag. Although there is a slight bias in favour of males in the bag of most species when sex ratios in the population are equal (largely because they present a bigger target), this is slight enough to discount in considering the major differences that occur through differential migration of the sexes. The average proportion of males in the adult bag has been 46% for Mallard, 57% for Teal and 70% for Wigeon. The results of duck wing surveys are considered in more detail on p.524.

Shooting surveys

One of the basic requirements of population management is the effective monitoring of the harvest and of the population from which that harvest is taken so that shooting regulations can be sensitively controlled. In North America, hunting regulations are very flexible and are changed annually in an attempt to tailor the harvest to the changing numbers of ducks in the autumn—the "fall flight". In Europe hunting regulations are less easily controlled and in most countries consist of a regulation of shooting practices and the imposition of a close season. Nevertheless, in many countries, notably Denmark and the USSR, comprehensive records are kept of the harvest levels. A resource harvesting division of IWRB coordinates the monitoring effort in an attempt to achieve a harvest estimate for the whole of the north-west European flyway and a rational pattern of shooting regulations among the many countries involved.

An annual shooting survey was initiated by the BASC in 1979-80 – the first attempt to assess the British shooting kill. A questionnaire was sent to a randomly selected sample of a quarter of the BASC membership. The questionnaire asked for information on the

Table 3. *The number of duck wings collected from Great Britain for the WAGBI/BASC Duck Production Survey from its inception in 1965. Data from annual reports of the survey published in the organisation's Yearbook and magazine. Reports compiled by J.G. Harrison, A. Allison, H. Boyd and J. Harradine.*

Season	Mallard	Teal	Wigeon	Others*	Total
1965-66	155	325	265	} 191 (1965-66 to 1967-68)	} 3827 (1965-66 to 1967-68)
1966-67	} 1511 (1966-67 to 1967-68)	} 742 (1966-67 to 1967-68)	} 638 (1966-67 to 1967-68)		
1967-68					
1968-69+	555	372	331		
1969-70+	1111	594	668		
1970-71	879	601	475	197	2152
1971-72	910	557	797	184	2448
1972-73	1313	913	1215	280	3721
1973-74	1491	963	1572	406	4432
1974-75	1180	752	1115	235	3282
1975-76	2176	1540	1814	429	5959
1976-77	1540	859	1452	402	4253
1977-78	1603	1086	1466	457	4612
1978-79	1043	1090	1068	386	3587
1979-80	1495	884	1079	302	3760
1980-81	2012	1163	1435	417	5027
Total	18974	12441	15390		

* Largely Pintail, Shoveler, Gadwall, Pochard, Tufted Duck and Goldeneye with very small numbers of other species.

+ Mallard, Teal and Wigeon only available for 1968-69 and 1969-70.

number of shooting outings, their timing and location, and a breakdown of the bag in terms of species. The survey covered birds other than wildfowl but the majority of the quarry was of ducks, geese and waders. The sample was large as a proportion of BASC members, but only some 60,000 of the estimated 160,000 wildfowl shooters in Britain are members of BASC. Nevertheless, as long as something is known of the behaviour of non-BASC shooters, reasonable estimates of the kill of the more common quarry species can be obtained. Data on the less common ones, of course, suffer from great sampling problems especially if, as with grey geese, their distribution is localised and relatively few hunters are likely to take a large proportion of the total bag.

Data from these annual surveys, which are just becoming available, should not only help us to assess the impact of shooting on British wildfowl populations but also provide information necessary to the understanding of population dynamics in our more common ducks and geese.

Lead poisoning research

Deaths of wildfowl resulting from the ingestion of lead pellets, either from shotgun cartridges or in the form of anglers' weights, are widespread and in many of the United States the use of lead shot for wildfowl shooting is banned. In Britain the occurrence of shot in wildfowl gizzards has been recorded in connection with feeding studies (422, 582), but no systematic attempt was made to assess the extent of the problem until 1979. Then a project initiated by the RSPB, BASC and the Wildfowl Trust began and a report was produced two years later (360).

Traditionally one of the main methods of estimating the proportion of birds affected is to record the proportion carrying ingested shot in their gizzards. Pellets which have been shot into the gut are distinguished from ingested ones by the lack of erosion of their surface and the entry hole can usually be found. All of the information gathered prior to 1979 concerned south-east England, but the recent survey attempted to obtain samples from all parts of Great Britain. The guts were collected mainly by wildfowlers and usually all the viscera including heart, liver, etc. were sent in. In addition to scanning the gizzard for pellets, samples of liver, where lead accumulates following ingestion, were also analysed. In all some 2,600 guts and 1,400 liver samples were examined during the two years of the survey, and these came from a reasonable scatter of sites throughout the country.

Lead also tends to accumulate in bones and, whereas liver lead is rather transitory, the build-up in bones continues throughout the bird's life. Pieces of wing bone from ducks sent in to the duck production

Fig. 5. The proportion of Mallard wings collected from Kent which were juveniles (solid line) and the September index of Mallard numbers in south-east England (dashed). The point in brackets is based on a small sample of sites.

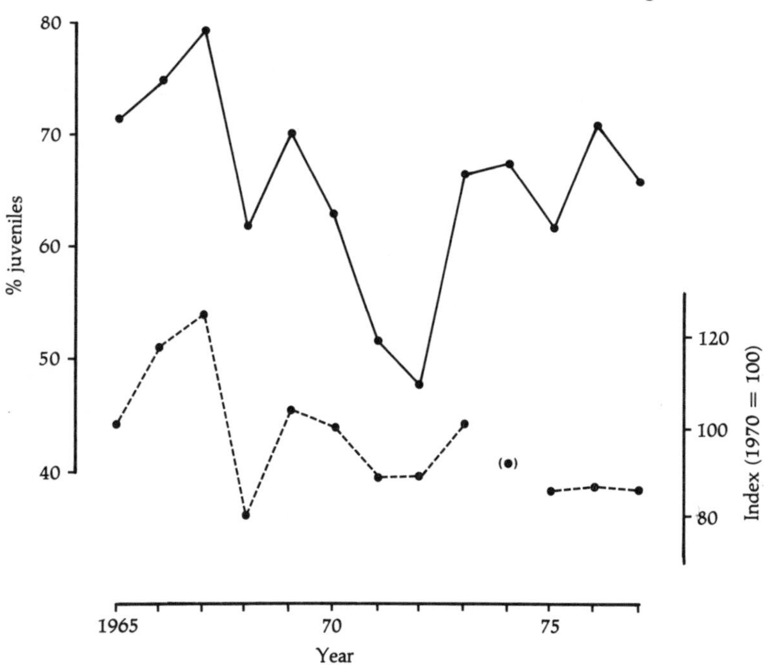

survey were also analysed to give an indication of the birds' lifetime exposure to lead from all sources. A total of over 1,800 bone fragments were analysed.

All these analyses suffer from the drawback that the samples are collected opportunistically and for only a limited part of the year. It could be argued that ducks are more likely to ingest lead after the shooting season when they move to feed in areas previously heavily shot. On the other hand, ducks suffering from lead poisoning are more vulnerable to the gun than healthy ones so that the incidence of poisoning in birds in the bag would be expected to be higher than in the wild (45). The results of such surveys should thus be treated with caution, but in all probability estimates of incidence during the shooting season provide an under rather than an overestimate of that in the population at large. The results and implications of this and other studies of lead poisoning are treated in more detail in Part IV.

Analysis of diet

Stomach analysis has provided the main technique of assessing the diet of wildfowl; indeed many of the early studies of feeding habits were carried out by interested wildfowlers. Following extensive surveys in North America, the Wildfowl Trust appointed a biologist in the mid 1950s to conduct a countrywide survey of wildfowl foods (421). Our understanding of wildfowl habitat and food requirements improved greatly and the sections on food and feeding habits in Part III rely heavily on the results of this work. The main method of collection was through wildfowling clubs and individuals who retained the viscera of shot ducks and sent them in to the Wildfowl Trust. Detailed instructions on techniques of removal and storage of samples were given to contributors (241), but many boxes of unpreserved guts found their way to the Slimbridge Post Office in rather poor condition and the survey was not popular with postal staff!

The details of the analysis technique (425) will not be described here, but a few points are relevant to the interpretation of the results, which in many cases are presented without comment in the species accounts. The most serious problem with the technique is that different foods can pass along the guts of ducks at different rates, which means that the analysis of food from the gizzard in particular may represent a biased sample of the food in the gut as a whole. Soft-bodied animals tend to pass through the gizzard very rapidly whereas hard items such as seeds stay there for a long time. In an analysis of Wigeon gizzards from the Ouse Washes, for example, a check on this was made using faecal analysis, and it was found that gizzard analysis would have exaggerated by more than five times the importance of seeds in the diet (453). Some of the bias was removed by considering only the contents of the oesophagus, but this severely restricts the sample since this part of the gut is often empty. Many animals have identifiable hard parts which remain in the gizzard after the softer portions have been digested and these can be used to estimate the number of animals consumed (425).

The survey suffered from the disadvantage of all opportunistic sampling schemes in that coverage of the country was not representative. Protected species are not, of course, represented and neither is any species outside the open season. Some were collected under licence specifically for research (428), but most of the gaps had to be filled by special studies using different techniques. Nevertheless, gut analysis has given us an invaluable guide to the feeding requirements of wildfowl, and where more detailed special studies have been conducted these have tended to confirm earlier findings or amended them only slightly. Knowledge of diet has proved extremely valuable for managers developing man-made habitats for wildfowl (see p.543). A small booklet was also produced to guide wildfowlers and others who wanted to improve or create wildfowl habitat (603).

This brief review has indicated that shot birds have provided us with a great deal of background material on which to base our conservation decisions. The need for surveys of feeding habits and lead poisoning has now been catered for, while the use of wing analyses to monitor production has severe limitations. The chief contribution in information gathering that today's wildfowlers can make is to continue the surveys of their own activities and to pass on their results, so that we can understand more fully the relationship between bird numbers and the harvest rate. Considerable progress has already been made in this direction, but much more needs to be done. As the various accounts in Part III illustrate, our wildfowl have been faring well over the last two decades with only slight control over the activities of hunters in comparison with regulations in force in North America. We cannot, however, assume that this situation will prevail for ever, and if a change comes we should be in a position to know why, and how, the situation can be corrected.

Part II

The survey of wildfowl habitat and distribution

Introduction

The data from the National Wildfowl Counts and the International Censuses have been used in many different ways in this book, the aim throughout being to furnish a sound scientific basis for the conservation of both birds and habitat. Their use to examine trends in numbers, to assess British populations and to map wildfowl distribution is described in Part III. Part IV discusses the ways in which counts are used to establish criteria for recognising sites of international importance and for assessing the value of individual areas in the national context.

The survey of wildfowl habitat and distribution which follows is concerned with the wetlands within each district, the populations of wildfowl occurring on them, and the changes which have taken place over the past 20-30 years. Apart from the sections written by the main authors, G.V.T. Matthews wrote "The Warwickshire Avon, middle Severn and Teme Basins", "East and central England", "North-west England", and "Teesdale, Tyneside and the Borders"; M.A. Ogilvie wrote "The inner Solway Firth", "Inland Dumfries", "The Stewartry and Wigtown", "The Clyde Basin" and "South-east Scotland"; M. Smart wrote "The lower Severn Vale"; and G.M. Williams and A. Henderson wrote "North Kent".

The survey has been divided into nine chapters, corresponding to the principal water basins or "faunal regions". The continued proliferation of man-made waters, often close to watersheds, has rendered the divisions between the regions less distinct than in the First Edition, but this is still considered the most logical system for a review of wildfowl distribution. The regional boundaries are shown in Fig 60 (p.339), and an enlargement of the appropriate part of this map is presented for each region. The locations of places and geographical features mentioned in the text are shown on 43 individual area maps.

The results of the counts at each significant resort are summarised to provide, in a single set of figures, a measure of the numbers of the various species that are likely to be found there in the course of a normal winter. These indices of abundance are set out in tables at intervals through the text, and are accompanied by a second set of figures showing the highest count for each species in the period under review.

The indices are calculated in two different ways, one method being used for the ducks, the other for the geese and sometimes for the swans. The first assesses the size of the "regular" population occurring during the months when the species are most plentiful, the second is based on the mean of the highest single count in each season, or in some species on the mean of the annual counts in a given month. The aim in both cases has been to provide as realistic a figure as possible, and to minimise any likely sources of bias. The detail of the methods and the reasons for their use are defined more fully below. The use of the maximum counts is also discussed.

The *regular* population is derived from the mean of the three highest counts in each of the seasons for which adequate data are available. For example, if 12 seasons are involved the mean will be based on a total of 36 high counts. For a season to qualify, counts must have been made in at least five of the six months between October and March, this being the period during which the numbers are usually largest. If this requirement is met, the September count may also be considered for inclusion as one of the three highest counts (though it cannot be accepted as one of the requisite five seasonal records).

In the course of the analysis, separate sets of means were taken for the 10 seasons 1960-1969, for the 13 seasons 1970-1982, and for the 23 seasons 1960-1982. If the means in the first period were markedly different from those in the second, the former were rejected, and only the more up-to-date results included in the table. In such cases the extent of the change is discussed in the text. If the results for the first and second periods are roughly the same, the means are drawn from the two combined. The number of seasons on which the means are based is shown for each site, together with the date of the first and the last. For example, "8 seasons 1968-1979" implies that the results are based on 8 of the 12 seasons within that time-span. This information is included as a guide to the reliability of the figures: the longer the run of data, the less bias introduced by exceptional records.

Although less than half of the data collected throughout the season are utilised, the method has several advantages. Sufficient scope is allowed for the cancelling out of chance fluctuations, and account is taken of the length of time during which large flocks were present. The means from a site which supports large numbers for several months each winter will be substantially higher than those from a site holding similar numbers for only a few weeks; the relative importance of the two resorts is thus reflected in the indices. Before the method was finally adopted, a number of other possibilities were tried and found to be less satisfactory. One such approach was to take the mean of all the available records. This was rejected because of the wide fluctuations which occur in the course of each season, and because of the bias which would be introduced by inequalities in the amount of data available for the various months; there was also the likelihood of including a proportion of unrepresentative counts, resulting from extremes of hard weather and other abnormal factors. Another possible approach was to use the mean of the single highest count in each season (a course which has in fact been adopted for the geese and swans). The disadvantage of this, when dealing with the ducks, is that too many data are left unused, and too much emphasis is placed on exceptional occurrences. The selected method is less subject to these shortcomings, and seems to provide the best compromise.

The *mean annual peak*. This alternative measure of local abundance is based either on the mean of the highest single count in each of a number of seasons, or on the mean of the numbers recorded in the same month of successive years, the month being the one in which the species is likely to be most plentiful. It is used primarily for the geese, the migratory swans and the sea ducks, all of which are difficult to count, because of the remoteness of their haunts or because of their day-time dispersal to feed in places well away from water. In most districts the flocks can only be covered effectively by specially planned surveys, undertaken either from the air or by mobile teams on the ground. These intensive searches for individual species are costly in time and money, and are normally restricted to a single occasion each winter, when an effort is made to conduct a census at all the known resorts. A good deal of information on the numbers in other months may also come to hand, through the efforts of individual observers or from other sources, such as the county bird reports. Even so the data generally are less extensive than they are for the ducks, and are better suited to this modified treatment.

The method also allows for the changes in distribution which take place in the course of each season. Most of the geese and swans have traditional arrival points, which they use for a while every autumn; they then move elsewhere to their winter quarters, and in due course move again to the assembly areas from which they eventually leave on their spring migration. Many of the resorts are thus used by large numbers for a month or two only, but may well be of special importance to the species during that period. In view of this, the "mean annual peak" is likely to provide a more realistic measure of the value of a site than the "regular" level, which tends to give a falsely poor impression.

In many parts of Britain the local flocks of geese and swans have increased substantially over the past 20-30 years (or more occasionally have shown a decrease). At sites where the changes are especially marked, the recent annual peaks are presented individually, together with the mean of the annual peaks in each preceding period of given length. If the flocks on several neighbouring resorts are closely interrelated, the results for the individual sites are accompanied by a similar set of figures for the group as a whole, the latter being based on the highest synchronised totals in the seasons under review. Any information on the numbers in other months is included in the text.

The *maximum* population. The figure for this is shown in the tables in brackets after the "regular" figure for each species at each site. It is defined as the largest single count recorded during the period under review, whether or nor it occurred in one of the seasons contributing to the index of "regular" numbers. By itself the figure is of no great value since it often depends on some exceptional circumstances; it nevertheless provides a useful guide to the limitations of a site, and when read in conjunction with the "regular" index, gives a good indication of the frequency with which large numbers occur. If a species is present in strength during most of the winter the "regular" figure amounts to about half the maximum; a ratio of 1:5 reflects a less stable population subject to considerable fluctuation, either each season or from year to year, and a ratio of 1:10 implies that large numbers are present only occasionally. Where the counts are too erratic for a regular figure to be calculated, the maximum only is shown.

For many sites the data are too scanty to permit the use of an index and the maximum figure may even be misleading. In this event a general assessment is included in the text, based on whatever data are available, and on the comments of local observers. The information may also be summarised in tables, the details of which are defined in the individual captions. Most of the sites concerned are either unimportant (which is why no regular observer can be found), or are covered only on the dates prescribed for the international censuses, when a special effort is made to cover as many places as possible.

Throughout Part II, reference is made to "internationally important" concentrations and "the Ramsar Convention". The main criterion for international importance used here is that a site should regularly hold 1% of the individuals in a population of one species or subspecies of waterfowl. For a citation and explanation of the full criteria see p.527. The Convention on Wetlands of International Importance Especially as Waterfowl Habitat was drawn up at Ramsar, Iran, in 1971, with the aim of encouraging the conservation of wetlands on an international scale. The implementation and progress of the Ramsar Convention are discussed on p.537.

The complete Wildfowl Count data to 1981-82 has been analysed and incorporated in Part II. Some updating has been possible up to summer 1984, when the manuscript was submitted.

South-west England and southern Wales

Cornwall

The choice of Land's End as the starting point for this survey has much to commend it: the district is far removed from the east coast with its constant traffic of migrant wildfowl, and few parts of Britain are more self-contained. In short, Cornwall is a cul-de-sac, lying at the end of a flyway and lacking any special quality which might attract large flocks of wildfowl. This account of the scattered resorts serves, therefore, as a prelude, affording some measure of perspective to the more important areas that follow.

The north Cornish coast

Three estuaries only are located along the 160km of coastline between Land's End and Hartland Point, those of the Hayle near St Ives, the Gannel at Newquay and the Camel at Padstow. Elsewhere there is nothing but cliff and occasional stretches of sand and dune, exposed throughout to the full force of the Atlantic gales. The wildfowl are thus confined to the sheltered inlets, the only exception being the Common Scoters which appear offshore in late summer and throughout the winter months. The largest numbers of scoters are usually seen between August and November, when the birds are on passage. Most of the records at this time have come from St Ives Head and Godrevy Point, and refer to flocks of 50-100 moving westward down the coast. On one occasion in November 1959 at least 500 were recorded in the course of a four-hour watch.

The estuary of the Hayle has for many years been kept as a virtual sanctuary, and is seldom

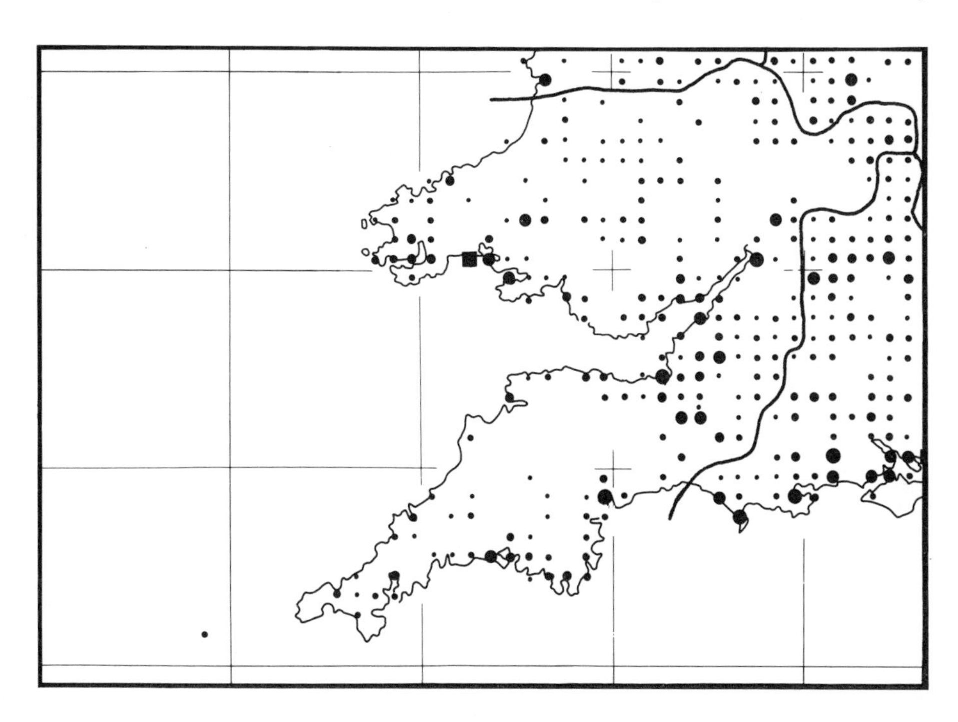

Fig. 6. South-west England and southern Wales. Total wildfowl and regional boundary (see p.339).

disturbed except by bait-diggers and fishermen. The main resort is on the upper basin, a shallow sandy expanse of 71ha with some strips of saltmarsh at the edge. At low water the whole of this is available as a feeding ground, except at one side where a bunded pool retains the tide and provides a permanent roost. Further downstream the tideway is dominated by Hayle town, while towards the mouth the channel cuts starkly through dunes and open beach. In winter the upper basin has for many years supported a steady total of at least 300 ducks between October and February, and a peak of 600 or more in December or January (Table 4). The only change since the counts began in 1959 has been a slow decline in the number of Teal, and a corresponding increase in the flock of Wigeon. Mallard have been recorded in only five of the seasons under review, the largest count being one of ten.

The short and narrow estuary of the River Gannel, although superficially quite attractive, is subject to a good deal of casual disturbance and seldom harbours many birds. In the 8 seasons between 1969 and 1976 the largest counts amounted to 50 Mallard, 4 Teal, 16 Wigeon and 5 Mute Swans, and

Fig. 7. Cornwall and the Isles of Scilly.

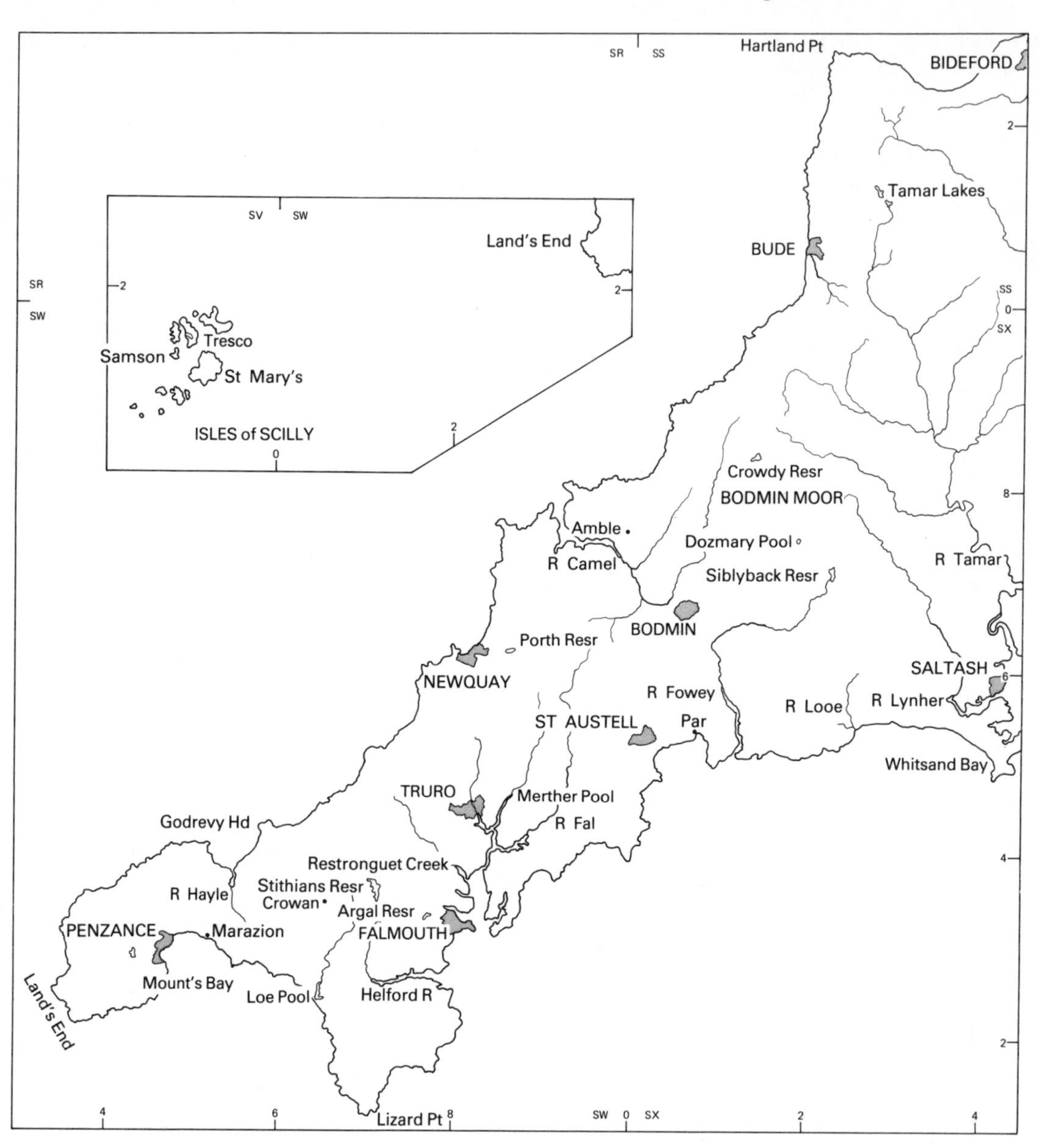

on 27 count dates out of 45 there were no ducks at all.

The Camel Estuary is the largest of the north Cornish inlets, and also the most important, though less so than it was some 20 years ago. The lower half is mostly choked with sand and often much disturbed, but further upstream the more secluded inner basin has extensive mudflats, backed in places by broad belts of saltmarsh. At Trewornan, on the northern bank, the saltings are up to 500m wide, and a further feeding ground is available on the 17ha of the Walmsley Sanctuary, a short way up the Amble Valley.

The Sanctuary, named after its donor, was founded in 1939 and confirmed by Statutory Order in 1948. At that time the valley was subject to frequent flooding, and one of the purposes of the Sanctuary was to encourage the flock of European White-fronted Geese which had recently taken to wintering thereabouts. Prior to 1932 only 20 individuals had been recorded in the course of 13 years, but thereafter the numbers increased to 24 in 1939 and to as many as 200 in 1945. Between 1950 and 1971 the annual peaks varied between 50 and 150, and in 1963, in a spell of hard weather, up to 2,000 were present for a short while. Since then the flock has diminished, and now seldom occurs unless the weather elsewhere in southern England is severe. This rapid decline is attributed partly to the general decrease in the British winter population, and partly to a drainage scheme, completed in 1964, which now prevents the valley flooding.

In the Sanctuary itself the water level is controlled by sluices, and parts of it can be flooded at will to provide a feeding ground for ducks and waders; there is also a small man-made pool which adds to its attractiveness. The counts on the reserve and on the neighbouring stretch of the estuary have been made over different periods, and are summarised separately in Table 4. From general reports it seems that the two sites together are now supporting a normal winter level of about 100 Wigeon, 50 Teal, 200-250 Shelducks and 20-30 Mallard. The current estimates are much the same as those of 20 years ago, but some of the less numerous species have apparently decreased. The small flock of Pintail, which used to reach an annual peak of 40-50 in the early 1960s, is now reduced to less than 10, and the Mute Swans have diminished from 50 or more to about 25.

In addition to the three north Cornish estuaries there are several lakes and reservoirs within 5-10km of the coast. The reservoirs have mostly been built during the past 20 years to meet the demands of the tourist trade, and the birds which use them are undoubtedly new to the district. The waters include Porth Reservoir near Newquay (15ha, flooded 1960), Crowdy Reservoir near Camelford (47ha, flooded 1973), Tamar Lake near Bude (21ha, flooded c.1901) and the adjoining Upper Tamar Reservoir (33ha, flooded 1975). The results of the counts at Porth and Crowdy are shown in Table 4. The most notable feature here is the influx of 1,200 Wigeon which took place at Porth in January 1982 – a prime example of an otherwise minor resort being used as a haven by hard-weather migrants. Both reservoirs are currently

Table 4. *The regular (and in brackets the maximum) numbers of ducks and swans recorded at six sites in north Cornwall.*

a) Hayle Estuary — 23 seasons 1960-1982
b) Camel Estuary, Dinham – Trewornan — 18 seasons 1960-1982
c) Amble Marshes (Walmsley Sanctuary) — 12 seasons 1960-1982
d) Porth Reservoir, near Newquay — 11 seasons 1960-1982
e) Crowdy Reservoir, near Camelford — 4 seasons 1977-1982
f) Tamar Lake and Upper Tamar Reservoir — 8 seasons 1975-1982

	a	b	c	d	e	f
Mallard	0 (10)	3 (21)	45(300)	47 (200)	120(321)	141(340)
Teal	136(450)	15 (97)	97(425)	42 (160)	38(100)	99(325)
Wigeon	447(900)	255(1420)	171(800)	0(1200)	18(110)	98(545)
Pintail	0 (8)	9 (53)	1 (30)	0 (0)	0 (2)	0 (5)
Shoveler	0 (5)	0 (6)	1 (8)	0 (12)	0 (1)	3 (12)
Tufted Duck	0 (8)	0 (15)	0 (0)	28 (90)	2 (6)	42 (90)
Pochard	0 (2)	0 (5)	0 (0)	24 (125)	6 (18)	60(155)
Goldeneye	2 (9)	3 (13)	0 (1)	1 (6)	1 (3)	6 (12)
Shelduck	27 (59)	100 (222)	6(200)	0 (1)	0 (0)	0 (0)
Mute Swan	8 (19)	17 (40)	2 (6)	0 (3)	0 (0)	4 (9)

undisturbed, but their use for recreation is now being considered, and their future value may be much reduced.

The much older reservoir, Tamar Lake, is protected by a Sanctuary Order, dating from 1950, and is seldom disturbed except by fishermen. Most of the pool is fairly shallow, and in places there are marshy areas at the edge which provide both food and cover. In the 15 seasons between 1960 and 1974 the lake on its own held a regular winter level of 30-40 Mallard, Teal and Tufted Ducks, 25-30 Wigeon and Pochard, and 2 or 3 Gadwall, Shoveler, Goldeneyes and Mute Swans. A dozen of the less common species were also reported on odd occasions, mostly in ones or twos. With the flooding of the adjacent upper reservoir in 1975, the joint population has increased to more than twice the earlier level, and totals of above 500 ducks are now recorded in most winters (Table 4).

The marsh at Maer, on the outskirts of Bude, is another site of local interest. In 1981, the only full season for which data are available, it held an autumn level of 60 Teal and 35 Wigeon, the latter increasing to a cold-weather peak of about 230 in the middle of January. Up to 15 Mallard and 6 Shoveler were also present in the early part of the winter.

Southern Cornwall and the Isles of Scilly

The wildfowl resorts along the Channel coast comprise a further series of scattered outposts, more frequent than those in the north, and more varied in character, but still of somewhat limited importance. To the west of the Lizard the coast itself is inhospitable, and the great majority of the ducks are located inland, on the marsh at Marazion, on the Loe Pool near Helston, and on the reservoirs at Drift above Penzance and at Stithians near Redruth. Further to the east the sheltered estuaries of the Helford River, the Fal, the Fowey and the Looe contain a certain amount of foreshore and salting, but here again the neighbouring lakes and reservoirs are often more important. The only major coastal resorts are on the estuaries of the Lynher and Tamar, which together with the Tavy open into the great land-locked harbour of the Hamoaze, astride the Devon border.

The main centre in Scilly is the Great Pool on Tresco, which has long been maintained as a private reserve. Although separated by more than 50km from the nearest mainland resorts, the wintering flocks of Mallard, Shoveler and Pochard are appreciably larger than on any of the Cornish lakes (Table 5). Another feature is the flock of Gadwall, which originated from birds introduced in the 1930s, and which now totals 70-80. Elsewhere in the islands, there is often a flock of 20-30 Pochard on Porthellick Pond, St Mary's, and similar numbers of Shelducks are centred on the Samson Flats. The other species, apart from those listed, are mostly scarce and irregular.

The reservoir at Drift near Penzance and 30ha of fresh marsh at Marazion together hold a winter peak of about 200 ducks, and another 3-400 are often present on the Loe Pool, a short way to the east (Table 5). At Marazion the marsh is now largely overgrown

Table 5. *The regular (and in brackets the maximum) numbers of ducks and swans at six freshwater sites in Scilly and south-west Cornwall.*

a) Great Pool, Tresco — 4 seasons 1969-1974
b) Drift Reservoir, nr Penzance — 11 seasons 1967-1977
c) Marazion Marsh — 21 seasons 1960-1982
d) Loe Pool — 21 seasons 1961-1982
e) Stithians Reservoir — 12 seasons 1965-1982
f) Siblyback Reservoir — 4 seasons 1976-1979

	a	b	c	d	e	f
Mallard	260(400)	8 (35)	17 (77)	117 (600)	49(137)	35 (50)
Gadwall	40 (75)	2 (15)	0 (6)	1 (19)	1 (14)	0 (4)
Teal	45 (85)	26(200)	24(120)	65 (400)	50(140)	15 (30)
Wigeon	4 (28)	18 (90)	3 (46)	47(1000)	88(300)	40(120)
Pintail	0 (0)	0 (5)	0 (14)	0 (4)	0 (3)	0 (0)
Shoveler	21 (62)	2 (9)	11 (44)	10 (56)	1 (9)	0 (1)
Tufted Duck	1 (4)	27 (70)	4 (28)	63 (300)	24 (81)	4 (7)
Pochard	65(170)	35(140)	6(104)	44 (150)	32(131)	15 (30)
Goldeneye	0 (1)	1 (3)	0 (5)	3 (11)	2 (11)	1 (2)
Mute Swan	13 (20)	3 (8)	3 (11)	4 (19)	6 (15)	0 (0)

with reeds and is subject to a fair amount of disturbance from walkers, and the counts over the past 20 years have dropped by nearly half, while those at Drift have shown a corresponding increase. The reservoir, covering 26ha, was flooded in 1961, and by agreement with the Water Authority is now maintained as a Cornwall Naturalists' Trust reserve.

The Loe Pool, with an area of about 60ha, is the largest natural lake in Cornwall, and for many years it, too, has been kept as a virtual sanctuary. At the lower end it is separated from the sea by a shingle bank, and the shoreline there is rather barren, but further inland there are stretches of marsh and reed-bed, lying in a deepset valley and sheltered by hanging woods. In recent years the numbers of several species, including Teal, Wigeon, Pochard and Tufted Duck have been somewhat lower than they were in the 1950s and the early 1960s, but the decrease has been more than offset by the inflow of birds to the new 110ha reservoir at Stithians, some 12km inland. Stithians was flooded in 1965, and within 5 years was attracting a regular mid-winter peak of 4-500 wildfowl, a figure which has since settled back to a stable level of about 250 (Table 5). As in other parts of Cornwall the influxes of hard-weather migrants are sometimes a great deal larger: in January 1982 the Loe Pool and Stithians together held more than 1,100 birds, and in 1963 the Loe Pool alone had 2,500.

Totals of up to 250 ducks have also been recorded on Argal Reservoir (26ha) and the adjoining College Reservoir (15ha) on the outskirts of Falmouth. Occasional counts over the past 15 years suggest a normal winter peak of 40-50 Tufted Ducks (maximum 110), 30-40 Pochard (max 180), and perhaps a few Teal and Wigeon (max 20-30 of each). The majority of these are usually found on the smaller of the two lakes, which dates from 1906, and has a mean depth of only 1.7m; at Argal, which was flooded in 1940 and enlarged in 1961, the mean depth is now 4.7m, but the water level varies and at times there may be quite extensive shallows.

Some of the smaller lowland waters hold a further scattering of birds, amounting in all to perhaps another 200. The Swan Pool in Falmouth has 20-30 Mallard and 30-40 Tufted Ducks; Tory Pond near Stithians has a wintering flock of 20-30 Teal, and Crowan Reservoir near Cambourne at times has up to 20 Pochard. Odd parties of Mallard may also occur on the sea around St Michael's Mount, and 20-30 Common Scoters are regular offshore.

At Falmouth the sweep of the open coast is broken by the broad deep- water harbour of Carrick Roads, and by the complex of creeks and estuaries which join it from either side. The harbour itself holds a notable winter gathering of 40-50 Goldeneyes and 60-70 Mergansers, but is otherwise of little value for wildfowl, being devoid of foreshore and often much disturbed by sailing. The creeks and estuaries are rather more attractive, but even here the numbers of most species are normally quite small. Most of these sites are typical of the drowned valleys which occur at several points along the Channel coast, being narrow and deepset, and full of mud and silt. Disturbance by leisure craft is less severe than in the Roads, especially at low water, but the narrowness of the tideway and the absence of extensive saltmarsh makes them vulnerable to even minor intrusions. In all there are more than a dozen inlets, the largest and most important of which are Restronguet Creek, the Truro and Tresillian Rivers, and the upper estuary of the River Fal from Polgerran up to Ruan Lanihorne (Table 6). Some sizeable flocks are also found on the small freshwater pool near Merther, which was once a side-arm of the Tresillian River, but is now cut off from the tide by a causeway. The estuaries in general are rich in invertebrates, and thus well suited to Shelducks and Teal, but, except for the Fal, they lack the saltmarsh to attract large flocks of Wigeon.

The estuary of the Helford River, a short way to the south of Falmouth, is normally devoid of ducks, being deep and steep-sided, with little or no foreshore except in the side-creeks. It is also much disturbed by the oyster fisheries, especially in the winter months. Further to the east there are regular records of 100-130 Mallard on the freshwater lagoon at Par, of 30-40 Mallard and 20-30 Mute Swans on the estuaries of both the Fowey and the East and West Looe, and of parties of 20-30 Common Scoters at various points along the coast, notably in Falmouth Bay and Whitsand Bay. Whitsand Bay is also used at times by flocks of several hundred Wigeon, presumably as an alternative to the neighbouring centres on the Lynher Estuary and St John's Lake.

The inland waters in this part are much less attractive than those in the north and west of Cornwall, and all but a few are virtually deserted. This applies in particular to the numerous china clay workings around St Austell, where the high turbidity of the water precludes all life. Elsewhere the ground rises steadily to the central spine of granite moorland; the rivers are short and swift, and standing water is scarce except for a scattering of acid upland pools. The only large lakes are Dozmary Pool (16ha) and Siblyback Reservoir (58ha) on the southern slopes of Bodmin Moor. The former is a natural lake, lying in open moorland at an altitude of 270m, and has a normal winter peak of about 40 Mallard and Teal, 20-30 Pochard and 5-10 Wigeon and Tufted Ducks.

The reservoir, lying at 220m, was completed in 1968 and almost from the outset was developed as a leisure centre, with sailing, fishing and canoeing continuing throughout the year. Despite this, it attracts an annual peak of at least 100 ducks, which is possibly as much as the food supply will allow (Table 5).

Devon

Topographically the scene in Devon has a great deal in common with that in Cornwall: the coast is still cliffbound for much of its length, the interior is dominated by a granite dome of moorland, and inland waters are scarce in many districts. The estuaries, however, are much more productive than their Cornish counterparts, and in places the wildfowl flocks can be numbered in thousands rather than hundreds.

The Tamar Estuary and the south Devon coast

After the rather meagre numbers on the south Cornish coast, the improvement at the Devon border is all the more remarkable. First comes the big composite estuary of the Tamar and its tributaries, with a population of above 2,000 ducks, followed in the space of 25km by a further five resorts with a total of 2-3,000 between them; then, beyond Start Point, there are three smaller centres holding 7-800 birds, and beyond that again the estuary of the Exe has a population of at least 5,000, and often 6-7,000, wildfowl.

The main resorts in the Tamar complex are along the Cornish side of the channel, on the muddy inlets of St John's Lake and the Lynher River, which lie only 2km apart and together contain more than 550ha of foreshore. Both areas are used extensively by Wigeon, the flocks moving freely from one to the other. Their numbers have declined, however, from a regular total of about 4,000 in the 1960s to little more than 2,000 since (Table 6). Teal have also decreased from 200 to 120, and the flock of 40-50 Pintail which used to frequent the Lynher has now virtually disappeared. The decline in the Pintail is in keeping with the trend at most of the other resorts in the south-west peninsula, but this scarcely applies to the Wigeon and Teal, whose numbers elsewhere have been generally buoyant. In their case the decline could perhaps result from improvements further east enabling the birds to winter there.

In the upper part of the complex, the two tidal areas of the Tamar and Tavy contain some further expanses of muddy foreshore, backed in places by a fringe of saltmarsh, and at Lopwell on the Tavy a 16ha reservoir has been created by damming the head of the estuary. This freshwater pool, dating from 1957, is less disturbed than the tideway and is often used as an

Table 6. *The regular (and in brackets the maximum) numbers of ducks and Mute Swans on the principal estuaries and lagoons of the south Cornish and south Devon coast (Lizard Point—Teignmouth).*

	Seasons	Mallard	Teal	Wigeon	Shelduck	Mute Swan
FAL						
Restronguet Creek	20:60-79	27 (95)	37(200)	2 (50)	50(150)	15 (40)
Truro River	8:60-73	7 (17)	115(350)	0 (150)	155(390)	11 (25)
Tresillian River	18:60-82	40(150)	25(165)	1 (22)	80(300)	6 (16)
Merther Pool	7:67-73	120(210)	70(150)	40 (120)	4 (85)	0 (0)
Upper Fal Estuary	5:73-81	65(225)	80(280)	280(1010)	85(250)	4 (11)
TAMAR						
St John's L & Lynher	10:73-82	86(412)	117(235)	2066(4200)	310(510)	40 (80)
Tamar	7:70-82	99(250)	9 (37)	11 (27)	83(318)	4 (11)
Tavy & Lopwell Dam	7:70-80	40(135)	70(260)	12 (85)	10 (25)	5 (10)
Plym	10:60-82	105(163)	10(158)	7 (210)	42(150)	3 (17)
SOUTH DEVON						
Yealm & Kitley Pool	9:73-82	210(462)	101(290)	34 (403)	77(145)	12 (38)
Erme	15:60-82	90(280)	79(360)	33 (129)	5 (25)	6 (20)
Avon	17:66-82	48(100)	25(100)	18 (150)	13 (40)	44 (66)
Milton Ley	10:69-81	45(170)	375(500)	55 (450)	0 (2)	0 (4)
Kingsbridge	17:60-82	14 (70)	51(500)	1172(3500)	143(500)	47(110)
Slapton Ley	23:60-82	163(600)	15(294)	49 (400)	0 (6)	17 (80)
Dart	10:68-82	221(417)	9 (60)	28 (155)	38 (95)	21 (32)
Teign	20:60-82	11 (69)	9 (77)	12 (355)	17 (95)	14(110)

alternative roost and feeding ground. In the 1960s the area as a whole held a regular total of 2-300 Wigeon, 150 Mallard and 100 Teal and Shelducks, but here again the Wigeon have decreased and now total only 20-30 (Table 6). The parties of 20-30 White-fronted Geese which used to frequent the Tamar meadows near Weir Quay have also ceased to come. The increased use of power-boats, and the recent draining of Lopwell Dam for several months each winter are among the probable reasons.

The adjoining sector between Landulph and the Tamar Bridge may also hold a few small flocks of ducks. Some occasional counts in the early 1970s suggested a normal winter level of 40-50 Mallard (max 160), 15-20 Teal (max 125) and upwards of 100 Shelducks (max 320), but the numbers since then have possibly decreased. Another small group is centred on the tidal basin of the River Plym above the Laira Bridge. Although flanked to the west by the streets of Plymouth, the eastern side is still largely unspoiled,

Fig. 8. Devon and west Somerset.

and for many years the area has attracted a steady winter peak of 150-200 birds (Table 6).

To the east of the Tamar the drowned valleys of the rivers Yealm, Erme and Avon and the Kingsbridge Estuary are spaced at intervals of 5-10km along the cliffbound coast. All of them are narrow and deepset, and extend inland for 4-5km. At the seaward end the shores are sandy and of little value, but further upstream they have attractive stretches of mudflat and salting, and both the Yealm and the Erme have pools of fresh or brackish water adjacent to the channel. There is also a sizeable tract of fresh marsh at Milton Ley, midway between the Avon and the Kingsbridge. In winter the fields here were often flooded for weeks at a time, and used extensively by Teal and other dabbling ducks, especially in hard weather (Table 6). Flooding is now less frequent following a drainage scheme in 1982 and there are signs that the area has become less favourable.

The Avon Estuary is more sandy than the others, and has suffered considerably from the loss of a former feeding ground. This comprised some 25ha of low-lying pasture, which reverted to saltmarsh during the Second World War and was not reclaimed until 1956. At that time the estuary was holding regular flocks of about 100 Wigeon and 60 Teal, compared with the current levels of 20-25. The main feature now is the flock of Mute Swans, which has recently regained its post-war strength of 50-60 after several years at half that level.

The Kingsbridge Estuary is the largest of the four inlets, and the only one which holds a regular peak of above 1,000 ducks. On the lower reaches around Salcombe the channel is deep and narrow, and much disturbed by sailing, but further upstream the tideway broadens into a shallow muddy basin, with a series of creeks on either side. The disturbance here, although much less severe than at the mouth, has been increasing steadily for many years, and the numbers of wildfowl are now smaller than they were in the 1950s and the early 1960s. The Wigeon, in particular, have decreased from around 2,000 to a current level of about 1,000 in the past 8 years, and Mute Swans from 80 to 50 (Table 6). The Shelducks, on the other hand, have shown a marked increase. Mallard are never much in evidence by day, but at dusk the birds from Slapton Ley nearby may at times flight in to feed in more substantial numbers. The estuary has regular flocks of a wide variety of other wildfowl, including 10-20 Tufted Ducks, Goldeneyes, Red-breasted Mergansers and Brent Geese. The Brent first arrived in 1978, and there are occasional hard-weather records of White-fronted Geese, with up to 250 in January 1982.

The freshwater lagoons and marshes at Slapton Ley are of special interest botanically and physiographically, and have long been maintained as a nature reserve by the Herbert Whitley Trust, and by the Field Studies Council who use the site extensively for education and research. The two pools, one of which is overgrown with *Phragmites*, are separated from the sea by a narrow shingle bar, and stretch for almost 4km along the coast. The total area is about 100ha, a third of which is occupied by reed-beds. The water, which is nowhere more than 3m deep, is rich in nutrients brought down by the streams from the farmland behind, and supports a wide diversity of aquatic plants and animals. In recent winters the Lower Ley has held a steady peak of about 400 ducks, or possibly rather more if allowance is made for the Mallard and Teal concealed in the reeds (Table 6). The only obvious changes since the counts began in 1948 have been an increase in Pochard, and a reduction in the early autumn peak of Mallard, which used to amount to about 500 but has since declined to 150. The numbers of Wigeon have also diminished, a reflection no doubt of the trend on the Kingsbridge Estuary. The appearance of small flocks of Ruddy Ducks is another recent development. These presumably stem from the strong feral population which is now established in Avon and throughout the Midlands. The first record at Slapton Ley was in the cold winter of 1978-79 when 7 were present for a while in February; this was followed by some further reports of ones and twos in the course of the next two seasons, and by an influx of no less than 80 in the cold spell of January 1982. An exceptional gathering of 75 Gadwall was also present at this time.

A short way to the south of Slapton the similar, but very much smaller lagoon at Beesands Ley attracts a regular population of 20-30 Pochard and Tufted Ducks (max 60-70), and 5-10 Mute Swans and Mallard. Parties of up to a dozen Gadwall and 5-6 Shoveler are sometimes recorded as well.

On the next 50km of the coast, from Slapton north to Exmouth, the only resorts of note are the estuaries of the Dart and Teign, and a few small ponds nearby. The tidal reaches of the Dart, although stretching inland for more than 15km, are mostly narrow and steep-sided, and are subject to a good deal of disturbance by pleasure craft, especially on the lower half. The majority of the birds, except for Shelducks and Mute Swans, are thus confined to the more secluded upper stretch above Stoke Gabriel (Table 6). Mallard, totalling at least 150, are present here throughout the winter months, and further flocks of 100 or more have at times been reported on the park lakes in Paignton, a little to the north. Tufted Ducks also occur regularly on the river above the Totnes weir, the numbers ranging from 20-30 to upwards of 100 in some seasons.

The estuary of the Teign has a total area of about 350ha, the greater part of which is dry at low water. On the upper reaches, where most of the ducks are located, the foreshore is muddy with a little saltmarsh, and a sizeable stretch along the southern shore is kept as a sanctuary by the Devon Wildfowlers' Association and the Devon Trust for Nature Conservation (485). Elsewhere the flats are mostly sand or shingle, and ducks are scarce except near the mouth, where the channel is flanked by mussel beds and attracts a regular flock of 8-10 Goldeneyes and about 20 Mergansers.

The numerous small pools and claypits along the valley of the Teign around Kingsteignton may also hold some quite large flocks of birds. A single count on Stover Lake in November 1969 produced a total of 200 Mallard, and in 1971 a run of records from the New Cross ponds nearby showed a winter level of 35 Mallard (max 45) and 15 Tufted Ducks and Pochard (max 20-30). The other pools have not yet been surveyed.

Further inland, along the southern slopes of Dartmoor, there are several upland reservoirs, most of which have neither the food nor the shelter to harbour many birds. The most productive of them are the chain of three at Hennock, near Bovey Tracey (46ha), and the ones at Fernworthy near Chagford (31ha) and at Burrator near Yelverton (61ha) (Table 9). The others, at Meldon near Okehampton and at Venford near Ashburton, are less productive and are virtually deserted, and the same is no doubt true of the Avon Dam and the various natural pools and mires.

The estuary of the Exe and the east Devon coast

The great majority of the wildfowl gatherings described so far have been of purely local interest, and even the largest have had little more than regional significance. In contrast the estuary of the Exe stands out as one of the major winter strongholds in the northern half of Europe. Its special interest lies in the large concentrations of Wigeon and Brent Geese, the numbers of both amounting to upwards of 1% of the north-west European population. The Exe also holds a normal peak of around 1,200 Mallard and 500 Teal, and small but regular flocks of half a dozen other species (Table 7).

The distribution of the flocks over the 1,000ha of foreshore is influenced partly by the variations in the habitat, and partly by the presence of a Bird Sanctuary along the greater part of the western shore. The sanctuary, dating from 1934, includes the whole of the area between high and low water mark, from Turf in the north to the Langstone Rock, some 8km downstream. It also covers the spit of sand and dune at Dawlish Warren, bringing the total area of the sanctuary to above 400ha. The Warren itself is further protected by a Local Nature Reserve, created in 1979. The effect of this has been to draw the birds away from the more heavily developed eastern bank, and to concentrate them westwards of the main channel.

The Mallard and Teal are centred mainly on the upper estuary, where the flats are muddier and the water less saline than further downstream. Their favourite area is the stretch along the western shore between Powderham and Topsham, half inside and half outside the refuge. In the absence of any large expanse of saltmarsh, the feeding grounds are either on the tideline or on the low-lying meadows along the valleys of the Exe and Clyst. The Teal may also resort to the pool and stream in Powderham Park.

The Wigeon, which usually reach a mid-winter peak of at least 5,000, are located for the most part on the lower estuary, either on the Cockle Sand between Exmouth and Lympstone, or in the bight between Starcross and Dawlish Warren. The tideway here is bounded abruptly on either side by the stone pitchings of the railway embankment, but food is plentiful on the beds of *Zostera* and *Enteromorpha*, which cover more than 250ha of the intertidal flats. In the event of disturbance the sea is the nearest and safest retreat, but at times in rough weather the flocks move upstream to the head of the estuary, and feed on the Exminster marshes nearby. Parties of 2-300 and more

Table 7. *The regular (and in brackets the maximum) numbers of ducks and swans on:*

Exe Estuary, east Devon — 23 seasons 1960-1982
Axe Estuary, east Devon — 23 seasons 1960-1982
Taw – Torridge Estuary, north Devon — 9 seasons 1974-1982

	Exe	Axe	Taw – Torridge
Mallard	910 (2014)	11 (65)	208 (515)
Teal	359 (1218)	11 (60)	291 (674)
Wigeon	3599 (7000)	146 (560)	580 (1495)
Pintail	40 (152)	0 (2)	0 (4)
Shoveler	7 (44)	0 (4)	9 (29)
Pochard	3 (50)	0 (1)	2 (16)
Tufted Duck	4 (25)	0 (0)	1 (20)
Goldeneye	16 (90)	0 (0)	2 (9)
Eider	9 (140)	0 (0)	14 (102)
Merganser	57 (108)	0 (4)	6 (16)
Shelduck	215 (640)	9 (75)	268 (502)
Mute Swan	104 (300)	16 (73)	86 (139)

have also been recorded on the Stoke Canon meadows to the north of Exeter.

The other species likewise have their favourite stations, the Pintail on the Cockle Sand, the Goldeneyes on the river near Topsham and the Mergansers on the channel off Starcross. The Mute Swans range more widely, and move from place to place according to the time of year. From April to the end of June they feed predominantly on the young saltmarsh grasses along the Exe and Clyst above Topsham, and on the estuary of the Otter, a short way to the east. They then resort to the *Zostera* beds below Lympstone, where they stay until December. Later on, they move again either to the river near Exeter or to the floods on the Exminster marshes (210). Over the past 30 years their numbers have varied considerably: between 1963 and 1971 they decreased from a regular peak of 2-300 to no more than 90, but have since built up to about 140. The Pintail and Shelducks have also declined, the former from a regular level of 70 to about 35, and the latter from 380 to 240. Apart from these few setbacks, the populations of all species in the estuary have either maintained their earlier levels or tended to increase.

The Brent Geese, in particular, have gone from strength to strength. These belong to the Dark-bellied race, whose total population has shown a sevenfold increase over the past 20 years. In the early 1930s the flock on the Exe had at times amounted to about 300, but thereafter it diminished, and by the early 1950s the winter peaks had dropped to less than 30. A slight improvement then began, and this continued with growing momentum until the late 1970s, by which time the flock had increased to about 2,000 (Table 8). At first the increases here kept pace precisely with the upward trend in the population as a whole; in recent years, however, the estuary has held an increasingly large proportion of both the British and the European total. From the counts elsewhere it is clear that most of the initial expansion of the Brent population was absorbed by the prime resorts in the eastern counties, but that after a time these reached capacity; the surplus birds were then forced to make increasing use of the outlying haunts, such as the Exe, which must hitherto have been much under-used. In common with the other south coast resorts, the Exe is essentially a mid-winter centre, the numbers building up from a few in October to a January peak, followed usually by a rapid decrease. Throughout their stay the geese both roost and feed on the lower estuary, their distribution being governed by the availability of *Zostera*. At present the flocks are not unduly disturbed, but several recent proposals, currently in abeyance, have been a continuing cause for concern. The most serious of these is the plan for a large marina on part of the Cockle Sand, an area which is used extensively as one of the two main feeding grounds. Meanwhile, the increase in windsurfing at Exmouth, where up to 100 surfers have been seen crossing the

Table 8. *The peak numbers of Brent Geese,* Branta b. bernicla, *recorded on the Exe Estuary, 1978-1982, and the means of the annual peaks and the maxima in each of the ten preceding 3-year periods. The two right-hand columns show the percentage of the British and world population occurring at this one resort. The percentages prior to 1978 are the highest recorded in each 3-year period.*

	Mean annual peak		Maximum	Percent of total population	
				British	World
1948-50	45		55	0	0
1951-53	27		29	0	0
1954-56	45		53	0.6	0
1957-59	68		81	0.8	0
1960-62	87		118	0.8	0.5
1963-65	87		101	0.9	0.5
1966-68	126		141	0.9	0.5
1969-71	322		350	1.5	0.9
1972-74	523		650	2.0	1.2
1975-77	1110		1300	2.7	1.2
1978		1575		2.5	1.1
1979		2400		3.2	1.5
1980		1945		2.9	1.3
1981		1700		2.8	1.5
1982		1400		1.5	0.7

Zostera beds at high tide, is a worrying development. The spread of *Spartina* in the bight at Dawlish Warren is another insidious threat.

The Canada Goose is another species whose numbers have greatly increased, both on the estuary and throughout the lower basin of the Exe. The population stemmed originally from a stock of 11 birds released at Shobrooke Park, near Crediton, in 1949. Since then their numbers have increased to around 400, and pairs are now breeding freely on many of the pools and reservoirs in the central and eastern parts of the county. In July and early August the bulk of the birds assemble at Shobrooke, where gatherings of above 300 are recorded regularly; they then tend to split into smaller groups, which drift erratically towards the coast. The peak numbers on the estuary are normally present in December or January, when counts of 2-300 are now an annual occurrence. The feeding grounds are either on the Exminster flood-meadows or more often in Powderham Park, which is less disturbed. A regular winter peak of 50-150 is also reported a little to the east, either on the lake at Bicton or on the nearby marshes of the lower Otter. The reservoirs at Hennock are another regular resort, which has at times held up to 100, and in recent years the estuaries of the Yealm and Erme have begun to attract an autumn flock of 40-50, reflecting an extension of the westward range.

The Eiders and Common Scoters, which occur on the sea off Exmouth and the Warren, are a further point of interest, the flocks being amongst the largest on the Devon – Cornish coast. Over the past 30 years the Eiders have increased from a few birds occasionally in the 1950s to a current peak of 25-30. In 1975 a flock of 100-140 was present for most of the winter, but this was exceptional and shows no sign of recurring. Elsewhere along the coast there are often parties of 5-10 Eiders around Hope's Nose, and in Start Bay and Plymouth Sound. Scoters are present throughout the year, the numbers increasing to a mid-winter peak of 3-400. Although centred off Exmouth, the flock ranges widely and sizeable groups are likely to be found anywhere between Dawlish and Sidmouth, and as far east as Seaton. Flocks of up to 50 are also reported off Hope's Nose, at the northern point of Torbay, and a further group of 100 or more occurs on the sea around Slapton.

The rest of the East Devon coast from Exmouth to Lyme Regis is flanked by cliffs along most of its length, the only inlets being the estuaries of the Otter and Axe, both of which are unimportant. In the seven seasons 1976-1982 the Otter held a regular winter level of 24 Shelducks (maximum 43), 8 Mute Swans (max 19) and a similar number of Mallard and Wigeon. Canada Geese were also present, especially in the last three seasons, when a flock of 30-40 was recorded in most of the winter counts. These are presumably the same birds as those which at other times are found on the nearby lake at Bicton in company with 15-20 Mallard and half a dozen Tufted Ducks.

The estuary of the Axe, which extends inland for about 2km, is rather more attractive. The muddy channel is itself quite narrow, but is flanked, especially on the western side, by a sizeable tract of fresh and brackish marshland, which provides both a feeding ground and a buffer against casual disturbances. The counts of ducks, extending back to 1948, have shown remarkable stability, the peak numbers amounting regularly to about 200 in a normal winter, with an increase to perhaps 400 in the colder spells (Table 7). The pressures on the estuary are at present slight, but there have in the past been proposals for the development of a large marina on the Seaton marshes, and similar threats may well arise in future (485).

Mid and north Devon

The 80km of the north Devon coast from Hartland Point to the Somerset border are cliffbound throughout, except for the stretch around Bideford and Barnstaple where the twin estuaries of the Taw and Torridge combine to provide the only major resort for wildfowl in the northern half of the county. In all recent winters the population here has totalled at least 1,000 ducks, and has at times approached 2,000 (Table 7). The majority of these are normally spread along the channel of the Taw, which is muddy and bordered to the south by some notable stretches of saltmarsh. The shallow sandy basin at the junction of the two rivers also attracts some smaller groups of ducks, and at night the grazing marshes to the north provide a further feeding ground, extending to about 400ha. The sewage farm by the Taw below Barnstaple is another linked resort, which is used at times by up to 100 ducks, mostly Wigeon, Teal and Shelducks. The narrow channel of the Torridge is much less attractive than the Taw, being flanked for most of its length by the waterfronts of Bideford and Appledore on the one side, and by the road and railway on the other. The sand and shingle beaches of the lower estuary are equally unsuitable; although flanked to the north by the Braunton Burrows NNR, the tideway here is much disturbed by sailing, and this will intensify when the marina planned at Appledore is brought into effect. At the mouth of the estuary and at several other points along the coast there are regular groups of up to 20 Eiders and 60-100 Common Scoters, the latter occurring throughout the year. The main resorts, apart from Bideford Bay, are at Baggy Point and Ilfracombe to the north, and at Hartland Point to the west.

The inland waters in the north are mostly too

small and too acid, and often too exposed, to be of any value. Apart from Tamar Lake on the Cornish border (p.31), the only sites of possible interest are the 17ha reservoir at Wistlandpound and the neighbouring lake at Arlington Court, some 10km to the north-east of Barnstaple (Table 9). The reservoir, flooded in 1956, has at times held totals of 150-200 ducks, but the recent peaks have dropped to less than 50. A certain amount of boating takes place, but not enough to cause undue disturbance. The estate at Arlington Court is controlled by the National Trust, and has long been maintained as a private reserve. The pool, although covering less than 3ha, is shallow and sheltered, and the numbers of Mallard and Teal are often rather more than those on the reservoir.

Somerset and Avon

In contrast to the pattern further west, the majority of the wildfowl in Somerset and Avon are located inland, notably on the Mendip reservoirs near Bristol, and in and around the floodplain in the centre of Somerset. A sizeable population is also dispersed over the many minor lakes and reservoirs which occur in nearly all districts. The main resorts on the coast are in Bridgwater Bay and on the stretch around the mouth of the Yeo near Clevedon.

Despite the widespread loss of habitat resulting from the drainage of the Somerset Levels in the 1960s, the area as a whole continues to be one of the major strongholds for wildfowl in the south-west region of England and Wales. This success is attributable partly to the creation of the National Nature Reserve at Bridgwater Bay in 1954, partly to the protection afforded on the numerous private lakes, and more especially to the enlightened planning which has allowed the important populations on the reservoirs to co-exist with a wide range of leisure activities. The Mendip reservoirs, in particular, provide a model solution to the conflict between wildfowl and recreation.

West Somerset and Bridgwater Bay

The westernmost districts of Somerset are dominated by the high ground of Exmoor and by the ridges of the Brendon, Quantock and Blackdown Hills; standing water is scarce and most of the coast is steep and rocky. The only notable resorts are the two small stretches of freshwater marsh near the coast at Porlock and Minehead, and the two upland reservoirs at Wimbleball, near Dulverton, and at Clatworthy, some 6km to the east. Flocks of 50-100 ducks have also been recorded on the lake at Cothelstone in the Vale of Taunton Deane, and on Luxhay Reservoir and the pools around Churchstanton in the Blackdown Hills.

The marsh at Porlock comprises about 100ha of wet pasture, with some freshwater pools at the seaward end and a shingle bank beyond. The data for recent years are scanty, but, except for a decrease in Shoveler, it seems that the present numbers are much the same as they were in the 1950s and early 1960s (Table 10). The Minehead marshes and the adjoining foreshore at Dunster Beach have also retained their earlier value, despite the imposition of the holiday camp, which opened in 1962. This involved not only the outright loss of 60ha of grazing, but also a major revision of the drainage to obviate further flooding. Rather surprisingly the development had no apparent effect on the wildfowl numbers; in fact the present flocks of Wigeon and Teal (Table 10) are substantially larger than they were before the camp was built. The only major loss has been the wintering flock of 50-100

Table 9. *The regular (and in brackets the maximum) numbers of wildfowl at:*

a) Burrator Reservoir — 8 seasons 1972-1982
b) Fernworthy Reservoir — 11 seasons 1969-1982
c) Hennock Reservoirs (3) — 9 seasons 1970-1980
d) Wistlandpound Reservoir — 9 seasons 1972-1982
e) Arlington Court Lake — 7 seasons 1972-1979

	a	b	c	d	e
Mallard	52 (100)	12 (50)	25 (55)	37 (200)	30 (155)
Teal	22 (50)	21 (45)	4 (25)	11 (43)	45 (100)
Tufted Duck	9 (26)	3 (15)	10 (35)	5 (25)	2 (10)
Pochard	8 (18)	3 (30)	25 (80)	16 (54)	2 (15)
Goldeneye	1 (5)	0 (2)	0 (2)	2 (7)	0 (3)
Goosander	5 (16)	1 (5)	0 (2)	0 (2)	0 (0)
Canada Goose	0 (0)	0 (6)	22 (100)	0 (2)	0 (0)

White-fronted Geese which in common with the groups elsewhere in south-west England has now virtually disappeared.

The two upland reservoirs at Clatworthy (53ha) and Wimbleball (150ha) are fairly recent developments, the former dating from 1961 and the latter from 1979. Both of them lie at an altitude of about 230m and have mean depths of above 10m, and both are used for fishing and sailing. Even so, the flocks of several species are sometimes quite large, especially at Wimbleball, where the initial attraction of the newly flooded basin has not yet started to wane (Table 10). The main interest at Clatworthy is the flock of Canada Geese, which started to appear in the late 1960s, having spread, perhaps, from the well-established group at Shobrooke Park, some 30km to the south. A

Fig. 9. Somerset and the Severn Estuary.

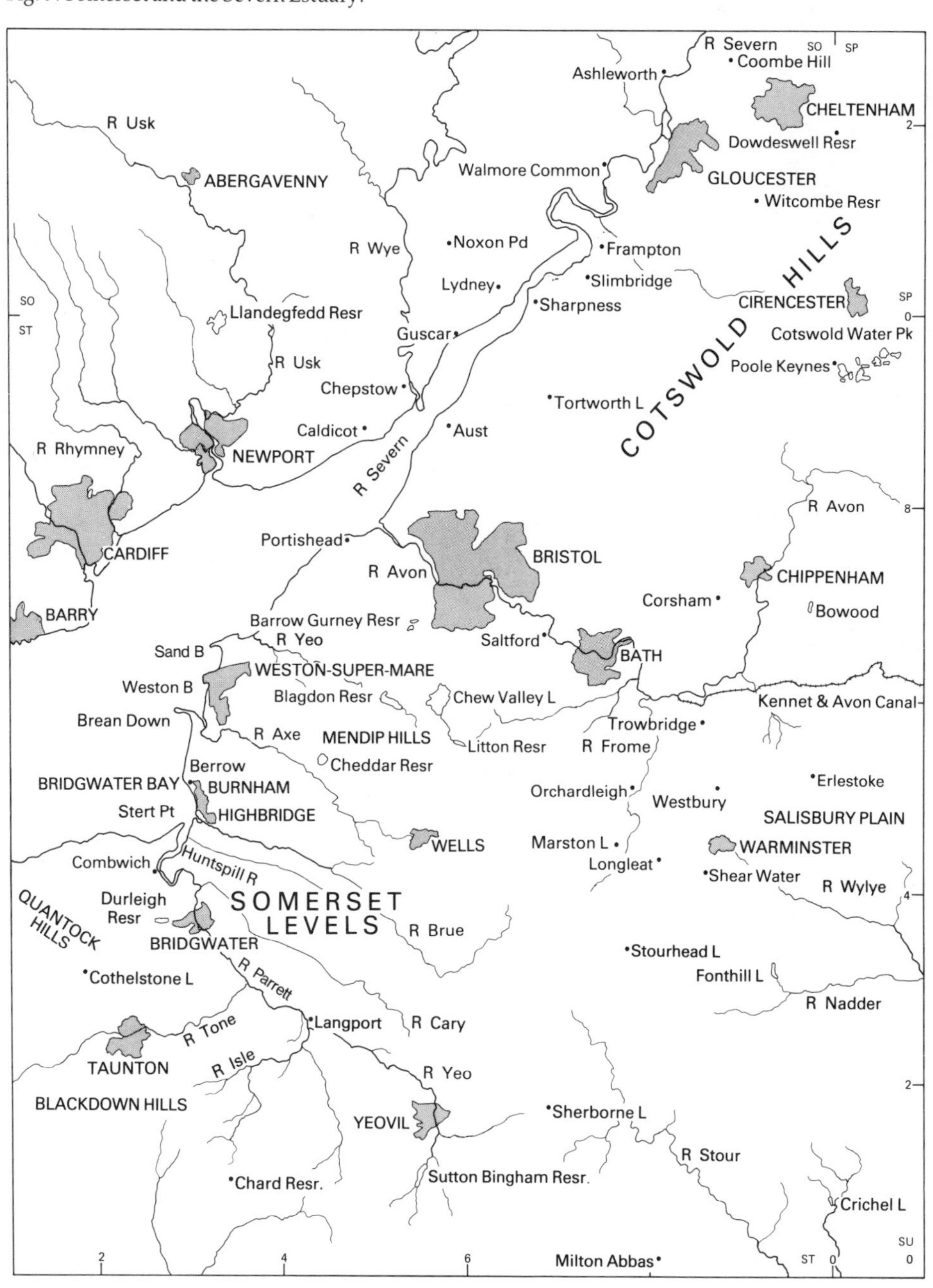

pair first bred in 1969 – the first in Somerset for nearly 70 years – and the flock has since increased to at least 110 (550). The primary centre for Canadas is still at Clatworthy, but in winter the flock often moves to Wimbleball, and sometimes to Minehead. Parties of 10-20 also appear from time to time around the Vale of Taunton Deane, and odd pairs have recently begun to nest at several of the local pools.

To the east of the Quantock Hills the land falls away to the great alluvial plain of the Somerset Levels, and the low-lying coast is flanked by a vast expanse of intertidal mud and sand, varying in width from 2 to 5km. At the southern end, around the Parrett Estuary, the flats support extensive beds of *Spartina*, with smaller stretches of sea-washed turf along the inner edge. Further to the north, from Burnham to Sand Bay, the open beaches are backed by dunes and the saltmarsh is restricted to two small stretches near Berrow and in Weston Bay, around the outfall of the River Axe. The southern, and by far the most important, sector is protected by the Bridgwater Bay NNR, which came into being in 1954, and which has since been enlarged by the acquisition of Fenning Island (a peninsula of saltings no longer a true island) in 1958 and the major extension of the Parrett Estuary agreed with the Crown Estate Commissioners in 1974. It is also included in the British list of important wetlands attaching to the Ramsar Convention (p.538). In its present form the reserve takes in the whole of the intertidal shore from Lilstock in the west to the Parrett channel in the east, including Stert Island, the saltmarsh on Fenning Island, and a strip along the western bank of the river upstream to Combwich. A further strip covers 2.5km of the eastern bank opposite Fenning, and continues inland along the lower 8km of the Huntspill River, bringing the total area to 2,460ha. In some of the outlying parts of the reserve a limited amount of shooting is allowed by permit, but this in no way affects the two main aims, to foster the flocks on the mudflats and saltings, and to protect the flight lines leading inland. Improvements to the bird interest on Fenning Island saltings have been achieved by sensitive management of the sward and by the construction of lagoons; hides have also been provided along the edge of the reserve.

Bridgwater Bay is of special interest for its summer assembly of moulting Shelducks, this being one of the few sites in Britain where such gatherings are known to occur. The major part of the British population migrates, so it seems, to the west German coast to join the huge concentrations of moulting birds at the mouth of the Weser and elsewhere (p.538). The Somerset gathering was first noted in 1950 (466), and was studied intensively in 1959, when frequent counts were made from both ground and air (174). In the course of 20 sorties flown between May and November of that year the observers identified four distinct influxes, the first of which occurred in July and was merely a passage of birds, on their way presumably to the German coast. This was followed by a second influx in early August, when the first moulting birds began to appear. While these were still flightless the third and largest wave arrived, bringing the September total to 3,400. Then in October a fourth small group appeared and started to moult a full month after the others had regained their power of flight. Thereafter the numbers dropped to the normal winter level of less than 250. Further observations in 1960 revealed a

Table 10. *The regular (and maximum) numbers of wildfowl at five sites in west Somerset.*

a) Porlock Marsh — 15 seasons 1950-1967
b) Dunster Beach and Minehead Marsh — 14 seasons 1960-1973
c) Wimbleball Reservoir — 4 seasons 1979-1982
d) Clatworthy Reservoir — 9 seasons 1972-1982
e) Cothelstone Manor Lake — 9 seasons 1972-1982

	a	b	c	d	e
Mallard	85 (250)	105 (240)	162 (291)	35 (120)	25 (100)
Teal	165 (400)	50 (185)	46 (71)	0 (15)	4 (40)
Wigeon	70 (190)	515 (1100)	64 (120)	0 (8)	0 (0)
Shoveler	8 (75)	6 (25)	0 (0)	0 (0)	0 (8)
Pochard	0 (0)	1 (6)	49 (132)	13 (55)	2 (10)
Tufted Duck	0 (0)	1 (10)	25 (43)	7 (30)	8 (26)
Goldeneye	0 (0)	0 (3)	8 (25)	0 (0)	0 (0)
Shelduck	4 (25)	12 (40)	0 (5)	0 (0)	0 (0)
Canada Goose	0 (0)	6 (30)	50 (91)	34 (115)	3 (16)
Mute Swan	0 (2)	9 (16)	0 (0)	0 (0)	0 (2)

maximum of only 2,000 birds, but the timing and relative size of the waves was the same. The influx in August was thought to consist of yearlings and non-breeding birds and the one in September of birds which had bred in the current season. The later peak in October was supposedly made up partly of late breeding birds and partly of the "nannies" left behind in the earlier migration to care for the "creches" of young. The origins of these various groups have not yet been established, but judging from the winter numbers the majority must come from areas outside the Bristol Channel.

Following the hard winter of 1962-63 the summer influxes of Shelducks were greatly reduced, and for the next 10 years or so the peaks were seldom much above 1,000. Since then the numbers have increased to around 2,000 and have at times approached 3,000, though the size of the true moulting flocks is not exactly known. There has also been a marked increase in the autumn and winter population, and in the early summer passage (Table 11). Apart from this the pattern of movement continues as before.

Wigeon also use the reserve in some strength, both as a roost from which to exploit the floods inland, and as a feeding ground, their favourite day-time resorts being the grass marsh on Fenning Island and the flats around Steart Point. At night they move up the Parrett to exploit the wet meadows of the Pawlett Hams and also fly up to 10km inland to marshes bordering the Huntspill River (433). Over the past 20 years the winter peaks have varied from the apparently normal level of 1-2,000, which currently obtains, to above 5,000, sometimes to 10,000 in the seasons between 1967 and 1972 and once, in 1968, peaked at 12,000. Mallard and Teal were likewise abundant at that time, the former reaching an annual peak of 3-4,000, and the latter 700-2,000. In most of the other seasons, before and since, the levels were 1-2,000 and 50-100 respectively. This sudden rise and fall, which also occurred in Pintail and Shoveler, was presumably linked to some extent with the growing curtailment of flooding inland, and with the recent encroachment of *Spartina* on the coastal feeding grounds.

The long-term decline in the local flock of White-fronted Geese is another source of sadness. During the mid 1950s, in the early days of the reserve, some 5-700 were recorded annually from January to March, roosting on the flats around Stert Island, and

Table 11. *The regular (and maximum) numbers of wildfowl on four stretches of the Somerset and Avon coast between Hinckley Point and Sand Point.*

a) Steart (Bridgwater Bay NNR) — 20 seasons 1960-1982
b) Parrett Estuary at Huntspill — 7 seasons 1955-1962
c) Axe Estuary and Weston Bay — 8 seasons 1960-1969
d) Sand Bay — 8 seasons 1955-1962

	a	b	c	d
Mallard	1175 (4000)	95 (360)	110 (285)	135 (290)
Teal	202 (2150)	115 (400)	20 (80)	5 (30)
Wigeon	1363 (12000)	520 (2000)	25 (145)	0 (100)
Pintail	33 (300)	5 (90)	0 (0)	0 (3)
Shoveler	26 (400)	0 (0)	0 (0)	0 (6)
Shelduck	1133* (2900)*	295 (1030)	120 (250)	120 (270)

* Based on winter counts (October—March inclusive) in 12 seasons, 1970-1982; the corresponding figures for 8 seasons, 1960-1968, were 560 (2200). The means of the highest monthly counts of Shelducks at Steart in the 8 years 1955-1962, the 6 years 1963-1968 and the 8 years 1975-1982 were as follows:

	Jan	Feb	Mar	Apr	May	Jun	Jul	Aug	Sept	Oct	Nov	Dec
1955-62	146	364	480	260		755	1840	1760	1930	1010	355	365
1963-68	(50)	(75)	(105)	(120)		545	935	620	915	815	565	125
1975-82	635	475	398	(400)	(515)	(1630)	2149	1394	1803	1541	1399	960

Notes: Figures in brackets are based on short runs of data, i.e. no counts in 2 or more years.
Compiled largely from records in *Somerset Birds*.

flighting inland to feed over a wide area of the flooded Somerset Levels. By 1960 their numbers were already starting to decrease, and by 1970 the annual peaks had dropped to less than 100. Since then their visits have become irregular and brief, and totals of more than 40 are now rare, except in the hardest of winters. This continuing fall, although hastened perhaps by the drainage inland, is symptomatic of the general decline of the species in Britain (p.369), and differs in this respect from the decrease in dabbling ducks.

Despite these recent, disappointing trends the Bridgwater Bay reserve is still a site of international importance, the present flocks of Shelducks amounting at times to as much as 2% of the north-west European population. Its future value will depend on the outcome of the Severn Barrage proposals, the implications of which are discussed on pp.483-5. Almost all the current plans, which are many and varied, are bound to affect the tidal range to a greater or lesser degree, and the Shelducks, being tied to the foreshore, will suffer accordingly. Most of the other species are less vulnerable, and some might even benefit from a change in the present regime.

The records from the coast to the north of the Parrett mouth are far from complete, but it seems that the present numbers are probably much the same as the earlier figures contained in Table 11. Sand Bay and Weston Bay are both dominated by the holiday front of Weston-super-Mare, and the wildfowl are mostly relegated to the small, but more secluded corners near Sand Point and around the outfall of the Axe. The Berrow Flats, which stretch for more than 7km from the Parrett channel northwards to Brean Down, are probably of little value for wildfowl, except perhaps as an alternative site for some of the Shelducks from Steart; small flocks of Mallard may also occur, mostly at the northern end below Brean Down. The Parrett Estuary, from Steart upstream to Pawlett Hams, is a good deal more attractive as a sheltered roost and feeding ground, and has at times been known to hold substantial flocks of Wigeon, Shelducks, Teal and Mallard. Many of these birds are no doubt the same as those recorded on Steart Flats.

The Somerset Levels

To the south and east of Bridgwater Bay the floodplains of the Parrett, the Cary, the Brue and the Axe extend inland for up to 30km, and cover in all some 68,000ha. Between and within these four main basins there are ridges and spurs of slightly higher ground, which restrict and retard the drainage, and form, in effect, a series of shallow saucers. This is especially so in the Parrett basin, which is joined from the south by the tributary streams of the Tone, the Isle and the Yeo, each with its own floodplain (Fig 10). The natural drainage is also hampered by the bank of clay, about 8km wide, which flanks the length of Bridgwater Bay, and which is 2-3m higher than many of the inner Levels.

Prior to 1940 the low-lying "moors" were often flooded for long periods throughout the winter, in places to a depth of several metres. The farming was thus restricted to about 5 months of summer grazing, and even that was sometimes curtailed by flooding in the wetter seasons. Since then the drainage has been progressively improved, and the periods of winter flood are now reduced from months to weeks, or even days. The summer floods have greatly reduced, allowing an extra six weeks of grazing in both spring and autumn.

The first of the recent series of drainage projects was in 1940, when the new Huntspill River was dug to take the water from the southern levels of the Brue directly to the sea instead of through the river. This was followed in the late 1960s by the straightening of the River Axe, by the deepening and widening of the King's Sedgemoor Drain, which carries the Cary in its lower reaches, and by similar work on the Huntspill cut. Large new pumping stations were also installed on the Axe at Hixham in 1972, and in the upper basin of the Parrett at Huish Episcopi in 1963, at Middleney in 1964 and at Long Load in 1977.

These developments, and a number of lesser projects, have succeeded in stabilising the water table to such an extent that several groups of landowners are now undertaking their own supplementary drainage schemes with a view to converting from pasture to arable farming. The areas involved were quite small, but the successful practice is now becoming widespread. Over most of the Levels the beds of peat and alluvium are potentially as fertile as those in the Cambridgeshire Fens, but the present arterial drainage is hardly capable of carrying the extra flow, and the cost of pumping would certainly increase dramatically. Other obstacles are the unsuitability of the drove-ways for wheeled traffic, the small size and fragmented layout of many of the farms, and the possible conflict between arable and dairy farmers (the former want low levels throughout the year, and the latter the ditches well-filled to provide their herds with water and to stop them from straying) (370).

The effect of the present volume of drainage on the winter flocks of wildfowl has been much more pronounced than appears at first sight. A few areas, notably in the north, are now completely dry, but elsewhere the total prevention of flooding is impossible to achieve except at exorbitant cost. After prolonged rain or a rapid snow melt, several of the moors are still liable to heavy flooding, and when this occurs (as in 1981-82) the numbers of ducks are as large as ever (Table 12). The difference lies in the length of their stay. In the 1950s, when flooding was common from January to March, the big concentrations used to assemble as soon as the shooting was finished, and often remained for 6 or 8 weeks. They are now rarely present for as much as a fortnight.

The figures in Table 12 should therefore be treated with caution; they act as a guide to the relative value of some of the main resorts, but cannot be taken as a measure of the changes in usage, although the recent data are much more reliable in this respect. At a guess, the total reduction in usage might be as much as 80% in the past 25 years.

The Bewick's Swan is the only species whose numbers have actively increased during the period of drainage, from a total of 20-30 in the early 1950s to a present level of around 400. As in the Cambridgeshire Fens, the tradition of regular wintering dates from the mid 1950s, when substantial flocks arrived in Britain from the Netherlands during spells of cold weather (390). At that time the Somerset Levels were the only regular centre in the western half of England, pre-dating the neighbouring resorts at Slimbridge and elsewhere by upwards of 10 years. The main area in Somerset is now the southern moors around Langport, although sizeable flocks still visit Tadham Moor, which still contains some moderate stretches of wet pasture. Depending on the state of the floods, the flocks either roost on the moors or retreat at night to the neighbouring reservoirs of Durleigh (to the west)

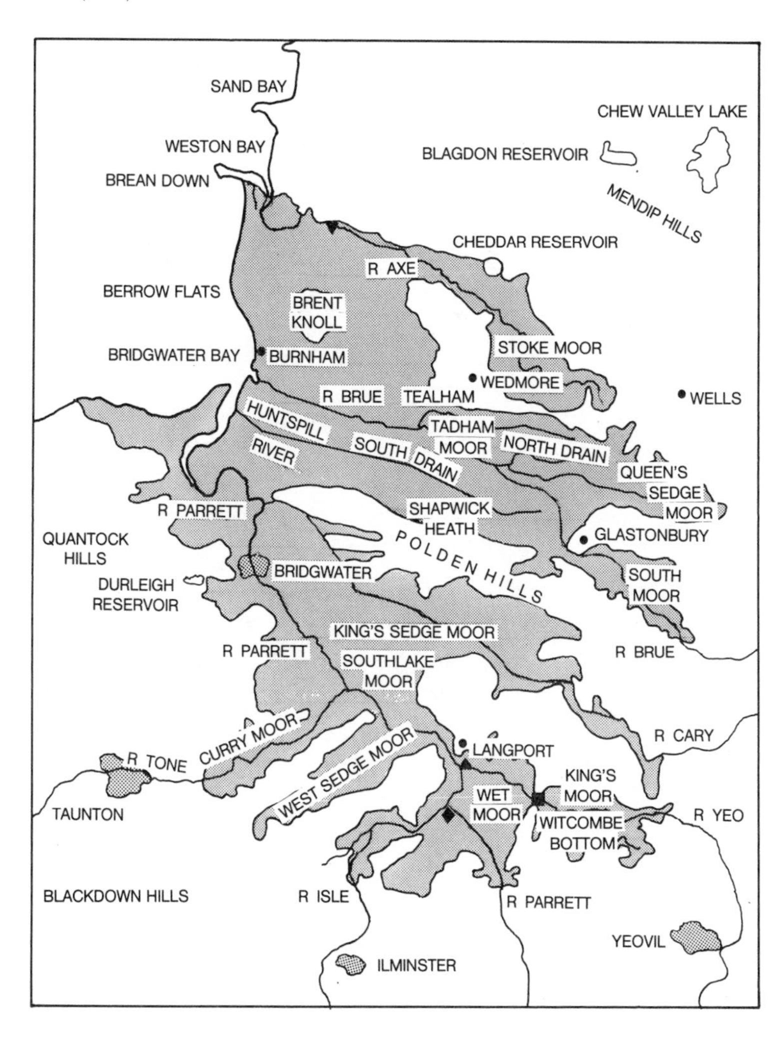

Fig. 10. The Somerset Levels. The shaded area is land below 50m. The major pumping stations are marked as follows: triangle pointing downwards – Hixham (opened 1972); pointing upwards – Huish Episopi (1963); diamond – Middleney (1964); square – Long Load (1977).

Table 12. *The highest counts of six selected species of ducks and swans at various sites in the Somerset Levels in each of seven 4-year periods, 1954-55 to 1981-82, together with 1982-83, when a full survey was undertaken for the RSPB, but when flooding only occurred in early February. The figures in brackets do not include data for one or more of the seasons in which extensive flooding probably occurred; a blank = no information. Most of the records were provided by D.E. Paull, or are drawn from the annual lists in* Somerset Birds. *Data from the RSPB surveys of 1976-77 (Round 1978) and 1982-83 (Murfitt and Chown 1983) are also included.*

	1954-57	58-61	62-65	66-69	70-73	74-77	78-81	82
TEAL								
King's Moor		(0)		500	1000	1100	(400)	89
Witcombe Bottom						1100	1250	150
Wet Moor	200	800	1000	40	500	350	1400	58
West Moor					(300)	(20)	1250	4
West Sedgemoor				(50)		350	3100	150
King's Sedgemoor		(250)				(385)		
Southlake Moor	(50)		(200)	100		(40)	160	10
Curry Moor	200		(100)	200	(40)	120	70	287
Tealham/Tadham Moors	6	0			75	165	150	81
WIGEON								
King's Moor		(60)		1550	4000	2000	(870)	0
Witcombe Bottom						400	1000	70
Wet Moor	1000	5000	3800	300	2200	1000	13000	200
West Moor		(8)	1500			(60)	1300	0
West Sedgemoor					(120)	20	10000	4
Southlake Moor	(1000)		(300)	500		(20)	22000	0
Curry Moor	275		(250)	20	260	(5)	2000	90
Tealham/Tadham Moors	60	0			400	75	1000	80
PINTAIL								
King's Moor		(100)		100	250	145	(135)	0
Wet Moor	3000	1000	130	60	150	200	(40)	0
West Sedgemoor					200	140	40	4
Southlake Moor	(200)		(50)	50		(200)	75	0
Curry Moor	165		(2)	0	(13)	27	20	64
SHOVELER								
King's Moor		(0)		180	450	40	(40)	0
Wet Moor	120	675	60	30	100	100	(150)	9
Southlake Moor	(100)		(0)	150		(100)	20	0
MUTE SWAN								
King's Moor		(105)		95	100	20	40	22
Witcombe Bottom						5	52	15
Wet Moor	265	165	85	85	155	100	60	30
West Sedgemoor						50	45	87
King's Sedgemoor		(23)			(70)	(60)		44
Southlake Moor	(100)		(130)	120		(102)	110	14
Curry Moor	100	50		(4)	(30)	120	45	45
Tealham/Tadham Moors	80	30		(20)	70	40	72	105
BEWICK'S SWAN								
King's Moor		(0)	(11)	208	324	89	(6)	7
Wet Moor	56	88	31	110	358	273	380	120
West Sedgemoor				(37)	(295)	64	82	195
Southlake Moor	(36)		(20)	100		(55)	245	0
Curry Moor	52		(60)	87	72	78	220	99
Tealham/Tadham Moors	52	4	(14)	17	79	73	65	3

or Sutton Bingham (to the south). They also roost on Southlake Moor, which is often flooded intentionally to enrich the fields with silt, and to provide a harvest of ducks.

Hitherto, the Bewick's Swans have not been over-troubled by the changes arising from drainage, and may well continue to find sufficient splashy pasture for their needs. The other species have fared less well, and the present adverse trends show no real sign of abating. Apart from Southlake and Tadham, the only moors which still flood frequently are Wet Moor, King's Moor and Witcombe Bottom (now largely drained) to the south of Langport, and West Sedgemoor a short way to the north. In the Cary Valley the huge expanse of King's Sedgemoor, some 4,000ha in all, has been partially flooded in less than one year in three in the past three decades, and Stoke Moor, in the valley of the Axe, is now completely dry.

The outlook, however, is not wholly black. At West Sedgemoor a block of more than 200ha in the centre of the 1,200ha basin has recently been bought by the RSPB with the aim of re-creating some of the conditions which obtained prior to the drainage schemes of 1942 and 1961. Possible plans include the provision of a permanent pool to replace an earlier mere, and the control of water levels to provide the optimum habitat for breeding and wintering waterfowl. Most of the fields are already surrounded by low banks of spoil from the ditches, and with minor amendments could be flooded at will with water pumped from the drainage channel or rhynes. This could be done on each field separately, without affecting the levels elsewhere, and would open the way to a system of controlled rotational flooding, which has long been envisaged, but rarely put into practice (331).

The widespread extraction of peat in the middle valley of the Brue is another development which may in time be of benefit to wildfowl and other marshland birds. The industry is centred on the two areas of raised bog at Shapwick and Westhay, the former covering a total of about 2,400ha and the latter about 900ha. Planning permission has already been given for extraction on 1,100ha, and well over 600ha is currently being worked, mostly in the neighbourhood of Shapwick. This digging will eventually leave a series of nine pools, some 2-3m deep and lying on a bed of clay or gravel. Although the peat itself is acid, the water draining from the limestone scarp to the east is strongly alkaline, and the pools are likely to be rich in plant and animal life. At least two of them will probably be kept as a nature reserve; another two are ear-marked for water supply, and two more might possibly be used as a temporary holding ground for excess water from the river Brue. The rest would be available for recreation, so long as funds could be found for the building of approach roads and other capital developments. Failing that they might perhaps be added to the reserve, or left to develop by natural succession from open water to reed-swamp and carr. So far as the ducks are concerned the prime requirement would seem to be a large and undisturbed roost from which to exploit whatever feeding grounds remain in the area round about. The long-term value of the pools will thus depend to a great extent on the practicalities of management.

Otherwise, one of the few examples of permanent standing water on the moors is a man-made 8ha pool at Ashmead Wildfowl Reserve, Witcombe Bottom, which, when there is no flooding elsewhere, proves highly attractive to Teal (363).

The need for substantial stretches of water to serve as roosts and alternative feeding grounds is illustrated by the large concentrations of ducks and swans which frequent the lakes and reservoirs adjacent to the Levels. The most notable gatherings are at Durleigh and Sutton Bingham to the west and south of the Levels, and on the Mendip reservoirs to the north. Sherborne Lake and Chard Reservoir are also used, mainly by Mallard (Table 13), and the numerous claypits in the coastal strip may at times attract a scattering of several hundred ducks. The pits include the Local Nature Reserve at Screech Owl Ponds near Bridgwater, now largely overgrown with reeds (Table 13), the pool at Combwich by the Parrett Estuary, the group of 24 small pits at Chilton Trinity nearby, and as many again between Highbridge and Burnham.

Durleigh Reservoir, covering 31ha, is conveniently placed within 10km of the coastal reserve at Steart and 15-20km of the floods along the middle and upper valley of the Parrett. The site is protected by a Sanctuary Order and, being fairly shallow with natural banks, it provides an admirable retreat for birds disturbed from the moors. This is apparent in the sudden influxes which occur in most seasons, giving rise to temporary peaks of 1,500-2,000 wildfowl. Most of the common species, such as Mallard, Teal, Pochard and Tufted Ducks, have increased substantially in recent years, and Pintail and Shoveler still occur regularly, though the peaks are now much smaller than they were in the 1950s. The reservoir is also used by Bewick's Swans as one of their principal roosts. These first started to come in the 1950s, in flocks of up to 20, and have since increased to a normal peak of 50-100, and at times to over 250. The Mute Swans, on the other hand, have shown a marked decline. Prior to 1965 a moulting flock of 60-80 birds was present regularly each summer; their numbers then dropped

Table 13. *The regular (and maximum) numbers of wildfowl at six inland sites in the western and southern surrounds of the Somerset Levels.*

a) Durleigh Reservoir near Bridgwater	11 seasons 1970-1982
b) Hawkridge Reservoir near Bridgwater	6 seasons 1963-1976
c) Screech Owl Ponds near Bridgwater	3 seasons 1971-1973
d) Sutton Bingham Reservoir near Yeovil	13 seasons 1970-1982
e) Sherborne Lake, Dorset	5 seasons 1960-1966
f) Chard Reservoir	11 seasons 1972-1982

	a	b	c	d	e	f
Mallard	526 (2100)	40 (140)	40 (140)	237 (505)	365 (900)	524 (750)
Gadwall	8 (30)	0 (0)	0 (2)	3 (27)	3 (13)	0 (0)
Teal	360 (1800)	4 (30)	37 (60)	195 (696)	145 (200)	8 (61)
Wigeon	70 (350)	0 (0)	0 (0)	617 (2700)	4 (25)	2 (18)
Pintail	27 (135)	0 (0)	0 (0)	26 (140)	0 (0)	0 (0)
Shoveler	20 (98)	0 (0)	7 (13)	87 (305)	25 (90)	6 (21)
Tufted Duck	90 (380)	28 (60)	0 (6)	163 (340)	20 (30)	15 (51)
Pochard	131 (410)	54 (270)	0 (25)	125 (585)	30 (150)	30 (81)
Goldeneye	2 (11)	0 (3)	0 (0)	2 (11)	0 (1)	0 (5)
Goosander	0 (12)	0 (19)	0 (0)	0 (2)	0 (0)	0 (9)
Canada Goose	7 (52)	0 (2)	0 (0)	0 (34)	0 (0)	0 (5)
Mute Swan	4 (14)	0 (5)	3 (4)	10 (26)	10 (22)	1 (9)

Note: The regular and maximum levels in the period 1960-1969 differed as follows from those shown above:
Durleigh (10 seasons): Mallard 265 (710); Teal 260 (800); Wigeon 30 (65); Pintail 2 (15); Shoveler 10 (30); Tufted Duck 45 (120); Pochard 80 (400).
Sutton Bingham (10 seasons): Mallard 380 (690); Wigeon 445 (1840); Tufted Duck 115 (210).
Chard (8 seasons): Mallard 238 (595).

Table 14. *The regular (and maximum) numbers of wildfowl on the Mendip reservoirs. The regular totals are for the four sites combined in the periods 1960-1969 (9 seasons) and 1970-1982 (13 seasons). The final column shows the highest total count in the period 1960-1982; maxima marked* were recorded prior to 1970. Includes data from Avon and Somerset* Bird Reports.

	Cheddar 23 seasons	Blagdon 23 seasons	Chew Valley 23 seasons	Barrow Gurney 22 seasons	Regular totals		Maxima
					60-69	70-82	60-82
Mallard	391 (980)	380(1104)	803(3093)	293(589)	1620	1770	2950
Gadwall	24 (216)	12 (163)	41 (285)	2 (19)	30	100	235
Teal	259(1400)	496(1850)	578(2064)	30(172)	935	1427	3670
Wigeon	72 (580)	214 (560)	596(1325)	3 (55)	995	717	1765*
Pintail	11 (88)	14 (88)	14 (119)	0 (5)	31	36	125*
Shoveler	36 (164)	61 (244)	169 (622)	9 (73)	215	270	920*
Pochard	522(1800)	317(1790)	298(1410)	81(306)	1485	851	3710*
Tufted Duck	124 (495)	277 (770)	246 (700)	93(302)	515	640	990
Goldeneye	11 (45)	14 (38)	17 (72)	3 (14)	23	54	97
Goosander	0 (40)	2 (58)	17 (64)	0 (24)	4	33	120
Smew	0 (4)	1 (6)	1 (8)	0 (1)	2	2	10*
Ruddy Duck	4 (105)	31 (415)	58 (600)	2(160)	7	150	758
Canada Goose	0 (20)	0 (16)	24 (130)	0 (0)	3	43	130
Mute Swan	18 (102)	15 (48)	29 (113)	3 (28)	35	63	155*
Bewick's Swan	2 (31)	8 (98)	10 (141)	0 (8)	17	15	135

to about 40, and remained at that level until 1978, when a further fall began. Since 1980 the highest counts have amounted to only 14. The winter numbers have also decreased, from a former level of about 15 to less than 5 in recent years. This decline, which is noticeable at several other sites around the Levels, has no doubt stemmed from the loss of floods which used to support the birds in winter; it may also have partly resulted in the recent increase on the Chesil Fleet near Weymouth (p.74).

The two small reservoirs at Hawkridge and Ashford, some 5km to the west of Durleigh, are much less favoured. The former, dating from 1962 and covering 13ha, has a mean depth of almost 7m, and is much disturbed by sailing and summer fishing, and at times by sub-aqua training. It is used mainly by Mallard, Tufted Ducks and Pochard, the total amounting usually to about 100 (Table 13). Ashford Reservoir, with an area of only 3ha, holds a smaller total of 40-50.

The reservoir at Sutton Bingham, on the southern fringe of the Levels, was completed in 1956, and covers 57ha to a mean depth of 4.5m. Here again, the site owes much of its importance to its situation, being set within 15km of the floodplain of the Yeo and Parrett to the north, and within 25km of the major wildfowl centres at Chesil Fleet and Weymouth, on the Dorset coast. The T-shaped conformation of the water is another factor which has helped it to maintain its value in the face of considerable recreational pressure; the long, shallow, southern arm is kept free of boating and fishing. Some 1-2,000 Wigeon come to roost when the southern Somerset Levels are flooded, but those species which use the reservoir for feeding as well have been disturbed by illegal evening shooting in recent years; Shoveler have declined from 2-300 to less than 50 since the mid 1970s. Bewick's Swans use Sutton Bingham Reservoir as a roost less often than Durleigh, but have numbered 102.

North Somerset and Avon

The southern environs of Bristol, Britain's second largest city south of Birmingham, have received an injection of wildfowl interest this century parallel to that in the suburbs of London, if on a lesser scale, with the construction of several large new drinking water reservoirs at the foot of the Mendip Hills: Blagdon (178ha) in 1905; Cheddar (95ha) in 1938; and Chew Valley Lake (490ha) in 1953. Table 14 gives the numbers of wildfowl since 1960 at these waters and the much older Barrow Gurney Reservoirs (51ha), and shows that the Bristol Waterworks Company's policy of carefully balancing the varying recreational interests has been successful; the total duck population has maintained a level of 5-7,000 and changes in the numbers of individual species have probably been in response to factors outside the locality. The decrease in Wigeon has coincided with the great reduction in flooding on the northern Somerset Levels, while Pochard have declined – and Teal and Ruddy Ducks increased – in line with the national trends on reservoirs.

Chew Valley Lake, the largest reservoir in Britain until the construction of Grafham Water in 1964, has remained the most important of the Mendip sites for wildfowl. Summer sailing is restricted to 80ha in the northernmost part but nearly 300ha is available for winter (September – April) sailing, introduced in 1967, in the northern and central parts of the reservoir. In common with Blagdon and Barrow Gurney Reservoirs, Chew is heavily fished for trout. The southern arm, around Herriot's Bridge, is kept as a reserve; though this is small it is buffered by the surrounding non-sailing area.

Blagdon lies across the watershed, 2.5km to the west, at the head of the Yeo. Like Chew it has natural banks and extensive shallows, and only its smaller size restricts its capacity for wildfowl. Sailing is prohibited but shooting is undertaken several times each winter. Cheddar Reservoir, lying south of the Mendips, in the Axe Valley, is a round, artificial basin with raised banks and stone pitchings. It is sailed on all year, but its proximity to the Levels as well as to the other major reservoirs enables a large population of ducks to be supported. The three small basins at Barrow Gurney, a few kilometres south-west of Bristol itself, are also artificial and stone-faced throughout, and usually hold many less wildfowl than the bigger waters to the south, although they can sustain quite large numbers for short periods.

It was at Chew Valley Lake that, excepting a few isolated records elsewhere in the early 1950s, North American Ruddy Ducks were first recorded at large in Britain. A young bird was present in 1957, the first year that a substantial number escaped from Slimbridge, and up to five were present on the north Mendip reservoirs the following year. Breeding commenced at Chew in the early 1960s, but has been erratic since, the species having failed to take the foothold that it has in the west Midlands and now elsewhere (pp.171 and 460). In autumn and winter, however, there is a large influx of Ruddy Ducks onto Chew Valley Lake and Blagdon Reservoir; the total

reached 110 in 1974 and continued to rise sharply to a peak of 758 in 1982. Most of the Avon birds apparently arrive direct from their breeding grounds to the north. During the spring moult most of the Ruddy Ducks gather on one or other water before gradually departing north in March (274, 301, 602).

At Chew there is a notable late summer gathering of wildfowl: in recent years up to 1,880 Mallard, 365 Gadwall, 275 Shoveler, 425 Pochard, 700 Tufted Ducks, 110 Mute Swans and 115 Canada Geese have been found in July and August. Mallard, Gadwall (up to 29 pairs), Tufted Duck, Shelduck, Mute Swan and Canada Goose breed at the lake, with Pochard, Shoveler and Ruddy Duck in some years. At Blagdon and Cheddar the summer populations are generally much smaller, but Blagdon is important for moulting Tufted Ducks (maximum 770), while 1,600 Mallard were briefly present at Cheddar in July 1982 (550).

To the east of the Mendip reservoirs, and on either side of the Wiltshire border, there are numerous small pools scattered along the valleys of the Avon and Frome, the more interesting of which are shown in Table 15. The largest numbers of ducks are found on the bigger private waters, some of which are stocked with Mallard, while others provide good feeding conditions for Pochard and Tufted Ducks or Teal. In addition, the park ponds of Bath and Bristol hold several hundred Mallard between them, along with a few Tufted Ducks and Mute Swans.

The coast between Sand Point and the Severn Bridge (Table 16) contains nowhere of the importance of Bridgwater Bay, to the south, or the New Grounds, to the north – indeed the cliffs between Clevedon and Portishead, the industrial waterfront of Avonmouth and the rocks and gravel banks of Severn Beach render those stretches quite unsuitable for wildfowl. There are, however, several narrow belts of saltmarsh: at Woodspring Bay and the Yeo Estuary north to Clevedon; at Portbury and St George's Wharves; and at Northwicke Oaze Aust, immediately south of the Severn Bridge. There is also a large population of Mallard around the Royal Portbury Dock, while Shelduck total 3-400 from Sand Bay northwards

Table 15. *The regular (and maximum) numbers of wildfowl at various sites in east Somerset and Avon, and the adjoining parts of Wiltshire. If full data are not available, the maxima only are quoted; the left-hand column in this case shows the total number of counts to hand. The letter in brackets after the place name is the initial of the county in which it lies.*

	Seasons	Mallard	Teal	Tufted Duck	Pochard	Mute Swan
SOMERSET—WILTSHIRE						
Wincanton – Warminster – Frome						
Stourhead (W)	13:70-82	115 (205)	0 (0)	55(105)	7 (42)	5 (10)
Shear Water (W)	23:60-82	75 (195)	2 (50)	18 (93)	6¾47)	1 (5)
Longleat (W)	23:60-82	99 (350)	0 (5)	65(130)	25(115)	11 (23)
Orchardleigh (S)	5:60-70	55 (200)	7 (33)	40 (65)	45(160)	1 (4)
Mells Park (S)	1:68	85 (105)	35(110)	0 (2)	0 (0)	0 (0)
Marston L (S)	27 counts	(100)	(30)	(21)	(29)	(5)
AVON						
Radstock – Bristol – Bath						
Litton Reservoir	6:75-82	9 (75)	3 (25)	38 (75)	13 (75)	3 (20)
Chew Magna Reservoir	28 counts	(65)	(39)	(40)	(25)	(3)
R Avon: Saltford – Bath	20 counts	(115)	(9)	(15)	(25)	(27)
R Avon & Canal in Bath	14 counts	(110)	(0)	(4)	(1)	(30)
WILTSHIRE						
Frome – Chippenham – Devizes						
Heywood, Trowbridge	2:80-82	56 (100)	0 (4)	7 (10)	4 (7)	0 (0)
Mineholes, Westbury	16:66-82	9 (20)	0 (1)	3 (25)	5 (29)	3 (8)
Edington, Westbury	10:70-80	20 (80)	0 (10)	3 (12)	8 (30)	0 (2)
Erlestoke, Westbury	11:65-80	10 (85)	3 (22)	3 (24)	2 (30)	2 (11)
Drews Pond, Devizes	8:75-82	110 (205)	8 (30)	0 (1)	1 (5)	0 (2)
Bowood L, Calne	23:60-82	424(1600)	8 (40)	8 (30)	7 (38)	0 (3)
Corsham Pk L	23:60-82	87 (235)	2 (14)	14 (52)	20 (74)	2 (12)

between January and April (28); other species are abundant only after hard weather.

The lower Severn and the Warwickshire Avon

The English part of the Severn Valley and its tributaries is largely agricultural, but at the northern edge of this sector lie the massive conurbations and industrial complexes of the west Midlands. Inland, the wildfowl population is small and scattered among the few ornamental lakes and canal feeder reservoirs, though large numbers of Canada Geese, arrivals in the last twenty years, can be found in the northern part of the area.

Very substantial numbers of wildfowl, including internationally important concentrations of several species, are found on the estuarine parts of the Severn downstream of Gloucester. These concentrations have also given rise to satellite sites on temporary floodwaters and marshes round about.

The lower Severn Vale

The major wildfowl haunts in Gloucestershire west of the Cotswolds are all closely associated with the Severn (for the Thames catchment and the Cotswold Water Park see p.132). In the southern part of the county, the Severn has the character of a tidal estuary with sand and mudflats uncovered at low tide, and with feeding areas for wildfowl in the riverside saltmarshes and raised saltings and, inside the sea wall, on pastureland. The tidal Severn encompasses the New Grounds at Slimbridge, a site of international importance for waterfowl. Further north, essentially above the great "S" bend at Arlingham, the Severn flows through a lowland valley where marshes and winter floods are the main attraction for wildfowl.

The Severn Vale in Gloucestershire is notably lacking in open water and, especially in summer, is surprisingly dry away from the immediate vicinity of the river. The few permanent waters are artificial, being either former gravel diggings like Frampton Pools (much the most important site), small lakes in the grounds of country houses and public parks, or reservoirs created by damming small streams flowing down from the Cotswold scarp. The area between the Severn and the Wye (including the Forest of Dean) is particularly lacking in wildfowl sites.

The tidal Severn falls naturally into two basins, the first from the Severn Bridge to the constriction between Lydney on the west bank and Sharpness on the east, and the second from Sharpness to the Arlingham bend. While the two basins have much in common ecologically, they differ sharply in the number of wildfowl they support.

The seaward basin from Aust to Sharpness (a distance of some 18km) has low-lying fields on either side for most of its length. Wildfowl concentrate mainly in the Aylburton – Guscar area on the west bank, and though data from the Berkeley shore opposite are scanty, there must be considerable movements to and fro across the river, which is about 3km wide in most of this basin. However, the figures do not bear comparison with those from the inner basin, either in total number of wildfowl or in range of species (Table 18). Wigeon, Teal, Mallard and Shel-

Table 16. *The regular (and maximum) numbers of wildfowl at five sites on or adjoining the Avon coast between Sand Point and the Severn Bridge.*

a) Yeo Estuary—Clevedon — 17 seasons 1965-1982
b) Portishead—Avonmouth — 2 seasons 1981-1982
c) Avonmouth—Severn Bridge — 7 seasons 1976-1982
d) Avonmouth Sewage Farm — 5 seasons 1978-1982
e) Bristol Docks—Netham Weir — 4 seasons 1978-1981

	a	b	c	d	e
Mallard	35 (135)	854(1275)	80 (158)	42 (64)	165 (220)
Teal	30 (200)	200 (485)	118 (314)	20 (65)	0 (8)
Wigeon	15 (245)	77 (185)	154(1500)	0 (1)	0 (1)
Shoveler	0 (2)	16 (35)	0 (4)	17 (45)	0 (0)
Tufted Duck	0 (6)	0 (2)	0 (3)	2 (8)	0 (15)
Pochard	0 (12)	0 (4)	0 (6)	10 (27)	1 (5)
Shelduck	130 (280)	52 (100)	60 (126)	0 (2)	0 (0)
Mute Swan	1 (7)	7 (12)	0 (2)	0 (7)	40 (55)
Bewick's Swan	0 (15)	5 (12)	2 (19)	0 (0)	0 (0)

duck are the only regular ducks, and their usual numbers scarcely reach three figures, though immediately after the very cold spell of December 1981, 1,000 Wigeon were recorded at Aylburton (214). Yet Aylburton Warth is an extensive area of sheep-grazed, raised saltmarsh, immersed by the highest tides about half a dozen times a year, very similar in character to the Slimbridge "Dumbles". Lydney has its own New Grounds, an area of reclaimed grassland behind the sea wall, which, however, attracts only tiny numbers of Whitefronts in the coldest winters. The reason for the disparity between the two areas must lie in the differing shooting regimes. Slimbridge New Grounds have been relatively undisturbed for many years; wildfowling at Aylburton has always been less controlled, with greater and more constant disturbance, so that there is no sense of sanctuary and little likelihood of wildfowl establishing a tradition of visiting the site.

The inner basin, extending for 9km from Sharpness to Arlingham, has few suitable areas for wildfowl on its west bank, where cliffs fall sharply to the river, and low-lying fields occur only on the Awre Peninsula. The east bank, however, is occupied by the New Grounds of Slimbridge – a crescent-shaped stretch of land covering some 500ha between the Gloucester – Berkeley canal and the Severn. Some of the wildfowl regularly move outside this core area, notably to Frampton Pools just beyond the canal. The New Grounds are of international importance for Bewick's Swans, European White-fronted Geese and Wigeon.

The New Grounds are made up, as their name suggests, of reclaimed saltmarsh, as the sea wall was built progressively closer to the river. The Dumbles, an area of some 65ha of salting, remains, like Aylburton Warth, subject to flooding, and these are the only considerable areas of unreclaimed raised saltmarsh in Gloucestershire (and indeed in the whole of the Severn, apart from an area near the Wye Estuary and Fenning Island in Bridgwater Bay). The fields between the sea wall and the canal are mainly used for cattle and sheep grazing, though a few carry arable crops. The whole of the New Grounds has, for two or three centuries, been managed by the Berkeley estates as a shooting preserve, with infrequent and limited disturbance, which permitted wildfowl to build up a wintering tradition. The headquarters of the Wildfowl Trust was established here in 1946, and its enclosures cover some 40ha, including the Berkeley New Decoy. The shallow pools and captive collection, with its "free" grain supplies, provide an added attraction for certain species of wildfowl. The Trust has always aimed to provide its human visitors with extensive facilities for viewing wildfowl at close range without disturbing them. In recent years, after much research on food preferences of wildfowl, in particular the European Whitefront, the Trust has taken over agricultural control of about half of the New Grounds through a system of leases, and has negotiated non-disturbance agreements in other parts. It has therefore been able to adopt a management policy allowing grazing of the Dumbles and the fields just inside the sea wall by cattle and sheep during summer, but prohibiting winter grazing. As a result, not only is there sufficient winter forage for Whitefronts, Bewick's Swans and Wigeon, but disturbance from shepherding activities is dramatically reduced, and the birds, while still ranging elsewhere, tend to concentrate in a few fields (442).

Another major development of the last twenty years is the build-up in numbers of Bewick's Swans. The first few were attracted into the Trust enclosures in the 1950s by the grain provided for the captive birds; many more came in the colder winters of 1961-62 and 1962-63, and the calls of captive Whistling and Bewick's Swans, deliberately placed in an undisturbed pen, attracted still more (400). The tradition has gradually strengthened, so that several hundred now winter regularly, with higher peaks in cold winters such as 1978-79 and 1981-82 (Table 17). While the Bewick's do take grain, they have tended in recent years to use the pools in Trust enclosures partly as a roost, feeding in the surrounding fields by day or flying further afield to the low marshy land beyond the canal known as the Moors, across the river to Walmore Common or as far as Coombe Hill or Ashleworth (see below).

Many Bewick's Swans have been caught for measuring and ringing (using "darvic" leg rings engraved with a number – letter code legible at up to 200m) in a specially constructed swan trap, since copied at other Wildfowl Trust centres (p.14). The Old Berkeley Decoy at Purton, 3km to the south-west of Slimbridge, was abandoned after the excavation of the canal in the early 19th century, since it was felt that ducks would be loath to fly across the canal! The New Decoy at Slimbridge was constructed on a secluded pond in a wood 1km from the river. During the 19th century, many thousands of ducks were caught for sale on the commercial market, and when the Wildfowl Trust was established catching continued for ringing purposes (p.10).

The main claim to fame of the New Grounds is, however, as a goose haunt. The numbers of European Whitefronts wintering in Britain have decreased in recent years, partly because of the succession of mild winters in the 1960s and 1970s, and partly because of the new goose habitat in the Netherlands, where there have been spectacular increases (p.367). In the cold winters of 1978-79 and 1981-82 peaks of 5,100 and 4,500 were a reminder of old times, though with heavy snow covering most of the New Grounds in both years, the majority of the Whitefronts left the area, some moving to Somerset and Cornwall, though returning rapidly after the thaw. The New Grounds remain, however, by far the most important British haunt of the European Whitefront, generally holding half of the British total (Table 17).

With the excellent viewing facilities, it is hardly surprising that stragglers of every species and subspecies of goose on the British list are recorded, one or two Lesser Whitefronts, Bean, Pinkfeet and Brent being seen almost every year. The flocks of Pink-footed Geese, which used to occur in October and November, are now a thing of the past; during the 1930s there were up to 1,200. Between the late 1940s and early 1960s the number varied between 50 and 130, all these birds being, as shown by ringing recoveries, of Icelandic origin. Nowadays the occasional Pinkfeet which occur among the Whitefronts at Slimbridge are thought to be of continental origin. Feral flocks of Greylag (up to 150) and Canada Geese (up to 110) occur on the New Grounds; both species breed at Frampton Pools and flocks move almost daily between the two sites. While Greylag numbers remain constant throughout the year, there is a small influx of Canadas in winter, from the Midlands to judge by ring sightings.

The Wigeon is the most numerous of the ducks at the New Grounds and has increased considerably in recent years; the regular population was 1,457 (maximum 3,000) in the 1960s and 2,913 (8,000) between 1978 and 1982. It appears that the cold winters of 1978-79 and 1981-82 drove more Wigeon to this part of western Britain, and they seem to have returned in the intervening and subsequent mild winters, taking advantage of the grassland management policy of the Wildfowl Trust. Wigeon are not attracted to the grain provided for collection birds in the enclosures, and neither are Teal which occur in numbers on autumn passage, with several counts of 1,000 or more in September and October. Most other ducks, however, find the free food supplies hard to resist; Mallard, Pintail and Shoveler come in appreciable numbers, and from feral beginnings the Gadwall population has reached in excess of 200 birds. Tufted Ducks and

Table 17. *Numbers of White-fronted Geese and Bewick's Swans at the New Grounds, Slimbridge, Gloucestershire. Data derived from Ogilvie (1969 and 1970), articles in* Wildfowl News *by E.C. Rees, and* Gloucestershire Bird Reports.

Period/ season	Whitefront			Bewick's Swan	
	Maximum	Mean	% of mean British total	Maximum	Mean
1946-49	4200	3600	42.6	0	0
1950-54	5000	3900	45.6	7	2
1955-59	5000	4340	55.9	16	10
1960-64	4500	3780	49.0	56	32
1965-69	7600	6120	59.0	404	273
1970-74	6000	4280	58.3	411	305
1975	2900		71.6	315	
1976	4000		66.6	354	
1977	2700		53.0	291	
1978	5100		53.7	610	
1979	2100		42.0	390	
1980	3000		53.3	412	
1981	4500		65.1	576	
1982	3000		53.3	285	

Pochard are also attracted by the feeding; these two species move back and forth a great deal between Frampton Pools and Slimbridge, but, while the numbers observed at Frampton have remained remarkably stable over the last twenty years (at a regular 100-150 of each), there has been a spectacular increase at Slimbridge, from a handful in the 1960s to a regular (and maximum) 295 (540) Tufted Ducks and 380 (1,020) Pochard from 1970 to 1982.

With the continual movement between Slimbridge and Frampton Pools it is convenient to treat the latter with the New Grounds. The deep water at Frampton often attracts small numbers of diving ducks rarely observed elsewhere in Gloucestershire, such as Goldeneye, Scaup, Long-tailed Duck and the three sawbills. There are also regular Frampton records of Red-crested Pochard and Ruddy Duck of Slimbridge origin, both in winter and as breeding birds; indeed the 1964 nest at Frampton was the first breeding record of Red-crested Pochard in Britain. The oft-quoted record of Scaup nesting at Frampton in 1959 has proved to be due to a misidentification of Tufted Ducks (563). Since then Tufted Ducks have rarely nested regularly at Frampton, probably because of the predation of the ducklings by the numerous pike (*Esox lucius*). They do, however, breed in small numbers within the Wildfowl Trust enclosures, notably in the decoy.

The Severn Vale must in centuries past have provided much damp marshy habitat, ideal for wildfowl. The Severn still floods extensive areas between Gloucester and Tewkesbury in most years, though the waters are usually too deep to provide food for ducks; furthermore the drainage system allows water to run off rapidly, leaving few marshes that hold numbers of wildfowl for any time.

Downstream of Gloucester, Elmore Marsh, once a major winter wildfowl haunt, was drained in the early 1960s and is no longer subject to flooding. Walmore Common, west of the river, is mainly significant as a satellite site for New Grounds Bewick's Swans. Some swans which roost at Slimbridge fly the 11km to Walmore each day to feed; in mild weather they may stay to roost on floodwater at the Common or nearby Rodley. Mallard, Teal and Wigeon are regular in small numbers (Table 18). Up to 150 Bewick's have also occurred on grassland at the Arlingham passage, about 5km north of the New Grounds.

Above Gloucester, the pools and brickpits along the Severn are of little interest for wildfowl, and the two main sites are the low-lying sumps which retain water longest – the Long Pool near Coombe Hill Canal, east of the river, and the Duckeries at Ashleworth on the west (Table 18). There are few data from Tewkesbury Ham or from the Avon Valley north of Tewkesbury, but it does not seem that these areas hold many wildfowl.

It would appear that twenty years ago there was a regular evening flight of ducks coming to feed in the

Table 18. *The regular (and maximum) number of wildfowl at six sites in Gloucestershire.*

a) Severn, Aylburton to Guscar — 7 seasons 1972-1982
b) New Grounds, Slimbridge — 12 seasons 1960-1982
c) Frampton Pools — 19 seasons 1960-1982
d) Walmore Common — 4 seasons 1976-1979
e) Coombe Hill — 11 seasons 1960-1982
f) Ashleworth/Hasfield Ham — 7 seasons 1976-1982

	a	b	c	d	e	f
Mallard	17 (41)	1494(2775)	68(235)	46(250)	37(350)	385 (850)
Teal	16 (60)	527(1700)	75(500)	67(300)	58(350)	545(1015)
Wigeon	114(380)	2255(8000)	1 (41)	76(600)	91(500)	975(3000)
Pintail	0 (4)	202 (657)	2 (42)	5 (30)	5 (90)	13 (100)
Shoveler	0 (11)	86 (240)	14(100)	6 (30)	3 (15)	51 (130)
Gadwall	0 (0)	106 (440)	27(125)	1 (4)	0 (2)	17 (50)
Shelduck	14 (31)	63 (130)	0 (20)	0 (4)	0 (0)	0 (3)
Pochard	0 (0)	258(1020)	123(400)	0 (0)	3 (30)	3 (30)
Tufted Duck	0 (10)	206 (540)	158(400)	0 (5)	3 (23)	1 (9)
Mute Swan	0 (1)	4 (54)	11 (30)	4 (13)	3 (27)	2 (17)
Bewick's Swan	3 (21)	189 (610)	1 (25)	46(140)	10(125)	5 (35)
Greylag Goose	0 (0)	49 (148)	21(120)	0 (0)	0 (0)	0 (0)
Canada Goose	0 (13)	34 (110)	32 (90)	0 (3)	0 (6)	0 (2)

Ashleworth and Coombe Hill area from the estuary, 20km away. Nowadays this traditional evening flight seems to have died out, and there is a permanent nucleus of wintering ducks based at Ashleworth. This site and Coombe Hill are complementary, and only 5km apart, but their relative importance has varied considerably in recent years. Until the end of the 1960s, the main concentration of ducks was at the Long Pool and on floodwater alongside the old canal at Coombe Hill; during this period Ashleworth and Hasfield Hams were regularly shot over and the ensuing disturbance prevented any concentration of ducks building up. From 1968 onwards, however, the Duckeries were acquired by a private owner who invited the Gloucestershire Trust for Nature Conservation to manage the area as Ashleworth Ham Nature Reserve (41ha), with shooting rights over neighbouring fields at Hasfield Ham. As a result, the numbers of surface-feeding ducks have steadily increased at Ashleworth and Hasfield; the numbers of Wigeon and Teal have been particularly impressive, Wigeon often exceeding 1,000 and Teal reaching this figure several times in early spring. At Coombe Hill, on the other hand, increased shooting pressure and consequent disturbance has led to a relative decrease in importance. Both sites attract small numbers of Bewick's Swans from Slimbridge, but the Whitefronts, once recorded regularly, are now only occasional visitors.

The other permanent waters for which count data are available each hold at most 100 Mallard, with a few Pochard and Tufted Ducks and occasional Mute Swans, other wildfowl being something of an event. Dowdeswell Reservoir near Cheltenham (9ha) and Witcombe Reservoir near Gloucester (16ha) are managed as reserves by the Gloucestershire Trust for Nature Conservation. No count data are available for the two small waters in the Forest of Dean, Noxon and Cannop Ponds. Noxon has been remarkable, however, as a regular haunt, one of the most southerly in Britain, of small numbers (maximum fourteen) of Whooper Swans, which may move to Cannop or Walmore Common if Noxon freezes, but very rarely venture across the river to join the Bewick's at the New Grounds (563). In recent winters, disturbance from recreational interests has apparently deterred the Whoopers from visiting Noxon.

The two major wildfowl habitats in the lower Severn Vale, the tidal estuary and the inland marshes, are both under threat, in the medium to long term, of major ecological change. If the Severn barrage is ever built (p.483), what is at present the Gloucestershire Severn could have a major change in tidal regime. The effect on waders and birds totally dependent on mudflats would be more severe than on most wildfowl; even so, many wildfowl would be forced to adapt to this fundamental change in ecological conditions.

Change in the drainage regime is also the major threat to the few remaining inland marshes. While there is little prospect of eliminating winter flooding completely north of Gloucester, there are plans to reduce the incidence and duration of flooding. Changes to drainage systems will tempt more and more farmers to grow arable crops (and to claim damages when the inevitable flood happens!). Plans exist to drain Walmore Common, while the fields around Coombe Hill may well become drier as the result of a drainage and floodwater relief scheme. At Ashleworth, improvements in the runoff of water from agricultural land were proposed, but after a Public Enquiry it was decided that the nature reserve should be exempted from this scheme and water supplies guaranteed.

The Warwickshire Avon, middle Severn and Teme Basins

The tract of country drained by these three river systems is not of much interest for wildfowl, though there is an occasional small passage of White-fronted Geese and Bewick's Swans on their way to more westerly haunts. The gently rolling countryside is intensively cultivated and includes the fruit-growing Vale of Evesham. There is a distinct lack of standing water, which would have been remedied by proposals, in the 1960s, for a bunded reservoir in the Upton-on-Severn area. This is now unlikely to materialise.

The Warwickshire Avon

In Stratford itself the Avon used to be graced by a large flock of Mute Swans, but this has virtually disappeared, due to lead poisoning from fishing weights (p.490). At Bancroft Bank, on the southern outskirts of the town, up to 353 Mallard have been counted, although data are only available for one season. The pool at Wootton Wawen, 10km to the north, has a good variety of wildfowl but in small numbers, and two dammed stretches of the river, Charlecote Park Lake with regular (and maximum) counts of 181 (300) Mallard and 14 (26) Canada Geese, and Warwick Castle Park Lake with 99 (375) Mallard, 29 (110) Teal, 15 (83) Wigeon (the only regular flock in the area), 16 (54) Pochard and 26 (97) Tufted Ducks, are rather more important, although both Mallard and Teal have decreased at Warwick since the 1960s. To the east of this stretch of the Avon, the few existing waters carry little, except for Chesterton Pools, which have been enlarged by the excavation of a further half-dozen trout lakes, and now hold 247 (600) Mallard, 37 (135) Pochard, 36 (81) Tufted Ducks and 98 (253) Canada Geese. Ufton Fields is a Local Nature Reserve, but carries only a few ducks, including up to 14 Mallard, 10 Pochard and 17 Tufted Ducks.

The Leam tributary, flowing from the east, has, at Leamington Spa, where it has been impounded to form a lake, 182 (237) Mallard, 10 (27) Tufted Ducks and 22 (30) Canada Geese, but the Grand Union Canal nearby holds only a handful of Mallard and Mute Swans. The small Leamington Reservoir has 10 (22) Mallard, 12 (57) Pochard, 14 (84) Tufted Ducks and 17 (182) Canada Geese, but the much larger (300ha) Draycote Water Reservoir further upstream carries a much more substantial population (Table 19). Besides the seven species listed, 13 other wildfowl have been reported, including, most exceptionally for this area, 34 (75) Goldeneye and 9 (30) Gadwall. The reservoir was first flooded in 1968 and in the following two seasons had "first flushes" of up to 1,000 Mallard, 1,500 Teal, 1,000 Wigeon, 27 Shoveler, 500 Pochard and 175 Tufted Ducks. The dabbling ducks have since declined in the usual way for new reservoirs. Draycote has also become heavily used for fishing and sailing; with little natural shoreline and no islands, size alone is its major asset from the point of view of wildfowl. A large (mostly Black-headed) gull roost, exceeding 100,000 birds, is present in winter. The reservoir is important in a regional context but further increase in recreation, if allowed, would almost certainly erode its value. The small canal reservoir at Napton, to the south, has a regular population of under 50 wildfowl, mainly Pochard and Tufted Ducks.

Around the Avon itself, north of Warwick and south of Coventry, there is a group of four waters, three of which are of value. The complex of gravel pits at Ryton-on-Dunsmore holds regular (and maximum) numbers of 45 (58) Mallard, 23 (38) Pochard, 56 (64) Tufted Ducks and 40 (56) Canada Geese. Brandon Grounds (or Brandon Marsh), just east of the outskirts of Coventry, is a varied site where a mining subsidence has been augmented by gravel pits and the whole complex is maintained as a reserve by the Warwickshire Conservation Trust. Teal are the most important wildfowl (Table 19), but there are many other ornithological interests and a varied breeding bird population. A little to the north, on the tributary Sowe, is the Coombe Country Park. This has a large lake with a good variety of wildfowl (Table 19),

Fig. 11. The middle Severn and the Warwickshire Avon.

Shoveler being particularly notable, and reaching 180 in the 1960s. Another feature of interest is the large heronry on the main island.

Further up the Avon, nearing its origin in Northamptonshire and lying just west of the watershed to the Welland Basin, are few, rather small, waters. The small lake in the grounds of Stanford Park holds only 25-50 Mallard and Tufted Ducks and just over a hundred Canada Geese. Stanford Reservoir, 2km north-east, is very much larger at 68ha (Table 19), and is of importance. Flooded in 1935, it has a long history of careful documentation by Rugby School. Most species of ducks have increased in recent years; 10 seasons of counts in the 1960s gave 349 (800) Mallard, 19 (107) Teal, 71 (255) Wigeon, 20 (175) Shoveler, 21 (69) Pochard and 17 (49) Tufted Ducks. Canada Geese were rare then, but there were 12 (44) Goosanders, diminishing to 6 (58) in 1970-1982. In all, 24 wildfowl species have been reported. There is a little coarse fishing, but other sports are excluded (and concentrated on Draycote Water). The Rugby Joint Water Board (now amalgamated into the Severn Trent Water Authority) have dedicated Stanford Reservoir for wildfowl conservation and protected it with a by-law under the 1968 Countryside Act. The Northamptonshire and Leicestershire Naturalists' Trusts have established a joint nature reserve on the reservoir in the Blower's Lodge Bay.

Finally the two canal reservoirs of Sulby and Naseby have smaller numbers of wildfowl; the combined regular (and maximum) numbers for the four seasons, 1979-1982, for which we have good data, are 81 (310) Mallard, 19 (40) Teal, 34 (100) Wigeon, 38 (80) Pochard and 48 (84) Tufted Ducks. Both waters have also had a good sprinkling of other species, notably Shoveler, Mute Swans and Canada Geese.

The middle Severn

Taking the division between middle and upper Severn as the line from Wenlock Edge to the Wrekin, around Ironbridge, the middle river has little to offer. Of the few waters between Tewkesbury and Worcester for which we have data, only the Beckford gravel pits, with up to 300 Mallard and 35 Canada Geese, are of any note.

Between Worcester and Stourport, the small pools at Witley Court to the west held up to 150 Mallard and a few other ducks in the 1960s, but there are no recent counts. The Great Pool at Westwood Park, 10 km to the east and just west of Droitwich, is more notable with 191 (450) Mallard, 14 (100) Teal, 12 (40) Shoveler, 85 (200) Pochard, 30 (80) Tufted Ducks and 18 (108) Canada Geese. This site lies alongside the tributary Salwarpe which, as that name implies, has a higher than usual salinity, derived from salt deposits in its upper reaches. Near Upton Warren there are several mining depressions which are distinctly saline, together with two freshwater pools, one a gravel pit, an area now managed as a Worcestershire Nature Conservation Trust reserve. Fluctuations in the flash pools expose tracts of mud which are attractive to waders, 30 species of which have been recorded. While 19 species of wildfowl have been counted, only seven (Table 19) are present in substantial numbers. With up to 18 Ruddy Ducks here and 16 at Westwood we are in the Midlands breeding stronghold of that newly established species, though few remain here to winter.

Table19. *Regular (and maximum) numbers of wildfowl recorded at sites in the Worcestershire Avon (a – d) and middle Severn (e,f) basins.*

a) Draycote Water, Rugby 7 seasons 1970-1982
b) Brandon Grounds, Coventry 5 seasons 1971-1982
c) Coombe Pool, Coventry 5 seasons 1971-1982
d) Stanford Reservoir, S Kilworth 13 seasons 1970-1982
e) Upton Warren, Droitwich 8 seasons 1970-1982
f) Bittell Reservoirs, Bromsgrove 13 seasons 1970-1982

	a	b	c	d	e	f
Mallard	666(1600)	47(150)	422(840)	693(1750)	165(290)	207(375)
Teal	48 (400)	155(300)	7 (50)	41 (406)	47(100)	86(265)
Wigeon	130 (525)	24(100)	17(100)	61 (240)	3 (14)	15 (67)
Shoveler	2 (10)	19 (60)	45 (95)	99 (480)	48 (85)	6 (26)
Pochard	207(1200)	26 (71)	49(146)	92 (246)	67(210)	32(158)
Tufted Duck	398(2000)	44(106)	35 (80)	186 (900)	27 (62)	107(200)
Canada Goose	0 (2)	21(245)	55(237)	55 (144)	28 (62)	125(330)

A major waterway connection between this section and that of the upper Trent is provided by the Worcester/Birmingham Canal. This roughly parallels the Salwarpe and then continues north-east. Just beyond Bromsgrove and barely 10km from Upton Warren are the Bittell Reservoirs (Table 19), very close to the Longbridge car plant. The reservoirs together cover 63ha and are surrounded by rough pasture providing grazing for the Canada Geese and Wigeon (240).

At Stourport the Stour enters the Severn from the north. Up its length and mainly to the east there are several small waters, of which only Captains Pool, Kidderminster, with 61 (150) Mallard, 14 (43) Pochard and 36 (75) Canada Geese is of any importance. Hurcott Pool, also on the outskirts of Kidderminster, has 20-50 Mallard and Tufted Ducks.

On the Severn itself, above its junction with the Stour, there are no areas of importance for wildfowl. It is not until a few kilometres short of Bridgnorth that one finds two places of interest. Chelmarsh Reservoir on the west bank holds 174 (342) Mallard, 13 (37) Pochard, 28 (56) Tufted Ducks and 88 (187) Canada Geese, while Dudmaston Pools support very similar numbers of those species. Both sites also have a good variety of other ducks in small numbers. Being barely a kilometre apart there will clearly be much exchange between the two waters and with Corner Wood Lake nearby, which has a record of 94 Tufted Ducks. To the north of Bridgnorth there is only the Bog at Cranmere, which can hold 35 Mallard and Canada Geese; Willey Estate lake with 82 (190) Mallard, 16 (32) Tufted Ducks and 13 (48) Canada Geese; and Trench Pool, Telford, with 31 (88) Pochard and 22 (69) Tufted Ducks. There are areas of floodlands around the Severn near Bredon and Upton-on-Severn and several other sites which perhaps carry substantial numbers of wildfowl but precise data are lacking.

The Teme Basin

The Teme enters the Severn from the west, below Worcester. Rising near Newtown, Montgomery, it is joined from the north by a series of tributaries and together they drain a large area of southern Shropshire. A few sections have some floodwater and hold numbers of wildfowl. The section of the Teme between Ludlow and Knighton, including the wet pastures near Wigmore, Leintwardine and Brampton Bryan, has up to 500 Mallard and 470 Canada Geese. Apart from a scattering of other dabbling ducks and Tufted Ducks, and a few Goosanders and Mute Swans, the remainder of the upper Teme Valley is of little interest to wildfowl.

On the lower reaches, Stanford Court Lakes held regular (and maximum) numbers of 108 (220) Mallard, 18 (40) Pochard and 34 (70) Tufted Ducks between 1960 and 1972, but more recent counts are lacking. There is nothing of note then until above Ludlow where Steadvallets Lakes held 210 (450) Mallard and 12 (15) Teal and a handful of Tufted Ducks in the 1960s. Near Bishop's Castle, the lakes at Walcott Hall (once the home of a famous collection of exotic waterfowl) counted in the 1980s, gave only tens of Mallard, Tufted Ducks and Mute Swans, but the Canada Geese at 219 (500) are numerous as they are in other parts of the west Midlands.

South Wales

This sector includes the industrial areas of South Wales and the valleys of the Usk and Wye (including the old county of Herefordshire). The coastline is of considerable interest to wildfowl as are those lowland reservoirs which are protected from excessive recreation. In the uplands, however, wildfowl populations occur in isolated groups and many of the reservoirs, fed from acid and wooded uplands, are of little interest to wildfowl.

The coast and coalfield

The 130km of coast between Chepstow and Mumbles Head is divided into four distinct zones: the mudflats, saltmarshes and grassy levels of the Gwent shore, the cliffs between Cardiff and Bridgend, the dunes and marshes of the east shore of Swansea Bay, and the now fully developed surrounds of Swansea itself.

The Welsh shore of the Severn Estuary, running the length of Gwent (formerly Monmouthshire) and extending to Lavernock Point, just beyond Cardiff, differs in several ways from the English side (pp.51-4). The mudflats, saltmarsh and grazed marshes are more extensive, and there are no cliffs or headlands equivalent to those which interrupt the coast of south Avon. The wildfowl (Table 20) are more diverse but not apparently more numerous (though they are probably always underestimated because of counting difficulties), and at no point approach the size of the flocks at Slimbridge. The most important areas are the vicinities of Collister Pill, near Undy, Peterstone Wentlooge, and Rhymney Estuary and Great Wharf, between the mouths of the Usk and Rhymney. The latter stretch includes a long, though *Spartina*-covered strip of saltmarsh and a particularly intensive system of ditches behind the sea wall. The largest variety of ducks in the Severn Estuary is found here, including frequent small groups of Scoters offshore and numerous Shoveler on the shoreline, but no species reaches nationally significant numbers. In recent years several hundred Pochard and Tufted Ducks have occurred, having apparently established a tradition of wintering on the estuary following hard-weather movements from inland waters in early 1979 (see also p.62). Since 1979 the shooting rights at Peterstone Wentlooge have been jointly controlled by the Gwent Trust for Nature Conservation and the Caerphilly Gun Club, and a small but crucial no-shooting area is maintained around Peterstone Pill. Caldicot Moor, on Caldicot Level, was once important as a feeding, roosting and breeding site, but its value was greatly reduced by drainage in 1966. The Nedern Brook area a few kilometres east is, however, still prone to flooding and

Fig. 12. South Wales.

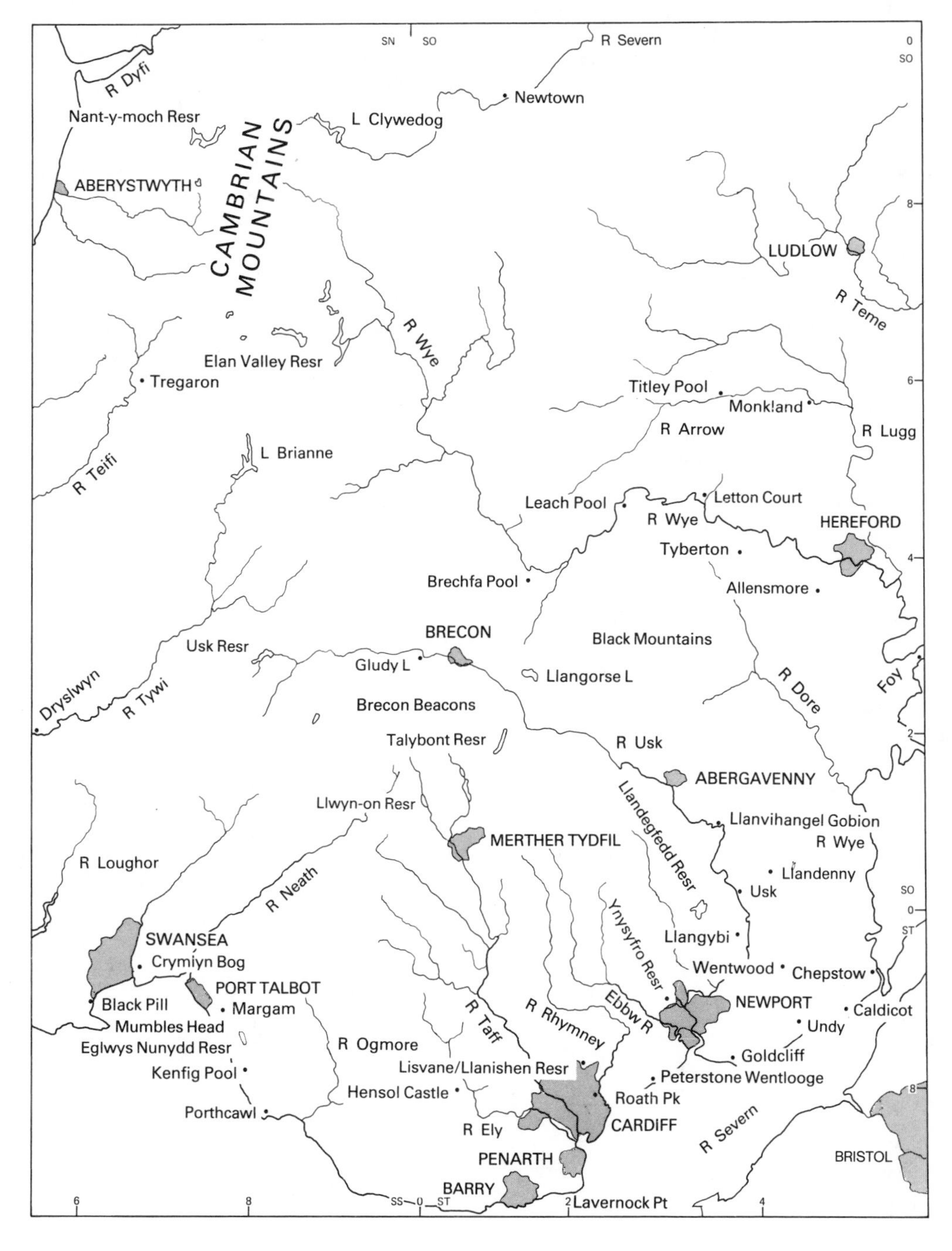

small numbers of White-fronted Geese may appear, while up to 130 Teal, 20 Shoveler and 70 Bewick's Swans have been recorded in recent years. The Usk Estuary is of little importance, its shores having been long since reclaimed for industry, but the ash ponds of Uskmouth Power Station and the saltmarshes to the east have both, in the past, attracted good numbers of wildfowl, and Newport Docks have held up to 21 Pochard. On the confluent estuaries of the Taff and Ely, precariously sandwiched between Cardiff and Penarth, industrial encroachment has gradually reduced the feeding ground available for ducks, but there are still areas secluded enough for roosting, and Pochard occur when inland waters are frozen.

From Lavernock Point to the Ogmore Estuary, the 40km of the Glamorgan coast are cliffbound and the numerous tiny bays are of no importance to wildfowl. The timber pools at Cadoxton, close to the shore east of Barry, however, attract one of the largest gatherings of Mute Swans in South Wales, reaching 42 in November 1972. Bewick's Swans are regular (maximum 10) and Whoopers occasionally visit. Ducks are few; the presence of 61 Pochard, 23 Tufted Ducks and 6 Goldeneyes in January 1970 was apparently exceptional. The disused quarries at Cosmeston, just south of Penarth, are managed as a Country Park by South Glamorgan County Council. Public access is unrestricted and recreational pressure is probably responsible for the lower than expected wildfowl numbers, 30 Mallard, 12 Pochard and 7 Mute Swans being present in October 1982. At the small boating lake at Cold Knap, Barry, 33 Mute Swans were found in January 1970. The reclamation of the Thaw Estuary for a power station and the diversion of the river in 1957 removed the only interesting site for wildfowl on the remainder of the coast, until the Ogmore Estuary. The latter is of some interest, especially in hard weather. Recent peak counts have included 104 Mallard, 42 Teal, 80 Wigeon, 48 Goldeneyes and 23 Canada Geese.

West of the Ogmore Estuary the cliffs give way to 45km of sand and dune as far as Mumbles Head. However, most of the stretch is now built over. The area from Margam Moors (once a major centre for ducks and White-fronted Geese) to the mouth of the Tawe at Swansea is occupied by a great array of industries, including the vast steel works and the town of Port Talbot. Westwards to Mumbles Head the land is completely residential. Between the steel works and the thriving holiday resort of Porthcawl, however, the coast is largely unspoilt.

The dunes of Kenfig Burrows became a Local

Table 20. *The regular (where sufficient data are available) and maximum numbers of wildfowl on the following sectors of the Welsh Severn, 1970-1981:*

a) Wye Mouth
b) Collister Pill
c) Goldcliffe
d) Uskmouth Power Station Ponds
e) St Bride's Wentlooge*
f) Peterstone Wentlooge*
g) Rhymney Mouth/Great Wharf
h) Taff/Ely Estuary

	a	b	c	d	e	f		g	h	
	Max	Max	Max	Max	Max	Reg	Max	Max	Reg	Max
Mallard	50	300	110	56	690	253	1150	90	82	234
Teal	48	40	220	140	305	142	630	450	33	270
Wigeon	400	450	60	0	0	20	175	20	0	1
Pintail	13	40	0	0	0	5	18	149	0	5
Shoveler	0	20	10	0	0	63	157	4	0	0
Pochard	0	16	0	4	0	16	260	409	2	60
Tufted Duck	0	0	0	2	0	56	500	402	0	0
Scaup	0	0	0	0	0	5	20	11	0	0
Common Scoter	0	18	10	0	0	2	22	0	0	0
Shelduck	50	650	156	0	120	227	650	2600	170	320
Bewick's Swan	0	29	0	0	0	0	20	0	0	0

* Data primarily from *Gwent Bird Reports*.

Nature Reserve in 1978, and the 28ha freshwater pool within them (Table 21) was made a statutory Sanctuary. Before the Second World War Kenfig Pool held up to 2,000 ducks, but immediately thereafter the population dropped dramatically, possibly as a result of war-time disturbance, and the process continued as the recreational use of the area increased, until the early 1970s. Nowadays, only fishing and canoeing are permitted, the former entailing the use of no more than two, non-motorised boats. Although shooting is still allowed on the northern dunes, away from the pool, and has some impact, and there can be considerable disturbance from visitors to the waterside, access to which is unrestricted, the numbers of wildfowl are recovering well. Furthermore, the perimeter of the pool is now protected from sand encroachment by a band of *Phragmites* and dense scrub-carr.

Some wildfowl are known to occur at the tiny mouth of Afon Cynffig, on the northern boundary of the Kenfig LNR, but exact records are lacking. Eglwys Nunydd Reservoir, however, on the other side of the main railway line on the approaches to the steel works, is well documented (Table 21). What started as a colliery subsidence pool in the 1930s, ultimately reaching 10ha, was transformed into an 89ha reservoir in the early 1960s, to provide a permanent water supply to the steel works. The numbers of dabbling ducks were surprisingly unaffected by this change, and have recently declined, due to the high water levels maintained by the steel works, but diving ducks showed a marked increase as soon as the reservoir was constructed. Disturbance here is greater than at Kenfig Pool, to which the birds readily move. Sailing was transferred from the Pool to Eglwys Nunydd in the 1970s, and is carried out throughout the year, while fishing also involves the use of a motor-boat, which causes considerable disturbance.

The construction of the steel works just after the Second World War engulfed the Morfa Pools on the old Margam Estate, where many hundreds of duck had formerly been present, but an 8ha reservoir remains, where 26 Tufted Ducks and 2 Mute Swans were recorded in February 1980. Margam Docks, sandwiched between the works and Port Talbot town, sometimes hold several hundred Pochard and Tufted Duck, but have not been formally surveyed. Large numbers of White-fronted Geese occurred on Margam Moors in the past but industrialisation drove them away. The three lakes in Margam Country Park, on the hillside 1km from Eglwys Nunydd, hold a few wildfowl most of the time, the maximum from a handful of counts since 1968 being 52 Mallard, 104 Teal, 28 Wigeon, 25 Gadwall, 4 Pochard, 14 Tufted Ducks, 2 Mute Swans and 2 Canada Geese. The lakes probably assume a greater importance when the reservoir and Kenfig Pool are disturbed, and Gadwall have been seen to move between the park and Kenfig Pool.

The narrow, polluted lower reaches of the Neath, opening into the largely reclaimed Baglan Bay, produced, during monthly counts in the early 1970s, a few Mallard, Teal, Wigeon, Pintail, Common Scoters and Shelducks, amounting at any one time to no more than 65 birds. At the small canal pool 4km upriver, no more than eleven wildfowl, mainly Mute Swans, were found in monthly counts during 1977-78. The remaining area of Swansea Bay, including the vast, but

Table 21. *The regular (and maximum) numbers of wildfowl at Kenfig Pool and Eglwys Nunydd Reservoir.*

	Kenfig Pool			Eglwys Nunydd Reservoir		
	1956-59*	1960-69	1970-82	1951-59	1960-69	1970-82
Mallard	0 (4)	5(28)	15 (63)	61(200)	14 (80)	28 (68)
Teal	4(10)	6(45)	44(169)	83(225)	55(300)	15 (85)
Wigeon	2(13)	11(37)	7 (50)	87(450)	9 (42)	2 (24)
Shoveler	4(20)	4(26)	24(190)	11 (24)	2 (10)	23(118)
Gadwall	0 (2)	2(15)	15 (52)	0 (0)	2 (10)	7 (54)
Pochard	5(28)	11(93)	93(350)	25 (84)	239(870)	157(380)
Tufted Duck	9(22)	14(50)	57 (80)	32 (64)	131(430)	121(301)
Goldeneye	6(13)	6(16)	11 (38)	0 (3)	8 (28)	13 (50)
Mute Swan	8(18)	7(23)	4 (11)	4 (22)	2 (8)	4 (18)
Whooper Swan	8(15)	10(25)	3 (11)	0 (0)	1 (10)	0 (5)
Bewick's Swan	2 (8)	1 (3)	1 (7)	0 (0)	1 (2)	1 (3)

* No complete data before 1956.

industrially hemmed-in Crymlyn Bog, is of no account, excepting one part: the sands around Black-pill on the west shore. Although numerous counts there since 1970 have found that only Teal (15), Wigeon (38), Common Scoter (12), Scaup (2) and Shelduck (9) are regular, after hard weather larger numbers can be found. Up to 196 Teal, 71 Common Scoters, 28 Scaup, 29 Brent Geese and 32 Shelducks have been recorded.

Inland of the coastal strip the only wildfowl resorts of any importance in Glamorgan are Lisvane and Llanishen Reservoirs, Roath Park Lake, Cardiff (Table 22), and Hensol Castle Lake, 14km to the west. The adjacent reservoirs of Lisvane and Llanishen, on the northern outskirts of Cardiff, are both 10.5m deep. Lisvane is much the smaller, but is more sheltered and holds far more wildfowl, mainly because it is not used for sailing, whereas Llanishen is. The artificial lake in Roath Park, in the centre of the city, is nowhere deeper than 4.25m and is frequently drained to reduce the risk of flooding. Despite disturbance on its surrounds it holds many more Mallard than the reservoirs, though fewer diving ducks. Teal were once plentiful, reaching 200 in the late 1950s, but since the mid 1960s they have been scarce. Mute Swans exceeded 100 in the 1950s, but gradually declined, until virtually disappearing in the 1970s. They used to moult on the reservoirs and move to the lake for the rest of the winter. Pochard travel readily between the reservoirs and Roath Park. In early 1979, when these and most other inland resorts were frozen, large numbers of both Pochard and Tufted Duck moved to the Severn. Since then a regular night-time exodus has apparently occurred to the estuary, even in mild weather, from the Cardiff waters (255) (see also p.59).

Hensol Castle, a hospital since 1930, has a 14ha lake, irregular counts of which since the mid 1960s have produced maxima of 170 Mallard, 40 Teal, 4 Wigeon, 2 Shoveler, 87 Pochard (212), 36 Tufted Ducks and 2 Mute Swans. A count on the tiny Llyn Yoy, 250m away, in January 1967 found only 3 Mallard and 37 Teal. At the same time, the nearby lake at Talygarn held just 5 Mallard, and the ponds at Sant-y-Nyll, towards Cardiff, 2 Mallard and 20 Teal.

Gwent contains only two significant inland areas: Llandegfedd Reservoir and the floodplains of the Usk Valley. Llandegfedd Reservoir (Table 23) lies 3km east of Pontypool. Its 230ha were flooded in 1964, following the damming of Sor Brook, a tributary of the Usk. At first, only Mallard were found in significant numbers, but in the late 1960s Teal, Wigeon, Pochard and Tufted Ducks increased greatly, and in the early 1970s Goosanders appeared in strength. All these species reached a peak in the early 1970s and have since, for reasons that are not clear, levelled off or declined. The control of shooting at Peterstone may have attracted birds there rather than to Llandegfedd, while the laying of a bypass pipeline at the reservoir during the 1979-80 winter caused some disturbance. Alternatively, it may be that the usual pattern of increase followed by stabilisation or decline shown by lowland dam reservoirs took especially long to occur in this case.

Llandegfedd provides a fine example of balance between conservation and recreational interests. Except for sub-aqua training in a small part of the south-eastern corner, no recreation is allowed from November to February inclusive, but for the rest of the year sailing and fly-fishing are permitted virtually throughout the water, with canoeing and rowing in parts. This regime has been maintained by committees, comprising all interested parties, which liaise with the Welsh Water Authority. Continuing pressure for winter sailing has been resisted, indeed proposals for the site to be designated an LNR have been shelved following assurances from the Welsh Water Authority

Table 22. *The regular (and maximum) numbers of wildfowl at Roath Park Lake and Llanishen and Lisvane Reservoirs.*

	Roath Park Lake		Llanishen/ Lisvane Reservoirs	
	1960-69	1970-82	1960-69	1970-82
Mallard	249(417)	217(409)	15(118)	77(223)
Teal	3 (16)	0 (30)	2 (22)	0 (6)
Pochard	25(117)	63(400)	86(189)	70(260)
Tufted Duck	57(460)	19(540)	173(320)	281(567)
Goldeneye	0 (1)	0 (8)	0 (3)	1 (5)
Mute Swan	36 (79)	2 (13)	3 (44)	1 (4)

Table 23. *The regular (and maximum) numbers of wildfowl at Llandegfedd and Ynysyfro Reservoirs.*

	Llandegfedd			Ynysyfro
	1964-68*	1970-75	1976-82	1974-82
Mallard	312(1120)	760(1300)	492(1700)	3 (13)
Teal	79 (220)	119 (250)	91 (130)	1 (9)
Wigeon	188 (500)	582(1000)	547(2000)	0 (20)
Pochard	201 (400)	138 (300)	83 (260)	14 (50)
Tufted Duck	67 (100)	83 (170)	22 (105)	57(117)
Goldeneye	3 (9)	5 (24)	4 (19)	1 (7)
Goosander	1 (6)	12 (51)	9 (27)	0 (0)
Mute Swan	6 (11)	2 (7)	4 (13)	1 (5)

* 1969 data incomplete.

that the interests of the overwintering wildfowl would be fundamental to their policy on any future proposals for the site. The shallowest and most important part of the reservoir is the northern end – where, in 1970, a birdwatching hide was built, controlled by the Gwent Ornithological Society – but the southern, dam end is often used for roosting, especially by Mallard (377). In addition to the species tabulated, Bewick's and Whooper Swans (up to 38 and 25 respectively), together with a few Mutes, have been known to use the reservoir as a roost from feeding grounds in the lower Usk Valley, 6km east.

The small reservoirs at Ynysyfro (9ha), immediately north-west of Newport (Table 23), and Wentwood (17ha), nestling between two hills beneath the forest of the same name (Table 24), are of little importance. Both are heavily disturbed, and Ynysyfro is split in two by a farm track.

Of the freshwater areas with irregular coverage shown in Table 24, by far the most significant are the Usk Valley floods, which, though not extensive, are regular and prolonged in occurrence. The principal stretch affected by the floods is between the towns of Usk and Newport, with the birds centred at Llangybi Bottom, which is the most important site for Bewick's Swans in Wales. Their average annual maximum for 1972 to 1982 was 40, with a peak of 68 in December 1981. The flock, which has contained up to 5 Whooper Swans, occasionally moves to Llandenny on the Olway Brook, 8km north-east, and roosts together with the Mute Swans at Llandegfedd Reservoir (228). At Llanfihangel Gobion, halfway between Usk and Abergavenny, Teal have their county stronghold, and up to 10 Goosanders are regular. Smaller numbers of both species occur 10km upriver at Llanwenarth.

The upper Usk and Wye

In westernmost Gwent (Table 24) and the northern half of Glamorgan the land rises steeply, and in the Brecon Beacons a height of almost 900m is attained. The numerous reservoirs and pools in this area hold few wildfowl, but on either side of the Usk Valley, which breaks in beyond the Beacons, lie two important resorts, 6.5km apart: Talybont Reservoir and Llangorse Lake (Table 25). Llangorse (153ha) is lower-lying than Talybont, more productive, shallower and bigger, being the largest natural lake in South Wales. However, it holds many fewer wildfowl, since it is subject to recreational pressures of all varieties throughout the year. Although the site is privately owned, there are almost no restrictions on recreational activities, which involve the whole of the water area. In autumn the lake is virtually untenable for wildfowl in day-time, while even in mid-winter, when sailing and power-boating are at their minimum, the small numbers of Whooper and Bewick's Swans which feed in the area use Talybont alone to roost (329, 597, 599). When disturbed, most ducks stay on the periphery of the lake, but some apparently fly to the Usk rather than the more distant Talybont Reservoir (599). A surprising feature at Llangorse was the summer gathering of non-breeding Mute Swans, which aver-

Table 24. *The maximum numbers of wildfowl at inland resorts in Gwent, 1970-1981. Data mostly from* Gwent Bird Reports. *(For Llandegfedd and Ynysyfro see Table 23.)*

	Mallard	Teal	Wigeon	Shoveler	Pochard	Tufted Duck	Mute Swan
USK FLOODS							
Llangybi Bottom	1000	100	600	9	23	8	33
Llanvihangel Gobion	250	350	11	1	16	8	23
Llanwenarth	0	30	15	6	70	30	0
LOWLAND WATERS							
St Pierre Lake	5	18	0	12	21	0	0
Wentwood Reservoir	7	0	15	6	70	30	3
Llanyrafan Boating Pond	0	0	0	0	4	22	0
Tredegar House Lake	0	3	0	1	1	0	0
Pant-yr-eos Reservoir	0	0	0	0	0	12	0
Llanfoist Lake	0	22	2	0	0	0	0
UPLAND WATERS							
Pen-y-fan Pond	0	8	0	0	46	22	10
Cwmtillery Reservoir	0	0	0	0	4	2	0
Semtex Pond, Brynmawr	0	0	0	0	7	2	0
Garnlydan Reservoir	0	0	8	0	28	20	0

aged 38 between 1971 and 1976 (86); the number has now, however, dwindled to 10-15 birds. The greatest overall threat to the lake probably stems from its over-enrichment with nitrates and phosphates, a process which (though inevitable in the long term) has been hugely accelerated in this case. Since 1961 all of its fifteen species of underwater plant have disappeared. Although the high level of recreation is thought to have contributed indirectly to the process of eutrophication, its major cause was removed in 1981, when the sewage works at Llangorse village, which had been pumping 200,000 litres of treated effluent a day into the lake, were modernised. Other causes remain, however, including enrichment from surface drainage (probably the main cause) and a roost of up to 10,000 Black-headed Gulls (147, 597). Canada Geese have recently become established and increased from 1-2 pairs in 1978 to 6-8 pairs in 1982. In winter up to 90 birds have been counted. Tufted Duck is the only species of wildfowl to have shown a recent decline in winter, but there may soon be a time when the food supply at Llangorse is too small to support a wildfowl population even of the present dimensions.

Talybont Reservoir (131ha) is surrounded by steep, wooded hillsides, but has an attractive marshy area at its southern end, where the Caerfanell River enters. This part is much the most favoured by feeding ducks, and was designated a Local Nature Reserve in 1975, managed jointly by the Brecnock Naturalists' Trust, the Brecon Beacons National Park Authority and the Welsh Water Authority. The only sporting activity permitted at the reservoir is game fishing, the close season for which covers almost the entire winter. Goosanders, in very small numbers at Llangorse, are attracted to Talybont by its plentiful trout (597) and probably also use the water as a roost, feeding on nearby rivers.

Within the Brecon Beacons National Park there are a further 15 reservoirs and a number of small pools. The following are included in Table 25: Usk Reservoir (on the Dyfed boundary, near the source of the river); Llwyn-on Reservoir (10km south-west of Talybont); and Gludy Lake, Brecon. The remainder are of little interest, mainly because of their high altitude (329). Small parties of Mallard occur along the upper Usk itself. One hundred were reported from the Crickhowell area in the 1960s, roosting on a pond on the Glanusk estate.

For the 88km from its source (close to that of the Severn) to the English border the Wye is closely lined by mountains and hills. Amidst some of the highest lie the Elan Valley Reservoirs (Craig Goch, Pen-y-garreg, Garreg Ddu, Caban Coch and Claerwen), which, despite covering over 400ha, hold very few wildfowl. Pochard, Tufted Ducks, Goosander and Whooper Swans are occasional, but a record of 100 Mallard in 1982-83 was exceptional. The only other water of interest along this stretch is Brechfa Pool, a Brecknock Naturalists' Trust reserve on the slope of the valley, 12km short of Hay-on-Wye. Fifty Teal are regular

Table 25. *The regular (and maximum) numbers of wildfowl at the principal Brecknock lakes and reservoirs.*

a) Llwyn-on Reservoir	2 seasons 1970 and 1972
b) Talybont Reservoir	12 seasons 1970-1982
c) Llangorse Lake	9 seasons 1970-1982
d) Gludy Lake	1 season 1970
e) Usk Reservoir	5 seasons 1970-1982

	a	b	c	d	e
Mallard	90(140)	294(670)	61(250)	118(170)	91(220)
Teal	0 (10)*	59(165)+	70(220)	3 (20)#	43 (80)
Wigeon	0 (2)*	18 (90)+	25(220)	0 (1)	6 (20)+
Shoveler	0 (0)	5 (30)	4 (20)	0 (0)	0 (0)
Pochard	5 (20)#	138(440)+	51(190)	4 (5)	6 (25)*
Tufted Duck	5 (8)	87(400)+	44(150)	18 (30)#	7 (15)+
Goldeneye	0 (11)#	6 (13)	1 (12)#	0 (0)	0 (4)*
Goosander	1 (3)*	14 (38)	0 (47)	0 (0)	3 (20)
Mute Swan (winter)	0 (1)*	1 (4)	12 (25)	1 (1)	0 (0)
Whooper Swan	2 (5)*	4n(12)	2 (7)	0 (0)	0 (4)
Bewick's Swan	0 (0)	1 (13)	1 (18)#	0 (0)	0 (0)

* From Massey (1975).
+ From Massey (1976).
# From *Breconshire Birds*.

there and up to 30 Wigeon and 34 Bewick's Swans have been found, with small numbers of other species (511). Up to 26 Teal, 10 Bewick's Swans and 16 Goosanders have been seen on the river nearby at Llyswen (86).

In the old county of Herefordshire the Wye meanders through a broad, low-lying vale of farmland in which are numerous pools and ornamental waters. These have been poorly covered by the counts, but it appears that none is of more than local importance. A total of over 100 wildfowl (chiefly Mallard, Teal, Wigeon, Pochard and Tufted Duck) has been recorded at Leach Pool, Letton Court, Tyberton and Allensmore, together with the outlying lakes at Titley, north-east of Kington, and Eastnor Castle, below the southern edge of the Malverns near Ledbury. The attraction of the region is increased in mid-winter by flooding along the Wye and its tributaries. The most important such areas are along the Arrow at Monkland, the Lugg near Hereford and the Wye around Foy. Although exact counts are lacking, each apparently holds moderate numbers of Mallard, Teal and Wigeon, with Bewick's Swans at Monkland (511) and a record of 100 White-fronted Geese on the lower Lugg. At Monmouth, 15 Mute Swans were recorded on the river in April 1976 (228). From there to its mouth, at Chepstow, the Wye forms the border between Gwent and Gloucestershire. Here the valley is narrow, wooded and steep-sided, and of little significance for wildfowl.

West Wales

Most of the wildfowl resorts in West Wales are coastal; inland standing water is scarce and rivers narrow and fast flowing. Several new reservoirs have been built but these are largely unproductive and fed from impoverished upland catchments. On the coast the few attractive estuaries are separated by stretches of cliff, sandy or shingly shore.

Gower and old Carmarthenshire

The south shore of the Gower Peninsula is cliffbound, apart from a short stretch of dune and saltmarsh in Oxwich Bay. The bay, together with the adjoining freshwater marsh, shallow pools and scrubland behind, was designated a National Nature Reserve in three stages between 1963 and 1973. Though the peak wildfowl population is normally about 150, up to 768 Mallard, 930 Teal and 71 Shoveler have been recorded.

The north side of the peninsula constitutes the southern, and by far the more important, shore of the Burry Inlet. Here the rivers Loughor, Lliw and Llan converge into the most significant wholly Welsh estuary for wildfowl. Whereas the north shore is mostly built up, the south (Table 26) is unspoilt, and it is here, on the large expanses of saltmarsh and mudflats, that the great majority of the wildfowl occur. Their numbers were declining, presumably because of pressure from wildfowling, until a series of conservation measures in the mid 1960s halted the trend. In 1966 a 770ha NNR was declared at Whiteford Burrows, a peninsula at the mouth of the Inlet. On its leeward side, and at the west end of the adjacent Llanrhidian Marsh, a no-shooting agreement with the West Glamorgan Wildfowlers' Association means that the most important area of foreshore is free from disturbance. To the east, where much of the land is owned by the National Trust, the shooting is strictly controlled. In 1969 3,436ha were designated a statutory Bird Sanctuary, although most of this area is too far offshore to be of practical benefit (320).

Further increases in dabbling ducks in the Inlet occurred during the late 1970s and early 1980s. The numbers of Pintail now frequently exceed the internationally important level (of 750). It is one of the few species to be found in strength on the inner estuary, above Penclawdd. The Brent Geese, whose westernmost stronghold in Britain this is, are mainly found off Whitford Point. There also small numbers of sea ducks are found. Eiders, present in all seasons since at least the last century, have increased markedly in recent years, though breeding has not yet been recorded. Scoters are surprisingly few considering the proximity of the huge Carmarthen Bay flocks (see below).

European Whitefronts declined from 3-400 in the mid 1940s to about 100 in the 1950s and early 1960s, and have not been recorded regularly since.

On the north shore of the Burry Inlet suitable feeding exists in small areas at Pembrey Burrows, Cefn Padrig, Penrhyn Gwyn and the upper Loughor, but disturbance is high and only Mallard and Shelducks are common, although the Pintail and Brent Geese occasionally move to this side. In some places shooting has recently been more tightly controlled and the numbers of wildfowl are apparently increasing.

The main threat to the Burry Inlet may come from a plan, currently shelved but with continuing calls from recreational interests, to build a barrage across the narrow opening of the Loughor River at Loughor town. The resultant silting of the Inlet as a whole would destroy much of the feeding ground for wildfowl.

To the west, the broad sweep of Carmarthen Bay, particularly on its eastern shore, contains a vast area of sandy shallows which, although unattractive to most other species, hold the largest gathering of

Common Scoters in England and Wales. Counts from boats in March of 1973 and 1974 recorded 20,000 and 25,000 respectively, while aerial surveys by the RSPB between 1976 and 1978 found up to 12,000. These and other records (Table 27) have shown that Carmarthen Bay is of major importance throughout the year. Since 1978, spring and winter counts from land have indicated that there may have been a reduction in numbers. Mass strandings of feathers, however, have suggested that the area is still an important moulting ground. The scoters feed on the benthic invertebrates which abound on the sandy sea bed. Although not venturing into the outflow currents of the rivers, on high tides the scoters come fairly close to the shore, but at extreme low tides many may disappear to as yet undiscovered localities (358). Generally they are found around the five fathom line, and are spread throughout the sweep of the bay, with the flocks packed more tightly at low tide (512).

Although the oil terminals in nearby Milford Haven have an exemplary record in avoiding and containing spillages (p.492), the Carmarthen Bay scoters were shown to be vulnerable to oil pollution by several incidents in the 1970s. The most serious

Fig. 13. West Wales.

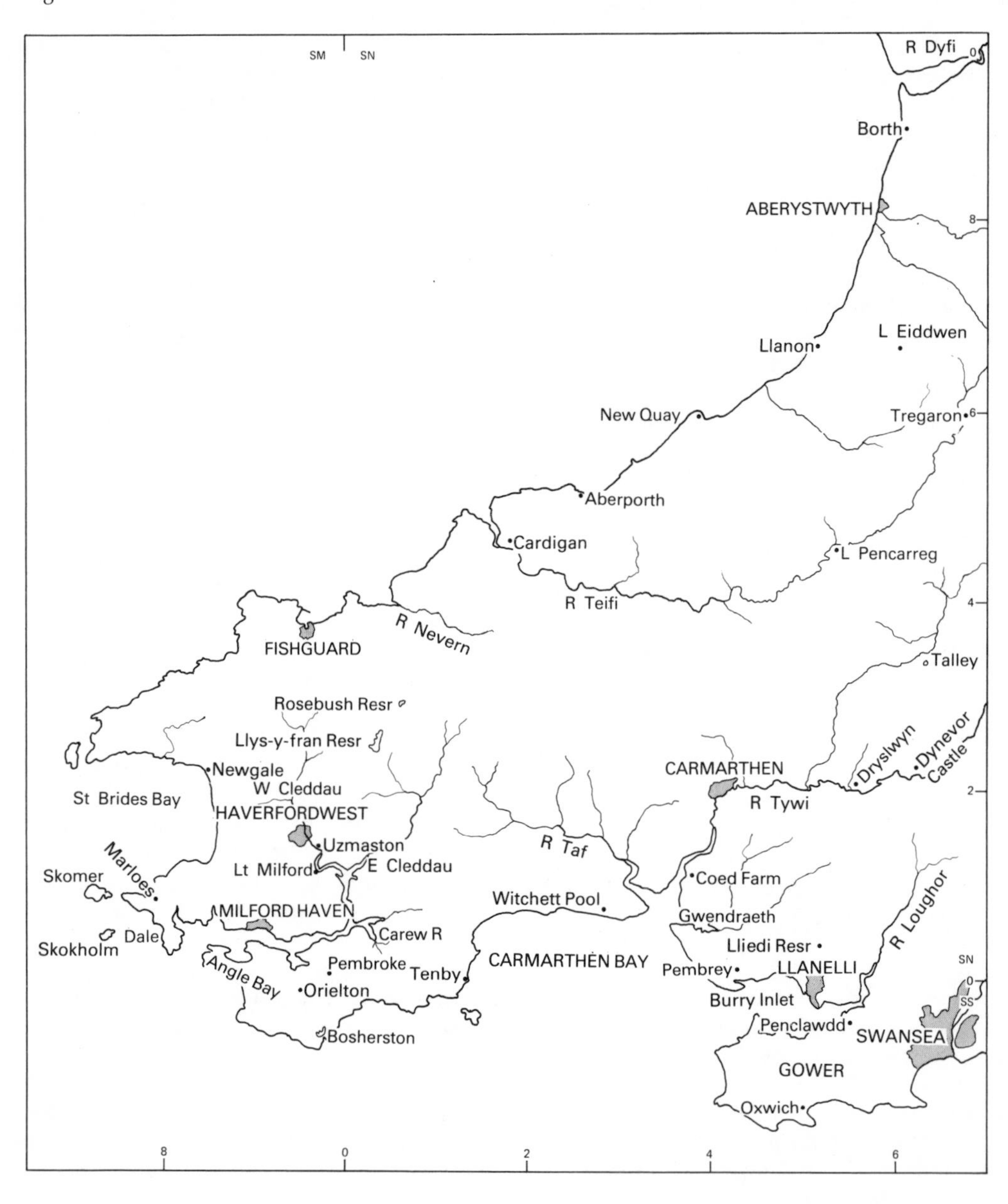

occurred in 1973, when over 300 dead or dying scoters were stranded on the shore between Freshwater East and Pembrey in late December/early January (562). In another serious incident in 1976, over 250 scoter corpses were found between Swansea and Tenby, and between 500 and 1,250 were thought to have perished (358). In both cases nearly all the birds washed ashore were oiled, but the source of the oil was unknown. In the latter instance, however, it appeared not to emanate from sources in the Burry Inlet or the bay itself.

The narrow estuaries of the Gwendraeth, Tywi and Taf converge in the north-eastern corner of Carmarthen Bay. They provide a large area of sheltered mudflats, but the feeding grounds themselves are few and the numbers of wildfowl disappointing. Several stretches of river are cliffbound or have hills just behind, while the firing range at Pendine, on the north shore of the bay, causes a great deal of disturbance. The only important areas in the complex are the south shore of the Gwendraeth (Morfa Cydweli and Morfa Coedbach), where maxima of 1,174 Mallard, 400 Teal, 2,000 Wigeon, 288 Shelducks and 51 Mergansers have been recorded during irregular counts since 1968; the outermost west shore of the Taf, which includes a wide saltmarsh protected by Ginst Point and has held up to 6,500 Mallard, 300 Wigeon, 120 Pintail and 189 Shelducks; and the vicinity of Coed Farm, 2.75km up the Tywi. At Coed Farm a WAGBI reserve was established in 1966, incorporating 18ha of saltmarsh between the railway line and the river. Shooting, initially heavy, gradually ceased and the numbers of ducks rose accordingly – Wigeon to 800 by early 1976, from a previous peak of 200. The construction of a 1.75ha freshwater pool during the 1970s increased the attractiveness of the area. A flock of 1,700 Wigeon was present in January 1979 and many other species are now regular in small numbers (58).

The 12ha Witchett (or Whityer) Pool (Table 29) lies 3km west of the mouth of the Taf, amidst the Pendine Gun Range. The few counts available since 1970 show the continuance of a good variety of wildfowl, but in small numbers. Of most interest are the Shoveler, of which 90 were found in January 1974, and scoters, small groups of which regularly seek refuge from Carmarthen Bay, often including some oiled birds. Flocks of 31 scoters in January 1974 and 17 a year later included, respectively, 3 and 2 Velvet Scoters, suggesting that these may be present in the flocks offshore. Up to 50 Scaup have also occurred on the pool, although no gatherings are known on the South Wales coast.

The narrow strip of pasture that fringes the

Table 26. *The regular (and maximum) numbers of wildfowl on the Burry Inlet (south shore).*

	1957-61	1962-66	1967-71	1972-76	1977-82
Mallard	108 (400)	26 (110)	33 (105)	51 (138)	109 (230)
Teal	76 (300)	24 (120)	128 (500)	448(2657)	565(1390)
Wigeon	1313(2500)	995(2250)	1527(2200)	1434(5596)	2056(4359)
Pintail	744(1500)	407 (700)	342 (600)	325(1679)	684(2535)
Shoveler	139 (350)	9 (100)	3 (25)	11 (76)	16 (90)
Scaup	0 (0)	1 (18)	2 (23)	0 (2)	1 (4)
Shelduck	429 (950)	262(1000)	318 (800)	273(1092)	826(1588)
Eider	6 (36)	12 (62)	11 (48)	43 (148)	92 (241)
Common Scoter	0 (0)	0 (0)	0 (0)	5 (26)	10 (92)
R-b Merganser	1 (5)	0 (1)	2 (10)	9 (45)	16 (33)
D-b Brent Goose	9 (35)	14 (29)	9 (33)	57 (352)	225 (475)

Table 27. *The numbers of Common Scoters in Carmarthen Bay. (From Moyse and Thomas 1977; RSPB 1978; R. Lovegrove)*

	1971	1973	1974	1975	1976	1977	1978
Jan/Feb	5000				6700	5000	3100
March		20000	25000		430	1800	
April/May						260	
June	700					1100	
July/Aug/Sept		5000	16000	6000	10600	2300	12000
Oct/Nov/Dec				2600	6700	4100	5700

lower valley of the Tywi probably attracts small numbers of wildfowl throughout its length, but there is one particularly important area – around Dryslwyn, 15km above Carmarthen. The flooded meadows here hold, in addition to the species listed in Table 29, one of the largest flocks of European White-fronted Geese in Britain. This was at its peak in the late 1960s, when the total briefly reached 3,000. Since then, as at Slimbridge, the next most westerly stronghold (100km away), there has been a decline; the average peak for 1971-1982 was 615, including a brief resurgence to 1,100 during the cold weather of early 1979. They are suspected of roosting in the Burry Inlet, although only a handful of Whitefronts have been seen there since the early 1960s.

The water-meadows and oxbow lakes at Dynevor Castle, 5km upstream from Dryslwyn, were, together with the adjacent woodland, established as a reserve in 1979 by the West Wales Naturalists' Trust, who had leased the shooting rights since 1975. Up to 190 Mallard, 400 Teal, 500 Wigeon and occasional Whitefronts have been recorded there. Above Llandeilo, however, the pasture is probably too rough to provide good feeding.

The two lakes at Talley, 10km north of Llandeilo, are encircled by hills and hold only a few Teal and a handful of diving ducks, although Pochard have one of their few Welsh breeding haunts here.

In 1972, a new reservoir of 210ha, Llyn Brianne, was flooded amidst the mountains on the Powys/Dyfed border, near the source of the Tywi. A series of counts there the following winter produced maxima of only 32 Mallard, 27 Wigeon and 4 Pochard, and numbers have remained low. Goosanders have recently started to breed in this area.

Elsewhere in the old county of Carmarthenshire the only waters of any size are the upper and lower Lliedi Reservoirs, just north of Llanelli, which cover 27ha (Table 29).

South Pembrokeshire, Preseli and Ceredigion

To the west of Carmarthen Bay the coastline is again cliffbound and of little interest to wildfowl, until it is broken by the long tidal arm of Milford Haven, which is lined with small estuaries and leads into the complex Cleddau Estuary (Table 28). The area is surrounded by rich farmland (although the shores of the Cleddau are wooded in many parts), and the whole system contains the feeding and roosting grounds of some 4,000 wildfowl. The most important stretch is the Western Cleddau below Little Milford, which, in combination with the adjacent outflows of the Sprinkle and Llangwm Pills in the main arm of the Cleddau, holds the greatest variety of ducks, although the Pembroke River holds the bulk of the Wigeon. Most of this area became a statutory Bird Sanctuary in 1970. A series of complete counts from the Gann Estuary to Uzmaston in 1982 and 1983, including some

Table 28. *The regular (and maximum) numbers of wildfowl in the following sectors of Milford Haven:*

(a) Gann Estuary (Dale Roads) (4 seasons: 1970-81)
(b) Sandyhaven Pill (5: 70-82)
(c) Angle Bay (4: 70-82)
(d) Pembroke Estuary (2: 75-82)
(e) Cresswell/Carew Rivers (2: 70-82)
(f) Daucleddau (Castle and Beggars Reaches); Garron Pill (1: 81)
(g) Eastern Cleddau (2: 81-82)
(h) Little Milford—Llangwm (10: 70-82)
(i) Uzmaston—Haverfordwest (1: 82)

	Mallard	Teal	Wigeon	Pintail	Goldeneye	R-b Merganser	Shelduck	Mute Swan
(a)	1 (10)	17 (45)	74 (270)	0 (0)	7(22)	0 (2)	4 (18)	8(21)
(b)	5 (32)	0 (12)	11 (43)	0 (0)	0 (2)	0 (1)	0 (15)	7(24)
(c)	31(254)	2 (23)	370(2200)	0(11)	0 (1)	0 (3)	41(300)	0 (4)
(d)	18 (54)	68 (300)	1054(1400)	3 (5)	2 (3)	0 (2)	267(441)	12(23)
(e)	22 (60)	227 (455)	138 (248)	0(11)	0 (6)	0 (1)	110(316)	2(13)
(f)	53(133)	139 (270)	336 (584)	1 (8)	1 (9)	0 (3)	151(194)	7 (7)
(g)	26 (40)	173 (400)	19 (75)	0 (0)	0 (1)	9(20)	12 (23)	0 (0)
(h)	146(569)	359(1500)	454(1436)	7(38)	5(47)	10(24)	165(358)	1 (7)
(i)	42 (51)	297 (524)	127 (200)	0 (0)	3(30)	0 (2)	18 (34)	0 (2)

areas with little previous coverage, recorded up to 1,700 Teal, 2,070 Wigeon and 1,200 Shelducks.

In the early 1960s three large oil terminals were constructed on the shores of Milford Haven, two on the north side, either side of Milford Haven town, and one on the opposite bank at Rhoscrowther. This area now constitutes one of the biggest oil ports in Britain. After initial problems, with several spillages in the first few years after the opening of the terminals, the system of preventing and controlling spillages was soon improved; nowadays the passage of tankers through the Haven, intensive though it is, causes no problems for wildfowl. Control of the vessels outside the Haven is harder, however, and there are still occasional spillages off the South Wales coast, such as those involving the Carmarthen Bay scoters described in the previous section (see also Fig 160).

In the Castlemartin Peninsula, to the south of Milford Haven, there are three freshwater resorts of interest, whose populations are detailed in Table 29: Pembroke Mill Pond, Orielton Pools and Bosherston Lakes. The Upper Mill Pond at Pembroke is separated from the Pembroke River estuary by a railway embankment. A reserve of the West Wales Naturalists' Trust (WWNT) was established there in 1979. The main pool at Orielton, 2km south of the Pembroke River, was built as a decoy in 1868. It was revived as a ringing station in 1934 after years of neglect, and continued to operate until the early 1960s. Nowadays the lake is owned by a fish farm. Bosherston Lakes (or Lily Ponds), three coastal marl lakes near St Govan's Head, lie in the Stackpole Estate, which was acquired by the National Trust in 1977. The lakes form part of Stackpole NNR, established under a management agreement with the Trust in 1981. In winter, the water level rises to form one continuous lake, but with different aquatic communities in each part (373).

The main freshwater sites immediately north of Milford Haven are Marloes Mere (which is leased by the WWNT from the National Trust), the two pairs of tiny reservoirs at Bicton and Castle Farms, St Ishmaels (Table 29), and a small reservoir at Thornton, just north of Milford Haven town, where 500 Teal were reported in the late 1960s. However, this is now drained.

The two major islands to the south of St Brides

Table 29. *The regular (and maximum) numbers of wildfowl at the following sites in Dyfed:*

(a) Witchett Pool (3 seasons: 1968-70)
(b) R Tywi: Dryslwyn (Av of 29 Nov – Feb counts, 66-82)
(c) Lliedi Reservoirs (6 seasons: 68-81)
(d) Pembroke Mill Pond (3: 75-82)
(e) Orielton Pools (5: 60-67)
(f) Bosherston Lakes (4: 79-82)
(g) Marloes Mere (2: 81-82)
(h) Bicton/Castle Farm Reservoirs (9: 73-82)
(i) Llys-y-fran Reservoir (6: 72-82)
(j) Rosebush Reservoir (11: 72-82)
(k) Llyn Pencarreg (1: 69)
(l) Cors Tregaron (13: 70-82)
(m) Llyn Eiddwen (1: 66)

	Mallard	Teal	Wigeon	Shoveler	Pochard	Tufted Duck	Goldeneye	Mute Swan
(a)	71(200)	35 (150)	20 (106)	9(90)	11 (39)	12 (60)	1(15)	3 (6)
(b)	135(348)	152(1050)	794(2500)	0 (0)	4 (49)	3 (22)	0(11)	2(24)
(c)	19 (65)	3 (31)	0 (0)	0 (0)	17 (46)	18(100)	0 (4)	0 (1)
(d)	48 (89)	20 (54)	0 (5)	1 (8)	5 (13)	43 (59)	0 (3)	28(48)
(e)	27 (70)	78 (250)	32 (190)	15(75)	8 (43)	5 (23)	0 (3)	1 (5)
(f)	41(182)	16 (74)	2 (9)	1 (5)	110(452)	13 (50)	4 (6)	2 (9)
(g)	32(106)	35 (95)	106 (208)	14(20)	6 (32)	0 (2)	0 (0)	0 (0)
(h)	9 (65)	20 (78)	33 (320)	1 (7)	32(150)	10 (41)	2(10)	1 (8)
(i)	30 (90)	16 (70)	3 (124)	0 (5)	60(255)	39(100)	4(14)	0 (0)
(j)	43(130)	34 (67)	0 (16)	0 (4)	5 (45)	5 (15)	0 (6)	0 (0)
(k)	58 (86)	5 (21)	7 (11)	0 (5)	3 (16)	5 (9)	1 (8)	1 (2)
(l)	223(380)	190 (380)	74 (215)	2(12)	4 (12)	3 (10)	2 (4)	3(11)
(m)	5 (12)	5 (13)	0 (30)	0 (0)	8 (60)	1 (3)	0 (0)	0 (0)

Bay, Skomer (a 294ha NNR) and Skokholm (97ha), are both WWNT reserves, mainly owing to their famous sea-bird colonies. They also contain small ponds which are surprisingly attractive to wildfowl at all seasons, though especially at passage times. At Skomer 1,200 Mallard and 320 Teal were found in autumn 1981, and at Skokholm, 3.5km south, up to 25 Mallard, 170 Teal, 51 Wigeon, 28 Shoveler and (offshore) 97 Common Scoters have been recorded. Of particular interest is the flock of 20 or more apparently wild Barnacle Geese which has wintered on Skomer in recent years. No rings have been observed on these birds, which suggests that they belong to the Greenland population, rather than the Spitsbergen group or the fully-winged flock at Slimbridge.

The rest of the coast of West Wales, encompassing the Districts of Preseli (the remainder of old Pembrokeshire) and Ceredigion (formerly Cardiganshire), is predominantly cliffbound or rocky, holding little apart from Mallard (which are in fact surprisingly abundant). However, the picture is enlivened by several small bays and estuaries and a stretch of flatter coast towards Aberystwyth. On the east shore of St Brides Bay, 300 scoters were counted off the sands of Broad Haven, in the south-east corner of the bay, in January 1972, and up to 1,000 are often seen at Newgale Sands, in the north-east corner. Insufficient records are available to gauge the seasonal trends of these flocks.

The Nevern Estuary, opening into Newport Bay, supports a regular population of 59 Mallard, 2 Teal, 11 Wigeon and 6 Mute Swans, while the Teifi Estuary, 10km up the coast, together with the tidal Rosehill Marsh, just upstream of Cardigan, holds several hundred Mallard, Teal and Wigeon and up to 30 Mute Swans. Most of these occur on the marsh and the adjacent Pentood Meadows (outside the reserve). The meadows have been increasingly prone to flooding, due to the gradual silting up of Rosehill Marsh (610), and this, with stricter control of shooting in recent years, has made the area more attractive to wildfowl.

Between New Quay and Aberystwyth the coast is much less rugged, but the lack of estuaries or any significant foreshore feeding grounds, and the proximity of the hills, mean that the wildfowl are few. Off Borth there is a regular flock of over 100 Common Scoters, and a few, rarely up to 100, are seen off the coast at Aberporth. A count of 20 Teal and 160 Wigeon off Llanon in January 1982 was exceptional, and followed a period of severe weather throughout Britain.

At the head of the Teifi valley lies the great raised bog of Tregaron, where an NNR of 792ha, now known as Cors Caron, encompasses all the wetland habitat of the area. Here, apart from the species shown in Table 29, whose numbers have remained stable for the last twenty years, the largest flock of Greenland White-fronted Geese in England and Wales could once be found. In 1962-63 a total of 500 (rather more than the usual flock of 2-300) was present until the birds were driven away by snow cover at the end of December. Many apparently moved to the coast, where food was short and they were subjected to heavy shooting, and only 200 returned with the thaw. For the remainder of the 1960s no more than 100 were found at Tregaron, and thereafter, for reasons that are not clear, they disappeared almost entirely (516). Cors Caron also has the most southerly regular flock (20-50 birds) of Whooper Swans in Wales.

There are no freshwater resorts of comparable importance in the remainder of Ceredigion or northern Pembrokeshire, which are largely mountainous. Counts from Llys-y-fran and Rosebush Reservoirs, in the southern slopes of Mynydd Preseli, Llyn Pencarreg, by the Teifi below Lampeter, and Llyn Eiddwen, amidst Mynydd Bach, 10km north-west of Tregaron, are given in Table 29. Apart from these Trefeiddan Pool, St David's, Ffynone Lake, near Boncath, 8km south-east of Cardigan, and the Aberffrwd and Nant-y-moch Dams, to the east of Aberstywyth, each support between 50 and 100 dabbling ducks, mainly Teal. As many as 44 Pochard were once recorded at Nant-y-moch.

South and south-east England

This section covers the length of the Channel coast from the Devon/Dorset border eastwards to the North Foreland, and extends inland for some 50km to the watershed of the Thames and Kennet Basin. Within this area the main geological features are the chalk, the heathlands, and the Wealden Clay. The chalk uplands, which dominate the northern boundary, extend from the Dorset coast near Abbotsbury north-eastwards to Salisbury Plain and continue through Hampshire to the Hog's Back and the North and South Downs. Because of its porous nature, the upland chalk is virtually devoid of standing water, but the rivers flowing from it are rich in wildfowl foods and support a major part of the inland population. The most important rivers are the Hampshire and Dorset Avon, the Test, the Itchen, the Arun and the western Rother, all of which have extensive floodland and numerous associated pools. The heathlands and the Weald are generally of little value for wildfowl, except in parts of East Sussex and Kent, where some of the newer reservoirs and gravel pits are favoured by substantial flocks. Elsewhere there are few large sheets of water, and most of the pools are either too acid or too confined to harbour many birds.

The coast, although flanked for most of its length by cliffs or open beaches, has several inlets of outstanding value, notably the Chesil Fleet and Poole Harbour, the Solent, and the group of shallow land-locked harbours at Portsmouth, Langstone, Chichester and Pagham. Brent Geese and Shelducks are especially numerous, the concentrations in several places being of international importance; Wigeon and Teal are also plentiful, and Mallard and Shoveler are common and widespread both on the coastal marshes and inland. Other notable occurrences include the

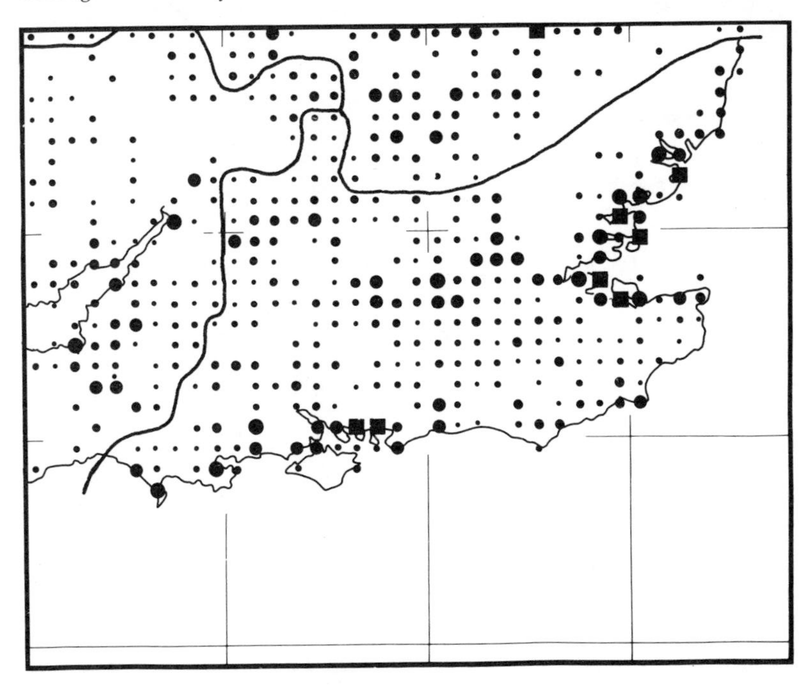

Fig. 14. South and south-east England. Total wildfowl and regional boundary.

flocks of White-fronted Geese and Bewick's Swans in the Avon Valley, the Dorset population of Mute Swans, and the passage of sea ducks along the coast.

Most of the coastal resorts, except the Chesil Fleet, have already suffered in some degree from urban or industrial development, and from various forms of disturbance or pollution. This is a trend which is certain to continue, and which calls for vigilance and compromise. Statutory reserves are now established at a dozen sites, either on or close to the coast, and another six are maintained by the voluntary bodies. The ones most important for wildfowl are the National Nature Reserve on the North Solent Marshes (660ha); the Local Nature Reserves at Christchurch (59ha), Calshot (49ha), Newton Marsh, Isle of Wight (120ha), Titchfield Haven (85ha), Farlington Marsh (119ha) and Pagham Harbour (440ha), and the RSPB reserves and NNRs in and adjoining Poole Harbour (631ha), Langstone Harbour (554ha) and Dungeness (483ha). The statutory Bird Sanctuary on Radipole Lake in Weymouth is also managed and wardened by the RSPB.

Southern England

Dorset and the Hampshire Avon

The wildfowl on the Dorset coast are confined almost exclusively to the Chesil Fleet and Poole Harbour, and to a few much smaller sites nearby. Elsewhere there are long stretches of cliff and shingle bank, separating the two main groups from one

Fig. 15. Dorset and the Hampshire Avon.

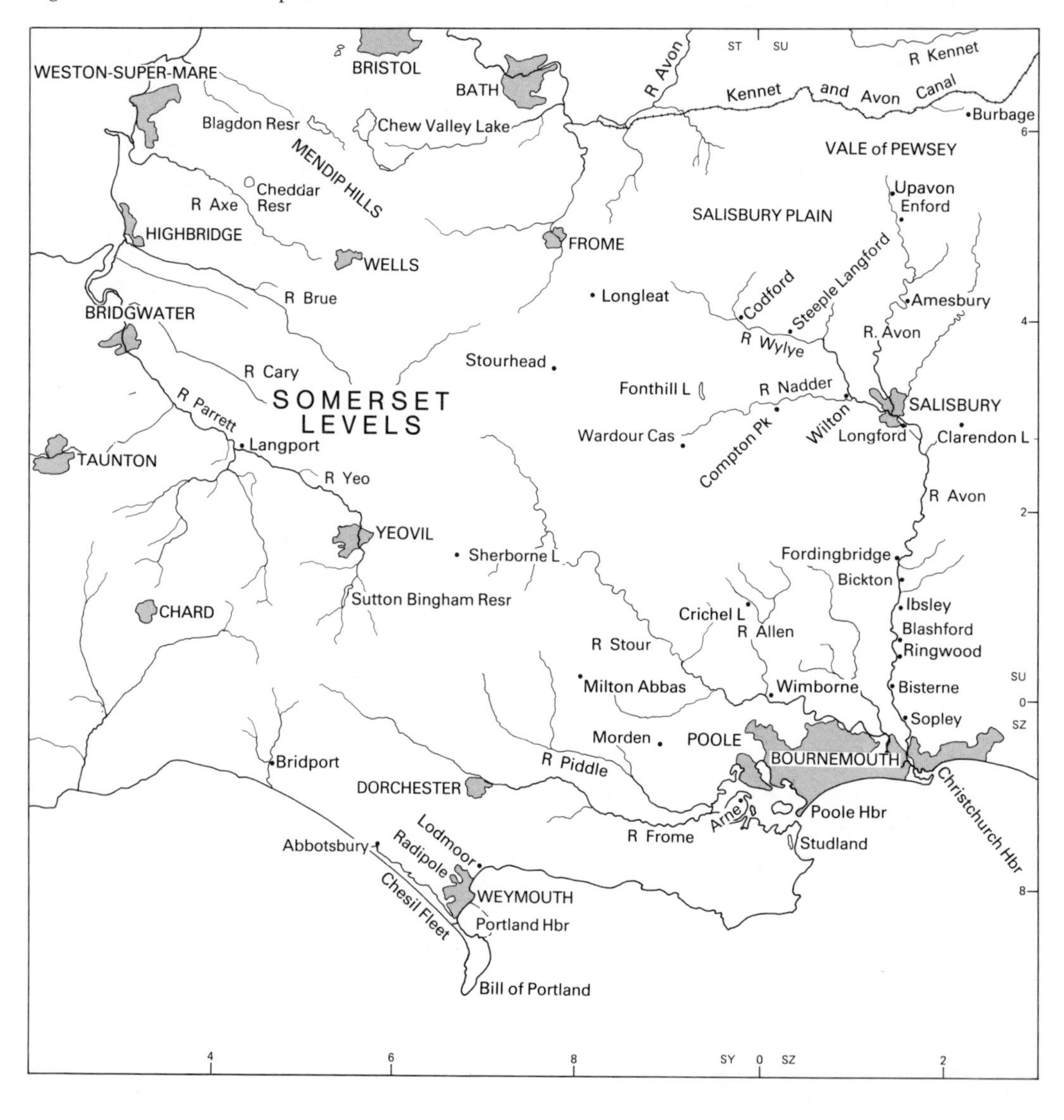

another and from the neighbouring flocks on the Devon and Hampshire shores.

The Chesil Fleet is a tidal lagoon, running parallel to the sea for 14km and covering about 480ha. At the eastern end it opens into Portland Harbour through a narrow channel, but otherwise it is divided from the sea by the high pebble ridge of the Chesil Bank. Because of the constricted entrance the tidal range is very much smaller than outside, amounting to less than 15cm at the upper end, and, although the depth is mostly under 2m, there are no large areas of intertidal shore. The salinity of the water also decreases towards the upper end, where the Fleet is joined by several small streams. *Ruppia* and all three species of *Zostera* are plentiful, and provide an abundance of food for the regular winter gatherings of 5-6,000 ducks and up to 1,000 Mute Swans. Brent Geese also occur in small but increasing numbers.

Wigeon are especially numerous with a normal winter peak of 4-5,000, and most of the other dabbling ducks, including Gadwall, Pintail and Shoveler, are present in fair numbers. Pochard, Tufted Ducks and Goldeneyes occur regularly, mostly on the brackish upper reaches, and Mergansers are common both on the lower Fleet and in Portland Harbour, the late winter total amounting to above 200 (Table 30).

The Brent Geese started to winter regularly in 1978, and now reach a January or February peak of 2-300 with an exceptional influx of 628 in December 1982. In the previous nine years of regular counts they were seen in four seasons only, twice as stragglers and twice in flocks of 90-120 (Table 36). This earlier scarcity, and the present lowish level, is hard to explain, in view of the ample food; perhaps the smallness of the tidal range restricts their feeding pattern.

Mute Swans have been established on the Fleet for at least 600 years, the earliest reference to them being in 1393. At that time the herd was already being managed on behalf of the Lord of the Manor as a much prized source of food, and the same medieval practices continue to this day, though the young birds are no longer sent to the table. Unlike those elsewhere in Britain, the swans at Abbotsbury nest colonially within an area of about 1ha, the number varying from 19 pairs in one recent season to more than 90 in another. As soon as the cygnets are hatched a proportion of them, usually 70-100, are placed in seven pens, each with a pair of adults, and are fed by hand until their release in mid-September. In this way the losses of young, which would otherwise be heavy, are greatly reduced, and the resident population is maintained at a buoyant level of around 500. At the start of the breeding season the residents only are present. In June and July the herd is augmented by birds from elsewhere arriving to moult, and further

Table 30. *The regular (and maximum) numbers of wildfowl at the five main sites in Dorset. The figures under the place names show the numbers of seasons for which full data are available, and the period in which these seasons occurred.*

	Chesil Fleet 11:1972-82	Radipole Lake 13:1970-82	Poole Harbour 13:1970-82	Christchurch Harbour 12:1970-82	Crichel Park Lake 11:1970-82
Mallard	172 (382)	159(420)	856(1750)	37 (90)	427(1500)
Gadwall	38 (175)	3 (23)	27 (120)	3 (14)	14 (48)
Teal	186 (777)	180(580)	1058(1787)	79 (350)	10 (51)
Wigeon	3718(10200)	15(205)	980(3550)	77(1000)	12 (47)
Pintail	78 (270)	2 (20)	169 (525)	4 (32)	0 (8)
Shoveler	71 (255)	87(236)	89 (215)	10 (64)	8 (25)
Scaup	7 (20)	2 (10)	23 (185)	1 (7)	0 (0)
Tufted Duck	130 (740)	353(775)	338 (679)	6 (27)	64 (125)
Pochard	331 (1440)	310(870)	189 (683)	8 (144)	45 (115)
Goldeneye	90 (210)	1 (4)	73 (315)	13 (35)	0 (1)
Merganser	74 (210)	0 (1)	162 (535)	2 (10)	0 (0)
Shelduck	69 (180)	8 (60)	1620(2700)	56 (123)	0 (0)
Canada Goose	0 (60)	0 (0)	315 (622)	0 (2)	22 (61)
Mute Swan	827 (1240)	59(151)	55 (419)	see text	5 (16)

Notes: Chesil Fleet includes Abbotsbury; Poole Harbour includes Brownsea Island Lagoon and Little Sea, Studland.

For Brent Geese, see Table 36.

influxes occur from September onwards, giving rise to a mid-winter peak of 8-900, and recently more than 1,000 (468). The origin of many of these visiting swans has not yet been determined: up to 100 are known to come from Radipole Lake nearby, and another 50 or so may perhaps be local birds from the pools and streams inland; the rest, totalling 100-250, and occasionally up to 350, are apparently drawn from a much wider area, beyond the county boundary. The ringing results so far have shown a small amount of interchange between the birds on the Fleet and those on the Exe to the west, but not with those on Poole and Christchurch Harbours to the east.

A wide range of species has been recorded in Portland Harbour, largely sea ducks, though only Red-breasted Mergansers occur in other than tiny numbers. Their regular number is 73 but as many as 200 have been recorded. Only a few scoters winter but thousands occur on spring passage at Portland Bill, no doubt the same birds that are recorded elsewhere along the south coast.

Radipole Lake is set in the middle of Weymouth, some 3km from the lower end of the Fleet. At one time it was part of the Wey Estuary, but in 1920 the inner basin was cut off from the sea by a causeway and sluice, and became a shallow freshwater lake, covering about 100ha. Since then the area of open water has been greatly reduced by silting and by the spread of *Phragmites* over much of the upper part. Despite this the lake continues to attract a winter peak of 1-2,000 ducks and a summer gathering of more than 100 Mute Swans. These are mostly located on the central and upper reaches including the reed-beds, which provide a series of secluded bays. The lower part is virtually devoid of food plants, and often much disturbed by boating and the like. In the other, more important sections the birds are protected by a Sanctuary Order, imposed in 1948, and by the RSPB reserve, established in 1978, which covers much the same area. Under the present regime the marsh is wardened and managed, and several species have increased, notably Shoveler and Pochard, both of which have doubled in numbers since the 1960s (Table 30).

The swans, on the other hand, have steadily decreased. Prior to 1967 there were normally 2-300 in September, 80-100 through the winter, and 180 in March. Since then the autumn counts have dropped to about 65, the winter level to less than 15, and the spring return to around 40. Summer counts throughout the 1970s have shown that some 20 pairs attempt to breed each year, but the number of cygnets fledged is seldom more than 10 (468). The lake is also used as a moulting place by swans from elsewhere, bringing the total in July and August to perhaps 100, a level which is certainly much lower than it was some 20 years ago. As soon as the moult is complete, both the local and the visiting birds begin to move away, mostly to Chesil Fleet, where they mingle with the flocks already there. If the two groups are taken together, it emerges that the recent losses at Radipole have been largely offset by the gains on the Fleet. For more than 50 years the Abbotsbury estate has organised an annual autumn count on both resorts, the results of which are shown in Fig 16. During this time the totals have varied from a maximum of 1,580 in 1929 to a mere 400 in 1947, when

Fig. 16. The number of Mute Swans on the Fleet, 1930-1982. From Perrins and Ogilvie (1981).

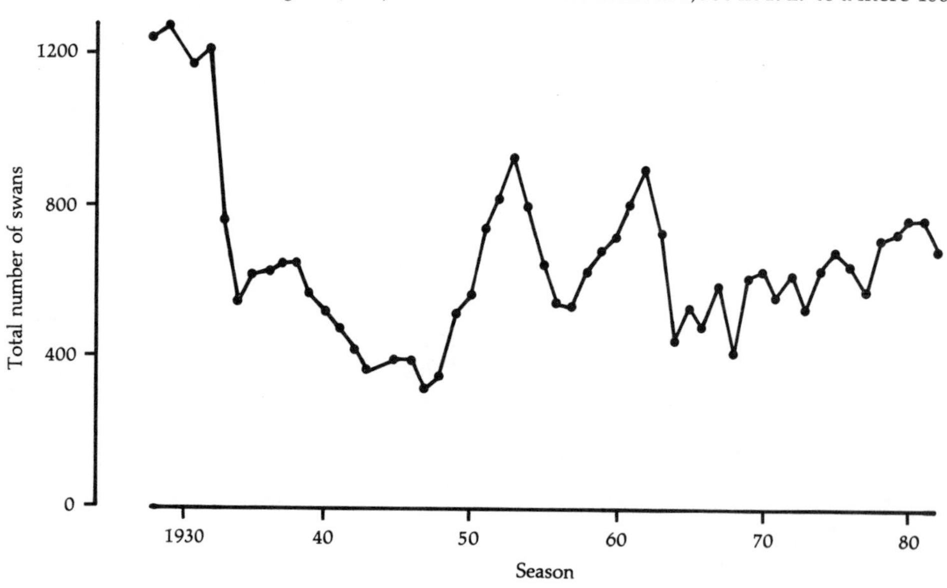

the cumulative effect of the *Zostera* failure in the 1930s and the subsequent war-time disturbance was capped by a winter of unusual severity. Since then there have been two periods of rapid increase, each followed by an equally steep decline, the decreases following the two hard winters of 1953 and 1962. Thereafter, the numbers have again shown an upward trend, but, perhaps because of the wane at Radipole, the rate of increase has been much slower and less sustained.

A short way to the east of Radipole, the coastal lagoon and fresh marsh at Lodmoor has also been acquired by the RSPB. At one time this 50ha site was used by several hundred dabbling ducks, mostly after the close of the shooting, but in 1956 the water level was lowered, and the numbers thereafter declined. In recent years the area has held a normal peak of 50-100 Teal, 20-30 Mallard and Shelducks, and up to a dozen Mute Swans; there have also been records of up to 100 Shoveler and 4-500 Wigeon, but only in the hardest winters. With suitable management and wardening the levels in future may well be improved.

Poole Harbour, with an area of more than 3,000ha and a current winter peak of 5-7,000 wildfowl, is one of the six prime centres on the Channel coast. As at Chesil, the shallow tidal basin is almost completely land-locked, the only access to the sea being a narrow channel at the eastern end. In consequence the tides within the harbour are restricted to about 2m, and are subject to a double ebb and flow. Most of the deeper, northern half is dominated by the urban spread of Poole and Parkstone, but apart from this the shores and surrounds are virtually unspoiled. The southern side, in particular, is shielded from development by the National Nature Reserves on Studland and Godlingston Heaths (631ha), which reach to the water's edge and include some intertidal land, and by the RSPB reserve on the Arne Peninsula, which covers another 530ha of the heath adjoining the shore. Brownsea Island, in the centre of the harbour, is also well protected, being owned by the National Trust, who lease 100ha to the Dorset Naturalists' Trust as a nature reserve.

Despite the smallness of the tides, the harbour has nearly 1,000ha of muddy foreshore and another 1,000ha of saltmarsh, including some of the largest and oldest *Spartina* beds in Britain. There are also some extensive reed-beds, and a few small patches of *Zostera* on the sandy stretches near the mouth (492). The southern shore, between Arne and Studland Heath, is especially attractive, with numerous sheltered creeks and bays, enclosed by a series of islands and outlying saltings. The western end, around the outfall of the Frome and Piddle, is more open and estuarine in character, and is also favoured as a feeding ground – much more so than the northern shore including Holes and Lychett Bays. In addition to these tidal areas, there are several freshwater pools nearby, which are used extensively by birds from the harbour, both as feeding grounds and as day-time retreats. The most notable of these are the Little Sea in the Studland Heath reserve, the marshy lagoon on Brownsea Island and a newly constructed shallow freshwater lake on the north shore of the Arne Peninsula. Some of the lakes to the west around Morden used also to carry several hundred ducks, but no recent counts are to hand.

The wildfowl in the harbour, including those on the Little Sea and Brownsea Island, have increased remarkably over the past 30 years. Mallard, Wigeon, Shoveler and Shelduck have doubled in numbers since the 1950s; Tufted Duck, Pochard and Merganser have increased threefold, and most of the other species, including Teal, Pintail and Goldeneye have shown substantial gains. Gadwall and Brent Geese have also become established during this time, and the flock of Canada Geese, introduced in 1956, now reaches a mid-winter peak of around 400 (Table 30).

The gatherings of Shelducks, which by January often top 2,000, are important internationally, amounting each year to between 1 and 2% of the north-west European population. The recent success of the Brent Geese is also of interest, especially from the local viewpoint. Prior to the wasting of the *Zostera* in the 1930s, they were said to be abundant in the harbour, but by 1950 the flocks had almost disappeared, and for the next 20 years the monthly counts only once reached 25. Then in the early 1970s they started to return; by 1980 they had reached 500 and in 1982 they exceeded 1,000. They now appear in strength from November onwards, the largest numbers being in January or later (Table 36).

To the east of Poole Harbour, the seafront of Bournemouth extends for nearly 15km to the next, much smaller inlet of Christchurch Harbour, at the mouth of the Avon and Stour. Here again the tidal basin is almost completely enclosed, and is flanked to the north by the town. Apart from a few small stretches of foreshore, the only area of value is the 60ha of fresh marsh near Stanpit, which is now a Local Nature Reserve. Considering its size, the area attracts a surprisingly wide range of species, though the numbers are mostly quite small (Table 30). The main features are the large summer gatherings of Mute Swans, and the regular occurrence of Bewick's Swans, and Brent and White-fronted Geese.

The Mute Swans, which come here to moult, reach a peak in July of about 350 (maximum 451 in August 1982). They are drawn for the most part from the areas upstream along the Avon and its tributaries, to which they return in the course of the autumn, leaving a regular winter level of 10-40. They may also be linked with the flock in Poole Harbour. The White-fronted Geese and Bewick's Swans, both of which began to winter regularly in 1977, are presumably offshoots from the well-established groups in the Avon Valley between Ringwood and Fordingbridge. The geese are normally present from December to February, and over the past six seasons have attained a mean annual peak of about 300 (max 500). The Bewick's Swans occur mostly in parties of 5-10, but up to 30 are on record. A few Whooper Swans have also been seen from time to time. The Brent Geese, like those at Poole and on the Fleet, began to appear in sizeable flocks in 1978, and have since been present regularly from November to March each year. The late winter peak is usually close to 100, but it has been as low as 15, and as high as 220.

The inland parts of Dorset are a great deal less important than the coast. Standing water is scarce in nearly all districts, especially on the chalk and limestone, and for the most part the wildfowl are scattered in smallish groups along the rivers and adjacent meadows. The only major resorts are Crichel Park Lake, in the Allen Valley to the north of Wimborne (Table 30), and Sherborne Park Lake on the head waters of the Yeo, which is taken in context with the Somerset Levels (p.48). At Crichel the flocks of Mallard and Teal have decreased substantially since the 1960s, the former from a regular level of about 940 to scarcely more than 400, and the latter from 50 to 10. The diving ducks, on the other hand, have held their own, and both Gadwall and Canada Geese have shown a slight increase. Most of the other inland records date from the early 1960s. At that time the lake at Milton Abbas held a normal level of about 30 Mallard, Pochard and Tufted Ducks (maxima 60-100), and the stretches of the Stour and Allen to the west and north of Wimborne together held about 100 Mallard, 50-60 Teal and up to 20 Mute Swans. Small flocks of Mallard, and a few Mute Swans and Teal, were also found at several points in the upper valleys of the Frome and Piddle.

The floodplain of the lower Avon, lying partly in Dorset but mainly in Hampshire, ranks equal in importance with the coastal sites at Poole and Chesil, though the flocks are spread more widely. In the 30km stretch between the estuary at Christchurch and the Wiltshire border, there are several related resorts, which together hold a winter peak of around 5,000 ducks, 800 White-fronted Geese and at least 100 Bewick's Swans. Mute Swans and Canada Geese are also present in substantial numbers.

The two most important sections lie between Sopley and Bisterne villages to the south of Ringwood, and between Blashford and Bickton to the north. Both are subject to prolonged winter flooding, and at Blashford a string of gravel pits, extending for 3km along the eastern side of the valley, provides a permanent roost and a supplementary feeding ground. Several of these pools are already ear-marked for sailing and other activities, but the digging continues and there is hope that at least two pits may in future be reserved for birds. The ones most used at present are Ellingham Lake and Ivy Lake near Rockford, and the continually expanding Mockbeggar Lake, which is being excavated on the site of Ibsley airfield (124).

The adjoining water-meadows between Blashford and Bickton, lying mainly on the Somerley estate, are reasonably well protected from disturbance, and provide the prime resort for the White-fronted Geese and Bewick's Swans. The geese began to winter here in 1940, and by the early 1950s were occurring regularly in flocks of 6-700 (127). This was followed by a further increase, bringing the total to about 1,000 in the mid 1960s, and to upward of 1,500 in the seasons around 1970. Since then the annual peaks have steadily decreased, in keeping with the trend elsewhere in Britain, the only exceptions being in the cold winters of 1978-79 and 1981-82 when totals of 1,400-1,500 were again recorded (Table 31). Although feeding mainly on the Blashford section of the meadows, they sometimes move to the areas south of Ringwood between Bisterne and Sopley, and up to 300 birds are sometimes seen on the Stanpit Marsh near Christchurch. When at Blashford, the geese normally roost on the flooded meadows, but occasionally turn to one or other of the pits at Ellingham. They are usually present from December to early March.

The Bewick's Swans prefer the Ibsley and Harbridge meadows, a little to the north of Blashford and the area around Sopley, but have recently become regular at other points along the lower valley, especially when their favourite fields are too deeply flooded. The flock first started to winter regularly in the 1960s, and is now of international importance, with an average peak of about 125 over the past 7 seasons (Table 31). The main flock arrives in November, and remains in strength until early March, the largest numbers being around the New Year.

The Ibsley and Harbridge meadows attract in addition a winter gathering of 100-150 Mute Swans, and similar numbers are often reported on the stretch to the north of Fordingbridge. In summer the majority depart for the moulting areas at Christchurch, and possibly Poole, leaving only the immature birds and a scatter of breeding pairs. The species is also common on the meadows at Sopley, and at several points along the Avon and Wylye Valley to the north and west of Salisbury (Table 32).

The Canada Geese, based mainly on the gravel pits, have increased from a maximum of 150 in 1973 to about 350 in 1981, and seem now to have reached a fairly stable level. In 1978 there were 17 breeding pairs, of which 9 were on Spinnaker Lake near Rockford, and 72 goslings were successfully reared (124). The largest counts are usually in autumn, when the numbers at Poole are also approaching their peak. Elsewhere in the Avon basin the flocks are small and erratic, the nearest other groups being in the Test Valley, 16km to the east, at the mouth of the Beaulieu River, some 25km east, and in the Kennet Valley more than twice as far to the north.

The meadows and pits between Blashford and Bickton are also used extensively by ducks. Wigeon are especially numerous, with a normal peak of around 2,000, and the flocks of Pintail and Shoveler are often the largest in Hampshire. Gadwall, too, have become established during recent years, mainly on the Ellingham pits, which are also favoured by the diving ducks, including Goldeneyes. Tufted Ducks are present throughout the year, and 5-10 broods are raised in most seasons (max 13 in 1979).

The Sopley meadows lack the advantage of an alternative roost and feeding ground close by (although the Wigeon sometimes fly north to Ivy Lake when disturbed). They nevertheless attract a regular mid-winter total of 1,500-2,000 ducks, and in several recent seasons have held flocks of 50 or more Bewick's Swans. The largest numbers are usually found when the floods at Blashford and Harbridge are too deep for the ducks to feed there. There is, in any case, a good

Table 31. *Above: the regular (and maximum) numbers of wildfowl at five resorts in the Avon Valley between Sopley and Salisbury. Below: the seasonal maxima of European White-fronted Geese and Bewick's Swans on the Avon water-meadows, Sopley – Fordingbridge (including data from Clark (1979) and Hampshire Bird Reports).*

(a) Sopley – Bisterne, water-meadows 5 seasons 1978-1982
(b) Blashford – Bickton, meadows and gravel pits 7 seasons 1976-1982
(c) Fordingbridge – Wiltshire border, water-meadows 5 seasons 1978-1982
(d) Longford Castle, river and meadows 23 seasons 1960-1982
(e) Clarendon House Lake 3 seasons 1980-1982

	a	b	c	d	e
Mallard	500 (950)	215 (500)	70(145)	95(270)	122(250)
Gadwall	3 (12)	46 (110)	0 (0)	0 (5)	7 (20)
Teal	348(1300)	176 (900)	9 (40)	5 (30)	8 (15)
Wigeon	647(2500)	1459(3700)	31(250)	0 (40)	38(105)
Pintail	5 (55)	54 (190)	0 (2)	0 (1)	0 (0)
Shoveler	14 (50)	123 (430)	0 (3)	0 (3)	5 (12)
Tufted Duck	9 (30)	161 (380)	17 (40)	20(200)	32 (70)
Pochard	45 (200)	198 (370)	3 (10)	10 (90)	13 (40)
Goldeneye	0 (2)	15 (46)	0 (0)	0 (0)	0 (0)
Canada Goose	12 (61)	195 (350)	3 (30)	0 (47)	2 (5)
Mute Swan	28 (75)	95 (140)	130(190)	11 (40)	2 (5)

	Whitefronts	Bewick's Swans		Whitefronts	Bewick's Swans
1968	1550	32	1976	830	138
69	1550	17	77	750	118
70	1400	26	78	1400	100
71	1000	26	79	400	97
72	1200	34	80	350	156
73	850	35	81	1500	183
74	420	52	82	290	173
75	800	114			

deal of movement between the two stretches, which are less than 10km apart.

The middle valley of the Avon, between Fordingbridge and Salisbury, is relatively unimportant, the only major feature being the flock of Mute Swans which winters at the southern end, below the Wiltshire border. The largest gatherings, totalling 140 or more, occur in spring and autumn when the birds are progressing to and from their summer quarters on the coast at Christchurch; the normal winter level is 80-100. Totals of 50-100 Mallard, and a few Wigeon, Teal and diving ducks are also recorded regularly, and further to the north, near Salisbury, there are slightly larger numbers on the river by Longford Castle and on the lake at Clarendon House (Table 31).

Above Salisbury the valleys of the Avon and its tributaries, the Nadder and Wylye, are hemmed in by the dry chalk downland, which occupies the greater part of Wiltshire northwards to the watershed. The wildfowl are thus confined almost exclusively to the rivers and water-meadows, and to the scattered lakes and gravel pits along the valley floor. Most of the records, excepting those from the main resorts, were gathered in the mid and early 1960s, but the present pattern of numbers and distribution is no doubt much the same as then. Mallard and Mute Swans were reported, often in some numbers, on nearly all the stretches visited, together with a sprinkling of Teal and a few outlying flocks of diving ducks (Table 32). The most notable centres are the gravel pits at Steeple Langford, with an annual peak of 3-400 ducks, and Fonthill Lake near Tisbury with 4-500. Several hundred dabbling and diving ducks are also said to occur on the pools at Wardour Castle near Donhead.

In its uppermost reaches, to the north of Salisbury Plain, the Avon is supplemented by the Kennet and Avon Canal, which crosses the Vale of Pewsey *en route* from Hungerford to Bristol. Here again there are regular records of 1-200 Mallard and a dozen Mute Swans, mostly on the stretch between Pewsey and Wootton Rivers, and further small flocks are known to occur on the ponds around Burbage and elsewhere.

The Solent and the Hampshire – Sussex harbours

The coast to the east of Christchurch Harbour is again backed by cliffs and narrow pebble beaches, and by belts of housing, stretching for more than 10km to the Hurst Point shingle spit at the western entrance to the Solent. The sandy heathland of the New Forest, extending inland from the coast for some 20km and

Table 32. *The regular (and maximum) numbers of the five most common species of wildfowl at various places in the Avon basin above Salisbury. The figures in brackets after the place names show the number of seasons on which the results are based: * means that the counts were made in the 1960s. The lengths of the stretches of river are shown in km.*

		Mallard	Teal	Tufted Duck	Pochard	Mute Swan
NADDER VALLEY						
Wilton – Barford, 4km	(3)*	160(215)	10 (20)	0 (0)	0 (0)	7(13)
Hurdcott House Lake	(1)*	100(250)	10 (15)	0 (0)	0 (0)	1 (2)
Compton Park Lake	(6)*	70(200)	0 (1)	7 (20)	10 (35)	2 (3)
Fonthill Lake	(11)	170(400)	25(110)	93(165)	65(157)	6(11)
WYLYE VALLEY						
Wilton – Wishford, 4km	(1)*	170(190)	0 (0)	0 (0)	0 (0)	40(50)
Steeple Langford GPs	(9)	130(320)	3 (20)	60(130)	80(160)	35(80)
Bathampton House 1km	(4)*	45(125)	1 (5)	0 (0)	0 (0)	15(70)
Wylye – Bapton, 2km	(4)	25 (70)	0 (6)	0 (0)	0 (0)	6(12)
Bapton – Codford, 2km	(6)*	120(230)	10 (35)	0 (0)	0 (0)	6(16)
Codford – Sherrington, 1km	(3)	10 (25)	2 (6)	0 (2)	0 (0)	8(20)
AVON VALLEY						
Sarum – Amesbury, 14km	(3)*	620(820)	30 (80)	3 (15)	0 (7)	25(55)
Enford – Upavon, 4km	(1)*	30 (35)	0 (0)	0 (0)	0 (0)	4 (5)
Upavon – Woodbridge, 2km	(1)	130(200)	0 (6)	4 (10)	0 (2)	6 (7)
KENNET – AVON CANAL						
Pewsey – Wootton Rivers	(4)*	155(255)	0 (0)	0 (2)	0 (0)	0(10)
Savernake – Gt Bedwyn	(2)*	30 (55)	0 (0)	0 (0)	0 (0)	11(13)

Note: Fonthill Lake also held 10 Shoveler (max 55) and 20 Gadwall (max 90).

occupying the greater part of the area between the Avon Valley and Southampton Water, is likewise devoid of any suitable resorts, except perhaps for a few small pools and boggy valley bottoms.

At Hurst, however, the character of the coastline changes, and for the next 60km there are frequent shallow bays and expanses of muddy foreshore, which together attract a current peak of 30-40,000 wildfowl. The resorts here fall into three main groups, the first comprising the marshes of the Western Solent between Hurst and the Beaulieu Estuary, the second the mudflats of Southampton Water, and the third the complex of harbours between Gosport and Pagham on either side of the Sussex border. Some of the inlets along the northern coast of the Isle of Wight are also of value, and are no doubt linked with the sites on the mainland shore. In the area as a whole there are nearly 6,300ha of intertidal mud and sand, and a further 3,000ha of saltmarsh, 85% of which is dominated by *Spartina*, the rest by *Puccinellia* and Sea Purslane (*Halimione portulacoides*) (596).

The principal sites in the Western Solent are the Keyhaven and Pennington Marshes in the lee of the Hurst Peninsula; the Sowley shore and adjacent freshwater lake to the east of Lymington; the area around the mouth of the Beaulieu River, including the freshwater pools on Needs Ore Point; and the Newtown Estuary on the Isle of Wight (Table 33). At the western end, around Keyhaven, the mudflats are upwards of 1,500m wide, but become increasingly narrow and eventually reduce to less than 100m to the east of the Beaulieu River. For the most part they are covered with dense growths of *Spartina*, though these are now showing signs of die-back, and are being replaced by *Enteromorpha* which is much preferred as a wildfowl food plant. Substantial beds of the narrow-leaved *Zostera noltii* are also established, both on the lower flats and in the saline lagoons along the sea wall, and in places there are fair amounts of *Salicornia* and Sea Aster (*Aster tripolium*). At Sowley and on the

Fig. 17. Hampshire and West Sussex.

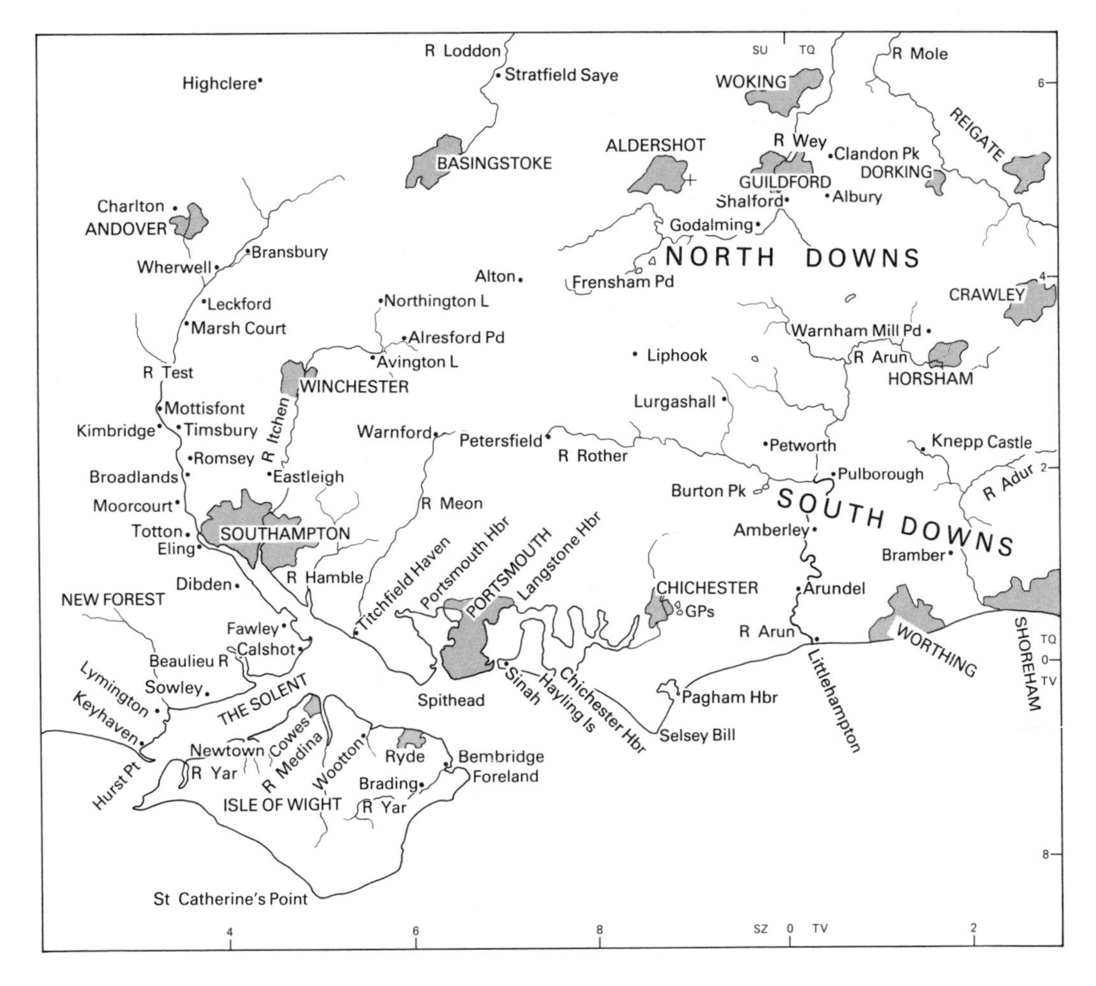

Beaulieu Estuary the freshwater pools adjoining the shore are a valuable adjunct to the saltmarsh, and are used extensively as a day-time retreat. Sowley Pond is especially favoured and often holds 3-400 Teal (max 800) and 1-200 Mallard (max 500), with smaller numbers of Wigeon and diving ducks. The pools on Needs Ore Point are noted chiefly for the flock of Canada Geese, which was brought here in 1964 and now totals more than 100.

Although the estuaries of the Lymington and Beaulieu Rivers are popular sailing centres, the disturbance on the Solent shore is not unduly great, and is further controlled by the establishment of two substantial reserves. The larger of these, the North Solent NNR, was declared in 1980 and covers 660ha of the marsh and foreshore between Hurst and Lymington. The other, covering 160ha, includes the greater part of the freshwater marsh and the adjoining shore at Needs Ore Point, and is managed by the Hampshire and Isle of Wight Naturalists' Trust by agreement with the owner.

On the opposite shore, in the Isle of Wight, the tidal reaches of the Newtown River are owned by the National Trust, and are further protected by a Local Nature Reserve covering 120ha. At the seaward end the harbour is enclosed by a shingle spit, pierced only by a narrow channel, behind which the tideway branches into two main arms and a number of smaller bays and creeks. Between the two arms lies a promontory of former fresh marsh which, following the breaching of the wall in 1954, has developed partly into saltmarsh and partly into a level expanse of mud, densely clad with *Enteromorpha*. In the two main channels there are further areas of saltmarsh, including a certain amount of *Spartina*, and some sizeable beds of *Salicornia* and Sea Lavender (*Limonium vulgare*) (492). The estuary as a whole is still remarkably unspoiled, and disturbance is minimal, thanks to the National Trust which controls all mooring and other harbour rights (366).

Encouraged by these three reserves, and by the recent spread of *Zostera* and *Enteromorpha*, the wildfowl of the Western Solent have increased substantially over the past 20 years to a current peak of at least 6,000 birds. The Dark-bellied Brent Goose, in particular, has repossessed its former stronghold, abandoned in the 1930s, and now occurs throughout the district in flocks totalling up to 2,000 (Table 36). Teal are also plentiful, rather more so than Mallard or Wigeon, and several other species, including Pintail and Shelduck, occur in moderate numbers (Table 33).

The other estuaries in the Isle of Wight are less commodious and more disturbed than Newtown Harbour, and hold many fewer birds. The West Yar, at Yarmouth, has a normal peak of about 150 ducks, consisting mainly of Teal and Wigeon, with perhaps a dozen Mallard and Shelducks; the Medina, upstream from Cowes, carries up to 100 Wigeon, 100 Shelducks and 40-50 Mute Swans, and Wootton Creek, with the freshwater Mill Pond at its head, has at times held 200 Mallard and 25 Teal and Pochard, but totals of less than 50 are probably more usual. On either side of Wootton Creek the coast facing on to Spithead is bounded by extensive sandflats, especially to the east of Ryde. *Zostera* is plentiful along the outer edge, and in recent years the area has attracted occasional flocks of 50-100 Brent Geese. These have been seen at several scattered points, but the data are too meagre to

Table 33. *Regular (and maximum) counts from five sites in the Solent, 1970-1982. Figures under the site name are number of seasons: period of counts.*

	Keyhaven/ Pennington 13:1970-82	Sowley Marsh 6:1972-82	Needs Ore Point 11:1970-82	Newtown I of W 11:1971-81	Brading I of W 13:1970-82
Mallard	24 (130)	11 (32)	314 (600)	7 (47)	211(577)
Teal	237(1300)	167(580)	363(1200)	326(1000)	153(499)
Wigeon	95 (480)	189(450)	422 (900)	321 (583)	12 (64)
Pintail	8 (100)	0 (5)	8 (120)	79 (160)	0 (10)
Shoveler	5 (40)	0 (2)	11 (29)	0 (74)	15 (60)
Pochard	14 (150)	0 (0)	3 (45)	0 (0)	60(128)
Tufted Duck	1 (24)	0 (0)	1 (11)	0 (0)	29 (49)
Goldeneye	7 (24)	0 (0)	1 (7)	5 (11)	0 (4)
Merganser	19 (40)	2 (10)	5 (13)	14 (44)	1 (4)
Shelduck	303 (700)	52 (93)	73 (140)	290 (484)	50(102)
Canada Goose	0 (2)	0 (0)	72 (143)	43 (120)	1 (23)
Mute Swan	22 (81)	3 (8)	2 (6)	5 (10)	19 (28)

establish any pattern. Parties of 20-30 Mergansers, increasing at times to 50 or more, are also common along this stretch, a favourite centre being in Osborne Bay. The numbers of other species are not known.

A short way to the east, near Bembridge, the estuary of the East Yar and the adjoining freshwater marshes along the lower river support a further total of 4-500 ducks, making this the second most important centre in the island (Table 33). Most of the birds, except the purely estuarine species, are normally found on the reclaimed marshes, which extend inland for nearly 3km, as far as Brading, and are subject to regular flooding, lasting in some years for two or three months. In addition there are several freshwater pools, which provide a permanent resort for both dabbling and diving ducks. The estuary itself, now greatly truncated, is too confined to be of value, except perhaps round the outer banks. Some quite large flocks of ducks have also been reported on the open coast near Bembridge, in the reefs around the Foreland. Two counts only are to hand, but they seem to suggest a fairly regular usage. In January 1974 the area held 65 Mallard, 35 Wigeon and 10 Shoveler, and in January 1980 it had 180 Mallard, 100 Wigeon and 26 Shoveler.

Elsewhere in the island there are no large lakes or ponds, and most of the remaining coast is steep and inhospitable. Common Scoters, sometimes totalling several thousand, are seen on passage from St Catherine's Point, moving eastwards from mid-March to May, and westwards in much smaller numbers during August, but very few occur at other times. Totals of up to 100 Mergansers have also been seen moving along the southern coast in spring, but apart from this the numbers of all species are negligible.

The busy tideway of Southampton Water, although overlooked by long stretches of urban and industrial development, retains a number of attractive areas, which together hold an annual peak of 3-5,000 ducks and 7-800 Brent Geese. On the lower half of the western shore the extensive mudflats between Calshot and Hythe are dominated by the power station at Fawley and by the adjoining oil refinery, with its deep-water jetties and array of storage tanks and other works, stretching in all for about 6km. This particular development has in fact proved beneficial, because of the restrictions placed on access to the shore. The flocks have also benefited from the Local Nature Reserve, established in 1979 on 50ha of the marsh and foreshore in the lee of Calshot Spit. In consequence the numbers of both ducks and Brent Geese are now the largest in the district (Tables 34 and 36). Wigeon and Shelducks have increased substantially since the 1950s, Teal have maintained their earlier abundance, and Tufted Ducks and Pochard have been drawn increasingly to the new works reservoir adjacent to the shore. The Brent Geese, which started to winter on the Calshot flats in 1975, have also increased and now occur regularly in flocks of around 400. The only cause for concern is the possibility of a major oil spill, which could be disastrous in these narrow waters.

Above Hythe the former expanse of Dibden Bay is occupied by a series of embanked lagoons, which in turn are being infilled to a depth of 3-4m with the mud dredged from the fairway to Southampton docks. In the early stages, when the mud is soft and overlaid with salt or brackish pools, the pans are used extensively by Shoveler, Teal and Shelducks, but are much less suitable for Wigeon than the foreshore in its natural state. Later on, when the pans are full, the mud dries out and develops into firm fresh pasture-land, which may in time be suitable for ploughing. At

Table 34. *Regular (and maximum) counts from five sites in or near Southampton Water in the period 1970-1982. Figures under the site name are number of seasons: period of counts.*

	Calshot – Fawley 13:1970-82	Dibden Bay 6:1977-82	Eling – Totton 5:1978-82	Hamble Est /Netley Shore 5:1978-82	Titchfield Haven 13:1970-82
Mallard	166 (350)	8 (35)	81(260)	5 (10)	125 (255)
Teal	933(3000)	121(800)	73(254)	139(325)	380(2140)
Wigeon	88 (450)	5 (50)	115(380)	0 (1)	558(2110)
Shoveler	18 (43)	35(130)	1 (7)	0 (2)	19 (74)
Pochard	78 (175)	0 (0)	2 (28)	0 (0)	3 (25)
Tufted Duck	107 (220)	0 (0)	5 (34)	0 (30)	3 (19)
Goldeneye	0 (2)	1 (6)	2 (12)	20 (98)	0 (2)
Shelduck	334 (710)	165(500)	13 (55)	25(106)	11 (27)
Canada Goose	0 (1)	0 (0)	66(295)	0 (0)	12 (64)
Mute Swan	5 (9)	3 (21)	19 (57)	14 (40)	8 (50)

Dibden there are four such pans, begun at intervals of about 10 years between 1940 and 1967, and covering a total of 180ha (216). Only the most recent of these, covering 65ha, is now wet enough for wildfowl, and before long the whole of the bay will have been reclaimed except for a narrow strip of foreshore at the outer edge. Meanwhile the site still holds a normal peak of 2-300 ducks, and occasionally up to 1,000.

More than 200ha of foreshore have also been reclaimed in the Test Estuary, upstream from Dibden, initially in the 1930s for the Western Docks, and again in the 1970s for the new container terminal. The Marchwood power station on the other side is likewise built on reclaimed land, leaving only the narrow strip of Eling saltmarsh on the upper half. This is still used regularly by 40-50 Teal and 100-150 Wigeon, and another 2-300 ducks, mostly Mallard and Teal, are found above the Totton Bridge, where a part of the adjoining freshwater is controlled by the Hampshire and Isle of Wight Naturalists' Trust. The estuary also attracts an autumn gathering of 40-50 Mute Swans, and is often used by the flock of up to 140 Canada Geese which winters by the Lower Test.

The mudflats on the eastern side of Southampton Water are relatively narrow and the wildfowl are mostly confined to the Hamble Estuary, and more especially to the fresh marshes at Titchfield Haven in the lower Meon Valley. The only notable occurrences elsewhere are the newly established flocks of Brent Geese and the small groups of sea ducks which winter on the open tideway.

The Brent Geese began to winter off the Hamble Estuary in 1975, when they also appeared at Calshot, and by 1978 had spread to two more points along the shore, the one off Weston and Netley, above the Hamble outfall, the other lower down towards Hill Head. A good deal of interchange no doubt takes place between these sites, none of which is more than a few kilometres distant from the next, but each is occupied throughout the season, and seems to be used by fairly constant numbers. As in the western Solent, the spread of the species has been prompted by the massive increases in Chichester and Langstone Harbours, which in turn reflect the rapid growth of the Dark-bellied Brent over the past 10-15 years.

The Hamble Estuary is long and narrow, and with moorings for about 3,000 yachts has neither the space nor the solitude required for a major assembly of ducks. It nonetheless attracts a regular winter peak of 2-300 Teal, and at times holds 50-100 Shelducks. A few Mute Swans occur as well, and further groups totalling 20-30 are found along the open shore off Netley and Weston, in company with 10-20 Goldeneyes.

At Titchfield Haven the former estuary of the River Meon is separated from the sea by a wall and sluice, built in 1611, and the river now flows through a narrow freshwater pool, about 1.5km long, with marshy meadows on either side. Winter flooding is frequent and prolonged, and gatherings of more than 1,000 ducks are often present for two or three months on end (Table 34). In the late 1960s the numbers of all species were greatly reduced, following the dredging of a new channel, but they have since regained their earlier level, thanks largely to the creation of a Local Nature Reserve on 85ha of the choicest stretch. In addition, the Hampshire and Isle of Wight Naturalists' Trust maintains its own reserve on 28ha of the meadows next upstream.

Offshore in the Solent the mussel beds around the Meon outfall attract one of the largest flocks of Eiders on the Hampshire coast. These first started to appear in the 1950s (574), and in 1969 reached a record level of about 130. Since then the annual peaks have varied between 5 and 80, but 10-20 is more usual (235).

To the east of Southampton Water the great tidal basins of Portsmouth, Langstone and Chichester Harbours, and the smaller inlet at Pagham, are important internationally as a wintering ground for 8-10,000 ducks and 16,000 Brent Geese. The three main harbours, although divided by Portsea and Hayling Islands, are joined in their upper reaches by tidal channels, and form a single ecological system. All three have narrow entrances, opening into broad expanses of shallows, with a fan of creeks and waterways stretching inland for 6 or 7km. The extent of the habitat is shown in Table 35. Pagham Harbour, lying 8km to the east of the main group, is similar in conformation, but covers less than 300ha in all.

The human impact on the harbours is unavoidably severe in a populous district such as this, but much has recently been done to lessen the effects by sympathetic planning and by the creation of reserves.

Table 35. *Area in hectares of the main intertidal habitats in Portsmouth, Langstone and Chichester Harbours at mean low water, 1975. From Tubbs (1977).*

	Portsmouth	Langstone	Chichester
Mudflats	776	1320	1298
Sand	0	72	164
Spartina marsh	173	216	611
Saltmarsh	0	48	42

Note: Since 1975 the *Spartina* marsh has been progressively reduced by die-back, and parts of it are now replaced by open mud.

At the same time, a great store of information has been assembled on the ecology of the Solent in general (367), and Langstone Harbour in especial detail (479). Studies have also been made of the wildfowl food resources, and of the feeding habits of the common species (595).

Portsmouth Harbour, in particular, is beset by a ring of towns, and so is Langstone to a lesser degree. Both have lost substantial tracts of foreshore during recent years: at Portsmouth about 200ha were reclaimed around 1970 for roads and industry, and at Langstone more than 70ha have been infilled with rubbish, a process which is still continuing. The increase in sewage effluent is also causing changes in the foreshore vegetation, to the detriment of some wildfowl, and the benefit of others (p.84). Chichester Harbour is still largely unspoiled, and, being designated an Area of Outstanding Natural Beauty, is likely to remain so. Its only possible detraction, now that the airfield at Thorney is closed, is the great intensity of sailing at all stages of the year. In 1973 the number of craft based here was established at 5,800, compared with 3,000 in Portsmouth Harbour and 1,400 in Langstone (479). The present totals are certainly much higher. The disturbance, however, is mainly confined to the larger channels, and except at high water the flats and shallows are inviolate.

Reserves are now established in three of the four harbours, Portsmouth being the exception. At Pagham a Local Nature Reserve, dating from 1964, covers the whole of the harbour and part of the surrounding marsh, the area amounting in all to 384ha. The reserve includes a visitors' centre and a nature trail, and is wardened by the Pagham Wildfowlers' Association, who have the right to shoot there from September until mid-October. Sailing and canoeing are restricted to people living locally, and are banned completely from October to the end of April. In Chichester Harbour, the LNR, covering 364ha, was declared in 1975, and includes nearly the whole of the Thorney Channel and the flats below high water mark on either side. The channel itself has long lines of moorings and is open to sailing throughout the year, but the fairway is narrow and the traffic fairly restricted. Shooting is controlled by the Chichester Harbour Wildfowlers' Association, who warden the reserve and maintain a substantial part as a sanctuary. The Farlington Marshes at the head of Langstone Harbour are also protected by a Local Nature Reserve, set up in 1971. The site comprises a track of damp reclaimed pastureland, 119ha in extent, and includes a series of fresh and brackish pools, some of which are fringed with extensive reed-beds. The primary interest is botanical, but the area also provides a valuable retreat for the wildfowl and waders which frequent the adjoining stretches of the harbour. Public access is restricted to the sea wall, which surrounds the marsh on three sides, and to a smallish section at the landward edge; otherwise the place is wholly undisturbed. In addition, a reserve covering more than 550ha of the adjacent foreshore was acquired by the RSPB in 1978. Within this area there are several low islands, surrounded by the only sizeable stretches of saltmarsh remaining in the harbours. These provide an important roost and feeding area and are used extensively by dabbling ducks and waders. Before the reserve was created, the shooting in the harbours had for many years been leased and strictly controlled by the Langstone Wildfowlers' Association, an arrangement which is still in being in parts of the reserve. There are no restrictions on sailing, except that landing on the islands is forbidden.

The creation of this network of reserves has undoubtedly fostered the recent increase in the number of wildfowl wintering in the district. There have also been some major improvements in the food supply, which have allowed the harbours to accommodate much larger flocks of certain species. The most notable events over the past 20 years have been the steady decline of the once invasive *Spartina*, and the rapid spread of *Zostera* and *Enteromorpha*.

By the late 1950s the *Zostera*, which had formerly flourished throughout the Solent, was reduced by disease to one small intertidal bed in Langstone Harbour, and to another four localities below low water mark. The patch at Langstone then started to expand: by 1968 the two narrow-leaved species, *Z. angustifolia* and *Z. noltii*, were both increasing rapidly, and by 1974 they had spread to the other harbours as well. When the beds were last surveyed in 1979 they covered a total of 280ha in Langstone Harbour, 130ha in Chichester and 20-40ha at Portsmouth. *Zostera* is also present at Pagham to a lesser extent (479, 596).

The increase in *Enteromorpha* began at about the same time, and it now occurs abundantly over large areas of the intertidal mud, including those left vacant by the die-back of *Spartina*. Recent studies have shown that the spread and density of the beds is related to the volume of treated sewage discharged into the various channels. In Langstone Harbour, in particular, the flow of effluent has increased fivefold since 1959, and in places the growth is now so dense that other forms of plant and invertebrate life are in danger of being smothered. Dense growths are also found in the other harbours, but the beds are less extensive, and on some of the more sheltered flats the *Enteromorpha* is replaced by the Sea Lettuce (*Ulva*).

The Brent Geese, in particular, have responded

to these changes with a massive increase, the local trend being generally in keeping with the growth in the world population. In 1951 the harbours – and the Solent – were virtually devoid of geese, except for a flock of about 300 at Langstone. By the spring of 1965 the total had reached 3,000, and thereafter it climbed to 10,000 in 1972-73 and to 16,000 in 1978-79. At the same time the birds began to spread away from their earlier enclave, and flocks now winter regularly at a dozen other south coast sites between the Exe and Pagham. The progress of their spread and increase is summarised in Table 36. The figures shown are the highest number counted at each site in successive seasons, and runs of seasons, since 1963. In many cases they are drawn from the county bird reports or other sources, and are not confined to any given dates. The totals for the various groups of sites are also maxima, and are based as far as possible on synchronised counts. There are, however, some disparities, and the figures in brackets should be treated with caution. This applies especially to the totals for Southampton Water, where the sites are closer together and the distribution less stable than elsewhere.

All four of the Hampshire – Sussex harbours are identified as areas of international importance for Dark-bellied Brent Geese, the total numbers in the group amounting regularly to at least 11% of the world population. The other prime resorts are the Exe Estuary, and the sites in the Western Solent taken as a group.

When they first arrive in the harbours in October and November, the geese feed by preference on *Zostera*, and continue to do so till Christmas, by which time the supplies are depleted, not only by grazing but by normal winter die-back. They then turn to *Enteromorpha*, and also flight inland to feed on the neighbouring farmland, where they take both grass and cereals. This adaption to inland feeding, which dates from the early 1970s, may well have contributed to the species' recent success, though there seems in fact to be sufficient food on the foreshore to support the local flocks throughout the winter. At first the number was too small to matter, but latterly as many as 11,000 geese have been counted on the fields around the harbours, mostly within 200m of the shore. This has led, inevitably, to complaints of agricultural damage, the usual grievance being that the birds are puddling the ploughland or filching the early bite. The various methods of dealing with such problems are discussed in full on pp.501-6; in this case it has been suggested that some of the existing reserves on the foreshore should be extended to include convenient areas of grassland, and that scaring devices be used elsewhere to protect the more valuable crops and leys. Substantial flocks of Brent Geese have also taken to feeding on the playing fields near Langstone Harbour, where they do no harm to the turf (596).

The Shelducks have been less well suited by the changes in vegetation, especially in Langstone and Portsmouth Harbours. By the early 1970s the growth of *Enteromorpha* was in places becoming so dense that it started to smother the invertebrate fauna, which had previously abounded on the open mud. This curtailment of the Shelducks' staple diet resulted in a sharp decrease in the size of the wintering flocks. In Langstone Harbour the regular level dropped from about 1,850 in the 1960s to 1,040 in the 1970s and early 1980s, and similarly in Portsmouth Harbour the annual peaks of about 1,000, which used to occur in the late 1960s, have dwindled to less than 200 in most recent years (see also Fig 158). This decline has to some extent been offset by a modest increase in Chichester and Pagham Harbours, and the totals for the four inlets now reach an annual peak of about 5,000 (or around 4% of the north-west European population), compared with 6-7,000 in the period 1968-1971, and less than 4,000 between 1974 and 1978.

The harbours are also used extensively by other wildfowl, but except for Brent and Shelducks, the gatherings are nowhere of international importance (Table 37). Most of the common species, especially amongst the dabbling ducks, have increased substantially since the 1960s. Teal have doubled in numbers at Langstone and Pagham, and in Chichester Harbour have increased tenfold. Pintail and Wigeon have also shown sizeable gains, and so have Mergansers on a smaller scale. The extent of these increases, and the value of the complex as a whole, is well illustrated in Table 38, which sets out the annual maxima on the four sites combined.

In addition to the tidal basins, there are several freshwater pools nearby which form part of the same complex. The most important of these are the gravel pits to the south and east of Chichester, which lie within 5km of both Pagham and Chichester Harbours (Table 37). Over the past 30 years the area of open water here has steadily increased, and the wildfowl numbers have closely followed the same trend (p.510).

Table 36. *The maximum numbers of Dark-bellied Brent Geese recorded at their main resorts on the Channel coast in each of the seasons 1978-1982, and in each of the five preceding 3-year periods. The totals for the groups of sites, and for the region as a whole, are the highest synchronised counts within each period. The date after the place name is the season in which the species began to winter regularly (after Tubbs and Tubbs 1982). A blank = no data.*

		1963-65	66-68	69-71	72-74	75-77	78	79	80	81	82
Exe	pre-1946	100	140	350	650	1300	1580	2400	1950	1700	1400
Chesil Fleet	1975			95	0	120	140	120	310	220	630
Poole Harbour	1972	2	10	60	200	400	540	440	490	680	1240
Christchurch Hbr	1978			0	0	3	220	80	10	20	110
DEVON – DORSET TOTAL		100	140	360	850	1410	2250	2980	2480	2530	2800
Keyhaven	1965	80	85	350	500	800	800	940	920	1000	1600
Sowley					0	450	170	290	200	400	550
Beaulieu Est	1973	0	3	5	50	280	80	270	270	230	460
Newtown Est, I of W	1969	20	90	250	550	700	680	890	700	750	845
WESTERN SOLENT TOTAL		80	150	390	665	1300	1450	1980	1780	2180	2990
Calshot	1975			0	45	400	310	550	430	350	390
Dibden/Weston	1975					190	20	40	260	80	100
Hamble	1975					200	200	350	210	370	180
Hill Head	1978				36	0	250	190	210	70	1450
SOUTHAMPTON WATER TOTAL				45	400	650	920	880	740	1590	1450
Portsmouth	1968		200	140	360	1640	1560	2450	1480	3320	855
Langstone	pre-1951	1240	2500	4080	6080	6480	5530	6420	7400	6190	7540
Chichester	1953	2000	2630	3300	7390	6350	8140	9500	7090	8630	10550
Pagham	1964	130	190	310	700	1500	1500	2700	1500	1860	4710
HARBOURS TOTAL		3000	4940	7560	13170	13820	16730	20620	16130	16680	22943
CHANNEL COAST TOTAL		3000	5170	8100	14660	16950	20660	26500	20270	21870	29500
1% of British population		175	190	240	410	490	630	750	665	600	925
1% of world population		270	315	410	845	1190	1420	1660	1470	1170	2025
% of world population in Harbours		11.1	15.7	18.4	15.6	11.6	11.8	12.4	11.0	14.3	11.3
% of UK population in Harbours		17.1	26.0	31.5	32.1	28.2	26.6	27.5	24.3	27.8	24.8

Table 37. *Regular (and maximum) counts of wildfowl other than Brent Geese in the four harbours and in Chichester Gravel Pits in the period 1970-1982.*

	Portsmouth	Langstone	Chichester Harbour	Pagham	Chichester GPs
Seasons	11	13	9	8	8
Period	1970-82	1970-82	1970-82	1970-82	1970-81
Mallard	123(331)	70 (200)	321(2400)	197 (443)	230(487)
Teal	50(200)	670(1300)	1214(3253)	250(1363)	42(133)
Wigeon	40(300)	1010(2000)	618(1201)	165 (730)	6 (42)
Pintail	3 (30)	75 (210)	120 (215)	112 (300)	2 (24)
Shoveler	10 (75)	70 (215)	11 (46)	15 (41)	100(261)
Pochard	60(146)	0 (30)	14 (78)	3 (13)	263(465)
Tufted Duck	15 (60)	2 (10)	23 (130)	11 (28)	292(400)
Goldeneye	35(115)	65 (120)	63 (225)	3 (14)	1 (4)
Eider	0 (0)	0 (13)	6 (27)	11 (100)	0 (0)
R-b Merganser	15 (70)	60 (195)	45 (97)	13 (68)	0 (1)
Shelduck	145(670)	1041(2950)	2379(4552)	563 (800)	12 (40)
Canada Goose	0 (40)	0 (15)	6 (36)	22 (129)	50(148)
Mute Swan	20 (60)	15 (45)	58 (126)	12 (29)	26 (42)

The other pools comprise the Sinah Gravel Pits on Hayling Island, close to the south-east corner of Langstone Harbour, and Baffins Pond in one of the Portsmouth parks. Both sites attract a wide variety of wildfowl, including estuarine species and occasional sea ducks, but except perhaps for Mallard, the numbers are seldom very large (Table 39). The Canoe Lake in Southsea is also of interest for its flock of 50-60 Mute Swans.

Away from the coast, standing waters in this region are few, other than a number of lakes and floodlands in the narrow valleys of the Test, Itchen and Meon. Data for the common species are given in Table 39 for 21 sites in these valleys. On the Test, a chalk river famous for its trout fishing and for its luxuriant aquatic vegetation, the numerous small waters hold a few breeding ducks (including Pochard) as well as the wintering populations. Kimbridge Lake used to hold upwards of 100 ducks but in recent years has been periodically drained and is now devoid of wildfowl. Some 2-300 Canada Geese are associated with the ornamental lakes and gravel pits, and a feral flock of Greylags has been established. This flock is centred on Wherwell Priory and now numbers 150-200, the highest count for the area being 213 on Bransbury Common, nearby, in November 1981.

The Itchen meanders through a valley topographically similar to that of the Test but rather less rural, and is likely to suffer further from the proposed extension of the M3 motorway to the south of Winchester. The waters around and downstream of Winchester (the first four Itchen sites in Table 39) have very little other than Mallard. Upstream, however, four impoundments of the river and its tributaries, associated with parks and country houses, hold substantial populations. As well as those listed in Table 39, the waters hold small numbers of a wide variety of other species, including up to 30 Shoveler at Alresford and Northington and a sprinkling of feral and exotic species. On the Meon the only area of note is at Warnford Park, again associated with impoundments of the river.

South-east England

The Arun and upper Wey Basins (Table 40)

Eastwards of Pagham Harbour, the chalk ridge of the South Downs – initially 10km inland – gradually nears the coast, reaching it 40km to the east at Brighton. In between there are two small estuaries – those of the Arun (at Littlehampton) and Adur (at Shoreham) – holding between them a few hundred dabbling ducks and smaller numbers of Shelducks and Mute Swans. Three January counts of the Adur below Bramber in the late 1960s recorded up to 100 Mallard and 71 Mute Swans.

The narrow outlets of the Sussex rivers make their lower reaches still readily prone to flooding, particularly in view of the lack of flat land to the north. The lower Arun Valley contains the most important flood-meadows in the region. Of particular interest are the Bewick's Swans, centred between Pulborough and Amberley. From 1975, when a large increase occurred, to 1982 their average seasonal maximum was 84, with a peak of 116 in February 1979. Large numbers of Mute Swans have been present in the same area since at least the mid 1950s, and have shown little change during this time. They usually peak in autumn, then apparently move a few kilometres south to the floodlands between Amberley and Arundel. These floodlands are less important for Bewick's Swans, although 61 were found in January 1981. Otherwise,

Table 38. *The maximum numbers of wildfowl in the four Hampshire/Sussex harbours combined in each winter since 1976-77. The totals at the foot of the table include Brent Geese and all other species.*

	1976	77	78	79	80	81	82
Mallard	571	635	980	1044	940	1345	1372
Teal	1987	2073	2278	3959	3665	4904	3465
Wigeon	1263	1932	2435	2187	2706	3896	1610
Pintail	276	252	457	404	391	675	532
Shoveler	130	107	151	147	149	238	155
Goldeneye	183	183	154	172	179	190	127
R-b Merganser	149	117	196	194	227	318	256
Shelduck	3081	2528	5967	3978	4850	7380	3500
Mute Swan	87	65	54	148	125	187	127
Canada Goose	65	69	30	7	141	149	51
Total wildfowl	20795	19641	26566	30987	27165	32142	31963
(Month)	(J)	(D)	(J)	(J)	(J)	(J)	(J)

the nearest regular wintering grounds for this species are on the Hampshire Avon (85km west) and Walland Marsh, Kent (92km east).

When there is sufficient flooding the lower Arun attracts in addition some of the largest inland concentrations of Teal, Wigeon and Shoveler in south-east England. Pulborough Levels and Amberley Wildbrooks are the main areas, but almost as many of some species can occur between Amberley and Arundel. North of Amberley are the winter feeding grounds for the Canada Geese which breed and moult at Petworth Park and form the bulk of the north-west Sussex subpopulation. Canada Geese were introduced to Petworth between the Wars and, despite occasional breeding control, had reached 500 by December 1977 and top 700 nowadays. There is evidence that the lower Arun receives birds, directly or indirectly, from the other centres of this subpopulation – Knepp Castle Lake and Warnham Mill Pond. These two localities did not receive introductions, their Canada Geese having presumably spread from Petworth (275).

Amberley Wildbrooks were threatened with pump drainage in 1978, but grant aid was refused by the Minister of Agriculture after a public enquiry. In 1981 the Sussex Trust for Nature Conservation established a reserve of 43ha at Waltham Brooks, at the north-west end of the Wildbrooks. The drainage already undertaken has concentrated 95% of the ducks on this reserve; elsewhere floodwater can be removed very quickly and the effect of drainage on the site as a whole is considerable.

By the river at Arundel itself the Wildfowl Trust has a 22ha reserve, opened to the public in 1976. It comprises a collection of over a thousand captive wildfowl and a wild area with reed-beds, fields and a scrape. As at the other Wildfowl Trust centres, the ponds created for captive birds attract many wild ones; they provide the most important concentration of Pochard in Sussex. Swanbourne Lake, in the grounds of Arundel Castle, opposite, has increased in import-

Table 39. *The regular (and maximum) numbers of six common species at various freshwater sites in the Test, Itchen and Meon Valleys, and on Portsea and Hayling Island in the period 1970-1982. The number of seasons from which the regular figures have been calculated is in brackets after the site name.*

		Mallard	Teal	Wigeon	Pochard	Tufted Duck	Mute Swan
TEST							
Moorcourt Lake	(7)	21 (40)	5 (25)	0 (2)	45 (87)	88(150)	14(47)
Broadlands	(5)	777(998)	118(316)	158(479)	11 (82)	44(129)	25(64)
Timsbury	(13)	49(120)	15 (60)	13(110)	49(110)	44(134)	10(44)
Mottisfont Abbey	(9)	74(218)	3 (17)	4 (45)	19 (48)	25 (81)	16(45)
Marsh Court	(12)	146(450)	33 (85)	43(173)	5 (40)	5 (24)	9(30)
Leckford	(6)	354(600)	17 (68)	3 (33)	24 (45)	76(170)	29(67)
Wherwell Common	(3)	34(110)	6 (50)	16 (73)	0 (0)	0 (1)	8(16)
R Anton, Andover	(2)	133(147)	0 (0)	0 (0)	0 (0)	0 (0)	4 (6)
Rooksbury/Charlton GP	(5)	100(200)	2 (9)	0 (2)	39(134)	82(160)	15(25)
St Mary Bourne	(3)	25 (91)	0 (0)	0 (0)	1 (4)	14 (31)	1 (5)
ITCHEN AND MEON							
Eastleigh GP and SF	(4)	14 (50)	6 (24)	0 (0)	0 (0)	0 (0)	0 (0)
Winchester: Itchen Meadows	(2)	48 (79)	0 (0)	0 (0)	0 (0)	0 (1)	3 (3)
Sewage Farm	(1)	18 (32)	1 (1)	1 (2)	4 (7)	35 (43)	2 (2)
Winnall Mere	(3)	76(128)	2 (10)	1 (2)	8 (25)	32 (59)	6(20)
Avington Lake	(11)	144(520)	50(200)	11 (65)	0 (6)	3 (18)	1 (4)
Northington Lake	(12)	352(650)	32(120)	1 (9)	14 (50)	18 (58)	5(10)
Arlebury Lake	(5)	32 (60)	3 (14)	0 (0)	3 (9)	11 (31)	6(24)
Alresford Pond	(12)	129(410)	98(300)	3 (40)	35 (92)	18 (82)	7(21)
Warnford Park	(8)	24 (94)	3 (18)	0 (1)	7 (15)	15 (39)	5(13)
PORTSEA/HAYLING							
Baffins Pond	(13)	171(480)	0 (0)	0 (0)	4 (12)	13 (66)	8(58)
Sinah Gravel Pits	(13)	66(120)	0 (8)	0 (0)	49(147)	49(109)	4 (9)

Note: Most waters at times hold Canada Geese; the most important are Timsbury (max 160), Marsh Court (315) and Leckford (115).

ance with the creation of the Arundel Reserve. Long the home of some 200 Mallard and small numbers of other species, it now holds flocks of Gadwall (of introduced stock), Shoveler and Tufted Ducks.

Along the western River Rother before it joins the Arun at Pulborough, are the small ornamental lakes of Burton and Petworth Parks. Further upstream, the pond to the south of Midhurst held 110 Mallard when counted for the only time, in January 1967.

The upper valleys of the Arun and Wey (the latter's lower reaches, north of Guildford, being covered in the Lower Thames Basin section) encompass the remainder of north-west Sussex, the bulk of rural Surrey and the corner of Hampshire to the east and north of Alton. This area, heavily wooded and rising in parts to almost 300m, contains numerous (mainly ornamental) lakes, the most important of which are included in Table 40, but no waters of over 35ha. Partly due to stocking for shooting, Mallard are very numerous, but apart from Canada Geese and Mandarin Ducks other species are relatively uncommon. Mandarins are spreading due to colonisation from north Surrey, some deliberate introductions, and escapes from wildfowl collections. The Sussex population was recently put at 150-175 birds (276). Some of the largest gatherings of Mandarins ever recorded in Britain have been in the vicinity of Godalming, amounting on one occasion to 300 in a single flock.

The main arm of the Adur rises south of Horsham and serves only one enclosed water of note: Knepp Castle Lake at Shipley. Wigeon find the surrounding fields here attractive, while Canada Geese are again numerous, mainly in winter, when they move in from the Arun Valley (275). The lower Adur is still subject to flooding between Steyning and Bramber, and can hold several hundred dabbling ducks in wet winters.

Table 40. *The regular (and maximum) numbers of wildfowl on 26 major resorts in the Arun and Upper Wey basins, 1970 to 1982. The figures in brackets after the place names show the number of seasons on which the summaries are based.*

		Mallard	Teal	Wigeon	Gadwall	Shoveler	Pochard	Tufted Duck	Canada Goose	Mute Swan
LOWER ARUN										
Swanbourne Lake	(13)	265 (441)	0 (20)	0 (2)	17(134)	7 (62)	14(106)	44(148)	1 (13)	7 (13)
Wildfowl Trust, Arundel	(6)	459(1003)	123 (391)	41 (63)	34 (74)	31 (70)	213(386)	157(239)	79(216)	12 (28)
Arundel – Amberley	(4)	153 (250)	287 (600)	63 (279)	0 (2)	1 (5)	0 (1)	4 (14)	34 (79)	3 (61)
Amberley Wildbrooks/ Pulborough Levels	(13)	101 (400)	331(1000)	312(3000)	1 (10)	56(400)	0(120)	0(120)	142(700)	58(126)
WEST ROTHER										
Burton Ponds	(8)	29 (81)	0 (4)	0 (5)	0 (6)	3 (15)	21 (59)	58(119)	5(111)	6 (14)
Petworth Park Lake	(6)	122 (205)	0 (1)	0 (0)	1 (7)	2 (7)	5 (24)	18 (70)	266(600)	3 (19)
Lurgashall Millpond	(2)	74 (95)	7 (11)	0 (1)	0 (0)	0 (0)	6 (14)	5 (11)	58(150)	1 (4)
Heath Pond, Petersfield	(13)	68 (126)	0 (4)	0 (0)	0 (0)	0 (1)	0 (12)	1 (9)	36 (94)	1 (4)
The Wilds, Liss Forest	(2)	31 (54)	11 (17)	0 (0)	0 (0)	0 (0)	2 (11)	9 (23)	5 (16)	1 (2)
UPPER ARUN										
Warnham Mill Pond	(8)	63 (270)	12 (40)	0 (1)	0 (0)	1 (2)	24 (50)	13 (27)	46(196)	3 (10)
Broome Hall, Coldharbour	(3)	32 (65)	4 (26)	0 (1)	0 (0)	0 (0)	0 (1)	1 (2)	19 (45)	0 (0)
UPPER WEY										
Forest Mere, Liphook	(2)	9 (32)	5 (16)	0 (0)	0 (0)	0 (0)	7 (32)	22 (46)	28 (81)	0 (0)
Frensham Ponds	(13)	56 (114)	0 (5)	0 (2)	0 (0)	1 (5)	49(250)	54(320)	17 (72)	10 (31)
Clandon Park	(7)	35 (320)	2 (10)	0 (2)	0 (0)	0 (1)	6 (32)	24 (43)	6 (25)	4 (9)
Thursley Ponds	(8)	35 (72)	6 (24)	0 (0)	0 (0)	0 (0)	0 (16)	9 (25)	0 (2)	2 (10)
Cutmill Ponds	(9)	143 (264)	2 (18)	0 (0)	0 (0)	0 (5)	1 (7)	4 (15)	2 (28)	2 (14)
Enton Ponds	(8)	10 (44)	7 (34)	0 (0)	0 (0)	0 (0)	11 (52)	35 (73)	6 (51)	9 (26)
Busbridge, Godalming	(1)	31 (60)	0 (6)	0 (0)	1 (2)	0 (0)	0 (1)	18 (23)	17 (37)	0 (0)
Broadwater, Godalming	(8)	74 (143)	0 (0)	0 (0)	0 (0)	0 (0)	1 (29)	0 (4)	0 (4)	1 (5)
Winkworth Arboretum	(9)	51 (122)	2 (16)	0 (0)	0 (1)	0 (0)	0 (4)	21 (57)	7 (63)	0 (0)
Eastwater, Bramley	(8)	111 (214)	0 (12)	0 (0)	0 (0)	0 (0)	3 (19)	14 (26)	9 (90)	0 (0)
Tangley Mere, Chilworth	(8)	59 (160)	0 (0)	0 (0)	0 (0)	0 (0)	5 (19)	1 (8)	15(136)	0 (0)
Albury Ponds	(7)	49 (63)	0 (8)	0 (0)	0 (0)	0 (0)	0 (18)	0 (6)	8 (52)	3 (7)
Shalford Water-meadows	(3)	42 (211)	0 (0)	0 (0)	0 (0)	0 (0)	0 (0)	0 (4)	28(132)	1 (5)
ADUR										
Knepp Castle Lake	(12)	115 (350)	11 (50)	67 (280)	0 (4)	5 (32)	36(410)	7 (30)	48(122)	4 (30)
Bramber – Henfield	(3)	29 (48)	86 (250)	42 (200)	1 (4)	8 (30)	0 (1)	0 (1)	11(200)	20 (35)

East Sussex and south Kent (Table 41)

To the east of Brighton, 30km of cliff culminates in the 150m high promontory of Beachy Head. This stretch of coast is interrupted by two small outfalls. That of the Ouse, with the adjacent Tide Mills Pools at Newhaven, holds a good variety of species, but in very small numbers. Some 7km east, the Cuckmere, with its flood-meadows, freshwater meanders and sheltered Haven, which is a favourite roost of dabbling ducks, is more important.

Off Beachy Head itself a large eastward passage of sea ducks occurs each spring. This can also be observed elsewhere, notably off Dungeness, the next headland to the east, and is part of the general movement through the Straits of Dover and the English Channel noted in the previous and following sections. The great majority of birds involved are Common Scoters, of which up to 13,300 have been counted in a day off Beachy Head and 27,000 off Dungeness, which is more intensively watched, being the site of a permanent Bird Observatory and reserve (299, 561). Both these counts referred to early April. In spring 1979 a total of 56,000 Common Scoters were recorded off Dungeness (299). Daily peaks of 2-300 Velvet Scoters, Eiders and Mergansers have passed in April and May. The autumn numbers are much smaller (for reasons that are unknown); no more than 570 Common Scoters and few of the other species have passed Dungeness in a day in that season.

The coast of Sussex and south and east Kent also holds its own concentrations of Common Scoters, although these have declined since the 1950s. The main areas are offshore from Littlehampton (maximum 145), Pevensey (180), Pett Levels (1,000), Rye Harbour (1,100), Dungeness (2,000) and South Foreland/St Margaret's Bay (500). Flocks may be present throughout the year, but the highest numbers are

Table 41. *The regular (and maximum) numbers of wildfowl at the following resorts in E Sussex/S Kent, 1970 to 1982.* (* = *data from* Kent Bird Reports.)

Area	Site	Seasons
OUSE:	(a) Newhaven Harbour	(1 season)
	(b) Glynde Levels	(12)
	(c) Barcombe Mills Resr	(4)
	(d) Possingworth Park	(2:pre-1970)
	(e) Ardingly Resr	(5)
CUCKMERE:	(f) Cuckmere Haven	(13)
	(g) Arlington Resr	(11)
PEVENSEY–HASTINGS:	(h) Pevensey Levels	(2)
	(i) Alexandra Pk, Hastings	(4)
	(j) Asten Valley, Filsham	(1)
ROTHER:	(k) Darwell Resr	(11)
	(l) Powdermill Resr	(4)
	(m) Pett Levels	(9)
	(n) Rye Harbour/Pits	(9)
	(o) Walland Marsh	(2)
	(p) Shirley Moor	(2)
	(q) Dungeness	(9)

	Mallard	Teal	Wigeon	Shoveler	Pochard	Tufted Duck	Shelduck	Canada Goose	Mute Swan
(a)	18 (21)	8 (15)	0 (0)	0 (6)	0 (0)	0 (7)	13 (14)	0 (14)	0 (2)
(b)	100 (230)	203 (420)	697(1750)	9 (32)	0 (0)	0 (5)	0 (3)	6 (76)	15 (47)
(c)	132 (620)	13 (94)	109 (903)	2 (22)	22 (77)	43 (77)	0 (0)	30(188)	4 (13)
(d)	64 (121)	1 (4)	1 (2)	0 (0)	7 (11)	4 (9)	0 (0)	0 (0)	0 (0)
(e)	100 (160)	2 (15)	0 (1)	0 (2)	42(118)	16 (72)	0 (0)	8 (17)	1 (5)
(f)	93 (254)	67 (176)	53 (623)	2 (12)	3 (21)	13 (61)	21 (55)	2 (24)	29 (79)
(g)	248 (800)	5 (30)	629(3300)	20 (67)	37(220)	21(100)	0 (3)	76(400)	2 (21)
(h)	192 (300)	194 (260)	186 (220)	14 (27)	125(208)	56 (90)	9 (12)	11 (21)	148(240)
(i)	208 (332)	0 (0)	0 (0)	0 (1)	11 (39)	4 (8)	0 (0)	0 (0)	10 (15)
(j)	43 (45)	223 (420)	39 (70)	6 (8)	0 (0)	0 (0)	2 (5)	0 (0)	0 (0)
(k)	222 (711)	47 (180)	25 (97)	1 (7)	27(141)	47(188)	0 (3)	22(135)	5 (13)
(l)	69 (230)	22 (70)	2 (8)	1 (9)	7 (19)	25 (54)	0 (0)	0 (2)	4 (15)
(m)	133 (450)	68 (300)	48 (200)	13 (66)	14 (23)	10 (20)	6 (34)	6(100)	19 (37)
(n)	212 (409)	58 (225)	50 (183)	18 (52)	232(465)	112(234)	35(122)	45(180)	20 (52)
(o)	316 (500)	102 (150)	806(3000)	4 (20)*	9 (31)	20 (68)	4 (22)	0 (0)	137(200)
(p)	87 (180)	60 (120)	6 (18)	0 (12)	0 (0)	0 (0)	2 (5)	21 (54)	61(112)
(q)	985(3284)	120 (510)*	73 (762)	90(300)	217(950)	153(450)	9 (76)	46(135)	16 (74)

usually in the autumn. It is not known whether the different areas have separate concentrations or whether, as for example in North Wales, flocks are moving about between sites in response to food availability.

Inland, the lower reaches of the Ouse and Cuckmere are still prone to flooding, but counts are mostly lacking except from the permanent waters and the tidal Cuckmere. Glynde Levels, however, between the two main rivers along the tidal Glynde Reach, are well covered, and rival the Arun floods in importance for dabbling ducks, thanks mainly to a flooding scheme operated by the Sussex Ornithological Society on 6ha (544). These levels act as the main feeding ground for the south-east Sussex subpopulation of Canada Geese, which emanated from introductions at Herstmonceux Castle and Bentley Wildfowl Collection at Halland, and goose flocks now use most waters in the area (275). The 63ha Arlington Reservoir, built alongside the Cuckmere River in 1971, apparently acts as a roost when the levels are flooded, although the large number of Wigeon present at times suggests that other feeding areas may be involved. During hard weather the surrounds of the reservoir itself are used; in early 1982 the Wigeon fed on winter wheat alongside. Barcombe Mills, a 16ha holding reservoir on the Ouse above Lewes, is less important, being subject to frequent complete draw-down. When the Glynde Levels are dry in mid-winter, however, it acts as an alternative resort, as in several seasons around 1970, when up to 1,000 Wigeon used the reservoir.

Records are lacking for the upper Ouse Valley, which contains numerous ornamental waters, except from the 73ha Ardingly Reservoir (just inside West Sussex). This was flooded in 1978, after damming the Ouse a few kilometres above Haywards Heath. For its first three winters the reservoir held about 100 diving ducks, mainly Tufted, but now little occurs apart from Mallard and Canada Geese. To the east lies Ashdown Forest, the main watershed between the North and South Downs. In the forest's gently undulating southern approaches the eastern Rother has its

Fig. 18. East Sussex and Kent.

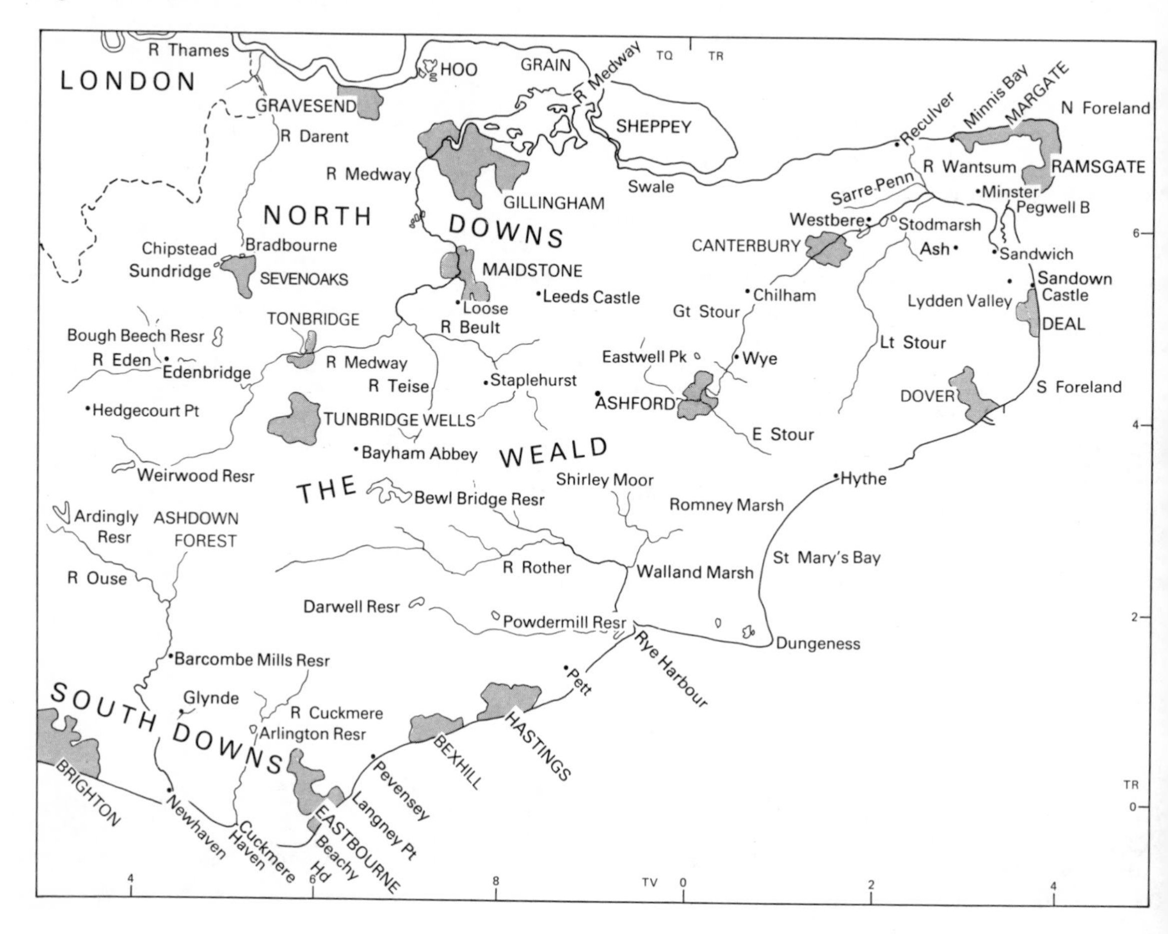

source. There are no sizeable waters in this basin (apart from Wadhurst Park Lake, for which no records are available) upstream of the 69ha reservoir of Darwell, above Robertsbridge. Some 8km to the east, Powdermill Reservoir (23ha) drains into the Brede Level, a western arm of the Walland Marsh complex. Both reservoirs are subject to heavy disturbance from shooting and fishing, which restrict and cause fluctuations in the numbers of wildfowl. Furthermore, Darwell is relatively deep, at an average of 15.5m.

At Beachy Head, and beyond Eastbourne, the South Downs give way to a coastal plain of low-lying pasture land, broken only by the low cliffs of Bexhill and Hastings (where the Upper Greensand reaches the coast) and the shingle beds at Langney Point and (much more extensive) Rye and the Dungeness Peninsula. To the west of the greensand lie the Pevensey Levels and to the east the Pett Levels. Both comprise reclaimed land intersected by dykes and subject to patchy flooding in winter, but owe their importance partly to their proximity to relatively sheltered seas, where Mallard and Wigeon, in particular, roost. There is a small reserve of the Sussex Trust for Nature Conservation on Pevensey Levels.

Almost contiguous with the Pett Levels, the sands and 80ha of pits (worked out in the 1970s) at Rye Harbour, at the mouth of the eastern Rother, attract a wide variety of species in moderate numbers. Some 85ha of shingle beach alongside were designated a Local Nature Reserve in 1970, with a resident summer warden (mainly to safeguard the nesting terns). The northernmost pit, by Camber Castle, is used for water-skiing, but those nearer the sea are undisturbed.

The lower reaches of the Rother flow through the western edge of the huge expanse of reclaimed land comprising the Rother Levels themselves, Shirley Moor, Walland Marsh and Romney Marsh. Like Pevensey and Pett Levels this area, formerly marsh, is now mainly pastureland intersected by numerous dykes. Conversion to arable land since the Second World War, and particularly in recent years, has further reduced the attraction for wildfowl. Romney Marsh is nowadays almost devoid of wildfowl, except in the severest weather, but Walland Marsh, Shirley Moor and Wittersham Level have retained some importance. Even there the number of ducks is high only after periods of heavy rainfall or cold continental weather, as in January 1982, when Walland Marsh held over 4,000 ducks, including 3,000 Wigeon. Since then, however, a further 40ha of this area have been converted to arable land. Swans have increased considerably in recent years, despite the drainage operations, and a regular wintering flock of over 50 Bewick's Swans has been present at Brookland on Walland Marsh since 1976, reaching a maximum of 182 in January 1982. In March 1983 a total of 343 swans was counted there – 200 Mute and 143 Bewick's. Shirley Moor and Wittersham Level are occasionally used by the swans, which, in common with the dabbling ducks, roost at Dungeness.

The 2,500ha shingle peninsula of Dungeness abuts Walland and Romney Marshes and is most famous for the Bird Observatory and two nuclear power stations on the point itself. For wildfowl the interest lies in the 166ha of lagoons scattered throughout the peninsula from Midrips, by the eastern end of Rye Bay, to Lydd Beach on the opposite side. Some of the lagoons, such as the Oppen Pits, were originally salt, but as the shingle built up they were cut off and became fresh; others were dug for gravel, but all have been worked at some time, and some still are. A 483ha part of the outermost peninsula, including most of the pits, comprises the RSPB's oldest reserve, established in 1930. The whole peninsula attracts 1,500-2,000 wildfowl each winter (excluding the offshore sea ducks mentioned earlier) and contains one of the largest gatherings of Mallard in south-east England. In the cold spell of early 1979 up to 19 Smew were present on the pits, part of the small invasion of this species to Britain which occurred at the time (119). Since then Dungeness has held the largest gathering of Smew in Britain each winter, the seasonal peaks for 1979 to 1982 being 14, 17, 22 and 10. Canada Geese colonised Dungeness from a small flock introduced onto the Royal Military Canal at Hythe in 1956. The small Rye Harbour population, itself an offshoot of the same introduction, now also uses the area as an autumn gathering ground (275). Twenty-one Canada Geese from Sevenoaks (see next section) were released at Lydd Pits in 1964 to start another flock (538). The breeding population of Tufted Ducks on the Dungeness pits has increased rapidly in recent years, reaching 65 pairs in 1980, although these produced only 81 young (299).

The abandoned 37km long Royal Military Canal forms the northern perimeter of Romney Marsh. In January 1971 its extremities were counted: the western end, from Appledore to Warehorne (owned by the National Trust), held just 14 Mute Swans, and east of West Hythe there were 168 Mallard and 20 Mute Swans.

The broad sweep of St Mary's Bay from Dungeness to Sandgate, with its narrow sands, and the rock-strewn and cliffed coast beyond are unsuitable for shore-based wildfowl, except perhaps as a roost (150 Mallard were counted in St Mary's Bay in January 1971).

South-east Surrey and west Kent (Table 42)

The upper valley of the Mole, south-east of Dorking, is much flatter and more open than that of the Wey, but this south-eastern quarter of Surrey contains little standing water. The slopes of the North Downs themselves, however, have several sites of interest. Buckland Sand Pit, 2km west of Reigate, has held up to 73 Mallard, 18 Teal, 2 Pintail, 8 Pochard and 1 Mandarin during two January counts. The lake in the grounds of Gatton Hall and the sand pits at Holmethorpe are only a kilometre apart, and much interchange takes place between them, especially as a result of disturbance from shooting at Gatton Park (where Mallard are reared and released in large numbers) and from fishing and general activities at the pits.

The small sand pits at Godstone were converted to emergency water supply reservoirs in 1977. The Bay Pond close by is run as a private nature reserve, but is plagued by mink and now holds just one or two Mallard and Mute Swans.

Just inside the perimeter fence of Gatwick Airport (now in West Sussex) there is a tiny (under half a hectare) lagoon which receives purified non-human effluent from the airport. Shallow and well-fringed with vegetation, the pool is protected by a 3m fence and has attracted up to 88 Mallard and 4 Mute Swans in recent Januarys.

To the south-east the land rises and becomes thickly wooded again as it approaches the Weald of Kent. To the north-east the Sevenoaks Weald, part of the Lower Greensand belt, acts as a buffer to the upper Darent Valley, along which, immediately north of Sevenoaks town, lies an important chain of gravel pits: Sundridge, Chipstead and Bradbourne. The last-named is the site of the famous Sevenoaks Reserve, the establishment and extensive management of which is described on p.543 and in full detail by Harrison (244) and the Annual Reports of WAGBI/BASC. The reserve, now managed by the Jeffery Harrison Memorial Trust, covers 45ha, most of which is occupied by four closely adjoining lakes, two much smaller, shallower and newer than the others. The number of wintering ducks has stabilised at about 300, while both Canada and Greylag Geese exceed 200 in most years. Canada Geese were introduced to Sevenoaks in 1956. This population has now spread throughout west Kent and as far as Weirwood Reservoir (see below), but most gather at Sevenoaks each autumn. Greylag Geese have been released on the reserve on several occasions, starting in 1961, but were much slower to increase and spread. They, too, show a marked autumn peak there. In recent years introductions of the Eastern, pink-billed race have, unfortunately, been made. Hand-reared Mallard and Gadwall have been released in large numbers on the reserve, though the latter are relatively scarce in winter.

At the western end of the complex is Sundridge Ballast Pit, which, from 1956 until it was worked out and its ownership changed in 1966, was run as a reserve by WAGBI. No management of any kind was carried out, so that it acted as a control for the Sevenoaks Reserve. Its numbers of wildfowl increased during this period, but on a much smaller scale than at Sevenoaks (248). After Sundridge was worked out, recreational disturbance increased and the duck numbers dropped considerably; fishing is particularly intensive.

Between Sevenoaks and Sundridge, the worked-out Chipstead (or Longford) Pit is also heavily disturbed. Six January counts since 1966 have produced maxima of only 90 Mallard, 4 Tufted Ducks, 1 Mute Swan and 45 Canada Geese.

The southern approaches of the Weald are traversed by two main rivers, the Medway and its tributary the Eden. In addition to several important ornamental lakes (Table 42) each valley contains a reservoir of over 100ha. To the north of the Eden lies Bough Beech (117ha), completed in 1970. Much the most numerous species there, apart from occasional large gatherings of Canada Geese, has always been the Mallard. In the first few autumns after flooding the reservoir held about 500, but nowadays 300 are more normal. A further peak occurs in mid-winter and can be substantial after hard weather, the all-time maximum of 771 having been in January 1979. Other species are in part deterred by the considerable depth of the water (maximum 23m). In the north-west corner there is a reserve of the Kent Trust for Nature Conservation, extending to 18ha.

The 114ha Weirwood Reservoir, flooded in 1955, straddles the West Sussex/East Sussex border near the source of the Medway (and lies just 7km from Ardingly Reservoir, on the upper Ouse). Though only 75m above sea level, Weirwood's position in heavily wooded country just north-west of the Ashdown Forest perhaps restricts the numbers of wildfowl using it. The initial influx of wildfowl, common to new reservoirs of this type, did not occur, but in the late 1950s and early 1960s over a thousand ducks, with a maximum of 782 Mallard, were present at times. Since then all the main species have shown a gradual decline, the start of which coincided with the introduction of winter sailing in early 1962, although the hard winters of 1978 and 1981 brought temporary recoveries. The western end of the reservoir is kept free of disturbance.

Four kilometres west of Bough Beech Reservoir – at Edenbridge – is one of the few remaining old-style sewage farms in the country, at which up to 39 Mallard and 20-30 Teal, Shoveler, Pochard, Tufted Ducks and Canada Geese were counted in 1982. In March 1978 as many as 66 Tufted Ducks were present (299).

The low-lying fruit-growing country along the Medway between Tonbridge and Maidstone, and along its tributaries the Teise and Beult, contains numerous farm ponds which would hold a negligible population were it not for the rearing of Mallard for shooting and the spread of Canada Geese from their surrounding centres. As it is, several of the ponds hold some 50 Mallard and 30 Canada Geese even in January. Three of the larger lakes in this region are excluded from Table 42 by a lack of recent count data; Eridge and Ashurst Parks near Tunbridge Wells held 50-100 Mallard and a few Teal in the 1950s; Leeds Castle 200 Canada Geese in autumn 1978 (299).

The rivers themselves are apparently of little importance for wildfowl, although recent data are largely lacking. A series of counts on the Beult just below Staplehurst produced maxima of 25 Mallard and 50 Teal. The Loose Valley (Table 42) consists of a

Table 42. *The regular (and maximum) numbers of wildfowl at the main resorts in SE Surrey and W Kent, 1970 to 1982. + = pre-1970 data. * = data from* Kent Bird Reports. *The figures in brackets after the place names show the number of seasons on which the summaries are based.*

		Mallard	Teal	Pochard	Tufted Duck	Canada Goose	Mute Swan
UPPER MOLE							
Dorking Mill Pond	(4)	101 (135)	0 (0)	0 (0)	5 (10)	2 (9)	2 (5)
Reigate Priory	(13)	146 (300)	0 (0)	1 (7)	7 (34)	1 (16)	7(16)
Earlswood Lakes	(5)	66 (158)	0 (1)	1 (8)	1 (7)	9 (16)	3(23)
Gatton Park	(9)	440(1200)	0 (6)	27 (42)	25 (68)	90(260)	3 (8)
Holmethorpe	(11)	47 (122)	13 (85)	27 (84)	32 (69)	78(294)	4(18)
Ifield Mill Pond/ Tilgate Lake	(2)	76 (112)	0 (0)	4 (12)	8 (16)	12 (17)	5 (9)
UPPER DARENT							
Sevenoaks Reserve	(3)	176 (370)*	37 (63)	22 (61)	83(155)	90(300)*	1 (9)
Sundridge Pits	(7)	41 (84)	0 (1)	11 (36)	18 (58)	9 (85)	1 (4)
EDEN							
Godstone Bay Pond	(9)	13 (82)	0 (3)	5 (19)	2 (8)	4 (54)	2 (8)
Leigh Place	(4)	38 (57)	0 (1)	0 (1)	11 (28)	22 (53)	6(12)
Hedgecourt Pond	(8)	12 (36)	0 (0)	1 (9)	0 (8)	5 (21)	6(21)
Godstone Reservoirs	(6)	20 (60)	0 (1)	5 (15)	15 (22)	16(123)	3 (8)
Hever Castle Lake	(7)	230 (452)	1 (5)	23 (93)	18 (41)	17(124)*	3 (9)
Bough Beech Resr	(10)	455 (841)*	67(341)*	71(255)	104(189)*	120(480)*	n9(27)
MEDWAY							
Mote Park	(6)	117 (250)*	0 (0)	8 (47)	3 (40)	58(170)	3 (7)
Hall Place, Leigh	(9)	23 (75)	0 (0)	23(159)	21 (44)	23(107)	4(23)
Loose Valley	(5)	51 (67)	0 (4)	0 (0)	0 (2)	1 (4)	2 (4)
Haysden GP	(2)	66 (117)	3 (25)	14 (55)	12 (25)	89(205)	1 (2)
Weirwood Reservoir	(13)	136 (285)	45(150)	84(330)	70(156)	36(157)	4(10)
TEISE							
Bayham Abbey	(1)+	33 (60)	14 (25)	0 (0)	0 (1)	0 (0)	8(10)
Bewl Bridge Resr	(5)	501 (736)	53(300)*	137(700)*	140(300)*	323(900)	19(33)

Note: Other species.

Wigeon: Sevenoaks 3(30); Hever Castle 1(4); Bough Beech Resr 16(125); Weirwood Resr 50(120); Bewl Bridge Resr 184(500)*.

Shoveler: Holmethorpe 2(18); Sevenoaks 3(11); Bough Beech Resr 16(69); Bewl Bridge Resr 3(12).

Gadwall: Sevenoaks 6(29); Bough Beech Resr 2(12); Bewl Bridge Resr 9(30).

Goldeneye: Sevenoaks 1(4); Bough Beech Resr 6(12); Bewl Bridge Resr 5(17).

Greylag Goose: Sevenoaks 311(474); Hedgecourt Pond 5(35); Bough Beech Resr 2(12); Hall Place 10(48); Haysden GP 85(140); Weirwood Resr 1(9); Bewl Bridge Resr 1(3).

series of linked ponds either side of the village of Loose, just above Maidstone.

In 1975 a 312ha reservoir – Bewl Bridge – was flooded along a tributary of the Teise on the Kent/East Sussex border, in the heart of the High Weald. Its initial peak of wildfowl has so far been maintained, apart from Pochard and Tufted Duck, which numbered 700 and 300 respectively early in 1977 (299), and it currently ranks as the most important (as well as by far the largest) reservoir in Sussex or Kent. A wide range of recreational activities is carried out, but from the outset 51ha have been designated a Water Authority Reserve, free from disturbance and wardened by the Sussex Trust for Nature Conservation. Although the reservoir's maximum depth is 29.5m, there are extensive shallow areas, particularly in the reserve, and the variety of species is high at all times. As at Dungeness (40km south-east) Smew have established a stronghold here. During the national influx of January 1979, 17 occurred on the reservoir, and 2-9 have been present in subsequent winters. Large numbers of Canada Geese began to appear in 1979; in 1982 there was a remarkable autumn gathering of 900.

In 1975 and 1976 unprecedented totals of, respectively, 2,641 and 3,536 Brent Geese were recorded migrating overland in Kent. They were moving either westwards along the Thames or south or south-west over the heart of the county, many reaching the coast at Rye Bay (245). 1,400 passed over Bewl Bridge Reservoir between 5th and 13th November 1982, coinciding with abnormal movements in East Anglia but also suggesting that the overland route may have been maintained.

In contrast few Brent Geese stay for the winter along the coasts of Sussex – east of Pagham Harbour – and south Kent, in the absence of any large estuaries or suitable areas of foreshore. The record of 285 at Pett in March 1978 was exceptional, the wintering flocks which became established there and at such places as Newhaven Harbour and Cuckmere Haven having remained at 20-30 despite the recent large increases further north and west.

East Kent (Table 43)

The Great Stour rises in the weald of Kent, flows south-east to Ashford, where it is joined by the East Stour, then bends through 90° to break through the Downs near Wye. Thence it flows to Pegwell Bay, the valley at first being narrow and lined with orchards and woods, with a few small gravel pits. Downstream of Canterbury, the river becomes tidal and the valley widens into an area of gravel pits and large subsidences resulting from the now disused Chislet colliery. The lower part of the valley, including the Sarre Penn, the River Wantsum, the Minster and Ash Levels and, south of Sandwich, the Lydden Valley, is now mainly arable. About 17km short of its mouth, the Great Stour is joined by the Little Stour. Virtually the whole of east Kent drains into this system.

There is little enclosed water, except alongside the river itself, where by far the most important area consists of the colliery subsidences of Stodmarsh NNR and the adjoining lagoons and marshes towards Westbere. These reed-fringed lakes and reed-beds were formed in the 1930s and total some 300ha, the NNR at Stodmarsh amounting to 163ha. A non-breeding flock of Mute Swans gathers each summer on one of the subsidence lagoons or nearby gravel pits; between 1979 and 1983 it averaged 186 (299).

Immediately to the west, the disused gravel pits at Westbere and Fordwich include about 70ha of open water, with moderately well-developed fringing vegetation. Table 43 gives maximum figures from Kent Ornithological Society records for Westbere – Fordwich, and for the smaller gravel pit complex at Seaton in the Little Stour Valley. In addition to the sites in Table 43, Stonar Lake, a worked-out gravel pit of 18ha by the river immediately north of Sandwich town, held an average maximum of 134 Pochard and 167 Tufted Ducks between 1978 and 1982 (299).

Full surveys of the Stour valleys have been undertaken, as the East Kent Lowlands Survey, each January or February during the period 1978-1982 (253). On average the following totals have been found away from the Stodmarsh – Westbere area (including the coast from Reculver to Minnis Bay, and Pegwell Bay to Sandown Castle): Mallard 1,481, Teal 1,128, Gadwall 8, Wigeon 256, Pintail 3, Shoveler 37, Tufted Ducks 198, Pochard 109, Goldeneyes 15, Shelducks 159, Canada Goose 36, Mute Swan 224. Most of the Teal (on average 573) and Mute Swans (averaging 120) are on Ash and Minster Levels, and about half the Mallard (average 762) on the Wantsum Marshes and adjacent coast. The Wantsum held 50 Bewick's Swans in January/February 1981 and, together with Minster and Ash, 18 Bewick's Swans and 10 Whoopers in February 1979. In that month, following the hard weather which had affected all of north-west Europe, there was an exceptional influx of 1,290 European White-fronted, 80 Bean and 92 Barnacle Geese to the Stour marshes.

The largest concentration of breeding wildfowl in east Kent is in the Stodmarsh – Westbere area. Averages of three surveys (1977, 1979 and 1981) give totals of 154 pairs of Mallard, 24 of Shoveler, 6 of Gadwall, 10 of Pochard, 11 of Tufted Ducks and 18 of Shelducks in this area. In addition 13 pairs of Mute Swans and 2 pairs of Garganey are worthy of note.

Elsewhere, notable figures are 20 pairs of Tufted Ducks on the small gravel pits upstream of Canterbury, especially at Chilham, and 27 pairs of Shelducks in the Pegwell Bay/Sandwich Bay area.

The mouth of the Stour links Pegwell and Sandwich Bays, immediately south of Ramsgate. This is the only significant area of intertidal mudflats in east Kent. Pegwell Bay (Table 43) has been devalued over the years by industrial development, a rubbish tip and a hoverport (not currently in use). Pegwell Bay, Sandwich Bay (the location of an important bird observatory) and the hinterland of the Lydden Valley have been counted as a whole for the East Kent Lowlands Survey; the averages of the five years' counts included 269 Mallard, 108 Teal, 95 Wigeon, 18 Shoveler, 48 Dark-bellied Brent Geese, 163 European White-fronted Geese, 78 Greylag Geese, 13 Mute Swans and 150 Shelducks. Some of these species may, however, be absent except in hard weather, when numbers of many wildfowl can be much higher. For example, 1,860 Wigeon, 500 Whitefronts and 780 Shelducks were present in January 1979. Brent Geese are scarcer than at equivalent localities elsewhere in the south-east, perhaps partly because the area is bypassed by the overland movements described earlier, but disturbance and the shortage of *Zostera* may also be important.

The remainder of the east Kent coast is cliff-bound and unsuitable for wildfowl, other than sea ducks. St Margaret's Bay (between Deal and Dover) holds, in addition to the Common Scoters mentioned earlier (under East Sussex and south Kent), small numbers of Eiders, rarely as high as 50. Both species, and also Dark-bellied Brent Geese, are observed on passage in spring and autumn.

North Kent

The coastline of north Kent is one of contrasts: the rocky shoreline of the Isle of Thanet, east of Whitstable, at one extreme, and the soft shores of the North Kent Marshes in the west at the other. Because of the range and extent of habitats it provides and because of its situation, ideal to receive continental migrants, the area has long been recognised as one of outstanding value for wildfowl.

The open coast

The coastline of the Isle of Thanet, where the North Downs meet the sea, is made up of chalk cliffs, with wave-cut platforms, rock pools and occasional mussel beds below. Apart from small numbers of Mallard and Wigeon, the wildfowl are limited to scattered flocks of sea ducks (Table 44). A passage of moderate to large numbers of wildfowl, however, is visible from Fore-

Table 43. *The regular (and maximum) numbers of wildfowl at the following localities in the Stour Valley, 1970 onwards:*

a) Eastwell Park (5 seasons)
b) Naccholt Claypits (2)
c) Chilham (Bagham Pond)/Shalmsford St GPs (3)
d) Wye floods (1)
e) Chartham GP (2)*
f) Fordwich/Westbere GPs +
g) Stodmarsh NNR/Collard's Lagoon (13)
h) Seaton GPs +
i) Pegwell Bay (12)

	a	b	c	d	e	f	g	h	i
Mallard	235(1500)	22(65)	38 (69)	34 (83)	22(50)	(320)	773(3550)	(60)	63(250)
Teal	0 (11)	0 (0)	1 (6)	37(104)	0(20)	(240)	255 (690)	(100)	0 (80)
Wigeon	0 (1)	0(10)	0 (0)	40(120)	0 (1)	(620)	93(1250)	(0)	0 (50)
Shoveler	0 (3)	0 (0)	2 (6)	0 (3)	0(80)	(20)	79 (300)	(6)	0 (5)
Gadwall	0 (1)	0 (0)	0 (7)	0 (0)	0 (2)	(62)	10 (75)	(64)	0 (1)
Pochard	7 (92)	5(20)	44(146)#	0 (4)	9(30)	(160)	94 (460)	(64)	0 (1)
Tufted Duck	32 (42)	1(15)	62(162)#	0 (0)	3(16)	(350)	71 (277)	(116)	0 (0)
Greylag Goose	0 (1)	0 (0)	0 (2)	0 (0)	0 (2)	(24)	8 (27)	(0)	0 (2)
Canada Goose	15 (48)	0 (0)	15 (43)	0 (0)	0 (4)	(109)	13 (65)	(75)	0 (0)
Mute Swan	1 (4)	1 (7)	8 (26)	0 (7)	2(10)	(184)	23 (91)	(116)	0 (0)

Note: Other species.
Shelduck: Stodmarsh/Collard's Lagoon 4(17); Pegwell Bay 39(80); Westbere (16); Seaton (16).
Dark-bellied Brent Goose: Pegwell Bay 13(150).

* Pre-1970 data.
+ Maxima only.
# Data from *Kent Bird Reports*.

ness, especially during strong north-westerly gales in autumn and severe weather in January. In the period 1978-1982, this passage included European Whitefronts (maximum 440, March 1979), Dark-bellied Brent (2,015, Oct 1980), Wigeon (1,118, Dec 1981), Eiders (310, Dec 1982), Common Scoters (743, Sept 1981) and Velvet Scoters (240, Jan 1982). Associated with the passage are movements of other species – divers, grebes, kittiwakes and auks. An exception to this general pattern is the Common Scoter, which exhibits a sustained migration from July to late October/early November, only visible with northerly winds.

Towards Minnis Bay, in the west, the cliffs grad into low-lying agricultural land of the north Chisle and Plumpudding Island Marshes, which form a lin between the north coast and the valley of the Rive Stour, at one time part of the Wantsum channel separating the Isle of Thanet from the mainland. Ver little grazing marsh still exists, with much of the are progressively underdrained and converted to arable i recent years. Here the coastline consists of sand an mud beaches interspersed with rocks. A greate variety of wildfowl are regularly attracted to this area, but numbers are again small – although numbers o dabbling ducks can be sizeable.

Between Reculver and Whitstable, the coastlin consists of cliffs composed of soft unstable Londo

Fig. 19. North Kent, Essex and Suffolk.

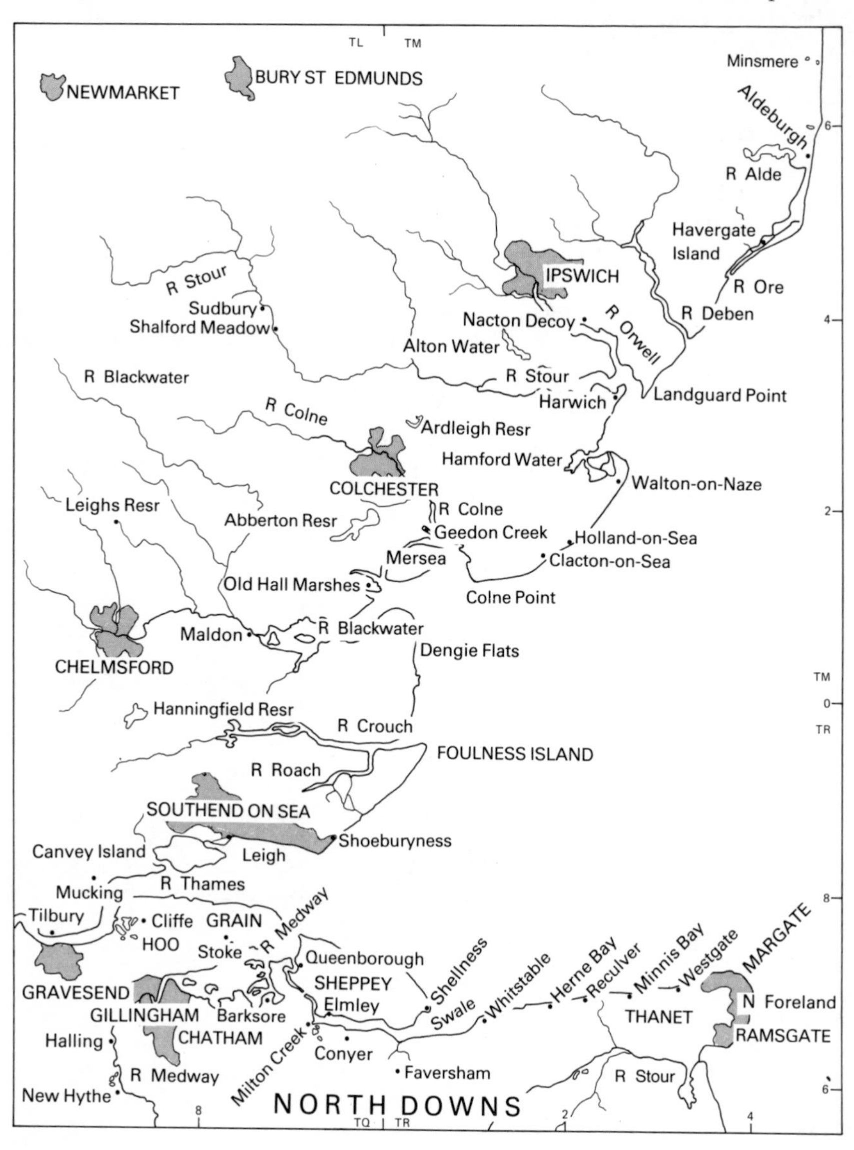

Table 44. *The regular (and maximum) numbers of wildfowl on the north Kent coast between North Foreland and Herne Bay.*

a) North Foreland to Foreness — 5 seasons 1967-1973
b) Walpole Bay and Palm Bay — 2 seasons 1967-1968
c) Westgate to Birchington — 3 seasons 1967-1970
d) Minnis Bay to Reculver — 4 seasons 1967-1981
e) Herne Bay — 1 season 1967

	a	b	c	d	e
Mallard	9 (90)	0 (0)	40 (61)	109(457)	9(26)
Teal	0 (0)	0 (0)	0 (0)	40(110)	3 (4)
Wigeon	0 (0)	4(18)	0 (2)	17(140)	0 (0)
Shelduck	0 (5)	1 (3)	2 (8)	6 (13)	1 (2)
Pochard	0 (0)	0 (0)	0 (0)	6 (56)	0 (0)
Tufted Duck	0 (0)	0 (0)	0 (0)	2 (23)	0 (0)
Scaup	0 (1)	0 (0)	0 (0)	0 (4)	0 (0)
Eider	15 (70)	7(16)	0 (0)	8 (30)	0 (0)
Common Scoter	16(380)	3 (8)	37(134)	10 (66)	1 (2)
Bewick's Swan	0 (0)	0 (0)	0 (0)	3 (28)	0 (0)
European Whitefront	0 (0)	0 (0)	0 (0)	4 (27)	0 (0)
Dark-bellied Brent	0 (3)	0 (0)	3 (6)	6 (53)	0 (0)

Clay, fronted by clay beaches. The principal wildfowl interest of this coastline is the sea duck population occurring during autumn and winter (Table 45).

Table 45. *The regular (and maximum) numbers of selected sea ducks off the north Kent coast, 5 seasons 1977-1981, from Kent Ornithological Society records.*

	Thanet	Reculver
Eider	49(191)	24(100)
Common Scoter	98(270)	22 (91)
Red-breasted Merganser	6 (15)	4 (19)

The North Kent Marshes

West of Whitstable the landscape changes totally, rocky shorelines being replaced by the soft muds, saltmarshes and enclosed pasture and arable land of the North Kent Marshes.

Geographically, the North Kent Marshes can be delimited as the area between Whitstable in the east and Gravesend in the west, below the 12.6m (25ft) contour lines. Three discrete areas can be identified, both on a geographical and an ornithological basis: the Swale and Isle of Sheppey in the east (13,000ha); the Medway Estuary in the central areas (9,000ha) and the South Thames Marshes in the west (8,000ha).

The topography of the area is produced by the London Clay, which outcrops as the series of plateaux and low hills which form the spine of the Hoo Peninsula, the Isle of Sheppey and parts of the south Swale. To the north of the Hoo Peninsula, and in the Medway and Swale estuaries, the clay is overlain by extensive deposits of river gravels and alluvial silts and muds, by the action of the Thames and Medway.

The drainage infrastructure of the 19th century remains basically the same today, although many localised changes have been made to permit industrial developments or the conversion of grazing marsh to arable. In addition, sea walls have been improved throughout, both as a result of the breaches caused by the disastrous 1953 surge tide and as part (in 1980-1983) of the Thames Tidal Barrier Scheme. In several areas, attempts to enclose land have been abandoned. During the 18th century, for example, much of the south Medway was enclosed fresh marsh until the sea walls fell into disrepair and were breached. Victorian attempts at reclamation eventually failed; a series of ornithologically rich saltmarsh islands now remain. Both here and elsewhere in north Kent the erosion of established saltmarsh is occurring at the same time as the accretion of silt results in the production of new low-level saltmarsh, often close by.

A succession of localised drainage improvement schemes have permitted the widespread conversion of grazing marsh to arable. Between 1935 and 1982, the net area of grazing marsh fell from 14,750ha to 7,675ha – a reduction of 48% (Table 46). Until 1968 losses to urban and arable land-uses were relatively small and about equal. After 1968, however, the total losses of grazing marsh accelerated (Table 47).

The ornithological importance of the North Kent Marshes has been apparent since the mid 1850s (209).

It is clear that the process of enclosure and drainage affected the birds dramatically, as saltmarsh turned to fresh marsh and fresh marsh became drier. Despite these changes the area remains ornithologically rich and extremely important for wintering and breeding wildfowl and waders (Table 48), no doubt because of the juxtaposition, diversity and extent of the habitats present – open water, mudflats, saltmarsh, shell beaches and grazing marsh.

The north Sheppey coastline consists of unstable cliffs of London Clay, with little wildfowl interest. To the south, the Swale channel separates the Isle of Sheppey from the Kentish mainland, extending for 20km from Queenborough in the west to Shellness in the east. To the north and south of the channel extensive areas have been enclosed, up to 4.5km inland on Sheppey and 2.5km on the mainland.

The eastern Swale is of particular importance for wildfowl, the mudflats and associated *Zostera* beds of the Faversham – Whitstable area (designated as a Local Nature Reserve) and the saltmarshes of the north shore between Windmill Creek and Shellness being highly attractive feeding areas. The mudflats have a very rich invertebrate community, in terms both of quantity and species diversity. At high tide the open sea off the Leysdown – Shellness shore is often used for roosting by Brent Geese and Wigeon. The Shellness and Harty saltings are also important roosts. Of the enclosed marshes, those of the Isle of Sheppey are of greater importance than those of the Kentish mainland, with the remaining grassland mostly damper and less intensively grazed than that of the south Swale. Flooding is minimal, although the area is intersected by a network of creeks – which are the centre of attraction. Of special interest are the Swale NNR, just west of Shellness, and the RSPB reserve at Elmley.

In the First Edition Gillham and Harrison wrote "sheep and cattle are grazed over the levels and in all probability no major change has been made in the husbandry for several hundred years". The same does not hold today, as large areas have been drained – particularly on Sheppey – and the process may well continue. By and large, the area has avoided major industrial development – although port development at Ridham Dock in the west Swale, warehousing facilities inland of Queenborough and dredgings dumping at Rushenden on Sheppey have intruded onto the grazing marsh. Although the area is popular for recreation, the estuary has escaped major incursion to date. Marina proposals for Milton, Conyer and Oare Creeks, and Harty have all been rejected, to the credit of Swale Borough Council, the local planning authority. The detrimental impact of the changes in agriculture has been buffered by the establishment of the RSPB reserve at Elmley Marshes in 1975. Here the RSPB has leased 282ha of Spitend; the shooting rights are also held over a further 1,068ha of Elmley Island and the adjacent foreshore. Management of the reserve has involved dam construction and the pumping of water onto the grazing marsh to create a flooded "mere" in winter, and to keep water levels high enough to keep the grassland wet during summer. Cattle grazing of moderate intensity has produced a sward structure extremely attractive to both wintering and breeding wildfowl (Table 50). Certainly wintering dabbling ducks and European Whitefronts have responded to management over a remarkably short period of time (Table 49), and Elmley Marshes now qualify as internationally important for Wigeon.

The River Medway, downstream of the ports of Rochester and Chatham, is a complex of tidal creeks, mudflats, saltmarsh and grazing marsh, extending

Table 46. *The reduction in the area of grazing marsh in north Kent, 1935 to 1982 (in hectares). (After Williams* et al. *1983.)*

	1935	1968	1979	1982
Grazing marsh	14750	12250	8200	7450
Area converted to grazing marsh	0	200	275	225
Total area of grazing marsh	14750	12450	8475	7675
Net loss of grazing marsh per annum	70	361		267

Note: 725ha of land in the Sheerness area of the Isle of Sheppey was not surveyed in 1979.

Table 47. *Area of land-uses to which grazing marsh was converted 1968 to 1982. Percentages are given as a proportion of the total grazing marsh identified in 1935. (After Williams* et al. *1983.)*

	1968		1979		1982	
	area (ha)	%	area (ha)	%	area (ha)	%
Saltmarsh	125	1	175	1	175	1
Urban	1150	8	1450	10	1925	13
Arable	1225	8	4200	28	5200	35
TOTAL	2500	17	5825	39	7300	49

Note: 725ha of land in the Sheerness area of the Isle of Sheppey was not surveyed in 1979.

Table 48. *The regular numbers of selected species of wildfowl on the Swale and the Isle of Sheppey, Medway Estuary and South Thames Marshes in six periods, 1952 to 1982, with the maxima for 1977-82. Includes data from Kent Ornithological Society records.*

	1952-56	1957-61	1962-66	1967-71	1972-76	1977-82 (Max)
THE SWALE AND THE ISLE OF SHEPPEY						
Mallard	796	838	758	694	574	1771 (3764)
Teal	1368	933	355	221	324	2482 (5000)
Wigeon	1844	1310	2316	3250	2476	8129(20000)
Pintail	91	53	43	25	28	184 (341)
Shoveler	36	82	104	54	202	307 (650)
Gadwall						53 (104)
Shelduck	447	477	826	625	579	1592 (4158)
Pochard						202 (563)
Tufted Duck						61 (140)
Mute Swan						87 (150)
European Whitefront	144	350	461	804	660	1091 (2700)
Dark-bellied Brent	79	59	105	490	719	1003 (1848)
MEDWAY ESTUARY						
Mallard	180	200	355	843	640	827 (1503)
Teal	575	613	1035	2129	3617	1365 (5000)
Wigeon	535	688	2284	5454	5765	2156 (5500)
Pintail	38	33	143	444	625	286 (607)
Shoveler	91	53	43	25	28	96 (279)
Shelduck	410	552	814	1908	1896	2397 (4507)
Dark-bellied Brent	52	53	109	330	692	812 (2654)
SOUTH THAMES MARSHES						
Mallard	1154	1532	966	236	167	1070 (1750)
Teal	1400	1632	402	125	324	824 (1705)
Wigeon	2728	3027	2116	2532	974	1156 (5185)
Pintail	197	236	458	97	77	93 (227)
Shoveler	128	329	86	137	126	251 (390)
Gadwall						43 (100)
Shelduck	2638	5367	3527	1095	554	1257 (5185)
Pochard						452 (1730)
Tufted Duck						404 (809)
Mute Swan						151 (211)
European Whitefront	570	460	634	850	670	380 (1300)

Table 49. *Regular numbers of wildfowl at Elmley Marshes RSPB reserve since its acquisition in 1975. Includes data from RSPB.*

	1975-77	1978	1979	1980	1981	1982	Maximum 1975-1982
Mallard	711	533	567	1065	1610	2876	3744
Teal	722	1600	1567	2534	1777	1738	4000
Wigeon	2878	2200	3667	4430	10933	9464	20000
Pintail	40	2	106	180	189	250	323
Shoveler	143	129	164	239	294	336	600
Gadwall	15	30	35	29	49	88	104
Shelduck	508	77	1067	426	713	703	2000
Pochard	34	17	51	129	85	120	217
Tufted Duck	8	8	16	13	40	19	118
Mute Swan	20	12	42	44	37	116	137
Bewick's Swan	2	0	3	7	8	36	45
European Whitefront	269	583	418	829	797	669	1800
Dark-bellied Brent	144	61	41	78	40	208	800

18km to Sheerness in the east and about 8km at its broadest. It is of outstanding value for wintering wildfowl, and ranks as internationally important for Shelducks. Most of the north shore of the estuary rises steeply into the ridge of the Hoo Peninsula, and there is little grazing marsh, except near Stoke. Of particular interest are Stoke Saltings and Ooze, a complex of intertidal mudflats and small saltmarsh islands attractive to Mallard, Teal and Wigeon. Much of this side of the estuary is highly industrialised, with a power station and petrochemical plant centred on Kingsnorth and a larger complex which, since 1930, has expanded inexorably to occcupy most of the southern half of the Isle of Grain. Since the early 1970s, Stoke Saltings and Ooze have also been under the threat of reclamation – first from Greater London Council (GLC) proposals for the disposal of London's refuse, and now from waste pulverised fuel ash (PFA) from Kingsnorth Power Station. Planning permission was given during 1982 to dump PFA on 17ha of saltings and mudflats in the westernmost part of the area: it is to be hoped that this will be the last incursion.

South of the Medway channel lies a vast complex of mudflats, saltmarshes and freshwater grazing marsh. The central Burntwick – Milfordhope – Greenborough islands and mudflats are particularly attractive to dabbling ducks, both for roosting and feeding. The enclosed freshwater grazing marshes of Chetney also provide useful feeding areas especially for Wigeon.

Since the mid 1970s, the populations of most dabbling ducks in the Medway Estuary have fallen – although the numbers are still much higher than during the 1950s (Table 48). There is no clear reason for this decline; although considered for major industrial development on several occasions in recent years, the south Medway is still largely unspoilt. Possibly the provision of more attractive habitat at Elmley has drawn birds there. The erosion of the saltmarsh islands in the Burntwick – Greenborough complex may also have contributed. While these islands still offer good feeding, the ability of the area to provide secure, sheltered and undisturbed roosting sites may have been impaired. The conversion to arable of over 40% of the Medway's main block of grazing marsh at Chetney may also have had a detrimental impact. The wildfowl of the estuary continue to face very real threats from the reclamation of the saltmarsh by dredgings dumping (the Barksore Peninsula is being reclaimed, and the drainage of Nor Marsh is currently under consideration); increased disturbance from waterborne recreation; loss of the remaining areas of grazing marsh; and the continuing possibility of industrial development.

The extensive enclosed grazing marshes and arable land of the South Thames Marshes extend from the Isle of Grain in the east to Gravesend in the west – a distance of about 23km. At the broadest point the Marshes extend about 3km inland. On the seaward side there is little saltmarsh except in Higham Bay, but the area is fringed by the extensive mudflats, up to 1.5km in width, of Blyth Sands and the Roas Bank. There are also a number of mineral workings, including clay and chalk pits at Cliffe and sand pits on the Isle of Grain. Although the mudflats provide an important feeding area for Shelducks, the principal interest is centred on the grazing marshes, which support good numbers of European Whitefronts and dabbling ducks (Table 48). Undoubtedly, however, wildfowl numbers have declined severely since the publication of the First Edition where the South Thames Marshes were described as "the most outstanding wildfowl haunt in Kent, holding at least as many dabbling ducks as the Medway and Swale combined". Much of the decline can be attributed to the progressive drainage of the key areas of St Mary's, Halstow and Cooling Marshes – a process which looks set to continue to erode the remaining areas of grazing marsh. Although plans for the large-scale industrial development of Cliffe Marshes – in the shape of an oil refinery (near the present large refinery) and dredgings dumping grounds – have not yet been implemented, only fragments of the ornithological importance remain. Fortunately, plans to enclose Blyth Sands and the Roas Bank as part of the Thames Tidal Barrier Scheme were dropped on expenditure grounds, or the decline in wildfowl numbers might have been more severe.

The grazing marshes of north Kent, with their high water levels, dense network of fleets and cattle or sheep-grazed pasture, attract a diverse and abundant breeding wildfowl population, the numbers of Shelducks, Shoveler and Pochard being particularly important. Because of the extensive area and the practical difficulties of surveying breeding wildfowl, firm data for most species are not available. Using the results of localised surveys and unpublished Kent Ornithological Society and RSPB records, however, reasonable estimates can be made (Table 50).

There is evidence that for several species there have been marked fluctuations over the years. This is especially true of Shoveler, which respond rapidly to wet conditions on the marshes. For example, in the very wet season of 1975, 20 pairs nested on Chetney Marshes compared with only one in 1973 (576). In a study of 121.4ha of Cooling Marsh, 3 pairs of Shoveler were nesting on grazing marsh during the dry spring of 1974 (247). In 1975, after the area had been drained,

15 pairs were recorded. The spring of 1975 was, however, so wet that not only was standing water present on the marsh, but also on the arable parts. Increasingly efficient drainage and improved sea-defence works have meant that high densities of breeding Shoveler are now unlikely to occur away from nature reserves. Pochard increased to reach a peak during the 1960s, but have probably declined since (576), with a shift away from areas converted to arable at that time. Garganey have decreased from 13-18 pairs in the early 1960s to 9 pairs in 1975 and only 1 or 2 since 1980; while this may be partly associated with climatic change, habitat changes also may be implicated.

It is clear, therefore, that if the breeding wildfowl populations of the North Kent Marshes are to be maintained, this can only be achieved by safeguarding the remaining areas of grazing marsh – through reserve purchase and the application of the provisions of the Wildlife and Countryside Act (1981) to the full.

Inland north Kent

Since the inland boundary of the north Kent region has been taken as the North Downs, its extent is limited. There are few areas of open water except in the Medway Valley. Even fewer are of significance for wildfowl (Table 51).

Upstream of Rochester bridge the River Medway has only been irregularly covered by counts. The figures available, however, suggest that the river supports few wildfowl. An exception is the marshes between Halling and Eccles, which attract small numbers of dabbling ducks and also flocks of Mute Swans. Additional information suggests that the area is also important for Shelducks: two counts in March and December 1981 averaged 102 birds (Kent Ornithological Society). Over 20 pairs were estimated to have bred in 1982, and 30 and 46 young were reared in 1979 and 1980 respectively.

A little further upstream, on the west side of the river, is the most important inland site for wildfowl in this sector, the Leybourne – New Hythe Gravel Pits, a complex of about 15 pools covering about 80ha. Although most of the pits are mature, with good scrub growth on their margins, the majority are steep-sided and deep. As a result, they regularly attract good numbers of Tufted Ducks and Pochard only. Table 51 probably understates the numbers of diving ducks using the gravel pits for recent years. Of the nine counts between October 1980 and March 1983, five produced over 200 Pochard and six over 200 Tufted Ducks.

Essex and east Suffolk

The coastal plain of Essex and Suffolk is broken up by a succession of broad estuaries, the difficulty of bridging which caused the main road and railway to be built well inland, preventing the development of most of the shore (383). To this day, despite its proximity to London, the coast is largely unspoilt, and the parts where additional protection is afforded, either directly or, as in the case of the Ministry of Defence's ownership of Foulness, incidentally, amount to a virtual wilderness.

The estuaries of Essex, together with the Orwell in the south-east corner of Suffolk and the reservoirs at Abberton and Hanningfield, comprise one of the most important complexes of habitat for wildfowl in

Table 50. *Estimated total breeding wildfowl populations for the North Kent Marshes based on Scott (1978), Williams (1979), Taylor* et al. *(1981) and unpublished Kent Ornithological Society and RSPB records.*

Mallard	c.500 pairs
Teal	15-25 pairs
Wigeon	1-2 pairs breed irregularly
Pintail	1-2 pairs regularly
Shoveler	70-90 pairs
Garganey	1-2 pairs
Gadwall	5-10 pairs
Shelduck	200-500 pairs
Pochard	30-50 pairs
Tufted Duck	15-25 pairs
Mute Swan	c.50 pairs
Greylag	feral flock of c.50 pairs

Table 51. *The regular (and maximum) numbers of wildfowl at two sites in inland north Kent.*

a) Burham and Eccles Marshes, River Medway
5 seasons 1967-1973
b) Leybourne and New Hythe Gravel Pits
17 seasons 1961-1980

	a	b
Mallard	12 (30)	132(312)
Teal	26(113)	5 (40)
Shoveler	0	1 (12)
Shelduck	1 (11)	6 (55)
Pochard	0 (1)	102(270)
Tufted Duck	0 (16)	68(260)
Goldeneye	0 (2)	1 (6)
Mute Swan	34 (70)	22 (59)
Canada Goose	0	30 (98)

north-west Europe, holding a regular population of approximately 60,000 wildfowl, including some 20% of the world population of Dark-bellied Brent Geese. Several individual estuaries, as well as Abberton Reservoir, are of international importance in their own right.

Dark-bellied Brent Geese (see Table 52, which gives the recent numbers in the whole Thames estuarine basin) have their main autumn gathering ground at Foulness Island, where 25-30% of the peak British population concentrates between early October and late November. They feed mainly on the 325ha *Zostera noltii* beds, probably the largest in Europe (492), and roost on the vast Maplin Sands and the farmland on Foulness. A large proportion of the geese move readily to Canvey Island and Leigh Marsh, particularly when there is heavy activity at Shoeburyness military range, and the Foulness – Leigh area has therefore been combined with Foulness in Table 52.

The threat that a third London airport would be built on Maplin Sands, when it was first seriously proposed in the late 1960s, prompted a detailed study of the numbers, movements and habits of Dark-bellied Brent Geese at Foulness and throughout their range. The studies were originally funded by the Department of the Environment, but even after the shelving of the airport and associated seaport development plans it has been possible to finance a continuing study from a variety of sources. The programme, involving extensive marking and ringing for several years, has complemented and expanded upon the data from the regular wildfowl counts, the organisation of which on the Essex and Suffolk estuaries was itself funded from November 1972 to September 1975 by a contract between the Natural Environment Research Council (NERC) and the BTO, as part of the research into the likely impact of the proposed airport (56).

A series of papers and reports has resulted from the recent studies of Brent Geese in Essex (e.g. 47, 121, 122, 123, 521) covering most aspects of their ecology and behaviour (see also pp.388-93). In December the number of geese at Foulness/Leigh drops sharply, mainly because of the depletion of the *Zostera* stocks, remaining at a few thousand for the rest of the winter.

In mid-winter the Blackwater Estuary and Hamford Water become the main centres in Essex, while birds from Foulness also move in force to the south coast, turning at the same time to their secondary food source, *Enteromorpha*. It was in Essex that Brent Geese were first observed feeding in large numbers within the sea walls; 3,000 were inland in 1962-63, but the habit did not occur again until 1969-70. In 1973-74 9,000 out of an Essex/Suffolk total of 16,000 were observed inland; by 1975-76 it had risen to 14,500 out of 19,200. Field-feeding is now regular on all parts of the coast, often in some numbers as soon as early October, but most intensively in mid-winter.

Other migratory goose species are rare in Essex, except when hard weather forces them across the North Sea, such as in January and February 1979, when perhaps 1,000 European Whitefronts, 100 Bean Geese, 100 Pinkfeet and 60 Barnacle Geese were present in the county (177).

Table 54 (p.107) sets out the numbers of the principal ducks, plus Canada Geese and Mute Swans, along the coast from Canvey Island to the Orwell, including Abberton and Hanningfield Reservoirs.

The north shore of the Thames Estuary proper, below Tilbury, is very different from the south and much less important for ducks. Above Canvey Island the deep-water channel lies close to the north shore, restricting the intertidal area to a narrow strip between East Tilbury and Stanford le Hope, incorporating Mucking Flats. A series of counts for this area in the mid 1970s suggested a regular (and maximum) population of 300 (338) Mallard, 169 (297) Shelducks, 6 (15) Pochard, 32 (47) Tufted Ducks and 9 (16) Mute Swans. Since then maxima of 90 Pintail, 100 Shoveler, 50 Gadwall and 34 Tufted Ducks have been reported (177). Downriver, the coast as far as Shoeburyness is entirely built up, apart from small but vital areas at Hole Haven, Canvey Point and Two Tree Island/Leigh Marsh NNR. The vast petrochemical refineries and storage depots at Coryton, Thames Haven and Canvey Island give way to the conurbations of Canvey Island and Southend. The Brent Geese are confident enough to utilise any available feeding area, even immediately adjacent to the industrial and residential areas, but other species are restricted to the creeks to the north and west of Canvey Island.

The 12,000ha of flats and saltmarsh of Foulness and Maplin Sands (the largest continuous area of intertidal flats in Britain), amount to a virtual refuge, public access to the island being strictly controlled by the military authorities and disturbance from the firing ranges being only slight. Ducks can find relatively little prime feeding area, however, compared with the estuaries to the north. Wigeon are plentiful on the sands, but other species are mainly concentrated in the dykes, fleets and creeks around New England and Havengore Islands (56). Shoveler increased considerably in the 1970s and now have one of their largest east coast gatherings here.

A major roost of Common Scoters is present for most of the year off the Maplin Sands. In February 1976 7,000 were counted, but this was apparently exceptional, no more than 1,000 having been found in subsequent winters (177). They possibly feed at Buxey Sands, the shallows due east of the Dengie coast (215). This area may well also be used by the smaller flocks of scoters which are present elsewhere along the Essex coast, notably off Clacton, where up to 800 have been found, the peak there usually occurring during the first four months of the year (177).

Along the west and north shores of Foulness Island run the outermost reaches of, respectively, the rivers Roach and Crouch. These long, narrow estuaries, which converge 5km from the sea, are treated together in Table 54, but most of the wildfowl occur on the saltings and fresh marsh of the upper Crouch Estuary, especially on Bridgemarsh Island and Hydemarsh (56).

The coast between the mouths of the Crouch and Blackwater – the Dengie Flats and Marshes – is of major importance in its own right. Its strategic position makes it particularly valuable at passage times, and its remoteness means that human disturbance is slight and the winter wildfowl numbers are also high. The extensive saltings – one of the biggest expanses in Essex – provide excellent feeding for dabbling ducks and Shelducks.

The shallows off Sales Point, at the mouth of the Blackwater, are the main roost for the dabbling ducks in the Crouch – Colne stretch (215). The same area is favoured by a flock of 1-200 Eiders (the largest in Britain south of Walney Island in Cumbria) which frequents the outer Blackwater Estuary all year round. If undisturbed the Eiders may use the Tollesbury shore, on the north side, but in summer, when boating is particularly intensive, they prefer the seas off Bradwell and the Dengie Flats.

The Blackwater is the largest and perhaps most important estuary wholly in Essex. Although most of its saltmarshes were reclaimed long ago, it still has a superb diversity of shoreline habitats. As well as Brent Geese, which by the New Year usually outnumber those at Foulness and Leigh, Shelducks are present in internationally important numbers. Most wildfowl favour the inner north shore between Osea and Northey Islands, where the discharges from Maldon sewage works have resulted in a luxuriant growth of *Enteromorpha* (485), although Mallard prefer the outer north bank between Goldhanger and Tollesbury (56). A gathering of about 100 Mute Swans has been

Table 52. *The maximum numbers of Dark-bellied Brent Geese recorded at their main resorts in the greater Thames Estuary in each of the seasons 1978-79 to 1982-83, and in each of the five preceding 3-year periods. The totals are the highest synchronised counts within each period; the percentages refer to the season in which that count occurred. A blank = no data; brackets = incomplete data.*

	1963-65	66-68	69-71	72-74	75-77	78	79	80	81	82
Swale	150*	500*	1040*	680	1434	683	1274	567	1010	1848
Medway	110*	400*	550	716	146	1064	1002	844	850	2654
Foulness/Leigh	7074	6816	13943	15868	15593	12488	20469	17758	19961	18208
Roach				26	550	2392	4288	0	10	1100
Crouch	200	165		800				3120	3550	5059
Dengie	700	900	3000	2800	2800	1600+	3500	1100	914	(610)
Blackwater	3232	3008	5015	6117	8097	13287	12498	9167	9003	5820
Colne	1034	1348	2785	2033	2640	2450+	1900+	1000	1106	2500
HamfordWater/Naze	700	840	1580	3500	7245	4000	8200	4500	4000	8000
Stour	770	540	450	414	430	538	800+	686	558	1535
Orwell			375+	540	830	575+	400+	800	+1212	1000
FOULNESS/LEIGH – ORWELL										
Total	8969	11415	18317	20790	23700	25080	30914	26534	28019	28350
Month	Dec 65	Jan 68	Nov 70	Nov 73	Nov 76	Jan	Jan	Nov	Nov	Jan
% of UK pop.	51	61	77	50	49	40	41	40	47	31
% of world pop.	33	37	45	25	22	18	19	18	24	14

* Data from Harrison (1972).
+ Data from Essex and Suffolk *Bird Reports*.

present at Fullbridge, Maldon, at the head of the estuary, since at least the early 1960s, attracted by the spillage from grain wharves; the peak apparently occurs in mid-winter, suggesting a movement from the moulting area at Abberton Reservoir (see below).

Immediately north of the Blackwater Estuary, the Colne and its numerous associated channels have their outlet. The 8km between Salcott Channel, at the mouth of the Blackwater, and the Colne itself are occupied by Mersea Island, with the Strood and Pyefleet Channels to its north, and the broad Mersea Flats to the south. The most important parts of the complex for wildfowl are the largely inaccessible saltmarshes along several of the side-creeks, notably Old Hall Marsh, and Fingringhoe and Langenhoe Marshes, either side of Geedon Creek. The Mallard prefer the more open Mersea Flats, while the Goldeneyes and Mergansers occur in the Colne's main channel, along with occasional passage Long-tailed Ducks (up to 40) and Pochard. Sea ducks are often present in strength at the mouth of the Colne: up to 420 Common Scoters off East Mersea, 45 Eiders off Colne Point in autumn, and in severe weather up to 220 Scaup (56).

The Hythe, the uppermost tidal reach of the Colne between Rowhedge and Colchester, has held a sizeable flock of Mute Swans since about the mid 1950s (620), coinciding with an increase in grain shipments to Europe. The counts show a regular (and maximum) Mute Swan population from 1970 to 1981 of 118 (170), with a slight decrease since the mid 1960s, but otherwise very little change either during the season or year to year. Counts of other species in the same area between 1972 and 1974 indicated a population of 163 (260) Mallard, 4 (21) Gadwall, 9 (17) Pochard, 26 (37) Tufted Ducks and 39 (44) Canada Geese.

Immediately south of Colchester, just a few kilometres from both the Blackwater and Colne estuaries, lies the older and by far the more important of the two main Essex reservoirs – Abberton. Completed in 1940, this 490ha water has, since regular counts began in 1949, ranked as much the most important reservoir in Britain for wildfowl; only the relatively new Rutland Water can approach its average maximum of 12,115 ducks, geese and swans between 1978 and 1982 (Table 53). Abberton owes its importance mainly to its proximity to the coast and to its protection as a wildfowl site by the Essex Water Company (recreation is restricted to coarse fishing and general access to the banks is by permit only; the reservoir became a statutory Bird Sanctuary in 1968). Its primary role is still as a roost for the local estuarine population. Increasing numbers of Teal, Wigeon, Goldeneyes and other ducks use the reservoir for feeding as well, however, reflecting the trend in the country as a whole (p.515) and contributing to a continued increase in the total number of birds using the water since the mid 1970s. Even Brent Geese used the area for their entire daily cycle in November and December 1983, a maximum of 165 roosting on the reservoir and feeding on adjacent winter wheat, until scared off because of their refusal to use the local pastures for feeding!

Other factors making Abberton so attractive for wildfowl are its open aspect, the predominant shallowness of the water (the mean depth is under 5m, and above an island 100m from the perimeter the feeding conditions for diving ducks are ideal), the

Table 53. *The regular (and maximum) numbers of wildfowl at Abberton Reservoir in seven periods, 1949-53 to 1979-82. (NB: September to March only; for moulting figures see text.)*

	1949-53	1954-58	1959-63	1964-68	1969-73	1974-78	1979-82
Mallard	926(1367)	1320(3243)	1996 (4005)	3201(7380)	2953(5392)	2793 (4760)	2909(8600)
Teal	1745(4600)	2777(8609)	4846(12221)	1456(2953)	1080(2709)	1301 (5237)	1308(3370)
Wigeon	1863(5000)	1613(3790)	1641 (5321)	921(2341)	1940(3635)	3473(11574)	3166(5725)
Pintail	69 (260)	84 (205)	107 (453)	44 (262)	84 (405)	33 (198)	24 (73)
Shoveler	99 (544)	202 (403)	283 (1387)	169 (414)	358 (657)	365 (858)	322 (612)
Gadwall	1 (5)	8 (20)	12 (21)	16 (39)	39 (136)	135 (248)	143 (280)
Pochard	1070(3870)	1404(2224)	1077 (2512)	948(2804)	880(1591)	877 (1756)	510 (865)
Tufted Duck	308 (546)	250 (418)	306 (623)	281 (548)	964(2657)	1189 (1982)	1270(2670)
Goldeneye	125 (278)	240 (365)	205 (397)	309 (730)	267 (408)	359 (526)	403 (610)
Goosander	10 (30)	33 (67)	42 (87)	37 (69)	33 (114)	18 (51)	26 (38)
Smew	38 (84)	33 (90)	21 (52)	15 (28)	6 (16)	6 (8)	3 (5)
Shelduck	16 (80)	10 (70)	21 (126)	13 (41)	16 (65)	21 (61)	32 (117)
Mute Swan	not counted	265 (477)	206 (485)	44 (179)	43 (95)	50 (138)	52 (184)

abundance of both plant and invertebrate life and the substantial shoreline. The reservoir is divided into three parts by road causeways and the largest, eastern section is entirely concrete-faced, but the two small upper pools (totalling 73ha) have natural banks.

In these western sections the most successful duck ringing station in the country was established in 1949. Since 1954 it has been operated by the Wildfowl Trust and supported by the Essex Water Company. Between 1949 and 1982 74,291 ducks were ringed in the cage traps at Abberton – 44% of the entire catch at Wildfowl Trust stations (p.8). In the 1950s and 1960s the autumn provided the largest counts and catches. Nowadays a further, often greater peak in numbers occurs in mid-winter, but, as at other ringing stations, the largest catches are still made in autumn. Teal and Shoveler remain predominantly passage birds, but Mallard, rather than Teal, nowadays form the bulk of the catch. Very large numbers of Teal occurred at Abberton during the influx into Britain of the late 1950s and early 1960s, but not during the recent national increases (p.409).

In recent years the autumn flocks have consisted partly of ducks remaining from the massive moulting concentrations of July and August. These became established in the mid 1960s, initially comprising mainly Pochard, but now Tufted Ducks moult in equal numbers, and several other species have become important. Between 1970 and 1982 the late summer gathering amounted to, on average (with the maximum in brackets): 1,491 (2,310) Mallard, 143 (366) Shoveler, 66 (148) Gadwall, 2,144 (3,096) Pochard, 2,196 (3,389) Tufted Ducks and 167 (346) Mute Swans. The swans have moulted in large numbers at Abberton for many years but, as with the autumn and winter flocks on the Stour and Orwell (see below), declined markedly in the 1970s. They rose again in the 1980s, however, to 346 in 1982 and 427 in 1983. A flock of up to 450 Canada Geese also occurs in August.

In about 1960, Bewick's Swans became regular visitors to Abberton, though in markedly fluctuating numbers, peaking in mid-winter at 20-50. In the late 1960s they began commuting between the reservoir and Old Hall Marsh on the outer Blackwater Estuary, 8km south. Curiously, this change coincided with a sharp drop in numbers to under 20. In the mid 1970s, however, the flock increased to nearly 50, then in the cold weather of the 1978-79 winter to 145, followed by 80, 57 and 100 respectively from 1979-80 to 1981-82. The day to day fluctuations continue to be marked, and there is clearly a strong element of passage involved.

Abberton has always been one of the main British strongholds of the Smew, although their numbers have declined greatly, in concert with those in Britain generally (p.453). Not surprisingly, most of the Smew ringed in this country have been caught at Abberton, indeed the trap where the first one was caught has been known ever since as the "Smew Trap"!

About 30km south-west of Abberton, and 6km from the head of the Crouch Estuary, lies the 365ha Hanningfield Reservoir. Constructed 15 years after Abberton, Hanningfield is much less important for wildfowl, being further from a major estuary and having a less open approach, but, with an average maximum of 1,860 wildfowl between 1978 and 1982, it still ranks among the top 20 reservoirs in Britain. It was declared a statutory Bird Sanctuary in 1978, and must also benefit from being the only sizeable water directly between the Lee Valley, north London, and the east coast.

Hanningfield's most notable features are a large autumn passage of Gadwall and Pintail and a moulting flock of Canada Geese. Canada Geese were introduced to the reservoir in 1959 (273), and between 1978 and 1982 the moulting flock averaged 220. Moulting ducks are also numerous, though not to the same extent as at Abberton: up to 600 Pochard, 1,590 Tufted Ducks and 150 Gadwall have been noted at Hanningfield in late summer since 1978 (177). Trout fishing is allowed in the northern half of the reservoir, involving up to 25 dinghies, but disturbance is not great. A small number of walking permits have also been issued (273).

Not suprisingly, with such a vast array of oustanding habitat to be covered on the estuaries and two big reservoirs, the lesser inland resorts of Essex and south and east Suffolk have received relatively little attention. (Note, however, that Heybridge and Chigborough Gravel Pits, at the head of the Blackwater Estuary, are included in the count data for the estuary itself.) The most important of the smaller freshwater sites in Essex is probably Ardleigh Reservoir (56ha), flooded in 1970. Though quite well placed – between the heads of the Colne and Stour Estuaries – it is subject to heavy recreational pressure, notably sailing and fly-fishing. Since 1980, when monthly counts were first made, it has held regular (and maximum) numbers of 210 (480) Mallard, 15 (32) Teal, 3 (20) Wigeon, 24 (54) Shoveler, 35 (125) Gadwall, 11 (250) Pochard, 53 (179) Tufted Ducks, 19 (85) Mute Swans and 26 (87) Canada Geese. Of note are the autumn gatherings of Gadwall (often the largest in the county) and Shoveler.

Records from the numerous gravel pits in the coastal plain of Essex are available from the Essex Bird Reports; they included the following maxima between

1976 and 1981: Chigborough, Maldon: 80 Teal, 145 Shoveler, 98 Gadwall, 270 Pochard, 140 Tufted Ducks, 32 Goldeneyes, 113 Canada Geese; Heybridge, Maldon: 60 Shoveler, 75 Gadwall, 238 Pochard, 225 Tufted Ducks, 200 Canada Geese, 75 Mute Swans; Ulting, Maldon: 300 Mallard, 10 Gadwall, 50 Canada Geese; Tillingham, Dengie: 300 Mallard, 106 Pochard, 36 Tufted Ducks, 130 Canada Geese; Layer-de-la-Haye, 1km from Abberton Reservoir: 250 Shoveler; Rowhedge, Colchester: 22 Gadwall. The records show how such sites can benefit from their proximity to major estuaries or reservoirs, and in turn how they provide valuable supplementary resorts for these complexes.

To the west, Leighs Reservoirs, 6km south-west of Braintree (and completed in 1967), have held up to 207 Mallard, 122 Pochard, 137 Tufted Ducks and 111 Canada Geese in recent years (177), while to the north, Shalford Meadow, in the Stour Valley below Sudbury, on the Essex/Suffolk border, held a regular (and maximum) 22 (65) Mallard, 4 (35) Teal, 12 (37) Tufted Ducks, 4 (9) Mute Swans and 28 (71) Canada Geese between 1970 and 1982. The scattering of small lakes and ponds elsewhere in the interior of Essex and Suffolk has received no coverage latterly, but counts from a few places in the 1950s and early 1960s gave very small returns.

The sandy coast between Colne Point and the Naze has the most urban development in Essex. Not surprisingly, wildfowl are scarce, apart from the Common Scoters mentioned earlier. Holland Brook still floods occassionally just before it reaches the coast at Holland-on-Sea. In the cold weather of late December/early January 1981-82 the area held 6,500 Wigeon, along with 80 Mallard, 50 Teal, 35 Pintail and 17 Shoveler; in the mild winter of 1982-83 the maxima were only 40 Mallard, 110 Teal and 25 Wigeon, although 17 White-fronted and 4 Bean Geese were briefly present in early December.

In the lee of the Naze Peninsula lies the shallow, muddy basin of Hamford Water, containing a number of islands, 940ha of flats and 1,200ha of saltmarsh (492). Its innumerable creeks, many highly secluded, support a large and varied wildfowl population. In hard winters massive numbers can occur for short periods: in early January 1979, 25,300 Wigeon and 12,000 Shelducks were present, as well as 215 Mergansers, while the official count of 10th January 1982 recorded 29,155 wildfowl of 23 species, including 950 Pochard, 280 Scaup and 52 Eiders. As a rule, dabbling ducks, apart from Mallard, which are largely restricted to the mouth, favour the extensive saltings in the north and west of Hamford Water. Shelducks prefer the innermost estuary. An unusual feature is the autumn passage of 2-300 Pochard on the flooded meadows of Bramble Island (56). Horsey Island, the largest in the basin, is notable as the favourite feeding area of the Brent Geese and one of the first places where field feeding became established.

The saltmarsh on the north side of Hamford Water extends up the coast, bordered by a narrow strip of sand and shingle, almost to Harwich. Beyond lie the confluent estuaries of the Stour and Orwell, the former straddling the Essex/Suffolk border. The Stour Estuary is the larger and more important, containing several sheltered bays, which compensate for its sharply sloping margins. Saltmarsh is scarce and mainly restricted to the outer, more disturbed end, but the shallow, productive upper reaches (where sailing is difficult) are ideal for dabbling ducks. In contrast to most estuaries to the south, however, and for reasons that are not evident, the wildfowl of the Stour have declined in the last 20 years; of the species in Table 54, only Mallard, Teal, Pintail, Shoveler and Canada Goose had a higher regular population from 1970 to 1982 than between 1960 and 1969, and the increases in these species were small. The biggest losses have been the autumn gatherings of 5,000 or more Wigeon and 800 Mute Swans (the latter attracted by waste grain from the mill at Mistley). Most remarkably Brent Geese showed no increase during the 1960s and 1970s, although they have now (Table 52).

In 1977 a 158ha reservoir was completed between Tattingstone and Holbrook (roughly halfway between the upper estuaries of the Stour and Orwell), known when under construction as Tattingstone Reservoir but since its opening as Alton Water. At the time of flooding it was too enclosed by woods and orchards, as well as heavily disturbed, for wildfowl to take advantage of its large size and strategic position, and in its first winter (1977-78) it held no more than 177 Mallard, 20 Teal, 30 Wigeon, 102 Shoveler, 6 Gadwall, 185 Pochard, 278 Tufted Ducks and 6 Mute Swans. Many of the woods have now been cleared and the natural banks and many bays have proved attractive to a wide variety of wildfowl. Few counts are available, but maxima of 150 Teal, 158 Shoveler, 630 Pochard, 500 Tufted Ducks and 43 Goldeneyes have been reported. When shooting is heavy on the nearby estuaries, large numbers of Wigeon seek refuge at Alton and a few Bewick's Swans are frequently recorded.

Although wholly in Suffolk, the Orwell Estuary is linked much more closely, in character as well as importance, with those of Essex, and so is included in Table 54. The lower estuary differs from the Stour in having a deeper channel and consequently larger numbers of diving ducks. Dabbling ducks are also present (Pintail being especially notable), but at the upstream end the grain wharves of Ipswich Docks attract large numbers of Mute Swans, though, as elsewhere in the region, many fewer than in the 1960s.

Some 1.5km north of the Orwell Estuary, secluded among the woodland of the Orwell Park estate, lies Nacton Decoy, the last commercial duck decoy in Britain until its conversion to a ringing station in 1967 (p.12). Its value in this capacity stemmed, as at most decoys, from its attraction to dabbling ducks on autumn passage (although there was also a mid-winter influx of Teal in most years), but the regular numbers declined between 1969-75 and (in brackets) 1976-82 as follows: 591 (135) Mallard, 58 (45) Teal, 68 (37) Wigeon, 80 (50) Pintail, 36 (22) Shoveler, 16 (7) Tufted Ducks. The parallel decline in the number of ducks caught caused the Wildfowl Trust, who had run the decoy since 1968, to give up its lease in 1982.

North of Landguard Point as far as Lowestoft, where the present sector ends, the coast of Suffolk (Table 55) is typified by long open beaches of coarse

Table 54. *The regular (and maximum) numbers of wildfowl on the estuaries and major reservoirs of Essex (including the Orwell Estuary, Suffolk), 1970 onwards. The figure in brackets after the place name is the number of seasons on which the "regular" populations are based. Maxima include counts made away from the official dates, from* Essex Bird Reports. *Totals are based on synchronised counts. For Brent Geese see Table 52; for detailed data from Abberton Reservoir see Table 53.*

		Mallard	Teal	Wigeon	Pintail	Shoveler	Gadwall	Shelduck
Canvey Island	*(13)	47 (140)	32 (190)	691 (2332)	2 (18)	2 (13)	0 (0)	84 (362)
Foulness	(13)	339 (771)	285 (828)	1159 (3875)	10 (63)	67 (363)	2 (21)	518 (1355)
Roach/Crouch	(7)	441 (1056)	943 (2058)	794 (3500)	30 (113)	56 (302)	4 (50)	425 (1162)
Hanningfield Resr	(10)	658 (1415)	363 (1800)	160 (560)	62 (400)	96 (353)	50(167)	17 (59)
Dengie Flats	(8)	1047 (2130)	301 (2000)	689 (3000)	78 (564)	4 (25)	0 (3)	318 (752)
Blackwater	(12)	748 (1817)	560 (2000)	1284 (3347)	78 (483)	28 (149)	15 (75)	1834 (3679)
Abberton Resr	(13)	2819 (5900)	1189 (5237)	2967(11574)	50 (405)	366 (852)	110(300)	23 (117)
Colne	+(6)	535 (1020)	210 (763)	107 (500)	7 (25)	20 (136)	1 (8)	1048 (1860)
Hamford Water	#(13)	368 (2800)	1189 (6500)	1169(25300)	270(1450)	11 (110)	2 (17)	1564(12000)
Stour	(10)	1158 (3207)	113 (391)	2258 (4502)	362 (940)	12 (45)	0 (9)	1754 (2891)
Orwell	(5)	282 (529)	98 (350)	609 (2000)	118 (307)	31 (75)	0 (2)	561 (2000)
TOTAL								
1960-69		5282 (8030)	3102(11661)	10903(20961)	498(1070)	328(1434)	19 (40)	5026(10187)
1970-82 (excl 75)		6702(10914)	4363(10742)	9506(26336)	884(2032)	542(1138)	160(357)	7211(18614)

		Pochard	Tufted Duck	Goldeneye	R-b Merganser	Mute Swan	Canada Goose
Canvey Island	*(13)	0 (9)	0 (2)	0 (2)	0 (1)	4 (13)	0 (0)
Foulness	(13)	63 (331)	47 (214)	1 (6)	8 (32)	68 (203)	73(241)
Roach/Crouch	(7)	21 (120)	28 (155)	1 (11)	5 (18)	12 (59)	27 (85)
Hanningfield Resr	(10)	237 (515)	425 (769)	40 (270)	0 (0)	11 (33)	74(201)
Dengie Flats	(8)	1 (20)	0 (2)	8 (80)	16 (55)	1 (10)	0 (0)
Blackwater	(12)	84 (238)	79 (227)	210 (799)	59 (113)	99 (203)	40(182)
Abberton Resr	(13)	729 (1756)	1182 (2670)	342 (610)	0 (4)	50 (184)	42(450)
Colne	+(6)	63 (300)	41 (250)	107 (306)	45 (105)	33 (138)	0 (3)
Hamford Water	#(13)	84 (2000)	35 (300)	30 (150)	34 (215)	13 (39)	18(105)
Stour	(10)	5 (475)	16 (347)	48 (177)	0 (3)	220 (353)	23(140)
Orwell	(5)	52 (320)	56 (112)	43 (80)	6 (170)	110 (400)	0 (15)
TOTAL							
1960-69		1306 (2819)	688 (1179)	479 (890)	78 (223)	914(1257)	34(112)
1970-82 (excl 75)		1059 (2073)	1697 (3305)	690 (1250)	126 (364)	656(1093)	232(663)

Notes: The table covers September to March only; for details of moulting concentrations see text.
The 1975 counts were too incomplete to contribute to the overall totals.
* Canvey Point, Tewkes Creek, Leigh Marsh and Two Tree Island.
+ Incl. Fingringhoe and Langenhoe Marshes, Flag and St Osyth Creeks; excl. The Hythe.
# Incl. The Naze.

sand and shingle, backed in places by low bluffs and in others by large expanses of grass marsh and reed-bed. The terrain is extremely exposed and rainfall low, but by way of compensation the area has long been split up among a number of large estates, each of which has rigorously controlled access, so that the marshes and estuaries have acted in many cases as virtual sanctuaries, even if no official protection exists. These estates are slowly being broken up, however, and are likely to be increasingly open to disturbance in the years to come. Further inland the distribution of all species is cut short by the belt of dry sandy heathland which backs most of the coast.

Some 6km to the north and east of the Orwell lies the estuary of the Deben, which is similar in length to those further south, but narrower and, owing to several other factors, much less attractive for wildfowl. In its lower reaches saltmarsh is severely restricted by the heavy scour of the tide; upstream the estuary is broader and shallower. The shore, to which the heathland breaks through in most parts, is relatively steep, while disturbance from sailing and shooting is heavy.

Northwards to Aldeburgh the coast is dominated by the great shingle spit of Orfordness, which deflects the channel of the Alde from its former mouth near Aldeburgh to a course lying parallel to the shore and extending down to Shingle Street. The channel is known as the River Ore from east of Orford to its mouth. Behind the shingle spit, and protecting the river from a great deal of casual disturbance, lies a broad belt of what used to be all grazing marsh, intersected with numerous ditches and dotted with occasional flashes of winter flood. Regrettably, much of this has now been converted to arable. Even more remote are the Lantern and King's Marshes, which lie on the inner edge of the spit, between the river and the sea. Access is prohibited here, so there are no count data. Immediately downriver, stretching to the mouth of the Butley Creek, is the most important individual resort in the area – Havergate Island. A National Nature Reserve since 1954, the island is owned and managed by the RSPB, who have created six shallow lagoons, mainly for the sake of the famous breeding colony of Avocets. The salinity and depth of the lagoons are controlled by means of wells, sluices and dykes (513). A good variety of wildfowl are attracted to the island, which is largely bordered by saltmarsh, but the huge concentrations of Teal and Wigeon which appeared at times in the 1950s no longer do so.

The Havergate populations are closely linked with those on the river and adjoining marshland and the whole should perhaps be treated as one site. Table 55 therefore includes two entries for this area: one for Havergate Island alone and one ("River Ore/Butley") for the island in combination with three other count-points nearby: Boyton Marshes, on the mainland side of the River Ore/Butley junction; Stonebridge Marshes, on the west shore of the Butley Creek, 2km north; and the creek itself. The regular numbers of wildfowl in this area are boosted by cold-weather immigrants, such as the exceptional 10,000 Wigeon at Boyton in late December 1981. The Alde Estuary (where the broad shallow basin between Iken and Barber's Point represents the only extensive intertidal zone in the area) has been kept separate, no full counts having been made there since 1972. A summering population of over 350 Shelducks has been reported (485). Sudbourne Marshes, downriver, have not been formally counted since 1959 and 1960,

Table 55. *The regular (and maximum) numbers of wildfowl at the main resorts on the Suffolk coast north of Felixstowe, 1970-1982.*

	Deben Estuary (5 seasons)	Havergate Island (10)	River Ore/Butley (incl. Havergate) (9)	Alde Estuary (3)	Minsmere (13)	Blyth Estuary (9)
Mallard	168 (635)	311 (965)	616(2502)	405(750)	340 (822)	109 (900)
Teal	292 (820)	318 (905)	437 (985)	32(100)	458(1000)	22 (129)
Wigeon	728(6039)	501(3470)	1404(4070)	256(500)	424 (727)	499(2000)
Pintail	176 (352)	48 (320)	54 (320)	47(105)	3 (18)	0 (7)
Shoveler	0 (10)	74 (160)	83 (176)	0 (4)	110 (420)	2 (16)
Gadwall	0 (0)	20 (77)	27 (82)	0 (11)	83 (300)	0 (0)
Shelduck	639(1388)	112 (250)	167 (320)	250(425)	66 (168)	573(1094)
Goldeneye	24 (58)	0 (6)	2 (14)	0 (21)	4 (24)	1 (11)
Mute Swan	53 (132)	3 (10)	25 (50)	10 (21)	14 (25)	6 (29)
Canada Goose	0 (24)	0 (20)	40 (180)	0 (40)	145 (320)	10 (85)
D-b Brent Goose	0 (450)	13 (500)	28 (500)	0 (8)	13 (72)	0 (48)

when they produced regular (and maximum) returns of 550 (1000) Mallard, 123 (250) Teal, 163 (350) Wigeon, 6 (12) Pintail, 6 (10) Shoveler and 12 (20) Shelducks. Numbers at least as high as this are still probably present, and White-fronted Geese are still regular, with 1,000 in the hard weather of 1978-79, 100 in 1981 and 310 in 1982. A single count of the Ore between Boyton Hall and its mouth at North Weir Point/Shingle Street found 20 Mallard, 2 Teal, 40 Wigeon and 1 Common Scoter. A flock of 40-50 Bean Geese has become regular in the Orfordness area in the last few years, at different times being seen in different marshes.

North of Aldeburgh the coast, though mainly flat, is interrupted in several places by low, sandy cliffs, where the heathland breaks through. Between Aldeburgh and the mouth of the Blyth, the fresh marshes at Thorpeness, Sizewell, Dunwich and Walberswick are believed to hold fair – perhaps in places large – numbers of dabbling ducks, but counts are lacking. The RSPB reserve at Minsmere Level has, however, been covered regularly since its inception in 1948. Both Minsmere and Walberswick were once estuaries, but were finally enclosed in the early 19th century, acting as grazing land until reflooded in 1940 as a war-time defence. In 1953 severe sea water flooding necessitated the repair of the shingle and sand embankments, greatly reducing the salinity of the marshes. At Minsmere, the rapid spread of *Phragmites* divided the shallow flood into separate meres, and the control of reed has been the most intensive management operation on the reserve (29, 513). The excavation of a shallow, 18ha lake, known as The Scrape, has partly compensated for any reduction in the freshwater meres from the point of view of wintering wildfowl, but the reserve's fame still stems from its remarkable breeding populations of marshland birds, including some 20-30 pairs of Teal, Shoveler and Gadwall. In winter Minsmere has a regular flock of about 100 White-fronted Geese (559).

The estuary of the Blyth, lying just north of Walberswick Marshes, is embanked along its lower reaches, but some 2.5km inland opens out into a shallow basin, similar to that on the Alde, but full of breached bunds left from relics of former reclamation (485).

The remainder of this sector, the coastal plain between Southwold and Lowestoft, contains three small brackish broads with adjacent fresh marsh, all of which have been poorly covered by wildfowl counts. Easton Broad held a regular 49 Mallard, 17 Teal, 9 Wigeon and 18 Mute Swans between 1958 and 1960, and Covehithe Broad had peaks of 85 Mallard, 24 Teal, 3 Shoveler and 3 Scaup in 1969. Neither site has much standing water left, however, so there are no substantial wildfowl numbers nowadays. At Benacre Broad, clearly the best locality in this stretch, maxima of 500 Mallard, 2,000 Teal, 500 Wigeon, 28 Pintail, 24 Shoveler, 32 Gadwall, 375 Pochard, 160 Tufted Ducks, 20 Goldeneyes, 40 Scaup and 2 Mute Swans have been recorded between 1966 and 1982. The Broad is of such interest that the NCC have signed an agreement with the owner; it is to be declared an NNR. Kessingland Level, immediately north of Benacre, holds one of the largest flocks of Canada Geese in east Suffolk, amounting to some 500 throughout the year, and these often appear at Benacre, 700 being counted there in September 1982. Bewick's Swans are regular at Kessingland (average peak 1970-81 of 78), and large numbers of Wigeon congregate in hard weather. Further south, gatherings of 150 Canada Geese occur at Minsmere and Aldeburgh. The other Suffolk centres for Bewick's Swans are at Havergate Island/Boyton (average peak 88), Sudbourne/Aldeburgh (74), Minsmere/Dunwich (55) and Southwold/Walberswick (61); there is, however, considerable interchange between these places and, as with the Abberton/Old Hall Marshes flock (see above), a strong element of passage (559).

Sea ducks are abundant off the coast between Orford Ness and Lowestoft Ness. Common Scoters are present all year, but are most numerous in summer, when they gather off Minsmere/Walberswick. Over a thousand were found in the 1960s, occassionally moving as far north as Gorleston, but more recently 2-600 have been more usual, although flocks of 1,000 were present at times in 1980, 1981 and 1982. Small parties of Velvet Scoters frequently occur, while in summer and autumn 30-40 Eiders are regular off Dunwich and Benacre (559).

The outstanding importance of the coast of Essex and Suffolk has been recognised by the establishment of a number of reserves. Among wetland sites, there are National Nature Reserves at Leigh Marsh (257ha), Orfordness – Havergate (225ha) and Walberswick (514ha). The Walberswick/Minsmere complex was designated under the Ramsar Convention (p.540) in 1976, as was Abberton Reservoir (the only reservoir in the United Kingdom to have been so recognised) in 1981. The RSPB have reserves at Stour Wood and Copperas Bay (289ha), Havergate Island (108ha) and Minsmere (595ha), while the local voluntary bodies manage a number of other sites.

Nevertheless, the proportion of major wildfowl habitat in this sector which is under any form of protection (apart from the meagre safeguards given under SSSI designation) is very small, and the fragmented ownership of the Essex coast makes it particularly vulnerable to isolated development. The principal single threat – the construction of an airport and/or seaport on Maplin Sands – which would destroy the most unspoilt area in this region, has been shelved, but was one of the alternatives considered at the recent Public Enquiry into the building of a third London airport at Stansted, Essex. However, a process remains whose total effect could ultimately be as serious: the proliferation of sailing, winter water-skiing and other forms of recreation. Blindell (56) regarded this as the greatest threat to the coast

Fig. 20. The London area.

between the Thames and Orwell in the mid 1970s, noting that 14 new marinas were under construction or proposed in September 1975. The Eastern Sports Council (164) recorded that the number of moorings and dinghy parks on the Suffolk Coast increased as follows between 1968 and 1972: on the Stour Estuary by 80%, Orwell 180%, Deben 110%, Alde/Ore 80%, remainder 50%. They stated, with regard to Essex and Suffolk: "Certain estuaries already seem to have reached saturation point as far as recreational facilities are concerned, but there appear to be possibilities for development in other estuaries, particularly with regard to improved access and mooring facilities". Furthermore, they suggested an increased recreational use of Abberton and Hanningfield Reservoirs – easily the most important inland sites in the region for wildfowl. When they wrote, the estuaries of the Blackwater, Colne and Crouch, and much of Hamford Water, were already heavily used for sailing. Since then, the situation has deteriorated further. The outlook may, however, be brighter: the Eastern Sports Council has been holding consultative meetings again recently and acknowledged that there is now less demand for increased yachting facilities. The Port Authorities are also against such expansion, and Essex County Council are taking a refreshingly balanced view of the problem.

The conversion of grazing land to arable, with the associated lowering of the water table and the conversion of reedy channels to canalised drainage ditches, has also been a significant factor in many parts. Apart from the nuclear power stations at Bradwell and Sizewell, industrial development has been confined to the petrochemical refineries and storage depots on and around Canvey Island. Extensive though they are, these developments have had little or no adverse effect on the local wildfowl as yet, but the potential for a major pollution incident remains.

The Thames Basin

The catchment of the Thames and its tributaries covers about 13,000sq km. The lower part of the valley, separated from the upper basin by the Chiltern Hills and Lambourn Downs, is largely built up and traversed by numerous roads and railways. The upper reaches of the river run through open countryside and small villages. In the late 1950s, apart from the river itself and around London, waters were rather small and few and far between. In the last two decades, however, the explosion in gravel digging and the creation of more reservoirs has increased the water area in the region considerably, and the wildfowl, especially the inland diving ducks and Canada Geese, have responded by greatly increasing their numbers. The growing demand for water for recreation means that this requirement competes with the need for infilling for waste disposal and that some of the area excavated for gravel remains unfilled; recreational pressures do, however, have an adverse effect on some pits and reservoirs.

The lower Thames Basin: London

The vast sprawl of London may seem an unlikely centre for wildfowl, as well as a difficult area to define precisely. Yet within a circle of natural boundaries there lies an extensive complex of prime habitat for wintering ducks, holding a significant proportion of the British population of several species. The boundaries are: to the north and north-east, the edge of the chalk downland of Hertfordshire and Essex; to the east, the start of the true estuary of the Thames at Tilbury; to the south, the North Downs; to the south-west and west, the Surrey and Berkshire heathland; and to the north-west, the first slopes of the Chilterns. Apart from the exclusion of the North Downs, this definition corresponds closely to a circle with a 20-mile (32km) radius centred on St Paul's Cathedral, as used by the London Natural History Society.

The River Thames

The Thames, which winds for roughly 100km across the region, constitutes London's only important natural habitat for wildfowl. It is tidal below Teddington, well to the west of Central London. The 40km stretch between London Bridge and Tilbury (though nowhere more than 1km wide) is generally known as the inner Thames Estuary. In this unlikely setting, bordering the abandoned dockyards and warehouses of the East End, the new housing estates of Thamesmeade, the power stations at Dartford and Thurrock and the modern docks and container depot at Tilbury, the most dramatic story in the modern ornithology of the London area, and one of the most exciting in Europe, has unfolded. Although a recent decline in most species has spoilt the picture slightly, the fact remains that in the space of ten years the inner Thames Estuary was, to borrow the word used by Harrison and Grant (247) in the title of their definitive work on the subject (on which much of the following account is based), "transformed" from a condition of gross pollution, rendering it incapable of supporting more than a few birds, to the status of international importance for wildfowl. Table 56 charts these developments.

In the period between 1945 and 1950, when

frequent observations were made, the inner estuary held few ducks, and above Dartford hardly any (95). In the late 1950s a programme was introduced by the Port of London Authority and the London County Council (the latter being superseded by the Greater London Council in 1964) to clean the river by means of new large-scale sewage treatment works, through which all industrial and domestic waste had to be channelled. The scheme was a success, and by 1972 66 species of fish had been found in areas previously devoid of them. An increase in wildfowl occurred as early as 1961 and 1962, when up to 500 Teal, 200 Wigeon, 100 Pochard and 70 Shelducks were found between Barking and Swanscombe (222). At the time these influxes were attributed to hard weather, but the increases (and further movement upriver) continued in the following, mild winters. Then in 1968-69 remarkable flocks occurred, and over the next few years most species increased further, Shelduck and Pochard attaining levels of international importance. Each winter the large numbers penetrated further upriver (partly at the expense of the lower sections), and in 1973 nearly 2,000 Pochard were found above Dartford – several hundred of them in the Upper Pool of London – a level which would have been unthinkable ten years previously. The upsurge was helped by the favourable conditions which prevailed at the Surrey Docks for a few years between their abandon ment at the end of 1970 and their becoming intolerabl disturbed and, in 1980, finally being reclaimed fc industry. At their peak in the early 1970s, Surre Docks (inside the bend of the river formed by th Upper Pool and Limehouse Reach) held up to 12 Mallard, 800 Tufted Ducks and 1,700 Pochard, all c which used the area as a roost from their feedin grounds on the river (2, 207, 221).

A research programme organised by WAGBI i the early 1960s, under licence to the NCC, made th suprising discovery that the main food of the wildfow involved in these influxes was *Tubifex* worms (25C 567). Some species had never before been known t take this food – indeed Pochard actually took t dabbling in order to obtain the worms.

Ironically, *Tubifex* thrive where pollution i worst, and had probably always been plentiful in th Thames mud. The presence of these worms perhap provides the key to both the sudden increase and th subsequent decline in wildfowl on the inner Thame Estuary. It is likely that the cleaning up of the rive served to create conditions which the ducks coul tolerate so that they could exploit this vast, untappe food resource. As the level of oxygen in the wate become too high for the worms to bear, so they died of (as shown in surveys by the Thames Water Author

Table 56. *Seasonal maxima of wildfowl on the Thames between Tilbury and London Bridge, including Rainham Marsh. 1960-1975 data from Harrison and Grant (1976). Brackets indicate incomplete data; n.c. = not counted.*

	Mallard	Teal	Wigeon	Pintail	Shoveler	Pochard	Tufted Duck	Shelduck	Mute Swan
Max 1945-50*	n.c.	55	30	0	0	n.c.	n.c.	12	9
1960-67	n.c.	500	200	60	n.c.	700	14	600	48
1968	500	450	200	130	9	1500	400	1000	71
69	700	1000	150	350	2	2500	800	3000	200
70	900	1500	200	367	n.c.	2000	400	1500	300
71	1174	730	65	275	n.c.	4000	458	2600	160
72	1490	800	100	485	n.c.	2600	300	929	420
73	1445	800	25	231	n.c.	2471	300	868	616+
74	1300	430	26	196	34	581	200	629	245
75	1179	650	41	70	15	175	81	932	325
76#	(200)	(300)	(20)	(100)	(150)	(88)	(5)	(162)	(82)
77#	n.c.	(600)	n.c.	(80)	(80)	n.c.	n.c.	(130)	(71)
78	1072	4000	260	300	51	5835	249	260	94
79	1080	2000#	30	130#	200	200	100	245	74
80	820	3000	27	260	300#	225	55	640	50
81	720	2000#	75	100#	100#	266	147	1125	25
82	780	1800#	1	80#	53	61	93	2020#	16

* From Burton (1974).
+ June 1973.
# From Essex and London *Bird Reports*.

ity), but there was clearly a time-lag between the elimination of pollution in particular areas and the disappearance of the *Tubifex*. This helps to explain the steady upriver shift of several species of wildfowl. The lower parts of the river were the first to receive the benefit of the reduced pollution and the earliest to be colonised by the birds. Similarly, it was here that the *Tubifex* first died out, presumably causing the birds to move upstream. The worms were badly affected by the low rainfall between 1974 and 1976, which resulted in the greatly increased salinity of the river, a condition unacceptable to at least one of the species involved (in the exceptionally dry summer of 1976 the flow of water over Teddington Weir was virtually nil).

The improved oxygenation of the Thames water has resulted in a proliferation of *Enteromorpha*, which supplements the diet of Mallard and Mute Swans, and which has enabled the Mallard to maintain their numbers while other species have declined. The Mute Swans, previously scarce on this part of the river, were first attracted to the grain and other foods spilt from the wharf at Silvertown, adjacent to the secluded Bow Creek, and when this closed in 1975 they took to the Mollasine Wharf at nearby Greenwich. Presumably the swans, like the birds feeding principally on the *Tubifex* worms, needed the general cleaning up of the river to be able to take advantage of this longstanding food resource. Their subsequent decrease is harder to explain, since none of the factors contributing to the recent decline in the other species in this area can have significantly affected the Mute Swan. Lead poisoning cannot, however, be ruled out as a likely cause.

Besides the loss of the *Tubifex* worms there are two other probable reasons why the huge concentrations of ducks on the lower Thames were shortlived. Firstly, the reclamation of the foreshore between Margaret Ness and Cross Ness, known as Woolwich Bay, in the 1970s deprived the birds of one of their favourite feeding grounds. This development was decided upon as part of the Thamesmeade New Town development, before the area became important for wildfowl. Between 1969 and 1973 this 3km stretch had held up to 900 Mallard, 1,720 Teal, 350 Pintail, 1,000 Pochard, 800 Tufted Ducks, 330 Mute Swans and 1,600 Shelducks. Secondly, although the Thames Barrier itself at Silvertown, completed in 1982, is unlikely to pose a problem to wildfowl populations (its gates being fully submerged except during flood-tides) its construction over the previous ten years caused considerable disturbance along one of the most important parts of the river. The associated improvement in the sea wall defences has resulted in the destruction of several major areas of saltings, notably Aveley Marshes, between Rainham and Purfleet, which were the main site for Wigeon and also held breeding Shelducks.

Despite these recent setbacks, the inner Thames Estuary is still of far greater importance to wildfowl than it was at any previous time for which records are available. Its present importance is probably as much as can be expected in view of the apparent extinction of the supplies of *Tubifex* and the virtual elimination over the years of the available foreshore (which was never very extensive). The main centres for wildfowl in the area are now the riverside gravel pits and, in particular, Rainham Marsh. On the north shore between Dagenham and the former Aveley Marshes, Rainham is the main dumping ground for mud dredged from the river. The settling lagoons, with their profusion of Sea Aster (*Aster tripolium*), have of late proven increasingly attractive, especially for Teal. The recent gatherings of several thousand Teal shown in Table 56 have occurred entirely at Rainham Marsh, and render this site of international importance. Around Dartford, on the south shore, the pools at Littlebrook Power Station have carried regular (and maximum) populations of 181 (382) Mallard, 11 (45) Teal, 22 (42) Shoveler, 166 (950) Pochard and 67 (115) Tufted Ducks, while the gravel pits of Darenth (now a sports centre) and Brooklands hold between them 15-40 Mallard, Tufted Ducks and Canada Geese.

Although other factors (such as the reduction in heavy shipping caused by the steady closure of the inner London docks) may have helped, the remarkable achievement of the Port of London Authority, the new Thames Water Authority and the London County and Greater London Councils in eliminating serious pollution from the Thames represents one of conservations's greatest modern gains.

During the national Mute Swan census of April/May 1956, a total 744 non-breeders were found on the London Thames: 59 from Greenwich to Westminster; 408 between Westminster and Kew (including the major section above Putney shown in Table 57); and 277 from Kew to Staines (138). In December of that year the sinking of an oil barge at Battersea, then their main locality in Central London, resulted in the death of 243 swans. By the 1961 census the numbers on the stretches listed above had declined to 39, 214 and 174 respectively, a total of only 427 (139). Since then a further massive decrease has occurred above Putney and, although no separate figures are available, there is no reason to suppose that the London Bridge – Putney stretch has not suffered the same trend. The large flocks which were present on the inner Thames Estuary in the late 1960s and early 1970s (see above) did not apparently penetrate the heart of London. The decline on the river above

Central London was thought to have been caused largely by lead poisoning, although some element of the normal fluctuations found (140) may have been involved (420). The recent survey also cast doubt on whether this decline was caused by a shift downriver to the scene of the influxes of the late 1960s, as the greater part of the reduction upriver occurred well before then. The decline in man's industrial activities along the river in Central London may have contributed to the disappearance of swans from there, as they had thrived on the scraps provided. A survey of the Thames above Richmond in January 1983 found 54 Mute Swans (and 36 Canada Geese) along the 26km between Richmond and Chertsey (199 – Table 66). In July 1983 the same area held 28 Mute Swans and 74 Canada Geese. Between Richmond and Grays, 4km upstream of Tilbury, an area not surveyed in January, there were 291 Mute Swans including 20 cygnets.

As a breeding species the Mute Swan has declined in the London area as a whole from 188 pairs in 1956 to 32 in 1981, although there was no significant reduction in the number of pairs after 1974 (420).

In neither 1956 nor 1961 were any breeding pairs of Mute Swans found below Hammersmith, presumably because of the lack of suitable sites (139), and the paucity of breeding Mallard on the river through Central London was attributed to the same reason (143). Ducks have apparently rarely occurred in numbers in this stretch, however; during weekly surveys throughout 1951-1953, monthly averages of no more than 3 Mallard were found between Waterloo and Blackfriars Bridges, and 2 between Vauxhall and Lambeth (142). Occasional Tufted Ducks were recorded, but the flock of 61 along with 16 Pochard and 6 Mallard at Waterloo in December 1952, following a cold spell, was exceptional. A count from a launch in February 1973 revealed only 253 Mallard between London Bridge and Putney (317). A total of 119 Mallard, 135 Pochard and 107 Tufted Ducks was counted on the same stretch the following winter (250). Occurring as it did during a series of exceptionally mild years, this survey may indicate an upriver extension of the influx on the inner estuary, but a lack of subsequent records for this central area makes it impossible to be sure.

Good numbers of Mallard occur on the Thames above Putney (Table 57), along with occasional examples of the other species found so profusely on the adjacent reservoirs and gravel pits. Notable are Tufted Ducks, which can be abundant on the river when the enclosed waters are frozen, and Teal, which are common between Putney and Barnes, around Barn Elms Reservoir (see below). A sharp decline in Mallard in the late 1960s, following a period of

Table 57. *The regular (and maximum) numbers of wildfowl on four stretches of the Thames between Putney and Teddington in three periods, 1954-1982.*

a) Putney – Barnes (excl. 1957)
b) Barnes – Kew (excl. 1977-82)
c) Kew – Richmond (excl. 1980-82)
d) Richmond – Teddington (excl. 1979-82)

	a	b	c	d
MALLARD				
1954-59	321 (836)	40 (149)	681 (1015)	335 (545)
60-69	699 (1790)	252 (656)	490 (1323)	328 (445)
70-82	311 (593)	83 (110)	213 (350)	168 (318)
TEAL				
1954-59	7 (58)	1 (15)	16 (30)	0 (2)
60-69	31 (181)	0 (1)	4 (35)	0 (0)
70-82	21 (69)	0 (1)	0 (0)	0 (0)
TUFTED DUCK				
1954-59	0 (2)	2 (6)	0 (2)	0 (5)
60-69	19 (201)	2 (12)	0 (2)	2 (11)
70-82	6 (38)	0 (0)	1 (6)	1 (4)
MUTE SWAN				
1954-59	259 (379)	6 (28)	49 (77)	28 (64)
60-69	124 (280)	9 (28)	9 (35)	9 (25)
70-82	5 (23)	1 (3)	5 (18)	3 (8)

increase, coincided with the influx onto the inner estuary, but a count of 980 between Teddington and Walton in January 1975 (317), the only time this stretch has been fully surveyed, suggested that there had been no such exodus higher up the river.

Enclosed waters in London

For most of this century the great majority of London's wildfowl have occurred on reservoirs and gravel pits, which now comprise a total of some 3,600ha. There are two main groups of reservoirs – one along the River Lee and the other following the Thames between Molesey and Staines – and several isolated waters, of which Barn Elms, Brent and Hilfield Park are the most important. All the principal reservoirs, except Brent, are for drinking water, and are bunded throughout, with artificial banks.

The first reservoirs, built in the early 19th century, were the canal-feeders of Brent, Stoke Newington, Ruislip (23ha) and Aldenham (28ha), all in the north-west of the region in what was then open country, together with Lonsdale Road, Barnes. For a time these were the only important water-bodies away from the Central Parks, but by the end of the century many of the smaller drinking water reservoirs of the Walton, Hampton/Kempton and Walthamstow groups, together with Barn Elms, had been constructed. Then, in 1902, came the first of the "giants" – Staines. A further 650ha in the Thames and Lee Valleys had been added by 1925. There was then a lengthy pause, but 850ha have appeared since the Second World War, including the large Queen Elizabeth II, King George VI, William Girling, Wraysbury and Queen Mother Reservoirs. Table 58 lists the principal reservoirs in each group and shows their overall importance for wildfowl in the last thirty years.

Following the construction of each of the pre-war drinking water reservoirs there was a distinct time-lag before the appearance of large numbers of ducks, and the subsequent increases were not steady, but in a series of jumps. The largest apparently occurred in the 1930s, during a period when no new basins were constructed, suggesting a link with the numbers and distribution in Britain as a whole or a

Table 58. *The principal London reservoirs, in order of construction within each group, showing size, depth and general importance to wildfowl.*

Name (No. of waters where more than one)	Year of completion	Area (ha)	Normal depth (m)	Total wildfowl (Mean annual maxima)		
				1950-59	1960-69	1970-82
LEE VALLEY						
Stoke Newington (2)	1834	17	6	121	644	1110
Walthamstow (11)*	1866-1903	139	3-10	649	524	1506
King George V (2)	1913	172	7-9	578	396	729
William Girling	1951	135	13	320	495	920
NORTH-WEST						
Brent (Welsh Harp)	1835	61	8(max)	210	141	364
Hilfield Park	1955	47	11	508	754	640
THAMES (BARNES)						
Lonsdale Road	1838	6	6	711	601	140
Barn Elms	1897	35	4	1787	2488	1266
THAMES (WEST)						
Walton (4)+	1877-1907	96	6-13	989	700	448
Hampton/Kempton (4)#	1898-1906	40	6-13	542	489	707
Staines (2)	1902	172	9-12	1478	1146	1636
Island Barn	1911	49	9	557	417	424
Queen Mary (Littleton)	1925	286	12	1033	1154	978
King George VI	1947	142	16	1104	1557	1220
Queen Elizabeth II	1962	128	18	-	1364	661
Wraysbury	1971	203	15	-	-	838
Queen Mother (Datchet)	1975	192	23	-	-	see text

* Banbury, Lockwood, Low and High Maynard, Nos. 1-5 Walthamstow, E and W Warwick.
+ Chelsea, Lambeth, Knight, Bessborough.
# Kempton Park W and E, Stain Hill W and E.

redistribution within north-west Europe. In contrast, the post-war reservoirs received an almost immediate influx upon their completion, the wildfowl populations elsewhere in the area being by then firmly established at their modern level.

The total of 7,191 Tufted Ducks counted on the London reservoirs in December 1979 comprised 17% of those recorded in the whole country in that month. Generally, London's Tufted Ducks move around between sites from year to year, and no doubt the complex of habitats is essential to maintain present numbers (Table 59).

The gravel pits of London cover almost as big an area as the reservoirs – roughly 1,700ha of wet workings in 1979 – but are of less importance, mainly because the relatively small size of the individual pits makes them prone to freezing and means that the wildfowl on them are more susceptible to disturbance (including continuing extraction in many cases). Nevertheless, the pits in total hold some 2,500 Tufted Ducks, 1,000 Pochard and good numbers of dabbling ducks and Canada Geese.

There was a dramatic increase in the number of gravel pits around London during the 1950s and 1960s (in 1954 their total area was only about 100ha) but the general reduction in road-building caused by the economic recession and the decline in motorway construction has meant that few new complexes have been created since then. Continued excavation and expansion of existing workings has barely kept pace with the areas filled in to meet the increased demand for waste disposal. The appearance of new pits and reservoirs around them facilitated increases in wildfowl at most of the traditional resorts, whose birds had a greater choice of alternative waters. Tufted Ducks, as in other parts of the country, have benefited in particular.

The north-eastern extremity of the region, i common with most of rural Essex, is largely dry, whil few data are available for the gravel pits and lake scattered between there and the rivers Thames an Lee. They are, however, believed to act as alternativ feeding grounds for wildfowl from the inner Thame Estuary. The River Rom and adjacent gravel pit together with the new flood control lake at Dagenhar Chase, 5km inland from Rainham Marsh, hold up t 150 Teal (154).

From its entry into the present sector near War to within 10km of its mouth on the Thames at Cannin Town, the River Lee (or Lea) is lined by an almos continuous cluster of gravel pits and reservoirs currently comprising about 870ha (Table 60). Ban fishing is allowed on nearly all waters, but sailing o only a few, including Rye Meads, the southern half o King George V Reservoir and Banbury Reservoir Walthamstow. Sailing was not permitted on any o London's drinking water reservoirs until the earl 1970s. The Lee Valley became a statutory Regiona Park in 1967. The gravel pits and associated water occupy the upper valley and now total 370ha. The hold over 1,500 Tufted Ducks and moderate number of several other species. The purification works at Ry Meads, opened in 1956 (with 23ha of lagoons) together with the 5ha RSPB reserve at Rye Hous Marsh and the adjacent Stansted Abbots pits provide a particularly interesting complex of habitats, including a small remnant of water meadow (1). Amwell, a littl way to the north, has recently become a major centr for Mute Swans, while the Cheshunt pits, at th southern end of the complex, have a notable autumr passage of Gadwall.

The reservoirs, covering 500ha, lie to the sout of the pits, in the inner London suburbs. Although n new waters have been built since 1951 there has been a

Table 59. *Tufted Duck: annual maxima at main London sites. Brackets indicate incomplete data.*

	1973	74	75	76	77	78	79	80	81	82
LEE VALLEY										
Stoke Newington Resr	1696	2077	427	488	417	1034	290	278	252	146
Walthamstow Resr	790	890	1400	810	550	1225	885	994	1037	820
King George V Resr	223	500	650	498	280	390	216	2000	350	300
BARNES										
Barn Elms Resr	1450	470	630	390	320	340	190	380	469	230
WEST										
Queen Mary Resr	850	750	500	300	259	360	541	1252	1147	213
Staines Resr	158	120	86	325	180	1700	4000	327	500	665
Wraysbury Resr	341	309	483	478	498	1246	422	1358	893	(30)
Wraysbury GPs	699	576	561	656	848	1411	909	1528	1343	1512

marked increase in the numbers of wildfowl in recent years (Tables 58 and 60). Smew, though, have virtually disappeared, in keeping with the trend throughout the country and particularly in London. At their peak in the mid 1950s over 100 Smew were regular in the London area, Brent Reservoir (see below) being the most favoured site. By the mid 1960s the total had declined to 50 and by the early 1970s to 30, with Stoke Newington Reservoir the main centre. Either Brent or Stoke Newington, the two reservoirs closest to the centre of London, often held the largest group of Smew in Britain. During this time, London followed the trend in the country as a whole, but recently, however, while the national total has stabilised at a very low level, the main centre has shifted to Dungeness, Kent, and a nadir of four was reached in London in 1980. During periods of hard weather on the Continent, influxes of Smew to London still occur; in the cold spell of early 1979 at least 84 were present and there were 22 in 1981-82.

The two small basins of Stoke Newington Reservoir lie 2km west of the main Lee Valley group, just 6km from St Paul's Cathedral. For several years they held the largest flock of Tufted Ducks in the London area (Table 59). Wood Green Reservoir, 3km north-west, carried average winter maxima of 23 Mallard, 4 Shoveler, 142 Tufted Ducks, 124 Pochard and 5 Smew between 1970 and 1976.

Walthamstow Reservoirs contain a number of wooded islands, which, apart from a large heronry, also support a notable collection of breeding ducks, including 30-40 pairs of Tufted Ducks and Pochard. Tufted Ducks breed in strength at several other waters in the Lee Valley, and in the London area as a whole may now total 4-500 pairs.

Canada Geese have established a major stronghold at Walthamstow (from which they move freely up the Lee Valley and elsewhere). The species was first brought to London in the 17th century but expansion did not take place until it was successfully introduced to Hyde Park/Kensington Gardens in 1955. The lakes of these and other Royal Parks (Table 61) represent much the most significant wildfowl haunts in Central London. All are, of course, subject to enormous human pressure around their perimeters, but it is largely because rather than in spite of this aspect that they are so interesting, if only to a limited number of adaptable species. The public provide a constant source of food through the scraps they throw to the birds, while the pinioned "ornamental" wildfowl, managed for the public to enjoy, act as "decoys" for the wild birds. As well as the species tabulated,

Table 60. *The regular (and maximum) numbers of wildfowl at the Lee Valley reservoirs and gravel pits, 1960-1982. (Includes data from* London Bird Reports *and reports of the Lee Valley Project Group.) The figures in brackets after the place names show the number of seasons with regular counts and the period they cover. For Tufted Ducks see also Table 59.*

		Mallard	Teal	Shoveler	Pochard	Tufted Duck	Goldeneye	Mute Swan
Stoke Newington Resr	(21:60-82)	25(153)	0 (1)	2 (53)	89 (433)	489(2077)	0 (3)	5 (14)
Walthamstow Resr	(23:60-82)	114(450)	13(109)	48(335)	279(1639)	519(1640)	2(11)	16 (73)
Wm Girling Resr	(22:60-82)	222(535)	125(450)	58(233)	14 (168)	203 (726)	16(48)	2 (34)
King George V Resr	(21:60-82)	66(200)	2 (22)	13(120)	135 (650)	278(2000)	18(44)	2 (15)
ALL RESERVOIRS:								
	1960-69	213(542)	36 (76)	24(105)	222 (911)	762(2149)	23(43)	23 (76)
	1970-82	419(977)	200(527)	161(356)	658(2021)	1724(3147)	42(59)	20 (38)
Cheshunt GPs	(16:67-82)	167(355)	12(115)	16 (70)	77 (150)	339 (848)	1 (4)	11 (29)
Holyfield Marsh GP	(2:73-81)	170(565)	13 (50)	2 (8)	28 (61)	114 (216)	0 (4)	3 (8)
Nazeing GPs	(7:60-75)	81(250)	1 (6)	1 (19)	115 (397)	330 (678)	0 (6)	7 (14)
Broxbourne GPs	(14:62-82)	68(231)	13 (50)	24 (98)	75 (300)	260 (607)	1 (7)	6 (39)
Rye Meads	(14:60-82)	227(507)	20(200)	15 (70)	17 (164)	193 (625)	0(10)	16 (50)
Stanstead Abbots	(5:73-82)	94(450)	17 (98)	6 (16)	39 (86)	129 (241)	0 (1)	5 (11)
Amwell GPs	(2:81-82)	209(450)	3 (5)	3 (12)	30 (47)	151 (239)	1 (3)	39(102)

Note: The following additional species were regular at one or more sites:
Wigeon: Wm Girling (regular 2/maximum 40); King George V (1/18); Rye Meads (1/13); Amwell (3/25).
Gadwall: Cheshunt (16/155); Broxbourne (1/5).
Goosander: Walthamstow (1/33); Wm Girling (11/100); King George V (7/60).
Smew: Stoke Newington (6/41).
Canada Goose: Walthamstow (38/278); Cheshunt (6/72); Amwell (6/22).

almost anything else can turn up, especially in hard weather. Tufted Ducks breed in large numbers, thanks partly to introductions. In St James's Park 30-50 pairs are generally present, but breeding success is low. Ringing has shown that the winter population of Tufted Ducks in the Royal Parks is largely immigrant, there having been a number of recoveries from Scandinavia and eastern Europe. The numerous suburban parks also contain lakes and ponds, and Table 61 includes those at which over a hundred wildfowl have been recorded. The more rural waters, notably Effingham and Little Bookham Ponds, on the edge of the North Downs, are stocked with Mallard reared for shooting, which probably augment the numbers in the region as a whole in early winter.

In the north-west suburbs there are four sizeable reservoirs. Ruislip and Aldenham (the latter a Country Park) are used so intensively for water sports and general recreation that their value for wildfowl is minimal. Hilfield Park, next to Aldenham, on the other hand, is kept free from disturbance and between 1970 and 1982 held regular (and maximum) popula-

Table 61. *The regular (and maximum) numbers at the London area lakes at which over 100 wildfowl have been recorded. Where regular counts have not been undertaken, the maxima only are given. (Includes data from* London Bird Reports.)

	No. of seasons (Years)	Mallard	Pochard	Tufted Duck	Mute Swan	Canada Goose
NORTH/CENTRAL						
Wanstead Park		(125)	(1)	(40)	(2)	(150)
Eagle Pd, Snaresbrook		(76)	(10)	(30)	(10)	(30)
Grovelands Pk, Southgate		(583)	(2)	(6)	(0)	(0)
Broomfield Pk, Palmers Green		(268)	(0)	(4)	(2)	(0)
Pymmes Pk, Edmonton		(172)	(0)	(28)	(12)	(0)
Alexandra Park		(37)	(600)	(1)	(0)	(0)
Finsbury Park	1 (55)	142(150)	0 (0)	13 (34)	7 (7)	0 (0)
Clissold Pk, Stoke Newington		(280)	(0)	(62)	(8)	(0)
Regent's Park		(690)	(138)	(158)	(6)	(55)
Serpentine/Longwater	21(60-82)	330(490)	19 (88)	146(295)	3(10)	87(287)
Round Pond, Kensington Gardens	21(60-82)	104(220)	7 (50)	93(250)	4(20)	20(160)
St James's Park	1 (81)	141(153)	32(101)	124(365)	0 (0)	0 (0)
Waterlow Park	1 (55)	142(175)	0 (0)	1 (21)	4 (7)	0 (0)
Hampstead/Highgate	18(60-82)	221(440)	3 (23)	40 (91)	0(10)	0 (21)
WEST						
Clapham Common	2(65-66)	100(110)	5 (13)	29 (38)	0 (6)	0 (5)
Wimbledon Common	5(57-67)	124(309)	0 (1)	3 (34)	1 (3)	0 (0)
Wimbledon Park	21(60-82)	101(200)	9 (50)	33(120)	2 (9)	0 (78)
Kew Gardens	4(60-63)	84(136)	1 (3)	64(160)	0 (2)	0(210)
Pen Ponds, Richmond Pk	23(60-82)	136(370)	14 (82)	53(154)	3(11)	6 (43)
Bushy Park		(300)	(5)	(12)	(4)	(7)
Home Pk, Hampton Ct		(115)	(0)	(12)	(1)	(0)
Fulmer Mere, Gerrards Cross	14(69-82)	34 (95)	3 (17)	6 (21)	1 (5)	3 (19)
SOUTH						
Kelsey Park, Beckenham	15(60-77)	144(320)	1 (7)	25 (54)	3 (7)	0 (15)
South Norwood	10(61-79)	59(117)	2 (17)	1 (7)	0 (6)	0 (5)
Danson Park, Welling	14(60-77)	75(123)	6 (30)	6 (24)	2 (8)	0 (1)
Bourne Hall	3(60-62)	155(204)	0 (0)	0 (0)	1 (2)	0 (0)
Carshalton Grove	6(57-64)	133(160)	0 (1)	0 (12)	2 (4)	0 (0)
Grange Park	14(48-64)	96(232)	0 (0)	0 (1)	1 (4)	0 (0)
Effingham	6(76-81)	405(834)	18 (42)	31 (51)	7(13)	45(110)
Little Bookham	6(76-81)	120(204)	2 (9)	9 (25)	5(13)	6 (25)
Fetcham	3(76-79)	31(130)	0 (25)	1 (14)	2(10)	0 (0)

Note: The following additional species were regular at one or more sites:
Teal: Fulmer Mere (regular 34/maximum 95).
Gadwall: Pen Ponds (9/74).
Shoveler: Regent's Park (maximum 60); St James's Park (maximum 24); Wimbledon Park (maximum 33).

tions of 102 (337) Mallard, 36 (116) Teal, 12 (55) Wigeon, 7 (29) Shoveler, 146 (617) Pochard and 271 (585) Tufted Ducks; Mallard and Teal have declined but Tufted Ducks have increased since the 1960s. A Local Nature Reserve was declared there in 1969.

Brent Reservoir, better known, on account of its shape, as the Welsh Harp, lies midway between Hilfield Park Reservoir and Hyde Park. Unlike Hilfield Park it has natural banks, but is heavily used for sailing and canoeing and has been the venue for the World Water-Skiing Championships. One small arm of water is, however, inaccessible to boats, and the reservoir is able to retain a small significance to wildfowl (37). Indeed, counts undertaken when there has been no sailing have partially reflected the increases in Pochard and Tufted Ducks which have occurred in London as a whole. Between 1970 and 1982 the Welsh Harp held regular (and maximum) populations of 63 (211) Mallard, 5 (24) Shoveler, 9 (39) Gadwall, 76 (490) Pochard and 134 (318) Tufted Ducks. Smew reached a peak of 100 in the mid 1950s and were still regular in the 1970s, but are now rare.

At Barnes, well to the east of the main complex of Thames-side waters, lie two of the oldest reservoirs – Barn Elms and Lonsdale Road (Table 62). Barn Elms, divided into four basins by grass-covered concrete causeways, has always held one of the most varied duck populations in the region. Gadwall long had their only London breeding site there, having escaped from the captive colony at St James's Park in the 1930s, but they no longer nest at Barn Elms – and rarely elsewhere in the region. Barn Elms is the only London reservoir where boat fishing is allowed, and a sharp decrease in Pochard and Tufted Ducks in the mid 1970s coincided with an intensification of this activity. Between 1974 and 1982 the regular numbers were only 52 Pochard and 274 Tufted Ducks compared with 609 and 803 respectively in the previous nine-year period. These changes may, of course, have been part of the normal movement of these species around London – see Table 59. Lonsdale Road, adjoining the river 1km west, is much less important, but in the late 1950s and early 1960s often held more than a thousand Mallard, and since the mid 1970s has regularly been used by the Barn Elms Gadwall.

At the western edge of this sector, the Thames and Colne Valleys house 2,300ha of reservoirs and gravel pits (Table 62). The reservoirs, lying, as in the Lee Valley, in the south of the complex, follow the course of the Thames; the gravel pits, now comprising over 1,000ha, run as far north as Rickmansworth. On the reservoirs most species have increased substantially over the last twenty years, but Mallard have declined to almost the same extent as on the Thames (see above), and Goosanders have never repeated the exceptional influx of early 1963, when over 800 were present. The cold spell of January 1979 brought 250 and that of early 1982 no more than 280 Goosanders to the area.

Exceptional numbers of Pochard and Tufted Ducks occurred at Staines Reservoir around 1979 following the drainage and reflooding of the southern half for maintenance: 5,000 Pochard and 4,000 Tufted Ducks were present for some weeks. During such operations, which are increasingly frequent at Staines and elsewhere, large expanses of shallow water and mud are produced which attract substantial numbers of dabbling ducks, especially Teal. These species are otherwise largely restricted to the reservoirs with grassy or gently sloping banks, or which are adjacent to open land. Island Barn Reservoir fulfils all these requirements, while Staines and King George VI are close to Staines Moor, which, although drained, has attracted up to 500 Wigeon in recent years (317). London's Shoveler were virtually confined to the Thames-side reservoirs until the late 1960s, for reasons that are unclear, although the inflow pipes from the Thames may have provided a source of food. Certainly this group of reservoirs has long held one of the largest concentrations of Shoveler in Britain.

Queen Mother (formerly Datchet) Reservoir is the newest and third largest of all London's reservoirs, but recreational activity has been so intense ever since its completion in 1975 that there has been little opportunity for a large regular wildfowl population to build up, and few counts have been made. However, Shoveler and Gadwall are apparently often numerous, and there are records of up to 120 and 92 respectively, along with 86 Mallard, 160 Teal, 60 Wigeon, 52 Pochard, 264 Tufted Ducks, 7 Goldeneyes and 65 Goosanders. In times of hard weather, notably in December 1981 and January 1982, or when sailing activity is temporarily reduced, important concentrations are understood to have been present, though no precise figures are available.

The most important gravel pits in the Thames/Colne complex are those lying amidst the reservoirs, notably the Shepperton, Thorpe and Wraysbury groups. The abandoned pits of Thorpe Water Park are almost untenable to wildfowl in summer because of heavy disturbance from water sports, but for most of the winter they are closed to recreation and assume a considerable importance. The Wraysbury group is greatly disturbed by recreational and industrial concerns, but its birds (which have recently included the largest gathering of Tufted Ducks in the London area) are able to seek refuge on the smaller pits and adjacent reservoirs when necessary. Sailing is also permitted –

Table 62. *The regular (and maximum) numbers of wildfowl at the Thames/Colne Valley reservoirs and gravel pits, 1960-1982. The figures in brackets after the place names show the number of seasons with regular counts and the period they cover. For Tufted Ducks see also Table 59. For moulting Pochard and Tufted Ducks see text. (Includes data from* London Bird Reports.)

		Mallard	Teal	Wigeon	Gadwall	Shoveler	Pochard	Tufted Duck	Golden-eye	Goos-ander	Mute Swan	Canada Goose
RESERVOIRS												
Barn Elms	(22:60-82)	102 (763)	19 (315)	19 (75)	38(130)	34 (138)	341(3000)	550(2700)	1 (6)	0 (11)	5(39)	11 (92)
Lonsdale Rd	(21:60-82)	144(1238)	0 (105)	0 (45)	19(110)	5 (76)	16 (101)	67 (540)	0 (2)	0 (1)	3 (9)	0 (0)
Island Barn	(23:60-82)	49 (144)	196 (550)	1 (40)	0 (40)	5 (110)	16 (150)	97 (266)	3 (20)	16(140)	0 (3)	0 (37)
Queen Elizabeth II	(22:60-82)	188 (506)	194 (458)	13(200)	0 (32)	122 (570)	3 (56)	78 (365)	5 (27)	12(429)	0 (3)	0(125)
Walton	(22:60-82)	29 (167)	13 (184)	27(120)	0 (47)	19 (194)	42 (592)	276 (557)	20 (80)	28(314)	1(12)	0 (80)
Hampton/ Kempton	(21:60-82)	104 (306)	15 (198)	20 (85)	12 (68)	43 (339)	107 (400)	257 (524)	5 (28)	0 (19)	7(74)	0(410)
Queen Mary	(14:62-82)	171 (923)	4 (40)	3 (36)	0(102)	89 (535)	76 (450)	539(1455)	35 (95)	28(188)	2(40)	0 (70)
Staines	(22:60-82)	131(1130)	71(1000)	73(350)	0 (30)	41 (683)	460(5000)	539(4000)	9 (61)	20(130)	1(14)	0(303)
King George VI	(19:60-82)	358 (860)	383(1102)	51(308)	5 (39)	106 (660)	37 (749)	164(1110)	14 (85)	21 (96)	1 (8)	0 (11)
Wraysbury	(11:71-81)	96 (498)	17 (150)	0 (80)	4 (45)	60 (720)	29 (417)	366(1500)	18 (64)	7 (59)	0 (1)	0 (40)
COMBINED	1960-69	1756(3591)	913(1470)	207(570)	22 (53)	110 (241)	913(2606)	2532(5467)	49 (77)	144(835)	10(41)	0 (3)
	1970-82	700(1258)	886(1637)	166(500)	122(218)	563(1227)	1132(5131)	2655(5557)	109(214)	90(279)	21(85)	80(446)
GRAVEL PITS												
Shepperton	(2:81-82)	113 (256)	5 (10)	0 (3)	1 (2)	12 (47)	70 (116)	468 (522)	8 (13)	2 (6)	23(89)	145(269)
Thorpe	(5:78-82)	109 (228)	13 (36)	18 (70)	18 (52)	10 (30)	64 (246)	430 (893)	1 (2)	4 (31)	19(36)	75 (14)
Wraysbury	(15:68-82)	192 (381)	32 (187)	5(120)	10 (65)	12 (86)	126 (333)	730(1528)	20 (44)	9 (61)	14(38)	104(570)
Old Slade	(16:65-82)	89 (191)	16 (61)	0 (5)	13 (84)	13 (51)	93 (235)	177 (502)	0 (8)	0 (5)	11(40)	8 (77)
Maple Cross*	(21:60-82)	115 (358)	5 (150)	0 (55)	0 (62)	0 (126)	41 (270)	192 (432)	4 (21)	0 (80)	29(72)	53(350)
Hampermill	(4:67-70)	36 (120)	0 (1)	0 (0)	0 (0)	0 (5)	45 (220)	40 (77)	0 (1)	0 (0)	10(14)	0 (4)

Notes: In the frozen conditions of January and February 1963 the distribution of Pochard and Tufted Ducks on the Thames-side reservoirs was particularly abnormal, because of the varying degrees of ice on the different waters; maxima of 2370 Pochard and 3130 Tufted Ducks were recorded on the new Q E II Reservoir, and 2052 Tufted Ducks on the Walton group. That season's data for these two species have therefore been omitted from the figures for individual reservoirs.
Smew were counted regularly at Hampton/Kempton Reservoirs (regular 2/maximum 17); Staines Reservoir (5/30); Shepperton GPs (1/10); Wraysbury GPs (3/36).
* Maple Cross, Royal Oak, West Hyde, Helicon, Troy Mill, Pynefield, Springwell, Stockers and Batchworth.

all year round – at Island Barn and Queen Mary Reservoirs, but there, too, the presence of disturbance-free waters nearby has helped prevent any apparent declines so far. At Queen Mary, the Water Authority severely restricts the entry of boats into the eastern half and strict discipline ensures an effective refuge even at weekends. Casual access is prohibited at all the reservoirs except Staines, which is crossed by a (well-fenced) public causeway.

The lower Colne Valley (an unofficial Regional Park) is almost continuously lined by gravel pits. In addition to those listed in the table there are many which have received irregular count coverage. By far the most important of these is the 60ha Harefield Moor or Broadwater Pit, where maxima of 1,200 Mallard, 85 Wigeon, 120 Shoveler, 800 Pochard, 700 Tufted Ducks, 25 Goldeneyes, 4 Smew and 355 Canada Geese have been reported (317). Sailing is undertaken there, but only on the northern half.

Mute Swans have only recently appeared in numbers on London's enclosed waters, despite the presence of large gatherings on the River Thames. When these flocks virtually disappeared in the 1960s and early 1970s there was no increase on the reservoirs or gravel pits. More recently, however, large numbers have been recorded at Hampton/Kempton Reservoirs (74 in November 1980), Queen Mary Reservoir (58 in summer 1980), Harefield Moor Pit (55 in November 1976), Tilehouse Pit (75 in November 1976), the Maple Cross group (up to 72 in winter; 52 in late summer) and in the Lee Valley, at Amwell Gravel Pits.

The old-style sewage farm at Perry Oaks, by the western edge of Heathrow Airport, is still of considerable importance, up to 350 Mallard, 40 Teal, 20 Wigeon, 100 Shoveler, 50 Pochard and 100 Tufted Ducks having been seen in recent winters. The proposed fifth terminal building at Heathrow would destroy this site. The purification works at Maple Cross, amidst the gravel pits, held up to 400 Mallard and 130 Teal in the late 1970s (317) but there are no more recent data. At both these localities access is severely restricted.

In recent years the London area has become of major importance for moulting Tufted Ducks. Synchronised counts in 1979 and 1980 revealed the following numbers (nearly all being males) on the reservoirs and principal gravel pits: 1979 – 5,298 (July) and 6,608 (August); 1980 (poorer coverage) – 1,939 (August) (419). Many of these birds are known to have been flightless and it is probable that most were moulting. The largest total located in the July and August counts carried out at most reservoirs between 1948 and 1957, for the National Wildfowl Counts, was only 645 in 1957. Although no more coordinated summer counts were undertaken until 1979 the indications from various isolated records are that there was a steady increase throughout the intervening period. As with the wintering flocks, the moulting birds move around the waters between seasons; Walthamstow, Walton, Queen Mary, Wraysbury, King George VI and Staines Reservoirs have all held over 1,000 in late summer in the last ten years. The Walton Reservoirs, however, have consistently proved more attractive to moulting than to wintering Tufted Ducks. Between 1975 and 1980 they held an average of 877 in July/August, compared with an average September – March peak of 299.

In the south of the region, away from the Thames, there are few notable wildfowl resorts, apart from the ornamental lakes and ponds mentioned earlier. A record of 200 Teal at Beddington Sewage Farm, near Croydon, in November 1980 (317) suggests that this site has retained some attraction, despite modernisation. Along the Mole in rural Surrey, Mandarin Ducks are common, their spread being described in the next section. Elsewhere in the London area Mandarins breed only in Epping Forest, though they may turn up almost anywhere. Ruddy Ducks have been reported in most years since 1965, but breeding was not confirmed until 1980 (317).

Although few wetland sites in London have statutory protection, the very nature of most of the wildfowl haunts has made them largely safe from destruction or spoilation. A potential exception to this arose in 1982 when a plan to extract gravel and clay from the northern basin of Staines Reservoir, which would have deepened the water enough to reduce its attraction to wildfowl drastically, was averted largely because of the site's SSSI status. The proposal, which arose from the construction of the M25 motorway nearby, served as a warning that, should the general economic situation improve, some of London's time-honoured havens for wildfowl may quickly come under threat. Meanwhile, much the biggest problem stems from the continuing proliferation of water-based sports. These have apparently had little overall impact as yet, although the increase in some species could well have been greater in the absence of recreational pressures. Any further expansion in recreational activity in the region should only take place with the greatest care, ensuring that in all areas a large proportion of undisturbed waters remains.

The lower Thames Basin: the Home Counties and the Vale of Kennet

To the north and south of Greater London the basin of the Thames is bounded by the rolling chalkland of rural Hertfordshire and by the ridge of the North Downs, between Maidstone and Guildford. In the southern part the area up to the regional boundary has already been covered in the previous section; to the north, however, there are several groups of lakes in the upper basins of the Lee and Colne, which lie beyond the limits of the London area.

The first of these groups, around Hertford and Welwyn, is closely linked with the gravel pits along the middle reaches of the Lee between Cheshunt and Broxbourne. The two main centres are the shallow ornamental lakes at Panshanger and Brocket Park, both of which hold a regular peak of 2-300 ducks (Table 63). Flocks of about 100 Mallard are also found on the river pools at Digswell and in Woodhall Park and parties of 20-30 have from time to time been reported on some of the smaller pools nearby.

At the head of the Lee Valley, the lake at Luton Hoo supports a further population of 2-300 ducks, and in recent years has become the centre for a new and flourishing group of Canada Geese. These were introduced in the early 1970s, and after a slowish start have increased to upwards of 100. At first their numbers were the same throughout the winter, but latterly the resident flock has been augmented by a

Fig. 21. The Thames Basin and the Vale of Kennet.

late-autumn influx, probably from Brocket Park, where a similar group, now totalling 20-30, was introduced at about the same time.

Further to the west, near Chesham, the lake in Shardeloes Park provides a similar centre for quite large flocks of ducks (Table 63). In addition there are several smaller sites nearby, each holding a regular flock of 20-30 Mallard, and a peak of 50-100. Parties of up to 40 Teal are also found on some of these, notably on the marshy ponds at Little Missenden, the sewage farm near Chesham, and the lake in Latimer Park.

To the west of London, there are records of wildfowl from upwards of 60 lakes and gravel pits in the 30km square between Slough, Reading, Hook and Guildford. Within this area the landscape varies from sandy heathland with extensive fir-woods in the eastern and central part, to farmland with coverts of hardwood in the north and west. The dividing line between the two is generally clear cut, although in places the heath is interspersed with salients of richer land, notably along the Wey and Blackwater which drain from the chalk ridge at the southern limit of the region. These variations are reflected in the size and distribution of the wintering flocks, the farmland pools being much the more favoured. The numerous lakes and pits are thus divisible into a series of convenient groups. These comprise the lakes in Windsor Great Park, the gravel pits along the Thames and Blackwater, the pools along the Wey between Woking and Guildford, and the ornamental lakes and dams in the north-east parts of Hampshire.

The lakes in Windsor Great Park, although lying within 10km of the reservoirs and gravel pits along the western fringe of London, have little in common with their larger and much deeper neighbours. In all there are more than a dozen pools, set either in the park or close by, the surroundings varying from open pasture to wooded heath, with thickets of rhododendrons reaching in places to the water's edge. The largest and most favoured of the pools are Great Meadow Pond (13ha) and Virginia Water (53ha), both of which attract a peak of at least 250 ducks, and in some years up to 500 (Table 67). Totals of several hundred birds have also been recorded on the lakes at Ascot Place and Sunninghill, and some of the other pools, such as Obelisk Pond and the Rapley Lakes, near Bagshot, may at times hold 50-100. The rest have mostly less than 25.

The park itself is perhaps best known for the population of Mandarin Ducks, which became established in the early 1930s, following the spread of full-winged birds from a private collection near Cobham, and possibly from the London parks. At first they were centred mainly on Virginia Water, but later spread to many of the neighbouring pools between Windsor and Bagshot. By 1950 their numbers had increased to an estimated 400, and another 150 were thought to be based around Cobham. Since then their spread has continued westwards to the Loddon Valley, and southwards along the Mole and Wey as far as the Sussex border (542). The size and distribution of these local groups is governed by their specialised requirements. The most favoured sites are secluded pools or streams, surrounded by hardwoods, such as oak, sweet chestnut and beech, these trees providing both the requisite nest-holes and the nuts and acorns which form the staple winter diet (527). An overhanging fringe of rhododendrons or other shrubs in which the birds may perch and roost is also needed. Because of their retiring habits, and the frequent movement of the birds from pool to pool, the monthly counts give no real indication of the numbers occurring at each site or of the totals on the various groups of sites. From time to time, however, there are records of substantial gatherings on one or other of the main resorts, which provide a basis for a rough assessment (Table 64). On the Windsor – Bagshot group of lakes the population seems to have maintained a fairly constant level since 1965, and probably since 1950. The population around Cobham has also remained at about the same level, or

Table 63. *The regular (and maximum) numbers of wildfowl at five of the main resorts to the north of Greater London. The figures in brackets after the place names are the number of seasons on which the summaries are based. Sites marked * were covered mainly in the 1960s.*

		Mallard	Teal	Shoveler	Tufted Duck	Pochard	Canada Goose	Mute Swan
Panshanger, Herts	(7)*	248(450)	31(90)	0 (8)	1 (2)	0 (0)	0 (0)	1 (3)
Brocket Pk, Herts	(10)	142(300)	1 (7)	0(16)	61(143)	13 (41)	15 (46)	6(12)
Woodhall Pk, Herts	(2)	107(350)	0 (1)	0 (0)	26 (64)	4 (10)	16 (28)	4(10)
Shardeloes, Bucks	(11)	119(400)	13(34)	4(47)	23 (58)	30(300)	0 (28)	1 (6)
Luton Hoo, Beds	(9)	180(385)	20(45)	2(15)	40 (75)	20 (40)	35(365)	7(11)

possibly increased a little. In 1969 there were 55-60 pairs along the 25km stretch of the Mole between Esher and Mickleham, compared with the earlier estimate of 150 individuals in a rather smaller area (542). From this it seems that the population in the two initial centres is now at its upper limit, but some of the new outlying resorts may well provide bases for further expansion. Winter flocks of 20-30 have recently been found at Bearwood Lake and Stratfield Saye in the Loddon Valley, 150-200 have been seen on several occasions at Virginia Water, and as many as 300 were seen in a single flock near Godalming in 1981 (p.88).

The lakes around Windsor and Ascot are also used by one of the many groups of Canada Geese which are now established throughout the region. Some 20-25 pairs are currently breeding in the area, mostly towards Ascot, and in July a moulting flock of about 200 birds assembles either on Great Meadow Pond or one of the other pools nearby. In autumn these disperse, and probably move eastwards to the reservoirs and gravel pits between Wraysbury and Shepperton where 4-500 are present in winter. The numbers remaining in the park are usually less than 100, but are sometimes augmented by visiting flocks (Table 65).

Further to the south the tracts of heathland between Bagshot and the Hog's Back are virtually devoid of ducks, except perhaps for occasional flocks of 20-30 Mallard on some of the scattered pools on Pirbright and Whitmoor Commons. In contrast, the resorts along the Wey and Blackwater, on either side of the heath, have gained in value during recent years, and are now supporting sizeable numbers of several of the common species. On the eastern side by the Wey, the sand pits at Sendmarsh and Papercourt Farm hold a normal total of 2-300 ducks (Table 67), and another 80 or so, mostly Mallard, use the smaller pits at Send

Table 64. *The maximum number of Mandarin Ducks recorded at various sites around Windsor Great Park between 1975 and 1982, and in the two preceding 5-year periods. The figures in brackets are taken from short runs of data; a blank = no information.*

	1965-69	1970-74	1975-82
Virginia Water	70	110	112
Obelisk Pond	84	53	53
Great Meadow Pond	45	72	65
Cow Pond	58		
Sunninghill Pk	70	92	23
Titness Pk		154	
Ascot Place		(34)	
Swinley Brick Pits	60		
Rapley Lakes	24	(16)	

nearby. Canada Geese are also present in fair numbers, especially in August and early September, when they reach a normal peak of 90-120, increasing to 400 in 1981. The winter flocks, although smaller and less regular, are likewise showing signs of increase. A short way to the south, the lakes in Clandon Park are occupied by droves of several hundred Mallard, mostly hand-reared, and by regular parties of 20-30 Tufted Ducks, with perhaps a few Pochard and Teal. The water-meadows along the Wey at Stoke Park, Sutton Place and Send are also used by Mallard and Teal, the latter occurring at times in flocks of up to 40.

The birds in the Blackwater Valley are attracted mainly to the gravel pits between Eversley Cross and Aldershot, which have doubled in size in the past decade to a current total of 25ha. Six separate groups of workings are now scattered along this 20km stretch, and several other lakes are available nearby. The most favoured of the gravel pits, each holding a regular total of at least 100 ducks, are those between Sandhurst and Yateley, which date from the 1940s, the newer workings at Eversley Cross, Frimley and Ash Vale, and the 10ha lake at Badshot Lea (Table 67). The lake in the grounds of the Royal Military Academy, Sandhurst, is also used by 80 Mallard, and the pits at Mytchett and Aldershot hold another 60 between them. The other lakes at Hawley, Frimley and Mytchett are apparently of little value.

Canada Geese are recorded at many points along the valley, the largest numbers being at Yateley and Eversley, on the fringe of the urban development. The breeding population, centred mainly on Yateley, now totals 20-30 pairs, compared with 8-10 in the early 1970s, and the winter numbers stand at about 250 (Table 65). Gatherings of 2-300 moulting birds have also been reported in July, but latterly these summer numbers have decreased to 100-150, while those on the neighbouring centres to the west have steadily increased.

These next resorts, lying mostly within 10km of the Blackwater Valley, are set in the belt of fertile farmland between Fleet, Odiham and Stratfield Saye, and are used by a total of several thousand birds. Mallard and Canada Geese both attain an autumn peak of 1,500-2,000, and flocks of up to 50 Teal and diving ducks are common on the larger pools. Most of the more important sites, including those at Dogmersfield, Elvetham, Bramshill and Stratfield Saye, are ornamental lakes in the grounds of estates or institutions, and are virtually free from disturbance. Fleet Pond, although flanked by houses on one side, is also well protected, and has gained considerably in value since the declaration of a Local Nature Reserve in 1977.

The concentrations of Canada Geese are espe-ially notable, the numbers in this one small area mounting at times to 6 or 7% of the national total. The rime resorts are the lakes at Elvetham and Bramshill, vith a current output of 2-300 goslings a year, and the arkland pools at Stratfield Saye, which provide an utumn centre for 1,500 birds. These start to assemble ı August, following the moult, and are present in trength until October or November, when they start) disperse to their winter quarters. This tradition egan to develop in the late 1960s, at a time of rapid xpansion, and now involves at least a third of the opulation in the lower Thames and Kennet Basin. ater on, from November through to March, the flocks reak up into wandering bands, and the distribution varies from month to month depending on the food supplies.

Table 65 attempts to trace this changing pattern, both here and in the neighbouring districts, including the western parts of London. The two left-hand columns are concerned with the breeding and moulting distribution, and are based on the census undertaken in July 1976, this being the only recent source of detailed summer data (405). The figures show the number of goslings at each site, and the total numbers of geese, including parents, non-breeding birds and young. The number of pairs has been omitted, because by July, when the counts were made, the broods had often coalesced into larger groups, and the individual families could no longer be identified. The

'able 65. *The numbers of Canada Geese recorded in western London and the lower Thames and Kennet basin in July 1976 columns 1-2) and the mean of the monthly counts in four autumn and winter periods 1978-1982 (columns 3-6). The means in he two-monthly periods (columns 4-6) are based on the higher of the two counts in each season. The July total (column 2) ncludes parents, non-breeding birds and young. Figures in brackets are based on meagre information; a blank = no data.*

	July 1976		Means 1978-82			
	Goslings	Total	Sept	Oct – Nov	Dec – Jan	Feb – Mar
LONDON – MAIDENHEAD – GUILDFORD						
London: Parks/Thames Reservoirs			305	320	360	170
Colne Valley GPs	15	56	130	150	80	110
Wraysbury – Shepperton	51	240	350	400	370	290
Windsor – Maidenhead	(6)	224	270	150	70	55
Effingham – Clandon – Send		(40)	75	115	35	110
Area total	72	560	850	960	770	530
Highest count			1180	1060	1170	625
NORTH-EAST HAMPSHIRE						
Yateley/Eversley GPs	77	236	65	165	135	235
Frimley – Aldershot – Fleet	9	45	35	105	140	80
Tundry Pond/Potbridge	7	9	110	120	150	150
Elvetham Pk	101	270	145	225	450	255
Bramshill Lake and Pits	3	50	80	110	40	50
Stratfield Saye/Country Pk	22	154	1400	1125	790	335
Area total	219	764	1835	1585	1360	950
Highest count			1910	2000	1800	1310
MAIDENHEAD – NEWBURY						
Medmenham/Aston/Henley	34	94				
Twyford GPs			225	(125)	(50)	40
Sonning GPs	0	240				
Henley Rd GPs, Reading	34	364	250	(230)	195	175
Woodley GPs/Bulmershe L	20	158	75	190	135	130
Theale/Burghfield GPs	22	197	(125)	(155)	85	270
Aldermaston/West Canal GPs	39	58				
Thames: Purley			4	85	165	75
Area total	149	1111	580	480	450	505
Highest count			740	600	750	760
Regional total	440	2435	3150	2825	2420	1950
Highest count			3580	3070	2805	2700

four remaining columns show the means of the September counts in the seasons 1978-1982, and the corresponding figures for the periods October – November, December – January and February – March. In the paired months the means are based on the higher of the two counts in each season, or, where several sites are taken together, on the higher of the two synchronised totals. Many of the figures are almost certainly too low, especially some of the area and regional totals. Except in Hampshire, the runs of data are often marred by gaps, and several sites of probable importance have been neglected altogether. These omissions have perforce been ignored, and the data used as they stand, with no attempt at interpolation.

Despite these shortcomings, the table gives a fair idea of the interrelationship between the various resorts, and of their role as breeding, moulting or wintering centres. It also reveals an apparent decrease in the population over the winter, amounting in all to more than 1,000 birds. This is explained partly by losses through shooting and natural mortality, but mainly by the progressive dispersal of the flocks on to the smaller lakes and marshes, many of which lie outside the scope of the regular counts. There is also evidence of a movement to areas outside the region, especially to the south, but the numbers involved are probably quite small.

Over the past 30 years the population in the region as a whole has shown a steady increase, somewhat steeper than the national trend (p.382). In 1953 the total for the lower basin was estimated at 100-180, the main centre being at Englefield, to the west of Reading (57). By 1968 the level had risen to at least 570, and by 1976 to a minimum of 2,470 and possibly some hundreds higher (405). Since then the autumn counts have increased still further, to a total of around 3,500 in each of 1980, 1981 and 1982. On these occasions the numbers at several sites were unrecorded, and the true total now may well be nearly 4,000.

Feral flocks of Greylags, Snow Geese and Barnacle Geese are also established at Stratfield Saye, and stragglers from here are common enough on many of the other waters round about, with a few breeding pairs at Bearwood Lake and at Burghfield Gravel Pits. The Greylags were introduced in the late 1960s, and by 1981 had increased from 10 to 130. Four or five pairs are now breeding regularly, and 15-20 goslings are recruited each year. The other two species total 20-30 apiece, and have lately shown signs of rapid increase. In 1981 there were three broods of Barnacle Geese and four of Snow Geese, from which totals of 12 and 16 young were successfully reared.

The gravel workings by the Thames and Kenne provide a further series of resorts, extending intermittently for nearly 50km, from Windsor to Newbury an beyond. Over the past 30 years the number of pits ha grown from 10 or so in 1949 to upwards of 50 in 1981 and the area of water has increased from less than 10 to above 600ha. Many of the larger pools are now use for sailing, fishing or water-skiing, but in mos localities the birds disturbed from one area are able t find alternative roosts and feeding grounds on th other pits nearby. There are also several lakes an marshes, and some quiet stretches of the river, whicl are known to hold ducks, and may well be used a convenient day-time retreats.

The results of the counts on some of the mor important centres are summarised in Table 67. Thes begin in the east with the scattered pits at Bray Maidenhead and Marlow, of interest mainly for diving ducks, and continue after a gap of 15km with the mucl more extensive group on the eastern outskirts o Reading. The main resorts here, together holding a least 1,000 ducks, are two large workings on th Oxfordshire bank at Sonning Eye and Caversham (Henley Road) and the complex of pits at Twyford Woodley and Dinton Pastures in the lower Loddo Valley. The lakes at Bearwood and Whiteknights an the South Lake at Bulmershe also hold 50-100 duck apiece, and the water-meadows at Wargrave are ofte used as a feeding ground by the local flocks of Canad Geese, the numbers at times amounting to above 500.

In addition to the species listed, the Loddon an Caversham pits hold an annual total of 1-200 Wigeon and in recent years have attracted a growing numbe of Gadwall. The latter were first recorded in strengt in 1969, when up to 100 appeared, but this wa exceptional and for the next nine seasons the total wa seldom more than 10. Then in 1978 they increased t around 30, and from 1980 to 1982 reached peaks o about 120. These concentrations are present only from November until February, and are not apparently o local origin, though the increase may perhaps be linked with a similar trend on the West London pits and reservoirs, which began at about the same time. Their main resort is on the pits in the Dinton Pastures Country Park, where some of the smaller pools are kept as reserves.

The Mute Swans which used to frequent the Thames at Reading and elsewhere have also taken to using the pits as their principal winter stronghold. Totals of 75-85 are now recorded regularly from September through to March, though as many as 163 have occurred, and numbers tend to be higher in hard weather. The largest flocks are usually at Caversham in early autumn, and thereafter in the Loddon Valley, the pools at Dinton Pastures being especially favoured (515). Similar shifts from the river onto standing water have apparently occurred at several other points along the Thames, notably at Oxford, Dorchester and Abingdon/Radley and in the western parts of London. One of the reasons is said to be the increase in power-boats, which have the effect of stirring up mud, and destroying the plants on which the swans feed (408). At the same time many of the newer pits are losing their initial bareness, and becoming more attractive. The results of surveys in 1983 of the Thames from its source to Richmond and Grays are given in Table 66.

On the western fringes of Reading a further population of up to 1,000 ducks is centred around the gravel pits at Burghfield and Theale, which now stretch for nearly 6km along the lower Kennet, and cover a total of about 280ha. The preponderant species are Tufted Duck and Pochard, which together reach a peak of 6-700, and flocks of up to 300 Teal have recently begun to winter regularly. Several other species, such as Mallard, Shoveler, Mute Swan and Canada Geese (as many as 500 in moult), are also present in fair numbers, but the totals generally are a good deal smaller than those in the Loddon Valley at Twyford, Dinton and Stratfield Saye. Wigeon and Gadwall are scarce and irregular.

Table 66. *The number of Mute Swans on the Thames and its backwaters according to surveys on 9th January 1983 (from French 1983) and 10th July 1983.*

River section	Swans Jan	Swans July
Source – Cricklade	37	75
Cricklade – Lechlade Br	85	85
Lechlade Br – Tadpole Br	56	36
Tadpole Br – Newbridge	56	35
Newbridge – Swinford Br	11	21
Swinford Br – Folly Br	22	22
Folly Br – Abingdon Br	9	13
Abingdon Br – Clifton Hampden	31	27
Clifton Hampden – Fairmile Hosp	35	22
Fairmile Hosp – Whitchurch Br	18	25
Whitchurch Br – Caversham Br	29	16
Caversham Br – Henley Br	20	49
Henley Br – Marlow Br	15	25
Marlow Br – Maidenhead Br	21	35
Maidenhead Br – Bells of Ouzeley	33	15
Bells of Ouzeley – Chertsey Br	45	8
Chertsey Br – Hampton Ct	11	3
Hampton Ct – Richmond	43	25
Total on whole stretch	577	536

Note: The 62km between Richmond and Grays were also covered in July, holding 291 Mute Swans. The grand total of 827 for that month included 152 cygnets.

Elsewhere in the neighbourhood there are several smaller sites, which supplement the feeding grounds on the deeper and less secluded pits nearby. These include the lagoons at Manor Sewage Farm, near Reading, and Cranemoor Lake, near Theale, both of which have held peaks of 50-100 Mallard and similar numbers of Teal, although the sewage farm has dried up and become less important. Totals of 1-200 Mallard are also reported on the Thames around Purley, and late in the winter there are often parties of up to 100 Canada Geese both there and at Cranemoor.

Further to the west, towards Newbury, the resorts are more scattered, and the numbers of most species are relatively small. The main centres, each holding a peak of 1-200 ducks, include the old 14ha pit at Aldermaston, long kept as a private reserve, and the recent workings at Thatcham, now covering more than 40ha. The lakes at Donnington Grove and Benham Park, on the western outskirts of Newbury, are also used by fair-sized flocks of Mallard and diving ducks, and another group of 120 Mallard (max 240) is centred on the lakes at Highclere, a short way to the south (Table 68). Several of the other outlying resorts, including the lakes at Ewhurst, Sydmonton and Hampstead Marshall, hold further flocks of up to 50 Mallard and perhaps a few diving ducks, though the latter are generally scarce except near the course of the Kennet. Canada Geese occur from time to time on most of the larger pools, and are probably increasing, but their visits are still somewhat erratic and flocks of more than 30 are unusual.

In its middle reaches the Kennet meanders through long stretches of marshy meadow, splitting in places into parallel channels and enclosing a network of quiet ditches and waterways. This is especially so between Newbury and Hungerford, where several sections of the marsh are at least 500m across. The wildfowl here have not yet been fully surveyed, but the total for the 12km stretch could well amount to several hundred. In the seasons round 1970 there were several records of 80-100 Wigeon and 40-50 Tufted Ducks on the river near Marsh Benham, and similar numbers of Tufted Ducks were present regularly on Eddington Mill Pond, near Hungerford. Mallard were not much in evidence, but are probably more numerous in the central section, near Kintbury, where the marshes are broad and secluded.

Above Hungerford the dry chalk ridges of the Wiltshire Downs close in on either hand, and wildfowl are scarce, except on the river and on Wilton Water, some 10km to the south. The two main centres on the

Kennet are the broad ornamental pools at Ramsbury and Chilton Foliat, both of which support substantial flocks of Mallard (Table 67). The stretch at Chilton Foliat also holds up to 100 Tufted Ducks, and parties of 30-40 Wigeon appear for a while in the course of most seasons. Elsewhere in the valley there are regular flocks of 50-100 Mallard on the cress beds at Axford, and as many again are spread along the 5km reach between Clatford and Silbury, to the west of Marlborough.

The reservoir at Wilton Water was constructed in the late 18th century to supply the Kennet and Avon Canal in its passage over the Crofton summit, a function which it still fulfils. Although covering less than 5ha, it is shallow and fairly secluded, and, despite its isolated position, attracts a regular peak of 100-150 ducks. The canal also supports a few Mallar and Mute Swans, but the numbers are a great de smaller than those on the sections further west, on th other side of the watershed.

Canada Geese breed freely and are common i winter at several points along the upper Kennet an on Wilton Water. They first started to appear in th mid 1960s and have since increased to an autumn tot of about 150. In the first instance they must presun ably have come from the strongholds to the eas beyond Newbury, but they now form an independer group, which in turn may be linked with the flocks t the north of the downs around Swindon.

The less important sites in the lower Thame and Kennet basin are listed in Table 68.

Table 67. *The regular (and maximum) numbers of six selected species of wildfowl on 26 major resorts in the lower Thames and Kennet basin. The figures in brackets after the place names show the number of seasons on which the summaries are based. The counts were made between 1970 and 1982, except at sites marked * for which data from the 1960s have been included.*

		Mallard	Teal	Shoveler	Tufted Duck	Pochard	Mute Swan
A:WINDSOR – BAGSHOT – GUILDFORD							
Virginia Water	(12)	253 (709)	0 (2)	0(10)	55(203)	31(131)	3 (9)
Gt Meadow Pond	(12)	129 (403)	5 (21)	20(73)	41(103)	29 (71)	3 (11)
Papercourt GPs, Woking	(13)	114 (258)	2 (12)	5(28)	63(151)	72(137)	8 (30)
B:BLACKWATER – ODIHAM – STRATFIELD SAYE							
Yateley/Eversley GPs	(7)	204 (280)	3 (10)	4(15)	227(310)	67(130)	32 (56)
Frimley/Mytchett/Ash	(6)	142 (225)	39 (70)	0 (4)	90(130)	39 (65)	14 (25)
Badshot Lea	(9)	24 (114)	0 (2)	0 (0)	36(258)	29(197)	4 (20)
Fleet Pond LNR	(13)	129 (350)	12 (40)	2(14)	19 (65)	44(125)	2 (8)
Dogmersfield/Tundry Pd	(7)	186 (446)	26 (90)	4(15)	35 (85)	43 (90)	4 (8)
Bramshill/Elvetham	(9)	192 (420)	21 (50)	0 (3)	6 (22)	5 (20)	0 (8)
Stratfield Saye	(12)	750(1660)	37(225)	1 (8)	37 (84)	33(105)	3 (9)
Wellington Country Pk	(6)	35 (84)	2 (10)	0 (2)	12 (21)	43(145)	11 (34)
C: MAIDENHEAD – READING							
Bray GP	(3)	24 (61)	15 (45)	5(15)	50(134)	65(124)	1 (4)
Maidenhead GPs	(7)	33 (55)	0 (2)	0 (1)	59(190)	37 (82)	4 (15)
Marlow GPs	(13)	7 (44)	0 (8)	0(12)	127(246)	73(254)	11 (22)
Sonning GPs*	(6)	85 (265)	0 (4)	0 (1)	58(153)	24(102)	3 (10)
Caversham GPs*	(7)	116 (286)	1 (14)	3(38)	155(480)	74(178)	6 (45)
Twyford/Dinton GPs	(6)	210 (460)	43(130)	34(70)	189(280)	95(270)	67(140)
D: READING – NEWBURY							
Burghfield GPs	(10)	112 (644)	5 (50)	16(60)	201(500)	260(675)	6 (32)
Theale GPs	(11)	48 (305)	62(342)	3(20)	120(279)	111(380)	12 (40)
Aldermaston Main GP*	(3)	45 (194)	2 (35)	6(45)	38 (52)	40 (98)	4 (6)
Thatcham Marsh GP	(2)	21 (37)	3 (7)	4(10)	52 (64)	79(124)	8 (16)
E: NEWBURY – MARLBOROUGH							
Donnington Grove*	(7)	91 (159)	9 (36)	0 (1)	28 (63)	33(103)	3 (8)
Benham Pk Lake	(4)	81 (171)	2 (15)	0 (0)	82(260)	56(122)	6 (18)
Chilton Foliat	(13)	91 (362)	2 (31)	0 (0)	67(173)	6 (33)	6 (37)
Ramsbury Manor	(12)	365 (800)	0(382)	0 (1)	13 (61)	2 (12)	3 (26)
Wilton Water	(13)	70 (240)	6 (36)	0 (0)	41(100)	14 (34)	11 (27)

The upper Thames Basin

To the north of Reading and Newbury the lower valley of the Thames and Kennet is bounded by the long chalk ridge of the Chiltern Hills and the Berkshire Downs, and by the narrow defile of the Goring Gap, through which the river breaches the escarpment. Above this feature the Thames and its many tributaries fan out again through the broad bowl of the upper basin, which stretches westwards for 70km to the brink of the Severn Vale. In the west the main stream is joined by the Coln, the Leach, the Windrush and the Evenlode, all of which drain from the Cotswold ridge; in the north the Cherwell rises in the Northamptonshire uplands, close to the head of the Nene and Ouse, and meets the Thames at Oxford; to the east the Thame has its source at the foot of the Chiltern Hills near Tring. For the most part the watershed is well defined, except in the north between Linslade and Bicester, where the basins of the Thames and the Fenland Ouse are separated by a scarcely perceptible ridge, rising in places to less than 100m, and affording an easy route for birds between the two regions. The closeness of this link is further strengthened by the many man-made lakes along the Ouse, which attract large gatherings of ducks to that side of the boundary (p.155).

Viewing the region as a whole, the two prime areas for wildfowl are the environs of Oxford, within 20km of the city centre, and the stretch between Lechlade and Cirencester, containing the Cotswold Water Park. In both these districts there are frequent clusters of gravel pits, interspersed with occasional lakes and reservoirs, which provide in each case a closely knit complex of habitats for upwards of 5,000 ducks. Elsewhere a further 2-3,000 ducks are dispersed over the many smaller resorts which are scattered throughout the region. Canada Geese are also numerous and widespread. Mute Swans have declined less markedly than in the lower reaches of the Thames (Table 66).

The chief resorts near Oxford are the groups of gravel pits at Dorchester and Stanton Harcourt, each covering 40-45ha, the lake in Blenheim Park (48ha), and the two adjacent reservoirs at Farmoor, constructed in 1966 and 1974 on the site of some earlier workings (total water area 174ha). The building of the reservoirs made a significant improvement in the local situation for wildfowl, especially because of their proximity to the gravel pits, providing additional and alternative refuges for birds using those sites. The largest gatherings are at Stanton Harcourt, where the ducks have increased from a maximum of 200 or so in the mid 1960s to a mean peak of 1,520 between 1974 and 1982. The other three sites hold 6-800 ducks apiece (Table 69). Some gravel pits in the area have been filled with fly-ash or refuse and there are threats to those at Abingdon/Radley and Stanton Harcourt. The area of water in these complexes may well decrease in future years.

The pits at Dorchester and Stanton Harcourt have also become the main autumn centre for a

Table 68. *The total numbers of wildfowl recorded regularly at a further 23 sites in the lower Thames and Kennet basin (excluding Canada Geese and Mandarin Ducks). The figures after the place names are the highest totals recorded at each site. Except at sites marked *, the counts were made exclusively between 1970 and 1982; "d" indicates a predominance of diving ducks. The sites are grouped by districts as in Table 67.*

Regular totals 25–50		51–100		101–200	
A:Sunninghill Pk, Ascot	120	Obelisk Pd, Windsor	100	Thames:Windsor	250
				Send GPs, Woking	150
B:Potbridge Pds, Odiham	50	RMA Lakes, Sandhurst	90		
C:		Bearwood L, Winnersh	160	South Lake, Woodley	175
		Whiteknights L, Reading*	150		
D:Aldermaston Court L	70d	Manor Sewage Fm, Reading*	125	Highclere L, Newbury	240
Marlston Pds, Thatcham*	70	Cranemoor L, Theale	180		
Ewhurst L, Kingsclere	120	Thames:Purley	210		
Sydmonton Ct, Kingsclere	50				
E:Benham GPs, Newbury*	70	Eddington Mill Pd	130d		
Kennet:Benham	140d	Kennet:Axford	270		
Hampstead Marshall L	80	Kennet:Silbury–Clatford	130		

thriving group of Canada Geese. These first colonised the district in 1970, following increases in the Reading area, and have since doubled in numbers every 2-3 years, from an initial level of 50-60 to a total of at least 900 in 1981 and 1982. The largest individual counts, all made in September or October, have come from Dorchester (700), Stanton Harcourt (500), Abingdon/Radley (230) and Standlake (200). Later on in the season, when the autumn gatherings have dispersed, a regular flock of 2-300 (with a maximum of over 500) frequents Port Meadow, near Oxford, and wandering bands are likely to appear erratically on many of the other pools throughout the district.

In addition to the four prime centres there are several places, such as the pools in Eynsham Park and the pits near Abingdon/Radley, Cassington, Standlake and Hardwick, which attract a regular peak of at least 250 ducks, and another four or five resorts hold upwards of 100 each.

The riverside floods and meadows are also used by transient flocks of ducks and swans. At one time, before the gravel pits were dug, these areas ranked as major strongholds, but over the years the extent and duration of the winter flooding has been steadily curtailed due to Water Authority control of rivers, and the present flocks are often relatively small compared with those on the new and populous pools nearby. The largest flocks in recent years have been on Port Meadow, by the Thames at Oxford, and on the floodplain of the Cherwell near Somerton and Aynho. The first holds a regular late-winter flock of 1-200 Wigeon, and is used occasionally by parties of up to 100 Teal and 20-30 Shoveler. A few Mallard and Mute Swans are also present, and Canada Geese are now regular from November until January (see above).

The Somerton meadows, although reduced in value by the dredging of the Cherwell in 1971, are still subject to brief but regular bouts of flooding, and continue to attract an annual peak of 3-400 ducks (Table 69). Teal, in particular, are well represented and

Table 69. *The regular (and maximum) numbers of seven selected species of wildfowl on 22 of the main resorts in the upper Thames basin. The figures in brackets after the place names are the numbers of seasons on which the summaries are based. The counts were all made during the period 1970-1982, except at sites marked *, where the data relate to the 1960s.*

		Mallard	Teal	Wigeon	Shoveler	Tufted Duck	Pochard	Mute Swan
A: OXFORD GROUP, SOUTH								
Dorchester GPs	(13)	301 (607)	19 (94)	85(286)	10 (33)	133(225)	107 (195)	31 (78)
Sutton Courtenay GPs	(11)	216 (450)	75(450)	0 (8)	2 (10)	79(180)	15 (45)	9 (24)
Abingdon/Radley GPs	(11)	63 (170)	14 (65)	0 (20)	0 (4)	87(170)	66 (170)	14 (35)
Farmoor Reservoirs	(13)	203 (800)	10(200)	125(380)	27(110)	188(650)	94 (480)	8 (26)
Stanton/Hardwick GPs	(11)	451(1000)	39(196)	21(118)	13 (97)	346(560)	503 (700)	27 (50)
B: OXFORD GROUP, NORTH								
Wolvercote/Cassington	(13)	45 (170)	40(120)	6 (65)	2 (20)	103(210)	22 (56)	11 (55)
Eynsham Lakes	(13)	337 (690)	21 (80)	0 (5)	0 (10)	19(150)	2 (14)	0 (0)
Blenheim Pk Lake	(13)	285 (676)	14 (60)	0 (2)	5 (26)	156(580)	125 (292)	9 (21)
Cherwell: Somerton	(8)	66 (200)	74(324)	32(400)	0 (5)	0 (2)	0 (9)	14 (30)
Middleton Stoney*	(5)	160 (460)	25(100)	15(200)	0 (2)	10 (35)	10 (75)	0 (2)
Boarstall Decoy	(5)	160 (370)	70(385)	0 (15)	2 (10)	0 (6)	0 (2)	0 (0)
C: AYLESBURY – BICESTER – BANBURY								
Wotton Underwood	(11)	153 (349)	17 (85)	3 (34)	4 (11)	26 (55)	31 (102)	3 (7)
Eythrope Pk	(11)	201 (350)	28 (78)	8 (58)	0 (10)	6 (18)	5 (22)	1 (7)
Tring Reservoirs	(13)	220 (563)	79(300)	28(195)	92(177)	207(331)	154 (600)	16 (42)
Calvert Brick Pits	(12)	182 (448)	1 (9)	27(132)	0 (3)	39 (74)	46 (118)	1 (8)
Grimsbury Reservoir	(7)	87 (266)	0 (1)	4 (32)	0 (4)	12 (72)	32 (108)	2 (10)
D: COTSWOLD								
Buscot Lake/Reservoir	(12)	325 (510)	5 (50)	5 (80)	0 (2)	34(110)	19 (90)	0 (2)
Cotswold Water Pk East	(10)	317 (650)	64(220)	98(400)	3 (15)	390(550)	909(1687)	42 (90)
Cotswold Water Pk West	(8)	623(1200)	164(360)	272(750)	7 (25)	477(800)	978(1610)	108(140)
Bourton GPs	(7)	111 (300)	4 (75)	9 (50)	0 (2)	106(200)	93 (200)	1 (15)
E: SWINDON								
Coate Water/Queen's Pk	(12)	256 (490)	10 (91)	0 (50)	6 (45)	10 (40)	34 (100)	3 (13)
Stanton Lake	(7)	166 (250)	4 (12)	0 (0)	0 (2)	10 (19)	3 (10)	0 (5)

the flocks are often the largest in the county. There is also a regular flock of 5-15 Whooper Swans, and Bewick's Swans occur at times in small numbers, although they have once topped 100. The only major decrease has been in the numbers of Wigeon, from a former level of 2-300 to less than 70 nowadays. The Pintail and Shoveler, which used to occur in parties of up to a dozen, have likewise become increasingly scarce.

The floodplain of the River Ray at Ot Moor is another area which was once a major stronghold. The meadows here are set in a saucer of land some 3-4km across, the whole of which was a permanent swamp until the start of the 19th century. Since then a succession of drainage schemes have robbed the site of much of its interest. A certain amount of flooding still occurs in some seasons, and at times may be quite extensive, but even when conditions seem ideal there are seldom more than 150 ducks. In the seasons 1965-1976 the regular (and maximum) levels amounted to: Mallard 25 (60), Teal 50 (250), Wigeon 5 (65), and Mute Swan 5 (25). Up to 15 Pintail and 30 Shoveler were also seen.

On the rising ground nearby the 18th century decoy pool at Boarstall was formerly dependent on the Ot Moor floods for its working lead of ducks, and is still a favourite day-time retreat for several hundred Mallard and Teal (Table 69). In the 1880s, some 50 years after the initial drainage at Ot Moor, the decoy was still taking an average of 800 ducks a year, and as many as 2,500 in one exceptional season (462). This continued on a lesser scale until 1925 or thereabouts, and by 1963 the place had long since fallen into disrepair. The decoy was then leased by WAGBI and restored to working order, the catch being kept for breeding stock in connection with their duck rearing scheme (p.518). During this period, which lasted until 1979, the pool was heavily fed to maintain a good lead of birds, and this may perhaps account for the dearth of ducks on Ot Moor. The decoy and the surrounding woodland is now owned by the National Trust, and is managed as a reserve by the Berks, Bucks and Oxon Naturalists' Trust (BBONT).

Further to the east, between Boarstall and the Chiltern ridge, the main points of interest are the ornamental lakes at Wotton Underwood and Eythrope Park, and the reservoirs at Weston Turville and Tring. The Calvert Brick Pits to the north are also well favoured, and are usefully placed within reach of the Ouse Valley lakes around Buckingham (Table 69).

The 20ha lake at Wotton has several secluded creeks and islands, and although much disturbed by anglers in boats, is able to hold some hundreds of ducks for the greater part of the winter. It is also the local centre for a new and flourishing group of Canada Geese. These were introduced in 1974, and by 1978 had increased from 23 to their present, stable level of 80-100. At first they remained in the area throughout the winter, but latterly they have started to wander, and the recent parties of 20-30 at Eythrope and Tring have no doubt stemmed from here. Some 2-300 ducks, mostly Mallard, are also found at Eythrope, on the quiet stretch of the Thame which runs through the park, and on the landscaped arm which branches from it.

The canal-feed reservoirs at Tring and Weston Turville are set at the foot of the Chiltern escarpment, about 3km apart, and the water drawn from the chalk is rich in nutrients, encouraging the growth of plants and invertebrates. At Tring there are four adjacent pools with a total area of 77ha, and a maximum depth of around 6m, though this is subject to considerable draw-down (492). The banks, which are mostly natural and flanked by a belt of reed-swamp, were declared a National Nature Reserve in 1955. This arrangement, which in effect protects the water as well as the surrounds, has benefited both the wintering and the breeding population of wildfowl. Amongst the species which nest here regularly are Mallard, Shoveler, Tufted Duck and Pochard, and occasionally a pair of Teal. Ruddy Ducks have also bred, beginning in 1965, but are not yet fully established (542). The winter wildfowl population, with a current peak of 6-800, is substantially larger than it was in the 1950s. Pochard and Tufted Ducks have both trebled in numbers, and Shoveler have increased from a regular level of less than 20 to about 100 in recent seasons. There are also regular records of 5-10 Gadwall and Goldeneyes, and of feral Greylags and Canada Geese in parties of 20-30 (max 40-50).

The 9ha pool at Weston Turville (an SSSI) is equally attractive, with natural banks and extensive reed-beds, but is much disturbed by sailing and fishing. Even so it often holds 100-150 ducks. In the seasons 1970-1982 the regular population comprised 12 Teal (max 42) and 20-30 apiece of Mallard, Shoveler, Tufted Ducks and Pochard (max 60-90).

At the Calvert brickworks, near Steeple Claydon, there are two separate claypits, covering 11 and 18ha. Since 1978 the smaller of these has been maintained by the County Naturalists' Trust (BBONT) as a "Jubilee" nature reserve, and the numbers of ducks have risen from a mean peak of 200 or so in the early and mid 1970s to about 500. The other larger pool is used for sailing, and is usually deserted, at any rate by day.

To the north of the Oxford complex there are several outlying resorts in the upper Cherwell Valley, around Banbury, but none is of any great importance. The lakes at Astrop, near King's Sutton, and Farnborough Park each support a regular level of 125 Mallard; Grimsbury Reservoir on the outskirts of Banbury has a similar but rather more varied population (Table 69), and another six resorts, comprising the canal reservoirs at Byfield, Clattercote and Wormleighton, and the pools at Edgcote, Farthinghoe and Canons Ashby, each hold a regular total of 40-80 ducks (maxima 100-200). Mallard are predominant, except at Clattercote, where the bulk of the birds are Pochard and Tufted Ducks, and Wormleighton, which can hold substantial numbers of Teal (up to 110) when water levels are low, and which has a regular population of 27 (max 80) Tufted Ducks. All other species are scarce or absent.

The Cotswold Water Park, in the western part of the region, is a new and exciting development, which has steadily gained in importance over the past 20 years. It comprises two separate stretches of low-lying land along the Thames and Coln, amounting in all to 5,700ha. Throughout these two sections, one centred on Fairford, and the other on South Cerney, there are beds of alluvial gravel overlying a layer of impervious clay, and the water table is seldom more than 60cm below the surface.

The digging of gravel began here in the 1920s, and went on sporadically until the mid 1950s; the rate of production was then greatly increased, and by 1983 no fewer than 96 separate pits had been dug, with a total area of 842ha. The eventual area, envisaged in the plans, will be twice as great, but not perhaps for another 20 years. The abandoned pits, which are seldom more than 6m deep, are rich in calcium dissolved from the limestone gravel, and develop into marl lakes, with a characteristic abundance of aquatic plants and invertebrates. Belts of willow and alder have also grown up around many of the older pools, providing both shelter and a screen from passers-by (264).

The purpose of the park is to serve as a leisure centre, and many of the larger pools are now allocated for sailing or fishing, or for pursuits, such as water-skiing, which require a space of their own. Plans have also been made for a nature reserve, covering upwards of 100ha, with facilities for education and experimental management (135). However, this ambitious and much needed project has not yet been realised, and for the time being the only formal reserves are two small pools, administered by the Gloucestershire Trust for Nature Conservation. Several pools in private ownership are also kept as refuges, and others again are hard to reach, and virtually free from interference. In consequence the birds disturbed from the active pits are able to find alternative roosts and feeding grounds close by, even on the busiest days.

Because of the frequent movement of the flocks the population in each of the two sections of the park is best considered as a whole, and the records are combined accordingly (Table 69). In the eastern sector there are currently 25 pools, disposed in three distinct groups around Fairford, Dudgrove and Lechlade. The total area of water is 220ha, and the population reaches a normal peak of 1,600-1,800 ducks. Regular totals of 3-400 ducks are also found on the lake and reservoir in Buscot Park, a short way to the east. The lake, which is much the more favoured of the two, was acquired by the National Trust in 1956, and for many years has served as a virtual sanctuary.

The western part of the Water Park has a total of 71 pits, with an area of 622ha. These lie in one continuous chain, stretching for more than 7km, from Poole Keynes in the west, past Somerford and Ashton Keynes, and on to South Cerney and Cerney Wick. The sewage farm at Shorncote is also a part of the complex, and the records from there are included in the western sector totals (Table 69). As in the eastern sector, the pits are mostly steep-sided, and the wildfowl population, currently reaching 3,000, is largely composed of Mallard and diving ducks.

Pochard are especially numerous, and continue to increase both here and on the eastern pits. In the park as a whole their numbers have risen from a regular level of 1,550 between 1973 and 1976 to 2,100 between 1977 and 1981, and 2,400 in 1982. Tufted Ducks breed freely in both parts of the park, and their winter numbers have increased from a regular total of 680 in the earlier period to a current level of about 930. The breeding population now totals at least 40 pairs, and sometimes as many as 80, but less than a quarter of the 2-300 ducklings hatched each year appear to survive to fledging. Parties of Goldeneyes, totalling 20-30 (max 50), are also recorded regularly, mostly on the newer pits, and in recent years a feral flock of 15-20 Red-crested Pochard has become established in the western park around Poole Keynes. One or two pairs are now breeding annually, but the number of fledglings is again disappointingly small.

The dabbling ducks are less well suited to the pits, many of which are steep-sided, and useful only as roosts. Mallard are common enough, with a regular total of nearly 1,000, but the individual gatherings are mostly quite small, and rarely exceed 200. The Wigeon and Teal are usually found on the pools around Ashton Keynes and South Cerney, and on Shorncote sewage farm, which provides a convenient and secluded feeding ground. They also favour the

Lechlade and Dudgrove pits in the eastern part of the park, but are generally scarce elsewhere. Gadwall first started to appear in the early 1970s, probably from Slimbridge, and have since increased to a regular winter total of 20-30, with occasional peaks of up to 45. Their prime resort is around Poole Keynes, but a few may at times occur on some of the other pits, including those to the east. In summer they move elsewhere, except perhaps for an odd bird or two, and they seldom return in strength until October or November. There are no reports of breeding.

The Water Park also supports a substantial breeding and wintering population of Mute Swans and Canada Geese, and a small, but regular, flock of feral Greylags. The swans increased considerably in the early 1970s and in recent years have reached an autumn total of 150-200. Later on in the winter the flocks disperse, probably to the floods along the Cotswold streams, and the numbers drop to less than 50. Some 10-15 pairs breed annually and together rear 25-35 young (214).

The Canada Geese are descended from the local flock of 30-40, which was introduced at Buscot at some stage prior to 1953 (57). By the early 1970s odd pairs were nesting annually within the park, mostly on the eastern pits, and in 1976 a flock of 50 was present at Dudgrove during August and September. Since then their numbers have increased to above 200, and 8-10 pairs are now breeding regularly, giving an annual recruitment of 30-40 young. The Dudgrove and Fairford pits are still the main resort in summer and early autumn, but thereafter the birds disperse, and in recent years the western pits have begun to attract a wintering flock of 60-100.

The Greylags first appeared in ones and twos in the early 1970s, mostly in late winter, and in 1974 a pair stayed on to breed. Since then odd pairs have bred on at least three more occasions, and a flock of up to 25 is now present throughout the year on the pits around South Cerney. From the rather variable numbers, it seems that some may occasionally move elsewhere, but the alternative sites have not yet been discovered.

To the north and west of the park the land rises to the dry limestone plateau of the Cotswold Hills, and the wildfowl are strictly confined to the deepset, and often marshy, valleys. The largest numbers, amounting to about 300, are centred on the gravel pits at Bourton-on-the-Water, and several hundred more are scattered in numerous small parties along the length of the Thames and its tributaries. Counts on the 15km stretch of the Windrush between Witney and the Gloucestershire border have shown a normal winter total of at least 100 ducks, with occasional peaks of above 200 in times of extensive flooding. There are also records of 50-100 ducks on the Sherborne Brook, a short way upstream, and of similar numbers on the

Table 70. *The total numbers of wildfowl (excluding Canada Geese) recorded regularly at a further 29 resorts in the upper Thames basin. The figures after the place names are the highest totals recorded at each site. Except at sites marked *, the counts were made exclusively in the period 1970-1982; "d" indicates a preponderance of diving ducks. The sites are arranged by districts, as in Table 69.*

Regular totals					
25–50		51–100		101–250	
A:Cote GP, Bampton	160	South Hinksey Pool	110	Dorchester Lake	185
				Appleford GP*	375
				Standlake GPs	410d
B:Witney GP	50	University Pk, Oxon	150	Port Meadow, Oxon	430
Windrush:Ducklington	70	Shotover Pk, Wheatley	150		
Cornbury Pk, Charlbury	60	Otmoor	390		
Bunker's Hill Quarry	140				
C:Aynho Pk	70	Farthinghoe Lake	100	Weston Turville Resr	210
Clattercote Resr	140d	Edgcote Pk	205	Astrop Pk	295
Canons Ashby	175	Byfield Resvr	155	Farnborough Pk	380
		Wormleighton Resvr	125		
D:Windrush:Sherborne	90	Windrush:Witney–Burford	250		
Coln:Bibury	65	Leach:Eastleach	90		
E:		Lyden Pool, Swindon	190		
		Shaftesbury L, Swindon	120		

Coln at Bibury, and at two separate points on the Leach. The great majority of these are Mallard, except at times in the Windrush Valley, where transient flocks of 30-40 Teal and Wigeon have been reported in several recent seasons. Totals of 20-30 Mute Swans (max 45) and a few Tufted Ducks and Canada Geese are also recorded regularly, mostly from December onwards.

Further to the south, around Swindon, there are several small lakes and man-made pools, which together hold 5-600 Mallard and a sprinkling of other species. The largest gatherings, each totalling more than 150, are on Coate Water, Stanton Lake and Queen's Park Lake, and several other places, such as Shaftesbury Lake and the Lyden lagoon, have regular flocks of 50-100 birds (Table 70).

Coate Water was enlarged to 49ha in 1974 by the excavation of a new lake on its eastern side, and is now a nature reserve. Three or four pairs of Tufted Ducks and Canada Geese are now breeding regularly, and the latter are sometimes reported in flocks of more than 100. As in the Cotswold Water Park and the Kennet Valley, some 15km to the north and south, the geese here have doubled in numbers in recent years, and it may well be that the flocks are interrelated. Teal were also plentiful at one time, occurring regularly in flocks of over 100, but in 1963 the water level was raised and stabilised, and they lost the muddy shallows in which they used to feed. Following this their numbers fell abruptly, and flocks of more than 10 are now rare. The Mallard were much less affected, and the diving ducks, being suited to the deeper water, have shown a slight increase.

East and central England

Norfolk

Norfolk is well situated to receive migrants from the east, and in hard weather is of special importance. Its diversity of habitats includes the harbours and the saltmarshes of the north coast, considerable stretches of which are safeguarded by the National Trust and by Ramsar designation (p.538). Inland the marshes and fens along the Yare and around Breydon Water have all but disappeared and threats to the Broads from enrichment and human pressure continue. Few inland sites in this important wetland area are entirely shielded from these effects.

Norfolk Broadland

From Lowestoft northwards to Sheringham the coast has long open beaches backed by low cliffs, sand-dunes or town frontages and is thus of little interest to wildfowl. The Lothing Lake in Lowestoft once had up to 150 Mute Swans, but there are no recent counts. Oulton Broad, behind the town, is largely sterilised by disturbance, particularly sailing, and has only up to 60 Mallard, 25 Teal, 30 Pochard and 63 Tufted Ducks.

The course of the Waveney River from its source to Diss has little of interest to wildfowl, though Frenze Hall just east of Diss carried up to 490 Mallard and 46 Teal in the early 1960s. On its eastward and northward course to Breydon Water, between Lowestoft and Great Yarmouth, the river receives water from Fritton Decoy (now a reservoir for Lowestoft), which used to hold a thousand dabbling ducks and had a productive decoy pipe. Pochard were also caught in drop nets. There are no recent data on wildfowl numbers and catching is understood to have ceased. There are several other waters and areas of wet grassland along the Waveney Valley but unfortunately data are completely lacking.

Fig. 22. East and central England. Total wildfowl and regional boundary.

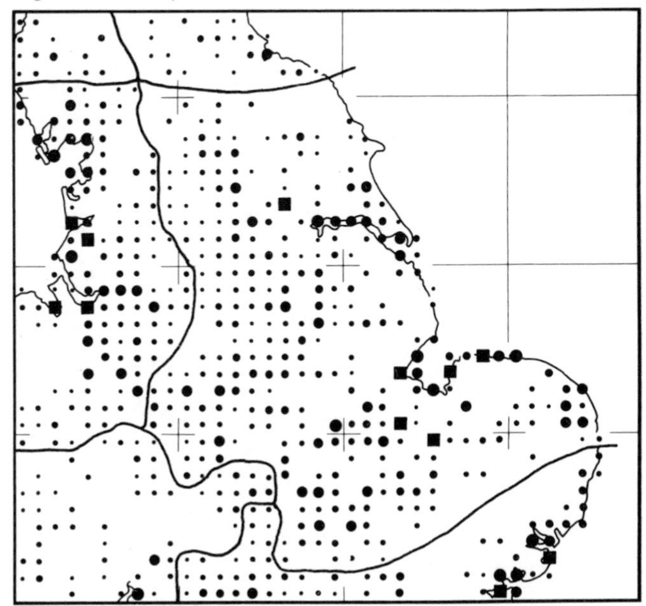

Breydon Water, the combined estuary of the Waveney, Yare and Bure, lies behind Great Yarmouth. The tidal flats of this 453ha LNR (Table 71) carry quite substantial flocks, particularly of Wigeon and Shelducks, though Wigeon only reach the thousands in severe weather. There is a pre-moult migration gathering of Shelducks, in June and July. Most species have held their numbers or increased, except the European White-fronted Geese, which are now sporadic. These used to frequent the adjacent Halvergate Marshes, along with Pink-footed Geese, in flocks of several thousands. The Pinkfeet all but disappeared during the period of general withdrawal from eastern England in the 1960s and 1970s. The Whitefronts have declined as more and more of the wet splashy marsh has been drained to provide dry pasture or even arable crops. Two-thirds of the Halvergate "triangle" has recently been threatened by further drainage, but a significant advance was made when goverment drainage subsidies were refused on conservation grounds; moreover agreement appeared to have been reached with private owners not to drain and plough at their own expense (378). However, this did not entirely have the intended result, since several owners expressed a renewed wish to convert their land to arable. The outcome has been the setting up by the Countryside Commission, in 1985, of the Broads Grazing Marsh Plan, providing an agreement for safeguarding the "core" area; the remainder has been the subject of management agreement under the Wildlife and Countryside Act (1981). The future of the Halvergate Marshes now looks much brighter. Flocks of Mute Swans, some Bewick's Swans (which largely graze winter wheat) and Wigeon are now the main users of the grassland. A threat to Breydon Water itself, and the whole Broads ecosystem – a proposal for a tidal flood barrage at Yarmouth – was abandoned in the mid 1970s as uneconomic, but a surge barrier to prevent flooding on major surges was in the offing until April 1984, when that too was abandoned for financial reasons. In the meantime, however, much of the land had been "improved" in response to the demand for greater agricultural returns.

The drained marshes extend up the Yare as far as Rockland St Mary, but there, on the opposite bank, is the 243ha RSPB reserve of Strumpshaw Fen, where

Fig. 23. The Norfolk Broads.

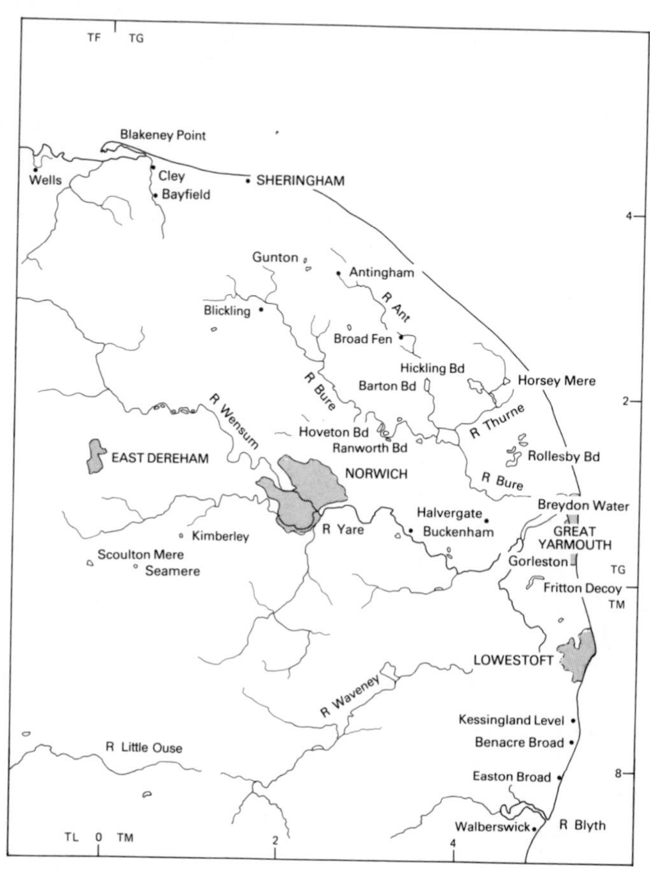

intensive management is endeavouring to restore and maintain some of the original habitat. Embankments prevent the nutrient-enriched river from flooding into the reserve and old water areas are being re-opened. Nearby in the Buckenham Marshes the only sizeable flock of Bean Geese in England received full protection under the 1981 Wildlife and Countryside Act. In the 1960s the maximum numbers in the Yare Valley fluctuated between 25 and 51, and in the 1970s between 66 and 141, the last figure being reached during the hard weather of early 1979, when the numbers in Norfolk as a whole were over 400. There was another sharp increase in the cold winter of 1981-82 and in the last two winters numbers have exceeded 300 (p.365). The numbers are now at the highest recorded since 1939 – partly thanks to control by the RSPB of excessive disturbance. A few hundred Bean Geese were normal in the area in the 1920s and 1930s, exceptional influxes, related to severe winters, being 5,000 in 1927 and 1,000 in 1936. Wigeon numbers have also increased recently on these marshes, to a maximum of 7,500 in 1982. At Cantley Sugar-beet Factory and neighbouring marshes just to the east up to 871 Teal, 350 Wigeon and 68 Bewick's Swans were reported in the early 1970s.

In Norwich itself, the University Broad has up to 128 Mallard and 22 Tufted Ducks, both of which also breed there and at the Bawburgh pits 3km upstream. Sites upstream and just south of the Yare include the lake in the grounds of Kimberley Park, on the River Tiffey, with about 400 Mallard and 20-50 Teal, Shoveler, Gadwall, Pochard and Tufted Ducks. Seamere, a circular pond on a small tributary near Hingham, has up to 300 Mallard and little else. Scoulton Mere, 5km to the west, has regular (and maximum) counts of 180 (300) Mallard, 17 (25) Teal, 12 (24) Pochard and 20 (24) Canada Geese. On the River Wensum, which joins the Yare at Norwich, there are gravel pits at Taverham, some 5km upstream, which have held regular (and maximum) numbers of 77 (107) Mallard, 22 (34) Pochard, 12 (25) Tufted Ducks and 11 (17) Goldeneyes. Ten kilometres upriver are the gravel complexes of Lenwade and Lyng, supporting a similar range of species in similar numbers, but each also has records of Canada Geese, with a maximum of 123 at Lyng.

The importance of the Yare basin as a whole to breeding birds has been assessed recently (506). In 1979, 49 breeding pairs of Mute Swans (and 240 non-breeders), 433 Mallard males, 11 probable pairs of Gadwall, 3 pairs of Garganey, 34 male Shoveler (and 7 broods) and 7 male Pochard (and 3 broods) were found. The lower Yare and Waveney and the south side of the lower Bure are of especially high ornithological interest and most at risk from drainage schemes.

It is in the catchment of the Bure and its

Table 71. *Regular (and maximum) numbers of wildfowl recorded at the four most important sites in Broadland.*

a) Breydon Water i) 10 seasons 1960-1969
ii) 13 seasons 1970-1982
b) Ranworth/Cockshoot Broads i) 6 seasons 1960-1969
ii) 4 seasons 1971-1982
c) Hickling Broad 8 seasons 1971-1982
d) Horsey Mere 8 seasons 1971-1982

	a i	a ii	b i	b ii	c	d
Mallard	57 (150)	120 (425)	779(1220)	334 (780)	738(1400)	683(2000)
Teal	53 (280)	88 (360)	101 (350)	42 (210)	870(2027)	161 (600)
Wigeon	496(1600)	705(3650)	197 (500)	276(3000)	144 (752)	185 (600)
Pintail	58 (154)	95 (186)	0 (2)	0 (3)	5 (16)	0 (2)
Shoveler	11 (100)	4 (16)	106 (200)	32 (150)	224 (800)	18 (100)
Gadwall	0 (5)	1 (22)	0 (2)	18 (65)	44 (196)	31 (150)
Shelduck	379 (600)	373 (580)	0 (1)	5 (12)	0 (0)	1 (2)
Pochard	1 (16)	2 (32)	12 (138)	51 (140)	96 (220)	14 (50)
Tufted Duck	12 (150)	5 (48)	6 (50)	63 (140)	101 (230)	13 (150)
Goldeneye	3 (19)	19 (70)	0 (12)	2 (5)	25 (60)	5 (9)
Mute Swan	65 (128)	63 (142)	7 (30)	1 (8)	50 (150)	7 (17)
Whooper Swan	0 (5)	1 (7)	0 (0)	0 (1)	22 (88)	12 (50)
Bewick's Swan	18 (91)	71 (229)	1 (13)	0 (70)	20 (141)	18 (120)
Greylag Goose	0 (1)	9 (46)	100 (227)	71 (130)	10 (52)	3 (15)
White-fronted Goose	132 (500)	29 (176)	1 (19)	0 (1)	1 (17)	14 (60)
Bean Goose	16 (56)	58 (103)	0 (0)	0 (0)	0 (0)	4 (50)
Canada Goose	1 (7)	1 (12)	16 (40)	16 (70)	56 (136)	3 (17)

tributaries, the Ant and Thurne, running from the north into Breydon Water, that most of the Broads are to be found. The lakes originated as pits dug to remove peat for fuel, mainly from the 11th to 13th centuries. With changes in land and water levels these became flooded and started on the slow succession back to land through siltation and plant growth, the loss of open water speeding up as the areas became shallower. The Nature Conservancy (368) recognised 42 Broads which in the 1880s had 1,220ha of open water. This had shrunk to 685ha by the 1950s, and at that time 14 Broads were less than 1m deep. Of the remainder, only 16, with a water surface of 383ha, were open to navigation. At one time the encroachment of vegetation was held in check by commercial reed-cutting, but this has now become uneconomic. Among the "natural" retardants of vegetation are the South American Coypus (*Myocaster coypus*) which escaped in the 1930s and 1940s from Nutria fur farms and became established in Broadland. They cleared substantial areas of vegetation (on Ranworth Broad open water increased by 22% in 17 years) but also became agricultural pests, which led to campaigns for their elimination. It is apparent that intensive management, including reed-cutting, mowing, grazing, coppicing, mud-pumping and ditching, will be necessary to "freeze" portions of the Broads at one or more stages in their succession.

These shallow, slow-moving waters are particularly open to pollution, from the runoff of agricultural fertilisers and sewage effluent from the local communities and the seasonal influxes of holiday-makers. Improved drainage of soils in the neighbourhood may also result in the release of ferric hydroxide which is particularly destructive to aquatic invertebrates. In the last twenty years many Broads have lost all or most of their waterweeds and a diverse invertebrate fauna has been replaced by worms and chironomid larvae. The traditionally clear water is now turbid with phytoplankton. Most of these changes result from eutrophication, and the polluting inputs must clearly be rigorously controlled if the ecosystems are to be saved. Recent outbreaks of botulism, killing many hundreds of ducks, are indicative of the unhealthy state of affairs. Additionally the disturbance by the vastly increased numbers of holiday-makers with their boats and other paraphernalia causes a severe impact which can only be made tolerable to wildlife by rigid zoning arrangements whereby a few sanctuary areas are retained. At least boat traffic has been exonerated from major blame for the increased water turbidity (related rather to phytoplankton numbers), but the wash from craft undoubtely causes bank wear and vegetation damage (357). The excavation of flight ponds, heavily fed, has also been significant over recent years.

In the face of so many inimicable developments it is not surprising that the general impression is of a reduction in the wildfowl populations. Unfortunately detailed series of counts are only available in a few instances.

Of the Broads associated with the Lower Bure, South Walsham, Hoveton Little and Decoy have regular totals of under 50 wildfowl. The counts for the Ranworth/Cockshoot Broads (Table 71) reveal a considerable decline in Mallard, Teal and Shoveler; among the dabbling ducks only Wigeon (feeding away from the Broad) have increased along with the newcomer, Gadwall. Rather suprisingly the diving ducks have also increased. The Greylag flock was introduced at Ranworth in 1935 and totalled 3-400 at the end of the 1950s; its subsequent decline is unusual. These two Broads, along with Woodbastwick Fen Decoy and Hoveton Great Broad, comprise the 451ha Bure Marshes NNR, which is also listed as a 412ha Ramsar Site. Ranworth/Cockshoot have been owned since 1949 by the Norfolk Naturalists' Trust, the pioneer of the voluntary County Naturalists' Trust movement. The Trust was formed as long ago as 1926, and now holds a wide range of valuable reserves in the county. Ranworth is now the home of their Broadland Conservation Centre. Recently Cockshoot Broad has been cut off from nutrient-rich river water and much of its phosphorus-laden sediment removed. Its subsequent development will be followed with interest. Nearby, Burnt Fen Broad had regular (and maximum) counts of 77 (95) Mallard, 22 (25) Teal and 19 (30) Tufted Ducks in the early 1970s. Salhouse Broad, 2km to the west, held substantial goose flocks in the 1960s, with regular numbers of 172 Greylag and 42 Canada Geese, but these have also declined, to 59 and 23 respectively. Wroxham Broad nearby has 10-40 Mallard and Tufted Ducks regularly, but up to 400 Greylag and 150 Canada Geese. Upstream on the Bure there is little of interest until, near Aylsham, the lake on the National Trust property of Blickling Hall (Table 72). The Canada Geese derive from the Norfolk population focused on Holkham and there are also up to 30 Egyptian Geese. Also feeding into the upper reaches of the Bure are the lakes of Gunton Park, only 7km inland from the coast at Cromer (Table 72). The Gadwall at Gunton Park are particularly outstanding and, since the large numbers are only present in September and October, the probability is that the lakes are being used as a moulting site for much of the county's population. There has been a significant decline in 1983-84, however, due apparently to increased disturbance, but further research is needed (see also Rutland Water p.159). Only 4km away, though actually on the Ant headwaters, Antingham Lake has many fewer wildfowl.

The Ant turns into the Bure from the north and has a range of marshes and Broads along its lower reaches between Ludham and Stalham. The most important is Barton Broad, a reserve of the Norfolk Naturalists' Trust, though with a navigable channel through it and subject to the full range of disastrous impacts summarised earlier. However, sewage (especially phosphate) input is being reduced. The enrichment of the water no doubt explains why it supports, despite its relatively large size (about 135ha), fewer wildfowl than might be expected for a water in this region. A private ringing station was started at the 4ha Crome's Broad at How Hill (Table 72), which has adjoining marshland, in 1936 by the Messrs Boardman. Between 1954 and 1965 they caught, and ringed on behalf of the Wildfowl Trust, over a thousand Mallard and a hundred Teal, mostly in late winter. The Mallard gave an unusually high proportion of overseas recoveries. Three other counted sites nearby hold only small numbers of wildfowl, but further upstream, near Stalham, Broad Fen had good numbers in the 1960s (Table 72), but there are no recent counts. At the beginning of the century Sutton Broad was almost entirely open water equipped with its own Freshwater Laboratory. Now its 170ha form one of the richest areas of vegetation in Broadland (492) and only a narrow navigation channel remains.

At Thurne, the tributary of that name runs into the Bure from the north-east and supports the other Ramsar Site in Broadland, the Hickling Broad/Horsey Mere/Heigham Sound/Martham Broad complex of 892ha. Hickling Broad is a National Nature Reserve, part owned and part leased by the Norfolk Naturalists' Trust; Horsey Mere is a National Trust property, managed as a nature reserve. Despite unrestricted boating (but strictly controlled access to the marshes) Hickling Broad supports a substantial wildfowl population (Table 71) and is the most important site in the area for Teal and Shoveler. Its Whooper Swans are also exceptional and Bewick's Swans often exceed 100 on return passage or after severe weather overseas. There are very few counts from the 1960s, but those of the 1950s indicate that there has been a drop in the Mallard, then with regular (and maximum) levels of 1,150 (3,000), Tufted Ducks 225 (500), Goldeneyes 55 (90) and especially Mute Swans 300 (400). In the 1920s the then owner organised massive Coot shoots, the birds being accused of destroying aquatic plants. Many thousands of Coots were killed but plant eradication (and through that reduction in the numbers of Coot and Pochard) has been accomplished far more effectively by over-enrichment with phosphates and nitrates leading to algal blooms and turbidity. The alga *Prymnesium parvum*, which is part of the summer bloom, is toxic to fish and outbreaks cause many deaths. In hot summers botulism takes a toll of water birds. The massive winter roost of Black-headed Gulls is a major cause of this enrichment. The gull roost has been smaller in the early 1980s and waterweeds are making a comeback – hopefully to be followed by a recovery of the numbers of wildfowl.

Horsey Mere, lying within a couple of kilometres of the coast, is distinctly saline and also has an exceptionally high calcium level. It has more submerged vegetation than Hickling and, although a fifth of its size, carries a not dissimilar population of wildfowl (Table 71). Limited counts from the 1960s indicate that dabbling ducks were then fewer, with regular (and maximum) counts of 306 (600) Mallard, 30 (80) Teal and 66 (150) Wigeon, but diving ducks more frequent – 100 (250) Pochard and 100 (150) Tufted Ducks. Martham Broad, 2km to the south, is of little interest to wildfowl, but has achieved conservation fame in that government money has been forthcoming to arrange that water contaminated with ochre and salt

Table 72. *Regular (and maximum) numbers of the commoner species in six of the less important sites in the Broadland and in Bayfield Park in north Norfolk. The numbers of seasons in the period 1970-1982 is given in brackets after the site name. * indicates early 1960s counts.*

	Mallard	Teal	Wigeon	Shoveler	Gadwall	Pochard	Tufted Duck	Greylag Goose	Canada Goose
Blickling Lake (6)	74 (300)	3 (21)	1 (18)	3 (21)	0 (2)	4 (22)	22(63)	1 (5)	131(354)
Gunton Park (9)	260 (700)	178(450)	3 (35)	19 (59)	228(630)	7 (28)	8(30)	97(250)	137(457)
Antingham Lake (2)	98 (210)	28(104)	0 (1)	0 (0)	11 (37)	0 (7)	1 (3)	0 (0)	0 (4)
Barton Broad (4)	159 (470)	59(270)	26(300)	3 (11)	3 (13)	36(187)	87(78)	55(103)	42(130)
Crome's Broad (2)	364 (550)	218(270)	88(200)	76(120)	35 (50)	28 (40)	23(30)	79(140)	47 (70)
Broad Fen (4)*	577(1200)	59(100)	6 (25)	8 (16)	0 (4)	0 (0)	0 (0)	24 (55)	83(200)
Bayfield Park (9)	141 (350)	28(120)	2 (10)	1 (8)	3 (15)	1 (8)	11(45)	18(115)	52(200)

will be pumped and channelled away downstream, because this is part of a Ramsar Site and international responsibilities are involved and have been respected.

The remaining group of Broads lies on the small tributary Muck Fleet to the south, entering the Bure from the north-east near Stokesby. In sequence there are Filby, Rollesby and Ormesby Broads. These are much deeper (up to 3m) than other Broads, and together serve as a stand-by water supply reservoir. They are used for angling and sailing and apparently carry few wildfowl, though counts are only available for Rollesby, which in 1982 had up to 67 Mallard, 60 Teal, 18 Pochard, 34 Tufted Ducks, 27 Greylags and 10 Canada Geese.

The north Norfolk coast

From Sheringham, 40km westwards to Hunstanton, the north Norfolk coast is a nearly continuous strip of prime wildfowl habitat. This was recognised by its designation in 1976 as a Ramsar Site of International Importance, comprising 5,559ha of inter-tidal sands and muds, shingle, sand-dunes and saltmarshes. From east to west are included the marshes of Salthouse, Cley, Blakeney, Holkham and Scolt Head.

It was to purchase the Cley Marshes that the Norfolk Naturalists' Trust was formed in 1926. Their 176ha are still owned and managed by the Trust; along with the Salthouse Marshes (81ha) they are covered by statutory Bird Sanctuary provisions. The wildfowl populations (Table 73) were higher in the 1960s than in the 1950s, but have since declined quite markedly for no obvious reason except perhaps increased recreational disturbance. Only Pintail, Canada Geese and Dark-bellied Brent Geese have increased markedly (in line with national and regional trends) and the numbers of Wigeon have also risen. The two geese reflect upsurges in their populations as a whole and are best considered on a county basis later, because there is movement, at least of Brent, between the various harbours. Blakeney Point (445ha) is a property of the National Trust (as are the Morston and Stiffkey Marshes to the south) and is managed as a nature reserve. Counts are limited but in 1982 there were up to 340 Mallard, 165 Teal, 800 Wigeon, 270 Pintail, 360 Shelducks, 90 Goldeneyes, 27 Eiders, 10 Long-tailed Ducks, 28 Red-breasted Mergansers, 17 Greylags and 3,200 Dark-bellied Brent Geese. Bayfield Park, 4km inland on the banks of the Glaven, is included in Table 72.

The unreclaimed saltmarshes between Blakeney and Wells and the reclaimed marshes west to Bur-

Table 73. *Regular (and maximum) numbers of wildfowl at three sites on the north Norfolk coast.*

a) Salthouse/Cley i) 2 seasons 1962-1965
ii) 8 seasons 1971-1982
b) Holkham Lake i) 3 seasons 1965-1969
ii) 9 seasons 1970-1980
c) Scolt Head i) 2 seasons 1961-1969
ii) 13 seasons 1970-1982

	a i	a ii	b i	b ii	c i	c ii
Mallard	490 (850)	286 (600)	850(1250)	605(1300)	305 (450)	526(1000)
Teal	758(2000)	582(1168)	3 (20)	8 (50)	67 (400)	30 (450)
Wigeon	2250(4500)	2848(5000)	252 (800)	131 (420)	623(3000)	1079(6000)
Pintail	2 (10)	105 (300)	1 (5)	0 (3)	0 (0)	0 (0)
Shoveler	135 (250)	68 (250)	11 (23)	25 (50)	0 (0)	0 (0)
Gadwall	60 (120)	22 (80)	0 (2)	6 (32)	0 (0)	0 (0)
Shelduck	237 (400)	77 (185)	0 (2)	2 (20)	521 (700)	633(1200)
Pochard	0 (0)	0 (0)	13 (128)	25 (75)	0 (0)	0 (0)
Tufted Duck	31 (180)	2 (21)	0 (2)	28 (100)	0 (0)	0 (0)
Goldeneye	53 (250)	2 (9)	1 (6)	6 (31)	36 (100)	53 (75)
Common Scoter	0 (0)	21 (500)	0 (0)	0 (1)	0 (0)	0 (0)
R-b Merganser	21 (80)	0 (1)	0 (0)	0 (2)	1 (2)	17 (84)
Mute Swan	10 (20)	10 (18)	0 (2)	19 (143)	0 (0)	0 (0)
Bewick's Swan	11 (40)	1 (14)	0 (0)	0 (11)	0 (0)	0 (0)
Greylag Goose	7 (17)	9 (70)	24 (40)	87 (300)	0 (0)	7 (100)
White-fronted Goose	5 (18)	2 (15)	25 (85)	0 (2)	0 (0)	0 (0)
Pink-footed Goose	0 (2)	1 (11)	0 (1)	0 (3)	0 (0)	200(3000)
Canada Goose	56 (200)	151 (240)	774(1725)	428(1500)	0 (0)	0 (0)
Brent Goose	14 (85)	516(1800)	0 (0)	0 (0)	600(1300)	1623(4000)
Egyptian Goose	0 (0)	0 (4)	34 (84)	19 (98)	0 (0)	0 (0)

nham Overy form a fine range of habitats, the majority of which are included in Holkham NNR (3,925ha), owned by the National Trust. Prior to the Second World War this area supported large numbers of Pink-footed Geese feeding on reclaimed pasture at Holkham and fields inland to Fakenham. This flock subsequently disappeared, but since 1980 the geese have returned, up to 6,000 being recorded flighting between sugar-beet fields south of Brancaster and roosting grounds at Scolt Head. A smaller flock of over 700 wintered in 1983-84 on the Holkham Marshes, roosting on offshore sandbanks. The reclaimed pasture at Holkham is also a regular wintering ground for European White-fronted Geese, with maxima increasing from 50-80 in the 1960s to 150-250 in the late 1970s and early 1980s.

The origin of the latest upsurge of Pinkfeet in north Norfolk has given rise to some speculation, since they appear from their timing (late arrival and early departure) and habits to be largely independent of those on the Wash. Circumstantial evidence suggests that they could be Spitsbergen birds, coming over the North Sea, rather than part of the Icelandic stock. Some have been seen arriving from the north-east and their departure in February, long before the Wash geese leave, also suggests a different origin. Added evidence is the presence of European Whitefronts, Bean Geese and Barnacles (up to 45 in 1983) among the flocks. It is true that in recent years, as the Spitsbergen Pinkfeet have increased in number, some thousands, present in continental Europe in autumn and spring, have been unaccounted for in December and January despite thorough searches. This all fuels speculation, but we cannot be sure until we have some direct evidence from marking.

Counts along the Blakeney – Wells stretch have been scattered and incomplete. Those at Wells in 1982 returned maxima of only 35 Mallard, 40 Shelducks, 60 Mute Swans and 2,000 Brent Geese. Just inland is Holkham Park, whose lake has been counted for many years (Table 73). There has been a general decline in wildfowl numbers at Holkham but the reasons for this are unclear. Contrary to the national trend the Canada Goose has also decreased. This indicates successful control measures, for they had been increasing from 440 in 1953 to 2,000 in the early 1960s, but the Holkham flock is still the biggest assembly in the county. The north Norfolk total was 1,700 in 1976, with 140 on the Broads and 250 on Breckland waters (405); the Breck birds have been shown by ringing to have connections with Holkham. The main assembly at Holkham is for moulting, much breeding taking place at scattered sites some distance away. The other feral bird for which Holkham is well known is the Egyptian Goose. This was fairly widely introduced in Britain in the 18th century but is now largely confined to Norfolk. A number of places have regular winter counts of Egyptian Geese but usually in single figures; even the maxima only reach double figures at Gunton Park (38), Blickling Lake (30), Wroxham Broad (13), Barton Broad (12) and Salhouse Broad (11). Counts at some other times have reported higher figures. There were 140 at Beeston in 1973 (542) and the Norfolk Bird Report for 1979 says that the species was reported from 29 localities, with 78 birds at Flitcham – Hillington in November, 91 at Burnham Overy in January and 200 at Holkham in July (presumably a moulting assembly). A considerable number can sometimes also be found at Narford Lake, near Narborough. The population of Egyptian Geese is thus probably holding its own at the 3-400 level.

West of Holkham Marshes is another NNR, that of Scolt Head, a 737ha island owned by the National Trust and the Norfolk Naturalists' Trust and leased to the Nature Conservancy Council. In general its wildfowl numbers (Table 73) have increased rather in line with the decreases at Salthouse/Cley, so there may have been a shift along the coast for species such as Shelduck, Goldeneye and Red-breasted Merganser. A small number (up to 50) of Eiders are present most of the year at Scolt Head. The Brent Geese were the subject of a detailed ecological study in the 1950s (490). This is therefore an appropriate place to consider the north Norfolk population of Brent Geese as a whole. The numbers of geese between north Norfolk and the Humber since 1972 are given in Table 74. The numbers in north Norfolk have increased in line with that of the world population (p.243) and have continued to constitute between 4 and 5% of the population, reaching 11,860 (out of 202,000) in January 1983. There does not appear to be any limit imposed by habitat available to the birds in Norfolk as yet. Nevertheless, the Brent here, in common with those in other parts of England, feed as much as 10km inland, though they have been much less studied than in Essex and Hampshire (p.84). The autumn of 1982 was extraordinary in that not only had a remarkably good breeding season (50% young) boosted the population to a record level, but unusual wind combinations brought many geese to Norfolk that would not normally winter here. As a result 25,000 were seen streaming south along the coast just north of Great Yarmouth on just one day, 5th November.

At Titchwell the RSPB have established a reserve of 206ha on an interesting area which had been reclaimed for arable crops but whose sea-defences were breached in the storm surge of January 1953. Natural vegetation had recolonised the area and the habitat was deliberately made more diverse by building a sea wall across the reserve in 1979. This created a brackish marsh, a freshwater marsh and a freshwater reed-bed. Management is keeping these different habitats in their most productive stages. Counts to 1980 had given up to 200 Mallard, 450 Teal, 72 Wigeon, 29 Shelducks, 12 Eiders, 200 Common Scoters and 500 Brent Geese. The Pinkfeet inhabiting this section have been discussed above.

The last coastal marsh in this stretch is that at Holme where the Norfolk Naturalists' Trust has a 162ha reserve, and where there is a bird observatory. Counts in the 1970s at Thornham and Holme gave respective regular figures of 48 and 57 Mallard, 11 and 60 Teal, 8 and 38 Wigeon, 40 and 40 Shelducks, and 192 and 204 Brent Geese. Holme also records up to 153 Canada Geese and both places have odd records of 150-250 Common Scoters, and 70-150 Pink-footed Geese.

To summarise, this stretch of coast is unusually well protected by ownership or agreement through governmental and non-governmental agencies. It is not subject to the onslaught of disastrous features that threaten the very existence of the Broads, but recreational pressure is very high and uncontrolled in places. The effect of this pressure on waders and terns has been emphasised (485) and it must have an impact also on breeding ducks. It is unfortunate that for a well bird-watched area there was, until recently, such a lack of information on wildfowl numbers.

The Wash

This great bay has 23,700ha of sand – silt flats and 2,600ha of saltmarsh (492), the largest such system in Britain. It incorporates the estuaries of the East Anglian Ouse (not to be confused with the Yorkshire river), the Nene and the Welland flowing in from southerly directions, and the Witham from the north-west. The eastern and part of the southern shores are in Norfolk, the remaining southern and western shores are in Lincolnshire. Over the centuries there has been piecemeal reclamation by empoldering and about 470sq km of agricultural land have been abstracted in this way. In the past, new marsh had developed outside the enclosing sea walls, but now the stage appears to have been reached where any further reclamation will result in a net permanent loss of intertidal habitat, since evidence indicates that the low water mark has not moved appreciably for many years. The Lincolnshire County Council has published a detailed subject plan summarising the conflicts of requirements for recreation, conservation and agriculture and making suggestions to resolve them (316). A moratorium on planning permission for reclamation works on the Lincolnshire coast of the Wash has been declared until 1986. This welcome step was the subject of a Public Inquiry in 1983 and was upheld by the Inspector conducting the Inquiry on behalf of the County Council.

In the early 1970s it was proposed to construct a

Table 74. *The maximum numbers of Dark-bellied Brent Geese in north Norfolk (Salthouse – Thornham), the Wash and the Humber in each season 1970 to 1982 (earlier data too incomplete). The totals for north Norfolk and the Wash are the highest synchronised counts.*

				North Norfolk and the Wash		
	North Norfolk	The Wash	Humber	Total	% of UK pop.	% of world pop.
1970		2500				
71		2450				
72	2180	3460		5340	18.9	10.3
73	3860	5930		7850	19.1	9.2
74	2580	3860		5350	16.1	7.5
75	6180	8780		12910	25.8	11.7
76	4740	7800	580	11620	24.1	10.7
77		6420	1280			
78	6100	9090	1310	15180	24.0	10.7
79	8000	11390	1110	14170	17.4	8.5
80	5500	17000	400	22500	33.4	15.3
81	7000	8690	230	15120	25.2	12.9
82	11860	24500	680	19860	21.4	9.8

series of bunded water reservoirs at the southern end of the Wash. The likely effects of such a development on the ecology of the Wash were made the subject of detailed study (118) and this had the advantage that it was possible to fund the gathering of much new information on bird numbers and distribution. The loss of feeding grounds to all or any of the proposed reservoirs would be serious, but for the present the scheme appears to be in abeyance (see also p.483).

Full monthly count coverage has not been achieved on most shores since the 1960s, so the figures in Table 75 for this and other stretches are averages of the annual maxima in the years 1972-1982 and (in brackets) the overall maxima during this period. The wide ranges between average and overall maximum numbers indicate that there is considerable movement between the different sectors. The eastern shore from Hunstanton to Heacham is open with some low cliffs and only the sea ducks are present in any numbers. The Eiders (and other sea ducks, which include, apart from those in the table, up to 100 Long-tailed Ducks) are now fewer than they were at the height of the sudden expansion of their winter range in the 1950s (575). Smaller parties occur along the north coast to Scolt Head and in the stretch from Heacham to North Wooton. This has fringing shallow lagoons, gravel pits and sizeable areas of fresh and saltmarsh, making it the most populous and varied wildfowl resort on the Wash. It is therefore most fortunate that the RSPB were able to establish their Snettisham Reserve of 1,316ha there in 1975. The Mallard population is notable and the Shelduck numbers here alone rate double the criterion for international importance. This part of the coast is also the Wash headquarters of Pink-footed Geese. Following hard-weather movements their numbers have greatly increased over those for the 1960s, when the regular (and maximum) numbers were 418 (1,348). This marks a reversal of the

Fig. 24. The Wash, Fenland and Breck.

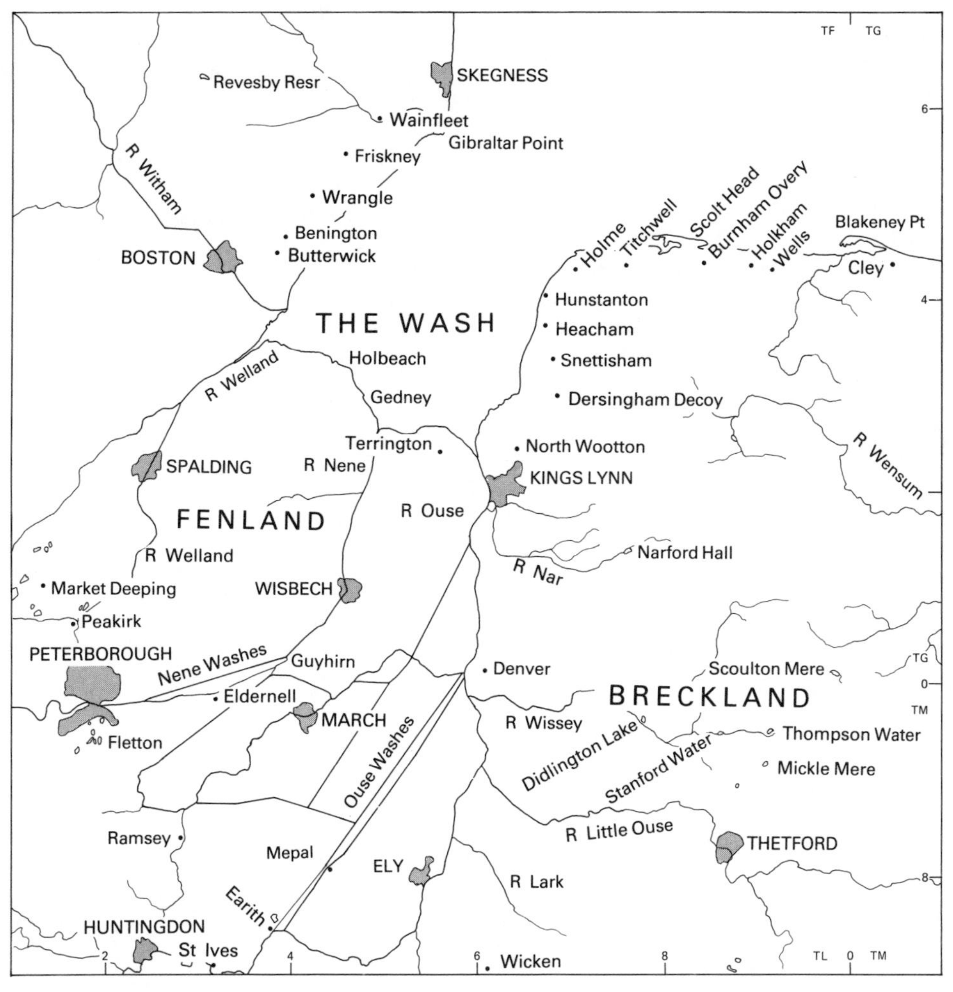

trend away from eastern England also noted on the Humber (p.162). There has also been a shift towards the eastern side of the Wash compared with the position in the 1950s.

Inland from Snettisham is Dersingham Duck Decoy, restored by private enterprise in 1966 (307). In association with the Wildfowl Trust, nearly 1,500 ducks, mostly Mallard and Teal, were ringed in a year there and in traps in an adjacent creek. However, the struggle to maintain water in the pool, standing now above the shrunken fen, became too much and catching was again abandoned by 1971. There are no recent counts from the decoy, nor from Sandringham, where the lakes held regular (and maximum) numbers of 144 (200) Mallard and 70 (150) Teal in the early 1960s.

The mouth of the River Ouse has relatively small numbers of wildfowl, but the large expanse of Terrington Marshes has more substantial populations. Wildfowl are well spread over the marshes, though the Brent Geese favour the more eastward areas. Across the mouth of the River Nene the Gedney Marshes (a particular target of the reclaiming agencies) are somewhat less important. The adjacent Dawsmere Marshes are very extensive and heavily used as a bombing range, but still support considerable flocks of wildfowl. Numbers build up again as we approach the mouth of the River Welland through the marshes of Holbeach St Matthew and St Marks, while the river mouth itself, included in the Holbeach counts in the table, often has large numbers of dabbling ducks and geese. This south-western part of the Wash is certainly the second major focus for the Shelducks and Brent Geese.

The southern end of the west coast at Kirton has much smaller numbers of dabbling ducks, but the 826ha saltmarsh is extensively used for grazing by Wigeon and Brent Geese. The marshes below the mouth of the River Witham at Frampton, and those to its north as far as Wainfleet, hold substantial concentrations (Table 76). Brent Geese are numerous in the whole area, which is the focus from which the species has spread to other parts of the Wash. The ducks and Pink-footed Geese are commonest to the south, between Frampton Marsh and Butterwick, the Goldeneyes (which take advantage of waste and washings) and Pinkfeet being found mainly off the mouth of the Witham. There is a Lincolnshire and South Humberside Trust for Nature Conservation Reserve (271ha) on Frampton Marsh, which harbours one of the largest Black-headed Gull colonies in Britain (22,000 pairs in 1974, though fewer now). Gibraltar Point, at the northern extremity of the Wash, housed Britain's first statutory Local Nature Reserve, declared in 1952 and managed by the Lincolnshire and South Humberside Trust for Nature Conservation. An NNR of over 400ha was declared there, on the same basis, in 1984. Large numbers of wildfowl occur sporadically around the Point, as indicated by the large disparity between the averages and maxima. The Pink-footed Geese have deserted the area of Croft Marsh behind Skegness, where they peaked at 5,000 in the early autumn during the 1950s.

Table 75. *Average annual maxima (11 seasons 1972-1982) and (in brackets) overall maxima in the counted sectors in the east and south of the Wash.*

a) Hunstanton to Heacham
b) Heacham to North Wootton
c) Ouse Mouth (6 seasons only)
d) Terrington Marshes
e) Gedney and Dawsmere Marshes
f) Holbeach Marshes/Welland Mouth

	a	b	c	d	e	f
Mallard	0 (0)	1737 (3790)	111(360)	592(1235)	171 (308)	356 (646)
Teal	0 (0)	148 (250)	7 (15)	267 (710)	16 (59)	115 (500)
Wigeon	0 (0)	1323 (5040)	30 (80)	1975(9070)	356(1100)	1074(2322)
Pintail	0 (0)	630 (2690)	10 (50)	153 (494)	6 (26)	39 (300)
Shelduck	0 (0)	3797 (7680)	303(753)	4137(8620)	940(2308)	3363(5696)
Pochard	0 (0)	81 (117)	0 (0)	0 (6)	0 (0)	4 (40)
Tufted Duck	0 (0)	99 (157)	0 (5)	1 (12)	0 (0)	1 (10)
Goldeneye	31 (50)	54 (71)	3 (7)	4 (19)	10 (60)	93 (200)
Eider	69 (350)	30 (144)	2 (5)	6 (38)	2 (25)	6 (40)
Common Scoter	743(2000)	70 (306)	2 (10)	10 (67)	0 (0)	0 (0)
R-b Merganser	30 (51)	36 (65)	4 (10)	8 (58)	2 (4)	1 (5)
Pink-footed Goose	0 (0)	3432(10500)	134(800)	114 (800)	178(1000)	1425(2222)
Brent Goose	367 (600)	765 (2300)	47(260)	40 (260)	1090(1895)	1477(6100)
Bewick's Swan	0 (8)	41 (92)	3 (20)	5 (20)	0 (5)	0 (0)

Over such a vast area as the Wash it is obviously difficult to ensure synchronisation of counts on more than a few occasions. But sufficient has been achieved by the diligence and dedication of counters in recent years for definitive statements to be made regarding peak counts for the Wash as a whole. The area has already been shown (485) to be of international importance for eight species of waders. Four species of wildfowl – Shelduck, Dark-bellied Brent Goose, Pink-foot and Wigeon – also fulfil the criterion. For the last seven seasons from 1976 the average (and maximum) peaks for Shelduck have been 14,532 (20,296), far exceeding the 1% level (1,250). The Wash is also of great importance for breeding Shelducks, cautiously estimated by the RSPB at about 500 pairs, while 1,000-1,500 flightless Shelducks have been found (92, 93), making the Wash one of the few areas in Britain where moulting concentrations of this species occur. For Brent Geese (see Table 74) the average (and maximum) peaks have been 12,472 (24,497). Much of their feeding is on the extensive saltmarshes, although *Enteromorpha* is an important intertidal food and there is some farmland feeding. Between 1963 and 1975 the Wash held 7.0 to 10.1% of the then much smaller world population (418), and since then this has varied from 6.1 to 12.1%. As with north Norfolk there is therefore no sign of a decline in the percentage held, which would have indicated that the area's holding capacity had been reached. The Wash is notable in that Brent Geese stay here later than at any other site in Britain, fattening up for their long migration north-eastwards. In the early 1970s increasing numbers, up to 2,000 in 1974-75, stayed on the Wash in late April and many well into May (105). Nowadays spring flocks have increased further – to about 3,000 in 1983. The Pink-footed Goose, with an average (and maximum) peak of 4,395 (6,493), is a species for which the Wash has regained international importance, holding between 4 and 8% of the Iceland/Greenland population in all except one year, when there were as few as 2%. Wigeon have largely moved inland, and are only just present in internationally important numbers on the Wash, with 5,299 (13,634). This was due to an out-of-line maximum in 1976 and without that year we would have 3,910 (8,698). Taking all wildfowl together the Wash has held the impressive average (and maximum) peak totals of 38,846 (53,745) birds since 1976.

The Wash as a whole, and especially its south-east and south-west shores, merits every effort to ensure protection from wholesale agricultural reclamation or usurpation by bunded reservoirs. The construction of these waters would not be absolutely disastrous to Mallard, Wigeon and Pink-footed Geese, which could use them as roosts and feed on farmland as they do already. Such structures would, however, rob Shelducks and Brent Geese, as well as many waders, of a major part of their feeding areas, and the Shelduck has shown less adaptability than the Brent in adjusting to habitat loss.

The Fenland and its fringes

The south-east fringes of the Fenland

The Great Ouse River above King's Lynn is joined successively by the Nar, Wissey, Little Ouse, Lark and Cam/Granta and it is the basins of these tributaries which are covered in the present section. The eastern boundary is that of the Breckland; the southern is the chalk uplands between Newmarket and Royston.

The heaths and grasslands of the Breck cover some 7,650ha and comprise a great diversity of habitats within quite a small compass. These range

Table 76. *Average annual maxima (1972-1982) and (in brackets) overall maxima for eight sectors in the western part of the Wash.*

	Mallard	Teal	Wigeon	Shelduck	Goldeneye	Pink-footed Goose	Brent Goose
Kirton Marsh	17 (55)	14 (50)	95 (500)	89 (315)	0 (1)	65 (250)	149 (700)
Frampton Marsh	132(660)	35(120)	1298(6000)	266 (550)	38(260)	17 (78)	1354(3600)
Witham Mouth – Butterwick	63(177)	62(300)	636(3000)	713(2585)	75(320)	186(1739)	2350(5320)
Benington	29 (70)	8 (17)	52 (135)	478(1150)	5 (15)	12 (70)	718(1050)
Wrangle	13 (30)	4 (21)	48 (420)	198 (521)	0 (1)	0 (9)	903(1760)
Friskney	12 (40)	1 (10)	22 (73)	53 (112)	0 (0)	0 (4)	368 (950)
Wainfleet	56(500)	4 (20)	55 (222)	258 (419)	0 (5)	54 (520)	998(2460)
Gibraltar Point	356(850)	56(224)	96 (500)	111 (400)	3 (11)	243 (350)	279(1550)

from open sand-dunes, through dry heathland, chalk grassland and grassland heath to conifer plantations. In the present context the most interesting places are the rivers and their dammed lakes, the gravel pits and the Breckland meres. The meres are small shallow bodies of water lying in glacial sands and gravel, and fed not by streams but by upwelling ground-water, variations in which give rise to random fluctuations in water level. The level varies over a range of several metres, following that of the ground-water, but lags behind by different periods in each mere. The waters dry out occasionally in the summer months and would be at risk if there were any greatly increased abstraction of water from the underlying chalk.

Along the Nar, a scatter of five gravel pits around Narborough only hold small numbers of wildfowl, but Narford Hall Lake has held a diverse and important assemblage. Gadwall numbers have remained high into November, when the birds on Gunton Park Lake (p.138) have vanished. Narford indeed was the site of the original introduction of this species, a pair being released there in 1850. Count data from this and other less important waters in the Wash and Fenland area are given in Table 77.

The main course of the River Wissey passes Didlington, which has a large lake studded with islands providing breeding sites for ducks and Canada Geese. A tributary of the Wissey, upstream of Didlington, runs through the extremely busy Stanford Practical (i.e. Military) Training Area, where it is impounded to form two large shallow areas of standing water – Thompson Water and Stanford Water. Counts on the latter and on the several small meres within the Training Area have not always been complete, but do provide a fair picture of the populations (Table 77). There appears to have been some reduction since the 1960s, possibly because of greater disturbance. Micklemere, near East Wretham, remains the main centre.

The Little Ouse river and its branches have little to attract wildfowl in any numbers. The park lakes at Euston and Shadwell, near Thetford, held substantial numbers of ducks, particularly Mallard, in the 1950s but there are no recent data.

The River Lark is interesting limnologically and has a number of nesting pairs of Tufted Ducks and Gadwall, as well as scattered wintering diving ducks. In its vicinity are two waters – the relatively new gravel pits at Lackford, 10km north-east of Bury St Edmunds, and the long pool in Livermere Park, 8km north of Bury – which are, no doubt, interdependent as far as their wildfowl are concerned. There are few counts, but the gravel pits have in recent years held as many as 270 Pochard, 295 Tufted Ducks and 440 Canada Geese; Goosanders and small numbers of other ducks are regular. Livermere, as well as having a number of wintering duck species, was the first Suffolk site for breeding Ruddy Ducks and is notable for its Mallard (920 in August 1981) and Canada Geese (up to 1,400).

The Cam/Granta basin is somewhat more productive of wildfowl waters. There is a group of gravel complexes near Cambridge of which Waterbeach Pits are the most important (Table 77). This group includes the Landbeach Marina, which is subject to very high pressure from recreation; the waterfowl are predominantly found at the northern end west of the airfield. The nearby Milton complex holds less than a hundred ducks, but to the south in the suburbs of Cambridge, Romsey Town Pit is of greater value (Table 77). A number of other pits are developing as promising wildfowl sites, especially the newly excavated areas at Histon and Hauxton, which are still being dug and gaining in value. Cambridge Sewage Farm, haunt of generations of undergraduates, was modernised out of usefulness for wildfowl in the late 1960s. An 8km stretch of the river between Cambridge and Milton (Baits Bite Lock) north of the city holds substantial numbers of Mallard, and this species is also legion along the Backs. The former fens at Fulbourn and Chesterton rarely flood now, but attract Mallard and Teal when they do. To the south a new reservoir at Babraham is worthy of inclusion in Table 77, but five other gravel or cement pits and pools around Barrington, although counted, carry few birds.

The Fenland

The flat arable expanse of the Fenland, a checkerboard of canals and ditches covering some 3,400sq km, is the drained part of the Wash basin. The outer section, north and east of Downham Market, Guyhirne, Crowland and Billingham, is composed of silt, laid down at successive coastlines of the Wash and empoldered by banks piecemeal from as early as Roman times. Inland a vast marsh with a few islands of higher land, as at Ely and March, built up great deposits of peat. These were difficult to drain as they lay slightly lower than the outer siltlands and it was not until the 1600s that Dutch engineers under the leadership of Vermuyden began the massive works which were completed by the mid 1800s, when the last great meres at Soham and Whittlesey disappeared. The courses of the main fenland rivers – the Great Ouse, Nene, Welland and Witham – have been so perturbed and supplemented by the drainage operations that there is no point in tracing their courses successively inland. Instead we may concentrate on the main features from the wildfowl point of view, the confusingly-named Washes. Straight, embanked canals were cut through the marshes and water pumped into them from the lower-lying land alongside. To allow for seasonal floods two such canals were cut parallel and the land between left as floodplains (washes), into which sudden surpluses of water could be directed and held for later removal. Predominant are the Ouse Washes (known to engineers as the Hundred Foot Washes – a singularly inappropriate name for an internationally important site) running from Earith to Denver, and the Nene Washes from Peterborough to Guyhirne. The Welland or Cowbit Washes near Crowland have been drained and ploughed.

The Ouse Washes extend for 30km and the two drainage cuts, the Old and New Bedford Rivers (named for a patronising earl, not in relation to the town), lie about 1km apart for most of their length.

Table 77. *The regular (and maximum) numbers of the principal species at 38 lesser resorts in the Wash catchment.*

		Mallard	Teal	Wigeon	Pochard	Tufted Duck	Mute Swan	Canada Goose
FENLAND AND FRINGES								
Narford Lake	(1 season:82)	760(2000)	69(142)	1 (4)	183 (215)	138(165)	67 (72)	152(339)
Didlington Lake	(6:73-79)	159 (450)	18 (49)	0 (2)	7 (25)	18 (58)	7 (30)	141(233)
Stanford PTA	(8:72-82)	294(1500)	103(400)	0 (60)	22 (60)	86(200)	22 (48)	30 (80)
Waterbeach GPs	(10:70-82)	76 (343)	2 (10)	3 (30)	113 (366)	102(382)	12 (32)	5 (23)
Romsey Town GP	(5:78-82)	6 (12)	0 (0)	0 (0)	83 (150)	18 (48)	0 (0)	0 (2)
Cam:Cambridge – Milton	(11:72-82)	728 (912)	11 (47)	0 (2)	0 (1)	0 (6)	12 (30)	0 (1)
Babraham Reservoir	(5:78-82)	118 (240)	0 (0)	0 (0)	0 (12)	3 (14)	0 (0)	0 (0)
Welland:Spalding – Boro' Fen	(4:79-82)	64 (101)	11 (38)	5 (12)	20 (83)	74(211)	207(269)	1 (8)
Coronation Channel	(2:71-80)	268 (375)	0 (0)	0 (0)	0 (1)	1 (4)	19 (31)	10 (21)
Mepal GPs	(9:70-82)	24 (290)	0 (25)	0 (25)	182(1034)	68(240)	1 (9)	0 (0)
Tanholt GPs	(1:82)	74 (85)	30 (65)	1 (1)	7 (13)	22 (24)	2 (3)	0 (0)
Baston Fen	(7:68-82)	89 (315)	123(375)	58(250)	0 (10)	11 (44)	6 (23)	0 (8)
Grimsthorpe Lake	(9:60-82)	462(1000)	38(250)	0 (16)	8 (87)	17 (70)	2 (23)	316(550)
Denton Reservoir	(2:81-82)	128 (250)	0 (5)	11 (60)	6 (30)	9 (70)	1 (7)	36(265)
Revesby Reservoir	(1:82)	87 (100)	0 (6)	0 (0)	5 (12)	5 (72)	0 (1)	40(100)
West Ashby GPs	(1:82)	61 (90)	36 (42)	0 (0)	9 (18)	38 (50)	6 (10)	70 (82)
Whisby GPs	(2:79-82)	440(1029)	95(259)	55(167)	115 (258)	76(159)	12 (25)	30 (87)
GREAT OUSE								
Fen Drayton GP	(3:77-80)	19 (80)	10 (39)	6(106)	58 (120)	93(206)	8 (27)	0 (7)
Meadow Lane GPs	(9:61-80)	26 (170)	3 (23)	7 (65)	90 (290)	106(380)	11 (44)	0 (4)
Marsh Lane GPs	(3:78-82)	28 (40)	0 (10)	0 (1)	44 (125)	43 (75)	6 (9)	9 (34)
Fenstanton GPs	(1:82)	233 (300)	63 (70)	0 (6)	125 (300)	57 (85)	6 (6)	34 (80)
Wyboston GPs	(10:60-73)	110 (350)	28 (75)	3 (50)	66 (337)	90(290)	8 (21)	0 (2)
Barkers Lane, Bedford	(14:60-82)	48 (166)	55(200)	7 (80)	23 (239)	24(198)	5 (13)	0 (0)
Southill Lake	(12:60-82)	379 (630)	66(166)	1 (6)	55 (125)	23 (60)	0 (10)	16(140)
Stewartby Lake	(4:70-82)	277(1000)	6 (40)	2 (15)	25 (215)	33(114)	7 (30)	0 (70)
Coronation Clay Pit	(3:77-80)	220 (600)	3 (22)	4 (10)	31 (95)	19 (50)	2 (6)	30(120)
Radwell GPs	(5:77-82)	136 (320)	33(140)	35(200)	41 (100)	49 (86)	11 (45)	24(112)
Emberton GPs	(11:71-82)	142 (253)	0 (2)	3 (90)	23 (81)	74(144)	8 (24)	26(153)
Newport Pagnell GPs	(17:60-81)	27 (157)	3 (40)	0 (22)	17 (55)	104(336)	4 (11)	0 (17)
Tongwell Lake	(4:78-82)	23 (68)	0 (0)	0 (0)	14 (48)	34 (65)	13 (35)	0 (0)
Mount Farm GP	(3:80-82)	45 (123)	0 (0)	0 (42)	30 (41)	31 (92)	4 (12)	6 (26)
Woburn Park Lakes	(1:74)	178 (286)	0 (14)	0 (3)	8 (26)	44 (65)	2 (11)	48 (61)
Hyde Lane GPs	(10:71-80)	4 (17)	0 (0)	2 (39)	10 (29)	80(163)	14 (62)	0 (2)
NENE								
Deene Lake	(15:62-82)	128 (535)	12(200)	0 (18)	21 (66)	28 (58)	2 (9)	35(200)
Oundle GPs	(4:71-82)	10 (210)	0 (2)	0 (4)	18 (38)	45(115)	4 (13)	18 (86)
Ringstead GPs	(3:74-82)	76 (212)	7 (23)	46(156)	60 (234)	64(200)	21 (51)	2 (18)
Sywell Reservoir	(4:79-82)	14 (79)	1 (6)	3(100)	23 (180)	44(175)	4 (9)	5 (54)
Castle Ashby Lakes	(4:79-82)	88 (226)	0 (0)	0 (0)	1 (16)	11 (43)	7 (15)	0 (2)

The land between belonged to a plethora of owners and securing it for conservation has been an exemplary effort by three non-govermental organisations, the Cambridgeshire and Isle of Ely Naturalists' Trust (CAMBIENT), the RSPB and the Wildfowl Trust. From the mid 1960s these bodies have been purchasing land piecemeal and together they now hold about 1,200ha, well over half the total area. This means that conservation interests carry as much clout as do agricultural ones when decisions as to the future of the Washes are discussed. It also led the Government to designate the site under the Ramsar Convention in 1976, the only instance in the original list where designation was of land without National Nature Reserve status, i.e. not protected under the law from development. It does have SSSI status but this is not in itself sufficient to prevent changes, the ownership by conservation bodies being a much better safeguard. The flood control requirements of the Anglian Water Authority nevertheless still have ultimate priority and the Authority have the power to control the water regime on this wholly artificial wetland area.

The New Bedford River is embanked on both sides (the span from bank to bank giving the alternative name, Hundred Foot, although the normal river width is barely a third of that) and only spills on to the floodplain with unusually high water levels. It carries all the normal flow of the Great Ouse from Huntingdon and beyond. The Old Bedford to the west is embanked on its outer side only and its flow is controlled by sluice gates at Earith. These are opened when the Ouse reaches a drawmark level, which is lower in winter than in summer, and the Old Bedford then rises and spills over into the washes. The flood creeps across from the western to the eastern side, which is 1m or so higher. Ideal shallow water feeding conditions are thus provided along a great frontage until the water extends from bank to bank, sometimes up to 2m deep. It then still provides feeding opportunities for diving ducks, but the dabblers and grazers are forced out on to the retaining embankments or onto the arable land, while continuing to use the floods as a roost. Flooding of course depends on the rainfall and soil water deficit in the watershed but may occur three or four times a winter for periods of two or three weeks. As the waters subside, the shallow pools and waterlogged pasture again provide ideal temporary feeding conditions (582).

In recent years, river siltation and other engineering problems have combined with a somewhat wetter climate to prolong the flooding, reducing the fluctuations in level which are so valuable to feeding wildfowl and, even more serious, extending the floods into the spring with disastrous consequences to breeding birds. For once, conservationists and agriculturalists are in accord in pressing for measures which will get the water off the washes by mid-April at the latest. The farming interest is to put cattle and sheep on to graze the washes, these being the only significant grasslands in the Fens. Grazing, controlled in intensity and duration, is also essential if there is to be both a close sward attractive to Wigeon in autumn and sufficient seeding to feed the dabbling ducks. The engineers, too, need the grazing, to prevent rank grass, rushes and scrub taking over and so impeding the water flow. The remarkable and varied flora of the area (588) demands that water be maintained in the network of ditches and so does the grazier, for whom they serve as fences to control the cattle. Until recently the ditches were the only nursery areas for breeding ducks; now they have been supplemented by lagoons of open water created when excavating material for screen banks by the Wildfowl Trust at Welney. Killing *Glyceria* with chemicals, by the RSPB at Welches Dam also produces shallow pools as the dead root systems contract. The ditches are kept topped up by letting in water from the New Bedford River at high tide. (Although the river is tidal to Earith the sea water does not penetrate very far and it is backed-up freshwater that enters the ditches, except in very dry summers.) In the drought year of 1976 the Water Authority installed a temporary dam on the New Bedford at Earith as part of Operation Rodeo (Reversal of Direction of Ely Ouse). Representations by the conservation interests, citing the newly ratified Ramsar Convention, resulted in agreement to install pumps to draw freshwater from the Old Bedford to top up the ditches. However, the drought then ended and the dam was removed.

Apart from maintaining suitable grazing and water regimes, the conservation bodies have been concerned to control disturbance, by preventing ingress and, at Welney, by building shielding banks and planting screens of willow trees. Birdwatching has been catered for by numerous hides ranging from the simple to the frankly luxurious. The Wildfowl Trust has also taken the more controversial action of baiting in wildfowl, particularly Bewick's Swans, to the lagoons in front of its observation hides, and extending the period of observation by flood-lighting in the evening. Shooting has been prevented over large parts of the Washes. The Wildfowl Trust rapidly acquired a complete block of some 300ha north of the Welney road and excluded shooting, while the RSPB purchases have coalesced into a similar no-shooting block south of the railway viaduct, with another nearing completion. Many other scattered washes have been purchased with a view to the nesting birds or, particularly in the case of CAMBIENT, the flora. The interspersal of shooting areas and sanctuaries concentrates the birds on the refuges from September to January. After the shooting season closes the birds spread out, taking advantage of food previously denied them and so probably extending the length of their stay (582).

The success of the conservation efforts is clearly shown in the numbers of wintering wildfowl set out in Table 78. Mallard had already reached a stable level in the 1960s after a threefold increase from a regular (and maximum) 1,160 (4,000) in the 1950s. Teal, which were then 820 (4,205), have continued to rise and their numbers now exceed the criterion for international importance for this species. Wigeon, which in the 1950s reached 5,680 (19,300) as part of the move inland away from the heavily shot coasts, have built up until this is the largest inland concentration in Britain and rivals the largest at any British site, that at Lindisfarne (p.207). Pintail have fluctuated around a level rather less than twice the 1950s figure of 775 (5,000). Shoveler have also stabilised after a great jump from a regular 70 and a maximum of 400, while the inland diving ducks have similarly increased from regular (and maximum) figures for the 1950s of 120 (570) Pochard and 20 (285) Tufted Ducks. The Gadwall and Shelducks are virtually newcomers, responding to the provision of standing water on the refuges. The Whooper and Bewick's Swans are much increased from the 1950s maxima of 25 and 677. The swans used to feed

exclusively on the washes, chiefly grazing soft grasses and rooting, but they have increasingly resorted to flighting to the arable fenland within 1km of the washes, returning there to roost. The commonest crops used are waste potatoes and winter wheat, but the birds have also taken to gleaning sugar-beet fields and even taking waste peas. The latest increase in Bewick's Swans means that at peak about a quarter of the entire population wintering in western Europe are present in what is by far the biggest gathering in Britain and, probably, anywhere.

The only group which might be better represented are the geese. Undoubtedly disturbance of this narrow site drove them away in the past and they may now return. The Greylags are presumably of feral origin; it is perhaps no bad thing that Canada Geese have not taken to the area, although one or two pairs have nested.

Table 78. *Regular (and maximum) numbers of wildfowl recorded at four sites in Fenland.*

a) Ouse Washes i) 4 seasons 1961-1969
ii) 13 seasons 1970-1982
b) Nene Washes 3 seasons 1980-1982
c) Wicken Fen 9 seasons 1960-1982
d) Fletton Brick Pits 11 seasons 1971-1981

	a i	a ii	b	c	d
Mallard	3416 (6800)	3833 (6469)	410 (829)	549(1250)	359(1100)
Teal	1442 (3000)	2444 (7570)	814(2283)	128 (800)	21 (94)
Wigeon	14604(36000)	24334(42500)	1056(2213)	281 (975)	8 (91)
Pintail	1296 (3050)	1325 (8450)	264 (677)	1 (76)	0 (0)
Shoveler	461 (1200)	383 (1080)	57 (176)	25 (120)	1 (8)
Gadwall	27 (60)	77 (371)	3 (12)	5 (38)	0 (2)
Shelduck	2 (19)	23 (174)	35 (83)	0 (1)	0 (0)
Pochard	1404 (3250)	1238 (5480)	184(1200)	23 (150)	141 (298)
Tufted Duck	323 (900)	336 (1165)	178 (895)	16 (75)	129 (369)
Goldeneye	18 (48)	4 (22)	2 (4)	0 (1)	1 (5)
Mute Swan	280 (433)	306 (621)	64 (88)	3 (13)	13 (23)
Whooper Swan	12 (46)	73 (223)	0 (4)	0 (0)	0 (0)
Bewick's Swan	474 (938)	1415 (2995)	209 (600)	0 (12)	0 (0)
Greylag Goose	1 (7)	48 (214)	10 (54)	0 (3)	7 (90)
White-fronted Goose	20 (140)	5 (60)	0 (0)	0 (2)	0 (0)
Canada Goose	3 (34)	1 (6)	0 (3)	6 (42)	56 (317)

The Ouse Washes are thus an outstanding wintering site, specifically of international importance for three duck and one swan species. As regards total numbers of wildfowl they regularly hold over 30,000 birds, with a maximum (in February 1981) of 54,610, which closely rivalled the highest count for Britain, at the Mersey marshes, of 56,330 (p.188).

While the wintering birds alone make the Ouse Washes quite outstanding, the breeding waterfowl also provide the largest and most varied inland concentration in Britain. Indeed it was the return of the Black-tailed Godwit *Limosa limosa* as a British breeding bird, nesting at Welney in 1952, which provided the initial impetus to land acquisition in the Washes. They remain the national focus for this species with 40-60 pairs. Ruff *Philomachus pugnax* also became established in the mid 1960s. In 1982 there were 326 pairs of Lapwings *Vanellus vanellus* on the Ouse Washes, and they and the Nene Washes together held 258 pairs of Redshank *Tringa totanus* and 738 drumming Snipe *Gallinago gallinago*, this being 37% of the total drumming birds recorded for lowland England and Wales (548). The ecology of breeding waterfowl on the Ouse Washes has been investigated in detail (583). In the 1970s Mallard pairs numbered 400-1,300, Shoveler 133-306, Gadwall 12-52 and Tufted Ducks 30-50. The last two, which first bred in 1953 and 1964 respectively, clearly benefited from the increased standing water provided by the conservation bodies. Up to 4 pairs of Pochard, 12 pairs of Shelducks and Garganey, 20 of Teal and 34 of Pintail were also reported in the period; Garganey and Pintail became established in the 1950s, the others in the 1960s. Records vary from 50-450 pairs of Coot *Fulica atra* and 120-600 pairs of Moorhens *Gallinula chloropus* nesting, depending largely on water conditions.

Compared with the fabulous Ouse Washes, the Nene Washes, 20km to the north-west, are less important, but they nevertheless support substantial numbers of wildfowl. Similar in construction and purpose to the Ouse Washes, they are 10km shorter, but in places up to 2.5km broad. They serve a smaller catchment area and flooding is less frequent and seldom prolonged. The dry 1960s encouraged farmers to plough up much of the washlands, leaving grassland mainly at either end. In recent, wetter years the gamble on the infrequency of floods has been less rewarding and unharvested crops of potatoes and carrots have attracted temporary concentrations of Bewick's Swans and Pintail. The High Washes, to the west of the Whittlesey – Thorney road and extending back towards Peterborough, have but few wildfowl. The birds are found mainly to the east of the road on the Low Washes (Eldernell and Guyhirne). Counts on the Low Washes from 1964 to 1979 showed little variation between the two decades and, compared with the more favourable 1950s, were quite small – the regular (and maximum) numbers being 98 (400) Mallard, 95 (300) Teal, 142 (500) Wigeon, 84 (500) Pintail, 35 (107) Mute Swans, 24 (400) Bewick's Swans and 24 (300) Pink-footed Geese. (The Pinkfeet have never regained the several thousands that used the Nene Washes as a late-winter resort in the 1950s.) Counts on the other washes to the east were formerly sporadic. However, in the 1980s wildfowl counts were obtained over all the washes east of the road. They indicate much higher levels (Table 78) than in the previous two decades. With the establishment of a 220ha RSPB reserve in 1983 the future looks rather hopeful in the light of the achievements of conservation management on the Ouse Washes. The nesting wildfowl on the Nene Washes vary widely in numbers, from 1,700 pairs (including 360 of Shoveler) in the wet 1981 to 340 pairs in the dry 1982 (106).

The ditches and canals of Fenland total many hundreds of kilometres and provide a temporary refuge for very substantial numbers of wildfowl feeding in the arable land around. Some idea of the numbers involved is given by a scatter of counts. The River Welland is regularly counted for 30km above Spalding (Table 77). The Mute Swans exploit the drained Cowbit Washes alongside the river. The 14km of the River Cam between Milton and the Great Ouse, including the washland west of Wicken Fen, has had up to 128 Mallard, 290 Teal, 575 Wigeon, 30 Shoveler, 12 Pochard, 75 Tufted Ducks and 60 Mute Swans. A 5km stretch of the Coronation Channel around Spalding is recorded in Table 77, and the Hilgay New Cut near Denver has held up to 750 Mallard, 40 Teal, 32 Wigeon and 30 Tufted Ducks. Certainly many thousands of wildfowl, especially Mallard, must be dispersed along the Fenland waterways. The BTO Breeding Atlas (542) recorded densities ranging from 1.5 to 3 Mallard per square kilometre of fenland.

Substantial bodies of water only exist where they have been excavated for gravel or some other purpose. At Wicken Fen (15km east of Earith) the National Trust has maintained a 295ha remnant of fenland in an approximation of its former undrained condition. In 1955 a 4ha mere was dug to improve the area for wildfowl, and dabbling ducks became and remain quite numerous (Table 78). There is much interchange between Wicken Fen and the nearby Cam Washes at Upware. The NNR at Woodwalton Fen west of Ramsey holds a roosting population of Mallard, at peak up to 1,100. Near Ely the Roswell pits at the sugar-beet factory carried up to 300 Teal at peak, but the factory closed in 1982. Just west of the Ouse Washes, Mepal Gravel Pits have substantial numbers of ducks, no doubt interchanging with the larger site, but the Earith pits hold fewer than 100. Associated with the Nene Washes are the brick pits at Fletton (Table 78) (though these are threatened with infilling) and nearby pits at King's Dyke and Drysides. The gravel pit at Tanholt is recorded in Table 77. At the seaward end of the Nene the sewage farm at Wisbech used to carry Teal and Shoveler in some numbers as well as many waders. In 1978, after a century of use, it was replaced by a modern processing plant and the lagoons only received the effluent of a canning factory. Efforts are being made to save a remnant of the old-style sewage fields as a man-made marsh.

Near the Welland at Peakirk is the ancient Borough Fen Decoy, whose history has been described in detail (130). The decoy is now run in conjunction with the nearby Wildfowl Trust collection at Peakirk. First mentioned in written records in 1670 the decoy was probably built 20-30 years earlier. In a 6ha wood the pond of 1ha has eight radiating screened and netted "pipes", an unusually complex construction. Ducks were enticed off the pond by food in the pipe or followed a small dog trained to run along its bank (p.9). The records of the size of the catch for the market are available from 1776 (129), but not the composition by species, the old decoymen being only interested in meat. From 1888, at least, Mallard and Teal made up most of the catch. Ringing started in 1947, the Wildfowl Trust taking over full operation in 1951. From 1947 to 1977 40,716 ducks were ringed, with a catch of 3,150 in 1967 probably as high as any in its history. In the 1960s the "lead" of birds using the pond was at regular (and maximum) levels of 482 (1,150) Mallard and 235 (935) Teal; between 1970 and 1977 this declined to 289 (960) and 153 (425); since then the numbers have fallen right away and catching is no longer worthwhile. The decoy is designated as an Ancient Monument and is being maintained as such. It would seem that now other and larger waters have appeared nearby and been protected from disturbance the ducks prefer these to the secluded little ponds of the many decoys, which were once the only apparent sanctuaries in the Fens (a similar fate appears to have befallen Nacton Decoy, Suffolk as described on p.12). Within 2km of Borough Fen Decoy, for instance, are Deeping St James Ballast Pits of over 40ha which are known to carry several thousand ducks at times, although winter counts only run to a maximum 625 Mallard, 380 Teal, 36 Wigeon, 32 Shoveler, 230 Pochard and 320 Tufted Ducks. Cage traps are still operated with considerable success by the Wildfowl Trust on an island in the pits, 1,252 ducks having been ringed in 1977 (including 908 Mallard and 263 Tufted Ducks).

Another group of gravel pits which probably attracts many of the ducks which used to go to Borough Fen is around Market Deeping (Table 79). At the Lolham and Maxey pits (Maxey is now largely deserted because of disturbance), the dabbling ducks have shown some decline, more than offset by increases in the diving species. At Tallington and Langtoft all ducks have increased over the 1960s figures, in line with those shown in the table for Baston Common, though the pits are now subject to considerable recreational disturbance. The nearby Baston Fen (Tongue End) reserve of the Lincolnshire Naturalists' Trust, comprises 27ha of controlled flooding on washland of the River Glen, a tributary of the Welland. It is frequented by small numbers of a dozen wildfowl species in addition to those listed in Table 77.

The north-west fringes of the Fenland

The Fenland continues to the north, mainly in the basin of the River Witham, which enters the Wash

Table 79. *Regular (and maximum) numbers of wildfowl recorded at four gravel pits near Market Deeping on the edge of Fenland.*

a) Lolham and Maxey 22 seasons 1960-1982
b) Tallington 8 seasons 1970-1978
c) Langtoft 8 seasons 1970-1978
d) Baston Common GPs i) 10 seasons 1960-1969
ii) 8 seasons 1970-1978

	a	b	c	d i	d ii
Mallard	82(340)	175(460)	77(225)	322(1200)	699(1310)
Teal	64(300)	4 (30)	24 (75)	38 (180)	93 (420)
Wigeon	2 (65)	8 (37)	0 (1)	2 (18)	73 (420)
Pochard	62(272)	303(822)	155(610)	114 (351)	404(1100)
Tufted Duck	61(165)	149(462)	86(128)	64 (115)	147 (265)
Mute Swan	27(217)	13 (57)	7 (42)	8 (30)	8 (18)
Greylag Goose	0 (2)	5 (35)	1 (7)	1 (8)	32 (116)
Canada Goose	3 (35)	7 (75)	19 (70)	3 (17)	87 (205)

at Boston, having flowed past Grantham and Lincoln, to which the peatland extends. The flat drained land is similar in character to the rest of the Fens, having but little standing water. It is cut off from the north Lincolnshire coast (p.159) by the higher ground of the Lincolnshire Wolds. To the west it abuts the Middle Trent basin.

About 20km south-east of Grantham, Grimsthorpe is a large isolated park lake whose main interest is for feral Greylag and Canada Geese. It is one of the major strongholds of Canada Geese in this region, perhaps representing, at peak, up to half the population of Cambridgeshire, Lincolnshire and Northamptonshire (405). Their virtual absence from the Fenland has already been noted. To the west, some 5km from Grantham, Denton Reservoir (Table 77) is within 5km of Belvoir and Knipton Reservoirs in the Middle Trent. The park lake at Denton has nothing of note, but the two pools nearby at Harlaxton have held 140 Mallard. The sewage farm on the southern outskirts of Grantham used to carry up to 500 Wigeon in the 1960s but is now of little interest. Counts are not available for the park lakes at Belton and Syston although they probably hold some numbers of ducks and Canada Geese. Rather further to the north-east the ponds at Ancaster have had up to 209 Mallard, 29 Pochard and 52 Tufted Ducks along with 108 Canada Geese, but the lake at Culverthorpe does not rate more than 42 Mallard, 24 Pochard and Tufted Ducks and 56 Canada Geese. There are several gravel pits around Sleaford, those at Scredington and Burton being unimportant, but those at Sleaford itself held regular numbers of 28 Mallard, 43 Pochard and 31 Tufted Ducks when last counted in the early 1970s.

Well to the east of Lincoln, Revesby Reservoir and the gravel pits at West Ashby on the banks of the River Bain to the north are reported in Table 77. The pits at the sugar-beet factory on the banks of the Witham at Bardney used to carry up to 328 Mallard, 230 Teal, 100 Pochard and 60 Tufted Ducks in the

Fig. 25. The Great Ouse, Nene and Welland.

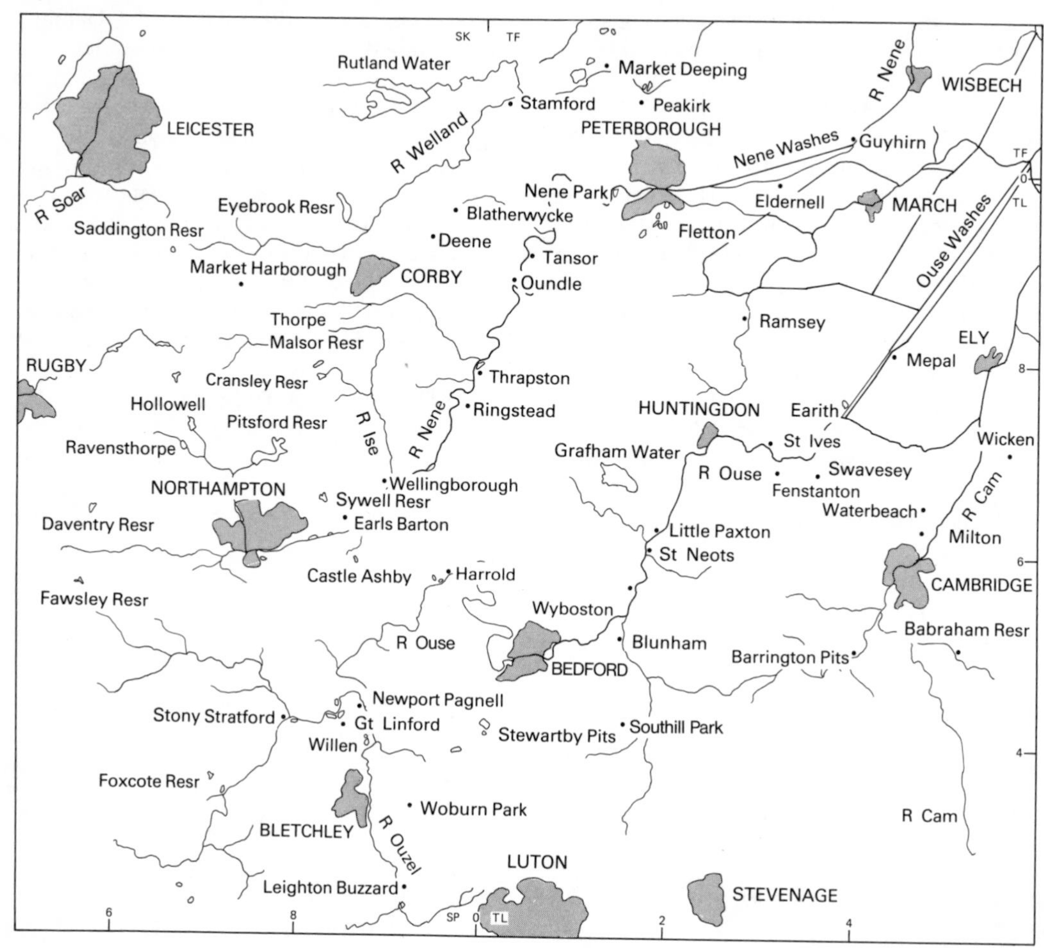

1960s, but there are no recent detailed counts, though the area is said to hold as many as 2-3,000 Mallard in early autumn.

To the north of Lincoln, Fillingham Lake has held up to 150 Mallard, 27 Pochard, 13 Tufted Ducks and 125 Canada Geese. Only 9km to the east, and no doubt closely associated in its wildfowl, is Toft Newton Reservoir, which has carried up to 55 Mallard, 60 Teal, 118 Wigeon, 328 Pochard and 365 Tufted Ducks. It lies at the headwaters of the River Ancholme which runs north into the Humber. Around Lincoln itself seven counted sites are of little interest, but there are three gravel or ballast pits of some note – the ballast pits at Lincoln, and Burton and Whisby gravel pits (Table 77). The Whisby pits to the south-west of the city are relatively undisturbed and act at times as a refuge for birds from other pits south of Lincoln. The maxima thus probably refer to such occasions. This welcome relief in an area otherwise rather devoid of wildfowl haunts lies less than 10km over a ridge of higher ground from the River Trent, and the many gravel pits that line its course.

The middle and upper basins of the Great Ouse, Nene and Welland

Away from the Fenland the courses of its three major rivers may be successively traced inland for up to 100km, with a catchment area of some 6,000sq km. All three rivers rise within 40km of each other in the uplands of Northamptonshire and flow on roughly parallel courses through gently rolling farmland. In the middle reaches the valleys are broad and the gradient slight, and extensive flooding used to occur. This has now been largely eliminated, but the floodland habitat has fortunately been replaced by the opening up of numerous gravel pits along the valleys. Many reservoirs have also been established on the low limestone hills above. These new habitats have largely been developed in the last 30 years, preserving and indeed strengthening the links between the wildfowl of the Wash and Fenland and those of the basins of the Thames to the south, the Severn to the west and the Trent to the north.

The Great Ouse

The Ouse and the riverside fields from Earith to Swavesey have carried up to 142 Mute and 106 Bewick's Swans. Between St Ives and St Neots a dozen gravel pits lie along the valley, mostly started to meet demands of airfield construction in the Second World War. Half this number have been counted irregularly and do not appear to be of great importance. However, the others carry fair numbers, particularly of Mallard, Pochard and Tufted Ducks. The counts for Fen Drayton 3km east of St Ives, Meadow Lane, St Ives, Fenstanton (the pits south of the A604) and Marsh Lane near Hemingford Grey are given in Table 77. The deep, water-gorge Stirtloe Pit, only 5km east of Grafham Water, once held up to 940 Mallard, 88 Pochard and 1,200 Tufted Ducks, but the use of the site has now declined to the extent that it is no longer counted. Little Paxton pits (Table 80) have been enriched by the flooding of Grafham Water only 5km away and provide an extensive complex – the most important gravel site in the area. There is undoubtedly much interchange with Grafham, although in recent years ducks have tended to favour the reservoir rather than the pits for roosting. The Gadwall on Little Paxton are now eighth in importance in Britain,

Table 80. *Regular (and maximum) numbers of wildfowl recorded on five sites along the Great Ouse.*

a) Little Paxton Gravel Pits 10 seasons 1972-1982
b) Grafham Water i) 4 seasons 1964-1969
ii) 13 seasons 1970-1982
c) Blunham Gravel Pits 10 seasons 1971-1982
d) Brogborough Clay Pits 8 seasons 1972-1982
e) Harrold/Odell Gravel Pits 11 seasons 1972-1982

	a	b i	b ii	c	d	e
Mallard	291(1150)	2174(5315)	1291(2920)	82(198)	219(1200)	398(766)
Teal	88 (385)	232(1000)	356(1425)	6 (56)	1 (13)	23 (80)
Wigeon	35 (450)	220 (640)	525(2060)	11(144)	1 (6)	15(109)
Shoveler	37 (280)	14 (100)	42 (150)	7 (21)	1 (35)	1 (4)
Gadwall	59 (300)	0 (2)	36 (157)	13(133)	0 (2)	1 (4)
Pochard	143 (380)	292 (630)	138 (474)	119(337)	85 (200)	42(200)
Tufted Duck	416(1210)	300(1250)	943(3050)	107(272)	38 (80)	79(206)
Goldeneye	13 (50)	20 (77)	96 (180)	4 (12)	1 (14)	4 (11)
Mute Swan	22 (92)	12 (39)	78 (196)	6 (30)	4 (18)	6 (14)
Greylag Goose	36 (290)	0 (0)	0 (1)	34 (77)	0 (0)	44(201)
Canada Goose	3 (40)	0 (0)	1 (11)	0 (5)	15 (224)	33(108)

measured by the average peak over the five years 1978 to 1982. Indeed in 1980-81 the maximum there was only exceeded by Gunton Park in Norfolk (p.138).

Grafham Water commenced flooding in December 1964, its water being pumped from the Ouse at Offord, and it was full by February 1966. The early history of this 738ha reservoir which, until surpassed by Rutland Water, was the largest lake in southern England, has been well described (132, see also p.544). Like other reservoirs it was particularly attractive to ducks soon after flooding had released the nutrients and seeds from the inundated meadows. More than 5,000 Mallard were counted on several occasions. The numbers then declined but recovered somewhat as lacustrine flora and fauna became established, and after the drought of 1976 (Table 80, see also Fig 171 p.512). However, the potential increase was checked by the great development of water-related recreation. Fishing is considered here to be a much greater disturbance factor than sailing, which is less continuous, tends to be in the deeper water and is banned altogether from a buoyed off area in front of the 150ha nature reserve (managed by Bedfordshire and Huntingdonshire Naturalists' Trust). An observation hide overlooks the reserve, and other management measures have been carried out therein. While many species have achieved some stability, the Tufted Duck, which had responded well to the increased maturity of the reservoir and reached their record peak of 3,050 in November 1979, have suddenly declined, to less than 200 in 1982-83. It is thought that this may be related to an unexplained crash in the population of chironomids, whose larvae are a major food item. In addition to the species in Table 80, Goosander are regular at Grafham but highly mobile, with a maximum of 26; Bewick's Swans occur in all years, with up to 33 in the early days; Shelducks are usually present, with a peak of 32, and have bred, as have Garganey. The Shoveler is a regular breeder, but Mallard and Tufted Ducks are the main nesting ducks; Mute Swans breed, but their particular use of the reservoir is as a moulting area. This habit developed in the early 1970s and a maximum count was of 268 in July 1983, the birds feeding on the extensive algae mats which appear in summer.

Between St Neots and Bedford the Ouse Valley has many further gravel pits. Those at Chawston, with up to 170 Mallard, and at Girtford and Sandy, with up to 100 Tufted Ducks, are of less importance, but Wyboston and Barkers Lane, Bedford (the counts from which include the sewage farm and which is now Priory Marina), are more productive (Table 77). Wyboston pits, however, have lost much of their former interest through increased water-skiing, sailing and fishing; parts have also been infilled with fly-ash. On the Ivel tributary coming from the south, the Blunham pits (Table 80) not only carry substantial numbers of ducks but are the scene of a successful new private ringing operation, under the auspices of the Wildfowl Trust. In four years from 1979 it had ringed 1,383 Pochard and 1,231 Tufted Ducks, becoming a major marking centre of diving ducks. The pits at Langford, Henlow and Arlesey, and Radwell Mill are much less notable, but the lake at Southill Park, 4km west of the Ivel at Biggleswade, is worthy of note (Table 77).

Ten kilometres south of Bedford, excavation by the brick industry has resulted in a number of pits around Stewartby. These are up to 25m deep and steep-sided, but form useful roosts for dabbling ducks; the presence of diving ducks is not easy to explain. Stewartby Lake, which suffers from considerable disturbance, and Coronation Clay Pit, 3km north-east are recorded in Table 77. Brogborough Pits, 5km south-west of Stewartby (Table 80), is the main resort in this area, though windsurfing takes place on half the lake. Five other pits nearby have regular totals of 100 or less.

As it passes through Bedford itself, the Ouse carries up to 315 Mallard and 31 Mute Swans, while the town park lake has up to 200 Mallard. In the valley to the north-west, a series of gravel pits have similar mixes of species. Those at Felmersham have a regular (and maximum) 48 (95) Mallard and 42 (100) Tufted Ducks; Radwell Gravel Pits are recorded in Table 77, while the major complex along this stretch is at Harrold/Odell (Table 80). The rather scattered Bedfordshire/North Buckinghamshire population of Canada Geese was estimated to be 210 in 1976 (405).

Around Newport Pagnell there is another major group of gravel pit complexes. Three sites, though counted, are of little import, but Emberton at Olney, the Newport Pagnell pits just north-east of the town and those at Tongwell just to the south are recorded in Table 77. The centrepiece is, however, provided by the Great Linford pits, some 300ha, mostly excavated in the 1970s and now nearly worked out. The considerable increase in numbers of wildfowl there (Table 81) is not only due to the expansion of the water surface, but owes much to the creative management of a 37ha reserve, established in 1971, financed by the gravel company and run by the Game Conservancy. They have made a "lacustrine oasis" (556) in the sterile "desert" of raw gravel pits. An intensive research programme has been aimed at finding ways of hastening maturity of the habitat and of increasing its value for nesting as well as wintering wildfowl. It is also a breeding site for the common species; in 1980 47

Mallard and 32 Tufted Duck nests were found. Observation hides have been established and an imposing research and education centre has been built overlooking the reserve.

The whole area between Newport Pagnell and Bletchley to the south has been completely perturbed by the establishment of the New Town of Milton Keynes. An interesting spin off for wildfowl was the building of a "balancing lake" of 68ha at Willen. This, and other similar sites under construction or planned, are designed to collect storm water, increased by the concretisation of so much countryside, and release it slowly, thereby controlling the flash floods that were a feature of the area. The difference in water level on the lake may be more than a metre. Such fluctuations can provide rich feeding opportunites for wildfowl and the lake certainly holds a fair population (Table 81).

The tributary Ouzel flows from the south into the Ouse at Newport Pagnell. At Bletchley, Mount Farm Gravel Pit (Table 77) has an unusual pre-eminence of Shoveler – a regular 47 and maximum 86. Woburn Park Lakes are notable only for Mallard and are also listed in Table 77. Further south again, Leighton Buzzard Sand Pits have up to 365 Mallard, 48 Pochard, 18 Tufted Ducks and 46 Canada Geese. Three more counted sites have small numbers, but the sewage farm at Dunstable, near the source of the Ouzel in the chalk hills, has up to 112 Mallard, 48 Shoveler, 30 Pochard, 49 Tufted Ducks and 26 Mute Swans.

Returning to the upper reaches of the Ouse, the Berks, Bucks and Oxon Naturalists' Trust reserve at Stony Stratford, with 20ha of wet meadows and water, is worth a mention, though numerical data are not available. The lake at Cosgrove Park has held up to 48 Mallard, 100 Canada Geese and 20 Mute Swans, but the gravel pits are unproductive. Those at Hyde Lane, Buckingham, have a suprisingly small number of Mallard. An unusual feature here was the regular group of 5-9 Whooper Swans which occurred in the late 1970s. The only major site in the area is Foxcote Reservoir. This small, 20ha water, first flooded in 1956, is a County Trust reserve; the population that became established in the 1960s has remained remarkably similar in later years (Table 81), only Shoveler reflecting in any way the general increase in wildfowl populations in the country as a whole. Probably the wealth of habitat that has been created downstream has short-stopped many migrants. Only a few wildfowl are recorded on the sites remaining before the source of the Ouse is reached. Stowe School Lake has up to 31 Mallard, 15 Teal and 20 Tufted Ducks; the lake at Brackley has 27 (58) Mallard and only single figures for other species. At its source the Ouse is within 10km of the River Cherwell and the upper Thames Basin.

The Nene

Like the Ouse, the Nene Valley has many gravel pits along its length, maintaining its value as a corridor of habitat for wildfowl. Just west of Peterborough, the pits at Ferry Meadows, Orton Longueville, were excavated with a view to the site becoming a public recreational area, the Nene Park (126). Extraction began in 1972 and some 50% of the area was left intact to preserve archaeological, woodland and meadowland features. The Park with its 50ha lake (Table 82) was opened in 1978. Recreational pressures have since then reduced the value of the area, particularly for diving ducks, Pochard and Tufted Ducks not exceeding 86 and 39 in 1982-83. The two geese, however, both reached a peak in that year. The establishment of a small, 13ha reserve in 1980 has helped to stabilise the

Table 81. *Regular (and maximum) numbers of wildfowl recorded on three sites along the upper valley of the Great Ouse.*

a) Great Linford Gravel Pits i) 4 seasons 1960-1968
ii) 12 seasons 1970-1982

b) Willen Balancing Lake 9 seasons 1974-1982

c) Foxcote Reservoir i) 6 seasons 1960-1968
ii) 13 seasons 1970-1982

	a i	a ii	b	c i	c ii
Mallard	367(860)	586 (975)	422(900)	502(1000)	530(1300)
Teal	33(100)	55 (266)	100(300)	33 (80)	22 (75)
Wigeon	84(700)	265(1000)	102(241)	27 (120)	25 (150)
Shoveler	2 (8)	16 (66)	61(180)	7 (15)	27 (95)
Pochard	26 (68)	134 (497)	148(655)	50 (96)	67 (192)
Tufted Duck	69(160)	190 (475)	161(441)	72 (130)	74 (183)
Goldeneye	0 (2)	2 (9)	21 (39)	1 (3)	1 (3)
Mute Swan	6 (10)	14 (50)	27 (70)	3 (12)	1 (8)
Greylag Goose	0 (0)	72 (163)	0 (4)	0 (0)	0 (0)
Canada Goose	1 (2)	53 (187)	55(209)	0 (0)	2 (27)

populations of dabbling ducks. On the Stibbington pits further upstream there are no recent counts, but Mallard reached 52 in the 1960s. The pits at Yarwell have carried 24 Mallard, 36 Tufted Ducks and 88 Canada Geese. This area also holds an increasing population of Mandarin Ducks, currently about 20 pairs, centred on the Nene Park Lake.

At Elton the Nene is joined by a tributary, Willow Brook. Although its loop to the north brings it to within 3km of the Welland, it is properly taken in the Nene context. Along its length it has been impounded to form a string of private park lakes of value to wildfowl. The lake at Apethorpe returned regular (and maximum) counts of only 36 (75) Mallard, 9 (18) Tufted Ducks and 17 (25) Canada Geese in the early 1970s. Specific counts are not available more recently but are known to be higher and to include some full-winged exotics. Further upstream Blatherwycke Lake (Table 82) carries large numbers of Mallard, though the 5,000 in September 1973 was exceptional. Nevertheless, 1,000 has been exceeded on four occasions since. The lake at Lynn Wood just to the north carries a few Mallard and Teal. Deene Lake (4km south) is not so populous as Blatherwycke but has good numbers of the common species (Table 77).

Returning to the Nene itself, we find a further series of gravel pits, with moderate numbers of Mallard, Pochard and Tufted Ducks and interesting variations in the other species. The relatively new gravel pits at Tansor already carry good numbers of the common species (Table 82). Some of the pits near Oundle have been turned into the Barnwell Country Park and have a much smaller population (Table 77). The pits at Thrapston, however, being excavated in the 1950s have had time to mature and, not being unduly disturbed, carry a substantial and varied wildfowl population (Table 82). The pits continue to be extended and now reach Thorpe Waterville 4km away. Particularly notable are the Wigeon, Goldeneyes and Goosanders. This body of water more than amply replaces the flooded meadows of Titchmarsh and the still-extant heronry which were favoured birdwatching sites in the 1920s and 1930s of schoolboys from Oundle, including James Fisher and Peter Scott. It is a nice coincidence that one of the many achievements of the latter was the saving from extinction of the Ne-ne Goose (of Hawaii and bisyllabic). The next pit complex at Ringstead is again newer, with less variety and numbers (Table 77). The Ditchford pits on the northern outskirts of Rushden are largely older and among their varied wildfowl are those listed in Table 82. The numbers of Teal, Shoveler and Gadwall are higher than in other waters downstream, and the birds are usually found on the newer, shallow pits near Higham Ferrers. The Mute Swans also find this an attractive site.

At Wellingborough the tributary Ise enters the Nene from the north. Associated with this, on the southern outskirts of Kettering, Wicksteed Park Lake had up to 300 Mallard in the 1960s, but there are no recent data. Just to the west of Kettering are two small hill reservoirs, Cransley and Thorpe Malsor, carrying regular (and maximum) levels of 10 (24) Mallard at Cransley and 42 (107) at Thorpe Malsor, together with about 10 Pochard and 35 Tufted Ducks apiece.

Upstream of Wellingborough the gravel pits along the Nene at Earls Barton carry a useful population; the pits have greatly increased in size in

Table 82. *Regular (and maximum) numbers of wildfowl recorded on six sites in the middle Nene Valley.*

a) Nene Park, Orton Longueville 8 seasons 1974-1982
b) Blatherwycke Lake 10 seasons 1970-1982
c) Tansor Gravel Pits 2 seasons 1973-1982
d) Thrapston Gravel Pits 7 seasons 1975-1982
e) Ditchford Gravel Pits 4 seasons 1979-1982
f) Earls Barton Gravel Pits 4 seasons 1979-1982

	a	b	c	d	e	f
Mallard	134(300)	578(5000)	301(600)	161(420)	269(450)	154(300)
Teal	113(325)	4 (90)	22(100)	45(170)	156(256)	5 (13)
Wigeon	24(300)	66 (300)	16 (37)	102(480)	36 (81)	94(225)
Shoveler	2 (20)	10 (100)	1 (6)	7 (18)	20 (60)	1 (8)
Gadwall	2 (10)	4 (22)	0 (7)	13 (50)	44 (77)	3 (8)
Pochard	103(350)	61 (400)	65(224)	179(500)	195(437)	89(194)
Tufted Duck	111(257)	71 (300)	70(100)	190(700)	219(420)	203(263)
Goldeneye	3 (10)	0 (0)	1 (5)	15 (40)	1 (5)	7 (17)
Goosander	8 (74)	4 (46)	1 (9)	43 (99)	0 (3)	0 (1)
Mute Swan	7 (29)	2 (8)	3 (6)	19 (43)	37 (86)	26 (47)
Greylag Goose	63(200)	32 (127)	11 (23)	1 (17)	0 (2)	6 (63)
Canada Goose	136(350)	158 (420)	96(169)	71(140)	1 (6)	6 (18)

ecent years (Table 82). A small tributary from the orth entering the Nene just upstream of Earls Barton s impounded to form the 32ha Sywell Reservoir. This as rather smaller numbers than the lakes at Castle shby 3km to the south of the river. Much more nportant is a complex of well-established pits (which nclude an "aquadrome") lying between the villages of ogenhoe, Little Houghton and Little Billing to the ast of Northampton. These are counted as two roups, Billing and Clifford Hill, but the counts are enerally made on the same day and so the regular nd maximum numbers can be combined without too nuch fear of exaggeration (Table 83). More Mallard are ound on the western (Clifford Hill), more Wigeon nd Greylags on the eastern (Billing) pits, but therwise the proportions are similar. In the southern utskirts of Northampton two gravel pits, Harding- tone and Stortons, have smaller numbers of up to 100 Iallard and Tufted Ducks. Hardingstone has rather ewer than Stortons but also has up to 19 Mute Swans.

Into the Nene at Northampton flows a tributary which, 10km to the north, is impounded to form the 00ha Pitsford Reservoir, completed in 1955. Much is iven over to sailing, fishing and other recreation, but, nost unusually, nearly half of the water, separated om the rest by a causeway and road, is a nature eserve managed by the Northamptonshire Trust for Iature Comservation. A fine raised hide gives xcellent views of the substantial numbers of wildfowl sing the area (Table 83). Feeding conditions are avoured by the fluctuating water levels in this hallower end of the reservoir. Most species have ncreased following the establishment of the reserve in ne 1970s, although Wigeon and Pochard no longer find the place so attractive. The numbers of Goldeneyes are notable as are those of Goosanders. Further to the west, but also impoundments of branches of the same tributary of the Nene, are two smaller reservoirs also carrying good numbers of wildfowl. These are Hollowell and Ravensthorpe (Table 83). At Hollowell (54ha), where there is fishing from bank and boat as well as sailing, Mallard numbers have dropped from the 1960s regular (and maximum) level of 420 (1,000) but those of other species have remained much the same, except that Tufted Ducks and Goldeneyes have increased from 57 (100) and 11 (24) respectively. There is a reserve on one arm of the nearby 114ha Ravensthorpe Reservoir where the numbers of wildfowl have increased from regular levels of 215 Mallard, 35 Teal, 8 Shoveler, 69 Pochard, 49 Tufted Ducks, 2 Goldeneyes, 2 Mute Swans and no Canada Geese in the 1960s. There is, however, considerable interchange between these two reservoirs, partly due to recreational disturbance at Hollowell.

Finally, in the headwater branches of the Nene near Daventry, there are two more reservoirs. Drayton is small and supports only up to 100 Mallard and 20 Tufted Ducks, other species being in single figures. Daventry Reservoir is both larger and carries more ducks (Table 83), at least on passage. Sailing is permitted and fishing allowed from the bank on one side. There is an observation hide, with a path all round the perimeter. This and other canal-feeder reservoirs in the area were constructed to maintain canal water levels over the hills between the Thames/ Avon and the Trent. Neither large nor deep they were given over to recreation at a time, before the 1960s,

'able 83. *Regular (and maximum) numbers of wildfowl recorded on five sites in the upper Nene Valley.*

) Billing/Clifford Hill Gravel Pits 4 seasons 1979-1982
) Pitsford Reservoir i) 4 seasons 1964-1969
ii) 9 seasons 1970-1982
c) Hollowell Reservoir 13 seasons 1970-1982
d) Ravensthorpe Reservoir 13 seasons 1970-1982
e) Daventry Reservoir 5 seasons 1972-1982

	a	b i	b ii	c	d	e
Mallard	888(1496)	488(1000)	862(1548)	191(1035)	284(1181)	233(635)
'eal	49 (155)	257 (600)	393(1185)	74 (240)	97 (555)	21(155)
Nigeon	134 (180)	1038(2000)	546(1700)	184 (384)	62 (240)	7(135)
hoveler	6 (8)	6 (100)	53 (175)	12 (54)	12 (65)	37(120)
adwall	6 (16)	0 (1)	14 (104)	1 (15)	0 (3)	0 (6)
ochard	220 (355)	487(3000)	284(1100)	137 (365)	197 (855)	60(368)
'ufted Duck	385 (709)	95 (300)	365 (985)	131 (245)	151 (275)	36(140)
oldeneye	2 (5)	13 (35)	37 (95)	31 (71)	10 (84)	1 (16)
oosander	8 (35)	15 (45)	21 (57)	3 (16)	1 (5)	1 (9)
Aute Swan	54 (106)	5 (60)	25 (64)	3 (17)	2 (8)	2 (11)
reylag Goose	19 (54)	0 (0)	0 (3)	0 (1)	0 (0)	0 (0)
anada Goose	67 (94)	0 (0)	89 (358)	47 (175)	56 (210)	9 (60)

when the hygiene considerations of the period made drinking water reservoirs effectively wildfowl refuges – a policy later sadly reversed. The lakes at Fawsley to the south have maxima of only 40-50 Mallard, Pochard and Tufted Ducks. They are much disturbed and the birds tend to move to Daventry.

The Welland

Unlike the Ouse and Nene, the Welland does not have a chain of gravel pits along its middle course. However, any loss of floodland habitat has been more than replaced by two large reservoirs. The first is Rutland Water, formed by damming the tributary Gwash, less than 10km upstream from its junction with the Welland at Stamford. Initially named Empingham Reservoir from the small village nearby, it was given its present title in memory of the small county it partly flooded and which was finally eliminated by boundary changes. Its great size, 1,260ha, makes it the largest man-made water in Britain, surpassing even the new Kielder Water (p.206), and nearly the size of the largest natural eutrophic lake in Britain, Loch Leven at 1,403ha (p.246). The advent of such an inland sea (it began flooding in 1975 and reached top level in 1979) has naturally had profound consequences for waterfowl in the region.

Rutland Water is not only outstanding in sheer size, but also for the careful planning for future use that occurred even before construction began. Right from the start the conservation interest was considered, along with the more obvious potential for waterborne recreation. Its shape, with two large arms separated by the high peninsula carrying the village of Upper Hambleton, was of considerable help in this respect. The nature reserve, run by the Leicestershire and Rutland Trust for Nature Conservation, supported by the Water Authority, is located on these two western arms (13, see also Fig 181). Sailing boats are not allowed access to 425ha adjacent to the 142ha nature reserve and even the low-profile fishing boats, present in summer only, are not allowed within 50m of the reserve shores, from which bank fishing is also excluded. Bunded lagoons were constructed in the southern arm and they and former fishponds in the other arm, in all some 80ha, are also kept boat-free. The three separate lagoons can have their water levels controlled independently of the draw-down of the reservoir and ideal feeding conditions for migrant waders are created. (This is now one of the most important inland wader sites in Britain, with up to 19 species being recorded in a day.) Since 1975 a full management programme has included the planting of bushes and reeds and the construction of islands in the lagoons, and of trails to 12 birdwatching hides. (In 1982 there were 8,000 visitors to the reserve and in 1983 13,000.) The effect of reserve creation is discussed in more detail on p.544.

The wildfowl recorded at Rutland Water (Table 84) are spectacular in numbers and variety, 37 species

Table 84. *Regular (and maximum) numbers of wildfowl recorded at three sites in the Welland basin.*

a) Rutland Water 8 seasons 1975-1982
b) Eyebrook Reservoir i) 10 seasons 1960-1969
ii) 13 seasons 1970-1982
c) Saddington Reservoir 9 seasons 1970-1982

	a	b i	b ii	c
Mallard	1707(2961)	831(2100)	884(1750)	27 (81)
Teal	931(2038)	184 (590)	353 (831)	15 (86)
Wigeon	2208(4518)	676(1600)	820(1750)	2 (32)
Pintail	22 (60)	2 (17)	2 (26)	0 (0)
Shoveler	254 (616)	17 (90)	27 (170)	4 (26)
Gadwall	168 (493)	0 (0)	2 (23)	0 (0)
Pochard	631(1556)	150 (310)	176(1260)	25(179)
Tufted Duck	1271(2380)	45 (125)	142 (419)	31(120)
Goldeneye	107 (219)	11 (22)	19 (52)	0 (1)
Goosander	23 (125)	34 (94)	23 (63)	0 (2)
Ruddy Duck	10 (60)	0 (0)	3 (43)	0 (4)
Mute Swan	49 (184)	2 (18)	4 (13)	2 (6)
Bewick's Swan	9 (76)	11 (52)	8 (69)	0 (0)
Canada Goose	41 (194)	6 (22)	82 (310)	2 (12)

of ducks, geese and swans having already been recorded in its short history. The initial flooding effect discussed under Grafham Water (p.154) was also present here though the creation of reserves modified future trends (Fig 171). Leaving aside the 1975 season, when flooding had only just commenced, we may compare regular numbers for the three seasons 1976 to 1978 with the four seasons 1979 to 1982. Mallard, with 2,119 versus 1,637, and Pochard, with 964 v 479, show the most marked fall away. In the case of Teal (1,165 v 931), Wigeon (2,290 v 2,021) and Shoveler (303 v 285) the effect is slighter. Considerable gains have occurred with Gadwall (61 v 285) and Rutland has already surpassed the peak numbers at Gunton Park, Norfolk (p.138), and must now be considered the main stronghold of this species in Britain. This was further confirmed in September 1983 when the peak leapt to a remarkable 796, and in December of the same year when 947 were counted. The average annual peak is rapidly approaching the international level of importance (namely 550) for this species. The ducks mainly dependent on animal food have also increased as the reservoir's fish and invertebrate fauna matures; Tufted Ducks (1,022 v 1,832) have established Rutland Water as their major site in England, surpassed in Britain only by Loch Leven (p.246). The peak was further raised in September 1983 to 2,404. Goldeneyes (75 v 155) and Goosanders (17 v 31) have also shown expected increases. The feral Ruddy Ducks appear to have established their most substantial easterly population, 15-60 being present through the season. Wildfowl are also beginning to use Rutland Water as a moulting site, as witness a mid-August 1983 count of 1,700 Mallard, 829 Teal, 200 Gadwall, 524 Shoveler (more than any winter count), 1,032 Pochard, 3,062 Tufted Ducks (again an all-time peak), 145 Mute Swans and even 13 Garganey. At least 50 Mallard, 25 Tufted Ducks and 14 Mute Swans breed in the reserve area alone, along with a few Teal, Shoveler, Gadwall, Shelducks, Pochard, Ruddy Ducks, Greylags and Canada Geese. Contrary to the general trend at the edge of its range, Garganey was confirmed as a breeder for the first time in 1983.

Barely 10km south-west of Rutland Water is another reservoir, formed by damming the Eyebrook tributary of the Welland. This is much smaller (164ha) but longer established (flooded in 1934) than Rutland Water. The numbers on Eyebrook Reservoir (Table 84) do not appear to have been diminished by the advent of its giant neighbour. Indeed, compared with the 1960s, the numbers of most species have increased in recent years; only Goosanders would appear to have moved over to the new water to some extent. Sailing has not been allowed, but this is an important fishing site, otherwise maintained as a reserve. The reservoir was built to supply water to steelworks in Corby; now that these have closed, its future is in some doubt.

Further upstream there are several small park lakes north of the Welland on which there are no recent data, but which held 1-200 Mallard in the 1960s: Blaston, Keythorpe, Rolleston, Noseley and Gumley. The only other water of consequence before the watershed west of Market Harborough is the canal-feeder, 18ha Saddington Reservoir, 7km north-west of the town. This has but small regular numbers (Table 84) probably due to disturbance, but the considerably higher peaks indicate its use by passage migrants.

The Humber catchment

The Humber drains a vast area of England, some 26,000sq km, or around 20% of the land area. It is formed by the confluence of the Trent and Yorkshire Ouse, with minor contributions by the Ancholme from the south and the Hull from the north. We consider first the coast on either side of the Humber, then the estuary itself and then, in a clockwise direction, the contributory rivers and the various waters associated with them.

The Humber coast and estuary

The Humber coast

To the north of the Wash, the Lincolnshire coast from Skegness to Grimsby is flanked throughout by dunes and open sandy beaches. Of interest are the numerous borrow-pits from which clay has been taken to strengthen the sea walls. Huttoft Bank Pits, 5km south-east of Sutton-on Sea (Table 85), provide a good example of the numbers and species mix to be found.

Just south of the Humber, much of the area between Donna Nook and Saltfleet is used as a bombing range, but the Lincolnshire Trust for Nature Conservation has a management agreement with the Ministry of Defence. To the south of Saltfleet, the dunes betweens Saltfleetby and Theddlethorpe constitute a 440ha National Nature Reserve. There is a substantial passage of dabbling ducks, and sea ducks are often seen, including 320 Common Scoters in 1979. The Humber Brent Geese have their centre in this area. Their numbers have shown a welcome increase in recent years, reaching a peak of 2,500 in 1983. The counts of Brent Geese in the Humber are given, with those for other sites from Norfolk northwards, in Table 74 (p.142).

The small number of sites a few kilometres inland, short of the Lincolnshire Wolds, hold notable numbers. Fulstow Fish Pits, 10km north of Louth, have a regular (and maximum) 27 (90) Mallard and 123

(215) Tufted Ducks. The 88ha Covenham Reservoir (Table 85) is only 2km south of the pits and no doubt associated with them; it has good numbers of both dabbling and diving ducks, including Goldeneyes. Some 15km to the west, the two ponds at Croxby, despite their largely wooded margins, hold regular (and maximum) flocks of 220 (375) Mallard and 53 (115) Teal.

North of the Humber, between Spurn and Flamborough Heads, low clay cliffs face onto sandy beaches and offshore banks of coarse shingle. The numbers of wildfowl are consequently small, though parties of dabbling ducks may briefly appear. Common Scoters assemble during the summer in the shallow waters of Bridlington Bay and several hundred may be seen in winter there or off Spurn Head. A few kilometres inland, just west of the village of Aldbrough, is the Lambwath Stream, which occasionally floods, holding up to 750 Wigeon, 650 Teal and 400 Mallard in most winters. Much more impor tant are the massive flocks of several species to b found on the 230ha Hornsea Mere, lying within kilometre of the sea (Table 85). This has for long bee noted as a breeding and wintering ground for ducks Hornsea Mere is a shallow (3.4m) kettle-hole in ric agricultural land and was already extremely eut rophic, with very high phosphate levels, in the earl 1970s (492). Dense algal crops occur and the paucity o the insect fauna might reflect the beginning of further deterioration. Some disturbance from boatin and angling occurs, but the site is an RSPB reserve. I the 1960s all species were considerably more plentifu than they had been in the 1950s, but in more recen years most have fallen back to the level of the 1950s though Shoveler and feral newcomers like Gadwa and the two geese have maintained an increase Besides the species tabulated, numerous other appear sporadically. Mallard, Shoveler, Gadwall Pochard and Tufted Duck breed regularly and Tea hatch in some years. The Canada Geese, which ar

Fig. 26. Humberside and the lower Vale of Trent.

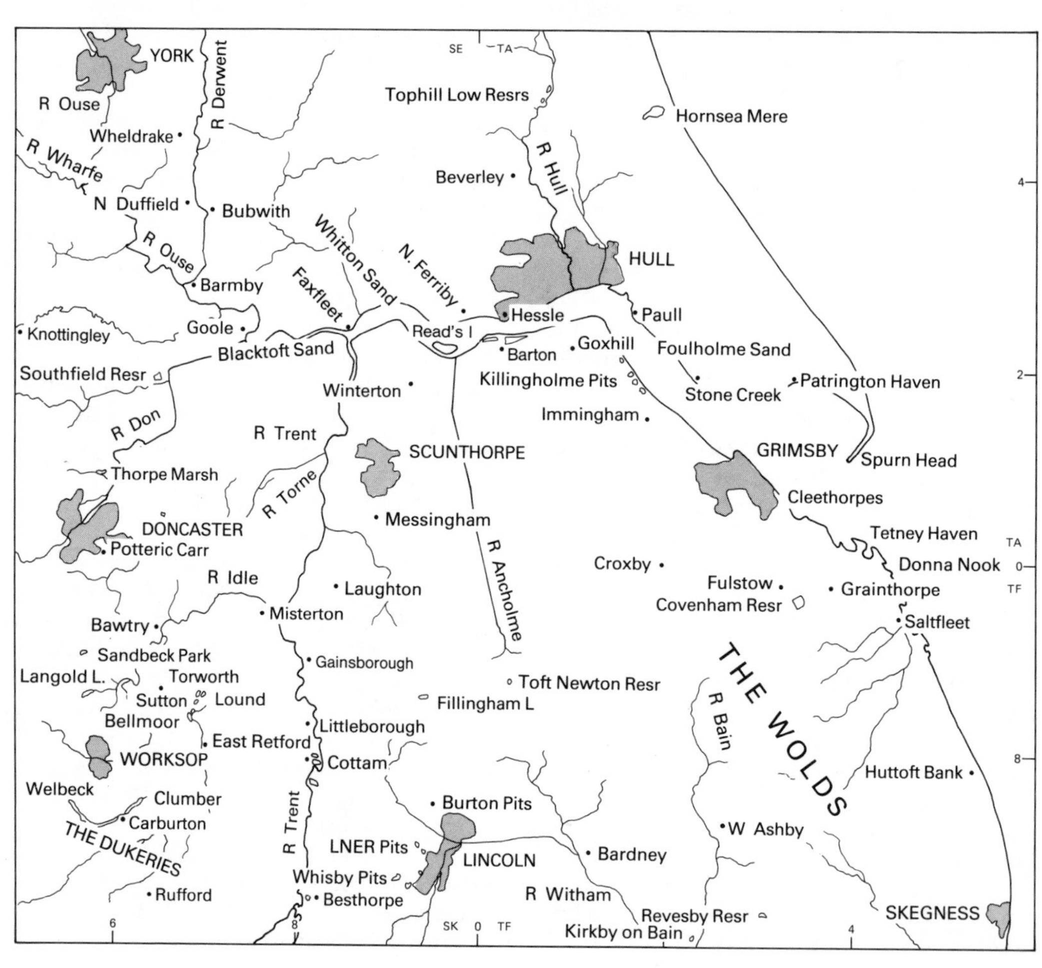

based on Hornsea and East Park Lake, Hull (see below), form a separate group from the main Yorkshire population and were estimated to total 190 birds in July 1976 (405). There have been no counts from East Park since the late 1960s, but recent maxima from Hornsea indicate no increase on the 1976 levels.

Other near-coastal sites in the Hornsea area include the large number of gravel pits in the Brandesburton complex, which, rather surprisingly, have small numbers of wildfowl, seldom exceeding 50, though coverage has been sporadic. Farther north near Bridlington standing water is scarce, but Burton Agnes Mere has recorded up to 105 Mallard. Beyond Flamborough Head up to Robin Hood's Bay the coast becomes increasingly rocky, with only small numbers of wildfowl offshore. Very few stretches of coast are surveyed here, but Cornelian Bay, Scarborough, has recorded 73 Scaup, 47 Eiders, and about 20 Mallard, Tufted Ducks, Goldeneyes and Common Scoters. Coastal freshwater does, however, attract the ducks; for example Hunmanby Mere near Filey, and Seamer Road Mere, Scarborough, hold 1-200 Mallard and small numbers of Pochard and Tufted Ducks.

The Humber Estuary

The Humber is a long, relatively narrow estuary. From Spurn Point to Trent Falls, where the Trent and Ouse rivers join to form the Humber, is a distance of 62km. Over much of this length it is considerably less wide than the 7.5km at Grimsby. The Humber Advisory Group and Hull University published a valuable set of papers concerning many aspects of the estuary's structure and dynamics (277). In the lower reaches thousands of hectares have been reclaimed from the estuary, around Sunk Island on the northern shore in particular. In the shelter of Spurn Head, however, some 2,400ha of mud and sandflats are largely untouched, though under threat with natural infilling. Invertebrate food resources are locally rich and this area is one of the most favoured by wildfowl. On the Kilnsea Clays at Spurn Point the highest numbers recorded since 1970 have included 1,020 Mallard, 45 Teal, 160 Wigeon, 253 Shelducks, 365 Scaup and 339 Dark-bellied Brent Geese. Comparable maxima for the clays from there to Patrington Haven are 5,250 Mallard, 200 Teal, 1,680 Wigeon, 1,000 Shelducks and 330 Brent. Up to 1,000 Mallard, 80 Teal, 350 Wigeon, 400 Pintail and 350 Shelducks are to be found on the next section, Patrington Haven to Stone Creek, while the shore between Stone Creek and Paull has been counted frequently enough to give the regular/maximum figures in Table 86. Most of the wildfowl here are on Foulholme Sand, where a small flock of moulting Shelducks was found in July 1978 (573). At Paull the built-up waterfront of Hull starts, and only relatively small numbers of wildfowl are found with any regularity between there and Hessle, which can be taken as the limit of the lower Humber, the estuary at that point being only 2.4km wide. The numbers of wildfowl have, however, increased recently, Teal reaching a maximum of 755, as many as in any other part of the Humber, in March 1981.

Table 85. *Regular (and maximum) numbers of wildfowl at three sites on or near the coast south (a,b) and north (c) of the Humber.*

a) Huttoft Bank Pits, Sutton on Sea 5 seasons 1963-1982
b) Covenham Reservoir 4 seasons 1971-1981
c) Hornsea Mere i) 9 seasons 1960-1969
ii) 11 seasons 1970-1982

	a	b	ci	cii
Mallard	184(400)	684(1250)	2459(5190)	1879(8000)
Teal	30(500)	6 (25)	401(1700)	188(1600)
Wigeon	28(400)	29 (113)	495(3000)	268(2100)
Pintail	2 (20)	2 (5)	6 (61)	10 (125)
Shoveler	6 (12)	1 (3)	39 (460)	78 (240)
Gadwall	0 (40)	2 (6)	14 (70)	93 (275)
Pochard	14 (60)	18 (62)	820(2150)	357(1525)
Tufted Duck	7 (30)	123 (199)	434 (728)	484(1035)
Goldeneye	1 (6)	68 (105)	185 (480)	123 (358)
Goosander	0 (1)	0 (2)	44 (147)	11 (218)
Mute Swan	5 (11)	1 (4)	41 (175)	41 (185)
Greylag Goose	0 (1)	0 (0)	0 (28)	112 (416)
Canada Goose	1 (5)	0 (1)	18 (55)	76 (161)

The Grainthorpe Basin, in the southern part of the outer Humber, is counted only irregularly, but annual peaks of 400 Wigeon and Brent Geese, 800 Shelducks and 60 Teal are not unusual, with higher numbers in severe weather. Tetney Haven (where there is an RSPB reserve of 126ha) records up to 50 Mallard, 200 Teal, 1,000 Wigeon and 1,200 Shelducks. Brent Geese also spread north to this area in flocks of 1-400 from their headquarters in the Saltfleet area.

The southern coast of the lower estuary has a much narrower foreshore, lined by farmland except for the built-up frontages of Cleethorpes and Grimsby, and the docks at Immingham and North Killingholme. Wildfowl, as would be expected, are relatively scarce, only Shelducks forming any concentrations, with maxima of 660 off Grimsby Docks and 1,530 between Grimsby and Immingham. Small numbers of sea ducks are seen regularly in the outermost parts of the estuary, and there is a sizeable passage of scoters flying westwards in autumn.

A feature of the southern coast is the number of small freshwater excavations (mainly borrow-pits) just inland. At Cleethorpes the two boating pools hold 1-200 Mallard; Fisons Lagoon at Immingham regularly has 60 Pochard, 20 Tufted Ducks and 30 Mute Swans, plus an autumn gathering of up to 50 Wigeon and 50 Shoveler; the Haven Pits at North Killingholme are used by refugees from the estuary of many species, and also support 40-90 Mute Swans. The marshy area near Goxhill, including a number of small pits, has held up to 360 Wigeon and 200 Pochard after hard weather, but otherwise carries little. These workings, together with those at Killingholme, support a number of breeding wildfowl, including Pochard and Shoveler. The pits between Barrow and Barton have been regularly counted in recent years, together with the adjacent foreshore; the results are given in Table 86.

The southern shore of the upper Humber, above Barton, contains 5,000ha of medium to low-level silt and sandflats, in a tidal range of 6m. It has similar species to those occurring elsewhere on the Humber but in no great numbers. Thus from Chowder Ness to South Ferriby there are 50-100 Mallard and occasional influxes of several hundred Teal, Wigeon or Shelducks, while South Ferriby Pits have small numbers of Teal and Shelducks. Only at Read's Island is there a sizeable concentration, amounting in 1982 to regular (and maximum) numbers of 438 (550) Mallard, 28 (61) Teal, 355 (400) Wigeon and 64 (100) Shelducks. Along the northern shore, substantial regular flocks of 280 Mallard, with up to 1,200 at peak, occur between Hessle and North Ferriby. This stretch also has up to 150 Shelducks. The area is, however, dominated by the assemblage of wildfowl on the Whitton Sand and adjacent saltmarsh. These lie within the statutory Humber Wildfowl Refuge, established in 1955 with the particular aim of protecting the roost of Pink-footed Geese, which had somewhat earlier moved there from the Read's Island flats, now rarely used. At the time the Humber was the most important English site for this species, exceeding the Wash and the Ribble, and having a regular autumn population of 6,000, with peaks up to 10,000. However, hardly had

Table 86. *Regular (and maximum) numbers of wildfowl at four major sites on the Humber, and on the estuary as a whole.*

a) Stone Creek to Paull — 7 seasons 1970-1982
b) Barrow/Barton Pits and foreshore — 3 seasons 1980-1982
c) Humber Wildfowl Refuge — i) 8 seasons 1961-1969; ii) 13 seasons 1970-1982
d) Faxfleet to Goole — 4 seasons 1969-1980
e) Humber complete: Donna Nook/Spurn to Goole — 9 seasons 1974-1982

	a	b	ci	cii	d	e
Mallard	423(1250)	65(950)	2815(7500)	2740(8140)	391(1200)	4560(8430)
Teal	84 (500)	9(340)	537(1200)	649(3163)	268(1100)	1040(3663)
Wigeon	333(2000)	6 (23)	541(2170)	2527(9600)	32(1250)	3900(9740)
Pintail	104 (400)	0 (10)	5 (21)	15 (154)	4 (33)	107 (800)
Shelduck	167 (800)	8(130)	108 (317)	284 (800)	47 (132)	1130(2408)
Pochard	0 (2)	71(250)	1 (11)	14 (400)	0 (5)	61 (600)
Tufted Duck	0 (3)	55(100)	16 (200)	6 (50)	0 (2)	44 (200)
Goldeneye	0 (0)	25 (48)	4 (43)	1 (60)	0 (4)	21 (60)
Mute Swan	0 (0)	18 (63)	1 (6)	3 (15)	1 (4)	22 (94)
Pink-footed Goose	0 (10)	0 (35)	1516(6000)	591(2826)	3 (35)	430(2020)
Canada Goose	0 (0)	42(150)	0 (5)	1 (12)	0 (0)	14 (150)

the Refuge been established than the geese started to decline. This was probably due to short-stopping in Scotland, through the greater amount of autumn feed available on barley stubbles, and to the preferential development of the Ribble concentration. The geese range far over the Yorkshire and Lincolnshire Wolds when feeding on stubble fields in autumn, flights of 30km being commonplace. In winter, feeding is generally closer on the reclaimed land. In January 1975, geese feeding near Coleby, 5km to the south, suffered mass poisoning through ingesting wheat treated with an organophosphorus pesticide (233). The grains were accessible on the surface following poor sowing in wet conditions. The bodies of 243 birds were recovered on the fields or on the roost. The total kill was undoubtedly higher and the Humber population slumped from 1,300 before the incident to 300 shortly afterwards. Only 473 Pinkfeet were counted at peak the following winter and it was February 1979 before numbers again touched the thousand mark. The protection afforded on the Refuge has, however, enabled Mallard and Teal to maintain their numbers and the Wigeon have increased fivefold.

Although no longer the Humber, the stretch of the Ouse from Faxfleet to Goole, some dozen kilometres, is generally considered as part of the estuary. Despite being narrow it has, in total, quite a useful population of wildfowl (Table 86). The counts include a small length of shore between the Trent and Ouse as well as the mouth of the Ouse.

The importance of the Humber is centred around the Whitton Sand and the 13km of mudflats west of Spurn Point, but the area as a whole can be considered as being of international importance. Since 1974 counts have been sufficiently extensive to give a good idea of the total number of wildfowl using the estuary. The regular total wildfowl population is 10,600, the maximum being 18,882 in January 1980. Wigeon, Shelduck and Brent Goose are individually of international importance, according to the average annual maximum. The rather lower regular figures in Table 86 show that the peak is short-lived. The species for which the Humber is most noteworthy, however, is the Mallard (Table 86); the Humber is easily the most important area for this species in Britain.

Despite these healthy numbers, the Hull University studies (277) pointed out that ten million people live in the catchment area, which produces 25% of the nation's industrial output and 62% of its coal. The inflow of sewage and industrial waste is considerable through the River Trent and the Ouse and a substantial effluent load is discharged directly into the estuary. The possibilities for deterioration of water quality are thus manifest. Nevertheless, the Hull studies concluded that dissolved oxygen rarely fell below 50% saturation in the middle and outer estuary, but it was very variable west of Brough. Moreover, the Humber is relatively unpolluted by heavy metals. Although background pollution may not have reached dangerous levels, the vulnerability of the long, narrow estuary to acute pollution was dramatically demonstrated when in the autumn of 1983 an oil-tanker holed itself on an Immingham jetty and released 6,000 tons of oil. Some 10,000 birds, mostly gulls, were contaminated to some extent, but fortunate combinations of tide and wind prevented a major disaster and only a few hundred birds, including Shelducks, were found dead.

The lower Vale of Trent

The marshes of the now-canalised River Ancholme, which enters the Humber south of Read's Island, have been drained and cultivated, and many of the small pools on the heathland of the Lincolnshire Ridge have been rendered valueless to wildfowl by afforestation. In the western districts thousands of hectares of marsh and mere have been reclaimed around the Isle of Axholme and on the Thorne, Goole and Hatfield moors. For a while the loss of natural resorts was offset by the practice of warping – the deliberate autumnal flooding of land near the rivers to allow suspended silt to settle on and fertilise the fields (331). Now the gradual raising of the land and the availability of artificial fertilisers has led to the abandonment of the practice and the wildfowl are centred mainly on scattered lakes and gravel pits, but make use of the agricultural land for feeding.

Where the Trent joins the Yorkshire Ouse to form the Humber, an RSPB reserve has been established on Blacktoft Sand. This comprises a total of 186ha including a tidal reed-bed with a saltmarsh fringe and artificial lagoons, mainly of interest to breeding birds but also holding large numbers of wintering Teal. Alkborough Flats, on the opposite shore of the Trent, record regular (and maximum) numbers of 18 (165) Mallard, 157 (215) Teal and 39 (86) Shelducks.

Further inland, Bagmoor Mines near Winterton, harbour 130 (209) Mallard, 27 (156) Teal and 19 (54) Tufted Ducks, while the Crosby Warren Gravel Pits on the outskirts of Scunthorpe have 289 (420) Mallard, 52 (70) Pochard and a few Tufted Ducks. The Twigmore Gull Ponds (or Black Head Ponds), 5km south-east of Scunthorpe, used to be a good example of heathland meres and in the 1950s supported a regular 240 Mallard, 95 Teal and 15 Shoveler. These levels were maintained through the 1960s, but more recently are down to 5-15. The pools have woodland surrounds and there has been a great deal of conifer planting; the decline in the ducks is, however, thought to have been caused by the increase in the number of other waters nearby. Messingham sand quarries (Table 87) have a varied and quite substantial wildfowl population. Ten kilometres north of Gainsborough, Laughton Lake held up to 1,250 Mallard, 600 Teal and 80 Shoveler in the 1960s, and the nearby floods of Corringham Scroggs up to 600 Mallard and 80 Teal. The Laughton Forest has been extensively planted with conifers since the Second World War and there are numerous pools which still contain Mallard and Teal, but there is no recent information from Corringham.

Just upstream of Gainsborough, the Bole Ings, including in the counts sometimes-flooded sites at the West Burton Power Station, have in mid-winter recently held up to 200 Mallard and Teal, and 17 Bewick's Swans, but the Lea Marsh, just across the river, has not been counted. The Trent itself at Littleborough carried up to 900 Mallard, 1,300 Teal and 500 Wigeon in the early 1970s, with occasional influxes of other species, but wildfowl numbers, like floods, have been low in the past few years. The lagoons of Cottam Power Station (Table 87), almost on the river bank, provide an extensive water area which holds substantial numbers of both dabbling and diving ducks. The rest of the useful wildfowl habitat in the Trent Valley up to the environs of Newark is mainly provided by flooded gravel pits. The pool at Ossington, 5km from the river, and the pits at Cromwell just to the south-east, have regular populations of under 100 wildfowl; those at Indies Farm, near Spalford, a regular (and maximum) 161 (202) Mallard, 74 (82) Teal and 29 (32) Tufted Ducks. The pit complexes at Besthorpe/Girton and at South Muskham on the outskirts of Newark (Table 87) are, however, much more interesting. The Besthorpe complex has seen considerable increases in dabbling ducks since the 1960s, when only 200 Mallard and 100 Teal and Wigeon were regular, whereas diving ducks have shown little change. Six other waters near Newark have been counted, but have little of consequence, though a riverside site – East Stoke Backwater – has records of up to 700 Mallard.

Just north of Gainsborough the Trent is joined by the River Idle from the west and we may trace the tributary's course, which for much of its length roughly parallels the Trent. Between Misterton and Bawtry there are areas of grassland which have served as flood relief sites, the Misson Washlands. The number of wildfowl there varied widely with the extent of flooding, but in the period 1966 to 1980 there were up to 309 Mallard, 800 Teal, 900 Wigeon, 80 Pintail, 45 Shoveler, 26 Gadwall, 10 Shelducks, 1,489 Pochard, 170 Tufted Ducks, 84 Mute Swans, 15 Whooper Swans, 165 Bewick's Swans and 19 Canada Geese. The fluctuating conditions prevented the total reaching international significance for Bewick's Swans and, sadly, a new flood prevention scheme, now in operation, will prevent the return of their former glory. There is a small area beside the Idle where the

Table 87. *Regular (and maximum) numbers of wildfowl recorded at six artificial sites in the lower Trent and Idle Valleys.*

a) Messingham Sand Pits 7 seasons 1976-1982
b) Cottam Power Station 9 seasons 1971-1982
c) Besthorpe/Girton GPs 13 seasons 1970-1982
d) S Muskham/N Newark GPs 10 seasons 1973-1982
e) Torworth GPs 4 seasons 1970-1974
f) Sutton/Lound GPs 11 seasons 1970-1982

	a	b	c	d	e	f
Mallard	43(165)	659(1400)	471(1350)	112(409)	71(125)	232(600)
Teal	70(150)	137 (500)	134 (480)	75(320)	115(250)	43(310)
Wigeon	0 (0)	241 (700)	266(1000)	63(212)	5 (30)	20(211)
Shoveler	10 (44)	9 (36)	9 (72)	1 (12)	12 (32)	12 (38)
Pochard	55(149)	120 (380)	43 (130)	66(226)	6 (15)	117(300)
Tufted Duck	46 (80)	107 (237)	92 (400)	94(250)	24 (65)	152(264)
Goldeneye	2 (15)	6 (16)	3 (14)	1 (8)	0 (1)	11 (24)
Goosander	0 (3)	1 (11)	21 (62)	1 (14)	0 (0)	0 (5)
Canada Goose	26(106)	4 (33)	25 (206)	26(185)	1 (4)	13 (70)

Nottinghamshire Trust can hold some floodwater, but this site is hardly tenable without the more extensive washland floods. The demise of the area is illustrated by the counts for 1981-82 when, despite some flooding, in five of the seven counts there were fewer than 50 wildfowl, with a maximum of only 216, compared to a usual peak previously of 1-2,000. In 1982-3 there was no flooding at all and an average of less than 50 wildfowl (maximum 82) in the five counts November – March.

There are eight waters with counts around Bawtry and East Retford of which the gravel pits at Torworth and the large complex of Sutton/Lound just north of Retford are recorded in detail (Table 87). Torworth is remarkable in that Mallard are not the most numerous species. The small pools have recently been amalgamated into two lakes by the Nottinghamshire County Council and one of these is set aside for birds, so this site should increase its wildfowl interest. Of the remainder, Sandbeck Lake, an impoundment of a small stream at Sandbeck Hall, 8km west of Bawtry, holds up to 260 Mallard and 64 Tufted Ducks. The gravel pits at Ranskill are very close to the complexes at Torworth and Lound and, no doubt, share wildfowl with them. These pits have held up to 260 Mallard, 120 Teal, 22 Shoveler, 27 Pochard, 60 Tufted Ducks and 23 Mute Swans and, in one year at least, had 47 Bewick's Swans. Five kilometres north of Bawtry the several pits around Finningley and Blaxton attract combined peaks of 200 Mallard and 180 Teal.

The Idle and its branches farther south drain the area of Nottinghamshire named with Victorian grandeur as the Dukeries. Strictly this comprises the demesnes of Worksop Manor, Welbeck Abbey, Clumber House and Thoresby Hall. The noblemen in their heyday sought to diversify this remnant of Sherwood Forest by building weirs on the rivers to form long, picturesque lakes which are of considerable attraction to wildfowl. The largest, and the one attracting the most wildfowl, is Welbeck Great Lake, an impoundment of the River Poulter and Millwood Brook which stretches for more than 5km. Table 88 gives details, Rufford being included because of its proximity – indeed these lakes are all within 10km of each other. Worksop Manor has several small pools and lacks recent data, but has carried up to 400 Mallard and 50 Canada Geese; all the lakes are of especial importance to these species. The Dukeries form the headquarters of the north Nottinghamshire population of Canada Geese, thought, on considerable ringing evidence, to be a distinct subpopulation and estimated to number 1,300-1,400 birds (612), an increase from 850 in 1976 (405). This total includes many smaller sites not included in the Wildfowl Counts.

The middle Vale of Trent

The Trent forms a natural route for wildfowl across the heart of England, the loss of its floodlands having been offset by the proliferation of gravel excavations and reservoirs. This section covers the Trent from Newark to Burton and its tributaries: the Devon and Soar to the south and the Derwent and Dove to the north. Derbyshire, Nottinghamshire and Leicestershire contribute to the area.

The Devon river joins the Trent at Newark, having amalgamated with the Smite a little to the south. The source of the Devon is close to that of the Eye which runs west to join the Wreake and the Soar before running north to enter the Trent at Nottingham. The associated waters thus form an arc which borders on the Fenland/Wash region, and have close links with waters along the Nene and Welland. Following this arc round from the east, we come first to empoundments on the Devon. That at Belvoir has been assiduously counted over 23 years but yields smallish regular (and maximum) numbers: 89 (300)

Table 88. *Regular (and maximum) numbers of wildfowl recorded at five artificial lakes in the Dukeries district of Nottinghamshire.*

a) Welbeck Great Lake 13 seasons 1970-1982
b) Clumber Park Lake 13 seasons 1970-1982
c) Carburton Dams 13 seasons 1970-1982
d) Thoresby Lake 13 seasons 1970-1982
e) Rufford Lake 12 seasons 1970-1982

	a	b	c	d	e
Mallard	512(1276)	551(1014)	158(620)	320(750)	135(281)
Tea	10 (100)	1 (4)	1 (15)	55(196)	0 (0)
Pochard	108 (364)	20 (57)	26(162)	50(148)	7 (32)
Tufted Duck	65 (330)	108 (251)	134(356)	58(210)	8 (31)
Mute Swan	8 (35)	25 (48)	9 (51)	11(125)	0 (5)
Canada Goose	300 (800)	266 (802)	145(770)	44(200)	44(400)

Mallard, 48 (170) Wigeon, 32 (106) Tufted Ducks and 67 (190) Canada Geese. Knipton Reservoir is likewise well counted but of limited value, with 215 (900) Mallard, 12 (140) Wigeon, 23 (100) Tufted Ducks and 15 (200) Canada Geese. The River Smite is of little interest, though 200 Mallard and Teal have been noted on flooded meadows near Langar.

By the River Eye, the ornamental pools at Stapleford Park have up to 155 Mallard and 85 Canada Geese, but little else. The gravel pits at Kirby Bellars near Melton (Table 89) and Frisby have similar populations, though those at Frisby are smaller, with regular (and maximum) numbers of 44 (200) Mallard, 11 (80) Teal, 16 (88) Pochard, 48 (108) Tufted Ducks and 17 (37) Canada Geese. At Wanlip, where the Eye, now the Wreake, joins the Soar, the largest gravel pit complex in this area is found in the triangle between the Soar, Wreake and the Grand Union Canal (Table 89), and the numbers here have shown considerable increases since the 1960s. The flocks of Gadwall are unusually large for the district. Near Leicester, three man-made waters – Groby Pool, Cropston Reservoir and Swithland Reservoir (Table 89) – have all shown considerable increases since the 1960s, of the same order as those at Swithland. Groby and especially Swithland have regular populations of the feral Ruddy Duck, though it is as yet a predominantly upper Trent bird. Blackbrook Reservoir is of less importance, with a regular (and maximum) 158 (400) Mallard, 31 (68) Teal, 20 (110) Pochard and 40 (80) Tufted Ducks, and the dabbling ducks there have actually declined in the last 20 years. Thornton Reservoir also has a limited number of species, 112 (350) Mallard, 39 (190) Pochard and 29 (69) Tufted Ducks, and other waters in the area are of little interest to wildfowl.

Returning now to the River Trent we trace its course upstream from Newark to Burton. Gravel pits again form the major habitat, beginning with the large

Fig. 27. The middle and upper Vale of Trent.

complex between Hoveringham and Bleasby (Table 90). These have shown considerable increases for some species since the 1960s (when only 25 Teal and Wigeon and 55 Tufted Ducks were regular), despite the fact that infilling with fly-ash has continued to erode the area of water. The Gunthorpe pits are smaller and have fewer birds in the same species mix. Stoke Sewage Farm, 1km west of the river, which in the early 1960s regularly supported 57 Mallard and 103 Teal, has been rendered valueless by modernisation. On the south-eastern outskirts of Nottingham the Holme Pierrepont Gravel Pits (Table 90) with the Netherfield pits north of the river and the Colwick Country Park waters to the west, constitute a substantial complex. The pits at Netherfield are of rather little importance but those at Colwick are also listed in the table. The Holme Pierrepont complex has been extensively developed for recreation, including a 2km Olympic rowing course, but the large size and substantial number of waters in this complex mean that some value for wildfowl is still maintained.

In the south-west of Nottingham, Attenborough Gravel Pits (Table 90) have received a much greater degree of nature protection than any other complex in the region and the population of wildfowl has greatly increased. The upward leap in Shoveler, Pochard and Canada Geese numbers is particularly noteworthy. A 100ha nature reserve was established in 1965 and extensive management and planting undertaken. The history and management of this area has been recorded in detail (11). The River Erewash was breached in the course of extending the pits, in winter

Table 89. *Regular (and maximum) numbers of wildfowl at five sites north of Leicester.*

a) Kirby Bellars Gravel Pits 8 seasons 1975-1982
b) Wanlip Gravel Pits 12 seasons 1970-1982
c) Groby Quarry Pool, Leicester 10 seasons 1970-1982
d) Cropston Reservoir 10 seasons 1970-1982
e) Swithland Reservoir i) 8 seasons 1960-1969
ii) 13 seasons 1970-1982

	a	b	c	d	e i	e ii
Mallard	94(250)	163(418)	213(850)	431(1500)	245(415)	393(765)
Teal	32(200)	80(360)	7 (90)	101 (260)	38(120)	89(725)
Wigeon	52(330)	192(700)	10(150)	156 (450)	81(370)	140(564)
Shoveler	4 (18)	16 (55)	9 (41)	17 (130)	4 (17)	33 (90)
Gadwall	11 (94)	26 (86)	0 (1)	0 (0)	0 (0)	9 (54)
Pochard	37(107)	75(154)	48(140)	28 (70)	89(228)	122(435)
Tufted Duck	56 (95)	232(454)	49(120)	47 (130)	55(200)	109(319)
Goldeneye	2 (9)	3 (18)	0 (3)	8 (37)	8 (22)	16 (43)
Ruddy Duck	0 (1)	1 (6)	12 (46)	8 (94)	0 (0)	21(105)
Mute Swan	23 (60)	54(115)	5 (9)	3 (14)	2 (8)	8 (27)
Canada Goose	6 (31)	72(222)	35 (82)	9 (115)	8 (49)	14 (96)

Table 90. *Regular (and maximum) numbers of wildfowl at six excavated sites in the middle Vale of Trent.*

a) Hoveringham/Bleasby Gravel Pits 13 seasons 1970-1982
b) Holme Pierrepont Gravel Pits 12 seasons 1970-1982
c) Colwick Country Park 12 seasons 1970-1982
d) Attenborough Gravel Pits i) 10 seasons 1960-1969
ii) 13 seasons 1970-1982
e) Swarkeston Gravel Pits 13 seasons 1970-1982
f) Drakelow Power Station 13 seasons 1970-1982

	a	b	c	d i	d ii	e	f
Mallard	132(548)	200(580)	84(244)	427(1281)	748(1791)	77(334)	459(1500)
Teal	71(195)	16 (53)	4 (20)	27 (114)	63 (220)	20(125)	83 (351)
Wigeon	60(280)	5 (42)	22(125)	126 (706)	123 (747)	11 (71)	7 (40)
Shoveler	4 (18)	5 (38)	5 (40)	4 (18)	74 (247)	1 (10)	12 (70)
Pochard	134(253)	48(131)	69(195)	42 (208)	324 (840)	26 (86)	68 (258)
Tufted Duck	135(230)	133(224)	113(308)	148 (442)	229 (506)	103(203)	223(1000)
Goldeneye	9 (30)	5 (21)	4 (24)	1 (4)	12 (50)	8 (28)	6 (28)
Mute Swan	9 (24)	14 (27)	13 (37)	10 (25)	14 (33)	23 (78)	11 (26)
Canada Goose	111(199)	22(120)	24(150)	5 (28)	98 (229)	40(108)	312 (840)

1972, thus introducing polluted river water to the relatively pure water in the pits. Undoubtedly there have been significant changes in water chemistry and increases in algal crops, leading to a degree of impoverishment in macrophytes and invertebrates. However, the high incidence of flooding effectively flushes out the ecosystem and has helped to maintain the biological importance of the pits. The diving ducks are the most likely to reflect any deterioration in the aquatic habitat. The Pochard, a plant feeder, has in fact remained steady with regular numbers of 320 for 1970 to 1975 and 327 for 1976 to 1982. The Tufted Duck, however, which is mainly an animal feeder, has declined sharply, its regular numbers for these two periods being 321 and 149, bringing it back to the 1960s level. Long-term, insidious change is the likely prospect, and the pressure for recreational use of sites on the edge of a city of half a million people is obvious.

On the outskirts of Nottingham, in the Erewash Valley, is a group of mine subsidence flashes which, though individually having rather small numbers of ducks, nevertheless combine to form a valuable complex in the local context.

The city parks in Nottingham are mainly of interest to Mallard, but Wollaton Park also supports a handful of Pochard and Tufted Ducks. The river at Trent Bridge used to attract a non-breeding flock of up to 75 Mute Swans but this has disappeared following the decline of the species related to lead poisoning (p.490). At the junction with the Soar at Redhill Lock, the Trent carried up to 100 Mallard and several hundred Teal and Wigeon in the 1960s, but there have been no counts since. The Trent floods are much more under control than they used to be and upstream to Burton only a few hundred dabbling ducks, mainly Teal and Wigeon, now occur, along with 2-300 Canada Geese. There have been no counts since 1971 on the formerly prolific (up to 1,000 dabbling ducks) section of river and margins from Burton to Walton. The riverside gravel pits, such as those at Swarkeston, south of Derby (Table 90), again provide the more important habitat. At Drakelow Power Station, with its associated gravel pits and lagoons, the Central Electricity Generating Board has developed a well-managed wildfowl reserve (Table 90), much favoured by Mallard, Tufted Ducks and Canada Geese. It is a focal point for the south Nottinghamshire subpopulation of the Canada, estimated to have been 380 birds in 1976 (405) and about 550 in the early 1980s (612) and separated, on ringing evidence, from the more northerly group around the Dukeries. The numbers of ducks at Drakelow have increased considerably and steadily since the late 1960s, with total wildfowl rising from a regular 100 in the late 1960s to 500-800 in the early 1970s, 1,000-1,500 in the late 1970s and close to 2,000 in the 1980s. The Branston Gravel Pits, just across the Trent, undoubtedly share the same Canada Geese (up to 400) and ducks, with regular (and maximum) counts of 48 (200) Mallard, 39 (150) Pochard and 106 (332) Tufted Ducks, all considerable decreases since the 1960s.

Just south of the Trent are a pair of reservoirs, Staunton Harold (flooded in 1966) and Foremark (flooded in 1977), where the effect of recreational activities has been studied in a comparative fashion (609, p.497). The zoning of sailing to leave a winter refuge area on each reservoir has been largely successful in maintaining their value to wildfowl despite the fact that there is considerable informal recreation through the winter (Table 91). The Goosander flocks, with the maximum reached in February 1979, are notable for this region. Melbourne Pool, only 1km from Staunton Harold, no doubt benefits from this proximity; whereas regular numbers are low,

Table 91. *Regular (and maximum) numbers of wildfowl at five reservoirs in the middle Vale of Trent.*

a) Staunton Harold 13 seasons 1970-1982
b) Foremark 5 seasons 1977-1982
c) Church Wilne 12 seasons 1971-1982
d) Ogston 13 seasons 1970-1982
e) Moorgreen i) 9 seasons 1960-1969
ii) 13 seasons 1970-1982

	a	b	c	d	e i	e ii
Mallard	412(929)	366(580)	180(388)	323(459)	237(600)	205(396)
Teal	22(100)	36(113)	78(339)	137(256)	135(400)	189(420)
Wigeon	95(272)	89(385)	536(964)	97(194)	1 (16)	3 (15)
Shoveler	12 (81)	14 (92)	10 (37)	0 (4)	5 (19)	6 (19)
Pochard	52(202)	54(154)	88(251)	76(357)	20 (98)	31(132)
Tufted Duck	106(292)	68(129)	197(342)	21 (46)	27(100)	27 (89)
Goldeneye	5 (18)	8 (21)	17 (41)	2 (10)	0 (1)	0 (1)
Goosander	15 (67)	39(161)	0 (5)	1 (9)	0 (1)	0 (0)

maxima have reached 240 Mallard, 109 Wigeon, 34 Gadwall, 42 Pochard and 101 Tufted Ducks.

The gravel pits at Clay Mills, by the confluence of the Trent and Dove north of Burton, have held their numbers since the 1960s, now having a regular (and maximum) 29 (115) Mallard, 49 (300) Teal, 53 (190) Pochard, 86 (215) Tufted Ducks and 24 (150) Canada Geese. The greater abundance of Teal versus Mallard is unusual, but the site is the last water to freeze in cold periods, presumably because of the inflow of warmer water from an adjoining sewage works; it is therefore of considerable importance in hard winters. The Eggington pits north of the river have regular populations of around 200 wildfowl; Willington, only 1km away, 150; and Hilton, 5km to the west, 100. Further upstream along the Dove, the lake at Sudbury Hall, a National Trust property, has a number of the commoner species, including up to 480 Mallard, 130 Tufted Ducks and 70 Canada Geese. Of nine counted sites scattered around the northward line of the Dove only Calwick Abbey Lake, just over the border in Staffordshire, is of interest, with up to 140 Mallard and 45 Teal, but Osmaston Park near Ashbourne has regular (and maximum) counts of 36 (129) Mallard, 11 (60) Teal, 27 (59) Tufted Ducks, and 35 (320) Canada Geese for which it is a traditional moulting site. The high ground to the north-west has little to offer. Near Leek, of a group of four reservoirs only Tittesworth is of interest with regular (and maximum) returns of 154 (290) Mallard, 21 (52) Teal, 101 (201) Wigeon, 56 (270) Pochard and 36 (72) Canada Geese.

At the junction of the Trent with the Derbyshire Derwent just south of Nottingham, the 32ha Church Wilne Reservoir, built in 1971 (Table 91), has substantial numbers, especially of Wigeon and Tufted Ducks, despite its steep banks and regular perimeter. The flock of Goldeneyes is also notable. There are several not very important waters nearby and northwards on the line of the Derwent. Five have less than 100 wildfowl regularly, but others worth detailing in Table 92 are Elvaston Country Park on the south-eastern fringes of Derby, the British Celanese Pools, 2km away on the banks of the Derwent at Spondon, the adjacent Derby Sewage Works, Markeaton Park Lake on the other side of Derby centre and Locko Park Lake. Shipley Country Park encompasses the two reservoirs of Shipley and Mapperly, just west of Ilkeston. The heartland for the Canada Geese is Kedleston Park Lake, just north-west of Derby, where the ornamental lakes have been formed by the damming of a tributary of the Derwent. The species was introduced here early this century and there were regular numbers of over 700 and a maximum of 1,200 in the 1960s. In contrast to the national proliferation, the small increase indicates that a measure of control has been contrived. Indeed there was a drop from 890 in 1968 to 570 in 1976 for the Derbyshire subpopulation, which has links with both the south Nottinghamshire group and the west Midlands group (405). Kedleston hosts a great variety of wildfowl other than those listed, though in rather small numbers.

Above Belper, the Derwent runs deeper into the foothills of the Peak District and counts along the river have not been productive. However, east of Bakewell the valley opens out and the tributary Wye downstream of the town carries up to 400 Mallard and 30 Teal. The lakes at Chatsworth House east of the river have good numbers of Mallard. Most waters around the upper reaches of the Derwent are of no great interest, and even the large Ladybower, Derwent and Howden Reservoirs are not very attractive. Of much greater importance is the empoundment of the tributary Amber to form Ogston Reservoir near Clay Cross (Table 91) which has improved on its early

Table 92. *Regular and maximum numbers of the common species at the less important sites in the Derwent Valley, from 1970 to 1982. (Number of seasons (max 13) in brackets after site name.)*

		Mallard	Teal	Pochard	Tufted Duck	Canada Goose	Mute Swan
Elvaston Country Park	(7)	17 (62)	1 (5)	3 (9)	5 (14)	0 (2)	1 (4)
British Celanese Pools	(4)	87(160)	138(300)	34 (87)	94(252)	0 (0)	1 (3)
Derby Sewage Works	(4)	68(139)	276(350)	0 (0)	0 (0)	0 (0)	2 (4)
Markeaton Park Lake	(3)	114(186)	0 (0)	1 (6)	5 (16)	42 (192)	5 (8)
Locko Park Lake	(13)	54(164)	1 (15)	9 (31)	22(170)	42 (370)	1 (3)
Shipley Country Park	(7)	109(181)	7 (27)	87(212)	44 (81)	16 (72)	7(16)
Kedleston Park Lake	(12)	143(350)	15 (40)	21 (64)	77(194)	855(1800)	3(13)
Chatsworth Lakes	(10)	363 (74)	9 (22)	1 (15)	22 (48)	2 (12)	0 (2)
Ladybower, Derwent and Howden Resrs	(8)	33 (96)	10 (60)	13 (37)	2 (12)	0 (0)	0 (0)

promise and is the most important site in north Derbyshire.

Around Chesterfield, on the upper reaches of the Rother (p.174), thirteen counted waters are in the under-50 category and mention need only be made of the 200 Mallard on Queen's Park Lake. We are then left with a group of waters scattered between Nottingham and Mansfield and so verging on the Dukeries (p.165). The most important is the most southerly, Moorgreen Reservoir (Table 91). The long run of counts shows remarkably little change in numbers between the 1960s and the more recent data. Nine other waters in the group have under fifty wildfowl, and the only resorts of note are Newstead Abbey Lakes, with 30-40 Mallard, Pochard, Tufted Ducks and Canada Geese; Salterford Dam, with 40 Mallard and 20 Tufted Ducks; and Kings Mill Reservoir on the outskirts of Mansfield, with 80 Mallard, 35 Pochard and 20 Tufted Ducks.

The upper Vale of Trent

This section covers the Trent, no longer pre-eminent as a wildfowl haunt, from Burton to its source near Newcastle-under-Lyme, together with its tributaries: the Mease, Tame, Anker and Blythe from the south, Sow from the west and Blithe from the north. The southern boundary along the Clent Hills includes the greater part of the metropolitan West Midlands, otherwise the counties involved are Leicestershire, Warwickshire and Staffordshire.

The Mease is of little interest; there are no floods and very few associated standing waters. The Tame, which flows into the Trent shortly upstream of the Mease, just east of Alrewas, leads to some important waters, although up to its junction with the Anker there are 15 counted sites, mostly to its west, none of which have more than 50 wildfowl regularly. On the Anker itself there is a series of mining subsidences, Alvecote Pools near Polesworth (Table 93), with a nature reserve operated by the Warwickshire Nature Conservation Trust. The reed-fringed shallow pools and the open marsh have attracted a wide variety of birds (190 species) and the largest and deepest of the waters – Pretty Pigs Pool – is particularly attractive to diving ducks. Over the period of the counts there have been few major changes in regular numbers though Shoveler have increased from 4 in the 1960s to 21 in the 1970s and Tufted Ducks from 55 to 145. There is also a moulting flock of Mute Swans, numbering 200 at times. Further upstream a group of waters on the southern outskirts of Nuneaton – Arbury Park, Astley and Seeswood Pools – carry a combined maximum of 200 Mallard, 60 Teal and Tufted Ducks, 25 Pochard and 105 Canada Geese.

Upstream on the Tame a large amount of wildfowl habitat has been created by gravel extraction and by the recently completed settling lakes, built by the Severn Trent Water Authority at Lea Marston. Middleton Hall Pool (in fact an older, ornamental lake) carries around 50 wildfowl, but much more important is Kingsbury Water Park (Table 93), despite considerable recreational pressures, inevitable with a city of a million people a few kilometres away. However, without the recreational demand the land could well have returned to agriculture. The numbers of diving ducks are outstanding. These are reflected in the records for the settling lakes and pools at Coton nearby, where the increase has also been recent, and for the seasons 1980 and 1981 the regular (and

Table 93. *Regular (and maximum) numbers of wildfowl at five sites around Birmingham.*

a) Alvecote Pools, Polesworth 20 seasons 1960-1982
b) Kingsbury Water Park 7 seasons 1976-1982
c) Ladywalk NR, Coleshill 7 seasons 1976-1982
d) Packington Hall 9 seasons 1970-1982
e) Chillington Hall 17 seasons 1960-1982

	a	b	c	d	e
Mallard	139(359)	298 (513)	491(900)	932(4200)	365(820)
Teal	100(215)	140 (300)	182(400)	74 (250)	20 (90)
Wigeon	42(100)	37 (93)	92(178)	40 (450)	8 (37)
Shoveler	13 (80)	62 (132)	30 (70)	3 (15)	29(209)
Gadwall	0 (7)	31 (68)	3 (8)	0 (0)	0 (1)
Pochard	155(331)	517(1674)	36(300)	37 (150)	30(195)
Tufted Duck	104(237)	416(1514)	128(700)	187 (300)	81(175)
Goldeneye	1 (3)	34 (68)	2 (7)	2 (20)	7 (17)
Ruddy Duck	2 (16)	4 (13)	0 (1)	0 (6)	0 (2)
Mute Swan	41(185)	44 (91)	4 (9)	14 (36)	1 (4)
Greylag Goose	0 (0)	43 (103)	5 (64)	94 (203)	0 (0)
Canada Goose	8(116)	329 (497)	41(125)	97 (350)	82(330)

maximum) numbers were 743 (2,800) Pochard and 785 (1,600) Tufted Ducks. The reasons for the upsurge are probably associated with the settling lakes, which are little disturbed, and the adjacent well-vegetated gravel pit at Marston, which no doubt provides good feeding. However, the increase in water areas in the district will undoubtedly result in some redistribution from older established sites like Ladywalk Nature Reserve (Table 93), though the warm water from Hams Hall Power Station would still give it the advantage of ice-free conditions in hard weather. Operations at the power station, and consequently the outflow of water, have now been reduced. Nearby Shustoke Reservoir is less frequented though still at times holding good numbers, up to 45 Mallard, 164 Wigeon, 496 Pochard, 930 Tufted Ducks and 31 Goldeneyes. Two sites are counted but are of little consequence where the Tame turns west to plunge into the conglomeration of Birmingham and is joined from the south by the Blythe. In the Blythe Valley the lakes at Packington Park are of note (Table 93); the massive Mallard flocks are heavily supplemented by captive reared stock and the Greylag, as well as the Canada Geese, are of feral origin.

Birmingham draws most of its water from mid-Wales and is therefore, in common with Manchester, not ringed by sizeable reservoirs. Ten small waters, reservoirs and park ponds, within its conurbation, hold very small numbers of wildfowl, though the pools in Sutton Park carry a regular 207 (maximum 400) Mallard, 58 (116) Tufted Ducks and 42 (86) Canada Geese. To the north and west in the outskirts of the city is a line of canal reservoirs and gravel pits. Chasewater (Cannock Reservoir) has 28 (103) Mallard, 44 (159) Pochard, 230 (430) Tufted Ducks, 41 (93) Goldeneyes and 28 (50) Mute Swans. Numbers here have increased substantially in the last decade, especially Goldeneyes, which peak at around 80 each winter; this despite considerable pressure from water sports and recreation. Gailey Pools Gravel Pits support 115 (315) Mallard, 58 (231) Pochard, 146 (411) Tufted Ducks and 85 (257) Canada Geese. The Mallard and Pochard have decreased since the 1960s (regular counts of 207 and 115 respectively) probably through recreational pressure, sailing having been allowed from 1970 and trout fishing from the mid 1970s. Just over the western watershed are the pools at Chillington Hall and Patshull. Chillington (Table 93) has maintained its populations since the 1960s except in the case of Mallard (456 down to 315). Patshull, however, which held a regular (and maximum) 449 (725) Mallard, 35 (100) Pochard, 50 (140) Tufted Ducks and 141 (321) Canada Geese in the 1960s, became virtually deserted in the 1970s, with maxima of 30 Mallard and 32 Tufted Ducks in the last season of counts (1979). This was a consequence of greatly increased recreational disturbance and the establishment of a hotel complex. Undoubtedly the most important water in this group is the 75ha Belvide Reservoir (Table 94). This has long been the concern of the West Midland Bird Club; they rented the surrounding land to avert threats from water sports and established a formal reserve in 1977. Management in the form of building observation hides and islands has been undertaken. Nearly every species of wildfowl has increased in numbers and the Ruddy Duck has established this water as one of its main foci in the Midlands (the other being Blithfield). It is appropriate,

Table 94. *Regular (and maximum) numbers of wildfowl at three sites in the upper Vale of Trent.*

a) Belvide Reservoir i) 9 seasons 1960-1969; ii) 12 seasons 1970-1982
b) Aqualate Mere 21 seasons 1960-1982
c) Blithfield Reservoir i) 9 seasons 1960-1969; ii) 13 seasons 1970-1982

	a i	a ii	b	c i	c ii
Mallard	523(3100)	995(1520)	1445(2539)	1431(2549)	1311(2062)
Teal	63 (800)	123 (314)	98 (693)	641(1423)	436(1179)
Wigeon	63 (850)	49 (156)	211 (708)	1016(1842)	650(1330)
Shoveler	52 (230)	150 (570)	139 (520)	76 (227)	34 (230)
Pochard	76 (260)	121 (342)	67 (267)	143 (304)	295(1066)
Tufted Duck	70 (268)	276 (843)	38 (131)	247 (429)	111 (441)
Goldeneye	36 (70)	68 (139)	1 (14)	30 (46)	28 (58)
Goosander	10 (48)	3 (15)	1 (9)	46 (120)	51 (137)
Ruddy Duck	3 (25)	144 (425)	13 (137)	4 (17)	146 (630)
Mute Swan	3 (17)	6 (17)	1 (7)	6 (36)	4 (27)
Canada Goose	33 (300)	93 (350)	228 (592)	36 (125)	239 (680)

if surprising to traditional birdwatchers, that the Club should have adopted this recently established exotic escapee as its logo. The maximum numbers of Ruddy Ducks gathering on Blithfield and Belvide Reservoirs represent a high proportion of the national population. Thus in January and February 1981 they shared 800 birds between them, over two-fifths of the British population. Another species which is clearly excellently suited by conditions at Belvide is the Shoveler, which trebled its numbers between the late 1960s and early 1980s.

The Sow tributary joins the Trent below Stafford, but counts over two sections have not indicated large numbers of wildfowl. The pool at Brocton is part of an LNR of 49ha and has modest flocks of up to 60 Mallard, 25 Tufted Ducks and 55 Canada Geese. Farther upstream, Copmere, a deep kettle-hole like some of the Shropshire meres, carries substantial numbers of the commoner species, with regular (and maximum) counts of 347 (700) Mallard, 20 (30) Pochard, 107 (170) Tufted Ducks and 54 (94) Canada Geese. Just across the watershed, Knighton Reservoir has records of up to 22 Mallard, 27 Teal, 139 Wigeon, 25 Tufted Ducks and 26 Mute Swans. Far more important is Aqualate Mere (Table 94), a shallow lake with a reed fringe on about three-quarters of the perimeter and open parkland on the remainder, which is privately protected. The ecology of the lake is clearly highly suitable for dabbling ducks, but no less important has been the rigorous exclusion of disturbance; duck shooting occurs, but not frequently. The population of Mallard is at present higher than that at Blithfield and is indeed one of the highest for an inland site in Britain. Considering the mere is only some 60ha in area, the concentration is outstanding. The other species have also held their own. The Shoveler has increased as it has at Belvide, and for their size these two waters are of great importance for this species.

At King's Bromley, where the Blithe joins the Trent, the gravel pits hold substantial numbers of four species: 262 (388) Mallard, 48 (220) Pochard, 104 (302) Tufted Ducks and 222 (420) Canada Geese. There is little else along the line of the Trent, but two sites are of some interest, Trentham Park with up to 320 Mallard and Westport Lake in the northern outskirts of Newcastle/Stoke with 50-100 Mallard, Pochard and Tufted Ducks and up to 28 Mute Swans.

Less than 10km upstream the Blithe (not to be confused with the Blythe to the south) has been impounded to form Blithfield Reservoir, 324ha, which though bisected by a road forms the biggest contiguous water surface in the area. It is used by a great range of wildfowl, many in very substantial numbers (Table 94). It was flooded in 1952 and the numbers of

Fig. 28. The Don, Rother and Calder.

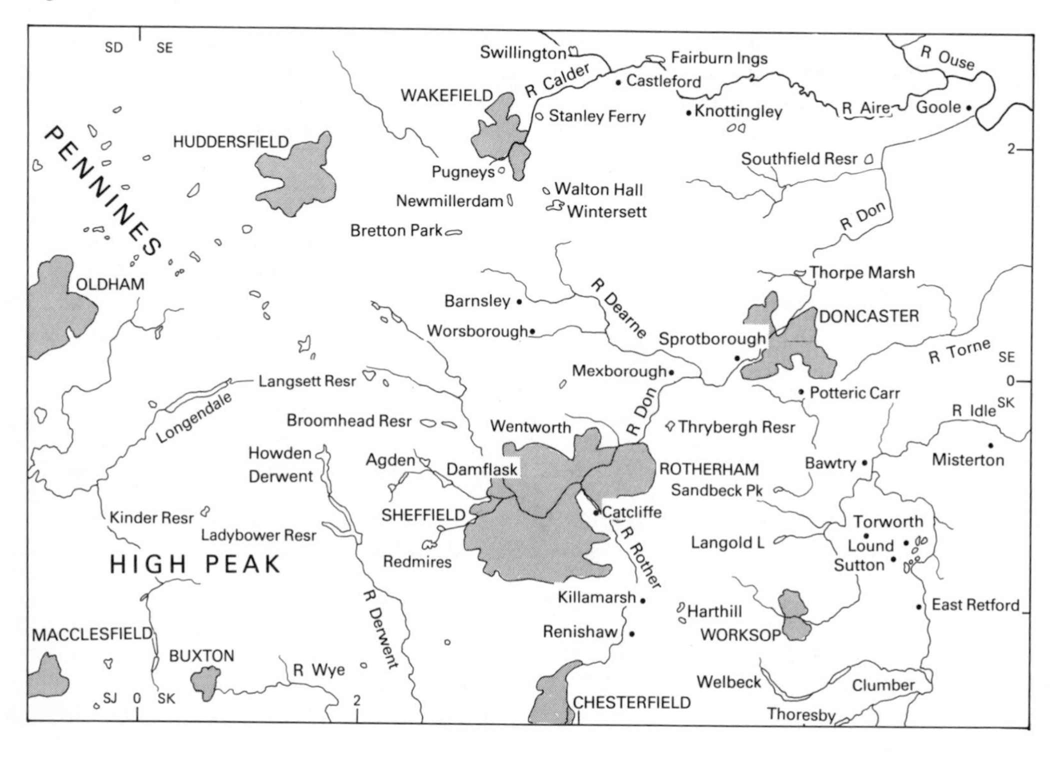

dabbling ducks regularly present in the 1950s (1,130 Mallard, 585 Teal, 945 Wigeon and 75 Shoveler) were surpassed in the 1960s, but have since fallen back. This probably reflects the normal slow deterioration of reservoirs formed by flooding farmland, for there is little disturbance from sailing in the northern shallower end, where most dabblers are found. Fishing from boats and from the banks does, however, proceed in this area. Sailing is confined to the southern end and is extensive, but is probably not the reason why the diving ducks (225 Pochard, 245 Tufted Ducks, 40 Goldeneyes and 60 Goosanders in the 1950s) have declined or barely held their own. More likely reasons are changes in water quality and a succession of autumns with considerable draw-down at the reservoir, affecting in particular the food supply for the Tufted Duck, which has suffered the most severe setback. The flock of Goosanders is still outstanding for the region and the importance of the reservoir to Ruddy Ducks has already been discussed. The Canada Goose flock is now second in size in the area to that on Kingsbury Water Park, but Blithfield has been a major focus for this expanding species for a long time. Extensive ringing of the geese in the west Midlands has suggested considerable interchange between the various "sub-units". The south and east Staffordshire group should be considered together with those in north Worcestershire, Warwickshire and east Shropshire. An overall total of nearly 2,500 was estimated for this population in summer 1976 (405), and numbers have increased in many waters since.

The Don Basin

The River Don enters the Yorkshire Ouse at Goole via the canalised Dutch River. The area covered in this section lies predominantly in South Yorkshire, overlapping into Derbyshire. Some 10km up the Don and 1km to the west of the channel, Southfield Reservoir, which is linked with the adjacent floodplain of the River Went, where winter flooding is regular, harbours a good mixed population (Table 95). Thorpe Marsh by Doncaster Power Station holds 27 (82) Mallard, 57 (200) Teal, 36 (76) Pochard, 55 (112) Tufted Ducks and an unusual number of swans, 16 (32) Mute and 10 (28) Whooper. Potteric Carr, on land owned by British Rail near a major junction south-west of Doncaster, has substantial dabbling duck populations, including Shoveler (Table 95). Mining subsidences (flashes or ings) are a feature of the area. Denaby Ings, next to the river near Mexborough (Table 95), again attract Shoveler, while Wath Main Ings, 5km to the west, have a regular (and maximum) 112 (400) Mallard and 76 (250) Teal, but little else. Sprotborough Flash, a long pool close to the Don just south-west of Doncaster, has less than 100 wildfowl regularly and the other nearby flashes have rather small numbers. All these waters are managed as Yorkshire Wildlife Trust reserves.

At Mexborough the River Dearne joins the Don, but its course from the west is not studded with wildfowl resorts. Worsborough Reservoir south of Barnsley holds a regular (and maximum) 134 (326) Mallard, 47 (108) Pochard and 20 (72) Tufted Ducks, and Carlton Marsh nearby favours Teal, with up to 200. Six small reservoirs to the west of Barnsley on higher ground are all of little interest, each holding less than 50 wildfowl regularly.

The Don flows through Sheffield and the city parks carry quite a lot of wildfowl, especially Mallard. Hillsborough has up to 130; Porter Valley a regular 80 (based on one season – 1977); Graves Park holds up to 50 ducks, while Wentworth (Table 95) has larger numbers and variety. In addition to the ducks, Wentworth used to hold 50 Canada Geese regularly, being one of the main sites for the small South Yorkshire population, but the flock has now declined to very few birds. This decline is said to have been due to increased pressure from angling. On the moors to the west and north-west of Sheffield, there are count data on 14 reservoirs but duck populations are low

Table 95. *Regular (and maximum) numbers of wildfowl at six sites in the Don basin, Yorkshire.*

a) Southfield Reservoir 2 seasons 1966-1982
b) Potteric Carr, Doncaster 4 seasons 1975-1982
c) Denaby Ings/Flash, Mexborough 1 season 1982
d) Wentworth Park, Sheffield 20 seasons 1963-1982
e) Thrybergh Reservoir 11 seasons 1970-1981
f) Catcliffe Flash 9 seasons 1973-1982

	a	b	c	d	e	f
Mallard	70(100)	745(1300)	148(288)	198(500)	232(400)	64(173)
Teal	39 (70)	278 (480)	150(204)	21(300)	89(235)	16(102)
Shoveler	0 (0)	43 (120)	39 (75)	1 (7)	3 (20)	1 (4)
Pochard	94(120)	32 (92)	35 (85)	65(204)	35(140)	45(128)
Tufted Duck	52 (65)	9 (14)	27(100)	58(148)	24(138)	27 (70)

despite the large size of some of the waters. Of the few exceptions Broomhead south of Stockbridge has held up to 620 Mallard and Langsett 190. At Agden Reservoir Mallard have increased from 10-30 in the early 1970s to 1-200 since. Redmires Reservoir in the Hallam Moors is made up of three separate lakes and is notable for a small flock of moulting Tufted Ducks, numbering 172 in August 1975.

Near Rotherham, Thrybergh Reservoir (Table 95) carries both dabbling and diving ducks in some numbers, while the pits at Aldwarke Toll had up to 160 Teal in the late 1970s, but the site was destroyed in 1978. The River Rother flows into the Don in Rotherham and Catcliffe Flash – now effectively a reserve – 5km upstream on the Rother (Table 95) is also of interest, mainly for the diving ducks. A few other sites in the lower reaches of the Rother have small numbers of ducks, but farther south Harthill Reservoir, 4km east of the river, has regular (and maximum) counts of 62 (407) Mallard, 11 (45) Pochard and 21 (80) Tufted Ducks. The development of the Rother Valley Park near Killamarsh, mainly for recreation but with a nature reserve area, may augur well for the future. Only 45 Teal have been reported at Killamarsh, but farther upstream the lakes in Renishaw Park have 50 (210) Mallard, 23 (38) Teal and 14 (27) Tufted Ducks. The park is also the local stronghold for wintering Gadwall, with a number present all winter and a peak of 52 in January 1983. The floods at Breck Farm near Staveley hold up to 325 Mallard and 450 Teal. Four other sites nearby are of marginal interest; the upper Rother is dealt with in another section (p.170).

The Aire Basin

To the west of Goole the low-lying arable plain around the River Aire in South Yorkshire continues inland to West Yorkshire, and only where the ground rises towards the Pennine foothills do sizeable stretches of water occur. The scene there is dominated by industry and by the northern fringes of the coalfield, but around Leeds and Wakefield, in particular, resorts are many and varied.

Between Goole and Knottingley there are few reports of wildfowl numbers along the Aire Valley, but up to 300 Mallard and Teal have been counted at Birkin, though this area is now virtually drained, and up to 320 Mallard, 44 Teal and 180 Pochard on Gale Common, which has also now disappeared under dumped fly-ash. The most important of the ings (flooded mining subsidences) occur in a strip some 500m wide and stretching for 10km westwards along the north bank of the Aire from Brotherton to Swillington, though the eastern end near Brotherton is filled in with fly-ash. Mining and tipping of waste render some of the remainder unattractive but the amenities of Fairburn and Newton Ings have largely been preserved by the declaration of a Local Nature Reserve and, since 1976, the operation of an RSPB reserve over 275ha of shallow lakes, marshy depressions and flood pools. Of principal interest are the numerous wintering wildfowl (Table 96), but all the common species are also found nesting. Upstream, Mickletown Ings north-west of Castleford are frequented mainly by diving ducks, with regular (and maximum) numbers of 43 (103) Pochard and 124 (198)

Table 96. *Regular (and maximum) numbers of wildfowl at five sites in the Aire/Calder basin (including Bretton Park, which is close to the Calder but drains into the Dearne).*

a) Fairburn Ings 8 seasons 1975-1982
b) Wintersett Reservoir 5 seasons 1972-1982
c) Pugney Water, Wakefield 3 seasons 1980-1982
d) Bretton Park 13 seasons 1970-1982
e) Coniston Cold Lake 10 seasons 1970-1982

	a	b	c	d	e
Mallard	994(2600)	56(144)	87(186)	172(322)	101(430)
Teal	384(1600)	2 (51)	16 (62)	6 (43)	11 (91)
Wigeon	27 (68)	1 (28)	5 (30)	1 (7)	32(108)
Shoveler	135 (276)	2 (22)	2 (7)	0 (6)	3 (30)
Gadwall	22 (64)	1 (3)	0 (0)	0 (3)	0 (0)
Pochard	238 (424)	43(120)	345(774)	27(125)	43(230)
Tufted Duck	334 (528)	108(513)	89(215)	53(176)	32 (91)
Goldeneye	30 (84)	17 (60)	17 (28)	2 (20)	3 (8)
Goosander	10 (31)	0 (2)	0 (0)	0 (6)	1 (8)
Mute Swan	72 (105)	5 (18)	18 (50)	3 (10)	5 (14)
Whooper Swan	41 (87)	2 (11)	1 (3)	0 (0)	0 (3)
Canada Goose	90 (190)	1 (7)	79(202)	134(247)	10 (45)

Tufted Ducks, but also 23 (93) Teal. Swillington Ings, a complex of flashes adjacent to the Aire, have recovered some of their importance, lost in the 1950s, and now have regular (and maximum) counts of 168 (342) Mallard, 102 (251) Teal, 10 (52) Wigeon, 47 (324) Pochard, 114 (312) Tufted Ducks and 18 (36) Mute Swans. The area is subject to further open cast mining and if, as is planned, the site is restored to stretches of water, its value may well eventually increase. Another dozen or so sites in the environs of Leeds and Wakefield have been counted but have regular wildfowl populations of less than 50.

Following the Calder tributary towards Huddersfield, and mainly to the south, lie a scattering of waters which carry small numbers except for Bretton Park Lakes (which are in the Dearne drainage but are included here because of their proximity), Wintersett Reservoir and Pugneys (Table 96). An impoundment of a tributary, the Newmillerdam holds up to 200 Mallard, 28 Teal, 137 Pochard and 301 Tufted Ducks despite its wooded surround. Pugneys is of especial interest being a recent (June 1979) flooding of a coal extraction site on an earlier gravel working. The site was landscaped to form a Country Park by the National Coal Board, largely for recreation but with a small reserve lake.

Within 15km of Huddersfield lie to the west nine, and to the south ten waters, mostly reservoirs, where counts do not regularly exceed 50 wildfowl in all. Farther north, between the Calder and Aire, a cluster of sixteen small waters, likewise hill reservoirs with small regular populations, spreads westward from Bradford. Back towards the Aire, Farnley Balancing Reservoir between Leeds and Pudsey is surrounded by roads and built-up areas but nevertheless has maxima of 52 Mallard, 36 Pochard and 80 Tufted Ducks, while Reva Reservoir on the edge of the Rombalds Moor near Guiseley has up to 116 Mallard. Although not very supportive of wildfowl, Airedale is one of the few easy routes through the Pennine chain from the Humber to Morecambe Bay, and a passage of scoters is sometimes noted. The numbers of ducks recorded on the reservoirs probably do not reflect the valley's true importance to wildfowl. It is probably in

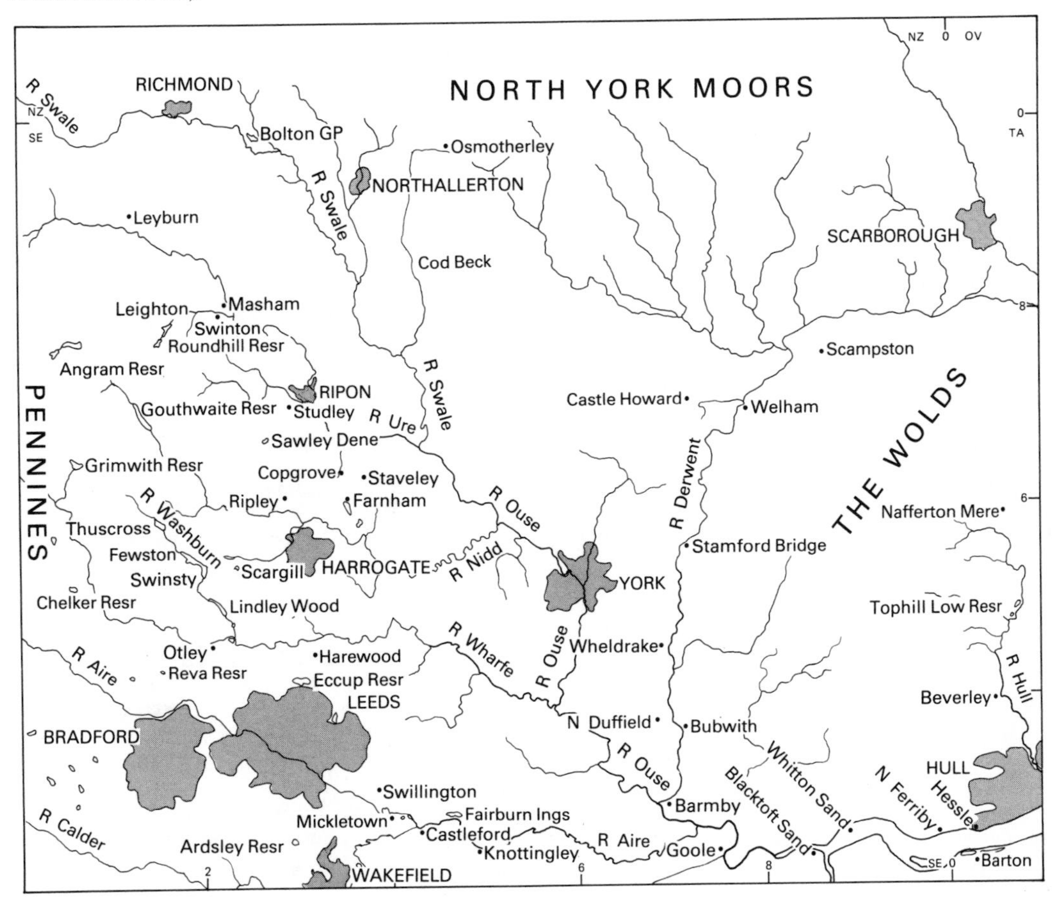

Fig. 29. The Yorkshire Ouse Basin (Wharfe, Nidd, Ure, Swale and Derwent).

relation to this that a scatter of sites right up at the headwaters of the Aire and on the watershed to Ribblesdale have quite high maxima, thus Long Preston records 77 Wigeon and the Ribble Marshes, near Settle, 90. Coniston Cold Lake (Table 96) attracts quite substantial regular populations as well, while Eshton Tarn has up to 120 Mallard, 40 Teal and 20 Tufted Ducks. Malham Tarn, owned by the National Trust, is managed as a reserve and houses a Field Studies Centre. Malham is the highest marl lake in Britain (at an altitude of 380m); yet it has recorded as many as 54 Mallard, 18 Teal, 48 Wigeon, 39 Shoveler, 214 Pochard, 104 Tufted Ducks, 6 Goldeneyes and 14 Goosanders.

The Yorkshire Ouse, Derwent and Hull Basins

Above its confluence with the Aire the Yorkshire Ouse is joined by the Derwent (p.177), draining from the Cleveland Hills and the North York Moors, and then by four further tributaries draining from the west, the Wharfe, Nidd, Ure and Swale. The area covered, which in contrast to Airedale is generally rural, lies almost entirely in North Yorkshire, though fringing into Humberside.

Again, there are few waters of interest along the Ouse until the ground begins to rise to the west in Wharfedale. Roundhay Park and Golden Acre Park (with the adjacent Adel Dam, which is a Yorkshire Wildlife Trust Reserve) on the outskirts of Leeds have recorded up to 277 and 84 Mallard respectively, but of considerably more interest to the north of that city are Eccup Reservoir and Harewood Lake (Table 97). Both waters have shown remarkably little change in regular numbers through the long runs of counts, though Canada Geese have increased at Harewood, from 152 (1960-69) to 383 (1970-82). At Eccup Reservoir the regular Goosander population has increased from 21 to 51. Near Otley, the gravel pits at Knotford Nook have held up to 104 Mallard, 93 Pochard, 59 Tufted Ducks and 248 Canada Geese, and Farnley Lake up to 80 Mallard, 57 Teal and 30 Tufted Ducks, but a more interesting water hereabouts is Lindley Wood Reservoir formed by the damming of the Washburn, a tributary of the Wharfe (Table 97). Although counts have been irregular since the 1960s, the indications are of no great change, except an increase in Canada Geese. The string of large reservoirs farther up the Washburn Valley – Swinsty, Fewston and Thruscross – holds a total of some 200 Mallard and 100 Canada Geese. Fifteen Teal and Tufted Ducks were regular on Fewston in the 1960s when counts were frequent. Closer to Harrogate, Beaverdyke and Scargill Reservoirs are much smaller and are also mainly used by Mallard, respective maxima being 54 and 310, and Canada Geese with up to 46 and 123. Beaverdyke again had a regular presence of Teal and Tufted Ducks. To the west up Wharfedale, Chelker Reservoir just south of Bolton Abbey has had up to 130 Mallard and 22 Goosanders, while Grimwith Reservoir on the Dibb tributary to the north has recently been trebled in size, but, presumably because of its altitude, has records of only 34 Mallard, 30 Teal and 17 Wigeon; 110 Canada Geese have mounted there.

Upstream on the Ouse around York there are few significant concentrations of wildfowl, although the University Lake has had over 300 Mallard joining tame birds there. The lower reaches of the Nidd, which runs northwards to join the Ouse to the north-west of the city, have half a dozen sites of little importance. However Hay-a-Park and Farnham Gravel Pits near Knaresborough have an interesting mix of species (Table 97). Ripley Castle Lake, farther up the dale, has smaller numbers of all species except Teal (Table 97). The upper reaches of the Nidd are

Table 97. *Regular (and maximum) numbers of wildfowl at six sites in Wharfedale and Nidderdale.*

a) Eccup Reservoir, Leeds 21 seasons 1960-1982
b) Harewood Lake 18 seasons 1961-1982
c) Lindley Wood Reservoir, Otley 10 seasons 1960-1969
d) Knaresborough Gravel Pits 4 seasons 1976-1982
e) Ripley Castle Lake 11 seasons 1960-1977
f) Gouthwaite Reservoir 11 seasons 1960-1982

	a	b	c	d	e	f
Mallard	915(1853)	589(3000)	345(1200)	128(243)	42(132)	343(1930)
Teal	23 (100)	14 (90)	38 (132)	0 (8)	19 (50)	167 (400)
Wigeon	9 (41)	2 (20)	7 (23)	4 (14)	3 (32)	43 (107)
Pochard	10 (95)	24 (123)	18 (63)	84(187)	4 (15)	28 (125)
Tufted Duck	48 (113)	42 (98)	22 (67)	126(213)	28 (50)	20 (65)
Goldeneye	7 (18)	5 (29)	4 (11)	12 (20)	0 (2)	15 (34)
Goosander	37 (132)	0 (2)	1 (6)	23 (47)	0 (6)	2 (10)
Canada Goose	70 (380)	319 (580)	5 (27)	164(493)	138(330)	54 (143)

dominated by the empoundment of the river to form Gouthwaite Reservoir (Table 97). The numbers of wildfowl here have remained largely unchanged since the early 1950s, the autumn passage of Teal being particularly notable. There was, however, a period of high autumn Mallard numbers in the 1960s, while a small regular flock of Whooper Swans disappeared in the early 1960s. Three neighbouring sites are of little general import for wintering birds, though Angram and Scar House Reservoirs harbour moulting Canada Geese.

The Ure, above its junction with the Swale near Boroughbridge, runs past Ripon, round which is a cluster of a dozen counted sites, all except two being of little consequence. Studley Park Lake and Ripon Racecourse Gravel Pits have respective maxima of 119 and 76 Mallard and 134 and 270 Canada Geese. However, counts in the 1960s showed more variety at Copgrove Lake, with a regular (and maximum) 166 (385) Mallard and 15 (38) Tufted Ducks, and Sawley Dene Lake, with 233 (610) Mallard, 21 (150) Teal, 11 (42) Pochard and 31 (100) Tufted Ducks as well as 23 (86) Canada Geese. In the 1970s Staveley Gravel Pit, a lagoon created by subsidence after rehabilitation of gravel workings in the early 1970s, had up to 320 Mallard, 60 Teal, 22 Shoveler and nearly 300 Canada Geese as well as small numbers of a wide variety of other species. Counts of the Ure itself near Ripon have turned up no less than 20 species of wildfowl, but no great numbers of any except 900 Canada Geese, which moult in the Givendale area near the racecourse.

We are indeed in the heartland of the feral Canada Goose in central Yorkshire. The population in July 1976 was estimated to be 1,550, plus 920 away on moult migration in the Beauly Firth (405). This was one-eighth of the total for Britain as a whole at that time. Undoubtedly they have increased since (in 1979 they were estimated at 2,600), in line with the national trend, for the scattering of waters through parkland is very suitable habitat for them. Efforts by humans to control breeding and competition for nest sites are the probable cause of a growing tendency for pairs to disperse to breed in remote moorland as well as on park lakes (206). Since 1967 a very active group of ringers has annually rounded up several hundred geese at 18 sites in Yorkshire and marked them, at first with metal rings and then with large plastic rings whose code may be read at a distance (p.15). By 1973 nearly 60% of the summer population was so marked. Major catching sites (50 or more caught in one year) included Harewood Lake, Farnham Gravel Pits, Swinsty, Grimwith, Gouthwaite, Angram, Scar House and Roundhill Reservoirs, and Ripley, Studley and Swinton Lakes. Studies of their migrations (606) and mortality and other population parameters (579) have been published, giving valuable insights into the performance and behaviour of this feral species in Britain.

Proceeding up the Ure we reach a group of sites around Masham. Nosterfield Gravel Pit has had up to 500 Mallard and a sprinkling of other species; Marfield Pond up to 167 Mallard, 77 Teal, 40 Pochard, 38 Tufted Ducks, 15 Goldeneyes and 320 Canada Geese. Swinton Park is notable for Canada Geese mainly in the summer and there are sometimes as many as 200 Mallard in the several lakes. Leighton and Roundhill Reservoirs, although lying at an altitude of around 200m, have records of 19 wildfowl species but only three regularly in any numbers – regulars (and maxima) of 1,117 (3,500) Mallard, 24 (92) Goosanders and 114 (420) Canada Geese. Beyond Leyburn the river enters Wensleydale proper, much better known for cheese than for wildfowl. Apart from a record of 164 Mallard on Locker Tarn, the only concentration of any note is on Semerwater, a high altitude lake (around 250m) on the tributary Bain. Here are found 20-40 Mallard, Teal and Wigeon, a few diving ducks and a flock of about 30 Whooper Swans.

In the lowlands, the other branch of the Ure/Ouse, the Swale, also runs through country devoid of wildfowl concentrations, except for the gravel pits at Bolton-on-Swale, near Catterick. These had recorded up to 69 Mallard, 15 Wigeon and 83 Pochard in the 1970s, but exceptional concentrations of 1,300 Mallard and 700 Wigeon were present in December 1983. Osmotherley Reservoir, which has at peak 90 Mallard, although lying in the Cleveland hills is connected to the Swale, being on its tributary, Cod Beck. Forcett Park, farther north beyond Richmond, however, lies within 5km of the Tees and is considered in the Teesdale section.

The main southward drainage of the Cleveland Hills is via the Derwent, which joins the Ouse near Goole. Its lower reaches are extremely notable for the low-lying grassland adjacent to the river, known locally as "ings". (These are natural in origin and not to be confused with mining subsidences given the same title.) They traditionally serve as floodrelief areas, and the main concentration of ings land lies south of Stamford Bridge, the most important being between Wheldrake and Bubwith. Five blocks amounting to 300ha are recognised as a site of international importance to wildfowl, and the SSSI was extended in 1981 to form a continuous 783ha area. In the northern part, Wheldrake Ings is owned by the Yorkshire Wildlife Trust. At the time of going to press, efforts are in train to safeguard the whole area's future more securely because of a succession of potentially detrimental proposals. A threat of pumped drainage at North Duffield Carrs was effectively ended, after a protracted fight by conservationists, by the Minister of Agriculture's decision in 1984 to refuse grant aid for the scheme. This threat followed the construction of a barrage across the mouth of the Derwent at Barmby in 1975 to prevent the intrusion of salt water at high tide. Floods since have been sporadic, but as they have always been so there is no evidence that the barrage has caused a change. Pumping in the interests of agriculture was, however, proposed and the fear was that the opportunity might be taken permanently to lower the water table and plough out the grasslands, wholly undesirable in this important area. The latest threat stems from an application to the High Court by the "Derwent Trust" (an offshoot of the Inland Waterways Association) to try to prove an unfettered right of navigation along the whole length of the river, despite the rights above Sutton Lock (just north of Wheldrake) having being extinguished by Parliament in 1935. To raise money to oppose the case (which if successful could cause a huge increase in boat traffic and associated development through the main Ings area) the "Yorkshire Derwent Appeal" has been launched by a notable amalgamation of conservation, farming, landowning and angling bodies.

The number of birds present on the Derwent Ings is, of course, strongly dependent on the extent of the floods. Thus the markedly higher regular numbers in the 1970s and early 1980s, as compared with the 1960s (Table 98) reflects, as well as improvements in coverage, the establishment of the reserve at Wheldrake Ings and its management, chiefly the retention of floodwater. This should have less effect on the maxima, so there appears to have been a real increase in Mallard, Shoveler, Pochard and Tufted Ducks. Several species are now regular that were not present before, such as Gadwall, Shelduck and Greylag, the geese almost certainly being of feral origin. A wide variety of other species of wildfowl are also recorded. Nevertheless the Derwent Ings reach the species criterion of international importance only for the Bewick's Swan. In most recent years, however, the total wildfowl criterion (10,000 ducks, geese, swans) has been exceeded. The four seasons 1979 to 1982 gave regular total counts of 10,336 with a maximum (31st January 1982) of 19,169 wildfowl. Occasional records of Teal and Wigeon have also exceeded their qualifying levels, especially in recent years. Clearly it is highly desirable to safeguard the area's future as a key wetland in the region, which, of necessity, requires a satisfactory flooding regime. A relatively small area is, in fact, involved compared with the extensive floods that used to occur in Yorkshire and Humberside (488). In 1625 virtually the whole of the Vale of York was under water. As well as the winter interest, the Derwent Ings are also important for breeding wildfowl, with most British breeding ducks being recorded regularly, including over 100 pairs of Mallard, 40 of Shelduck, 50 of Shoveler and 3-7 of Garganey.

There are few sites of interest on the upper reaches of the Derwent. Near Malton, Welham Park Lake has up to 287 Mallard, and Scarpston Lake up to 98. A scatter of sites up to 15km to the west of Castle

Table 98. *Regular (and maximum) numbers of wildfowl at three sites on the Yorkshire Derwent (a,b) and Hull Rivers (c).*

a) Lower Derwent Ings i) 2 seasons 1960-1961
ii) 6 seasons 1976-1982
b) Castle Howard, Malton 17 seasons 1962-1982
c) Tophill Low Reservoirs 8 seasons 1970-1982

	a i	a ii	b	c
Mallard	1350(2300)	2830(8142)	991(1850)	441(1360)
Teal	908(3000)	1801(4450)	54 (195)	347 (800)
Wigeon	3160(5500)	3659(8250)	4 (100)	224 (520)
Pintail	61 (165)	51 (171)	0 (3)	0 (4)
Shoveler	31 (90)	80 (250)	19 (111)	39 (105)
Gadwall	0 0	2 (26)	0 (5)	12 (50)
Shelduck	0 (3)	15 (74)	0 (2)	2 (7)
Pochard	202(1500)	410(3115)	19 (74)	176 (300)
Tufted Duck	23 (124)	128 (500)	11 (41)	510 (780)
Goldeneye	59 (120)	15 (56)	3 (28)	38 (136)
Mute Swan	5 (15)	8 (39)	0 (1)	7 (47)
Whooper Swan	9 (35)	10 (34)	0 (6)	0 (2)
Bewick's Swan	51 (278)	104 (263)	0 (6)	0 (8)
Greylag Goose	0 (0)	78 (240)	46 (172)	6 (43)
Canada Goose	0 (1)	82 (326)	125 (245)	1 (13)

oward are of minor interest. The Canada Goose flock Castle Howard Lake (Table 98) appears on the nging evidence to be a fairly discrete sub-group ther than a part of the central Yorkshire population.

Eastward beyond the low Yorkshire Wolds is e last river draining into the Humber, the Hull, ompleting the arc of the present section. The Hull is ow extensively canalised and the fens of Holderness, hich once stretched along it for more than 30km from everley to Barmston, have been reclaimed. There is ttle to suggest today that this was once one of the reat strongholds for wildfowl in the north-east of ngland. The only substantial numbers of ducks are n the few pieces of open water left or created. East ark Lake, on the outskirts of the city of Hull, may old up to 500 Mallard, but otherwise there is little on ecord until Tophill Low Reservoirs alongside the river ear Brandesburton. These form the major wildfowl resort in the valley, with large populations of both dabbling and diving ducks (Table 98). Further upstream, Nafferton Mere, near Great Driffield, has had up to 70 Mallard and 180 Tufted Ducks.

The Humber catchment has, in the last few decades, lost much of its floodlands and the remaining few pieces are under threat. The creation of large expanses of gravel pits and reservoirs has provided some compensation, but these new habitats are generally important only for Mallard, inland diving ducks and introduced species. The washland species, notably swans, Wigeon and other dabbling ducks, are undoubtedly severely depleted compared with the past when flooding was regular along the Trent, Ouse and Derwent. The fight to save the Derwent Ings must succeed if a mere remnant of this past glory is to be retained.

North-west England and North Wales

North Wales

Apart from on Anglesey, where there are several inland marshes, pools and reservoirs, most of the wildfowl in North Wales are on estuaries and shallow roosts. In mainland Gwynedd the ground rises sharply to the mountain ranges and the many upland lakes are poor and acid, and hold very few wildfowl. The coast of Clwyd also has very sparse wildfowl populations, but the freshwater marshes of the River Clwyd form a locally important haunt for ducks.

Mainland Gwynedd

The three largest estuaries in the northern half of Cardigan Bay – the Dyfi, Mawddach and Glaslyn/Dwyryd – have much in common. In each case the northern shore is steep and rocky; the estuary is sandy and supplied from an impoverished upland catchment, with a strip of low-lying marsh on the southern side; the lower reaches are sheltered by long spits of

Fig. 30. Northern England and North Wales. Total wildfowl and regional boundary.

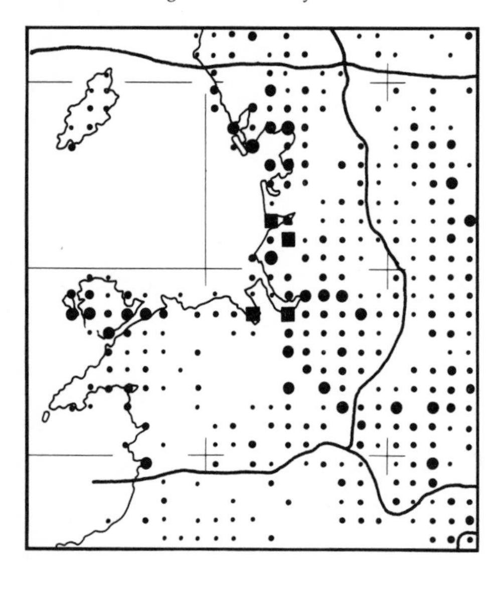

dune overlapping from the south; and in the case of the Dyfi and Mawddach there are areas of raised peat bog just behind the coast. The outfalls of the Dysynni, 8km north of the Dyfi, and the Artro, at the south end of Tremadog Bay, show some of these features on a small scale (Table 99).

South of the Dyfi Estuary (within Ceredigion District) is the famous Borth Bog, or Cors Fochno, a National Nature Reserve and one of the original thirteen sites designated by the United Kingdom on its ratification of the Ramsar Convention (p.540). The vicinity of the bog has long supported a small wintering flock of Greenland White-fronted Geese, now the largest and one of the few in Wales. Their numbers have generally been within the range 50-100 during the last 20 years, reaching peaks of 130 in 19[illegible] and 1970, and 108 in 1981. The geese usually roost on the mudflats and nowadays prefer to feed on the *Spartina* and grass marshes along the south shore, especially in the NNR west of Ynys-hir RSPB reserve – rather than on the bog itself. They have also been found roosting and feeding at Bugeilyn, a tiny lake in the Montgomery mountains 20km to the east (511), and at other hill bogs and lakes in the Plynlimon area. The Dyfi flock amounted to over 500 in the 1950s but records suggest that these were of the European race, possibly linked to the Severn Valley group. However, reports since before 1960 of Whitefronts staying into April demonstrate that at least some Greenland birds occurred. In the last 20 years the geese have been entirely from Greenland, with the exception of the very occasional individual of the European race in severe weather.

The dabbling ducks are concentrated at the head of the Dyfi Estuary, on the Ceredigion shore, where the Wigeon in particular find ideal feeding on the saltmarsh at Ynys-hir and elsewhere.

The land-locked estuarine lagoon of the Broad Water, at the foot of the Dysynni, has rarely been counted except in January. In addition to the species shown in Table 99, up to 50 Common Scoters and 1[illegible]

Table 99. *Cardigan Bay north. The regular (and maximum) numbers of wildfowl at the following sites. The figures for sites marked * are January averages (and maxima).*

a) Dyfi Estuary 12 seasons 1970-1981
b) Broad Water, Towyn* 7 seasons 1970-1982
c) Mawddach Estuary* 5 seasons 1978-1982
d) Artro Estuary* 6 seasons 1977-1982
e) Glaslyn/Dwyryd Estuary 10 seasons 1970-1981
f) Afon Wen Pools/Coast 5 seasons 1973-1978
g) Pwllheli Harbour 5 seasons 1974-1978

	a	b	c	d	e	f	g
Mallard	853(1505)	15(101)	134(263)	47 (56)	450 (900)	51(160)	132(250)
Teal	468(1300)	0(152)	10 (41)	227(450)	183 (523)	10 (60)	5 (39)
Wigeon	1646(5506)	255(497)	376(500)	229(500)	646(1380)	61(500)	79(230)
Pintail	160 (360)	0 (4)	6 (22)	0 (0)	76 (218)	0 (2)	0 (2)
Shelduck	111 (245)	21 (67)	32 (54)	15 (43)	176 (323)	1 (7)	32 (72)
Goldeneye	20 (40)	0 (2)	0 (0)	0 (2)	10 (21)	2 (10)	3 (10)
R-b Merganser	46 (109)	1 (11)	26 (41)	3 (5)	18 (103)	16 (65)	10 (52)
Mute Swan	0 (3)	0 (5)	2 (4)	4 (16)	11 (25)	5 (9)	15 (19)

Fig. 31. North Wales.

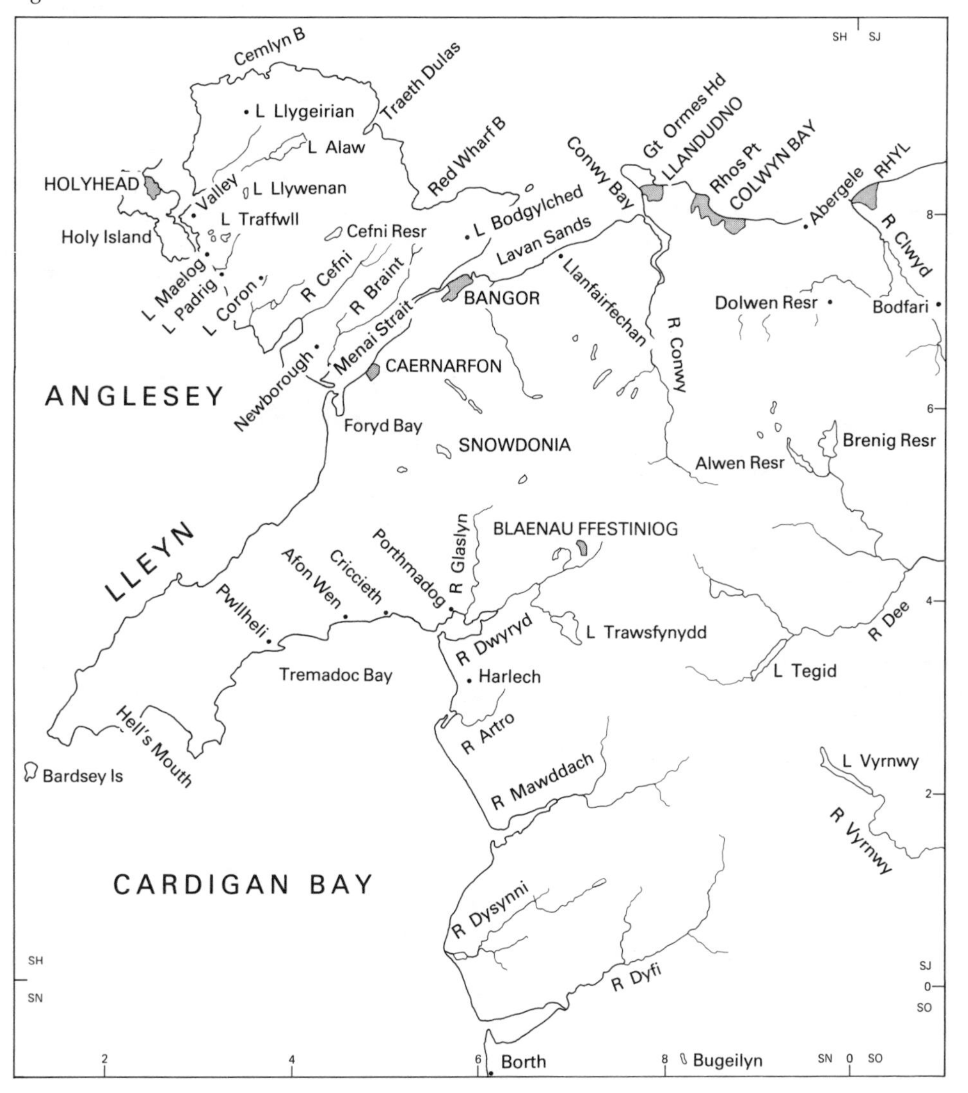

Eiders have been seen offshore (108), the latter representing a recently increased summer gathering.

The Mawddach Estuary is narrower than that of the Dyfi and the greater scour of the tide leaves less mud over the sand. The saltings, most of which have been reclaimed, and the feeding grounds on land are severely limited. As a result the wildfowl are relatively few, with no species exceeding local importance.

The great dunes of Morfa Dyffryn and Morfa Harlech (created NNRs in 1962 and 1958 respectively) extend for almost 20km to the north, broken only by the airfield, several new holiday camps and the small estuary of the Artro. Sizeable flocks of scoters are regularly seen off this stretch. Beyond lie the confluent estuaries of the Glaslyn and Dwyryd. Traeth Mawr, the "big estuary" of the former river, has been largely reclaimed, although the cob pool to the landward side of the man-made causeway at Porthmadog has recently held up to 200 Teal, 250 Wigeon, 70 Pintail and 20 Whooper Swans. The remaining basin, Traeth Bach, the "little estuary" of the Dwyryd, with its extensive sheep-grazed saltings, usually holds most of the ducks and has retained its importance despite the expansion of the holiday trade nearby.

Drainage at Llyn Ystumllyn, 3km west of Porthmadog, formerly the chief wildfowl haunt of the area and now controlled by BASC, has vastly reduced its importance. Only some 50 Mallard and 70 Teal now occur.

The Lleyn Peninsula is mostly unsuitable for wildfowl. The north shore is entirely rocky and for much of its length rises sharply to the adjacent hills. The south shore lacks any sizeable inlets, apart from Hell's Mouth (or Porth Neigwl) where 250 Mallard were counted in December 1967, probably disturbed from the nearby marshes along the River Soch. Apart from this area only the pools behind the railway at Afon Wen and the small harbour of Pwllheli are of interest to wildfowl (Table 99). Off Criccieth, however, 900 Common Scoters and, in summer, 80 Eiders have been found (108). The scoters are probably the same birds as those sometimes occurring on the opposite side of Tremadog Bay. The scoter flocks, often accompanied by a few Scaup, Goldeneyes and Mergansers, move along the north coast of Cardigan Bay and may regularly reach 1,000 birds in total.

The inland populations of mainland Gwynedd are mostly insignificant. On the Lleyn Peninsula there is little standing water, and in Meirionydd and Snowdonia the numerous lakes are mostly too high. A nearly annual series of comprehensive January surveys has produced, since 1967, combined maxima of only 291 Mallard, 71 Teal, 130 Wigeon, 345 Pochard, 122 Tufted Ducks, 18 Goldeneyes, 10 Whooper Swans and 20 Mute Swans. The most important waters are Llyn Tegid, or Bala Lake (at 445ha the largest natural lake in Wales), where 377 ducks, mostly Mallard, Teal and Wigeon, were counted in January 1982, and the

Table 100. *The regular (and maximum) numbers of wildfowl at the following waters on Anglesey:*

a) Llyn Rhosddu 2 seasons 1978-1979
b) Cefni Reservoir 5 seasons 1976-1981
c) L Coron 10 seasons 1970-1981
d) L Maelog 2 seasons 1976-1978
e) L Traffwll 3 seasons 1976-1981
f) L Penrhyn 3 seasons 1976-1981
g) L Dinam 2 seasons 1976-1981
h) L Llywenan 1 season 1976
i) L Llygeirian 3 seasons 1976-1981
j) L Alaw 8 seasons 1970-1981
k) Bulkeley Memorial Lake 1 season 1976

	Mallard	Teal	Wigeon	Shoveler	Pochard	Tufted Duck	Goldeneye	Greylag Goose	Canada Goose
a)	12 (130)	49 (96)	0 (0)	1 (3)	19 (47)	11 (16)	0 (0)	0 (0)	1 (2)
b)	295 (677)	114 (555)	16 (280)	6 (20)	90(350)*	160(299)	5 (14)	0 (0)	0 (1)
c)	163 (840)	102(1110)	280(1120)	19 (50)	10 (49)	34 (69)	3 (12)	83(350)*	44(260)
d)	24 (41)	8 (32)	13 (38)	5 (20)	61(135)	46(134)	8 (16)	29(175)	0 (2)
e)	502(1060)	86 (227)	285(1040)*	120(368)*	47(200)*	55(200)*	12 (19)	195(540)	137(600)*
f)	31 (77)	10 (50)	0 (0)	68(410)*	45(106)	44 (64)	12 (23)	0 (43)	1 (5)
g)	19 (32)	49 (291)	6 (50)	41(110)	9 (40)	7 (50)	2 (58)	0 (6)	0 (2)
h)	35 (84)	88 (250)	78 (250)	54 (85)	7 (11)	19 (65)	4 (5)	76(250)	14(150)*
i)	32 (76)	57 (200)*	5 (60)	12 (29)	17 (27)	53 (71)	2 (9)	0 (2)	0 (58)
j)	519(1132)	144 (465)	758(2104)	10 (58)	75(297)	104(500)*	17(107)	97(594)	55(290)
k)	30 (210)	19 (150)	81 (492)	33 (44)	1 (10)	30 (70)	0 (1)	2 (4)	16 (50)

* From *Cambrian Birds*.

514ha hydroelectric reservoir at Trawsfynydd, now the site of a nuclear power station and a Central Electricity Generating Board nature reserve. This lake is used for cooling water from the power station and is less likely to freeze than other fresh waters; Mallard, Pochard and Tufted Duck have all reached 100 here.

Anglesey and the North Wales coast

At the southern end of the Menai Straits lie two interesting resorts: Foryd Bay, on the mainland, and Traeth Melynog (the Braint Estuary), on Anglesey (Table 101). Both are surrounded by dune systems and only Wigeon and Shelducks are present in numbers for any length of time. The inner parts of Foryd Bay are now largely choked with *Spartina*.

The great, partly afforested dune system of Newborough Warren NNR, on Anglesey, contains several small pools, of which Llyn Rhosddu on the edge of the dunes (Table 100) is much the most important. The Cefni Estuary beyond, known as Malltraeth Sands, is an important roost for dabbling ducks, notably Pintail. The cob pool behind the sea wall and the adjacent marshes often hold several hundred ducks, Teal finding the spot particularly attractive (Table 101). Upstream the river is canalised and runs through the Malltraeth Marshes (Cors Ddyaga), the reclaimed upper estuary of the Cefni. Here there are some suitable feeding grounds for wildfowl, but bird numbers are rather small, probably partly due to heavy shooting pressure. Since 1970 no more than 110 Mallard, 180 Teal and 20 Shoveler have been seen, together with occasional Bewick's Swans, the largest number being 16 (108), and 60 feral Greylags. Near the head of the river, in the heart of Anglesey, lies Cefni Reservoir, built in 1950. This is a sheltered lake surrounded by conifer plantations, where a particularly interesting autumn passage of Teal occurs, exceeding 500 in 1976 (Table 100). They are probably attracted by the shallows resulting from the large summer draw-down. Here, too, occurs the only sizeable flock of Mute Swans on Anglesey at present, amounting to a regular 14 and maximum 32 between 1976 and 1981. Whooper Swans have reached 6 and 29 respectively in those winters. There is a public hide at the edge of the lake, but disturbance from shooting is slight, being restricted to six days a year. The reservoir is crossed by a little-used railway line.

Llyn Hendref, 4km to the west, was a mid-winter roost for several hundred Wigeon in the 1960s, but drainage in 1970 reduced the water area by three-quarters. The few counts since then have recorded only a handful of ducks, except in August 1976, when 115 Mallard and 125 Teal were found.

Inland of the south-west Anglesey dunes there are a number of shallow lakes (Table 100), of which the most important is the 25ha Llyn Traffwll. Llyn Coron used also to be important but a decline occurred in all the main species during the 1970s. The regular populations of Mallard, Teal, Wigeon and Tufted Ducks decreased, respectively, as follows between the

Table 101. *The regular (and maximum) numbers of wildfowl at the following sites on the coasts of Anglesey, north Gwynedd and Clwyd:*

a) Foryd Bay 2 seasons 1971-1976
b) Traeth Melynog 3 seasons 1976-1979
c) Cefni Estuary, incl. Malltraeth Cob 3 seasons 1976-1979
d) Inland Sea/Beddmanarch Bay/Alaw Estuary 3 seasons 1976-1981
e) Cemlyn Bay/Lagoon 8 seasons 1972-1981
f) Traeth Dulas 2 seasons 1973-1974
g) Red Wharf Bay 7 seasons 1971-1980
h) Traeth Lafan 6 seasons 1970-1980
j) Conwy Estuary 4 seasons 1971-1974
k) Clwyd Estuary 3 seasons 1970-1976

	Mallard	Teal	Wigeon	Pintail	Shelduck	Goldeneye	R-b Merganser	Mute Swan
a)	12(100)	7(200)	273(1450)	6 (24)	100(217)	6 (16)	11 (22)	21(38)
b)	157(370)	179(830)	860(2000)	54(111)	204(542)	2 (4)	5 (12)	0(25)
c)	30 (82)	31(600)*	280 (873)	115(400)*	220(350)*	5 (15)	6 (14)	10(14)
d)	102(500)	74(123)	947(1600)	10 (29)	280(370)	66(110)	88(210)	9(37)
e)	151(347)	142(500)	158 (500)	0 (1)	5 (10)	19 (32)	16 (33)	3 (9)
f)	2 (6)	14 (28)	53 (100)	0 (0)	16 (23)	0 (1)	9 (18)	0 (0)
g)	6 (17)	7 (85)	149 (290)	0 (0)	39 (78)	0 (1)	9 (67)	0 (0)
h)	457(860)	83(188)	565 (944)	12 (72)*	250(548)	38(160)*	see text	3(21)
j)	116(280)	60(132)	91 (320)	4 (9)	66(560)	3 (11)	1 (9)	2 (3)
k)	62(274)	17 (84)	8 (40)	0 (2)	12 (72)	2 (14)	1 (8)	10(20)

* From *Cambrian Birds*.

periods 1960-1969 and (in brackets) 1970-1981: 325 (163), 178 (102), 549 (280), 66 (34). There was a reduction in aquatic vegetation on the lake, due to increasing algal growth (601), and it was suspected that this may have caused the disappearance of a non-breeding herd of some 50 Mute Swans which occurred in the mid 1960s. The same factor may have occasioned the roughly simultaneous decline in wintering ducks, and, although Mallard have maintained the level of their annual peak, the largest numbers are now present only for a short time in early autumn, when they can feed on the adjacent stubble fields (498). Pintail, well established on the nearby Cefni Estuary, sometimes exceed 100 at Llyn Coron in autumn, but are rare at other times.

Llyn Padrig is the smallest of the coastal lakes, at 6ha, but its closeness to Llyn Coron (3km) enables birds disturbed from there to find speedy refuge, and maxima of 240 Teal, 800 Wigeon and 24 Shoveler have occurred on Llyn Padrig.

Llyn Maelog is only half a kilometre from the coast, but is subject to disturbance from the nearby village and the Valley airfield. The Valley lakes – Traffwll, Penrhyn and Dinam – are subject to some aircraft disturbance but they hold the largest numbers of Shoveler in North Wales and provide the stronghold for Anglesey's feral populations of Ruddy Ducks, Canada Geese and Greylags. Ruddy Ducks have bred there since 1978, while in the hard weather of January 1982 the Valley lakes held 76 of the 101 to reach Anglesey from their Midland strongholds. In October 1978 540 Greylag and 375 Canada Geese were counted on Llyn Traffwll (108), the Greylags probably representing the entire Anglesey population and the latter comparing to an estimated 4-500 Canada Geese on the island (321). Both species, however, are highly mobile and likely to turn up in numbers anywhere within Anglesey (Table 100).

The shallow straits between Holy Island and the Anglesey mainland are of considerable importance, especially the northern complex of the Inland Sea (between Four Mile Bridge and the Stanley Embankment), Beddmanarch Bay and the contiguous Alaw Estuary (Table 101). The few full counts made to the south of Four Mile Bridge, in the late 1960s and one in 1983, produced maxima of only 21 Mallard, 23 Teal, 77 Wigeon, 40 Shelducks, 6 Goldeneyes and 2 Red-breasted Mergansers. In Cymyran Bay, at the southern entrance to the straits, however, 250 Common Scoters were found in December 1980 (108).

The west shore of Holy Island is mainly rocky, but Trearddur Bay, just across the central isthmus of the island, serves as an alternative resort for Goldeneyes from the Inland Sea area; 50-100 are usual, though 158 were found in Janaury 1978.

Llyn Llywenan, 4km inland from the head of the Alaw Estuary, holds a good variety of species, sometimes in large numbers, but a paucity of counts other than in January makes the impression of a decline since the 1960s (when over 1,000 ducks were sometimes found) difficult to confirm. Llyn Llygeirian, 10km to the north, is most notable for its flock of Gadwall, the largest on Anglesey, which amounts to a regular 26.

The 314ha Llyn Alaw was formed by damming the river 7km upstream in 1966. It is shallow (under 5m on average), and became the principal inland site for wildfowl on Anglesey shortly after its completion; it has since maintained that status. In addition to the species in Table 101, Mute and Whooper Swans are regular in small numbers, the latter reaching 38 in February 1976.

The north and east coasts of Anglesey are rocky and for the most part hold few wildfowl. The main interest is in the concentration of Common Scoters off Red Wharf Bay, a flock which reached 1,800 in December 1976. Velvet Scoters are sometimes seen in the group, with 30 in February 1977 (108). This scoter flock may be linked with those of up to 1,100 which occur at times along the remainder of the North Wales coast, especially between Llanfairfechan and Penmaenbach, and in Colwyn and Abergele Bays, their distribution possibly being connected with the changing availability of bivalve molluscs. In 1981 these molluscs appeared to be unusually sparse in Red Wharf Bay, coinciding with an absence of scoters from that area and their presence in strength further east (497).

Three inlets along Anglesey's north and east coast are included in Table 101: Cemlyn Bay, Traeth Dulas and Red Wharf Bay itself. At Cemlyn Bay most of the ducks are found on a brackish lagoon, divided from the sea by a shingle bar, or in a small inlet in the coast nearby. The lagoon has long been run as a private sanctuary and is now a reserve of the North Wales Naturalists' Trust which leases it and the bar from the National Trust, which also owns the headland and much of the surrounding farmland. Traeth Dulas is small and sandy, but protected from the bay by a grassy spit. Its wildfowl, however, are few, perhaps deterred by heavy shooting pressure and the fairly steep rise of the ground to either side. Red Wharf Bay is the largest on Anglesey, but too sandy and exposed to be of much importance for wildfowl.

The interior of eastern Anglesey is short of standing water. Beaumaris Reservoir, at Pen-y-Parc, near the entrance to the Menai Straits, has held up to

50 Tufted Ducks (108) but little else. Bulkeley Memorial Lake (Llyn Bodgylched), 2km north (Table 100), is of interest for its wintering flock of White-fronted Geese, which averaged 43 between 1970 and 1979, with a peak of 72 in January 1973. The few close observations made confirm that this flock comprises birds of the European race, whose nearest stronghold is in the Tywi Valley, 150km south. Whitefronts began wintering at Bodgylched in 1962, having possibly moved from the south-west of the island, where they were found in similar numbers in the late 1950s and early 1960s, but were once much more abundant.

Parts of Anglesey have, in recent years, received occasional flocks of Pink-footed Geese displaced from Lancashire, the largest being a group of 500 at Alaw Reservoir in February 1982. These records may become more common, especially in hard winters, if the concentrations in Lancashire are maintained.

For most of their length the Menai Straits hold few wildfowl, due to the lack of suitable bays. Between Caernarfon and Bangor, where the January counts have been fairly complete, no more than 280 Mallard, 50 Teal, 60 Wigeon, 110 Shelducks, 25 Goldeneyes and 30 Mergansers have been found. From the northern entrance of the Menai Straits to Great Ormes Head stretches Traeth Lafan/Conwy Bay (Table 101). Most of the wildfowl of the bay occur in its south-western corner, occupied by the vast Lavan Sands (2,300ha) and the Ogwen Estuary and Bangor mudflats (300ha); most of this area was declared an LNR in 1979. Particularly notable are the Red-breasted Mergansers which have gathered to moult off the Lavan Sands every summer since at least 1965, averaging 283 between 1971 and 1980, with a peak of 441 in August 1973. They are present from late June to late October and usually feed at Llanfairfechan, at the eastern end of the sands, drifting into roosting rafts off Afon Aber, a few kilometres west, at high tide (148). By January there are usually fewer than 10 Mergansers in the area, although 55 were found to the east of Llanfairfechan in February 1977. Groups of 3-600 Mallard also moult on the Sands, while Shelducks have a pre-moult gathering of 300-350 between April and June (148). In winter the Ogwen Estuary is the favoured area for most species, while Bangor Harbour acts as a refuge for Tufted Ducks during heavy frosts, up to 140 having been found.

The south-east corner of Conwy Bay comprises the estuary of the River Conwy, with its sandbanks and mussel beds and, above the railway bridge at Conwy town, mudflats and an increasingly extensive *Spartina* marsh. It is this upper section which supports most of the Wigeon and Shelducks. The latter reach a spring peak, as in the west of the bay (485). Other ducks are rather scarce (Table 101).

From Great Ormes Head to the entrance of the Dee Estuary the coast, though largely flat, has no large inlets and has been heavily developed for the holiday industry. The only significant concentrations of wildfowl are the scoters, mentioned earlier, the Mergansers, which occasionally exceed 80 off Rhos Point, and a fairly regular gathering of Scaup in Abergele Bay, which reached an exceptional 196 in December 1973 (267). This flock had probably moved from the mouth of the Dee, where such numbers are normal. There are also occasional records of flocks of up to 200 Pink-footed Geese from various sites on the coast and on the Clwyd in recent years, since the great upsurge in Lancashire.

The Clwyd Estuary (Table 101) has a good variety of ducks but in paltry numbers, although 30-40 Common Scoters are consistently present on the sea in mid-winter. The Vale of Clwyd, especially the flood-meadows between Bodfari and Ruthin, 10-20km upstream, was once the haunt of Pink-footed and White-fronted Geese, the former sometimes reaching 200. Both species are now rare except in hard weather. Up to 2,000 Mallard and 2,000 Wigeon are believed to occur still in this area, but recent data are inadequate (511).

In the hills to either side of the Vale there are numerous lakes and reservoirs, some quite large, but few wildfowl visit them. The 372ha Brenig Reservoir, constructed in 1977, has, despite its size, held regular populations of only 141 Mallard, 23 Teal, 27 Pochard, 4 Tufted Ducks, 4 Goldeneyes and 6 Goosanders, and seems to be undergoing the decline usually associated with new reservoirs after their first few winters. The nearby and much older Alwen Reservoir is apparently almost devoid of wildfowl. Both lie at over 350m. Much lower and nearer the coast, Llyn Helyg and Ysceifiog Reservoir, on the eastern slopes of the Clwydian Range, held average peaks in the 1960s of, respectively, 151 and 15 Mallard, 56 and 12 Teal, 152 and 1 Wigeon, 21 and 0 Shoveler, 22 and 9 Pochard and 36 and 29 Tufted Ducks. Subsequent records are sparse, but there were similar concentrations at least up to the late 1970s. The upper Dolwen Reservoir, on the opposite side of the Vale, was counted from December 1978 to March 1979, holding peaks of 41 Mallard, 330 Wigeon, 38 Pochard and 72 Tufted Ducks, and remaining unfrozen even during the severe cold of that winter.

North-west England

The north-western districts of England are divided into three distinct sectors. The first comprises the great lowland plain of Cheshire and north Shropshire with its numerous meres and a short but valuable stretch of coastline around the estuaries of the Mersey and Dee. Further north is the industrial area of south Lancashire and the reclaimed fenland around the mouth of the Ribble. Lastly there is the sandy expanse of Morecambe Bay with its series of small estuaries, and a scattering of inland waters including the southern part of the Lake District.

The Dee and Mersey Estuaries, Cheshire Plain and upper Severn

In contrast to the mountainous districts of North Wales, the greater part of Cheshire and north Shropshire is occupied by a belt of level farmland, lying at less than 100m and extending inland fo upwards of 60km. Of this about half drains north wards to the Mersey and the Dee, the remainde southwards to the upper reaches of the Severn Throughout its length the watershed between the tw systems is scarcely perceptible. An arbitrary division i made unrealistic by the many meres and ponds whic feature prominently everywhere. The area has there fore been treated as an entity, the southern boundar being drawn to include the whole of the upper Severn from the point where it leaves the hills at Welshpool t the entrance of the gorge at Ironbridge.

The estuary of the Dee, greatly reduced in siz by reclamation schemes in the past two centuries, has in the last twenty years, been threatened by furthe

Fig. 32. Cheshire and Shropshire (upper Severn).

reclamation, road bridge, barrage and water reservoir schemes (p.480), and there are plans, currently being appraised, to extend the marina at West Kirby. The barrage proposals appear to have receded for the present (p.483), but they have had the advantage of bringing into being a very active Dee Estuary Conservation Group, which, together with the NCC, produced a wide-ranging research review (372). The wildfowl tend to concentrate in three zones: the lower saltmarsh between Burton and Heswall and between Mostyn and the Point of Air; the upper marsh and pools around Burton and Denhall; and the outer estuary between Caldy Blacks and Hilbre Island (619).

In the 19th century Barnacle Geese came to the Dee in thousands, but around the turn of the century they were replaced by Pink-footed Geese. These in turn decreased to the odd party until recent years when regular flocks have been seen briefly, while White-fronted Geese, never numerous, hardly occur now. Wildfowling and other disturbances as well as habitat loss have undoubtedly caused the decline of the Dee as a goose resort. The situation is improving, however, following the establishment of the RSPB reserve of 2,040ha on Gayton Sands. The ducks have remained very numerous, but have shown wide fluctuations; Pintail occur consistently in internationally important numbers. A massive increase took place in the present century until, by the 1950s, the normal peak varied between 2,000 and 3,500 birds, sometimes reaching 5,000. Numbers then fell sharply to between 500 and 1,000 in the early 1960s, but have since built up again, to a maximum of 11,265 in November 1983. The high numbers now tend to remain through to January, whereas earlier there was more of an autumn passage. Table 102 shows that along with Pintail several other species increased during the 1970s, Mallard, Teal, Wigeon and Shelducks being especially notable.

Shelducks are also present in internationally important numbers. They breed in small flocks along the coast, but those which assemble in late June and early July represent an influx from elsewhere. Summer counts from 1957 to 1971 on the Dee and elsewhere in the north-west region gave an average of 1,252 adults and 34 young between 1957 and 1967 (6). This was well exceeded in 1968 when a total of 2,705 adults and 13 young was recorded. In spite of several surveys there was no definite evidence (flightless birds or quantities of feathers) that the Dee is a moulting area like Bridgwater Bay and other estuaries until 1983, when at least 75 adults were moulting. A moult migration from Cheshire is more usual and has been well studied (5), and many more birds than the summer population pass through on their way, presumably, to Heligoland Bight.

The estuary of the Mersey, just across the Wirral Peninsula, is quite different in character from that of the Dee. Along the lower reaches the tideway is at first narrow and deep, and dominated by the waterfronts of Liverpool and Birkenhead. Further inland the estuary becomes broad and shallow, with large areas of intertidal mud extending along the whole length of the southern shore. Disturbance in this part is less intense, due partly to the dangerous softness of the

Table 102. *Changes in the regular (and maximum) numbers of wildfowl at the Dee and Mersey Estuaries.*

a) Dee Estuary i) 5 seasons 1965-1969 ii) 11 seasons 1970-1982*

b) Mersey Estuary: Ince/Stanlow Banks i) 10 seasons 1960-1969 ii) 12 seasons 1970-1982

	a i	a ii	b i	b ii
Mallard	928(1956)	1802(3750)	238 (510)	1238 (2440)
Teal	89 (492)	741(2710)	2090(5150)	10370(35000)
Wigeon	266(1022)	595(2015)	1591(3200)	4087(15200)
Pintail	722(2000)	3111(7360)	927(3900)	8509(18450)
Shoveler	12 (59)	18 (45)	4 (53)	22 (140)
Shelduck	1670(2160)	3148(7315)	213 (660)	4112(12170)
Pochard	44 (225)	19 (85)	0 (2)	3 (41)
Tufted Duck	9 (50)	12 (64)	0 (2)	1 (8)
Goldeneye	22 (43)	17 (51)	0 (0)	0 (4)
Scaup	70 (325)	125 (383)	0 (0)	0 (2)
Common Scoter	30 (220)	7 (43)	0 (0)	0 (0)
R-b Merganser	17 (52)	16 (52)	0 (0)	0 (8)

* All time maxima of 4950 Mallard and 11265 Pintail were recorded on the Dee in 1983-84.

mud, but mainly to the Manchester Ship Canal, which skirts the edge of the marsh and limits access.

The Ince and Stanlow Banks are the focal point for ducks and support a remarkable assemblage (Table 102). Wigeon, Teal, Pintail and Shelducks exceed the levels of international importance – the last three by a huge margin. The striking concentrations of Teal noted in the 1950s were maintained during the 1960s and then increased prodigiously in recent years. Maxima jumped to 9,750 in 1972 and in following winters to 13,700. Numbers were then lower for a few years but were back at 12,870 in the cold January of 1979. However, they increased further, to 17,400 and 25,850, in the next two mild winters before rocketing to 35,000 at the start of the severe weather in December 1981. A real population shift as well as the impact of extreme climatic conditions would seem to be involved. Certainly the peak in 1982-83, another mild winter, was still 26,100. An improvement in the conditions of the area itself is also suggested by the parallel increases, again mainly in the 1970s, in Mallard, Wigeon, Pintail and Shelducks. The first three all reached their maxima in 1980-81, which was not a hard winter. During the last decade a good deal of effort has been put into reducing pollution in the Mersey, but a Department of the Environment document (158) still stigmatised it as the most polluted estuary and river system in the United Kingdom. In 1979, for instance, there was substantial mortality among waders due to poisoning by organic lead from an industrial process (p.489).

The initial build-up of wildfowl was ascribed to a marked improvement in the state of the river, evinced by increased growth of food plants (4). The establishment of a sanctuary from Stanlow Point to Mount Manisty by the Merseyside Naturalists' Association was also instrumental. Hopefully the studies by Liverpool University into the ecology of the estuary started in autumn 1984, may shed further light. Regrettably these have been prompted by a plan to reclaim Ince and Stanlow Banks for petrochemical development. If this goes ahead a site of outstanding international importance will be lost. There is one other sad note in this extraordinary success story: the European Whitefronts which used to be a feature of the area have now disappeared and the only geese seen nowadays are a few Pinkfeet in hard weather.

Other sites around the Mersey Estuary hold substantial numbers of wildfowl, though for many the counts have not been sufficiently frequent to do more than provide maximum figures (Table 103). Two have been well enough covered recently to give full data (Table 104). Fiddlers Ferry Lagoons are associated with the power station and are freshwater. Woolston Eyes are a series of deposit grounds for the Manchester Ship Canal, embanked reservoirs into which canal sludge is pumped, covering in all some 400ha. Other deposit grounds at Frodsham and Weaver Valley hold but few ducks during the day-time counts, but these areas of soft mud undoubtedly are important as night-time feeding places. The dabbling ducks at Woolston and Fiddlers Ferry may well interchange with the much greater concentrations on the Ince Banks, but the diving ducks are additional.

To the south and east of the Mersey Estuary, and 15-25km inland, lies a ring of meres whose duck population seems closely linked with that on the tidal water. Pre-eminent amongst these is Rostherne Mere near Knutsford, a National Nature Reserve and a Ramsar Site, a status merited as much for the limnological interest as for its value to wildfowl and other birds (252). In fact its use by wildfowl has fallen away from the times in which it supported the largest inland concentration of Mallard in England (607). Their regular numbers have been halved and indeed

Table 103. *The maximum numbers of wildfowl recorded on five sites in the Mersey Estuary 1960-1979 (not including Ince Marshes).*

	No. counts	Mallard	Teal	Wigeon	Pintail	Shelduck
Dungeon Banks	5	250	0	20	29	1200
Weston Marshes	24	29	240	540	48	12
Runcorn Sands	15	438	521	150	103	2
No Mans Land	5	430	1100	500	120	6
Moss Side	7	305	582	10	116	25

ıe data presented in Table 105 give too optimistic a icture, for since 1976 the peak number of Mallard has een below a thousand. Teal, despite the upsurge on ıe Mersey, have also declined here though less ramatically than the Mallard, whereas the numbers f Pintail and Shoveler, and especially the diving ucks, have increased. The lake is up to 30m deep and ıe dabbling ducks must feed mainly on the shallower ıeres and agricultural land, so some change there ıay have a bearing. Increased noise from motor ehicles and aircraft is also a possible factor but the ther habitat changes are probably more important. Iearby Tatton Mere (Table 105) has experienced an ıcrease in Mallard, though not in the numbers lost to ostherne. Both waters have had an increase in anada Geese, but only in proportion to that species' verall rise. The deepness of Rostherne means that it is ıe last of the meres to freeze in severe weather, and ınuary 1982 produced a gathering of 76 Ruddy ucks, now firmly established as a feral breeder in the istrict.

Table 104. *The regular (and maximum) numbers of wildfowl recorded at two artificial sites on the Mersey Estuary.*

a) Fiddlers Ferry Lagoons 4 seasons 1975-1982

b) Woolston Eyes 4 seasons 1979-1982

	a	b
Mallard	452 (881)	266 (831)
Teal	321 (1080)	1521(4590)
Wigeon	23 (85)	0 (2)
Pintail	12 (45)	165 (782)
Shoveler	41 (110)	249 (516)
Shelduck	19 (104)	4 (16)
Pochard	265 (890)	319 (765)
Tufted Duck	23 (356)	76 (219)
Ruddy Duck	0 (0)	14 (104)
Mute Swan	0 (2)	7 (17)
Canada Goose	11 (37)	2 (14)

'able 105. *Changes in the regular (and maximum) numbers of wildfowl recorded at two neighbouring Cheshire meres.*

) Rostherne Mere i) 7 seasons 1960-1969 ii) 12 seasons 1970-1982

) Tatton Mere i) 2 seasons 1960-1964 ii) 6 seasons 1974-1980

	a i	a ii	b i	b ii
1allard	2055(3000)	1209(3000)	175(402)	350(520)
eal	469(1000)	372(1300)	40(160)	17 (56)
Vigeon	37 (150)	49 (150)	15 (50)	0 (3)
'intail	0 (2)	71 (320)	0 (0)	1 (9)
hoveler	11 (50)	73 (300)	3 (14)	11 (89)
ochard	125 (600)	279(1300)	41(146)	81(156)
'ufted Duck	29 (80)	81 (750)	89(162)	186(362)
;oldeneye	2 (16)	7 (16)	9 (20)	21 (31)
.uddy Duck	0 (0)	8 (76)	0 (0)	0 (0)
:anada Goose	34 (210)	134 (450)	91(286)	259(734)

'able 106. *Regular (and maximum) numbers of wildfowl recorded on six north Cheshire meres and pools.*

) Oulton Mere 12 seasons 1965-1978

) Great Budworth Mere 16 seasons 1960-1982

) Radnor Mere 6 seasons 1977-1982

d) Redesmere 7 seasons 1976-1982

e) Withington Sand Pits 6 seasons 1976-1981

f) Farmwood Pool Gravel Pit 7 seasons 1976-1982

	a	b	c	d	e	f
1allard	114(400)	95(633)	400(648)	265(670)	318(905)	426(800)
'eal	3 (60)	8 (79)	182(312)	3 (23)	2 (9)	19 (83)
Vigeon	7(100)	3 (38)	4 (18)	1 (16)	4 (27)	4 (19)
hoveler	3 (20)	5 (40)	28 (75)	5 (35)	7(120)	2 (10)
'ochard	70(400)	19(170)	19(101)	12 (46)	8 (18)	41(107)
'ufted Duck	64(250)	23 (91)	36 (61)	62(212)	39(148)	50(121)
'uddy Duck	0 (0)	0 (4)	25 (41)	4 (14)	0 (4)	11 (32)
:anada Goose	18 (93)	14(147)	105(232)	170(500)	8 (87)	28(205)

Scattered over 15km to the south-west are a number of meres which have often been well counted but which, apart from Mallard, hold only a few ducks. The counts for Great Budworth and Oulton Meres are set out in Table 106, but Tabley Mere, Marston Flash, Petty Pool, Oulton Mill Pond, Nunsmere, Cuddington, Little Budworth, Oakmere and Acre Nook Quarry only rate regular double figures for a few species, while the others are even less important. Further east, between Wilmslow and Macclesfield, are other waters of which Radnor Mere and Redesmere, along with the pits at Withington and Farmwood, are important enough to set out in Table 106. It is clear that, although close to the Mersey, these inland waters do not intercept the large numbers of Teal an Pintail passing to the estuary. The Ruddy Duck tenc to frequent the more easterly of the waters, althoug the species continues to spread in the north-we Midlands. Around Macclesfield a group of a doze small reservoirs holds mainly Mallard but in no gre number, 20-40 apiece, with a sprinkling of oth species. Mallard also occur over the Derbyshire bord on the hill reservoir at Fernilee in the Goyt Valley which had a regular count of 168 and a maximum 343 between 1962 and 1982. Similar numbers occur o Todbrook Reservoir, but many fewer on the other tw reservoirs in this group – Errwood and Combs.

In central Cheshire, in the Winsford – Crewe Sandbach triangle, a number of "flashes", permaner shallow subsidences caused by underground sa extraction, provide useful wildfowl habitat. A scatte ing of counts reveals an interesting mix of specie (Table 107). Railway Farm Flash, south of Elton, ha few counts, but maxima of 64 Mallard, 130 Teal, 5 Wigeon and 32 Tufted Ducks have been recorded South of Crewe, Doddington Pool, Shavington Park Combermere, Bar Mere and Cholmondeley Park forr a chain of important waters, particularly for dabblin ducks and Canada Geese (Table 108). The Dee aroun Aldford has produced occasional counts of up to 1,50 Mallard, 926 Teal, 600 Wigeon, 440 Pintail and 25 Canada Geese, though not all at the same time, and maximum of 265 Wigeon is reported from Tattenhal Ponds. Associated with these are several waters wit small regular populations of 10-50 Mallard, Teal Pochard or Tufted Ducks. Rather more substantia numbers are to be found on Bache Pool, with regula (and maximum) numbers of 86 (142) Mallard and 2 (40) Tufted Ducks, and on Norbury Meres, with 20

Table 107. *Regular (and maximum) numbers of wildfowl recorded on four salt-extraction subsidences in central Cheshire.*

a) Elworth Flashes 7 seasons 1960-1976
b) Warmingham/Crabmill Flashes 2 seas. 1977-1978
c) Elton Hall/Watch Lane Flashes 4 seas. 1977-1982
d) Winsford Bottom Flash 4 seasons 1973-1979

	a	b	c	d
Mallard	16 (91)	21(60)	268(400)	24 (78)
Teal	116(540)	24(50)	437(970)	8 (28)
Wigeon	14(100)	21(73)	136(508)	4 (30)
Pintail	3 (25)	0 (0)	35 (52)	0 (1)
Shoveler	9 (35)	1 (1)	25 (68)	1 (8)
Pochard	43(180)	1 (1)	39(105)	152(277)
Tufted Duck	15 (48)	1 (2)	42 (69)	56(180)
Mute Swan	11 (25)	4 (6)	9 (25)	5 (20)

Table 108. *Regular (and maximum) numbers of wildfowl recorded on five waters in south Cheshire/Shropshire.*

a) Doddington Pool 6 seasons 1960-1979
b) Shavington Park 18 seasons 1960-1981
c) Combermere 20 seasons 1960-1981
d) Bar Mere 5 seasons 1960-1981
e) Cholmondeley Park 2 seasons 1980-1981

	a	b	c	d	e
Mallard	135(450)	1176(2000)	289(826)	37(250)	289(434)
Teal	12 (59)	57 (200)	33(261)	17 (62)	18 (41)
Wigeon	12 (42)	116(1000)	46(450)	35(118)	11 (35)
Shoveler	19 (81)	56 (200)	6 (40)	20 (76)	90(204)
Pochard	59(252)	56 (200)	11 (56)	5 (15)	14 (33)
Tufted Duck	48(138)	105 (550)	41(117)	14 (41)	65(110)
Goldeneye	2 (7)	1 (4)	5 (30)	2 (10)	7 (10)
Goosander	6 (42)	15 (50)	0 (5)	0 (0)	9 (23)
Ruddy Duck	1 (11)	1 (14)	14(115)	12 (55)	14 (20)
Mute Swan	15 (56)	3 (27)	2 (32)	0 (1)	2 (7)
Canada Goose	14(123)	225(1600)	75(536)	111(430)	275(550)

300) Mallard, 51 (150) Teal, 22 (70) Shoveler and 48 (184) Canada Geese.

A few kilometres further west another group of meres includes Ellesmere and a dozen smaller pools, the most imporant of which are Blakemere, Newton Mere, Cole Mere, Crose Mere and Sweat Mere. This Ellesmere group has been counted as such for many years with the results shown in Table 109. Most of the dabbling ducks use these somewhat deep and steep-sided lakes mainly as roosts, flighting out to the farm ponds and marshy hollows of the surrounding countryside. Hanmer Mere, although across the Welsh border in Clwyd, is clearly associated with this group, having a similar mix of species and not inconsiderable numbers for the one water. Bettisfield Pool and Marsh nearby have only produced 20 Teal.

Around Oswestry, Aston Hall has lost the sizeable flock of up to 300 Mallard present in the 1960s; it was drained for a year in the early 1970s and now only supports single figures of several species, as do Oswestry Ponds. Oerley Reservoir, counted between 1967 and 1982, has regular (and maximum) numbers of 33 (130) Mallard, similar counts of Canada Geese and 19 (80) Tufted Ducks. West of Shrewsbury there are resorts that hold sizeable flocks if not too much disturbed: Marton Pool, Fenemere and Polemere. Table 109, gives an indication of the main species frequenting the area. Sherwardine Pool had up to 170 Mallard, 25 Wigeon, 33 Shoveler, 6 Pochard, 28 Tufted Ducks and 48 Canada Geese in 1982, but there are no data from Isle Pool, Alkmond Park Pool or the several other waters in the vicinity of Shrewsbury which look promising. Counts on the River Severn from Coton Hill to the Weir (Shrewsbury) have given regular (and maximum) numbers of 223 (296) Mallard, but little else. To the east of Shrewsbury several waters support mainly Mallard, with around 50 wildfowl regularly, such as Mere Pool, Weeping Cross; Bomere, Shomere and Betton Pools; Acton Burnell; Pitchford Hall; Sundorne Pools and Alderham Park Pool. Venus Pool near Cound, a flooded dip caused by the collapse of a road culvert, provides an interesting wetland. The early flocks of up to 500 ducks in 1961, soon after flooding, have not been maintained and regular counts now yield less than 100, though 200 Teal were recorded in January 1982. Another interesting site is the Allscott sugar-beet pits near Walcot, with 65 (234) Mallard, 154 (450) Teal, up to 50 Tufted Ducks and 330 Canada Geese. These waters, it should be noted, are only some 20-30km from the big Staffordshire resorts at Aqualate and Belvide (p.171) and some interchange with them no doubt takes place.

Upstream from Shrewsbury, the area of the confluence of the River Severn and the River Vyrnwy holds considerable numbers of wildfowl when there is extreme flooding. Thus in January 1982, 423 Mallard, 1,125 Teal, 502 Wigeon and 42 Pintail were seen (144). However, in the following milder winter the maxima were, respectively, 440, 680, 123 and 13 (145). The importance of the floodwater is emphasised by the demise of areas further upstream through agricultural improvements, as in the low-lying meadows which flank the Severn and Camlad between Chirbury and Welshpool. The substantial flock of European White-fronted Geese, which used to peak at 1,000 to 1,500, has virtually disappeared. Ducks still come to the valley and roost on the hill lakes, but only in small numbers. Thus in January 1976, coverage of fifteen

Table 109. *Regular (and maximum) numbers of wildfowl recorded on waters in Shropshire and the associated Hanmer Mere in Clwyd.*

a) Ellesmere group — 17 seasons 1960-1982
b) Marton Pool, Baschurch — 10 seasons 1968-1981
c) Fenemere — 10 seasons 1968-1981
d) Polemere — 4 seasons 1978-1982
e) Hanmer Mere — 2 seasons 1981-1982

	a	b	c	d	e
Mallard	751(1923)	261(700)	25(170)	116(252)	395(600)
Teal	30 (390)	32(150)	8 (70)	45(113)	4 (30)
Wigeon	109 (377)	57(320)	32(140)	4 (15)	17 (61)
Shoveler	11 (44)	16 (40)	3 (22)	4 (20)	24 (43)
Pochard	62 (177)	26 (70)	10 (52)	1 (5)	20 (70)
Tufted Duck	244 (573)	43 (85)	29 (58)	3 (11)	37 (52)
Goldeneye	31 (72)	2 (9)	3 (9)	0 (0)	3 (4)
Goosander	3 (22)	7 (25)	4 (31)	0 (0)	13 (24)
Ruddy Duck	29 (143)	1 (14)	2 (13)	0 (0)	34 (71)
Mute Swan	10 (47)	1 (6)	1 (7)	1 (5)	1 (4)
Canada Goose	393 (923)	25 (78)	32(140)	15 (26)	113(310)

resorts associated with the upper Severn totalled only 89 Mallard, 78 Teal, 49 Tufted Ducks, and a handful of Pochard, Goldeneyes, Mute Swans and Whooper Swans; while another nine resorts associated with the Vyrnwy, including the 450ha Lake Vyrnwy itself (an RSPB reserve), gave totals of 155 Mallard and 37 Tufted Ducks and small numbers of the other specie

Although the region covered by this sectio nowadays has only a few truly wild geese – a flock 20-40 Greenland Whitefronts in the hills abov Newtown – it contains one of the major breedin populations of the feral Canada Goose. A late summ census in July 1976 gave totals of 1,589 in nort

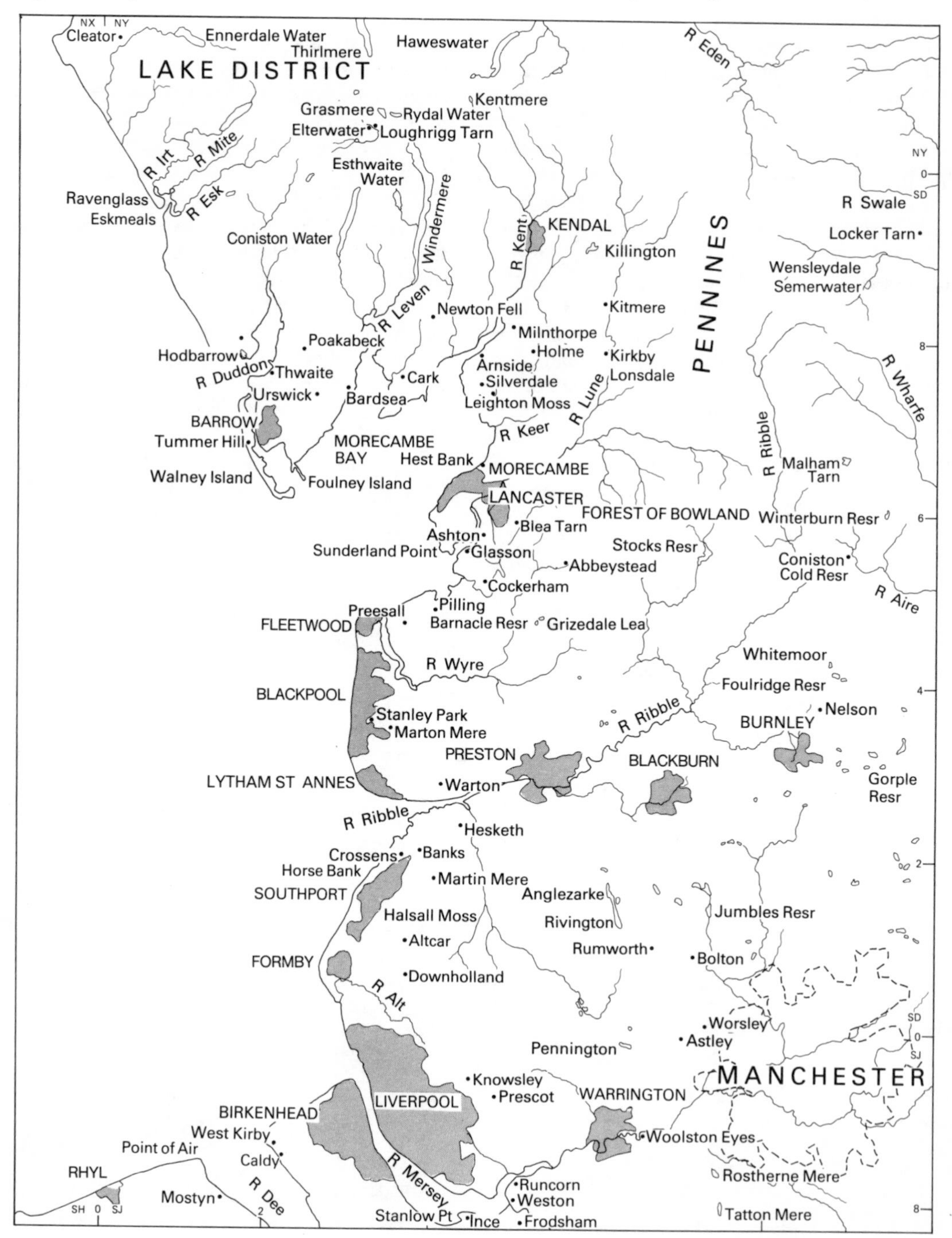

Fig. 33. Merseyside, Lancashire and south Cumbria.

Shropshire and south Cheshire, 666 in north Cheshire and 355 in south-west Shropshire and north Powys – the divisions representing fairly discrete subpopulations (405). Together they represented 13.5% of the then national total and they have in all probability increased in step with it (p.382).

Another, more recent, colonist which has taken to Cheshire and Shropshire, favouring the small reedy meres for nesting sites, is the North American Ruddy Duck. In 1975, when some 50 nesting pairs were estimated for England, 30 of them were in these two counties (274). The Mallard and Tufted Duck are also widespread breeders and the Shropshire meres have seen a marked increase in breeding Shoveler.

The Ribble Estuary and sites inland

From the Mersey northwards to Blackpool, much of the coast is either built over or flanked by dunes and bare sandy beaches, the only exception being around the estuary of the Ribble between Southport and Lytham St Annes. Here the soil is finer, and siltation of the estuary, hastened by embankments, has led to the formation of salt and freshwater marshes which provide extensive feeding grounds for wildfowl. On the other hand, the whole of the fenland behind Southport has been reclaimed and cultivated, including the 1,200ha of marsh and water of the ancient Martin Mere. The rich soil is devoted almost entirely to the growing of arable crops, including substantial acreages of brassicas, onions and carrots.

This agricultural development has been one of the factors involved in the massive increase of Pink-footed Geese in Lancashire in recent years. From the peaks of 4,500 in the 1950s, they reached 9,000 in 1965 and 15,000 in 1973, and each winter since 1974 the peak has exceeded 16,000; in the severe weather of January 1982 numbers soared to 36,580, more than 40% of the national total. Over the past thirty years the British Pinkfoot population has more than doubled, but that for Lancashire has increased sixfold. The situation is now studied in detail, with a series of coordinated surveys nine times each winter since 1977 (196).

In 1956 a Sanctuary Order was made to cover the Horse Bank opposite Southport, where the majority of the Pinkfeet then roosted. The extraction of sand and movement of lorries, by night as well as day, began in the early 1970s and has continued unabated. Recently the goose roost has moved upriver to the intertidal areas of Great Brow, George's Brow and Banks Sands. These are now protected within the Ribble Marshes NNR. From the Ribble the geese fly inland to feed on Martin Mere, Halsall Moss and Plex Moss, and, less frequently, they travel further south to the Altcar area, up to 20km away. Altcar is also served by a smaller roost of up to 4,000 geese on Taylors Bank, off Formby. To the north there is a largely separate unit based on a roost of 5-6,000 geese on Pilling Sands at the mouth of the Lune Estuary. This has been protected since 1963 by a Sanctuary Order, and the geese from it feed in the Fylde area. An unusual feature of the Lancashire goose situation is the formation of field roosts on dry inland sites, such as those at Downholland Moss and Altcar Withins. These are in undisturbed areas with good all-round visibility and are used year after year. The Wildfowl Trust Refuge at Martin Mere, Burscough, also serves as a roost at times for 10,000 or more geese, and the inland sites have an equal footing with the intertidal sites for substantial periods. Although the feeding areas of the geese cover in all some 260sq km, certain very limited parts are of overwhelming importance; nearly half of the total goose-days are recorded on just 8% of the area (196). This naturally leads to agricultural problems but, equally, enables crop rotation to be planned which can minimise damage. Certainly cooperation with agricultural interests is essential if this internationally important concentration (regularly 21% of the world population) is to be maintained.

The Ribble area was the scene of two impressive conservation initiatives during the 1970s. Early in the decade the Wildfowl Trust purchased 147ha of the former Martin Mere at its eastern, less well drained, end. Besides the setting up of a captive collection of wildfowl with associated ponds on 16ha, the provision of public viewing facilities, and excavation of an 8ha mere at the sump of the ancient lake, the rest of the area was managed for wildfowl by rotational flooding and selective grazing. Shooting controls were established over a wider area. As well as the Pinkfeet, the Refuge is now frequented by a wealth of wildfowl, as set out in Table 110, and by many thousands of waders.

If Martin Mere demonstrated what could be achieved by a voluntary body (with some government support), the other initiative was governmental, although strongly supported by the voluntary bodies, especially the RSPB. In 1978 the most important areas of the Ribble Estuary, 1,000ha of the Banks and Crossens saltmarshes and 1,200ha of Banks Sands, were bought by a Dutch entrepreneur with the intention to embank and drain most of the area for agricultural use. The exciting year-long battle, described by the NCC in its 5th Annual Report (376), resulted, following parliamentary and ministerial interventions, in the Government purchasing the whole site for over 1.7 million pounds, an unprecedented investment in wetland conservation. In

1980, 120ha of the Hesketh Out Marsh was added under a Nature Reserve Agreement and the Ribble Marshes NNR was scheduled. Controlled wildfowling is permitted on part of the reserve, but adequate areas of grass marsh, and Banks Sands, are kept as sanctuaries. Further upstream, part of Hesketh Marshes was embanked in 1981, emphasising the economic pressure for marsh destruction even in these days of agricultural surpluses in the European Economic Community. Longton Marsh to the east, and Warton Marsh on the north bank, are shot over, but the wildfowling associations operate sanctuary areas on both.

The Ribble is especially important for wader populations (485), but substantial numbers of wildfowl are also recorded. Recent counts, together with those from the 1960s and early 1970s (224), are set out in Table 110. The estuary is internationally important for its Bewick's Swans, Pink-footed Geese, Wigeon, Teal and Shelducks, while Martin Mere holds this status for Pinkfeet, Pintail and Teal. Of course, there is considerable interchange between the two sites, only 10km apart. This is especially obvious in the case of the Bewick's Swan, now with a combined total comfortably exceeding internationally important numbers. Indeed, in 1983-84 the increase continued, with an unprecedented 336 being counted at the two sites combined. Some of the decreases in the Ribble counts (e.g. Mallard and Teal) are probably accounted for by the development of Martin Mere. In the 1960s and early 1970s, 483 pairs of Mallard bred in the Ribble area, producing about 800 young, but wildfowling associations were releasing 3-5,000 hand-reared Mallard each autumn (224). As a breeder the Shelduck is also widespread, perhaps reaching a total of a hundred pairs. Several other coastal sites in this area are worth noting. Hesketh Park Lake, Southport, was harbouring up to 1,100 Mallard in the early 1960s, but more recent counts (224) suggested only 250.

The Common Scoter, by contrast to most of the species mentioned, has shown a major decline in the area. Recorded in tens of thousands in the late 1930s, only a few hundred are now occasionally seen.

The Alt Estuary is of some importance for ducks, with maxima in the period 1973-1982 of 123 Mallard, 83 Pintail, 242 Shelducks, 275 Pochard, 105 Tufted Ducks and 17 Goldeneyes. A single count on Seaforth Dock Pools, Liverpool, gave 21 Mallard, 200 Pochard, 42 Tufted Ducks, 16 Goldeneyes and 4 Scaup. Inland, when flooded, Formby and Downholland Mosses have held substantial numbers of dabbling ducks.

Elsewhere in south Lancashire, Merseyside and Greater Manchester the lowlands are heavily developed, especially in the triangle between Liverpool, Manchester and Blackburn. Standing water is plentiful, however, and comprises park ponds, colliery subsidences and drinking and industrial reservoirs. Many of the reservoirs are high up in the Pennines but the lower ones can hold significant numbers of ducks. The various waters may be considered in somewhat artificial groupings, associated with one or two of the more important ones.

Knowsley Park, Liverpool, and the nearby Prescot Reservoirs hold substantial numbers of ducks

Table 110. *Regular (and maximum) numbers of wildfowl recorded at two major sites in Lancashire.*

a) Martin Mere, Burscough 7 seasons 1974-1982
b) Ribble Estuary i) 14 seasons 1960-1973 (Greenhalgh 1975)
ii) 4 seasons 1974-1982

	a	b i	b ii
Mallard	1526 (4000)	1890 (2750)	979 (1557)
Teal	3329(10000)	996 (2615)	2004 (5274)
Wigeon	649 (2000)	2645 (6600)	6173(13823)
Pintail	1254 (4500)	1146 (4700)	917 (4719)
Shoveler	65 (156)	20 (76)	25 (150)
Gadwall	165 (260)	18 (51)	0 (3)
Shelduck	31 (96)	1587 (2937)	1546 (2218)
Pochard	38 (124)	0 (0)	25 (63)
Tufted Duck	33 (86)	0 (6)	5 (32)
Common Scoter	0 (0)	0 (50)	13 (50)
Whooper Swan	18 (76)	0 (34)	16 (41)
Bewick's Swan	57 (200)	0 (137)	117 (220)
Pink-footed Goose	4756(15500)	0(14000)	2240 (9715)

(Table 111), but the dabbling ducks have decreased while the diving ducks have increased, compared with counts in the 1960s, when, for example, regular Mallard were 244 at Knowsley and 25 on the reservoirs. Knowsley Park was formerly a stronghold for feral Canada Geese. They have since spread to lakes and flashes between Liverpool and Bolton and in 1976 the south Lancashire subpopulation totalled 280 (405).

Pennington Flash, near Leigh, holds good numbers of Mallard and Pochard (Table 111), but seven other counted waters nearby hold less than 50 ducks apiece. Lightshaw Hall Flash is worth noting, with regular (and maximum) numbers of 54 (410) Mallard and 105 (420) Teal, while diving ducks are better represented on Pearson's/Scotsman's Flashes, with regular flocks of 113 Pochard and 128 Tufted Ducks. Further east, Shore Top Reservoir, Radcliffe, supports 142 (260) Mallard but little else. Only Astley/Worsley Waters, which hold 53 (184) Mallard and 93 (260) Teal, are worthy of note in the vicinity. Further east again, beyond Oldham and on over the Derbyshire border, among half a dozen waters with small numbers, the Longendale Reservoirs have held up to 204 Mallard and 47 Tufted Ducks.

Further north, between Bolton and Blackburn, are three groupings of waters. Anglezarke, Yarrow and Rivington Reservoirs, north of Horwich, have largely wooded shores and are of interest to Mallard only, with a combined regular total of about 200. Rumworth Lodge Reservoir, which holds a regular 10-40 Mallard, Teal, Pochard, Tufted Ducks and Whooper Swans, is of some interest, as are Doffcocker Lodge and Delph Reservoirs, each of which has 100 ducks regularly. Associated with them are, however, many smaller waters with less than 50 birds each.

Around Burnley are several reservoirs with regular numbers mostly in single figures, but Whitemoor Reservoir, just north of Colne, has substantial counts (Table 111). The adjacent Foulridge Reservoirs have a wide variety of wildfowl, but most occur only occasionally or in small numbers. There are, however, regularly 70 Mallard, 20 Pochard and Tufted Ducks and 50 Canada Geese. Apart from 200 Mallard at Victoria Park, Nelson, the numerous other waters in the locality have very small flocks of wildfowl.

Table 111. *Regular (and maximum) numbers of wildfowl at four inland sites in the Merseyside/Manchester area.*

a) Knowsley Park, Liverpool 12 seasons 1970-1982
b) Prescot Resr., Liverpool 10 seasons 1970-1982
c) Pennington Flash, Leigh 9 seasons 1972-1981
d) Whitemoor Reservoir, Colne 12 seasons 1970-1982

	a	b	c	d
Mallard	123(323)	291(596)	189(732)	234(430)
Teal	54(270)	11(125)	32(137)	23(335)
Wigeon	10 (70)	5 (41)	4 (58)	25 (80)
Pochard	57(232)	43(375)	158(534)	9 (38)
Tufted Duck	69(150)	95(278)	59(157)	15 (38)
Canada Goose	43(180)	14(227)	1 (17)	32(110)

To summarise, of the many resorts in this area 73 have been visited, but 51 have held less than 50 ducks regularly, five have 50-100, and only 17 are worth setting out in any detail, and most of these rely on Mallard for the major portion of their interest.

Morecambe Bay and south Cumbria

From Lytham St Annes to Fleetwood the coast is sandy and largely built over. Of the few waters in this stretch, the lake in Stanley Park, Blackpool, is of little interest, but Marton Mere, just to the west, has assumed some importance since its enlargement in the early 1970s to its present 11ha, despite the establishment of a caravan park on the south bank. The mere now holds 2-300 Mallard at peak, and there are up to 75 Teal, and maxima of 100 Shoveler, 250 Pochard and 70 Tufted Ducks, with smaller numbers of other species. The only other feature of interest in this area is an isolated record of 275 Common Scoters off the south promenade at Blackpool.

Between Fleetwood and Walney Island lies the vast expanse of Morecambe Bay, which includes the estuaries of the Wyre and Lune in the south and the Kent and Leven in the north. Morecambe Bay is second only to the Wash in Britain in the area of intertidal estuarine flats; it contains 15,800ha, with about 1,200ha of marsh fringing them. While some parts are of more value to wildfowl than others it was not judged practical by the *Nature Conservation Review* (492) to divide up the system and the whole area was included in the boundaries of one Grade I site. It was as a whole, too, that the estuary was threatened by various proposals for tidal barrages (there is a large tidal range of 11.5m) and reservoir impoundments in the late 1970s (p.480), which led to the illuminating Morecambe Bay Feasibility Studies. However, such a large area is best reviewed for wildfowl in sections around the contributory estuaries. In 1968 and 1969 monthly counts were undertaken in most sectors (224, 517), but since then only the RSPB reserve on the east shore and Walney Island have been covered that frequently. Elsewhere, the counts have generally been in January only.

The mouths of the Wyre and the Lune, together

with the intervening 10km of sand and salting, form a major wildfowl resort in their own right. As on the Ribble, the coastal marshes provide excellent grazing and are backed by broad tracts of reclaimed fenland largely devoted to arable crops. In the 1960s, regular (and maximum) winter populations were estimated at 1,000 (3,900) Mallard, 180 (430) Teal, 1,360 (4,100) Wigeon, 105 (240) Pintail, 34 (71) Shoveler and 200 (290) Shelducks (224). January counts between Pilling and Cockerham between 1970 and 1982 have given maxima of 1,390 Mallard, 70 Teal, 3,300 Wigeon, 10 Pintail and 20 Shoveler, all rather lower, but the Shelducks, peaking at 930, indicate an increase. Scattered mid-winter counts on the Wyre Estuary have produced maxima of 600 Mallard, 1,100 Teal, 24 Wigeon and 190 Shelducks. These, together with some rather doubtfully amalgamated maxima from within the Lune Estuary of 515 Mallard, 50 Teal, 350 Wigeon and 175 Shelducks, suggest a rather lower dabbling duck population than in the 1960s. Just inland, a few counts indicate up to 350 Mallard on the half-dozen small gravel pits at Preesall, 32 at Glasson Dock on the Lune, and 50 at Ashton Hall. Goldeneyes appear regularly on the Lune, with a maximum of 65, and there have also been up to 47 Red-breasted Mergansers. A reserve on the Wyre covering Burrows Marsh and Barnaby's Sands is operated by the Lancashire Naturalists' Trust. The Sanctuary Order on Pilling Sands protecting the roost of 5-6,000 Pink-footed Geese has already been noted (p.193). The Fylde area inland does not hold geese before the New Year, but by March half of the Lancashire Pinkfoot population are feeding there (196).

Moving north along the eastern shore of Morecambe Bay a fair number of counts are available from the 1970s and up to 1982. About Heysham, from Sunderland Point to Hest Bank, maxima of 420 Mallard, 1,060 Wigeon, 330 Shelducks, 200 Goldeneyes and 170 Red-breasted Mergansers have been recorded. More regular counts are available around the Keer Estuary, from Morecambe to Arnside, and these are summarised in Table 112. January counts are also available for separate segments of this stretch, Hest Bank – Keer Estuary – Jenny Brown – Arnside, which in some cases are added to produce the maxima given in the table. An unusual record was a count of 120 Bewick's Swans in March 1978, probably stopping on spring migration.

The RSPB now owns land and freehold rights over 2,485ha of Morecambe Bay between Hest Bank and Silverdale (two-thirds being sandflats). They also own 130ha of Leighton Moss at the northern end of the marsh. The Moss was under the plough for nearly 80 years, but was allowed to reflood in 1917. Now a reed-swamp with meres, willow and alder carr, it is managed to suit wildfowl, with considerable increases

Table 112. *Regular (and maximum) numbers of wildfowl recorded around Morecambe Bay.*

a) Coast Morecambe – Arnside 4 seasons 1974-1981
b) Leighton Moss i) 6 seasons 1964-1969
ii) 12 seasons 1970-1982
c) Leven Estuary (Cark) 2 seasons 1970-1975

	a	b i	b ii	c
Mallard	413 (777)	370(867)	560(1800)	156(400)
Teal	27 (112)	334(460)	813(1550)	78(237)
Wigeon	489(1360)	51(300)	13 (65)	256(423)
Pintail	57 (280)	13 (38)	32 (403)	143(460)
Shoveler	14 (44)	187(350)	150 (537)	4 (10)
Gadwall	0 (0)	4 (10)	18 (34)	0 (0)
Shelduck	605(1585)	0 (0)	1 (12)	428(660)
Pochard	0 (0)	5 (9)	20 (75)	0 (0)
Tufted Duck	0 (5)	14 (38)	21 (60)	0 (0)
Goldeneye	42 (199)	4 (11)	11 (26)	14 (22)
Scaup	0 (3)	0 (0)	0 (1)	24 (41)
R-b Merganser	48 (247)	0 (0)	0 (1)	9 (18)
Goosander	1 (7)	0 (0)	2 (15)	1 (2)
Mute Swan	3 (12)	11 (18)	7 (13)	0 (2)
Greylag Goose	87 (405)	0 (0)	6 (42)	11 (65)
Pink-footed Goose	64 (650)	0 (0)	0 (0)	0 (0)

in most species (Table 112). Haweswater, 2km to the north, held up to 70 Mallard and Teal in the 1960s but there are no recent counts.

In the summer a notable feature of the Keer Estuary is the gathering of Shelducks prior to their moult migration to the Heligoland Bight. During July 1949, 4,000 were counted leaving to fly over the Pennines, though only 1,055 were seen on any one day (133). An even greater through-put was indicated by more recent summer counts, which gave an average total July count for Morecambe Bay for 1964 to 1968 of 1,074, with a maximum, in the last year, of 1,789. (These figures did not include the Lune flock of a maximum of 278 Shelducks.) Considerable numbers of Shelducks breed around Morecambe Bay and the species has shown a tendency to colonise inland sites. Mallard have been estimated to total 1,100 breeding pairs, producing 4,100 young (517). Another 1,400 artificially reared young Mallard were also said to be released. Otherwise the only duck with major breeding output is the Eider colony at Walney.

The estuary of the Kent, lying only a few kilometres to the north of Leighton Moss, is cut in two by a railway viaduct. Below this point the bay is broad and sandy, and used by wildfowl mainly as a roost, though there are grass marshes at Silverdale. Above the viaduct the estuary narrows somewhat, but with large areas of reclaimed pasture on either side and a sizeable stretch of saltmarsh near Arnside. For 1968 and 1969, regular (and maximum) populations for the estuary as a whole from Arnside round to Kents Bank of 280 (420) Mallard, 46 (79) Teal, 210 (450) Wigeon and 90 (140) Shelducks were indicated (517). There was also a regular flock of 120 Greylag Geese, then thought to be true migrants and the southernmost in England, afforded sanctuary on Brogden Marsh. These birds are now, however, largely a feral group, and were so as far back as the early 1960s. January counts from 1970 to 1979 indicate similar maxima for the estuary of 565 Mallard, 117 Teal, 370 Wigeon and 289 Shelducks. There were 244 Greylags in 1974, but this flock has now moved to the Lune, and in recent years has largely wintered on the RSPB reserve on Morecambe Bay.

The Leven Estuary is similar in conformation to that of the Kent, the upper narrow reaches being again divided by a viaduct from the broad sandy bay around the mouth. The marshes, however, are relatively small and the ducks are believed to flight well inland. For the whole estuary, from Humphrey Head Point round to Bardsea, regular (and maximum) populations of 410 (673) Mallard, 50 (149) Teal, 370 (749) Wigeon and 680 (2,566) Shelducks were estimated for the 1960s (517). Later complete counts are lacking but for five years in the 1970s counts were made from Cark on the eastern side (Table 112), indicating little general change.

The coast from Bardsea down to Rampside is less hospitable to wildfowl but 100 Mallard, 748 Shelducks, 3 Goldeneyes and 10 Mergansers were counted in January 1983. Around Foulney Island, mid-winter counts (1967-1982) have given maxima of 25 Mallard, 15 Teal, 620 Wigeon, 287 Shelducks, 12 Goldeneyes, 3,300 Eiders, 650 Common Scoters and 19 Mergansers. The massive flock of Eiders is not wholly distinct from those at Walney Island. In January 1982 2,400 were counted at Foulney and 2,000 at Walney, and several combined counts of 4-5,000 have been made. The build-up of this concentration, based on a local breeding population, has been remarkable, increasing more than tenfold from the maximum of 400 in 1959. The first breeding pair on South Walney Nature Reserve (operated by the Cumbria Trust for Nature Conservation) was recorded in 1949; the colony had grown to 300 pairs by 1972 (542) and more recently to over 500 pairs. It remains by far the most southerly breeding colony of Eiders in Britain. Walney is also remarkable for our largest breeding colony of Lesser Black-backed Gulls *Larus fuscus* and Herring Gulls *L. argenatus*, estimated at 25-30,000 pairs each (485). Despite the Barrow dockyards at the head of the South Walney inlet, the island itself is relatively unspoiled, with sand-dunes and a series of small marshy bays. The concentrations of Common Scoters, up to 1,600, seen in the Januarys of 1967-1969 have not been reported since, but flocks of up to 500 are regular. Tummer Hill Marsh is well favoured by Wigeon and Shelducks, with maxima of 1,000 and 300. Teal are particularly plentiful with 500-1,000 regularly and peaks of up to 1,500. A large variety of other wildfowl are recorded at Walney Reserve, including at times 300 Pink-footed Geese. In Barrow itself the 80ha expanse of Cavendish Dock, which acts as a cooling basin for Roosecote Power Station, is useful to diving ducks. It is relatively deep, the warmed water never freezes, and it supports a rich growth of aquatic plants. The indications are, however, that its value to wildfowl is declining. In 1951 flocks of 1,000 Wigeon, 600 Pochard and 150 Tufted Ducks were reported. In the 1960s (Table 113) these numbers were maintained, but in recent years there has been a marked decrease. Mute Swans favour the Dock for moulting, with up to 112 recorded. Ormsgill Reservoir, also in Barrow, supports about 20 Mallard and Pochard, up to 60 Tufted Ducks and 30 Red-breasted Mergansers. The north end of Walney (Table 113) is much more attractive for dabbling than other ducks and, being smaller and more land-locked than the southern inlet, lacks the regular sea ducks.

While, as far as wildfowl are concerned, Morecambe Bay is really a series of small estuaries, for conservation purposes an estimate for the wildfowl significance of the overall area would be useful. The results of three surveys of the area are shown in Table 114, which indicates that the bay is internationally important at least for Wigeon and Shelduck. Teal numbers are also high and it seems likely that better count coverage would raise the total to over the qualifying 2,000.

The estuary of the Duddon, debouching into the Irish Sea immediately to the north of Walney, affords another 470ha of intertidal sand, backed by 300ha of fine marshes, with many splashes and gullies, along most of its length. Published data indicate regular (and maximum) numbers for the 1960s of 410 (704) Mallard, 320 (976) Teal, 740 (1,500) Wigeon and 230 (470) Shelducks. The flock of 150 Greylags recorded at that time was based on introductions by WAGBI in the 1960s (517). January counts are available for 1967 onwards, while in 1982 the maxima for September – April were 776 Mallard, 50 Teal, 630 Wigeon, 158 Pintail, 694 Shelducks, 14 Mergansers, 316 Pinkfeet and 7 Greylags. The January average (and maximum) population has been 1,035 (2,045) Mallard, 388 (810) Teal, 971 (1,842) Wigeon, 422 (525) Pintail, 381 (651) Shelducks and 136 (157) Greylag Geese. The first three species were boosted by a single exceptional count in 1972, so the general picture is one of little change.

Close to the Duddon Estuary lie mining subsidences much favoured by ducks. Hodbarrow Pond, one of the few waters on the north bank, has regular January populations of 340 Mallard, 370 Teal, 70 Pochard, 80 Tufted Ducks, 40 Goldeneyes and 20 Red-breasted Mergansers. The Mergansers gather in a large post-breeding flock, possibly to moult, a maximum of 292 being recorded in the summer of 1980 (53). Mining at Hodbarrow, and the pumping out of water, ceased in 1968. Mining has also ceased on the south side of the estuary, but the Thwaite Flat and Roanhead Ponds are less favoured and less well counted, with maxima of 82 Mallard, and less than 30 Wigeon, Pochard, Tufted Ducks and Goldeneyes.

The coast north of the Duddon has a good variety but small numbers of ducks. From Millom to Eskmeals Point the maximum numbers recorded are 250 Mallard, with 20-40 Teal, Wigeon and Shelducks. The Common Scoter is the most numerous of the diving ducks, with a peak of 130, but there is a wide variety of other species, including as many as 100 Pochard, 70 Scaup and 60 Mergansers. The small Esk Estuary, formed by the confluence of the Irt, Mite and Esk, is mainly sandy but with small areas of saltmarsh. The maximum numbers of wildfowl recorded have been 520 Mallard, 204 Teal, 1,453 Wigeon, 142 Shelducks, 12 Pochard, 38 Goldeneyes, 89 Mergansers and 108 Greylag Geese. The long dune peninsula at the mouth of the Irt is notable for one of the largest breeding colonies of Black-headed Gulls in Britain and is now a Local Nature Reserve (Ravenglass). From Drigg Point to Whitehaven 64 Scaup and 168 Common Scoters have been recorded at peak, together with a handful of Mallard.

Lying between Morecambe Bay and the Pennines there are a number of sites of interest: among these are Scorton Gravel Pits, on the banks of the Wyre north of Garstang, which have carried up to 600 Mallard, 35 Pochard, 60 Tufted Ducks and 43 Mute

Table 113. *Regular (and maximum) numbers of wildfowl recorded in the Barrow area.*

a) Walney Island South 8 seasons 1960-1981
b) Cavendish Dock, Barrow i) 7 seasons 1963-1969
ii) 3 seasons 1970-1981
c) Walney Island North 2 seasons 1970-1980

	a	b i	b ii	c
Mallard	119 (280)	6 (54)	4(158)	237 (600)
Teal	455(1400)	2 (31)	0(400)	207 (450)
Wigeon	364(3170)	630(1190)	149(800)	507(2000)
Shelduck	178(1094)	0 (0)	0 (4)	236 (600)
Pochard	3 (40)	401 (850)	79(130)	1 (5)
Tufted Duck	0 (2)	319 (700)	46(262)	4 (21)
Goldeneye	23 (170)	23 (72)	6 (10)	6 (18)
R-b Merganser	6 (45)	5 (18)	8 (42)	22 (42)
Mute Swan	1 (3)	38 (101)	10 (31)	0 (0)

Swans. Wyresdale Park Lake nearby has a maximum of 50 Mallard and 21 Tufted Ducks; Barnacre and Grizedale Lea Reservoirs together hold at peak 250 Mallard; and Abbeystead Reservoir, an impoundment on the Wyre itself, some 8km north-east of the gravel pits, holds up to 120 Teal. Blea Tarn and Langthwaite Reservoirs, just south of Lancaster, barely rate 50 ducks. To the east, lying high (180m) above the Ribble Valley and abutting Gisburn Forest, Stocks Reservoir (Table 115) has quite substantial dabbling duck populations as well as Goldeneyes and Goosanders.

The Lune Valley from Lancaster to Kirkby Lonsdale is broad and quite attractive to ducks. The whole length in January 1982 gave 275 Mallard, 147 Teal, 315 Wigeon, 40 Goldeneyes, 61 Goosanders, 250 Greylag Geese and 395 Canada Geese. There is little above Kirkby Lonsdale except the Rigmaden area which had, at peak in 1982-83, 277 Mallard, 209 Teal, 183 Wigeon, 29 Goldeneyes, 37 Goosanders, 20 Mute Swans, 340 Greylags and 320 Canada Geese.

Between the upper Lune and the M6 motorway lie a number of waters, of which Killington Reservoir, on the edge of the motorway and overlooked by its service area, is most notable (Table 115), particularly for its Goosanders and as a focus for moulting Canada Geese. In July 1981 a flock of 170 was present, but the introduction of sailing will probably reduce its use by geese in future. Nearby Lilymere regularly holds only about 50 wildfowl, but 5km further south, Wyndhammere and Kitmere support a combined regular population of 100 Mallard, 10-20 Goosanders and a good variety of other species in small numbers. Other tarns in the vicinity carry rather small flocks of Mallard. The mill pond at Holme, hard by the motorway, is an exception, with up to 98 Mallard, 22 Teal and 25 Shoveler.

On the River Kent, counts upstream to Kentmere have produced 356 Mallard, 21 Goldeneyes and 10 Goosanders, while three ponds near Milnthorpe have each recorded maxima of 150-200 Mallard and very little else.

On the Leven, counts between Backbarrow and Newby Bridge (the outlet of Windermere) have produced up to 45 Mallard, 16 Pochard, 17 Tufted Ducks and 60 Goldeneyes. Associated waters are Bigland and Mere Tarns, with negligible numbers of ducks, and Urswick Tarn, with up to 44 Mallard and 25 Tufted Ducks. Harlock, Poakabeck and Pennington Reservoirs grouped above Ulverston carry combined maxima of 50 Mallard, 37 Tufted Ducks and 10-20 Teal, Wigeon, Pochard and Goldeneyes.

The English Lakes, for all their size and number, are relatively unimportant to wildfowl, being mostly too deep and steep-sided to support any large concentrations. The lakes draining northwards are not considered here and will be dealt with in a later section (p.212).

The great length of Windermere, which as a statutory "public highway" is free of all shooting, does in total support quite a number of ducks (Table 115), but little enough for its 1,480ha. It is, however, the third most important site for Goldeneyes in England. Some remain all summer and, since the conditions are

Table 114. *Estimated numbers of wildfowl in the whole of Morecambe Bay according to three surveys.*

a) Data from the 1960s for the Wyre and Lune Estuary (Greenhalgh 1975) and the remainder of the bay (Ruxton 1973) – regular only.
b) Average count in best month, 1969-1974 (Prater 1981).
c) Average (and maximum) of 15 January wildfowl counts 1967-1981.

	a	b	c
Mallard	1850	3060	2390(2594)
Teal	790	950	1365(1618)
Wigeon	3600	4120	6427(6724)
Pintail		560	572 (625)
Shelduck	1500	6640	4058(4472)
Pochard			521 (660)
Tufted Duck			391 (518)
Goldeneye		390	269 (288)
Common Scoter		660	1223(1600)
Eider		630	2963(5000)
Red-breasted Merganser		330	268 (331)

Table 115. *Regular (and maximum) numbers of wildfowl recorded at four sites inland from Morecambe Bay.*

a) Stocks Reservoir, Slaidburn 4 seas. 1970-1982
b) Killington Reservoir, Sedbergh 7 seas. 1963-1982
c) Windermere* 16 seas. 1967-1982
d) Esthwaite Water 17 seas. 1960-1982

	a	b	c	d
Mallard	217(429)	96(550)	1118(1967)	67(195)
Teal	150(254)	3 (32)	4 (10)	13 (62)
Wigeon	289(534)	10 (50)	8 (100)	0 (0)
Pochard	50(107)	3 (65)	198 (591)	17(104)
Tufted Duck	5 (32)	11 (38)	378 (512)	57(198)
Goldeneye	11 (22)	4 (12)	242 (296)	16 (29)
R-b Merganser	2 (14)	0 (9)	56 (209)	0 (9)
Goosander	6 (80)	16 (84)	4 (15)	1 (5)
Canada Goose	109(181)	33(139)	0 (6)	3 (18)

* Mean annual maxima.

apparently suitable, are expected to breed before long. Red-breasted Mergansers, not known to breed in England before 1950, reached 16 pairs, producing 120 young annually, between 1969 and 1976. There was a spectacular fall in young after 1977, following a disease which killed off 98% of the lake's Perch (*Perca fluviatilis*) (15); in mid-winter only a handful of Mergansers are present. Associated waters to the east of Windermere include Simspon Ground Reservoir, near the outlet, with regular (and maximum) counts of 47 (104) Mallard and 34 (72) Teal; Newton Fell Reservoir with a maximum of 79 Mallard; and four other tarns with small numbers. West of the north end of Windermere another four tarns are equally unattractive, including, regrettably in view of its name, Drunken Duck Tarn. To the north, Grasmere, Elterwater, Loughrigg Tarn and Rydal Water are a little more productive. Each, however, holds less than a hundred wildfowl, mainly Mallard (up to 50 regularly), with smaller numbers of Pochard, Tufted Ducks, Goldeneyes and Goosanders and a sprinkling of other species.

Some 10km south, near Hawkshead, lies Esthwaite Water (195ha) which has been regularly counted for many years and has, as well as the species listed (Table 115), 20 Greylag Geese regularly. It occasionally has quite substantial numbers of wildfowl, as befits this richest of the Lakes, with a mean depth of only 6.8m (492). Coniston Water, although covered for many years, has not been counted frequently enough to calculate regular figures; the maxima recorded are 301 Mallard, 302 Pochard, 111 Tufted Ducks, 24 Goldeneyes, and 10-20 Red-breasted Mergansers, Goosanders and Greylag Geese. Considering it is about four times the size of Esthwaite, the paucity of wildfowl interest is apparent.

Furthest west, within 15km of the coast, lies Ennerdale Water, which apart from Wastwater is the least productive of the English Lakes. Small wonder its maximum counts are 90 Mallard, 20 Teal, 10 Goldeneyes and 18 Goosanders. Congra Moss, a small lake despite its name, is of little interest, though Cleator Gravel Pits, less than 5km from the coast, are of more note, with up to 52 Pochard and 15 Tufted Ducks.

Near Millom, by the Duddon Estuary, lies the Haws Reserve, the main site of a successful attempt by WAGBI to introduce breeding Greylags into Cumbria. Between 1959 and 1969 284 birds were released at Haws and four other sites (171). The stock was obtained from Galloway, from a flock originally reputed to have been taken from the Outer Hebrides (p.319). The feral birds now appear to be breeding in the wild and there is ringing evidence of a summer moult migration to the quieter lochs of Galloway. The population, fortunately for local farmers, does not appear to be increasing unduly and is still in the low hundreds. The largest assembly reported has been 380 in early October some 12km from Haws.

WAGBI were also responsible for introducing Canada Geese into Cumbria. In the 1976 census a total of 360 was estimated (405) and this has probably increased, in line with the continuing national upsurge, to at least 500. Feral Gadwall were also introduced, and the species now breeds in several places. Mallard breed throughout the region, as do smaller numbers of Teal, Wigeon, Shoveler, Tufted Ducks, Mergansers and Goosanders.

The Isle of Man

The Isle of Man is almost equidistant from England, Wales, Scotland and Ireland. Despite an area of 588sq km, with 121km of coastline, the island is of little importance to wildfowl, being composed of Cambrian rocks, with small areas of Carboniferous Limestone in the south-east, mid-west and north. Much of the land is steep with few water areas, other than some small reservoirs on the Cambrian rocks. The only places of real interest are a few stretches of coast, mostly in the south, and the low-lying district of Ayre in the north of the island.

Marshy areas near Ballaugh, including the Curraghs, hold some breeding Mallard and Teal and larger flocks of Mallard winter there and on the northern, shingle coasts. Small groups of dabbling ducks are found in the pond – stream area south of the Sandhills of the Point of Ayre. Glascoe Dub, for

Fig. 34. The Isle of Man.

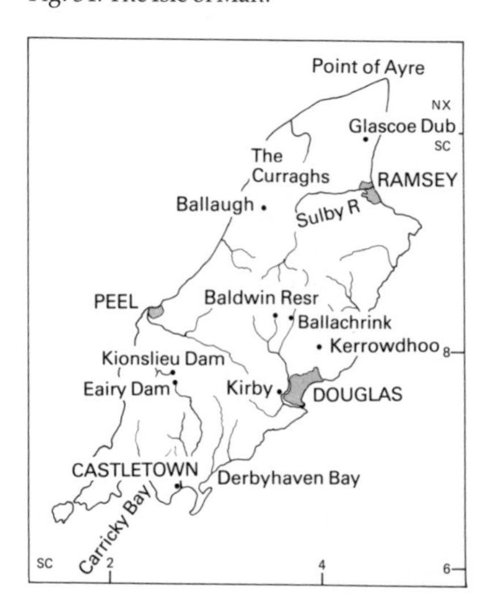

instance, has a maximum of 161 Mallard, 136 Teal, 100 Wigeon, and 10 Shoveler and Pochard.

Inland, Baldwin Reservoir regularly supports 50 Mallard and maxima of 90 are recorded there and at nearby Ballachrink Pond. Kionslieu Dam and Eairy Dam have long runs of counts, but hold at most a hundred Mallard and 50 Teal between them. Near Douglas, Kerrowdhoo Reservoir and Kirby Park regularly have only 10 and 22 Mallard and a sprinkling of other species.

On the south coast, Carrickey Bay may hold up to 110 Mallard, 20 Teal, 140 Wigeon and 70 Shelducks (which breed along the coast and on the Calf of Man). Castletown Bay has only a few ducks in its western half, but maxima of 122 Mallard, 46 Teal, 265 Wigeon and 37 Shelducks on its eastern stretches. Ten seasons (1973-1982) of counts of the eastern shore in combination with Derbyhaven Bay, on the other side of the narrow spit which leads to the Langness peninsula, gave regular (and maximum) numbers of 279 (480) Mallard, 110 (300) Teal, 114 (220) Wigeon and 23 (45) Shelducks, making this much the most important area on the whole island.

Very small numbers of Whooper and Bewick's Swans are recorded on the Isle of Man, likewise of Greylag and Canada Geese, Pintail, Pochard, Tufted Ducks, Goldeneyes and Red-breasted Mergansers and occasional birds of several other species. A few Mute Swans breed and are resident.

Teesdale, Tyneside and the Borders

The eastern border regions of England and Scotland present sharp contrasts. The narrow bands of lowland in the south of the area are heavily industrialised, and much of the natural wildfowl habitat has been lost or rendered untenable by disturbance. The presence of the coalfield, however, has, through subsidence, provided a considerable amount of wildfowl habitat. Distinct pockets of farmland remain, moreover, providing feeding grounds within easy reach, and many of the lowland pools support sizeable numbers of several species. Further to the west there are resorts along the foothills of the Pennine chain, and beyond that again a zone of remote moorland, dotted with reservoirs and tarns. In the northern districts the lowlands are entirely agricultural and it is here that the great majority of the wildfowl are to be found.

Teesdale

The coast north of Robin Hood's Bay is largely cliffbound and rocky, with only narrow inlets, such as that of the Esk at Whitby to the north, until Teesmouth is reached. However, Scaling Dam Reservoir, 15km west of Whitby and lying hard by the A171 main road, has maintained its early promise after its flooding in 1957 (Table 116). Lockwood Beck Reservoir, a little further along the road to the west, is mainly of interest for Mallard, with a regular figure of 180 in 20 years of counts. Haverton Hole just to the west of Teesmouth has held only up to 30 Tufted Ducks, and Longnewton Reservoir, 15km further west, is equally unattractive with up to 60 Tufted Ducks. Wynyard Park Lake, an impoundment of a small stream just north-west of Teeside, supports regular (and maximum) numbers of 396 (930) Mallard, 15 (151) Pochard and 19 (130) Teal, and also 129 (237) Greylag Geese. The geese are present in the late summer and are doubtless feral birds resulting from WAGBI-sponsored introductions in the region. In the 1960s there were at least 227 releases at 12 sites in Yorkshire, Durham and Northumberland (171). Some 10km further north, Hurworth Burn Reservoir (Table 116) has an interesting range of ducks and also up to 130 Greylag Geese; Crookfoot

Table 116. *Regular (and maximum) numbers of wildfowl recorded at four Teesdale sites.*

a) Scaling Dam i) 7 seasons 1963-1969
ii) 11 seasons 1970-1982
b) Teesmouth i) 7 seasons 1960-1969
ii) 13 seasons 1970-1982
c) Hurworth Burn Reservoir 17 seasons 1961-1982
d) Forcett Park, Gainford 4 seasons 1967-1970

	a i	a ii	b i	b ii	c	d
Mallard	486(750)	442(1484)	818(1750)	570(1280)	134(376)	347(750)
Teal	100(400)	126 (225)	126 (406)	593(1788)	41(200)	29 (95)
Wigeon	46 (85)	47 (180)	106 (340)	277 (677)	55(339)	139(245)
Pintail	0 (2)	0 (1)	15 (35)	13 (46)	0 (3)	0 (1)
Shoveler	0 (2)	0 (1)	1 (3)	21 (71)	1 (6)	4 (28)
Shelduck	0 (2)	0 (1)	1854(4445)	1666(3500)	0 (11)	1 (5)
Pochard	28 (59)	40 (180)	12 (148)	178 (810)	39(194)	25 (48)
Tufted Duck	21 (32)	23 (104)	3 (19)	33 (166)	11 (44)	15 (28)
Goldeneye	13 (27)	16 (38)	8 (33)	99 (440)	5 (18)	4 (19)
Scaup	0 (0)	0 (2)	14 (82)	8 (61)	0 (7)	0 (0)
Mute Swan	0 (0)	1 (3)	9 (23)	10 (18)	1 (10)	4 (10)

Reservoir has regular populations of only 81 Mallard, 14 Teal, 11 Pochard and 17 Greylags.

Teesmouth itself has been the scene of a long-running battle of attrition between reclamation, development and conservation (p.478). Reclamation began in earnest in the late 1930s and by 1967 only 655ha of the original 2,600ha remained, namely the important Seal Sands. Infilling behind a porous wall reduced this to a 149ha remnant by 1976. Conservationists fought tooth and nail to retain this last piece and, after fluctuating fortunes, the present threat effectively ended in 1984, when Parliament failed to renew the Port Authority's powers to complete the reclamation of Teesmouth. Despite its shrinking area Teesmouth actually holds more ducks than it did in the 1950s, when it supported regular numbers of 290 Mallard, 45 Teal, 110 Wigeon and 300 Shelducks. Table 116 details the populations in the 1960s and more recently. It is certainly remarkable to view these substantial flocks hemmed in by wirescapes and petrochemical complexes. The Shelducks are particularly surprising, still being present in internationally important numbers. Counts separated by only a few days often fluctuate by several hundreds, indicating

Fig. 35. Teesdale and Tyneside.

continual waves of migrants passing through, but substantial numbers are consistently present from October to January. Durham University has closely monitored the effects of successive infillings on Shelducks and waders (185).

On the coast outside the estuary, from North Gare to Hartlepool, regular numbers of wildfowl have not exceeded a hundred birds despite the fact that this stretch must be rated as one of the most favourable south of Sunderland, but occasional substantial flocks of Scaup (up to 1,063), Common Scoters (361) and Eiders (93) have been noted.

Inland the Tees meanders in its middle course through a broad vale, but the distribution of wildfowl is restricted by the lack of substantial water-bodies. Forcett Park Lake, 5km south of Gainford, had an exceptional population in the 1960s (Table 116), and may still have but there are no recent counts. Three small waters have been counted west of Barnard Castle but hold well under 50 birds altogether. The large reservoirs on the Balder and Lune tributaries of the Tees are not very productive; five hold regular populations of around 50 birds apiece. Only Hury Reservoir is of note with regular (and maximum) numbers of 103 (451) Mallard, 38 (132) Wigeon, 13 (31) Canada Geese and a scatter of other species. Near the headwaters, Cow Green Reservoir, site of a famous (botanical) conservation battle, lies at an altitude of about 500m and has less than 20 wildfowl regularly.

Weardale

The coast is uninteresting from the point of view of wildfowl between Hartlepool and Sunderland, at which point the River Wear debouches. Slightly upstream of the Sunderland conurbation the Wildfowl Trust's Washington Refuge, opened in 1975, has created a fragment (40ha) of artificial wetland, with numerous pools and lagoons, a habitat largely absent in this built-over land. More typical are the adjacent Barmston Ponds, small mining subsidences somewhat irregularly flooded, which hold regular flocks of 48 Mallard, 32 Teal, 13 Tufted Ducks and a few other species. Washington Waterfowl Park counts are given in Table 117. The picture is now complicated by some refuge-bred birds remaining full-winged, but Pintail, Shoveler, Gadwall, Pochard, Tufted Duck and Goldeneye, are all increasing.

Between Washington and Durham, and up along the River Wear, there are 17 count points with regular numbers in the under-50 category. Only Low Butterby Ponds, 4km south of Durham, and Brasside Ponds, are worthy of note. Butterby Ponds regularly have 103 Mallard, 25 Teal, and a few Tufted Ducks and Goldeneyes; Brasside much less. It is not until Witton-le-Wear that we find two sites with quite substantial numbers and variety of wildfowl (Table 117). Low Barnes at Witton is a nature reserve owned by the Durham County Conservation Trust and kept undisturbed. Lambton Ponds are within a kilometre of this site and no doubt to some extent share the same wildfowl. Up at the headwaters, five small reservoirs are counted, of which the most important are Smiddyshaw with a regular 100 Mallard and Tunstall with 50, the others having a regular population of less than 50 wildfowl.

Only 4km across the watershed is an important fairly high altitude (220m) reservoir of substantial size (400ha), Derwent Reservoir, formed by impounding the Derwent, a tributary of the Tyne. The data (Table 117) show the wide variety and substantial numbers of

Table 117. *Regular (and maximum) numbers of wildfowl counted on four Weardale/Derwentdale sites.*

a) Washington Waterfowl Park — 3 seasons 1976-1978
b) Low Barnes, Witton — 4 seasons 1971-1978
c) Lambton Pools, Witton — 2 seasons 1977-1978
d) Derwent Reservoir — i) 2 seasons 1968-1969; ii) 10 seasons 1970-1982

	a	b	c	d i	d ii
Mallard	288(450)	255(605)	50 (96)	568(950)	356(709)
Teal	114(158)	64(125)	33 (43)	61(123)	112(289)
Wigeon	57 (74)	37(220)	176(236)	110(155)	285(465)
Pochard	3 (13)	35 (80)	13 (48)	64 (99)	65(346)
Tufted Duck	1 (4)	26 (44)	31 (52)	10 (17)	26(109)
Goldeneye	3 (11)	5 (15)	4 (6)	26 (34)	16 (57)
Goosander	0 (0)	10 (23)	3 (18)	26 (64)	19 (48)
Mute Swan	0 (3)	7 (28)	19 (38)	0 (0)	0 (0)
Whooper Swan	0 (4)	0 (6)	8 (36)	1 (6)	0 (4)
Greylag Goose	0 (0)	1(116)	12 (21)	0 (0)	7 (41)

wildfowl using this reservoir. Moreover, apart from some decrease in Mallard, there has been no general fall away from the early years after flooding (in 1965) as often happens with reservoirs, and this despite the introduction of sailing. The setting aside of a substantial area as a nature reserve at one end seems to have produced an acceptable compromise. Mallard, Teal and Tufted Ducks have bred there; Wigeon also are breeding sparsely but are widespread in upper Weardale and in Teesdale.

Tyneside

The Derwent does not join the Tyne until Blaydon, where the Shildon Ponds, despite being hemmed in by a vast conurbation, hold a good variety of wildfowl (Table 118) and in 1980 became a Local Nature Reserve of 14ha. At the mouth the Tyne's lower reaches are occupied by shipyards, docks and industrial development. Several wetland resorts like the Jarrow Slakes have been virtually eradicated. The coast on either side is more attractive; Whitburn Bay to the south has recorded 157 Mallard, and to the north the stretch to Seaton Sluice has had as many as 600 Mallard, 96 Scaup, 105 Eiders and 49 Goldeneyes. A few kilometres inland the Northumberland Wildlife Trust reserve at Holywell Pond, 2km south-west of Seaton Sluice, carries quite substantial numbers of ducks (Table 118) and is notable for a regular flock of Whooper Swans. Gosforth Park Lake (Table 118) once had good numbers of wildfowl but its importance has been virtually eliminated by drainage. (In 1983 the maxima were just 8 Mallard and 30 Teal.) Dinnington Colliery Pond (Table 118) on Seaton Burn (also known as Big Waters), however, remains important; it is a reserve of the Northumberland Wildlife Trust. Several other small ponds hold a scattering of ducks, but Killingworth Mere was infilled in 1964. Although rather further to the north, just beyond Newbiggin, we may mention here the subsidence ponds of Cresswell and Linton (Table 118), some of the last of the coalfields. The numbers of wildfowl here have apparently increased considerably in the 1970s, but the cause is unknown.

The River Tyne from Heddon to Hexham holds several hundred Mallard, small parties of Goldeneyes and Goosanders and occasional Mute and Whooper Swans. The Whittledene Reservoirs, north of the river, have shown marked increases, particularly in their regular numbers, in recent years (Table 119).

Upstream of Hexham the Tyne splits into its North and South branches. Above the southern arm, hard by Hadrian's Wall, there is a cluster of shallow loughs (Table 119). Although lying at an altitude of around 300m they are on the pass from east to west coasts and receive sudden influxes of migrants, including an autumn passage of diving ducks. Broomlee Lough is the most important, but substantial numbers are found on Greenlee Lough – the largest of the group – and Grindon, which has a single record of 39 Bean Geese.

Associated with the North Tyne is a group of waters lying at lower altitude and consequently having less erratic populations (Table 119). Hallington Reservoirs, in contrast to Whittledene 15km to the south-east, have shown a marked decline in recent years, only Greylag and Canada Geese increasing. Nearby Colt Crag Reservoir, however, has shown little evidence of change over twenty years despite, or

Table 118. *Regular (and maximum) numbers of wildfowl recorded at five near-coastal sites in Tyneside.*

a) Shildon Ponds, Blaydon 4 seasons 1976-1982
b) Holywell Pond, Seaton 18 seasons 1960-1982
c) Gosforth Park Lake 13 seasons 1960-1982
d) Dinnington Pond 20 seasons 1960-1982
e) Cresswell/Linton Ponds and sea 12 seasons 1970-1982

	a	b	c	d	e
Mallard	107(155)	215(470)	203(706)	92(500)	153(465)
Teal	67(112)	100(475)	111(410)	45(300)	135(445)
Wigeon	1 (5)	18(110)	23(100)	5 (50)	108(465)
Shoveler	6 (20)	2 (36)	22 (75)	5 (35)	4 (45)
Pochard	90(252)	42(145)	50(223)	38(200)	25(100)
Tufted Duck	54(210)	56(170)	48(160)	36(250)	44(121)
Goldeneye	0 (0)	7 (30)	1 (8)	0 (3)	6 (42)
Mute Swan	4 (10)	5 (14)	2 (18)	5 (40)	19 (60)
Whooper Swan	0 (0)	59(170)	1 (14)	2 (35)	33(125)
Greylag Goose	0 (3)	9 (83)	0 (0)	0 (4)	5(100)

Note: Apart from (c) all these sites are colliery ponds.

perhaps because of, being drained (for trout management) in 1961-62 and 1968-69. Sir Edward's Lake, Capheaton, less than 10km north-east of the reservoirs, has also maintained its wildfowl numbers in the last two decades, although the larger goose numbers are recent. This group is the headquarters of the feral Canada Goose population in Northumberland, which was estimated at only 40 in 1976 (405); it clearly has increased since. Another species which has shown a marked increase hereabouts is the Goosander. The first nest was found in Upper Coquetdale in 1941 and North Tynedale was colonised in 1956 (336). In 1967-68 there were an estimated 35 pairs in Northumberland, in 1975 130-150. Breeding now occurs in virtually every river system and there were an additional 10-20 pairs in County Durham in 1975. Ringing of ducklings netted on the rivers has shown a dispersal north and west, with some evidence of return to the natal area for breeding (337).

Also associated with the North Tyne resorts are six small waters scattered west of Morpeth to the fringes of the Cheviot Hills. Ryal, Little Swinburne, Belsay Castle, Bolam and Angerton Lakes have regular wildfowl totals of less than 100 apiece. Only Sweethope Loughs at the headwaters of the River Wansbeck have larger numbers, with regular (and maximum) counts of 158 (450) Mallard, 56 (125) Tufted Ducks and 31 (122) Canada Geese.

Further north and east, in the valleys of the Wansbeck and Coquet, another varied group of waters includes Rayburn Lake, with 100 Mallard and smaller numbers of Teal, Wigeon, Pochard and Tufted Ducks; Fontburn Reservoir, with 1-200 Mallard and occasional parties of Teal and Greylag Geese; an three other counted waters which hold less than 10 regularly.

Further west, at the head of Redesdale, Catc leugh Reservoir, impounded on the River Rede affords little interest to wildfowl, with regular num bers of 45 Mallard, 13 Teal and 7 Whooper Swans. Th great stretches (1,086ha) of the new Kielder Wate have yet to record more than 84 Mallard, 50 Pochar and 23 Goldeneyes, presumably because of their hig altitude, steep banks and enclosed surrounds.

The Northumberland coast

Apart from counts associated with the Cress well/Linton Ponds (Table 118), the coast north fro Seaton Sluice has little to offer apart from scattere sometimes sizeable parties of Mallard and Eiders and more occasionally, Goldeneyes, Long-tailed Duck and Common Scoters. The 6ha Coquet Island lies 1kr off the coast at Amble and is an RSPB reserve. Here i the most southerly breeding colony of the Eider on th east coast, numbering about 450 pairs. Northward parties of Eiders are frequently encountered an increase to many hundreds as the main breedin colony on the Farne Islands, some 30km north, i approached. This colony, centred on the islands o Inner Farne and Brownsman, has a venerable histor of 1,300 years, back to the time of St Cuthbert whos cooing "doves" the Eiders were. Persecution in th Second World War reduced the colony to 130 pairs, but under the protection of the National Trust a massive increase has occurred, to 1,699 nests in 1981, with over 4,000 birds being recorded between Bead-

Table 119. *Regular (and maximum) numbers of wildfowl recorded at five inland sites in Tyneside.*

a) Whittledene Reservoirs, Harlow Hill
i) 10 seasons 1960-1969
ii) 8 seasons 1970-1982
b) Hallington Reservoirs
i) 10 seasons 1960-1969
ii) 11 seasons 1970-1981
c) Colt Crag Reservoir, Throckington
20 seasons 1960-1981
d) Sir Edward's Lake, Capheaton
20 seasons 1960-1982
e) Loughs by the Wall (Broomlee etc.)
12 seasons 1971-1982

	a i	a ii	b i	b ii	c	d	e
Mallard	177(441)	253(417)	252(1100)	104(360)	411(1200)	208(600)	216(508)
Teal	12(110)	33 (90)	26 (106)	12 (75)	14 (80)	18(127)	48(124)
Wigeon	17(120)	58(191)	52 (180)	23(120)	27 (140)	23(100)	237(467)
Pochard	29(109)	68(300)	10 (60)	4 (60)	6 (24)	26(132)	49(190)
Tufted Duck	98(350)	146(570)	33 (93)	24(135)	55 (220)	46(156)	55 (92)
Goldeneye	4 (9)	23 (90)	11 (41)	7 (25)	8 (28)	12 (29)	33 (67)
Goosander	5 (16)	12 (40)	11 (32)	12 (74)	9 (54)	0 (3)	9 (32)
Whooper Swan	0 (3)	1 (22)	5 (28)	0 (7)	5 (48)	6 (41)	36(101)
Greylag Goose	0 (0)	45(460)	0 (6)	20(260)	6 (200)	1 (18)	33(254)
Canada Goose	0 (0)	0 (0)	7 (44)	30(104)	24 (90)	4(100)	0 (0)

ell and Lindisfarne. The colony has been the subject f considerable study, and ringing has shown movements northwards as far as the Forth (33).

As well as the Eiders, several hundred Common coters are scattered along the coast from the Coquet Budle Bay. The Coquet Estuary itself is notable for a ock of up to 47 Mute Swans. A long series of counts 8 years) covering Embleton Bay and Newton Haven, gether with the small Newton Pool (a National Trust eserve), returned regular (and maximum) numbers of 1 (314) Mallard, 48 (170) Teal, 13 (150) Wigeon and a rge variety of other species in small numbers.

Lindisfarne National Nature Reserve covers the ast sands and mudflats from Budle Bay to Cheswick, gether with the sand-dunes of Holy Island (or indisfarne) itself. Declared in 1964, it was extended in 966 to its present 3,278ha. Arrangements were made r restrained wildfowling to continue over much of oly Island Sands and Fenham Flats (the most nportant area for wildfowl) under a permit system – ery appropriate as the initial pressure for an NNR, in part at least, came from the local wildfowling association. The benefits of such protection are easily seen (Table 120) by comparing numbers recorded during the 1960s with those of more recent years. Nearly every species has shown a substantial increase except the two swans. The Whooper Swan reduction is surprising, though it may be linked with the increasing cultivation of winter cereals in the Tweed Valley, but Lindisfarne remains a major stronghold for this species. The decline in Mute Swans is in line with that for many parts of England, but it is difficult to pinpoint the reason in this area, though lead poisoning may be a factor. The drastic decrease has at least allayed one-time fears by the wildfowlers that the swans were limiting the feeding possibilities for Wigeon. The numbers of Wigeon have escalated dramatically, although the area was already historically famous for its flocks, threatened, however, by overshooting before the establishment of the NNR. Lindisfarne now has the largest assemblage of this species in Britain, having in recent years consistently exceeded the total for the Ouse Washes, the next most important site. The other species for which the area was traditionally

Fig. 36. The Borders.

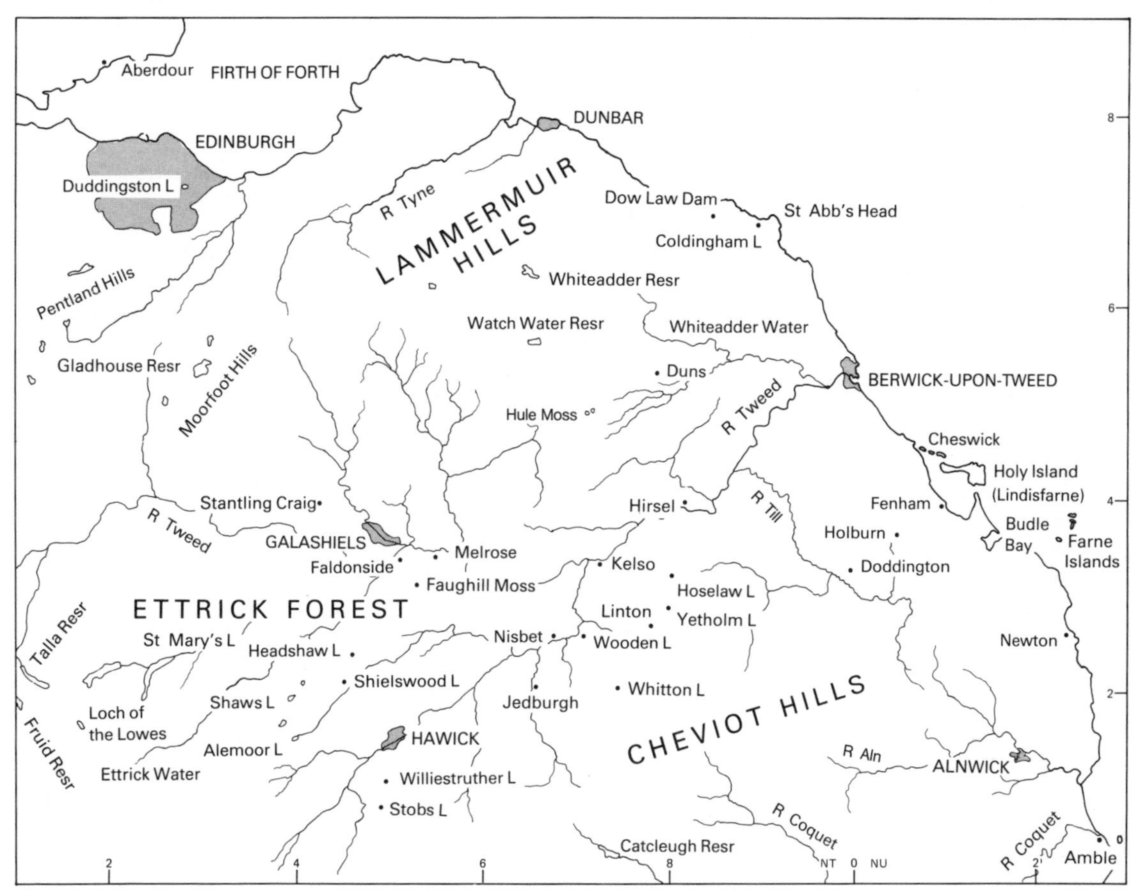

known is the Brent Goose, for which complete data are given in Table 244. The peaks for the last twenty years only are given here in Table 121. (In many cases the maxima occurred between the monthly wildfowl counts.) The Brent are overwhelmingly of the Light-bellied race (*hrota*), which breed on Svalbard (Spitsbergen) and winter mainly in Denmark. Influxes are therefore related to the severity of the winter on the continent. Recently it has been pointed out that the maximum influx recorded by Abel Chapman in March 1886 was 20,000 not 30,000 as generally quoted (338). Even so, present numbers are still but a shadow, though the Svalbard population does appear to be regaining numbers slowly. The Dark-bellied race (*bernicla*), which used to contribute the majority of the more static population of between 1,000 and 2,000 Brent, still has not returned in force despite its great increase elsewhere (p.389). It is unusual for more than a score of these birds to be seen at Lindisfarne nowadays.

Lindisfarne passes the criteria for international importance for Whooper Swans, Light-bellied Brent Geese and Wigeon, and also in respect of the total number of ducks, geese and swans – there were over 47,000, including 40,000 Wigeon, present in October 1982, the highest ever. Besides the wildfowl listed in Table 121, small numbers of Shoveler, Scaup, Velvet Scoters, Goosanders, Bewick's Swans, Pink-footed Geese and (stopping briefly on autumn migration to the Solway) Barnacle Geese are regularly seer emphasising the wide variety of species using th area.

The Greylag Geese at Lindisfarne are mostl genuine immigrants from Iceland, though sma numbers present in the summer are presumably c feral origin. The geese are also found roosting o Holburn Moss, 5km inland (Table 121), and th records certainly overlap to some extent. The regula grey goose surveys in early November recorded total of 1,430, 4,130 and 3,850 Greylags in Northumberlan in 1980, 1981 and 1982; however, maxima are generall reached in January. Only very small numbers c Pink-footed Geese are found in Northumberlanc Also apparent from Table 121 is the fact that Wigeo have largely moved from Holburn Moss to favour th now-protected Lindisfarne. Wildfowl counts are als available from several small lakes between the Mos and the coast, but none are of particular note. Som 7km south-west near Doddington, however, the wid plain of the River Till (in fact a tributary of the Tweec held up to 250 Mallard, 350 Teal, 220 Wigeon and 350 Greylags in the late 1970s.

The coast north to St Abbs Head is not of muc interest except for the estuary of the River Tweed, wit maxima recorded there in the 1970s of 82 Mallard, 3 Wigeon, 25 Tufted Ducks, 350 Goldeneyes (mostl immature males), 27 Red-breasted Mergansers, 22 Mute Swans and 43 Greylag Geese. The Mute Swan

Table 120. *Regular (and maximum) numbers of wildfowl recorded at two neighbouring sites in north Northumberland.*

a) Lindisfarne NNR i) 9 seasons 1960-1968; ii) 12 seasons 1970-1982

b) Holburn Moss i) 10 seasons 1960-1969; ii) 11 seasons 1970-1982

	a i	a ii	b i	b ii
Mallard	224 (1000)	747 (1400)	124 (250)	276 (650)
Teal	90 (500)	494 (1200)	105 (300)	188 (550)
Wigeon	8400(17000)	19792(40000)	2216(8000)	282(2000)
Pintail	8 (75)	39 (81)	5 (24)	22 (85)
Shelduck	301 (850)	576 (1000)	0 (0)	0 (0)
Pochard	9 (100)	24 (91)	0 (0)	1 (12)
Tufted Duck	5 (100)	14 (73)	1 (5)	1 (5)
Goldeneye	7 (70)	26 (70)	1 (17)	0 (3)
Eider	19 (2000)	2156 (4800)	0 (0)	0 (0)
Long-tailedDuck	0 (40)	120 (463)	0 (0)	0 (0)
Common Scoter	0 (3505)	432 (1270)	0 (0)	0 (0)
R-b Merganser	4 (25)	46 (100)	0 (0)	0 (0)
Mute Swan	336 (517)	46 (137)	0 (0)	0 (0)
Whooper Swan	231 (635)	156 (410)	0 (4)	0 (0)
Greylag Goose	85 (500)	1362 (4500)	68 (450)	1593(4500)
Pink-footed Goose	23 (350)	9 (71)	7 (190)	6 (95)
L-b Brent Goose	388 (2756)	647 (2170)	0 (0)	0 (0)
D-b Brent Goose	0 (0)	7 (39)	0 (0)	0 (0)

peak in the summer when Lindisfarne is practically devoid of them. It was therefore thought that these were the same birds. However, in contrast to Lindisfarne, the numbers on the Tweed have not declined, with August peaks of 431, 683, 482 in 1979, 1980 and 1981. The Tweed could well, therefore, be drawing on the more successful Scottish breeding populations. The species only breeds in scattered pairs through Northumberland.

The Tweed Basin

The Tweed runs inland between the Cheviot and Lammermuir Hills having, until it is joined by the Teviot at Kelso, low-lying agricultural country on either side, good feeding grounds for dabbling ducks and grey geese. Counts on several stretches of the river itself have produced mainly Mallard, but also Wigeon, Goldeneyes, Goosanders, and Mute and Whooper Swans in some numbers. Close to the river at Coldstream is a major Mallard roost on Hirsel Lake (Table 122). Their numbers have fluctuated erratically over the years (possibly because many are reared locally) and are now at a generally lower level than in the 1950s. To the north, above Greenlaw, the small loch on Hule Moss forms the main roost of the Pink-footed Geese in the district, with 1-4,000 present throughout the winter. There are also 1-200 Mallard and a few other ducks. Duns Castle Lake holds over 100 Mallard and Tufted Ducks, but the only other areas of note to the north of the Tweed are Watch Water and Whiteadder Reservoirs, high up in Lammermuir. Watch Water has been increasingly adopted by Greylags, with up to 1,000 present in December 1980. It also carries Mallard, Tufted Ducks and Goosanders, with maxima of 350, 87 and 40. Whiteadder Reservoir, 7km north, just inside Lothian region, is much larger and somewhat more important for ducks, holding regular (and maximum) levels of 208 (380) Mallard, 42 (95) Teal, 107 (330) Wigeon, 11 (40) Pochard, 38 (90) Tufted Ducks and 15 (86) Goosanders, but only 67 (250) Greylags. The main roost of the Greylag Goose is south of the Tweed on Hoselaw Loch, with an occasional presence on nearby Yetholm Loch (Table 122). These two lochs also carry useful numbers of a good range of other species. The Mallard, Shoveler and Goosanders have all greatly increased on Hoselaw, their 1960s regular numbers being respectively 190, 27 and nil. Yetholm has shown a like jump in Mallard and Shoveler from 69 and 6, Mallard having exceeded a thousand in two recent autumns. Linton Bog, 2km away in the neighbouring hills, has up to 250 Mallard and 220 Teal. The relative proportions of the two grey geese are apparent in the annual November surveys, collated on a county basis. Greylag/Pinkfoot numbers in 1981, 1982 and 1983 were: for Berwickshire (including Hule and Watch

Table 121. *The maximum number of Light-bellied Brent Geese recorded at Lindisfarne, 1963 to 1982.*

1963	450	1973	1710
1964	1000	1974	300
1965	2760	1975	550
1966	500	1976	780
1967	900	1977	1000
1968	1500	1978	1540
1970	650	1980	700
1971	700	1981	1800
1972	400	1982	600

Table 122. *Regular (and maximum) numbers of wildfowl recorded at seven sites in the Tweed Basin.*

a) Hirsel Lake, Coldstream 7 seasons 1960-1981
b) Hoselaw Loch 8 seasons 1970-1982
c) Yetholm Loch 7 seasons 1970-1982
d) Wooden Loch, Jedburgh 10 seasons 1970-1982
e) Whitton Loch 8 seasons 1970-1982
f) Headshaw Loch, Selkirk 14 seasons 1961-1982
g) Faughill Moss, Melrose 10 seasons 1970-1982

	a	b	c	d	e	f	g
Mallard	874(4000)	405(1500)	407(1200)	88(450)	172 (550)	43(300)	78(280)
Teal	19 (150)	20 (100)	44 (224)	0 (5)	10 (107)	2 (15)	1 (14)
Wigeon	32 (146)	118 (400)	72 (250)	19(100)	53 (500)	21(280)	92(200)
Shoveler	11 (42)	54 (154)	22 (100)	0 (3)	1 (27)	1 (25)	0 (0)
Pochard	13 (100)	34 (153)	57 (175)	4 (15)	5 (45)	6 (50)	8 (70)
Tufted Duck	35 (68)	49 (120)	44 (135)	61(135)	11 (55)	8 (32)	18 (60)
Goldeneye	13 (40)	25 (72)	33 (65)	12(110)	2 (25)	3 (12)	6 (42)
Goosander	7 (120)	12 (114)	5 (20)	0 (4)	4 (17)	4 (55)	25 (80)
Greylag Goose	0 (0)	1278(4100)	12 (700)	0(160)	74(2500)	0 (0)	0 (0)

Water) 820/1,000, 200/1,320 and 580/3,340, for Roxburghshire (including Hoselaw) 3,950/440, 1,510/120 and 3,850/1,800. The grey goose surveys of March 1982 and 1983 found Greylags (1,200 and 1,790) almost all in Roxburghshire and Pinkfeet (1,800 and 500) only in Berwickshire.

Following the River Teviot south-westerly along its dale, counts on the river itself have a similar mix as on the Tweed. At Nisbet records show up to 675 Mallard, 71 Teal, 220 Wigeon, 90 Tufted Ducks, 150 Goldeneyes, 50 Goosanders, 104 Mute Swans, 192 Whooper Swans and 950 Greylag Geese (mainly after the turn of the year). The Whooper Swans are especially notable, reaching 100 in most winters, and occurring in smaller numbers on other stretches of the Teviot and on a number of other waters in the vicinity. Some of these birds doubtless move to Ploughlands Pond above Nisbet. Five other waters in this area north of Jedburgh have regular populations of well under 50 wildfowl. Apart from Wooden Loch and Whitton Loch, further east (Table 122), there are no waters with notable wildfowl numbers.

Further along Teviotdale, around and mainly south and west of Hawick, is a scattering of small waters of which most are unimportant. Mention need only be made of Williestruther Loch, with regular (and maximum) counts of 36 (120) Mallard, 8 (24) Pochard and 21 (62) Tufted Ducks, while Stobs Loch, 4km south, has had maxima of 138 Mallard, 25 Teal and 34 Tufted Ducks.

On the higher ground westward to Ettrick Water there have been determined efforts to gain data on every water. However, of eighteen sites covered each have fewer than 50 wildfowl regularly, and the encroachment of forestry plantations in this area, lying mostly above 300m, is not making it more attractive to wildfowl. Mention may be made, however, of Alemoor Loch, with maxima of 290 Mallard and 20-40 Teal, Pochard, Tufted Ducks, Goldeneyes and Goosanders; Hellmoor Loch, with up to 160 Mallard and 43 Tufted Ducks; Shielswood, with 200 Mallard; and Shaws Lochs, with 290 Mallard, 50 Wigeon and 40 Goosanders. Headshaw Loch has a long run of counts (Table 122).

To the west again, St Mary's Loch, although sizeable, is over 40m deep; not surprisingly it supports only 55 (180) Mallard and a scattering of other species. The smaller, conjoint Loch of the Lowes carries even less, 30 Mallard at most. Still further west are Talla Reservoir and Fruid Reservoir, near the source of the Tweed and the watershed to the Clyde and the Annan Rivers. The two reservoirs carry small numbers of Mallard, Goldeneyes and Goosanders, but Fruid often has Wigeon, with as many as 21 present in September. These, like other early ones recorded in the district, are probably local-bred, the Ettrick Forest being one of the main centres of breeding (albeit in small numbers) for this species.

Returning to the River Tweed, between Melrose and Galashiels, there is a cluster of small waters, mostly to the south of the river. Most of these hold very little, but the loch on Faughill Moss (Table 122) carries more and is unusual in that the Mallard is not the major duck species. Again up to 90 Wigeon are present in September. The Goosanders must clearly be fishing on the Tweed, 4km away. Indeed many more of these lochs may be used, unnoticed, for roosting; observers report that the birds arrive and depart in the dusk and that as many as 227 have been seen congregated on one loch. Faldonside Loch nearby interestingly has no Goosanders, although only a kilometre from the river, but it has regular numbers of 82 Mallard and 12 Wigeon.

Somewhat separate from this group and north-west of Galashiels, Stantling Craigs Reservoir lies at 240m and supports only Mallard (1-200) and Teal (12) regularly in any numbers. The important waters lying to the north-west, in the Moorfoot and Pentland Hills, are considered in the section on the Forth (p.241).

North Cumbria and south-west Scotland

The Solway Basin

The Solway Basin, lying partly in England and artly in Scotland, is deservedly famous for its large nd widespread population of geese, for its great xpanses of foreshore and merse, for its popularity as wildfowling centre, and for the various conservation ıeasures, both formal and informal, which may have elped to correct the earlier trend towards excessive hooting. Over the past half century the changing attern of disturbance has had a marked effect on the ›cal distribution of the geese, but taken as a whole the eak numbers within the region are now larger than ver before. Whooper Swans are also plentiful, but in ommon with the ducks the concentrations at any one lace are seldom outstandingly large.

Although the firth is the primary centre of ıterest, the numerous inland waters and marshes are equently used in preference to the coast, and ›gether support a sizeable population of breeding nd wintering wildfowl. Many of the more important ites are now protected to a greater or lesser degree, ither by the owners or by one of the voluntary bodies, ıch as the RSPB, the Scottish Wildlife Trust and the Cumbria Trust for Nature Conservation. Statutory eserves are also established on Siddick Pond, near Vorkington, on the Castle and Hightae Lochs, ochmaben, and on the merse and foreshore at Caerlaverock, by Dumfries.

Inland waters of north Cumbria

The inland waters on the English side of the firth ave much in common with those in south Lakeland, nd carry a similar population of breeding and vintering ducks. Greylags and Whooper Swans also vinter in some numbers, and there is in addition a eral population of several hundred Greylags and Canada Geese.

The lakes, of which there are 40 or more, range ı size from pools of a few hectares to the great xpanses of Ullswater (890ha), Thirlmere (330ha), Bassenthwaite (530ha) and Derwent Water (540ha). Most of the larger ones are typical glacial troughs, and are much too deep and too steep-sided to be of value, except perhaps as occasional roosts or summer moulting sites. The only exceptions are Bassenthwaite and Derwent Water, both of which have a mean depth of less than 6m, and are flanked in places by small but attractive stretches of marsh. The smaller lakes, covering 10-30ha, are generally more productive, especially those in the basins of the Lune and Eden, and in the lowlands to the east of Carlisle. The pools and floodplain of the River Eden are also used by ducks, and are the main resort of the wintering Greylags and Whooper Swans.

The wild Greylags, as opposed to the feral stock, are confined for the most part to a 10km stretch of the Eden Valley to the east of Penrith, and to another less important area between Crosby and Aglionby, on the eastern outskirts of Carlisle. The main centre around Culgaith was probably adopted sometime in the 1950s, and by 1965 was holding a regular flock of 3-400. The numbers then rose sharply to 900 in 1969, and to 1,800 in 1972. Since 1975 the peak counts, which are mostly in January, have ranged between 1,300 and 2,200. The feeding grounds are spread along the low-lying farmland between Temple Sowerby and Langwathby, the favourite area being around Staingills and Watersmeet. The roost is normally on the shingle banks in the river, but up to 800 have at times been found on Whins Pond above Penrith. The flock at Aglionby usually reaches a mid-winter peak of about 250, and is probably linked with the coastal roost on Rockcliffe Marsh, some 10km to the west. There is also a record of several hundred Greylags on the gravel pits near Longtown, presumably from the same group.

All the other records of Greylags occurring inland are believed to refer to the feral birds introduced by WAGBI in the 1960s and 70s (p.378). These are now breeding freely on several of the larger lakes, including Derwent Water and Crummock Water, but are less widespread than in the southern districts of Lakeland, and have not yet become established in the northern and eastern parts of the county. The present winter total in north Cumbria is estimated at about 200 birds, the great majority of which are centred on the marsh at the southern end of Bassenthwaite. A few wild Greylags may also occur here from time to time, giving rise to temporary peaks of up to 350. In summer the flock at Bassenthwaite disperses, and some of the birds may possibly leave the district altogether, moving either to the southern parts of Lakeland or to join the moulting flocks in Galloway. The only local assembly at this time is on Derwent Water, where 50-60 are normally present in June and July.

The Canada Geese were introduced into Lakeland in 1957, and, with further introductions, th stock has since increased to 140 in 1969, 360 in 19 (405) and above 500 in 1981. In the northern distric they are now well established as a breeding species i several parts of Allerdale and Eden, but are st uncommon in the coastal plain and in the district Carlisle. Counts of adult birds on the three ma moulting centres in July 1981 produced a total of 36 of which 40 were on Crummock Water, and 160 c both Thirlmere and Haweswater (53). At the san time a flock of 170 was present on Killington Reservo in south Lakeland. In winter the largest gathering

Fig. 37. North Cumbria, Dumfries and the inner Solway Firth.

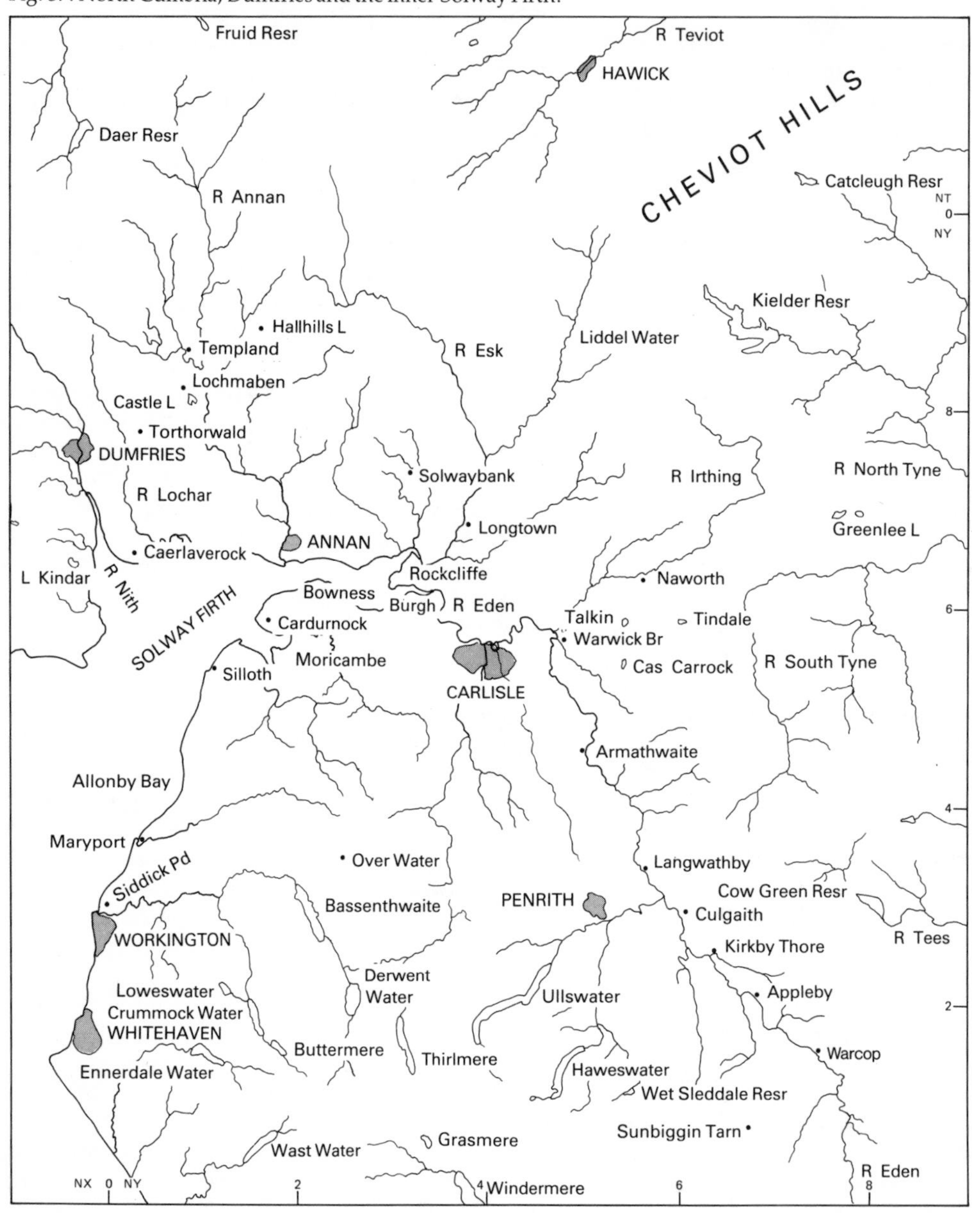

amounting at times to 200 or more, are found in the Eden Valley to the east of Penrith, in the area frequented by the Greylags. Another flock of 60-70 winters regularly on Derwent Water, and occasional parties of 10-20 are not uncommon elsewhere.

The Whooper Swans have two major resorts in the Eden Valley, one on the middle reaches around Culgaith and the other on the haughlands near Carlisle. The herds in the middle valley usually total about 50 or 60, and are present in strength from November to March. They have several favourite feeding areas, spread over 15km of valley floor, the most important of which are at Langwathby, Watersmeet and Kirkby Thore. Occasional parties of 20-30 have also been reported near Temple Sowerby, at Crackenthorpe near Appleby, and at Warcop still further upstream. The group near Carlisle usually totals between 80 and 120, and seems to have no connection with the flocks in the upper valley. The feeding grounds are confined almost exclusively to the low-lying land between Warwick Bridge and the M6 motorway, especially near Aglionby, Crosby and Linstock. The peak numbers are usually present in January and February, but sizeable flocks of up to 110 sometimes linger to the end of March, or into early April (53).

Whooper Swans also occur with fair regularity on at least eight of the lakes and tarns, and on some of the coastal marshes as well, usually in parties of 5 or so, but sometimes in groups of up to 20. During the course of the winter the flocks appear to roam from place to place, and the total numbers involved are probably not more than 50. The larger groups of 10 or more are mostly found on Rockcliffe Marsh at the head of the firth (where the Carlisle group sometimes roosts), on Siddick Pond near Maryport, on Wet Sleddale Reservoir near Shap, and on Tarn House Tarn in upper Lunedale. Other resorts include Bassenthwaite and Derwent Water, Sunbiggin Tarn and Greenside Tarn in Lunedale, Tindale Tarn by Brampton, and the coastal marshes round Moricambe.

Mute Swans are resident on many of the larger lakes and lowland tarns, but the numbers at any one site are seldom more than 5, or occasionally 10. The only major gathering is the summer moulting flock of 50-60 which assembles on the Longtown Gravel Pits, and seems to be increasing (53). A summer flock of about 40 has also been reported at Siddick Pond. The largest winter counts have come from the coastal marshes at Rockcliffe (max 32) and Moricambe (14), from the River Eden between Armathwaite and Warwick Bridge (20), and from Bassenthwaite (12), Talkin Tarn (12) and the Longtown Gravel Pits (15).

The winter population of ducks is disappointingly small and prosaic, considering the great extent of open water. Table 123 summarises the results of the counts on a sample of 21 lakes and reservoirs in the three administrative districts of Allerdale in the west, Eden in the east and Carlisle in the north. The figures are based for the most part on short runs of data, and some of the detail may perhaps be awry, but the general pattern of numbers and distribution is substantially correct. Bassenthwaite, already noted for its feral Greylags, is the principal centre for both dabbling and diving ducks, and is the only lake which often holds totals of more than 1,000 wildfowl. The next most important are Derwent Water, Haweswater and the group of gravel pits near Longtown, all of which have held occasional peaks of above 500 ducks. Most of the other lakes support less than 250 birds, and have no particular features of note, except perhaps for Siddick Pond, in the industrial outskirts of Workington, and some of the pools in the east of the county. The pond at Siddick, which is now a Local Nature Reserve, is a flooded colliery subsidence, 11ha in extent, with a shallow marshy shoreline and a sizeable fringe of reed-bed. Its interest lies not so much in the size of its population, which seldom exceeds 350, as in the variety of species occurring in the course of most seasons. These include a wintering flock of up to 20 Whooper Swans and a regular autumn gathering of 20-30 Shoveler; one or two Gadwall are often present as well, and the odd Pintail, Scaup and Common Scoter at times seek shelter from the nearby coast.

In the east of the county the moorland pools at Sunbiggin Tarn and Tarn House Tarn have an isolated summer population of 10-20 Gadwall, the largest in the region, and at Longtown the autumn peak of Shoveler is seldom less than 30, and has exceeded 90.

Mergansers and Goosanders also occur on many of the lakes and rivers, the former during summer only, the latter throughout the year. Both species first started to breed in Lakeland in the early 1950s, and, after a period of rapid increase, are now established at a steady level throughout the northern districts. In winter the Goosanders are widely dispersed, usually in groups of up to 5, or sometimes 10-15. The largest counts have been on Derwent Water (30), Ullswater (35) and Talkin Tarn (42); totals of 30-50 have also been recorded on several occasions on the 12km stretch of the River Eden between Armathwaite and Warwick Bridge. The Mergansers winter in flocks of 20-30 at several points along the coast, but are rarely found inland.

The inner Solway Firth

The middle and upper reaches of the Solway Firth form a large complex of interdependent resorts.

The outer boundary is a matter of choice, but for the purposes of this survey it is taken as running from Maryport, on the Cumbria coast, to the outflow of Southwick Water, on the border of the Nithsdale and Stewartry Districts of Dumfries and Galloway, a line which includes virtually all the intertidal sandflats. These amount to over 10,000ha; there are also more than 3,300ha of saltmarsh. Most of the marsh, especially on the Cumbrian shore, is covered only by the highest tides, and is summer-grazed to a short sward which provides an important feeding ground for geese. The largest areas are at Rockcliffe and Burgh around the estuaries of the Eden and Esk (1,400ha), at Caerlaverock between the channels of the Nith and Lochar (760ha), at Kirkconnell on the estuary of the Nith near Glencaple (240ha), and in Moricambe around the outfall of the Wampool and Waver (930ha). On the landward side there are further feeding grounds for geese on the arable land which occupies the greater part of the coastal lowlands, notably in the stretch between Annan and Dumfries.

The Solway has traditionally been a wildfowling centre, and shooting pressure over much of it is high and in some areas excessive. The creation of the Caerlaverock National Nature Reserve in 1957 provided the first complete non-shooting area, though with a permit shooting zone within the reserve; the sanctuary is in the course of being extended to cover the saltmarsh on the north side of the Lochar. Local wildfowling clubs try to exert some control, but the freedom to shoot in Scotland below high water mark makes any such regulation very difficult. On the English side, there has been rather better success, while private landowners also restrict the shooting in some areas. Overall, the pattern of use of the estuary by the wildfowl is much influenced by the shooting pressure, and it is reasonable to suggest that for some species substantial increases could come about if the

Table 123. *The regular (and maximum) number of the six commonest species of ducks on a sample of 21 lakes in the north Cumbrian districts of Allerdale, Eden and Carlisle. Most of the counts were made between 1976 and 1982. The figure in brackets after the place name shows the number of seasons in which regular counts were made. A blank = insufficient data.*

		Mallard	Teal	Wigeon	Pochard	Tufted Duck	Goldeneye
ALLERDALE							
Loweswater	(2)	138(235)	14 (28)	0 (0)	7 (14)	5 (10)	7 (9)
Crummock Water	(0)	(55)	(0)	(0)	(43)	(20)	(17)
Derwent Water	(4)	195(389)	7 (25)	7 (22)	110(236)	70 (87)	21 (34)
Bassenthwaite	(4)	451(903)	9 (56)	93(300)	125(514)	297(867)	51(117)
Over Water	(2)	5 (43)	2 (6)	0 (13)	11 (23)	9 (20)	4 (13)
Thirlmere	(0)	0 (52)	(0)	(0)	(17)	(1)	(5)
Siddick Pond	(2)	72 (93)	55(120)	5 (14)	42 (55)	70(150)	13 (35)
Tarns Dub	(0)	(250)	(60)	(0)	(0)	(0)	(0)
Martin Tarn	(2)	18 (34)	1 (3)	1 (3)	7 (12)	2 (6)	0 (0)
EDEN							
Ullswater	(2)	183(348)	11 (57)	1 (10)	6 (35)	13 (42)	3 (12)
Haweswater	(1)	270(460)	18 (32)	0 (1)	1 (6)	0 (3)	2 (28)
Wet Sleddale Resr	(2)	95(212)	21(126)	0 (20)	0 (5)	1 (3)	0 (2)
Whins Pond	(3)	142(411)	25(110)	65 (80)	18 (80)	21 (46)	3 (13)
Sunbiggin Tarn	(3)	30 (80)	47(114)	7 (90)	3 (10)	3 (12)	2 (9)
Tarn House Tarn	(5)	31(130)	6 (30)	27 (80)	13 (38)	13 (27)	1 (4)
CARLISLE							
Thurstonfield L	(4)	30 (75)	12 (22)	17(160)	14 (32)	35 (75)	1 (3)
Longtown GPs	(4)	233(695)	100(260)	94(167)	73(115)	99(120)	34 (82)
Talkin Tarn	(3)	7 (26)	0 (0)	27(150)	7 (18)	25 (53)	10 (46)
Castle Carrock Res	(3)	40 (95)	3 (15)	25 (69)	9 (26)	31 (65)	10 (32)
Naworth Lake	(4)	75(200)	6 (32)	0 (6)	3 (9)	11 (30)	8 (14)
Tindale Tarn	(4)	122(260)	7 (25)	2 (13)	12 (23)	14 (41)	9 (18)

Note: Occasional counts at the following sites failed to show totals of more than 50 ducks:
Allerdale: Mockerkin Tarn (9 counts), Buttermere (8).
Eden: Brothers Water (9), Greenside Tarn (3), Chapel Rigg Tarn (2).

shooting and its associated disturbance could be reduced.

The winter populations of no less than four of the species in the inner Solway are internationally important. These are Barnacle Goose, Pink-footed Goose, Pintail and Shelduck. In addition there are nationally important concentrations of Shoveler and Scaup. The inner Solway flats and marshes were listed as a Grade I site in the *Nature Conservation Review* (492), and the area has been proposed for addition to the list attaching to the Ramsar Convention. Certainly there is a need for some statutory designation which will protect the area against long-standing plans for a barrage, and more recent intentions to reclaim areas of saltmarsh, as well as providing local controls on the amount of shooting.

Barnacle Goose

The tremendous increase in this population since the war has been well documented (406, 444, 451); the peak annual totals are summarised in Table 124. The birds involved comprise the whole of the population breeding in Svalbard (Spitsbergen). They arrive at Caerlaverock in the last few days of September and stay on the Solway until the end of April or the beginning of May. Even in periods of severe weather there is no evidence of them moving out of the area. The population growth has been linked with successive protection in Britain, Svalbard and Norway, during the 1950s and 1960s, of particular importance being the establishment of the Caerlaverock NNR in 1957 (451). The establishment by the Wildfowl Trust of its Eastpark refuge in 1970, and the subsequent protection from disturbance and management of the farmland, together with the establishment of breeding sanctuaries in Svalbard in 1972, have ensured continued growth. Illegal shooting is still a significant mortality factor, but one which may reduce after the proposed extension of the Caerlaverock NNR. Poor breeding seasons in recent years have contributed to the apparent levelling out in the population, but there is probably potential for some further growth (444).

The two main haunts for the geese, of roughly equal importance, are at Caerlaverock, on the saltmarsh (known locally as merse) of the NNR and the adjoining farmland (the birds roosting on the very extensive mudflats of Blackshaw Bank), and at Rockcliffe at the head of the firth. Use of Rockcliffe is slight until after the end of the shooting season, but thereafter the entire population may transfer there for prolonged periods, often right through until the spring departure. More use has been made of Caerlaverock in spring, particularly in the most recent years. The farmland and small area of merse between Southerness and Southwick Water, on the west side of the Nith, is also used by the Barnacle Geese, sometimes by up to half the population, but more usually only 1-2,000. Elsewhere in the inner Solway the birds are irregular, though occasionally in large flocks, as at Burgh Marsh, just across the Eden from Rockcliffe.

When the population was smaller most of its

Table 124. *Numbers of Barnacle and Pink-footed Geese wintering in the inner Solway Firth. The Barnacle Goose figures are the means of the peak count each winter, usually made soon after their autumn arrival. The November figures for Pinkfeet derive from the anual censuses; the March figures from the less frequent censuses, with additional counts where available.*

a) North side of inner Solway, from Southwick Water to Annan, and including Lochmaben area
b) South side of inner Solway, from Rockcliffe to Moricambe

Period	Barnacle Goose	Pink-footed Goose			
		November		March	
		a	b	a	b
1946-50	470 (700)	2000(2000)			
1951-55	866(1200)	2827(4000)			
1956-60	1590(2800)	2982(7000)			
1961-65	3510(4250)	4270(7500)	632 (900)	4580 (7220)	4015 (9000)
1966-70	3760(4200)	2090(3700)	990(3083)	3681 (6000)	5938 (8874)
1971-75	4890(6050)	1222(1900)	173 (470)	4847(10333)	10917(17585)
1976-80	7880(9050)	1065(3380)	106 (374)	4776 (8000)	11000(12000)
1981	8300	2205	0	8597	1750
1982	8500	1860	0	12200	3700
1983	8400	2617	23	20850	3950

feeding was done on the extensive saltmarshes at Caerlaverock and Rockcliffe. Now, there is much feeding in autumn on barley stubble, up to 3km inland from Caerlaverock, as well as the maximum use of the specially managed grassland of Eastpark Farm. Other pastures, on both sides of the Lochar Water, are also important, especially in mid-winter.

Pink-footed Goose
The inner Solway has traditionally been a very important haunt for this species, though the very high numbers mentioned in the First Edition (an autumn peak of 15,000 and up to 10,000 present at other times through the winter) seem to have been based on a single, and very atypical, season. Complete November counts date back to 1961, with some information for the northern side back to 1950 (Table 124), and these show that there has been a decline in autumn levels, particularly during the late 1960s and early 1970s, though with some signs of a recovery on the northern side in the last few years. Numbers at this time of year are considerably affected by the available food supply in the regular autumn haunts in eastern Scotland.

The main roosts for the Pinkfeet lie on the Blackshaw and Priestside Banks between the channels of the Nith and the Annan Water, and on the extensive sandflats off the Rockcliffe saltmarsh. The main feeding areas are on the farmland on either shore and on the saltmarshes, but also extend inland, for example up Nithsdale and Annandale, for up to 20-25km. In recent years the Pinkfeet have begun to use Castle and Hightae Lochs, Lochmaben, as roosts, especially the larger Castle Loch, from February onwards. A record 9,000 were counted roosting at Castle Loch in mid-March 1984. Pinkfeet are regular, often in very large numbers, on the west shore of the Nith, both at Kirkconnell Merse (though here rarely much before the end of the shooting season, the peak numbers being in spring) and behind Southerness and along to Southwick Water.

Pintail
There is an early-winter population of this species which peaks at around 2,000 birds, making this one of the six most important sites in the country. The birds are centred on the mouth of the Nith, feeding either on the Caerlaverock NNR and adjacent farmland, or between Airds Point and Southerness, but particularly round Carse Bay. Complete counts are few, not least because the birds frequently feed by day on the farmland, for example on autumn stubble, and so are not visible to the counters on the shore. Nevertheless, between 1974 and 1979 the peak count, occurring between November and January, varied from 1,470 to 2,969 with a mean of 2,061. Since then the maximum has been below 1,000, but coverage has been nowhere near complete. The birds sometimes arrive in numbers as early as mid-September (1,500 in September 1977), while major departures occur as early as February. Elsewhere in the Solway the numbers of Pintail are small, with a mean of less than 100 for all other haunts

Table 125. *Regular (and maximum) numbers of ducks and swans in seven areas of the inner Solway Firth, and mean annual peaks for the entire estuary.*

a) Maryport to Beckfoot 5 seasons 1971-1976
b) Moricambe 5 seasons 1971-1976
c) Cardurnock to Port Carlisle 5 seasons 1971-1976
d) Rockcliffe 5 seasons 1971-1976
e) Annan Mouth to Priestside 8 seasons 1970-1982
f) Caerlaverock NNR 13 seasons 1970-1982
g) Carsethorn to Southerness 8 seasons 1970-1982
h) Entire inner Solway, 1971 onwards

	a	b	c	d	e	f	g	h
Mallard	62 (89)	232(522)	602 (633)	270(608)	82(240)	499(1727)	317 (680)	2343(3712)
Teal	11 (25)	52 (84)	20 (53)	53(329)	20 (60)	319 (820)	26 (60)	668 (902)
Wigeon	13 (32)	607(812)	2367(4600)	449(848)	11(370)	894(3000)	114 (350)	3832(6041)
Pintail	28 (49)	78(191)	2 (9)	17 (19)	18(117)	483(2150)	144(1400)	2061(2969)
Shoveler	5 (12)	11 (25)	22 (33)	9 (23)	10(342)	15 (80)	12 (30)	193 (354)
Goldeneye	0 (2)	5 (9)	21 (35)	60 (93)	59 (83)	3 (8)	0 (0)	220 (370)
Scaup	165(214)	186(220)	510 (958)	16 (48)	417(876)	0 (6)	490 (800)	1095(1505)
Common Scoter	13 (50)	14 (43)	0 (8)	0 (0)	0 (2)	0 (1)	0 (2)	72 (107)
R-b Merganser	17 (32)	23 (33)	32 (57)	20 (38)	0 (8)	4 (17)	16 (40)	120 (179)
Shelduck	48(125)	352(438)	318 (539)	142(374)	52(470)	156 (450)	445(1500)	2268(2379)
Mute Swan	4 (5)	7 (14)	10 (13)	21 (32)	0 (3)	17 (42)	0 (0)	99 (244)
Whooper Swan	1 (4)	0 (14)	1 (3)	6 (18)	0 (0)	42 (129)	0 (0)	164 (218)
Bewick's Swan	0 (0)	0 (0)	0 (0)	3 (8)	0 (0)	21 (78)	0 (0)	22 (80)

ombined. Further details of Pintail numbers in lifferent parts of the inner Solway will be found in 'able 125.

helduck

'he number of Shelducks in the inner Solway, eaking at a little over 2,000, also puts this estuary into he top half-dozen in Britain. The birds are quite well pread throughout the area, with the main concentraions in Moricambe and from there all the way to Rockcliffe on the south side, and at Caerlaverock and Carse Bay on the north (Table 125). Overall the mean f the highest complete counts in the period 1971-1976 vas 2,268, with a peak of 2,379. There is some ndication that the north side holds more than the outh, but the fluctuations between months are quite onsiderable, suggesting much movement within the stuary. Numbers build up to a maximum in January t the earliest, with March or April seeing the largest iumbers in some years.

hoveler

'he true status of this species in the inner Solway is ot very clear. A flock which peaked at 342 was resent between the mouth of the Annan and 'riestside Bank in the first part of the 1971-72 winter. n the years on either side, for which admittedly there re rather few counts, the maximum in this stretch has ever been more than 150, and often under 100. Elsewhere in the Solway, up to 55 Shoveler are regular t Caerlaverock (presumably from the Priestside to Annan group) and another 30-60 are scattered along he south side (Table 125). The mean of five annual ounts for the whole of the inner firth (all between September and November) in the period 1971-1976 vas 193, with a maximum of 354.

Scaup

Carse Bay, on the west side of the Nith, holds a regular vintering flock of this species of between 400 and 700, vhile the area around Powfoot, east of the Lochar, is ilso important (Table 125). What are presumably some f the same birds are also seen at Caerlaverock and in he stretch to the south of Carse Bay. On the south side f the Solway, Scaup occur particularly between Maryport and Silloth, in Moricambe, and off the stretch from Cardurnock to Bowness. The Solway Scaup have already arrived in substantial numbers by October, and stay at least until April. Such coordinated counts as there are suggest that there may be novement between the two sides of the firth. The nean of four annual peak counts in the period 1970-1976 was 1,095, the maximum 1,505. In 1982-83 the Solway Firth held up to 1,244 Scaup, more than any other British haunt, and over a quarter of the number counted in Britain.

Other wildfowl

Taking this section from south to north, the stretch from Whitehaven to Maryport is rather exposed, and though there is a strip of mud and sand all along the coast, there are no rivers and few streams of any size reaching the sea. Consequently the numbers and variety of wildfowl are small, with regular (and maximum) counts of no more than 23 (100) Mallard, 7 (22) Teal and up to 50 Common Scoters. The next section, reaching to Beckfoot, is of more interest, and the figures are set out in Table 125. The Scaup have already been discussed, while there are a few Common Scoters and some Red-breasted Mergansers offshore.

The main channel sweeps close past Silloth and so reduces the area's attractiveness to wildfowl, and only small numbers occur along this stretch. Then comes the wide, shallow and fertile bay of Moricambe. As Table 125 shows, this holds reasonable numbers of several species, including Mallard, Wigeon and Pintail, though there is little doubt that if a sizeable sanctuary area were created, the bay would be able to keep substantially larger numbers. Small flocks of Greylags and Pinkfeet occur here; the Pinkfeet occasionally roost in some numbers in the spring, using the area as a secondary roost to Rockcliffe.

The section from Cardurnock round to Port Carlisle is the most important in the firth for Wigeon, one of the better stretches for Mallard, and the prime area for Scaup on the south side (Table 125). The very extensive sandflats are backed by dunes and saltmarsh and are occasionally visited by both Pinkfeet and Barnacle Geese.

Counts from Burgh Marsh are unfortunately not very complete, but sufficient to indicate that Wigeon can number up to 400. This area is quite heavily shot over, in contrast to Rockcliffe Marsh, across the channel of the Eden. At Rockcliffe the shooting is in private hands and limited in frequency. Even so some species, like the geese, rarely use the marsh in large numbers until the season has closed in February. Outside the open season the marsh is a reserve of the Cumbria Trust for Nature Conservation, which provides a warden. Table 125 shows that this is a good area for dabbling ducks, and also visited by all three species of swan.

On the north side, there is a long stretch from the mouth of the Sark by Gretna to Annan with a fairly wide stretch of intertidal mud but no backing saltmarsh. It is also heavily shot and apart from a twenty-year old record of 1,000 Mallard and 2,000 Wigeon, has not since held more than 250 of either, together with 150 Teal and 125 Shelducks. Up to 50 Goldeneyes and 100 Scaup are recorded regularly, particularly around the mouth of the Annan.

The stretch from the mouth of the Annan to the River Lochar, including the Priestside Bank, is important for the flocks of Shoveler and Scaup already discussed (Table 125). Shooting is particularly intensive over Priestside and the numbers of dabbling ducks are low. The mudflats are used as a roost by Pinkfeet though it is not possible to separate this from the Blackshaw Bank next door.

The Caerlaverock National Nature Reserve takes in all the extensive intertidal flats lying between the rivers Lochar and Nith, together with the merse. In addition, there is the Wildfowl Trust reserve at Eastpark, covering adjoining farmland and including artificial pools. For Mallard and Teal, as well as for the Pintail and the Barnacle and Pink-footed Geese discussed above, this is the outstanding site on the Solway (Tables 124 and 125), reflecting the value of the reserve in reducing shooting and disturbance. From the NNR's inception in 1957, there has been carefully regulated permit shooting on part of the merse.

The merse is grazed in summer and provides suitable feeding conditions for the Barnacle Geese in winter, and also encourages the Wigeon. The Mallard and Pintail regularly feed on autumn stubbles at Eastpark and on neighbouring farms.

Artificial pools at Eastpark are regularly baited with grain, and this has attracted all three kinds of swan to make this their winter quarters. Prior to 1971, when the first pool was dug, there was a flock of Whooper Swans at Islesteps (see below) while Bewick's Swans were no more than a vagrant. In the years since, the Whooper flock at Eastpark has built up to its present level of over 100, while a regular wintering flock of over 50 Bewick's Swans has become

established. Mute Swans, too, have grown from a casual visitor to the present 30-40. The Whooper Swans are still present at Islesteps in roughly their former numbers (though with a recent slight increase to 140), and while ringing has shown that there is a limited amount of interchange, there are now two wintering flocks, both over 100, where previously there was only one.

Kirkconnell Merse, on the west bank of the Nith, is heavily shot and ducks are scarce.

The coast from Airds Point to Southerness is chiefly notable for the Pintail, Scaup, Shelducks and geese discussed earlier. Here, too, there are up to 200 Mallard and 500 Wigeon.

Greylags are found almost throughout the Inner Solway, using the mudflats as a roost, and feeding on adjacent farmland. However the more important sites for this species are on inland lochs, and the estuary flocks, totalling no more than 1,000 at peak, will be put into that context later (p.222).

Table 125 contains regular and maximum levels for all the common species for the whole firth, based mainly on a small series of complete counts in the mid 1970s.

Inland Dumfries

The main interest in the eastern half of Dumfries, away from the Solway, lies in the cluster of lochs round Lochmaben. The largest, Castle Loch, together with Hightae Loch just to the south, comprise a 137ha Local Nature Reserve, declared in 1962. The numbers of wildfowl they carry, together with figures for nearby Mill and Kirk Lochs, are shown in Table 126, plus those for Hallhills Loch, about 8km to the north-east. Both dabbling and diving ducks are present in good numbers, together with some Goldeneyes and Goosanders. As mentioned in the previous section, Castle and Hightae Lochs are sometimes used by considerable numbers of Pinkfeet for roosting, particularly in the spring. There is also a Greylag roost here of up to 1,900 birds, which will be treated in more detail in the next section (Table 130).

Since 1981, counts have been made along the River Annan, particularly at the series of oxbows between Hightae and Templand and some associated floods. Combined populations of around 5-700 Mallard, 4-500 Teal and 200 Wigeon have been found, which are probably additional to the birds on the lochs.

To the east of Lockerbie, up to the Cumbria border, there are few places of wildfowl interest, the Longtown Gravel Pits having been dealt with on p.213. The pool at Solwaybank, 10km north of Gretna, holds regular (and maximum) numbers of 120 (400)

Mallard and 11 (40) Teal, and up to 240 Mallard have been counted on floodwater at the confluence of the Esk and Liddel Water. Otherwise none of the other small natural waters or reservoirs in the area, such as the pools at Springkell and Blackwoodbridge or the reservoirs at Torbeckhill and Purdomstone, holds more than 50 wildfowl of all species combined.

Close to the Solway shore just west of Annan lies Kinmount Loch, holding up to 400 Mallard and 40 Teal, and also the origin of a long-established flock of Canada Geese which now wanders quite widely in the area. There were about 50 in the late 1960s, increasing to 70 in 1976 (399, 405), but in the last few years a peak of about 200 has been reached. This has usually been in August or September, with rather lower figures through the winter, suggesting a possible late-summer gathering of birds from further afield. What are presumably some of the same birds are quite regular at the Lochmaben lochs.

Other small waters between Annan and Dumfries include Dormont Lake, with regular (and maximum) counts of 30 (60) Mallard and 24 (99) Wigeon, and Manitou Pond, Rammerscales (maximum counts of 20 Mallard, 13 Teal and 14 Wigeon). Ironhurst Loch, in the Mabie Forest which has been planted over much of the Lochar Moss, held no more than 47 (75) Mallard, 4 (85) Teal, 12 (46) Wigeon and 10 (21) Tufted Ducks when counted during two seasons in the early 1970s, though a single count in January 1967 produced 750 Mallard, 170 Teal and 150 Wigeon. It is also occasionally used as a roost by some hundreds of Greylag and Pink-footed Geese. The Lochar Moss itself is drastically changed and much reduced in size by forestry. Although geese flight into adjoining fields, their use of the actual Moss is virtually unknown, as is the other wildfowl interest. The only information is that 78 Mallard and 142 Teal were counted in the southern part in February 1983.

There are rather more counts, though none since 1973, for the upper part of the Lochar Moss, above Torthorwald. Here maxima of 140 Mallard, 130 Teal, 60 Wigeon and 350 Pintail were found in the early 1970s, though comparing with higher numbers (up to 500 Teal and 300 Wigeon) in the 1960s. This area has also held a few hundred Greylags, but with the development in recent years of the old airfield where they used to feed, these are now much less regular in occurrence.

Of some small pools to the north of Dumfries, only Rosehill Pond close to the Nith holds more than a handful of birds, with regular (and maximum) counts of 54 (170) Mallard and 39 (130) Teal. The Nith itself is not prone to prolonged flooding and areas along its length attractive to wildfowl are few. Near Keir 290 Mallard have been counted, as well as 300 Greylags. There are counts, though only from the 1960s, of 175 (600) Mallard on Kettleton Reservoir and Morton Loch, just north of Thornhill, where up to 12 Whoopers have also been seen. At least one other site on the river held up to 500 Mallard and 100 Teal in the mid 1960s, with Black Loch, Sanquhar, carrying 100 Mallard and 70 Teal in the same period.

The Stewartry and Wigtown

To the west and south-west of Dumfries a number of large lochs lie amidst the rolling country between the valleys of the Nith and the Dee. All are more or less disturbed by shooting and other recrea-

Table 126. *Regular (and maximum) numbers of ducks and swans at five lochs near Lochmaben, Dumfries and Galloway.*

a) Hallhills Loch 3 seasons 1971-1975
b) Kirk Loch 10 seasons 1970-1982
c) Mill Loch 7 seasons 1970-1982
d) Castle Loch 9 seasons 1970-1982
e) Hightae Loch 9 seasons 1970-1982

	a	b	c	d	e
Mallard	107(247)	17 (90)	4(27)	36(250)	14 (34)
Teal	9 (67)	1 (9)	1 (7)	9(150)	5 (22)
Wigeon	8 (35)	4 (27)	1(13)	93(283)	64(300)
Pochard	16 (56)	20 (77)	1(11)	25(205)	2 (12)
Tufted Duck	23 (56)	51(171)	7(38)	36(164)	13 (34)
Goldeneye	0 (5)	4 (12)	2(11)	16 (57)	2 (6)
Goosander	10 (27)	0 (6)	0 (0)	32(115)	5 (34)
Mute Swan	0 (2)	7 (35)	4 (7)	4 (26)	6 (22)
Whooper Swan	1 (10)	0 (0)	0 (0)	5 (14)	6 (32)

tion, or by the proximity of roads, but all hold useful populations of both dabbling and diving ducks (Table 127). Loch Kindar is perhaps the least favoured, despite being close to the estuary, and there is little to choose between the others. The fact that they are close together makes it very probable that on being disturbed from one loch, the birds will quickly settle on one of the others. Greylag Geese are regular at Milton and Auchenreoch Lochs, less numerous at Loch Arthur, and only occasional visitors to Loch Kindar. (Counts from the first three are summarised in Table 130.)

Other waters in the area hold few ducks, with Glenkiln Reservoir, to the north of Crocketford, having a peak of 144 Mallard, and Fern Loch, near Dalbeattie, 220 Mallard but little else. The group of lochs south of Dalbeattie is of little importance.

The coast south of Dalbeattie includes the quite large shallow bays of Rough Firth, Orchardton, and Auchencairn. There is a single count of as many as 3,500 Wigeon, but the average is very much lower at around 500, with similar numbers of Shelducks. Up to 300 Scaup sometimes appear, presumably part of the Inner Solway flock.

The next break in the otherwise rocky coast to the west is Kirkcudbright Bay, running back several kilometres. It has extensive sand and mudflats, with a little saltmarsh, but numbers of wildfowl are not very great, perhaps because of shooting disturbance. Recent counts have revealed regular (and maximum) numbers of 57 (150) Mallard, 12 (40) Teal, 193 (350) Wigeon and 121 (275) Shelducks. Up to 20 Mute Swans are also regular. The small flocks of Greylags and Pinkfeet, common in the 1960s, seem to have abandoned the bay.

The Dee Valley, running inland from Kirkcudbright Bay, has a long stretch of great wildfowl interes between Bridge of Dee and New Galloway. Th Threave estate, just west of Castle Douglas, is owne by the National Trust for Scotland and maintained b them as a wildfowl refuge. This includes marshes, th river and islands lying in it. The principal interest is i the geese (see below) but up to 4,000 Wigeon an several hundred Mallard and Teal have been re corded. Above Threave the Dee has been dammed fo hydroelectricity to form a 12km-long narrow reservoir Loch Ken, with a deeply indented shoreline, man marshy places along the margins and several smal islands, together with a more extensive marsh at th north end. The RSPB has two reserve areas, on beside Loch Ken and the other in the marshe downstream, while the whole area is considere eligible as a Ramsar Site (p.539).

The wildfowl counts from Loch Ken (Table 128 show the wide variety of birds present in som numbers, with the excellent mixture of both dabblin and diving ducks. The loch is an important centre o Whooper and Mute Swans, the former often to b found grazing on the marshes and adjoining fields The best of the smaller lochs in the area is Carling wark, beside Castle Douglas, which despite som disturbance carries good populations of diving duck and Mute Swans (Table 128).

While there are counts from several of the loch in the Dee Valley and its environs, others are littl known, though none are likely to be of grea importance for wildfowl. Earlstoun, Carsfad an Kendoon Lochs upstream from Loch Ken are a hydroelectric reservoirs and subject to considerabl short-term variations in water level, which thu greatly reduces their wildfowl value. The regular (an maximum) counts of 33 (132) Mallard at Earlstoun an

Table 127. *Regular (and maximum) numbers of ducks and swans at five lochs between Dumfries, Crocketford and New Abbey.*

a) Loch Kindar 4 seasons 1970-1977
b) Loch Arthur and Lochaber Loch 4 seasons 1970-1982
c) Lochrutton Loch 4 seasons 1970-1977
d) Milton Loch 8 seasons 1974-1982
e) Auchenreoch Loch 9 seasons 1971-1982

	a	b	c	d	e
Mallard	173(290)	274(500)	72(180)	37(225)	76(212)
Teal	4 (14)	41(110)	10 (52)	4 (38)	15 (56)
Wigeon	170(480)	128(260)	239(650)	53(210)	5(165)
Shoveler	0 (4)	5 (20)	1 (5)	0 (1)	0 (14)
Pochard	2 (36)	46(140)	3 (13)	28(230)	13 (80)
Tufted Duck	49(150)	85(120)	47(190)	118(265)	84(403)
Goldeneye	1 (8)	11 (30)	8 (24)	14 (43)	11 (36)
Goosander	0 (0)	4 (10)	0 (0)	1 (19)	19 (86)
Mute Swan	2 (3)	6 (10)	4 (13)	24 (86)	18 (60)
Whooper Swan	0 (4)	2 (10)	1 (5)	5 (27)	3 (98)

ble 128. *Regular (and maximum) numbers of ducks and ans at Carlingwark Loch and Loch Ken, Stewartry.*

Carlingwark Loch 13 seasons 1970-1982
Loch Ken 9 seasons 1974-1982

	a	b
allard	63(250)	297(545)
al	18 (57)	371(540)
igeon	27(107)	579(869)
ntail	0 (0)	83(188)
oveler	9(100)	0 (0)
chard	48(390)	36(156)
fted Duck	79(170)	37 (69)
ldeneye	22 (52)	54 (94)
osander	0 (3)	25 (63)
ute Swan	28(131)	63(101)
hooper Swan	3 (12)	84(146)
wick's Swan	0 (3)	4 (67)

Fig. 38. The Stewartry and Wigtown.

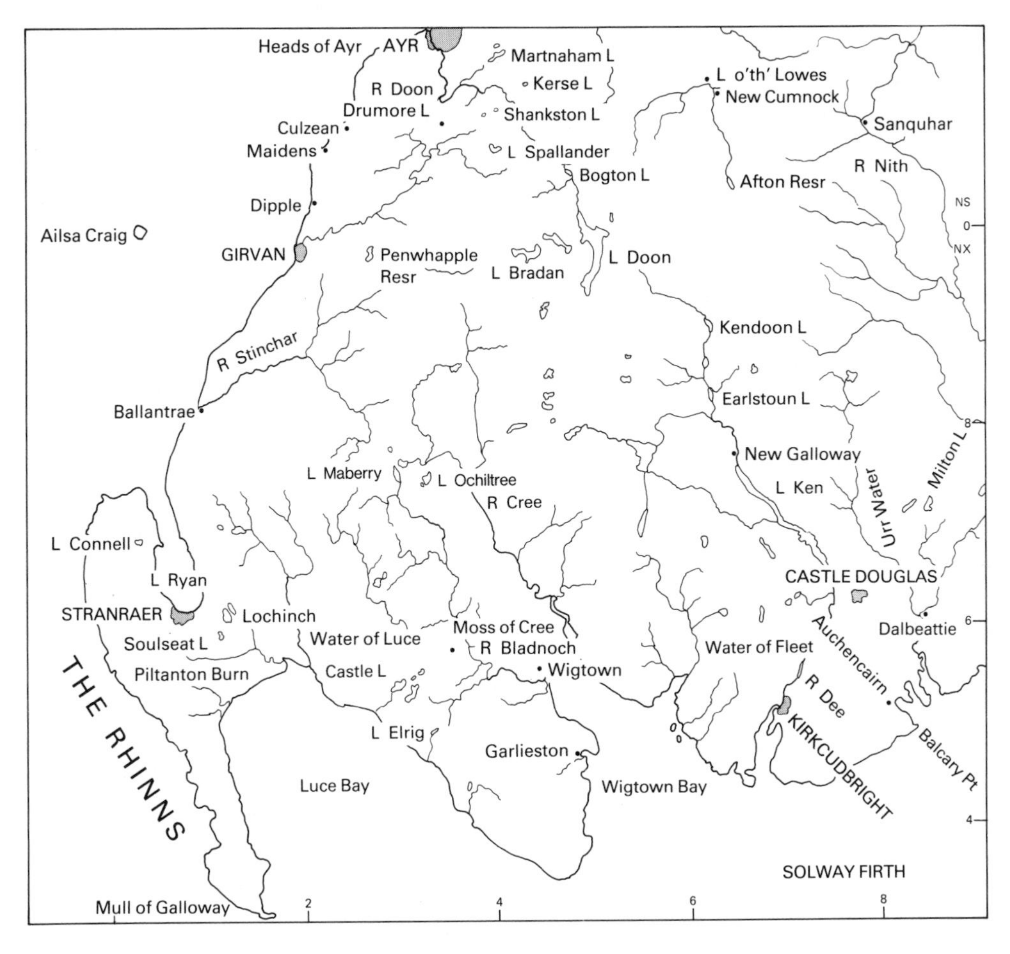

14 (32) Pochard at Kendoon represent the highest totals for any species. Further south Bargatton Loch has had maxima of 80 Mallard and Teal, while Stroan Loch was for several years a regular roost for the Greenland Whitefronts, though they seem to have deserted it more recently.

Loch Ken and the area below the dam round Threave are of particular importance for geese. Greylags are the most numerous, but because of their relative scarcity the Greenland White-fronted and Bean Geese rate higher. The Bean Geese comprised for a long time one of only two regular flocks of this species in Britain, the other being in Norfolk (p.137). As Table 129 shows the pre- and immediately post-war population of around 400 dwindled gradually through the 1950s and 1960s until in recent years it seems to be barely surviving. Whether there is any connection between this flock and the one which now seems more regular (p.244) in the Carron valley is not known. The geese formerly confined themselves mainly to a small area of pasture south of Threave, but when this was ploughed in the 1950s, a circumstance which may well have contributed to the decline, they tended to move more into the Threave area itself. The restoration of their original feeding ground to pasture did not bring about any recovery in numbers and it is difficult to see what additional protection or management could help this tiny flock far out on the fringe of the population's range.

The Greenland White-fronted Geese also feed around Threave, though their principal feeding grounds are more traditionally on fields on either side of the loch. They formerly made considerable use of the area at the head of Loch Ken but seem to have abandoned this in recent years. It has been suggested that the slow decline in numbers recorded during the 1960s and 1970s (Table 129) could have been caused by the birds being driven away by Greylags. Certainly the latter are now seen in greater numbers (Table 130) and occupy some of the areas once used by Whitefronts. However, it is more likely that the increased shooting

Table 129. *Average of annual peak counts, together with recent seasonal maxima, of Bean Geese and Greenland White-fronted Geese, at their haunts in Galloway. Data mainly from* Scottish Bird Reports.

(a) Bean Geese at Threave, Castle Douglas.

Period	Average (and maxima)
pre-1940	400-500 (500)
1945	400
1946-50	no counts
1951-55	214 (240)
1956-60	131 (190)
1961-65	99 (150)
1966-70	25 (60)
1971-75	48 (65)
1976-80	26 (82)
1981	38
1982	30

(b) Greenland White-fronted Geese at Loch Ken, Bladnoch Valley, and near Stranraer. Updated from Ruttledge and Ogilvie (1979).

Period	Loch Ken	Bladnoch Valley	Stranraer
1962-64	360(530)		0 (0)
1965-67	425(475)	41(50)	102(119)
1968-70	303(370)	22(26)	168(200)
1971-73	368(400)	51(62)	188(200)
1974-76	327(360)	38(66)	272(300)
1977-79	250(260)	55(55)	293(300)
1980	260		360
1981	330	0	480
1982	320	19	380

which has accompanied the increase in Greylags h had an adverse effect on the Whitefronts, throu mortality and disturbance. The slight recovery in t last two years may, one must hope, be a reflection the complete protection given to Greenla Whitefronts by the 1981 Wildlife and Countryside A

Table 129 also gives the numbers of Greenla Whitefronts at the other two haunts in Galloway, the Bladnoch Valley near Wigtown, and near Stra raer. The counts from the Bladnoch Valley are le complete than for the other two sites but there see to be not one but two separate small flocks involve one based on the Bladnoch Valley and roosting Clugston Loch not far away, and the other which seen around Moss of Cree and Wigtown Bay. T Stranraer flock was apparently not recorded befo 1965-66 and has since grown considerably in numbe It is quite possible that its growth is linked with t decrease at Loch Ken, but there is no positive eviden for this. The birds may suffer, like those at Loch Ke from the heavy shooting of Greylags in the area, b on the whole are reasonably secure. Both this flo and the one at Loch Ken qualify as of internation importance.

The Greylags of Dumfries and Galloway a summarised in Table 130. Those of the lochs to t west of Dumfries have already been mentioned. T species is only counted thoroughly during annu autumn censuses, but less complete March censuse plus other counts, suggest that mid-winter and spri numbers are higher than in autumn. This is certain true at the more easterly sites of the Lochmaben loc and Caerlaverock NNR, as well as further west. T increase that has taken place around Loch Ken sin 1961 is well shown in the table, though not parallel by any similar increase around the Ken – Dee marsh at Threave.

The Greylag picture further west in Galloway confused by the presence of considerable numbers introduced birds. Greylags of Outer Hebridean sto were originally released at Lochinch, Stranraer, about 1930, with a further release a few years later ne Mochrum, Wigtownshire. They have since sprea widely across Galloway, nesting on many of t inland lochs as far east as Dumfries. In 1971 t population was estimated at around 1,000 adults pl 300 goslings reared that summer (622). There ha been no complete surveys since, and it is known th eggs and birds from the Stranraer lochs have bee taken extensively for many years, mainly to start oth flocks of feral Greylags elsewhere in Britain, but als with the aim of keeping the Stranraer population check. There is no comprehensive information f more recent years, but counts of 800 birds at Lochin mid-September 1978 (before the Icelandic Greylags ve arrived in Britain for the winter) and 100 on Loch en in September 1981 suggest that the population is least maintaining its early 1970s level, and may well larger. With the presence in the area from October wards of some thousands of Greylags from Iceland is not then possible to separate the feral population. ne lochs just east of Stranraer (Table 130) hold the ost Icelandic birds, which feed in the surrounding rmland, while from time to time some of them move ross Loch Ryan and base themselves on Loch onnell. Apart from the sites shown in the table, from few tens to 2-300 Greylags, and often breeding pairs, ay also be seen on many other lochs in Galloway, for ample Castle, Mochrum, Woodhall and Earlstoun ochs.

The Greylags are mostly found on freshwater, ough they also roost on the estuaries, especially igtown Bay, from time to time. The Pinkfeet of alloway, on the other hand, are more or less nfined to the coast for roosting and most of their eding. The principal Pinkfoot haunt is Wigtown Bay d the Moss of Cree. The geese are rarely seen before ecember, and the peak numbers do not appear until bruary or March, after the end of the shooting ason, though earlier arrivals in hard weather have en recorded. Mid-March counts in 1982-1984 proiced an average 8,150 (max 9,500), a very considerle increase on the average 1,100 (2,500) from five arch counts in the period 1963-1967. Intervening, d not necessarily complete, counts suggest that the crease is very recent, with a previous maximum of 600 in spring 1978.

Wigtown Bay, with its extensive sands and saltmarsh, is also important for other wildfowl (Table 131). Wigeon and Pintail are both quite numerous and there is a small but regular flock of Shoveler present in the early part of the winter. The much smaller Fleet Bay to the east, has held up to 170 Mallard and 240 Wigeon, while there are also records of up to 300 Common Scoters. Along the coast to the west the two small bays of Garlieston and Rigg hold modest numbers of ducks, but the muddy estuaries of the Piltanton Burn and Water of Luce are more attractive, with a good flock of Pintail, and also Mergansers. The flock of 190 Mergansers referred to in Table 131 occurred in September 1982 (540). Up to 1,300 Common Scoters have been seen in Luce Bay in August – September, but whether they occur at other times of year except in small numbers is unknown. Like scoter flocks elsewhere in Britain their likely location some distance offshore precludes more than casual observation.

Loch Ryan, although large and well sheltered, lacks any extensive intertidal flats, though those along the southern shore attract a good flock of Wigeon, while Scaup, Eiders and Mergansers are all present in some numbers (Table 131). The rest of the Wigtownshire coast is mainly steeply sloping, with wildfowl interest low, apart from the only Solway breeding Eiders, particularly round the north part of the Mull of Galloway.

The lochs of Wigtownshire are, in the main, not particularly attractive to wildfowl (272). The majority are acid, and set in peat moorland or among hills. Castle and Mochrum Lochs support some diving

able 130. *Average (and maximum) numbers of Greylag Geese at haunts in south-west Scotland counted in the annual ovember censuses, and regular (and maximum) numbers derived from all available counts.*

Lochs at Lochmaben (9 seasons in regular figure)
Caerlaverock NNR (16)
Lochs Arthur, Lochrutton, Milton and Auchenreoch (12)
) Loch Ken (15)
e) Threave, Castle Douglas (0)
f) Wigtown Bay (12)
g) Bladnoch Valley (0)
h) Loch Ochiltree (0)
i) Lochs at Stranraer (19)

eriod	a	b	c	d	e	f	g	h	i
OVEMBER CENSUSES									
961-65	188 (351)	260 (400)	47 (50)	332 (510)	500(550)	175 (350)	206(500)	27 (54)	4135(5000)
966-70	217 (340)	191 (355)	134 (550)	290 (575)	470(500)	45 (100)	269(530)	21 (62)	1540(2500)
971-75	111 (277)	293 (696)	421 (649)	705 (990)	460(600)	14 (28)	510(704)	30 (82)	2380(3000)
976-80	30 (145)	204 (689)	181 (362)	564(1070)	300(500)	189 (513)	200(250)	29(102)	1720(2000)
981	0	310	250	700	100	141	80	24	1800
982	80	722	123	792	600	0	0	18	2600
983	270	28	320	1250	400	49	172	27	3100
LL AVAILABLE COUNTS									
961-82	649(1900)	256(1500)	267(1630)	725(1700)		82(1500)			1300(5000)

ducks (Table 132), but for their size the numbers are small, as is also true of Elrig Loch to the south. The tiny Castlewigg Loch, closer to Wigtown Bay, carries regular (and maximum) flocks of 43 (132) Mallard, 35 (74) Teal and 20 (68) Wigeon, but the larger pools at Bambarroch, to the north, seem to hold no more than a handful of Mallard.

There are few counts from the inland group of lochs (Ochiltree, Maberry and Dornal), or the group of Ronald, Heron and Black Loch, but such information as there is suggests very low numbers of wildfowl, apart from some breeding Greylags; Loch Ochiltree is a moulting place for up to 300 geese. Whitefield Loch, not far from Glenluce, holds no more than 30 Mallard, 20 Teal and 20 Tufted Ducks, but the Stranraer lochs are much more interesting. Soulseat and Magil Lochs are shallow and fertile and have both dabbli and diving ducks in some numbers (Table 132). Bla Loch and White Loch, at Lochinch, are set in parkla and somewhat disturbed, but have a good variety ducks, and there is also a flock of up to 70 Cana Geese. Finally, there are counts from two lochs on t Mull of Galloway, Lochs Connell and Dunskey (Tab 132). Loch Connell is much the better and, as well the ducks, has a regular flock of Whooper Swans.

The Clyde Basin

Throughout the Clyde region there is a wi diversity of landscape and habitat. In the south, lar

Table 131. *Regular (and maximum) numbers of ducks and swans in Wigtownshire estuaries.*

a) Wigtown Bay 6 seasons 1970-1975
b) Garlieston and Rigg Bays 3 seasons 1967-1969
c) Piltanton and Luce Estuaries 4 seasons 1970-19
d) Loch Ryan 10 seasons 1970-19

	a	b	c	d
Mallard	28 (146)	36(106)	58(263)	38 (240)
Teal	11 (48)	43(301)	15 (85)	2 (43)
Wigeon	591(2000)	163(392)	93(158)	1224(2200)
Pintail	221 (506)	2 (15)	187(480)	0 (2)
Shoveler	12 (43)	1 (7)	0 (1)	0 (8)
Scaup	0 (8)	0 (0)	0 (0)	148 (474)
Goldeneye	0 (17)	0 (0)	6 (23)	50 (312)
Eider	0 (0)	0 (0)	0 (0)	52 (228)
Common Scoter	0 (0)	0 (0)	5 (60)	2 (12)
R-b Merganser	2 (14)	0 (0)	21 (62)	112 (294)
Shelduck	95 (500)	43(123)	35 (67)	19 (80)
Mute Swan	7 (14)	0 (10)	1 (4)	42 (74)
Whooper Swan	19 (48)	0 (6)	0 (4)	5 (27)

Table 132. *Regular (and maximum) numbers of ducks and swans at six freshwater sites in Wigtownshire.*

a) Castle and Mochrum Lochs 2 seasons 1971-1975
b) Elrig Loch 3 seasons 1961-1971
c) Soulseat Loch and Loch Magillie 13 seasons 1970-1982
d) Black Loch and White Loch, Lochinch 4 seasons 1976-19
e) Dunskey Loch 1 season 1967
f) Loch Connell 4 seasons 1966-19

	a	b	c	d	e	f
Mallard	39 (62)	52 (82)	75 (300)	281(500)	65(76)	291(500)
Teal	10 (22)	38(135)	26 (198)	61(100)	40(86)	82(260)
Wigeon	55(103)	7 (41)	458(1000)	100(600)	12(19)	91(460)
Pintail	0 (0)	0 (0)	0 (3)	11 (24)	0 (0)	0 (0)
Pochard	42(141)	25(104)	13 (73)	50(500)	1 (2)	32(350)
Tufted Duck	62(127)	15 (60)	77 (187)	4 (35)	22(24)	35(200)
Goldeneye	23 (42)	7 (22)	1 (13)	11 (16)	2 (3)	3 (8)
Mute Swan	0 (2)	0 (0)	4 (13)	0 (0)	1 (2)	3 (10)
Whooper Swan	0 (0)	3 (17)	0 (2)	0 (0)	4 (4)	34 (53)

areas of good farmland occur over much of Ayr and Lanark. The central part of the region is completely dominated by Glasgow and its industrial satellites, with nearly half the human population of Scotland concentrated within 40km of the city centre. The north is mountainous, with fjord-like sea lochs extending far inland, themselves arms of the Firth of Clyde. This great estuary is 50km wide at its mouth, between the Mull of Kintyre and the Ayr coast, and contains many islands, both large and small. The central districts are linked with adjoining parts of the Forth region, thus affording an easy route between the east and west coasts of Scotland. Except for a narrow coastal strip, there is a more obvious high ground barrier to the south, between the Clyde and the Solway.

The majority of the wildfowl of the region are located in the south and central lowland districts, where food is plentiful and numerous local centres are provided by the reservoirs and lakes around the industrial and urban developments. The mountainous north and the long, narrow sea lochs are less suitable. Some stretches of the inner Firth of Clyde are particularly productive, while others are rocky and steeply sloping.

Greylag Geese are quite widespread in the south and central areas, but nowhere in very large numbers, though formerly so on Bute. Whooper Swans are plentiful in a few places. A wide variety of ducks occurs in fair numbers, but they are rarely abundant except on the inner Clyde, which provides the only extensive areas of estuarine habitat in the region. There are also some sea ducks in certain stretches.

Reserves for wildfowl are less numerous than in most other parts of Scotland, but a good deal of protection is afforded by landowners, including water authorities. On the other hand the scatter of generally small haunts through the region does not include any outstandingly large concentrations of wildfowl in urgent need of protection, though doubtless reserve status would increase the numbers in several cases.

Ayrshire

The southern part of the Ayr coast, from the mouth of Loch Ryan to the Heads of Ayr, is mainly cliff

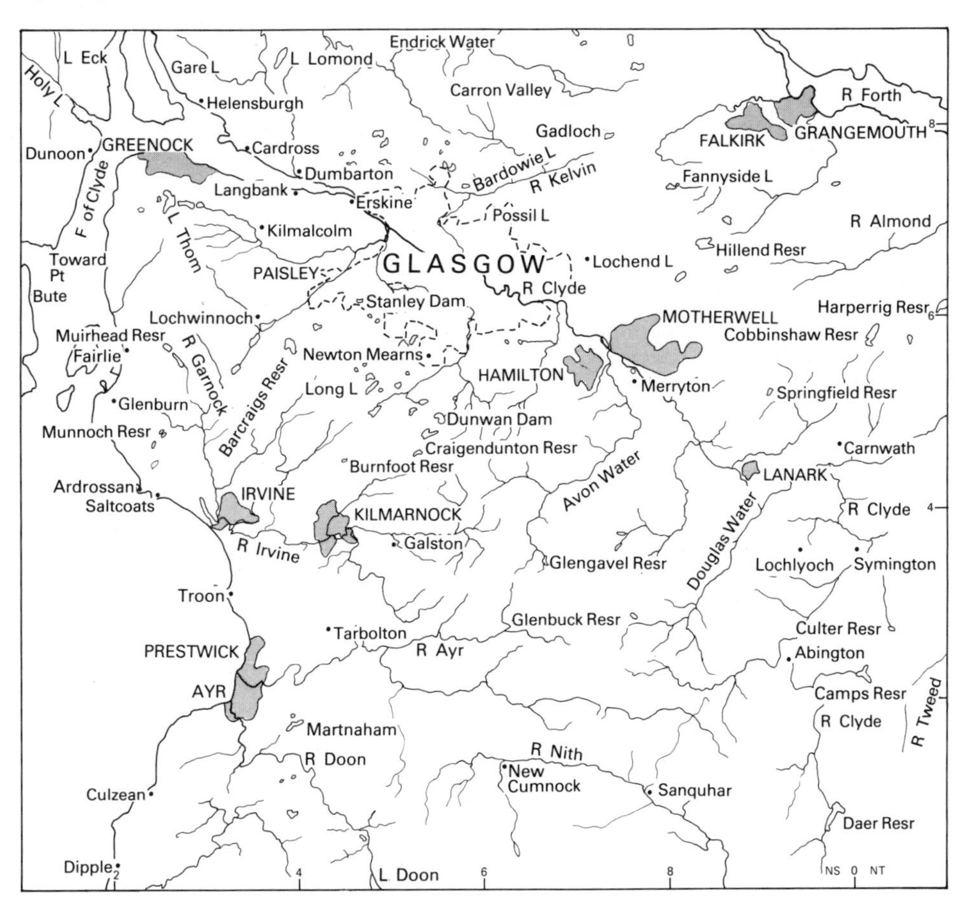

Fig. 39. Ayrshire and Clydesdale.

or low, rocky shore and, apart from Eiders and a few small parties of Mallard and some Wigeon at the outfalls of the occasional burns, is devoid of wildfowl. The only exceptions are the Stinchar Estuary at Ballantrae, Turnberry Bay and the strip of coast south to Dipple, and Maidenhead Bay at Maidens.

The Stinchar Estuary is quite small and from the few available counts holds no more than 80 Mallard, up to 25 Wigeon and 14 Mute Swans. The sandy intertidal flats of Turnberry Bay are bounded by a number of rocky outcrops, providing feeding possibilities and shelter for sea ducks. Only Mallard, with a regular 216 (maximum 533) are important among the dabbling ducks, but the 286 (525) Goldeneyes and 257 (530) Eiders are of more significance. In addition there is a late summer gathering of between 1,500 and 2,000 moulting male Eiders. In the last four winters, Red-breasted Mergansers have peaked at no more than 30, though there was a count of 112 in September 1978. The occasional appearance of up to 100 Scaup suggests that they, too, may often be out of sight offshore, or may be the flock from Loch Ryan. Shelducks peaked at over 100 in the early 1970s, but are now more usually around 50. Maidenhead Bay, just to the north, is much less important, with a maximum of 120 Mallard.

Inland there are several groups of waters lying in the low, rolling hills and the valleys of south Ayrshire. East of Girvan the Penwhapple Reservoir holds moderate numbers of the commoner ducks (Table 133) and the pond at Cairnhill beside the Water of Girvan has had regular (and maximum) numbers of 68 (140) Mallard and 41 (90) Teal in two recent seasons. In the low hills between the valleys of the Girvan and the Doon are several small lochs and reservoirs, the more important of these for wildfowl being shown in Table 133. Culzean Pond, also known as Swan Lake, is in a Country Park and although bounded at one end by a public area is secluded and sheltered at the other with good amounts of emergent vegetation. At none of the rest does any duck species regularly reach three figures, even when amalgamating the counts for the three close neighbours of Lochs Shanskton, Barnshean and Croot. Bogton Loch, near Dalmellington, is a colliery subsidence and there are shallow marshes and pools downstream. However, shooting disturbance is high.

Upstream from Dalmellington are some sizeable hydroelectric reservoirs, of which the largest are Lochs Doon and Bradan. They have boulder shores and peaty moorland surrounds, and are thus of little or no interest to wildfowl. Other waters in this area include Craigdow Loch, not far from Mochrum Loch, holding up to 70 Mallard and 65 Teal, and the Loch Spallander Reservoir with regular numbers of 10-20 Mallard and Tufted Ducks.

The next group of lochs to the north include the trio of Martnaham, Fergus and Snipe, south-east of Ayr, together with Kerse, Belston and Trabboch, spread around the town of Drongan. The figures for all these lochs are set out in Table 134. The Martnaham group seems to carry fewer birds than might perhaps be expected from its size and position.

About 15km to the east there is another collection of waters lying in the shallow valley between Cumnock and New Cumnock. They are actually in the catchment area of the Nith, which runs down to the Solway past Dumfries, but are close to the watershed with the Ayrshire streams and fall more naturally into this region. Mining subsidence has helped to create some of them, making for shallow water and marshy areas alongside. Loch o'th' Lowes is the most attractive but all carry useful numbers of ducks, as well as some Mute and Whooper Swans (Table 135).

Eight kilometres up the valley of the Afton

Table 133. *Regular (and maximum) numbers of ducks and swans at inland sites in south Ayrshire.*

a) Penwhapple Reservoir — 6 seasons 1972-1982
b) Culzean Pond — 11 seasons 1971-1982
c) Mochrum Loch — 11 seasons 1971-1982
d) Drumore Loch — 13 seasons 1970-1982
e) Blairbowie Farm Flash — 12 seasons 1970-1981
f) Shankston, Barnshean and Croot Lochs — 13 seasons 1970-1982
g) Bogton Loch — 6 seasons 1971-1982

	a	b	c	d	e	f	g
Mallard	104(280)	399(650)	63(164)	56(250)	34(163)	80(330)	47(160)
Teal	17 (55)	16 (51)	14 (50)	41(200)	59(333)	13(100)	27(126)
Wigeon	18(110)	1 (14)	10(115)	38(252)	29(152)	44(170)	9 (52)
Pochard	10 (40)	3 (12)	3 (8)	3 (16)	1 (20)	15 (60)	16 (57)
Tufted Duck	21 (50)	26 (89)	22 (49)	12 (29)	0 (0)	38 (80)	24(150)
Goldeneye	1 (7)	1 (3)	1 (7)	1 (3)	0 (0)	6 (24)	14 (56)
Whooper Swan	0 (2)	0 (0)	0 (7)	0 (0)	3 (19)	1 (12)	12 (36)

Water, south of New Cumnock, lies the Afton Reservoir, first counted in 1977-78 when it held up to 400 Mallard and 100 Teal. In September 1978 there were no less than 2,500 Mallard present, with 1,700 still there in October. This had dropped to under 400 in December, and the following winter no more than 500 appeared at peak. Since then the numbers have dropped even further and a maximum of 21 appeared in 1982-83. Teal numbers have stayed between 20 and 100. The reasons for this massive occurrence of Mallard are obscure, but may have had something to do with the particular water level or some short-lived but very abundant food supply. The reservoir is long and narrow and set in a steep-sided valley, and so at first sight is not a very likely place for large numbers of wildfowl.

The Greylags of south and central Ayrshire are difficult to census. The regular early November counts revealed an average of 180 in the 1960s, increasing to 240 in the 1970s, and topping 350 in each year from 1980 to 1983, with a maximum 490 in November 1983. The principal roosts are Drumore Loch and Blairbowie Farm Flash, and the Martnaham group of lochs. The numbers increase slowly through the winter, with probable totals of 1,000-1,500 during February to April, though complete counts are few. At this time up to 200 are fairly regular around New Cumnock. Not only are the geese occurring in fairly small flocks in the region, but they also use temporary floods for roosting, thus compounding the counting problems. During periods of exceptionally cold weather, Greylags may be driven into Ayrshire from other parts of Scotland. Thus during the early 1982 cold weather there were an estimated 3,500-4,000 geese there. Greenland Whitefronts are recorded in the area from time to time, but no regular wintering haunt has been found and it seems probable that they are birds moving to and from sites in Galloway. Similarly Pinkfeet are only irregular visitors.

From Ayr north-eastwards to Kilmarnock and beyond there are several mainly small lochs and reservoirs. Fail Loch and Loch Lea near Tarbolton have each held up to 40 Mallard, while Teal at Fail averaged 90 (maximum 125) during three seasons' counts in the late 1960s, perhaps involving the same birds as were counted on the Garrochburn, 2km to the east, in the same period (up to 50 Mallard and 117 Teal). The River Irvine marshes near Galston are probably too disturbed to be of much value, and the only counts, in the late 1960s, revealed no more than 70 Mallard and 90 Teal.

Of the waters just north of Kilmarnock, neither North Craig Reservoir nor Craufurdland Loch carry more than a handful of wildfowl, but Burnfoot Reservoir, just to the east of Stewarton, is quite attractive, with 59 (275) Mallard, 12 (50) Teal and 33 (230) Pochard. Small numbers of Greylags occasionally appear. Finally, in this area, Craigendunton Reservoir, to the east of Burnfoot and close to the watershed, has held 40 Mallard and 20 Teal.

The coast of central and northern Ayrshire has several good stretches of sand and mudflats, as well as small river estuaries and outfalls providing food for sea ducks. The wildfowl counts from the Doon Estuary, just south of Ayr, to Fairlie, near Largs, are summarised in Table 136. The River Doon flows into a small sandy bay at Doonfoot and wildfowl use the area quite regularly, though disturbance is often great. Around Ayr the outfalls support good numbers of, particularly, Goldeneyes, as well as the small flocks of

Table 134. *Regular (and maximum) numbers of ducks and swans at inland waters in central Ayrshire.*

a) Martnaham, Fergus and Snipe Lochs 7 seasons 1971-1982
b) Kerse Loch 6 seasons 1977-1982
c) Belston Loch 6 seasons 1977-1982
d) Trabboch Loch 1 season 1965

	a	b	c	d
Mallard	86(281)	39(180)	46 (97)	13(15)
Teal	30(127)	3 (28)	6 (21)	7(12)
Wigeon	28 (85)	0 (0)	10(100)	5(21)
Pochard	31(172)	4 (26)	9 (40)	21(45)
Tufted Duck	66(158)	11 (27)	22 (60)	14(16)
Goldeneye	29 (82)	1 (4)	6 (15)	0 (1)
Mute Swan	2 (6)	0 (1)	1 (2)	2 (2)
Whooper Swan	6 (21)	5 (62)	1 (7)	0 (0)

Table 135. *Regular (and maximum) numbers of ducks and swans at inland waters in western Ayrshire.*

a) Black Loch, New Cumnock 6 seasons 1976-1982
b) Loch o'th' Lowes 3 seasons 1979-1982
c) Creoch Loch 6 seasons 1976-1982
d) Knockshinnoch Lagoons 5 seasons 1976-1980

	a	b	c	d
Mallard	14(30)	118(254)	35(105)	40(111)
Teal	17(40)	91(150)	23(100)	69(153)
Wigeon	6(25)	65(132)	30 (96)	1 (7)
Pochard	2(13)	25 (39)	14 (61)	1 (6)
Tufted Duck	8(26)	75(150)	33 (93)	10 (54)
Goldeneye	2 (7)	14 (30)	10 (25)	0 (3)
Mute Swan	0 (2)	9 (13)	1 (2)	9 (15)
Whooper Swan	11(22)	3 (17)	0 (1)	3 (17)

Scaup and Eiders. The stretch either side of Troon is especially noteworthy for the number of Mallard, which do not reach their peak until mid-winter, and for moulting Eiders, with up to 1,500 present in late summer. Disturbance from walkers along the beaches can, however, often be quite serious.

The best mudflats and virtually the only salt-marsh on this coast are found in the estuary of the Irvine and Garnock Rivers. The area is quite badly disturbed by aircraft, bait-diggers and others, but the numbers of birds are important, particularly the Eiders, mainly offshore, as well as the Wigeon, Teal and swans. Eiders again feature strongly in the northern two sections which are a mixture of rock, sand and mud. The Fairlie flats by Hunterston encompass a considerable area of intertidal sand and hold the largest numbers of Shelducks on the coast as well as a useful flock of Wigeon. Greylags occasionally roost in small numbers, generally less than 200. Whooper Swans were formerly regular here, with 10-30 in the 1960s, but have moved away, perhaps to one of the reservoirs inland. There has been considerable reclamation and development in this area in recent years and some decline in wildfowl numbers.

Winter wildfowl counts from Great Cumbrae Island, just opposite Fairlie and Largs, show comparatively few birds, with maxima of 150 Mallard, 88 Wigeon, 28 Mergansers and 80 Eiders, though there are good numbers of Eiders breeding round this and Little Cumbrae Island. Some indication of the total Eider population in the outer Firth of Clyde comes from the late summer counts of between 1,500 and 2,000 moulting males along the coast from Girvan to Turnberry, and 1,000 to 1,500 between Ayr and Saltcoats. These are separate from the even larger numbers which have been counted in the inner Clyde (see below).

Just inland from Ardrossan and West Kilbride there are a number of reservoirs, of which the best for wildfowl are Muirhead and Camphill up the Rye Water valley, with regular (and maximum) counts of 100 (155) Mallard, 35 (84) Teal and 23 (37) Pochard. Most of the others are smaller and carry fewer ducks. Glenburn has held up to 80 Mallard, Mill Glen 50 Mallard and 80 Tufted Ducks, and Munnoch 60 Mallard, 40 Pochard and 120 Tufted Ducks, but regular levels at all of these are very much lower. Kilbirnie Loch is just on the watershed between the River Garnock and the Black Cart Water, and data are limited to a single count in the 1960s when no ducks were seen. Barr and Castle Lochs at the head of the Black Cart Water will be dealt with in the section on Renfrew. Greylags are regular in small numbers at Munnoch Reservoir, averaging only 14 in recent years, but with an occasional peak of over 200 in late winter.

Clydesdale

The greater part of southern and western Lanark is hilly and lacks open water resorts for wildfowl. The large Daer Reservoir lying at 300m in the extreme south held no more than 30 ducks of all species when counted in the late 1960s. The Camps, Cowgill and Culter Reservoirs in the hills east of Abington are little

Table 136. *Regular (and maximum) numbers of ducks and swans along six stretches of the Ayrshire coast, Firth of Clyde.*

a) Doon Estuary (Greenan to Ayr)	12 seasons 1970-1982
b) Ayr to Pow Burn, Prestwick	9 seasons 1972-1982
c) Pow Burn to Barassie, Troon	13 seasons 1970-1982
d) Irvine Bay and Garnock Estuary	7 seasons 1971-1982
e) Ardrossan to West Kilbride	11 seasons 1971-1982
f) Hunterston to Fairlie	8 seasons 1972-1982

	a	b	c	d	e	f
Mallard	363(600)	25(103)	429(1110)	32(430)	2 (17)	10 (60)
Teal	1 (5)	0 (6)	0 (3)	126(420)	1 (16)	0 (5)
Wigeon	67(200)	0 (2)	0 (8)	264(650)	41(164)	141(578)
Tufted Duck	4(100)	13 (93)	2 (40)	3 (60)	0 (0)	0 (0)
Goldeneye	71(200)	125(386)	28 (277)	23(145)	6 (28)	1 (9)
Scaup	3 (48)	35(192)	4 (77)	1 (33)	0 (0)	6(150)
Eider	7 (45)	26(100)	57 (320)	205(850)	89(632)	84(500)
R-b Merganser	2 (10)	6 (18)	4 (26)	10 (22)	10 (37)	3 (27)
Shelduck	3 (24)	0 (0)	2 (16)	39(110)	11(120)	61(155)
Mute Swan	9 (30)	1 (7)	0 (3)	12 (31)	2 (18)	1 (7)
Whooper Swan	0 (1)	0 (3)	0 (5)	17 (61)	0 (0)	0 (18)

better, Cowgill having held maxima of 85 Mallard and 14 Teal. This and Culter are roosts for Pinkfeet, as is Lochlyoch Reservoir on the other side of the Clyde. The geese seem to feed regularly in the farmland close to Lochlyoch, and along the Clyde between about Symington and Lanark. Their numbers are usually in the range 500-1,000, but occasionally rise to over 2,500. There have certainly been some instances when birds roosting at West Water or Baddinsgill Reservoirs in the Pentland Hills (p.243) have fed in this part of the Clyde Valley, and this may be the origin of the larger flocks. Up to 500 Greylags are also regular in the area though their roost or roosts are not known with certainty. Springfield Reservoir on the north side of the Clyde may be one of them; Pinkfeet also roost here sometimes.

The small waters in this area are relatively unimportant for wildfowl. There are single counts of 230 Teal at Quothquhan Loch, south of Carnwath, and of 105 Mallard at Bowmuir Loch nearby, but the large White Loch held no ducks when checked in the early 1970s. A flock of Whooper Swans is regular in the Symington to Lanark stretch of the Clyde, numbering up to 30, occasionally more.

Two large reservoirs lie in the headwaters of streams draining eastwards into the upper Clyde: Glenbuck on the Douglas Water and Glengavel on the Avon. Both hold fair numbers of Mallard, as well as of diving ducks and Goosanders (Table 137). Other waters in this area, though, are at too high an altitude to provide much interest for wildfowl, or are surrounded by peat moorland, or both.

Between Lanark and Motherwell the Clyde runs in a narrow valley, but at Motherwell and Hamilton it broadens out and meanders through a wide flat valley with numerous marshy pools. This area, known as Hamilton Low Parks or Bothwell Brig, has undergone enormous changes in recent years, starting with the building of the M74 motorway through the middle of the valley in the late 1960s. Subsequently the marshy pools on the east side downstream of Motherwell have been embanked and turned into a sailing and recreation area. However, the pools on the west side are still mostly in a natural state and the same applies to the Merryton Ponds upstream of Motherwell. Although placed amidst a heavily populated district, these waters are still attractive to wildfowl (Table 137) and it may be that the construction of the motorway has actually made human access more difficult to some of them. As the table shows, the numbers of Mallard, Teal and the diving ducks on the Low Parks have increased in recent years, though the Wigeon have declined considerably. The flocks of Whooper Swans, which occur in many waters in the region, and in substantial numbers in Hamilton Low Parks and Stanley Dam, are also of considerable interest.

Renfrew

On the rising ground to the south and south-west of Glasgow there are about 30 reservoirs and lochs scattered between Newton Mearns and Lochwinnoch. The majority are set in moorland at altitudes up to 200m, and are too exposed and bleak to harbour many birds. The numbers of the commoner species on 11 of these, centred within about 9km of Newton Mearns, are set out in Table 138. For many of them there are only relatively infrequent counts available but they probably represent the true picture well enough, though it would be useful to have further records to confirm the high Mallard counts at Lochgoin and Dunwan Dam.

The reservoirs above Paisley are little more productive with the exception of Stanley Dam (Table 139). The Whooper Swans roost here but feed down on the lowland between Renfrew and Erskine where they have sometimes come into mild conflict with the authorities at Glasgow Airport. Close to the head of the Black Cart Water lie the two lochs of Barr and the shallow and attractive Castle Semple at Lochwinnoch. There has been a reserve and study centre of the RSPB here since 1979, and this is much the best freshwater wildfowl locality in the whole south Clyde area (Table 139). In addition to the ducks and swans, a wintering

Table 137. *Regular (and maximum) numbers of ducks and swans at four waters in Lanark.*

a) Glenbuck Reservoir 6 seasons 1970-1977
b) Glengavel Reservoir 4 seasons 1974-1977
c) Merryton Ponds 3 seasons 1974-1976
d) Hamilton Low Parks 11 seasons 1971-1982

	a	b	c	d
Mallard	190(360)	363(940)	28 (80)	352(1262)
Teal	2 (8)	3(141)	50(200)	181 (600)
Wigeon	0 (4)	0 (20)	67(170)	75 (430)
Pintail	0 (0)	0 (0)	6 (23)	7 (30)
Pochard	84(420)	55(450)	4 (22)	122 (340)
Tufted Duck	151(462)	69(220)	33(104)	259 (704)
Goldeneye	2 (14)	0 (11)	3 (6)	20 (57)
Goosander	17 (40)	21 (36)	0 (2)	4 (26)
Mute Swan	2 (16)	2 (8)	5 (13)	8 (24)
Whooper Swan	1(126)	0 (0)	37 (70)	26 (69)

Note: The regular and maximum levels for Hamilton Low Parks in the period 1960-1969 (5 seasons counted) differed from the above as follows: Mallard 137(300); Teal 152(360); Wigeon 298(1200); Pochard 26(71); Tufted Duck 94(209); Goldeneye 2(19).

flock of up to 500 Greylags is present. The nearby Barcraigs Reservoir is less attractive but has a small population of Canada Geese, which were first noted in the 1960s, and now number about 60, though there was a count of over 150 in 1976. They are sometimes also seen at Castle Semple Loch.

The group of small waters around Kilmacolm, near the Clyde coast, have rarely been counted. Single counts of 490 Pochard and 510 Tufted Ducks at Leperstone Reservoir, and 295 Pochard and 385 Tufted Ducks at the next-door Auchendores Reservoir (though not on the same date), suggest that they would repay further visits. The major Gryfe and Loch Thom Reservoirs above Greenock are apparently not attractive to wildfowl.

Dunbarton and north Lanark

North of the Clyde, and to the east and north of Glasgow, there are a number of lochs, mining subsidences and reservoirs. The best of these are listed in Table 140, with the Shoveler on Gadloch and Possil Loch of some interest. Apart from Endrick Mouth, Loch Lomond (see below), this is the most northerly regular haunt of the species on the west side of Scotland. Unfortunately Gadloch (or Lenzie Loch) has been opened up for recreation in the last few years and the numbers of wildfowl have already fallen. In particular the formerly regular Greylag flock, which used to roost on the loch and feed in the surrounding fields, has transferred to the Balmore Haughs, 4-5km to the north-west. It averages 400 during the winter, but often has a November peak of 600. Infrequent counts on other waters in this area have revealed up to 115 Mallard, 150 Pochard and 90 Tufted Ducks on Bishop Loch, 67 Pochard on Johnston Loch, 70 Mallard and 30 Tufted Ducks on Antermony Loch, and 65 Mallard, 86 Tufted Ducks and 38 Goldeneyes on Bardowie Loch, but it is not known how representative these numbers are. Many of the waters are very disturbed, often lying in public parks. Tannoch Loch, lying just below Mugdock Reservoir, has had maxima of 144 Mallard and 20 Tufted Ducks, while of the various hill reservoirs between Bearsden and Dumbarton, Greenside and Burncrooks have had counts of 90 and 60 Mallard respectively, the others failing to reach even these modest totals.

Separated from the inner Clyde by the Kilpatrick Hills and the length of the River Leven is Loch Lomond, the largest freshwater lake in Britain and one of the original thirteen areas designated by the United Kingdom under the Ramsar Convention. The narrow, northern end is set between steep-sided mountains and is essentially acid and poor in nutrients. This is in considerable contrast with the broader, much shallower, southern end, particularly around the mouth of the main tributary feeding the loch, the Endrick Water. This flows through agricultural land and brings nitrogen into the loch, while some limestone is present in the area, which also helps to raise the pH of the water.

The mouth of the Endrick, and an adjoining area of low-lying marshes and carr woodland together with five of the islands in the loch, in all some 416ha, is a National Nature Reserve. The wildfowl interest is high, and the details for the ducks and swans are set out in Table 141. In the last few years one or two pairs of Whooper Swans have bred in the area having escaped from a local wildfowl collection. Greylag

Table 138. *The regular (and maximum) numbers of five species of ducks at eleven sites in southern Renfrew. The figures in the left-hand column show the number of seasons for which full data are available, and the period during which these seasons occurred, or, if full data are not available, the total number of counts to hand and the period. In the latter case only the maxima are quoted.*

Site	Seasons	Mallard	Pochard	Tufted Duck	Goldeneye	Goosander
Lochgoin	7 counts: 1968-72	(650)	(0)	(30)	(0)	(1)
Dunwan Dam	7 counts:1968-72	(250)	(8)	(32)	(2)	(0)
Bennan Loch	2 counts:1968-69	(105)	(0)	(6)	(1)	(0)
Lochcraig	1 count:1968	(83)	(0)	(0)	(0)	(0)
Black Loch	6 seasons:1967-72	6 (12)	26 (40)	10 (40)	5 (5)	2(12)
Brother Loch	6 counts:1971-72	(35)	(6)	(53)	(2)	(3)
Long Loch	7 counts:1967-71	(88)	(57)	(27)	(75)	(12)
Harelaw Dam	6 counts:1967-76	(18)	(7)	(87)	(13)	(12)
Craighall Dam	12 counts:1960-75	(45)	(56)	(61)	(2)	(15)
Walton Dam	3 seasons:1961-72	0 (6)	32(123)	24 (55)	0 (2)	4(27)
Balgray Resr	1 season:1972	3 (7)	2 (16)	79(105)	4(10)	5(29)

Geese are regular, a flock of 5-800 being present throughout the winter. For a short time in the late 1970s the numbers rose to between 2,000 and 3,000, but they have since returned to the lower level. Greenland Whitefronts also winter here, having first appeared in 1962-63, when 6 were counted. A slow but steady increase has brought the flock up to a regular 100-120. A small number of Bean Geese, up to 20, occurred in the early 1960s but have not been seen since. Up to 40 Canada Geese are now resident. Several species of ducks breed in and around the reserve, including Shoveler and up to 35 pairs of Shelducks (52).

Counts of wildfowl elsewhere on Loch Lomond are fragmentary, but a complete survey of the west shore in January 1968 discovered no more than 98 Mallard, 29 Teal, 40 Wigeon, 26 Pochard, 31 Tufted Ducks and a handful of Goldeneyes and Mergansers. Up to 50 Mallard and 10 Goldeneyes and Mergansers occur around Ardlui at the very northern end.

The inner Firth of Clyde

This section covers the Clyde Estuary downstream from the Erskine Bridge to Greenock and Dunoon, together with the long arms of Gare Loch, Loch Long, Loch Goil and Holy Loch.

Although heavily industrialised along much of its length, the estuary of the Clyde above Greenock and Helensburgh contains a number of stretches of suitable foreshore including some very extensive intertidal mudflats. Together these support a winter population of several thousand wildfowl, principally intertidal feeders such as Mallard, Wigeon and Shelduck, and offshore feeders including Goldeneye and Eider.

In recent years the counts along the entire estuary below the Erskine Bridge and up the length of the adjoining sea lochs have been fairly comprehensive. The pattern of wildfowl distribution is now fairly well established; the most important areas of the estuary are the stretches between the bridge and Langbask on the southern shore, and either side of Cardross on the northern side. Elsewhere much of the shoreline is dominated by the waterfronts of Greenock, Port Glasgow and Dumbarton, or interrupted by rocky sections. The RSPB has a 231ha reserve covering an area of mudflats on the southern shore, but there is considerable overall disturbance in the estuary, particularly from indiscriminate wildfowling. Plans for reclamation, marinas and further industrialisation have all been put forward in recent years.

The sea lochs contribute rather little to the total

Table 139. *Regular (and maximum) numbers of ducks and swans at three sites in central Renfrew.*

a) Castle Semple and Barr Lochs 8 seasons 1970-1980
b) Barcraigs Reservoir 2 seasons 1971-1972
c) Stanley Dam 4 seasons 1971-1976

	a	b	c
Mallard	328 (680)	177(241)	98(239)
Teal	111 (190)	6 (18)	0 (2)
Wigeon	252 (639)	29 (98)	0 (3)
Shoveler	22 (83)	1 (4)	0 (0)
Pochard	424(1700)	2 (11)	69(175)
Tufted Duck	328 (920)	39 (96)	57 (82)
Goldeneye	48 (98)	8 (11)	7 (12)
Mute Swan	17 (38)	0 (2)	1 (3)
Whooper Swan	25 (68)	2 (5)	41(117)

Table 140. *Regular (and maximum) numbers of ducks and swans at seven sites to the east and north of Glasgow.*

a) Lochend Loch 2 seasons 1972-1973
b) Woodend Loch 2 seasons 1972-1973
c) Hogganfield Loch 4 seasons 1973-1980
d) Gadloch 11 seasons 1970-1981
e) Possil Loch 2 seasons 1971-1972
f) Craigmaddie and Mugdock Reservoirs 6 seasons 1971-1976
g) Fannyside Loch 2 seasons 1981-1982

	a	b	c	d	e	f	g
Mallard	130(250)	48(203)	247(713)	232(600)	144(200)	351(907)	113(188)
Teal	0 (0)	1 (7)	2 (48)	70(190)	80(220)	0 (3)	41(120)
Wigeon	0 (2)	34 (66)	2 (17)	90(320)	25 (41)	27 (75)	0 (1)
Shoveler	0 (0)	0 (0)	0 (0)	9 (55)	35 (83)	3 (29)	0 (0)
Pochard	24 (65)	43 (70)	18 (67)	43(151)	3 (8)	12 (36)	29 (41)
Tufted Duck	111(167)	101(134)	73(155)	78(166)	57 (75)	122(604)	7 (23)
Goldeneye	0 (1)	2 (5)	9 (16)	1 (6)	1 (2)	8 (44)	9 (16)
Mute Swan	62(119)	7 (15)	11 (29)	8 (27)	2 (2)	0 (7)	0 (0)
Whooper Swan	17 (43)	7 (43)	0 (1)	20 (93)	11 (26)	0 (0)	0 (2)

numbers of birds, despite some exposed sand and mudflats at their heads (231). The exception is Holy Loch which is shallow and well protected. The counts for the four most important stretches of the inner Clyde are shown in Table 142, together with the figures for the whole estuary and for Holy Loch. The number of Eiders stands out, especially when taken with the totals for the outer firth, along the Ayr coast (see above). Some at least of the birds may be the same, as the highest counts in the estuary come one to two months after the peak in the outer firth, which is perhaps the major moulting area from which the birds move into the estuary. Shelducks and Mallard are also present in important numbers and heavily dependent upon the extensive mudflats (232).

The few freshwater lochs on the north or west side of the Clyde are of little interest to wildfowl, lying high up in the hills, or in steep-sided valleys. Loch Eck, the largest, typically holds less than 50 Mallard and Teal, plus some Goldeneyes and Mergansers.

The section of the Firth of Clyde lying between Dunoon and Cloch Point at the north end, and Toward Point and Skelmorlie at the south, is bounded by mainly low, sometimes rocky shores, with a road running along much of the coast. The only exceptions are Lunderston and Inverkip Bays on the east side and some small patches of exposed mud and rock on the west. The predominant wildfowl of this section are the Eiders, with counts of up to 780, together with up to 135 Goldeneyes and 130 Mergansers.

Fig. 40. The Firth of Clyde and Loch Fyne.

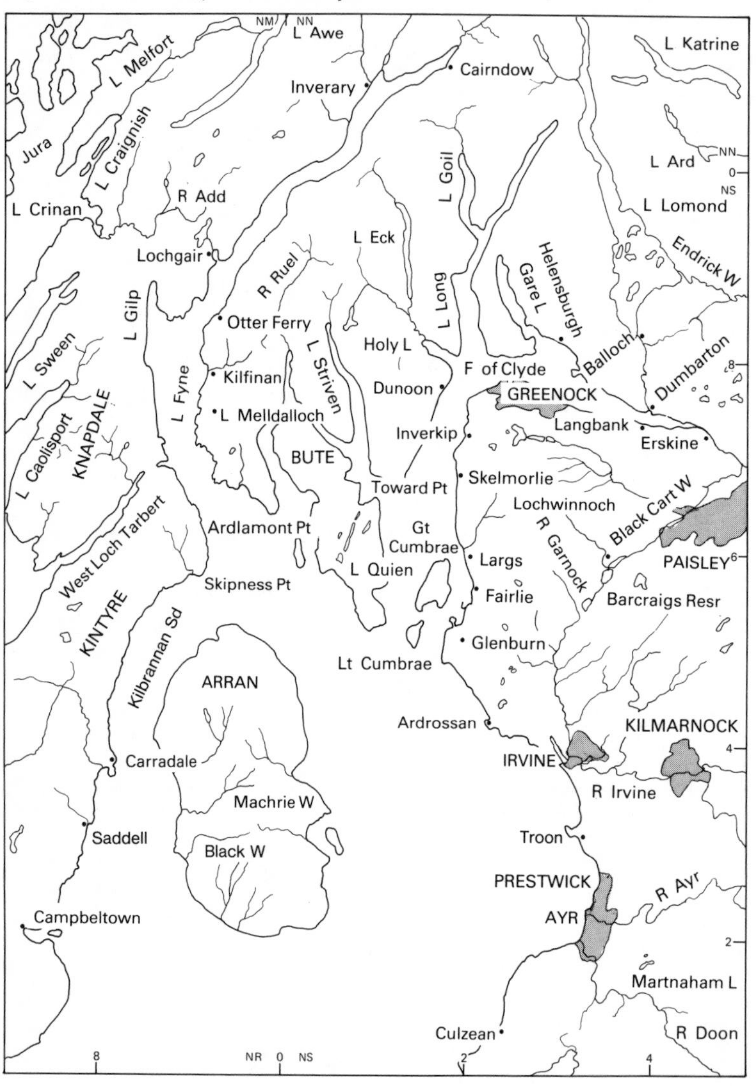

Bute, Loch Fyne, Arran and east Kintyre

The island of Bute contains a rich variety of wildfowl habitat, with several quite shallow and fertile lochs, set in good quality farmland, and a much indented and sheltered coast, some of which is rocky and inaccessible. The five principal lochs are counted in three groups, and these are set out in Table 143. It is quite probable that there is movement between the lochs, according to food availability and disturbance, so that the total populations are less than the sum of the counts.

The island has long been a regular wintering haunt for Greylag Geese. They increased from about 1,000 in the late 1940s to 6,000 in the early 1960s. At that time complaints of damage to crops led to successful efforts on the part of the main landowner to halt and then reverse the trend through a campaign of scaring and shooting. From an average early November count of 5,100 in the mid 1960s (maximum 6,300 in 1965) the level fell to an average 1,700 (2,620) in the early 1970s and to no more than 750 (1,370) in the late 1970s. In the last three years there has been some recovery to around 2,000. Counts at other times of the winter suggest that there are further influxes, often linked to cold weather on the mainland of Scotland further east, to a maximum of around 3,000.

A small flock of Greenland Whitefronts also winters on the island. Counts are incomplete but since 1967 they have been reported in most years, with maxima of 48 in 1969-70, 50 in 1973-74 and 70 in 1982-83. Up to 54 Bean Geese were also reported in several winters between 1968 and 1972.

The coastal areas of Bute are populated by good numbers of Eiders, with breeding colonies on small islands off the north coast. A survey round the entire coast in May 1982 revealed a total of 1,414 Eiders, not including ducklings (491). The largest numbers were round the north coast and in the south-west. Up to 300 Wigeon have been seen in Kilchattan Bay in the south-east and Scalpsie Bay on the west coast; the latter may be the same flock as occurs on Loch Quien, close by.

Loch Striven, running north from Bute, carries very few ducks, mainly Mergansers and a few Eiders scattered along its shores. Loch Riddon, and the 3km long mudflats of the Ruel Estuary at its head, is much more attractive and, as well as several hundred Eiders to be seen around its mouth, there are regular (and maximum) counts over four years in the early 1970s of

Table 141. *Regular (and maximum) numbers of ducks and swans at Endrick Mouth NNR, Loch Lomond, 1971-1979 (9 seasons).*

Mallard	345(667)	Goldeneye	16(37)
Teal	303(500)	R-b Merganser	2 (7)
Wigeon	502(647)	Goosander	1(14)
Shoveler	21(100)	Shelduck	5(20)
Pochard	13 (35)	Mute Swan	8(47)
Tufted Duck	25 (56)	Whooper Swan	27(56)

Table 142. *Regular (and maximum) numbers of ducks and swans in the inner Firth of Clyde and in Holy Loch.*

a) South shore—Woodhall to Erskine 5 seasons 1972-1982
b) North shore—Dumbarton to Cardross 5 seasons 1972-1982
c) North shore—Cardross to Helensburgh 5 seasons 1972-1982
d) Gare Loch 5 seasons 1972-1982
e) Inner Firth of Clyde—total, excluding Holy Loch 5 seasons 1972-1982
f) Holy Loch 5 seasons 1972-1981

	a	b	c	d	e	f
Mallard	362 (512)	132 (263)	19 (61)	106(143)	376(1400)	96(188)
Teal	7 (15)	8 (100)	0 (18)	0 (0)	15 (100)	20(127)
Wigeon	19 (67)	188 (289)	17 (205)	0 (0)	186 (291)	229(447)
Pintail	117 (167)	0 (2)	0 (2)	0 (0)	56 (167)	0 (2)
Pochard	70 (136)	0 (0)	0 (0)	1 (11)	22 (136)	0 (0)
Tufted Duck	128 (266)	0 (44)	0 (6)	0 (0)	20 (298)	0 (0)
Goldeneye	97 (136)	30 (46)	4 (130)	107(173)	287 (535)	19 (34)
Scaup	26 (201)	70 (105)	0 (141)	0 (0)	38 (306)	0 (0)
Eider	1218(3200)	15 (380)	1133(1500)	200(994)	2121(3600)	173(308)
R-b Merganser	3 (32)	10 (18)	0 (26)	9 (43)	43 (100)	17 (33)
Shelduck	428(1091)	1106(1500)	21 (237)	0 (1)	301(2330)	2 (12)
Mute Swan	11 (20)	46 (56)	3 (32)	3 (8)	29 (56)	15 (41)

66 (120) Mallard, 120 (277) Teal, 197 (264) Wigeon and 22 (71) Mergansers. The upper reaches of the estuary, including an area of saltings, are particularly important for the dabbling ducks (230).

The great length of Loch Fyne, in excess of 50km from its mouth at Ardlamont Point to its head at Cairndow, has a mainly low, rocky coast, broken by a few bays with intertidal mudflats. On the east side Kilfinan Bay holds up to 40 Eiders but is more important for the small flock of Greylags which is found here. The birds, numbering about 150 (maximum 320), feed around the head of the bay, sometimes roosting there but more usually flighting to Loch Melladoch, 5km south, where up to 50 Pochard have been counted. A similar number of Greylags is also found at times feeding on the Ardlamont Peninsula, and it is believed that these may be the same group of birds.

To the north of Kilfinan Bay the only other distinctive feature of the east side of Loch Fyne is the long spit of intertidal sand and shingle at Otter Ferry. Up to 20 Mallard, 30 Wigeon and 15 Goldeneyes have been seen here. Eiders are scattered along the whole length of Loch Fyne, a count of 1,022 between Otter Ferry and Inverary, about 30km to the north, in January 1969 giving some indication of how numerous they are. Mergansers, too, are to be seen frequently from the road running along the west side.

The bays at Inverary and Loch Gair on the west side hold few ducks, apart from a few Mallard, and some Eiders and Mergansers, but Loch Gilp is more important. The count figures include the birds in Loch Crinan on the west side of Knapdale, but it is thought that Crinan does not hold more than about 30 Mallard and 100 Wigeon. The overall counts include regular (and maximum) levels of 113 (470) Mallard, 170 (500) Wigeon, 8 (31) Goldeneyes, 35 (145) Mergansers and 15 (37) Mute Swans. South from there to the mouth of Loch Fyne at Skipness Point there are no counts, but Eiders and Mergansers are evident from the coast road, with small numbers also present in East Loch Tarbert.

The island of Arran is generally too mountainous and lacking in standing freshwater to be of much interest to wildfowl other than sea ducks. Up to 50 Mallard and 45 Wigeon have been counted in the Black Water Valley where there is also a regular wintering flock of Greylags. These, which at the time of the November censuses generally number between 150 and 250 (maximum 500), are also found feeding in the Machrie Water Valley to the north. There are very few counts from the coasts of Arran but those there are confirm the expected presence of small flocks of Eiders and Mergansers (up to 81 at Machrie in summer 1982 (540)), together with a few Mallard, Teal and Wigeon in the bays. Eiders breed sparsely; 37 pairs reared 80 young in 1982 (540).

The east side of the Mull of Kintyre across the Kilbrannan Sound from Arran has a low, rocky shore, with small bays at Carradale and Saddell, and the larger indentation of Campbeltown Loch. The two bays hold no more than a few Mallard and Wigeon, with Eiders scattered along the intervening coast. Campbeltown Loch has only been counted twice, in mid-January each time, when it held maxima of 24 Goldeneyes, 150 Eiders, 15 Mergansers and 14 Shelducks. The wildfowl of the broad Machrihanish Valley behind Campbeltown are included in the account of Argyll and the Inner Hebrides.

Table 143. *Regular (and maximum) numbers of ducks and swans on the lochs of Bute.*

a) Greenan Loch 4 seasons 1978-1982
b) Lochs Quien, Fad and Dhu 7 seasons 1970-1982
c) Loch Ascog 4 seasons 1978-1982

	a	b	c
Mallard	202(505)	548 (990)	444(600)
Teal	172(300)	364 (950)	106(300)
Wigeon	186(550)	467(1228)	573(750)
Shoveler	0 (0)	5 (30)	0 (0)
Pochard	5 (20)	43 (115)	0 (20)
Tufted Duck	10(100)	128 (230)	0 (25)
Goldeneye	16 (70)	36 (119)	75(100)
R-b Merganser	0 (2)	19 (40)	0 (0)
Goosander	0 (2)	8 (29)	0 (4)
Mute Swan	3 (7)	8 (16)	1 (6)
Whooper Swan	0 (0)	2 (9)	0 (2)

South-east Scotland: Forth and Tay

The First Edition drew attention to the importance of this area to geese and ducks, particularly the Pinkfoot, the Greylag, and the diving and sea ducks. However, it was suggested that some of the then very favourable conditions might turn out to be only transitory. For the flocks of ducks which depended almost solely on the outpourings of untreated effluent from the City of Edinburgh, and the mussel beds this encouraged, the statement has proved all too percipient. The numbers of geese and some other ducks remain high, however, still benefiting from the combination of large firths and estuaries, many inland lochs and reservoirs, and ample feeding opportunities in the shallow waters of the coast or inland on the fertile farmland.

Forth

The outer Firth of Forth

The coast north-westwards from St Abb's Head has little to offer wildfowl until one reaches the estuary of the Tyne, near Dunbar. Small numbers of Eiders and Shelducks breed along the coast, while a handful of counts at Dunglass and Thornton Loch have revealed maxima of 167 Eiders, 28 Goldeneyes and 4 Common Scoters. Up to 35 Mallard and the occasional Mute Swan complete the winter picture.

Fig. 41. The Borders and central Scotland. Total wildfowl and regional boundaries.

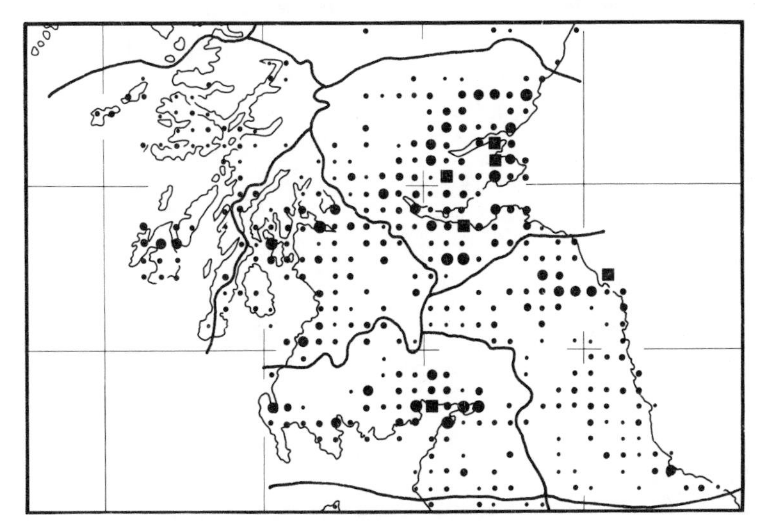

Tyninghame Estuary, with the associated sandy flats of Belhaven Bay and adjacent offshore reefs, supports a moderate wintering population of quite a wide variety of species (Table 144), most of which have shown no change in numbers since the early 1960s. However, the Pinkfeet, which were thought to be increasing at that time, have declined again, and only small numbers now appear even after the end of the shooting season. Up to 500 Eiders are present in the estuary or offshore in the early summer together with about 150 Shelducks. Both species breed freely along the coast.

The next stretch of the firth, round past North Berwick to Aberlady Bay, is also comparatively rocky, but here supports rather more breeding Eiders, with up to 200 pairs in the Tantallon – Scoughall area, as well as a very large moulting concentration in late summer. In addition to the mainland nesters, there are further breeding Eiders on the offshore island of Fidra, which had 200 nests in 1974-1976 (113). The moulting concentration regularly reaches at least 3,000 birds, with this number present from late June until early September. There was an apparent increase to a peak of 7,000 in September 1975, and 6,000 in the following two summers, but no more than 4,000 have been reported since (540). The Eiders are found mainly off Gullane, and round the point to Eyebroughty and Fidra. The fluctuations in numbers may be connected with the changes further west in the firth, round the Edinburgh sewage outfalls, though the flocks wintering there (see below) were never as great as the peak moulting totals. These moulting flocks may well include birds from further north. Gullane Bay holds several hundred Common Scoters through the summer, occasionally up to 2,000, and these presumably moult there, as does a flock of up to 200 Mergansers. Common Scoters have wintered in Gullane Bay, too, with a peak of 2,500 in October 1973, but since the mid 1970s fewer than 1,000 have remained. The decline may be linked in some way with the cut off of the Edinburgh sewage, though the Scoters only rarely

Fig. 42. The Firths of Forth and Tay.

ent that far west. Up to 300 Velvet Scoters and 160 ong-tailed Ducks used also to be found; these, too, ave declined recently.

Aberlady Bay, with its extensive areas of sand nd mud exposed at low tide, is the most important abbling duck and Shelduck haunt on the south side f the firth. It has been a Local Nature Reserve since 952; the control of wildfowling through a permit stem has meant that both the ducks and the roosting inkfeet (which are dealt with below in the context of e Lothians reservoirs) regard the area as secure. ecreational use of the area by birdwatchers, walkers nd golfers, particularly on the east side, has in-eased in recent years and may be causing some isturbance, especially at weekends. Up to 300 pairs of iders breed within the Nature Reserve, in the ssociated dunes or further inland. Some of the Eiders re nesting as much as 3km from the sea (286). In ddition there are 40-45 pairs of Shelducks.

West from Aberlady Bay, the coast becomes ore and more built up, and the wildfowl interest entres round the sea ducks. The only other occurr-nces in the 25km stretch to Muirhouse, in west dinburgh, are one or two flocks of up to 150 Mallard, nd a group of 10-40 Mute Swans between Mussel-urgh and Leith. The latter have declined from a gular level of 100-150 in the mid 1960s.

The status and decline of the once great flocks of caup, Goldeneyes and Eiders on the south side of the rth, particularly centred round the sewage outfalls om Edinburgh between Leith and Musselburgh, ave been thoroughly documented in published apers (112, 115, 353), as well as in a series of npublished reports. The birds were very largely dependent for food upon the discharges of raw sewage, untreated except for maceration. This provided not only a rich food source in its own right but also the nutrient to support enormous mussel beds, while the substrate held dense concentrations of worms and other invertebrates. The building of an efficient treatment works in the mid 1970s has had a dramatic effect on the Scaup and Goldeneyes, but less so on the Eiders, as can be seen in Table 145. A word of caution about the counts: there are obvious difficulties in assessing the size of such large flocks of ducks, coupled with natural winter hazards of poor visibility and rough seas. In the earlier years, particularly, there were some occasions when counts covered only part of the area; the possibility of duplication also exists. Nevertheless, the picture that emerges is of an increasing number of all three species using the area from the early 1960s, having previously increased through the 1950s. While the Scaup have virtually disappeared, a remnant population of Goldeneyes remains, and the Eiders have managed to sustain a flock little smaller than at the beginning of the period.

While these three species made up the great majority of the ducks in the area, a few hundred each of Common Scoters and Long-tailed Ducks, and up to 100 Velvet Scoters and Mergansers also occurred. Most of these have also disappeared since the building of the treatment works, as have the Pochard and Tufted Ducks which used to come here to feed from their roosts on the Edinburgh lochs (see p.241).

To the west of Edinburgh the south shore of the firth as far as the bridges bounds extensive areas of intertidal mud and sandflats on either side of the mouth of the River Almond. Dabbling ducks reappear, with a regular (and maximum) 570 (1,500) Mallard, 5 (64) Teal and 12 (180) Wigeon, as well as 40 (120) Shelducks. Goldeneyes are the most numerous

able 144. *Regular (and maximum) numbers of wildfowl* (a) Tyninghame Estuary and (b) Aberlady Bay in the othians. Counts available from Tyninghame for 23 seasons the period 1960-1982, and from Aberlady Bay for 17 asons in the period 1962-1982.*

	a	b
lallard	308(900)	211(1000)
'eal	103(300)	24 (104)
Vigeon	363(850)	320 (737)
;oldeneye	15 (57)	3 (20)
caup	1 (22)	3 (45)
ider	106(475)	51(1000)
helduck	65(137)	118 (230)
Vhooper Swan	3 (36)	15 (50)
lute Swan	44(104)	2 (9)

For details of goose numbers see Table 152.

Table 145. *Regular (and maximum) numbers of Scaup, Goldeneyes and Eiders on the south side of the Firth of Forth, between Musselburgh and Granton. 3-year periods up to 1980.*

Period	Scaup	Goldeneye	Eider
1960-62	2200 (6000)	505(1080)	incomplete
1963-65	7400(14000)	1430(2550)	1860(3050)
1966-68	12700(40000)	1470(3010)	1650(3250)
1969-71	10160(20000)	1910(3850)	2800(4710)
1972-74	13420(20820)	2750(4230)	5140(8450)
1975-77	3470(10280)	1560(2570)	2780(3660)
1978-80	555 (2250)	670(1225)	1270(1830)
1981	51 (114)	428 (468)	2300(2800)
1982	0 (0)	458 (643)	2100(2650)

of the diving ducks, with 84 (500), and a few Scaup, Eiders and Mergansers also occur. The Goldeneyes and Scaup have declined in recent years, suggesting their dependence upon the sewage outfalls further east.

From the bridges eastwards along the north shore of the firth, rocky stretches predominate, and the exposed tidal flats are generally much smaller than their counterparts on the southern shore. From Inverkeithing to Aberdour, the only wildfowl of note are the wintering Mergansers. There are few counts, but up to 100 seem regular, while 425 were counted in December 1977. It is possible that the birds are normally further out in the firth, around the many small islands and skerries, and only occasionally come close enough inshore to be counted. Up to 30 Shelducks and 80 Eiders also occur. No ducks seem to breed on Inchcolm, but there were about 30 pairs of Eiders, 8 of Mallard and 1 of Shelduck on Inchmickery in 1976, and 100 pairs of Eiders on Inchkeith in 1975 (113). In 1982 the following numbers of Eider nests were found: Craigleith 66, Fidra 85, Inchkeith 24 and Inchmickery 42 (540).

From Aberdour to Kinghorn there is little change in the shore or the wintering wildfowl. The sands east of Burntisland have rather limited invertebrates present and so support few birds. Mergansers (up to 200), and Eiders (up to 50) occur there, while limited counts in the late 1960s suggested that 3-400 Mallard were regular; however, these seem to have reduced to no more than 50 in recent years. There is considerable disturbance along the shore and in the bays.

The 35km stretch of coast from Kinghorn to Buckhaven lacks any extensive intertidal sands or much in the way of sheltered bays. Its wildfowl interest is limited to the outfall of a sewer off Kirkcaldy

Table 146. *Regular (and maximum) numbers of ducks in Largo Bay (Fife) in three four-year periods from 1970 to 1981.*

	1970-1973	1974-1977	1978-1981
Mallard	275 (400)	278 (440)	265 (630)
Wigeon	336 (622)	6 (18)	6 (7)
Pochard	631(1800)	157 (300)	262 (700)
Tufted Duck	328 (500)	276 (800)	280 (838)
Scaup	760(3265)	969(1600)	1978(2750)
Goldeneye	335 (650)	386 (500)	482 (783)
Eider	700(3265)	1417(3500)	1408(2748)
Long-tailed Duck	88 (300)	231 (349)	212 (539)
Common Scoter	550(1855)	483 (811)	319 (640)
Merganser	14 (50)	28 (50)	63 (171)

Harbour. Here a regular (and maximum) flock of 5((1,615) Goldeneyes winters, together with up to 1: Eiders and 10 Mergansers. Scattered records of up 850 Pochard and 400 Tufted Ducks perhaps refer birds from the south side of the firth. No more tha 200 of either species have been seen in the last fe years.

Beyond Buckhaven the coast swings east in great arc round Largo Bay. Exposed tidal flats ar limited but there are good populations of inverte brates, including mollusc beds, as well as a number sewage outfalls. Much of the bay is backed by dune and links. The numbers of ducks counted in the ba since 1970 are set out in Table 146, separated int four-year periods. During this time, which has see the collapse of the huge Scaup and Goldeneye flock on the opposite shore of the firth at Edinburgh, th numbers of Scaup here have shown a marke increase, presumably reflecting the movement of least a small part of the Edinburgh Scaup flocks to th new locality. The Goldeneye total has moved up onl slightly, however. Interestingly, the numbers Pochard and Tufted Ducks have not changed in th same way. The Pochard total had already droppe before the virtual disappearance of the Edinburg flock (p.241); the Tufted Duck numbers have remaine fairly stable. It is believed that these birds commut from Kilconquhar Loch, where similar changes hav taken place. Largo Bay is also important for Eiders; a well as the wintering flock, up to 500 are present spring and summer, suggesting perhaps as many a 100 breeding pairs in the vicinity. The Long-taile Duck and Common Scoter flocks feed especially o Methil Docks and Levenmouth. The Scoter flocks als include some Velvet Scoters, normally about 5(though an exceptional 550 were recorded in Marc 1972 and 200 in April 1982. The reason for th disappearance of the Wigeon is not known. Scattere earlier counts gave totals of up to 1,000 in the lat 1960s. The final stretch of coast in this section, t Fifeness, is nearly all rocky and lacks shallows. Sma numbers of Mallard can be found scattered along th shore, totalling no more than 200 in all. Simila numbers of Eiders also occur, with several pair breeding, particularly between Elie and Pittenweem There are also breeding Eiders on the Isle of May where nearly 300 nests have been counted. About 5 birds stay around the island in the winter. The othe wildfowl interest of the May is the regular passage Barnacle Geese seen flying over in the autumn Sightings of leg rings on birds which have lande show that these geese belong to the Spitsberge breeding population and are heading for their Solwa wintering grounds.

The inner Firth of Forth

The inner Firth of Forth, above the bridges, has en divided into seven main sections. The numbers the commonest ducks and the swans are set out in able 147.

On the south shore as far west as Bo'ness there e good muddy stretches, rich in invertebrates, posed at low tide. The most important ducks are allard, Wigeon and Mergansers. The Wigeon con- ntrate in Blackness Castle Bay and peak in the first alf of the winter. The next section west to Grange- outh Docks includes an extensive mudflat up to 2km ide, with a fair-sized river, the Avon, flowing across There has been considerable pollution over the ars from the industrialisation around Grange- outh, which includes an oil refinery and chemical orks. Nevertheless, there is abundant invertebrate una in some places, though little or none in areas fected by effluent. The Mallard flock is based around e entrance to the docks, while the Mergansers tend scatter along the northern edge of the Kinneil udflats. They are part, or sometimes all, of the inner orth wintering population which is normally to be und somewhere in mid-channel opposite Kinneil.

The Kinneil area is the most important centre for wintering Shelducks on the Forth and additionally holds a summer moulting flock of about 500 birds. The latter were first discovered in 1976 and make this one of only a handful of Shelduck moulting sites in the country (91). Some pairs breed in the neighbourhood, and upwards of 100 juveniles occur in late summer.

The Skinflats section of the inner Forth, from Grangemouth Docks to Kincardine Bridge, is mostly taken up by a large mudflat bounded on its inner edge by saltmarsh. The numbers of wildfowl using the site have declined in recent years (Table 147), the result perhaps of excessive shooting. Certainly the former roost of more than 1,000 Pinkfeet has virtually disappeared, except for a few in late winter after the end of the shooting season. The peak numbers of Shelducks here probably represent at least some of the birds from the Kinneil area. Skinflats is the most important site on the Forth for Pintail, the birds using particularly the area just west of Grangemouth Docks.

From Kincardine Bridge to Alloa, the mudflats on either side of the river become much narrower, but are still backed by saltings up to 200m wide. Kennet- pans, on the north shore, is the most favoured area within this section. Again it can be seen from Table 147

able 147. *Regular (and maximum) numbers of ducks and swans at seven sites in the inner Firth of Forth.*

) Queensferry to Bo'ness 8 seasons 1973-1982
) Bo'ness to Grangemouth 8 seasons 1969-1976
) Grangemouth to Kincardine Bridge (Skinflats) 7 seasons 1969-1975
) Kincardine Bridge to Alloa (Kennetpans) 7 seasons 1969-1977
) Alloa Inches (Tullibody) 8 seasons 1970-1979
Torry Bay 7 seasons 1972-1979
) Cultness 6 seasons 1970-1977

	a	b	c	d	e	f	g
Iallard	177(400)	400(1000)	325(1100)	163(365)	240 (700)	250(930)	27(139)
eal	46(350)	15 (128)	88 (275)	29(320)	419(1092)	30(173)	175(543)
Vigeon	351(700)	20 (250)	55 (190)	53(207)	2 (47)	214(498)	155(530)
intail	1 (6)	30 (140)	105 (290)	0 (1)	82 (236)	0 (3)	0 (0)
ochard	3 (45)	0 (0)	6 (42)	10(233)	21 (263)	0 (4)	0 (2)
ufted Duck	30(149)	0 (0)	1 (7)	25(638)	18 (400)	0 (0)	36(109)
oldeneye	18 (68)	0 (0)	11 (60)	234(748)	89 (500)	16(129)	2 (13)
Ierganser	86(239)	250(1500)	110 (280)	1 (8)	1 (5)	234(600)	16(220)
helduck	18 (47)	1000(2562)	857(1814)	40(150)	44 (160)	37(127)	53(327)
Iute Swan	0 (0)	0 (0)	1 (3)	2 (13)	35 (114)	3 (17)	0 (1)
Vhooper Swan	0 (0)	0 (0)	1 (10)	2 (10)	8 (50)	0 (0)	0 (0)

ote: The regular and maximum levels in the period 1960-1968 differed significantly from the above as follows:
rangemouth to Kincardine Bridge: Mallard 799(1646); Wigeon 178(482); Merganser 246(2000).
incardine Bridge to Alloa: Mallard 519(1500); Teal 121(640); Wigeon 380(1000); Pochard 45(615); Tufted Duck 20(2300); Goldeneye 274(760); Mute Swan 17(89).
lloa Inches: Teal 827(4330); Wigeon 52(253); Mute Swan 217(408); Whooper Swan 33(113).
ultness: Wigeon 262(1285); Merganser 6(43).

that the numbers of several species have declined in recent years and again disturbance seems to be the main cause. All the dabbling ducks are well below former levels, but exceptional numbers of Tufted Ducks still occur during cold weather, when the inland waters are frozen.

Two islands break the river at Alloa, the Alloa Inch, which is farmed, and the reed-covered Tullibody Inch. In between are soft mudflats with some invertebrates; again the area is heavily disturbed in the shooting season. The peak Mallard count occurs in August; the numbers then fall to no more than 100 in mid-winter before building up again in the spring, and there is a summering population of 2-600. The site used to be of international importance for Teal, but their numbers are now much reduced, partly by disturbance, and partly because the discharge of waste from the Cambus distillery was stopped in 1965. This also affected the numbers of Goldeneyes and Mute Swans. The decline of the latter has continued, and in the last five years no more than 10 have been counted. Pintail are very irregular, but the largest counts are invariably after the end of the shooting season. They concentrate in the vicinity of Tullibody Inch.

The north shore of the firth between the Kincardine and Forth Bridges has several areas of mudflats, the largest in Culross and Torry Bay. These have been greatly encroached upon by the construction in the early 1970s of large ash-settling pans for the nearby Longannet Power Station. Initially filled with sea water, they are now nearly full of ash. There are more mudflats in Ironmill Bay but very much smaller in extent. At the eastern end of this section there is a freshwater marsh, at Cultness. Tipping of waste has limited the natural area quite considerably in rece years, while encroaching reeds have reduced the op water. Wigeon in particular have decreased and bo Mallard and Teal have declined in the last few ye (Table 147). The Mergansers between Longannet a Cultness are part of the same population which counted at Kinneil, and it seems that the peak for t whole inner Forth is probably between 1,500 and 1,7 birds, 3-4% of the estimated north-west Europe population. Small numbers of Shelducks also bre along the north shore.

Inland waters of the Lothians

The natural inland waters of the Lothians a few and mainly small, but the increasing demand f water by the inhabitants and industries of Edinbur and its environs has led to the creation of about reservoirs, mainly in the Moorfoot and Pentland Hi to the south of the city, as well as in the Lammermu to the south-east. Many of these waters lie 2-300m more above sea level and so often freeze over winter. Some of them, however, are extreme important to roosting geese in the autumn and sprin and also hold good numbers of a wide variety ducks.

There is a cluster of lochs inside St Abb's Head which Coldingham Loch is the largest and the best f ducks, with a regular (and maximum) 95 (23 Mallard, 18 (77) Teal and 45 (80) Tufted Ducks. Milla Moss has held up to 400 Wigeon, but irregularly, whi Mire Loch holds a small population of diving specie including 45 (167) Tufted Ducks and 10 (29) Go deneyes.

Just to the north is Dow Law Dam, a far

Table 148. *Regular (and maximum) numbers of wildfowl at six sites in the City of Edinburgh. See also Table 149.*

a) Duddingston Loch 23 seasons 1960-1982
b) St Margaret's Loch 23 seasons 1960-1982
c) Lochend Loch 22 seasons 1960-1982
d) Dunsappie Loch 22 seasons 1960-1982
e) Figgate Pond 23 seasons 1960-1982
f) Blackford Pond 12 seasons 1960-1982

	a	b	c	d	e	f
Mallard	103 (255)	133(236)	181(330)	60(140)	103(230)	138(209)
Teal	14 (48)	0 (0)	0 (2)	0 (0)	0 (2)	0 (0)
Wigeon	1 (10)	3 (18)	1 (25)	8 (41)	2 (45)	0 (1)
Shoveler	7 (54)	0 (1)	0 (8)	0 (5)	0 (6)	0 (0)
Pochard	2800(8200)	15 (82)	34(550)	8 (24)	8 (87)	0 (3)
Tufted Duck	261 (800)	60(370)	79(280)	27(212)	62(470)	17 (45)
Greylag Goose	44 (240)	17(262)	0 (3)	6(165)	2 (65)	0 (0)
Mute Swan	2 (9)	3 (16)	3 (7)	2 (8)	4 (10)	2 (7)

Notes: Counts at Craiglockhart Pond in 3 seasons between 1968 and 1971 found 7(14) Mallard and 21(31) Tufted Ducks.
Counts at Inverleith Park Pond in 1960 revealed 43(45) Mallard and 3(4) Mute Swans.

impoundment which attracts rather more birds: 96 (400) Mallard, 80 (390) Teal and 46 (150) Wigeon. This is also a regular Greylag roost, with 2-300 in most recent years, and a peak of 570. They feed mainly within a few kilometres of the dam, to the south and south-east, and sometimes roost on Coldingham Loch.

A series of small lochs and reservoirs lies along the northern flank of the Lammermuirs. Stobshiel and Hopes Reservoirs are in steep-sided valleys at an altitude of about 300m. Stobshiel holds very small numbers of Mallard and Teal, and little else. Hopes Reservoir is a little better, with regular (and maximum) numbers of 21 (400) Mallard, but less than 10 of any other species on a regular basis. It has been used as an occasional Greylag roost, with 300 present in November 1976, but no more than 150 since at the time of the censuses. At a lower level are the Donolly, Thorters and Pressmennan Reservoirs, and the much smaller pools of Lammerloch, Quarryford, Bara and Danskine. The larger reservoirs can have up to 100 Mallard on them, together with a few Tufted Ducks and Pochard. Of the smaller waters, only Quarryford Pool is at all important, with 100 (310) Mallard, 61 (250) Teal and 4 (26) Tufted Ducks. In December 1978 there was an exceptional count of 1470 Teal, presumably an effect of the prevailing cold weather. The Pool is used by Greylags as an irregular roost, with a maximum of 400 reported.

Close to the north coast, a little inland from North Berwick, are the Balgone Ponds, ornamental lakes in sheltering woodland, with a regular flock of 100 Mallard (max 431), 5 (19) Teal, 5 (16) Pochard and 12 (39) Tufted Ducks. The valley of the Peffer Burn, a short way to the west, is the feeding place of a flock of Whooper Swans, regularly 30-40, occasionally rather more, and sometimes with a few Bewick's Swans. These probably roost at Aberlady Bay, and perhaps also at Tyninghame.

Within the city limits of Edinburgh are a number of small ponds of which the largest, Duddingston Loch (a statutory Bird Sanctuary), has until recently been outstanding for a day-time roosting flock of Pochard, easily the largest in Britain. The birds fed at night in the Firth of Forth and were, it became apparent, as dependent upon the sewage outfalls at Leith and Seafield as the sea ducks. With the completion of the sewage works the flock virtually disappeared, while the numbers of Tufted Ducks, though never as great, also dropped sharply. Table 148 sets out the regular and maximum counts of wildfowl at the various ponds for the full period 1960-1982, while the diving duck numbers at the four most important waters are shown in more detail in Table 149. Although flocks of diving ducks on the waters other than Duddingston were always quite small they, too, have declined since the commissioning of the sewage treatment works in 1978. Some of the birds visiting the various ponds may well be the same; this is certainly true of the feral flock of about 200 Greylags, which is based on Duddingston Loch, but often moves to the other ponds round Holyrood Park.

The reservoirs and other smaller waters to the south of Edinburgh divide into two groups, those predominantly in the Moorfoot Hills, and those in the Pentlands. The main sites in the Moorfoot range are the reservoirs at Rosebery, Edgelaw, Gladhouse and Portmore, which all lie below the north-western scarp of the hills. A little to the east there is the small natural pool at Fala, lying on a moorland plateau (where no recent counts have been made). Table 150 sets out the numbers of ducks found at these five sites. Glad-

Table 149. *Regular (and maximum) counts of Pochard and Tufted Ducks at four sites in the City of Edinburgh during 1981 and 1982, and seven preceding 3-year periods.*

	Duddingston Loch		St Margaret's Loch		Lochend Loch		Figgate Pond	
Period	Pochard	Tufted Duck	Pochard	Tufted Duck	Pochard	Tufted Duck	Pochard	Tufted Duck
1960-62	1240(2500)	386(800)	16(34)	90(370)	145(550)	73(230)	18(87)	130(235)
1963-65	2530(3500)	408(780)	15(35)	40 (64)	10 (37)	45(134)	13(20)	149(470)
1966-68	3790(8000)	256(302)	20(42)	57(138)	45 (90)	105(280)	6(12)	53(110)
1969-71	3510(6000)	242(450)	8(20)	48(195)	18 (40)	115(277)	3 (8)	31 (56)
1972-74	6410(8200)	305(450)	5(10)	107(163)	11 (18)	146(240)	5 (9)	53 (87)
1975-77	3730(7400)	301(400)	11(18)	59 (82)	9 (18)	87(195)	13(26)	45 (81)
1978-80	175 (350)	48 (70)	7(12)	21 (27)	5 (14)	15 (30)	3 (9)	15 (25)
1981	42 (60)	62 (67)	6(18)	28 (44)	5 (6)	25 (43)	0 (0)	17 (21)
1982	24 (34)	99(117)	2 (4)	42 (64)	4 (5)	20 (24)	0 (0)	17 (25)

house, which is the largest and most important of them, was declared a Local Nature Reserve in 1979, and although there is as yet no obvious sign of any benefit from this it has certainly saved the site from any damaging development. Gladhouse, Portmore and Fala are all major goose roosts, particularly for Pinkfeet, but the first two also hold Greylags (see Table 152). The largest numbers of geese occur in the autumn, but the flocks may build up again in the spring after rather lower winter populations. Gladhouse is the most important single goose roost in the region, though now being challenged by West Water Reservoir (constructed in 1966), on the southern slopes of the Pentland Hills (see below). Fala Flow acts as a relief roost for the geese, somewhere they can go and sit if disturbed from the feeding fields during the day, and where they can also spend the night if Gladhouse is disturbed. The Pinkfeet roosting at Aberlady Bay also belong to the same population as those in the Moorfoot Hills. The feeding areas include all suitable farmland between the hills and the coast.

Also in the general area of the Moorfoot Hills are the very small pools at Blackcastle, Rosebery (close to the reservoir) and Rosslynlee. Only the last named, a small reservoir, has consistently held a peak of more than 100 ducks, with regular (and maximum) counts of 14 (85) Mallard, 4 (30) Teal and 4 (82) Wigeon.

At the north end of the Pentlands, close to the southern outskirts of Edinburgh, lies a further group of six small reservoirs. Loganlea, deep in the hills, is virtually useless for wildfowl, while Bonally, Clubbiedean and Torduff are too heavily disturbed. Occasional counts have revealed no more than 30 Mallard and 30 Tufted Ducks on any of them. Harelaw, with 45 (126) Mallard, no doubt benefits from being just downstream from Threipmuir Reservoir, while Glencorse is the best of the six, with a regular (and maximum) level of 68 (151) Mallard, 2 (30) Teal and 17 (57) Tufted Ducks.

The numbers of ducks and swans occurring at the five most important reservoirs in the Pentland Hills are set out in Table 151. They all hold a good

Table 150. *Regular (and maximum) numbers of ducks on five waters in the Moorfoot Hills.*

a) Fala Flow 9 seasons 1960-1969
b) Rosebery Reservoir 13 seasons 1970-1982
c) Edgelaw Reservoir 13 seasons 1970-1982
d) Gladhouse Reservoir 13 seasons 1970-1982
e) Portmore Reservoir 13 seasons 1970-1982

	a	b	c	d	e
Mallard	156(1000)	174(442)	256(620)	337(1150)	117(250)
Teal	22 (100)	44(540)	17(150)	139 (425)	22 (75)
Wigeon	2 (12)	54(181)	43(200)	158 (305)	10(102)
Pochard	1 (7)	1 (18)	1 (14)	7 (30)	28 (79)
Tufted Duck	2 (16)	7 (47)	6 (18)	79 (205)	66(175)
Goldeneye	1 (5)	3 (11)	1 (4)	20 (55)	33 (75)
Goosander	0 (0)	1 (6)	0 (4)	15 (73)	4 (15)

Table 151. *Regular (and maximum) numbers of ducks and swans on five waters in the Pentland Hills.*

a) West Water Reservoir 12 seasons 1970-1982
b) Baddinsgill Reservoir 10 seasons 1970-1982
c) Cobbinshaw Reservoir 12 seasons 1970-1982
d) Harperrig Reservoir 13 seasons 1970-1982
e) Threipmuir Reservoir 13 seasons 1970-1982

	a	b	c	d	e
Mallard	259(1210)	83(255)	107(450)	164(800)	284(552)
Teal	86 (320)	13(137)	23(100)	38(200)	19(121)
Wigeon	11 (135)	1 (10)	18(100)	3 (50)	3 (38)
Pochard	1 (7)	0 (1)	21 (85)	1 (9)	13 (31)
Tufted Duck	20 (41)	3 (15)	98(200)	19 (64)	33 (93)
Goldeneye	5 (25)	1 (3)	20 (50)	2 (12)	7 (23)
Goosander	4 (13)	2 (10)	8 (50)	3 (18)	0 (5)
Mute Swan	0 (1)	0 (0)	2 (9)	0 (8)	1 (5)
Whooper Swan	1 (11)	1 (9)	8 (41)	2 (15)	9 (23)

ariety of birds, and the waters in the two natural roupings of Baddinsgill and West Water, and obbinshaw, Harperrig and Threipmuir, lie close nough together for birds to move from one to another avoid the disturbing effects of shooting. Typically l these reservoirs hold their peak numbers in the first alf of the winter and are often deserted later on, specially in cold weather.

These five resorts, together with Crosswood eservoir, are important goose roosts, particularly in ne case of West Water for Pinkfeet, while all of them old at least some Greylags (Table 152). Baddinsgill nd West Water have long acted as alternate roosts for e Pinkfeet, though West Water has become the vourite in recent years, and is now one of their most nportant sites in south-east Scotland. Often nearly eserted in mid-winter, there is renewed use in the pring, though by smaller numbers of geese.

Crosswood is used by some ducks, notably a gular (and maximum) 25 (241) Mallard and 45 (400) eal, while Morton Reservoir a little way to the north as carried 46 (160) Mallard but less than 20 of any ther species. This seems to be true of other small aters in the general area, such as Penicuik Park, larfield, Whim and Macbiehill Ponds, all south of enicuik. A little further south, the floods along the Lyne Water near West Linton have held up to 12 Mallard and 75 Teal.

The Forth Valley

In its upper reaches the Firth of Forth lies in a broad valley, bounded on the north by the steep escarpment of the Ochil Hills, and on the south by the foothills of the Pentlands. The valley continues beyond Stirling for a further 30km, gradually changing from rich arable and pasture farmland to the extensive peat areas and forestry plantations of Flanders Moss. A number of the larger lowland lakes and neighbouring hill reservoirs are of some interest for their wintering ducks, while substantial goose flocks and some Whooper Swans occur, particularly above Stirling.

On the south side of the upper firth are Linlithgow Loch, with Beecraigs and Lochcote Reservoirs just to the south and Loch Ellrig to the west (Table 153). Numbers have held up well at Linlithgow despite the building in recent years of the M9 motorway close to the northern shore, and increasing recreational use. The Mallard, in particular, probably feed on the estuarine flats and use the loch as a roost.

There is a scattering of smaller waters in this area, including Dundas Loch, the pools at New Liston,

able 152. *November counts of Pink-footed and Greylag Geese at their principal haunts on the Moorfoot and Pentland servoirs, together with Fala Flow and Aberlady Bay—five-year averages (and maxima) 1961-1980 and annual counts 981-1983. n.c. = no count.*

INKFEET	Aberlady	Fala Flow	Gladhouse	Portmore	West Water*	Baddinsgill	Cobbinshaw
961-65	703 (1380)	729(2430)	4197(5500)	25 (40)		2776 (5500)	81(270)
966-70	1740 (2750)	989(4000)	3460(7100)	3 (15)	2586(4030)	580 (1800)	14 (55)
971-75	2540 (3500)	553(2600)	3233(7420)	824(3570)	3760(5500)	286 (700)	102(370)
976-80	6246(11930)	95 (470)	5980(9600)	172 (420)	3286(7000)	2779(10000)	37(185)
981	4510	5000	3500	0	12340	0	200
982	5165	1068	1050	70	9240	0	0
983	3240	1700	11900	0	3600	0	2000

REYLAGS	Gladhouse	Portmore	Crosswood	Cobbinshaw	Harperrig	Threipmuir
961-65	320 (500)	360 (710)	n.c.	0	333 (900)	564(1050)
966-70	1100(2350)	286 (700)	n.c.	20 (40)	406(1800)	385 (640)
971-75	224 (680)	358(1000)	84(175)	95(200)	186 (350)	26 (110)
976-80	270 (450)	0 (0)	8 (40)	5 (22)	362 (500)	2 (11)
981	235	0	160	28	1950	0
982	40	80	27	25	675	0
983	625	0	0	1430	550	300

West Water constructed 1966.

Carmelhill, Auldcathie and Glendevon Farm, and Bangour and Cockleroy Reservoirs, as well as the Union Canal past Linlithgow. From the occasional counts carried out at these sites, the numbers of wildfowl are small, with none except Dundas Loch (250 Mallard) holding more than 80 Mallard and 50 Tufted Ducks regularly. Flocks of up to 30 Whooper Swans are reported in the area from time to time but there is no regular wintering centre.

Further west, in the vicinity of Falkirk, are some more waters of only limited importance, including Callendar Loch with regular (and maximum) counts of 110 (200) Mallard, and Carron Dam with 155 (300) Mallard, 35 (74) Teal and 79 (143) Pochard. Up to 22 Mute Swans also occur at Carron Dam. The nearby Drumbowie and Little Denny Reservoirs are little used by ducks, while to the south, just into Strathclyde, lie Black Loch, with 67 Mallard, 21 Teal and 200 Tufted Ducks recorded on a single January count, and Fannyside Loch, for which counts in 1981 and 1982 produced a regular (and maximum) 113 (188) Mallard, 41 (120) Teal and 29 (41) Pochard.

In the hills to the south-west of Stirling lie a number of sizeable reservoirs, of which the Carron Valley Reservoir (370ha) is easily the largest and the most important for wildfowl; Loch Coulter and North Third Reservoirs also carry useful numbers (Table 154). There has been a single January count from the Touch Muir Reservoirs in the same hills, when there were 300 Mallard, 90 Teal and 70 Pochard, suggesting that they have some potential value for wildfowl. There are no counts available from the equally remote Earlsburn and Buckieburn Reservoirs.

In the First Edition it was noted that 3-400 Greylag Geese had been found feeding on the south side of the Carron Valley Reservoir, but since then there have been no reports of more than 40. Five Be
Geese had also been seen there in January 1959, a
the observer commented that many more could ha
been overlooked in such broken country. The perci
ence of this remark was shown as recently as Februa
1981 with the discovery of a flock of 73 Bean Gee
feeding mainly on the overgrown southern sho
with records in the two subsequent winters of up
100. The area they favour is difficult of access, a
how long they have been occurring there, or inde
the persistence of the flock through each winter, a
unknown. With the virtual disappearance of t
Castle Douglas flock of Bean Geese (p.221), this s
may hold the only regular flock of this species
Scotland.

On the north side of the river, north and east
Stirling, three waters in particular hold useful conce
trations of several species of wildfowl. These a
Peppermill Dam, near Kincardine, Gartmorn Da
Alloa, and Airthrey Loch in the Stirling Universi
campus (Table 155). Gartmorn Dam, the most impo
tant of these, has shown recent increases in sever
species, especially among the dabbling ducks. I
status as a Country Park, complete with birdwatchi
hides and wardens, has undoubtedly contributed
this success.

The Whooper Swans occurring at Gartmo
Dam certainly include birds feeding in the Devo
Valley between Stirling and Alva (257), as well as so
which also roost mainly on the Forth at Tullibod
where there has also been a decline. The Devon Valle
flock peaked at around 130 in the early 1970s. Mo
Whooper Swans winter along the River Forth west
Stirling, with 30-50, occasionally up to 100, bei

Table 153. *Regular (and maximum) numbers of ducks and swans at four sites in the south-east of the Forth Valley.*

a) Linlithgow Loch 13 seasons 1970-1982
b) Beecraigs Reservoir 7 seasons 1970-1982
c) Lochcote Reservoir 3 seasons 1969-1978
d) Loch Ellrig 2 seasons 1980-1982

	a	b	c	d
Mallard	506(2540)	56 (61)	120(145)	40 (73)
Teal	0 (3)	1 (13)	29 (41)	86(200)
Wigeon	3 (34)	28 (55)	40 (50)	0 (0)
Pochard	179 (401)	45 (77)	51 (80)	3 (15)
Tufted Duck	299 (585)	221(600)	155(216)	27 (40)
Goldeneye	18 (43)	1 (10)	0 (0)	0 (0)
Mute Swan	11 (27)	1 (2)	2 (7)	0 (0)

Table 154. *Regular (and maximum) numbers of wildfowl a three reservoirs on the south side of the Forth Valley.*

a) Loch Coulter Reservoir 7 seasons 1970-1982
b) North Third Reservoir 6 seasons 1970-1982
c) Carron Valley Reservoir 6 seasons 1970-1982

	a	b	c
Mallard	88(500)	177(765)	216(483)
Teal	4 (28)	94(680)	242(744)
Wigeon	5 (26)	4 (40)	13 (73)
Pochard	0 (5)	1 (15)	16 (62)
Tufted Duck	18 (87)	32 (87)	24 (77)
Goldeneye	6 (15)	1 (6)	13 (75)
Goosander	2 (7)	3 (25)	3 (8)
Whooper Swan	0 (2)	1 (6)	5 (15)
Greylag Goose	9(180)	0 (0)	4 (39)
Bean Goose	0 (0)	0 (0)	6 (73)

und anywhere between the outskirts of Stirling and anders Moss, often in two or three separate groups. nese birds roost on the river and its shingle spits (85).

Small numbers of Greylags occur from time to me at both Gartmorn Dam (up to 150), and eppermill Dam (c.50), but neither site is a regular oost.

In the upper stretches of the Forth Valley there re five more waters holding both numbers and ariety of ducks: Lochs Watston and Mahaick, either de of Doune, the Lake of Menteith on the north side f Flanders Moss, and Lochs Venachar and Achray in ne adjacent Teith Valley west of Callendar (Table 56).

None of the sites has been counted very egularly. Loch Watston is quite small but well neltered, and is used as an occasional goose roost (see elow). Pinkfeet as well as Greylags are attracted nere, though more usually only as a day-time rinking and bathing place, the geese preferring to oost at Loch Mahaick. This loch, lying at over 200m bove sea level, not infrequently freezes in mid-vinter, so that it is most used by ducks and geese in ne autumn and early winter months. The decrease in ne numbers of ducks, especially dabbling species, at ne Lake of Menteith in the 1950s has continued, robably for the same reasons of increased recreation-l use and associated disturbance. Part of the low-ying area between Lochs Achray and Venachar was nade a WAGBI reserve in the late 1960s. A feral flock f over 200 Greylags has become established, while esting baskets and habitat management have been imed at increasing the general attractiveness of the rea for wildfowl.

The other waters in the upper Forth Valley are f limited value to wildfowl, being either rather small, like Loch Laggan and Muir Dam near Kippen, or large and inhospitable, such as Lochs Ard and Lubnaig.

Only small numbers of geese, either Pinkfeet or Greylags, are found in the valley below Stirling, formerly a stronghold, and then mainly in the spring. However, the numbers above Stirling are quite substantial (Table 157). As already mentioned, Loch Watston is only an occasional roost, mainly used by birds feeding in the immediate vicinity of the loch. Usually the birds feeding at this end of the valley, and as far east as Bridge of Allan, flight to Loch Mahaick up in the hills. On occasion, however, Mahaick is deserted and the geese, instead, roost in the middle of Flanders Moss. The Greylags flight either to the Lake of Menteith, or to Loch Rusky, though the latter is not used as much now as it used to be. There are rather few counts other than in November, but these suggest smaller numbers in the middle of winter followed by a spring peak as large as or even larger than the autumn one. For example over 4,000 Pinkfeet were recorded in the Flanders Moss area in March 1982, and 8,700 in March of 1983 and 1984. Variations between years, whether autumn or spring, are dictated as much by food availability on the farmland as by overall changes in population size (p.360).

Inland Fife and Kinross

On the north side of the Firth of Forth the region of Fife is mainly rolling arable land, though there are some industrial centres, especially in the south-west. The industry includes some mining which has produced several flooded subsidences as well as encouraging the building of reservoirs. There are thus a

'able 155. *Regular (and maximum) numbers of ducks and wans at three sites in the north-east Forth Valley.*

) Peppermill Dam 10 seasons 1970-1981
) Gartmorn Dam i) 8 seasons 1960-1969
ii) 12 seasons 1970-1982
) Airthrey Loch 9 seasons 1971-1982

	a	b i	b ii	c
Iallard	28 (84)	318(653)	781(1531)	132(367)
'eal	21(143)	34(120)	128 (380)	0 (4)
Vigeon	2 (78)	39 (72)	96 (592)	0 (0)
'ochard	17 (56)	87(245)	108 (493)	5 (30)
'ufted Duck	47(121)	337(734)	284 (807)	137(454)
;oldeneye	8 (20)	23 (49)	35 (81)	0 (5)
Iute Swan	5 (20)	17 (74)	10 (63)	6 (15)
Vhooper Swan	3 (26)	48(237)	7 (67)	0 (0)

Table 156. *Regular (and maximum) numbers of ducks and swans at four sites in the upper Forth Valley.*

a) Loch Watston 3 seasons 1970-1982
b) Loch Mahaick 2 seasons 1971-1972
c) Lake of Menteith 3 seasons 1978-1982
d) Lochs Venachar and Achray 2 seasons 1970-1979

	a	b	c	d
Mallard	101(210)	61 (96)	91(185)	182(280)
Teal	16 (66)	46(139)	6 (14)	81(170)
Wigeon	22 (75)	5 (15)	4 (13)	78(120)
Pochard	21(100)	4 (7)	16 (55)	28 (78)
Tufted Duck	26 (44)	12 (29)	75(160)	31 (83)
Goldeneye	3 (29)	1 (7)	29 (44)	35 (65)
R-b Merganser	3 (23)	0 (0)	0 (0)	27 (40)
Goosander	0 (2)	4 (9)	3 (14)	16 (28)
Mute Swan	9 (22)	2 (2)	2 (6)	10 (17)

considerable number of freshwaters in the region, many of them attractive to wildfowl, and none lying more than about 20km from the firth. Although the administrative boundary between the regions of Fife and Tayside puts the old county of Kinross into the latter, geographically it lies with Fife, being separated from the Earn Valley to the north and west by the Ochil Hills.

From a wildfowl standpoint the whole of Kinross and western Fife is dominated by Loch Leven, covering some 14sq km, yet with two-fifths of this area no more than 3m deep. A National Nature Reserve since 1964, the loch is renowned for its wildfowl, and is one of the original 13 sites designated by the United Kingdom under the Ramsar Convention. It holds the largest concentration of breeding ducks in the country, and is an extremely important wintering site for a wide variety of ducks, geese and swans.

The loch is the largest natural eutrophic lake in Britain. It is set in rich agricultural land, and the increasing use of nitrogen fertilisers in the surrounding area, coupled with sewage effluent from Kinross and Milnathort, have raised the pH of the water over recent years. This has resulted in severe algal blooms which have greatly reduced the submerged vegetation. The blooms were at their worst in the late 1960s and more recently have become less severe, allowing the return of some aquatic vegetation. The invertebrate fauna has undergone massive changes over the same period. Much of the information on the loch comes from a detailed study carried out as a project under the International Biological Programme, extending over six years from 1966. The study has been comprehensively written up as a volume in the *Proceedings of the Royal Society of Edinburgh* (7, 310, 356).

Regular winter wildfowl counts did not start at Loch Leven until 1966, but there is a complete seri available since then. The counts of the main species ducks, geese and swans are set out in Table 158. Th show that the loch is of international importance f the two grey geese, and of national importance f several of the duck species as well as Whooper Swa The changes in numbers over the years reflect well th known changes in the food supplies. Thus the gre drop in the Pochard numbers can be linked with th loss of the submerged vegetation. The Tufted Duc and Goldeneyes are more dependent upon invert brates, and although both declined for a time, th now also seem to be recovering. Goosanders ma have suffered because of a drop in fish stocks – Lo Leven is renowned for its Brown Trout (*Salmo trutt* fishing. Whooper Swans declined at a time when th aquatic vegetation was disappearing, but were al often found feeding out on the farmland. They ha not reappeared in more recent years, so there may b other reasons for their reduction. The winter numbe of Mute Swans have stayed fairly stable, but moulting flock of up to 500 non-breeders in la summer dwindled as the vegetation died off and no numbers less than 50. Ringing of these birds in the mi 1960s indicated that some were coming from as f away as Northumberland and Durham. More rece catching of the moulting flock in Montrose Basi Angus, suggests a similar origin, and it seen reasonable to suppose that the Basin has replace Loch Leven as the region's moulting site.

The breeding ducks of Loch Leven were th subject of detailed research in the period 1966-19 (380). During these years the estimated totals of pai were 500-600 Tufted Ducks, 400-450 Mallard, 25- Gadwall and Wigeon, 11-13 Shelducks and up to each of Shoveler and Teal. Recent figures (540) a

Table 157. *Numbers of Pink-footed and Greylag Geese counted during November censuses at haunts in the upper Forth Valley—five-year averages (and maxima) 1961-1980 and annual counts 1981-1983.*

Period	Loch Watston Pinkfeet	Loch Mahaick Pinkfeet	Loch Rusky, Lake of Menteith, and Flanders Moss	
			Pinkfeet	Greylags
1961-65		1710(1725)	350 (650)	280 (554)
1966-70			1480(2300)	620(1000)
1971-75		1690(2500)	1075(1750)	795(2195)
1976-80	1700(2020)	1130(2050)	580(1440)	680(1140)
1981	0	1100	2500	400
1982	0	0	5240	470
1983	0	1300	2425	229

ithin the same range, though there were 40 pairs of adwall in 1982. The great majority of the ducks nest St Serf's Island (42ha), near the eastern end of the ch, with just a handful of pairs on the other islands round the edge of the water. About one-third of St rf's is covered with *Phalaris* reeds and tussocks of *eschampsia* grass, and it is this area which provides e cover for the nesting ducks. Full information on e distribution, density and success of nests, and on verall breeding success has been published (380). ccess is generally poor, largely because the size of e loch leads to considerable wave action in windy nditions, with a consequent lack of food-rich and eltered rearing areas for the ducklings.

The loch forms a major arrival point for the ink-footed Geese migrating from Iceland, and in ost years the counts are highest in the early autumn, lling to a lower, though greatly fluctuating level for e rest of the winter. The surrounding farmland rovides much feeding for these and the Greylag eese, though variations in harvest success and crops rown also cause variations in the numbers of geese nd how long they stay. Further details on the geese, nd on other aspects of waterfowl biology at Loch even, will be found in the publications already entioned and in Allison *et al.* (8).

In the area surrounding Loch Leven, and articularly to the south, there are a number of smaller aters, many of which hold interesting numbers of a ariety of waterfowl. The counts at nine of these aters, the two Craigluscar Reservoirs, Lochs Fitty, low, Ore and Gelly, and the Lomond Reservoirs (Ballo, Harperleas and Holl) are set out in Table 159. Of these, Loch Ore, a mining subsidence, is probably the most important as well as having the greatest potential. It is now a Country Park with greater freedom from shooting than it has ever had. Although the recreational use has increased, the wide natural history interest is also taken into account. In addition to the species shown, Loch Ore has a small population of Gadwall. All the sites act as occasional roosts for Greylag Geese, usually no more than 250, and for smaller numbers of Pinkfeet. When weather or disturbance interferes with the normal roosting pattern of the geese at Loch Leven, the Lomond Reservoirs can sometimes hold 1,000 or more of both species.

Other waters in this area include Town Loch, Dunfermline, holding regular (and maximum) numbers of 22 (80) Pochard and 44 (112) Tufted Ducks. Much larger numbers of Pochard have occasionally

able 158. *Regular (and maximum) numbers of wildfowl counted at Loch Leven, Kinross, 1966-1982.*

	1966-69	1970-73	1974-77	1978	1979	1980	1981	1982
lallard	2258(2882)	1737(3400)	1272(2200)	1320(1600)	1800 (2726)	1854(2337)	2220 (3666)	1550 (2200)
eal	412 (865)	379(1200)	188 (450)	740(1000)	660 (328)	390 (669)	1144 (2063)	580 (862)
Vigeon	823(1272)	793(1200)	742(1000)	670 (825)	817 (1000)	1144(1612)	1047 (1159)	920 (1400)
intail	2 (6)	20 (55)	8 (25)	13 (20)	27 (79)	31 (63)	11 (28)	21 (40)
hoveler	73 (400)	300 (640)	225 (700)	62 (100)	150 (293)	168 (431)	233 (696)	36 (60)
;adwall	16 (38)	17 (30)	29 (70)	48 (80)	75 (162)	99 (208)	60 (175)	97 (169)
ochard	785(1353)	1135(2450)	548(1600)	301 (353)	350 (660)	290 (310)	502 (760)	1003 (1160)
ufted Duck	669(1933)	1541(4000)	828(1500)	1400(2000)	3175 (4500)	2250(4273)	2490 (4560)	2500 (3455)
;oldeneye	313 (727)	389 (750)	140 (400)	83 (155)	102 (130)	144 (164)	205 (281)	317 (350)
;oosander	26 (112)	48 (183)	56 (140)	25 (44)	36 (73)	21 (25)	33 (64)	25 (40)
helduck	7 (20)	7 (14)	11 (24)	4 (9)	11 (18)	11 (20)	8 (22)	8 (14)
inkfooted Goose	4041(8506)	5125(9000)	5755(7250)	6830(7000)	10500(13460)	7190(9780)	8520(12640)	10210(10640)
;reylag Goose	2536(4846)	2408(4700)	2808(4000)	2223(4200)	2670 (2900)	2900(5615)	2000 (3000)	2030 (2500)
lute Swan	16 (42)	10 (26)	19 (65)	25 (58)	34 (73)	30 (30)	56 (89)	62 (97)
Vhooper Swan	164 (358)	221 (428)	71 (242)	84 (120)	44 (90)	92 (150)	51 (81)	61 (140)

been seen here, up to 2,800 in 1967-68, at a time when there were large flocks on the Forth and on Duddingston Loch, Edinburgh. The lochs at Lumphinnans Farm, Ornie and Camilla, and the reservoir of Balado are all unimportant for wildfowl. Arnot Reservoir, just east of Loch Leven, is a little better, with 111 (250) Mallard, 51 (130) Wigeon and 27 (74) Tufted Ducks.

To the north of Loch Leven, in the Ochil Hills, and midway between the loch and the Earn Valley, is Glenfarg Reservoir. There are few duck counts, with a maximum of 239 Mallard and 22 Tufted Ducks, but the reservoir is a more or less regular Greylag roost, holding up to 1,500 in November, though more usually 250-750. These birds appear to be part of the Loch Leven population and to use the roost in preference to Loch Leven when they are feeding to the north of the loch.

Close to the Firth of Forth between the bridge and Kirkcaldy are a number of small pools an reservoirs, separated from the waters already discus sed by a low ridge of hills, and therefore more close related to the firth than to the larger inland waters Beveridge Pond, south of Kirkcaldy, with regular (an maximum) counts of 122 (229) Mallard, 31 (24 Pochard and 146 (330) Tufted Ducks, and Ottersto Loch, Aberdour, with 71 (130) Mallard, 86 (50 Pochard and 52 (140) Tufted Ducks, are the mos important of them. The other sites of Cullaloe an Stenhouse Reservoirs, Kinghorn and Raith Lochs, an Mill Dam, rarely carry double figures of more than on of these three species.

There are comparatively few waters in th

Table 159. *Regular (and maximum) numbers of ducks and swans at six sites in western Fife and Kinross.*

a) Craigluscar Reservoirs 6 seasons 1974-1981
b) Loch Fitty 13 seasons 1970-1982
c) Loch Ore 13 seasons 1970-1982
d) Loch Glow 7 seasons 1960-1966
e) Loch Gelly 7 seasons 1973-1982
f) Lomond Reservoirs 5 seasons 1970-1982

	a	b	c	d	e	f
Mallard	32(125)	44(350)	228(1030)	261(460)	13 (50)	198(615)
Teal	5 (67)	2 (34)	17 (131)	0 (0)	5 (9)	1 (6)
Wigeon	20 (68)	2 (54)	269 (913)	22(193)	0 (0)	44(143)
Pochard	12 (33)	66(300)	408 (801)	7 (60)	64(390)	42(358)
Tufted Duck	43 (66)	69(177)	256 (807)	15 (62)	134(351)	20 (38)
Goldeneye	2 (5)	6 (20)	6 (20)	3 (12)	11 (91)	2 (6)
Mute Swan	6 (15)	7 (32)	28 (120)	0 (0)	10 (23)	0 (2)
Whooper Swan	0 (2)	7 (25)	17 (74)	0 (6)	2 (21)	13 (97)

Note: The regular and maximum levels in the period 1960-1969 differ as follows from those shown above:
L Fitty (8 seasons): Mallard 39 (202); Wigeon 17 (114); Pochard 251 (650); Tufted Duck 114 (320); Goldeneye 16 (41).
L Ore (6 seasons): Mallard 16 (76); Teal 27 (81); Wigeon 14 (31); Pochard 113 (228); Tufted Duck 247 (405).
Lomond Reservoirs (5 seasons): Teal 20 (70); Wigeon 189 (650); Pochard 6 (24).
Infrequent counts at L Glow since 1969 give maxima of: Mallard 38; Tufted Duck 14; Goldeneye 11.

Table 160. *Regular (and maximum) numbers of ducks at six haunts in Fife.*

a) Carriston Reservoir 3 seasons 1970-1973
b) Clatto Reservoir 12 seasons 1971-1982
c) Lindores Loch 11 seasons 1972-1982
d) Kilconquhar Loch 6 seasons 1970-1981
e) Cameron Reservoir 13 seasons 1970-1982
f) Morton Lochs 11 seasons 1970-1982

	a	b	c	d	e	f
Mallard	287(450)	23(100)	209(476)	760(2650)	225(530)	324(1200)
Teal	14 (46)	0 (0)	22 (89)	68 (130)	52(209)	290 (950)
Wigeon	220(520)	23(110)	37(237)	22 (215)	60(384)	26 (254)
Pintail	0 (0)	0 (0)	0 (2)	1 (5)	3 (30)	6 (30)
Shoveler	0 (1)	0 (0)	0 (0)	29 (127)	10 (64)	2 (14)
Pochard	14 (31)	1 (10)	72(730)	300(1800)	52(167)	6 (22)
Tufted Duck	78(132)	22(120)	177(398)	344 (900)	125(754)	93 (230)
Goldeneye	12 (22)	0 (2)	25 (84)	116 (379)	31 (53)	1 (7)
Goosander	0 (4)	0 (0)	4 (46)	0 (1)	4 (26)	3 (41)

stern half of Fife, and the former floodplain of the len, running back from Guardbridge almost to Loch even, is now being well-drained and given over to tensive farming. To the north, the Firth of Tay is parated from the rest of Fife by a range of hills along s southern shore, and the only related freshwater site Lindores Loch, near Newburgh. This and five other aters in the region are the subject of Table 160. ilconquhar Loch formerly held occasional large ocks of Pochard, up to 2,000, but with the disappearance of the Edinburgh – Forth flock, the numbers have cently been well below 500. Both these and the ufted Ducks flight out regularly to Largo Bay. Several undred Greylags commonly roost here, and similar umbers are quite regular at Carriston Reservoir. Both tes can sometimes have 1-2,000 roosting for short eriods. Star Moss, near Carriston, is also used as an ternate roost by up to 1,200 Greylags and by regular ocks of 84 (280) Mallard and 30 (250) Teal.

Cameron Reservoir is very attractive to wildwl, though its recent opening up for some recreation may have an adverse effect. It is also a very important goose roost, especially for Pinkfeet (Table 161). They feed in the surrounding farmland, often close to the reservoir, but sometimes flight north to feed between the Eden and the Tay just west of Tentsmuir Forest. Their numbers have varied widely in recent years, depending on the amount of food available in the autumn, particularly on stubbles. Pinkfoot numbers drop off during the winter but rise to a spring peak of 2-3,000. On the northern edge of this area lie the Morton Lochs, artificial pools covering some 25ha, and a National Nature Reserve since 1952. After becoming very overgrown through the 1960s they were extensively dug out and reshaped, and are now beginning to mature again nicely.

Other waters in this area are mostly small and hold few ducks. Pitlour Pond, Strathmiglo, was counted in two seasons in the early 1970s, when it held 229 (325) Mallard and 41 (74) Teal, but there are no more recent records. Neither Coul Reservoir, near Glenrothes, nor Carnbee Reservoir, near Pittenweem, normally hold more than 100 ducks.

able 161. *Numbers of Pink-footed and Greylag Geese at ameron Reservoir, Fife, counted during annual November nsuses. 3-year means (and maxima) prior to 1980.*

ears	Pinkfeet	Greylags
971-73	4580(5000)	184(450)
974-76	2730(4000)	152(250)
977-79	5420(5750)	90(250)
980	3600	700
981	6150	2250
982	270	90
983	8000	1250

Tay

The Fife and Angus coast

The coast from Fife Ness to St Andrews is mostly low and rocky. Eiders are quite plentiful, a flock of Common Scoters is regular offshore, and Mallard, Teal and Wigeon all occur, particularly round the mouth of Kenly Burn (Table 162). Small numbers of Eiders and Shelducks breed. St Andrews Bay itself is notable for the flock of Common Scoters. Counts in the 1960s gave peaks of 4-5,000, a level which does not seem to have been reached in recent years, though

able 162. *Regular (and maximum) numbers of ducks on five stretches of the coast of Fife between Fife Ness and the irth of Tay.*

) Randerston to Boarhills — 7 seasons 1970-1977
) Kenly Burn to Buddo Rock — 4 seasons 1970-1974
St Andrews Bay — 2 seasons 1979-1980
d) Eden Estuary — 11 seasons 1971-1982
e) Eden Estuary to Tayport — 2 seasons 1981-1982

	a	b	c	d	e
Iallard	314(540)	242(450)	266 (930)	225 (800)	214(1100)
eal	1 (5)	32(108)	1 (4)	279 (750)	0 (0)
Vigeon	20(300)	11 (20)	2 (12)	1128(1742)	93 (257)
intail	0 (0)	0 (0)	2 (10)	26 (77)	0 (0)
oldeneye	4 (14)	4 (10)	6 (26)	24 (57)	2 (30)
ider	242(550)	142(250)	142 (340)	96 (427)	398(1969)
-tailed Duck	7 (25)	0 (2)	4 (50)	0 (4)	110 (600)
ommon Scoter	242(397)	2 (16)	283(3000)	0 (4)	792(1700)
-b Merganser	1 (5)	0 (4)	1 (8)	19 (78)	214(1000)
helduck	1 (10)	1 (4)	0 (2)	1321(1974)	3 (78)

there are obvious difficulties in counting ducks on the sea. The maximum generally seems to occur in the period November – January. Scattered counts at other times suggest that there is still a late summer moulting flock in the area though perhaps this, too, has declined from earlier peaks (540).

The Eden Estuary is heavily disturbed, both by aircraft from the adjacent Leuchars base and from wildfowling, though its declaration in 1977 as a Local Nature Reserve has had a beneficial effect. The numbers of dabbling ducks are holding their own, though well below the levels of the early 1960s when peaks of 4,300 Mallard and 2,100 Wigeon were recorded. Shelducks, on the other hand, have increased markedly, nearly doubling their numbers of 20 years ago (Table 162).

The stretch of coast north from the Eden Estuary is low and sandy, with extensive sandbanks offshore, particularly the Abertay Sands, off Tentsmuir Poi and in the bay between the Point and Tayport. Nea 500ha of the foreshore, including the Abertay Sand are within the Tentsmuir NNR. The principal wildfo interest of the area is the sea ducks (Table 16 including a flock of Common Scoters and go numbers of Eiders, Long-tailed Ducks and Re breasted Mergansers. Regular ground counts in t early 1980s have been backed up by less frequent aer surveys extending back to the mid 1970s. Tentsm Point and Abertay Sands also act as a roost for up 8,000 Pinkfeet which feed on the north side of the firt These birds may also use Buddon Ness for roostin

The Eiders are only a small part of a very lar wintering flock in the outer Firth of Tay. The best th could be said in the early 1960s was that there we thought to be many thousands present, but that r accurate counts were available. The first such cou

Table 163. *Numbers of Eiders counted in the outer Firth of Tay, 1968-1983.*

Date	Type of count	Number	Source
8 Oct 1968	Ground	10000	Pounder 1971
5 Jan 1969	Ground	10000	Pounder 1971
11 Oct 1969	Ground	16000	Pounder 1971
28 Oct 1970	Ground	15000	Pounder 1971
16 Nov 1970	Ground	8000	Birds of Estuaries Enquiry
13 Dec 1970	Ground	10000	Pounder 1971
Dec/Jan 1970/71	Ground	18000	Milne & Campbell 1973
2 Feb 1971	Ground	10000	Pounder 1971
Feb 1971	Aerial	8400	Milne & Campbell 1973
20 Feb 1971	Ground	12000	Birds of Estuaries Enquiry
6 Mar 1971	Ground	c.10000	Pounder 1971
12 Mar 1971	Ground	9000	Pounder 1971
13 Nov 1971	Ground	15000	Birds of Estuaries Enquiry
28 Feb 1977	Aerial	3120	Milne (352)
29 Nov 1977	Aerial	9360	Milne (352)
14 Dec 1977	Ground	11007	Campbell 1978
12 Jan 1978	Ground	10196	Campbell 1978
10 Mar 1978	Ground	1158	Campbell 1978
30 Nov 1978	Aerial	9015	Milne (352)
21 Oct 1979	Ground	15460	Wildfowl counts
20 Nov 1979	Ground	13900	Wildfowl counts
20 Dec 1979	Aerial	9500	Milne (352)
20 Dec 1979	Ground	10400	Wildfowl counts
14 Sep 1980	Ground	4700	Wildfowl counts
13 Oct 1980	Ground	13600	Wildfowl counts
30 Oct 1980	Aerial	11500	Milne (352)
3 Nov 1980	Ground	8000	Wildfowl counts
22 Jan 1981	Ground	12500	Wildfowl counts
27 Nov 1981	Aerial	8700	Milne (352)
4 Nov 1983	Ground	14100	Wildfowl counts

as in February 1962, and the peak population in the eriod 1955-1964 was estimated at around 10,000 birds 49). A series of ground-based counts was carried out om both sides of the firth in 1968-1971 (480), while oth ground and aerial surveys were done in winter 970-71 (352). The results of these, and all subsequent ounts, are set out in Table 163.

The 1968-1971 surveys suggest a peak population of up to 20,000; none of the counters thought that ther ground or aerial totals were likely to be omplete. Numbers were already high at the time of e October counts, and fell away in the spring after a id-winter peak. After a gap of several years, a series roughly annual aerial surveys began in 1976-77, and egular ground-based counts were also restarted. On e basis of the 1977-78 figures, it was suggested that ere had been a considerable decline in numbers nce the early 1970s (114). If this was so it proved to be nly temporary, because higher peaks were found in ubsequent years. Because of the difficulties involved surveying large flocks of ducks, it is not possible to etermine whether any changes have taken place in e last twenty years; in fact the numbers seem to have emained reasonably constant.

The north side of the outer Firth of Tay, from undee to Monifieth, is heavily built up, but not ithout wildfowl interest largely because of the resence of several sewage outfalls. The figures of able 164 show the wide range of species and the umbers of birds involved (481, 482). Buddon Ness, to e east, is heavily used as a firing and bombing range, ut despite this the sands at the point provide an alternative roost for the Pinkfeet which feed in the rolling farmland between Monikie and Letham, and east to Arbroath (see below). They also use the Abertay Sands on the south side of the firth, but little is known about their preferences, which may well be affected by prevailing weather conditions.

There are further counts available around Carnoustie (Table 164) but only scattered records north from there to Montrose. However, the wildfowl found at Carnoustie seem typical for the whole stretch, with small numbers of Mallard and Wigeon on the shore, and Eiders and Common Scoters on the sea. Aerial surveys have confirmed that neither of the sea ducks is especially common (352).

At the northern limit of the Angus coast lies Montrose Basin, the almost totally enclosed estuary of the River South Esk. The basin has always been badly disturbed, originally by military aircraft, then by excessive wildfowling; throughout the 1960s and 1970s it was unable to realise its potential. Then in 1981 a Local Nature Reserve was created, covering some 1,024ha, and taking in virtually the whole of the basin plus a small amount of land above high water, particularly at the western end. The Scottish Wildlife Trust appointed a warden, shooting was restricted to specified areas, and the numbers of wildfowl responded immediately (Table 164) and in the case of the Pinkfeet, dramatically. In the period 1961-1980, the annual November grey goose censuses revealed an average of only 200 Pinkfeet (max 950), and in several years there were none. This compares with the counts of 1,000 in November 1981, 6,130 in 1982, and 9,500 in

able 164. *Regular (and maximum) numbers of ducks and swans at four coastal haunts in Angus.*

) Dundee to Broughty Ferry — 11 seasons 1970-1980
) Broughty Ferry to Buddon Ness — 4 seasons 1970-1973
) Carnoustie to Westhaven — 3 seasons 1970-1972
d) Montrose Basin — i) 9 seasons 1970-1979; ii) 3 seasons 1980-1982

	a	b	c	d i	d ii
1allard	74(340)	33(150)	85(130)	132 (470)	242 (393)
eal	0 (1)	19 (23)	1 (4)	5 (47)	184 (555)
Vigeon	0 (13)	16 (27)	12 (59)	2062(4000)	3297(3860)
intail	0 (0)	0 (1)	0 (0)	21 (100)	69 (202)
ufted Duck	181(730)	38(420)	1 (3)	7 (19)	22 (40)
;oldeneye	121(460)	106(254)	8 (27)	14 (70)	26 (38)
caup	5 (80)	2 (10)	3 (12)	0 (3)	5 (19)
:ider	*	*	101(250)	507(1200)	1086(1700)
:ommon Scoter	0 (2)	13 (30)	26(100)	0 (100)	0 (0)
:-b Merganser	33(500)	30(130)	4 (11)	16 (200)	20 (46)
;oosander	38(450)	2 (18)	0 (1)	1 (20)	1 (3)
helduck	0 (0)	0 (3)	0 (4)	216 (480)	327 (446)
1ute Swan	14(110)	10 (20)	0 (1)	162 (229)	190 (245)

The Eiders in these two areas form part of the Outer Tay flock shown in Table 163.

1983. Furthermore, substantial numbers of geese have occurred in every month in recent winters. Greylags, too, have averaged over 1,000 each season from 1981 to 1983, compared with about 250 in the previous years. This species has, however, always been a more erratic visitor, and often uses an alternative roost at Dun's Dish, a short way inland. The basin has in recent years become an important moulting site for Mute Swans, with several hundred birds. This area seems to have replaced Loch Leven as a moulting ground for the region's swans.

The inner Firth of Tay, Strathearn and Strathallan

The Firth of Tay runs back inland some 30km from the bridges, dividing Fife from Angus and providing a link between the North Sea coast and inland Perth. Unfortunately there are few complete counts for the estuary and even these suggest that it falls well below its potential wildfowl carrying capacity, perhaps because of disturbance. The inner firth has been considered for some kind of conservation status for 15 years or more but so far nothing has come to fruition. The *Nature Conservation Review* rates it as a Grade I site (492).

The available counts suggest that Invergowrie Bay on the north side of the firth just west of Dundee is the single most important sector, with regular (and maximum) counts of 760 (1,400) Mallard, 44 (868) Tufted Ducks and 38 (140) Shelducks. Counts in the 1960s indicated higher totals of Mallard, up to 3,000, though still below the 3,700 maximum reported in the 1950s. It is not known whether this reduction is due to disturbance, some change in the food supply, or the reclamation that has been taking place on the Dunde[e] side of the bay. The Tufted Ducks occur only sporad[ically] ically in numbers greater than 100, and these additio[n]al birds have probably been temporarily displace[d] from the stretch between Dundee and Broughty Ferry.

Further upstream such counts as there ar[e] suggest mid-winter totals of another 500-1,000 Ma[l]lard, 250 Teal, 40 Wigeon, 70 Goldeneyes and 20 Shelducks. There was a paucity of information in th[e] early 1960s, although counts of over 1,500 Teal we[re] reported. The very extensive reed-beds running f[or] some 12km along the northern shore are still a[n] enigma, but it now seems unlikely that this cove[r] conceals large numbers of wildfowl, particularly wit[h] the current shooting pressure. There is undoubtedly [a] need, still apparent over 20 years after the Firs[t] Edition, for a thorough and detailed survey of the Firt[h] of Tay, its birds and the influences thereon.

In the 1950s there were also large numbers o[f] Pink-footed and Greylag Geese using the inner firth a[s] a roost. These have largely abandoned the area in th[e] last twenty years, preferring the greater security o[f] inland waters. Only occasionally, under conditions o[f] hard weather or some other extreme factor, do th[e] large flocks return, reaching a now rare total of 10,00[0] Pinkfeet and 2,000 Greylags. The former once fed o[n] the Carse of Gowrie but seem now to ignore it, eve[n] after the end of the shooting season. Up to 1,000 sti[ll] flight north-west over the Sidlaw Hills to the Wolfhill-Pitcur area, while others intermittently use th[e] Rhynd, between the Tay and the Earn. A few hundre[d] Greylags feed on the south side of the firth. Th[e] majority of the geese are now found on inland sites t[o] be described in other sections.

Counts from the Tay on either side of Perth ar[e] set out in Table 165. Downstream from the city, mainl[y] Tufted Ducks and Goldeneyes are found, presumabl[y] dependent on outfalls, as are those in the Lowe[r] Harbour. The stretch from Perth upstream to th[e] mouth of the River Almond contains a fair selection o[f] species, including a feral flock of Mandarin Ducks, descendants of those released in the 1960s, and sti[ll] probably dependent to a large extent on the nest boxes available in the release area just north of Perth.

To the west and south-west of Perth lie thre[e] major goose roosts, the Dupplin Lochs about 8km west of Perth, Drummond Pond near Crieff, and th[e] Curling Ponds, Carsebreck, south-west of Auchterar[der]. The first and last of these are used by substantia[l] numbers of both Pinkfeet and Greylags, while Drum[mond] Pond is solely a Greylag roost. The number[s] occurring at the time of the annual Novembe[r] censuses since 1961 are set out in Table 166. Betwee[n] them the three sites regularly hold over 10% of the tw[o]

Table 165. *Regular (and maximum) numbers of wildfowl on four stretches of the River Tay, either side of Perth.*

a) Luncarty to Almond Mouth Jan 1969 only
b) Almond Mouth to Perth 7 seasons 1972-1982
c) Perth Lower Harbour Jan 1982 only
d) Pye Road to Inchyra 3 seasons 1968-1970

	a	b	c	d
Mallard	97	285(516)	17	1 (2)
Teal	0	0 (2)	0	0 (0)
Wigeon	0	19(150)	0	0 (0)
Tufted Duck	0	38(169)	100	34(46)
Goldeneye	5	19 (38)	8	66(78)
Merganser	0	3 (11)	0	0 (0)
Goosander	0	7 (31)	4	1 (5)
Mandarin	0	29 (64)	0	0 (0)
Mute Swan	2	3 (13)	0	3 (5)

oose populations at this time of year and have held ubstantially more on occasions.

The two small lochs at Dupplin (together only 0ha) are set in a plantation which screens them from asual disturbance and, being on private land, access s strictly limited. Although some concern has been xpressed in recent years about the goose droppings ossibly having an adverse effect on the trout in the ochs through encouraging eutrophication, so far the wner has not only tolerated this massive roost, mong the largest gatherings of Pinkfeet in the ountry, but has positively encouraged them through areful protection. There was a peak count in November 1973 of 28,500, or approximately one-third of the ntire Iceland/Greenland population. The apparent eduction in the last few years is probably the result of series of clean harvests in the neighbourhood epriving the geese of the usual early autumn feeding n spilt grain, coupled with the upsurge in sowing of vinter barley which requires much earlier ploughing f the stubbles than do other crops. The main feeding reas are in the Earn Valley to the south and the Pow Vater Valley to the north. The numbers of geese ecline fairly rapidly as winter approaches but may uild up again in the spring, with recent mid-March ounts of 4-5,000. Greylags also roost here in some umbers, mainly feeding to the west between Dalreoch and Kinkell Bridges in Strathearn.

The Greylag flock roosting at Drummond Pond ncreased in size during the 1960s and early 1970s, to a naximum of over 13,000 in 1973 (Table 166). There ave been one or two high counts since, but in general he level has fallen somewhat, probably at least partly ue to farming changes. The geese feed in the rather roken country to the east of the pond, but also use reas of parkland to the south, and are often to be seen razing under the scattered trees. Occasionally, erhaps influenced by disturbance or bad weather, the Greylags will also roost on the nearby Loch of Balloch. There is also some movement between Drummond Pond and the Dupplin Lochs, and south to the Curling Ponds in Strathallan.

Although the Allan Water runs south-west to join the Forth near Stirling, the movements of the geese link the valley more closely with the Earn to the north, from which it is separated by a comparatively low watershed. Both Pinkfeet and Greylags roost on the two small Curling Ponds in considerable numbers (Table 166), often sharing the same loch, but generally feeding apart. The Pinkfeet flight principally on the south side of the valley, often up the slopes of the Ochil Hills to the limit of cultivation, while the Greylags stay in the valley bottom, extending south-west as far as Kinbuck. When disturbed by shooting at the lochs, both species move to the nearby moss, while the Pinkfeet occasionally roost on upland heather moor, or visit the Glendevon Reservoirs, high in the hills to the south-east.

Small numbers of Barnacle Geese are not infrequently seen among the Pinkfoot flocks around Dupplin and the Curling Ponds, usually scattered in singles or family parties, but up to 100 have been seen in a single flock. They are probably part of the Greenland population migrating with the Pinkfeet from Iceland, instead of following their own kind to their haunts off the west coast.

The principal haunts for ducks in Strathearn and Strathallan are Drummond Pond and the Curling Ponds, together with the Loch of Balloch. In addition, Bertha Loch, a little way to the north of Perth, fits in with this group. The numbers of wildfowl other than geese counted at these sites are shown in Table 167. The note to the table mentions an unusual concentration of Mallard and Teal at Drummond Pond in 1970 and 1971 which has not occurred since. There are no regular counts of ducks from the Dupplin Lochs but it

able 166. *Numbers of Pink-footed and Greylag Geese counted in November at three major haunts in Strathearn and trathallan, Perth. 5-year means (and maxima) prior to 1981.*

	Pinkfeet		Greylags		
Period	Dupplin Lochs	Curling Ponds Carsebreck	Dupplin Lochs	Curling Ponds Carsebreck	Drummond Pond
961-65	9580(14060)	1925(3000)	455 (780)	3975(5500)	3780 (6000)
966-70	13280(22000)	2550(3660)	1070(1925)	4010(8280)	8240(10242)
971-75	11730(28500)	3020(6270)	2380(3990)	4750(6500)	7890(13450)
976-80	10290(19550)	2390(5680)	790(1440)	2235(5050)	5660 (7000)
981	5000	3380	900	4050	3400
982	5010	4920	1840	2180	3945
983	5570	2840	1800	4310	4500

appears that some hundreds each of Mallard, Teal and Wigeon may occur there, together with smaller numbers of Pochard, Tufted Ducks and Goldeneyes.

The small lochs of White Moss and Keltie, in the Earn Valley near Dunning, hold some dabbling ducks, with more than 200 Mallard regularly at the former, and a maximum of 550 Mallard and 40 Teal at Keltie. Floods along the course of the Earn are quite frequent and attract up to 200 Mallard, 100 Teal and 500 Wigeon, particularly in the stretch around Dalreoch Bridge. There was formerly a regular wintering flock of Whooper Swans at Dalreoch, numbering up to 50, but these seem to have left the area, though about the same number, or a little less, can be found further downstream around Bridge of Earn.

The two Glendevon Reservoirs, lying at about 300m in the Ochil Hills, often freeze up in mid-winter. Both they and the nearby Glensherrup and Glenquey Reservoirs attract Mallard, with up to 700 on the Glendevons and about 50 on the other two, as well a small numbers of Tufted Ducks, mainly in the autum and spring.

As well as Drummond Pond and the Loch c Balloch, the Crieff area has five other waters fror which there are at least some counts of wildfowl. Loc Bonnybeg, just below Drummond Pond, has held u to 200 Mallard and 40 Teal, while Loch Monzievaird t the north is less attractive, with counts of only 3 Mallard and 4 Teal. The shores of this loch have bee developed in recent years, with holiday chalets an associated water recreation. Occasional counts at Loc Meallbrodden suggest it is not particularly used b wildfowl (max 132 Mallard, 4 Teal). Loch Buchanty i the Almond Valley is a little better, with maxima of 15 Mallard, 50 Teal and 200 Wigeon. Methven Loch downstream, held regular (and maximum) flocks of 6 (120) Mallard and 11 (34) Tufted Ducks in the 1960s but there are no more recent counts.

Loch Earn itself is too deep and steep-sided to b of much interest for wildfowl. The only availabl

Fig. 43. The upper Forth and Strath Tay.

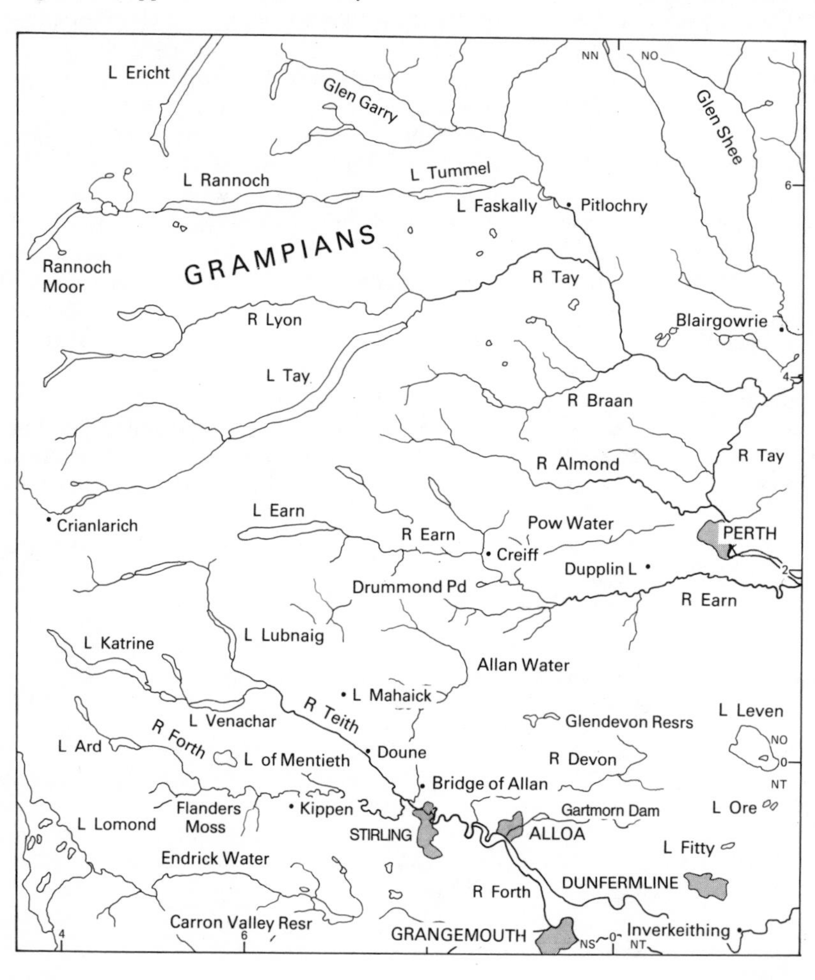

counts, from the St Fillans (east) end, are of no more than 140 Mallard, 90 Wigeon and 5 Tufted Duck.

Strath Tay

The River Tay, from its junction with the Almond north of Perth, northwards to its meeting with the Isla, is of little interest to wildfowl. The next stretch, however, where the Tay trends west – east beyond Meiklour is of considerable importance, especially for Greylag Geese, though also carrying some numbers of ducks. Here the river is broader, with exposed shingle banks and some low-lying often wet ground on either side. The lower reaches of the Isla, too, are used by wildfowl, while a few kilometres to the north of both rivers lie a number of important lochs, especially the groups of Monk Myre, Stormont and Hare Myre to the east, Marlee Loch and the Loch of Clunie west of Blairgowrie, and the Lochs of Lowes, Butterstone and Craiglush near Dunkeld. Most of these are used by roosting Greylags, and also hold many ducks. There are also some lochs on the south side of the Tay, of which King's Myre is the most important.

Table 167. *Regular (and maximum) numbers of wildfowl on four waters in south-western Perthshire.*

a) Bertha Loch 6 seasons 1974-1979
b) Drummond Pond 5 seasons 1978-1982
c) Loch of Balloch 5 seasons 1978-1982
d) Curling Ponds, Carsebreck 5 seasons 1970-1982

	a	b	c	d
Mallard	87(231)	220(1800)	27(80)	829(1830)
Teal	0 (15)	11(1370)	5(30)	62 (270)
Wigeon	22(107)	143 (290)	8(54)	348 (640)
Pochard	6 (18)	28 (74)	4(22)	62 (168)
Tufted Duck	128(295)	37 (79)	14(40)	117 (305)
Goldeneye	1 (11)	14 (37)	2 (6)	11 (38)
Goosander	0 (1)	22 (39)	1 (7)	4 (13)
Mute Swan	2 (7)	64 (136)	2 (4)	27 (68)

Note: In January 1970 there was a count of 1370 Teal at Drummond Pond, and in 1970-71 counts of 600 and 620. Since then the maximum count has been 37. Similarly the only two counts of Mallard over 1000 were in 1970-71. Since then the maximum has been 330.

The Greylags increased considerably during the 1960s, from the 3,000 reported in the 1950s, but have declined since (Table 168). The reasons for the change are not altogether clear, but include recent alterations in farming practice, particularly the change to winter barley; there has also been some increase in the amount of shooting and associated disturbance by the local landowners and farmers worried about possible agricultural damage. From the breakdown of the counts in the table it can be seen that the two main roosts are the shingle banks of the River Tay near Meikleour, and Stormont Loch, one of the least disturbed of the waters. There are considerable movements, even day to day, between the different roosts, in response to local conditions of feeding and disturbance. Although peak numbers of Greylags occur in the autumn, with some dispersal in early winter, some thousands of geese remain in the area throughout the winter, except in very hard weather, and into the spring. They feed on all suitable areas within a few kilometres of the various roosts, also flighting up the Isla Valley to Meigle or just beyond.

Pink-footed Geese were formerly present in the area around Wolfhill, south of Meikleour, with a regular 1-2,000 feeding there, and flighting over the Sidlaws to roost on the inner Tay. There have been very few records in the last ten years, but a smaller

Table 168. *Numbers of Greylag Geese roosting on the lochs and the River Tay in the Blairgowrie area, at the time of the annual November censuses. 5-year means (and maxima) prior to 1981. Totals based on synchronised counts.*

a) Loch of the Lowes, Butterstone Loch
b) Loch of Clunie
c) Marlee Loch
d) Stormont Loch
e) Monk Myre
f) River Tay, Kercock to Kinclaven
g) King's Myre, Old England Loch

Period	a	b	c	d	e	f	g	TOTAL
1961-65	65 (191)	1130(4000)	590(1600)	2710(4350)	1885(4380)	1430(3250)	0 (0)	7810(10000)
1966-70	0 (0)	3160(7750)	10 (63)	1350(3380)	1680(3100)	6120(7470)	220(900)	12540(15850)
1971-75	1410(2500)	495(1200)	360(1200)	865(3000)	1320(2500)	3340(4840)	530(900)	8320(10736)
1976-80	1600(5000)	330 (700)	275 (700)	1545(2800)	265(1100)	4170(6262)	270(800)	8455(12652)
1981	0	150	170	2500	0	805	0	3625
1982	0	0	0	4000	0	2175	0	6175
1983	0	820	0	50	410	3690	300	5270

flock, usually of less than 1,000, still feeds on the north side of the Sidlaws, around Pitcur, and roosts on the Tay, mostly to the west of Errol.

Also in the Meikleour area is a small but increasing population of Canada Geese. There have been a few breeding pairs in this part of Perthshire for 20 years or more, but at least some of the Meikleour birds appear to originate from the population from North Yorkshire, the non-breeders of which make a moult migration to the Beauly Firth near Inverness (p.277). The route of these birds can take them over this region, both on their way north in mid-summer and on their return in early autumn. Sightings of colour rings, and some recoveries, confirm that at least some birds ringed on the Beauly Firth have stopped off at Meikleour on their way south. Several of the earlier sightings of Canada Geese during the 1970s were in September, the main migration period. In recent years, however, the flock has built up to 2-300 in the autumn (540), and to rather fewer in the spring. Several pairs are now breeding in the area, so that further growth, independent of the migrating birds, seems very likely.

The principal duck haunts in the Blairgowrie district can be divided into still and running waters. In the enclosed waters, the counts from the eight most important sites are set out in Table 169. Stormont Loch, more eutrophic than the others and notable for Shoveler, as well as being one of the most important Greylag roosts, also carries the largest numbers and variety of ducks. This and the other two lochs in the eastern group, Hare Myre and Monk Myre, are shallower than the lochs further west. The Loch of the Lowes is a reserve of the Scottish Wildlife Trust, but being deep and tree-fringed is not used by large numbers of wildfowl.

There are also several smaller lochs in th vicinity, including Old England Loch, Lily Loch an Airntully Pool, on the south side of the Tay, and th White Loch, Rae Loch and Fingask Loch ne Blairgowrie. These normally hold totals of less tha 100 ducks each, though there is one exceptional recor of 720 Teal at Airntully in 1966. Similarly the waters i the hills to the north of Dunkeld, including Loch Ordie and Benachally, are mostly too deep an steep-sided to be attractive to wildfowl.

One stretch of the River Tay near Meikleour, an three of the River Isla towards Coupar Angus, hav been counted regularly and hold useful numbers o wildfowl (Table 170). The Wigeon find grazing on th river banks and adjacent wet areas, a habitat which i mostly absent around the lochs, and a few Whoope Swans are found regularly on the Isla.

A flock of 100 or more Whooper Swans wa regular in the Blairgowrie – Coupar Angus area in th 1960s, but following the deaths in 1968 of at least 4 from mercury poisoning, obtained from seed dre sings, the flock never recovered, and is now rarel more than 20 strong, often feeding up the Isla Valle towards Meigle.

Upstream from Dunkeld, the River Tay mear ders across a fairly narrow but flat valley in whic flocks of a few hundred Greylags are commor probably roosting on the shingle islands in the rive bed. Ducks are confined to small numbers of Mallard Goldeneyes and Mergansers. Counts from Loch Ta have revealed no more than 160 Mallard, 20 Wigeon and less than 10 of any other species, apart from up t 800 Greylags which may be regular in spring. Furthe up the valley beyond Killin are the two adjacent loch of Iubhair and Dochart. Despite their remote positio and the steep hills on either side, they both have quit

Table 169. *Regular (and maximum) numbers of ducks and swans on eight lochs near Blairgowrie.*

a) Loch of the Lowes 4 seasons 1970-1976
b) Butterstone Loch 4 seasons 1970-1976
c) Craiglush Loch 4 seasons 1970-1976
d) Loch of Clunie 11 seasons 1972-1982
e) Marlee Loch 11 seasons 1972-1982
f) Hare Myre 2 seasons 1972-1973
g) Stormont Loch 11 seasons 1972-1982
h) Monk Myre 7 seasons 1970-1976

	a	b	c	d	e	f	g	h
Mallard	41(200)	29(100)	9(40)	209(451)	74(231)	76(220)	151(835)	133(350)
Teal	0 (6)	1 (10)	1 (4)	5 (35)	1 (28)	33(142)	25(160)	31(100)
Wigeon	7 (80)	0 (30)	0 (0)	4 (38)	22(300)	0 (0)	52(356)	30(100)
Shoveler	0 (0)	0 (0)	0 (0)	0 (3)	2 (50)	0 (1)	13(100)	10 (70)
Pochard	4 (45)	2 (23)	2(10)	8 (33)	23(130)	2 (6)	15 (39)	6 (41)
Tufted Duck	14 (64)	10 (56)	4(22)	20 (58)	32(100)	12 (34)	26 (77)	24 (68)
Goldeneye	9 (37)	2 (20)	4(16)	4 (16)	6 (35)	0 (3)	4 (25)	0 (3)
Goosander	3 (18)	0 (17)	1 (6)	1 (5)	0 (6)	0 (0)	2 (18)	0 (6)
Mute Swan	2 (12)	2 (7)	0 (0)	1 (8)	3 (10)	0 (0)	2 (10)	2 (7)

productive shallow areas, and carry totals of up to 25 Whooper Swans, 45 Mallard, and 15 Wigeon, Pochard and Tufted Ducks.

Loch Faskally, formed by a hydroelectric dam on the River Tummel by Pitlochry, holds regular (and maximum) numbers of 78 (394) Mallard, 22 (41) Teal, 39 (55) Tufted Ducks and 28 (35) Goldeneyes, as well as smaller numbers of other species. A small flock of Greylags was introduced here in the 1960s, and has since increased to 220. Loch Tummel and the Dunalastair Hydro Reservoir further up the valley are neither of great value to wildfowl in winter, though up to 27 Whooper Swans occur regularly at Dunalastair. In summer, however, Dunalastair has a large and varied population, including breeding Wigeon, Teal and Tufted Ducks. Up to 20 Whoopers also occur at Loch Moraig near Blair Atholl, together with a hundred Mallard and 60 Pochard. Loch Rannoch has barren wave-swept beaches and holds no attraction for wildfowl.

The headwaters of the Tummel – Tay system, particularly Rannoch Moor and the Black Mount, were formerly an important breeding area for Wigeon but by the 1920s a substantial decline had already occurred. Surveys in the early 1970s suggested that the area is now virtually deserted, and that the population which used to breed here has moved to the lower-lying and more productive lochs and marshes some 50km east towards Glen Garry and Strath Tay (621).

Inland Angus

In the Sidlaw Hills, south-west of Coupar Angus, are a group of some nine lochs, mostly lying above the 200m contour. The best of these for wildfowl are Lochs Long, Lundie and Thriepley (Table 171). The others, including Lochindores, Laird's Loch and Ledcrieff Loch, rarely carry more than 100 wildfowl in total. Wigeon and Tufted Ducks have both decreased since the 1950s and 1960s. On the other hand, Greylag Geese, formerly of only irregular appearance, now use the lochs as a roost, feeding by day in the surrounding farmland. There are few counts before about 1970, but since then their numbers have risen from around 500 to between 1,500 and 2,300 at the time of the November censuses, though dropping to under 1,000 by the spring. Redmyre Loch, the southernmost of the group, has only been counted in 1967, when it held 150 Mallard, 45 Teal and 30 Wigeon. It is an occasional roost for both Greylags and Pinkfeet, the latter presumably those feeding around Pitcur to the north, stopping here instead of flying down to the Tay.

Further east, beyond Dundee, the Monikie Reservoirs hold useful numbers of Mallard and Tufted Ducks (Table 171), though the recent creation of a Country Park has opened the reservoirs and surrounds to recreation and there is much more disturbance now than formerly. The reservoirs also hold a small wintering flock of Greylag Geese (Table 172), though only since the late 1960s. Occasionally the large numbers of Pinkfeet feeding in the area also use the reservoirs as a roost. Nearby Crombie Reservoir holds fewer ducks (Table 171) but is used as an alternative roost by the Greylags and occasionally by the Pinkfeet.

The Pinkfeet feed mainly in the area bounded by Inverarity, Letham, Arbirlot and Monikie, a stretch of about 140sq km, consisting mainly of much open farmland with the large fields so favoured by this species. Apart from the occasional use of the Monikie

Table 170. *Regular (and maximum) numbers of ducks and swans on four stretches of the Rivers Tay and Isla near Meikleour and Coupar Angus, Perth.*

a) River Tay, Bloody Inches 2 seasons 1980-1982
b) River Isla, Bridge of Isla 4 seasons 1973-1976
c) River Isla, West Banchory 9 seasons 1973-1982
d) River Isla, Coupar Angus 9 seasons 1973-1982

	a	b	c	d
Mallard	45(138)	28(60)	66(300)	8 (46)
Teal	39(140)	1 (4)	3 (53)	0 (10)
Wigeon	138(400)	6(38)	57(340)	145(414)
Tufted Duck	0 (16)	12(20)	30(204)	37 (90)
Goldeneye	10(121)	6(13)	5 (18)	1 (5)
Goosander	2 (13)	0 (9)	2 (18)	1 (7)
Mute Swan	0 (0)	1 (5)	1 (7)	1 (6)

Table 171. *Regular (and maximum) numbers of ducks and swans at haunts in south Angus.*

a) Lochs Lundie, Long and Thriepley 3 seasons 1970-1972
b) Monikie Reservoirs 11 seasons 1971-1982
c) Crombie Reservoir 4 seasons 1971-1974

	a	b	c
Mallard	259(680)	378(1062)	233(650)
Teal	28(180)	2 (20)	2 (20)
Wigeon	143(260)	43 (120)	23 (70)
Pochard	6 (27)	37 (120)	0 (10)
Tufted Duck	107(155)	172 (392)	10 (39)
Goldeneye	14 (39)	16 (41)	0 (1)
Mute Swan	2 (3)	4 (10)	2 (4)
Whooper Swan	1 (4)	1 (10)	0 (0)

and Crombie Reservoirs, their main roost is on the Firth of Tay, either on the north bank around Buddon Ness, or on the other side on the sands off Tentsmuir Point. The counts in Table 172 show that 2-4,000 are regular, with higher peaks of up to about 10,000 on record. They are sometimes found outside their main feeding area, particularly along the shallow valley between Dundee and the Sidlaw Hills.

Immediately to the north of this Pinkfoot area are Rescobie and Balgavies Lochs (Table 173). The latter became a Scottish Wildlife Trust reserve in 1975. This has already had the effect of drawing birds from Rescobie, which is disturbed by boat fishing. Both lochs are used by geese for roosting, Rescobie by Pinkfeet which, perhaps surprisingly, seem completely separate from those feeding just to the south; instead they flight north to feed in the valley of the South Esk between Aberlemno and Brechin, or sometimes on the rolling countryside a little closer to the roost. Their numbers have been very variable over the last twenty years (Table 172). Small but increasing numbers of Greylags roost on Balgavies Loch, often feeding close by, but occasionally flighting a few kilometres to the north.

Just on the west side of the town of Forfar is the loch of the same name. Over the last twenty years it has gradually become increasingly disturbed by sailing and other forms of recreation, and by the

Table 172. *Numbers of Pink-footed and Greylag Geese counted at their major inland haunts in Angus in November censuses. (Five-year means 1961-1980 and annual counts thereafter.)*

a) Loch of Forfar and feeding area to west and south-west
b) Rescobie and Balgavies Lochs
c) Inverarity and Carmyllie—feeding area
d) Lochs of Lintrathen and Kinnordy
e) Monikie Reservoirs

	Pinkfoot			Greylag			
Period	a	b	c	a	b	d	e
1961-65	1575(2220)	75 (250)	2925(5000)	100 (500)	0 (0)	4085(6400)	0 (0)
1966-70	2250(3400)	50 (250)	1850(3555)	70 (280)	0 (0)	7650(9800)	30 (280)
1971-75	2110(3050)	1315(3150)	5900(8030)	800(1270)	140(500)	4950(7680)	60 (190)
1976-80	690(2270)	150 (600)	4680(7330)	135 (360)	165(480)	4480(7000)	400(1200)
1981	1190	0	2190	0	0	1070	0
1982	1460	290	2110	960	0	800	100
1983	0	36	3680	0	900	6200	300

Table 173. *Regular (and maximum) numbers of ducks and swans at six haunts in west and central Angus.*

a) Backwater Reservoir 13 seasons 1970-1982
b) Loch of Lintrathen 11 seasons 1970-1982
c) Loch of Kinnordy 9 seasons 1970-1981
d) Loch of Forfar 13 seasons 1970-1982
e) Rescobie Loch 13 seasons 1970-1982
f) Balgavies Loch 13 seasons 1970-1982

	a	b	c	d	e	f
Mallard	105(400)	1517(4500)	227(600)	388(720)	192(350)	133(450)
Teal	19(150)	104 (560)	69(300)	41 (75)	32 (65)	17 (55)
Wigeon	56(200)	241 (800)	13 (50)	168(478)	115(350)	21 (70)
Shoveler	0 (0)	2 (30)	9 (80)	31(160)	9 (40)	12 (55)
Pochard	17(100)	25 (150)	13(150)	79(160)	56(120)	22 (60)
Tufted Duck	23(100)	54 (240)	11 (60)	133(320)	80(230)	44(120)
Goldeneye	10 (50)	7 (20)	1 (13)	11 (20)	9 (18)	7 (20)
Goosander	10 (40)	4 (26)	0 (10)	0 (2)	2 (15)	1 (16)
Mute Swan	0 (0)	2 (15)	1 (7)	6 (24)	2 (12)	1 (6)
Whooper Swan	2 (20)	5 (67)	1 (15)	0 (3)	0 (4)	0 (0)

Note: Important differences in regular (and maximum) numbers in the period 1960-1969 are as follows:
Loch of Lintrathen (9 seasons): Mallard 2687 (5415); Teal 29 (95); Wigeon 938 (2450); Pochard 10 (76); Tufted Duck 209 (490)
Loch of Forfar (10 seasons): Wigeon 254 (1300).

encroachment of housing and industry. Nevertheless, it still attracts a variety of ducks including, notably for this area, a good number of Shoveler. The number of geese roosting on the loch has declined quite sharply (Table 172), though the position is obscured by the lack of records from the loch itself. There is a long series of counts of Pinkfeet feeding in the area between Forfar and Glamis to the south-west, and while some at least of these birds used to, and may still, roost on the Loch of Forfar it is thought that most fly down to the Tay, thus making them part of the Inverarity – Carmyllie group to the south-east. The occasional flocks of Pinkfeet coming into the Loch of Lintrathen to the north belong, presumably, to the same group. Greylags have always fluctuated in numbers at the Loch of Forfar, preferring the waters at Lintrathen and Kinnordy as roosts.

The Loch of Lintrathen is a large reservoir (c.120ha) with predominantly natural banks, well screened by plantations. A remarkable Mallard population built up here through the 1940s and 1950s giving regular (and maximum) levels of 1,620 (5,400) for the 11 seasons prior to 1960. This increase continued apace in the 1960s, reaching 2,685 (5,415) in that decade, but the level has since fallen to 1,517 (4,500) in the period 1970-1982 (Table 173). This last figure conceals an even more dramatic fall in recent years; comparing the period 1970-1975 with 1977-1982 (there being inadequate counts in 1976) the figures are 2,210 (4,500) and 830 (1,500) respectively. Other species, such as Wigeon and Tufted Duck, were also more numerous in the 1960s than in the 1970s but have not shown the recent sharp decline. The reason for the drop in Mallard, is far from clear, especially since the loch became a Scottish Wildlife Trust reserve in the early 1970s; perhaps there has been some fundamental change in the loch or its surrounds affecting the food supply. The loch became a Ramsar site in 1981.

The Loch of Kinnordy, not far to the east, is also a reserve, controlled by the RSPB since 1977. The loch is shallow and overgrown with weed and scrub. As well as wintering ducks (Table 173) small numbers of several species stay to breed, including Mallard, Teal, Gadwall, Tufted Duck and the Ruddy Duck in one of its most northerly outposts. Both Kinnordy and the Loch of Lintrathen are important Greylag roosts. The November census results for the two are added together in Table 172 as there is large-scale and frequent interchange between them, the birds disturbed at one roost at once setting off for the other. There has been some slight decline from a peak in the late 1960s, but there is also much fluctuation as the counts for 1981 to 1983 show. This is mainly due to availability of suitable stubbles at this time of year, both here and further north. The numbers in the spring are usually much lower.

In 1969 a large reservoir, the Backwater, was constructed a few kilometres upstream from Lintrathen. Lying in a steep-sided valley it lacks the attractive shallows of Lintrathen but is, nonetheless, used regularly by several of the common species (Table 173). In Glenisla, a little to the west, the Lochs of Shandra and Auchintaple hold small numbers of Mallard and Wigeon, and a few diving ducks.

North-east Scotland

The wildfowl wintering in the northern half of Scotland are isolated from the populations to the south by the massif of the central Highlands, and are mostly confined to the narrow lowland belt along the coast. Even there the populations of most species are separated into a series of independent groups by lengthy stretches of unsuitable terrain. Nevertheless, there are several areas of north-east Scotland in which the concentrations of wildfowl are of major international importance. The most striking occurrences are the gatherings of sea ducks along the Aberdeenshire coast and in the Moray Firth; the autumn and spring assemblies of grey geese in the Buchan and Easter Ross, and the concentrations of Whooper Swans on the Loch of Strathbeg and elsewhere. Wigeon are also abundant on the northern firths.

Deeside, Gordon and Buchan

The areas to the south of the Dee Valley are dominated by the foothills of the Grampian mountains, which reach in places almost to the sea. Lowland waters are scarce, and much of the coast from St Cyrus north to Aberdeen is steep and rocky, and of little consequence except at times for sea ducks. Small groups of Eiders breed at many points along the 50km stretch, and scattered parties totalling 4-500 are present regularly throughout the winter months. In early summer this resident population is augmented by the arrival off Gourdon of a moulting flock of about 2,000 birds; these remain in slowly dwindling numbers until September, and are gone completely by October, presumably to join the wintering flocks at the mouth of the Tay and in the Firth of Forth (349, 350, 352). There is also a summer gathering of Mergansers in St Cyrus Bay; 1,000 were recorded there in 1980, and 600 in 1981 (540). Other species are scarce, except for Mallard, which winter in smallish flocks at intervals along the coast; the recent aerial surveys suggest a total of about 400.

To the north of Aberdeen the cliffs give way to a long low sweep of dune and beach which stretches for

Fig. 44. Northern Scotland. Total wildfowl and regional boundaries.

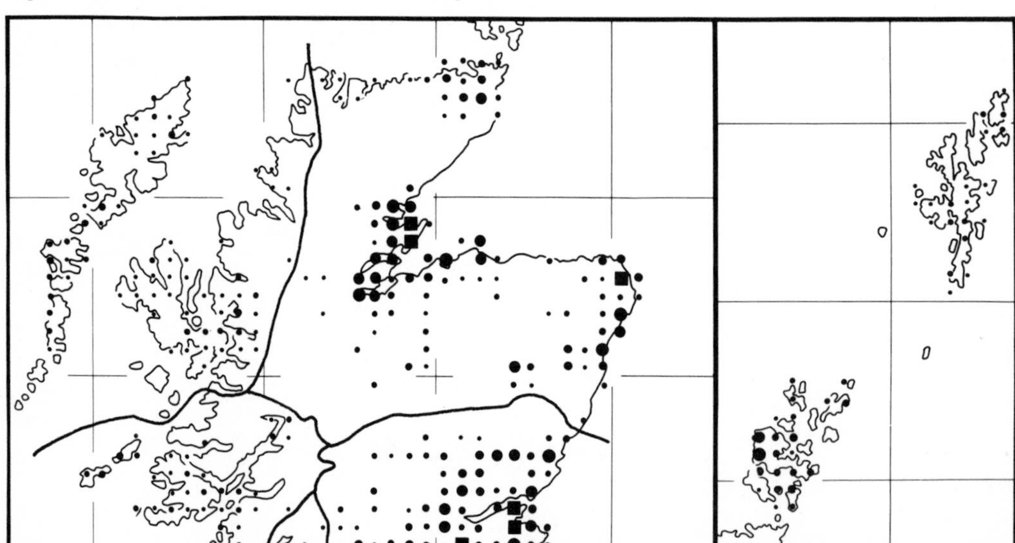

almost 20km to the outfall of the Ythan Estuary. Beyond this there are further stretches of cliff, extending for 10km to the links and beaches of Cruden Bay.

The most important centre on this part of the coast is on and around the Ythan Estuary, which runs inland for more than 6km, and contains two quite different types of habitat. On the lower reaches below the road bridge the channel is narrow and sandy with extensive mussel beds, and is bounded to the north by a tract of dune and heath, most of which is now included in the Sands of Forvie National Nature Reserve. Further upstream the estuary broadens into a shallow muddy basin, which attracts a fair number of ducks, and is also used as a roost by several thousand grey geese; for details see page 264.

The lower reaches of the estuary and the 700ha of dune and heath to the north are of special interest for the colony of Eiders, which is the largest in Britain. A ten year study of the population between 1960 and 1970 revealed an increase in the total numbers from about 3,000 in the early part of the period to 4,800 between 1964 and 1970. The estimated number of breeding pairs also increased from 1,200 in 1961-1963

Fig. 45. Deeside, Gordon and Buchan.

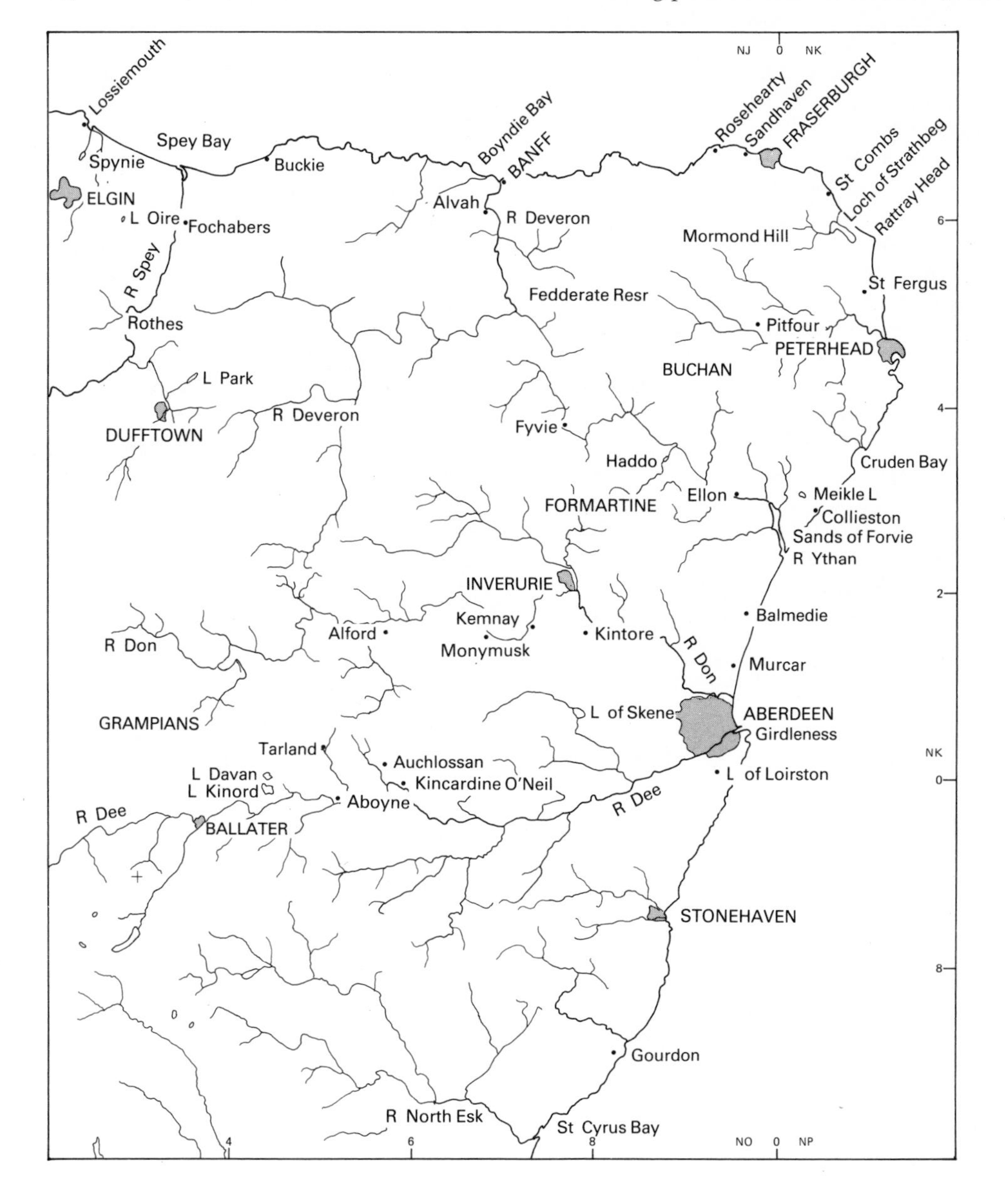

to 1,800 in 1965-1968, and to 2,000 in 1970 (350). The latest report, dated 1980, puts the total population at 6,000, but the breeding population is apparently unchanged: an estimated 1,900 pairs were present in 1976, 1,300 in 1977, 2,015 in 1978, 1,850 in 1979, 2,000 in 1981 and 2,100 in 1982 (540).

The fluctuation in the number of pairs is related to the rearing success in previous seasons. In most years the loss of ducklings between hatching and fledging amounts to 95% or more: this was so in seven of the ten years between 1960 and 1970. In the other three, the survival rate varied between 11.8 and 55.0%. The best season was in 1963 when 1,600 ducklings were reared. In the following year the total population showed a corresponding increase, and in the year after that, when the females reached maturity, the number of breeding pairs also showed a marked increase (350).

The large concentrations of Eiders are present around Forvie only during April and May: by mid-June most of the males have forgathered on the open sea, and these are followed a month later by the females, except those which remain to take care of the creches of young. On leaving the estuary the birds move southwards to a moulting centre off Murcar, midway between the Ythan mouth and Aberdeen, where they stay until September. The assembly here has increased greatly from 3,500 in 1962 (349) to around 10,000 in most recent years (540). Such numbers are far in excess of the resident population of 4,500-5,000 which winters along the length of the Gordon and Buchan coast (Tables 174 and 176). From the recoveries and resightings of birds marked at Forvie it seems that the bulk of the influx originates from the winter concentrations off the mouth of the Tay; there is also some evidence of a late summer movement southwards from the Moray Firth (33, 349).

Flocks of Velvet and Common Scoters also assemble on the sea off Murcar to undergo their summer moult. The latter usually totalled about 1,500 in the 1970s, but the peak was 2,200 in 1980 and 3,300 in 1981. They are present in strength from the end of June until early September, then decrease to a maximum winter level of about 400, spreading out in groups along the coast. The Velvet Scoters have a summer peak of up to 550, and a winter total of 15-30. Long-tailed Ducks are present regularly between October and April, and sizeable numbers of dabbling ducks and Goldeneyes occur on both the estuaries and the open coast (Table 175). The Goldeneyes feed at the sewage outfall at Girdle Ness and roost on the Don Estuary, where a maximum of 420 were counted in January 1982. Modifications now in progress to the sewage outfall may reduce the future value of this site.

To the north of Cruden Bay there are further flocks of Eiders along the shallow stretch of coast, which extends for more than 30km from Peterhead to the cliffs beyond Rosehearty (Table 176). As in other parts of north-east Scotland, the numbers wintering here have increased substantially in recent years, and now reach a level of around 3,000. In February 1971 an aerial survey of the coast from Banff to Collieston produced a total of 1,455 (353), compared with a count of 2,620 on a rather shorter stretch in February 1978. The main resort is at Rattray Head, where the first large flocks of about 1,000 began to appear in the early 1960s (353); the present level is upwards of 1,500 and there have been as many as 4,250.

Goldeneyes and Mergansers are present regularly at many points along the same stretch, and are sometimes joined by wandering groups of Long-tailed Ducks and Common Scoters. Dabbling ducks are scarce, except when the inland lochs are disturbed or closed by ice.

Apart from the fishing ports at Fraserburgh and Peterhead, and the North Sea gas terminal at St Fergus, which is probably harmless, there are no major coastal developments anywhere between Aberdeen and Banff. The only industrial hazard lies in the offshore traffic of tankers. The risks of pollution are

Table 174. *The average numbers of Eiders recorded between Girdle Ness and Cruden Bay in summer, autumn and winter, 1978-1982. A blank = no data.*

	August	Mid-Sept	Oct/Nov	Dec/Jan	Feb/March
Girdle Ness—Don		220	202	181	415
Don—Balmedie	8100	4409	196	319	317
Balmedie—Ythan		152	226	63	39
Ythan Estuary		1177	1213	930	641
Ythan—Cruden Bay		645	175	75	180
Total: Average		6603	2012	1568	1592
Maximum	9700	9646	2964	2153	1868

certainly no greater than elsewhere, though oiling incidents in 1982 and 1983 accounted for at least 700 and 1,000 birds respectively (see also p.491).

Away from the coast, the fertile lowlands of Aberdeenshire extend inland for upwards of 40km along the valleys of the Dee and the Don, and throughout Formartine and the Buchan. The farming is predominantly arable, with a high proportion of cereals, and the rolling terrain is scattered with lochs and ornamental waters. Most of these are used extensively by ducks, and often by geese and swans, and in general the populations of wildfowl are much less confined than in other parts of north-east Scotland.

Table 175. *The regular numbers of ducks, Girdle Ness—Cruden Bay, winter 1970-1982 (Ythan 1960-1982). The figures in brackets show the number of seasons for which adequate data are available. The totals below are based on synchronised counts, 1975-1977.*

		Mallard	Teal	Wigeon	Goldeneye	L-tail Duck	Velvet Scoter*	Common Scoter*	Merganser
Girdle Ness	(11)	2	27	0	68	0	0	0	2
Dee Est	(9)	109	0	0	27	0	0	0	0
Dee–Don	(4)	2	0	0	10	4	0	27	5
Don Est	(11)	119	77	1	40	1	0	0	0
Don–Balmedie	(5)	83	59	5	18	25	7	82	11
Balmedie–Ythan	(5)	55	21	4	20	28	13	196	22
Ythan Est	(15)	97	6	288	42	14	0	0	35
Ythan–Cruden Bay	(3)	710	7	138	18	123	1	187	22
Total:									
Regular	(3)	1088	223	452	238	208	27	404	85
Maxima	(3)	1505	357	1242	438	901	80	768	145

Note: The Ythan Estuary also has a regular population of 48 Shelduck (max 112), 52 Mute Swans (max 116) and 33 Whooper Swans (max 248). The Dee Estuary has 13 Goosanders (max 24).
* September records are omitted to avoid overlap with the moulting concentrations.

Table 176. *Peterhead Bay—Rosehearty, 1977-1982 (mostly 1977 and 1978); means of the three highest counts, each being in a different month. The figures in brackets show the number of different months in which counts were made.*

		Mallard	Teal	Wigeon	Goldeneye	L-tail Duck	Common Scoter	Eider	Merganser
Peterhead Bay	(3)	0	0	0	91	0	0	131	4
Peterhead–Rattray	(5)	125	18	124	41	121	27	1327	20
Rattray–St Combs	(5)	191	3	6	38	99	4	1182	30
Fraserburgh Bay	(4)	17	1	20	5	18	0	49	9
Fraserburgh–Sandhaven	(6)	33	69	15	38	0	0	1620	9
Sandhaven–Rosehearty	(5)	12	1	2	6	0	0	305	10
Highest monthly total		894	107	200	208	423	75	3383	72

Note: If large numbers were recorded in the same month of both 1977 and 1978, only the higher of the two is included in the mean.

One of the largest concentrations is located on the Loch of Strathbeg, which lies within 800m of the sea, just west of Rattray Head. The loch, now managed as a reserve by the RSPB, has an area of 200ha, and a mean depth of only 1-1.5m. On the seaward side it is flanked by a ridge of calcareous dunes, and the slightly brackish water is also rich in lime. The submerged vegetation includes an abundance of wildfowl food plants, but the shores are mostly bare and wave-washed except at the north-west end, where sizeable stands of *Phragmites* provide the only shelter. Immediately to the south the site of a war-time airfield is now occupied by an array of wireless masts, the tallest of which is 300m high. Contrary to expectations, this has not proved a hazard to birds in flight, but has provided a buffer zone to reduce disturbance; it is certainly much less objectionable than the treatment plant for natural gas which might otherwise have been sited here, instead of at St Fergus.

Table 177. *November counts of Greylags and Pink-footed Geese on the Loch of Strathbeg and on the Ythan Estuary (including Meikle Loch) 1980-1983, with corresponding means for earlier five-year periods. The figures in brackets are the maxima for each period. The counts of Greylags elsewhere in east Grampian are included for comparison (for details see Table 179). The figures are in thousands.*

GREYLAG November	Strathbeg	Ythan	Elsewhere
Mean 1955-1959	1.1 (1.7)	0.2 (0.4)	No data
Mean 1960-1964	1.8 (3.7)	0.7 (1.2)	0.4 (0.7)
Mean 1965-1969	2.5 (4.4)	3.4 (9.4)	0.5 (1.0)
Mean 1970-1974	3.4 (4.5)	3.4 (5.3)	1.7 (3.6)
Mean 1975-1979	7.0 (9.5)	1.1 (1.8)	4.0 (5.7)
1980	4.0	0.5	9.3
1981	4.8	0.7	9.7(−Haddo)
1982	9.6	0.5	16.3(−Haddo)
1983	3.6	0.7	14.3(+Haddo)

PINK-FOOTED GEESE November	Strathbeg	Ythan	Combined
Mean 1955-1959	0.8 (1.4)	0.4 (2.0)	1.3 (2.6)
Mean 1960-1964	1.3 (2.2)	1.2 (3.3)	2.5 (5.5)
Mean 1965-1969	2.3 (4.6)	10.6 (15.0)	12.9 (15.1)
Mean 1970-1974	2.2 (5.5)	4.8 (11.9)	6.8 (13.7)
Mean 1975-1979	5.6 (7.5)	6.2 (10.1)	11.8 (16.6)
1980	2.2	4.4	6.6
1981	5.9	7.3	13.2
1982	6.2	6.5	12.7
1983	2.9	10.7	13.6

The wildfowl population at Strathbeg is of major international importance, amounting at the peak to around 5,000 ducks, 660 swans and 11,300 geese. The diversity of species is also unusually great, reflecting the value of the loch as a staging post for the Scandinavian and Icelandic migrants which make their landfall in this easternmost corner of the Scottish mainland. No fewer than 33 species and subspecies of wildfowl have been recorded here in recent years, including 20 species of ducks and all three species of swans.

The loch is primarily important as a roost for Greylags and Pink-footed Geese, both of which have increased remarkably over the past 20 years, not only here but in other parts of Aberdeenshire. The Greylags, in particular, have extended their range and are now wintering regularly at five main centres and several lesser resorts throughout the east Grampian lowlands. Prior to 1939 they were limited to a few small flocks (50), and even in the early 1960s the total numbers in the county seldom amounted to more than 2,500. The first big increase to about 5,000 came in 1964, and this was followed in 1969 by a record jump to nearly 15,000. Since 1973 the autumn censuses have produced a fairly constant level of 10-13,000.

The distribution of the Aberdeenshire Greylags varies considerably from year to year, depending on the local food supplies and on the degree of disturbance at the individual resorts. The extent of these variations is apparent from the results of the November censuses set out in Table 177. The most striking features have been the apparent sudden decrease in the flocks in the Ythan Valley, dating from 1974, the overwhelming importance of the Loch of Strathbeg between 1975 and 1978, and the subsequent dispersal of some of the geese from there to other resorts in Aberdeenshire, notably on Deeside. This process of expansion is further reflected in the temporary appearance of large numbers in places where none had been reported previously. Amongst the several examples are the records of 1,100 on Fedderate Reservoir, near New Deer in November 1977, and of 3,000 on the loch at Fyvie in November 1979. In each case the flocks seem to have been exploring, unsuccessfully, for suitable roosts from which to exploit the untapped feeding grounds nearby. A major roost was discovered at Haddo House Lake between Ellon and Methlick in 1981 but may have held Greylags for some years previously. There were as many as 11,900 in December 1982, but the 1983 November census revealed only 4,600. Despite these recent signs of dispersal, the Greylags at Strathbeg are still of prime importance, especially in autumn when the flocks comprise at least 5% of the British – Icelandic population. Later on in the winter the numbers decrease to a level of 2-4,000, which continues well into March.

The Pink-footed Geese, unlike the Greylags, have failed to establish new centres, and are still largely confined to their two traditional resorts, the one on Strathbeg and the other in the Ythan Valley, some 30km to the south. During the course of the winter there is interchange between the flocks. This relationship is emphasised by the scarcity of the species elsewhere in the eastern parts of Grampian: except for a flock of 2-300 near Kemnay, the nearest concentrations are more than 80km away in Angus and Elgin.

The Pinkfeet in the Ythan Valley (and also the Greylags) roost either on the upper estuary, or more often on the Meikle Loch, which is only 3km away and is rather more secluded. The feeding grounds are widely spread, extending southwards to Balmedie and westwards to Ellon and beyond. The flocks at Strathbeg also forage over quite a wide area, ranging at times as much as 10km inland to the flanks of Mormond Hill. In October and November, at the time of the autumn peak, there are flocks of several thousand Pinkfeet at both of these resorts (Table 177). The numbers then decrease to a winter total of 2-3,000, the majority of which are normally found at Strathbeg. In March and April, prior to the spring migration, the flocks increase again, especially in the Ythan Valley, where peaks of 6-11,000 have been recorded in several recent years. There is also evidence of an increase at Strathbeg, perhaps to around 5,000.

In addition to the Greylags and Pinkfeet, the Loch of Strathbeg is known to attract another six kinds of geese, albeit in very small numbers. Barnacle Geese began to appear in the mid 1960s and now occur regularly in October and November, and sometimes in other months; each year since 1977 groups of up to 100 have been recorded, with a large flock of 530 at Strathbeg in October 1979. Greenland Whitefronts used to winter in the 1930s (50), but were absent, apparently, throughout the 1950s and the 1960s. They now appear annually in flocks of up to 5, mostly in February and March. The European race is also recorded in most years either here or in the Ythan Valley. Bean Geese have never been numerous, but continue to occur regularly in parties of up to 10 at one or other of the two resorts. There are also several recent records of Light- and Dark-bellied Brent Geese, mostly single birds.

The recent counts of ducks and swans at Strathbeg are summarised in Table 178. From an earlier run of data, between 1962 and 1968, it seems that the numbers of several species have changed substantially since the 1960s. Mallard decreased dramatically from a regular level of 2,500 in the 1960s to less than 500 in the early 1970s, but have since recovered to about 2,000, and Pochard have dropped from about 1,000 to less than 750. These losses are largely offset by the gains in other species: Wigeon

Table 178. *The regular (and in brackets the maximum) numbers of ducks and swans on five freshwater lochs in the Buchan and Formartine.*

a) Loch of Strathbeg — 11 seasons 1972-1982
b) Meikle Loch, Cotehill Loch and Sand Loch — 10 seasons 1970-1982
c) Pitfour Loch, Old Deer — 8 seasons 1969-1978
d) Haddo House Lochs — 1 season 1982
e) Fyvie Castle Loch — 4 seasons 1979-1982

	a	b	c	d	e
Mallard	1199 (2800)	504 (1150)	264 (1050)	255 (1300)	295 (602)
Teal	212 (850)	216 (1375)	10 (24)	61 (180)	1 (8)
Wigeon	798 (2300)	619 (1200)	48 (380)	2 (200)	0 (5)
Shoveler	5 (37)	3 (44)	0 (2)	0 (2)	0 0
Pochard	655 (2200)	33 (92)	3 (30)	12 (17)	0 (6)
Tufted Duck	911 (1950)	57 (228)	29 (75)	40 (131)	17 (43)
Goldeneye	200 (518)	28 (71)	2 (18)	3 (8)	0 (20)
Goosander	25 (109)	0 (4)	2 (9)	12 (16)	0 (0)
Shelduck	26 (95)	3 (18)	0 (0)	0 (2)	0 (0)
Mute Swan	307 (426)	1 (4)	15 (46)	12 (15)	1 (7)
Whooper Swan	255 (700)	8 (43)	6 (28)	0 0	0 0

Note: Loch of Strathbeg has also held Gadwall (max 5), Garganey (2), Pintail (10), Scaup (9), Eider (160), Long-tailed Duck (7), Merganser (83), Smew (up to 3 regularly), Bewick's Swan (6).

have increased from 600 to 800, Tufted Ducks from 380 to 910, and Goldeneyes from 75 to 200. The swans have also shown an upward trend.

Mute Swans are present in varying numbers throughout the year. In June and July the breeding population of up to 30 pairs is joined by a moulting flock of 2-300; the numbers then continue to increase to a late-summer level of 350-400. The latter is maintained until October, or sometimes November, when the flock slowly dwindles to a late-winter total of 50-100. The most notable change has been the increase in the late-summer peak from a mean 225 in the 1960s to 3-400 since 1974. The autumn dispersal has also been less rapid, and the mid-winter numbers are now a good deal higher than they were. The population in February and March has remained unchanged.

The Whooper Swans begin to arrive in September and reach a peak of about 300 in October and November; they then decrease sharply to a winter level of 50-100. Over the years the autumn peak has varied widely, and has sometimes been as high as 7-800, reflecting perhaps an accumulation of successive waves of migrants. There has been an underlying trend of increase recently; the autumn peak averaged 507 between 1978 and 1982, compared with 326 over the previous five years.

Bewick's Swans were unrecorded in Aberdeenshire prior to 1953 (41), but this was perhaps for want of looking. Between 1961 and 1971 they were seen at Strathbeg in nearly all years, normally in groups of four or five. Since then the records from Strathbeg itself have been less regular, but reports are coming with increasing frequency from other sites within the county. The largest recent count was one of 8 near Monymusk in December 1978.

The counts from a sample of the other freshwater lochs in north-east Grampian show a marked preponderance of Mallard, the regular levels amounting in most cases to several hundred. In addition to the records in Table 178, there are casual reports of up to 480 on Ladymire between the Ythan Estuary and Ellon, of 200-250 on Fedderate Reservoir, and of 100-150 on the dune slacks near Pittenheath and in the Ythan Valley above Fyvie. Teal are also widespread, usually in flocks of less than 50, but occasionally in quite large gatherings (maximum of 340 at Ladymire). Despite a recent increase, they seem to be much less abundant than they were in the 1930s (50). Wigeon are relatively scarce, except on the Meikle Loch, which provides an important day-time retreat for the flocks on the Ythan Estuary. The loch at Pitfour also used to hold some hundreds, but the recent development of a country club has greatly reduced its value.

Further to the south in Strathdon and on Deeside there are several major centres for ducks and geese, the most notable of which are the Loch of Skene, by Dunecht, and the twin lochs of Davan and Kinord, some 30km to the west near Dinnet. The haughs along the middle reaches of the Dee and Don are also used by Greylags, and many of the smaller lochs and marshes at times hold quite large flocks of ducks and swans.

The Greylags have increased tenfold over the past 15 years, and at least two of the concentrations may now be classed as internationally important. They have also established several new resorts in areas which formerly had none. These changes date mostly from the mid 1970s, and it may well be that

Table 179. *November counts (in thousands) of Greylag Geese in Strathdon and Deeside, 1980-1983, with corresponding means for earlier five-year periods. The figures in brackets are the maxima for each period. A blank = no data.*

November	Loch of Skene	Kemnay *	Kincardine O'Neil +	Lochs Davan and Kinord #
Mean 1960-1964	0.03(0.2)	0.4(0.7)		0.03(0.6)
1965-1969	0.3 (0.7)	0.2(0.3)		
1970-1974	1.4 (2.7)	0.1(0.2)		
1975-1979	1.2 (1.7)	0.1(0.1)	0.9(1.4)	2.2 (3.3)
1980	4.7		0.2	4.3
1981	5.7			4.0
1982	4.5		0.5	9.5
1983	1.7$			6.0

* No data since 1976.
\+ 800 in 1974.
\# No data 1963-1976.
$ Incomplete data.

further changes are in store. The relationship between the various groups is far from clear, but already there are signs of instability, the gains in some areas being offset by losses elsewhere. The records in Table 179 provide a fair indication of the course of events, although the detail is sometimes obscured by gaps and confusion in the data.

The two original roosts on the Loch of Skene and on the haughs of the River Don around Kemnay were established seemingly in the early 1960s, and for a while appeared to be discrete. By 1969 the group at Kemnay had increased to 275, the numbers remaining virtually unchanged from November until March. Thereafter the flock began to dwindle and become less stable, while the group at Skene (which may not be independent) continued to increase. Large numbers are also known to occur in the stretch of the Don Valley between Hatton of Fintray and Kintore; these are birds from the roost at Skene, 8km away, rather than a separate group. The information on them is confined to a few odd records from which it seems that the area has been used at least since 1961: 670 were recorded there in January 1968; 5,200 in October 1978 (540), and 1,440 in November 1979.

The spread of the Greylags along Deeside also seems to have occurred in the early or mid 1960s. The initial resort was probably on the Loch of Auchlossan, which had previously been drained, but had partly reflooded following the collapse of a culvert in 1960. At about that time there were rumours of a few geese appearing, but this was unconfirmed until 1969 when the first of a five-year run of counts revealed a regular late-winter flock of about 250. Shortly afterwards, in 1974, a flock of 800 was discovered on the riverside fields near Kincardine O'Neil, some 3km to the south, and this was followed in 1977 by the further discovery of a flock of 500 on Loch Kinord. Although separated by less than 20km, these two new groups have independent roosts and feeding grounds, and show no sign of daily interchange. Over the years, however, the flock at Kincardine O'Neil has declined, while the one at Lochs Davan and Kinord has continued to grow (Table 179).

Pink-footed Geese used to occur in two or three localities on Dee- and Donside, and for a while in the mid 1960s it seemed that they might succeed in establishing a major centre on the Loch of Skene. During the 1950s the numbers there had increased from the odd flock of 20 to a normal autumn level of 1-200, and in 1965 reached a peak of nearly 700. The flock then started to decline, and within six years had disappeared completely. The other resort, on the haughs of the Don near Kemnay, was adopted in 1967 and is probably still in use. When the site was last covered in 1976 it was holding a flock of 1-200 throughout the winter months. There were also rumours in the early 1960s of a similar group in the Howe of Alford, but this has never been confirmed.

The information on the ducks and swans is summarised in Table 180. The Loch of Skene, with an area of 119ha and a mean depth of 1.5m, is the largest and most favoured of the local centres, and the only one which normally holds a peak of more than 1,000 ducks. Over the past 30 years the numbers of most species have tended to increase, or have at least maintained a constant level. The only major change has been the marked decrease in Wigeon, especially in the autumn; in the 1950s the regular population amounted to about 900, compared with 130 in the 1960s and 70 between 1970 and 1982. The highest counts in the same three periods were 2,500, 1,130 and 280 respectively. Mute Swans have also decreased

Table 180. *The regular (and maximum) numbers of ducks and swans in Strathdon and Deeside.*

a) Lochs Davan and Kinord — 5 seasons 1970-1980
b) Loch of Skene — 16 seasons 1960-1982
c) Loch of Loirston — 19 seasons 1960-1982
d) River Don: Kemnay — 17 seasons 1960-1976
e) Loch of Auchlossan — 2 seasons 1970-1971
f) Loch of Aboyne — 5 seasons 1976-1982

	a	b	c	d	e	f
Mallard	310(674)	648(2235)	32(120)	75(128)	88(300)	10 (39)
Teal	23(147)	30 (214)	5 (41)	18 (36)	83(300)	1 (22)
Wigeon	281(511)	95(1128)	15 (92)	0 (5)	199(450)	80(322)
Pochard	13 (43)	87 (628)	0 (4)	0 (0)	31(120)	7 (36)
Tufted Duck	38 (66)	235 (830)	46(122)	0 (0)	22(100)	19 (38)
Goldeneye	26 (60)	107 (300)	7 (56)	5 (13)	7 (20)	2 (10)
Goosander	11 (32)	see text	1 (12)	9 (18)	0 (1)	2 (16)
Mute Swan	13 (39)	9 (28)	2 (11)	16 (38)	0 (1)	4 (8)
Whooper Swan	15 (47)	3 (26)	2 (41)	36(122)	43 (80)	1 (15)

from a former level of 60 to less than 10 in recent years. The loch is also used as a roost by about 100 Goosanders from November to January.

The Lochs of Davan and Kinord are situated at the western extremity of the Deeside lowlands, and now form part of the Muir of Dinnet National Nature Reserve (1,415ha, established 1977). They lie in a rather sterile tract of gravel and moraine, and are flanked by belts of reed-swamp and birch, and by stretches of bog along the eastern side. The total area of water is 134ha, and the mean depth about 1.5m. Most of the marshland vegetation is typical of the acidic conditions, but in places there are some richer areas of fen, especially to the west. The farmland by the River Dee and along the Tarland Burn may also provide a feeding ground for ducks as well as geese. Wigeon have increased considerably in recent years and are now breeding regularly, both here and on some of the smaller pools nearby. Several other species, such as Goldeneye and Mute and Whooper Swan, are likewise more numerous than they were in the 1950s and the early 1960s. The Mallard have probably decreased.

The Lochs of Aboyne and Auchlossan, some 10km to the east, have both been sadly degraded. The former has a camp site by the shore, and is used for water-skiing throughout the summer and early autumn, but still holds quite large flocks of wintering Wigeon so long as the water is open. The Loch of Auchlossan, after 200 years of unsuccessful drainage schemes, is now completely dry, except perhaps in the wettest of winters. Apart from the numerous reed-fringed ditches, the only trace of its former state is a small marshy patch in the middle. In the early 1970s the basin was still sufficiently flooded to attract a late winter total of 500-1,000 ducks and up to 80 Whooper Swans, but by 1972 the place was deserted, and it has not been reported on since. The Loch of Loirston, 1km south of Aberdeen, is notable as a roost for the Goldeneyes which feed by day on the nearby Don Estuary, and for occasional visits by sizeable flocks of Whooper Swans. The latter have their main south Grampian centre in the Don Valley at Kemnay.

The Moray Basin

The Moray Basin is the largest, and in several respects one of the most exciting of the Scottish faunal areas. Topographically it is a region set apart by the mountains along the southern and western boundaries. To the east, however, the way is wide open to incoming migrants; indeed the two converging arms of the coast form a funnel, concentrating the birds from a broad front onto the sheltered inner firths. In this central section the climate is relatively mild, and many of the immigrants are content to remain throughout the winter months. Wigeon are more numerous than in any other part of Scotland, and several other species are abundant, notably Greylags and Whooper Swans, and the sea ducks and sawbills. The breeding population is also large and varied, especially on Speyside, and in parts of Ross and Sutherland.

The funnelling effect of the outer firth is heightened on both the southern and northern shores by the long stretches of inhospitable terrain which face the open sea. To the west of Rosehearty, the Banffshire hills reach almost to the coast, which is flanked for upwards of 50km by a series of cliffs and rocky bays. Inland waters are scarce, and the coast itself is of interest only for the small groups of Eiders which are scattered at frequent intervals along the shore. In February 1971 an aerial survey of the 60km stretch between Fraserburgh and Buckie revealed a total of 1,045 birds, of which 470 were located to the west of Banff, and 575 to the east (353). Counts of the River Deveron in 1981 and 1982 suggested regular (and maximum) populations of 34 (60) Goldeneyes and 18 (42) Eiders on the estuary, and 134 (165) Mallard, 58 (93) Goldeneyes and 15 (22) Goosanders upriver towards Alvah. Boyndie Bay, immediately to the west, held 62 (103) Goldeneyes, 39 (57) Eiders, 2 (7) Long-tailed Ducks and 6 (8) Mergansers at the same time.

The only other notable resort between the Buchan lowlands and Strath Spey is Loch Park, in the hills between Dufftown and Keith. The loch, which was formed artificially in about 1860, is set in a narrow steep-sided valley, with fir plantations on either hand. The area is 14ha and the water, which is mostly less than 2m deep, supports an abundance of water-weeds and *Chara*. In most winters the loch is completely frozen by December, but earlier on it is likely to hold some 2-300 Mallard, and occasionally a few Wigeon and diving ducks. At one time it was also noted as an autumn staging point for Whooper Swans. These began to appear soon after the felling and replacement of the existing timber in the mid 1940s, and quickly increased to a peak of 200 in 1957 and 1958; the flocks then started to decline, and within ten years had virtually disappeared. The reason was almost certainly the development of the new plantations, which increasingly overshadowed the long and narrow water. A similar sequence of events is said to have taken place in the 1930s, when the previous plantings began to mature, but this is unconfirmed; there is also a record of swans occurring here in the 1870s, less than 20 years after the loch was created (259, 260).

Moray, Nairn and Speyside

At Buckie the hills retreat a little from the coast and the cliffs are replaced by long open stretches of dune and beach extending westwards for more than 60km to Fort George, at the entrance to the inner firth. The prime interest here lies in the offshore flocks of Long-tailed Ducks and Scoters, the details of which are discussed in a later section (p.274). Most of the other species are unattracted by the bare sandy bays, and greatly prefer the freshwater lochs in the coastal plain, and the sheltered inlets and saltmarshes at Findhorn, Culbin and Whiteness. Sizeable numbers of ducks and swans are also found on some of the lochs and marshes around the middle reaches of the Spey.

Fig. 46. The Moray Basin.

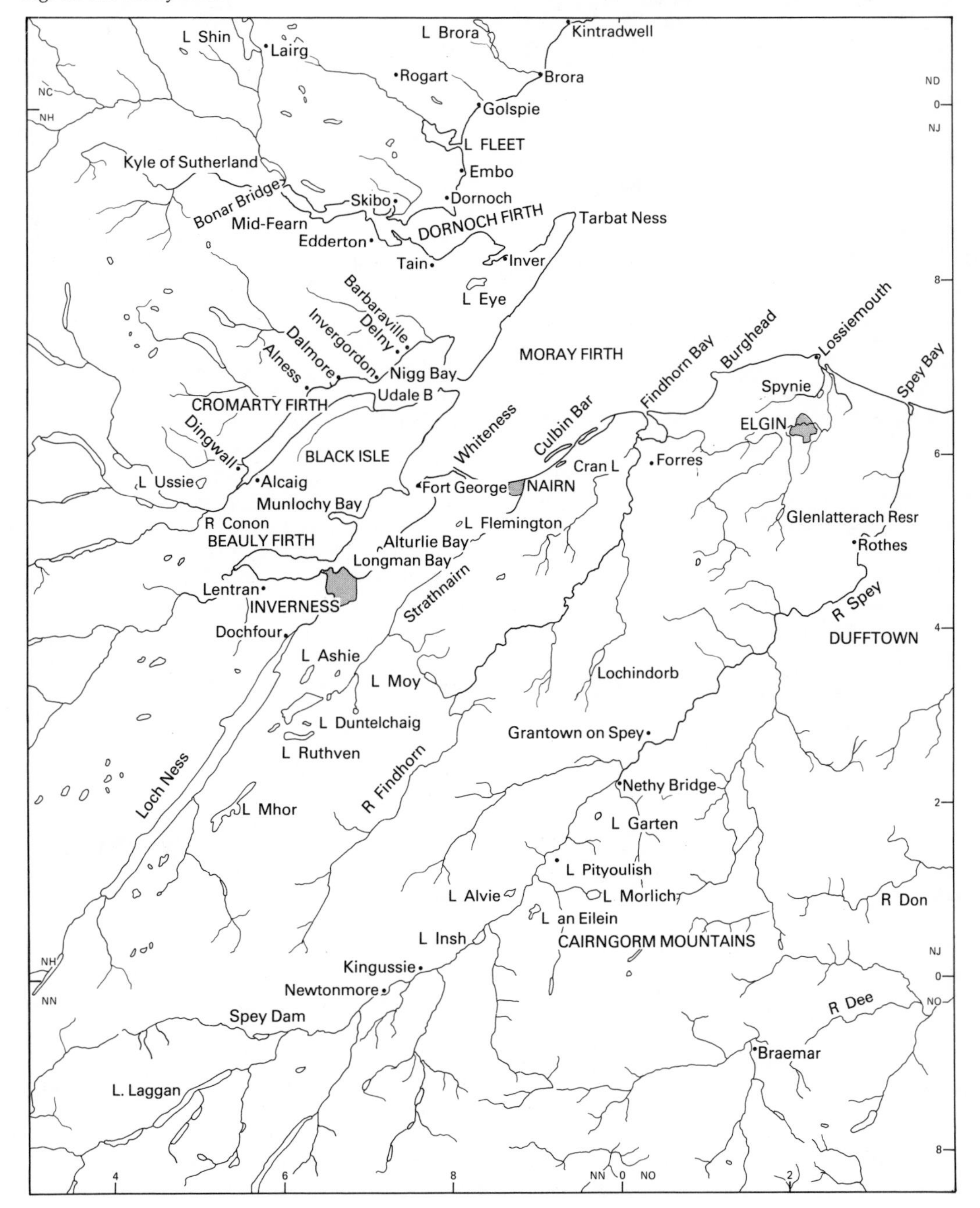

In the eastern parts of the region there are two major resorts at Spynie and on Findhorn Bay, and a number of secondary sites, the most important of which are the lochs near Lhanbryde. The 10km stretch of the coast between Lossiemouth and Burghead is also of substantial interest. Unlike the adjoining stretches of Spey Bay and Burghead Bay, the foreshore here is rocky, with several offshore skerries towards the eastern end. Mallard reach an autumn peak of 6-700, and up to 400 are present until March, mostly round the outer reefs. The reefs are also frequented by 1-200 Eiders, and as many again are scattered along the shore from Burghead eastwards to the outfall of the Spey. The largest numbers occur in mid and late winter when the flocks on the northern side of the firth around Dornoch have passed their autumn peak. Another feature is the flock of Goldeneyes which feeds around the outfall from the maltings at Burghead. Peaks of 4-500 were recorded here in January 1975 and 1978, and totals of 250-300 are known to have occurred in several other recent seasons. At night the birds either stay around the outfall or move along the coast to Findhorn Bay, depending apparently on the state of the tide (156, 359).

To the east of the rock coast, the two small estuaries of the Spey and Lossie have neither the space nor the food supplies to attract large gatherings of ducks. At Speymouth a flock of 150 Wigeon winters regularly in the sheltered inner harbour, and similar numbers of Goldeneyes are often found in the outer channel or on the sea nearby. Mute Swans also occur, both here and at Lossiemouth, but in common with the dabbling ducks the flocks are a good deal smaller than those on the inland lochs at Spynie and elsewhere (Table 181).

Table 181. *The regular (and in brackets the maximum) numbers of ducks and swans at three sites in east Moray.*
Speymouth 4 seasons 1970-1974
Lossiemouth 9 seasons 1970-1982
Loch of Spynie 12 seasons 1970-1982

	Speymouth	Lossiemouth	Spynie
Mallard	17 (58)	15 (110)	488 (1360)
Gadwall	0 (0)	0 (0)	0 (7)
Teal	25 (66)	0 (50)	88 (315)
Wigeon	140 (244)	94 (700)	213 (750)
Pochard	0 (0)	0 (70)	81 (380)
Tufted Duck	0 (21)	0 (20)	79 (194)
Goldeneye	80 (170)	28 (280)	2 (8)
Eider	3 (14)	60 (406)	0 (0)
Merganser	17 (36)	12 (40)	1 (9)
Goosander	1 (4)	2 (20)	13 (37)
Shelduck	5 (9)	3 (19)	1 (9)
Mute Swan	11 (23)	0 (23)	72 (148)
Whooper Swan	0 (0)	0 (0)	5 (26)

The Loch of Spynie has a long and chequered history extending back to the 16th century. Prior to that the greater part of the low-lying plain between Elgin and Lossiemouth was covered by a tidal inlet but shortly afterwards a shingle spit grew up across the mouth, enclosing a shallow, fertile lagoon, more than 650ha in extent. This remained until the early 1800s when the cutting of the Spynie canal reduced the surface area to little more than 30ha. Since then reed-beds have encroached considerably, and recent drainage schemes have done away with much of the marshy pastureland along the western side. Nevertheless, the loch continues to attract an annual peak of 1,500-2,000 Greylags and 1,000-1,500 ducks.

The Greylags, it seems, were greatly troubled by the earlier drainage, and by 1850 the large flocks of former years were reduced to occasional parties in spring (518). By 1900, however, the species was firmly re-established, and in recent years the flocks have shown a slow but fairly steady increase. They have also spread, presumably from here, to Findhorn Bay where a new and thriving centre was established in the latter half of the 1960s (Table 182). At Spynie the geese are normally present in strength from October until the close of the shooting season, when the numbers in Findhorn Bay show a marked increase.

Pink-footed Geese were also abundant at one time, especially in the 1930s, when flocks of many thousands used to occur in autumn and early winter, and again in spring. The occurrence of these concentrations dated apparently from the mid 1920s, and coincided with similar increases on the Beauly Firth and in parts of Easter Ross (50). The pattern of distribution then changed, and by the 1950s the flocks throughout the region were reduced to scattered remnants, most of which have since disappeared. At Spynie the process was probably hastened by the traffic of aircraft from the nearby station at Lossiemouth, which is said to have worried the Pinkfeet a great deal more than the Greylags. The parties of Greenland White-fronted Geese which used to winter in the 1930s have likewise disappeared completely.

Most of the other species have maintained a fairly constant level over the past 30 years. The only major change amongst the ducks has been a decrease in the numbers of Wigeon from a regular level of 700 in the 1950s to about 200 at the present time (Table 181). The Mute Swans, for which the loch has long been renowned, have also tended to decline. Prior to 1960 they used to assemble in strength in September or October, and increase to a winter level of about 110; the present flocks are much less stable, and the regular level has fallen to less than 60 during recent years. Mallard and Pochard have shown a slight increase. The loch also attracts a sizeable number of breeding ducks, including a pair or two of Shoveler and 2-3 pairs of Pochard.

Table 182. *November counts of Greylags on the Loch of Spynie and on Findhorn Bay, 1975-1982, with corresponding means for earlier five-year periods. The figures in brackets are the maxima for each period. The table also shows the means (and maxima) of the monthly counts at each site in seasons between 1975 and 1982.*

November	Spynie	Findhorn Bay
Mean 1955-1959	755 (1000)	0 (0)
Mean 1960-1964	1090 (1500)	0 (0)
Mean 1965-1969	660 (950)	390 (1560)
Mean 1970-1974	1030 (1780)	130 (400)
1975	1750	140
1976	1130	80
1977	1850	390
1978	1720	90
1979	1370	125
1980	860	160
1981	600	1200
1982	1650	560
Monthly means (and maxima) 1975-1982		
October	840 (2790)	146 (300)
November	1370 (1850)	340 (1200)
December	885 (1650)	270 (680)
January	1225 (2460)	380 (2000)
February	1025 (1500)	835 (1800)
March	945 (1900)	1755 (3000)

The populations on the neighbouring inland sites are rather more prosaic. A ten-year run of counts at Loch Oire and Loch na Bo, near Lhanbryde, shows a January mean of 130 Mallard (maximum 325), 60 Wigeon (maximum 315) and 25 Tufted Ducks (maximum 80), with perhaps a few Pochard and Teal. Mallard in flocks of 50-150 have also been recorded on the pond at Gordon Castle near Fochabers, on the Loch of Blairs near Forres, on the Spey at Rothes, and at Glenlatterach Reservoir in the hills behind Elgin.

At Findhorn Bay the twin channels of the Findhorn and the Muckle Burn debouch into a shallow land-locked harbour with extensive intertidal flats, amounting in all to nearly 700ha. The basin itself is important as a roost, but the food supplies are poor except along the southern shore where a sizeable tract of saltmarsh surrounds the outfall of the two rivers. *Salicornia* is abundant in this sector, and a small amount of *Zostera* is available in places (369). The adjoining arable land also provides a feeding ground, especially for geese. Elsewhere there are long stretches of barren sand or shingle, flanked to the east by the runways of Kinloss airfield, and to the west by the dour plantations of Culbin Forest.

As at Spynie, the main changes over the past 25 years have been a marked increase in the numbers of Greylags (Table 182), and a steady decline in the flocks of Wigeon. Between 1955 and 1964 the latter maintained a regular level of 1,400 and were present in flocks of 2,000 or more in five of the ten seasons; since 1971 the regular level has dropped to 523, and the annual peaks have only once topped 1,000 (Table 183). This decrease, which is not yet explained, contrasts with the small but perceptible gains in numbers of

Table 183. *The regular (and maximum) numbers of ducks and swans at five sites in west Moray and Nairn.*

a) Findhorn Bay — 12 seasons 1971-1982
b) Culbin lagoon, western half — 8 seasons 1970-1982
c) Cran Loch and Loch Loy — 19 seasons 1960-1979
d) Whiteness lagoon — 2 seasons 1979 and 1982
e) Loch Flemington — 13 seasons 1960-1982

	a	b	c	d	e
Mallard	107 (387)	45 (120)	157 (420)	70 (157)	7 (149)
Teal	26 (115)	25 (250)	68 (320)	4 (10)	13 (200)
Wigeon	523(1200)	168 (702)	69 (330)	148 (195)	299 (940)
Pintail	0 (2)	0 (6)	0 (5)	0 (0)	0 (0)
Pochard	0 (0)	0 (0)	1 (20)	0 (0)	1 (7)
Tufted Duck	0 (1)	0 (2)	26 (94)	0 (0)	29 (71)
Goldeneye	48 (199)	11 (80)	2 (17)	4 (7)	1 (12)
Eider	58 (320)	4 (101)	0 (0)	0 (0)	0 (0)
Merganser	25 (84)	34 (472)	0 (5)	48 (173)	0 (0)
Shelduck	58 (176)	70 (147)	1 (13)	10 (45)	0 (0)
Mute Swan	13 (39)	4 (29)	21 (88)	0 (2)	16 (63)
Whooper Swan	1 (14)	0 (1)	3 (23)	1 (0)	14 (184)

most of the other ducks, notably Mallard, Goldeneyes and Shelducks.

The advent of the Greylags in the late 1960s coincided with a temporary drop in the numbers at Spynie, and was no doubt prompted by the need for an alternative roost from which to exploit the unused feeding grounds on the arable land towards Forres and Nairn. The two resorts are in fact complementary, the largest numbers at Spynie ususally being in autumn and winter, and at Findhorn Bay in March and April (Table 182). Light-bellied Brent Geese were also plentiful up to 1939; the population then collapsed and virtually none have wintered since the war. The only recent records relate to September or early October when parties of 20-50 are sometimes seen briefly on passage, either here, or more often at Lossiemouth or Whiteness. A flock of 150 at Spynie in November 1974 was rightly described as amazing! (540).

To the west of Findhorn Bay the coast adjoining Culbin Forest is flanked by an offshore spit of sand and shingle, behind which lies a shallow tidal lagoon, some 6km in length and up to 1km in width. The lagoon itself has extensive stretches of intertidal mud and sand, and the shores, especially on the landward side are fringed with a belt of sea-washed turf interspersed with numerous pools and gutters. In the central and eastern parts the forest provides an effective shield against disturbance, and further protection is afforded by the reserve established by the RSPB. The counts from the western section are summarised in Table 183. The eastern end has not yet been covered effectively, but is said to hold similar numbers. A recent estimate suggests a total of up to 1,000 Wigeon on the lagoon and the adjoining stretches between Findhorn Bay and Nairn.

In addition to the coastal flocks there are often 3-400 ducks on the Cran Loch and Loch Loy, a short way inland on the western edge of the forest (Table 183). Greylags also come in to roost from time to time, either here or on the lagoon, but their numbers are erratic, and no clear pattern of usage has yet emerged. The flocks first started to come in 1964, and some at least have been recorded in every winter since. In most seasons the peak occurs in the latter part of the season, and varies between 100 and 450.

Another point of interest is the flock of Scaup which appears from time to time at scattered points along the Nairn and Moray coast. The occurrences are usually quite brief, and are seemingly unconnected with those elsewhere in the region. The principal records in recent years have been as follows: 84 at Burghead, February 1972; 400 off Culbin and Findhorn, winter 1972-73; 300 off Culbin, October 1973; 50 at Speymouth, January 1978; and 600 in Spey Bay, January 1979 (540). Some sizeable flocks of Mergansers have also been recorded on the sea off Findhorn and Culbin, mostly in late September and early October. Gatherings of up to 300 were present regularly in the early 1960s and similar numbers have been seen in several recent seasons. A count of 800 in October 1978 was presumably exceptional (540).

The tidal inlet at Whiteness is similar in formation to the one at Culbin, but less than a quarter the size. In 1972 about 100ha of the outer tideway was reclaimed and developed as an oil rig building yard, but the inner areas of saltmarsh were unaffected and continued to attract a winter peak of at least 500 ducks. The main change has been a decline in Teal and Wigeon, the regular (and maximum) numbers of which between 1965 and 1967 had been, respectively, 39 (160) and 511 (600). A flock of moulting Mergansers, 100-150 strong, forgathers in summer and early autumn on the seaward side of the bar. Goldeneyes are scarce, except off the outfall at Nairn, where up to 75 have been recorded in recent years.

Further flocks of Wigeon, often totalling several hundred, are found on Loch Flemington, a short way inland from Whiteness (Table 183). Their presence here was especially noticeable in the early 1970s, at the time of the Whiteness development, and the two resorts are doubtless closely linked. Greylags were also numerous hereabouts, notably in the mid 1960s, when the flocks increased from a normal level of 2-300 to a peak of 1,500. Several roosts were adopted at this time, including Whiteness, Loch Flemington and Alturlie Bay, but none was occupied for very long, and

Table 184. *Loch Insh and the Spey marshes, Kincraig—Kingussie: the regular and maximum numbers of wintering ducks and swans, 10 seasons 1973-1982, and the estimated number of breeding pairs in the seasons since 1975* (Scottish Bird Reports).

	Regular	Maximum	Breeding pairs
Mallard	320	695	c. 40
Teal	105	500	40-50
Wigeon	72	150	15-25
Shoveler	1	9	1*
Pochard	6	43	1*
Tufted Duck	37	75	30-40
Goldeneye	31	100	0
Goosander	5	16	0
Mute Swan	7	14	0
Whooper Swan	96	184	0

* A pair of Pochard bred in 1978, and a pair of Shoveler in 1979 (the first proven record).

by 1970 the flocks had ceased to winter regularly. The favourite feeding grounds, which are still used occasionally, were on the open farmland near Moss-side and Kildrummie, a little to the west of Nairn.

The valley of the Spey provides one of the few easy routes through the barrier of the central Highlands, and is used extensively by waterfowl on spring and autumn passage. In the middle reaches, at an altitude of 200-250m, there are numerous lochs and areas of marsh, the most important of which is the stretch of the floodplain between Kingussie and Loch Insh. Within this 10km section the river has a fall of less than 10m, and the flat valley floor is occupied by a tract of mire and sedge fen, 1,000ha in extent. Nearly half of the area is now either owned or leased by the RSPB, and is managed specifically for breeding and wintering birds. At the lower end, near Loch Insh (which itself covers 113ha), there are massive reed-beds interspersed with numerous small pools, and the whole is subject to frequent and extensive flooding. The main cause of this is the ridge of rock which spans the river bed below Loch Insh, and acts as a dam, retaining the sudden spates and regulating the flow to the lower river. Proposals have been made to remove the sill with the aim of reclaiming the marshes upstream, but this would be short-sighted folly; the catchment area above this point is 81,800ha, and without the natural pondage of the Insh Marshes the much richer haughlands downstream would be ravaged by flooding instead. The ornithological and botanical importance of the system would also be destroyed.

The value of the Insh Marshes as a centre for breeding and wintering wildfowl is apparent from Table 184. The main point of interest is the gathering of Whooper Swans, which starts to assemble in November and reaches a mid or late-winter peak of at least 120. Greylags also occur regularly from October to December and again in March and April. The peak numbers usually amount to about 250, but latterly have shown a major increase to 650 in November 1979, and to 975 in 1980. They normally roost on the floods in preference to Loch Insh, and feed on the pasture-land towards Kingussie. The Whooper Swans feed mainly on the seeds of *Carex*, which is abundant throughout the middle reaches of the marsh.

The information on the other Speyside lochs is restricted to a few disjointed records dating mainly from the early 1960s (Table 185). For the most part the autumn and winter numbers appear to be quite small, the only notable exception being at Loch Garten, which is now owned by the RSPB. A further flock of Greylags, up to 1,000 strong, occurs here regularly between October and December, and smaller numbers are sometimes present for a while in spring (e.g. 650 in March (540)). The loch itself is used solely as a roost, the feeding grounds being on the haughlands of the River Spey on either side of Nethy Bridge. Whooper Swans, in bands of 80-140, are also reported from time to time either on the loch or on the farmland nearby. Unlike the geese, they are probably linked with the flocks on the Insh Marshes, and a good deal of movement may well take place along the line of the valley.

Greylags and Whooper Swans also occur in small numbers in the upper valley of the Spey, between Kingussie and Spey Dam, and a further flock of feral Greylags is now establised on Loch Laggan, a short way over the watershed. The Loch Laggan Greylags stem from a group of 20, introduced from Cumberland in 1968, which have since increased to a level ranging from 60 in 1975 to 70 in 1981. Allowing for interchange, the peak population on Speyside, including the Insh Marshes, is estimated at about 1,500 Greylags and 3-400 Whooper Swans.

The valley between Grantown and Laggan is renowned more especially as a breeding ground for ducks. Wigeon nest regularly on several of the larger lochs and on many of the smaller pools and marshy hollows which abound throughout the area. A recent estimate suggests a total of at least 50 pairs, and possibly as many as 100, over the area as a whole. Teal are even more numerous, and Tufted Ducks are common enough, with a total of perhaps 100 pairs, including those at Insh. Goldeneyes are also breeding regularly, but the sites are not divulged. The first record, and the first for Scotland, was obtained in 1970 (323). The initial breeding was stimulated by the provision of nest-boxes, and the ready acceptance of these has been largely responsible for the subsequent dramatic increase and spread. In 1980 eggs were laid at 23 sites and 17 were successful, hatching 165 young. The average annual figures for the subsequent three seasons were 26 broods from 44 sites, hatching 238 young (540). The survival rate of the ducklings is good, and, with the erection of nest-boxes continuing, a further expansion can be expected (see also Fig 134).

The lochs in the neighbouring valleys of the Findhorn and Strathnairn are mostly set in open moorland, and the numbers of wildfowl using them are normally quite small (Table 185). The only exception is Loch Ruthven, which lies at less than 200m in a shallow bowl of farmland, and seems to be rather more productive than the rest. Loch Ashie is also of interest as a summer site for moulting Pochard. These have been coming regularly for many years, and have slowly increased from a peak of 65 in the 1950s to as many as 150 in 1975 (540). Most, if not all of them are males and are presumed to come from abroad.

The sea ducks of the Moray Firth

The offshore waters of the outer firth, from Buckie round to Golspie, have long been quoted as the haunt of Long-tailed Ducks and scoters. Several of the earlier accounts refer to the presence of impressive numbers, but the records are mostly vague and unrelated, and the pattern of events on the firth as a whole has become apparent only in the past 20 years. As in other parts of Scotland the detailed study of the flocks was stimulated by the threat of oil pollution, stemming in this case from the proposed developments on the Cromarty Firth and from the exploitation of the Beatrice field, which lies within 50km of some of the prime resorts. The first specialised surveys were initiated by the RSPB in the mid 1960s, and were continued intensively for the next 10 seasons by R.H. Dennis and his team. In the winters of 1977 and 1978 an even more ambitious programme was undertaken, supported by the RSPB and the Nature Conservancy Council. Extensive surveys have also been carried out since 1981 by the RSPB with financial support from Britoil. Although the full results of these are not available, some figures have been received and these have been included where relevant.

The species such as Scaup, Eider, Goldeneye and Merganser, which occur as much on the inner firths as on the open coast, have already been discussed in their local context. The present section is concerned with the Long-tailed Ducks and scoters which winter solely on the outer firth, and are best considered as an independent group.

The great majority of the birds are found in areas where the sea bed is sandy, and the water less than 10m deep. This applies particularly to the scoters which often feed quite close inshore; the Long-tailed Ducks are sometimes farther out, but seldom beyond the 20m line. The favourite resorts are in Spey Bay and Burghead Bay, in the area off the Culbin Bar between Findhorn and Nairn, in the triangle between Nairn

Table 185. *Maximum numbers of ducks and swans recorded in autumn and winter in the basins of the Spey, Findhorn and Nairn, and in Stratherrick. The figures in brackets show the number of occasions on which counts were made; sites not covered since 1965 are marked *; records marked x are from* Scottish Bird Reports.

		Mallard	Teal	Wigeon	Pochard	Tufted Duck	Goldeneye	Mute Swan
STRATHSPEY								
Spey: Spey Dam – Newtonmore	(5)	276	6	21	0	0	9	27
L Alvie	(4)	75	9	3	22	5	18	1
Kinrara*	(1)	36	24	0	0	0	0	0
L an Eilein*	(6)	63	5	11	0	0	9	0
L Morlich	(8)	8	6	2	43x	40	45	2
L Pityoulish	(4)	5	17	0	2	6	13	5
L Garten, L Mallachie	(4)	475x	52	12	0	12	13	143x
L Dhu More, Dulnain Br*	(1)	190	5	20	0	0	0	15
FINDHORN BASIN								
Lochindorb*	(21)	80	11	63	33	6	0	14
L Belivat	(1)	100	50	0	0	0	1	0
L Moy*	(37)	191	10	4	2	38	44	7
UPPER STRATHNAIRN and STRATHERRICK								
L Ruthven	(55)	136	14	110	60	7	7	5
L a'Choire*	(35)	68	4	3	2	22	4	0
L Ceo Glais*	(37)	13	0	0	6	3	14	0
L Duntelchaig*	(33)	10	1	3	0	0	6	2
L an Eoin Ruadha*	(31)	27	0	0	13	9	3	0
L a'Chlachain	(41)	51	10	4	14	2	8	10
L Ashie	(48)	56	2	30	36	27	10	2
L Bunachton*	(3)	80	0	0	0	3	0	0
L Mhor	(5)	64	0	214	0	0	0	47

osemarkie and the mouth of the Cromarty Firth, and
 the stretch between Dornoch and the start of the
iffs beyond Kintradwell. In each case the 10m
ntour is at least 1km, and in some places as much as
km, offshore.

The distribution of the scoter flocks varies
nsiderably from year to year; a certain amount of
ovement may also take place in the course of the
ason, but in general the flocks once committed to an
rea are likely to remain there until spring. This is
rtainly so in the northern sector, which is set apart
y the stretch of cliff between Cromarty and Tarbat
ess.

Table 186 shows the highest annual counts in
ach of the five main sections, and also the highest
nown totals for the Moray Firth as a whole. The latter
re based on synchronised counts in the period
ctober to March (or October to May in the case of
ong-tailed Ducks), and except in some of the later
easons are patently incomplete. They nevertheless
rovide a rough-and-ready basis for the estimation of
inter numbers. The shifts in population between the
arious resorts are likewise apparent, despite the
umerous gaps. The occurrences in other months are
iscussed in the text below.

The Long-tailed Ducks are difficult to count,
ven from the air, and a sizeable allowance should be
nade for birds overlooked in rough water and in areas
ut of sight of land. In 1977-78 the probable peak was
stimated at about 10,000, and similar totals may well
ave occurred in previous years. Unprecedented
ounts of 20,000 and 14,000 in 1981-82 and 1982-83
540) indicated a massive increase in numbers, initially
erhaps in response to the cold weather in the former
vinter. The majority of the birds in the late 1970s were
eeding along the southern side of the firth, in Spey
ay and Burghead Bay, and more especially on the Riff
ank, between Nairn and the Black Isle. In each case
he flocks moved at dusk to a single central roost on
he sea some 3-4km offshore in Burghead Bay; the
ame area was also used in some of the following
easons, although the distribution of the day-time
locks had changed considerably (3, 359, 361). This
radition of congregating by night is by no means
ncommon, and is often a cause for concern. At this
articular roost as many as 7,000 birds may at times be
ssembled in an area of a few hundred hectares, where
n ill-disposed oil slick would doubtless cause havoc.
The flocks on the northern shore around Dornoch and
Golspie also fly to this Burghead Bay roost (361). The
irds are normally present from October onwards,
vith a peak in December or January. Large numbers
re also recorded in April and May, when the birds
end to concentrate prior to migration, usually near an
outfall of freshwater as at Speymouth, Findhorn or Loch Fleet. These gatherings, which are full of birds displaying, are believed to consist not of migrants on passage, but of birds which have wintered locally; the records at this time have therefore been taken into account in Table 186.

The Common and Velvet Scoters are normally found in a series of long loose flocks lying parallel to the shore, and at no great distance from it. Aerial surveys have shown that by day the great majority of the birds are in view from the land, and the counts in the various sectors are probably representative. The only difficulty lies in distinguishing between the two species; both of them are often present in the same flock, in a mingling of little separate groups, and the Velvet Scoters may at times be overlooked. At dusk the flocks swim a short way out to sea, but otherwise seem to remain in the general area of the feeding grounds. They thus retain their day-time scatter, and are rather less vulnerable to pollution than the Long-tailed Ducks.

The Common Scoters have varied greatly in number over the past 15 years or so. Very large gatherings, totalling 8-10,000 or more were recorded regularly between 1972 and 1977, but such numbers were unknown in the mid and late 1960s despite extensive searches. In the winters of 1979 and 1980 the flock of 6,000, which had spent the two previous winters in Spey Bay, was nowhere to be found, and the totals for the firth as a whole were thought to be less than 3,000. Recent RSPB/Britoil surveys have, however, again found very sizeable numbers. A total of 14,242 scoters (mixed species) were found in February 1982, 13,907 in March 1982 and 12,266 in March 1983. In November 1982 6,929 Common Scoters alone were counted. Similar periods of abundance and scarcity have been recorded in several other parts of Britain, and the general impression is one of constant change, the reasons for which are yet to be explained.

The wintering flocks of Common Scoters arrive in September and are present in strength from October until March. The peak numbers are normally found between November and January, but sometimes a month or two later. Substantial flocks of moulting scoters are also recorded from time to time, and these may well occur regularly: 1-2,000 were present off Embo in July 1968 and 1975, and 500 were off Speymouth in July 1978. A gathering of 8,000 off Portmahomack in early September 1971 was probably a mixture of summering birds and incoming migrants.

The Velvet Scoters have apparently followed much the same pattern of increase and decrease. There was a period of great abundance between 1973 and 1978, when totals of 2-3,000 were recorded

Table 186. *Peak counts of Long-tailed Ducks, Common Scoters and Velvet Scoters in five sectors of the Moray Firth, October - March, 1970-1982. A blank = no data; ( ) = counts made in one or two months only.*
Includes data from Dennis 1975, Mudge 1978, Allen 1979 and Scottish Bird Reports.

a) Spey Bay
b) Burghead Bay
c) Findhorn–Nairn
d) Nairn—Rosemarkie—Cromarty
e) Dornoch—L Fleet—Kintradwell
f) Highest synchronised total

	a	b	c	d	e	f
LONG-TAILED DUCKS						
1971/72	350 A	2350			1000 A	(2850)
1972/73	300 A	3000			2000	5000
1973/74	1500 A	7000			2500 A	7250
1974/75		3100	(210)		3000 A	3450 E
1975/76	1600	5000			(4000)	8100 E
1976/77	750		25			1200
1977/78	2155	1850	760	2400	1625	6800
1978/79	2600	300	460	6500	560	8500
1979/80	35	3500	(100)	(4500)	(1610)	6800
1980/81	338				500*	
1981/82		3000*	4000*		250	20000*
1982/83	555				320	14000*
COMMON SCOTERS						
1970/71		(1800)			(7000)	(7000)
1971/72	(370)	3520	320		(2850)	(6750)
1972/73	1000	7000			7000	10000
1973/74		14000	600			14000
1974/75	(1500)	6500	500		2900	7900E
1975/76	2800	(7000)			(1800)	8250
1976/77	1715		200		(200)	
1977/78	6075	2455	1040	180	670	9060
1978/79	6000	1330	410	90	640	7000
1979/80	45	2000	330	650	610	2700
1980/81	350				500*	
1981/82	3400*		1900*		1000	
1982/83	3000*				3000	6930
VELVET SCOTERS						
1970/71		(740)				(740)
1971/72		(84)			(500)	(500)
1972/73	(105)	180			660	(825)
1973/74		2000				2000
1974/75		375				
1975/76	(200)	2500			(150)	2500
1976/77	80				(0)	
1977/78	2100	1450	910	20	125	3400
1978/79	3000	104	144	0	40	3000
1979/80	70* A	(0)	(0)	(0)	400* A	(75)
1980/81	100* A					
1981/82	403				100	
1982/83	1200*					8035

Note: Columns a & e: Figures marked A refer to April or May (see text). Column f: Figures marked E include a estimate for one or more of the areas, based on the mean of the counts in the two months adjacent to the peak period. Velvet Scoter: In April 1979 a flock of 5-600 was present in Spey Bay.
* From *Scottish Bird Reports*.

regularly, mostly along the southern shore between Speymouth and Nairn. At that time the Moray Firth was supporting rather more than 1% of the estimated European population, and well over half of the probable total in Britain. High counts were also made in 1981-82, when a substantial proportion of the scoters counted in February and March (see above) were Velvets. When the RSPB/Britoil surveys were able to distinguish between the species in 1982-83, no fewer than 8,035 Velvet Scoters were recorded in March. There is no reliable information on numbers elsewhere in Europe at the time, but this total was undoubtedly highly significant both in the national and international context. The timing of the winter peak varies from year to year, occurring sometimes in October, sometimes in December or January, and sometimes in spring: in 1977-78 there were 2,000 or so from December to February, 3,500 in March, and 5-6,000 in the middle of April (359, 540). The counts for the early 1980s given above also indicate late-winter peaks. The flocks, although lingering at times into May, are normally absent in summer, and seldom return until early October.

The northern firths

The four drowned valleys of the Beauly and the Inner Moray Firth, the Cromarty Firth, the Dornoch Firth and Loch Fleet together provide one of the most important areas for wildfowl on the Scottish coast. All of them have extensive intertidal flats of sand or silt, with a wealth of *Zostera* and other food plants, and further feeding grounds are available on the low-lying arable and pastureland nearby. Some of the neighbouring freshwater lochs are also of major importance, notably Loch Eye and Loch Ussie. Wigeon are especially plentiful with an autumn total of 20,000 or more; Whooper Swans are numerous, and Greylags have increased greatly to a normal autumn peak of 25,000. The winter gatherings of sawbills and the summer concentrations of Canada Geese are also of outstanding interest. Considering the size of the firths, the area in which they lie is remarkably compact, encompassing no more than 50km from north to south and 30km from east to west. Most of the coastline is still unspoiled, except around Inverness and on parts of the Cromarty Firth, where a good deal of industrial development has already taken place and further projects are proposed. In order to offset these losses a number of reserves have been established during recent years: in the 1970s these included a National Nature Reserve over parts of Nigg Bay and Udale Bay in the Cromarty Firth; a Local Nature Reserve on Munlochy Bay in the Inner Moray Firth; a statutory Bird Sanctuary on Loch Eye; a Scottish Wildlife Trust reserve on the tidal basin of Loch Fleet, and an RSPB reserve at Nairn and Culbin Bar.

The Beauly Firth and the inner Moray Firth

The two broad basins of the Beauly and the inner Moray Firth extend in all for more than 25km inland, from the seaward entrance at Fort George to the head of the estuary beyond Kirkhill. In the middle reaches the narrows of the Kessock channel act as a link, and also as a break between the upper and the lower tideway – a division which is emphasised by the marked difference in character of the two sections.

To the east of Inverness the shoreline of the inner Moray Firth has much in common with the open coast beyond, but lacks the bonus of the flocks of sea ducks. The southern side is occupied for the most part by long stretches of barren sand and shingle, while to the north the hills of the Black Isle fall steeply to a narrow and rocky shore. The only sites of interest are the two main centres at Longman Bay and Munlochy Bay, and the two minor resorts in Alturlie Bay and around the mouth of the River Ness (Table 187).

Longman Bay has declined substantially in value since the 1960s, but still attracts an annual peak of several thousand ducks. At the western end the bay is flanked by a narrow strip of saltmarsh, and by a sweep of foreshore up to 1,000m wide and varying in texture from mud to sand and gravel. The mud in particular is rich in invertebrates, and the wildfowl foods *Zostera* and *Ruppia* occur here at some of the highest densities in the area. In recent years the rather limited feeding grounds have suffered considerable damage from the dumping of rubbish over parts of the saltmarsh and the adjoining shore; it also seems that the mud is slowly being replaced by sand (369). The building of the new A9 highway along the inner edge of the saltmarsh has been another source of disruption, but one which is unlikely to persist; in the longer term the embankment may well provide a useful screen between the foreshore and the planned developments nearby.

These adversities at Longman Bay have been accompanied by a steady decrease in the numbers of several species. Teal have declined from a regular level of 450 in the early 1960s to less than 60 in recent years, the flock of Pintail which used to reach an annual peak of above 200 has now virtually disappeared, and Mallard and Shelducks are both somewhat scarcer than they were. In most cases these local losses have been largely offset by corresponding gains in Munlochy Bay and elsewhere. Wigeon were also scarce for a while in the early 1970s but have since regained their earlier level of 2-3,000 – implying perhaps that they use the bay chiefly as a roost rather than a feeding ground. In November 1976, and again in December 1979, the counts reached a peak of about 6,000, compared with the previous maximum of just under 4,000 in 1964.

On the opposite side of the firth the narrow inlet of Munlochy Bay pierces the line of hills along the northern shore and opens into a shallow sheltered basin, some 3km in length and rather less than 1km in width. At low water the greater part of this dries out and provides an excellent feeding ground for nowadays 2-3,000 ducks (Table 187). *Zostera* and *Salicornia* are abundant on the flats, and *Enteromorpha* is plentiful towards the inner end. The mud also supports a fairly dense population of the snail *Hydrobia ulvae*, a favourite food of Teal and Shelducks. Along the southern shore there are wide stretches of saltmarsh, with a tract of wet pasture beyond, and further areas of saltmarsh are found at the head of the bay (369). The shooting on the estuary and the adjacent farmland has been restricted for the past ten years or more, initially through the efforts of the Black Isle and Mid Ross Wildfowling Club, and since 1975 by the establishment of a Local Nature Reserve. Under the present regime the number of guns is controlled by permit, and the disturbance is limited to three days a week.

The most notable change at Munlochy has been the recent increase in the numbers of roosting Greylags. Prior to the recent curbs on shooting their appearances were limited to occasional parties of up to 300. They then began to winter regularly and in 1977, following the increases elsewhere in the Black Isle, the bay developed into a major centre, with an annual peak of 2-4,000 birds (Table 188). The largest flocks are normally present in autumn or early winter, but a great deal of interchange takes place between here an the neighbouring resorts, and the day to day numbe are likely to fluctuate widely. Pink-footed Geese als occur in some numbers during the March and Apr passage. In the late 1940s, flocks of about 1,000 we recorded regularly, but they later declined and by 196 had virtually disappeared. Then in 1972 they began return, and have since regained their earlier level. Th smaller flocks which used to appear in the autumn n longer occur.

The ducks at Munlochy have been rather les successful. The only major increase has been in th number of Teal, from a regular level of 100 in the lat 1960s to more than 1,000 in recent years. Mallard an Shelducks have remained unchanged, or possibl decreased a little, but Wigeon, after a marked declin in the early 1970s, are now regaining their earlier leve (Table 187).

A short way to the east of Longman Bay, th little inlet of Alturlie Bay contains a modicum saltmarsh, which at times attracts a total of sever hundred ducks. A run of counts in eight of the season between 1958 and 1967 showed a regular level of 23 Wigeon (max 500) and 20 Mallard (max 120), an occasionally a few Teal (max 8), Shelducks (max 46 and Mute Swans (max 11). Counts in 1981-198 produced maxima of 300 Mallard, 344 Teal, 2,12 Wigeon and 180 Shelducks. Clearly the area ha considerably increased in importance for wildfow Greylags also used to roost here regularly, notably in the winters of 1965 and 1966, but, as at Whiteness an Loch Flemington nearby, they failed to establish lasting tradition (p.272). None have been seen in an of the recent November censuses, apart from a flock o

Table 187. *Inner Moray Firth and Beauly Firth: the regular (and maximum) numbers of ducks and swans recorded in:*
a) Munlochy Bay 7 seasons 1975-1982
b) Longman Bay 8 seasons 1975-1982
c) Ness Estuary 9 seasons 1973-1982
d) Beauly Firth 10 seasons 1972-1982

	a	b	c	d
Mallard	77 (200)	145 (321)	59 (235)	446 (840)
Teal	462 (1180)	77 (400)	53 (251)	345 (800)
Wigeon	1321 (5000)	2001 (6000)	0 (69)	973 (2425)
Pintail	0 (0)	23 (230)	0 (1)	69 (129)
Scaup	0 (0)	5 (107)	6 (44)	5 (28)
Tufted Duck	0 (0)	0 (0)	85 (803)	100 (800)
Goldeneye	11 (27)	18 (262)	207 (641)	35 (404)
Merganser	18 (120)	3 (40)	2 (23)	see text
Goosander	1 (20)	0 (0)	0 (1)	and Table 190
Shelduck	134 (230)	72 (248)	0 (11)	55 (153)
Mute Swan	4 (10)	20 (59)	7 (25)	15 (31)
Whooper Swan	0 (2)	0 (0)	0 (0)	20 (88)

0 in 1971, and another of 220 in 1980, although 60 vere present in March 1983.

The tidal reaches of the River Ness are of interest hiefly for the flock of several hundred Goldeneyes, vhich congregates around the mouth and in the Kessock channel, attracted by the effluent from the naltings and the city sewers. The flock may also stray t times into the Beauly Firth or into Longman Bay, vhere the numbers are otherwise quite small (Table 87). Mallard and Teal are normally scarce, but may at mes occur in flocks of 1-200, especially in stormy veather. Tufted Ducks are also recorded on occasions vhen the inland lochs are frozen, either here or on the djacent stretch of the firth off Clachnaharry. In anuary 1981 they totalled upwards of 800. The najority of these were probably from Loch Dochfour, ome 8km inland at the outfall of the river from Loch Ness. Occasional counts from there, and from the Abbar Water close by, suggest a winter peak of 350-550 'ufted Ducks, 100 Mallard and 30 Wigeon, with erhaps some Pochard and Goldeneyes and a few Mute Swans besides. Loch Ness itself is probably leserted except for scattered Goldeneyes, and the ame no doubt applies to most of the hill lochs to the vest of the Great Glen. One possible exception is Loch Laide which is known to hold several breeding pairs of Mallard and Tufted Ducks.

The Beauly Firth, in contrast to the Moray Firth, s wholly estuarine in character. The upper reaches, in articular, contain a great expanse of intertidal mud and sand, 1,000ha in extent, with belts of saltmarsh at the landward edge. The latter are broadest on the southern shore, in the bight between Lentran and Lovat Point, and are backed by a stretch of marshy reclaimed pasture extending inland to Kirkhill. At Lentran there are sizeable stretches of reed-bed, which provide a favourite moulting site for Mallard, and extensive beds of *Zostera* and *Ruppia* are located on the neighbouring flats. *Salicornia* is also plentiful to the west of Lentran, and intermittently in the sectors eastwards to Bunchrew (369). On the northern shore there are further beds of *Zostera* off Redcastle, and some patches of *Ruppia* near Tarradale and Spital Shore; there is also an expanse of saltmarsh, some 2.5km in length, between Tarradale and Milton, but this is frequently disturbed by shooting.

The lower reaches of the firth are a great deal less attractive. The foreshore here is relatively narrow, with little or no saltmarsh, and in many places the muddy sand is overlaid with heavy growths of seaweed. The only area of value is the little bay between Clachnaharry and South Kessock, which at times attracts a total of several hundred Mallard, Teal and Wigeon. Flocks of 150 Tufted Ducks also used to occur in hard weather around the outfall of the Clachnaharry sewer (the same birds probably as those which sometimes appear at the mouth of the River Ness), but the sewer has now been moved, and the numbers are greatly reduced.

The results of the counts along the full length of

Table 188. *November counts of Greylags in the Black Isle and on Loch Eye and the Dornoch Firth, 1975-1983, with corresponding means for three preceding periods. The figures in brackets are the maximum for the period. Updated from Boyd and Ogilvie (1972).*

	Beauly Firth	Munlochy Bay	Cromarty Firth	Loch Eye	Dornoch Firth	Total
Mean 1960-1964	205 (410)	205 (270)	170 (320)	205 (500)	50 (200)	940 (1240)
Mean 1965-1969	360 (700)	65 (260)	610 (2000)	1065 (1850)	6 (18)	2070 (2650)
Mean 1970-1974	650 (1040)	250 (500)	350 (855)	1150 (2100)	22 (100)	2070 (2990)
1975	750	225	500	4000	26	5500
1976	200	100	100	3000	28	3430
1977		2600		12500	30	15130
1978		3700		11500	7	15200
1979		2080		13100	125	15300
1980		9330		8300	500	18130
1981		11850		20800	250	32900
1982		5450		3500	175	9130
1983		850		3130	720	4700

Notes: The prevalence of Greylags in the Black Isle since 1977 has made it impossible to relate the flocks to individual roosts. An exceptional count of 4000 on Munlochy Bay in 1960 is omitted from the mean and the total.

the northern and southern shores, including Clachnaharry Bay, are summarised in Table 187. Except for the diving ducks and possibly the Mallard, the great majority of the birds are normally found on the southern side to the west of Bunchrew, and more especially in the bight between Lentran and Lovat Point. Over the past 20 years the numbers of several species have shown a steady upward trend. Teal increased from a regular level of 150 in the early 1960s to 300 in the late 1970s, Wigeon from 500 to 1,000, and Pintail from 20 to 60.

The flats in the upper firth are also the principal stronghold of the Greylags which feed on the Black Isle and on the reclaimed pastureland towards Kirkhill. These first started to winter regularly in the 1920s, and by 1939 were occurring annually in flocks of many thousands from November until April (50). Thereafter their numbers declined, and from 1960 until 1971 there were seldom more than 700; they then made a rapid recovery and in most recent seasons have reached a peak of 2-3,000, with 6,000 in 1979. The largest numbers are usually present in December or January, when the autumn concentrations at Loch Eye are starting to disperse. In the Black Isle as a whole the current mid-winter peak amounts to about 10,000.

The Pink-footed Geese which used to appear on the firth between February and April in flocks of several thousand are now greatly reduced, and the smaller autumn influxes have virtually ceased to come. In common with the Greylags, they increased enormously during the 1920s and early 1930s, but a few years later were already showing signs of decrease (50). Since 1970 the flock has alternated between here and Munlochy Bay, the latter being perhaps the more favoured. The current total for the two sites is usually between 500 and 800, but larger numbers are on record: 2,000 were present on the fields to the north of the Beauly Firth in March 1977, and 1,000 were found on Munlochy Bay in April 1979 (540).

The upper reaches of the firth are also of outstanding interest for the flock of moulting Canac Geese, which starts to assemble in May and remair until early September. Their remarkable increas from the initial report of 2 in 1947 to a total of 1,100 i 1981, is traced in Table 189. Extensive ringing of th flightless birds has shown that the great majori belong to the population which is based in central an northern Yorkshire; a few are also drawn from the ea and west Midlands, and two are known to have com from as far afield as West Sussex. The moulting floc is composed predominantly of non-breeding bird both adult and immature, and in general the increas can be related to the growth in the Yorkshir population. It has also been suggested that the size the moulting flock increases in years when th breeding success is low; this was certainly so in 198 when a week of cold weather at the end of Apr prevented many birds from nesting. The destructio of nests as a means of controlling the population an lessening agricultural damage may likewise help t explain the rapid increase in the moulting flock notably in the early 1970s. During their stay on th firth, the geese are centred on the flats off Lentran an at the mouth of the firth, coming to the saltmarsh t feed and to the outfall of the Moniack Burn for water

Table 189. *The numbers of Canada Geese moulting on the Beauly Firth, July 1947-1982. From Walker (1970) and* Scottish Bird Reports.

Year	Number	Year	Number
1947	2	1971	378
1950	18	1972	520
1951-52	50	1973	630
1955	22	1974	704
1961	40	1975-80	800-900
1962	120	1981	1100
1963-67	150-200	1982	817
1968-70	225-325		

Table 190. *Peak counts of Goosanders and Red-breasted Mergansers on the Beauly Firth, 1970-1982, and a note of some major gatherings of Mergansers elsewhere on the coast between Findhorn and Golspie during the same period. The letter after each record is the initial of the month, September—March, to which it refers. Spring and summer counts are not included. Records marked * are taken from* Scottish Bird Reports.

	Goosander	Red-breasted Merganser		
	Beauly Firth	Beauly Firth	Elsewhere	
1970-71	800*D	21*M	300*S	Findhor
			1520D	Cromarty Firt
71-72	452*J	23O		
72-73	370J	200*N	200*D	Off Loch Flee
73-74	800*J	250J		
74-75	1200*DJ	210J	325*S	Loch Flee
			275*N	Findhorr
75-76	570N	23S		
76-77	570N	21D	700*O	Tain Bay
77-78	350M	30J	800*O	Culbin Bar
			400F	Off Dornoch
78-79	600D	100M	415O	Cromarty Firth
79-80	900J	133F	495J	Cromarty Firth
80-81	1550*D	2250*D	290D	Cromarty Firth
81-82	1954D	1744D	400N	Riff Bank
			300F	Cromarty Firth
82-83	3189F	1200F	500S	Off Dornoch

Another feature of the firth is the great assembly
sawbills which comes to the tideway in winter, in
ursuit of the shoals of sprats and young herrings.
oosanders are more numerous here than at any other
sort in Britain, the peak in recent seasons amounting
at least 1,000, and sometimes to 2,000 or more (Table
90). The largest flocks occur between November and
nuary, and, in the absence of any sizeable counts
sewhere in the vicinity, are thought to comprise the
ılk of the regional population. Scattered records
om the 1960s confirm that the present numbers have
een coming for at least the past 20 years, but perhaps
gnificantly there is no hint of their presence in any of
ıe earlier accounts (36, 41, 50). The Mergansers are
uch less predictable. Large concentrations are re-
orded in most seasons in one or other of the northern
rths, but the location of the main body may change
om year to year, and often from month to month.
he Beauly Firth has perhaps had rather less than its
hare of the big flocks, but between 1980 and 1982 the
umbers were particularly impressive (Table 190). In
ddition to those listed there are records of 250 in the
vinter of 1955, 400 in 1956, 575 in 1962 and 200 in 1969.
he recent huge gatherings of sawbills cast an unusual
delight on the relationship between *Mergus* and
ıan. The taking of herring in Scottish waters was
anned in 1980 in an effort to conserve the North Sea
tocks, the ban being enforced by the threat of heavy
nes. This effectively halted the netting of sprats as
vell, at any rate in the Beauly Firth, since the shoals
vere often intermingled. The fish were therefore left
o the sawbills and to a host of Cormorants and auks,
ll of which rejoiced exceedingly: even the fish,
naybe, came out of it quite well. The ban was lifted in
984, however, and it will be interesting to see what
ffect this has.

Hitherto the Beauly Firth has suffered very little
legradation, and has been able to absorb some of the
npact of the adverse developments elsewhere. There
re now rumours of a plan to reclaim at least 1,600ha of
he foreshore, including the area used by the dabbling
lucks as a feeding ground, by the Greylags as a roost,
nd by the Canada Geese as their moulting site. Were
his to materialise, the effect on the wildfowl would be
lisastrous, and the ultimate losses even more far-
eaching. In its present state the firth provides a classic
xample of an estuarine ecosystem in which the
vaterborne nutrients from land and sea are combined
nd brought into intimate contact with soil, light and
ir. Such areas are amongst the most naturally fertile
n the world, so long as the system is uncurtailed and
llowed to function properly. In this particular
nstance the abundance of sawbills and other pre-
lators, such as Cormorants *Phalacrocorax carbo* and
auks, is convincing evidence of the high productivity
of the firth, and of its value as a nursery for fish.

The Cromarty Firth

Despite the industrial growth of recent years, the Cromarty Firth is still by far the most important of the northern firths. Taken as a whole it covers some 12,500ha and stretches inland for nearly 30km. On the lower reaches the sheltered deep-water channel is flanked by extensive intertidal flats, especially on the northern shore, and further banks of mud and sand are located at the head of the estuary, bringing the total area of foreshore to about 3,750ha. A detailed survey of the physiography of the firth, and of the wildfowl food resources, has been published by the Nature Conservancy Council (369), and much of this is summarised in the following account.

In recent seasons the firth has held an average autumn peak of about 10,000 wildfowl, the majority of which are centred on the lower reaches in Nigg and Udale Bays. Substantial gatherings are also found on the middle reaches in Alness Bay and around Dalmore, and further flocks occur at the head of the firth in Dingwall Bay and at the Conon mouth (Table 191). Geese are abundant on the neighbouring farmland, but the numbers roosting or feeding on the firth itself are relatively small.

Nigg Bay, in its present state, contains some 1,600ha of intertidal flats, the greater part of which is now included in the Cromarty Firth National Nature Reserve, declared in 1977. To the east, the bay and the adjoining areas of the firth are sheltered from the open sea by the ridge of hills, which rises abruptly to 150m on either side of the narrow entrance. Within the bay, the foreshore is predominantly sandy, with an occasional overlay of mud, especially on the outer banks. The broad-leafed *Zostera angustifolia* and the smaller *Z. noltii* are both abundant throughout the area, and dense beds of *Salicornia* are established along the northern shore on either side of the Balnagowan River. Stretches of saltmarsh are also located in this area and around the northeast corner of the bay, together with a small amount of *Ruppia*. The remainder of the eastern shore, from Nigg village southwards to Dunskeath was formerly flanked by a broad belt of shingle, most of which is now reclaimed for industry. At Dunskeath Ness an oil rig building yard has been in operation since 1972, and more recently the area to the north has been developed as a pipeline terminal, and as a storage and shipment centre for oil from the Beatrice field. In conjunction with this, a deep-water channel has been dredged along the length of the new waterfront to give access to sea-going tankers. The oil itself is untreated, except for the separation of gas.

A good deal of development has also taken place to the west of Nigg Bay in the stretch between Barbaraville and the naval base at Invergordon, which is still operational. The main enterprises here have been the opening of the Invergordon distillery in 1965, the construction of an aluminium smelter on 300ha of land behind Saltburn in 1970, and the establishment of a pipe-coating works on an adjoining site of 20ha in 1971. In each case the factories are set well back from the shore, and the only intrusion on the firth itself is the jetty and cableway which links the smelter (now closed) with the moorings in the deep-water channel.

The impact of the developments to date has been much less serious than was feared ten years ago. At that time there were proposals for a major refinery and petrochemical plant near Barbaraville, and for the reclamation of 250ha of the adjoining foreshore, the purpose of the latter being partly to provide more space for industry and partly to obtain a frontage on the main channel. Neither project has yet progressed, and it now seems unlikely that they ever will. The threat of oil pollution is nevertheless a source of constant anxiety. In the eight seasons between 1969 and 1976 there were no fewer than 26 minor spillages within the firth, which affected a total of well over 5,000 birds, including many hundreds of ducks, geese and swans. With the advent of the oil terminal at Nigg in 1978, the risks have greatly increased.

Udale Bay, on the opposite side of the firth, contains a further 530ha of sandy foreshore, the whole of which is now included in the new National Nature Reserve. Although much smaller than Nigg Bay, the wildfowl food supplies are unusually plentiful, an the two areas are rightly regarded as complementar sections of the one major site. *Zostera* is agai abundant and widespread, especially *Z. angustifoli* and extensive beds of *Salicornia* and *Ruppia* are locate on some of the upper levels. In the western corner the bay there are stretches of saltmarsh around th outfall of the Newhall Burn, with a tract of marsh pastureland beyond. The saltings here include sizeable quantity of *Spartina*, and further stands ar established along the muddy channel of the bur *Spartina* was introduced with a view to reclamatio but the climate is against it and the rate of spread ha hitherto been slow. As in Nigg Bay, the invertebrat fauna is diverse in species, but nowhere very dens there are, however, some fairly extensive mussel bed along the edge of the main channel, with a scatter smaller beds elsewhere. The bay as a whole is quit untouched by industrial development, and none envisaged in the current plans.

The foreshore in Nigg and Udale Bays internationally important for the concentrations Wigeon, which assemble in October and Novembe and remain in dwindling numbers until spring. Th size of the autumn peak has varied greatly over th past 15 seasons, from a total of above 12,000 in 196 (the year in which the counts began) to a trough 3,250 in 1970. Since then it has ranged between 5,00 and 8,500. The sharp decrease in the late 1960s is i keeping with the trends at most of the other resort within the Moray Basin, and cannot be attribute solely to the effects of industrialisation. It seems, i

Table 191. *Cromarty Firth: the regular numbers of ducks and swans in six sections of the firth, 1970-1982, and the regular and maximum numbers in the six sections combined, based on nine seasons of full synchronised counts.*

a) Udale Bay 10 seasons
b) Nigg Bay 13 seasons
c) Barbaraville—Invergordon 13 seasons
d) Dalmore Bay 13 seasons
e) Alness Bay 11 seasons
f) Upper firth above causeway 11 seasons

	a	b	c	d	e	f	Combined reg	Combined max
Mallard	168	262	201	46	77	302	870	1257
Teal	34	4	0	320	158	63	538	1600
Wigeon	2119	3176	17	355	1578	867	7754	15022
Pintail	0	281	0	0	0	0	279	600
Scaup	25	0	8	1	0	0	35	82
Goldeneye	63	4	273	11	3	14	343	577
Merganser	28	45	25	10	29	25	176	1518
Shelduck	54	162	8	56	96	43	413	526
Mute Swan	61	23	49	53	7	25	183	332
Whooper Swan *	4	15	76	43	19	14	see Table 193	

* The figures include birds feeding on the fields a short way inland from the firth.

ıct, that the numbers in 1966 were a good deal higher ıan usual. At that time the new distillery at ıvergordon was discharging large quantities of spent arley into the firth, and this for a while attracted an xceptional concourse of up to 5,000 Wigeon, 500 caup, 1,500 Goldeneyes and 5-600 of both Mute and Vhooper Swans. The largess was then greatly re- uced, and the flocks moved elsewhere. The Wigeon nd Scaup disappeared without trace, the swans went) other resorts nearby, and only the Goldeneyes emained, albeit in much smaller numbers.

Nigg Bay is of further interest for the growing ock of Pintail, now the largest in the north of Scotland Table 191), and for the flocks of Pink-footed Geese, vhich use the flats as their principal roost. The latter re usually present throughout the winter, the umbers increasing from 2-300 in November and)ecember to a March or April peak of 1-2,000 naximum 2,500). They feed either on the arable land) the north and west of the bay, or more often on the Iill of Nigg, where Bayfield Loch at times provides an lternative roost. Greylags, although abundant on och Eye and on the farmland to the north of Nigg, are eldom numerous on the bay itself. The odd flocks vhich happen to appear have mostly been flushed rom the fields nearby, and their visits are usually hort. There is certainly no evidence of a regular roost, r of widespread feeding on the saltmarsh.

On the other side of the firth an apparently eparate group of 2-300 Greylags roosts in Udale Bay nd forages southwards into the Black Isle. The birds oncerned are almost certainly linked with those on Aunlochy Bay and the Beauly Firth, and with the urther group of 2-300, which roosts on the upper firth round Alcaig.

The firth was also renowned at one time for the locks of Light-bellied Brent Geese, which used to vinter in Nigg and Udale Bays. By the 1930s their ormer numbers had already been reduced to less than ıalf by decades of overshooting, but flocks of 4,000 or nore were still occurring regularly, and, moreover, he *Zostera* on which they fed showed no indication of he wasting which affected it elsewhere (50). In the easons prior to 1939 it seemed that the decrease had eased, but within 10 years the firth was utterly leserted, and so it is to this day. One possible cause vas the siting of a bombing range astride the main oost, but the disappearance of the population was so omplete that it may perhaps have been overwhelmed y some much greater war-time disaster.

On the middle reaches of the firth, from nvergordon upstream to the road bridge at Ardullie, he wildfowl are concentrated into the area between)almore and Evanton, the primary centre being in Alness Bay. Elsewhere the foreshore is narrow with long stretches of rock or shingle, overlaid in many places with a heavy growth of seaweed.

Alness Bay is similar in many ways to the inlets of the lower firth. The foreshore, covering more than 500ha, is mostly sandy with a strip of saltmarsh on the landward side, and stretches of mud and extensive mussel scalps along the outer edge. *Zostera* and *Salicornia* are plentiful, the former at the eastern and western ends, and the latter along the length of the northern shore; *Ruppia* is also widespread across the central flats. At the western end, the saltings around the mouth of the River Glass are rather more extensive than elsewhere and include a small lagoon which attracts some quite large flocks of ducks, especially in rough weather. Another favourite retreat is at Alness Point, where a shingle spit encloses a shallow muddy creek, some 800m long and up to 150m wide. The spit itself is clad with whins, and supports a regular breeding population of 20 pairs of Shelducks (369).

The flat land at either end of Alness Bay is zoned for industry, and a sizeable area near Evanton has already been developed for various engineering works. These are fairly unobtrusive and have not yet detracted from the value of the bay. The much more damaging plan for a petrochemical plant has now been dropped, but there is still the possibility of other adverse developments in this somewhat sensitive area.

To the east of Alness Point there are further expanses of foreshore around Dalmore, which are complementary to those in Alness Bay. The beaches here are mostly shingle, especially towards the west, the only areas of sand being along the edge of the channel. These outer banks support a densish growth of *Zostera*, and further beds of both *Zostera* and *Salicornia* occur on either side of Dalmore pier. There is also a small, but favoured, area of saltmarsh around the Alness River mouth. In the 1960s the effluent from the Dalmore distillery was another major attraction, but as at Invergordon the amount of spent grain being discharged to the firth is now greatly reduced. This reduction has led to a notable decrease in the numbers of Mute Swans in the vicinity of the two outfalls (Table 191). In the winter of 1966-67 the flocks totalled 5-600 from November until January, compared with an average peak of 70, and a maximum of 95, in recent years. To some extent the losses here have been offset by an increase at Loch Eye, but the population in the district as a whole has undoubtedly declined, perhaps by as much as 40% in 15 years. One of the probable causes has been the frequent spillages of fuel oil from the naval station at Invergordon, notably in March 1972, November 1975 and October 1976, and from the

Dalmore distillery in January 1974. The casualties included more than 100 Mute Swans.

Whooper Swans appear at times on the Dalmore shore, but are found more often on the fields to the east behind Belleport, where they reach a peak of 50-100 in most years. The largest numbers are usually in December or January, when the big autumn gatherings at Delny and Loch Eye are starting to disperse (Table 193). The foreshore also attracts a winter peak of 5-600 Teal and 2-3,000 Wigeon, the former occurring mainly at Dalmore, and the latter in Alness Bay. The Teal, in particular, have adopted this as their main resort within the firth, and are showing signs of a slight increase. The Wigeon numbers have varied widely from year to year, but on the whole continue to maintain the prosperous levels of the 1960s. The Shelducks, however, are tending to decrease: in recent winters the local flock has declined from a mean peak of 210 to less than 150, while those elsewhere in the firth have maintained a fairly constant level of around 200.

The upper reaches of the firth contain another 1,000ha of intertidal flats, similar to those on the Beauly Firth, which at this point lies less than 10km to the south. On the northern side the shores of Dingwall Bay are flanked by sizeable stretches of saltmarsh, and by broad expanses of mud, giving way to sand towards the outer banks. There are also extensive areas of saltmarsh around the islands at the Conon mouth, and along the southern side between the islands and Alcaig. The foreshore here again consists of mud, but further to the east becomes increasingly sandy and overlaid with seaweed. The food resources include a dense sward of *Salicornia* over the greater part of Dingwall Bay, and some extensive beds of *Zostera* and *Ruppia*, especially on the southern shore. Apart from the waterfront at Dingwall, and the busy road and railway along the northern bank, the surrounds are still completely rural, and are likely to remain so. The new road bridge, completed in the late 1970s, is having no appreciable effect on either birds or habitat; except for a short length of causeway at either end, the tidal flow is unrestricted, and the likelihood of increased silting seems remote. A more serious threat has been the introduction of *Spartina*, with a view to reclaiming parts of Dingwall Bay, but as on the lower firth the rate of increase has been slow.

Despite the apparently favourable conditions, the numbers of wildfowl on the upper firth are relatively small (Table 191). The highest counts since 1966 have amounted to 750 Mallard, 380 Teal, 2,400 Wigeon, 63 Goldeneyes, 100 Shelducks and 85 Mute Swans. There are also records of up to 400 Goosanders, 200 Mergansers and 90 Whooper Swans, but these are exceptional. Greylags roost regularly on tl flats off Dingwall, and feed to the south towards tl Beauly Firth. They normally reach a peak of about 25 but up to 1,400 are on record.

Loch Ussie and Loch Eye

The freshwater lochs around the Cromarty Firth are r less important than the estuary itself. This appli especially to Loch Eye, which in recent seasons ha attracted some astonishing concentrations of up 38,000 geese, 7,500 ducks and 1,200 Whooper Swan Loch Ussie is also important, with a normal peak of least 2,000 ducks, and several hundreds more a likely to occur on the smaller pools and marshes roun about. The breeding population of Mallard, Teal an Tufted Ducks is likewise quite substantial.

The most productive lochs are located in th coastal belt of Old Red Sandstone, which stretche from Beauly to Brora and extends inland for up t 20km. The numerous hill lochs to the west are mostl dour and acid, and of little value at any time of yea The only notable occurrences are the groups Common Scoters which breed on several of the uplan lochs in Ross and northern Inverness-shire. The currently total 10-15 pairs, and appear to be increasin and extending their range.

Loch Ussie lies in a sheltered bowl in the hills some 5km to the west of Dingwall, and is closel associated with the mudflats of the upper firth. It ha an area of 85ha, a maximum depth of 10m, and surrounded by farmland to the north and by broa belts of conifers on the other three sides. The shore are fringed with rush and sedge, and are sheltered i many places by a screen of alder, birch and willow There are also several small islands, all heavil overgrown with trees. The loch is fed mainly b springs, and is moderately nutrient-rich (369). Wigeo are especially numerous, with an annual peak o 1-2,000, and both Pochard and Tufted Ducks ar present regularly, at times in flocks of several hundre (Table 192). Parties of 15-30 Whooper Swans occur i autumn and early winter, and up to 500 Greylags ar said to appear on occasion, presumably from the roos on Dingwall Bay. The feeding grounds of the Wigeo are probably on the upper firth, the loch being use mainly as a day-time roost. The breeding populatio of Mute Swans, Mallard, Teal and Tufted Ducks is als of interest, although many of the young birds ar likely to be taken by the numerous large pike, an possibly by the Herons *Ardea cinerea* which breed o one of the islands.

Another notable place for breeding ducks is o the heathland bogs and pools of Monadh Mor, at the western end of the Black Isle. These at one time ttracted a variety of species, including Pintail and hoveler, but drainage and afforestation in the area ound about has tended to reduce their value. The nly common species now are Teal and Mallard.

Loch Eye owes its importance to a combination f factors, notably to its position midway between the ajor resorts in Nigg Bay and on the Dornoch Firth, to s size (165ha) and shallowness (mean depth 1.2m), to e abundant food supplies within the loch and in the urrounding farmland, and not least to the protection fforded voluntarily by the riparian owners and tatutorily by a Sanctuary Order. The bed of the loch is arpeted throughout with prolific growths of *Pota-ogeton* and other submergent plants, including some f great botanic interest. The shores, however, are ather exposed and, except for a patch of reed-swamp t the eastern end, are mostly composed of wave-vashed stones and gravel. The surrounding land is redominantly arable, except to the west where the atchment includes a large proportion of woodland, eath and pasture.

During recent years the wildfowl of many pecies have increased dramatically, and the loch now anks as one of the most important freshwater centres n the whole of Britain. Its special value lies in the ecurity which it now affords to birds disturbed from e Dornoch and Cromarty Firths, less than 5km away n either hand. It has also become the principal tronghold for Greylags and Whooper Swans in the orthern half of Scotland.

The Greylags, in particular, have increased normously over the past decade, from a normal peak f about 1,000 in the late 1960s to above 10,000 in the easons 1977-1980. Then in 1981 their number in-reased again to a level of 20-25,000 between mid-October and mid-November, and to as many as 38,000 t the peak. This occurred at a time when the harvest n all parts of Scotland was more than usually uccessful, so that scarcely any stubble remained vhen the geese began to arrive – except near Loch Eye vhere 150ha of barley was unaccountably left uncut. he geese were quick to exploit this unexpected treat, nd for a while nearly half of the British – Icelandic opulation could be seen feasting in the one field. hese local developments, which are symptomatic of a eneral increase in the north of Scotland, are already ntroducing some of the problems which have long een a source of concern in the central Lowlands. At och Eye the big concentrations of geese are accused f enriching the water with their droppings and ncreasing the frequency of algal blooms. The fishing s deteriorating (as it did at Loch Leven in similar ircumstances), and the botanical interests, which epend on a delicate balance between the diverse communities, are in danger of upset. There are also rumbles of discontent from some of the neighbouring farms. Despite all this, it is highly desirable that Loch Eye and the other northern resorts should continue to harbour a reasonable share of the British grey geese; otherwise the concentrations elsewhere, already large enough, might well become a problem.

Parties of Greenland White-fronted Geese are often reported as well, and seem to be slowly increasing. Since 1975 the autumn peak has reached an average level of 65, compared with a mean of 40 over the previous ten years. The birds begin to arrive in October, and are present in strength from November to March, either here or on their alternative roost and feeding ground on Morrich More, less than 3km to the north. These are now the southernmost of the east coast resorts, and seem to be quite independent of the other centres in Caithness, some 80km to the north. Pink-footed Geese, from the roost in Nigg Bay, are common enough on the fields to the south and east, but they seldom repair to the loch itself.

The Whooper Swans have several major centres in the district, the chief of which is currently Loch Eye. Over the past 15 years the numbers at the individual sites have fluctuated widely, but the population as a whole has shown a marked increase (Table 193). In autumn, when the numbers are largest, the vast majority of the birds are normally concentrated into two small areas, the one around Loch Eye, the other on the farmland between Barbaraville and Invergordon, and more especially in the neighbourhood of Delny. These concentrations, often totalling 500 or more, persist until December when many of the birds move south, leaving a winter total of 1-200. The flocks then make increasing use of resorts elsewhere in the district, including the fields and foreshore near Dalmore, Alness and Dingwall, the freshwater pool at Loch Ussie, the Skibo Estuary near Dornoch, and the

Table 192. *The regular (and maximum) numbers of ducks and swans on Loch Ussie, 7 seasons 1972-1982, and Loch Eye, 15 seasons 1968-1982.*

	Loch Ussie	Loch Eye
Mallard	46 (150)	340 (1500)
Teal	9 (80)	371 (3000)
Wigeon	1200 (2100)	870 (6000)
Pintail	0 (0)	12 (150)
Tufted Duck	270 (1000)	96 (600)
Pochard	50 (250)	102 (550)
Goldeneye	4 (40)	15 (60)
Mute Swan	7 (17)	118 (265)
Whooper Swan	11 (30)	88 (1200)

saltings and farmland at Tarradale and Lovat Point at the head of the Beauly Firth. Loch Fleet and the mere at Mid-Fearn, near Bonar Bridge, are also used occasionally by flocks of up to 50, mostly in autumn and spring.

Loch Eye has also become the stronghold of the flocks of Mute Swans which used to forgather around the sewage outfalls at Invergordon and Dalmore. Since 1966 the regular numbers on the firth have dwindled from 280 (max 640) to the present level of 50 (max 95), while those at Loch Eye grew from 60 to 175 in the late 1970s, and in one year exceeded 250. The length of their stay at Loch Eye likewise increased. Prior to 1970 the big concentrations appeared only in September and October, the numbers from November to March being normally less than ten, but in the 1970s they stayed in numbers until late winter. In the early 1980s a decline set in, and no more than ten were present in 1982.

The increase in swans was accompanied by an equally striking rise in the numbers of several of the common ducks. The dabbling ducks, in particular, benefited from the swans' habit of pulling up much more of the bottom vegetation than they actually consume. The residue is then left floating on the surface, and provides the ducks with a large amount of food which would otherwise have been beyond their reach. It may be significant that the decline in the swans has been accompanied by a decrease in duck numbers in the 1980s.

The recent upsurge in numbers was especially noticeable in Teal and Wigeon, both of which have achieved a four or fivefold increase. The Wigeo which have twice reached a peak of 6,000, a apparently drawn from the nearby bays of tl Dornoch Firth, where the flocks have corresponding decreased. Teal, on the other hand, have bee showing a gentle upward trend at all their loc haunts; the gain of about 800 at Loch Eye can therefo be viewed as a genuine increase. Mallard hav regained their 1960 level of around 600, after a drop t 200 in the mid 1970s, and Pochard and Tufted Duc have both increased by more than 200 from an earli level of 50.

The Dornoch Firth and Loch Fleet

The Dornoch Firth, although similar in size to th Cromarty Firth and containing about the same amou of foreshore, has less extensive food supplies an harbours rather fewer wildfowl. The estuary nonetheless important internationally as an autum staging place for above 6,000 Wigeon; it is als renowned for its botanical and physiographical fe tures, and not least for the beauty of its setting. Thes values are enhanced from the conservation viewpoi by the virtual absence of industrial and other develo ments. The only obvious intrusion is the nav bombing range on the dunes at Morrich More, but th birds have grown used to this, and it seems to cause n great amount of harm.

The most productive area of the firth is in th 10km stretch of the lower basin between Ardmor Point in the west and Dornoch Point and Morric More in the east. Within this span there are two majc

Table 193. *Changes in the numbers of Whooper Swans at various sites around the northern firths. The table shows the mean of the annual peaks at each site in each of five 3-year periods, 1966-1980, together with the two years 1981-1982, and the corresponding means and maxima for the full 17 seasons. The right-hand column shows the highest synchronised total recorded in each period. The figures in brackets are based on incomplete runs of data; a blank = no records. The sites, and the months of greatest abundance, are as follows:*

a) Loch Eye — November
b) Delny—Invergordon — November
c) Dalmore — January
d) Alness—Dingwall — March
e) Beauly Firth — December
f) Lower Dornoch Firth — December
g) Loch Ussie — Nov/Dec
h) Highest synchronised totals — November

	a	b	c	d	e	f	g	h
Means 1966-68	15	235	60	35	(8)	6		525
1969-71	17	185	50	40	(8)	35		380
1972-74	330	195	25	135	60	65	25	500
1975-77	175	325	50	(30)	75	35	(22)	725
1978-80	630	305	185	30	45	13	(17)	1325
1981-82	250	(35)	75	55	35	19	7	305
1966-82:Means	235	250	75	50	(55)	30	(19)	645
Maxima	1200	750	315	130	90	90	30	1240

esorts on the southern shore, in Tain Bay and dderton (or Cambuscurrie) Bay, and a further two on he northern side, on the Dornoch and Cuthill Sands nd in the small secluded inlet of the Skibo Estuary. hese four sectors, although contiguous, are appreci- bly different in character, which adds considerably to he value and attraction of the basin as a whole (Table 94).

Tain Bay is the largest of the four, with almost ,000ha of intertidal flats. These vary in composition rom pure bare sand at the seaward end around White Jess to mud or muddy sand in the western half of the ay towards Ardjachie Point. The foreshore here, eing a good deal richer and less exposed than it is to he east, supports a widespread growth of *Zostera* and *nteromorpha*; there are also some extensive mussel eds on the outer banks off Tain. The main wildfowl nterest is the gathering of Wigeon which assembles in trength in September, and reaches a peak around the niddle of October. In the twelve seasons prior to 1977 he annual maxima attained a mean of 3,500 and on wo occasions topped 5,000, but in the late 1970s and arly 1980s the numbers throughout the autumn and vinter dwindled to little more than half their earlier evel. This decline was offset by the massive increase n the sanctuary at Loch Eye, some 4km inland, and it nay well be that many of the birds which were ecorded there by day, were continuing to feed on the ay by night. The situation was reversed, however, in 982-83, when 8,275 Wigeon were counted in Tain Bay n September while numbers in Loch Eye were down. There has also been a recent decrease in the counts of Mallard, Pintail and Shelduck in Tain Bay, for which no obvious reason can be found.

Edderton Bay embraces a further 270ha of intertidal mud and sand, and supports a strong and widespread growth of *Zostera*; there is also a small amount of saltmarsh at the inner edge. To the east the flats are separated from Tain Bay by the Meikle Ferry shingle spit, which stretches far out into the centre of the firth and provides an important element of shelter. Wigeon are again abundant, with a normal autumn peak of 2,500 and a winter level of around 500. The peak counts usually occur in the same month as those in Tain Bay, giving a mean total of 5,500 and a maximum of 8,500. Since 1978, however, the numbers here, like those at Tain, have decreased by half, and most of the recent totals have been below 3,000.

Another feature at Edderton is the flock of Scaup, which is normally present from October until March, the numbers increasing from 50 in autumn to a peak of 2-300 in mid or late winter. This is now the last stronghold of the large population which used to frequent the Littleferry and the Dornoch and Moray Firths in the early 1900s (345). The only other resort in current use is the lower Cromarty Firth where the flock of 50-100, which used to occur in the 1960s, has recently returned after an absence of almost a decade. From the monthly counts it seems that a good deal of interchange takes place between the two resorts, but this is probably the limit of their local movement; there is certainly no suggestion that the birds which winter here are the same as those which appear erratically on the Nairn and Moray coast (p.272). At both places the

Table 194. *Lower Dornoch Firth: the regular (and in brackets the maximum) numbers of ducks and swans at:*

a) Tain Bay 15 seasons 1965-1982
b) Edderton (or Cambuscurrie) Bay 16 seasons 1965-1982
c) Skibo Estuary and Lochs Evelix and Ospidale 23 seasons 1960-1982
d) Dornoch and Cuthill Sands 4 seasons 1977-1980

	a	b	c	d
Mallard	244 (600)	92 (350)	110 (650)	119 (350)
Teal	158 (833)	83 (800)	454 (1650)	5 (1200)
Wigeon	2180 (5500)	1573 (4500)	347 (1150)	2077 (12000)
Pintail	13 (90)	39 (170)	0 (26)	23 (200)
Scaup	0 (40)	206 (482)	4 (150)	0 (35)
Tufted Duck	0 (0)	2 (51)	79 (220)	0 (0)
Goldeneye	2 (18)	2 (23)	3 (13)	0 (10)
Merganser	22 (700)	9 (60)	1 (15)	35 (600)
Goosander	3 (100)	0 (30)	0 (24)	0 (0)
Shelduck	108 (232)	20 (60)	8 (50)	43 (105)
Mute Swan	5 (26)	20 (106)	19 (42)	0 (10)
Whooper Swan	0 (15)	2 (26)	16 (70)	0 (16)

flocks are associated with the effluent from distilleries.

A short way to the west of Edderton the 40ha inlet at Ardmore (counted on a few dates in each recent season) provides an alternative resort, which sometimes harbours several hundred ducks. The bay, it seems, is frequently disturbed, and in some recent counts the total has amounted to less than 20 birds. On several occasions, however, there have been flocks of over 100 Wigeon (max 430), and on two dates 2-400 Teal.

The Skibo Estuary, on the opposite shore, is a shallow land-locked inlet some 70ha in extent. Despite its small size it has several attractions which are lacking elsewhere, notably the shelter and seclusion afforded by the policies of the Skibo estate. Its value is also enhanced by the freshwater pools of Loch Evelix (15ha) and Loch Ospidale (7ha) which at one time were arms of the estuary, but are now divided from it by a narrow causeway. The pools and the estuary are, in fact, complementary, and together provide an autumn roost and feeding ground for 1,000-1,500 ducks. Teal are especially well suited and have more than trebled in numbers since the 1950s, to a current peak of 1,000-1,650 (Table 194). Tufted Ducks and Mute and Whooper Swans occur regularly, at times in fair numbers, and Greylags are increasing, both in numbers and regularity. Small parties of Greylags have always been at least occasional autumn and winter visitors to the Skibo Estuary, but in the late 1970s 60-70 were frequently reported in September, presumably from the feral flock at Spinningdale, a short way to the west. Since then, the numbers occurring in the remainder of the season have risen to at least 100 in all months, with 6-700 in autumn and spring. The area must be benefiting from the general increase in Easter Ross (Table 188).

To the east of Skibo the 700ha of the Dornoch and Cuthill Sands attract a regular autumn influx of at least 3,000 Wigeon (maximum 12,000) and a winter total of up to 2,500. There is also a flock of Teal, numbering 1,200 at peak, a regular autumn passage of 50-100 Pintail (exceptionally 200), mostly in October, and a winter population of 100 Mallard and 30-40 Shelducks (Table 194). On the landward side the bay is flanked by a narrow shingle beach, with heath and whin clad ridges of sand and gravel beyond. At Dornoch Point there are areas of saltmarsh, which provide a likely feeding ground, but apart from this the birds are restricted to the open, windswept flats. As at Tain, the sands are overlain in places with a slick of mud, and the food resources are similar to those along the southern shore.

The upper reaches of the firth, from Skibo westwards to the Kyle of Sutherland, are narrow and deepset, and the only large expanses of foreshore a around Dun Creich (150ha) and in the stretch betwee Wester Fearn and Bonar Bridge (270ha). The su roundings also change abruptly from the low-lyin arable land of the coastal plain to acid moorland, risin in places to above 300m. The numbers of wildfowl a normally quite small, and even at the peak are unlike to total more than 1,000. The most interesting sector the bay between Easter and Wester Fearn, where th railway embankment crosses the foreshore and e closes a brackish lagoon of about 16ha (Table 195). Th is used predominantly by Mute Swans and Tufte Ducks, both of which have reached a peak of 100 c more in most recent years. Whooper Swans an Pochard also occur from time to time in company wit several other species, and in 1980 a flock of up to 18 Greylags was present from November until Februar their first appearance in the area. The records from th adjacent foreshore up to Bonar Bridge and from th brackish basin of the Kyle of Sutherland are summa rised in Table 195. There are also a couple of coun from the foreshore below Dun Creich which sugge that it may be an area of some importance. I September 1976 it held 80 Mallard and 500 Wigeor and in January 1972, 376 Mallard and 53 Merganser On the latter occasion synchronised counts were mad in all four sectors of the upper firth; these produced total of 875 ducks, including 480 Mallard, 100 Wigeo and 50-100 Teal, Tufted Ducks, Mergansers and Mut Swans.

At the seaward end beyond Whiteness the firt opens into a shallow sandy bay, some 15km acros the shores of which extend from Tarbat Ness t Golspie. Except for the inlets of Inver Bay and Loc Fleet, the coastline is exposed throughout to the nort and east, and is of interest only for the sea ducks which at times amount to many thousands (pp.274-7).

Much of the southern shore is occupied by th dune system of Morrich More, which covers upward of 1,000ha and extends inland almost to Loch Eye. I comprises, for the most part, an open grassy plain traversed with pools and marshy hollows, and ha long been used as a low-level bombing range. Despit this activity the area is used by a wintering flock o about 100 Greylags and sometimes by 1,000 or mor during spring and autumn passage (369). The flock o 60 Greenland White-fronted Geese which winters o Loch Eye also uses the area as an alternative roost an feeding ground. Pink-footed Geese may possibl occur as well: in 1972 it was rumoured that up to 1,00 might be present for a while in spring, and perhaps i autumn, but this has not yet been confirmed (369).

The sandy flats adjoining Morrich More ar

devoid of vegetation, except along the eastern shore, where patches of saltmarsh, pioneered by *Salicornia*, have developed in the lee of the outer bars. The records from here, and from the pools in the dunes, are fragmentary because of the problems of access, but it seems that there may at times be several hundred Mallard and Wigeon, and up to 100 Shelducks. In Inver Bay, on the landward side of the dunes, the sand is mixed with silt and supports a certain amount of *Zostera* and *Enteromorpha* – enough to attract a regular mid-winter peak of at least 600 Wigeon (Table 195). The rest of the coast to Tarbat Ness is rocky and of little interest, except for the offshore flocks of sea ducks and occasional groups of Mergansers.

From Dornoch north to Golspie, and on to Brora, there are long stretches of open sandy beach, broken at Embo by an outcrop of rock, and at the Littleferry by the narrow entrance to Loch Fleet. Long-tailed Ducks and scoters abound, especially off Embo (pp.274-7), and a thriving colony of Eiders is centred on the dunes around Loch Fleet. Mergansers are also common, occurring usually in scattered flocks, but at times coalescing into larger groups. Counts on Loch Fleet and on the coast up to Golspie suggest a total of about 300, and further flocks are often present to the north and south. In February 1978 an unusually large gathering of 400 was recorded off Dornoch, and another 100 were located elsewhere. There are also regular summer records of 150-250 moulting on the sea off Brora, and a similar flock may perhaps occur off Dornoch.

The Eiders at Loch Fleet have increased remarkably since the mid 1950s, when they first started breeding on the Coul and Ferry Links. During the previous 50 years they had spread sporadically from their early stronghold on the Pentland Skerries, and occasional nests and broods had been found at several points on the north-east coast, notably near Berriedale, Helmsdale and Brora (345, 463). The winter population, which had been sparse, was also starting to increase: in 1938 there were "crowds in the sea off Brora", which was most unusual (41), and by 1960 an autumn flock of up to 500 was occurring regularly at the mouth of Loch Fleet. Since then the breeding colony on the nearby Links has increased and autumn and winter totals of 2,500-3,000 are now recorded annually. In autumn the bulk of the birds are concentrated in the stretch between Embo and Golspie but thereafter they scatter more widely and some move elsewhere, either to the north coast of Grampian or possibly further afield (359).

Loch Fleet itself is a shallow, land-locked basin, sheltered from the sea by the dunes round the narrow mouth, and covering in all about 650ha. The foreshore, which accounts for some 500ha, is predominantly sandy and in places supports an abundance of *Enteromorpha* (492). At the western end the tideway is terminated by the Mound Causeway, built by Telford

Table 195. *Upper Dornoch Firth, Inver Bay and Loch Fleet. The regular (and in brackets the maximum) number of ducks and swans on:*

a) Lagoon and bay between Easter and Wester Fearn 8 seasons 1966-1981
b) Wester Fearn—Bonar Bridge and Kyle of Sutherland 3 seasons 1979-1981
Inver Bay 5 seasons 1973-1982
Loch Fleet 14 seasons 1960-1981

	Upper Dornoch Firth		Inver Bay	Loch Fleet
	a	b		
Mallard	18 (110)	114 (300)	46 (450)	263 (625)
Teal	1 (10)	52 (98)	0 (15)	131 (460)
Wigeon	47 (200)	16 (35)	373 (1200)	987 (1700)
Tufted Duck	97 (320)	58 (128)	0 (4)	0 (1)
Pochard	18 (130)	1 (6)	0 (0)	0 (19)
Goldeneye	7 (120)	25 (47)	2 (8)	12 (37)
Merganser	4 (80)	2 (6)	3 (14)	38 (150)
Goosander	2 (16)	11 (43)	0 (0)	0 (3)
Shelduck	2 (27)	2 (9)	42 (95)	85 (160)
Mute Swan	70 (150)	25 (77)	9 (27)	2 (7)
Whooper Swan	11 (44)	4 (15)	0 (11)	3 (44)

Note: Eiders at Loch Fleet – see text.

in 1816 partly as a bridge and partly to reclaim the estuarine marshes, which at that time ran back to Rogart. The latter are now replaced in the upper half of the valley by good farmland, and in the lower half by a large freshwater swamp, overgrown for the most part by alder but still retaining some open pools. There is also a brackish lagoon of about 16ha just above the causeway. Although cut off from the estuary by sluices, the water level in the lagoon and fresh marsh varies by as much as 15-20cm in the course of each tide, owing to the impoundment of the river when the sluices are closed at high water. This tidal effect adds considerably to the value of the marsh as a feeding ground for Mallard and Teal, both of which are probably more numerous than the figures in Table 195 suggest. Over the past 20 years the numbers of all the common species, including Wigeon and Shelduck, have remained more or less unchanged, implying perhaps that the food supplies are being used to the full. Greylag flocks have recently increased during the autumn; more than 300 were present during the 1983 autumn survey. Throughout the same period a high degree of protection has been afforded, initially by the landowners and tenants, and latterly by the National Nature Reserve established on the Mound Alder-woods in 1966 and by the Scottish Wildlife Trust reserve, which dates from 1970 and covers the whole of the tidal basin.

Taking the northern firths as a whole, it seems that the great majority of species have adapted with surprising success to the annual fluctuations in food supplies and other changes and adversities of the past 15-20 years. Greylags and Eiders have prospered exceedingly, Whooper Swans and Teal have in-creased, and the Wigeon have just about regained their earlier level of 20-25,000, after a major decrease in the early 1970s. The other common species are also holding their own, with the exception of the Mute Swan, whose numbers continue to wane.

At the same time, the distribution of the flocks over the various coastal and inland resorts has changed considerably, in response to local pressure and to the protection now afforded by the new reserves. Some examples of this, and of the recent trends in population, are contained in Table 196, which provides a summary of the records in each of five 3-year periods since 1966 and the most recent two seasons. The figures shown are directly comparable, being based throughout on the same substantial sample of sites. There are, however, a number of other sites which have had to be omitted, because the runs of data are too short. A note of the more important of these is included in the table as a reminder that the true totals are likely to be somewhat higher than thos shown in the two right-hand columns.

The most striking event has been the emergenc of Loch Eye as the principal resort within the regior Since the declaration of the Sanctuary, it has becom the prime centre for Greylags, Whooper Swans an Teal, and has also drawn to itself the bulk of th Wigeon from the Dornoch Firth, and many of th Mute Swans from Invergordon and Dalmore. Anothe major change has been the desertion of Longman Ba by Teal and Pintail, and the adoption of the Beaul Firth and Munlochy Bay in its place. The Pintail on th Dornoch Firth have also decreased, while those on th Cromarty Firth have shown a slow but stead increase.

The new reserves at Munlochy Bay, the Cromar ty Firth, Loch Eye and Loch Fleet have undoubtedl contributed to the present stability of these popula tions. The changes in farming practice, especially th increased growing of winter barley, have also be nefited some species, notably the geese. Nevertheless there are still several major threats, especially o pollution and the possibility of extensive reclamatio for agriculture in the Beauly Firth. Hitherto, majo threats have been partly averted thanks to th constant vigilance of individuals and organisations. I is more than ever important that this vigilance b continued.

Caithness and the north coast

The coast from Brora northwards to Duncansby Head is dominated by long straight stretches of rock and cliff, broken in places by open sandy bays, bu lacking any major inlets which might provide a element of food and shelter. Regular records are sparse, but the inshore waters have been checked from the air, and it seems unlikely that any large assemblies have been overlooked. Eiders now breed a many points along the shore, but in autumn and winter they mostly move away, probably to join the flocks around Loch Fleet and in the Pentland Firth. The only other known occurrences are in Sinclair's Bay, where flocks of up to 200 Long-tailed Ducks and 120 Velvet Scoters are recorded regularly, the largest numbers being in early spring. Spring flocks of 30-40 Common Scoters have also been located here (prob-ably local breeders), and in winter the harbour at Wick has held up to 120 Goldeneyes (540).

The northern coast from Duncansby west to Cape Wrath is equally rugged, but much more indented, and in the eastern half in particular there are several sheltered bays, which at times hold quite large flocks of dabbling ducks and Goldeneyes (Table 197).

Table 196. *Changes in the numbers of selected species at various sites within the northern firths. The table shows the regular numbers at each site in each of five 3-year periods, 1966-1980, together with the two seasons 1981-1982; it also shows the mean annual peak for all sites combined, and the highest total within each period. The sites are grouped as follows, unless otherwise defined.*

a) Longman Bay, Munlochy Bay and Beauly Firth (south shore)
b) Upper Cromarty Firth: Conon Mouth, Shoretown, Dingwall Bay, Alness Bay, Dalmore Bay
c) Lower Cromarty Firth: Udale Bay and Nigg Bay west to Invergordon
d) Loch Eye
e) Dornoch Firth: Tain Bay, Edderton Bay, Skibo Estuary
f) Loch Fleet (no data for 1982)

WIGEON	a	b	c	d	e	f	Mean peak	Maximum
1966-68	3430	2900	7150	880	3920	1180	21900	25700
69-71	2230	1630	3650	240	3240	970	13200	14400
72-74	2470	3560	5290	190	5220	725	19400	21800
75-77	3610		240	4340	1130		19000	
78-80	3830	2290	5030	3380	1900	1040	17800	20500
81-82	4240	2530	7610	35	3230	1470	22300	24000

TEAL	a	b + c	d	e	f	Mean peak	Maximum
1966-68	390	450	50	520	90	1480	1500
69-71	430	600	100	790	100	2620	3420
72-74	540	400	160	530	140	1800	2600
75-77	490	580	230	620	160	2050	2200
78-80	630	540	970	650	170	3130	4800
81-82	990	490	560	1100	270	3970	5420

PINTAIL	a	b + c	d	e	f	Mean peak	Maximum
1966-68	175	140	10	40	0	465	610
69-71	105	225	5	50	0	425	500
72-74	95	235	15	110	0	535	660
75-77	65	280	20	60	0	475	485
78-80	75	300	20	0	0	400	450
81-82	70	390	70	6	0	545	640

Note: Not included above: Dornoch Sands (regular 25, max 100).

MUTE SWAN
Sites as above except b) = Upper Cromarty Firth less Dalmore
bc) = Dalmore—Invergordon—Barbaraville
c) = Udale Bay and Nigg Bay west to Barbaraville

	a	b	bc	c	d	e	Mean peak	Maximum
1966-68	40	30	280	130	40	30	505	700
69-71	30	30	195	95	90	30	375	380
72-74	20	45	135	70	140	40	375	490
75-77	25	35	60		150	30		
78-80	25	15	50	65	170	60	335	390
81-82	35	30	55	110	10	60	245	260

Notes: Not included above: Wester Fearn Lagoon (regular 60, max 170). Loch Fleet has few Mute Swans.

The places concerned are mostly within easy reach of the fertile freshwater lochs of the Caithness plain, and perhaps for this reason are used more extensively than the larger inlets to the west, such as Loch Eriboll and the Kyles of Tongue and Durness, which lie in a zone of acid moorland. Eiders and Long-tailed Ducks are also common and widespread, and some of the islands to the west have wintering flocks of Barnacle Geese.

The Eiders are much more numerous than Table 197 suggests, especially to the east in the districts closest to the long-established colonies on the Pentland Skerries and on Stroma, where 60-70 pairs are known to breed. In November 1973 a survey of the coast between Duncansby Head and Holborn Head, near Thurso, produced a total of 800, while further to the west a summer moulting flock of about 500 is recorded regularly in the bay off Bettyhill (540). On the north coast as a whole, the winter population is probably of the order of 1,000-1,500. The Long-tailed Ducks are reported mostly in April and May, when the flocks come inshore and forgather in a few favoured areas, often near an outfall of freshwater. Their chief resorts at this time are in Dunnet Bay, off Castlehill, and in Balnakeil Bay, near Durness, both of which hold peaks of about 250. The winter counts are usually a good deal smaller, but the birds are more scattered and many of them may be overlooked. The largest records for the period October – February come from Dunnet Bay (max 200), Thurso Bay (100), Balnakeil Bay (80) and Sandside Bay (75).

The Barnacle Geese are an offshoot of the scattered population which frequents the islands off the north-west coast between Skye and Kinlochbervie (p.309). Their spread to the north coast dates from 1938, when a flock took possession of Eilean nan Roan, off the Kyle of Tongue, following the withdrawal of the last of the crofters. By 1959 their number had increased to about 200, and a second flock of similar size was established on Eilean Hoan off the mouth of Loch Eriboll (Table 214 p.310). Since then the two flocks have together maintained a steady winter level of about 400, with occasionally as many as 660 in March, these being the largest numbers that the islands can support for any length of time. Occasional parties of 50-100 are also recorded on the Durness Peninsula, mostly in late winter and during the spring passage. The grazing here is a great deal better than in other parts of Sutherland, owing to the isolated outcrop of Durness Limestone. The effects of this are especially noticeable in the four freshwater lochs

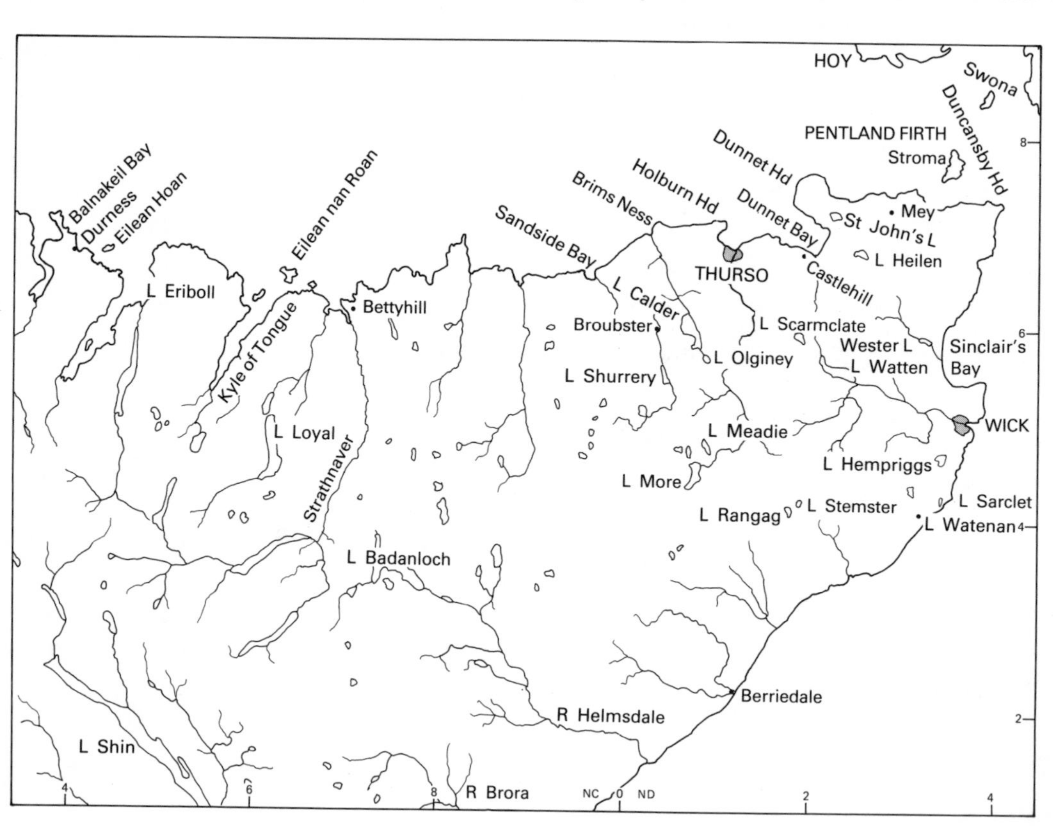

Fig. 47. Caithness and the north coast.

adjacent to Balnakeil village, which support a rich submergent vegetation, including an abundance of *Chara* and *Potamogeton*. Although lacking in shelter and marginal cover, they carry a sizeable breeding population of several species, including Tufted Ducks and Mergansers (492). The winter numbers are unknown.

The numerous freshwater lochs in the Caithness lowlands are also rich in wildfowl food plants, and are well positioned to attract and hold the autumn flow of incoming migrants. Like the lochs in the Mainland of Orkney, they are set on a bed of Old Red Sandstone, in an area of fertile farmland interspersed with tracts of moss and bog. Most of them are shallow, with mean depths of less than 2m, and except in the more exposed positions have a fair amount of marginal cover. In all there are more than a dozen major sites lying within a span of 30km, and covering a total of about 1,200ha. The sites concerned, although closely interrelated, are divisible into three convenient groups, one in the north near Dunnet Bay, another to the south and west of Thurso, and the third to the south and east along the valley of the Wick River. The northern group comprises Loch Heilen (77ha), St John's Loch (79ha), the Loch of Mey (25ha) and the Loch of Bushta (9ha), all of which lie within 8km of one another. The western group is centred on Loch Calder (341ha) and includes Loch Olginey (36ha) and the marshes at Broubster and Westfield; and the larger eastern group contains Loch Scarmclate (77ha), Loch Watten (376ha), the Lochs of Winless (7ha) and Wester (45ha) and Loch Hempriggs (88ha). These are only some of the more important sites; there are many smaller pools and marshes which may at times hold sizeable numbers of ducks and swans especially during the autumn passage. A few of the moorland lochs to the south are also of interest in autumn and winter, notably Loch Rangag (33ha) and Loch Meadie (29ha), but this is exceptional; the great majority of the upland waters in Sutherland and Caithness are of value only in summer, as potential breeding sites.

The lowland lochs are noted chiefly as a centre for Greylags and White-fronted Geese, and as an autumn staging point for Whooper Swans on passage. Pink-footed Geese, in flocks of several hundred, may also appear for a while in spring, and dabbling and diving ducks are numerous and widespread.

The Greylags have increased greatly over the past 20 years, both as a winter visitor and as a resident breeding species. In the late 1950s, after a century of steady decline, the native population had dwindled to an estimated 20-35 breeding pairs and 20-30 non-breeding birds. The former colonies in Caithness had by that time disappeared, and the birds were restricted to a final enclave around Loch Badanloch, in the eastern interior of Sutherland. The wild stock was then reinforced with feral birds brought in from elsewhere, and by the late 1960s at least 60 pairs were breeding regularly. These have since increased to more than 150 pairs, and many of the earlier sites are now reoccupied. The non-breeding birds have also increased, following a series of successful seasons in the 1970s. In recent summers the moulting flock on Loch Badanloch has amounted to at least 280, and the group as a whole may exceed 500. The two main breeding areas are in the western part of Caithness, to the north and south of Loch Calder, and in central Sutherland, in upper Strathnaver and around Loch

Table 197. *Maximum counts of ducks at various sites on the north coast of Caithness and Sutherland in the period September to mid-May. The figures in brackets show the number of occasions on which counts have been made. A blank = no data. Records marked * are taken from* Scottish Bird Reports.

		Mallard	Wigeon	Goldeneye	Eider	L-t Duck	Merganser	Shelduck
Gills Bay	(3)	80	50	10	100	1	1	0
Harrow Harbour	(2)	36	7	50	20	0	2	0
Dunnet Bay	(7)	3	220	150*	15	300*	4	5
Murkle Bay	(21)	40	2	5	20	30	5	6
Thurso Bay	(38)	63	360	78*	101	100	8	1
Brims Ness	(33)	50	60	1	180	3	2	2
Sandside Bay	(40)	100	350	4	58	75	1	23*
Kyle of Tongue	(19)	100	18	7	65	0	20	34
Loch Eriboll	(1)	3	0	1	37	0	8	0
Balnakeil Bay						250*		

Notes: Single counts at Scotland's Haven, Melvich Bay and Torrisdale Bay produced totals of less than 20 ducks. Most of the counts at Sandside Bay and Kyle of Tongue were made between 1948 and 1955; all other counts were made between 1966 and 1982.

Loyal and Loch Badanloch. They have also bred around Loch Shin and are suspected of having done so in several other districts (542).

Another feral flock is established on Loch Brora, some 25km to the south of Badanloch. This originated from two broods of wild Greylags introduced from the west coast in 1937 to replace the native colony which had died out some 70 years before. By 1952 the flock had increased to about 200, and was starting to spread to fresh localities. Their numbers were then reduced to 30-40, and have since remained at about or just above that level.

In autumn the Sutherland Greylags move northwards and eastwards to join the resident birds in Caithness, and remain with them until spring (540). Prior to 1973 it seems that these two groups comprised the sum of the autumn and winter population. The area then began to attract an increasing number of the migrants from Iceland, which had previously passed through without stopping. The largest flocks, which now total 4-6,000, arrive in late October and are present for perhaps six weeks (Table 198). The numbers then fall to a winter level of 800-1,000, rather more than half of which is accounted for by residents. In the course of their stay the flocks move freely between the three main roosts on Loch Calder, Loch Heilen and the Lochs of Scarmclate and Watten, none of which are more than 20km apart. They also roost from time to time at places such as Dale Moss and the Lochs of Scrabster, Wester and Winless, but seldom in large numbers.

The White-fronted Geese have only recently returned to Caithness after an apparent absence of almost a century (41, 50). In 1868 they were said to be not uncommon, but none were recorded thereafter until 1962 when a flock of 19 was discovered near

Table 198. *Peak counts of Greylags and Greenland White-fronted Geese in four areas of Caithness, 1978-1982, and the mean of the annual peaks (and maxima) in four preceding 3-year periods, based on Ruttledge and Ogilvie 1979, Stroud 1983 and regular wildfowl and grey goose counts. The totals in the right-hand column are conservative estimates from counts made on or about the same date. A blank = inadequate data. The areas comprise:*

a) Loch Calder, Loch Olginey and marshes at Westfield, Broubster and Dale
b) Loch Heilen, St John's Loch and Loch of Mey
c) Lochs Watten, Scarmclate and Toftingall
d) Lochs of Wester, Killimster and Winless

GREYLAG	Calder	Heilen	Watten	Wester	Total
Means 1966-68	170 (275)	30 (50)	(230)	(200)	250 (350)
Means 1969-71	135 (295)	255 (510)			340 (515)
Means 1972-74	535 (680)	545 (1000)	(900)		1070 (2150)
Means 1975-77	(4000)	1170 (2000)	(2000)		3520 (6000)
1978	405		4500	0	6750
1979			615		3250
1980	850	825	3050		4100
1981		3500		10	4300
1982					4140

GREENLAND WHITE-FRONTED GEESE	Calder	Heilen	Watten	Wester	Total
Means 1966-68	95 (115)	130 (150)		80 (140)	230 (260)
Means 1969-71	160 (200)	155 (250)	(150)	(300)	290 (320)
Means 1972-74	140 (200)	370 (615)	100 (150)	60 (95)	485 (730)
Means 1975-77	250 (315)	195 (300)	60 (90)	(90)	370 (400)
1978	370	280	130		640
1979					285
1980	220	180	250	0	410
1981					355
1982	246	160	160		464

Note: The 1982 and 1983 surveys recorded 224 and 91 Whitefronts respectively in November at Stemster, well to the south of the other centres, a concentration hitherto unrecorded.

Castletown. In the following seasons there were records of erratic but increasing numbers in autumn and spring, and in 1968 a flock of about 100 remained through the winter. Their progress since then is shown in Table 198.

The present winter population is made up of two distinct groups, one based on Loch Calder, the other on Loch Heilen and the valley of the Wick River. The Whitefronts at Calder roost either on the main loch or on one of the neighbouring pools, and feed on the adjacent marshes and farmland, notably at Broubster, Westfield and Knockglass. Except in hard weather, when they sometimes shift to the coast, they seem to be tied to this one small area, seldom moving more than 5km from the central roost. The group at Loch Heilen ranges more widely. For some reason the loch itself is not much favoured as a roost although the birds feed extensively on the nearby fields and marshes, and as many as half of them often resort to an alternative roost on Loch Meadie, some 26km to the south-west. In doing so they pass directly over Loch Scarmclate, and the flocks occurring there and on Loch Watten are presumed to be part of this group. The same may also be true of the flocks which occur from time to time on the Lochs of Winless and Wester, and elsewhere in the Wick Valley. In recent years the numbers at Scarmclate and Watten have tended to increase, while those at Heilen have declined following the drainage of a favourite feeding area on the marshy grassland to the west of the loch. In March 1983, 160 were present on the Loch of Mey, 5km north of Heilen (which is generally used after January), and these are likely to be the same birds. Apart from these local redistributions the flocks appear to be reasonably stable, and free from any serious threats. The shooting pressure was negligible even before protection, and the birds have adapted to feeding on farmland, thus largely offsetting the losses through drainage of their traditional mossland habitat (although a few do still resort to the remaining bogs at times). In the broader context, the success of the flocks here is some compensation for the sudden, and possibly related decrease in Orkney, which occurred in 1975 (p.327).

Whooper Swans have been found on nearly all the lowland lochs, at one time or another during the past 15 years. The largest flocks, totalling up to 350, are normally present for a month or so in autumn, but their distribution follows no set pattern, and places which hold big numbers in some years may be almost deserted in others (Table 199). This unpredictability, coupled with a shortage of data, has inhibited the study of trends and the assessment of total numbers, except in the past few years when a run of censuses has revealed a regular autumn level of 8-900. Isolated counts outside the areas listed include 212 at Scrabster and 204 at Hestigrew on the same day in November 1982, and 250 at Halkirk later in the month (540). By December the numbers are greatly reduced, but at least one flock of 1-200 is normally present until January, or later in a mild winter. The spring passage is less pronounced, and totals of more than 300 are probably uncommon. The only suggestion of a possible change in numbers comes from the Loch of Wester, where the present regular gatherings of 2-300 are a recent development, dating from the mid 1970s. Recent censuses of Whooper Swans have shown that the autumn concentrations in Caithness and in other parts of north-east Scotland are separate populations,

Table 199. *Maximum counts of Whooper Swans at their main resorts in Caithness, 1975-1981, and, in the left-hand column, the highest count in the period 1966-1974. The totals below are based on counts made on or about the same day. Records marked * are taken from* Scottish Bird Reports.

	1966-1974	1975	1976	1977	1978	1979	1980	1981
Loch Heilen	302	300*	60	250*	7	10	49	110
St John's L	34		170*		4	0	37	
Loch of Mey	54*			5	350*	1		
L Calder	373*	28	20	86	160	365*	72	
L Scarmclate	4	3	46*		20	18	67	21
L Watten	72	22	3	29	8	19	23	
L Winless/Reiss				261		140*	370*	
Loch of Wester	16			309*			250	322*
L Watenan						24*	48*	
Total	387				916*	781*		

Notes: The only available count for 1982 is of 150 at Loch Scarmclate.
The maxima at a number of less important sites are included in Table 201.

rather than a single group moving from one place to another (540).

Mute Swans are rather less numerous than they were in the early 1970s, when the population was expanding and totals of 100-125 were recorded regularly. Prior to that the normal level was 80-90, compared with 60-70 in the past few years. In winter they tend to concentrate on Lochs Scarmclate and Watten, or occasionally on Heilen, and are generally scarce elsewhere; in summer, however, there are breeding pairs on most of the lowland waters. As in Orkney, the population appears to be resident and wholly self-contained.

The winter population of dabbling and diving ducks is often substantial though the numbers at any one place are seldom outstandingly large (Tables 200, 201). In several of the seasons between 1966 and 1975 the counts from a sample of 10-15 sites produced January totals of 5-6,000, but the usage varies from year to year and the current level of 3-4,000 is perhaps more representative. The most favoured resort is Loch Watten, with a normal peak of around 2,000 ducks. Loch Heilen, St John's Loch and Loch Calder used to hold regular totals of above 1,000 each, but latterly the flocks have diminished to less than 500. The reason for this is perhaps a redistribution of the flocks to other parts of north-east Scotland, rather than an adverse change in the local conditions, of which there is no

Table 200. *The regular (and maximum) numbers of ducks and swans at eight freshwater lochs in Caithness. The figures below the place names show the number of seasons, and the periods, in which comprehensive counts were made. The regular figures for Whooper Swans tend to be misleading and only the maxima are shown (see text and Table 199).*

	L Heilen 12:1962-81	St John's L 9:1963-79	Loch of Mey 3:1977-79	L Olginey 11:1968-81	L Calder 14:1967-80	L Scarmclate 8:1974-82	L Watten 3:1979-81	L of Wester 2:1980-81
Mallard	367 (1000)	110 (500)	85 (240)	89 (305)	350 (935)	262 (1000)	870 (1350)	53 (88)
Teal	138 (400)	95 (600)	100 (200)	4 (20)	110 (380)	100 (500)	58 (150)	21 (50)
Wigeon	222 (800)	115 (600)	105 (250)	41 (200)	335 (1200)	116 (600)	830 (2700)	32 (66)
Tufted Duck	32 (200)	90 (220)	2 (12)	1 (20)	27 (115)	90 (250)	309 (523)	17 (42)
Pochard	32 (100)	120 (1200)	1 (8)	23 (80)	10 (75)	74 (300)	112 (310)	5 (17)
Goldeneye	91 (400)	90 (400)	4 (30)	3 (9)	23 (100)	2 (17)	37 (132)	21 (81)
Mute Swan	14 (64)	3 (13)	2 (8)	1 (7)	2 (4)	30 (65)	47 (112)	7 (11)
Whooper Swan	(302)	(34)	(7)	(21)	(160)	(67)	(72)	(261)

Table 201. *Maximum counts of ducks and swans at another 13 freshwater lochs in eastern and northern Caithness. The figures in brackets show the number of occasions on which counts were made.*

		Mallard	Teal	Wigeon	Tufted Duck	Pochard	Goldeneye	Whooper Swan
EASTERN								
L Rangag	(4)	60	0	0	5	200	2	0
L Stemster	(2)	22	0	0	0	0	3	0
L Watenan	(1)	2	0	0	3	0	0	48
L Sarclet	(6)	100	3	45	33	55	22	4
L Hempriggs	(5)	185	10	500	18	50	4	0
L of Winless	(21)	200	50	14	7	3	3	9
NORTHERN								
L of Bushta	(6)	450	0	65	0	0	0	0
Scrabster L	(5)	16	35	13	2	5	1	11
L'an Buidhe	(38)	100	25	65	10	15	0	13
Broubster Leans	(44)	75	30	35	10	10	0	37
Westfield Marsh	(26)	110	30	55	5	2	0	20
L Shurrery	(3)	12	0	14	0	0	1	15
L Saorach	(5)	6	0	2	0	0	0	9

Notes: Counts at L More (2), L Meadie (1) and L. Thormaid (5) produced totals of less than 15 ducks. Mute Swans were recorded only at Lochan Buidhe, max 2; L Shurrery, 2, and Broubster, 3.

obvious sign. The counts at Scarmclate showed a big increase in 1982, with over 2,000 ducks present in December, but recent data are insufficient to confirm whether a similar change has occurred elsewhere.

The results of the counts at eight of the main resorts are summarised in Table 200, and are followed in Table 201 by a note of the maxima recorded at another 13 less important sites. The latter should be treated with caution, being based for the most part on short runs of data, but it can perhaps be used as a guide to the numbers and distribution of the species. Mallard and Wigeon are common and widespread, and in bumper years may well total 3-4,000 each; the highest synchronised counts have in fact amounted to 2,500 and 3,500 respectively. The counts of Teal have on several occasions totalled from 500 to 900, but the flocks away from the main resorts are usually small and peaks of more than 1,000 seem unlikely. Pochard and Tufted Ducks are seldom reported in flocks of more than 100, except on St John's Loch and on Watten and Scarmclate. In recent years the counts of both species have tended to increase to a current level of 3-400, suggesting a true total of perhaps 500 each. From time to time, however, much larger gatherings appear, presumably from the neighbouring resorts in Orkney, which together hold several thousand diving ducks. Examples include flocks of 500 Tufted Ducks on Loch Watten in November 1971, 737 there in September 1982, and 1,200 Pochard on St John's Loch in November 1973. Occasional influxes of Goldeneyes have also been recorded on Loch Heilen and St John's Loch, notably in the early 1970s when totals of 5-600 were not uncommon on these two lochs combined. More usually the count on all the inland lochs amounts to less than 200.

The breeding population is large and diverse, and widely distributed over both the lowland and upland lochs. The commonest species are Mallard, Wigeon and Teal, followed by Tufted Ducks and Mergansers. Up to 4 pairs of Pintail and a few pairs of Shoveler are also recorded regularly, and Common Scoters are now well established on several of the hill lochs. The Scoter have been breeding hereabouts for at least 100 years, and appear to be increasing slowly. Their most successful season in recent years was in 1976 when 45 females were present on the lochs, and 65 ducklings were seen (540). In most seasons the number of pairs has varied between 10 and 20, but the number of broods has not exceeded six.

North-west Scotland

Argyll and the Inner Hebrides

Loch Fyne and the eastern coasts of Knapdale and Kintyre have already been described in the section on the Clyde (p.234). The present section covers the long and tortuous stretch of the Atlantic coast from the Mull of Kintyre northwards to Mallaig, and includes the three main island groups of Islay and Jura, Mull, and Tiree. The account has benefited considerably from the work and comments of D.A. Stroud. The outstanding features of the area are the great concentrations of Barnacle and White-fronted Geese which occur on Islay and elsewhere. Greylags also winter in some numbers, and Eiders and Mergansers are plentiful in many parts.

West Kintyre, Knapdale and Lorn

The main resorts in Kintyre are at Machrihanish, to the west of Campbeltown, and at Rhunahaorine, some 30km to the north; the Sound of Gigha and the sheltered waters of West Loch Tarbert are also of substantial value. Elsewhere the coast is open and exposed, with long straight stretches of beach and rock, from which the land rises steeply to crests of 3-400m along the watershed. The various hill lochs are still largely unsurveyed, but most of them are high and bleak, surrounded by forestry plantations and probably devoid of ducks, though some may serve as roosts for White-fronted Geese.

Table 202. *Peak counts of Greenland White-fronted Geese at Machrihanish and Rhunahaorine, 1980 to 1982, and mean annual maxima in the three preceding 5-year periods. The figures in brackets are the highest counts in the periods concerned. Summarised and updated from Ruttledge and Ogilvie (1979).*

	Machrihanish	Rhunahaorine
Mean 1965-1969	277 (375)	408 (520)
Mean 1970-1974	261 (400)	543 (700)
Mean 1975-1979	363 (450)	543 (627)
1980	450	755
1981	70*	570
Nov 1983	400	763

* An undercount.

Machrihanish and Rhunahaorine are of special importance for the flocks of Greenland White-fronted Geese, which have wintered regularly for many years. In recent seasons their numbers have tended to increase and the total for the two areas now amounts to about 1,000, in the region of 5% of the world population (Table 202). The Machrihanish flock of 3-400 feeds mainly on the low-lying arable land and pastureland to the south of the airfield, and roosts on Tangy Loch some 7km to the north, or alternatively on Lussa Loch nearby. The geese may also feed on the fields adjacent to the roost. Disturbance has not been a problem hitherto, but is nonetheless a constant threat: the airfield has a military as well as a civil role, and although the traffic is at present light, it could increase at any time; there are also plans, currently in abeyance, for an oil storage depot on part of the goose feeding grounds.

The Whitefronts at Rhunahaorine, first colonised in 1934, have increased substantially, especially in recent years and now total 7-800. The feeding grounds are on the triangle of arable and pastureland between the highway and the sea, and more occasionally on some of the rougher fields, arable land and bogs. At least four separate roosting areas have so far been identified: up to 100 move 6-8km eastwards to Loch Garasdale and the neighbouring hill lochs; others use the pools on the moss adjacent to the feeding grounds, and in calm weather they tend to stay on the open fields or move to the sea nearby. Disturbance is light, and the shooting well controlled.

Greylags winter regularly in at least five areas of Knapdale and Kintyre, the total numbers varying between 1,000 and 1,500. The main centres in Kintyre are at Machrihanish and Rhunahaorine, where the flocks are often mingled with the Whitefronts. Parties of 50-100 are also reported from time to time at Corran, on the southern shore of West Loch Tarbert. The resorts in Knapdale are at Carse, just north of West Loch Tarbert (whence come the birds at Corran), and on Moine Mhor (Crinan Moss), near Lochgilphead. Small flocks are reported from a few other sites from time to time, for example near Southend on the southern tip of Kintyre, but their status is uncertain because of the paucity of counts. The November counts in Table 203 give a fair indication of the winter numbers, except that some of the recent figures from Machrihanish and Rhunahaorine are demonstrably too low (the counts having been made before the birds had all arrived). In each case the late-winter counts were a good deal higher and more consistent. Allowing for this it seems that the levels of population have remained virtually unchanged at least since 1965, and probably since 1960.

The Sound of Gigha is remarkable for the large concentrations of Mergansers which forgather during May and June, and remain until the end of August. Counts in five recent summers have shown a mean of 1,400, with a maximum of 1,700 in 1978 (540). The numbers at other times of year are not recorded, but are probably much smaller. There are also some recent

Fig. 48. South-west Argyll, Islay and Jura.

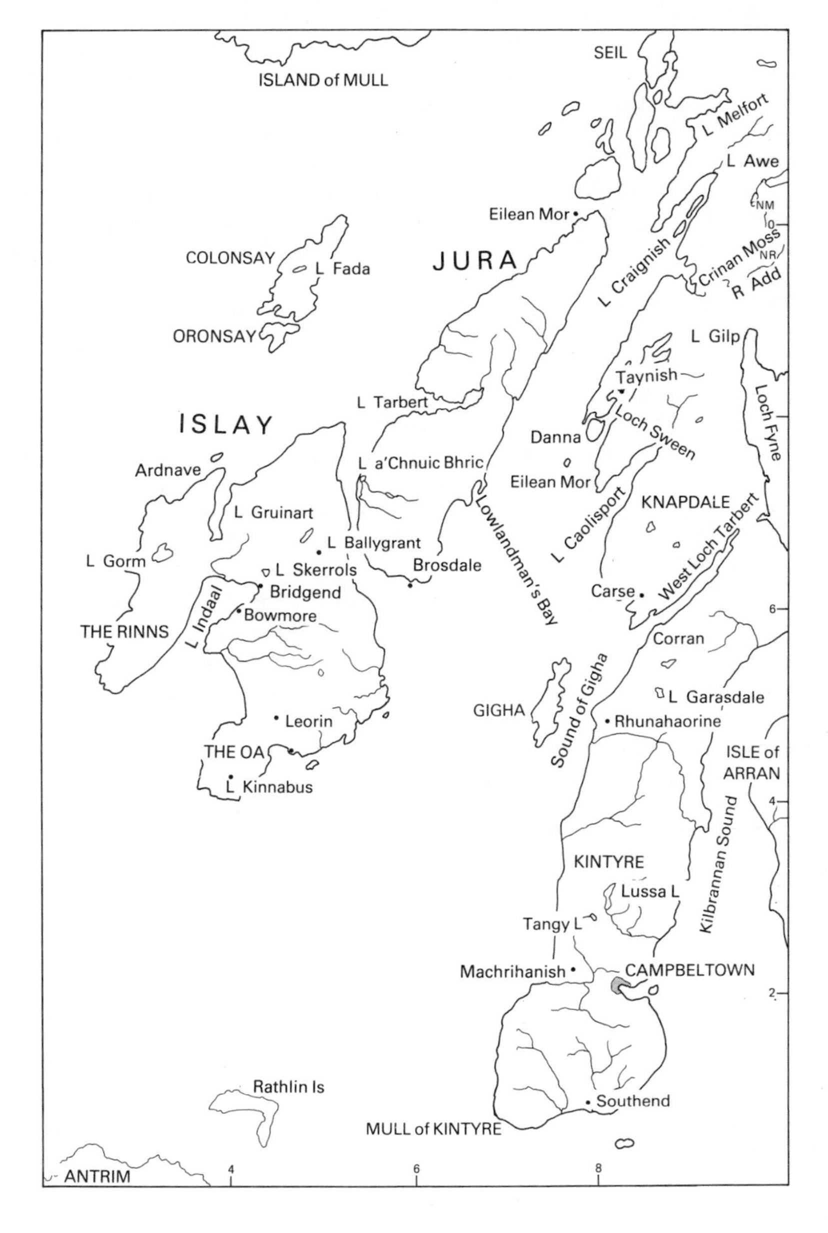

winter records of up to 80 Velvet Scoters, 200 Common Scoters and 300 Eiders.

A short way to the north the long narrow inlet of West Loch Tarbert holds a winter population of 20-60 Mergansers and 100-250 Eiders (rarely up to 1,000), and at times has quite large numbers of other ducks. The latter are usually centred on the sheltered shallows at the eastern end, especially in rough weather, but are sometimes scattered in little groups along the shore. The largest counts have totalled 400 Wigeon, 60 Mallard, 45 Teal, 60 Goldeneye and a few Mute Swans. Greylags occur from time to time, and parties of up to 20 Barnacle Geese have been seen, though much less frequently, on Eilean Traighe at the mouth of the loch.

Occasional counts from Tangy Loch and Lussa Loch in the hills above Machrihanish suggest a January total of perhaps 100 Mallard, 60 Pochard and 30-40 Teal and Wigeon. There is also a report of 166 Mallard near Machrihanish in August 1974 (540). These are the only inland records from Kintyre, and are probably much higher than those on the other freshwater lochs.

To the north of West Loch Tarbert the shores of Knapdale and the Firth of Lorn are sheltered from the Atlantic by the islands of Islay and Jura, and the many bays and sea lochs provide a series of resorts for sizeable numbers of ducks and swans. Greylags and Barnacle Geese occur regularly, and Greenland White-fronted Geese have also become established during recent years.

The main resort of the Greylags is on Crinan Moss, near Lochgilphead, where their numbers increased from about 100 in the late 1950s to a steady level of 3-400 throughout the 1970s (Table 203). The flock arrives in late October, and remains until March moving only a short way from the roost on the estuary to the feeding grounds on the adjacent moss and farmland. A small flock of 30-60 Greenland Whitefronts has re-adopted the site in the 1980s after deserting it in the previous decade.

The Barnacle Geese are based on the islands off the mouth of Loch Sween, notably on Eilean Mor, which provides both the roost and the principal feeding ground. Small parties also wander across to feed on Danna, and on the low-lying grassland towards Ulva and Taynish. In the 1960s they were thought to total about 70; the present estimate is 150-180, but 3-400 may at times pass through (Table 212 p.309). White-fronted Geese have also taken to wintering in this area during recent years. Prior to the autumn of 1977 they were seen only occasionally, but are now present regularly from October through to April. Their favourite feeding ground is on Danna, where there are usually 40-100 but occasionally up to 200. They also feed around the Ulva lagoons, in parties of up to 50, and on Eilean Mor and the neighbouring offshore islands.

The total number of ducks on the coast between West Loch Tarbert and Oban is estimated at about 1,500. Regular winter counts on six of the larger inlets show a combined level of around 1,250, and further small flocks no doubt occur at several other points (Table 204). Wigeon predominate, with an annual peak of 6-800, and in common with the other ducks have shown no sign of change in recent years. Eiders are not much in evidence in the inlets, but are possibly more numerous along the open coast. One of their favourite resorts is in Oban Bay, where up to 200 are sometimes seen. Mergansers are also rather scarce

Table 203. *November counts of Greylags at five sites in Knapdale and Kintyre, 1980-1983, and mean November counts in the four preceding 5-year periods. The figures in brackets are the highest counts in the periods concerned. A blank = no data. Summarised and updated from Boyd and Ogilvie (1972).*

	Crinan	Carse	Rhunahaorine	Machrihanish	Southend
Mean 1960-1964	272 (500)				
Mean 1965-1969	348 (527)	356 (400)	111 (205)	94 (116)	
Mean 1970-1974	299 (330)	416 (500)	104 (218)	102 (215)	43 (139)
Mean 1975-1979	347 (450)	435 (500)	193 (370)	123 (292)	
1980	405	100	200	170	
1981	350		40	0	
1982	426	160	120	24	
1983	437	50	225	245	

Notes: Late-winter counts (February to April) 1978-79 to 1982-83 have shown the following means (and maxima): Rhunahaorine 210 (390), Machrihanish 240 (590).
A count of 69 was made at Southend in March 1984.

except in late summer. In addition to the counts in the table there are isolated records of up to 15 Mallard and 15 Scaup on Loch Melfort, and of half a dozen Eiders and Mergansers on upper Loch Caolisport, the former in December, the latter in March.

Whooper Swans are plentiful in autumn, and quite large flocks are often present until late December, when they move elsewhere. The size of the autumn passage has fluctuated widely over the past 20 years, especially on Loch Sween, where up to 200 were reported in the 1950s. By 1961 their numbers were greatly reduced, and for the next 12 years they seldom totalled more than 20; they then increased to the present level of 40-50. Fluctuations such as this have occurred in several other parts of western Scotland, and probably reflect a change in distribution rather than a trend in population.

Mute Swans are present throughout the winter, occurring in substantial numbers on all the main resorts. The largest numbers are in October, when totals of 90-100 are recorded in most years; the winter level is usually 40-50, reducing to 30 in March.

The numerous inland waters of Knapdale and Lorn are much less favoured than the coastal bays. Many of them are set in open moorland at altitudes of 2-300m, and are almost certainly deserted. Those at lower levels hold a scattering of Mallard and Goldeneyes, usually in flocks of 5-10; small parties of Whooper Swans also appear in many places. The following winter maxima, collected at various times over the past 25 years, are a good deal larger than most other records: 80 Mallard at Loch Coille-Bharr, near Loch Sween; 18 Teal and 34 Wigeon on the lochs in Glen Lonan, near Oban; 25 Pochard and 24 Tufted Ducks on Loch Seil; 28 Goldeneyes on Loch Leathan, Kilmichael; 11 Whooper Swans at the north-east end of Loch Awe; and 21 Mute Swans on Loch Ederline, near Ford.

Islay, Jura and Colonsay

The wildfowl of the Inner Hebrides are rather more diverse, and in places a great deal more numerous, than those on the adjoining mainland. The importance of the individual islands varies enormously, depending on their topography and on the amount of disturbance from farming and other activities. The most valuable, or potentially valuable, areas are those in which there are outcrops of limestone, or extensive stretches of shell sand and machair. The geese, in particular, are attracted to these districts by the richness of the grazing, and the conflict with farming interests is in places severe.

This problem is nowhere more acute than on Islay, which currently supports a winter population of up to 20,000 Barnacle Geese and 3-4,000 White-fronted Geese, within an area of less than 50,000ha (Table 205). Thanks to the policy of the large estates the flocks have hitherto been shielded from excessive shooting disturbance, and this has allowed them to continue in such large numbers. The present shooting pressure is, however, a good deal more severe than it was in the seasons prior to 1976, when the open season for Barnacle Geese was extended from two months to five. The number of visiting wildfowlers has also grown, and this, coupled with the longer season, has effectively checked the earlier increase in the population of both species. The annual kill in the late 1970s and early 1980s was estimated at 1,500-2,000 Barnacle Geese and caused a decline in the island population, but more effective licensing control and the establish-

Table 204. *The regular (and maximum) numbers of ducks and swans recorded on six coastal sites between West Loch Tarbert and Oban.*

a) Loch Mhuirich and Loch of Keills (Loch Sween) — 24 seasons 1957-1982
b) Loch Crinan (Add Estuary) — 4 seasons 1970-1973
c) Loch Craignish — 8 seasons 1967-1974
d) Balvicar Bay, Seil Sound — 8 seasons 1967-1974
e) Ardencaple Bay, North Seil — 8 seasons 1967-1974
f) Loch Feochan — 3 seasons 1967-1973

	a	b	c	d	e	f
Mallard	27 (104)	67 (141)	23 (65)	18 (60)	13 (41)	35 (60)
Teal	27 (140)	44 (65)	0 (3)	0 (2)	12 (30)	21 (66)
Wigeon	161 (700)	234 (365)	196 (442)	125 (225)	29 (80)	59 (129)
Goldeneye	9 (83)	3 (10)	22 (59)	4 (16)	1 (4)	18 (26)
Eider	18 (70)	14 (56)	38 (89)	2 (12)	0 (4)	12 (22)
Merganser	10 (52)	41 (152)	26 (58)	9 (18)	3 (10)	14 (19)
Shelduck	10 (35)	17 (40)	15 (48)	0 (0)	0 (4)	0 (1)
Whooper Swan	17 (120)	0 (0)	22 (60)	12 (36)	0 (2)	0 (0)
Mute Swan	11 (50)	8 (13)	34 (56)	25 (49)	5 (23)	14 (26)

ment of an RSPB reserve on one of the main feeding areas will hopefully check this (see also p.387).

The Barnacle Geese belong to the Greenland population. Islay has always been a major stronghold, at least since 1870, and has recently become of paramount importance. Over the past 20 years the numbers on the island have almost trebled, and now comprise about 60% of the Greenland stock, and 20% of the entire world population. The principal roosts are on Loch Gruinart and Loch Indaal, both of which have held at least 10,000 geese in recent winters. The locations of the feeding grounds are shown on Fig 49. These are mostly on improved pastureland, the area of which has been greatly enlarged since the 1950s. The geese have also taken to feeding on the barley stubbles, which are often undersown with grass or clover. This extension and improvement of the feeding grounds has been largely responsible for the massive increase in population, and somewhat ironically has led to the recent complaints of excessive numbers.

The concentration of Greenland White-fronted Geese is also of vital importance, representing almost a quarter of the world population of 19,800. They too have been established on Islay for at least 100 years, and may well have spread from here to the other Hebridean islands. Between 1910 and 1939 they increased enormously (50), reflecting perhaps a redistribution of the flocks which appeared in South Uist at the turn of the century and later declined. In recent years their numbers have been buoyant, but have shown no sign of the increases noted at Rhunahaorine and other neighbouring resorts (Table 202). Because of their relatively small numbers and their wider distribution, they are less in conflict with the farming interests than the Barnacle Geese, and in some areas are given special dispensation. They were, nevertheless, affected by the increase in autumn shooting in the late 1970s, when the open season for Barnacle Geese began in September instead of December. This problem has now been largely overcome by the total protection of the Whitefront in Scotland.

Recent surveys and ringing suggest that the Whitefronts in Islay are divided into several distinct groups, each with traditional roosts and feeding grounds. The relationship between the groups is not yet clear, but a good deal of interchange no doubt takes place, depending on local disturbance. The average numbers of geese in the various areas on 12 occasions in 1982-83 and 1983-84 were as follows: The Oa, about 400; Leorin and the flats near the airfield, 715 (max 935); inland from Bowmore and Bridgend, 2,010 (2,355); Loch Gruinart and Ardnave, 290 (614); Loch Gorm, 440 (888), and the Rinns, 390 (660). In most of these areas the birds are usually scattered in parties of no more than 1-200 over a number of inland

Table 205. *Peak counts of Barnacle Geese, Greenland White-fronted Geese and Greylags in Islay, 1980 to 1983, and mean annual maxima in the six preceding 5-year periods. The figures in brackets are the highest counts in the periods concerned. A blank = no data. Summarised and updated from Boyd (1968), Boyd and Ogilvie (1972), Booth (1975), Ogilvie and Boyd (1975), Ruttledge and Ogilvie (1979).*

	Barnacle Geese	Whitefront	Greylag
Mean 1952-1954	5000 (9000)*		
Mean 1955-1959	4180 (7500)		
Mean 1960-1964	8460 (10400)	1567 (2400)*	372 (665)*
Mean 1965-1969	12780 (16800)	2904 (4700)	197 (300)
Mean 1970-1974	17440 (19400)	3118 (4180)	152 (210)
Mean 1975-1979	20900 (24000)	3560 (4210)	106 (165)
1980	20500	4850	71
1981	17000	3600	116
1982	14000	3880	65
1983	15000	4590	90

* Means based on 3 seasons only.

ɔosts and feeding grounds, but they sometimes oalesce into larger flocks. Roosts which regularly old large numbers include Feur Lochan, Eilean na 1uich Dubh, Loch Finlaggan, and Lochs Kinnabus nd Risabus on the Oa. The feeding grounds are ıostly on stubbles (in autumn) or rough pasture, but reas of improved grazing and bog are also used egularly (516).

Greylags, which are thought to be of the Outer Iebridean stock, have never been plentiful, and over he years have steadily declined. By the 1930s their ɔrmer resorts on the Leorin flats and the islets of the outh-east coast had already been abandoned, and the locks elsewhere were said to be greatly reduced (50). n the late 1950s, a temporary increase took place, and -600 were recorded in several seasons, roosting on ,och Indaal and feeding on the farmland north and ast of Bridgend. Thereafter the numbers fell sharply, nd have since continued to decrease (205). The rest of he island is now deserted.

Islay also holds substantial numbers of Scaup nd Eiders, and some fair-sized flocks of several other lucks. The Eiders are especially plentiful along the ast coast, in the Sound between Islay and Jura where large autumn and spring passage occurs. Very large ıumbers were reported there in the 1930s (41), and ecent records confirm that movements of 1-3,000 are till occurring regularly throughout the autumn and arly winter. Large flocks of moulting and wintering irds have also been located along the southern coast, presumably the same population. In 1969, after an unusually good breeding season, the numbers in Loch Indaal reached a peak of about 600, but were then cut back by an oil spill to less than 300 (412); since then the population has remained at the lower level. Elsewhere the species is widely distributed, occurring in several places in flocks of 100 or more. The breeding population is equally widespread.

The Scaup are restricted almost entirely to the shallow sheltered waters of Loch Indaal, their favourite station being around the sewage outfall off Bowmore. In the 1960s the late autumn gatherings often totalled 1,000-1,500, but in recent years the peak has rarely exceeded 6-700. They also resort to Loch Skerrols, a short way inland, mostly in small numbers but sometimes in flocks of 2-300. The birds are almost certainly of Icelandic origin, as are those on Lough Neagh and other Irish resorts.

Loch Indaal is also used by quite large numbers of dabbling ducks. These are mostly located on the flats of the inner bay between Blackrock and Gartbreck, where totals of 650-1,000 are reported every winter (Table 206). The remaining shoreline to Port Charlotte is less attractive, but may well hold another 50 Mallard, Teal and Wigeon and a regular small number of Shoveler. Mergansers are common in both

'able 206. *The numbers of ducks and swans on:*

,och Indaal, Blackrock – Bowmore, 1970-1982 (see ıote);
,och Skerrols, 4 seasons, 1970-1975.

		L Indaal		L Skerrols	
		Reg	Max	Reg	Max
1allard	(5)	148	279	50	403
'eal	(3)	140	470	58	204
Vigeon	(4)	457	800	87	250
·caup	(9)	971	1500	56	300
'ufted Duck		0	0	47	100
'ochard		0	0	149	320
;oldeneye	(5)	8	38	2	8
·helduck	(7)	33	81	0	0
1ute Swan	(5)	9	19	0	2
Vhooper Swan	(6)	13	53	0	0

Vote: Loch Indaal: some species were counted less requently than others; the figures in brackets show he number of seasons for which adequate data are vailable.

Fig. 49. Islay, showing the 60m contour and the feeding areas of Barnacle Geese.

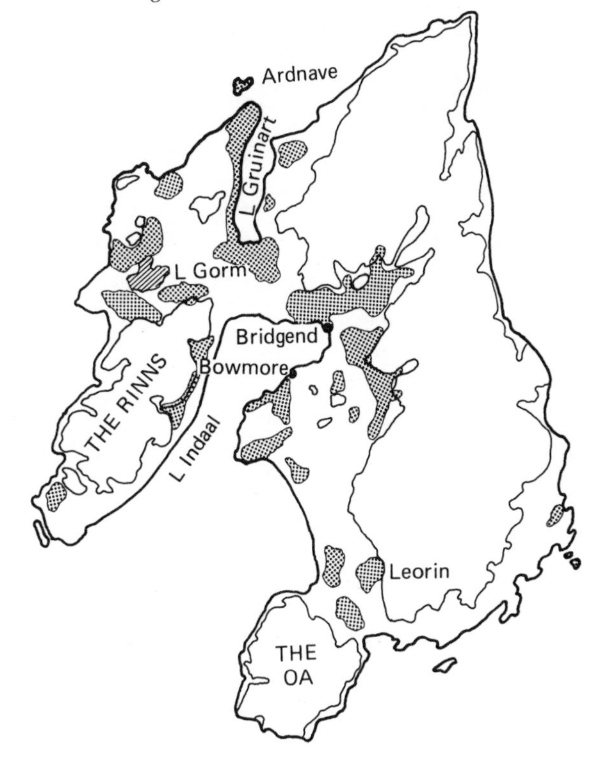

sectors, especially in July and August when up to 200 are sometimes present; the winter total is seldom more than 40. Shelducks reach a late winter peak of up to 80, and a further 150-200 occur on Loch Gruinart nearby; the total for the whole island is estimated at around 250. Loch Gruinart has also held up to 250 Mallard, 500 Teal and 540 Wigeon, but most of the counts have been a great deal smaller. In January 1974 a survey of the whole island produced a total of 660 Teal and 1,000 Wigeon (60).

The freshwater lochs of Islay support a further population of dabbling and diving ducks, but the numbers are mostly quite small. Loch Skerrols in the fertile lowlands attracts a regular autumn peak of about 500 birds (Table 206), and Loch Gorm, the largest of the inland waters, though usually almost devoid of ducks, has held 435 Teal, possibly from Loch Gruinart. The maxima elsewhere are listed in Table 207 and are probably representative, except that Loch Ballygrant is reported to hold rather larger numbers of diving ducks. An earlier count of 200 Tufted Ducks on Loch Allan in February 1969 is considered exceptional (60). Whooper Swans occur in some numbers during autumn migration, especially in the north of the island around Lochs Gruinart and Gorm. Totals of more than 110 have been recorded in early November of three recent seasons, but the normal peak is probably nearer 60-70. By mid-December the numbers are greatly reduced, and virtually none remain thereafter.

Summer surveys have shown that Mallard and Teal breed widely, the former occurring on nearly all waters up to 300m (340); Tufted Ducks have increased and are now well established in the northern part of the island, and Common Scoters have bred since least the early 1950s. Five pairs were recorded in 195 1965 and 1970, and 7 pairs in 1973 (60, 540).

Jura is ill-suited to nearly all species of wildfow Most of the interior is occupied by open moorlan rising in places to 800m, and the great majority of th lochs are known to be deserted; the coast is also stee and inhospitable over much of its length. Eiders a the only common species, occurring in places substantial numbers, notably around the Sound Islay. Surveys of most likely haunts on the island October 1981 and January 1982 gave totals of only Mallard, 15 Teal, 23 Wigeon, 6 Goldeneyes, Mergansers, 24 Eiders, 10 Shelducks and 5 Mu Swans. Somewhat larger numbers of ducks hav however, been found on occasion at two resorts on th west side: 66 Mallard and 17 Wigeon at Loch a' Chnu Bhric and 41 Mallard, 12 Teal and 35 Wigeon Whitefarland Bay nearby. The intertidal area at Loc Tarbert usually holds 20-30 Mallard, Shelducks an Wigeon and a few Goldeneyes and Mergansers. Up 80 White-fronted Geese winter regularly: about 50 Loch a' Chnuic Bhric, the remainder in a smal declining flock near Lowlandman's Bay. Barnac Geese also appear intermittently on the islets o Brosdale on the south coast and on Eilean Mor in th extreme north, but their numbers have always bee small and both of the flocks may now have di appeared (Table 212).

Colonsay is rather more attractive. The land mostly low-lying with quite large tracts of arable an pastureland, and the freshwater lochs are reasonab productive. The coastal waters are also of some valu

Table 207. *The maximum numbers of ducks and swans recorded at various freshwater lochs in Islay. The figures in brackets show the numbers of occasions on which winter counts were made.*

		Mallard	Teal	Wigeon	Tufted Duck	Pochard	Mute Swan
Kinnabus	(8)	21	1	4	0	0	4
Glenastle	(3)	20	30	7	16	40	0
Eighinn	(2)	50	0	100	0	0	0
Lossit	(4)	3	2	0	5	50	0
Ballygrant	(7)	4	9	0	28	50	8
Nan Cadhan	(2)	26	0	0	3	30	2
Finlaggan	(3)	7	21	2	18	23	0
Allan	(3)	13	0	0	19	45	8
Staoisha	(3)	13	0	0	37	64	0
Ardnave	(10)	19	164	101	35	51	1
Gorm	(8)	55	55	40	23	20	1

Note: Single counts from Lochs Muichard, Tallant, nam Ban and Ardnahoe in January 1971 produced a total of 30 Mallard and 11 Wigeon; Lochs Dubh, Airigh Dhaibhaidh, A'Bhogaidh, Corr, Laingeadail, na Lathaich and Feur Lochan were all deserted in January 1982.

specially in the south where the shallow channel etween Oronsay and the main island provides a izeable stretch of intertidal sand. The only adverse actor is the lack of shelter from the westerly gales, oupled perhaps with a recent increase in shooting.

Information on the ducks is limited to a two-year un of counts on Loch Fada in the mid 1950s. At that me the normal winter level amounted to 50 Mallard, 0 Teal, and 15-20 Wigeon, Pochard and Tufted)ucks. The coastal shallows around Oronsay also old a few Wigeon, and Eiders are abundant around ne islands and rocks off the southern and eastern oast; this, in fact, is one of the early centres of opulation from which the species spread to the areas urther south (40).

Geese of all species were formerly rare, but in ne 1940s or thereabouts the numbers began to icrease, and by the late 1960s more than 1,000 were aid to be wintering regularly. The increase was ttributed to the presence of a flock of Canada Geese, 'hich became established following the introduction f two breeding pairs in 1934. As a result of the itroduction (or so it is claimed) increasing numbers of ne Greylags, which had previously overflown on assage, were attracted to the island, and these were ater joined by substantial flocks of Barnacle and Vhite-fronted Geese (125). Table 208 sets out a ummary of the available data. Recent surveys suggest nat there has been a great reduction in the wintering umbers of all three species, but sizeable flocks of arnacles and Whitefronts may still occur from time to me – presumably *en route* to or from Islay, less than 0km to the south.

Tiree, Gunna and Coll

Tiree is the westernmost of the Inner Hebrides nd has much in common with the Atlantic seaboard f the Uists and Benbecula, some 80km to the north. 'he island, which covers 8,443ha, is mostly low lying vith wide expanses of shell sand and machair, and the numerous shallow lochs and wet machair provide an abundance of wildfowl foods. As in the Outer Hebrides the fertile farmland is divided into numerous crofts and small-holdings, and disturbance from the crofting communities imposes a marked restraint on the numbers and distribution, especially of geese. The western end of Coll has further areas of farmland and machair, but elsewhere the terrain is much rougher, with sizeable stretches of heather moor, bog and a good deal of naked rock. The land here is farmed in much larger units and the amount of disturbance is correspondingly less. In 1961 the human population on the 7,412ha of Coll was only 147, compared with 996 on Tiree. The Island of Gunna, covering less than 100ha, has no permanent dwellings, and serves as a virtual sanctuary. Grazing animals are, however, shipped to the island.

The islands together provide a stronghold for further flocks of Barnacle and White-fronted Geese, and for sizeable herds of Mute and Whooper Swans. Greylags also occur, but have never been numerous. After an increase during the 1940s, they reached a level of 50-100, which is apparently still maintained. In 1938 a pair bred on Coll for the first time in many years, and further records of several pairs breeding since the 1950s suggest that this is now a regular occurrence (81, 542).

The Barnacle Geese use Gunna as a central roost, and disperse by day to feed over a wide area of the farmland and machair along the northern side of Coll and Tiree. They also feed on Gunna itself, but in recent years the amount of stock being run on the island has been much reduced, and the longer, rougher turf is not so attractive to the geese as formerly. Other resorts include the Isle of Soa in Gott Bay, and Breachacha and the area around A'Chairidhe on the south-west coast of Coll. For 30 years or so their numbers have apparently been steady at about 400 (Table 212).

The White-fronted Geese first started to come to

'able 208. *Records of geese on Colonsay and Oronsay, 1962-1983. A blank = no count.*

	Greylag	Whitefront	Barnacle*	Canada	Source
962 April (Air)			230		Ogilvie & Boyd 1975
968 Winter level	c.600	c.200	c.400	65	Clark 1980
973 March (Air)	0	0	40		Ogilvie & Boyd 1975
975				44	*Scottish Bird Report*
978 March (Air)	0	0	45	0	M.A. Ogilvie
983 March	36	58	180	24	Ogilvie; Stroud 1983
983 October	9	75		18	Wildfowl Counts

See Table 212 for counts in other years.

Tiree in 1887, and by 1913 some 4-500 were appearing regularly. Their favourite resort was the marshy area of The Reef in the centre of the island, a locality which is still important despite the development of the airfield nearby. Coll was much less favoured and the numbers there were always small. In the early 1950s the total population was estimated at 50-150, with a maximum of perhaps 250 (66). Since then the numbers have greatly increased, especially on Coll, and totals of 6-800 are now recorded annually (Table 209).

Table 209. *Records of Greenland White-fronted Geese on Tiree and Coll, 1965-1978. Based on Ruttledge and Ogilvie (1979) and Stroud (1983).*

	Tiree	Coll
January 1965	0	
Winter 1966-67	200	
October 1969	28	
March 1970	244	
December 1973	63	
April 1978	0	610
Winter* 1982-83	433	343
November 1983	357	435

* Higher of two counts, November and March.

The occurrence of large herds of swans dat from the end of the last century. As in the Weste Isles, the numbers of Whooper Swans increas dramatically at this time, from a dozen in 1886 to abo 200 in 1898. The present level on Tiree is about 10 and has been the same for more than 30 years; t numbers on Coll are very much smaller. The first Mu Swans appeared on Tiree in 1898, and by 1911 th totalled 75. In the 1950s there were 50-100 in summe and 10-20 wintering (81). Since then the wint numbers have increased to 30-40, but the summ level has apparently declined; 75 were recorded in Ju 1973 (540), but only 28, including 7 nesting pairs, 1983. Bewick's Swans were formerly plentiful, occu ring regularly in flocks of up to 200 on the inlan waters of Tiree. In 1939 they were still fairly nume ous, but thereafter they followed the regional tren and by 1950 had disappeared completely (81).

Information on the ducks comes mainly fro the survey between 1952 and 1957 (66), and from a fe occasional counts on Tiree in subsequent seasons. Th results of the latter are summarised in Table 210, an probably give a fair indication of the total numbers the months in question. In each case they included th main inland resorts on Loch a' Phuill, Loch Bhasapo Loch an Eilein and Loch Riaghain, and the marsh and lochans of The Reef. The coastal areas were le

Fig. 50. Loch Linnhe, Mull and Tiree.

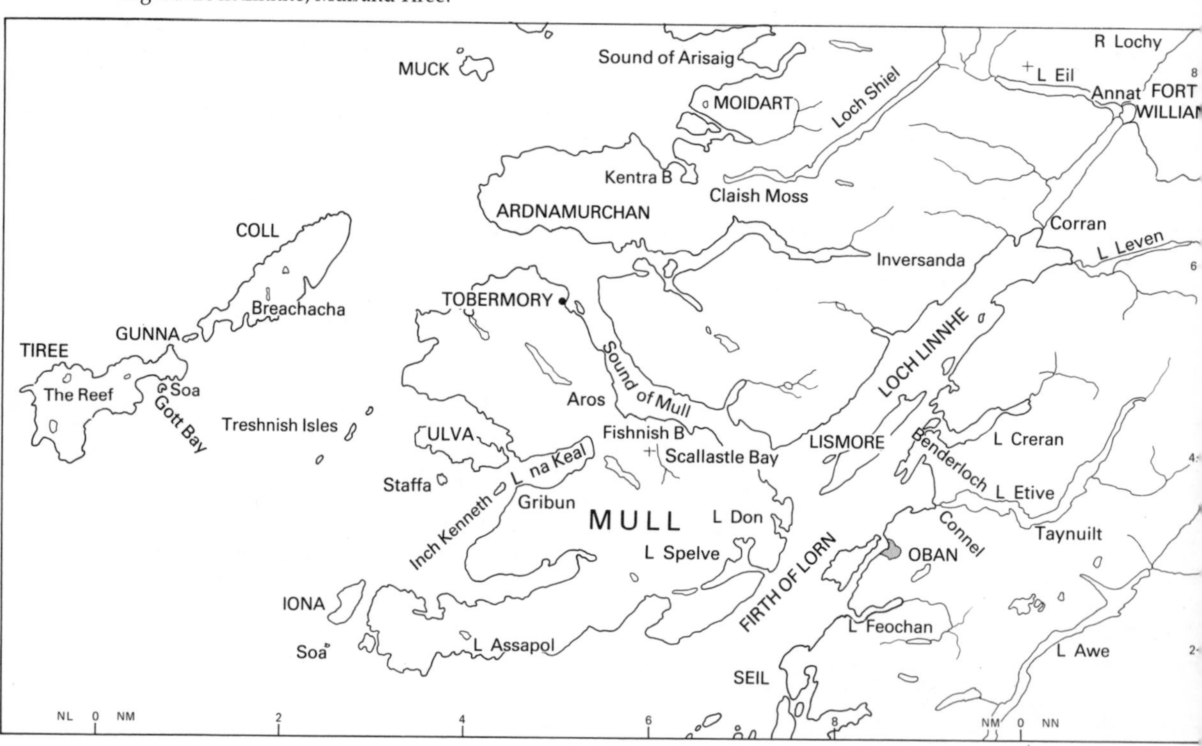

ell covered (hardly at all in the two most recent rveys), and the numbers of some of the maritime ecies are probably much higher than the figures ggest.

From Boyd's survey in the 1950s (66) and some lated counts at eight sites in 1982, it seems that most ecies are a good deal less numerous on Coll than on ree. Eiders are probably the commonest duck, curring in parties of 50 or so at many points along e north-west coast and in the Sound of Gunna. ocks of 200 or more are also found around Skerry-re, 17km off the south-west point of Tiree. Mergan-rs are another common species along the coasts of oll and Tiree, especially in late summer when therings of up to 200 are sometimes recorded. A ock of 150 in Gott Bay, Tiree, in July 1955 was mposed largely of immature birds. The winter umbers are usually quite small. The only count in the ound between Coll and Gunna – in November 1982 – ound 23 Wigeon.

Gadwall used to winter on Tiree in substantial umbers, arriving in autumn and remaining until late spring. They normally roosted on the sea, especially Balephetrish Bay, and flighted inland to feed, a abit which is virtually unknown in other parts of itain. By the 1940s, or perhaps before, they were owing signs of a marked decline, and the last corded gathering of any size seems to have been on och a' Phuill in January 1949. The birds were esumably of Icelandic origin, like most of those hich still appear in Ireland (279).

Lochaber and Mull

The wildfowl to the north of Oban and in Mull are confined by the mountainous terrain to a series of scattered resorts along the Atlantic coast, and around Loch Linnhe and the upper Firth of Lorn. In many places the land rises steeply from the shore to peaks of 1,000m or more, and with very few exceptions the freshwater lochs are dour and inhospitable. The commonest species are Eiders and Mergansers, with an occasional leavening of dabbling ducks and swans. Barnacle and White-fronted Geese occur in one or two localities.

The narrow deepset channel of Loch Linnhe cuts inland through the hills for more than 50km to Fort William at the foot of the Great Glen. The interest here lies mainly on the eastern shore, which is flanked by an outcrop of limestone along the greater part of its length. The effects of this are especially noticeable at the southern end in the low-lying farmland around the outfalls of Loch Etive and Loch Creran, and on the island of Lismore. There are several places hereab-outs, both on the coast and inland, which seem well suited for ducks, but the only counts are from the 10km of Loch Etive between Connel Bridge and Taynuilt, and Loch Creran. These have yielded similar numbers from both sites, of 20-50 Mallard, Wigeon, Teal, Goldeneyes and Mergansers, and about 100 Eiders. Loch Etive also has a flock of 20 Mute Swans, and up to 30 Shelducks are found on Loch Creran. Otherwise the main resorts are probably on the

able 210. *The numbers of ducks and swans recorded on Tiree.*

	Winter 1956-57*	January 1965	October 1969	March 1970	December 1973	November 1982	November 1983
allard	266	44	72	36	117	27	46
adwall	12	0	0	0	0	0	0
eal	205	156	14	23	20	116	59
'igeon	880	369	92	42	450	187	275
ntail	5	0	0	6	10	1	2
hoveler	20	0	9	5	0	11	20
ufted Duck	615	147	66	138	42	39	29
ochard	73	15	9	0	2	27	0
oldeneye	152	56	0	36	2	34	26
ider	154	179	274	100	60	0	0
ong-tailed Duck	7	30	0	12	0	0	0
1erganser	91	4	12	24	2	0	2
helduck	45	3	0	34	0	0	0
1ute Swan	30	29	78	41	44	10	5
Vhooper Swan	98	60	38	128	18	37	26

Highest monthly count, October – March.

freshwater lochs of Lismore, which are noted for the richness of their vegetation.

Lismore is also noted for the small flock of White-fronted Geese which seems to have wintered regularly for many years. In the 1950s they were said to total 30-40, and a record of 50 in April 1978 confirms that they have held their own. White-fronted Geese also occur on Eriska and the Benderloch peninsula adjoining Lismore, and are probably a separate flock. In the late 1950s and early 1960s they totalled between 25 and 75, with a maximum of 135 in the winter of 1964-65 (516). The recent censuses (558) found 20-60 on Eriska between November 1982 and February 1983, 150-170 in March 1983 (in two distinct flocks, suggesting a possible influx from Lismore) and 90 in November 1983. Goose droppings were found on Bernera Island, on the other side of Lismore, in November 1983.

The middle and upper reaches of Loch Linnhe and the tributary channels of Loch Leven and Loch Eil are all hemmed in by hills, and the shores are mostly steep and narrow. The only sizeable shallows are in the stretch around the outfall of the River Lochy between the tourist centre of Fort William and the pulp mills at Annat. Despite these developments the area still supports a small but varied population of up to 150 ducks, mostly Wigeon, Goldeneyes and Eiders (Table 211).

Substantial numbers of Eiders are also found on the neighbouring stretches of Loch Eil: up to 8 pairs have been recorded breeding and the autumn gatherings at times amount to nearly 200. Another group of

Table 211. *The regular (and maximum) numbers of ducks and swans at four coastal sites in Lochaber and Mull.*

a) Loch Linnhe: Fort William – Annat — 3 seasons 1970-1972
b) South-east Mull: North Loch Spelve and Loch Don — 5 seasons 1975-1982
c) North-east Mull: Scallastle Bay, Fishnish Bay, Forsa Mouth and Salen-Aros — 5 seasons 1975-1982
d) West Mull: Loch na Keal and Gribun-Inch Kenneth — 5 seasons 1975-1982

	a	b	c	d
Mallard	26 (60)	43 (96)	51 (93)	26 (74)
Teal	7 (20)	103(199)	19 (24)	3 (12)
Wigeon	65(110)	179(318)	32 (92)	3 (10)
Goldeneye	50(100)	2 (4)	4 (11)	17 (30)
Eider	36 (55)	8 (46)	106(276)	66(106)
Merganser	9 (17)	17 (51)	31 (73)	55 (78)
Shelduck	0 (1)	31 (71)	4 (13)	5 (17)
Mute Swan	20 (33)	6 (9)	5 (8)	0 (0)

100-150 occurs in late summer on the middle reach of Loch Linnhe, between Inversanda and the Corr narrows, and further parties of 10-20 are comm enough at many other points along the sho Mergansers are also plentiful, especially in August. least 250 are usually located off Inversanda at th time, and the total for Loch Linnhe and Loch Eil m well exceed 600. The winter numbers are apparen much smaller.

The coasts of Mull support another sizeab population of Eiders and Mergansers, interspersed places with flocks of dabbling ducks and Barna Geese. The largest gathering of Mergansers record is of 224 in the Sound of Mull in August 1982. T dabbling ducks – at their most plentiful in autumn are located mainly on the eastern side of the island, the bays and inlets opening onto the Firth of Lorn ar the Sound of Mull (Table 211). The inlets on th Atlantic coast may sometimes hold a few Mallar Goldeneyes and Shelducks, but most other specie apart from Eiders and Mergansers, are noticeab scarce. From the odd counts made at several poin along this stretch it seems that the numbers on Lo na Keal are a good deal larger than those elsewhe

The inland waters of Mull are even le favoured. Most of them hold a few Goldeneye, a sometimes some Mallard and sawbills, but the total any one place is seldom more than 20, and often le than 10. The only records of note come from the Ro of Mull, where Loch Assapol has held up to 29 Te and 70 Pochard, and Loch Poit na h-I 35 Pochard ar 30 Tufted Ducks. Whooper Swans are reported o passage, at times in some numbers, and a few a often present through the winter, notably on th Mishnish Lochs and Loch na Meal near Tobermor and less frequently on Loch Squabain.

The Barnacle Geese are centred exclusive along the western coast. Their main stronghold is o the Treshnish Islands, some 5-10km offshore, but the often wander to the other islets and to favoured are along the coast of Mull (Table 212). Amongst th places in regular use are the islands of Staffa, Iona ar Soa, the areas around the Sound of Ulva, and th Gribun and Inch Kenneth. The total numbers a usually between 300 and 600, but up to 850 are o record.

White-fronted Geese were formerly rare in Mu but now seem to occur regularly and in quite larg numbers in the Ross of Mull and Iona, the latter bein the scene of the earlier sightings (41) (Table 213).

On the mainland to the north of Mull anoth flock of White-fronted Geese has long been esta lished in the area between Kentra Bay and th southern end of Loch Shiel. In the early 1960s the were said to be less numerous than 20 years previously, but 60-70 were still appearing regularly each year (540). By 1970 they totalled at least 150; 250 were reported in the winter of 1974 and as many as 350 in 1977-78. Since then they have decreased; there were only 45 in 1982-83 and 44 in November 1983 (558). The feeding grounds are on Kentra Moss, to the west of Acharacle, and on Claish Moss to the east. The latter provides the main roost; it is also a site of great botanical interest, and on this account an area of 563ha was declared a National Nature Reserve in 1978. In 1981 Claish Moss was added to the list of sites designated by the United Kingdom under the Ramsar Convention.

Loch Shiel itself is deep and acid; on the only occasion of a winter count it held a total of 40 Pochard, 12 Whooper Swans and a few Tufted Ducks and Goldeneyes. Mallard and Teal are present, both in the breeding season and in winter, but the numbers are surely insignificant. Loch Morar and the other large inland waters in Lochaber are even less attractive, and may safely be ignored. Loch Lochy in the Great Glen is a possible exception, but has not yet been surveyed.

The Atlantic coast of Ardnamurchan, Moidart and Morar is of little value, except for Eiders and Mergansers. Geese are absent, and no appreciable gatherings of dabbling ducks have been seen, or are likely to be seen, in the areas which have so far been examined. The only known places of interest are Kentra Bay, which attracts up to 100 Mergansers, and Loch nan Cilltean, Arisaig, where Eiders first colonised the north-west mainland of Scotland (50).

The north-west Highlands and Skye

The coastline from Mallaig northwards to Cape Wrath is again deeply indented, with long stretches of rock and cliff broken only by a few small patches of saltmarsh around the heads of the sea lochs. The many hundreds of inland waters are also devoid of food and shelter, except in the one or two places where outcrops of limestone have encouraged a richer and more varied vegetation. The wildfowl are thus restricted to a few favoured localities and are nowhere very numerous. The most notable occurrences are the local gatherings of Eiders, and the flocks of Barnacle Geese which winter regularly on several of the offshore islets; there are also odd parties of White-fronted Geese, and some scattered groups of native and feral Greylags.

The Barnacle Geese, totalling about 900, are dotted at intervals along the length of the western coast. The largest flocks are located off the north-west coast of Skye, on Isay, Trodday and the Ascrib Islands, all of which are uninhabited, and serve as natural sanctuaries (Table 214). From these strongholds the birds are able to exploit the other islets round about, such as Staffin Island, Tulm Island and Eilean Beag, but this is probably the limit of their local movements; there is certainly no sign of any major interchange with the Hebridean populations on the Shiant Islands and in the Sound of Harris, some 20-30km to the west. Many of the islands off the mainland coast between Gairloch and Cape Wrath are also used regularly, though the numbers at any one place are usually quite small. The two northernmost centres on Eilean Chrona, near the Point of Stoer, and Eilean an Roin Mor, off Kinlochbervie, are probably the most important with a combined population of about 150. Roin Mor in particular is associated with two further centres on the northern coast of Sutherland, the one on Eilean Hoan, off Loch Eriboll, the other on Eilean nan Ron and the adjacent islands at the entrance to the Kyle of Tongue. These two sites mark the eastern limit of the normal winter distribution, and although lying in a different region, are best considered in the present context. Despite some local fluctuations, the popula-

Table 212. *Numbers of Barnacle Geese in Argyll and the Inner Hebrides: aerial censuses 1957-1983. A blank= not counted. From Ogilvie (1983).*

	Feb 1957	Dec 1959	Mar 1961	Apr 1962	Mar 1966	Mar 1973	Mar/Apr 1978	Mar 1983
Islay	3000	2800	5500	4800	8500	15000	21500	14000
Brosdale I (S Jura)	38	140	107	124	45	0	0	0
Eilean Mor (Knapdale)	35	0	14	44	196	110	436	210
Eilean Mor (N Jura)	0	10	4	0	16		0	0
Oronsay, Colonsay	16	0	0	230	18	40	45	180
Tiree, Gunna, Coll	420	25	380	484	534	143	390	619
Treshnish and Mull	317	299	470	390	795	419	610	620
Total, excluding Islay	826	474	975	1272	1604	712	1481	1629

tion in the region as a whole has shown a slight increase over the past 20 years, and is probably as large now as it ever has been.

The White-fronted Geese have also shown a small but welcome increase. Prior to 1939 they were said to be rare, occurring sporadically in Skye, but not apparently elsewhere (41, 50). By the 1950s a flock of 20 or so was wintering with fair regularity around Dunvegan and Loch Snizort, and this has since increased; 42 were recorded there in 1963, 71 in 1977, and 70-80 in the Novembers of 1978, 1982 and 1983. Two further centres have also been established: 15-20 wintered regularly on the mainland near Gairloch in the 1960s and early 1970s (540), though this area has apparently been deserted since drainage operations in

Table 213. *Greenland White-fronted Geese in the Ross of Mull (peak counts where more than one per season). Recent data from Stroud (1983).*

	Loch Assapol/ Bunessan	Iona/Fidden/ Loch Poit na h-I
1975	33 Jan	
1976*	45 Dec	90 Apr
1977		48 Mar
1978		29 Apr
1979		67 Apr
1980		Present
1981	26 Oct	60 Mar
1982	46 Nov	52 Nov
1983	23 Nov	36 Nov

* 134 in whole Ross of Mull, December 1976 (*Scottish Bird Report*).

about 1975, and another 15-25, increasing at times 40, are based on Broadford Bay in eastern Skye.

The population of native Greylags, althoug greatly reduced, is still ensconced in a few tradition resorts. The main centre is, and probably has alway been, on the Summer Isles, near the mouth of Loc Broom. At the turn of the century the population he is said to have totalled as many as 200, but by 1940 th numbers had dwindled to not more than 30 (149). Th present level is probably a little higher; in 1972 the were 15 breeding pairs (540), suggesting a total perhaps 50-60 individuals. In June and July th moulting birds assemble on the small inaccessib island of Glas-leac Beag, some 10km offshore. Th grazing here is much better than on the other island being much enriched by the droppings of the gre congregation of gulls (150).

Breeding Greylags have also been recorded i recent years in two parts of Skye, on Loch Maree an the lochs near Elphin, and in the area of Enard an Eddrachillis Bay (542). Some of these are no dout descendants of the birds which have been introduce at various times over the past 20-30 years. This i certainly so at Loch Maree where hand-reared birds c local origin were brought in by the Nature Conservar cy in 1958 to augment the wild stock, which by the had almost ceased to breed. The flock now tota upwards of 20, and 4-5 pairs are breeding regularly o the islands off Talladale, an area which has recentl been established as Loch Maree Islands Nationa Nature Reserve. Another feral flock, also of nativ origin, was established near the head of Loch Carro during the early 1950s. By 1960 the group totalled 80 o 90, and small parties were ranging over quite a wid

Table 214. *Numbers of Barnacle Geese in the north-west Highlands and Skye: aerial censuses 1957-1983. From Ogilvie (1983). A blank = no count.*

	Feb 1957	Dec 1959	Mar 1961	Apr 1962	Mar 1965	Mar 1966	Mar 1973	Mar/Apr 1978	Mar 1983
Isay (NW Skye)	151	130	140	395	420	380	297	290	250
Ascrib I (NW Skye)	193	0	122	204	308	272	132	140	172
Trodday (NW Skye)	182	60	108	47	264	236	143	94	255
Longa (Gairloch)	38	56	15	20	5	0		10	0
Eilean Furadh Mor (L Ewe)	21	0	0	11	0	0		0	0
Summer I (L Broom)	95	73	0	57		146		98	54
A'Chleit (Enard Bay)	52	37	33	0		33	0	49	0
Chrona, Handa (Stoer–Scourie)	121	172	100	9		55	96	121	75
Roin Mor (Kinlochbervie)	0	21	64	0		74	190	65	61
Eilean Hoan (Durness)*		214	180	244		425	6	220	0
Eilean nan Ron (Tongue)		199	179	135		130	350	339	255
Total	853	962	941	1122	997	1751	1214	1426	1122

* 250 on Eilean Hoan, November 1983.

rea. No news of them has been received since then, nd none were seen when the sea loch was surveyed n the winter of 1978.

In spring and autumn the resident Greylags are einforced by much larger numbers of migrants, but hese seldom stay more than a day or two. Substantial ocks of Whitefronts, Pinkfeet and Barnacle Geese are lso seen on migration, and in late September a few ight-bellied Brent Geese often appear in Broadford ay or thereabouts. A count of 150 in 1977 was larger han usual.

Whooper Swans are not uncommon, occurring om time to time on quite a number of the freshwater lochs, both in Skye and on the mainland. Their favourite resorts, each holding 10-15, are on Loch Suardal, near Dunvegan, and in Kintail, where the birds alternate between Loch Shiel and Loch a' Mholain. Elsewhere the flocks are smaller and less regular. In autumn and spring there are sometimes reports of 15-30 from odd points along the coast and inland, but no set pattern has yet emerged. A gathering of 116 near Gairloch in October 1976 was probably exceptional.

Most of the recent information on the ducks comes from the southern part of the region, where counts have been made on an extensive sample of coastal and inland waters. The most important areas here are along the sheltered eastern coast of Skye,

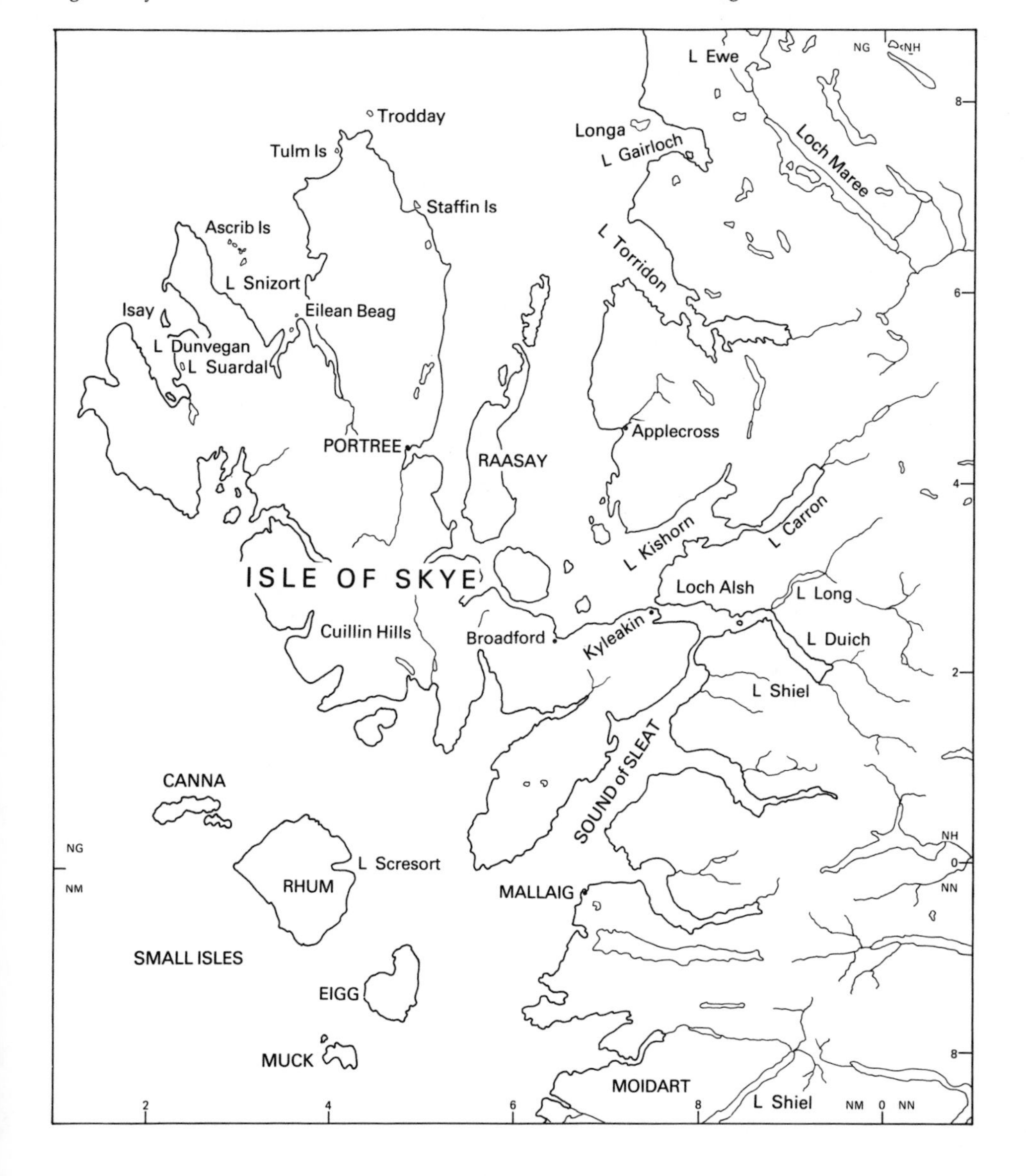

Fig. 51. Skye and Loch Alsh.

between Portree and Kyleakin, and around Loch Alsh and Loch Carron.

Eiders are plentiful throughout this stretch, their numbers amounting to at least 1,250; they are also scattered in much smaller numbers around the other coasts of Skye, and northwards to Loch Torridon. This is quite a recent development, stemming from the huge increases in the Outer Hebrides in the 1880s and 1890s. Prior to that the species was virtually unrecorded along the western coast of Ross and Sutherland. The colonisation began in the north around Cape Wrath and Handa, and from there spread southwards, reaching the Gairloch in 1912, and Loch Alsh in 1926 (40, 50). There are now two thriving colonies on Loch Alsh, on Glas Eilean and Eilean Ba which together hold at least 250 pairs. Up to 50 nes have also been found on Kishorn Island during rece years. The largest winter gatherings are located ne Kyle of Lochalsh and around the Strome Islands, at tl entrance to Loch Carron, where 5-600 have bee recorded regularly over the past 10 years. In April ar May the numbers decrease sharply, in common wi those at most of the other resorts nearby. This is part explained by the withdrawal of the breeding birds the nesting colonies in Loch Carron and Loch Alsh; tl immature and non-breeding birds may also mov away to a separate summer centre in the sea lochs east Skye. Table 215 shows the numbers of Eide recorded on various sections of the coast in tl autumn and winter of 1978-79 and the correspondir totals for some other common species.

Fig. 52. The north-west Highlands.

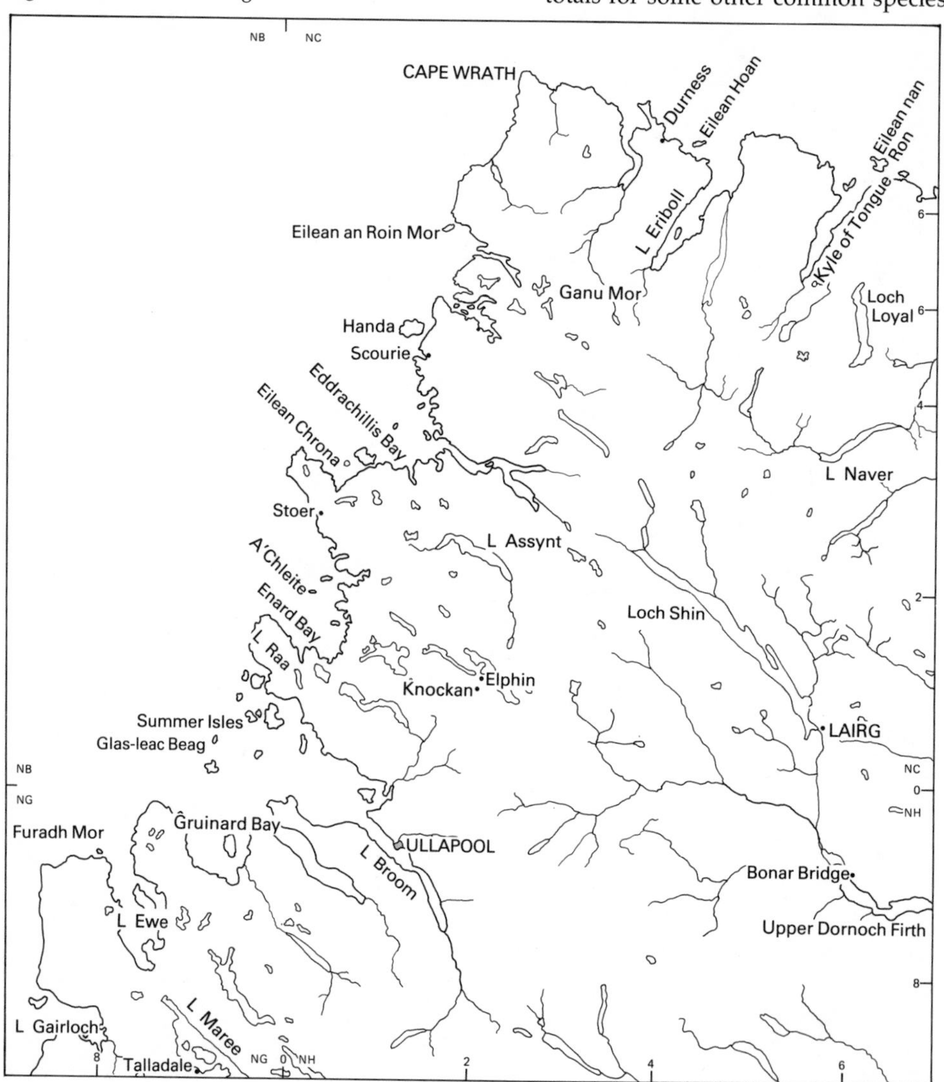

The Wigeon and most of the other ducks are normally found in the sheltered bays and sounds between Portree and Kyleakin, and on the inner reaches of Loch Alsh, Loch Duich and Loch Carron. The other coasts of Skye are seldom used except by scattered parties of Mallard and Mergansers, and occasionally by Shelducks. Mallard and Goldeneyes also occur on many of the inland waters, but their numbers are always small.

Despite the gaps in the cover it seems unlikely that any large gatherings have been overlooked. Most of the remaining coastline of Skye is beset with cliffs or rock, and can safely be discounted; Raasay is also steep and rugged, and is known from an earlier survey to hold only a few Mallard, Eiders and Mergansers (578).

The Small Isles, with a population of 5-600 Eiders, are rather more important. The largest flocks are on the north-east coast of Rhum, around Loch Scresort, and along the southern side of Canna. In the mid 1960s the latter held rather less than 100 breeding pairs, and the same was probably true of Rhum; Eigg and Muck each held 20-30 pairs. Ten pairs of Shelducks were also breeding regularly, mostly on Muck and Canna, Mallard and Mergansers were resident in very small numbers, and in winter a flock of up to 25 Barnacle Geese was often present on Muck (186). The last mentioned still occur.

The districts to the north of Torridon are still not fully surveyed. Eiders have increased greatly since the early 1900s, and now breed freely along the full length of the coast (40, 542). Winter gatherings of 1-200 occur off the Gairloch, but concentrations of this size are almost certainly atypical; in most areas the birds are dispersed in small groups and family parties. The only information on breeding numbers is from Handa, where 10-15 pairs rear a total of 10-20 young. Mergansers are resident in substantial numbers, and are probably the commonest breeding duck. Like the Eiders, they tend to be widely dispersed, but quite large gatherings may occur from time to time. Moulting flocks of 50-100 have been recorded in the Gairloch and Little Gruinard Bay, and Loch Torridon has held a similar number in November and December (540). Goldeneyes are much less common than in Skye, and Scoters and Long-tailed Ducks, although plentiful in the Outer Isles, are rare along the mainland coast.

Other species are scarce or absent, apart from Mallard, which are widespread in very small numbers, and Wigeon, sizeable parties of which can occur in the shallower bays and sea lochs. Between 50 and 100 Wigeon winter at Achnahaird Bay and the adjacent Loch Rad, 20km north-west of Ullapool.

The numerous inland lochs are mostly devoid of food and cover, and are virtually deserted. One of the few areas of interest is the stretch between Knockan and Loch Assynt, where outcrops of the Durness Limestone give rise to a small oasis of fertile land. No details are available, but several of the local lochs are believed to attract a variety of breeding and wintering wildfowl.

Table 215. *The numbers of ducks recorded in various parts of Skye, Lochalsh and Lochcarron, winter 1978-79. Data from Ellis (1979).*

	October	November -December	December	January -February	February -March
A: EIDER					
N and W Skye	76	26	28	28	39
Sound of Sleat	13	11	6	10	15
L. Alsh, Duich and Long	393	450	435	443	373
E Skye: Kyleakin – Portree	346	244	265	304	295
L. Carron and Kishorn	655	543	575	567	560
Applecross	87	56	63	74	89
L. Torridon	131	157	129	122	148
Total Eiders	1701	1487	1501	1548	1519
B: TOTALS FOR OTHER SPECIES					
Mallard	341	321	552	544	380
Teal	31	62	87	73	70
Wigeon	229	270	463	336	377
Goldeneye	0	22	45	94	98
Merganser	356	380	437	482	418
Shelduck	0	0	1	12	76

The Outer Hebrides

The problem of assessing the wildfowl populations of the Western Isles is still largely unresolved. Altogether the islands cover about 290,000ha, and stretch from north to south for more than 200km. Within this span the length of coastline is enormously increased by the ramifications of the numerous inlets and sounds, while in places the profusion of lochs is such that more than half the land is under water. Many districts are so inaccessible that aerial surveys are the only practical means of obtaining information, but even this approach is at times confounded by the wide dispersal of the flocks. Some parts are known to be much more attractive than others, but information on the actual numbers of birds is sparse and fragmentary. The dabbling ducks, in particular, are widely scattered and their distribution is subject to constant change. Even in the most favoured areas there are large concentrations, which might give some indic tion of total numbers. The main sources of informati on the wildfowl of the islands are the comprehensi report by J.W. Campbell (111) and some specialis studies of the geese, swans and sea ducks, dati mostly from the 1970s. A great deal of unpublish data and opinion has also been provided by Andrew, N.E. Buxton, W.A.J. Cunningham, Currie, N. Elkins and C.J. Spray.

Lewis and Harris

The northern island of Lewis and Harris is one the least suitable areas for wildfowl in the whole Britain. In both districts the interior is complete uncultivated, and the many hundreds of lochs are ac and bleak, without cover or food. The coast is equal inhospitable, being rock-bound or precipitous ov most of its length. In the few places where conditio are slightly less hostile there are small populations

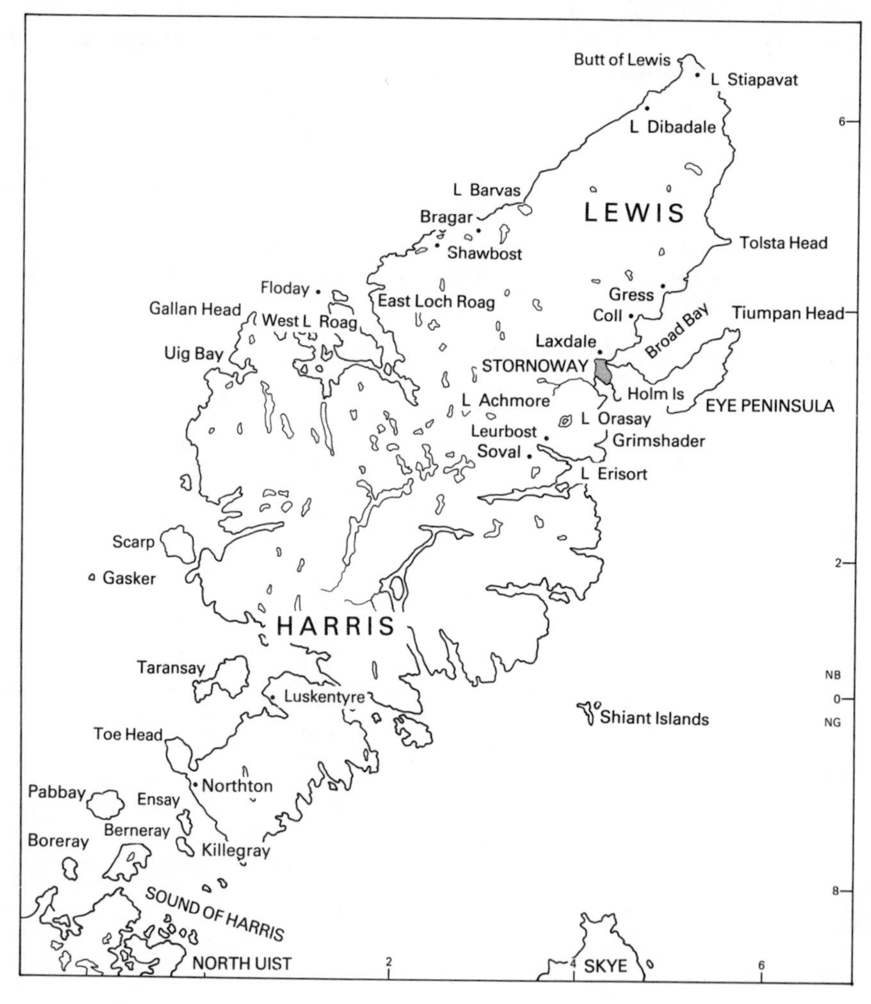

Fig. 53. Lewis and Harris.

vintering ducks, and maybe some geese and swans, ut these are isolated outposts, separated by long tretches of barren terrain. The breeding population is lso small and sparse.

In Lewis there are three known areas of note, och Stiapavat in the far north, the lochs between arvas and Shawbost on the north-west coast, and the oastal and inland waters around Broad Bay, near tornoway.

Loch Stiapavat is a shallow, fertile lake, lying on n isolated bed of calcareous shell sand, with a belt of nachair, bog and croftland round about. Although uite small, it supports a winter population of at least 00 ducks, and sometimes as many as 2-300. A few Greylags are also reported from time to time, mostly in March and April. The monthly averages in Table 216 re based on counts made at intervals between 1955 nd 1982. During this period the numbers have shown no sign of change. If the birds are disturbed they move o Lochs Dibadale and Drollavat, some 6km south-vest.

Between Barvas and Shawbost there are several ochs in the narrow strip of cultivated land which lanks the north-west coast. The largest of these, Loch Mor Barvas, has an area of 97ha and a mean depth of nly 1.3m. The surrounding land is reasonably fertile, ut the loch is rather exposed, and this may limit the numbers of birds. The normal winter population omprises 20-30 Mallard and 50-100 Wigeon. Green-and White-fronted Geese appear briefly on migration, ometimes in substantial numbers, and in recent years few have stayed throughout the winter months. This vintering flock was first reported in 1971, and has ince been present annually, the largest count being ne of 55; the 1982-83 survey reported between 23 and 32 from December to March (558). Some Greylags are often present as well, usually less than 10, but ometimes up to 30.

The neighbouring lochs around Bragar are used y further flocks of Wigeon and Teal, and less extensively by Mallard. Lochs Arnol, Ordais and Grinavat, and the lochan at Labost are usually the most favoured. Loch Ordais also holds up to 10 Whooper Swans, and similar numbers have at times een seen on Loch Mor Barvas.

A study of the Whooper Swans in Lewis and Harris in the winter of 1979-80 has done much to clarify their current status and to place the numerous records in perspective. Five main centres of population were identified, of which three were on the north-west coast of Lewis at Loch Stiapavat, Loch Ordais and Loch a'Bhaile near Tolstachaolais; the other two were on the east coast on the Coll marshes near Stornoway and on Loch Cromore, just south of the entrance to Loch Erisort. Odd birds and family parties were also recorded from time to time on a further 42 subsidiary resorts. Despite this multiplicity of sites, and the occurrence of some quite large gatherings on migration, the total population was relatively small. The peak count in late October amounted to 83, and the winter totals to less than 45 (100). The totals in 1980-81 were about the same.

The main resorts around Stornoway are on the shallow tidal waters of Broad Bay, the Branahuie Banks and the Laxdale Estuary, and on the several lochs and marshes which lie in the flanking belt of farmland. Holm Island at the entrance to Stornoway Harbour is also notable for the flock of Barnacle Geese which became established, so it seems, in the early 1960s. Between 1964 and 1971 the counts varied between 50 and 120, and similar numbers continued to come until 1979 when the place was deserted, in favour perhaps of an area with rather less disturbance. In the first instance the flock probably stemmed from the major centre on the Shiant Islands, 25km to the south, where 4-600 are recorded regularly, making it one of the three most important haunts for this species in the Outer Hebrides (Table 218).

Broad Bay, and the adjoining shallows of the Branahuie Banks, are used at times by quite large flocks of sea ducks. It seems, however, that the present numbers of Long-tailed Ducks and Common Scoters are a good deal smaller than those occurring in the 1960s. Mergansers have also decreased, although up to 140 still moult (99), and Velvet Scoters are scarcer than they were; there is now no sign of the flocks of up to 100 which appeared briefly in 1974 and 1976, and even the parties of 10 or so which used to occur regularly are no longer seen so frequently. The Eiders, on the other hand, have probably increased. This is certainly true of the flock of moulting males, some 200 strong, which assembles to the north of Tolsta Head, and is thought to include the bulk of the Broad Bay population. In the mid 1960s they totalled 75. Table 217 compares the current estimates of population, based on the observations of N.E. Buxton, with the maxima recorded in earlier years.

Table 216. *Loch Stiapavat: monthly averages.*

	Nov	Jan	Feb	Mar	Maximum
No. of counts	7	8	5	6	
Mallard	42	47	39	19	150
Teal	35	30	38	54	104
Wigeon	44	52	44	46	100
Whooper Swan	14	12	6	7	29

The Laxdale Estuary, at the south-west corner of Broad Bay, has a sizeable stretch of saltmarsh along its inner edge. This is used regularly by 1-200 Wigeon (maximum 330), and at times by up to 30 Mallard and Shelducks. It is also favoured by Greylags: in winter the flock of native birds often totals 30-40, and in spring and autumn there are sometimes transient flocks of migrants. A gathering of 100 in January 1979 was remarkable for the time of year.

A short way to the north another 50-100 ducks are normally found on the pools and fresh marshes near Upper Coll. Wigeon are present in flocks of up to 30 throughout the winter months, and at times exceed 100; Mallard and Teal have both topped 60, and on one occasion there were 20 Pochard. Whooper Swans, in parties of 10-15, are normally present throughout the winter, and much larger numbers are likely to occur for a week or two in autumn. As many as 100 appeared in October 1977, but this was exceptional; peaks of 30-40 are more usual (100).

Records from the neighbouring marshes at Gress and North Tolsta, and from the lochs on the Eye Peninsula, have all been relatively small. Loch Tiumpan has held up to 20 Mallard and a dozen Tufted Ducks and Whooper Swans; and 40 Wigeon have been found at Gress. Teal, in flocks of 15-30, are reported at several places, and are often the commonest species; their favourite resorts are Lochs Branahuie, Innis and Swordale, and the marsh at North Tolsta.

To the south of Stornoway a flock of up to 50 Barnacle Geese has recently taken to wintering on the islands at the entrance to Loch Erisort, an offshoot perhaps of the flock on Holm Island, some 10kr away. Loch Erisort also holds a few Mallard, Tea Goldeneyes and Mergansers, and further flocks c 10-20 Mallard are reported at Soval, Leurbost an Grimshader. Greylags and White-fronted Geese occu during spring migration, and are often quite nume ous, both here and at other points around the coas Pink-footed Geese are seen in flight, but seem to settl much less frequently. In April 1977 an aerial survey c the Outer Hebridean coast produced a total of 920 gre geese around Lewis and Harris, of which 450 wer identified as Greylags. Of the latter 50 were located o the Laxdale Estuary, 250 on the Shiant Islands and 15 at East Loch Tarbert; the rest were unidentified, an were found at various points along the Atlantic coas 20 on Scarp, 150 in Uig Bay, 250 at Gallan Head and 5 in West Loch Roag (351). There are also records of 20 Greylags and 420 Whitefronts near Leurbost in Apri 1974 (540), of up to 200 Greylags occurring regularly o migration at Loch Stiapavat (269), and of 100 Pinkfee around Loch Drollovat (1km from Loch Dibadale) i May.

Table 217. *The numbers of sea ducks recorded in Broad Bay and on the Branahuie Banks. (Includes data from Hopkins and Coxon 1979, Elkins 1974 and* Scottish Bird Reports.)

a) Estimated population, 1976-1982
b) Highest counts, 1976-1982
c) Highest counts, 1963-1976

	a	b	c
Long-tailed Duck	150-250	245	420 Dec 1964
			306 Nov 1974
Common Scoter	up to 25	22	220 Dec 1964
			150 Dec 1965
			120 Jan 1970
Velvet Scoter	up to 10		25 Dec 1965
			110 Dec 1974
			100 Apr 1976
Eider (winter)	c.250	250	360 Nov 1974
(moulting males)	200	200	75 Jun 1964
Merganser (autumn)	40-80	80	216 Sep 1964
(winter)	20-40	35	75 mid-1960s

The large west coast inlets of Loch Uig and Eas and West Loch Roag appear to be of little value, excep for birds on passage. A few Barnacle Geese winter o the islands around the mouth and along the coas southwards to Uig and Branish. The largest flocks totalling about 40, have been on Floday in February 1964, and on Eilean Molach in March 1965. Whoope Swans occur regularly in groups of 5-10 at severa points along the eastern side of East Loch Roag notably on Loch a'Bhaile near Tolstachaolais. Od parties of Greylags may also winter in the area. Th numbers of ducks are equally restricted: in the aeria survey of April 1977 the total for the three inlet amounted to only 2 Mallard, 27 Mergansers and 5(Eiders (351). The winter level is perhaps a little higher

The interior of Lewis is largely unsurveyed, bu even in the aggregate the wildfowl population must b wholly insignificant. Mallard and Goldeneye are th commonest species followed probably by Whoope Swan. The only known site of interest is Loc Achmore, with an autumn peak of 50-60 Pochard 50-60 Tufted Ducks and a dozen Mallard and Teal. I summer odd pairs of Mallard and Mergansers ar found in most districts, and a few Greylags still bree in some localities, notably on the island in Loc Orosay. In 1974 there were 5 breeding pairs on th loch, and 17 young were reared, but predation by fera mink has since become a serious threat; in 1976 ther were no young birds at all. A few isolated pairs als breed in the area south of Loch Leurbost. Th moulting areas are on Loch Orosay and on Tavay Mor, at the mouth of Loch Erisort (269).

Harris is even less suitable for wildfowl than Lewis. The interior is mostly mountainous, with several peaks of 500-800m, and the lochs, both upland and lowland, are dour and acid. Much of the coast is equally rocky or precipitous, and devoid of ducks apart from Eiders and Mergansers. Mallard and Teal occur in very small numbers, and are probably the only other breeding species; there is also a sprinkling of Whooper Swans.

The only area of note is the stretch of the south-west coast between the Sound of Taransay and Toe Head. The Sound itself is one of the main centres for sea ducks in the Western Isles. Eiders have recently increased, and up to 450 now moult (99), with 250-350 present in autumn. By December they start to move away, and the numbers in winter and spring are seldom more than 50. Common Scoters are probably less numerous than they were in the 1950s, but are still recorded in substantial numbers. The largest flocks, totalling 150-200, occur between September and December: they then decrease to a winter level of 80-120, very few being present in spring and early summer. The great majority of the autumn and winter birds are males (99), as was the case in late summer of 1959, when a remarkable 3-500 were present. Mergansers reach a peak of 50 between July and September, and similar numbers of Long-tailed Ducks are present during winter.

The adjoining inlet at Luskentyre is of little value, being choked with sand and virtually devoid of vegetation. Four or five Shelducks occur in late winter, and odd parties of up to 50 Wigeon are sometimes found near the mouth, and on the neighbouring strands, feeding on drifted *Zostera*. A few Brent Geese may also appear briefly during autumn migration; a flock of 75 in October 1980 was probably exceptional. Further to the south in the lee of Toe Head there are patches of saltmarsh and machair near Norton, which attract a few more Shelducks, and sometimes some Mallard and Greylags. Barnacle Geese occur regularly on Taransay and Gasker, the total for the two islands amounting usually to about 130 (Table 218). Small flocks have also been recorded from time to time on Scarp and the Norton machair. The distribution varies, and in recent years the numbers on Taransay have shown a marked decline; a former inhabitant, who visits the island daily, saw a maximum of 40 in the winter of 1979.

The Uists, Benbecula and Barra

At the Sound of Harris the topography changes, and substantial numbers of wildfowl are found in many places. The four main island groups of North Uist, Benbecula, South Uist and Barra all have their own characteristics, but their basic conformation is the same throughout. In each case the eastern coast is rocky and rises steeply from the deep water of the Minch to the moorland and peat bog which occupies the greater part of the interior. The attraction of this zone lies in the sea lochs which penetrate far inland and provide a unique system of salt and brackish feeding grounds. Closely associated with these are the myriad moorland lochs, many of which afford a modicum of food and cover. In contrast the Atlantic coast is shallow and shelves gently to broad stretches of strand, flanked by dunes and sheltered, especially in North Uist, by offshore reefs and islands. The dunes themselves contain frequent flashes of floodwater, and just inland there are large expanses of machair, the zone of fine rich turf typical of the calcareous shell sand on which it thrives. The influence of lime is equally apparent in the many shallow lochs and bogs which edge the machair, and provide attractive feeding grounds.

Table 218. *Numbers of Barnacle Geese in the Western Isles: aerial censuses 1957-1983. From Ogilvie (1983).*

	Feb 1957	Dec 1959	Mar 1961	Apr 1962	Mar 1965	Mar 1966	Mar 1973	Mar/Apr 1978	Mar/Apr 1983
Loch Roag	37	0	0	0	52	19	0	20	33
Loch Erisort	0	0	0	32	0	6	0	55	0
Shiant Islands	303	290	214	317	450	483	450	420	580
Taransay	201	15	120	7	0	120	125	0	0
Gasker	41	110	10	70	140	122	0	130	0
Sound of Harris	490	174	599	498	493	575	980	1330	1555
Monach Islands	330	480	519	860	750	1035	640	760	638
South Uist	200	110	250	0	23	0	0	0	0
Sound of Barra	238	86	452	415	392	360	336	455	375
Barra – Barra Head	223	49	142	171	289	443	80	154	371
Total	2063	1314	2306	2370	2589	3163	2611	3324	3552

This fertile strip along the west coast is the most populated area in the Western Isles, and human activities have had a marked effect on the distribution, especially of geese. Since about 1900 the large farms which were formed after the Uist evictions in the 1840s have gradually been divided into small-holdings, and the resultant increases in disturbance have made several former resorts untenable. A certain amount of disturbance has also stemmed from the construction of the airfield on Benbecula, from the rocket range on South Uist, and from the increase in tourists (and ornithologists) which has followed the linking of the islands by new roads and causeways. The advent of the oil industry and the development of wave energy schemes is another possible threat. On the other hand the evacuation of many of the smaller islands has allowed the geese to adopt new resorts, and the establishment of reserves has provided an added measure of security. The latter comprise the National Nature Reserves at Loch Druidibeg (1,677ha, established 1958) and on the Monach Islands (577ha, 1966), and the RSPB's reserve at Balranald (658ha, 1966).

Fig. 54. North and South Uist.

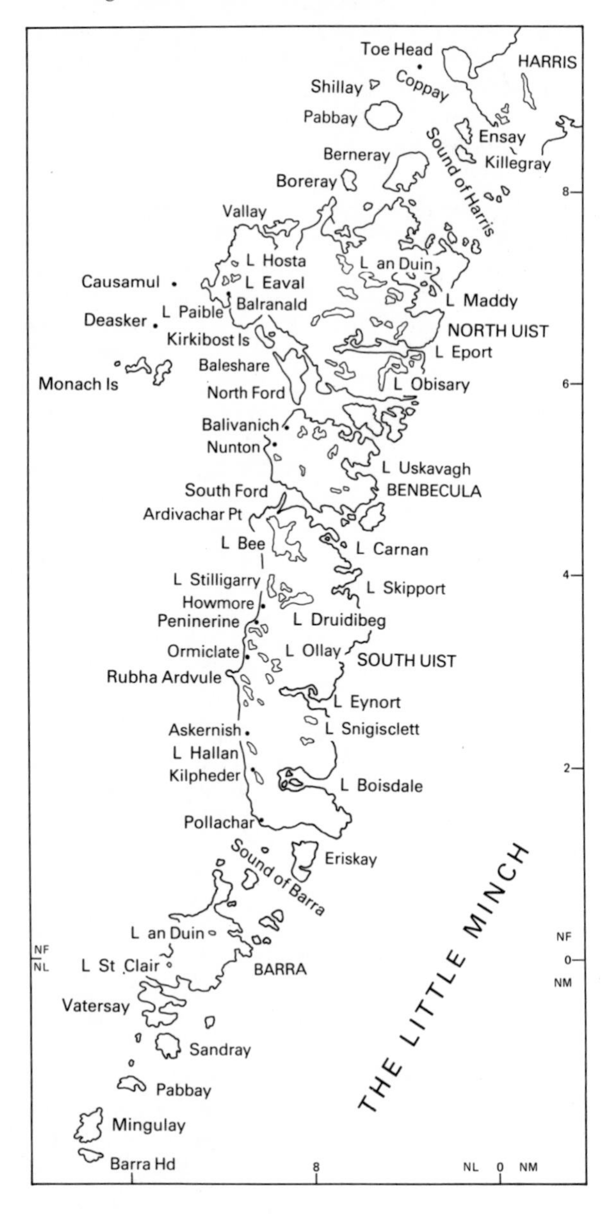

North Uist and the Sound of Harris

In North Uist the interest centres mainly on the coastal resorts and more especially on the complex of salt and brackish lochs formed by the incursions of Loch Maddy and Loch MhicPhail. These two systems, one opening eastwards to the Minch and the other northwards to the Sound of Harris are separated only by a narrow bridge of land, with a waste of uninhabited moor and peat bog on either hand. Their salinity ranges from pure sea water to near-fresh, and the many bays and channels provide an enormous quantity of food, notably *Zostera, Ruppia, Potamogeton* and *Chara*, most of which is available at any normal state of tide.

The Sound of Harris, with its numerous islands and skerries provides another large area of closely related habitat. The southern half, in particular, has extensive shallows, which continue westwards along the whole of the north coast of Uist. Among the several sources of food is the broad-leaved *Zostera marina* which flourishes at many points, though usually in rather deep water. Unlike those in other parts of Britain, the beds here were unaffected by disease during the 1930s; in 1959 the growth off Ensay and Killegray was just as abundant as it had been 20 years before. The islands, many of which carry sheep, provide good grazing for geese, and some have patches of bog and freshwater pools. These provide a welcome retreat for birds disturbed from the feeding ground along the North Uist coast. Loch Mor on Boreray is especially valuable in this respect, providing both food and security for quite large numbers of dabbling and diving ducks.

Wigeon are numerous throughout the area, and a good deal is known of the movements and feeding habits of the local flocks (110). Broadly speaking, their resorts fall into four closely related groups, each containing a variety of salt and freshwater pools. These comprise Loch Yeor and Loch an Duin at the head of Loch Maddy; Loch MhicPhail and Hornish; Loch Iosal an Duin and the shallows of the Sound of Harris to the east of Berneray; and Loch Mor, Boreray, coupled with the tidal flats around Lingay, Oronsay and Aird a'Mhorain. Within these groups the choice of feeding grounds is so large, and the dispersal of the flocks so general, that any form of census is virtually impossible. Some hundreds are known to use Loch Iosal an Duin and the Berneray shore, especially in October and November, and the total for the whole area may well be 1-2,000, perhaps more in some years. Further flocks, totalling several hundred, are scattered over the other islands of the Sound, and along the tidal inlets of Loch Eport and Loch Obisary, a short way to the south.

Teal are widely distributed in smallish groups throughout the north-east corner of the island, notably on Loch an Duin and Loch Iosal an Duin, and on the tidal channels of Loch Maddy, Loch Yeor and Loch MhicPhàil. The main centres of attraction are the stretches of ooze, which carry an abundance of the mollusc *Hydrobia ulvae*, a favourite item of food. Mallard are common enough, both inland and on the coast, but the numbers in any one place are invariably small; altogether they may total some hundreds. Eiders breed regularly and are plentiful in the Sound throughout the winter; up to 200 (females and immatures) moult in the Sound of Pabbay (99). Shelducks are resident in fair numbers, though scarce from August to October, and Common Scoters, a few hundred strong, occur at times on the deeper waters towards Toe Head, the same birds probably as those in the Sound of Taransay. Another feature is the marked movement of Mergansers which takes place each afternoon, as the flocks leave the Sound south-westwards for the open sea. Their winter numbers are far in excess of the breeding strength, suggesting a major influx of migrants, probably from Iceland. The area also attracts substantial numbers of Greylags, Barnacle Geese and Mute Swans.

The Mute Swans belong to the resident Hebridean population, which in recent years has totalled 750-1,000. Of these, perhaps one-third are located in North Uist. The largest numbers are centred on Loch an Duin and Loch Strumore, which together hold a total of at least 100 birds throughout the year. Moulting flocks of 30-50 birds are also found on Loch Grogary, and on Loch Bhruist in the island of Berneray. Elsewhere, the population is widely dispersed in pairs or small groups, mostly on brackish water or on the rich machair lochs along the western seaboard. The acid heather lochs in the central and eastern districts are very seldom used (552). Whooper Swans were also numerous at one time: in the 1930s, and again in the 1950s, there were said to be 2-300 wintering on Loch an Duin and thereabouts, but the place is now deserted. The only part of North Uist in which substantial numbers still occur is on the western machair, notably at Balranald where gatherings of up to 120 have been seen in recent years. They also frequent the areas around Hosta, Paible and Kirkibost but in smaller numbers. The current winter total for the island is probably between 100 and 150.

Barnacle Geese are widely distributed throughout the Sound of Harris, occurring on nearly all the islands at one time or another during the course of each winter. Being by nature nomadic, they are centred in no one area, often dividing into small groups and turning up in unexpected places. In general they prefer the smaller uninhabited islands, such as Shillay, Coppay and Ensay, but records of quite large numbers also come from Pabbay, Berneray and Oronsay, and from odd places along the northern and western coasts of North Uist. Besides these local movements there is evidence of interchange with the population further south, and with the scattered flocks along the Harris and Lewis coast. In the 1930s the total population in the Sound between November and February was estimated at 5-700. This tallies well with the results of the aerial censuses made at intervals between 1957 and 1966, mostly in March (Table 218). More recently the numbers in spring have increased, in common with those at several other Scottish resorts, and 750 were found in March 1983 (410). The winter numbers were estimated at 5-600 in January 1980. Another flock of 7-800 Barnacle Geese winters on the Monach Islands, some 16km to the west of North Uist. The islands, now uninhabited although grazed by sheep, have several hundred hectares of good grazing, and also some freshwater pools, so the geese are independent of the Hebridean mainland. Details of the numbers both here and in the Sound of Harris are included in Table 218.

The Greylags wintering in North Uist, and elsewhere in the Western Isles, belong almost exclusively to the native Hebridean population. Except perhaps for a brief period during the autumn and spring passage, there is no indication of reinforcement by migrants, and similarly no sign of any movement away from the islands by the resident stock. In the early 1970s the total breeding population was estimated at about 140 pairs (542), of which at least 50 were spread over the eastern districts of North Uist (269) and the neighbouring islands of the Sound. Except for a few pairs in Lewis and on the islands in the Sound of Harris, the remainder were concentrated in South Uist, mostly around Loch Druidibeg. Since then there seems to have been a notable increase; nesting birds are now found in nearly all districts of North and South Uist, and the Hebridean total is estimated at about 2-300 pairs (472). The number of individuals has increased equally rapidly. After a slight gain in the 1950s, the population became stable at an estimated level of 7-800 in the 1960s, but a large increase has occurred since. A partial census in 1982 accounted for 1,676 and the post-breeding total was estimated at some 2,000 birds. In winter a good deal of movement takes place within and between the islands, and the numbers at any one place are likely to vary almost from day to day. A count of 350 was made on Berneray in January 1982 (99). In North Uist there are two closely related groups, each totalling 1-200 birds. The first is centred on the Newton estate near Loch Maddy, and ranges from Loch an Duin over the northern machair and the islands of the Sound of Harris. The second is based in the southern part of the island, and wanders between the machair around Kirkibost and Baleshare, and the islands of the North Ford, at times straying southwards to Benbecula.

At one time a few Greenland White-fronted Geese and Light-bellied Brent Geese also wintered in North Uist. These wintering flocks no longer occur, though quite large numbers still pass through on autumn and spring migration. In the 1950s the Whitefronts were thought to total between 50 and 150, a level which had been maintained for the previous 50 years. The only major change during that period had been the temporary adoption of a new resort on Vallay in the late 1930s; apart from that the flocks were normally confined to a few favoured areas of moorland bog. The main resort of the Brent Geese was on Lingay Strand on the north coast. In 1936 about 100 were seen there, but shortly after their favourite feeding ground was reduced by erosion, and by 1938 the flock totalled less than 30. These continued to come for a few more years, but by 1950 were becoming increasingly scarce. Pink-footed Geese also appear briefly on migration, and parties of 10-15 are sometimes recorded in winter. Their regular occurrence dates from the 1930s, when the migration route through the Hebrides was first adopted. Past records of Bean Geese on Vallay should be treated with reserve; they were almost certainly either Pinkfeet or immature Whitefronts.

The rich machair lochs along the Atlantic coast of North Uist provide another important centre for breeding and wintering ducks. Information on the summer population is restricted to the Balranald reserve, where seven species of ducks are now nesting regularly. In recent years this area alone has held 25-30 pairs of Mallard, 15-20 pairs of Tufted Ducks and Eiders, 10-15 pairs of Shoveler and Shelducks, and 4-5 pairs of Teal and Gadwall, making 80-100 pairs in all. Wigeon have also bred in two of the past 15 years, a pair in 1977 and up to 3 pairs in 1978. The occurrence of Gadwall as a breeding species in the Western Isles is new development, dating from the 1960s. Since the they have bred annually, but the number of pairs ha not increased greatly (542). Autumn flocks of up to are now recorded at Balranald (269) and odd birds ar appearing with increasing frequency at other places i the Uists, and as far afield as Lewis. In 1982 a pair bre on Benbecula (540).

In winter the machair lochs are used mostly as night feeding ground. During the day there is often good deal of disturbance and the dabbling ducks, i particular, prefer the seclusion of the numerous ree which lie in a chain a few kilometres offshore. Th slightly larger islands of Causamul and Deasker ar both used regularly, and the day-time total of Mallar Wigeon and Teal on these and the smaller skerrie may well run to several hundred. Wigeon are als present regularly on the machair near Baleshare, th flock increasing from 80-100 in November to 180 i February (269). Shelducks are common at sever points along the shore, notably on Loch Paible, wher 50 or more are often present. Flocks of 30-40 are als found on the flats around Kirkibost and Baleshar and on the shallows of the North Ford. Pintail an Shoveler are scarce in winter, both here and in th other islands.

The diving ducks, being more tolerant c disturbance, remain on the machair lochs throughou the day, and are thus much more in evidence. Tufte Ducks have increased over the past 50 years, and ar now one of the commonest species. Recent autum and late-winter counts on a sample of the larger loch

have given totals of 1-300, which at a guess comprise rather more than half the local population. Scaup occur occasionally, mostly on fresh or brackish waters in preference to the sea. In February 1975 an unusually large gathering of 170 was recorded on Lochs Durasay and Leodasay, and another 45 were found on the east coast near Lochmaddy (234). A total of less than 10 is more normal. Pochard are found in parties of 20-30 on a few favoured lochs, such as Eaval and Hosta, but are otherwise scarce. The total for the island is probably 50-100.

Benbecula and South Uist

The majority of the wildfowl in Benbecula and South Uist are associated with the machair, and the many freshwater lochs along the western seaboard. The brackish lochs along the east coast of Benbecula and in the northern parts of South Uist are also of importance. The most notable features are the thriving colony of native Greylags, and the large populations of Mute and Whooper Swans. Dabbling and diving ducks are reputedly plentiful, and are said to be noticeably more numerous than they are further north. As in North Uist the crofting communities are responsible for a good deal of casual disturbance, but here again there are offshore reefs which provide a valuable day-time retreat. The geese are more vulnerable and have suffered several reverses over the past 80 years, not least from overshooting. The Greylags, being present throughout the year, have long been at odds with the crofters over damage to ripening crops, a problem which if anything has been exacerbated by the success of the nature reserve on Loch Druidibeg. Both they, and the wintering Barnacle and White-fronted Geese have also been affected, more obviously than other wildfowl, by the loss of grazing and the subsequent disturbances arising from the airfield at Balivanich.

Benbecula is the flattest and most drowned of all the Western Isles. Most of the western third is less than 10m above sea level, and except for Rueval (124m) the eastern part is not much higher. In the past ambitious drainage schemes have been attempted, the remains of which are marked by a network of ditches throughout the western plain. These waterways, known as "the Canal", have hitherto been a favourite feeding ground for ducks, the growth of cress and pondweed more than keeping pace with the manual clearance of the channels. With the introduction of mechanical dredging in 1980 this particular attraction will probably be lost, but the numerous associated lochs, which are equally valuable, will hopefully be unaffected. Most of them are shallow and fertile, with ample grazing round about, and some, such as Loch Torcusay near Nunton, are flanked by reed-beds which provide good cover for breeding and moulting ducks. The shallows of the North and South Ford, and the tidal ramifications of Loch Uiskevagh on the east coast are also used extensively by quite large numbers of Mallard, Teal and Wigeon. Loch Ba'Alasdair, on the southern side of the Uiskevagh complex, holds up to 200 Wigeon and 60 Mute Swans, and often some Scaup, Tufted Ducks and Whooper Swans.

About 50 pairs of Greylags were found on the islands and isolated peninsulas along the east coast of Benbecula in 1982 (472). Elsewhere, White-fronted Geese winter regularly in small numbers, usually on the bogs of Nunton Moor. Twenty years ago they were thought to total between 50 and 120; in the 1970s they were estimated at 25-50 (516). Thirty were counted in 1981 (99), but none were found during three counts in 1982-83 (558).

The wildfowl habitats in South Uist are similar to those in Benbecula, but on a much larger scale. The central and eastern parts of the island are in this case occupied by large areas of high moorland, rising in places to 600m. Much of the east coast is also unsuitable, except in the north around Loch Carnan and Loch Skipport, both of which hold groups of Mallard, Teal and Wigeon. The great majority of the wildfowl are thus restricted to the belt of machair, some 2km wide and 30km long, which flanks the length of the western seaboard. Within this sector there are more than 120 shallow lochs of varying sizes, all potentially valuable as resorts for breeding and wintering ducks.

The largest and most important centre is Loch Bee, in the north-west corner of the island. With an area of 673ha, and a depth of less than 1m in most places, it provides a prime feeding ground for a wide variety of water birds. Mute Swans are especially numerous, the population amounting to several hundred throughout the year. An estimated 250 Wigeon and 50 Goldeneyes are also present during winter. Information on the other ducks comes mainly from the surveys undertaken by D. Andrew, A. Currie and N.E. Buxton. Table 219 shows those inland sites at which a total of 100 or more wildfowl has been recorded. Twenty-five others have been covered individually. For comparison, the shooting records for South Uist and Benbecula for the seasons 1960-1978 (by permission of South Uist Estates) show an average annual bag of 183 Mallard, 60 Teal, 61 Wigeon, 8 Shoveler, 6 Pochard, 45 Tufted Ducks and 5 Goldeneyes.

The survey totals in Table 219, although far from complete, give a fair indication of the occurrence of the species, and their relative abundance. The dabbling ducks, in particular, are no doubt more plentiful than the figures suggest, because of their habit of retreating to the offshore reefs. This is confirmed by the shooting records, which contain an appreciably higher proportion of both Mallard and Teal.

The Mute Swans in the Uists and Benbecula (Table 220) have been studied in exceptional detail, firstly between 1971 and 1974 (287) and again since 1978 (552). The second survey, in contradiction to the first, has shown that the population is discrete, and is not at any time augmented by migrants from elsewhere. Comprehensive counts were made on foot and from the air at every likely resort throughout the three islands. The results showed a marked decrease in the population following the hard winter of 1979, when the losses of adults and young were unusually large. The population has since recovered, however, and 119 pairs reared 274 young in 1982 (540). Apart from this the numbers of adult birds remained more or less constant throughout the year. Loch Bee, and to a lesser extent Loch an Duin, emerged as centres of major importance, together holding some 60% of the total population. In each case the flocks consisted largely of non-breeding birds.

The absence of any large-scale movement of Mute Swans is confirmed by the recoveries of birds ringed either as cygnets or moulting adults. Only 8 out of a total of 551 birds marked in the Uists have been reported from other areas: one in Harris, 3 in Lewis, 3 in Tiree and one in Kintyre and later Northern Ireland. In each case the birds concerned were in their second year. A further 30 young and 43 adults were recovered locally, mostly within 10km of the place of ringing. The prime cause of death was collision with power-lines, especially during the hard weather when the birds, already in poor condition, were forced to move from place to place in search of food and open water (552).

Mute Swans were unrecorded in the Western Isles prior to the 1880s, when breeding pairs were introduced to Harris and North Uist. Within 50 years they were nesting commonly on Benbecula and South Uist, and the flocks on Loch Bee were already numbered in "hundreds". In July 1942 a total of 333 were counted there (41), which is roughly the present level. The paucity of the increase since then is attributed to a shortage of further nesting sites. The present breeding density of one pair per 500ha over

Table 219. *The regular (or, in brackets, maximum) numbers of ducks and swans at eleven resorts in the Uists and Benbecula. The figures after the place names indicate numbers of seasons and years. An asterisk in brackets after a place name denotes insufficient coverage for regular figures to be produced; the figure after the asterisk denotes how many individual counts have been made at the site. At the end of each island group, and at the foot of the table, the maximum totals recorded by D. Andrew (DA) during inland and coastal surveys in October 1970, 1971, 1975 and 1978, February 1968 and 1976 and March 1970 and 1973 are given.*

		Mallard	Teal	Wigeon	Shoveler	Pochard	Tufted Duck	Goldeneye	Whooper Swan	Mute Swan
SOUTH UIST										
L Hallan	(3:80-82)	49	13	0	0	3	33	6	24	13
L Aird an Sgairbh	(3:80-82)	29	24	34	0	0	25	0	7	2
East L Ollay	(*1)	(0)	(0)	(0)	(0)	(0)	(0)	(0)	(0)	(100)
L Druidibeg	(*1)	(50)	(0)	(0)	(0)	(0)	(10)	(0)	(0)	(3)
L Stilligarry	(*3)	(6)	(1)	(0)	(0)	(0)	(199)	(0)	(0)	(7)
L Bee	(*3)	(10)	(0)	(0)	(0)	(1)	(50)	(0)	(47)	(350)
October max, DA	(17 sites)	400	110	204	1	50	316	14	173	519
BENBECULA										
L Borve	(1:81)	18	11	19	0	0	3	45	5	6
L Fada/Mor	(1:81)	39	62	77	3	0	8	0	8	10
October max, DA	(4 sites)	245	120	206	23	40	52	1	12	95
NORTH UIST										
Balranald RSPB	(1:82)	135	141	113	11	17	43	1	9	11
L an Duin	(*4)	(9)	(12)	(97)	(0)	(0)	(0)	(3)	(0)	(70)
L an Sticir	(*2)	(12)	(0)	(86)	(0)	(0)	(0)	(0)	(0)	(13)
October max, DA	(14 sites)	205	176	594	14	60	153	4	98	322
OCTOBER max, DA, combined		727	380	1004	37	88	519	14	258	819
FEB-MAR max, DA, combined		282	187	410	16	131	492	56	309	662

the Uists and Benbecula as a whole is in fact double that in any other part of Britain. In contrast there is ample habitat for the non-breeding flocks, which now comprise a massive 65-70% of the population (552).

Bewick's Swans were once a common winter visitor, especially to the tidal channels of North Uist. In 1918 flocks of 100 or more were recorded, their favourite resorts being on the Newton estate around Loch an Duin, and on Loch Eport. They were also common on the machair in South Uist (41). Until then they were at least as numerous as Whooper Swans, but shortly afterwards the population slumped and by 1930 they had disappeared. This was by no means an isolated incident, but part of a major redistribution which affected the whole of the west of Scotland.

The Whooper Swan increased greatly as a winter visitor during the early part of the century, beginning around 1900 and continuing into the 1920s. At that time they were especially numerous in North Uist, with large flocks frequenting "every suitable bit of water". They were also abundant on their traditional resort on Loch Bee, where as many as 400 were recorded on occasion (41). Since then they seem to have suffered a major decline: the recent censuses by Spray and Andrew indicate a winter population of 2-300 in the Uists and Benbecula, more than half of which is normally located in the south. The largest groups are found on the flooded machair around Bornish, where more than 100 may occur, and in similar areas around Ormiclate, Peninerine and Loch Hallan. Loch Bee is no longer a stronghold; in the seasons between 1978 and 1980 the winter peaks did not exceed 28.

Loch Druidibeg, a few kilometres to the south of Loch Bee, is another site of major importance. The feature here is the colony of nesting Greylags, this being the largest remnant of the native British stock still occupying a traditional resort. Druidibeg itself is an acid heather loch of 248ha, with many secluded bays and islands, but not a great amount of food. So as soon as the young are hatched, the families move to the nearby machair lochs at Grogarry and Stilligarry, which all form part of the National Nature Reserve. After the creation of the reserve in 1958, the population increased slightly from the earlier estimate of 45-50 pairs, but soon became stable. In the early 1970s the level stood at 65-70 pairs, representing about one-third of the truly wild birds still breeding in Britain. The details in Table 221 are drawn from the intensive study undertaken between 1968 and 1972 (381). The extensive 1982 survey, however, indicated considerable redistribution, with only 38 nesting pairs and 240 individuals at the end of the breeding season (472). These represent 16 and 12% of the total Hebridean breeders and population respectively.

The main factors limiting the population are shooting and the frequent loss of clutches through predation and desertion. Of the 335 nests attempted during the five years of detailed study less than 200 were successful, and the annual production of young averaged only 150. The success rate depended largely on the weather and the date on which nesting could begin. In years when it started early a greater proportion of clutches was hatched, and more young were fledged per nesting pair. Similarly, within each season the birds which bred first laid larger clutches, deserted less often, lost fewer eggs to predators, and produced young which completed more of their growth within the period of optimum food supplies. During the same five-year period South Uist Estates recorded an average annual bag of 68 Greylags, but many more were undoubtedly taken by poachers. Since then the number shot by the Estate has increased to an average level of 85, with a maximum of

Table 220. *Numbers of Mute Swans recorded in North and South Uist and Benbecula. Cygnets of the previous summer are included as "full grown" from January onwards (Spray 1981).*

		Full grown	Cygnets	Total
1978	May	943		943
	June	948	267	1215
	August	889	218	1107
	December	845	156	1001
1979	February	820		820
	May	770		770
	July	753	216	969
	September	754	179	933
	December	738	152	890
1980	March	820		820

Table 221. *Numbers of Greylags breeding in Loch Druidibeg NNR between 1968 and 1972.*

a) No. of nests on L Druidibeg;
b) nests elsewhere in the reserve;
c) total no. of nests;
d) total no. of broods;
e) no. of young at 3-8 weeks old.

	a	b	c	d	e
1968	65	7	72	42	179
1969	53	8	61	30	116
1970	64	10	74	35	129
1971	45	15	60	36	137
1972	58	10	68	53	193

110 in 1975. This suggests that the annual production of young is offset almost entirely by the annual kill, although the spread of the population elsewhere has allowed the overall population level to increase. There is a call for licensed shooting of these geese during the close season to reduce alleged agricultural damage in parts of the Hebrides. The small size of the population and the existence of an open season, as well as sparse justification of damage claims, makes the granting of such licences inadvisable at present.

Loch Druidibeg is also used by non-breeding Greylags during the period of the annual moult. The flock, totalling 200-250, assembles in early June and stays until late July, when the birds move back to the machair, mostly between Howmore and the South Ford. A certain amount of the grazing in this area has been encroached upon by the rocket range, but apart from this the installation has had no real impact on the population. Loch Druidibeg lies 5km away, and neither the breeding nor the moulting birds have been affected.

In winter the flocks range rather more widely along the western machair, perhaps as far as Barra and the North Ford, but probably no further. All thirteen of the recoveries of birds ringed at Druidibeg have in fact been made within 28km. As in North Uist there is no sign of an influx of migrants, except on passage, and certainly no sign of emigration.

At one time the Greylag was more numerous in South Uist even than it is at present despite the recent recovery. In 1920 some 200 pairs were reckoned to breed in the island, and it seems that the numbers had been even higher (41). The reasons for the decline were partly the changes in land tenure and the growing persecution by crofters, and partly the overshooting which took place on the estate in the years prior to 1914. At that time the geese on South Uist were afforded no close season, and large numbers were shot as late as March and April. The only legal protection was a ban on the taking of eggs.

Barnacle Geese used to winter regularly on South Uist, but the flocks have long since moved elsewhere. In the early 1900s they were by far the commonest geese on the island, their numbers amounting to "perhaps 3,000 or more"; by the mid 1950s they had dwindled to about 500, with occasional peaks of 1,000, and a few years later they were gone. Their favourite resort was on the machair to the west of Loch Bee, in the area now partly occupied by the range. This, no doubt, was the final cause of their departure, but it cannot be blamed for the earlier decreases, which must have occurred sometime after 1937 (in that year the Barnacle was still described as *the* goose of South Uist (41)). One possible factor was disturbance by aircraft, especially during the war. In its latter stages, and possibly in the earlier stages as well, the decrease here was largely offset by increases at other resorts in the vicinity, notably on the Shiant and Monach Islands and in the Sound of Barra. Even more encouraging has been the recent increase in the total numbers wintering in the Western Isles, from 2,100 in 1957 to 3,300 in 1978 and 3,600 in 1983 (Table 218).

The status of the Greenland White-fronted Geese is less satisfactory. In common with those in North Uist and Benbecula, the flocks in South Uist have shown a marked decline, from a fairly stable level of 250 during the 1950s to the recent estimate of 1-200 (516) and the maximum of 75 counted in 1982-83 (558). They, too, have decreased on the machair to the west of Loch Bee, which used to attract the bulk of the population. In 1955 more than 200 were recorded there, and this was regarded as normal. At the same time there were 70 at Askernish, some 20km to the south, and a further 40 in Benbecula. In 1981 there were only 40 at Loch Bee (99) and in 1982, 60 (558). More recently there have been almost as many on the central and southern machair, mostly around Loch Hallan and Kilpheder where 27 were present in March 1981 and 20 in 1982-83. The associated roost is on Loch Snigisclett in the hills to the south of Loch Eynort.

Prior to 1887 Whitefronts were rare throughout the Western Isles; their numbers then increased dramatically, especially in South Uist, and within 12 years they were wintering there in thousands (50). In common with the Greylags they suffered a good deal from the heavy shooting in the early 1900s, and this may well have marked the start of their decline. In 1911 and 1912 the annual bags on South Uist estate amounted to 125, which is not far short of their total population at the present time. The progress of the decline, from thousands in 1900 to a few hundreds in 1950, is not apparently recorded. Berry (50) made no mention of it, so maybe they followed the same path as the Barnacle Geese, for reasons which were probably the same. The continuing decline since 1960 is attributed to factors other than mortality from shooting: over the past 20 years the bag has averaged only six a season, and has never exceeded sixteen.

Barra and the southern isles

The islands to the south of the Sound of Barra are important primarily for the flocks of Barnacle Geese, which are drawn by the good grazing and the virtual absence of disturbance. Many of the smaller islands in fact form natural sanctuaries, the steep shores and turbulent currents making human access almost impossible, except in rare periods of calm. The majority of the geese are normally found on Fuday and the islands in the Sound of Barra, but the distribution follows no set pattern and the flocks are often small and widely scattered. Altogether they total some 5-600 (Table 218). Barra itself is seldom used by the geese, and in common with the other islands has no expanse of freshwater habitat for dabbling and diving ducks. Loch St Clair, Loch na Doirlinn and Loch an Duin are known to hold a few Wigeon, Tufted Ducks and Whooper Swans, but these are the only lochs which seem at all attractive. The coastal waters are also of limited value, except for sea ducks.

Eiders are common, both in this southern sector and along the length of the Atlantic coast northwards to the Sound of Harris. The best available indication of their numbers comes from an aerial survey made in April 1977 (351). On that occasion there were 99 around Vatersay, 75 in the Sounds of Barra, Fuday and Eriskay, 155 along the west coast of South Uist (mostly to the north of Rubha Ardvule), 141 on the west coast of Benbecula, 192 around the Monach Islands and 47 in the Sound of Harris – a total of 709. Another 210 were located along the east coast and in Lewis and Harris, and more were no doubt present in the stretches which were not inspected, notably amongst the islands south of Vatersay, on the west coast of North Uist, and in Lewis to the north of Loch Roag and Tolsta Head. On this basis it was reckoned that the total breeding population in the Western Isles was unlikely to exceed 750 pairs, and was certainly less than 1,000 (351). The largest colony, estimated at about 100 pairs, is centred on the Monach Islands, which historically were the stronghold from which the species spread to the other islands and eventually to the western mainland (40, 269).

Long-tailed Ducks winter regularly at many points along the western coast. Their main stronghold is around the offshore reefs, but on calm days they tend to move inshore, and are then quite easy to observe. In the winters of 1971 and 1972 a series of counts between the Sound of Harris and the Sound of Barra produced a fairly constant total of 150-250 from October until April (89). The present numbers are thought to be a good deal higher, perhaps as much as 3-500. The largest flocks are usually recorded in South Uist, off Pollachar and more especially between Rubha Ardvule and Ardivachar Point, where upwards of 200 have been seen together. The main resort in North Uist is off Vallay; a rumour of large numbers in the Sound of Harris during late April (89) was confirmed by a count of 81 in April 1977.

Scoters are less in evidence, but are known to occur offshore, mostly in smallish groups, in areas where the depth of water is around 10m.

Orkney and Shetland

The two island groups of Orkney and Shetland are quite different in structure and character, and this is reflected in the composition of their wildfowl populations. The main features which they have in common, apart from the adversities of life at 60° N, are the scourge of frequent high winds and the constant threat of pollution from the traffic in oil, which has sprung into being since 1977. Both groups are of major importance for sea ducks, the former mainly for Long-tailed Ducks and the latter for Eiders. Orkney also has a large and varied population of dabbling and diving ducks.

Orkney

The Orcadian archipelago is made up of 90 islands, holms and skerries, less than a third of which have human inhabitants. The islands extend for 74km from north to south and for 61km from east to west, and have a total area of 97,250ha. The terrain consists for the most part of Old Red Sandstone overlaid with deposits of boulder clay. Although there are blocks of uplands, many of the islands are low-lying, with sluggish drainage, and pockets or tracts of peat cover occur in many places. The Ward Hill of Hoy (477m) is the highest point.

The climate, although boisterous, is mild enough, and the soil fertile enough, to support a prosperous farming community. The staple products are beef and dairy cattle; sheep are run on the rougher ground, and poultry and egg production is a major industry. In 1960 more than two-fifths of the land was under crops (15,000ha) or improved pasture (29,000ha); a further 28,000ha was classified as rough grazing (394). This intensive farming, although generating a certain amount of drainage and disturbance, and also giving rise to local pollution by silage effluent, is probably beneficial to wildfowl in providing an additional source of food, and in tending to enrich the lowland lochs.

The majority of the freshwater lochs in Orkney are situated on Mainland, mostly towards the western end. The largest and by far the most important of them is the Loch of Harray, with an area of 980ha and a mean depth of about 3m. At the southern end it is joined by culverts to the shallow Loch of Stenness (638ha), which in turn is joined to the sea by a narrow tidal channel. Taken together the two lochs provide an exceptional range of salt, brackish and eutrophic freshwater habitats, and the joint population of wildfowl is equally diverse. In addition to the species shown in Table 222 there are regular (and maximum) totals, mostly from Stenness, of 166 (386) Scaup, 102

(200) Long-tailed Ducks, 25 (150) Eiders, 29 (135) Mergansers and 10-15 Pintail. A flock of Greylags is also present throughout the winter, on and around the Lochs of Harray and Skaill. In the 1960s the regular level amounted to 15, and the highest count to 68; between 1970 and 1982 the corresponding figures were 130 and 525. Several of the other species on Harray and Stenness have shown similar increases: the Wigeon from a combined level of 670 (1,210) in the 1960s to 870 (1,840) in 1970-1982; the Pochard from 560 (3,000) to 2,010 (4,500), and the Tufted Ducks from 700 (1,250) to 1,715 (2,600). Mallard and Scaup have apparently decreased.

The other Mainland lochs, totalling about 900ha, are generally shallow, the mean depths ranging from 3m in the Loch of Swannay to 0.5m in the nearby Loch of Sabiston. Most of them have gently shelving surrounds, with extensive stretches of pebbles and stones along the water's edge; in several of them the bottom is also paved with flat stones, and virtually devoid of vegetation. The remarkable absence of silt is attributed to the wave action generated by the high winds, the turbulence in the shallow water being such that any debris is put into suspension and carried away in subsequent spates (364). Were this not so most of the lochs would have silted up long since; as it is, they continue to attract a sizeable number of ducks and swans.

The regular populations on 15 of the Mainland lochs, and on Echna Loch in the neighbouring island

Fig. 55. Orkney.

of Burray are summarised in Table 222. The table also shows the highest total recorded for each species on the 16 lochs combined. In several instances these group maxima were recorded on days when the cover was incomplete; the lochs for which data were lacking, and the species affected by this, are identified by an asterisk set against the relevant entry in the main part of the table.

The only major omission from the list of Mainland sites is the Loch of Wasdale (c.15ha). Some indication of its value is provided by a set of January records in the seasons 1969, 1970 and 1983, the means of which are set out in Table 223. The table also includes a summary of the occasional records obtained from a sample of the freshwater lochs on some of the other islands. These seem to follow the same general pattern of population, suggesting that the numbers on the lochs which have not yet been surveyed are probably similar.

In addition to the species listed there are reports of up to 100 Scaup on the lochs on Rousay, and of 20 Shoveler wintering in Sanday and North Ronaldsay. Both species have bred, the former at Isbister in 1973, the latter fairly frequently and at several different sites. Pintail also breed regularly on Mainland and on several of the other islands including Stronsay and Sanday. At least 10 pairs were reported in 1975, 1976 and 1977, compared wih only 1-3 pairs a few years earlier and later. Gadwall nested for the first time in 1976 but are not yet well established. Wigeon and Tufted Ducks breed regularly in small numbers and Shelducks are numerous and widespread; in 1977 there were more than seven nests on Papa Westray alone (540).

Greenland White-fronted Geese used to winter in numbers on Orkney, but have recently decreased. Between 1962 and 1975 upwards of 70 were recorded annually around the Loch of Tankerness, and another flock of 40-50 was established near Birsay in the north-west corner of Mainland. Since 1976 the flock at Tankerness has virtually disappeared; 15 were seen in November 1978, and 44 in March 1979, but none on any of the monthly visits. The numbers at Birsay and Tankerness have together amounted to no more than 80 between 1980 and 1983. The decrease at Tankerness is attributed partly to an increase in shooting, but mainly to the reclamation of the moorland which provided the main feeding ground. It may also be linked with the recent increase in Caithness (516).

Barnacle Geese appear in some numbers in autumn, and again in March and April. These are

Table 222. *The numbers of ducks and swans at 15 lochs on the Mainland of Orkney, and at Echna Loch on Burray: 1961-1982.*

A) the regular population at each site, and in brackets the number of seasons on which the figures are based,
B) the highest totals for each species on the 16 lochs combined, and the dates on which recorded.

		Area (ha)	Mallard	Teal	Wigeon	Pochard	Tufted Duck	Golden-eye	Whooper Swan	Mute Swan
A:REGULAR										
Stenness	(16)	638	222	80	352	28	296	124	15	109
Harray	(21)	981	196	54	544	1352	940	65	12	128
Clumly	(3)	c.20	55*	87	79	18*	13*	9*	10*	2*
Skaill	(17)	62	80	43	120	200	85	33	12*	3
Bosquoy	(19)	26	48	41	84	69	64	3*	4	23
Sabiston	(21)	31	116	45	80	5*	18	3	19	2
Isbister	(1)	34	20*	13*	350	0*	21	3*	1*	1*
Boardhouse	(20)	231	97	19	243	525	45	14	1	0
Hundland	(10)	112	55*	43*	41	170	31	5	2*	3
Swannay	(8)	244	37*	2*	2*	144	238	7	0*	0
Brockan	(2)	c.8	42*	15*	9*	3*	4*	13*	11*	5*
Kirbister	(10)	91	44	15	154	106*	37*	4	4*	0*
Tankerness	(17)	60	45	12	458	29	38	62	2	5
Ayre	(16)	c.6	6	0*	25*	23	20	9	7*	20
Graemeshall	(15)	c.8	8	5*	8*	9*	15	4	11*	4
Echna Loch	(15)	c.6	6	1	5*	37	38	5	1*	7
B:GROUP MAXIMA			1565	607	2709	5249	2812	542	133	382
			Oct 66	Oct 82	Dec 82	Dec 75	Jan 73	Mar 75	Jan 79	Dec 72

* These records not included in totals in B.

Spitsbergen – Solway birds on passage through northern Scotland, and marked birds are regularly seen. A separate group has taken to wintering on Orkney, their usual resorts being on the islets of Switha and Swona in the southern approaches to Scapa Flow. In recent years the flocks have continued to increase, to 400 in 1981-82 and 500 in 1982-83. The absence of marked birds makes it virtually certain that these wintering Barnacles originate from the Greenland population.

Whooper Swans are also more numerous in autumn than at other times, much more so than the tables might suggest. In mid-November 1976 a census of the Orkney population produced a total of 479, of which 70 were on Westray, 100 on Sanday, 95 on Stronsay and 144 on Mainland. Similar counts in 1981 and 1982 produced totals of 600 and 640, despite the fact that both were only partial surveys. These included 150 and 125 respectively on Shapinsay (500). A census of Mute Swans, undertaken on the same date in 1976, produced a corresponding total of 371. The population in this case is resident, and centred on the Lochs of Harray and Stenness, which together hold at least 30 breeding pairs. They also support a large non-breeding population which increases in winter to at least 200, and has reached 300 (35). On the occasion of the census there were 142 on Harray and Stenness.

Table 223. *Further records of ducks and swans on freshwater lochs in Orkney.*

a) L of Wasdale, Mainland: January mean, 3 seasons
b) North, Roos and Bea Lochs, Sanday: "Regular" numbers, 1980
c) L Trena, Lythe and Liddel, South Ronaldsay: maxima, 11 counts, 1973-1982
d) L of Wasbister and Scockness, Rousay: maxima, January – March 1975
e) 7 lochs in North Ronaldsay: combined maxima, October – March 1975/76
f) L of St Tredwell, Papa Westray: "Regular" numbers, 1980
g) L of Burness, Swartmill and Craig, Westray: maxima, 13 counts, 1981-1982

	a	b	c	d	e	f	g
Mallard	33	98	152	7	125	92	12
Teal	44	0	57	25	88	72	120
Wigeon	54	322	72	52	196	233	150
Pochard	9	246	177	0	5	9	2
Tufted Duck	6	89	57	64	12	14	45
Goldeneye	2	49	0	0	5	13	7
Whooper Swan	6	59	22	0	22	6	28
Mute Swan	1	34	14	17	9	13	8

As in the Western Isles the species increased greatly in the years between 1890 and 1940 (305), but has since maintained a fairly constant level.

The coastal waters of Orkney are remarkable for the great land-locked harbour of Scapa Flow, some 200sq km in extent, and for the numerous sheltered sounds between the northern islands of the archipelago. These enclosed waters are used extensively by Long-tailed Ducks and Eiders, and are greatly preferred to the open sea. The Long-tailed Ducks, with an estimated peak of about 6,000, are of special interest, the numbers here amounting to perhaps a third of the British winter population. The Eiders also total about 6,000, the winter numbers being rather higher than those in summer, when some of the males apparently move away to moult elsewhere.

One of the main resorts of the Orkney sea ducks is in Scapa Flow, which has recently become a major centre for the handling of North Sea oil. This development, which dates from 1977, is based on the island of Flotta, where the crude oil is brought ashore by pipeline and stored pending transfer by tanker to refineries elsewhere. The disruption caused by the installation itself and by the associated shipping is relatively unimportant, and has had no appreciable effect on the numbers and distribution of the ducks. A much more serious aspect is the potential hazard of pollution from ruptured pipelines or from accidental spillages during the transfer of oil from shore to ship. Several minor incidents have already occurred, and others will no doubt follow from time to time.

In 1974 the Nature Conservancy Council commissioned a four-year study of the birds in Scapa Flow, the aim being to establish a basis for future comparison, and to assess the extent to which the various species might suffer from pollution. The results of this work, undertaken by the RSPB, are summarised below (314). Table 224 shows the months in which the species are present in strength, and gives some indication of the total numbers; the true totals may well be somewhat higher.

The Long-tailed Ducks, with a peak of around 2,000, are much more vulnerable to pollution than other species. This is so particularly at night when most of the population resorts to the centre of the harbour, apparently in small parties scattered over a wide area. This critical roost lies a short way to the north and east of Flotta and the offshore loading terminals, in the lee of the prevailing wind, so that any spillage is likely to be swept directly through the roosting flocks. By day the birds disperse to feed in the shallower inshore waters of the Flow, entailing in some cases a flight of up to 12km (265). The chances of a day-time disaster are thus substantially reduced. The largest concentration of feeding birds, amounting at times to about 600, is located to the south and east of Graemsay, and is normally present from November until March; the bulk of the flock then moves to the eastern end of the Flow off the Barriers, Echnaloch Bay and inner and outer Water Sounds (314). The other feeding flocks are widely scattered and usually total under 100; the Loch of Stenness is probably the only site which often holds more than 150.

The Eiders vary in number from a summer level of about 1,000 to a winter peak of perhaps 2,000. They too are widely dispersed, and, unlike the Long-tailed Ducks, remain in scattered groups by night as well as by day. The largest wintering flocks, totalling 5-700, are centred around Graemsay and in the neighbouring Bay of Ireland, and are fairly well removed from the obvious sources of pollution. Most of their food in this area is made up of the razorshell *Ersis* sp, rather than mussels (314). In summer the moulting flocks are equally scattered, usually in parties of 20-30, and more occasionally in groups of up to 140. The majority of the birds at this time are either female or immature, the males comprising not more than 20% of the total (266). The breeding population has not been studied.

The small population of up to 100 Velvet Scoters is restricted to the eastern end of the Flow, to the area of Echnaloch Bay and outer Water Sound; a few are also found in Swanbister Bay in the latter half of the winter. The two main resorts are both downwind of Flotta, and at no great distance from it. The Goldeneyes and Mergansers are unlikely to suffer appreciably except in a major disaster; the Goldeneyes are mostly tucked away on the Loch of Stenness, and the Mergansers are so widely dispersed that only a few are at risk in any one place.

The sea ducks in the northern parts of Orkney have not yet been studied comprehensively, but enough is known from various surveys and reports to identify at least some of the main resorts. It is also clear that the numbers of most species are substantially larger than those in Scapa Flow. Fortunately the tideways here are shielded from the oil terminal by the barrier of Mainland, and the chances of pollution are probably no greater than those on most other stretches of the Scottish coast.

The information on Eiders relates mostly to the distribution of the moulting and non-breeding flocks. Throughout the summer these birds are widely scattered, usually in parties of 20 or so, but occasionally in groups of several hundred. Some of the largest gatherings, consisting mainly of adult males, have been found in the vicinity of Papa Westray, Wyre and Gairsay (c.400 each), and around Copinsay (c.300). The females and immature birds are normally dispersed in much smaller flocks. In summer 1976 a survey of 19km of the coast of North Ronaldsay produced a total of 580 birds, all in groups of less than 20; more than 85% of them were females or immature. As in other districts, the flocks were nowhere found on waters with a depth of over 20m (266).

The total number of Eiders in late summer is estimated at about 5,000, of which 1,000 are in Scapa Flow. The winter population is said to be somewhat larger; this is certainly so to the south of Mainland, and presumably to the north as well. The only definite counts date from the winter of 1973-74, when 400 were

Table 224. *The maximum number of sea ducks recorded in Scapa Flow, Orkney, 1974-1977 (including the Loch of Stenness). From Lea (1980).*

	Goldeneye	Long-tailed Duck	Velvet Scoter	Eider	Merganser
August	0	1	4	733	140
September	0	2	9	1181	170
October	65	864	47	1050	272
November	201	1284	46	1130	282
December	215	1243	87	1135	217
January	295	1620	88	1169	304
February	238	1912	83	1493	287
March	257	1390	94	1308	293
April	195	1390	97	753	158
May	5	293	15	995	160

Note: The true numbers are almost certainly higher than those shown; in February 1975 the total population was estimated at 2400 Long-tailed Ducks, 2000 Eiders and 350 Mergansers. The records of Goldeneyes and Velvet Scoters are probably representative.

located off Westray, and 1,100 in the neighbourhood of Wyre (540). Important breeding colonies are established on Papa Westray and Eynhallow, and many of the other islands have substantial numbers of nesting pairs.

Velvet Scoters occur in variable and possibly decreasing numbers. In 1975 the total Orkney population was estimated at 350, of which not more than 100 were in Scapa Flow (540). Parties of 5-10 are often present in Wyre and Rousay Sounds from January until April, and a flock of 80 was reported off Gairsay in March 1974 making this the most important area except for Scapa Flow. Common Scoters used to breed regularly, but this has ceased since 1958 and the species is now scarce and irregular throughout the year. The only recent record of any consequence refers to a flock of 50 in the Sound of Eday in January 1975.

Long-tailed Ducks are numerous and widespread throughout the northern sounds (Table 225), especially in places where the tide runs fast. The current estimate for Orkney of about 6,000, including upwards of 2,000 in Scapa Flow, is based on a survey undertaken in the winter of 1973-74, the results of which are set out in Table 225 (312). These have since been substantiated by several other records which are also noted; the count for Scapa Flow is included for comparison, and needs to be read in the context of Table 224.

Shetland

The Shetland archipelago comprises 117 islands, 16 of which are permanently inhabited. At the southern end of the group the isolated outpost of Fair Isle lies 45km from the nearest point in Orkney, and 37km from Sumburgh Head on the Mainland of Shetland. From there the chain of islands continues for another 112km to Muckle Flugga at the northern extremity of the British Isles. The total area of land, including Fair Isle and Foula, is 1,426sq km.

The geology of Shetland is complex, but the southern part comprises almost entirely the Old Red Sandstone of which Orkney is composed. The islands are studded throughout with numerous freshwater lochs, the majority of which are too acid and too bleak to harbour many ducks. The coast is equally broken, with many voes, or sea lochs, extending far inland. Even in Mainland, which accounts for two-thirds of the land mass, there is nowhere further than 5km from tidal water. The agriculture is much less prosperous than in Orkney: more than 90% of the land is classified as rough grazing and suitable only for sheep farming (394).

The freshwater lochs (Table 226) are of interest mainly for Whooper Swans, which arrive in strength in October and November, and remain in varying numbers throughout the winter months. Recent

Table 225. *Counts of Long-tailed Ducks in Orkney, winter 1973-74. (From Lea 1974.)*

Scapa Flow	1610		
N Mainland: Finstown – Deerness	300		
NW of Shapinsay	50		
Gairsay	340		
SE of Wyre	580		
Wyre and Rousay Sounds	480		
E of Egilsay	150		
W of Eday	50		
Sanday	350		
Eynhallow Sound	300		
Westray – Papa Westray	470		
Total	4680		
Additional records:			
North Ronaldsay	500, Feb 75;	500, Apr 80	
Wyre and Rousay Sounds	1634, Apr 75;	640, Mar 76;	500, Apr 80
Eynhallow Sound	264, Mar 75;	310, Feb 76;	400, Dec 80;
	356, Jan 81;	630, Jan 82	
Finstown	250, Apr 80		
Burray	400, Mar 80		
Inner Bay of Firth	350, Sep 80		

censuses have shown a November total of around 300, a figure which is probably close to the autumn peak; the actual counts amounted to 341 in 1976, 337 in 1977, 287 in 1978, 275 in 1979 and 273 in 1980 (540). In November 1982, however, there were 336 at the Loch of Spiggie, Dunrossness, alone – the second biggest concentration in Britain that season; the total for the archipelago then amounted to a record 486. The winter level is probably between 100 and 150, with perhaps a slight increase in spring. Large gatherings also occur regularly on the Loch of Brow in Dunrossness, the Lochs of Bardister and Kirkigarth near Walls, on Easter Loch in Unst, and at times on several other waters. In addition to those listed in Table 226 there are records of 25 or more from Sandness (West Mainland), Lochend (North Mainland) and Fair Isle. Counts from the spring are few, but an exceptional total of 568 Whooper Swans in March 1982 indicate that many do stop on passage.

As regards the freshwater ducks, the two main features are a small passage of Teal and Wigeon, and a wintering population of Pochard and Tufted Ducks amounting to several hundred of each. Only a small proportion of the lochs, particularly on Mainland, have ever been counted. Since 1970, however, the coverage has been sufficiently good for 17 waters to be included in Table 226, and it sems unlikely that any important concentrations of wildfowl have been

Fig. 56. Shetland.

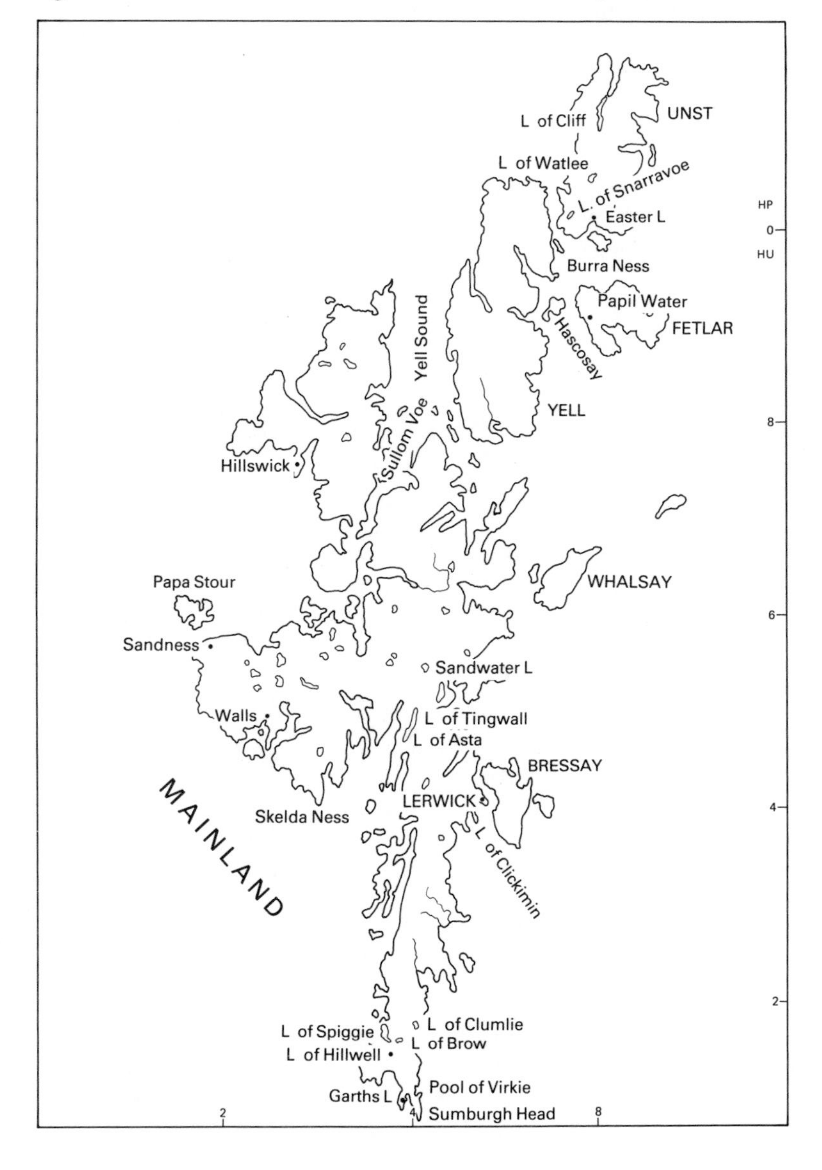

missed. The 88ha Loch of Spiggie, 8km from Sumburgh Head and only 100m from the sea at its north end, became an RSPB reserve in 1979. In recent years it has been increasingly important during autumn passage: in November 1981, 230 Greylag Geese were present, and a year later 336 Whooper Swans (see earlier) and 150 Goldeneyes. A few Greylags winter in Shetland, and a small return passage occurs in spring; Barnacles regularly overfly the islands in autumn on their way to the Solway, but other geese are rare. The numbers of breeding wildfowl on the lochs are small, but include some 25 pairs of Tufted Ducks and a few Wigeon, Shoveler and Shelducks. Up to seven pairs of Shelducks nested on the tidal Pool of Virkie until disturbance from the redevelopment of the adjacent Sumburgh Airport in the mid 1970s caused them to abandon the site.

The voes and sounds are too short of intertidal and shoreline feeding areas to hold many dabbling ducks, but sea ducks are abundant and have been well surveyed over the last ten years. These studies, covering all sea birds, were prompted by the impending advent of oil-related industry to Shetland, and followed, with commendable speed, the start of construction of the Sullom Voe terminal at Calback Ness and related developments. Between 1974 and 1978 the surveys were undertaken jointly by the Nature Conservancy Council and the Sullom Voe Environmental Advisory Group, and subsequently by the Shetland Oil Terminal Environmental Advisory Group as part of their wider monitoring programme. The resultant counts, mostly from a boat, have provided an excellent picture of the numbers, distribution and movements of sea ducks around Shetland, especially in the northern waters (303, 501).

Much the commonest species is the Eider, of which an estimated 15,500 were present around Shetland in August 1977, during an almost complete survey (266). This represented about a quarter of the entire British stock of Eiders. The main centre at the time was off Sumburgh, at the southern tip of Mainland, with 2-3,000, but 1,000 or more have also been found moulting off Skelda Ness, Whalsay, Bressay, Lang Ayre and Hillswick (Fig 57). There has, however, been a marked decline in the numbers of Eiders around Shetland in recent years. Comprehensive surveys between 1980 and 1983 found an average annual moulting population of only 8,850; in the main

Table 226. *The regular (and maximum) numbers of ducks and swans at various freshwater sites in Shetland, 1970 onwards. The figures in brackets after the place names show the number of seasons on which the results are based.*

	Mallard	Teal	Wigeon	Pochard	Tufted Duck	Goldeneye	Whooper Swan
SOUTH MAINLAND							
Garths Loch (6)	3(11)	4(23)	4 (33)	0 (7)	6 (40)	2 (7)	0 (2)
Loch of Hillwell (11)	5(24)	9(45)	4 (46)	0 (1)	6 (40)	2 (6)	12 (68)
Loch of Spiggie (11)	15(61)	1(10)	6 (51)	13 (80)	44(120)	32(200)	69(336)
Loch of Brow (11)	4(40)	0 (0)	9(181)	53(160)	46(142)	4 (20)	10 (73)
Loch of Clumlie (8)	6(28)	0 (3)	1 (12)	2 (40)	2 (20)	3 (8)	2 (12)
CENTRAL MAINLAND							
Loch of Clickimin (10)	30(55)	1 (4)	4 (46)	15 (70)	72(190)	13 (30)	2 (10)
L of Tingwall and Asta (6)	14(36)	3(14)	0 (0)	37(150)	67(170)	17(105)	1 (12)
Sandwater Loch (8)	5(14)	2 (8)	1 (9)	1 (23)	1 (9)	5 (15)	5 (15)
Benston Loch (7)	4(13)	1 (3)	8 (59)	10 (77)	3 (8)	6 (12)	4 (17)
Lochs of Bardister and Kirkigarth (4)	11(50)	0 (0)	5 (35)	0 (0)	46 (95)	1 (6)	27 (64)
PAPA STOUR							
Freshwater Lochs (1)	6 (7)	4(15)	42 (56)	0 (0)	0 (20)	0 (0)	4 (11)
FETLAR							
Papil Water (2)	15(30)	1 (2)	6 (13)	0 (0)	73(102)	19 (24)	13 (24)
Loch of Urie (2)	10(33)	16(28)	8 (15)	0 (0)	0 (0)	3 (18)	0 (0)
UNST							
Loch of Cliff (13)	20(85)	0 (0)	0 (4)	0 (0)	2 (67)	6 (55)	4 (12)
Easter Loch (9)	9(30)	0 (6)	3 (20)	6 (65)	17(100)	27 (90)	37 (86)
Loch of Snarravoe (8)	4(13)	0 (1)	4 (14)	7 (50)	83(310)	8 (18)	0 (10)
Loch of Watlee (1)	1 (2)	0 (0)	0 (0)	0 (0)	0 (0)	67(100)	0 (0)

vintering area, around the islands of Hascosay and 'etlar to the east of Yell, the total has decreased from i-6,000 in the mid 1970s (501) to only 1,300 in 1980 and i70 in 1982 (540). The bulk of the decline occurred luring the 1979-80 winter, when large numbers of lead Eiders were found. The birds were not oiled, and he cause of the mortality is unknown (258).

An estimated 1,500-2,000 Long-tailed Ducks vinter in Shetland (302). Up to 950 have been found in he Hascosay/Fetlar area, usually between Hascosay nd Burra Ness, Yell. Their numbers usually peak etween November and February, but notable gatherngs often occur in April and May (when the reshwater Loch of Spiggie can hold 30-50). Golleneyes and Mergansers are scattered all round the oasts, but rarely in large flocks, Goldeneyes being nuch more numerous on the freshwater lochs. A vintering population of 210-290 Mergansers has been

'ig. 57. Concentrations of Eiders in Shetland. Circles – noulting, triangles – wintering. From Hope Jones and Kinnear (1979) and *Shetland Bird Report*.

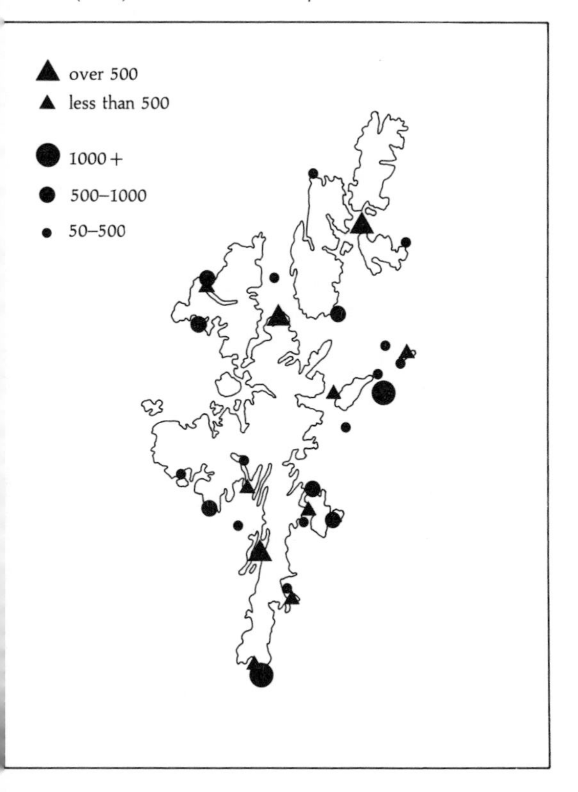

estimated (302); there are no recent records of the large autumn gatherings reported in the 1950s (600), but 50-60 regularly winter in Sullom Voe. Common Scoters are scarce and flocks of over ten unusual (though a few pairs often breed); about 20 Velvet Scoters formerly wintered in Sullom Voe, but since the *Esso Bernicia* accident in December 1978 (see below) no more than two have been seen (258).

As in Orkney, the construction of the oil installations had no serious direct effect on the wildfowl populations of Shetland, but the possible threat to the sea ducks from oil spillages is enormous. Since the opening of the Sullom Voe terminal in November 1978 there have been several incidents, including a particularly serious one just a month after the opening, when an accident during the berthing of the *Esso Bernicia* resulted in the spillage of 1,174 tonnes of fuel oil and the deaths of at least 3,700 birds, including 570 Eiders and 306 Long-tailed Ducks. An estimated 95% of the birds present in Sullom Voe at the time of the incident, and 75% of those in Yell Sound, were killed, including at least 540 Eiders, 270 Long-tailed Ducks, 33 Mergansers and 7 Velvet Scoters. The wintering numbers of Eiders, Mergansers and, to a lesser extent, Long-tailed Ducks have gradually recovered since the incident, but Velvet Scoters have been virtually absent (258).

The outlying islands of Foula and Fair Isle are important mainly for Eiders, 1-200 pairs nesting on each. On Foula the total numbers peak in July, and vary between 300 and 500 (201), and at Fair Isle there is an annual autumn gathering of 800-1,000. An autumn passage of geese occurs on Fair Isle, the maximum on any one day usually amounting to about 100 Greylags and 50 Pinkfeet and Barnacles (189).

Part III

Species accounts

Introduction

This part aims to summarise the available data on the numbers and distribution of each species of wildfowl in Britain. A brief description of the birds' behaviour and feeding habits is given first since this is very relevant to the interpretation of their distribution. There are a total of 48 wildfowl on the British List (549, and later reports in *The Ibis*); most are predominantly winter visitors. Thirty-one species occur here commonly, and these are fully treated in the following accounts. The remainder are either sporadic in occurrence or are usually in small numbers; they are considered in less detail, at the end of the accounts. The British List includes four species introduced to Britain and now well established, although two, the Egyptian Goose and Mandarin Duck, are in relatively restricted areas. The Canada Goose, on the other hand, is very widespread in England and Wales. It, and the Ruddy Duck, continue to increase in numbers and to expand their range (406, see also p.460).

Wildfowl are widely kept in captivity and several other species have been seen in the wild in Britain from time to time although recognised as originating from captive stock. Among these are the Black Swan (*Cygnus atratus*), Bar-headed Goose (*Anser indicus*), Ross' Goose (*Anser rossi*), Carolina or Wood Duck (*Aix sponsa*) and Ringed Teal (*Calonetta leucophyrus*). These are not likely to become widely established and recent legislation (p.539) makes further introductions more difficult.

Fig. 58. The monthly mean number of ducks, geese and swans counted in Great Britain 1976-77 to 1980-81, expressed as a percentage of the peak count.

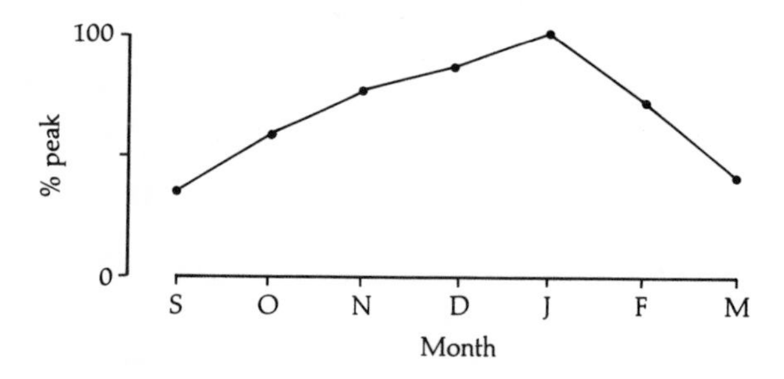

The information given here is the most up to date available and is gathered from published sources as well as the Wildfowl Count data and other Wildfowl Trust surveys. When the First Edition was published there were few quantitative data from outside the United Kingdom, but following the establishment of the International Wildfowl Counts in 1967 it is now possible to examine the numbers in Britain in the context of the flyway as a whole. The importance of sites can also be assessed using quantitative criteria on the proportion of British or flyway populations that they hold (22, 23). The importance of these criteria for wildfowl conservation is discussed more fully in Part IV.

Information on the occurrence of the scarcer species, and on breeding in those species which do so here occasionally, was obtained from regular accounts in *British Birds*.

The seasonal pattern of numbers

For species which are reasonably well covered a figure is presented which shows the pattern of numbers through the winter. The data used for this are the average monthly totals for the seasons 1976-77 to 1980-81, including four mild and one (1978-79) relatively hard winter. To make the figures more easily understandable the monthly totals are expressed as a percentages of those in the peak month (in most cases January). Because of differences in cover the same sites are not necessarily represented in each of the seven winter months. Ideally, the sample of sites should include only those which were covered in every month, but this would reduce the usable data to such an extent that the pattern revealed might no longer be representative. It therefore seems best to take the actual totals recorded each month, regardless of any variations in size of the sample. This is considered acceptable since the total number of sites covered each month is usually much the same (between 900 and 1,100), and it is normally the least important places which are covered the most erratically.

As an example the 5-year average of all ducks, eese and swans counted in each month is shown in ig 58. As might be expected there is a build-up in the utumn and a decline in the spring. The peak month is anuary, when we have received most of our immig- ants from continental Europe. There are, however, ifferences between the species; those originating rom Iceland tend to arrive early whereas those oming mainly from north-west Europe arrive later nd less predictably. Many species show annual dif- erences in pattern depending on overseas weather, nd these are shown where necessary.

The mapping of the British distribution

In the First Edition the species maps were lotted on a site by site basis, with a lower limit, below vhich a site was not included on the map. The limit vas sometimes as low as 5 (Bewick's Swan) but in ome species (Mallard, Wigeon) it was as high as 250, vhich meant that many areas where the species ccurred were not included. In this volume the maps re based on the number of birds found in each 10km quare. This has the great advantage that the lower imit can be set at 1 in all cases. It also avoids problems f delimiting sites, for example on estuaries or lengths f coastline counted as separate sectors. Some sites xtend over several squares and in these cases the total irds are allocated to the most important square (by he proportion of birds or of the site area).

A disadvantage of the system is that the dots are ot placed exactly on the site. However, with a 10km quare, the centre of the dot is no further than 7km rom the centre of the site. The largest dot has a liameter of 7km and in the vast majority of cases, at east for the larger or more important sites, the lack of recision is negligible. The extent of coverage of Britain's 10km squares is indicated in Fig 59. A dot is laced in the square if a count was made there at least nce in the 23 years 1960-1982.

Where a square is not covered, a triangle on the nap indicates that the square contains land at higher ltitude than 500m and an open square that it has no tanding water above 0.5ha in extent. In both cases the rea is unlikely to be an important one for wintering vildfowl.

The aim of the species maps is to illustrate the umerical distribution of species wintering in Great Britain. They are largely based on counts in the nonths September to March; where other data are ncluded mention of this is made in the text. Wintering eese are poorly represented in the Wildfowl Counts nd for them the maps are based largely on the regular November surveys of major autumn haunts, supplemented by data from the counts and by information from the literature. For some other species (such as sea ducks) counts are supplemented by data from special studies or from the literature.

Because of changes in the status of sites, data from five of the most recent years available (usually 1975-76 to 1979-80) have been used, but these have been updated where obvious changes occurred in 1981-82 and 1982-83. The wildfowl count data were analysed as follows:

a) for sites where the species occurred, the mean count in each month for the five years was calculated,

b) the site means were summed to give mean values for each 10km square which contained the species. The means for the square were rounded to the nearest whole number but, in order to eliminate very sporadic occurrences of single individuals, a mean of 1 was not accepted unless it was based on 3 or more counts. Thus a single individual seen once on a site with only 2 counts was excluded although it gives a mean of 0.5 rounded to 1,

c) the highest monthly mean for the square was selected and the dot size for mapping determined on this basis.

Since no assumptions could be made on changes in status of a site where counts were infrequent or had ceased, the site means were summed irrespective of their timing. For example if a square had two sites, one of which was counted twice in the 1960s, and had a mean of 200 ducks in January, and the other had complete coverage but yielded only 20 ducks, the square mean would be 220 ducks. Clearly this could lead to inaccuracies and the data were carefully examined to eliminate the most obvious anomalies. For example a square which had a single site had five counts in January with a mean of 59 and a maximum of 261 Whooper Swans. There was only one count in March, when 225 swans were present giving a "mean" of 225. When the full data were examined it became obvious that in only one winter had the site had a substantial number of Whoopers, 261 in January and 225 in March. In this case the mean of 59 based on the more complete month went forward for mapping. In another site counts were carried out in each of seven months in only one season. Whooper Swans were present only in January, when 9 were counted. This square had no other count-points and was given the smallest dot for that species (usually reserved for squares with means of 1-5 swans).

In practice most sites had recent counts and after

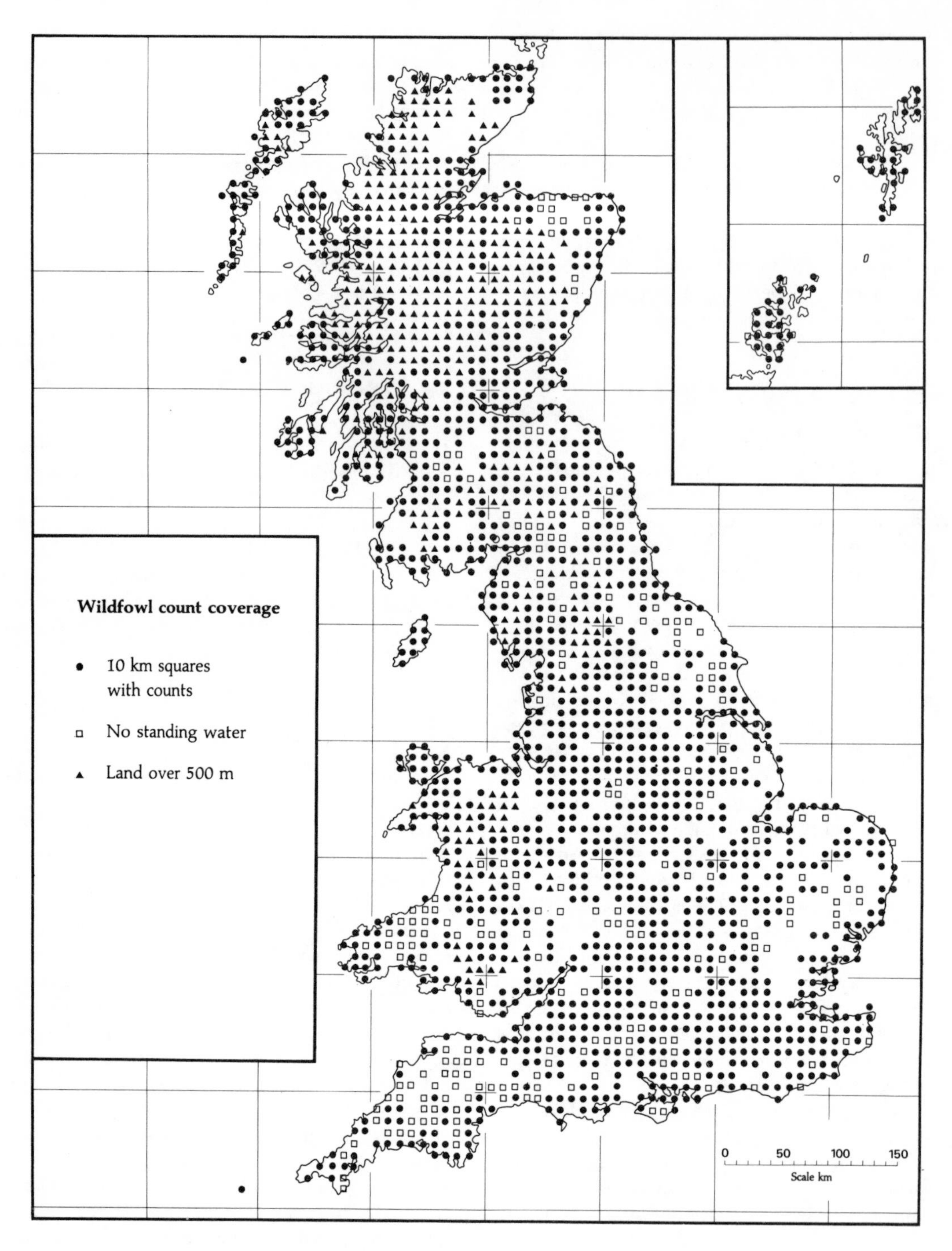

Fig. 59. Coverage of 10km squares in Great Britain by Wildfowl Counts, 1960-1982. Also indicated are uncounted squares with land above 500m (triangles) and squares with no standing water exceeding 0.5ha (open squares). Data on standing water from Sharrock (1976).

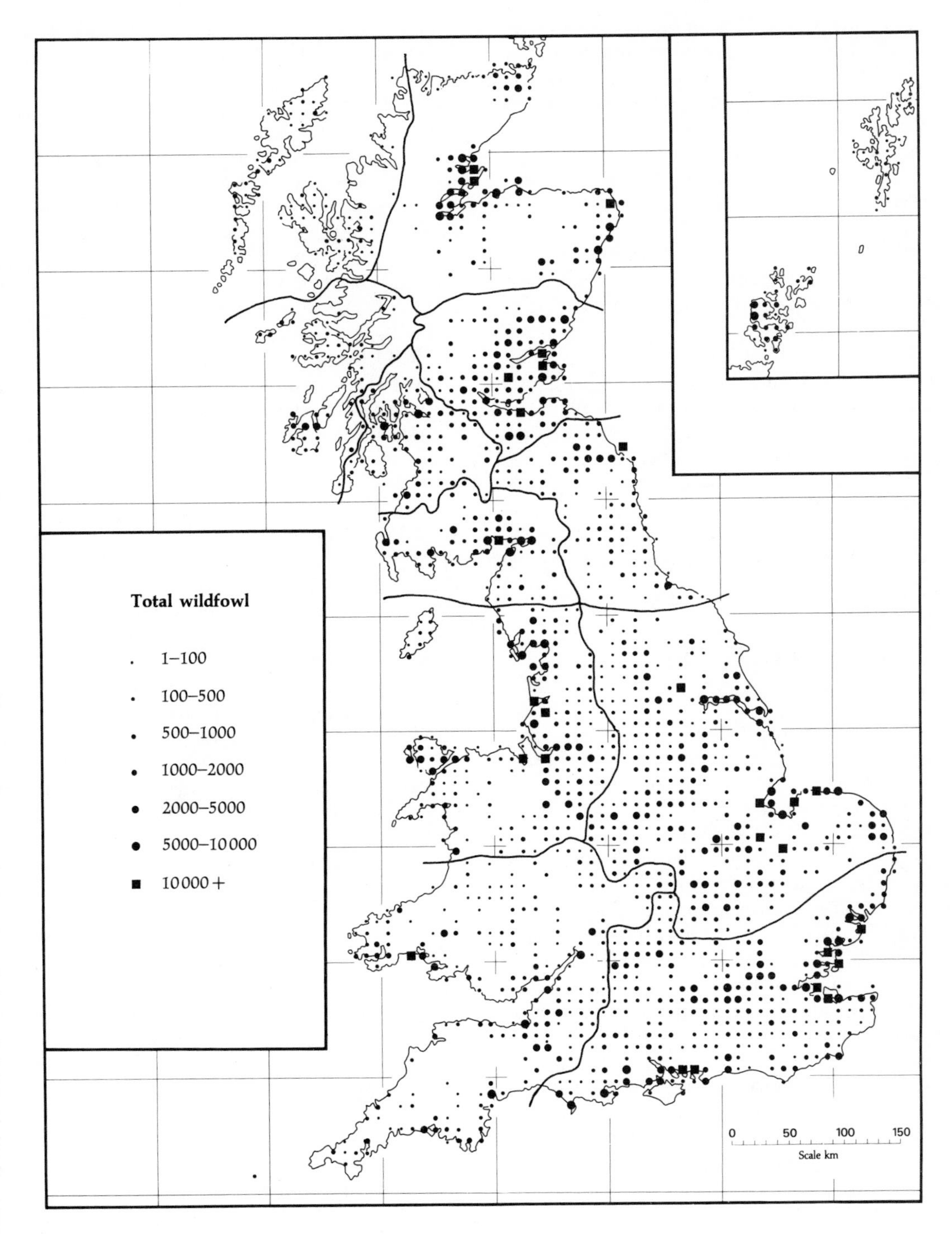

Fig. 60. The total number of wildfowl (ducks, geese and swans) counted in each 10km square in the National Wildfowl Counts, based on the average for the five seasons, 1978-1982.

obvious irregularities, such as in the examples given, had been eliminated the dots were a reliable indication of the average seasonal maxima in recent years.

At a small proportion of places some species were recorded as "present but not counted". In such cases the site was treated as if there had been no count on that date. If the counts were unsatisfactory over the whole period, and no other information was available, the smallest dot size was used.

Sporadic occurrences of a species outside its normal range were disregarded if there were good grounds for believing that the birds were escapes from captivity.

The numerical distribution of counted ducks, geese and swans in total is shown in Fig 60. In this and in the individual species maps square symbols are reserved for areas which qualify for international importance – i.e. holding at least 1% of the estimated mid-winter population of north-west Europe, or 10,000 wildfowl in total (22). The boundaries of the nine regions used in Part II are shown (see p.000), minor adjustments having been made to allow for the grid lay-out.

Breeding distribution and migration routes

The breeding distribution of all birds in the British Isles has recently been mapped (542). However, data of a quantitative nature are extremely sparse. Only two fully quantitative surveys have been carried out since the First Edition (598, 621) and these, as the authors admitted, were not comprehensive. The information on breeding in Britain given here is based on these and other sources in the literature.

Where a map is included, this is intended only to indicate the origins and migration routes of the populations wintering in Britain. For geese and swans the breeding area is often well known and the birds follow distinct and narrow migration routes and have regular staging areas. Ringing recoveries, sight records and counts along the flyway help to build up a picture of their normal movements. Ducks, on the other hand, migrate along a broader front or along unknown pathways which may vary from year to year. For many species the potential breeding area is vast and to a large extent unknown; our knowledge of their origins and migration routes must thus be based solely on ringing recoveries. In order to obtain a reasonable sample, recoveries of both British and the closely linked Dutch-ringed birds have been used. No recoveries have been excluded on the basis of age, sex or finding method, so that some of the bias (p.16) counterbalances. The recoveries are plotted on th map with an indication of the time of year, so that i some cases the progress of ducks to and from Britai can be followed. The interpretation of these maps not straightforward (p.16).

In some cases where British-ringed birds a recovered overseas the map also indicates the winte ing areas of British breeding stock or the eventu destination of birds on passage when ringed.

In general, the information on geese with the discrete populations and traditional habits is ve good, whereas for ducks it is rather inadequat Nevertheless, the extensive ringing carried out by th Wildfowl Trust and similar organisations oversea enables us to indicate the general areas from which ou wintering birds come and to determine their broa migratory pathways.

The analysis of trends in number

In the First Edition the trends in the populatior of the individual species were assessed by comparin the numbers recorded on a smallish sample of selecte sites in one particular season (the master year) wit the corresponding records in each of the other years i turn. The totals for the other years were the expressed as a percentage of the total in the maste year, to provide an annual index of abundance. Th disadvantage of the method was that too muc emphasis was placed on the turn of events in th master year: if a site was not covered in that particula season, it could never be included in the calculation and if the numbers of birds in that year wer abnormally large or small the indices for the othe years were correspondingly distorted. Another fau was that only a small proportion of the data wa employed.

The present method of analysis is more flexibl the results for each successive season are linked i turn with those for the preceding one, to provide directly comparable pair of totals. The one is the expressed as a percentage of the other, and since eac year is compared with both the one before and the on after, a continuous chain of comparisons is formed These are then converted into indices in the manne described below, and are related to a selected maste year, which by definition has a value of 100 (397).

The month selected for the comparison was no the same for all species but was the one in which mos individuals of that species were counted. In man cases the analysis was carried out on the data c several months and the one which gave an inde based on the greatest number of birds was selected. I most cases, because of completeness of cover as wel s because most species peaked then, January was the onth used.

The master season selected was 1970-71, which ad advantage of being near the middle of the period nder consideration and of not being in any way bviously exceptional.

The following procedure was adopted in each ase:

a) beginning in 1960-61 each pair of years was considered together, i.e. 1960 with 1961, 1961 with 1962, etc. The sites counted in the relevant months in both years were included in the paired sample and the total number of ducks in all of the paired sites in each year established,
b) the ratio of the two totals was calculated for each of the 22 pairs of years,
c) working backwards and forwards from the master year an index for a particular season was calculated by multiplying the index (which started at 100) by the ratio of the current year's to the preceding year's counts (after 1970-71) or to those of the subsequent year (before 1970-71).

To illustrate the method consider the following imple example, having selected the month under onsideration:

1) 10 sites were counted in January 1968 and 1969, when they contained respectively 800 and 1,000 Mallard,
2) 100 sites held 22,500 ducks in 1969 and 25,000 in 1970,
3) the figures for 1970 and 1971 from 50 sites counted in both years were 12,500 and 10,000,
4) there were only 20 paired sites in January 1971 and 1972 and these had respectively 15,000 and 10,000 Mallard.

Although this is an extreme case in coverage erms the use of all paired sites in all pairs of years has a great advantage. Table 227 can be constructed.

In order to give an assessment of how comprehensive and representative was the sample of sites on which the trend was based, the number of ducks and the number of counts included in the paired sample were expressed as percentages of the total number of that species counted. Thus if this percentage was 75% the index was based on three-quarters of all counted individuals of that species over the period.

Table 228, as an example of the analysis results, gives the data for Mallard in January in Great Britain. One obvious development is the great increase in the number of sites covered annually after 1967, when the International Wildfowl Counts began. This was a good reason for rejecting the first year as a master year. Since 1967 count coverage has remained relatively constant (except for 1968 when there was an epidemic of foot and mouth disease which made access to many areas difficult). With the increase in cover there was also an improvement in the number of ducks counted and in the number and proportion of birds included in the paired sample – an indication of better consistency from year to year. The fact that the proportion of birds included is higher than that of counts indicates that the more important sites are more regularly counted. This is much more striking for other species, which are considerably more concentrated than the ubiquitous Mallard. The proportion included overall is generally high; in this example more than four-fifths. Although the proportion of the overall British population which is on counted sites varies, we believe that the index is based on a sufficiently large and unbiased sample to give a reliable indication of the annual fluctuations in the numbers of British wintering wildfowl.

So that regional variations could be identified the index was first calculated for each of the 10 regions (see p.339) and these were grouped to give values for different areas and for Great Britain as a whole. Where regional differences do occur these are discussed.

able 227. *An example of the calculation of the trend index.*

Year		Paired	Paired counts			Index (1971=100)	
1	2	sites	Year 1	Year 2	Ratio*	Year	Index
968	1969	10	800	1000	0.8	1968	90
969	1970	100	22500	25000	0.9	1969	113
970	1971	50	12500	10000	1.25	1970	125
971	1972	20	15000	10000	0.67	1971	100
						1972	67

Note: The graph axes (Fig 62, etc.) are labelled according to the season (e.g. 1960 = 1960-61).
* Before 1971, the ratio is Year 1/Year 2, after 1971 Year 2/Year 1.

The geese, as for many other analyses, are exceptional in that good data on total population size are available in most cases. These are given rather than the Wildfowl Count indices, which would be inadequate for most species.

The assessment of the British population

The 5-year monthly means for the peak month as used for Fig 58 give the regular maximum counted, but this is an underestimate for the following reasons:

a) Most counts are of roosts in the day-time, whereas some species feed diurnally some way away from a site and would not be included in its counts. For this reason geese and some swans, dabbling ducks and Goosander are under-represented.

b) Of the 4,000 sites counted since 1960 only abo 1,200 are included each season. For most speci this results in the annual sample counted beir 70-80% of the total birds if all sites had bee included.

c) Some birds are on sites never covered by th scheme and we can only make informed gue ses, knowing the habitat requirements an dispersal of each species, as to the proportio missed. Thus we may count less than 50% of th Mallard population, which is widely dispersec and perhaps 70% of Red-breasted Merganser which are common in north-west Scotlanc where count coverage is poor. On the othe hand three-quarters of counted Scaup are o only 15 sites and no large concentrations ar missed, so we are confident that we cour practically all individuals of this species.

In order to obtain a total for all sites counted th

Table 228. *Analysis of trends of Mallard in Great Britain in January between 1960 and 1982.*

Year A	Year B	Year A counts	Total birds	Paired counts: Pairs	Paired counts: A	Paired counts: B	Ratio B to A	Index (1971=100): Year	Index (1971=100): Index	% Counts in pairs	% Birds in pairs
61	62	495	65258	378	56637	62605	90	61	78	76.4	86.8
62	63	541	77870	357	63161	53562	118	62	87	66.0	81.1
63	64	456	60502	366	55654	66310	84	63	74	80.3	92.0
64	65	564	87039	423	77343	75833	102	64	88	75.0	88.9
65	66	572	89425	431	74913	95087	79	65	86	75.3	83.8
66	67	573	107463	524	103736	97060	107	66	109	91.4	96.5
67	68	1225	160603	735	119151	109636	109	67	102	60.0	74.2
68	69	998	124056	777	108446	108014	100	68	94	77.9	87.4
69	70	1227	148190	884	114010	112045	102	69	94	72.0	76.9
70	71	1096	123015	836	106415	115750	92	70	92	76.3	86.5
71	72	1173	143892	847	123038	104462	85	71	100	72.2	85.5
72	73	1137	126216	901	113088	109362	97	72	85	79.2	89.6
73	74	1197	126488	924	105018	110220	105	73	82	77.2	83.0
74	75	1124	123123	878	109013	106717	98	74	86	78.1	88.5
75	76	1154	122919	853	103884	108563	105	75	84	73.9	84.5
76	77	1160	125245	899	110335	100766	91	76	88	77.5	88.1
77	78	1188	118333	944	107489	101459	94	77	81	79.5	90.8
78	79	1192	116676	920	97651	130732	134	78	76	77.2	83.7
79	80	1167	147386	934	132571	156836	118	79	102	80.0	89.9
80	81	1192	180177	967	166874	149379	90	80	120	81.1	92.6
81	82	1232	169763	968	151059	134544	89	81	108	78.6	89.0
82	83	1271	149275	1009	133400	143955	108	82	96	79.4	89.4
83		1469	174825					83	104	68.7	82.3

Notes: Number of sites included = 3524.
Assessment of comprehensiveness of the trend data:
The percentage of potential counts (sites x years) made is 27.1
The percentage of ducks counted included in paired counts is 86.6
The percentage of counts which are used in paired data is 76.4

-year monthly means used for mapping (see above) 'ere summed for each month, excluding those sites hich were consolidations of others. This total ıcludes a figure for each site whether it was recently ɔunted or not, and so the assumption is made that ɔnditions at that site have not changed since the last ɔunt was made. This may be rather a dangerous ssumption as some sites may have been abandoned y counters because they had declined in importance. Iowever, this seems a safer assumption to make than ıat the situation has changed in a particular direction. .s well as sites being abandoned others are forming nd take some time to be included in the scheme, so ıe errors are to some extent compensatory. The population estimate will in most cases be too low, since not all sites have counts in all months. For January, however, the peak for most species, this should be a reliable figure.

The total is corrected for the proportion missed completely to arrive at an estimate for the British population. As an example the total wildfowl means for Fig 58 range from a low of 249,000 in September to 707,000 in January. The peak population is likely to be well over a million after the various corrections have been applied. Since we accommodate birds in transit and since many ducks have been shot before the peak is reached the number of wildfowl which "use" Great Britain annually may well be double that figure.

Principal species

Mute Swan *Cygnus olor*

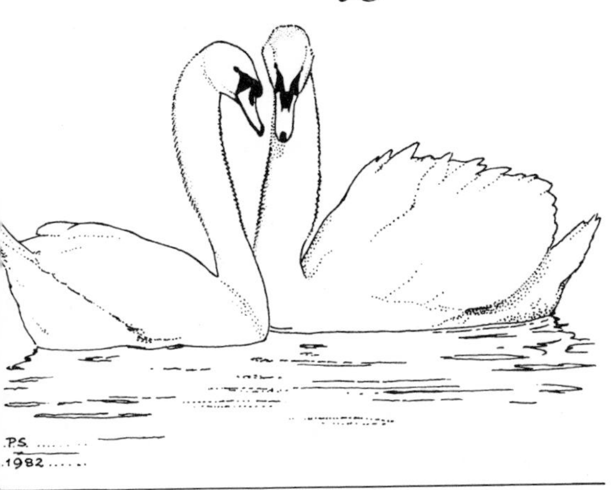

The Mute Swan is the largest and most common Eurasian swan. The male weighs up to 15kg and is very probably the heaviest flying bird. The species occurs around the Black and Caspian Seas and in eastern Asia and has been introduced to North America, South Africa, Australia and New Zealand, as well as much of northern and central Europe. The north-west European Mute Swans are split into seven more or less discrete groups, three in Britain, one in Ireland, and three in continental Europe (24). The Mute Swan is a resident or partial migrant and there is little or no interchange between groups, or even between different locations in the same group.

The Mute Swan is widespread on rivers, lakes and estuaries, often in close association with man. Outside the nesting season most birds are gregarious although some pairs remain on the nesting area and defend a territory throughout the winter. Wintering concentrations in Britain tend to be rather small; 91% of sites in our survey which held Mute Swans had less than 25 birds and 98.6% less than 100. In other countries, notably Denmark, where the numbers are much larger, coastal wintering flocks may be numbered in thousands.

The Mute Swan is not very agile on land and seldom strays far from water. Its feeding is usually confined to underwater and bank vegetation. Where it has been found on agricultural crops or grassland it has usually had access to fields from river or lake banks. The diet of Mute Swans has been studied in several countries (summarised in 447). Submerged aquatic plants, particularly the pondweed (*Potamogeton*), *Myriophyllum* and *Chara*, are overwhelmingly important. In brackish areas, such as the Chesil Fleet in Dorset, *Zostera* forms the main item, particularly the large-leaved *Z. marina*. On the Ouse Washes, East Anglia, roots and stolons of grassland plants are most important and fine grasses make up more than a third of the diet. The Mute Swans are much more aquatic than the Whooper and Bewick's Swans which winter there (445). On the Exe Estuary the species grazes saltmarsh grasses and succulents as well as taking *Zostera* and green algae from the mudflats and creeks (210).

The Mute Swan has long associated with man and in some places has been semi-domesticated. Large numbers gather in the non-breeding season in public parks or rivers used for public recreation, subsisting largely on bread and other titbits provided by the public. Others gather at the outfalls of breweries or distilleries and feed on waste grain. Mute Swans have been accused of damaging fish stocks by eating fish spawn, but where this has been investigated the allegation has been unfounded.

In the breeding season Mute Swans are territorial and aggressive, with a pair defending the whole of a small lake or stretch of river. In the Oxford area nests are usually between 2 and 3km apart and only exceptionally do two pairs nest within 100m (467). An average of 2km of river (including tributaries) is needed per breeding pair on the nearby Windrush, the most productive river in the area, with rich aquatic vegetation (30). In less suitable areas densities as low as 1 pair per 17km have been recorded.

By contrast, in some places Mute Swans breed colonially, with nests within a very few metres. The

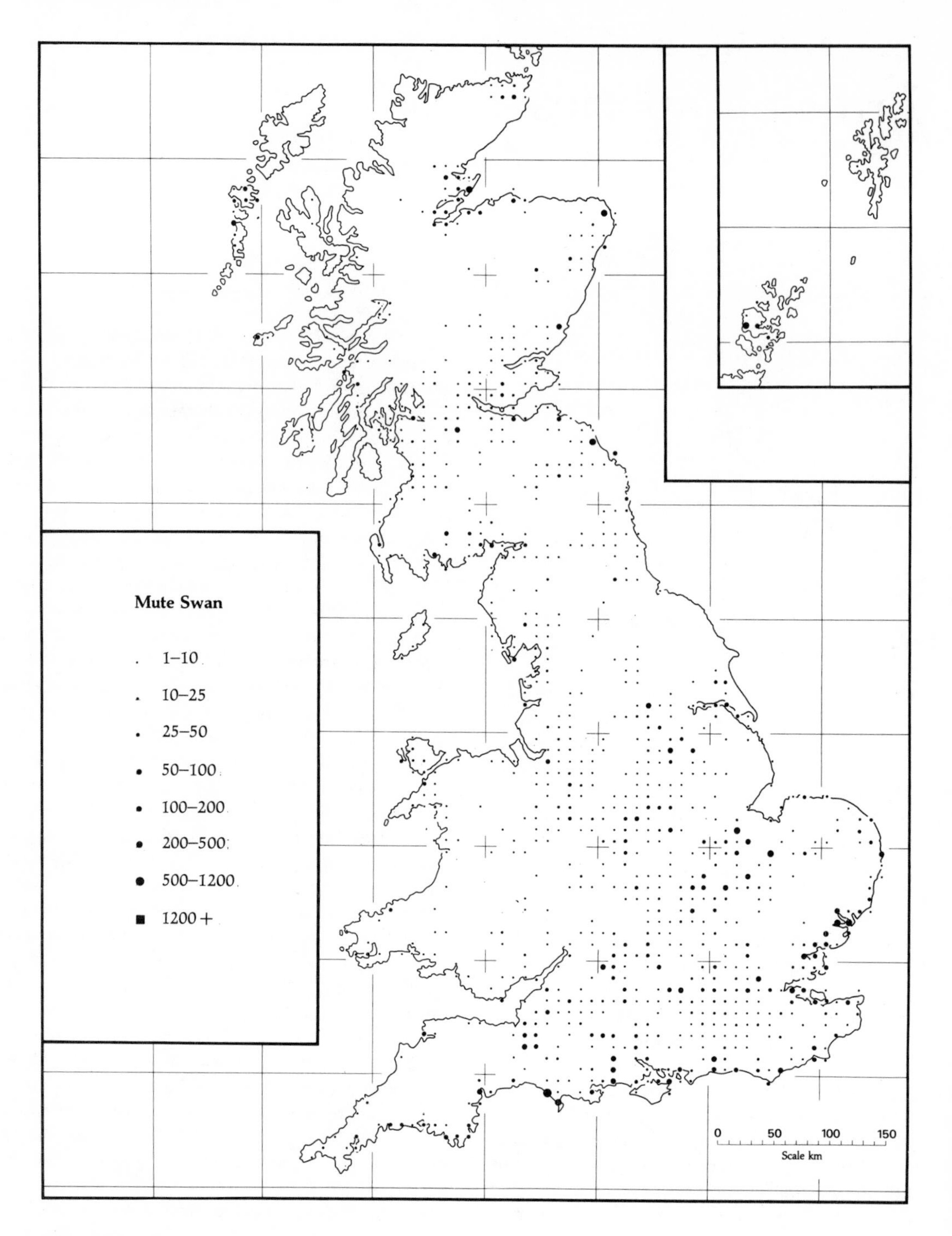

Fig. 61. Mute Swan – British distribution, September to March, as recorded by the National Wildfowl Counts.

nly important colony in Britain is at Abbotsbury on ne Chesil Fleet, and has been studied in detail (468). he colony was established artificially by monks and ates back at least as far as 1393. The swans were ncouraged to nest in the colony and their young enned up and fattened for the table. The tradition is till kept up although the cygnets are now released ather than eaten. In recent years the number of nests t the colony has varied from 19 to 100, and the rtificial rearing regime is responsible in large measure or the success of this local population. Colonial reeding is also common in Denmark, where abun-ant food is available on the shallow sea coast but safe esting sites are limited (298).

Non-breeding Mute Swans gather on lakes or stuaries to moult in May and June, while breeding birds moult on their territories, the female usually becoming flightless before the male (298). The flightless period lasts about 6 weeks (330), from mid-June till the end of July in the case of British non-breeders. The largest British moulting concentration, numbering 5-700 on average, is on the Chesil Fleet. In Denmark, moulting flocks can number up to 10,000.

There is very rarely any interchange between the Mute Swan populations of Britain and elsewhere (403). The numbers of swans in the various discrete groups delimited in Europe are given in Table 229. In addition to these the Caspian/Black Sea population, which is migratory, is estimated at 5-6,000 birds (403).

The Mute Swan is one of the most widely distributed of wildfowl in Britain (Fig 61), being recorded at more than 2,000 sites and in two-thirds of the 10km squares covered. Because of its sedentary nature the three British populations are more or less

'able 229. *Estimated numbers in the 7 European Mute ;wan groups. Figures for Great Britain modified from)gilvie (1981a), the remainder from Atkinson-Willes 1981).*

Group	Number
Scandinavian/Baltic	100300
Netherlands	5900
Central European	10100
England and Wales	13900
Scotland mainland and Orkney	2700
Outer Hebrides	1000
Ireland	5000
Total	139700

Table 230. *The estimated numbers of breeding and non-breeding Mute Swans in Britain in 1955, 1961, 1978 and 1983. Based on the survey reports by Rawcliffe 1958 and Campbell 1960 (reassessed by Ogilvie 1981a); by Eltringham 1963; by Ogilvie 1981a; and on the results for 1983 by Ogilvie (unpubl.). The totals in 1961 were not reliably estimated but were thought to be similar to those in 1955; the percentage of breeding and non-breeding birds in that year (including birds on territory but without nests) was based on a counted sample of 8848 birds.*

	Breeding	Non-breeding	Total
1955	7100-8000	12800-13600	19900-21600
1961	(29.4%)	(70.6%)	c.19000
1978	6230	11400	17630
1983	6300	12600	18900

Fig. 62. Mute Swan – trend in January numbers in Britain, 1960-1982. Diamonds linked by a dotted line are estimated summer census totals (right-hand scale) in 1961 and 1978 (Table 230).

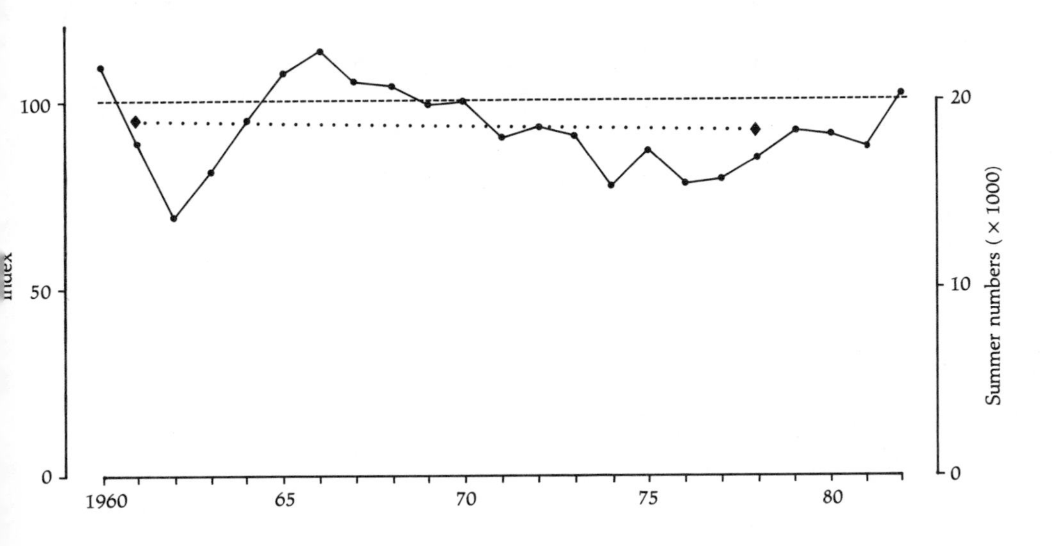

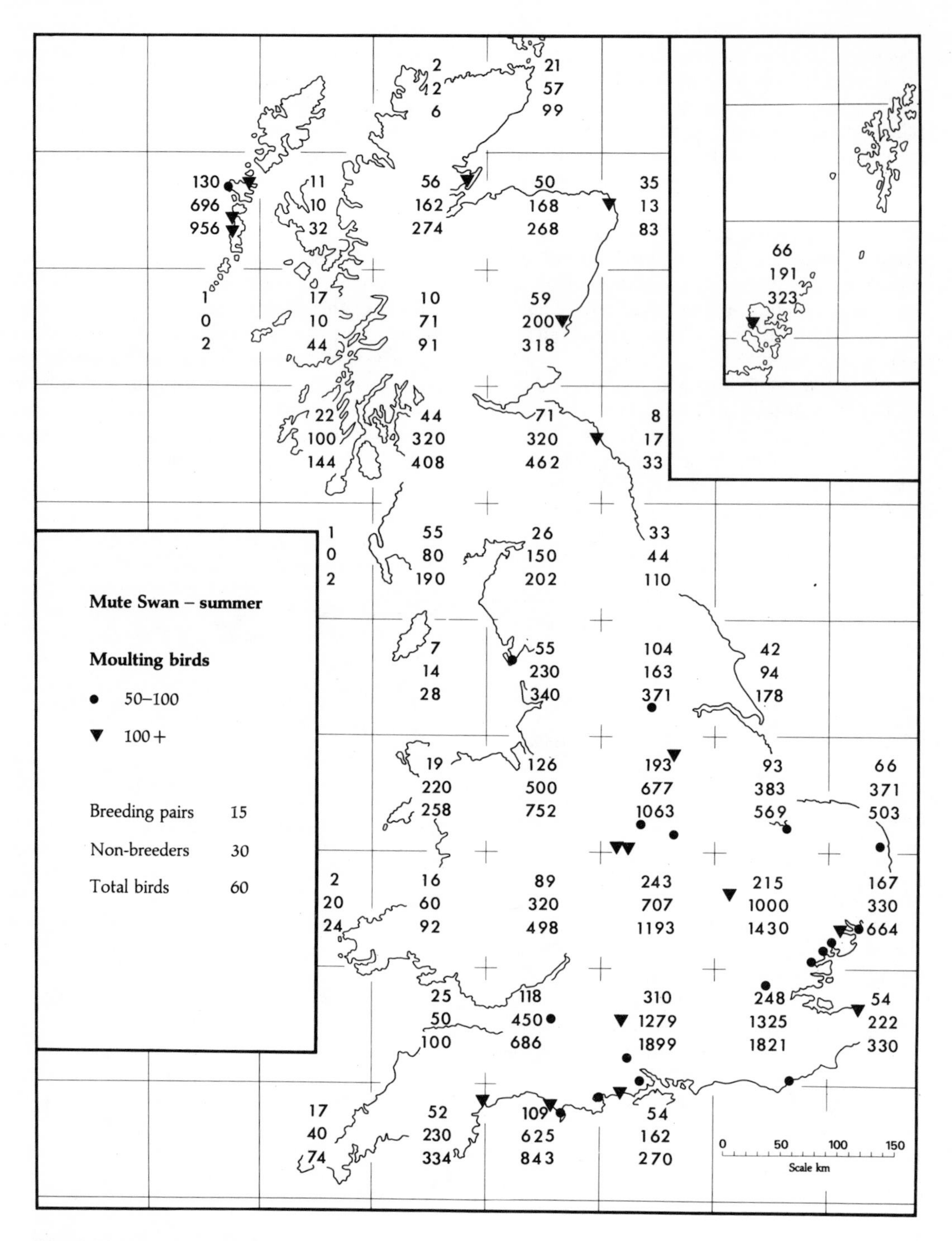

Fig. 63. The number of Mute Swans counted or estimated in the 1978 survey in each 100km square: breeding pairs (upper figure), non-breeders (centre) and total birds (bottom). Moulting concentrations are shown by dots (50-100 birds) and triangles (more than 100). From Ogilvie (1981a).

discrete. The Outer Hebridean population used to be thought to be partially migratory but has recently been found to be resident on the islands (552, p.322).

No sites in Britain qualify as internationally important (regularly holding 1,400 swans) and only one, the Chesil Fleet, approaches this. Between 1928 and 1980 the autumn gatherings there have fluctuated violently from 1,500 to as few as 400, usually because of food shortages and/or severe winters (468). The mean winter peak for the 5 years 1971-1975 was 824 and for 1976-1980 929, with a maximum of 1,238 in December 1980. Numbers fell thereafter, however, to 1,111 in 1981-82 and, following that hard winter, to a peak of only 890 in 1982-83.

Because of the Mute Swan's scattered distribution less than half of the British total is recorded in the winter counts. Since there are very few large concentrations and these are well covered, the index provides a reliable indication of trends (Fig 62). The most obvious feature is the trough in the early 1960s corresponding to the hard winters of 1961-62 and 1962-63, when large numbers were known to have died (76, 403). The lowest number appears to have been present in January 1963 and thereafter the population quickly regained its former level. This pattern is somewhat different from that obtained by Ogilvie (397, 403) using the same data, but the present analysis includes a greater proportion of counts. The steady decline in the late 1960s and early 1970s was again followed by an increase, but the hard winter of 1981-82 checked this temporarily. Mute Swan numbers have shown no overall trend over the period (see also below).

Alone among our native wildfowl, the Mute Swan population consists solely of residents, so that summer counts are relevant in assessing numbers and distribution. The BTO Atlas project (542) recorded Mute Swans in 1,622 of the 2,859 10km squares; of these breeding was confirmed in 1,411 squares and probable in a further 52. The distribution follows closely that in winter (Fig 61), with low frequency in the mountainous areas and a high density in the Outer Hebrides and in central and south-east England.

The numerical distribution of British Mute Swans is best established in spring, when breeding pairs are on territories and before non-breeders have gathered in moulting flocks. Four surveys have been carried out and the results are given in Table 230. No census was complete but the proportions which had to be estimated were relatively small, except in 1961.

These surveys confirm that the population has remained relatively stable over the 30 years, although fluctuations have occurred in the interval (see above). In contrast the numerical distribution of the swans changed considerably on a local scale between the 1955 and 1978 surveys. The number of swans estimated in each 100km square in the 1978 census is shown in Fig 63. Although on a county or 100km square scale the distribution was not much changed from that of previous censuses, on a more local level there had been a substantial reduction in the frequency of the Mute Swan. A quarter of the squares reported to have swans in 1968-1972 (542) had none in 1978 (408).

The declines have been concentrated in major lowland river systems, such as the Thames, Trent and Warwickshire Avon. This has been as a result partly of drainage and navigation works, but mainly of lead poisoning from fishing weights (219, 236). The mortality from lead poisoning has given cause for considerable concern, so much so that a governmental inquiry was set in motion in 1979 with the aim of identifying the causes of the local declines and suggesting ways of alleviating them. Since then, substantial progress has been made in finding non-toxic substitutes for lead weights, based on tungsten or stainless steel. Anglers have been slow to take them up, but the Government is now committed to banning the use of lead at the end of 1986 if a voluntary ban by the anglers has not proved effective by then. Although discarded weights will remain in the environment for many years after the active use of lead has been phased out, the prospects for the Mute Swans of lowland England are relatively bright.

Whooper Swan
Cygnus cygnus cygnus

The larger of the two species of migratory northern swans, *Cygnus cygnus* has two subspecies – *cygnus*, the Whooper Swan, and *buccinator*, the Trumpeter Swan of North America. There are four, apparently more or less discrete, populations: Icelandic, western Siberian, central Siberian (Black Sea wintering) and eastern Siberian. The vast majority of British Whooper Swans come from Iceland, with an unknown, but probably very small, number reaching the east coast from continental Europe.

In some areas, especially on migration, flocks can be large, but in Britain the wintering concentrations usually reach no more than a hundred or two. A survey in 1960 and 1961 (74) found that most flocks had less than 10 birds and that only 5% of 1,252 flocks encountered had more than 50 swans.

The Whooper Swan is not as agile on land as the Bewick's, but flocks are increasingly found feeding on pastures or arable fields. Potato eating became regular after the hard winters of the 1940s, and swans were also found in turnip fields after that time. While on water the birds move little and probably feed by day and night; when feeding on land they make one or two flights each day but do not normally wander further than 5km from the roost. At the Wildfowl Trust's reserves at the Ouse Washes, Martin Mere and Caerlaverock, Whooper Swans have readily visited ponds and taken grain and potatoes given as bait, feeding in very dense flocks in confined situations in company with Bewick's and Mute Swans. The Whooper Swan wintered in the past on shallow lakes, brackish lagoons and coastal bays, using its long neck to feed on underwater vegetation to a depth of up to a metre. Aquatic plants such as *Potamogeton*, *Myriophyllum*, Water Crowfoot (*Ranunculus*) and *Chara* are the traditional foods (447) and these are still important in freshwater lakes. *Zostera* and *Ruppia maritima* used to be most important on estuaries and the swans stil feed on *Zostera* on the mudflats at Lindisfarne. Tubers rhizomes and seeds are taken when available as wel as leaves and shoots.

On arable habitats stubble grain is the favourit autumn food, followed by potatoes, swede turnip and grass. In spring, grass and the shoots of winte wheat form the main part of the diet. Kear (295 implied that the arable feeding habit of Whoope Swans was acquired following a shift in distribution with more birds wintering in central Scotland, which was the most important area for arable (particularl potato) farming in Scotland. If this was true it seem that, once established, the habit spread rapidly so tha most British Whooper Swans now rely to a large exten on farmland for their feeding.

Whooper Swans are still very aquatic in summer, nesting on the shores of lakes rich in margina vegetation and feeding on submerged aquatic plants, algae and emergent vegetation. Moulting flocks in Iceland and Scandinavia are found on freshwater lakes or brackish lagoons and bays.

Most Whooper Swans leave Iceland during October, some migrating direct to Ireland or England. Others stop in Scotland for a time before moving on, while many winter there. Fig 64 shows the proportion of the November peak count that is recorded in Britain in other winter months. This is based on the total number of swans counted in each month and, because most important sites are included, can be regarded as giving a reliable pattern. It shows that 40% of Whooper Swans arrive by mid-October, but that by December the numbers are falling until only half of those present at peak are here in February. The swans have long been supposed to stay in Scotland *en route* to Irish wintering sites. There is another increase in March as the Irish birds return, and nearly all are again in Iceland by late April. The breeding area includes most of Iceland and the distribution is very scattered, with no more than one or two pairs occupying large lakes. There are also moulting flocks of non-breeders,

Fig. 64. Whooper Swan – monthly pattern of numbers in Britain, expressed as a percentage of the peak count.

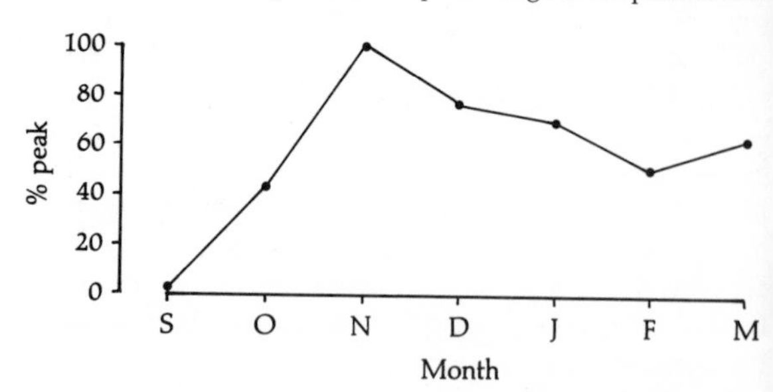

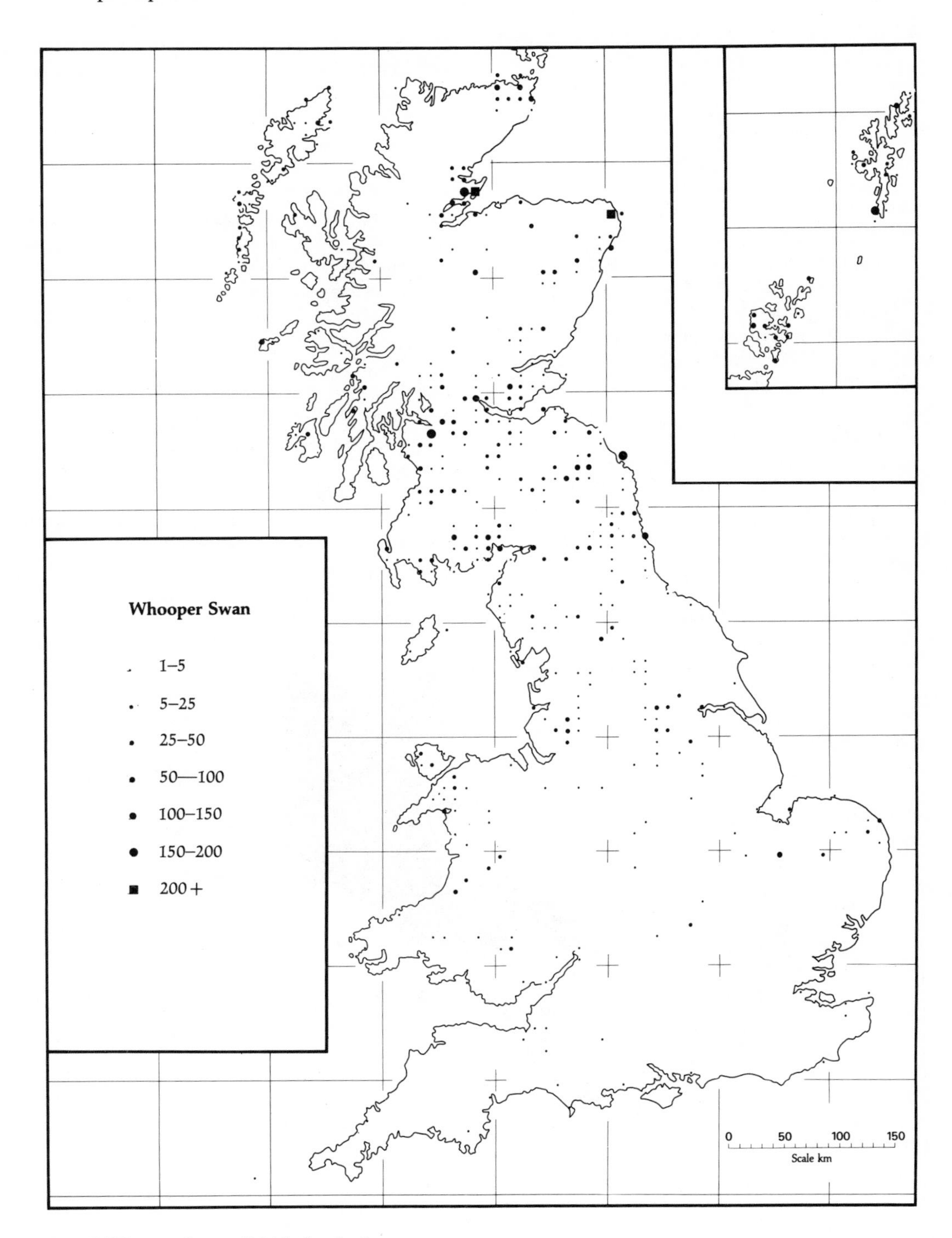

Fig. 65. Whooper Swan – British distribution.

numbering 200 or more, especially at Lake Myvatn and on the south-east coast.

The vast majority of British Whooper Swans have always been in Scotland, where the largest numbers are counted in November, when swans *en route* to Ireland, as well as Scottish wintering birds, are present. The recent distribution is shown in Fig 65; this shows a considerable number of differences from that given in the First Edition. The most obvious change is the lessening in importance of east central Scotland, which, in the early 1960s, held the largest concentrations. The Outer Hebrides also have many fewer Whooper Swans, whereas numbers have increased markedly in the Moray Firth area. There are many more sites in England holding Whooper Swans nowadays.

The Icelandic population is only partially migratory, with some hundreds, or even a thousand or more, remaining to winter on the south and west coasts and around some hot springs. Population counts are rare; the Iceland summer total was estimated as 5-7,000 in the early 1970s, with no increase in 25 years, but good coverage in the earl 1980s yielded a count of 9,000 and an estimate possibl reaching 11,000 (205).

The only census of the species in both Britai and Ireland was attempted in November 1979 (85) This resulted in totals of: Scotland 4,145, England an Wales 606, Ireland 2,014. There were no counts fron Iceland and the survey was incomplete even in Britair so the total of 6,765 is a minimum. The differenc between this and the population estimate of 9-11,00 illustrates the poor winter coverage in Ireland. If Fig 6 shows a true picture then up to 6,000 birds could be i Ireland in January and February. The main haunts i that country are in the north, at Lough Foyle, Loug Neagh, Lough Beg and Strangford Lough, but man other sites in the central and southern districts ar known to hold sizeable flocks, and the disperse population on the many hundreds of marshes an pools is probably substantial especially in wet periods

The continental population is wholly migratory breeding in Scandinavia and western Siberia an wintering in Denmark, the Netherlands and centra Europe (Fig 66). This group is numbered in the regio of 14-15,000 individuals (24).

The counts in the five most important sites i

Fig. 66. Whooper Swan – breeding range and migration routes.

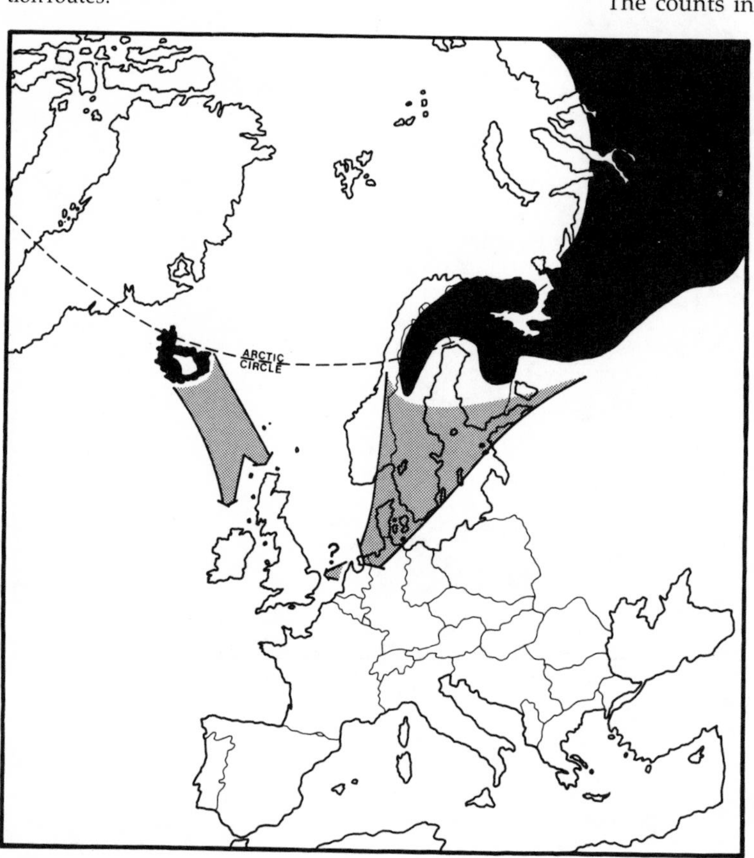

Britain are given in Table 231. The decline in Lindisfarne has corresponded with an increase in occurrence further south, particularly at Holywell Pond, Northumberland, in Yorkshire and at one or two sites in East Anglia, although the Holywell Pond flock is also now declining. It has been supposed that the Whooper Swans in East Anglia originate from the Scandinavian/Siberian population, and there was some evidence that this was the case when an influx of oiled Whooper and Bewick's Swans into the Ouse Washes followed an oiling incident in the Netherlands (445). In 1980, however, 46 Whooper Swans were marked with neck-collars in Iceland and resightings of 22 of these came from most parts of the British range, including four from the Ouse Washes (83). Thus at least some, and probably the majority, of the East Anglian Whooper Swans are Icelandic in origin.

In one of the few censuses of Great Britain, counts of 2,200 and 3,100 were made in November 1960 and 1961 respectively, and it was suggested that the British total probably exceeded 4,000 (74). This was said to be higher than in earlier decades, although information was sparse. The population in Scotland was put at 2,000 in the early 1970s, with 500 in England and Wales (403). The 1979 total of at least 4,751 in the most comprehensive survey since 1961 may not now be unusual.

Unfortunately because the species is so scattered the Wildfowl Counts coverage misses a considerable proportion; for example in the 1979 census only 2,989 (63%) of the 4,751 were located on regular count sites. Thus the national trends cannot be considered reliable on a year by year basis. However, a correlation on the whole 23-year run does show an upward trend ($r=0.78$ $P<0.001$), as would be predicted by the three estimates available. Apart from an apparent decrease in the middle of the decade there was no significant trend between 1960 and 1969, but thereafter from 1970 to 1979 the mean rate of increase was 6% a year.

Whooper Swans have consistently low breeding success; in 1948-1961 the average was 18% and the annual figures ranged from 6 to 26% (74). There are no indications of an improvement, indeed the 5% of 1979 was the lowest ever recorded. Thus annual mortality must be low and may have decreased in recent years to allow the recorded increase in numbers.

A few Whooper Swans remain in Britain through the summer, and odd pairs have bred sporadically. The BTO Atlas (542) quotes 6 records this century, the latest in 1947. Since then successful breeding has been reported at two different locations, one in 1978, the other in 1979; no fewer than three sites were occupied by breeding birds in 1980. Most of the records probably related to pairs in which one of the partners was injured and unable to return to Iceland; one pair included an escaped pinioned female.

Table 231. *Numbers of Whooper Swans on five internationally important sites. 5-year means 1960-1974 and seasonal maxima thereafter. A blank = no data.*

	Loch Eye/ Cromarty Firth*	Strathbeg	Lindisfarne	Holywell Pond
1960-64		291	360	
1965-69		481	266	25
1970-74	279	403	344	86
1975	344	440	206	110
1976	122	258	216	85
1977	693	248	267	157
1978	420	502	170	145
1979	1327+	495	179	155
1980	589	388	197	170
1981	324	519	93	84
1982	133	633	68	81

* Swans move freely between these adjacent sites and are counted sometimes on one and sometimes on the other. Both are internationally important in their own right.
+ From the November census (Brazil and Kirk 1981).

Bewick's Swan *Cygnus columbianus bewickii*

The smaller of the migratory northern swans, the Bewick's Swan is also the least abundant. The other subspecies of *columbianus* is the Whistling Swan *C.c. columbianus* of North America. This is distinguished by having much less yellow on the bill than the Bewick's Swan and is also slightly larger. Two races have been described from Eurasia, with the eastern *jankowskii* having longer, yellower bills than *bewickii*. Although slight differences may occur, these certainly do not constitute grounds for subspecific separation (183). There are, however, two discrete populations and only the western one will be considered here.

The Bewick's Swan winters on estuaries, lakes and river floods, roosting on the water and feeding either on submerged vegetation or on surrounding pastures. It walks well on land and is, of the European swans, the most suited to terrestrial feeding. Traditionally, once at a site the birds used to fly rather little, but recently, as they have increasingly resorted to feeding on arable land, regular dawn and dusk flights, and sometimes midday returns to the roost to drink and bathe, to a distance of 5 or 6km, are not unusual. The species has undoubtedly become much more land-based in recent years, with the reclamation of many of its traditional habitats in the Netherlands and drainage resulting in less flooding of meadows and floodplains.

On water, where they are relatively safe from predators, feeding flocks are well dispersed, usually with the families around the edges in the shallower water. On land, however, flocks are more compact, but still with the individuals well spaced out. In artificial situations as at the Wildfowl Trust's reserves at Slimbridge and on the Ouse Washes, where the swans are regularly fed with grain, very high densities, many hundreds of birds per hectare, are encountered, and there is a very high level of aggression in these flocks (532). On the Ouse Washes, Bewick's Swans feed for at least two hours after dark, whereas on surrounding farmland their feeding rhythm is more goose-like, with a peak in the morning and evening and a trough in the middle of the day (445).

On aquatic habitats, leaves and roots of pond weeds such as *Potamogeton* spp are the most important foods. Before the IJsselmeer was closed off from the sea, the large flocks of swans there fed on eelgrass (*Zostera*) leaves and rhizomes (88), and similar foods are recorded from Denmark. In the Netherlands pastures, mainly dominated by ryegrass, are the most common habitats nowadays and the diet there consists of soft grasses and clover (362). On the Ouse Washes the most common foods used to be the roots of Yellow Cress (*Rorippa palustre*) and soft grasses. Especially favoured were *Glyceria fluitans* and *Alopecurus geniculatus*. When the water level was high, swans grazed the surrounding banks or the shoots of the tall grass *Glyceria maxima*. Seeds also made up a small contribution (445). Nowadays Bewick's Swans at the Ouse Washes feed largely on agricultural crops in the surrounding fenland. Most common are stubble grain, potatoes and sugar-beet, all from harvested fields. The swans graze extensively on sprouting winter wheat in late winter and spring. In their main Irish haunt at the Wexford Slobs, arable feeding is also the rule, and waste carrots are taken as well as the crops mentioned.

It is evident that the Bewick's Swan has drastically changed the nature of its habitats and foods in this century and these new habits are still spreading. Many of these changes were forced on the birds by alterations in their traditional haunts, reclamation and drainage, but some, such as the recent change to arable farmland in Britain, appear to be voluntary. At least the behaviour has developed while the original habitat remains. Since farmland foods are easier to collect and often nutritionally more rewarding than wild plants, arable feeding may well become more widespread. Wet pastures and floodlands are everywhere coming under pressure from agricultural activities, so it is fortunate for the species that it can adapt so readily.

A great deal has been learnt about the movement patterns and behaviour of Bewick's Swans from studies carried out at Slimbridge and the Ouse Washes. The birds are individually recognisable by the pattern of black and yellow on the bill and many have been ringed; studies of return rates of individuals and their family history and behaviour have been carried out for many years. Bewick's Swans have very close pair and family bonds; no case of "divorce" has yet been recorded, but birds which lose a mate usually re-pair the following year. Families stay together throughout winter and spring and yearlings commonly associate with their parents, and that year's brood if there is one (179).

Once swans have adopted a wintering area they usually return there in following years. A bird which spends two winters at Slimbridge has a 55% likelihood of returning for a third. Yearlings, identified by the presence of some remaining grey immature feathers, are less loyal to the wintering site than adults (180). Although the feeding situation at Slimbridge is artificially crowded, the patterns of behaviour recorded there probably apply elsewhere. It is obviously advantageous to wildfowl, which occupy a localised habitat, to return to a successful feeding situation once found.

The western population of the Bewick's Swan breeds in western Siberia from the Yamal Peninsula (with scattered breeding westwards to the eastern White Sea) east to the Lena River (about 120° E). The Yamal is thought to be the densest breeding area but there is very little information. The migration route and breeding area, delimited with the help of recoveries of Slimbridge-ringed swans (969 ringed 1960-1979), are shown in Fig 67. There is a major autumn staging area in western Estonia, and the birds reach West Germany and the Netherlands during October. A few reach Britain in that month but they do not appear in numbers until November. The highest count is usually in January, but hard-weather movements giving rise to late peaks are not uncommon.

The return migration, along a similar route, begins in late February and March, and there are rarely Bewick's Swans in Britain in April. More is known of the spring migration route and most of the sightings of Slimbridge-ringed birds have been in April and May. There is a major staging area on the Elbe Estuary in West Germany, at various sites in East Germany and in Estonia (181). Long-distance movements from Slimbridge to the Netherlands and Germany have been reported as early as January, but in some cases the swans were pushed back by hard weather and the behaviour is thought to be the result of the artificial situation at Slimbridge. No such migrations are reported from Ireland, where most swans stay until mid-March (181).

The distribution of Bewick's Swans in western north-west Europe is always very restricted, especially in mid-winter. A number, up to 100 in recent years, winter on the Mediterranean coast at the Camargue,

Fig. 67. Bewick's Swan – breeding range and migration route. The stars indicate the major staging posts on migration.

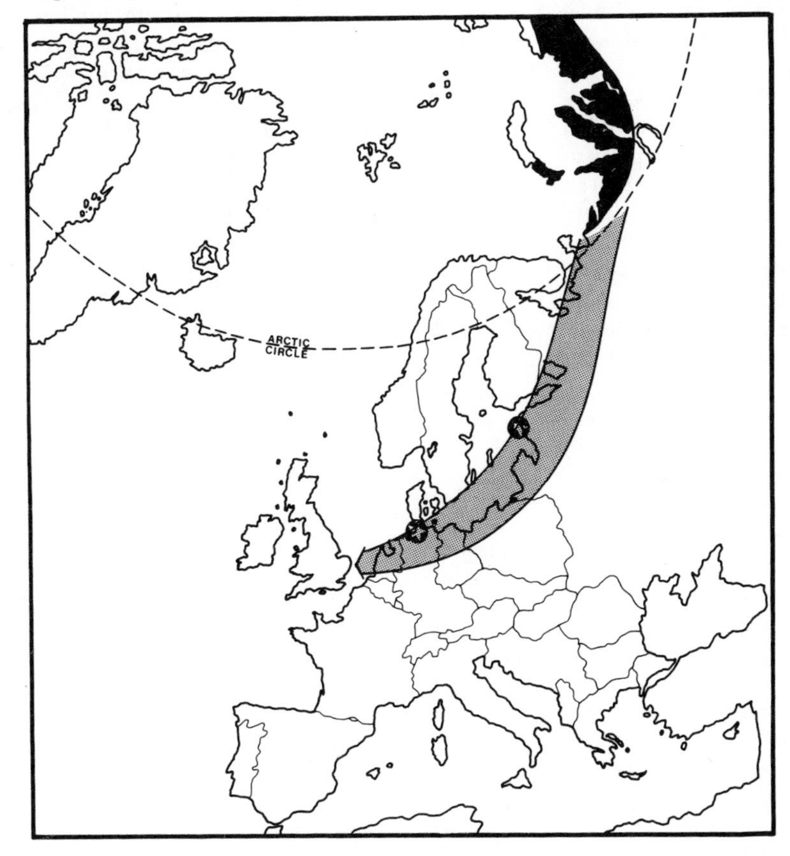

France, well away from the normal range, and this flock is still increasing. There have been a few sightings from North America both on the west and east coasts (183), but most have probably come from the eastern population.

The number of Bewick's Swans in the western population appears to have increased substantially in recent years. There were thought to be 6-7,000 in the early 1970s (22, 403), with a maximum of about 1,300 in Britain (400). Estimates of 9-10,000 were made for the mid 1970s (24, 362), and 12,000 for the late 1970s (529), when there were over 7,000 in the Netherlands and 3,000 in Britain. Preliminary results of the full census of Europe coordinated by the Dutch "Bewick's Swan Project '82-'84" suggest that the total population in January 1984 exceeded 16,000, more than 5,000 of which were in Britain.

The average breeding success is about 16% young annually, ranging from 7 to 23% in recent years, and the calculated survival of adults a minimum of 87% (179). If juvenile mortality is not much higher there is likely to have been 3 or 4% excess recruitment over mortality, allowing some growth in population. This is not, however, sufficient to account for the increase in numbers counted since the early 1970s, which must, therefore, have been due to improvements in count coverage (outside Britain) as well as the real rise in the population.

In the 19th and early 20th centuries the Bewick's Swan was rare in England and Wales but occurred in substantial numbers in north-west Scotland, particularly in the Outer Hebrides and on Tiree. The migration route passed over the Shetlands and western Scotland into Ireland (41). In the 1930s the numbers began to increase in England as the regular migration route shifted southwards. The numbers wintering in and passing through Scotland declined. Cold-weather influxes in 1938-39 and especially in 1955-56 (390) helped to establish a tradition of wintering in England, and thereafter significant numbers have been recorded in all years. The Scottish total dwindled to almost nothing until a tradition was built up in the 1970s at Caerlaverock on the north shore of the Solway Firth (peak up to 70 in recent years).

The present distribution (Fig 68) is largely restricted to England, with concentrations in the Midlands and the south-west. A total of 335 sites had Bewick's Swans at some time in the five winters used for mapping, but 302 of these had an average of less than 25 birds and many were satellites of major haunts. Eight sites qualify as internationally important, by far the most significant of which is the Ouse Washes, on the Cambridgeshire/Norfolk border, which had more than 3,000 for the first time in December 1983 (Table 232).

There has been a steady increase in the numbers on the Washes, and perhaps three factors have had an important influence. First was the cold winter of 1955-56, which brought 705 birds to the Washes and was important in the establishment of the tradition. Second was the establishment of the RSPB and Wildfowl Trust refuges in the late 1960s and early 1970s. This not only provided areas free from disturbance, but also more reliable floodwater and, at Welney, supplementary food. Not only did the numbers increase during this period but the swans also became more predictable, without the marked fluctuations of previous years. The third important factor was the change in feeding habits, from a complete dependence on the wet meadows throughout the winter to a predominance of feeding outside the Washes. This habit had begun in the early 1970s (445) but did not become the rule until after 1975-76; it meant that swans were dependent on the Washes only as a safe roost, which was always available on the refuges. The potential feeding area on arable fenland is very extensive and further increase in numbers and lengthening of stay could be accommodated.

Although numbers at other sites tend to be less predictable, a generally increasing trend can be seen (Table 233). The Nene Washes, near the Ouse Washes, have held 1,000 Bewick's Swans in two recent seasons, when birds have moved from the Ouse Washes from time to time. Since a safe roost there is dependent on floodwater the area must be considered as a satellite of the major site. The Wildfowl Trust grounds at

Table 232. *Average winter maxima of Bewick's Swans at the Ouse Washes for seven 6-year periods, 1939-1980, and seasonal maxima thereafter. From Nisbet (1955, 1959), Cadbury (1975) and Wildfowl Counts.*

Years	Mean maximum	Year of maximum (Jan or Feb)	Range
1939-44	6	1940	0-26
1945-50	13	1946	0-33
1951-56	226	1955	7-705
1957-62	238	1960	49-620
1963-68	527	1966	26-855
1969-74	976	1970	626-1278
1975-80	2048	1980	1257-2995
	Maximum		
1981	2842		
1982	2792		
1983	3364		

Slimbridge had only 10 Bewick's Swans, enticed by tame ones kept in enclosures, in 1960-61. The numbers built up rapidly to 200 in 1966-67, and in the 1970s and early 1980s the winter peak always exceeded 200 birds. Similarly the numbers have recently increased to internationally important levels at the Wildfowl Trust centre at Martin Mere. At Walland Marsh Bewick's occurred sporadically in small numbers before 1980, but counts were sparse. The Somerset Levels, particularly at Wet Moor, hold large numbers sporadically, and had a peak of 380 in January 1982. Both here and at the Derwent Ings, Humberside, where up to 200 birds

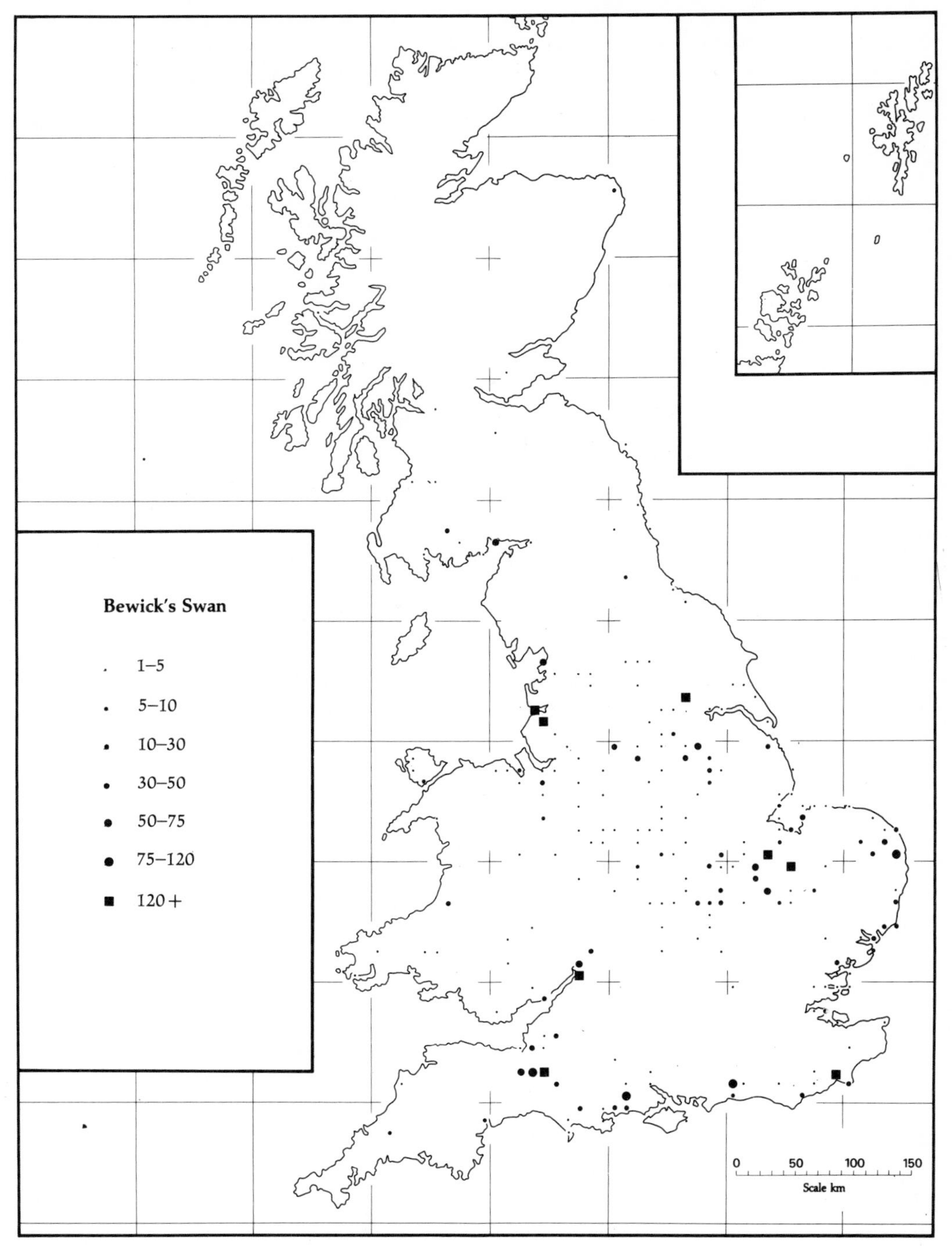

Fig. 68. Bewick's Swan – British distribution.

winter, the numbers largely depend on the availability of floodwater.

The trend in numbers in Britain as a whole is shown in Fig 69, where the recent upsurge is very evident. Because nearly all regular Bewick's Swan sites are well covered, total numbers in January are used rather than the index. Also shown are the January counts from the Ouse Washes. In four early years counts were not available, so the average of December and February totals was used. The patterns at the Ouse Washes and the remainder of the country have been similar, although the Washes now hold a greater proportion of the British stock than formerly.

The north-west European population of Bewick's Swans has increased markedly in recent decades and there has been an even sharper increase in those wintering in Britain. Whether their move to feed on agricultural land is a cause or a consequence of this increase in numbers is unknown. Whichever is the case, further increases are likely in the near future.

Table 233. *Numbers of Bewick's Swans at seven internationally important sites. 5-year means 1960-1974 and seasonal maxima thereafter.*

	Wet Moor, Somerset	Slimbridge, Glos	Walland Marsh, Kent	Nene Washes, Cambs*	Derwent Ings	Martin Mere, Lancs	Ribble Estuary+
1960-64	27	26		38	42	0	14
1965-69	48#	230		48	166	0	21
1970-74	174$	264		42	105	13	46
1975	65	280		0	180	18	76
1976	107	300		201	111	15	91
1977	235	290		30	111	5	103
1978	65	610		1000	100	21	213
1979	170	300		1010	187	58	96
1980	75	403	53	200	214	135	179
1981	380	580	182	222	24	154	267
1982	42	285	143	600	60	200	220

* Early counts from *Cambridgeshire Bird Reports*.
+ Early counts from Greenhalgh (1975).
# 4 years only, no count in 1965-66.
$ Mean of 2 counts only.

Fig. 69. Bewick's Swan – total number counted in Britain (solid line) and at the Ouse Washes (dashed), in January, 1960-1982. In the middle and late 1970s the increase nationally was at an annual rate of 16%.

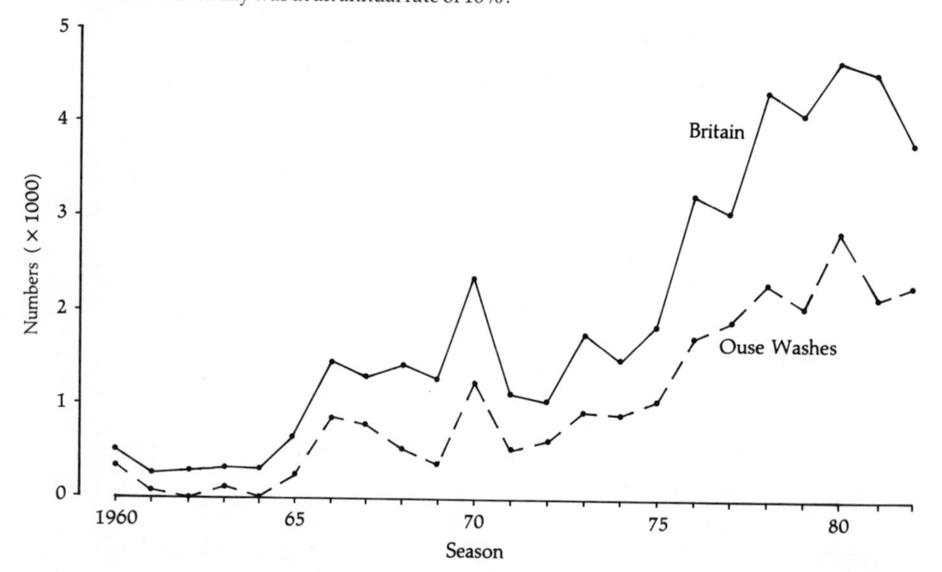

Pink-footed Goose *Anser brachyrhynchus*

Sometimes regarded as a subspecies of the Bean Goose, the Pinkfoot is obviously very closely related to it. There are several differences in appearance, and the Pinkfoot has a more coastal distribution and is more nomadic than its relative.

The main winter habitat used to be saltmarsh but in recent decades the species has increasingly moved inland to arable farmland, taking advantage of reservoirs and other freshwaters for roosting. Of 25 major Pinkfoot roosts in east central Scotland only 6 were coastal and those geese moved inland to feed. In other parts of Britain and in continental Europe coastal roosting is usual. Greylags often use the same Scottish roosts as Pinkfeet but the two species stay apart, and also show many differences in behaviour (382). Pinkfeet are more conservative in their choice of roost than Greylags, preferring sites larger than 20ha in area and more isolated. This intolerance of disturbance often means that Pinkfeet feed further from the roost, 15% of the feeding area being more than 10km away and 1% more than 20km. Most of the feeding is done by day although there is some moonlight feeding. On overcast nights Pinkfeet in Lancashire remain on fields, lit by reflected light from nearby conurbations.

Grain from barley stubbles constitutes the main food in autumn and early winter, followed by potatoes and an increasing proportion of grass. In late winter and spring, winter-sown cereals and grass are most important, though some grain can be gleaned from the surface of newly sown fields (379). Leaves and stolons of White Clover and other herbs are taken from grassland, as well as grass. Unharvested carrots were eaten in Lancashire in 1973-74 to the extent that farmers suffered losses, but greater vigilance by growers and an increase in the intensity of scaring stopped this becoming a habit and prevented further major incidents. The potential for further trouble remains, especially since numbers wintering in Lancashire continue to increase.

Feeding flocks of Pinkfeet frequently consist of several thousand birds, often densely packed. They are among the most wary of geese, keeping well away from hedges and other shelter that may hide potential predators. The farmland now used by the birds was not available before deforestation and Pinkfeet were then restricted to coastal saltmarshes (437). Their wariness and rather nomadic nature was adapted to this habitat, which was often transient and accessible to land predators. Saltmarshes are still important feeding areas in spring, and the reliance on November censuses for mapping purposes and the importance placed on agricultural feeding in relation to damage problems has resulted in an underestimation of the importance of coastal habitats.

Pinkfeet wintering in Great Britain breed in Iceland and east Greenland and represent the whole of that population. Another population breeds in Spitsbergen and migrates through Norway to Denmark, where it spends the autumn and spring, and to wintering grounds in the Netherlands and Belgium. This population was stable at between 10,000 and 18,000 in the 1960s and early 1970s but has recently increased to 30,000 (326). The breeding distribution and migration are shown in Fig 70.

Most British Pinkfeet nest in Thjorsarver, an oasis of vegetation in the central highlands of Iceland. In 1970 the area was estimated to hold 10,700 breeding pairs (300). This was thought to be about three-quarters of all pairs at that time. More recent information indicates that the importance of Greenland as a nesting area may have been underestimated and that other parts of Iceland have become relatively more important. Thjorsarver is nevertheless still vital to the well-being of the population.

In most breeding areas the nests are dispersed or in loose colonies but in Thjorsarver nesting density is high. Here the Arctic Fox population is small and the birds nest on low tundra or moraines rather than on cliffs, rock outcrops or islands as elsewhere. Non-breeders from Iceland migrate to east Greenland to moult, the first well-documented case of moult migration in geese (577). They return to the interior of Iceland in September and are joined by the breeding birds and their young. The earliest arrivals in Britain are in the first week of September but the main influx is in early October and the whole population has usually arrived by the middle of the month. The wintering period lasts until late April and a few geese remain into May.

Every year since 1960 a census of Pinkfeet has been carried out in early November. The distribution map (Fig 71) is based on the results of these censuses supplemented by data from the Wildfowl Counts and from the less frequent March censuses. Since many geese are missed from routine Wildfowl Counts the map is biased towards the autumn distribution (when a greater proportion is in northern and inland areas than later in the winter and in spring). Thirty-three 10km squares qualify as being internationally important for Pinkfeet. The main autumn concentrations are also shown in Table 234, which lists all sites with a 7-year average of more than 2,000 geese. Almost nine-tenths of British Pinkfeet are on only 12 sites in November, and most of these remain important through the winter.

Pinkfeet show major shifts in distribution according to feeding conditions in autumn, both on a local and regional scale. The most striking was th influx in Lancashire in 1976, a year when a clea harvest left little stubble grain in Scotland. Lancashir then held 37% of British Pinkfeet. The numbers ther dropped again in later years but remained at muc higher levels than in the 1960s (average 3,600). Two o the sites used predominantly in the spring are in th Solway, where 10-20,000 geese spend March an April. As the table shows, the numbers on many site change annually, chiefly in relation to disturbance an the availability of stubbles. One area not included i the table, Montrose Basin, has very recently come int prominence, holding 5,000+ geese following th establishment of a Local Nature Reserve in 1982.

The proportion of birds in various areas o Britain over 30 years is shown in Fig 72. This shows no only annual fluctuations in different areas but also th trends over longer periods. The most striking is th great increase in importance of east central Scotland i the 1950s at the expense of areas both to the north an south. In the late 1960s and early 1970s this region hel 50-60% of the population. The process was reversed t some extent in the late 1970s but eastern England ha not regained its former status. The early concentratio in Scotland has been attributed to the increase i

Fig. 70. Pink-footed Goose – breeding and wintering range and migration routes. The stars indicate the major staging posts on migration.

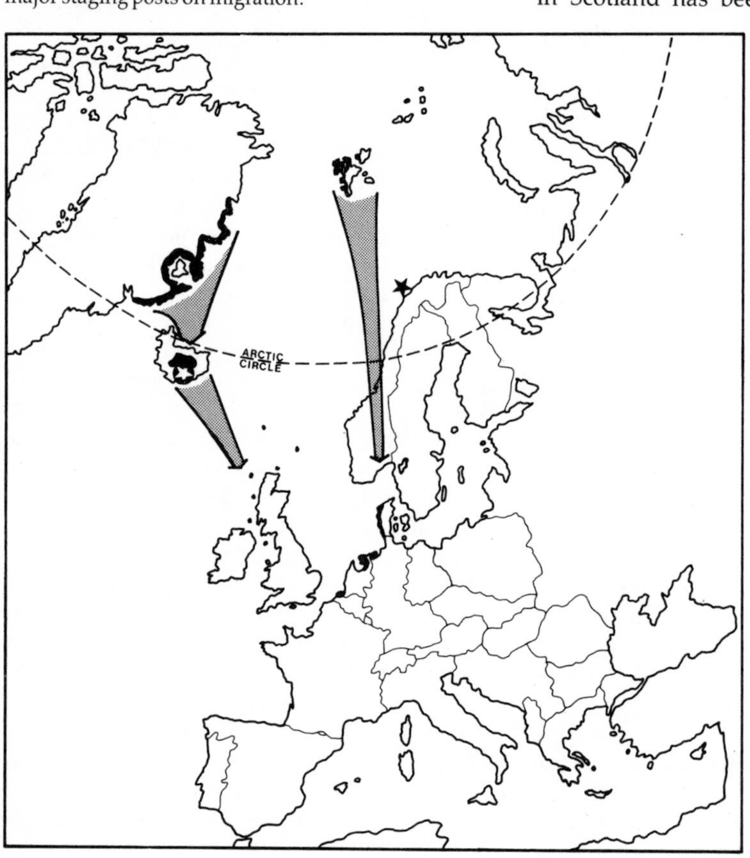

barley growing there (294) and other shifts are a response to changing agricultural patterns.

The way in which distributional changes occur, in view of the traditional habits of geese, is not well understood, although some clues may be found in an early analysis of ringing recoveries (61). Geese marked in the Forth area tend to be more mobile than those elsewhere so that the autumn gatherings contain geese with experience of many wintering areas. Inexperienced birds, including young geese, which show less attachment to the area of marking in subsequent seasons than do adults, follow those with

Fig. 71. Pink-footed Goose – British distribution.

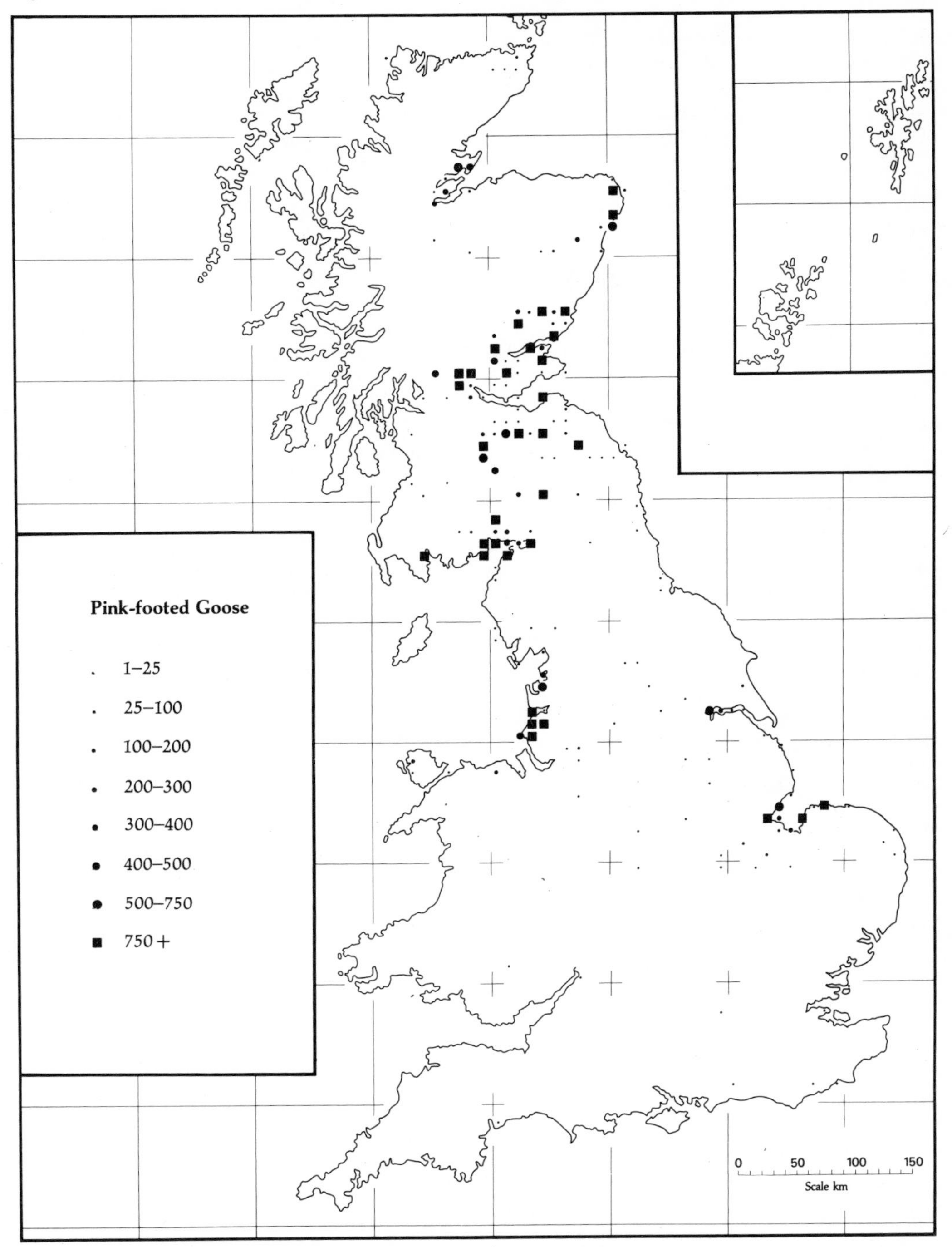

prior knowledge. The Pinkfoot has shown itself well adapted to exploit a patchy and rapidly changing food source, which makes it much less vulnerable to changing patterns of land use than goose species of more traditional habits.

There were about 30,000 Pinkfeet in Britain in 1950 and this had increased to 55-60,000 by the early 1960s (77,414). The numbers in 1977-1981, averaged 78,600. The increase was due to a lowering of the mortality rate rather than improved breeding success. The death rate, as estimated from counts and age ratio estimates, showed a rapid decline from 22% in 1960-1964, to 18% in 1965-1974. This was largely due to restrictions on shooting, particularly reinforced by

Table 234. *Sites or complexes which held on average more than 2000 Pinkfeet in the last 7 autumn censuses, and the total population at each census (data from M.A. Ogilvie). Also given are those sites which reach 2000 at other times of year, based on Wildfowl Count data.*

Census sites	1976	1977	1978	1979	1980	1981	1982	Mean
Loch of Strathbeg	6250	7500	6200	1400	2200	5900	6200	5094
Ythan Est/Slains L	2900	4113	7800	5850	4350	7300	5870	5455
Carsebreck Loch	630	920	3810	926	5680	3380	4920	2895
Arbroath Area	3160	645	6460	7330	5795	2190	2110	3955
Dupplin Loch	3344	14687	8010	5870	19550	5000	5012	8782
Cameron Reservoir	3000	5750	5000	5500	3000	6150	270	4095
Loch Leven	5000	7000	6500	4000	8750	6500	5800	6221
Gladhouse Reservoir	2000	5100	9600	8500	4700	3500	1050	4922
Westwater	0	2930	6500	0	7000	12340	9240	5430
Aberlady Bay	5800	5100	5200	3200	11930	4510	5165	5844
SW Lancashire*	26140	8684	5276	16260	8821	18240	18404	14546
Census total	71000	69000	78000	80000	95000	90000	89000	81700
% on above sites	83.2	94.1	90.8	86.0	86.1	83.3	72.0	83.3
Other sites (month)								
Hule Moss (O)		2300	3500		2000			(2600)
Inner Solway, North (F/M)	2500	8000	2600	3280	7500	8600	12200	6383
Inner Solway, South (F/M)			12000		12000	1750	3700	(7363)
The Wash (D/J)	7600	3880	2290	6450	3720	10500	5810	5750

* Peaks usually later than November. Overall peak was 36,580 in 1981-82 (see p.193).

Fig. 72. The proportion of the British population of Pink-footed Geese in various areas of Britain in early November. Note that the counts were incomplete before 1960 and proportions then are based on estimates from the few counts available. Based on Boyd and Ogilvie (1969) and Ogilvie and Boyd (1976).

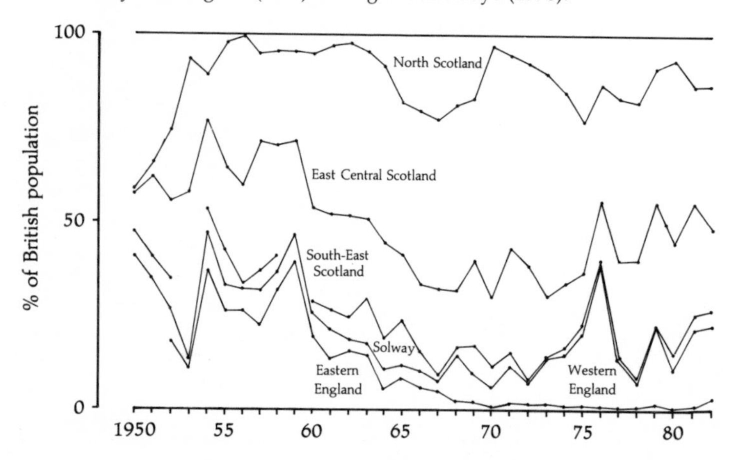

the ban on the sale of dead wild geese which came into effect in January 1968. Mortality fell again, to just over 10% in the late 1970s, but new legislation in the Wildlife and Countryside Act (1981) allows the shooting of wild geese, under licence, outside the shooting season in areas vulnerable to agricultural damage. The first two years of this legislation seemingly had little effect, however, with numbers reaching a record 101,000 in November 1983.

Mean annual recruitment rates (% young in November) were 25% in 1960-1964, 17% in 1965-1969, 21% in 1970-1974 and 11.7% in 1975-1979, and the average brood size was also declining. The most likely explanation for the poorer average breeding success seems to be a shortage of suitable nesting or rearing areas, a suggestion supported by the evidence of apparent over-exploitation of the main breeding area in Thjorsarver (203). The recruitment rate is still higher, on average, than the mortality rate, but since shooting pressure in spring is likely to increase, further growth in population, at least to any great extent, seems rather unlikely.

Bean Goose
Anser fabalis

The Bean Goose breeds across northern Eurasia from the highlands of Norway to Kamchatka in the east, and at least four races are recognised. In western Europe the western Bean Goose, *Anser f. fabalis*, from Scandinavia is much less numerous than the Russian or Tundra Bean Goose, *A.f. rossicus*. Populations are not easily delimited and the races are generally intermingled in autumn and winter. Nevertheless, there is still a reasonable degree of racial separation, both morphological and ecological (504).

As its name suggests the Bean Goose has long associated with man's cultivated crops and was probably the first European goose to adapt to arable agriculture. Its habitat in north-west Europe is still largely arable and its diet consists of stubble grain in autumn, potatoes and sugar-beet in mid-winter and winter wheat and grass in spring. It is often found in mixed feeding flocks with European Whitefronts. The Bean Goose withstands colder temperatures than most species, substantial numbers remaining in southern Sweden and East Germany to winter. The numbers in the Netherlands have increased recently, to 40-50,000 in normal seasons, with a maximum of 100,000 being counted during the very cold weather of January 1979.

During the 19th century the Bean Goose was a very common winter visitor to northern Britain and East Anglia, being the commonest species in many localities (50). A widespread decline began in the 1860s and 1870s until in the early part of this century only a few pockets remained. The present distribution (Fig 73) is restricted to two regular haunts and sporadic occurrences elsewhere. Bean Geese in the regular haunts and in Scotland are usually *fabalis* whereas those seen as ones or twos with White-fronted Goose flocks in southern England resemble *rossicus*.

The winter peaks at the two regular haunts are given in Table 235. Clearly the Threave flock is becoming less predictable, while that in Norfolk is increasing. During the 1970s occurrences outsid these areas were sporadic and rarely included mor than 10 birds together, with the exception a flock of 18 birds at Rutland Water in February 1977. In the har winter of 1981-82, however, there was a marke influx, with the Norfolk flock peaking at 329. Ther were at least five flocks of more than 20 birds in othe

Fig. 73. Bean Goose – British distribution.

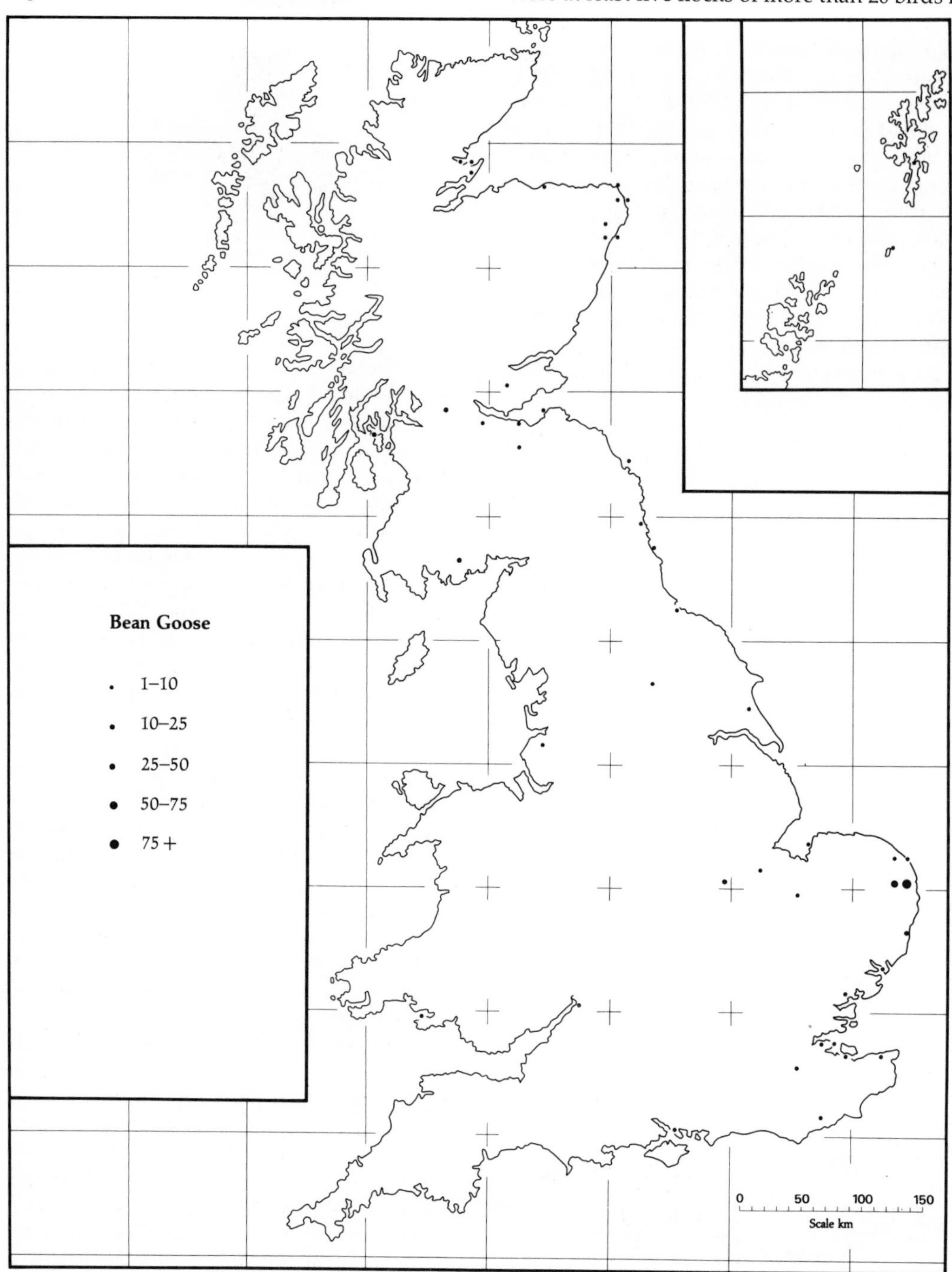

English sites, while in Scotland sightings were very numerous from October onwards, especially in the north-east and Northern Isles. The largest flock, apart from that in the Dee Valley, was of 25 which wintered at Munlochy Bay. Occurrences have become more frequent at Carron Valley Reservoir, Stirling, where there was a flock of 73 in February 1981.

With the exception of 1981-82, peak numbers in Britain are similar to those of 20 years ago; the species was given legal protection from shooting in the 1981 Wildlife and Countryside Act, despite the difficulty of distinguishing it from the Pinkfoot. Some of the feeding grounds of the Norfolk geese are protected by an RSPB reserve, and this remnant population may remain with us for some years to come.

Table 235. *Numbers of Bean Geese counted in the two sites where occurrences are regular, 1975-76 to 1982-83. Peaks are usually in January or February. From RSPB,* Scottish Bird Reports.

Season	Dee Valley, Threave, Dumfries & Galloway	Yare Valley, Norfolk
1975-76	60	102
1976-77	82	122
1977-78	0	103
1978-79	27	141
1979-80	40	155
1980-81	0	165
1981-82	38	329
1982-83	0	197

White-fronted Goose
Anser albifrons

Probably the most widely distributed of all geese, the Whitefront has an almost continuous Holarctic breeding distribution. It winters on both seaboards of North America, in southern and western Europe, in the Middle East and in eastern Asia. Two distinct races winter in Europe and two or three are recognised from North America.

European White-fronted Goose *Anser albifrons albifrons*

In western Europe most Whitefront roosts are on estuarine sandbanks, though with their increasing reliance on inland feeding areas many now roost on shallow lakes, some created by artificial flooding, especially in the Netherlands. The birds fly usually less than 10km to feed, although in continental Europe flights of 20-30km or even further are not infrequent. In suitable habitat flocks are very large, often many thousands of birds feeding together. Most foraging is done during the day but European Whitefronts are also found feeding on moonlit nights.

Traditionally wintering on coastal grasslands and inland floodplains, the European Whitefront is primarily a grazer. As its wintering areas have been drained and ploughed for arable crops it has, in common with other goose species, taken increasingly to frequenting arable land especially in autumn. In East Germany, though 45% of the autumn food consists of grasses, 29% is made up of grain from stubbles and the remainder of potatoes, couch grass rhizomes, clover and arable weeds (528). In the Netherlands and Britain, the winter diet is taken almost exclusively from pastures. On saltings the commonest species are Saltmarsh Grass (*Puccinellia maritima*), Red Fescue (*Festuca rubra*), Bulbous Foxtail (*Alopecurus bulbosus*) and Barley Grass (*Hordeum secalinum*). On inland pastures they favour Perennial Ryegrass (*Lolium perenne*), Meadow Grass (*Poa trivialis*) and Yorkshire Fog (*Holcus lanatus*). White Clover (*Trifolium repens*) stolons are an important item especially when the ground is wet (436). In spring, pastures and winter wheat fields are most commonly frequented. In the breeding season European Whitefronts occupy grassy and shrubby tundra, nesting on mounds or drier slopes. The nests are well dispersed and males are not markedly territorial.

The western population of the European White-fronted Goose breeds in north-western USSR and migrates through the Baltic in autumn to East and West Germany and finally to the Netherlands, Belgium and Britain. Some arrive at their final wintering areas as early as October, but in general, peak numbers are present in East Germany in October and November, the Netherlands in December and January and Britain in late January and early February.

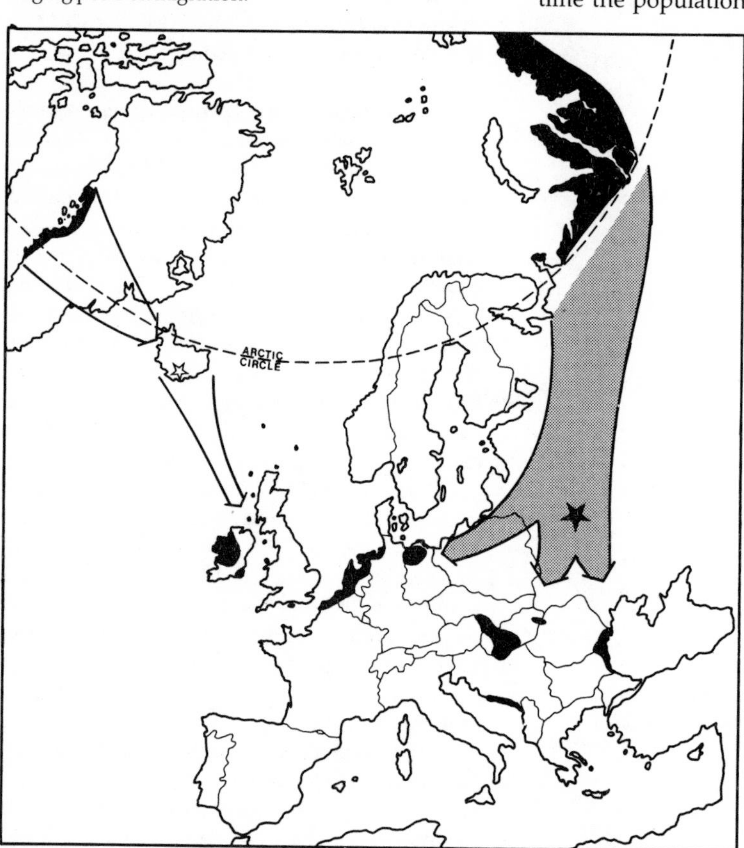

Fig. 74. White-fronted Goose – breeding and winter range and migration routes. The stars indicate major staging posts on migration.

Late periods of hard weather in the Netherland sometimes cause movements into Britain in lat February or even in March. The normal migratio routes and main wintering sites outside Britain ar shown in Fig 74.

The return migration begins in early March an very few remain in British haunts in the second half o the month. From the Netherlands the birds move i March almost due east to the Ryazan and Tula district of Russia, where they stay until late April or earl May; they then leave to the north for the breedin grounds. Recoveries of geese ringed at Slimbridge an in the Netherlands show clearly the pattern o movement and the different autumn and sprin migration routes.

The numbers of European Whitefronts through out the north-west European range have been moni tored by regular counts over the last 20 seasons, an the estimated total, together with the peak count i Britain, is shown in Fig 75. In the latter half of the 1960 the population entered a remarkable growth phase the numbers increasing from an earlier level o 50-60,000 to 100,000 in the mid 1970s. Since then the have topped 200,000 in several seasons. At the sam time the population has concentrated increasingly i he Netherlands, where about 90% of the total now vinters in most seasons. Exceptions occur in cold vinters, such as 1962-63, when 80% were in France for period, and 1978-79, when large numbers wintered n Belgium. In the exceptionally mild winter of 1977-78 38,000 geese (17% of the population) remained to vinter in East Germany (504).

The increase in population size is not due to an mprovement in breeding success; the proportion of oung birds in the autumn flocks has maintained an verage of over 30% throughout the 1960s and 1970s. t has been suggested that there has been a shift into orth-west Europe from central and eastern European ountries, chiefly Hungary, Rumania and the Black ea states. However, the increases in the Netherlands re consistent with the assumption that recruitment hrough breeding has been responsible and there is no vidence to support the idea of a massive redistribu- ion. Restrictions on hunting, resulting in a decrease in vinter and spring mortality, are considered to have een largely responsible (504). Spring shooting in the JSSR was banned in 1970, and in the early 1970s the Netherlands restricted shooting to between half an our before sunrise and 10 a.m. Increased food upplies in the Netherlands and East Germany, and he long series of mild winters since 1962-63, have also robably played a part. In cold winters the geese are ften driven as far as France where they encounter very high shooting pressure. Since the breeding success of the geese is undiminished, further increases are likely unless there are more changes in shooting regulations. There is now a mounting conflict with agricultural interests in the Netherlands and this may lead to steps being taken to reduce the numbers and disperse the flocks.

With the increasing concentration of European Whitefronts in the Netherlands the importance of Britain as a wintering area is much diminished. In the 1960s the peak counts in Britain represented 15% of the total; by the early 1970s this had fallen to 6% and in recent years has been below 3%.

The European Whitefront has a very localised British distribution, shown in Fig 76. Geese have been recorded in 200 sites, but only 33 have had a 5-year average of more than 25 birds and many of these are satellites of larger haunts. All these regular wintering areas are in the southern half of England and in Wales.

The coverage of Whitefront sites by the Wildfowl Count network is incomplete, but regular counts are made separately (65, 398, 406) and these are used to provide population totals and to assess trends. The census results since 1946-47 are given in Table 236. Numbers remained relatively stable through the 1950s and early 1960s but increased to more than 10,000 between 1967 and 1971. They then declined to the previous levels of around 8,000 and have further dwindled to 6,000 or less in 5 of the last 6 winters. The exception, 1978-79, was a hard winter in the Netherlands, which resulted in a late-winter dispersal from there into Belgium and Britain. The decline is a classic case of "short-stopping", where geese winter closer to their breeding areas following an improvement in feeding conditions.

Fig. 75. The estimated total of the north-west European population of European White-fronted Geese in 1960-1979, together with the peak number in Britain. From Philippona (1972), Ogilvie (1978), and Rooth *et al.* (1981).

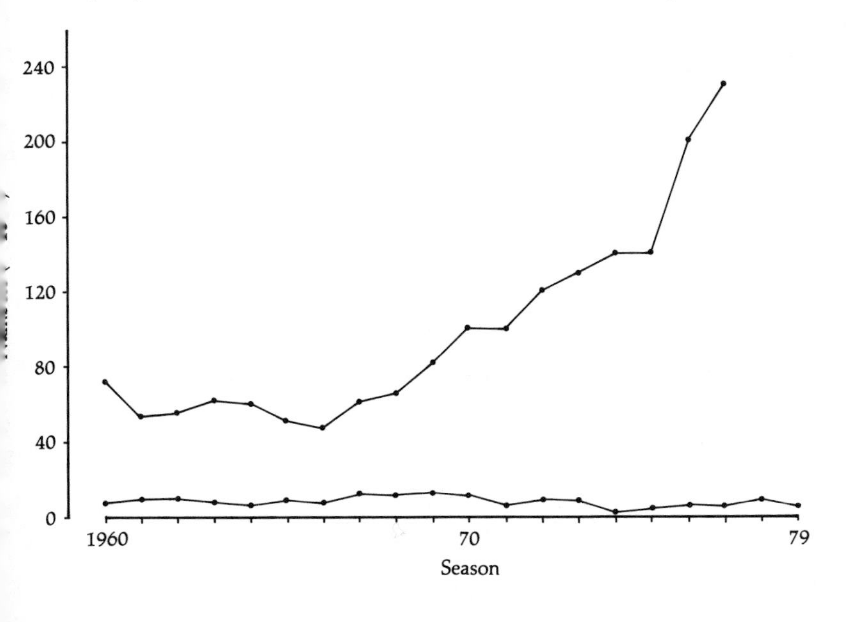

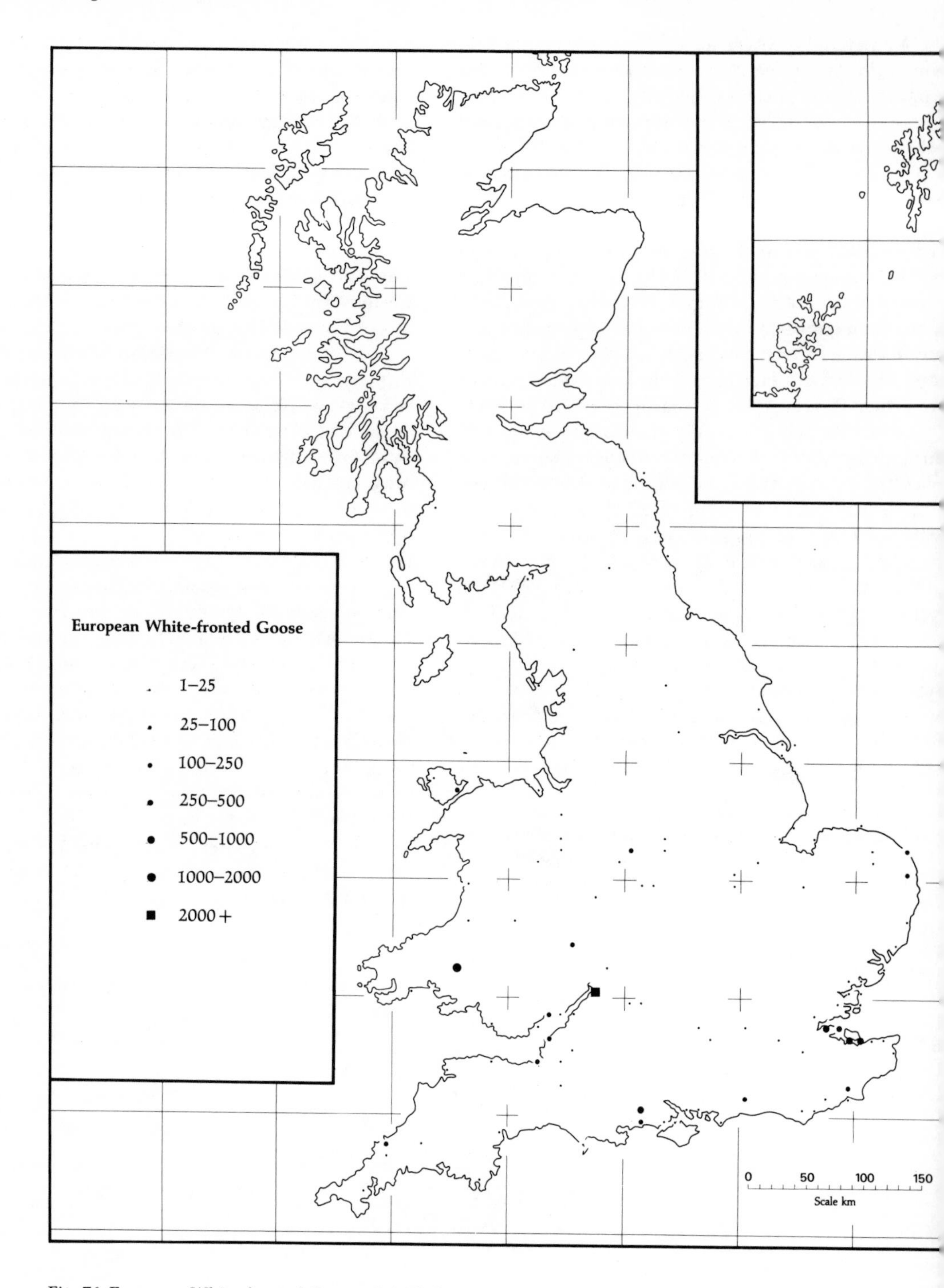

Fig. 76. European White-fronted Goose – British distribution.

The only site which qualifies as internationally nportant is the New Grounds, Slimbridge, on the anks of the inner Severn Estuary (Table 236). Only nce in the last 37 years has the number there been elow 2,000 – the present qualifying number for nternational importance – and the maximum was ,600 in 1968-69. In common with those at other British ites, the numbers have since declined, the lowest ver annual maximum of 1,450 being recorded in 974-75. Not surprisingly, the number at the New Grounds has a great influence on the national total. 'he correlation between them is close ($r = 0.87$) which hows that the variation is closely in parallel. Since the arly 1950s the Slimbridge flock has accounted for nore than 50% of Whitefronts wintering in Britain.

Among other important sites are the Tywi Valley, Dyfed (up to 3,000 in the late 1960s, now -500); the Avon Valley near Blashford, Hampshire (regularly 500-1,000); and a complex of sites on the Thames, Swale and Medway Estuaries, Kent (usually 500-1,200 with occasional peaks of 2,000+). Some further sites, especially in the south-west, such as the Camel Estuary, have occasional sizeable flocks, displaced from other haunts by hard weather.

Although there was an upsurge in the late 1960s and early 1970s the decline in the importance of the main haunts has continued and several major resorts have been abandoned altogether. Chief amongst these are Halvergate (Great Yarmouth) and Margam Moors (Port Talbot), which used to harbour 2,000 or more geese, the Severn/Camlad Marshes near Welshpool in mid-Wales, where over 1,000 geese used to winter, and Bridgwater Bay (Somerset), which held 500 geese in most winters in the 1950s. The Mersey Marshes, which used to hold more than 1,000 Whitefronts in the late 1940s, have also been abandoned. The range of the

Table 236. *Peak numbers of European White-fronted Geese in Britain and at the New Grounds, 1946-47 to 1982-83. In early years up to 30% of the British peaks were estimated annually (missing counts estimated on the basis of information from other years) but in the 1960s and 1970s this proportion has been below 10%. Peaks occur in January or February, but the year is that of the autumn e.g. 1946 = 1946-47 winter. From Ogilvie (1969, 1978) and Wildfowl Trust data published annually in* Wildfowl. *Numbers in brackets are the proportion of the British count which was at the New Grounds.*

Year	Britain	New Grounds	Year	Britain	New Grounds
1946	10500	4200	1965	8400	5500
1947	6600	3000	1966	7300	4200
1948	8400	3800	1967	12000	6700
1949	7400	3500	1968	11200	6600
1950	8900	3400	1969	13000	7600
1951	6300	2500	1970	11000	6000
1952	10000	4700	1971	6000	3400
1953	9000	5000	1972	9000	6000
1954	8600	3900	1973	8000	4500
1955	8400	5000	1974	2000	1500
1956	5600	3300	1975	4000	2900
1957	8400	4200	1976	6000	4000
1958	8000	5000	1977	5000	2700
1959	8400	4200	1978	9500	5100
1960	6800	3500	1979	5000	2100
1961	8600	4400	1980	5700	3000
1962	9800	3000	1981	6910	4500
1963	7400	4500	1982	5700	3000
1964	6000	3500	1983	5500	3400

5-year means	Britain	New Grounds
1946-49	8225	3575 (43%)
1950-54	8560	3900 (46%)
1955-59	7760	4340 (56%)
1960-64	7720	3780 (49%)
1965-69	10380	6120 (59%)
1970-74	7200	4280 (59%)
1975-79	5900	3340 (57%)
1980-83	6103	3475 (62%)

European Whitefront continues to contract; drainage and disturbance on many of its former haunts have helped to speed this process. In the absence of some drastic changes in the Netherlands in the very near future, the conclusion that the European Whitefront will cease to winter regularly in Britain in significant numbers seems unavoidable.

Greenland White-fronted Goose *Anser albifrons flavirostris*

The Greenland Whitefront occurs in western and northern parts of the British Isles. It is less gregarious than the European race, roosting on remote upland lakes and tarns, often among the bogs which form its feeding area. Because of the small size of these pools and the patchiness of the bogland habitats, both roosting flocks and feeding concentrations are small. This has changed to some extent as the species has resorted to improved grassland and arable land for feeding. Nowadays in the two main wintering areas roosting concentrations are larger than was traditional, though feeding flocks are usually small, exceptionally up to 700 birds. The remaining flocks on natural habitat can be found in small groups, often consisting of one or two pairs or families. Even in the larger flocks, the birds are well spaced and family integrity is well maintained. While on farmland the bird is diurnal, but a considerable amount of night feeding probably does take place in the safety of the centres of bogs or tarn sides.

The traditional food on bogs in Wales consisted of the roots and shoots of cotton grass (*Eriophorum*) (107) and the bulbils of *Rhynchospora alba* (478). On Islay bogs, *Eriophorum* is also important, and the diet elsewhere in the range on natural habitat is likely to be the underground parts of sedges, rushes and pondweeds. In spring Greenland Whitefronts in Wales move to graze on stream sides, and in many of the important British and Irish haunts the geese have relied increasingly on grasslands for their winter food. While many of these are unimproved, wet and rushy, drainage and improvement of grasslands, especially at the Wexford Slobs in south-east Ireland, has not deterred the geese provided the level of disturbance has not consequently increased.

On their main Irish haunt at the Wexford Slobs the geese are largely arable feeders, moving from stubbles in autumn to potatoes and sugar-beet in mid-winter and winter cereals in spring, with permanent and reseeded grasslands always well used. On these grasslands ryegrass is the main constituent of the diet and clover stolons are also very important. In parts of Scotland and Wales saltmarsh grazing is also usual. Nowadays the greater part of the Greenland Whitefront population feeds on farmland, but becaus the flocks are small and many boglands inaccessib the importance of their natural range may well b underestimated.

Although a few Greenland Whitefronts stop i Iceland in spring they do so for too short a time t make a substantial contribution to their feedin requirement on migration. In summer they fee intensively on marshes before and during laying eating a mixture of under- and above-ground plar parts. After hatching, families move to uplan lake-sides to graze grasses and sedges there (197). I autumn the geese stop in the southern and wester lowlands of Iceland feeding on the bulbils of Persicari (*Polygonum viviparum*), crowberries (*Vaccinium*) and o the shoots of horsetail (*Equisetum*) (204).

Like the European Whitefront, the Greenlan race is not markedly territorial, nests being widel dispersed in rather open grassy areas or in low willo scrub. Nests are sometimes on slopes or elevated sites not necessarily close to water but near marshy feedin areas (557).

Greenland Whitefronts breed between 64° an 73° N in the coastal tundra belt of west Greenland (Fi 74). They leave the breeding grounds in Septembe and stop in large numbers in Iceland before reachin Britain and Ireland in October, with the whol population in the wintering area by the end of th month. The wintering range is exclusively in th

Table 237. *Distribution of recoveries of Greenland White-fronted Geese ringed in Greenland and recovered elsewhere. These data include the 144 recoveries analysed by Boyd (1958). Other recovery details are as published by the Zoological Museum, Copenhagen, in* Dansk Orn. Forening Tiddskr., *1955-1979. Data collated by A. Fox and D.A. Stroud, who also provided unpublished information o Eqalungmiut Nunat (EN) ringing.*

Recovery location	Latitude of ringing			
	South of 69°N		North of 69°	Total
	EN	Other		
Iceland	2	1	20	23
Scotland	8	3	17	28
England	0	0	3	3
Wales	0	1	0	1
Wexford	1	9	123	133
Elsewhere in Ireland	0	42	19	61
Canada	0	1	2	3
Norway	0	1	0	1
Total	11	58	184	253

British Isles although there are sporadic occurrences in eastern North America. Substantial ringing in Greenland in the late 1940s indicated that birds in different parts of Ireland originated from different breeding areas. Only four of 69 recoveries from Wexford originated from south of 69° N on the breeding range whereas three-quarters of recoveries elsewhere were from the southern breeding range, a remarkably clear-cut result (66). There were insufficient recoveries from Scotland to come to a conclusion about their origins; more recoveries are now available and their distribution is shown in Table 237.

The remarkable and clear-cut split in Ireland is maintained in this analysis but the picture in Great Britain is less clear. The recovery pattern in Scotland is significantly different (Chi square) from that in both Wexford and the remainder of Ireland. In fact there is no significant difference in the Scottish pattern from that which would be expected purely by chance. It might be expected that birds recovered in Scotland while *en route* to Ireland would cloud the picture, but exclusion of the few October recoveries has no effect on the pattern.

Before ringing with clearly identifiable rings on the breeding grounds in 1979 (198) the picture appeared clearer, with the Scotland recovery pattern resembling that in Wexford, but the large proportion of Eqalungmiut Nunat birds (67°36′N) recovered in Scotland has changed this. In fact a much higher proportion of geese ringed in 1979 has been resighted in Scotland. However, since many of these were in large family groups, and the resighting probability is difficult to quantify, the resighted birds cannot be included in the recovered sample.

It remains true that there is a clear pattern in the wintering distribution of the Greenland Whitefront and that this is to some extent related to the breeding range. The fact that the locations of recoveries from the 1979 ringing are significantly different from those of birds ringed elsewhere south of 69° N (Chi square $P < 0.01$) may indicate that these differences may be on a smaller scale than was previously thought. More information is needed to clarify the position with regard to Scotland.

Ruttledge and Ogilvie (516) reported consistently higher breeding success in Wexford (mean for 12 years 20.2%) than in Islay (13.2%) and suggested that their breeding ranges were different. While this remains a possibility it is also possible that Wexford birds breed better as a consequence of having wintered there. Breeding success in geese is linked to feeding conditions in spring (review in 440), and this may well be better on the Slobs, where most of the foods are gathered from arable and improved pasture, than on the less intensively farmed Scottish island.

A low proportion of Icelandic recoveries originated from south of 69°N but this pattern was not significantly different from that which would have been expected by chance.

The return migration begins in late April and very few remain on the wintering grounds in May. Although there are several spring recoveries from Iceland (in May), the geese are not as consistent there in spring and apparently are in much smaller numbers than in autumn. Boyd (72) stated that "In spring most of the Whitefronts go straight from Ireland to Greenland..."; this makes the Greenland Whitefront one of the very few of the world's geese which do not have a significant intervening spring staging area, although some feeding occurs in Greenland before nesting begins. In 1979 the first geese arrived in the centre of the breeding range on 7th May and passage was complete by the 17th, with the peak of egg-laying on the 22nd (197).

Ruttledge and Ogilvie (516) discussed in detail the total numbers of White-fronted Geese in the Greenland population. For the 1970s they estimated a total of 14,300-16,600, compared with 17,500-23,000 in the 1950s. Since the numbers in Britain had increased, the decrease in Ireland had been sharp, from 13,000-17,000 to 8,000-9,000. They did not accept the suggestion that the British increase accounted in part for the Irish decline. Their interpretation seems correct in view of the recovery analysis above; it is the total outside the Wexford Slobs which has decreased while those at the Slobs have remained stable or increased. These birds apparently originate from a different part of the range from the Wexford/Scottish birds. This decline thus apparently represents a real reduction in that population segment, which may be due to changes on the wintering grounds (habitat loss, shooting) or on the breeding area. The first coordinated census in Ireland (392) and Britain (558) was carried out in spring 1983, and this yielded totals of 6,363 in Wexford, 3,025 elsewhere in Ireland and 7,282 in Britain. Similar surveys in 1983-84 and 1984-85 yielded population totals of 17,400 and 19,800 respectively. The British total indicates an increase of more than 40% over estimated totals in the 1950s.

The distribution in Britain (Fig 77) is restricted to the north and west, following the distribution of the traditional acid bog habitat. The subspecies has occurred sporadically as far south as the Scilly Isles but the most southerly regular sites are in mid-Wales. Although there are records from 59 10km squares, only in 35 are there regularly more than 25 birds, 10 of these on Islay.

As with other geese, records from the Wildfowl

Count scheme are not consistently good. The following paragraphs are based on the comprehensive review of the status of the Greenland Whitefront (516) and the 1982-83 survey (558), supplemented by information from the regular counts.

Nine or ten areas in Great Britain now hold more than a hundred Greenland Whitefronts regularly. Six of these which are well counted are listed in Table 238, together with the most southerly site, on the Dyfi Estuary in mid-Wales. The island of Islay, Argyll, is by far the most important, holding at least half the British

Fig. 77. Greenland White-fronted Goose – British distribution.

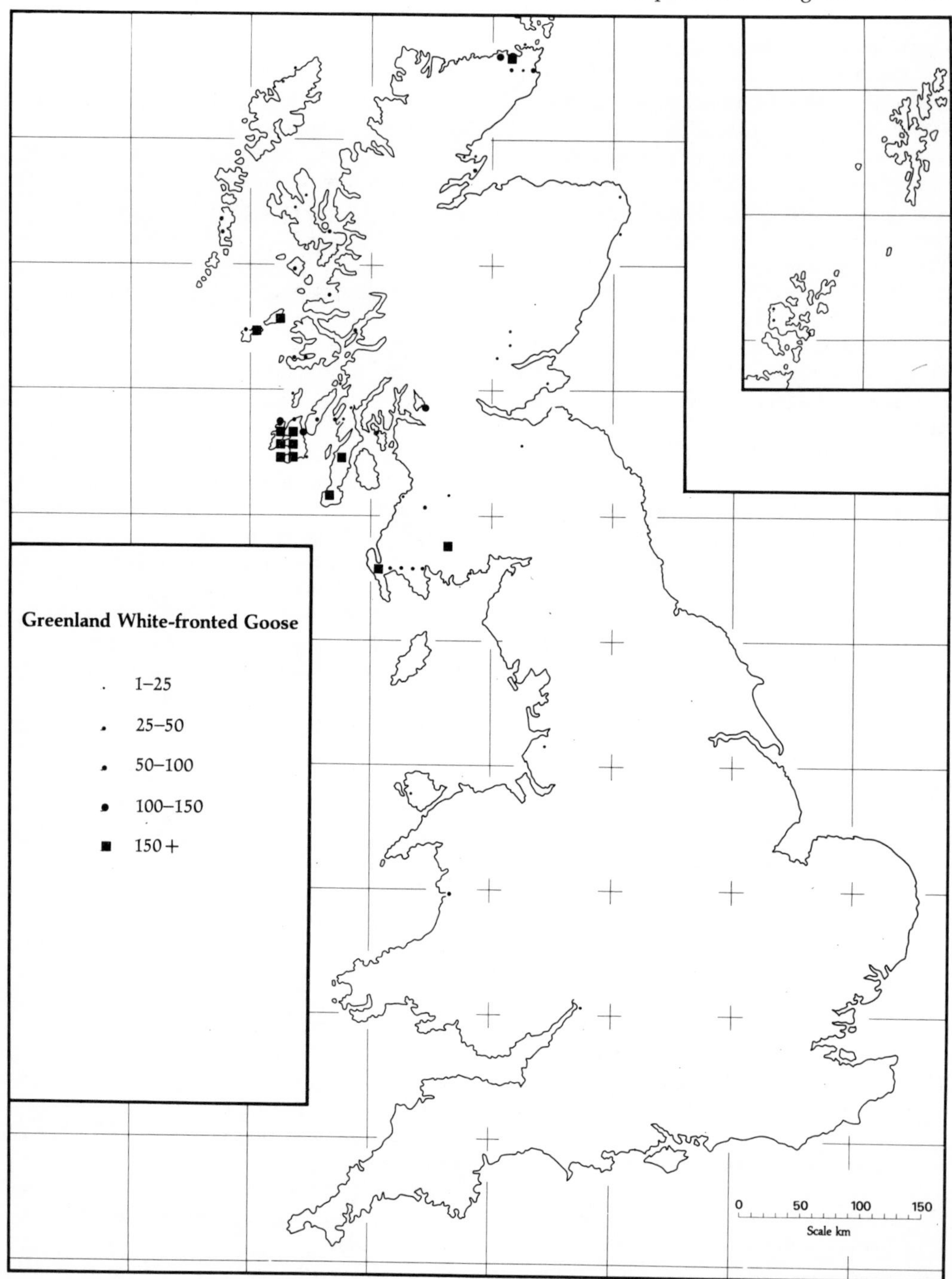

otal in recent years. The apparent increase in umbers since the early 1960s may be due largely to mproved count coverage. The highest ever count was nade in November 1980, following the best breeding eason for many years. Subsequent complete counts of the island later in the same winter, which included nany small groups in remote bogs, indicated a total of etween 4,900 and 5,000 birds.

As can be seen from the table all sites except Loch Ken, which suffered a slight decline, have held heir own or increased in importance. The two sites on he Mull of Kintyre – Rhunahaorine and Machrihan-sh – have become very important in recent years. An area of grassland near Stranraer, in Galloway, now egularly holds 250-350 geese, though this was not nentioned in the First Edition. Counting has im-proved in the last 10 years but the area has also ncreased in importance. The flock wintering on the Dyfi is small but holding its own.

The three remaining important sites are not often counted. The islands of Tiree and Coll were eported in the First Edition to hold 50-150, a decline rom 400-500 in the early part of this century. Recent counts include 610 in April 1978, but these were thought to have included birds on migration. The census yielded 715 in November 1982, and 763 in March 1983, so the area is obviously very important throughout the winter. Loch Shiel, Inverness-shire, held 30-100 in the early 1960s: this rose to 150-250 in the early 1970s and there was one count of 350. The maximum in 1982-83 was only 45. The Outer Hebrides formerly held 300-400 geese, and although counts are sporadic and incomplete the total there is probably around 100 nowadays (the maximum in 1982-83 was 101 in December).

Many other sites hold small numbers of Whitefronts regularly and in most of these the numbers have not changed. Three haunts have been deserted in the last 20 years, the most important of which is Tregaron Bog, Dyfed, which in the 1950s and early 1960s held 450-500 birds. Numbers fell to 83 in 1963-64 following the hard winter, when the geese dispersed and were either shot or died of starvation. There were sporadic occurrences in later years but by the early 1970s these were of only a handful of birds. Some may have moved to the Dyfi and there is still a

Table 238. *Peak counts of Greenland White-fronted Geese in the 7 British haunts which are regularly counted and which have held more than 100 geese at any time since 1970. Annual peaks 1970-1982 and average peaks for 4 periods since 1962. Figures in brackets are those obtained by adding the maxima on nearby sites, not necessarily counted at the same time. Updated from Ruttledge and Ogilvie (1979).*

Season	Caithness (total)	Islay	Mull of Kintyre	Loch Lomond	Vicinity of Stranraer	Loch Ken	Dyfi Estuary	TOTAL
1970	(278)	2000	(885)	40	160	330	130	3850
1971	(440)	3400	(788)	72	185	400	75	5350
1972	(350)	2580	(633)	79	200	365	64	4250
1973	(375)	4180	(668)	77	180	340	57	4300
1974	(450)	3430	(994)	97	300	340	36	5650
1975	(320)	4150	(978)	110	240	360	36	6200
1976	(550)	4210	1000	130	275	280	66	6500
1977	(550)	3300	1036	120	290	250	49	5600
1978	640	3380	730	110	290	240	40	5450
1979	400	2900	790	115	270	260	53	4720
1980	270	4300	1160	100	392	260	84	6570
1981	250	3300	650	124	480	330	104	5238
1982	460	3870	1280	118	380	305	77	6490*
5-year means								
1962-65		1475	337+	14		358	73	2600#
1966-70	255	3064	735	30	141	367	110	4700
1971-75	387	3548	812	87	221	361	54	5470
1976-81	482	3618	943	115	303	258	58	5780

* The census total 1982-83 (Stroud 1983) for the whole of Britain was 7282, so these sites in that winter held 89% of the total.
+ 3 years 1963-65 only.
# Assuming 2 sites not counted were similar to the next five years.

small flock in a remote hill area nearby. Up to 100 birds are said to have wintered in Morecambe Bay, Lancashire, and about 50 in Anglesey, Gwynedd, but both these sites are now deserted.

Only two species or subspecies of geese – the Hawaiian Goose (*Branta sandvicensis*) and the Aleutian Canada Goose (*Branta canadensis leucopareia*) – are less numerous than the Greenland Whitefront. Because of the recent decline and continuing loss of habitat, urgent conservation measures were recommended in the late 1970s to halt the decline (439, 516). The main recommendation was to remove the Greenland Whitefront from the shooting list at least for a period. Unfortunately the causes of the decline and particularly of the low average breeding success (less than 20% young annually compared with 30-40% for the other three Whitefront subspecies) are unknown. From the above account it is obvious that the situation is rather a complex one, with apparent subpopulations and different wintering conditions in different areas.

In recent years mortality for the population, judging by the relative stability of numbers and annual recruitment rate of about 19%, has probably been in the region of 20%, compared with 30-40% in the 1940s and 1950s (66). This, however, gives little buffer against adversity and, in the absence of information on other stages of the life cycle, the most useful immediate measure to increase numbers would be to reduce the winter mortality. If, say, annual mortality were reduced to 10% the population could grow to 25,000 without any increase in recruitment rate. Since mortality in Ireland is probably higher, protection there is more important than in Britain.

The subspecies has been removed from the shooting list in Scotland, in the Wildlife and Countryside Act (1981), following pressure from conservation bodies. Effective protection in Northern Ireland followed and in 1982 Eire prohibited the shooting of this goose for at least a 3-year period. The interest arising from the conservation discussions and from the expedition to Greenland in 1979 (198) has stimulated more work into the biology of the subspecies. We hope that the next 20 years will see an increase in the population to a healthier level and a better understanding of the life history of the Greenland White-fronted Goose.

Greylag Goose *Anser anser*

Our only native breeding goose, the Greylag i[s] in many places *the* wild goose and is also the ancesto[r] of the domestic goose. There are two subspecies, *anse[r]* of central and western Europe and *rubrirostris* to th[e] east with intermediates in east central Europe. Th[e] Greylags in Britain come largely from Iceland, th[e] native stock amounting to less than 3,000. There are n[o] differences in measurements between the two race[s] (335) other than bill size and no behavioural differ[-]ences have been described.

Greylag Geese winter both inland and on th[e] coast, roosting on estuarine sandbanks, lakes an[d] rivers. Large flocks frequently gather together at th[e] larger sites but Greylags also use small ponds, rive[r] islands and marshes, often only a few dozen bird[s] together. In central Scotland Pinkfeet and Greylag[s] regularly use the same roosts but keep apart. Greylag[s] leave the roost later than the Pinkfeet and fly shorte[r] distances to feed. 90% of Greylag feeding area i[n] central Scotland is within 5km of the roost and onl[y] 2% further than 10km away (382).

In Britain Greylag Geese feed almost exclusivel[y] on arable farmland and improved pastures. Typicall[y] cereal stubbles are used in autumn, followed b[y] potatoes, swede turnips and carrots in mid-winter. I[n] spring, sown grasslands and to a lesser exten[t] permanent grass and winter-sown cereals are impor[-]tant. The vast majority feed by day, mostly in flocks o[f] up to 100 birds (379). Greylags are less wary than othe[r] geese, come closer to human habitation and ar[e] fearless of farm stock. In the Netherlands thei[r] traditional habitat of *Scirpus* beds is still used to som[e] extent and here most of their feeding is nocturna[l] (624). In Britain little of this coastal habitat remains an[d] the inland fens and marshes have also largely bee[n] drained for agriculture.

Greylags are territorial when nesting but, lik[e] many grey geese, the males are often some distanc[e] from the nest and do not betray its position. Female[s] ely on camouflage to avoid detection and the nests are usually in cover, often heather or other shrubs. The clutch size of Hebridean Greylags averages about 5; peak of laying is in the second half of April and the goslings hatch in late May (381). Feral Greylags in south-west Scotland nest predominantly at the foot of trees or near rocks and on one island the density has been as high as 140 nests per hectare. Laying is in late March and early April and the geese lay on average 5.9 eggs (622). The migratory stock in Iceland lay in late May and their average clutch size is only 4 eggs (469). Scottish Greylags take their broods to the margins of productive lochs to graze and families merge to form large flocks. Non-breeders and immatures gather to moult in June and July and after the flightless period they and the breeding birds disperse to feed on surrounding farmland. The geese eat various marsh plants and moorland vegetation during the breeding and moulting season, but at other times grass and cereals are most important (379).

Many different populations have been separated in continental Europe although the breeding distribution is almost continuous. Greylag Geese breed as far north as 70° N in Norway. Birds from the Scandinavian/Central European population, which numbers 60-80,000 (504), migrate through Denmark and the Netherlands to winter largely in Spain (Fig 78). Icelandic birds arrive in Britain in late October, the whole population being here by the end of the month. The return migration starts in April and few remain into May. Hebridean Greylags are resident, although there may be short movements during the winter, and feral populations remain very close to their breeding areas throughout the year.

A complete census of the immigrant Icelandic population is attempted in early November each year and the map of their distribution (Fig 79) relies heavily on these counts. In November the birds are highly concentrated, feeding on stubbles largely in north-east and central Scotland. Thereafter they are more dispersed, but information is incomplete. Thus the distribution as plotted is biased towards the main autumn centres and shows a less dispersed pattern than the winter average. The dotted line shows the boundary of the range of the wild stock in winter, the birds to the south being exclusively feral. Some of the flocks to the north of the line are also feral, and those in

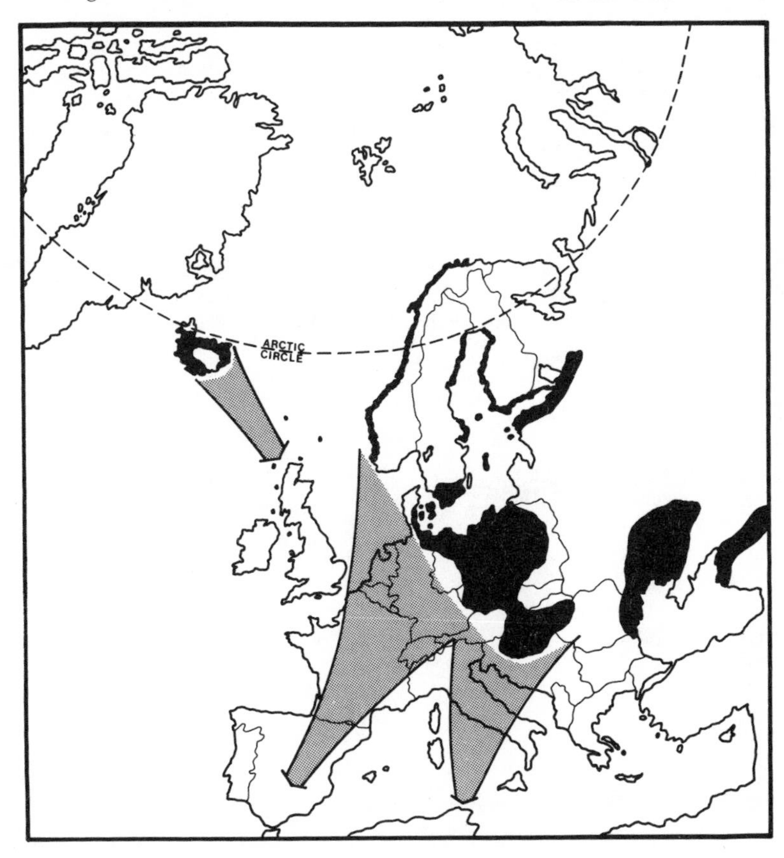

Fig. 78. Greylag Goose – breeding and wintering range and migration routes.

the north-west are resident breeders. No fewer than 46 roosts qualify as internationally important. The most recent counts from the 13 sites having a 7-year average of more than 2,000 birds are given in Table 239.

The table demonstrates well how concentrated the flocks are in November, with about 80% of British Greylags in only 13 areas. Annual fluctuations within sites are rather large and no doubt reflect the local availability of food, especially of stubble grain, the chief autumn diet. Over a longer time scale there have been marked shifts in distribution, as shown in Table 240.

Most notable is the increase in importance of north-east Scotland at the expense of east central Scotland which formerly held two-thirds of the stock. The numbers in north-east England have also markedly increased as the Solway has diminished in importance. The numbers in west Scotland have decreased, especially on the island of Bute, where they had increased rapidly to 5-7,000 in 1961-62 (293), but fell even more sharply following complaints of damage to the swede turnip crop and intensive disturbance of roosts and feeding grounds.

The first November census was carried out in 1960, when 26,500 Greylags were counted. Since then there has been a very marked increase, to an average of 65,000 in the 1970s and an all time maximum of 95,000 in 1981. Five-year running means of the breeding success of the population, expressed as the percentage of juveniles in autumn, and the total

Table 239. *Sites or groups of sites which have held an average of more than 2000 Greylag Geese in the last 7 autumn censuses, and the population total at each count. A blank = no data. Brackets = incomplete data.*

	1976	1977	1978	1979	1980	1981	1982	Mean
Caithness Lochs (consol.)	300	4240	6750	3250	3075	4300	4140	3722
Loch Eye	3000	12500	11500	13000	8300	20800	3500	10371
Beauly Firth/Black Isle	400	2600	3700	2080	9330	11000	4240	4764
Loch of Strathbeg	8250	5000	7000	5000	4000	4750	9600	6229
Loch of Skene	660	3050	950	1650	4700	5700	4500	3030
Davan and Kinord		506	2800	3300	4300	3980	9500	4064
Lintrathen and Kinnordy	2080	2000	7000	5750	4500	1070	800	3314
Blairgowrie Lochs (consol.)	9110	12650	6430	7140	9680	4140	6330	7926
Drummond Pond etc. *	4300	6000	11855	15000	12000	3440	3940	8076
Carsebreck Lochs	1370	1060	1780	1920	5050	4050	2180	2487
Loch Leven	3400	1100	1400	2900	5615	3000	2500	2845
Hoselaw Reservoir	1800	2963	2700	3200	2690	1700	730	2255
Holburn Moss +	1800	2500	3500	2800			1600	(2440)
Census total #	56000	67000	76000	81000	90000	95000	80000	77860
% on above sites	(65.2)	83.9	88.6	82.7	(81.4)	(71.5)	67.0	78.5

* The counts of over 8000 may have have been overestimates.
+ January counts (November mean 1780).
# Including 1000-2000 feral birds in south-west Scotland.

Table 240. *Average percentage of the Icelandic population of Greylag Geese in different parts of Britain in three-year periods between 1960 and 1982, as indicated by the November census results. Updated from Boyd and Ogilvie 1972, Ogilvie and Boyd 1976.*

	1960-62	63-65	66-68	69-71	72-74	75-77	78-80	81-82
NE Scotland	13.4	15.3	12.8	20.2	18.1	39.2	43.3	55.2
E Central Scotland	61.7	61.1	68.8	57.9	61.5	38.1	38.4	24.6
SE Scotland	2.1	5.4	4.0	7.0	5.1	5.6	7.7	5.5
W Scotland	9.1	10.1	9.9	6.2	6.2	5.9	5.2	4.3
Solway	13.2	7.2	5.4	9.3	7.7	7.7	3.1	5.2
NE England	0.4	0.7	0.5	1.0	1.5	3.5	2.4	5.5
Mean census total	33800	40200	57200	63400	71300	60700	82300	95000

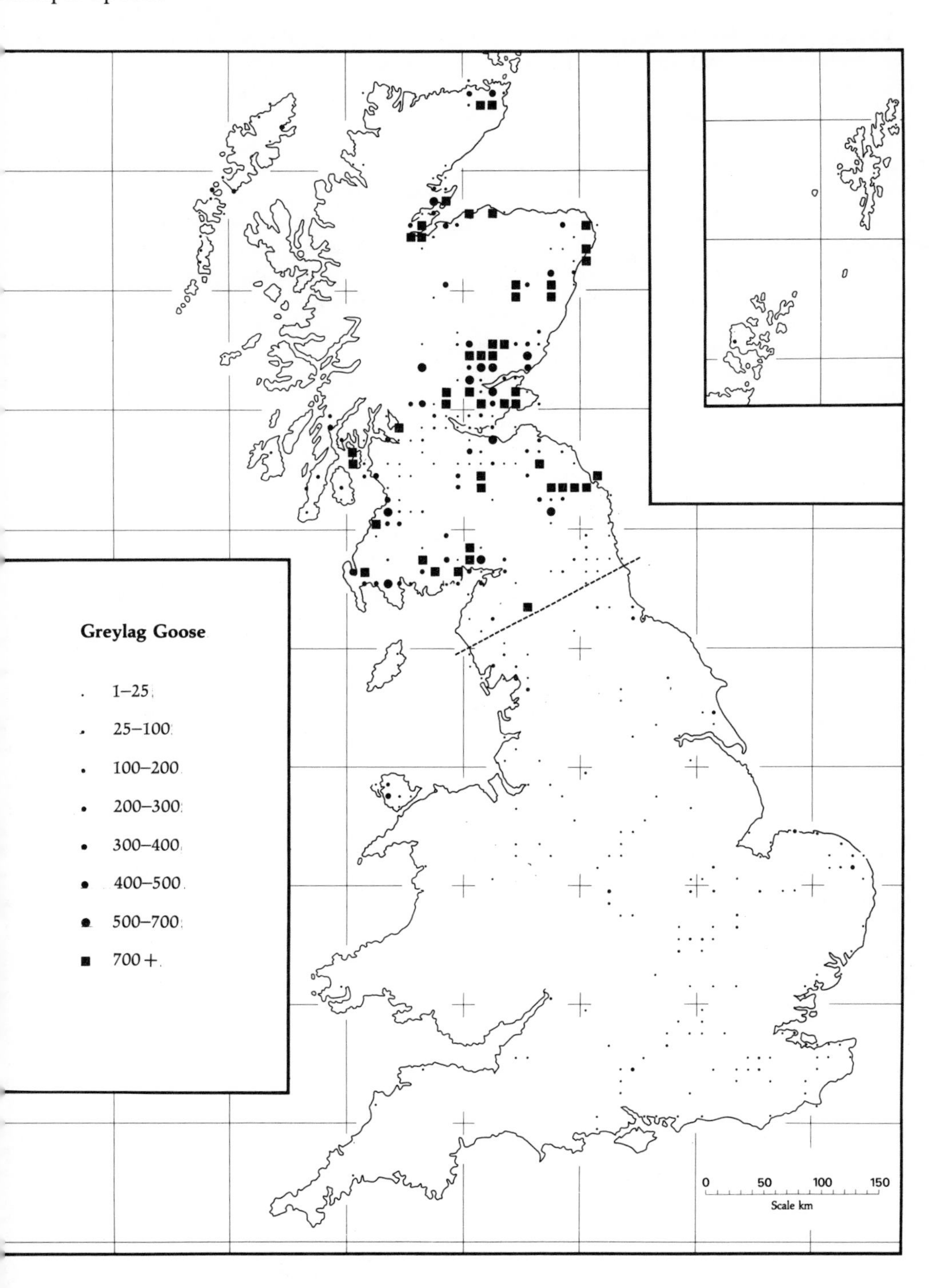

ig. 79. Greylag Goose – British distribution. The lashed line separates largely wild stock (to the north) rom almost exclusively introduced birds (to the outh).

population count are shown in Fig 80. The reduction in the percentage of young coincident with the increase in overall numbers in the 1960s is very striking. In the late 1960s and 1970s production only twice exceeded 25%, whereas it did so every year from 1958 to 1964. Thus although the total number of geese has increased, the annual number of young produced has not. In the four 5-year periods between 1960 and 1979 the average number of young arriving in Britain annually was estimated at 11,860, 9,620, 14,540 and 9,820. Competition either for food or for nesting sites was probably the limiting factor.

The increase in population in the last two decades is wholly attributable to a decrease in mortality. This was estimated as 22% in 1960-1964, 12.5% in 1965-1969 and 18.1% in 1970-1974 (414); similar calculations for 1975-1980 indicate only 11.1% annual loss. This low mortality, coupled with relatively good breeding seasons in 1978 and 1980, resulted in the upsurge in the population of the last few years. If this low death rate is maintained we can expect further increases in numbers in the 1980s.

The Greylag used to breed in the wild in the East Anglian fens, Lancashire, the Lake District and probably many other parts of Britain before the reed-marshes and fens were reclaimed for agriculture in the 19th century. By the early 20th century the species was restricted to north-west Scotland, but between 1930 and 1970 feral flocks were established many parts.

Most of the indigenous birds are on South Ui chiefly at Loch Druidibeg (p.323). The Hebride population has increased considerably in recent yea in 1982 1,676 geese were counted and further es mates indicated a total approaching 2,000 (472), wi perhaps up to 500 on the mainland and adjace islands in the eastern Minch. The total native stock between 2,500 and 3,000. The largest of the fe groups is in south-west Scotland, established arou 1930 by the release of young hatched from eg brought from Druidibeg. In 1971 the total number w estimated at 1,160, with 130 breeding pairs ((623), s also p.323). In the late 1960s and early 70s more th 1,000 Greylags, chiefly from eggs taken in south-we Scotland, were released by wildfowling clubs in ma other areas, especially in Caithness and Sutherland, Kent (numbers now over 400), and in Cumbria, ea Yorkshire, Anglesey and Bedfordshire. Some of th Kent releases were of pink-billed *rubrirostris* broug over from the introduced stock in Belgium rather tha from British stock. Introduced birds have increased i number in most areas but have spread rather little, least in the breeding season. The policy of widesprea release is not now encouraged, chiefly because complaints of damage to agricultural crops. The tot introduced stock probably numbers well over 3,0 birds. Their distribution is shown in Fig 81.

Fig. 80. Total numbers (solid line) and breeding success (% juveniles in autumn) of Icelandic Greylag Geese, 1960-1980, expressed as 5-year running means (mean at 1960 is of 1958-1962, that at 1978 is 1976-1980). From Boyd and Ogilvie (1972) and Ogilvie and Boyd (1976).

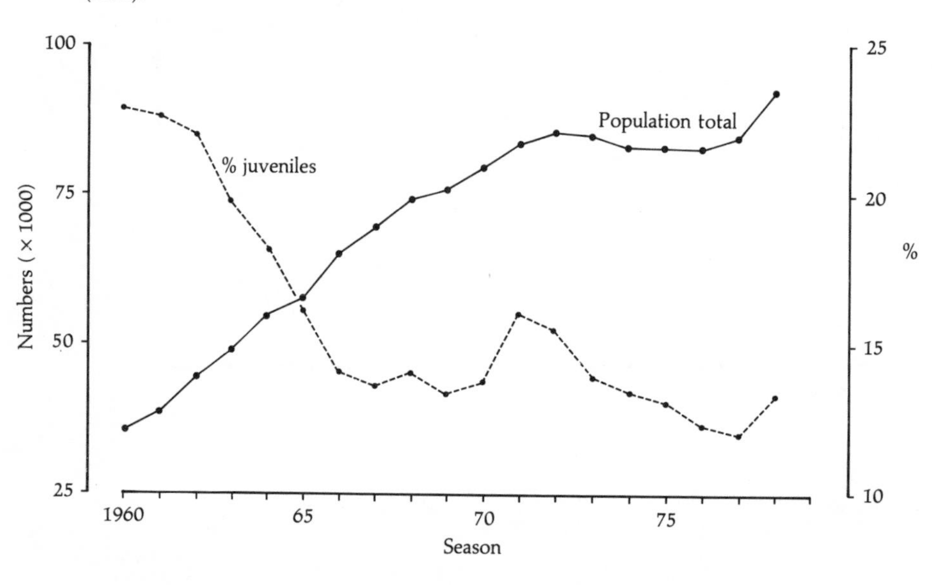

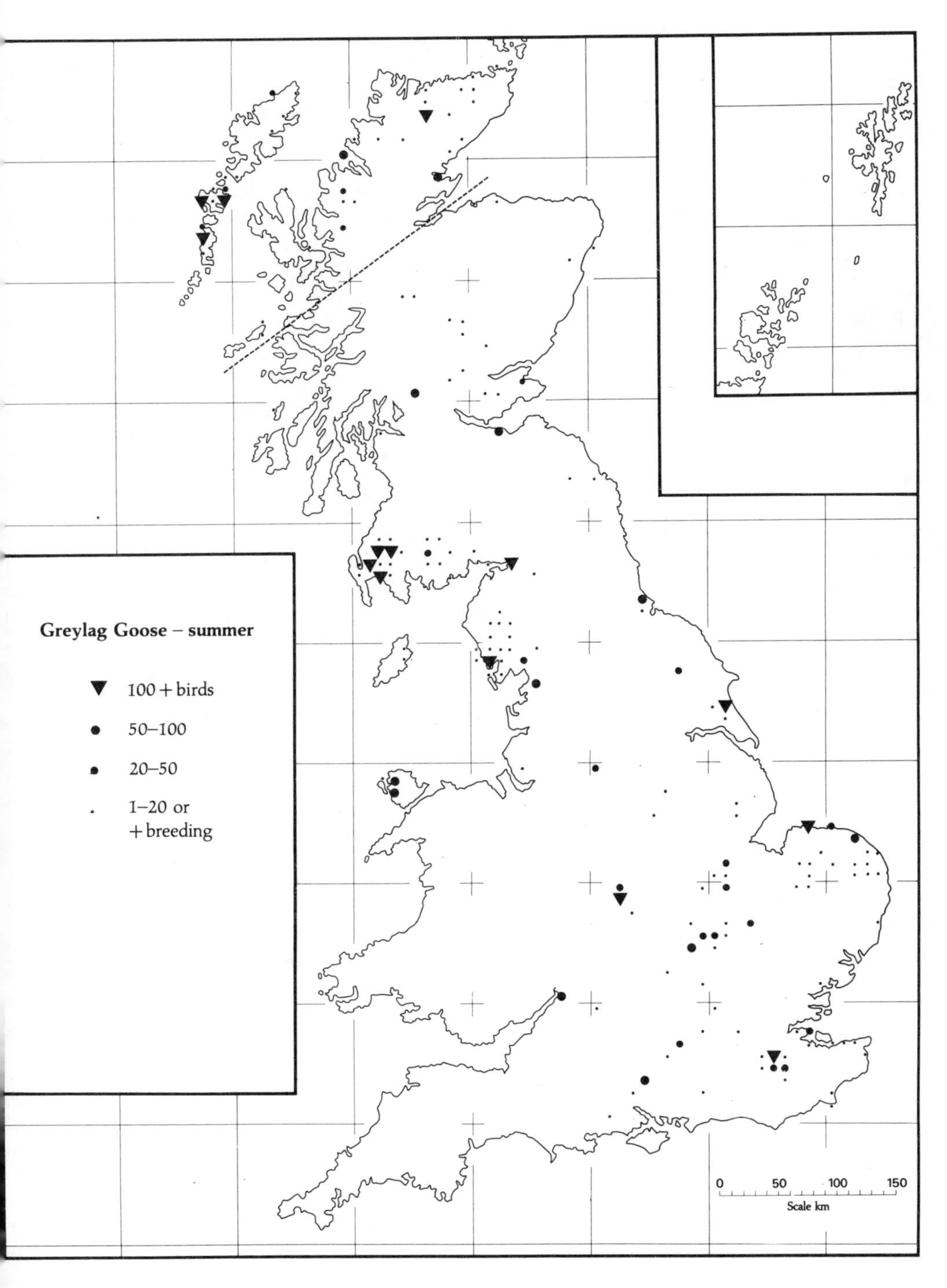

Fig. 81. Summer distribution of Greylag Geese in Britain. Based on:
a) Wildfowl Counts in the area south of the Icelandic population range, b) September counts within the range (on the assumption that migrants have not arrived in mid-September, c) Sharrock (1976), d) Young (1973), e) WAGBI reports on rearing and release programme (in WAGBI Annual Reports 1965-1973).
Where there is no information other than a positive breeding record from the Atlas the 10km square has been given the smallest dot.
The dashed line separates native (to the north) from the introduced stock.

Canada Goose
Branta canadensis

The most numerous goose in the world, the Canada Goose has been introduced from its native North America both to Europe and Australasia. The first introductions to Britain were in the second half of the 17th century, originally to adorn the grounds of large estates. The race to which they belonged is uncertain but present day British birds are as large as the largest races in North America. Until the middle of the present century they were restricted to a few areas, but in the 1950s and 1960s birds were caught up and transported to other suitable habitats. This was done partly because of agricultural problems and partly to provide shooting. The result was to speed up the dispersal of the birds considerably and to provide them with the opportunity greatly to increase their numbers.

The British geese are highly gregarious in winter, but, apart from the moult migration (see below), they rarely make journeys exceeding 100km, and remain in a series of independent groups. Feeding flights are short, no more than 5km, and in many cases the birds walk from the water to graze on nearby fields. Their foods have not been well studied but in most areas they concentrate on stubbles in autumn and on grass or cereals for the remainder of the year. Many flocks use public recreation areas, where they become very tame and accept food from the hand. Canada Geese will also feed on underwater vegetation and roots when available.

Male Canada Geese are territorial and very aggressive during the breeding season. Most nests are on islands in lakes, rivers or gravel pits, but in Yorkshire some birds have taken to nesting on open moorland (206). Pairs arrive on their breeding territories during March and some may lay in that month. Most birds lay in April and early May, the average clutch being 5-7 eggs. The broods are reared on the spring growth of grass in May and June, taking abo 10 weeks to fledge. The adults moult in June when t goslings are about a month old.

Moult migrations are common in geese and the direction is almost always northwards. Canada Gee were seen in small numbers on the Beauly Firt where none breed, from 1947 onwards, but it was n until 1963 that they were caught and ringed, esta lishing that the birds were moult migrants, originatir chiefly from Yorkshire, a distance of about 500k (155). The flock increased from 50 or less in the 1950s 300 in 1969 and around 1,000 in recent years. Abo half the migrants are immature birds (1 and 2 yea old) and the remainder older non-breeders. Eve when numbers are high, many immatures stay Yorkshire to moult (606). Birds from other, mo southerly, breeding areas, especially the Midland have also been found among the migrants in rece years.

The Canada Goose now occurs in well over 1,0 counted sites and 484 10km squares in Britain (Fig 82 Concentrations are usually rather small, with only sites holding more than 500 geese and nearly 70 averaging less than 25 birds. Smaller groups surrou major foci in the Thames Valley, the west Midlanc and Yorkshire. Table 241 compares the numbers different parts of Britain in 1953, 1968 and 1976, an indicates the growth and spread of the populatio during that period. Many of the southerly groups a based on gravel pit complexes and there is still muc room for expansion in this habitat, which continues

Table 241. *Numbers of Canada Geese recorded in three countrywide censuses in different areas of Britain. Based on the summarised data in Ogilvie 1977.*

Area	1953	1968	1976
South west and Hants	0	480	940
South east	50	660	890
Thames Valley	163	700	2790
Wilts, Glos	62	150	270
West Midlands	104	860	2230
Notts and Derbyshire	637	1350	1800
East Midlands	229	560	1200
East Anglia, Essex	725	1350	2450
Salop and Cheshire	901	1840	2620
Yorkshire (incl Beauly Firth)	575	1800	2810
North England	156	330	680
Wales	0	260	300
Scotland	194	100	140
Total	3796	10440	19120

Note: The Yorkshire totals include the associated moulting flock on the Beauly Firth.

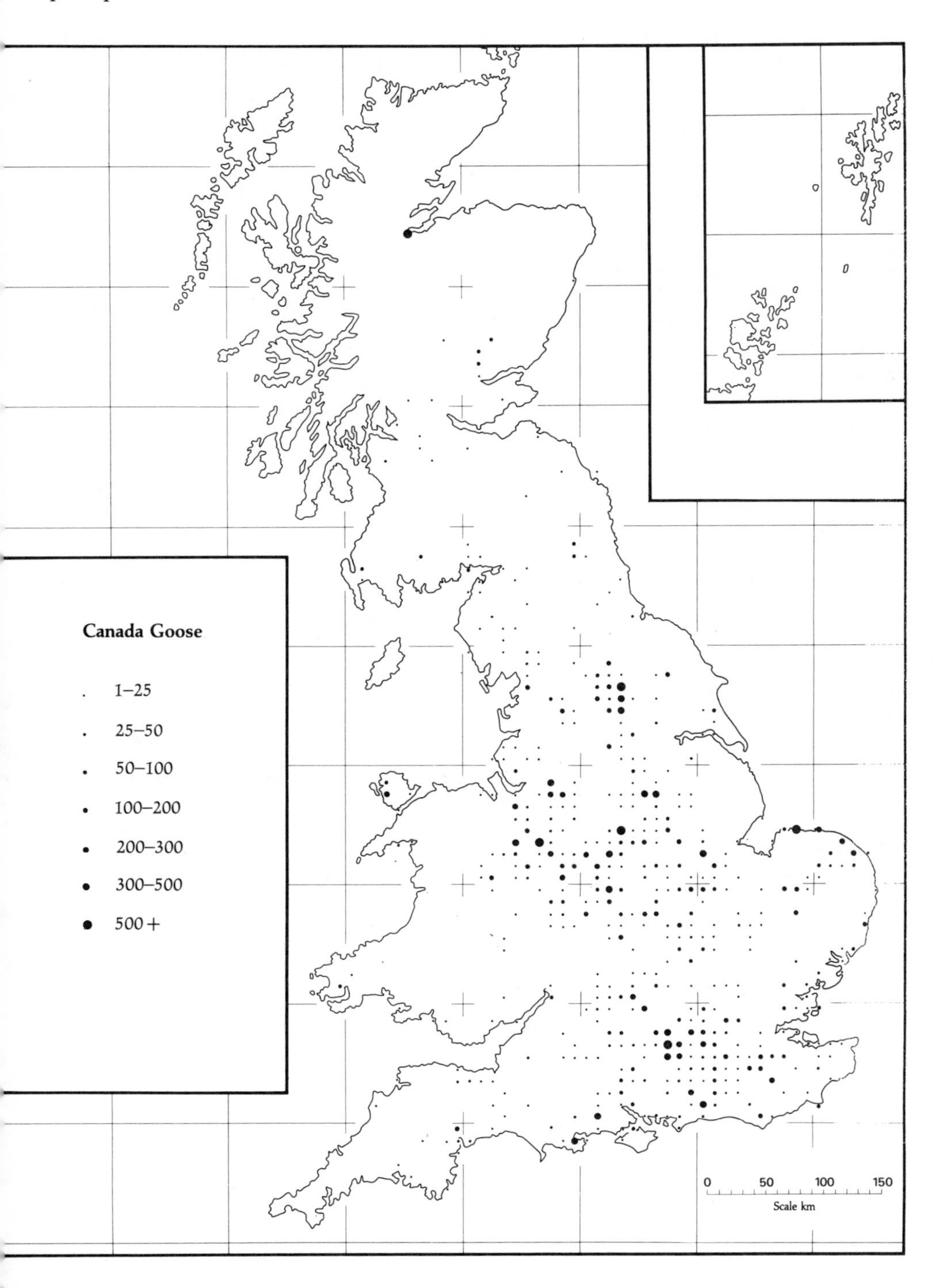

'ig. 82. Canada Goose – British distribution.

be created at a considerable rate. Of 43 groups or "subpopulations" recognised in 1976 (405) only 19 were established in 1953, and 32 in 1968. The fact that eleven areas have been colonised since the "transplanting" stopped indicates that natural range expansion is continuing.

The first organised census of Canada Geese was carried out in 1953 (57), when the total lay between 2,200 and 4,000 birds (the total in the table is arrived at using the maxima of the regional estimates (405)). Another census in 1968 (399) showed that, helped by transplanting, the numbers had increased to an estimated 10,500; by 1976 the total was more than 19,000 (405). More than half the birds are recorded in the monthly winter counts, and this enables us to examine the upward trend in more detail (Fig 83). The increase has been remarkably constant, at about 8% per annum, since the hard winter of 1962-63 when the numbers were somewhat reduced. The slope of the regression on 1962-1980 counts is almost identical to that of the line joining the two census totals. The apparent decrease in 1978-79 is attributed to the wide dispersal of the flocks during the hard weather.

A continuation of this trend would result in populations of 39,000 in 1985 and more than 50,000 by 1990. Such increases cannot go on indefinitely, but there seems little reason to believe that shortage of habitat will limit numbers in the near future. There are large parts of the country, especially in Scotlan which seem suitable but have not yet been colonise and new breeding habitat is continually being create Even in areas such as the west Midlands, where t geese are long established, the rate of increase is t same as in the national population.

Continued population growth would u doubtedly be of major concern to farmers, but as y there are only a few areas in which the numbers a large enough to cause substantial damage. In son places many acres of cereals adjacent to breeding moulting lakes have been completely destroyed l continual grazing by the geese (612). They also wa into standing cereals and strip the seeds. Damage pasture is more difficult to assess but since about fi of these large birds consume as much grass as a shee the losses can be substantial, especially in spring.

A concentrated and ruthless campaign cou probably eradicate the Canada Goose in Britain in few years, but most people would think this undesi able. Although wildfowlers were responsible f many of the early "transplants" the bird is not shot any great extent, except by the landowners in a fe localities. Most efforts at control are made during t nesting season by destroying the eggs, but this ineffective. The Wildlife and Countryside Act (198 allows the shooting of Canada Geese under licence the close season and this will give farmers t opportunity of taking more effective action. T maintain the population at its 1980 level an addition 2,000 birds would have to be culled each year. Becau of the birds' remarkable recovery capacity contr must be organised over all adjacent waters so that

Fig. 83. Canada Goose – trend in numbers in Britain, 1960-1982. To even out coverage irregularities the index is an average of that for January and September. The dotted line joins the two census points, 1968 and 1976 (diamonds).

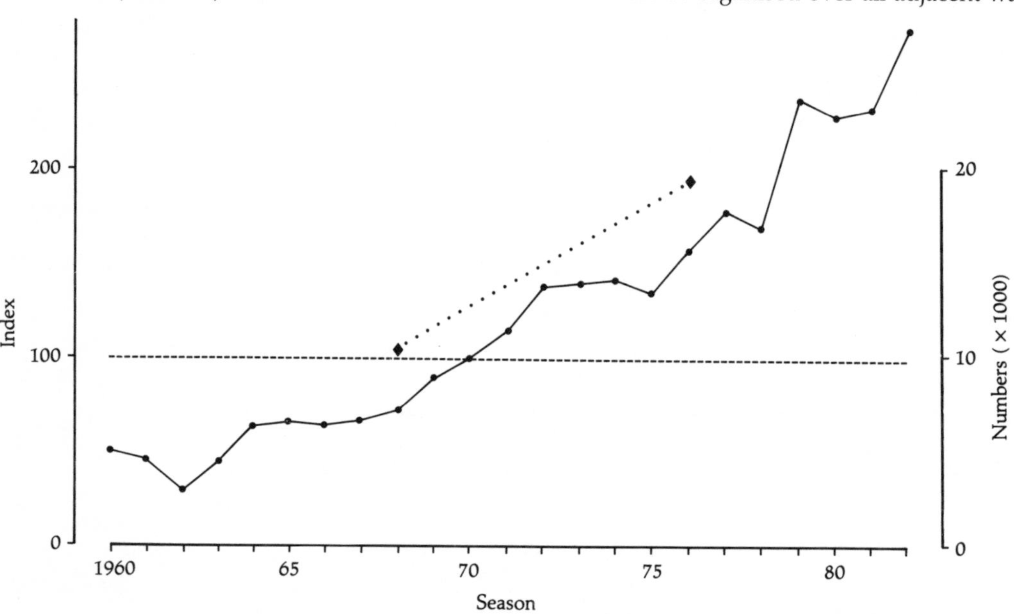

ffects the whole of a local population. Only coordin-ted action can prevent this introduced species from ssuming pest proportions.

In Europe the Canada Goose was introduced in Sweden in 1933 and later into Norway and Finland. The combined Scandinavian population now exceeds 0,000 birds following much faster growth than that in Britain. Their initial spread in Sweden was encour-ged artificially, but recent increases in the boreal orests of the Fennoscandian shield seem to reflect the bility of these geese to occupy habitats unsuitable for nost wildfowl (188).

A very few truly wild Canada Geese arrive in Britain as vagrants, usually with flocks of Greenland White-fronted or Barnacle Geese on the north-west oast of Scotland. These belong mostly to the smaller undra races, *hutchinsii* and *parvipes*, both of which reed in north-eastern Canada.

Barnacle Goose
Branta leucopsis

The Barnacle Goose breeds in three isolated areas – Greenland, Spitsbergen and western Siberia. Both the Greenland and Spitsbergen populations winter exclusively in the British Isles, on the western seaboards of Scotland and Ireland. The remaining, and largest, group from western Siberia winters in the Netherlands. Extensive ringing during the past 25 years has shown that the three populations are discrete even in winter (68, 440). Although some published weights suggest differences, a large sample of measurements indicates that birds of the three populations are identical in size (440).

Barnacle Geese are extremely gregarious, roosting in large flocks at high density, usually on estuarine sandbanks on uninhabited and fox-free islands. Their flight to the feeding grounds is rather short, usually no more than 5km. When travelling short distances they fly in dense bunches but on longer trips the characteristic skeins are formed. Feeding flocks are also closely packed and more easily put to flight than those of other geese.

Wherever they occur Barnacle Geese are never very far from saltwater; their traditional habitat in Britain is on the machair (grassland on shell sand) and on turf-topped rocky islands now mostly uninhabited, off the north-west coast, and some thousands of geese still occupy such sites. They are well adapted, with their short bills and rapid pecking action, to feed on the very short vegetation of so-called "*Plantago*-swards" on the sea-washed islands. Both on the Solway saltmarshes (448) and in Ireland, one of their chief autumn foods is the stolons of White Clover, which they grub out from among the grass. They also strip seeds of sedges and rushes from standing stalks and graze saltmarsh grasses such as *Festuca rubra* and *Puccinellia maritima*. Their habitats and foods in the

Netherlands used to be very similar but there they are largely dependent on inland grasslands nowadays.

Like other goose species in Britain, the Barnacle has come to rely increasingly on agricultural land. On its main British haunt on Islay, its diet is almost exclusively gathered from improved ryegrass pastures, and the same is now true of the Solway Firth and the Irish mainland. At Caerlaverock stubble feeding has become a regular autumn habit, and the flocks at Lissadell, Co. Sligo, have been known to take waste potatoes.

In the breeding season Barnacle Geese inhabit coastal tundra, nesting either on offshore islands or on rocky slopes or cliffs, inaccessible to Arctic Foxes. They rear their young on bog grasses and sedges and in Spitsbergen have been seen to visit offshore rocks to feed on algae exposed at low tide.

The migration routes of all three populations are well known (Fig 84) and the timing is relatively inflexible. The Siberian population starts its spring migration in April and stops for two or three weeks in the Swedish island of Gotland and in Estonia before leaving for the breeding grounds in late May. It leave the breeding grounds in late September, stops briefl on Gotland and in West Germany and stays in th Netherlands from mid-October to mid-April. Th population has grown from about 20,000 in 1959 (68) t an average level of more than 50,000 in the late 197((504).

The Greenland breeders leave the winterin grounds in April, and stop for about 3 weeks in th north-western valleys of Iceland. The return journe starts in mid-September, with a stop of about a mont in southern Iceland before proceeding to the Britis Isles in late October. The Spitsbergen flock leaves th Solway Firth in late April and stops in Norway, jus south of the Arctic Circle, for three weeks. It leaves th breeding grounds in mid or late September, stoppin for a time on Bear Island, 650km south of the breedin area. The flight of about 2,500km from there to th Solway is probably non-stop and the whole popula tion is in Britain by mid-October.

The wintering range of Barnacle Geese is mor restricted than that of any other European wildfow species. The Greenland population in Britain i confined almost entirely to the islands in wester Scotland, the scattered occurrences on the Scottis

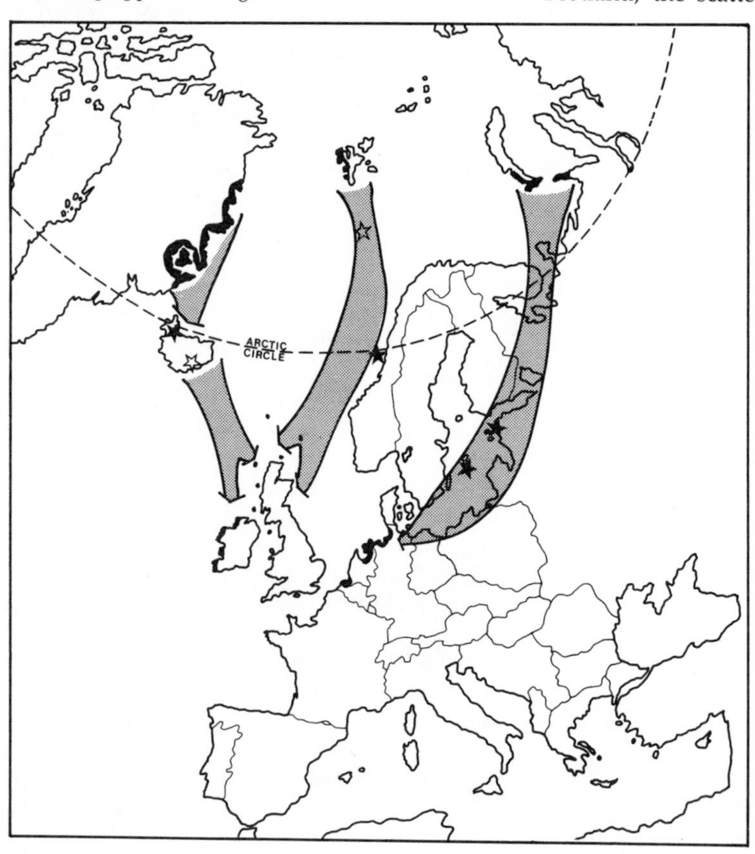

Fig. 84. Barnacle Goose – breeding and wintering range and migration routes. The stars indicate the major staging posts on migration.

mainland and in north-east England being mainly Spitsbergen breeders stopping briefly in October; the Spitsbergen population winters exclusively around the inner Solway Firth (Fig 85). Most records from southern England are of stragglers from the Netherlands, arriving either as a result of cold weather or as small groups among flocks of Whitefronts or Brents. The feral flocks which occur in several parts of England are not included on the map.

Counts of geese on the Solway and on Islay, and

Fig. 85. Barnacle Goose – British distribution.

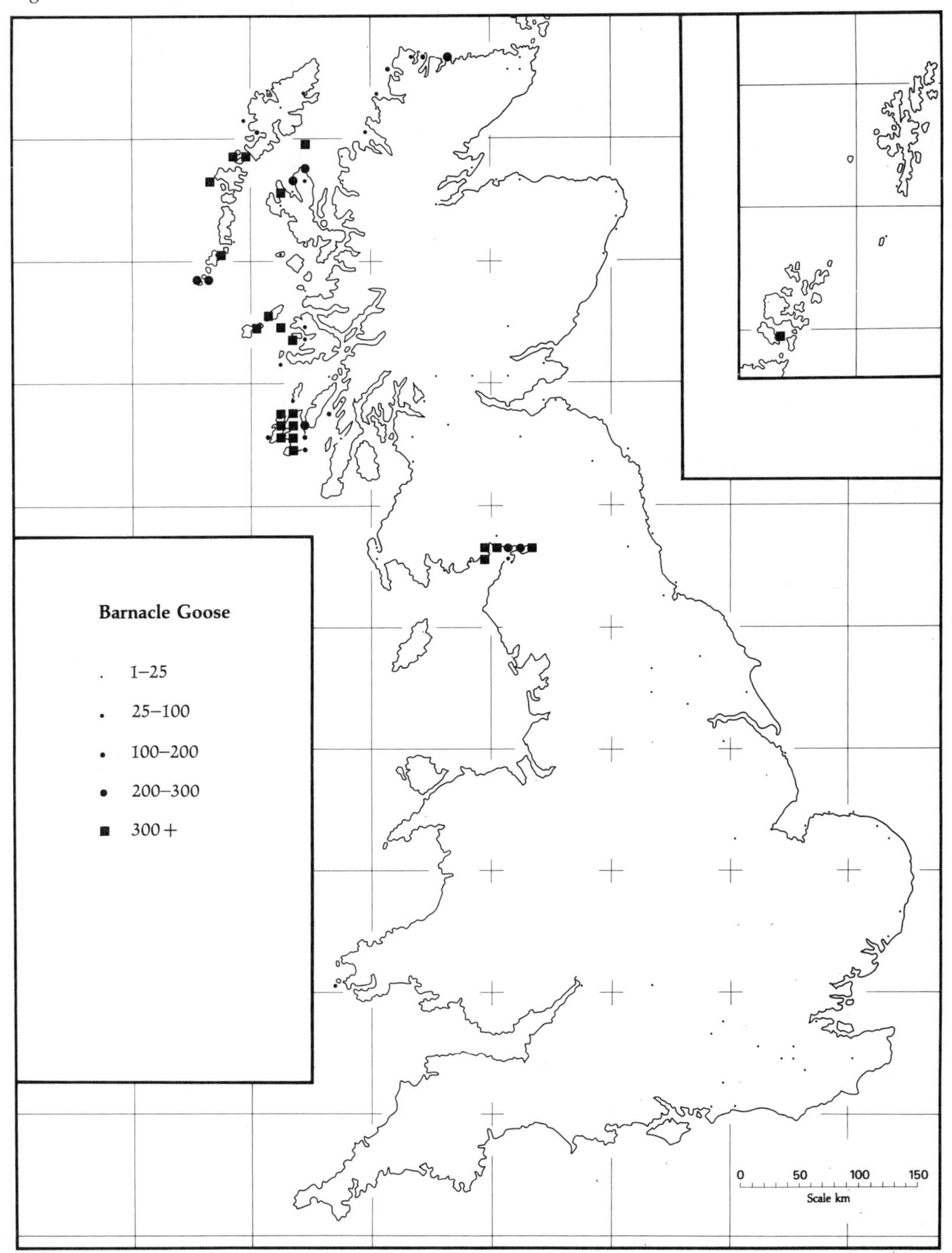

Table 242. *Greenland Barnacle Geese in their main haunts and population totals in six complete aerial surveys 1959-1983. Data from Boyd (1968), Ogilvie and Boyd (1975) and Ogilvie (1983).*

	Dec 1959	Apr 1962	Mar 1966	Mar 1973	Mar/Apr 1978	Mar 1983
Islay	2800	4800	8500	15000	21500	14000
Treshnish	299	390	795	419	610	620
Tiree/Coll	25	484	534	143	390	619
Sound of Barra	86	415	360	336	455	375
Isay (Skye)	130	395	380	297	290	250
Monach I	480	860	1035	640	760	638
Sound of Harris	174	498	575	980	1330	1555
Shiant I	290	317	483	450	420	580
Total Scotland	5506	9566	15103	19736	28056	20820
Total Ireland	2771	4404	4718	4398	5759	4432
Total population	8277	13970	19821	24134	33815	25252

Fig. 86. The numbers of Barnacle Geese in the Greenland population (upper line), on Islay (middle line) and the Solway Firth (lower line) from the early 1950s. The Greenland total for the first census was part estimated (Ogilvie 1978). The figures for Islay are maxima but numbers do not change much through the winter. Counts on the Solway were incomplete before 1970 and are presented as 5-year means.

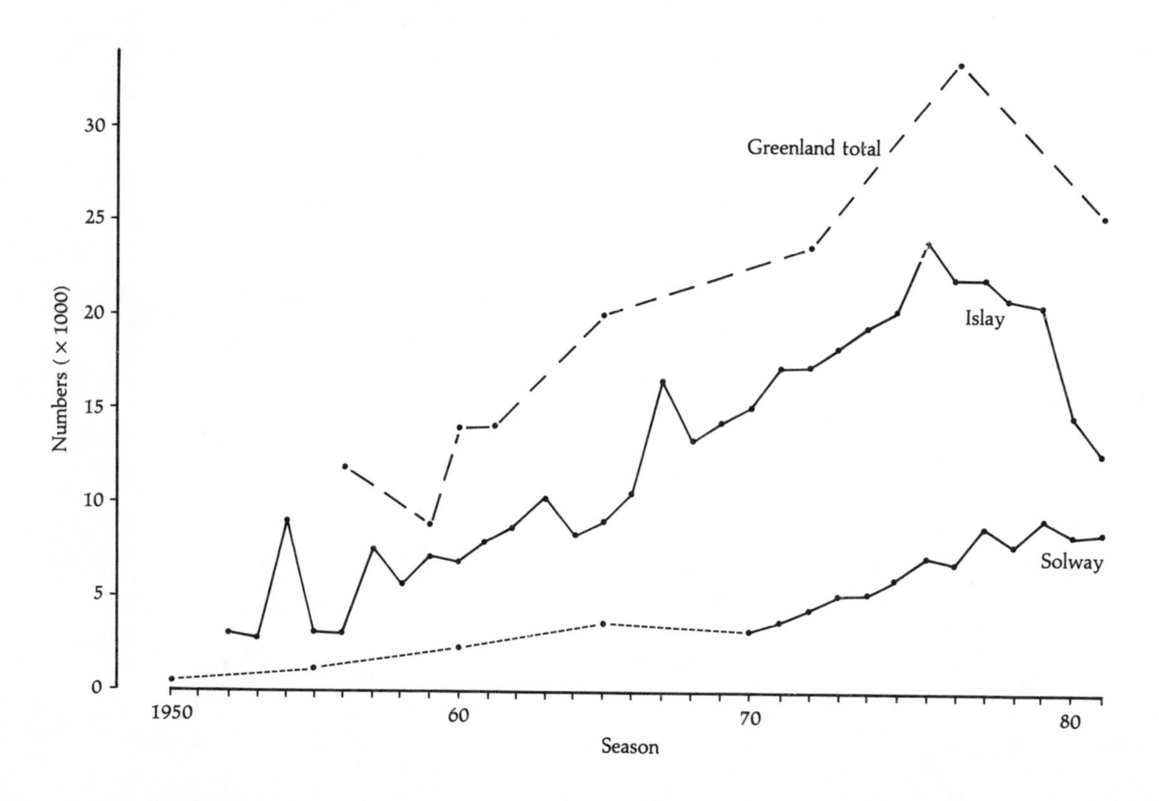

the results of periodical spring surveys of the whole Greenland population are given in Fig 86. Numbers on the Solway were precariously low after the war (only 300 counted in 1948) following a drastic decline. Former population levels are poorly known, though 6,000 were reported at the turn of the century (50). Since there were then few or no geese in some areas of Spitsbergen where they are now abundant, it may be that the majority of those recorded on the Solway in the early 20th century originated from Greenland. The Spitsbergen population increased gradually following protection from shooting in 1954 and the establishment of Caerlaverock NNR in 1957, which safeguarded a large portion of the feeding area. Concurrently conservation measures were taken in Norway and Spitsbergen, resulting in complete protection from shooting throughout the range by 1961-62, with breeding sanctuaries established in 1973. In the 1960s the number seemed to have stabilised at 3-4,000, but a further rapid increase took place in the 1970s, and numbers reached 10,500 in 1984-85. The increase was due to a lowering of the mortality rate rather than better breeding, and this probably resulted from better protection from illegal shooting, especially on the Solway where the Wildfowl Trust extended the disturbance-free refuge at Caerlaverock in the early 1970s (451). As the population increased, the other two main Solway haunts, around Southerness and on Rockcliffe Marsh, Cumbria, became much more heavily used, and by the late 1970s Rockcliffe Marsh rivalled Caerlaverock in importance (see also p.215).

The Greenland population has trebled since the late 1950s, the most notable increase being on Islay, which now holds about 20% of the world total (Fig 86). The records from most areas, except Islay, are infrequent, and come mainly from the six comprehensive surveys undertaken between 1959 and 1983 (Table 242). In Ireland, and in Scotland outside Islay, the numbers nearly doubled during this period, while those on Islay increased sevenfold from 2,800 to 21,500, before dropping back to 14,000. Concurrently the proportion of the population wintering on Islay rose from 34% in 1959 to 64% in 1977, and now stands at 55%.

Although the area of Islay is about 50,000ha, the great majority of geese are concentrated in the areas around the two adjacent roosts on Lochs Gruinart and Indaal, and are further restricted to the areas of improved grassland, nearly all of which lie below the 60m contour (Fig 49 p.303). This intense concentration of the feeding flocks has led to increasing demands by the farmers concerned that the numbers be drastically reduced. In 1977 the shooting season on the island was extended from two to five months, and efforts have since been made to disperse the geese by this and other means. The shooting of geese out of season is also permitted under licence where the damage to crops is excessive.

The dramatic decrease in the Islay flocks which has stemmed from this added mortality is now giving cause for concern. From the latest complete census in 1983 it seems that the losses here have not been made good by increases in other parts of the winter range; elsewhere in Scotland the numbers have changed only slightly and those in Ireland have decreased (410). It is also apparent that the geese which winter on Islay have a much greater breeding success than those on the west coast of Ireland (103, 406). The welfare of the Greenland stock is thus to a large extent dependent on the future of the Islay flocks.

The establishment of an RSPB reserve at Gruinart in 1983 is a welcome development, since it safeguards a large portion of the feeding grounds of the birds on one of the two main roosts. A careful watch must now be kept on the local trends and pressures, and if possible compromise must be achieved with the farming interests. In particular the issue of licences for shooting out of season must be more strictly controlled, to stop the geese being shot for sport, in the guise of crop protection.

Twenty years ago the world population of Barnacle Geese was 40,000 (68) and 40% of these were in the British Isles. That proportion is similar today, but the welfare of the species, with a world stock of 80-90,000, is much improved.

Brent Goose
Branta bernicla

The Brent Goose has a Holarctic distribution and three races are generally recognised. Of the two European subspecies, *Branta b. bernicla* has a single population breeding in western Siberia and *hrota* three populations, one restricted to eastern North America, one migrating to Ireland from breeding grounds in Canada and Greenland and a small Spitsbergen group wintering in Denmark and north-east England. The Pacific Black Brant, *nigricans*, migrates from its Alaskan, Canadian and east Siberian breeding sites to winter largely in California and Mexico.

The most maritime of geese, the Brent is rarely seen further than a kilometre or two from the sea. Most of its feeding has always been done on open flats of muddy estuaries or in shallow bays. Because of coincidences in the distribution of geese and *Zostera* the Brent Goose was once thought to rely almost entirely on that plant for its food. That was indeed the case in many places but stomach analyses have shown that other species, especially the algae *Enteromorpha* and *Ulva* and saltmarsh grasses, are also taken regularly (110).

The widespread and much publicised die-off of *Zostera* in the 1920s and 1930s resulted in a decrease in Brent numbers, and changed their distribution and probably their feeding habits. Nowadays *Zostera* is still the preferred food, but by mid-winter the supplies are usually exhausted, and the geese feed increasingly on *Enteromorpha* or grazed saltmarsh (121). In the spring staging areas in the Netherlands, saltmarsh grasses and other herbs form the bulk of the diet (165). This has probably always been the case, since the lush new

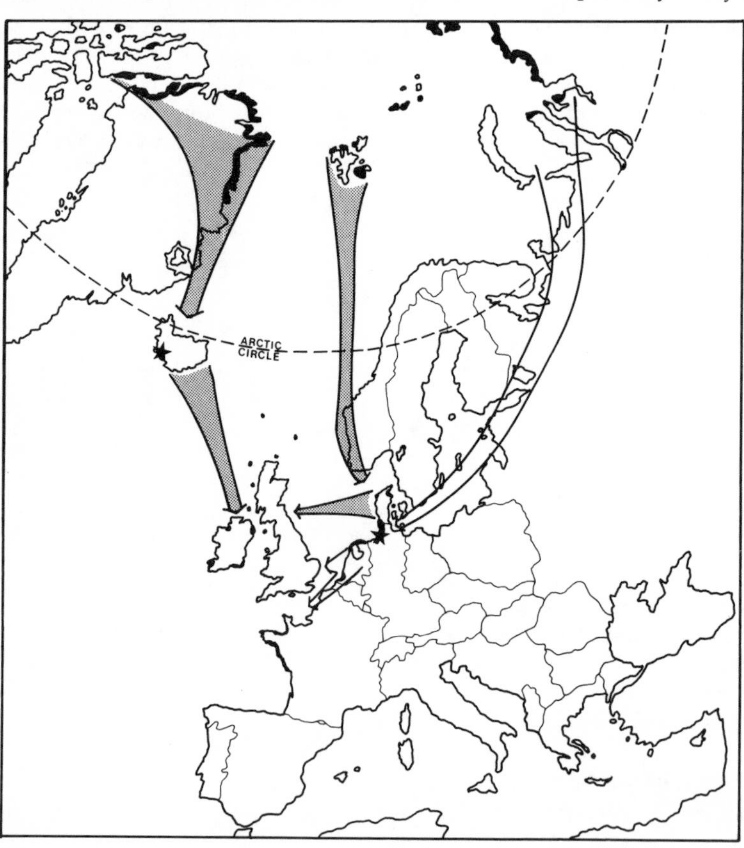

Fig. 87. Brent Goose – breeding and wintering range and migration routes. The stars indicate the major staging posts on migration. Shaded arrow – Light-bellied Brent. Open arrow – Dark-bellied Brent.

growth is produced before the *Zostera* regenerates (490).

Brent Geese are highly gregarious and since they became protected in Britain in 1954 they have become quite tame. They habituate to human activities through the winter and by spring move into small estuaries and creeks, sometimes very close to houses and harbours (456).

Brent Geese have probably always moved inland in search of food in small numbers in severe winters and Dark-bellied Brent have for many years used the Dutch "polder" grasslands in spring. In 1973-74, however, when the population was increasing rapidly, large numbers moved onto farmland without the stimulus of cold weather. The habit then became established, the numbers using inland pastures and winter wheat fields increasing to 14,500 in 1975-76 (520). By 1979-80 up to 40,000 were involved and the habit had spread to all the main resorts. Inland feeding has led to serious conflicts with farmers and there have been some reports of substantial damage (521, see also p.505).

Fig. 88. Total Dark-bellied Brent Geese counted in the whole population (solid line) and in Britain, 1956-1980. The numbers are the percentages of young birds in autumn flocks in Britain. From Ogilvie and Matthews (1969), Ogilvie and St Joseph (1976), St Joseph (1979a) and St Joseph (1981).

Dark-bellied Brent Goose *Branta bernicla bernicla*

The numbers and distribution of this goose have changed drastically in this century. Before the decline of *Zostera* in the 1930s large numbers used to winter in the Dutch Wadden Sea and Zuiderzee (now IJsselmeer). In Britain, Northumberland was a major stronghold and up to 30,000 (probably of mixed races) were counted near Holy Island in 1885-86 (26). The *Zostera* decline not only reduced numbers drastically but also caused major distributional shifts, with north-east Britain declining in importance in favour of south-east England and, presumably, France.

The present migration routes of the three populations of Brent Geese wintering in north-west Europe are shown in Fig 87, together with their distribution outside Great Britain. The Dark-bellied race breeds between 70° and 100° E on the seaboard of western Siberia, and probably on the Arctic islands of Severnaya Zemlya. The autumn migration begins in mid-August and the first geese are in the Baltic in mid-September and in the Netherlands and Britain later in the month. Numbers build up to a peak in November at the main gathering areas at Foulness and Leigh, Essex, from where they disperse to other parts of England and a few to France (519). The return migration begins in late February and early March and the geese spend April and May in the Dutch and

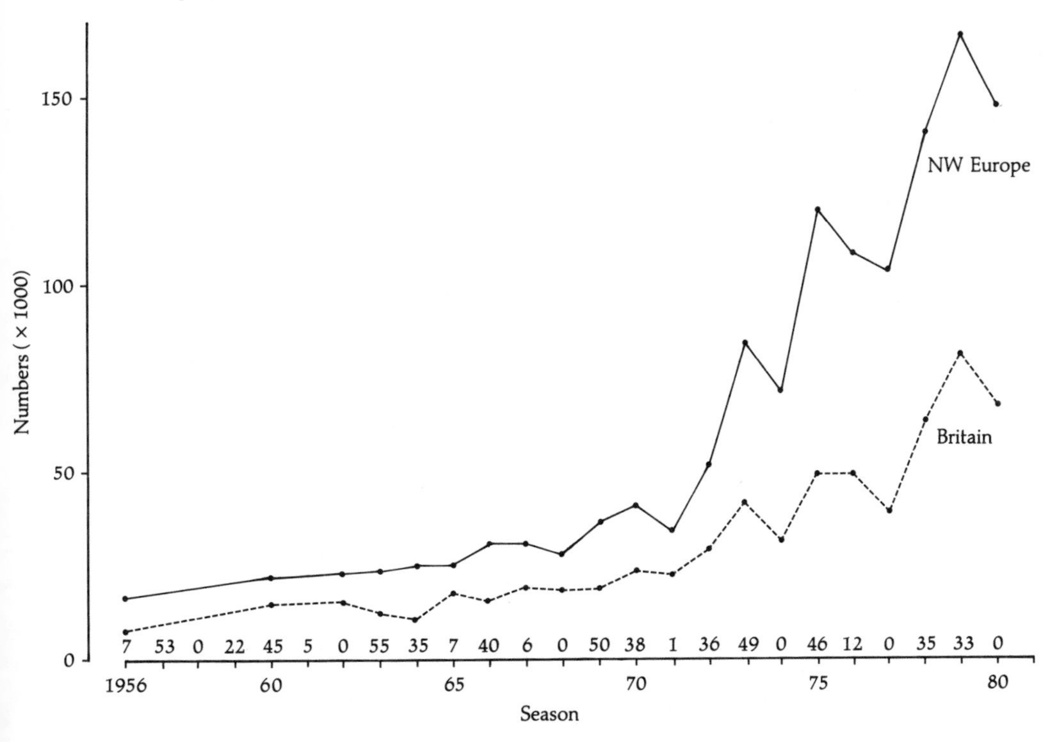

German Wadden Sea, returning to the breeding area in late May.

The first realistic estimate of population size was made in 1955-1957 (525) and yielded 16,500 birds. This was at most only a quarter of the numbers present at the turn of the century. Since then the population has recovered, at first gradually following protection in the Netherlands in 1950 and in Britain in 1954. Later it grew very sharply, after being protected in France in 1966 and Denmark in 1972 (Fig 88). The highest recorded total, of 166,000 in 1979-80, was 10 times as high as the 1955-1957 level. As the numbers increased sharply so did the percentage of the population which wintered outside Britain. In the 1960s the British peak only once amounted to less than half of the total, and in five years exceeded 60%; in the early 1970s the proportion was 55%, and in the late 1970s 45%. The main recipient areas were France and, in mild seasons, the Netherlands. The increasing trend has continued; there were 202,500 in 1982-83.

The distribution in Britain is shown in Fig 89; Table 243 lists the counts for the most important sites. As the population has increased in size so more geese have been found in the south and west and around the Wash. The Essex centres, although still holding the largest concentrations, have declined in relative importance, while Norfolk and the south coast harbours have increased in proportion. Whereas in the 1960s and early 1970s the latter two areas held 26% of the British population at peak, from 1976 onwards the proportion was 55-60%. There are only sporadic occurrences outside this very restricted range; the race was seen regularly but in very small numbers (average peak 7, max 18) in Northumberland in the 1970s.

The threat to establish the third London airport at Foulness in the early 1970s initiated a considerable amount of research into the habits and movements of Brent Geese, including a large marking programme. Among the important facts to be established were the nature of the autumn gathering places at Foulness and Leigh and the fact that British wintering geese spent the spring in the German part of the Wadden Sea, whereas those from France staged in the Netherlands. The geese were also shown to return not only to the same estuary but to the same part of it in subsequent years; geese caught together in winter are very likely to be seen together in the same area up to three years later. Predictably, return rates are higher to the final wintering areas than to the autumn gatherings at Foulness. Yearlings show a greater tendency to return than adults (519).

Dark-bellied Brent Geese, like other High Arctic breeders, show great fluctuations in breeding success from year to year and, as a consequence, wide fluctuations in numbers (Fig 88). In 5 of the 10 years 1960-1969 there were 30-50% young and in the other 5

Table 243. *Sites which have held an average peak of 1000 or more Dark-bellied Brent Geese in recent seasons and the months in which the maxima were recorded. Figures from Wildfowl Counts and* Norfolk Bird Reports. A blank = no count.

	1976-77	1977-78	1978-79	1979-80	1980-81	1981-82	1982-83	Mean	Month(s)
The Wash	7800	6420	9070	11390	17000	7000*	24500	11883	D/J
Scolt Head	1500	1800	2000	2000	2000	2000	4000	2214	J/F
Wells Harbour	1800	1000	2500	3500	2000		2000	2133	D/J
Cley/Salthouse Marshes	1800	2000	1500	1200	1000	1100	1800	1486	D/J
Hamford Water	6250	3310	4000	8200	4500	4000	8000	5466	J/F
Colne Est	2640	740	1000	790	1000	820	1450	1206	D-M
Blackwater Est	8100	7200	13300	12490	9170	9000	11500	10109	J/F
Dengie Flats	1800	2540	1070		1100	910	610	1338	N-J
Crouch & Roach Est		550	2400	4300	3120	3550	5050	3162	J/F
Foulness +	9700	9300	11100	15390	13760	14800	11400	12207	O/N
Leigh Marshes +	3600	4090	1100	5000	4500	5250	6800	4334	O/N
Pagham Harbour	1500	650	1500	2700	1500	1850	2960	1809	J
Chichester Harbour	n5700	5670	8140	9500	7090	8600	10550	7893	J/F
Langstone Harbour	6100	4970	5530	6420	7400	6200	7500	6303	J
Portsmouth Harbour	680	1460	1560	2450	1480	3300	860	1684	J
Exe Est	900	920	1580	2400	1950	1700	1400	1550	N-F
British total (January)	49000	38900	63200	81000	67400	53700	92600	63686	

* February count – no complete count in January when numbers may well have been higher.
+ Much interchange and overlap between these two sites.

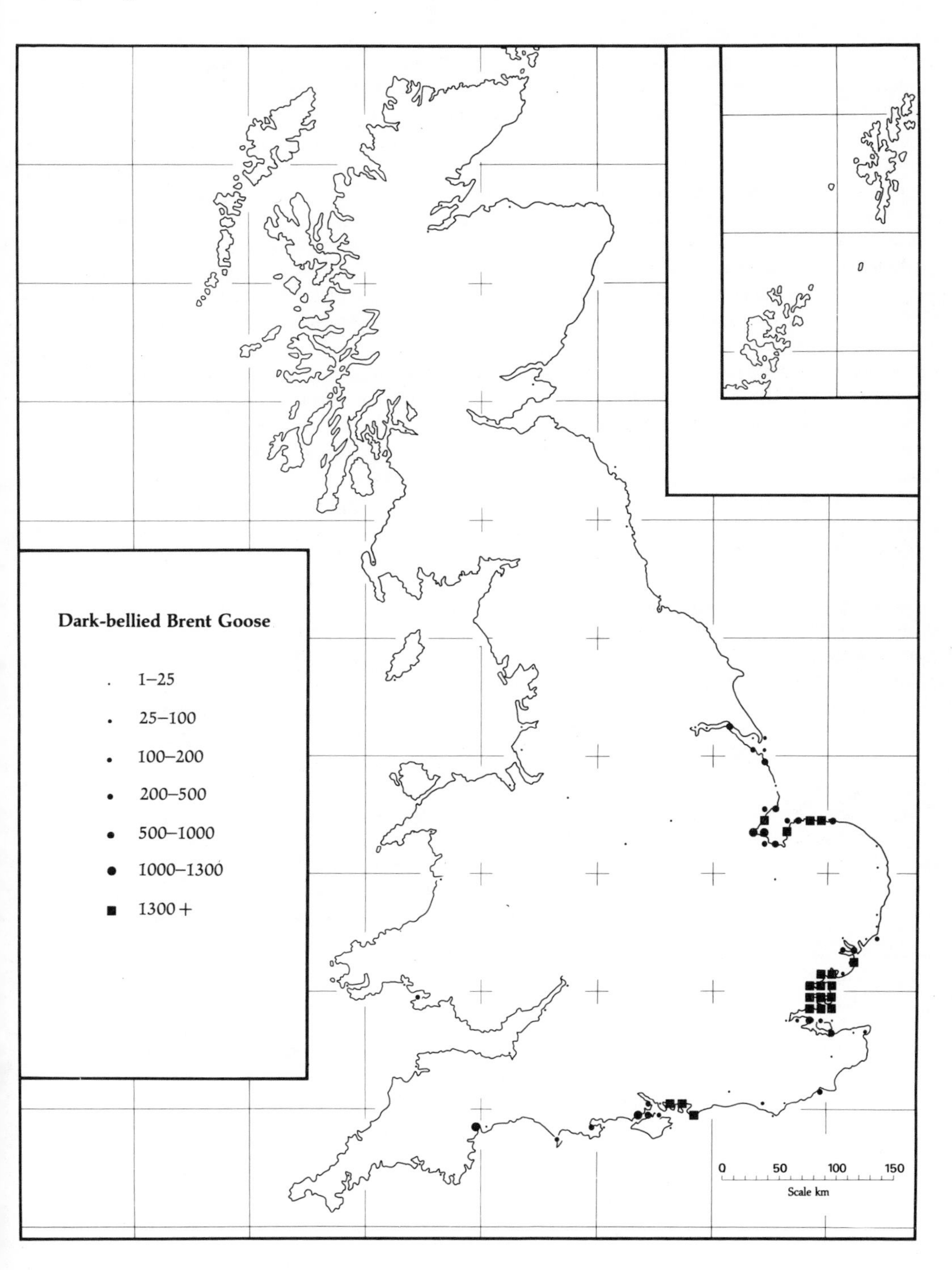

Fig. 89. Dark-bellied Brent Goose – British distribution.

less than 7%. In the 1970s 6 of the 10 seasons were "good", 3 poor and one intermediate. This high level of success, together with a lowering of mortality since the 1950s, is responsible for the rapid population growth (418, 521).

The fact that inland feeding has increased with the growth in population suggests that, at least in their present range, Brent Geese have exceeded the capacity of the estuarine habitat. Farmers in Britain have called for an alteration in their protected status, but an international group of scientists and administrators, meeting in Paris in 1977, decided that despite the recent increase the stocks were insufficient to permit the opening of a hunting season. Since then the problem has been eased by the Wildlife and Countryside Act (1981) which contains provisions for the shooting of Brent, under licence, in areas where damage occurs.

The key to the continued success of the Brent Goose is the protection of its estuarine habitat, but the threats are undiminished and many areas of the North Sea foreshore have been embanked in recent decades. A few well-managed reserves, strategically sited either on saltmarsh or on inland pasture, would help to prevent substantial agricultural damage and ensure that inland habitats continue to provide a "safety valve" for the geese (521).

Light-bellied Brent Goose *Branta bernicla hrota*

The population breeding in eastern Canada and Greenland winters almost exclusively in Ireland (Fig 87) but is regularly seen on migration on the west coast of Scotland, especially in autumn. Numbers normally fluctuate from 8-16,000 and the most recent census in January 1978 yielded 9,160. However, most, if not all, of the birds congregate at Strangford Lough, Co Down, in November and December, and numbers there averaged 12,300 between 1976-77 and 1981-82 (maximum 14,400 in November 1981).

The other, much smaller, population breeds in Spitsbergen and probably in small numbers in Franz Josef Land. It migrates along the Norwegian coast to Denmark, where many birds spend the winter. The rest cross the North Sea to their only remaining British stronghold on the NNR at Lindisfarne, Northumberland. In the early part of the century this population suffered a very serious decline, not only because of the *Zostera* die-off but also because of excessive shooting and disturbance in Britain and human interference in Spitsbergen. Up to 10,000 geese, nearly all Light-bellied, used to winter in the Moray Firth and the British total was probably at least 20,000. Up to 4,000 were still present in the Moray Firth in the 1930s but soon afterwards the area was deserted; the sporadic

Table 244. *Peak numbers of Brent Geese at Lindisfarne NNR, 1930-31 to 1982-83. Before 1960 most of the regular visitors were thought to be of the Dark-bellied race whereas the hard-weather influxes were Light-bellied. In the 1960s the vast majority were Light-bellied, as were all those listed for the 1970s, when the two races were distinguished. The month in which the maximum occurred is also given and severe winters are marked with an asterisk. A blank indicates no count.*

Winter	Maximum	Month	Winter	Maximum	Month	Winter	Maximum	Month
1930	30		1948	32	M	1966	500	J
1931	20		1949	12	J	1967	900	F
1932			1950	40		1968	1000	J
1933	500		1951	4	J	1969	1500	J
1934	500		1952	500	F	1970	650	J
1935	100		1953*	3500	M	1971	700	F
1936	400		1954*	2500	M	1972	400	J
1937	very few		1955*	2500	F	1973	1710	D
1938*	120		1956	1000		1974	300	D
1939*	5		1957	2000	F	1975	550	J
1940			1958	250		1976	780	D
1941*	2000		1959	810	F	1977	1000	D
1942			1960	2000		1978*	2170	J
1943			1961	420	F	1979	1540	J
1944			1962*	800	F	1980	700	J
1945			1963	450	J	1981*	1800	J
1946*	5000		1964	1000	J	1982	600	D
1947	52	J	1965	2760	F			

occurrences now are usually Dark-bellied birds. The flocks at Lindisfarne have, likewise, shown a massive decline. By the early 1950s the numbers were reduced to about 4,000 (525), and in 1966-67 to 2,500-3,000 (391). There is now a regular wintering flock of several hundred, reinforced on occasion by cold-weather influxes (Table 244). The birds are also arriving earlier, with peaks in December and January rather than in January and February. By the early 1970s the total Spitsbergen stock was down to about 2,000 and the count of 2,170 at Lindisfarne in January 1979 probably represented the entire population. In most other recent seasons the area has held about 40% of the stock, and is thus of prime importance.

Despite total protection throughout their range, the Spitsbergen Brent Geese are only increasing very slowly (in the early 1980s the numbers may have reached 3,000), and are probably one of the most vulnerable goose populations in the world. There have been suggestions that the expanding Barnacle population, which now nests densely on many of the Brent's former breeding islands, is contributing to the decline. Whatever the reasons, it seems likely that the group will do little more than maintain its numbers in the foreseeable future.

Common Shelduck
Tadorna tadorna

An almost exclusively coastal species, the Shelduck's world distribution is restricted to north-west Europe, parts of the Mediterranean and the Middle East. It used to be thought that there was no overlap between the populations and that the north-west European group could be considered discrete and closed (22). However, recent work has shown that Shelducks from the west Mediterranean migrate north in summer, probably to the moulting grounds in the West German Wadden Sea, where they mingle with moulting birds from the northerly population (608). Since pair formation probably does not take place during the moult the populations may still be relatively closed.

The majority of the north European birds are found in estuaries with extensive areas of mud or sandbanks exposed at low tide. Studies in the early 1960s showed that they rely for their winter food on the small snail *Hydrobia ulvae*, which lives near the surface of intertidal mud. All 46 of the stomachs examined contained this mollusc, which accounted for about 90% of the food present (428). Later detailed work on the Firth of Forth (94) and the Ythan Estuary (101) supported this conclusion. On the Ythan Estuary the adult Shelducks feed on *Hydrobia* in summer as well as in winter. On the Firth of Clyde, however, the birds show no preference for the snail where other foods are present. *Tubifex* worms, which occur there abundantly in the substrate, are, in fact, preferred to *Hydrobia*, and snails below 2mm in size are avoided (590).

Shelducks most commonly feed above the tide-line, either by moving the head from side to side with the bill just under the surface of wet mud (scything) or by dabbling in very shallow water or wet mud. They also feed while swimming, either by immersing the head and neck (head-dipping) or by upending in deeper water. The feeding rhythm is closely related to the tidal cycle, though the pattern varies from place to place. In the Firth of Forth feeding is much more intensive on a flowing than on an ebbing tide (94) whereas in the Clyde the pattern is similar in both situations, with very low feeding intensity at high and low tide (589). The birds move with the tide edge, taking advantage of the wet mud and the movement of prey into the upper layers. Some may also feed on the exposed mudflats at low tide, or on saltmarsh at high water.

The British breeding stock apparently accounts for the majority of our wintering birds and makes a substantial contribution to the flyway population. Shelducks nest in most coastal areas where there is suitable intertidal feeding, but the size of the British breeding stock has not been accurately determined. On the basis of evidence that rather less than half the Shelducks present in spring breed, the estimate of 50,000 birds in the British Isles between January and July has been used to estimate 12,000 breeding pairs, or 10,000 in Great Britain (621). Despite a rather large increase in numbers in the flyway and in Britain since the late 1960s, our estimate of mean numbers in March is 42,000, or about 10,000 pairs. This suggests that earlier estimates, from a smaller population, were too high but that the British stock may now be at those levels.

With increases in numbers there has also been an expansion in the breeding range, westwards and northwards from the centre around the southern North Sea, and in Britain increasing numbers are nesting inland, especially in the Fens. There have been regular breeding records from Warwickshire, more than 100km from the coast (542).

Shelducks nest in holes, usually rabbit burrows or other cavities in dunes and banks, but often in haystacks, buildings and a wide variety of other situations, usually under cover. Nests are often close together and where the density is high several females may lay in the same nest (271). The male retains a feeding territory which is used by the pair before and during incubation. Soon after hatching the young are taken to the estuary or, in inland sites, to the nearest standing water, where they feed on small invertebrates, chiefly *Nereis* worms, *Hydrobia* and crustaceans (101).

The broods remain discrete for the first week or two but later merge to form creches, sometimes containing up to 100 young. These are attended by a few adults, usually including the parents of some of the ducklings (141). Most non-breeding birds and parents leave the breeding area to moult.

The Shelduck's moult migration is remarkable both because breeding birds are among the migrants and because they gather in such massive concentrations. In mid-July Shelducks from all parts of north-west Europe move to the Heligoland Bight in the German Wadden Sea, where 100,000 or more gather, this being the vast majority of the flyway stock (218). The birds stay there at least until September and many do not return to their winter quarters until November or December. The conclusion, largely based on the seasonal pattern of numbers, that the return migration extended into January may have been mistaken (see below).

A much smaller moulting area was discovered at Bridgwater Bay, Somerset, in the early 1950s, and about 2,700 Shelducks were moulting there in 1959. Since these did not consist of local ducks, it was suggested that they originated from Ireland (174). Aerial and ground counts in 1976 confirmed that the moulting flock was substantially reduced. Although there was a peak of 2,000 on three occasions, the numbers were not above 800 for more than a very few days (404).

More recently moulting Shelducks have been found on the Firth of Forth, where 600-800 moulted in 1976 (91) and where summer counts have shown increases to 2,500-3,000 in 1979 and 1980 (540). Most of these probably originate from the Forth, with a few from elsewhere in Scotland (473). Another group, 1,000-1,500 strong, now moults in the Wash (92, 93); a few hundred at least occur in the Humber (573), and a further small flock of flightless birds has recently been found in the estuary of the Cheshire Dee. Clearly if suitable moulting sites exist in Britain it is advantageous for the birds to use them rather than make the journey to Germany. More groups probably exist and are likely to increase in importance as the habit spreads.

Figure 90 shows the pattern of Shelduck numbers in Britain on average and in a mild and a hard

Fig. 90. Shelduck – monthly pattern of numbers in Britain, expressed as a percentage of the peak count. Solid line – mean 1976-77 to 1980-81, dotted line – 1978-79 (relatively hard winter), dashed line – 1977-78 (mild winter).

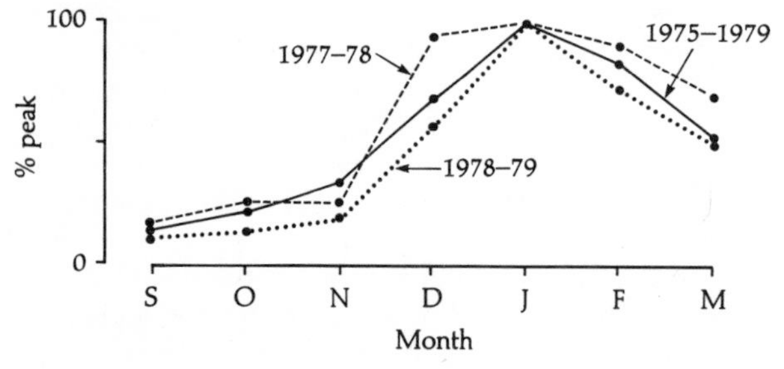

winter. The return of the moulting birds is gradual, but mainly takes place in November. On average during the late 1970s, 30,000 Shelducks were counted in March and 38,000 in December. The March count represents the British breeding stock, and since breeding birds do not disperse to their territories until April, there is no reason to suppose that the proportion counted is lower than in mid-winter. The December count, therefore, is at least equivalent to the British total and the increase to the January peak represents continental birds moving here to winter. On average this represents about a quarter of our

Fig. 91. Shelduck – British distribution.

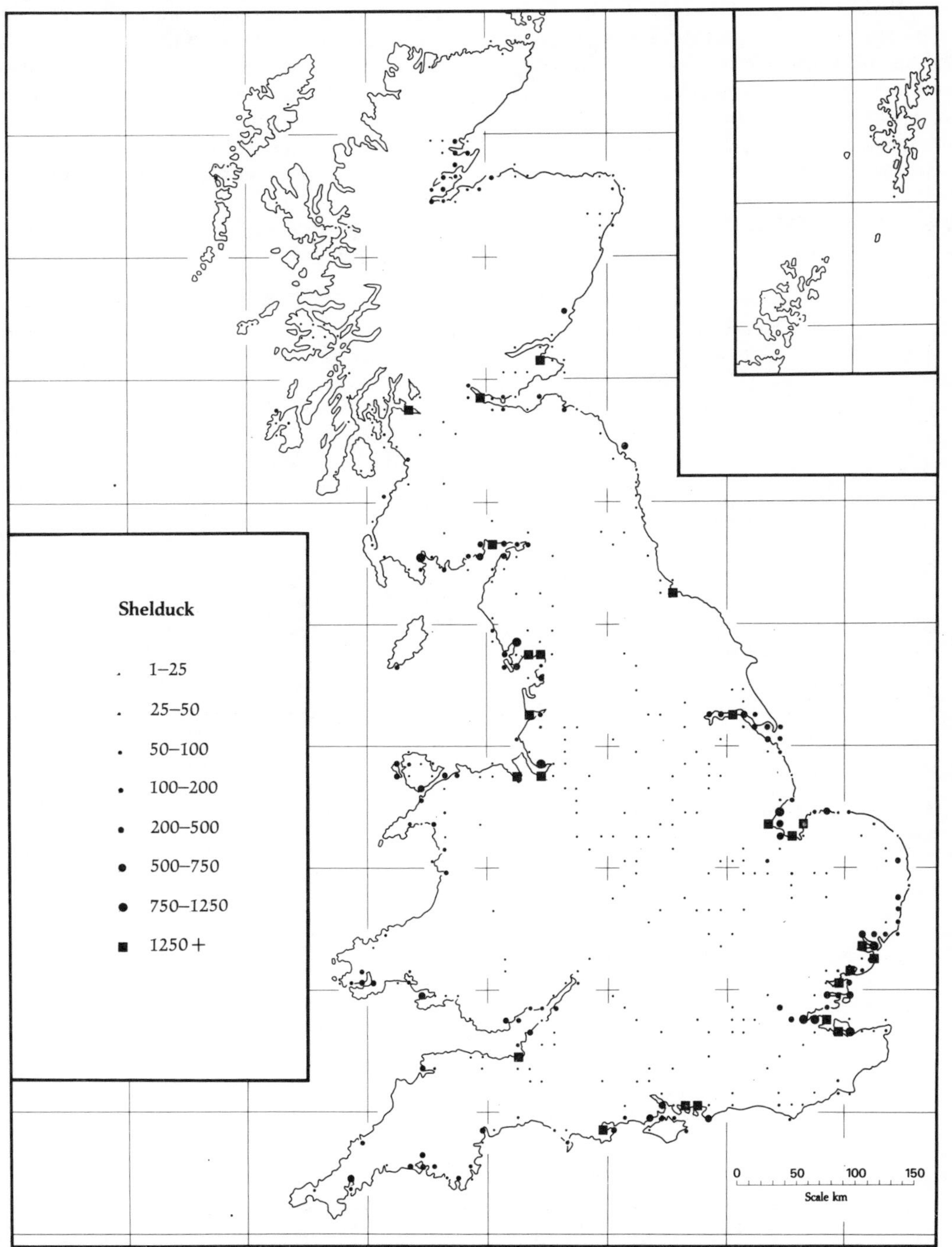

January total, or some 15-20,000 birds, rather more than has previously been thought.

The pattern clearly varies from year to year depending on the weather. In the very mild winter of 1977-78 only a few thousand Shelducks moved to Britain, whereas the hard-weather influx of 1978-79 almost doubled our total, from 38,000 in December to 68,000 in January.

The total flyway population is put at around 125,000 (22, 529). Outside Britain the main January concentrations are in the Netherlands, Denmark and France. There are very few in Iberia or in the northern Mediterranean and Adriatic; the southerly population, numbering 75,000, has its headquarters in the eastern Mediterranean and spreads westwards to Algeria and eastwards to Afghanistan and Pakistan (on inland sites). Some of the Mediterranean birds, at least from southern France, migrate to moult in the German Wadden Sea (608).

The vital importance of Britain for wintering Shelduck is apparent from Fig 91, which identifies no fewer than 25 internationally important concentrations. Most sites have peaks in December and January but in Bridgwater Bay and on the Cheshire Dee Estuary the highest counts are usually in September. The six sites regularly holding more than 2,000 Shelducks at peak are listed in Table 245.

Most sites showed an upsurge in numbers in the

Table 245. *Numbers of Shelducks on sites regularly holding 2000 or more. 5-year means 1960-1974 and seasonal maxima thereafter. The qualifying level for international importance is 1250, but all qualifying sites are too numerous to list (see Fig 91). A blank = no data.*

	Chichester Harbour	Medway Estuary	Hamford Water	The Wash	Dee Estuary	Mersey Estuary
1960-64	1215*	1255+	670			136
1965-69	2938	2140	1080		1798	422
1970-74	3092	2416	1163+	8826	3280	1640
1975		2700	2810	9690	2300	4260
1976	1500	2150	2860	12500	2530	4020
1977	1430		1230	9750	1090	3840
1978	3130	5150	1200	11750		7080
1979	2010	2150	1620	20050	4420	7380
1980	2750	3100	1200	19010	7320	11820
1981	4550	2500	5500	17230	3740	12170
1982	2260	1530	860	16950	5000	7110
Mean 1975-82	2519	2754	2160	14616	3771	7210

* 2 years only, 1960 and 1962.
+ 4 years only.

Fig. 92. Shelduck – trend in January numbers in Britain, 1960-1982.

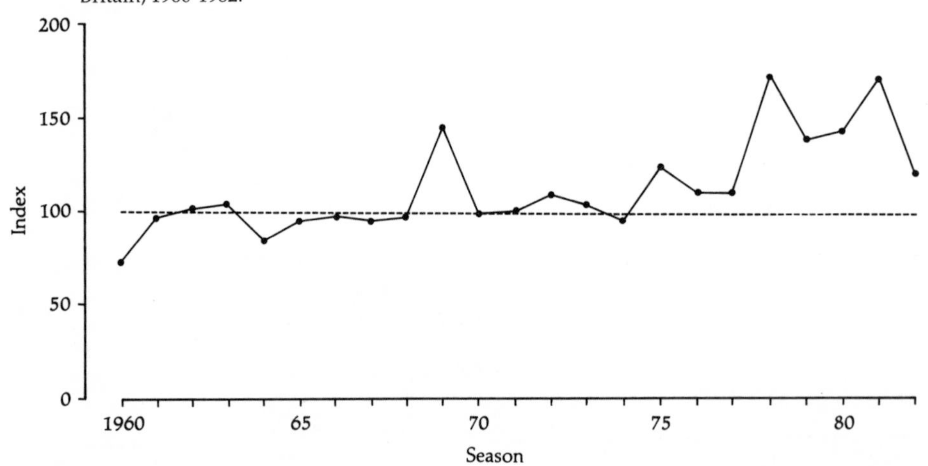

ate 1970s, with the Mersey and the Dee having shown pectacular increases. Few sites have declined in mportance, but the numbers at Morecambe Bay have pparently dropped from a maximum of 6,000 to ,000-1,500, though coverage is incomplete in most ears. At Teesmouth, Cleveland, most of the mudflats ave been reclaimed for industry, but Shelduck umbers fell only temporarily, from 3-4,000 to 1,500-,500, and have now regained their previous level p.479). Continued eutrophication of Langstone Harour has caused a spread of the alga *Entermorpha* over he mudflats to the detriment of the Shelduck's nvertebrate foods and the numbers have dropped rom 2-4,000 to 1,000-1,500 (p.488).

As might be expected from Table 245 the trend n Shelduck numbers in Britain showed a healthy ncrease in the late 1970s (Fig 92). The effect of the elatively cold winters of 1969-70 and 1978-79 is clearly shown, but even discounting these influxes the numbers have increased substantially over the rather stable levels of the 1960s and early 1970s. The reason for this is unclear, but Shelducks have undoubtedly benefited from the long run of rather mild winters from 1962-63 to 1980-81.

The trend conceals important regional differences; the index for Scotland has declined considerably, suggesting that numbers in the late 1970s were only 60-70% of those in the early 1960s. Unfortunately nadequate counts in such important areas as the Solway and Clyde make it difficult to make detailed comparisons. The changes in Scotland have been counterbalanced by large increases in England and Wales, particularly, and not surprisingly in view of the Dee and Mersey increases, in the north-west, where the numbers have more than doubled since the 1960s.

Comparison of the mean January count in the late 1970s (52,800) with the 5-year mean for all sites (76,000) suggests that 70% of Shelduck are counted each year. Since most sections of coastline, apart from those in north-west Scotland, where numbers are in any case very low, have been covered at one time or another, it seems unlikely that many Shelducks are missed completely. Thus the average numbers in Britain are between 75,000 and 80,000. The trend data ndicate that peaks have ranged from 42,000 in 1960-61 to 100,000 in January 1979. The Shelduck is one of the species most vulnerable to severe winters, so decreases might be expected in such circumstances, especially with the present very high population. Despite this, and in the absence of major developments on key wintering areas, the Shelduck population seems likely to remain in a healthy state for the foreseeable future.

Mandarin Duck
Aix galericulata

A native of far eastern Asia, the Mandarin Duck was introduced to Europe in the 19th century for its obvious ornamental qualities. The species is a member of the perching duck tribe Cairinini, of which we have no naturally occurring members, and its native habitat is temperate deciduous forests with sheltered ponds and streams. It is able to manoeuvre in enclosed situations and is well able to perch in trees. Its diet, gathered from the margins of ponds, consists largely of seeds and tree fruit, particularly acorns and beech mast, and some animal material. In summer it is more carnivorous, eating emerging insects on the water surface and from marginal plants (527).

Mandarins were breeding freely in captivity in Britain by the early 20th century and some were released into the wild in Surrey and Berkshire. The birds bred in a habitat similar to their native range and a viable population was soon established. The Mandarin nests chiefly in tree holes up to 10m above ground and also readily takes to elevated nest-boxes.

The British distribution (Fig 93) is still centred on the original release sites in Surrey and Berkshire, with scattered occurrences elsewhere. These are known to have occurred from escapes of local birds in collections or deliberate releases rather than a natural spread of the feral stock, which moves only very locally.

In 1952 the number in Britain was put at over 500 individuals (527) and the First Edition reported little change in the late 1950s. The Atlas survey suggested a total of 300-400 pairs based on known densities in some areas (542), which would be expected to result in a post-breeding population of 1,000-1,500. Unfortunately, because of their habitat and rather dispersed distribution, it is unusual for more than 100 birds to be counted in winter. There was, however, an exceptional concentration of 300 in Busbridge Lakes, Surrey, in March 1981.

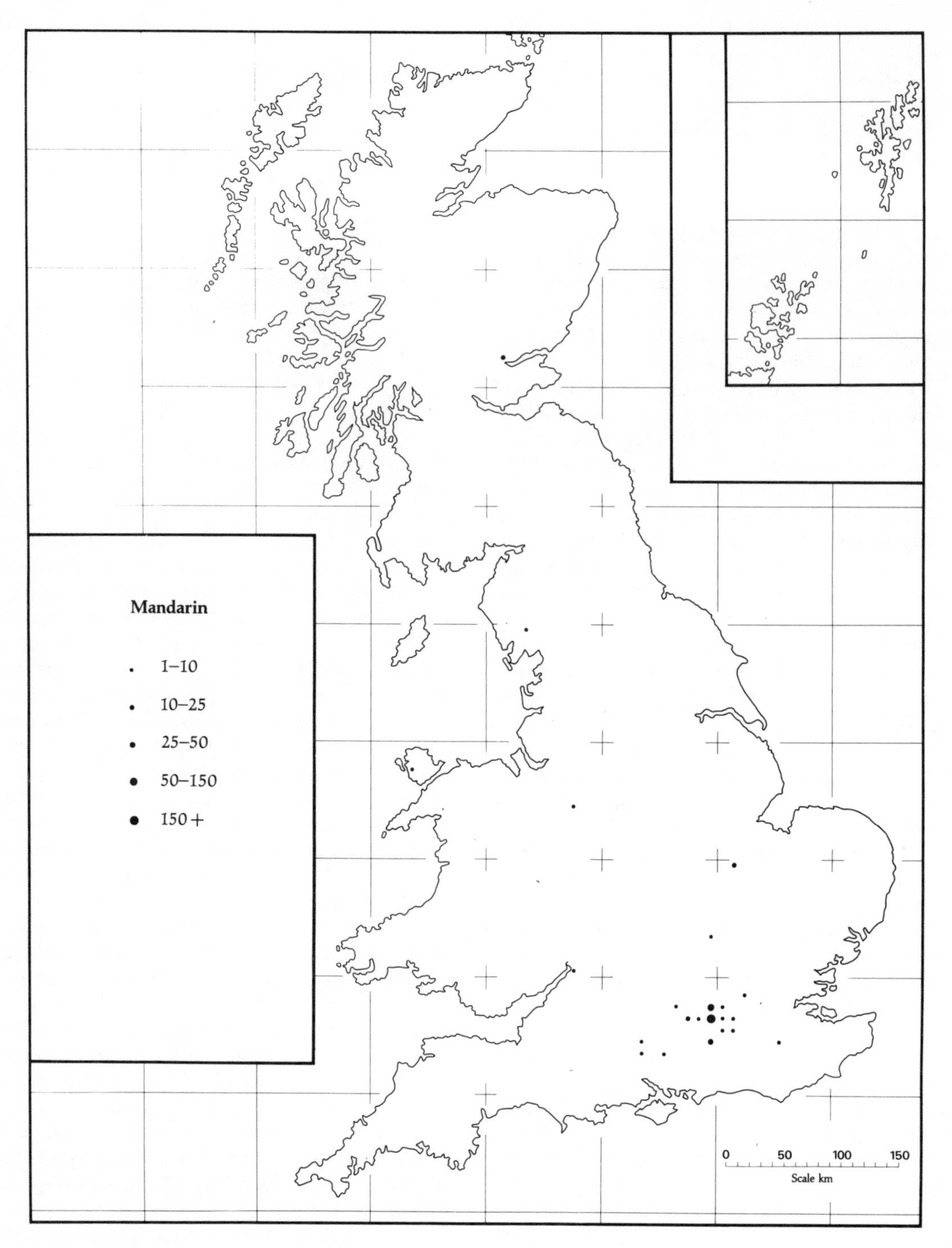

Fig. 93. Mandarin Duck – British distribution.

Wigeon
Anas penelope

One of our most numerous ducks, the Wigeon used to be very largely restricted to the coast, so much so that there was widespread concern over its future in the 1930s when stocks of *Zostera*, its main food on the mudflats, declined drastically following the wasting disease (50). The extent of the dependence on *Zostera* had, however, been overestimated, as studies of feeding habits indicated much more flexibility (213). The apparent decline in numbers in the 1930s was not maintained and the species continued to thrive and to move into new habitats.

A survey in 1975 showed that although 80% of Wigeon roosted on the coast nearly half of their food in Britain was gathered from freshwater habitats and only a third from their traditional mudflats (454). Because of increased development and reclamation on estuaries and the creation of new roosts inland on reservoirs and gravel pits, this trend is likely to continue, as it has in the Netherlands following the vast reclamation schemes carried out in recent decades.

Wigeon feed in tight, fast-moving flocks on grassland, pecking and walking very rapidly. Whereas their activity pattern used to be largely governed by the tide, they nowadays feed both by day and night in undisturbed areas and mainly at night where disturbed during the day. In some places they feed on mudflats at low tide and move to saltmarshes or inland when the water level is too deep. They are often in association with Brent Geese or swans, eating the leaves brought up and discarded by those wasteful feeders.

Apart from *Zostera* and the alga *Enteromorpha* the main coastal foods of Wigeon are fine grasses, particularly the saltmarsh grasses *Puccinellia maritima* and *Festuca rubra* (433). At the Ouse Washes, East Anglia, they prefer fine grasses to the coarser *Glyceria maxima* and also eat seeds and roots or stolons (especially stolons of White Clover) when available. They also fly out with Mallard to fen fields to gather stubble grain in autumn and to graze sprouting winter wheat in late winter and spring. Animal material is eaten only extremely rarely (453).

British wintering Wigeon originate from a very wide area, as shown by the recovery distribution (Fig 94), which indicates a similar pattern to that described in an early analysis (160). All Icelandic breeders, except the very few recovered in the USA and Canada, winter in north-west Europe, chiefly in the British Isles. The majority of recoveries are from Scotland and Ireland, but some Icelandic birds have been found in continental Europe and the USSR in later years, probably as a result of abmigration. Icelandic Wigeon are joined here and outnumbered by those breeding in Scandinavia and especially the USSR. There is a redistribution within Britain, presumably of Icelandic birds north – south and of continental ones east – west, during the winter (454).

Britain has a small breeding population, estimated in the late 1960s at some 350 pairs (621) and in the early 1970s at 300-500 pairs (542). The main breeding areas are in the north Pennines, the uplands of east and central Scotland and in the far north, although the largest concentration is in the lowlands, at Loch Leven, Tayside, where between 25 and 30 pairs breed annually (380). Wigeon nest there, together with several hundred ducks of other species, on St Serf's Island, which provides good cover. Brood rearing areas are not abundant, however, and rearing success is thought to be low. There are a few breeding records from the lowlands of England but few pairs are involved and breeding is rather sporadic. Although the number of Wigeon breeding in Britain has increased considerably this century they continue to make a neglible contribution to our wintering stocks.

The north-west European total is put at about half a million birds (22, 529), perhaps 100,000 originating from Scandinavia, a few tens of thousands from Iceland and the remainder from the north-western parts of the USSR. Another European group, wintering in the Mediterranean and around the Black Sea, is not entirely isolated from the north-west European one and similarly amounts to some half a million Wigeon.

As Fig 95 shows, British Wigeon are still largely concentrated on the coast, although one of the two most important sites – the Ouse Washes – is inland. A gregarious species, over a third of the Wigeon counted are on only 26 sites and 82% on 300, although the species has been recorded on over 2,000 (51%) of the waters counted. There are nine British sites which are internationally important for Wigeon and counts from these are given in Table 246.

It is no accident that many of these sites include reserves. There is no shortage of suitable habitat for Wigeon either on the coast or inland but their potential is not often realised while there is still disturbance, especially from shooting. A very good example is that of the Ouse Washes, always a good Wigeon site but one which did not hold large numbers regularly unt reserves were created and managed for the bird Before the late 1960s the number of Wigeon w largely dependent on the presence of floodwater an on shooting pressure. The creation of refuges reduce shooting, and some floodwater was provided b conservation bodies so that the site could be used at a times. The number of Wigeon using the Washe

Table 246. *Counts from the nine internationally important sites for Wigeon in Britain, 1977-1982. A blank = no data.*

	1977-78	1978-79	1979-80	1980-81	1981-82	1982-83	Mean
Cromarty Firth	2950*	6972	7200	10812	15002	9380	9873
Dornoch Firth	6724	4637	5416	3516	4026	8275	5432
Lindisfarne	21150	29500	22000	30000	25410	41000	28177
Ribble Estuary	1550*	4480	6640	6380	7242	13823	7713
Mersey Estuary	4060	3470	8040	15200	10800	9050	8437
Humber	5487	9569	7929	6340	5202	2692	6203
Ouse Washes	26532	17684	19340	26737	39368	28073	26289
Abberton Reservoir	3550	11574	4475	5725	5000	4070	5732
Elmley, Swale		5400	7000	5092	18500	14000	9998

* Incomplete counts, not included in the average.

Fig. 94. The origins of British wintering Wigeon as indicated by ringing recoveries. Triangles – May, circles – June, diamonds – July, squares – August.

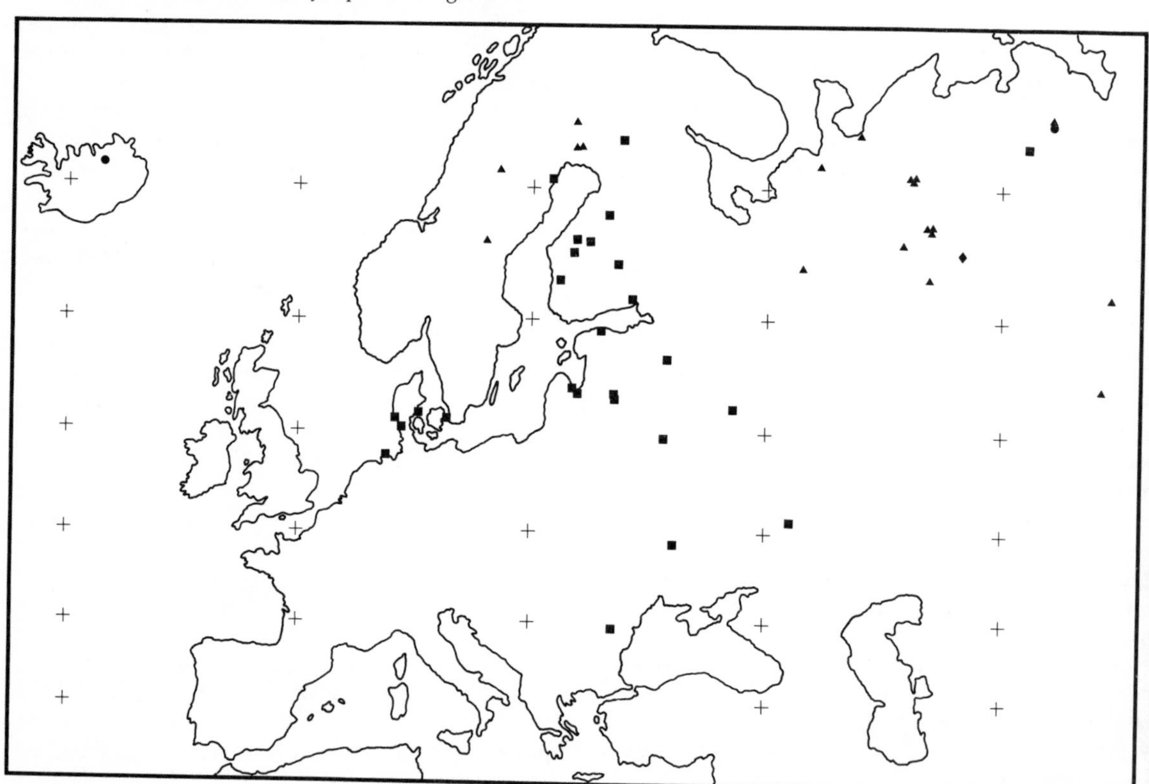

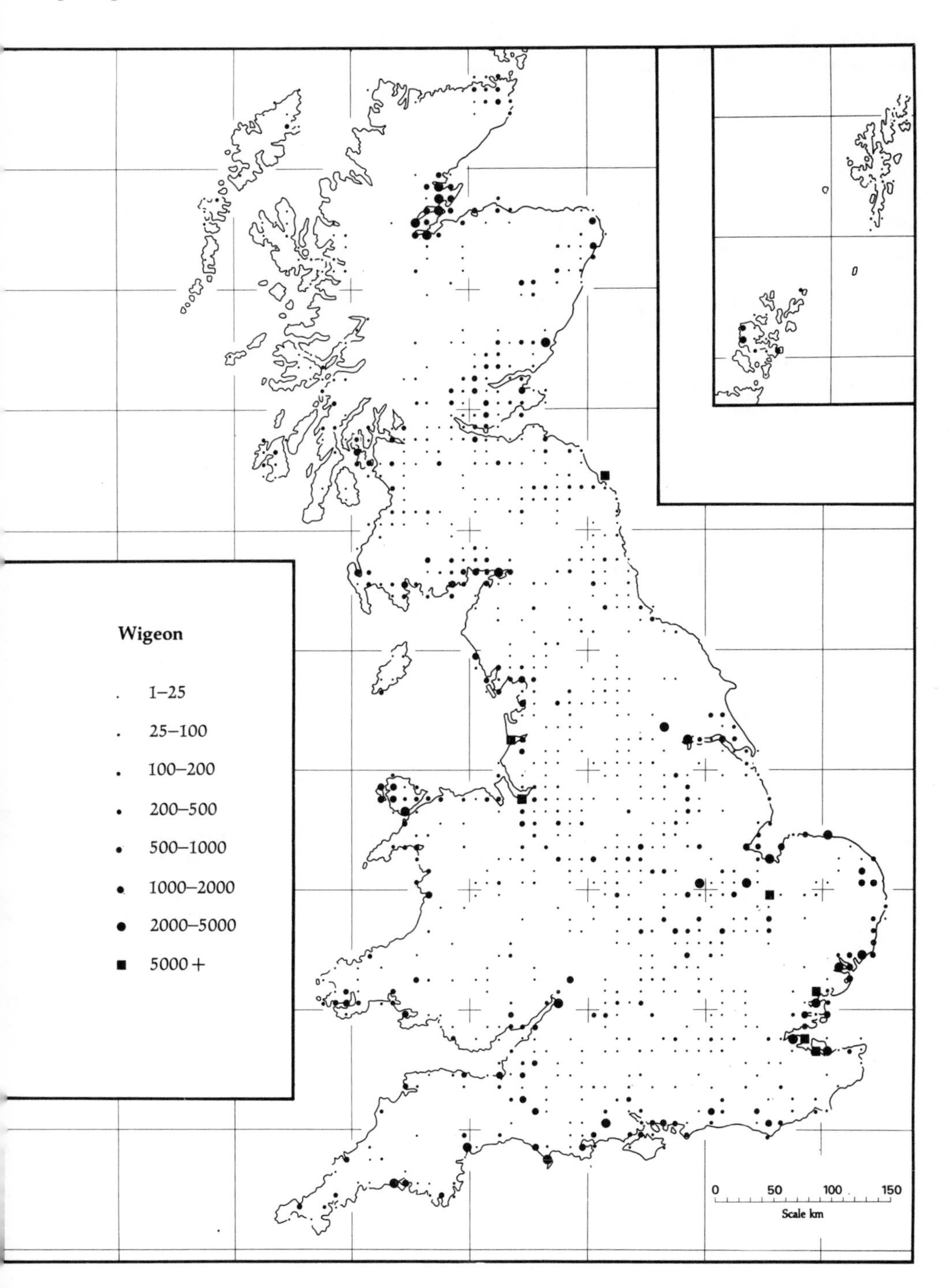

Fig. 95. Wigeon – British distribution.

increased dramatically, from an average of under 10,000 to 35,000 in the early 1970s. The wintering season also lengthened and high numbers were recorded much earlier and later in the winter. The usage of the site in terms of "bird-days" consequently increased more than fivefold (453). Numbers have declined slightly since the early 1970s but the Ouse Washes is still the first or second most important British site, depending largely on weather and flood conditions.

The maximum numbers of Wigeon are counted in Britain almost invariably in January (Fig 96), the average count for recent years being 161,000. The trend in January numbers since 1960 is shown in Fig 97. This shows an increase in the late 1960s followed by a decrease, though with a marked peak in the cold winters of 1978-79 and 1981-82. During the period of mild winters the trend in numbers in Britain probably reflects a real increase in the north-west European population. The cold-weather peaks here were, however, probably an indication of distributional shift

Fig. 96. Wigeon – monthly pattern of numbers in Britain, expressed as a percentage of the peak count.

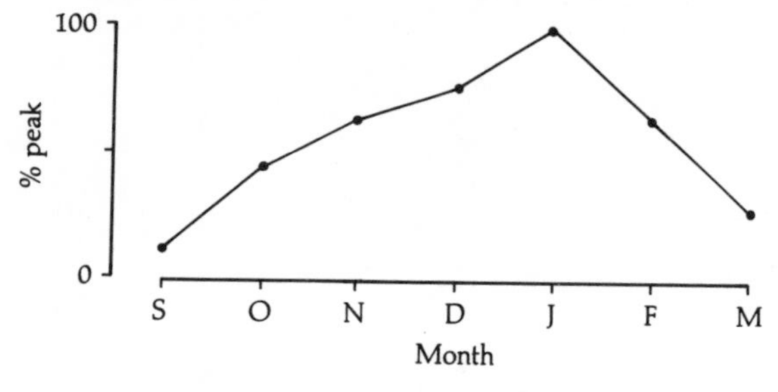

Fig. 97. Wigeon – trend in January numbers in Britain, 1960-1982.

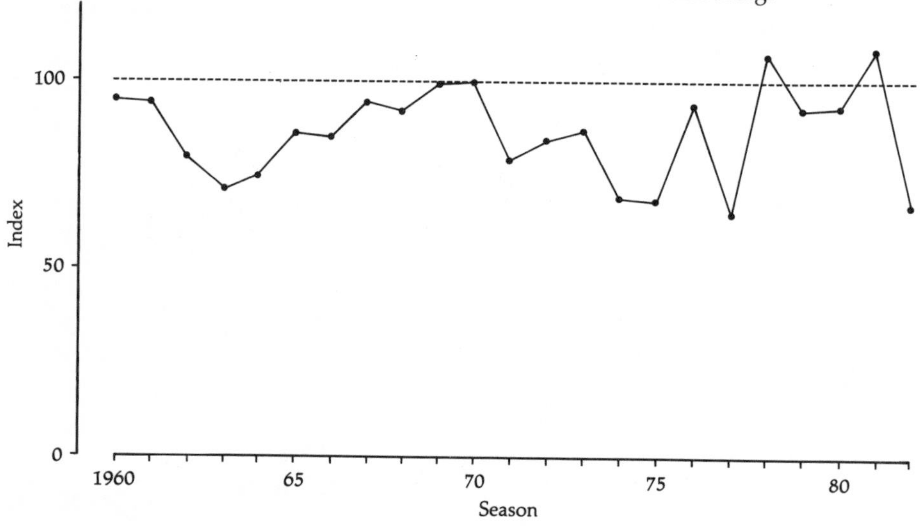

within northern Europe. If we look at only those site with complete coverage in 1978-79, for example, th numbers of thousands counted were: December 76 January 144 and February 100, compared with the mil season 1980-81: December 133, January 143 an February 136.

The probable size of the British maximum wa put in the First Edition at 250,000 and the sam approximate figure has been given subsequently (22 529). The five-year average for January given above i much less than this, as is the maximum eve counted – 209,800 in January 1982. Wigeon wer recorded on 2,022 of 3,973 sites counted over the 2 years 1960-1982, but 98% of the birds were on 84 sites. This indicates that, since an additional 2,800 site yield only 4,000 birds, it is highly unlikely that larg numbers are missed in the counts. Taking the averag count for the last five years covered at each site (th figure used for mapping), the January mean comes t 211,500. Although this figure is subject to errors it i unbiased and probably gives a reasonable estimate o average peak numbers. This means that, on average 76% of Wigeon are counted each January. Applying the total figure to the trend data we can say that the January population of Wigeon between 1960 and 1982 probably ranged between 168,000 (1981-82) and 265,000 (1978-79), with an average of about 200,000. This suggests that previous estimates were closer to absolute maxima than to averages or regular peaks.

While Wigeon numbers may have decline since the early part of this century they have held thei own in recent years; their spread into the newly available inland habitats and adaptability of feeding habits mean that they are under little threat at present The fact that most of their major resorts are reserves i also reassuring.

Gadwall
Anas strepera

The Gadwall is a southerly species, rarely breeding further north than 60° N and wintering mainly between 20° and 40°N in the Mediterranean, east Africa and Asia.

A freshwater species, 80% of Gadwall counted in Britain are inland. Like Wigeon, Gadwall are predominantly vegetarian in winter, inhabiting shallow lakes and floods with abundant submerged vegetation. In Germany more than 80% of the diet consists of floating and submerged plant parts taken from a swimming position, 15% by dabbling and 3% by diving (569). At the Ouse Washes (not typical wintering habitat) the vast majority of feeding is from water, by immersing the head and neck or upending, with a small amount of grazing on dry land. There the food consists predominantly of vegetative plant parts, especially leaves of the grasses *Glyceria fluitans* and *Agrostis* and submerged aquatic plants. Seeds and animal material are relatively unimportant although eaten regularly (582).

There is little information on food requirements elsewhere in Europe in more typical habitat, but in North America the leaves and stems of submerged aquatic plants are overwhelmingly important outside the breeding season (10). Even when present on the same habitat Gadwall apparently do not compete with Wigeon since they rarely walk out and graze, and, where possible, stay around permanent pools or ditches rather than move onto transient floodwaters.

Fig. 98. Gadwall – monthly pattern of numbers in Britain, expressed as a percentage of the peak count.

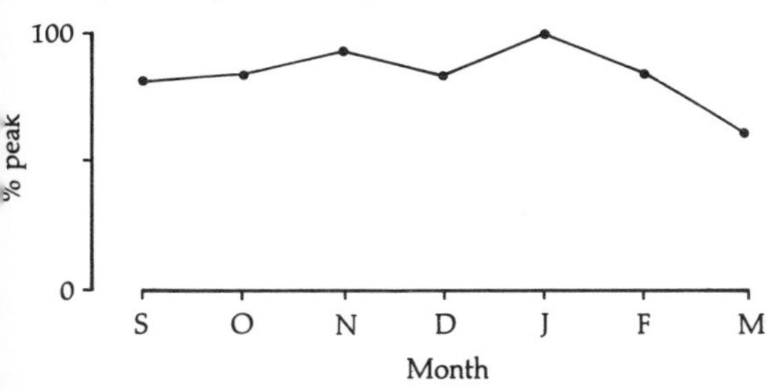

As Fig 98 shows, there are few more Gadwall in Britain at peak than there are in September and October when the count is of the local population, suggesting a rather low level of immigration. There is more movement than appears, however, since the local stock will have been reduced by January by mortality and emigration (see below). Ringing abroad indicates that our immigrants come from the Netherlands, Scandinavia, central Europe and Iceland. A substantial number of Gadwall have been ringed in Britain in recent years and a breakdown of their recovery locations is given in Table 247. About half the recoveries are of birds ringed, chiefly in late summer, at Abberton Reservoir, Essex, and most of the remainder are from Slimbridge, Gloucestershire. The small numbers of Gadwall at Loch Leven are long established and appear to behave differently (Table 247). There is no significant difference between ringing stations in either the proportion of overseas recoveries (27%) or in their distribution, so all the recoveries are considered as a single sample.

About 60% of the birds were ringed in their first winter and, as might be expected, most of the British

Table 247. *Recovery locations of Gadwall ringed in Britain (chiefly at Abberton Reservoir, Essex, and Slimbridge, Gloucestershire). Recoveries are split into same season (from July one year to June the next) and subsequent season reports.*

Area	Same season	Subsequent season	Total
Local (within 80km)	41	31	72
Other Great Britain	5	20	25
Ireland	2	2	4*
All British Isles	48	53	101
France	22	26	48
Spain/Italy	2	4	6
Netherlands	4	4	8
E & W Germany	0	6	6
Denmark/Sweden	2	2	4
Baltic Countries/USSR	0	6	6
All foreign	30	48	78
Total	78	101	179

* 3 of 8 recoveries of females ringed at Loch Leven, Kinross, were in Ireland, the other Irish recovery is of a juvenile ringed at Abberton and recovered in a subsequent season.

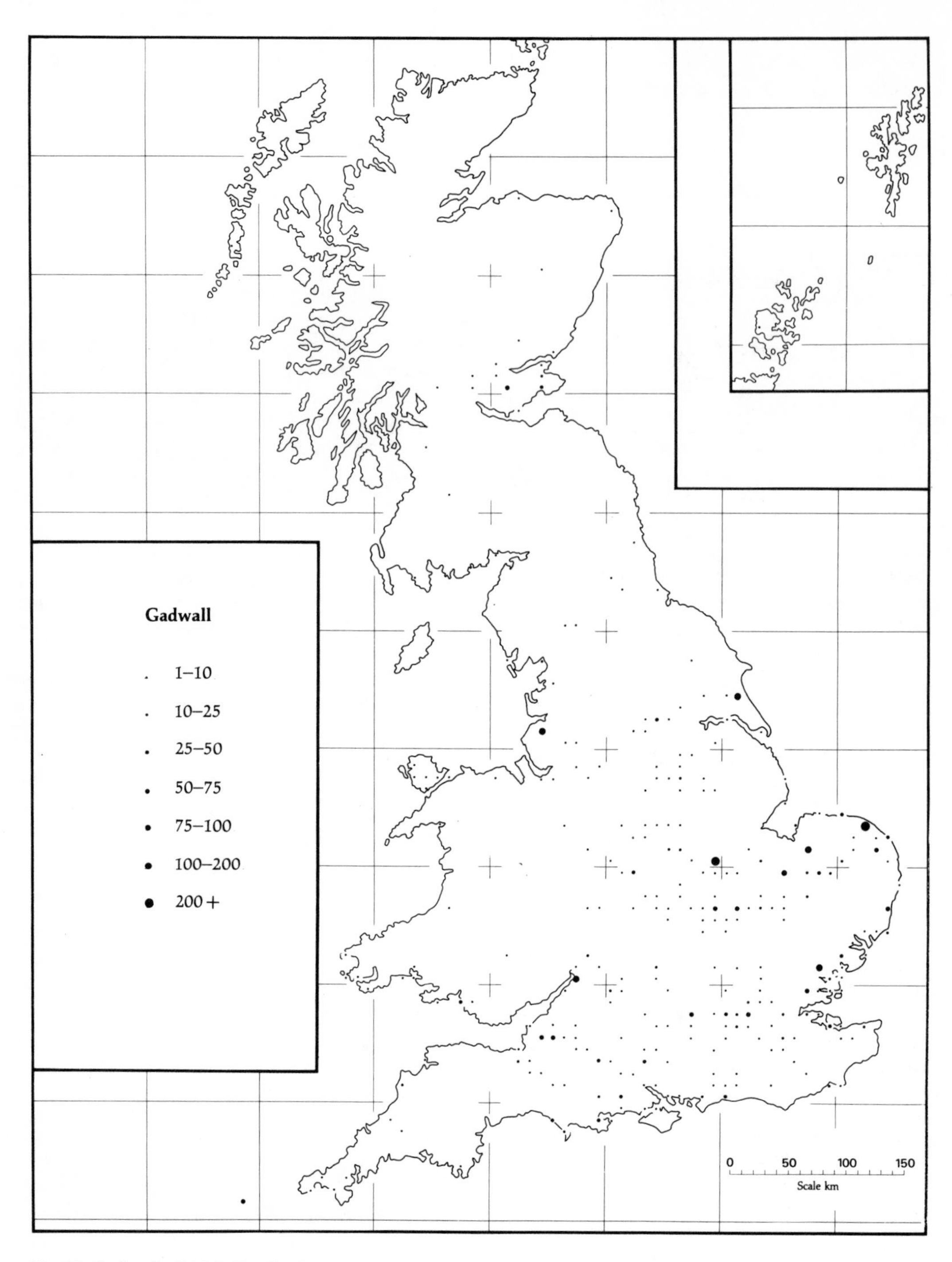

Fig. 99. Gadwall – British distribution.

recoveries in the same season were local, whereas reports from future years were more dispersed. There were, however, a surprising number of overseas recoveries during the season of ringing, most of which were juveniles. There was a significantly higher proportion of juveniles (48%) dispersing than adults (14%), and most of these moved southwards to France in autumn (most of their recoveries were in November). The likelihood of birds being shot in France is much higher than in Britain, so the proportion of overseas recoveries probably overestimates the emigrants. Nevertheless it seems likely that a substantial proportion, perhaps a fifth to a quarter, of British-reared Gadwall move south to winter. There are only a few recoveries in the same season from other parts of Europe.

France is the most usual overseas destination of birds recovered in subsequent years, and there is a significant difference in the proportion of males recovered there (76%) compared with elsewhere (47%). This lends some support to the suggestion that abmigration is partly responsible for the dispersal of Gadwall ringed in Britain (141). It also means that drakes originating from introduced British stock are entering the continental population.

Gadwall did not breed in the British Isles until a trapped pair were wing-clipped and released in Norfolk in about 1850. They spread rapidly in the Breckland, while further releases and escapes from captivity occurred in London, Gloucestershire, north-west England and elsewhere during the next 100 years. These resulted in a rapid increase in numbers and expansion of range. There are now substantial numbers breeding at the Ouse Washes, Minsmere (Suffolk), Martin Mere (Lancashire) and Slimbridge as well as the more traditional sites. Gadwall did not breed in Scotland until this century and the origin of the Loch Leven stock is unknown, although the species was spreading northwards and westwards in the early parts of this century (141). The Atlas survey had 88 squares in Britain and Ireland with breeding confirmed, the vast majority of these in south-east England. The British population was estimated at

Table 248. *Sites holding 200 or more Gadwall on more than one occasion in recent years.*

	New Grounds, Slimbridge*	Gunton Park, Norfolk	Rutland Water+	Martin Mere, Lancs#
1974	104	358		
1975	200	82	0	126
1976	100	27	20	100
1977	100	469	55	160
1978	100	480	135	260
1979	150	580	351	200
1980	250	503	141	200
1981	440	630	347	200
1982	250	427	493	200

* Captive bred stock released in late 1960s.
+ Flooding started in 1975, see also p.158.
# Captive bred stock released in mid 1970s.

Fig. 100. Gadwall – trend in numbers in Britain, 1960-1982, in October (dashed line) and January (solid line).

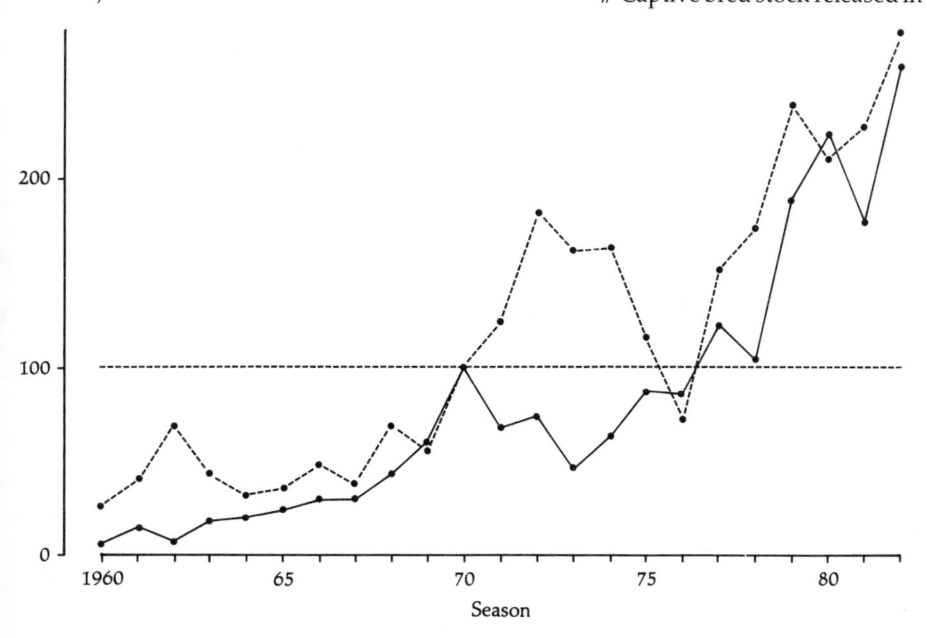

about 250 pairs in the early 1970s (542). The trend for October indicates a doubling of autumn numbers since then, so the breeding population is now probably at least 500 pairs.

Gadwall nests are not far from water, usually shallow lakes or ponds with abundant submerged aquatic vegetation. The colonisation and increase in numbers breeding at the Ouse Washes coincided with the provision of permanent water areas by conservation organisations (582). The spread of the species in eastern North America was also attributed to an increased availability of permanent freshwaters (256), but in the western Palearctic the range expansion may also be due to climatic amelioration.

The Palearctic population of Gadwall is centred on the Mediterranean/Black Sea area where there are about 50,000 birds. Most of the north-west European stock winters in Britain and the Netherlands and on the Atlantic coast of France. Numbers counted in these three areas in January 1980 were respectively 2,300, 1,200 and 1,750. With 3-400 in Ireland (279) and small numbers in Belgium and Germany the population seemed likely to be 6-7,000, compared with an estimate of 5,000 for the late 1970s (529). The British total was probably over 4,000 in October 1982 (3,820 counted), and 4,450 were counted in December 1984, so the increase continues.

The distribution of Gadwall in Britain is shown in Fig 99, which shows clearly the localised distribution around the release areas in East Anglia, London, Gloucestershire and Lancashire, as well as the scattering of birds in autumn in east central Scotland around Loch Leven. The species has, however, been recorded at no fewer than 740 sites. No waters in Britain exceed the qualifying level of 550 needed for international recognition, but because of the small north-west European population sites holding 200 or more are very significant. These are listed in Table 248.

Numbers at all these sites have increased recently as have those in Britain as a whole (Fig 100). Although the October and January trends are not well correlated they both show an enormous rise in numbers, especially in the late 1970s and early 1980s.

The average count in January 1975-76 to 1979-80 was 1,746; a comparison of this with the sum of means for all sites during the same period indicated that 89% of the Gadwall on all sites covered were counted annually. Since most flocks are accessible, few are probably missed. The British population has risen from about 520 in 1963-64 (following the hard winter of 1962-63); the total must surely soon reach 5,000.

Teal
Anas crecca

The smallest European duck, the Teal is also one of the most widespread, occurring both in brackish and freshwater habitats. The European subspecies *crecca* is smaller and lacks the vertical white stripe on the flanks of the American *carolinensis* race (p.410). In recent counts about half of the British Teal were found on sites classified as coastal (though many birds may only have been roosting on the coast).

European Green-winged Teal *Anas crecca crecca*

Teal are omnivorous and take a very wide variety of foods within a small size range. Plant seeds are overwhelmingly important, accounting for more than three-quarters of the food in Britain. Most seeds are small, chiefly *Polygonum* species, *Eleocharis* and buttercups (*Ranunculus*) in freshwater, and *Salicornia* and *Atriplex* on the coast. Since these seeds are the most abundant in the respective habitats the ducks probably take what is available. Chironomid larvae are the most abundant animal food and these are eaten together with small snails, when available (425, 582). On the Ouse Washes Teal have also been found to accompany Mallard to fenland fields to feed on stubble grain, a habit which is not uncommon elsewhere where the two species occur together. Most stubble feeding occurs in September and October, 10% of Teal collected in the shooting season had fed on arable land around the Washes (585).

In situations where several species of dabbling ducks co-exist, excessive competition between the omnivores is avoided because Teal specialise in seeds and animals intermediate in size between those taken by Mallard (which take larger items) and Shoveler. There are also differences in feeding locations and methods which help to separate the species (427).

In undisturbed situations Teal feed both by day and night (582) but where avian predators are numerous, such as at the Camargue, France, foraging is restricted to times when these are not active. Teal gather in very large flocks during the day to rest and preen (570). Typically Teal feed in very shallow water, walking slowly and dabbling with only the bill and head submerged. If shallow water is not available they will feed while swimming by immersing the head and neck or upending. However, their short neck and bill preclude them from using habitats where the water is more than 25cm deep (582).

Very large numbers of Teal have been ringed in the western Palearctic and up to the end of 1979 there had been more than 23,000 recoveries, 7,500 of these from Teal ringed in Britain. The origin of these birds is indicated by Fig 101 (ducks ringed in Britain and the Netherlands mostly between September and January). Apparently most of the Teal wintering in the British Isles breed in Scandinavia and in western parts of the USSR. Because the most common method of recovery is from shooting and spring hunting has long been prohibited there, the contribution made by Scandinavia is considerably underestimated, as shown for Dutch-ringed Teal (465). Teal migrate through Denmark and the Low Countries in September and October to give peak numbers in Britain in December and January. Some move on to France along broadly the same route in March and April.

Being small, Teal are more vulnerable to severe weather than any other European duck and would be expected to move more readily and over longer distances in hard winters. This is indeed the case: a preliminary analysis of cold-weather movements showed that large numbers of Teal from Britain move south to western France and Spain (409). In normal years there is a late-winter movement southwards; the February totals here are 70% of the December or January peak and the March ones only 40%. In hard winters movements are further and much larger numbers of birds are involved. Because of the high shooting pressures in the countries receiving our Teal the mortality is also very much higher than normal in severe winters.

The Teal is a widespread breeder in Britain, being confirmed to have bred in more than 700 10km squares in the early 1970s (542). Breeding density is, however, usually low; in the dense nesting colony at Loch Leven, Tayside, where other duck species nest in large numbers, only a handful of Teal breed (380). A

Fig. 101. The origins of British wintering Teal as indicated by ringing recoveries. Triangles – May, circles – June, diamonds – July.

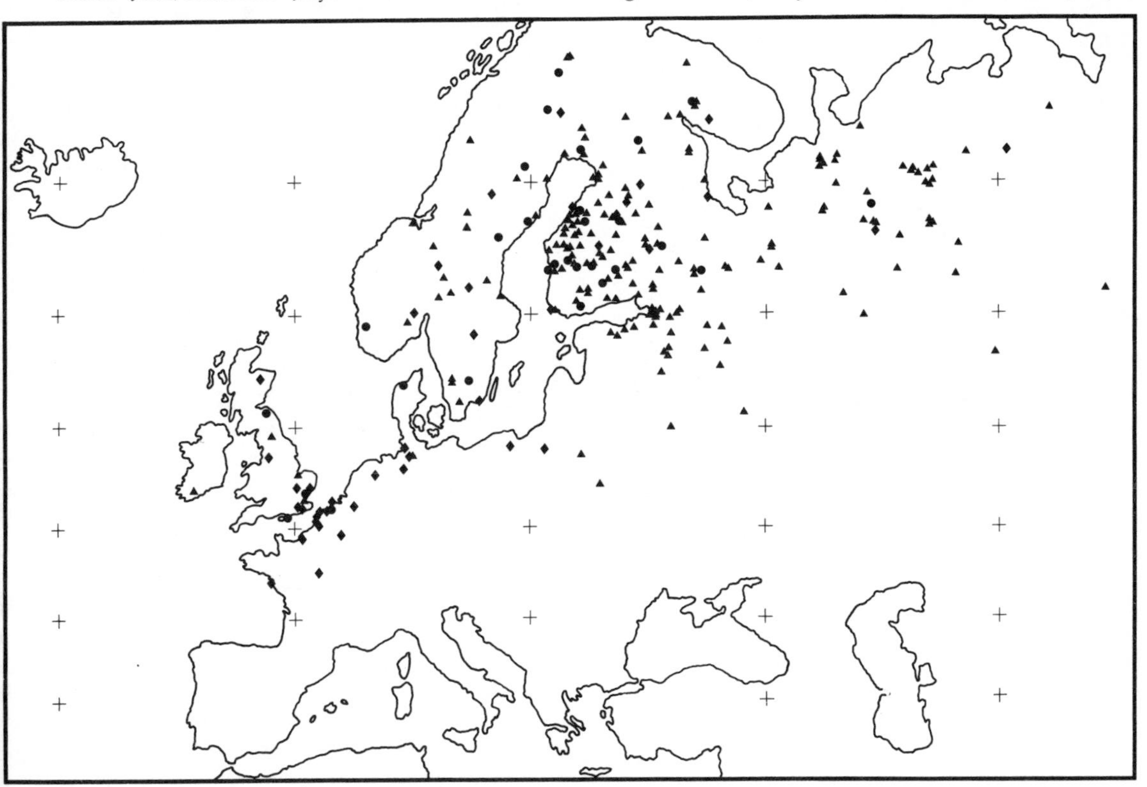

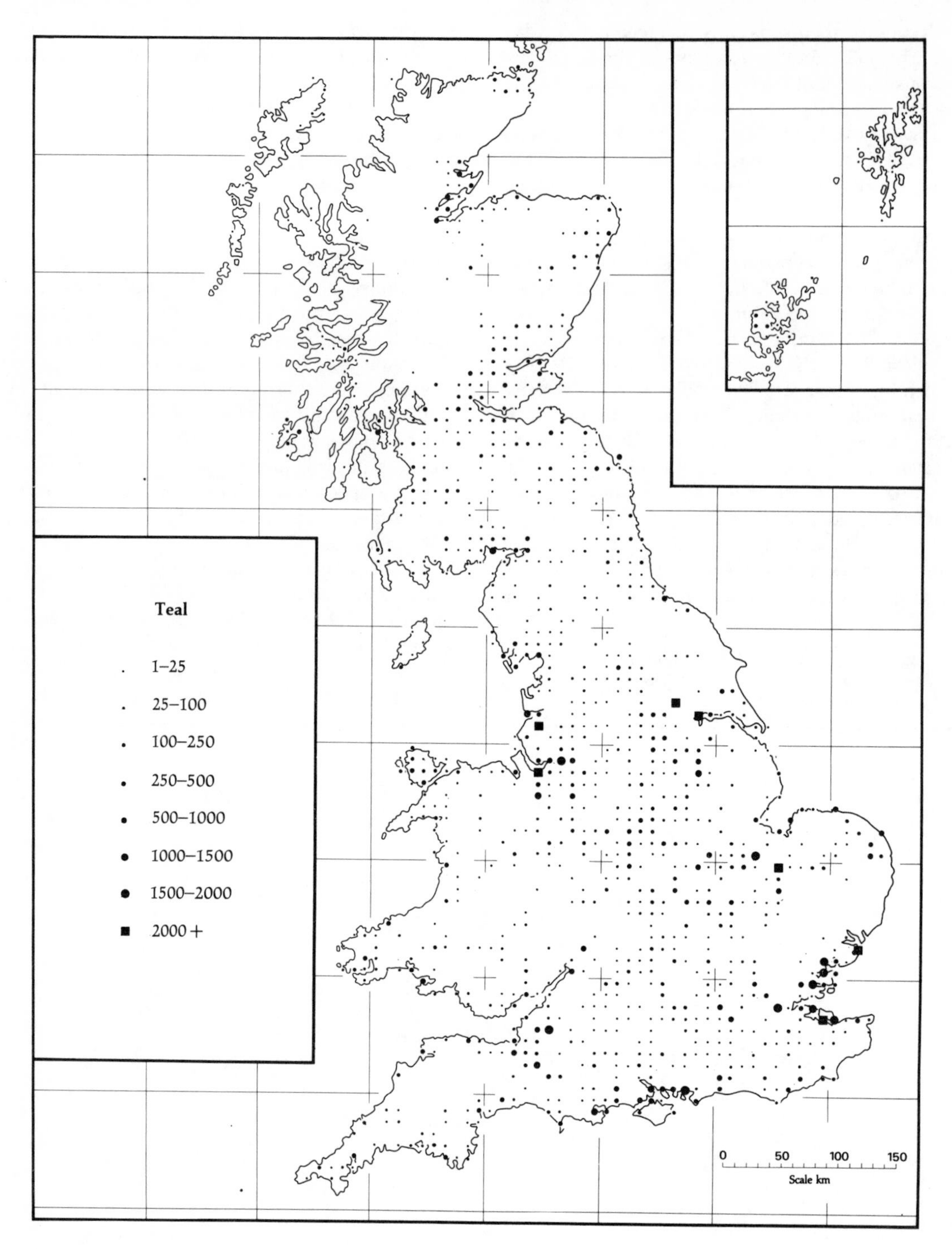

Fig. 102. Teal – British distribution.

vide variety of marshy habitats is used, with nests ısually in thick cover among tall grasses or shrubs. A ery approximate estimate of numbers in Britain and reland in the early 1970s was 3,500-6,000 pairs (542) nd if the density is similar throughout the range this uggests between 3,000 and 4,500 pairs in Great ritain. These might contribute 10-15,000 to the British vintering population.

The total north-west European stock in the late 960s and early 1970s was put at 150,000 birds (22). There followed a gradual increase which brought the total in the late 1970s to around 200,000 (529). The international counts suggest that between 40 and 50% of north-west European Teal were in Great Britain in January in most winters in the 1970s (all relatively mild). Another, much larger population breeding in central USSR winters in the Mediterranean – Black Sea area, with numbers around 750,000 (22, 529).

Fig. 103. Teal – monthly pattern of numbers in Britain, expressed as a percentage of the peak count.

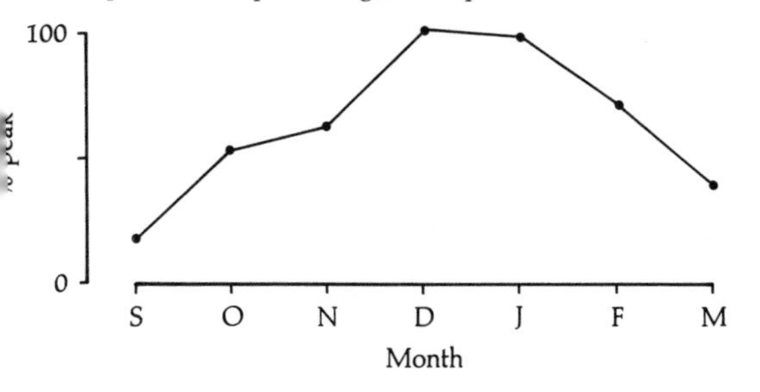

Fig. 104. Teal – trend in January numbers in Britain, 1960-1982.

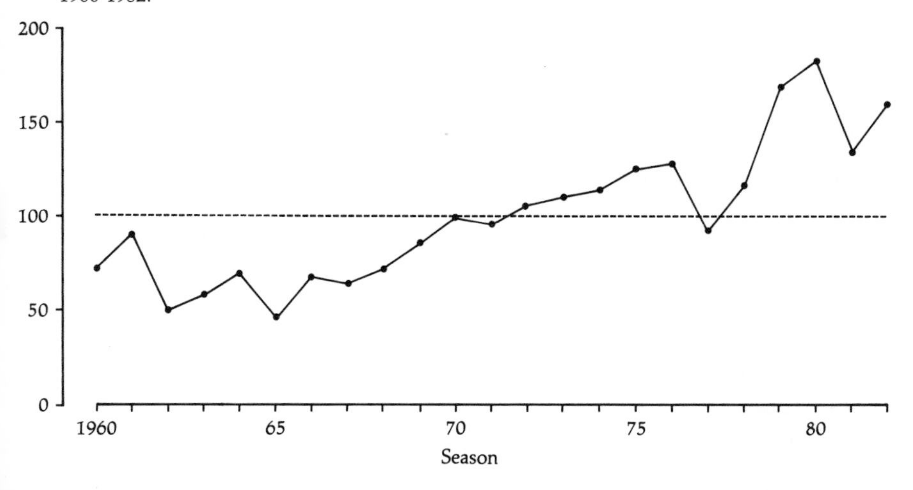

Teal are well distributed in Britain in winter, both on coastal and inland sites (Fig 102), and although there are a few spectacular concentrations the species is more dispersed than most ducks. Teal have been recorded on 61% of the sites counted, but nearly 40% of the birds are on sites holding less than 200 Teal. The numbers on the nine internationally important sites are given in Table 249.

By far the most important site at present is on the marshes of the Mersey Estuary at Ince. In the 1960s there were only two or three thousand Teal there each winter, but there was a spectacular increase in the early 1970s from 3,000 in 1970 to 13,700 in 1973. Since then the maxima have fluctuated between 5,000 and a peak of 35,000 (35% of the total British count) in 1981.

The seasonal distribution of counts (Fig 103)

able 249. *Peak numbers of Teal at the nine sites holding 2000 or more in Britain, 1977-1982. A blank = no data.*

	1977-78	1978-79	1979-80	1980-81	1981-82	1982-83	Mean
Mersey Estuary	5310	12870	17400	25850	35000	26100	20422
Martin Mere	5000	4000	3000	6000	6000	4000	4667
Ribble Estuary				2074	5274	4808	4052
Hamford Water	3000	2500	3780	4500	5400	2575	3626
Elmley Marshes		2000	2000	3000	2000	3000	2400
Ouse Washes	2755	1714	1874	2378	2970	4319	2668
Lower Derwent Ings	2000	420	2966	3682	4000	1183	2375
Abberton Reservoir	5237	997	1320	3370	1200	1850	2329
Chichester Harbour	1233	1636	1990	2760	3253	2235	2185

shows little difference between the means for December and January. On average January gives the more stable and slightly higher numbers so this is the month analysed for trends in Teal (Fig 104). Following a fall in the early 1960s the population has subsequently recovered. The gradual rise between 1967 and 1976 is related to the increase in the flyway population. The sharp drop in January 1978 is difficult to explain, but the decrease is not evident when December 1977 numbers are considered. The low January index may be due to redistribution within north-west Europe, though there was no severe weather in 1977-78.

It might be imagined that since a quarter of British Teal have been counted on the Mersey the trend for the country as a whole would be dependent on numbers there. This is not so, however, since the peaks there are not consistent, occurring in any month between November and February, often with relatively low counts in other months. For example in 1975 the peak of 13,000 was in November and only 7,400 were present in January. Since the peak in Britain seldom coincides, it seems that the Mersey attracts large numbers of Teal from other British sites to give an especially high total at some time during the winter; the very high numbers are seldom present over more than one monthly count.

The international data would suggest that the average peak number of Teal in Britain is around 100,000, the proportion of the population here, as well as the total number, having increased markedly since the early 1970s (22). The maximum ever counted in Britain is 102,000 (December 1981) and the sum of the January 5-year averages for all sites is 88,000, compared to a mean of 64,000 counted annually in the late 1970s. Thus, although Teal are dispersed, about 73% of those on sites ever covered are counted each year. It is difficult to estimate the proportion of the population missed completely, but this is probably substantial, even in counted sites, since Teal are liable to hide in vegetation or in small secluded pools. It seems likely that the annual peak is in the region of 100-200,000.

Teal numbers have probably always varied cyclically, with sharp declines being followed by a gradual recovery, though numbers in the early 1960s may already have been low following two severe winters in the mid 1950s. If we are to prevent too drastic falls in numbers in hard winters we must, as has been suggested (409), argue for shooting bans not only in countries which are suffering the severe weather (many of the Teal having already left Britain) but also in the recipient states, a move which is biologically sensible but politically will be difficult to achieve.

American Green-winged Teal *Anas crecca carolinensis*

Very similar in all respects to the Eurasian rac this subspecies is a regular vagrant to Britain. Th females are not distinguishable from our own Teal an all the records refer to males. There were only 1 sightings before 1957 but by 1982 a further 172 ha been reported, only a few of which were likely to b escapes from collections. No fewer than 17 were see in 1980, 13 in 1981 and 11 in 1982, and assuming a equal likelihood of females occurring here this subsp cies can now be regarded as a regular visitor.

Mallard
Anas platyrhynchos

By far our most numerous and familiar wildfowl pecies, the Mallard occurs throughout subarctic and emperate Eurasia and North America, breeding as far orth as 71°N in Norway and south to 35°N in North frica. The wintering range is mainly between 20 and 0° although some stay further north in Norway and celand. There are several subspecies, mainly in North nd Central America. Because of its importance as hooting quarry, the species is by far the most ntensively studied duck both in Europe and elsewhere; this account gives a brief summary of details elevant to our own stock, which is part of a large orth-west European population.

The adaptability of the Mallard contributes very argely to its success; it is found on all kinds of waters part from the open sea. In Britain about three-uarters of those counted are on freshwater habitats nd the remainder mainly on estuaries. The species is ighly tolerant of man and there are large numbers in arks and leisure areas, many relying almost com-letely on hand-feeding by the public.

The Mallard uses all the feeding methods used y any dabbling ducks to obtain its food, and its diet is xtremely varied. It walks well on land and feeds there y picking up large items such as cereal grains from he surface, by dabbling or grubbing in soft mud, by tripping seeds from standing stalks or by grazing like eese. In water, the commonest methods are upend-ng and immersing the head and neck, but food is also athered by dabbling in the shallows, pecking from he surface or even diving. The last-named activity is estricted to situations where the underwater food upply is rich (e.g. where fed artificially with cereal rains in water more than 30cm deep). The Mallard is nore awkward than the true diving ducks and emains submerged for only a few seconds (141, 569, 15).

In winter the diet consists mainly of vegetable matter in the form of seeds, either of marsh or aquatic plants, or, nowadays more commonly, waste cereal grains from stubbles. Mallard have long taken to stubble feeding in autumn, which is probably much more rewarding than foraging on the marsh, since the food is more abundant and more easily gathered. In some cases where unharvested cereals have been flattened by wind and rain Mallard will move into the fields, eat the grain and further damage the crop. In arable habitats the birds move onto potato fields and, recently, sugar-beet, eating the waste tubers and root fragments. Field feeding is a normal part of Mallard behaviour at the Ouse Washes, although the birds supplement the foods gleaned from the fen fields with a wide variety of seeds and animal material collected from the Washes (585). The commonest wild seeds eaten are those most readily available, especially *Eleocharis*, docks (*Rumex* spp), buttercups (*Ranunculus* spp) and persicarias (*Polygonum* spp). Animal material consists of a wide variety of slow moving invertebrates, chiefly beetles and chironomid larvae (582).

In more wooded areas tree fruits, especially acorns, and seeds of shrubs and marginal plants are important and are taken even though artificial feeding takes place (423). In brackish habitats seeds again predominate, with the larger ones being selected, with molluscs, crustacea and insects also being taken (426).

Clearly the Mallard is adaptable and opportunistic; there is little that limits its use of habitats provided they remain open in winter. The birds will stay in frozen areas for a few days to sit out short periods of severe weather; few winters in Britain cause long-distance hard-weather movements.

The Mallard is our commonest breeding duck, but there is also a sizeable proportion of immigrants in mid-winter. The wintering pattern (Fig 105) indicates a gradual increase through the autumn, and this is confirmed by ringing recoveries (333, 415). Shooting, of course, takes a considerable toll of Mallard in autumn, but this is more than compensated for by

Fig. 105. Mallard – monthly pattern of numbers in Britain, expressed as a percentage of the peak count. Solid line – 5-year average. Dashed line – 1977-78 (mild winter).

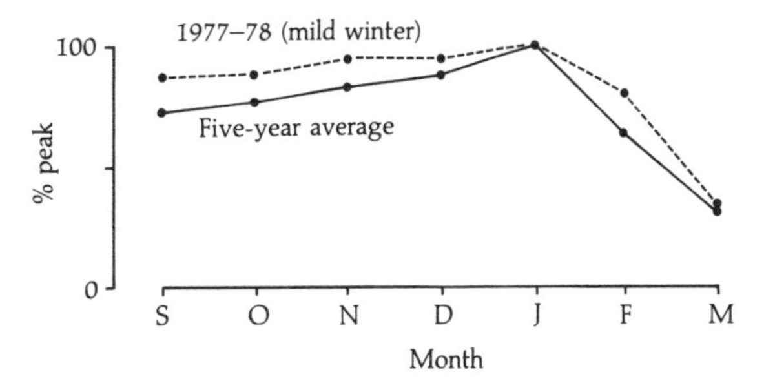

immigration. Estimating the proportion of immigrants in the British stock has been attempted using ringing recoveries (75). Of the number of winter-ringed birds recovered during the breeding season, 71% were in Britain and 29% abroad. Using the second season recoveries 32% were abroad. Despite the apparent close agreement between the estimates such calculations are fraught with difficulty, particularly the problems of regional variability in immigrant proportions and differences in recovery and reporting rates at home and abroad. The largest bias, leading to an underestimate in the proportion of migrants, is likely to be due to the fact that most ducks are ringed between August and October, when the proportion of immigrants is at its lowest. The ringing recoveries therefore suggest that more than one-third of British wintering Mallard are immigrants (see also below). These come from a very wide area (Fig 106) almost equally divided between three regions: Scandinavia and the USSR; Poland, Denmark and Germany; and the Netherlands, Belgium and France. There is a highly significant imbalance in the sexes of those recovered in Denmark, Germany and Poland (75%

Fig. 106. The origins of immigrant Mallard as indicated by ringing recoveries. Triangles – May, circles – June, diamonds – July.

males) through abmigration. There is also a pr ponderance of males in other areas, but this is not great (415).

The Mallard was shown by the Atlas survey be one of the most widespread British breeding bird breeding being confirmed in 85% of the 10km squar in Britain and Ireland (542). Nests are usually we dispersed, although there are high densities on son sites, e.g. 400-450 pairs at Loch Leven (380) an 400-1,300 pairs in and around the Ouse Washes (582

The breeding season is extended, starting February in the south, and eggs from re-nesters ma be found as late as July or August. A wide variety nest sites are used, most commonly in dense cov among trees, shrubs or tussocky grassland. Elevate sites in hollow trees or in the crowns of pollarde willows, up to 5m high, are also used (141, 380, 39 582). Mallard also take readily to artificial sites nest-boxes, platforms and specially constructed wic er baskets, and ducks have also been known to nest o buildings, sometimes at great heights. The nests a usually within a few metres of water but have bee found on hillsides with the nearest water 2km awa (542).

Mallard are very commonly managed by wil fowlers to increase shootable stocks, either by manag ing habitat and providing nest sites for wild birds or b

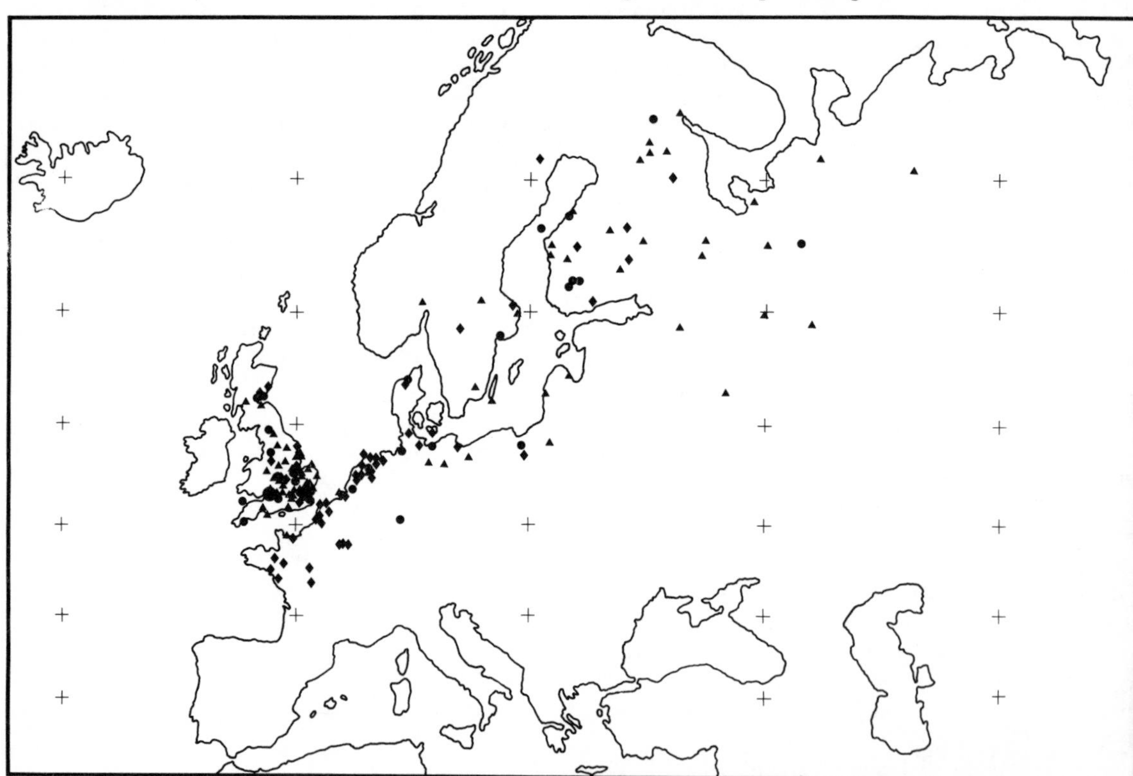

king eggs from the wild for hatching, rearing and lease. WAGBI (now BASC) began a scheme of xperimental release in 1954 and by 1968 109,000 [allard had been liberated. The practice spread uickly and is now widely carried out by shooters, hether or not in association with a recognised rganisation. Many of the released birds undoubtedly e shot early in the season, but many also integrate to the wild population and some move overseas in ter seasons (249, see also p.418).

There are an estimated 4-5 million Mallard in the estern Palearctic: a million and a half in north-west urope, a similar number in the Mediterranean/Black ea region and a million in the Middle East, chiefly ound the western Caspian (22). Although there are rge fluctuations in the number counted, there are no iscernible long-term trends. The species has a ispersed winter distribution in north-west Europe, ith only one concentration, in the IJsselmeer, the etherlands, exceeding 10,000 birds. Elsewhere in urasia the situation is different, with large concentra-ons more usual. Outstanding in the late 1960s and arly 1970s were January averages of 170,000 in the anube Delta and 120,000 in one site in the southern aspian.

In Britain the Mallard was recorded in 85% of e sites counted and in 1,334 10km squares (Fig 107). ecent counts from the five localities holding more an 3,000 birds regularly are given in Table 250.

These five sites have been recognised as having oncentrations of national importance (524). A few ther places, notably the Firth of Forth (max 6,066) and lartin Mere, Lancashire (4,000), occasionally hold rge numbers. In most major areas the counts are of roosting concentrations, the birds doing much of their feeding on nearby agricultural land.

The index of Mallard numbers shows no discernible trend either in September (the post-breeding British stock) or in January (the annual peak), as Fig 108 shows. Moreover, there is no correlation between the two indices; i.e. they vary quite independently of one another. This is not surprising since the January index includes immigrants, which also makes it much more variable than that of the British breeding population. There is a highly significant decline in the March index throughout the period. Since dispersal to nesting sites occurs in February, the March numbers are more related to the timing of breeding (the later this begins the more would still be in flocks in March) than to the pre-breeding population. The biological significance of this trend is unclear, but it seems unlikely that it is due to progressively earlier nesting through the 1960s and 1970s. There is a significant negative correlation between the index in March and that in the following September (i.e. September numbers are high when those in March are low). This agrees with the finding that breeding is more successful in early years; these years would show low March numbers because fewer Mallard would be in flocks.

The stability of the post-breeding population both in Britain and in north-west Europe as a whole suggests that there are density dependent factors operating to regulate the population. This density dependence can act in many ways on the productivity of the breeding stock. The main ones are through emigration from crowded areas, an increased proportion of non-breeders and reduced hatching and rearing success through interference and competition

able 250. *Numbers of Mallard on the five most important British sites. 5-year means 1971-1975 and seasonal maxima ereafter. A blank = incomplete data.*

	Abberton Reservoir	Ouse Washes	Humber	Lower Derwent Ings	The Wash
1971-75	3594	3583	5309		2640*
Maximum	4760	5450	7730		3487
1976	3220	4480	6920		1268
1977	4700	5400	4600	3000	1259
1978	4300	3460	5450	1350	1585
1979	4540	4460	8430	5903	2805
1980	2950	2880	6430	4436	5484
1981	2500	6260	4190	5985	4977
1982	5900	5547	6001	1559	4745
Mean 1976-82	4016	4641	6003	3706	3160

Mean of four years only.

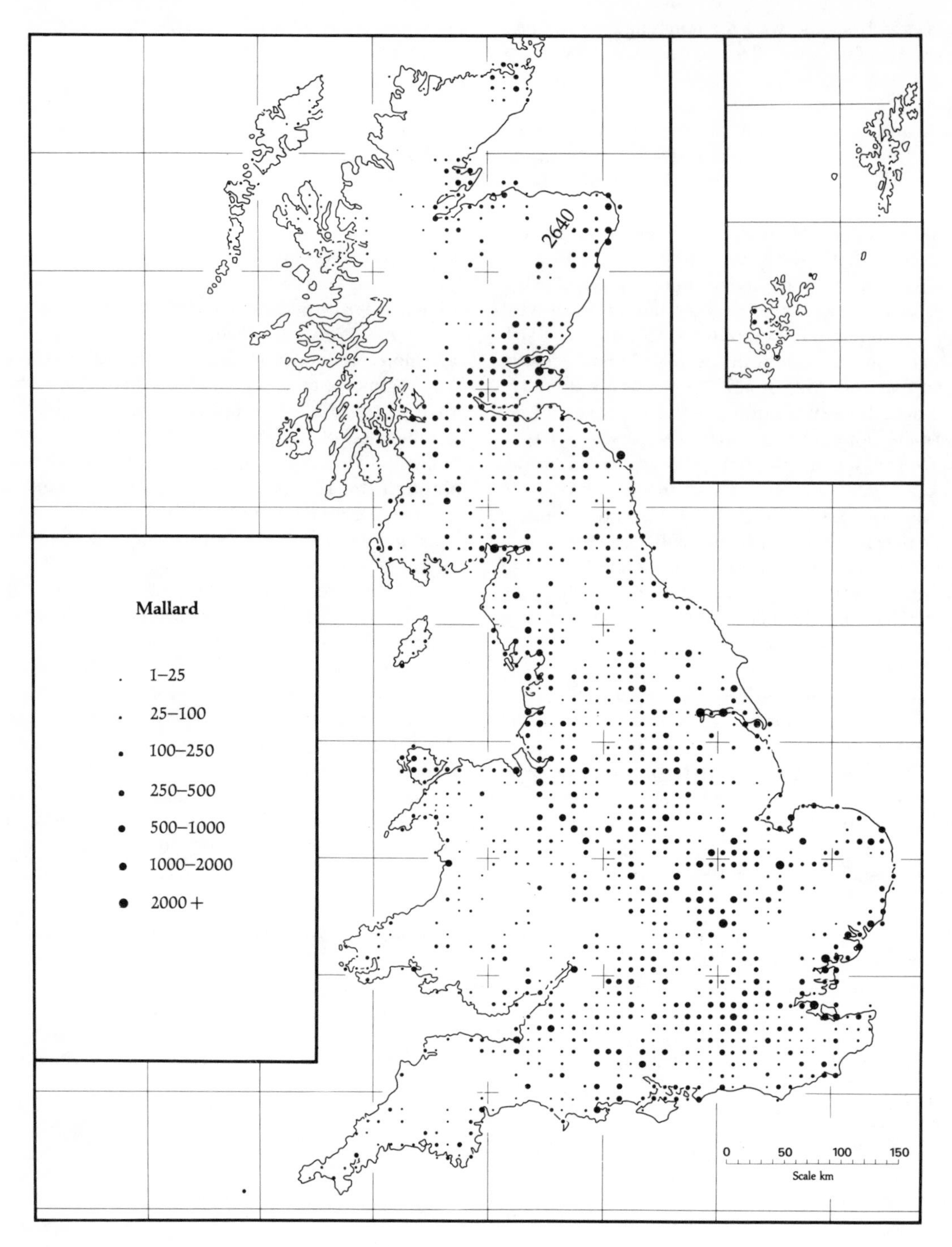

Fig. 107. Mallard – British distribution.

ı crowded habitats (163, p.525). Evidence from North ᵼmerica indicates that although shooting may ccount for a high proportion of birds, below a ıreshold level it is compensated for by natural ıortality (9). Thus, provided shooting pressure is not xcessive, it appears that our influence on the Mallard opulation either through its mortality rate or its reeding output has hitherto been slight.

Estimating the number of Mallard in the British opulation is fraught with difficulty, and very few ttempts have been made. In the First Edition the arlier estimate of 350,000 for the native population 191) was thought to be consistent with a guessed vinter total of 500,000. The number of breeding pairs vas put at 40,000 in the 1960s (20) but this was thought ɔ be a gross underestimate following the Atlas urvey, which suggested that the population could xceed 150,000 pairs in Britain and Ireland (542). Even sing a conservative measure of production of 1 oung per pre-breeding adult (one estimate (598) puts : at 1.8 young) the September population in Britain vould be in excess of 300,000; only 104,000 are ounted regularly. Recently the regular winter maxmum has been put at 300,000 (524) on the assumption hat about half the birds are counted in January. At the ame level of coverage this suggests a September opulation of 210,000, or the product of about 50,000 reeding pairs.

The summed 5-year means for all counted sites give a total of 146,000 for September and 190,000 for January, which means that 70 and 73% respectively of those on sites ever counted are included each year. The widespread distribution of the Mallard suggests that a substantial proportion are on sites which have never been surveyed. It seems likely therefore that the assumption that only half are missed (524) may be optimistic and the post-breeding population may be in excess of 500,000 ducks, or the product of about 100,000 pairs at 1.8 young per pre-breeding adult or about 130,000 pairs at 1.0 young. This broadly agrees with the estimate from the Atlas survey. Using the same method to estimate the January numbers the regular winter peaks are likely to be around 600-700,000.

Estimates from shooting organisations (237) suggest that about 700,000 Mallard are shot annually, and clearly these consist of immigrants as well as locally produced birds. The annual mortality rate of adults is 48% (70) and that of young birds greater. These figures are not consistent with the population estimates given above and the proportion of immigrants at 30-50%, but given that up to 400,000 Mallard may be released onto flight ponds and most of these will not appear in the counts the difference may be explained. It is, however, highly unsatisfactory, for such an important species as Mallard, that more accurate estimates of population size, turnover and the impact of the kill are not available (see also p.494).

Fig. 108. Mallard – trend in numbers in Britain, 1960-1982, in September (dashed line) and January (solid line).

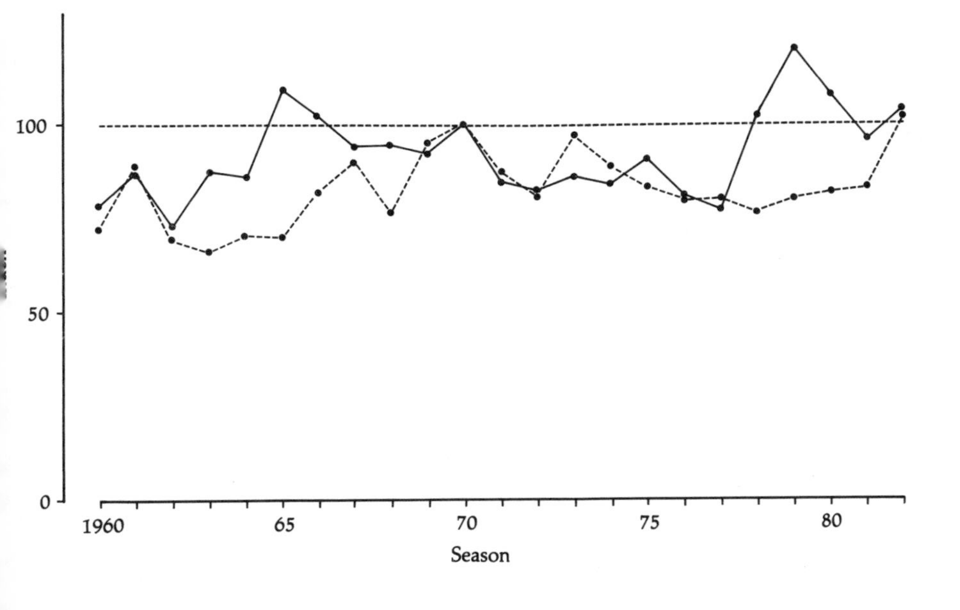

Pintail
Anas acuta

The Pintail breeds throughout northern parts of the Northern Hemisphere. In most places it is highly migratory, the birds breeding in north-west Europe wintering as far south as North Africa and the Nile Valley. In Europe its winter distribution is largely coastal although in North America it occurs equally inland.

Fig. 109. The origins of British wintering Pintail as indicated by ringing recoveries. Triangles – May, circles – June, squares – August.

There is rather little information on diet an habitat requirements but the Pintail is clearly a adaptable species. On estuarine habitat in Kent mo of the feeding is done in the intertidal zone, dabblin in the upper layers of mud or upending in shallo water. By far the most important food item here small snails of the genus *Hydrobia*, with seeds makin up most of the remainder (427).

Inland on the Ouse Washes, seeds are by far th most important food, accounting for more tha four-fifths of the diet. A wide variety is taken, the mo common naturally occurring species being *Eleochar palustris*. Animals, in contrast to the situation on th coast, make a very minor contribution (582). Unt recently field feeding was an unusual habit for Britis Pintail, but nowadays most of the diet is obtained i this way in many areas, as is also the rule in Nort America. Many of the Pintail wintering on the Ous Washes feed on nearby farmland, nearly alway accompanying flocks of Mallard. Cereal stubbles chiefly wheat, are the most frequented crops an when these are ploughed the birds move ont harvested potato fields, eating waste tubers from th surface, especially after frost (585). In recent years th habit of eating waste sugar-beet fragments fron harvested fields has become common near th Washes. When feeding on the arable land Pintail mak

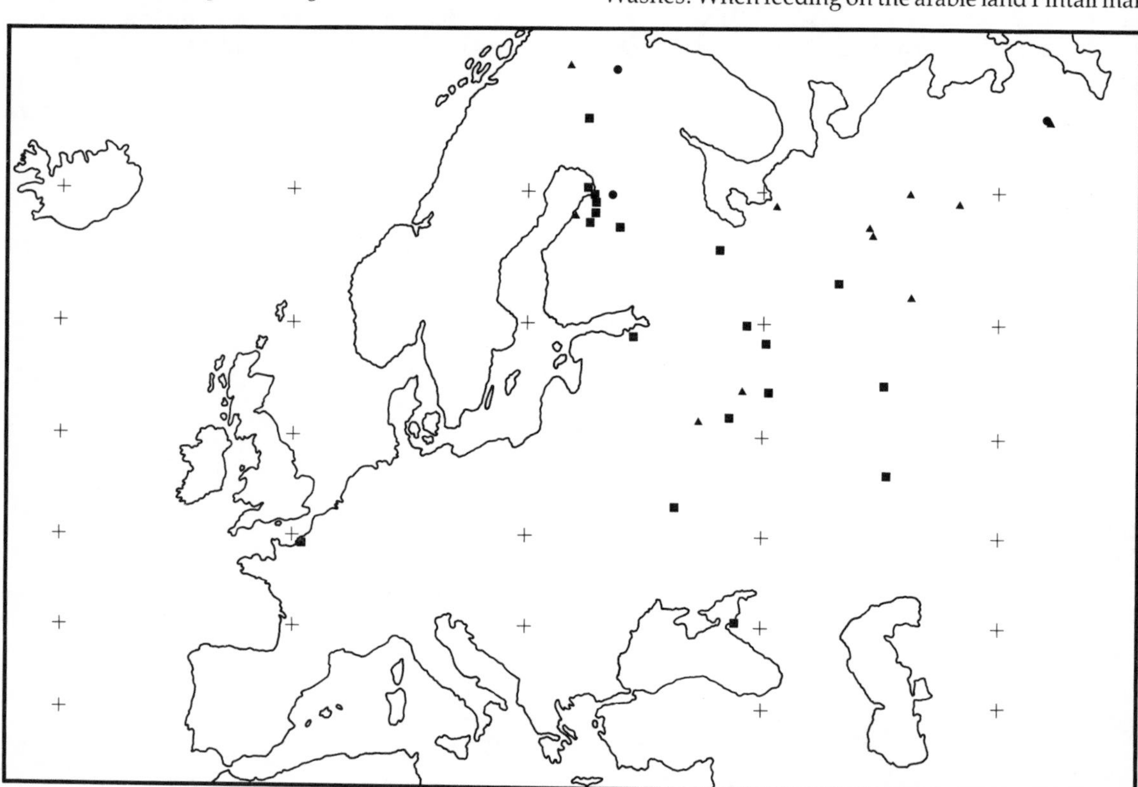

wo flights per day, one away from the roost on the Ouse Washes around dawn, returning just after unrise, and another to the fields in mid afternoon, eturning at dusk. Each feeding period lasts only an hour or two and the remainder of the day is spent esting and feeding on the Washes themselves. Flights o fields are more frequent in adverse weather and when the Washes are flooded. On the wet marshland Pintail feed largely from a swimming position either by immersing the head and neck or by upending. Their diet and habits are very similar to those of Mallard and the two species feed in very similar depths of water. Because of the considerable difference in the size of the sexes, female Pintail spend more time upending than do males (582).

From the few hundred recoveries of Pintail inged in Britain, the British stock is apparently part of that breeding in western Siberia and wintering in north-west Europe. Some birds also originate from Iceland, Scandinavia and other west European countries, though these are under-represented in the summer recoveries, which are mainly from shot birds. Autumn migration is via the Baltic countries and the Netherlands but in spring the route is more easterly through central Europe. British-ringed Pintail have been recovered as far east as 65°E (465, Fig 109).

Fig. 110. Pintail – monthly pattern of numbers in Britain, expressed as a percentage of the peak count.

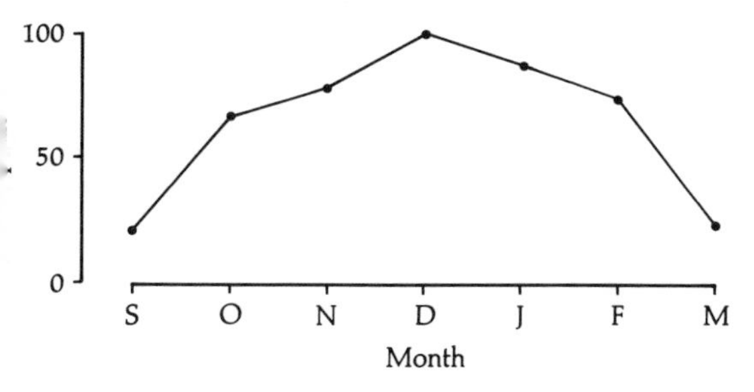

Some Pintail arrive in Britain in September and large numbers are here in October, though the peak is not reached until December in most years (Fig 110). Pintail are extremely mobile, a feature which makes them well adapted to use habitats which are temporarily available through flooding. This mobility causes major local changes in distribution and in the relative importance of wintering sites (22).

There is a very small British breeding stock but their wintering grounds are unknown. Breeding records are concentrated in the Cambridgeshire/Norfolk Fenland and north Kent, and scattered throughout the northern half of the country. Breeding was probable or confirmed in only 46 British and Irish squares in the five years of the Atlas survey, and there are probably less than 50 pairs in the British Isles in most years (542). Nests are usually in the open, often in marshland adjacent to shallow-water feeding areas.

The total north-west European population is estimated at about 75,000 (529), having increased from about 50,000 in the early 1970s (22). Typically in January they are concentrated in a very few major sites with a scattering elsewhere, usually near the coast. The more southerly and easterly populations are much larger, with the Black Sea/Mediterranean stock

Table 251. *Numbers of Pintail on the seven most important British sites. 5-year means 1960-1974 and seasonal maxima thereafter. A blank = no/incomplete data.*

	Caerlaverock	Ribble Est	Martin Mere	Mersey Est	Dee Est	The Wash	Burry Inlet
1960-64		280		360			632
1965-69	386*	374		2030	1260		560
1970-74	576	1900	1300+	8770	2840	103#	490
1975	2150	4000		9300	2420	236	304
1976	800	500	1000	9220	6000	487	1679
1977	1510	1800		15500		132	979
1978	1400	138	1400	8240		434	617
1979	1800	80	1200	10030	6700	550	726
1980	670	411	4000	18450	5510	1672	510
1981	70	1273	2000	11440	5395	2943	2426
1982	668	689	3700	13750	7360	1822	2535
Mean 1975-82	1134	1013	2157	11991	5564	1035	1222

* Mean of 4 years only.
+ None or very few (no flooding until 1973) in first 3 years, 4500 in 1973 and 2000 in 1974.
# Mean of 3 years only.

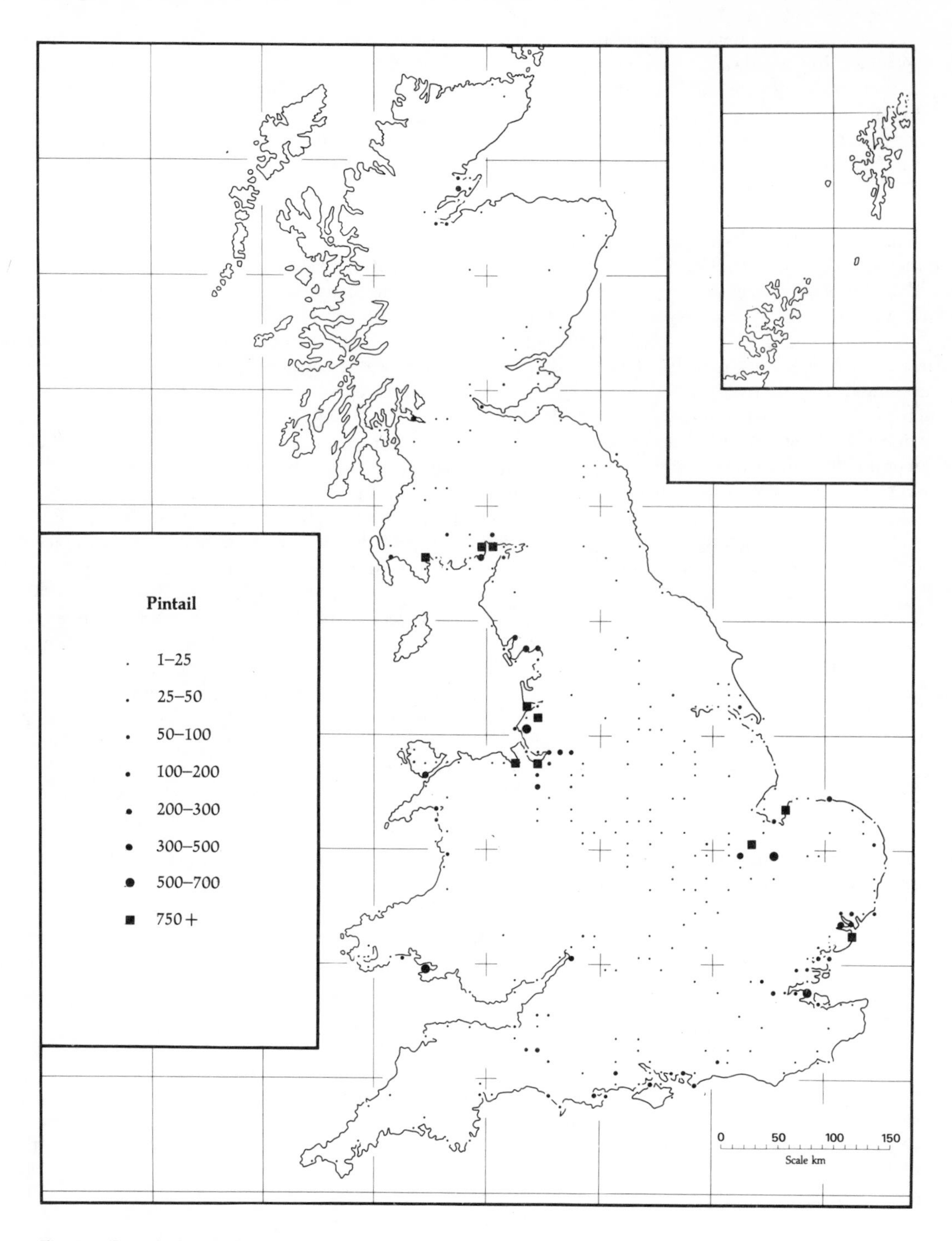

Fig. 111. Pintail – British distribution.

totalling a quarter of a million birds, and there are sometimes 200,000 south of the Sahara in Senegal and Mali (22).

The Pintail is one of our most concentrated wintering ducks, with over half the counted total in only six sites. Although Pintail have been recorded from time to time from more than 950 places, 85% of these have held less than 25 birds on average. The distribution is shown in Fig 111 and the seven most important sites are listed in Table 251. Because of large-scale movements, maxima occur at different times in different areas.

The marshes of the Mersey Estuary provide by far the most important site for Pintail in Britain. The increase in the importance of this area has been most dramatic, from a regular figure of only 70 in the First Edition. The first major influx occurred in the early 1970s, with fluctuating peaks in the later part of the decade. The count of 18,450 in November 1980 represented nearly three-quarters of the total Pintail counted in Britain in that month, which also produced the winter's peak. Ever since the Mersey has assumed such an important role, the numbers counted on the Dee have also been higher, the peak usually occurring at a different time to that on the Mersey. The Ribble Estuary has declined in importance, possibly as a result of the improvement of the habitat nearby at Martin Mere, where many birds roost. They feed on adjacent arable land.

The trend in Pintail numbers is shown in Fig 112, which indicates a remarkable increase in the early 1970s from what was a rather stable population. Although the Mersey was growing in importance at the time, the north-west European population as a whole was also increasing in parallel with that in Britain (22). The reason for the increase is unclear but is probably linked to good breeding seasons rather than to changes in mortality; there were no major changes in the protected status of the Pintail in the early 1970s.

Despite its highly localised distribution a comparison of the 5-year January mean on all sites (23,600) with the average count in January 1975-1979 (16,600) suggests that only 70% of Pintail are counted each year, a surprisingly low figure. It seems unlikely, however, that many Pintail would be missed altogether, so if we consider that the counts represent 70% of British Pintail the average number for 1975-1979 was 24,000, having ranged from 10,000 to 31,000 since 1960. There was another large increase in 1980, when 25,000 were counted in November, and in January 1983 (25,500 counted). The lowest winter maxima in recent years were recorded in 1977-78 (21,650) and 1980-81 (19,410), when the mild weather in northern Europe presumably caused large numbers to winter further east.

Although the two major habitats of Pintail are vulnerable to human interference and give cause for concern for the future (22), the habitat seems to have sustained the increases of the early 1970s. Perhaps the mobility of the species and its ability to use periodically a number of wintering sites give it the flexibility to overcome problems of local interference, at least in the short term. Long-term changes in habitat on a few of the major sites could, however, prove disastrous.

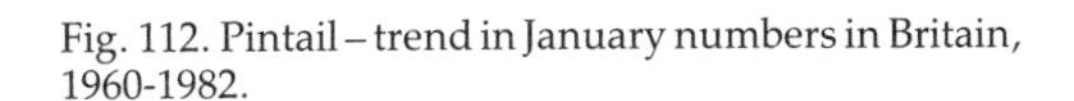
Fig. 112. Pintail – trend in January numbers in Britain, 1960-1982.

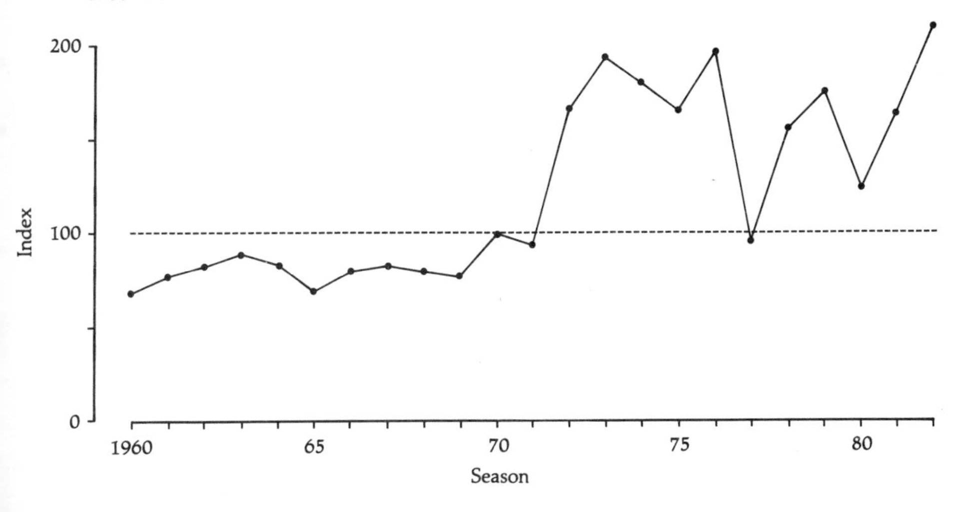

Garganey
Anas querquedula

The only wildfowl species which is a summer visitor to Britain, the Garganey breeds throughout Eurasia between 40 and 60°N and north to 65° in the Gulf of Bothnia. The wintering grounds of Palearctic migrants are in central Africa between the Equator and 20°N, although a very small number may be found in the Mediterranean.

The Garganey is a freshwater species, in winter frequenting rich, shallow lowland waters and river floodplains. In the Senegal Delta the diet consists almost entirely of seeds, chiefly wild grasses and marginal plants. Animal material is taken regularly in autumn soon after arrival but in winter is unimportant. The birds have recently taken to visiting rice paddies and eating the seeds, but the quantity taken is small and damage is negligible (594).

In spring and summer the species also inhabits freshwater, usually shallow floodwater or marshland bisected by ditches or other watercourses, sometimes in very transitory habitats where breeding is only possible in some years. The diet changes from a mainly vegetable to an animal one as the breeding season approaches. The most important items are snails, chironomid and other insect larvae, and various small, slow moving invertebrates (141, 153). As with other *Anas* ducklings, the young probably depend on emerging insects and other invertebrates from water or from vegetation on the margins.

Very few Garganey (average annual maximum about 10) are recorded during the Wildfowl Counts in March and September. Most of our immigrants arrive later than mid-March, sometimes in April and May, and most will have departed before the September count. Britain is on the edge of the breeding range; the numbers here are highly variable from year to year.

Garganey are found mainly in the south and east of Britain and breeding was confirmed in 46 10km squares during the Atlas survey. Nesting was sporadic, however, and it was thought unlikely that the total exceeded 70 pairs in any year (542). The Ouse Washes used to hold 25-35 pairs, by far the most important site in Britain. The numbers have declined there to much less than half of this and breeding is not recorded in some years. The total for Britain is probably no more than 50 pairs at present. The nest is usually in dense cover among grass clumps or emergent vegetation. The male remains in the vicinity of the nest throughout incubation, although he does not help with caring for the brood (141).

About 500 Garganey have been ringed in Britain, the vast majority at Abberton Reservoir, Essex, mainly in late summer. The distribution of the 61 recoveries outside Britain (there were also 4

Table 252. *Recovery locations of Garganey ringed in Britain (all but 3 at Abberton Reservoir) and recovered overseas.*

Country	Autumn (Sep – Nov)	Winter (Dec – Feb)	Spring (Mar – May)	Summer (Jun – Aug)	Total
France	4	2	15	2	23
Italy	3	2	15		20
Spain		1			1
Algeria	2				2
Belgium	1				1
The Netherlands				2	2
Germany			1	1	2
Austria				1	1
Hungary			1		1
Jugoslavia		1			1
Bulgaria			1		1
USSR	1		2	1	4
Turkey			2		2
Total	11	6	37	7	61

recoveries in Britain within a month of ringing) are shown in Table 252. It is clear that the main passage of Garganey both in spring and autumn is through France and Italy to North Africa.

All recoveries to the east were in subsequent seasons except the two summer recoveries from the Netherlands, both of which were juveniles ringed earlier that summer. The easternmost recovery is from Kazakhstan, 68°15′E, but European-ringed birds have been recovered as far east as 98° (141). These birds undoubtedly mix with those of the easterly groups in the wintering grounds and accompany them the following year.

There are no recoveries from the wintering area but this is not surprising since so few have been ringed and the wintering population is so vast (perhaps a quarter of a million birds). However, birds marked in other parts of north-west Europe have been recovered in the Senegal Delta and ducks ringed there have been recovered in most western parts of the breeding range (508). It seems, therefore, certain that our breeding Garganey winter in central Africa, perhaps towards the western end of the range.

Although British Garganey have declined considerably in the last 20 years, owing to a decrease in the main Fenland strongholds, their occurrence has always been sporadic, and with the degradation of many of their marshland areas it seems unlikely that they will breed here in substantial numbers in future.

Shoveler
Anas clypeata

An abundant species of southern Europe and north Africa, the Shoveler has a more southerly wintering distribution than most dabbling ducks. It occurs more commonly on freshwater than on the coast. The north-west European wintering population is very small in comparison with more southerly groups, and since there is much mixing (see below) it is not apposite to regard it as a discrete unit.

In Britain about 85% of Shoveler counted in November – the peak month – are found inland, the majority on reservoirs. There is little information on their food requirements, but a small sample from a saltwater site indicated a preponderance of animal food, chiefly *Hydrobia* snails which were abundant in the area. In brackish water seeds are most important but animals in the form of insect larvae and adults are also taken (427). The Shoveler is the most carnivorous of the dabbling ducks wintering at the Ouse Washes, with three-quarters of its diet consisting of animal material. Chief among these are freshwater snails of many different species, crustaceans and insect adults and larvae. Seeds are also important, of which *Eleocharis* spp are most commonly taken (582).

The bill of the Shoveler shows an adaptation extreme among ducks for filter feeding. The upper and lower mandibles have very prominent projections or "lamellae", which strain particles from water taken in through the tip of the bill. The Shoveler has by far the most pronounced and densest lamellae of all the northern ducks; the species is thus enabled to feed on the smallest particles, including plankton and other fine suspended material. On the Ouse Washes Shoveler take just as high a proportion of large items as do other species in autumn and early winter, but the advantage of filtering fine particles may be most important in spring and summer when plankton are more abundant (427, 582). Coastal and washland wintering sites are not entirely typical Shoveler habitats, and in other situations, such as in the deeper water of reservoirs and gravel pits, the filtering of small particles may be more important. Shoveler do not visit agricultural land in search of food, but in some situations where they are fed artificially they readily take grain, sometimes even diving for it in water up to 50cm deep.

Shoveler more commonly feed while swimming than other species of dabbling ducks, and less often upend. Groups of ducks can often be seen swimming in a line or in circles, each dabbling in the water disturbed by the movement of the one in front. The birds probably feed by day and night, at least in undisturbed areas (582).

The movements of British Shoveler are rather complex, since the native breeding population moves out in winter while we receive immigrants from north-west Europe (Fig 113). Most summer-ringed Shoveler are in France and Spain in mid-winter; none of 27 recoveries up to the end of 1960 were in Britain after November and several were in the Continent by October (395). This pattern has been maintained in recent ringing. Birds ringed here in autumn have been found later in the year in France and Spain, in subsequent years in summer in the western USSR and Scandinavia and on autumn passage in Denmark and the Netherlands (Fig 113). The results of Dutch ringing confirm that Shoveler move from there to Britain and Ireland and south-west Europe in January. Nearly all of these are in France and Spain during March (465). The pattern shown by the recoveries is thus relatively clear, with British breeders moving out in autumn to France and Spain, while immigrants from north-west Europe arrive in October and November. Many of these also move southwards later and most of the birds here in March stay to breed. Thus the bulk of our wintering birds originate from the USSR, where there have been recoveries as far east as 65°E.

The pattern of use of Britain by Shoveler (Fig 114) largely confirms the above, showing a pronounced November peak followed by a steady decline to February as the birds on passage move out. There is a sharp drop to March when most immigrants have left and the breeding population is returning. In the early 1950s the peak in numbers was apparently in February and March (17), and a February peak was also recorded in the 1960s (621). A comparison of the pattern in the 1960s and the early 1970s shows that a marked change has indeed taken place. Presumably the north-west European immigrants bypass Britain on the return journey, possibly as a result of the succession of mild winters (465).

Fig. 113. The origins of British wintering Shoveler as indicated by ringing recoveries. Triangles pointing downwards – April, pointing upwards – May, circles – June, diamonds – July, squares – August.

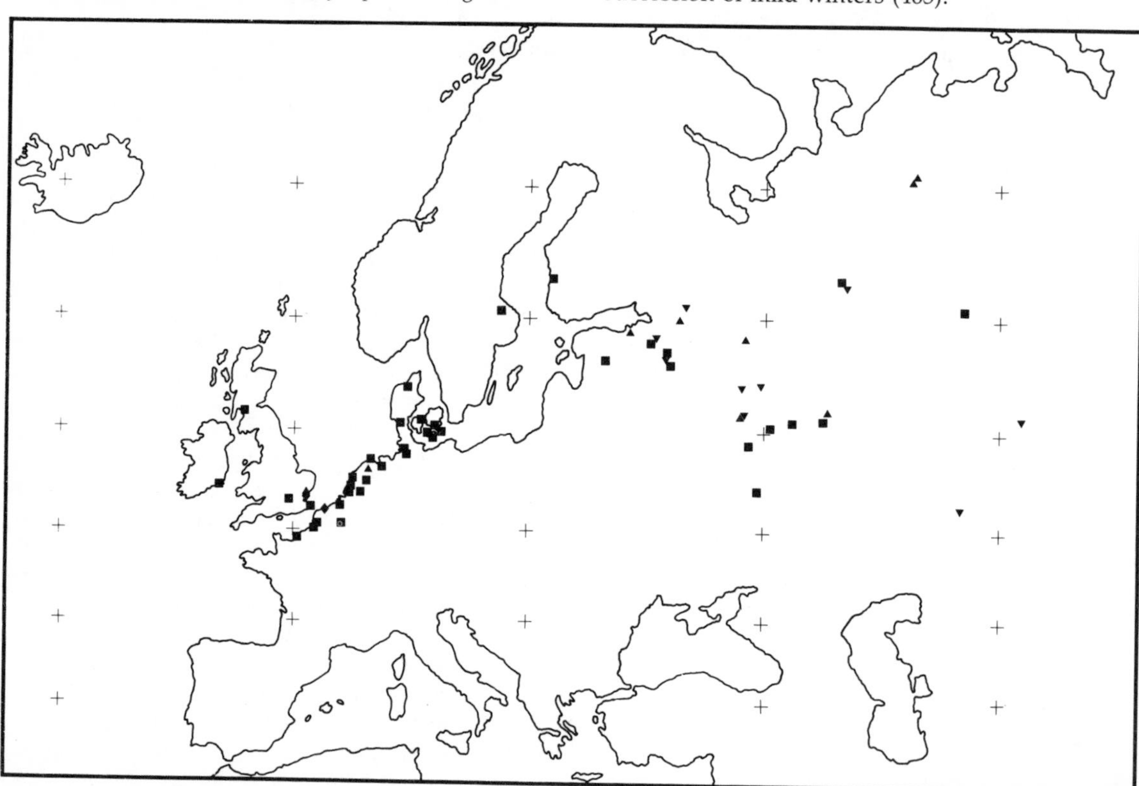

The size of the British breeding population was thought to be below 500 pairs in the late 1960s, with a marked concentration in south-east and east central England and a scattering northwards to Orkney (621). The Atlas survey confirmed the distribution but estimated the population to be about 1,000 pairs in Britain and Ireland, at least half in the Fens, East Anglia, and Kent (542). The most important site is the Ouse Washes, where the numbers have increased from 20-50 pairs in the early 1950s to 100-200 in the 1960s and 120-300 in the early 1970s, with a maximum of 306 pairs in 1975 (582). Since Shoveler occur in well-surveyed parts of the country the numbers counted are close to the population total. The present average of 5,000-6,000 birds in September suggests a breeding population of 1,000-1,500 pairs.

The breeding habitat is usually marshland adjacent to shallow open water, usually over rich calcareous soils providing abundant food for ducklings. The nest is usually in relatively sparse cover, barely concealed by tussocks of grass or sedge. Shoveler are unusual among dabbling ducks in being territorial, the male remaining close to the nesting sites and defending a territory well into incubation, though leaving before hatching.

Fig. 114. Shoveler – monthly pattern of numbers in Britain, expressed as a percentage of the peak count.

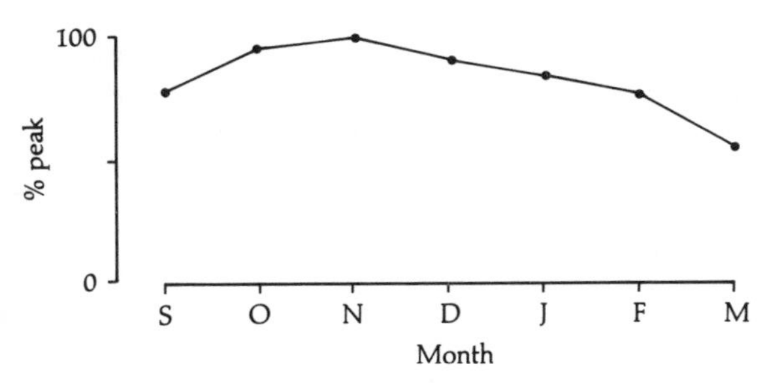

The north-west European stock was estimated at some 20,000 birds in January in the early 1970s (22), but since they form part of a much larger population, by then in more southerly areas, it is safer to regard the population as consisting of about 100,000 birds, only 20,000 of which remain to winter in north-west Europe (529). In January most are on freshwater or on brackish lagoons or bays, including a spectacular concentration of 20-40,000 on the Marismas of the Guadalquivir, south-west Spain. The population wintering in the Mediterranean and north Africa arrives there via an eastern route and numbers some 150,000 birds (529).

The distribution in Britain (Fig 115) shows the major concentrations in the south and east. The sites holding substantial numbers in northern England and Scotland do so usually in September and October. There are no sites qualifying for international importance, but the numbers at the four areas holding more than 300 regularly are given in Table 253. Three out of the four sites are reservoirs; nearly 40% of British Shoveler at the peak count are on this habitat. Rutland Water, Leicestershire, where flooding started in 1975, soon became a key Shoveler site. It may have drawn birds from the Ouse Washes, only 60km away, where numbers fell in the late 1970s following peaks of over 1,000 earlier in the decade.

Table 253. *Numbers of Shoveler on the four British sites holding over 300 regularly. 5-year means 1960-1974 and seasonal maxima thereafter. A blank = no data.*

	Abberton Reservoir	King George VI Reservoir, Surrey	Ouse Washes	Rutland Water*
1960-64	455	50	264	
1965-69	395	28	547	
1970-74	504	240	747	
1975	541	512	228	126
1976	649		479	231
1977	852		513	450
1978	328	311	212	471
1979	310	109	330	470
1980	281	488	411	316
1981	485	299	296	317
1982	612	539	685	443
Mean 1975-82	507	376	394	353

* Start of flooding in 1975.

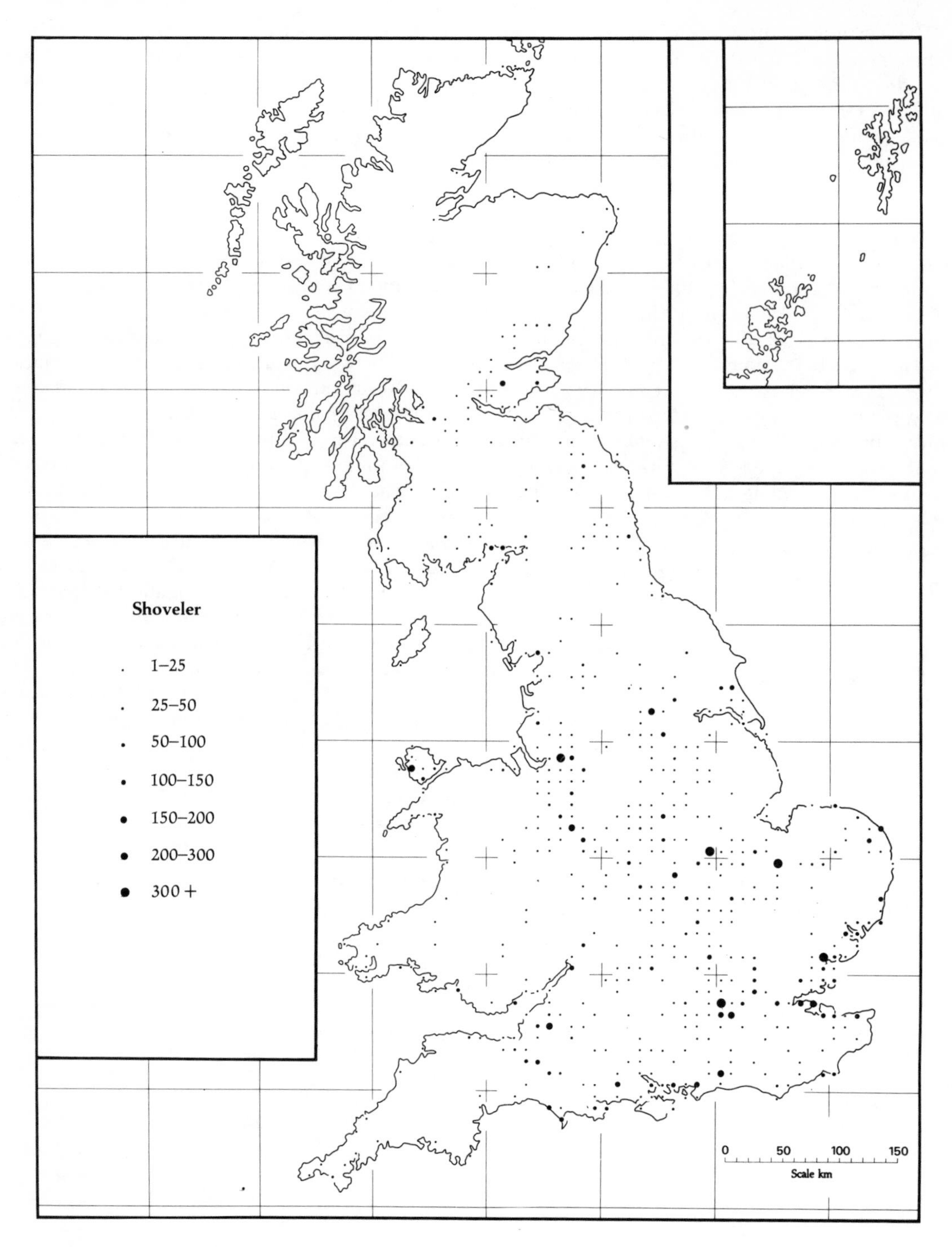

Fig. 115. Shoveler – British distribution.

The trend in numbers in October/November is shown in Fig 116. It appears that following a sharp decline in the first three years of the 1960s, particularly after the severe winter of 1962-63, the numbers have risen gradually and regularly (correlation coefficient for 1963-1979 $r = 0.894$) to high levels in recent years. The pattern is slightly misleading since the peaks in numbers in the early years were in February and March (see above), but it does reflect a real increase over recent seasons. This may in part be due to the larger contribution made to the numbers in autumn by the British breeding stock. If a substantial proportion now stays here until November this could account for all of the increase. The average numbers counted in November in the four five-year periods 1960-1964 to 1974-1979 were respectively 1,970, 2,780, 4,660 and 6,082. The increase has continued, the November average for 1980-1982 being 7,470. Numbers in the flyway as a whole (in mid-winter) increased between 1967 and 1976 in parallel with those in Britain (22).

The five-year mean count for November (total of all sites) is 8,200, compared with the average annual count in November of 6,100. Thus 74% of Shoveler are counted annually, and since they are mainly in the south-east, a well-surveyed area, the true total is unlikely to exceed 9,000 (i.e. about 10% missed completely). If we accept 9,000 as an average peak and relate this to the trend indices, the numbers in November have varied between 2,400 in 1963-64 and 11,500 in 1982-83. The numbers in February and March have declined drastically from the 8,000-10,000 quoted for the late 1960s (621). In the last five years, accepting 74% site coverage and 10% missed altogether, the average February population has been 6,900 and that in March 5,000.

Like some other birds, vulnerable in severe winters, the Shoveler has more than regained its former abundance since the setback in 1962-63. The species did not appear to suffer in the hard winter of 1981-82 – although numbers that season were low – and the highest ever count (7,690) was recorded in November 1982. Since the species' main wintering habitat – reservoirs – is one that is increasing in abundance there is no reason to suppose that the Shoveler is under imminent threat, at least in the northern part of its wintering area.

Fig. 116. Shoveler – trend in numbers in Britain, 1960-1982; average index for October and November (to even out irregular movement patterns).

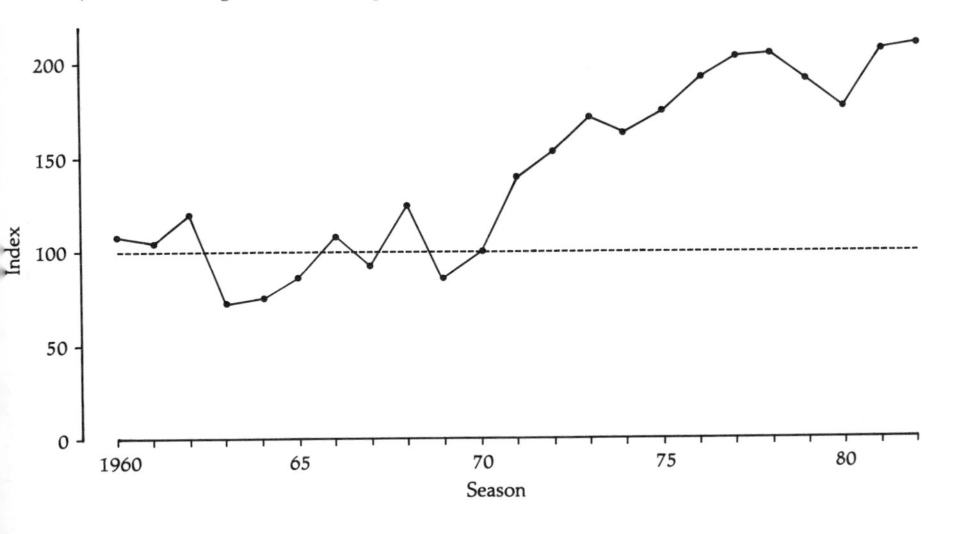

Pochard
Aythya ferina

A widespread and common wintering duck, the Pochard is very largely an inland species throughout its range. In Britain it is scattered in rather small concentrations although elsewhere single sites can hold very large numbers. Its breeding range extends over most of central Eurasia between 45 and 60°N, and its wintering grounds stretch from the extreme west of Africa to Japan.

The Pochard is a diving duck of rather shallow waters, rarely more than 3m deep. Food is gathered mainly from the bottom and the time spent submerged increases slightly with water depth. A large number of studies have indicated that depths of around 1m are preferred although diving has been recorded at depths varying from 0.3 to 3.4m. On average dives last 13-16 seconds, with a maximum of 25-30 seconds. During intensive feeding, the intervals between dives average about 5 seconds (304, 429). Although diving is by far the commonest feeding method Pochard sometimes upend in shallow water or feed with the head submerged like dabbling ducks. At the Ouse Washes, when floodwater is shallow diving accounts for only 55% of the feeding time (582).

Pochard feed both by day and night, though usually more intensively in daylight. There is no definite relationship between feeding intensity and time of day, and on average 7 hours a day (29%) are spent feeding (304).

The diet mainly consists of plant material in winter, chiefly the oospores of stoneworts *Chara* and *Nitella*, and the seeds of submerged aquatic plants, especially *Potamogeton* species. Shoots of these plants are eaten to a lesser extent, and tubers and rootstocks

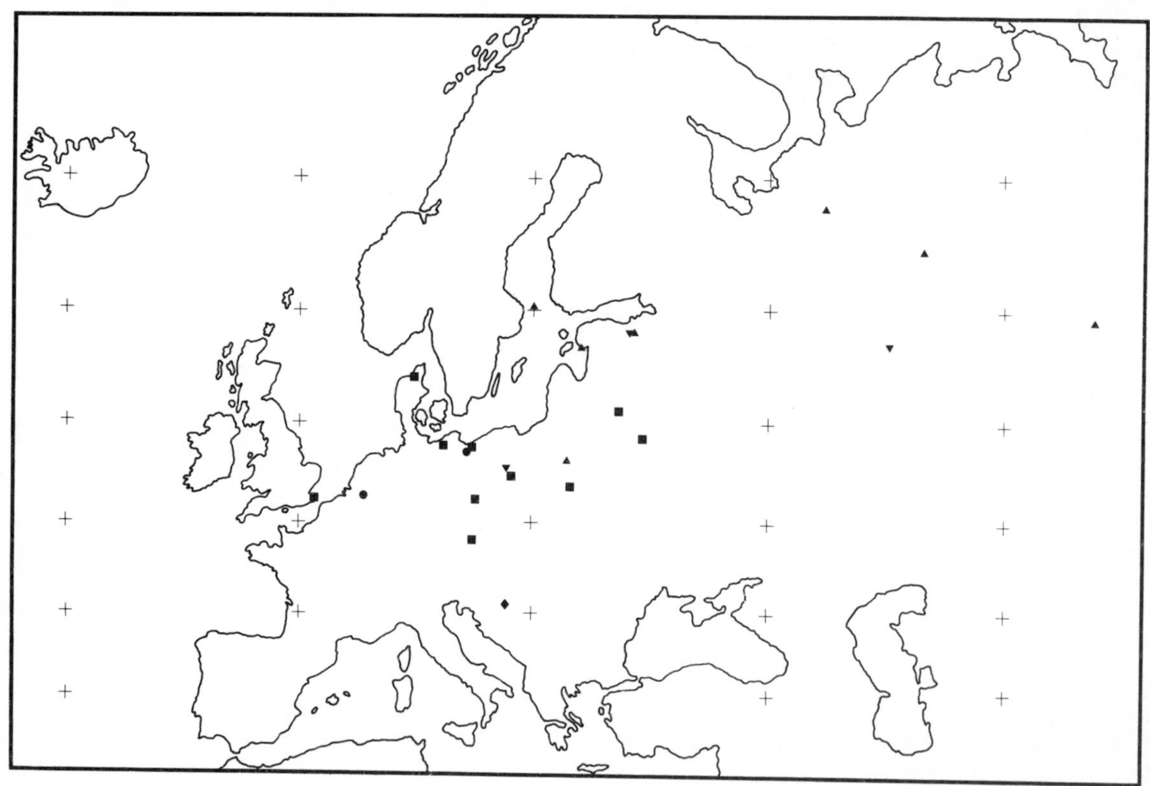

Fig. 117. The origins of British wintering Pochard as indicated by ringing recoveries. Triangles pointing downwards – April, pointing upwards – May, circles – June, diamonds – July, squares – August.

in some areas. Animal material generally consists of chironomid larvae and other small invertebrates, but rarely makes up more than 15% of the diet (429). The species had a more coastal distribution at the turn of the century, but a decline in *Zostera* and general deterioration of coastal habitats, together with an increase in suitable areas inland, particularly gravel pits and reservoirs, changed this situation. Nowadays more than 90% of Pochard counted in Britain are on freshwaters.

Britain has a small breeding population, with its centre in south-east England. Breeding is generally dispersed and sporadic, and in the late 1960s the total was estimated at about 200 pairs (621). The Atlas survey indicated rather larger numbers, perhaps as many as 400 pairs in Britain and Ireland, with confirmed breeding in 183 10km squares in Britain (542). The main centre is in the North Kent Marshes, where 30-50 pairs usually nest. The numbers in the Isle of Sheppey, first colonised in 1955, increased rapidly, to more than 20 pairs in the mid 1960s (270). The Norfolk Broads also have a sizeable population, including in recent years up to 50 pairs on the Filby/Ormesby complex just north of Great Yarmouth. The nest is usually on a platform of reed stems in a reed-bed surrounded by water, or in dense cover on the margins of lakes or creeks, never more than a few metres from the water's edge. Laying begins in May but most birds are incubating in June. Broods are reared in large stretches of open water with little cover and the young dive at an early age (270).

Table 254. *Numbers of Pochard counted on three British sites, 5-year means 1960-1974 and seasonal maxima thereafter.*

	Ouse Washes	Loch of Harray	Abberton Reservoir*
1960-64	728	1403+	
1965-69	2434	389	4560#
1970-74	2468	1839	1990
1975	400	3400	3096
1976	3055	2550	2319
1977	4897	3000	2877
1978	2948	3250	2282
1979	4706	1095	2380
1980	1203	1747	1250
1981	1310	1613	900
1982	1607	4500	2450
Mean 1975-82	2516	2644	2194

* Summer counts of moulting birds (July/August).
+ 4 years – 1962 missing.
# 2 years – 1968 and 1969 only.

British breeding birds make a minor contribution to the wintering population, however, even if all stay to winter, which is unlikely. Early ringing indicated that the immigrants came mainly from the Baltic countries and the USSR east at least to 61°. More recent ringing has confirmed this, and a Pochard has been recovered in Wales which was ringed at Barabinsk in the USSR (78°E). The pattern of recoveries (Fig 117) indicates a fairly continuous breeding range across the central USSR. Pochard travel as far as our furthest visitors among the ducks but they have a more southerly breeding range than other species.

About half the immigrant birds have arrived here by October, with peak numbers present in December and January (Fig 118), and the main exodus in February and early March. There is a large moulting concentration at Abberton Reservoir, Essex (Table 254), and small numbers elsewhere. The few recoveries from summer ringing suggest that the moulting birds originate from Britain, the Netherlands and Germany, but more information is needed to understand moult movements more fully.

The number of Pochard wintering in north-west Europe has been put at 250,000 (22, 529). The group is not entirely discrete, since ringing recoveries have shown that there is overlap between the breeding ranges of north-west European and Mediterranean birds and that some autumn-ringed Pochard in north-west Europe are found later in southern France and Spain. There are an estimated 750,000 wintering in the Black Sea/Mediterranean region and another 450,000 further east. In mid-winter there is a high concentration in the Netherlands and rather dispersed distribution elsewhere in north-west Europe. In the Black Sea/Mediterranean area the pattern is rather different, with a dozen sites each holding more than 8,000 regularly, and a spectacular concentration, averaging 215,000 in the late 1960s and early 1970s, in the Danube Delta.

The species is very widespread in inland Britain, having been counted on over 2,000 sites and occurring

Fig. 118. Pochard – monthly pattern of numbers in Britain, expressed as a percentage of the peak count.

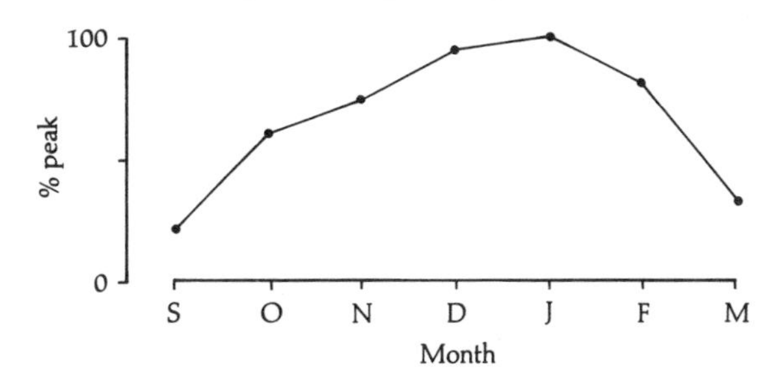

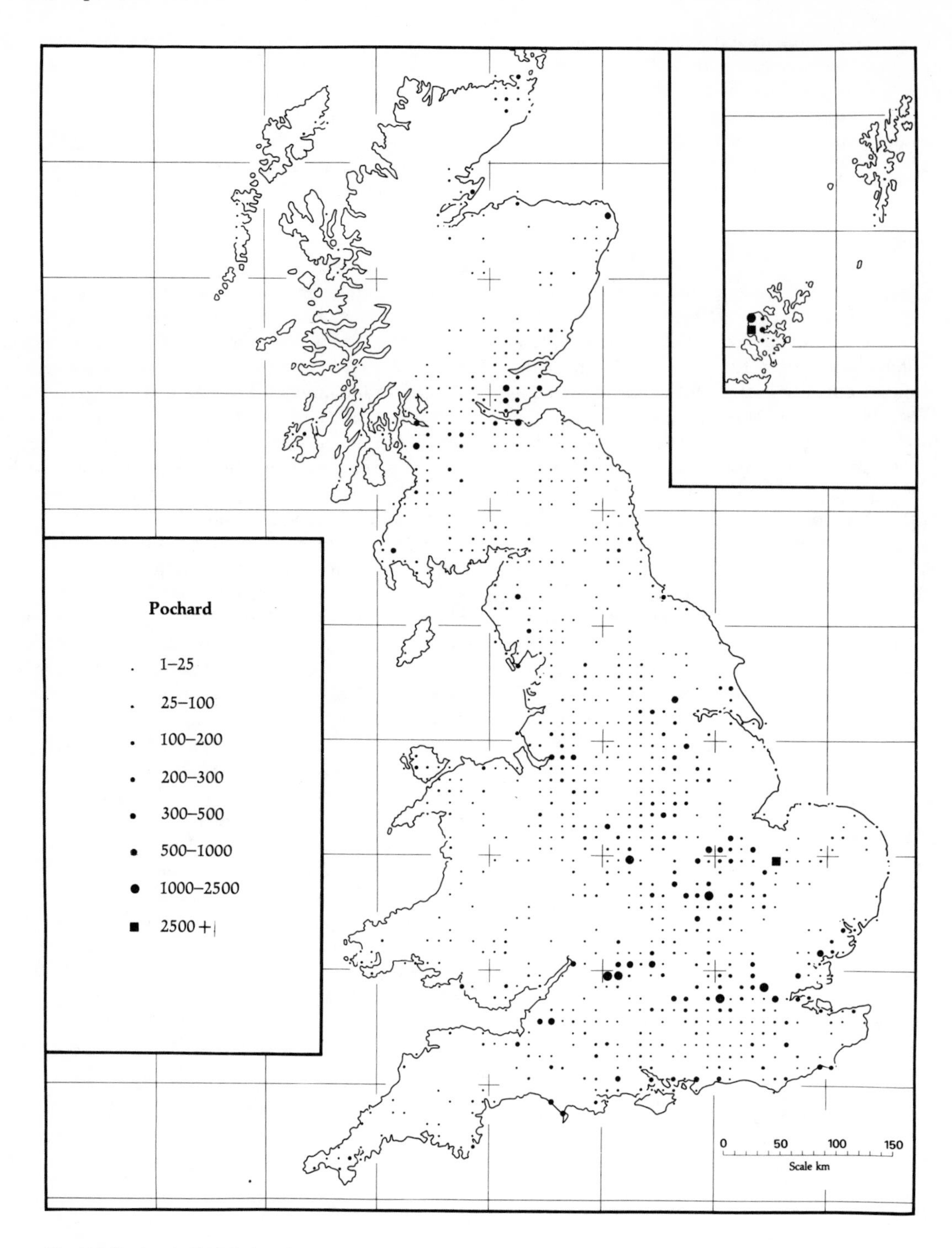

Fig. 119. Pochard – British distribution.

n nearly a thousand 10km squares (Fig 119). There are ıow only two internationally important wintering oncentrations in Britain (more than 2,500 birds), and ounts from these areas are given in Table 254, ogether with the numbers moulting at Abberton Reservoir.

The numbers at the Ouse Washes used to luctuate markedly in relation to the availability of floodwater, but the flocks have stabilised in recent years as the conservation organisations have provided permanent standing water and, to some extent, food. Duddingston Loch, a small water in Edinburgh, formerly held much the largest gathering of Pochard in Britain, peaking at 6-8,000 from the mid 1960s to mid 1970s. After 1977, when 7,400 were counted, there was a drastic decline, to only a few hundred in the next three seasons and less than a hundred since. The change is thought to be linked with the improvement in Edinburgh's sewage disposal (p.237). The

Fig. 120. Pochard – trend in September numbers in Britain, 1960-1982.

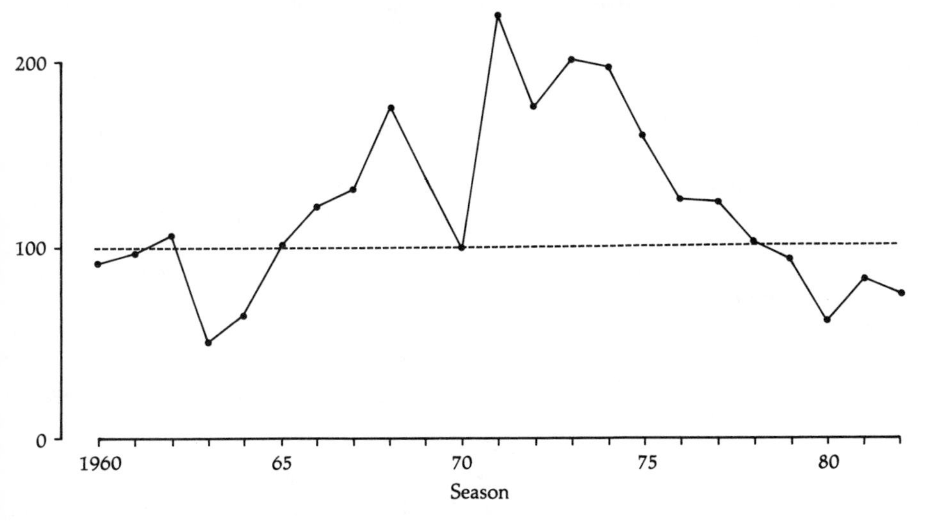

Fig. 121. Pochard – trend in mid-winter numbers in Britain, 1960-1982; average index for December and January (to even out irregular movement patterns).

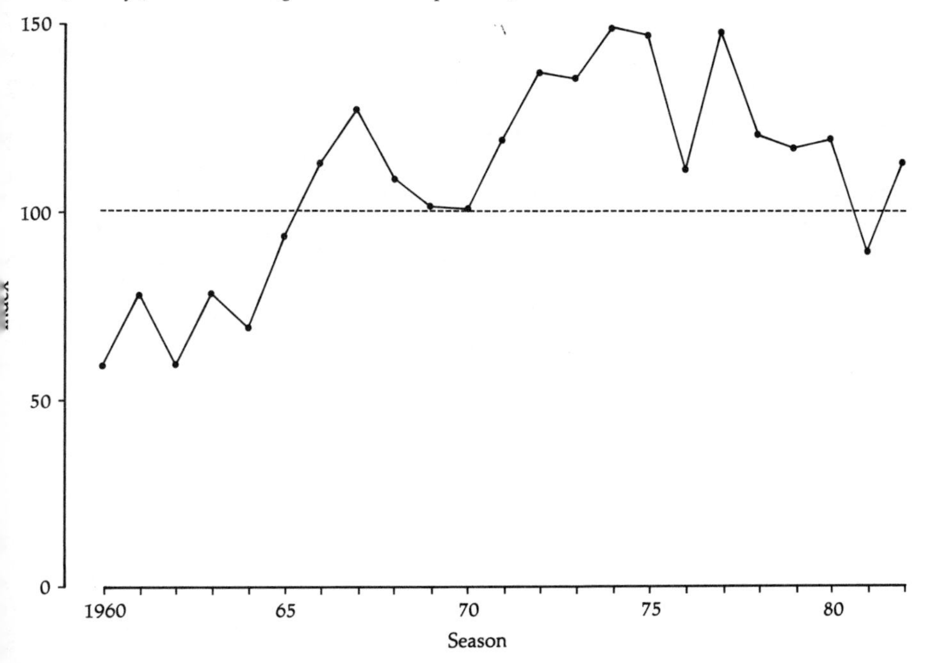

moulting flock at Abberton has also declined since the early 1970s, as have the numbers in Britain as a whole in September (Fig 120). Following a predictably sharp drop after the 1962-63 hard winter, numbers have fluctuated, presumably in relation to the arrival date of immigrants. There has, however, been a clear and major decline since the very high levels of the early 1970s. The index for south-east England, where most British Pochard breed, has changed even more remarkably, especially in recent years. The south-east now holds less than 14% of British Pochard in September, compared to between a third and a half previously. Thus the decrease both at Abberton and in the region in general may reflect a decline in the British breeding population since the early 1970s, but the movements of moulting birds are a complicating factor.

The British stock at its mid-winter peak rose sharply in the mid 1960s and again in the early 1970s, following a slight drop in the late 1960s (Fig 121). Since Duddingston Loch was such a major site and numbers there fluctuated so markedly the trend excluding that site was also examined. The result was the same, so that the increase at Duddingston was occurring in parallel with, rather than determining, the British trend. There was an overall decline in the late 1970s, but numbers rose again in 1980-81. There was a considerable increase in numbers in the flyway as a whole in the early 1970s but there is no recent information of sufficient completeness to indicate the present situation in the remainder of north-west Europe.

The average number of Pochard counted in January – the peak month – in 1975-1979 was 35,350 and the total 5-year mean for all sites 47,069. Half the ducks were on sites holding less than 200 and a third in concentrations of less than 100. It seems likely, therefore, that some Pochard are missed altogether (perhaps 10% – c.4,000 birds). Thus the probable average population is around 50,000, three-quarters of which are counted each January. This is considerably higher than the estimated population of 13,000-15,000 in the First Edition and higher than other recent estimates (524).

Pochard have increased their numbers in western Europe because they have taken advantage of the increasing amount of inland freshwater available as a result of gravel extraction, impoundment for drinking water and desalination following saltmarsh reclamation. The area of these habitats continues to increase and it seems likely that Pochard numbers will hold their own, despite occasional setbacks as a result of poor breeding or severe winters.

Tufted Duck
Aythya fuligula

Our commonest and most widespread divin[g] duck, the Tufted Duck is, like the Pochard, largely a[n] inland species, with about 95% of British stocks bein[g] counted on freshwaters. The species breeds acro[ss] Eurasia as far north as 70°N in Scandinavia and sout[h] to 45°N. Outside Britain it is often found in brackis[h] areas, particularly in the Baltic.

The Tufted Duck is primarily carnivorous, wit[h] animal material usually accounting for over 80% of th[e] diet. Molluscs are generally predominant, especiall[y] the Zebra Mussel (*Dreissena polymorpha*) and Spir[e] Snail (*Potamopyrgus (Hydrobia) jenkensi*) in freshwate[r]. The remainder of the diet is made up of a variety [o]f insects and plant material mainly in the form of seed[s] (424, 582). In brackish waters the food is almost a[ll] animal, with molluscs such as mussels, cockles an[d] crustaceans predominating (324). At Loch Leve[n], Kinross, chironomid larvae are overwhelmingly im[portant] especially in summer, but here large number[s] of molluscs are not available. Chironomids are filtere[d] from the bottom mud with an action similar t[o] dabbling while the bird holds itself at an angle of 45° t[o] the bottom (311). Molluscs, sometimes up to 25mm i[n] length, are presumably picked individually from th[e] bottom or from submerged plant leaves.

Nearly all the food is gathered by diving, th[e]

Fig. 122. Tufted Duck – monthly pattern of numbers i[n] Britain, expressed as a percentage of the peak count.

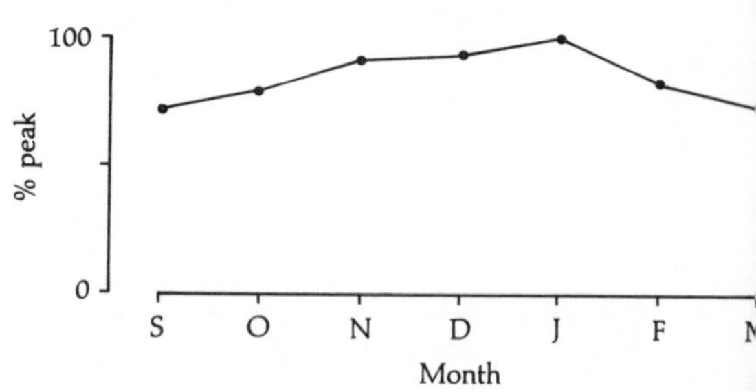

nly other feeding posture being upending. The referred depth is 0.6-2.5m but depths of up to 5m are ot unusual. The duration of a dive increases with epth but on average the birds remain submerged for 5-20 seconds, the normal maximum being 30-40 econds (195, 311). Males dive to greater depths than emales and consequently stay under for longer. This eads in some situations to a separation of the sexes, so hat feeding flocks may consist almost entirely of birds f one sex (615). During intensive feeding, pauses etween dives are about 10 seconds long and on verage feeding Tufted Ducks make 108 dives per our (195).

Tufted Ducks feed during the day and rest at ight in most places (141, 195, 582), although at Loch even and a nearby lake the ducks rest during the day nd feed at night (311). Diurnal feeders spend 61% of he day-time (7-8 hours) foraging (195).

Ducklings dive well as soon as they leave the est, but for the first week or two they pick insects rom the water surface as well as chironomid larvae, addis larvae and other items from the bottom (311).

Their need for animal foods restricts Tufted Ducks to rich lowland waters where the depth is no nore than 5m. For breeding, waters with rich bottom nud are required to provide the ducklings with bundant small larvae and emerging flies.

Fig. 123. Tufted Duck – trend in numbers in Britain, 1960-1982, in September (British post-breeding population – dashed line) and January (solid line).

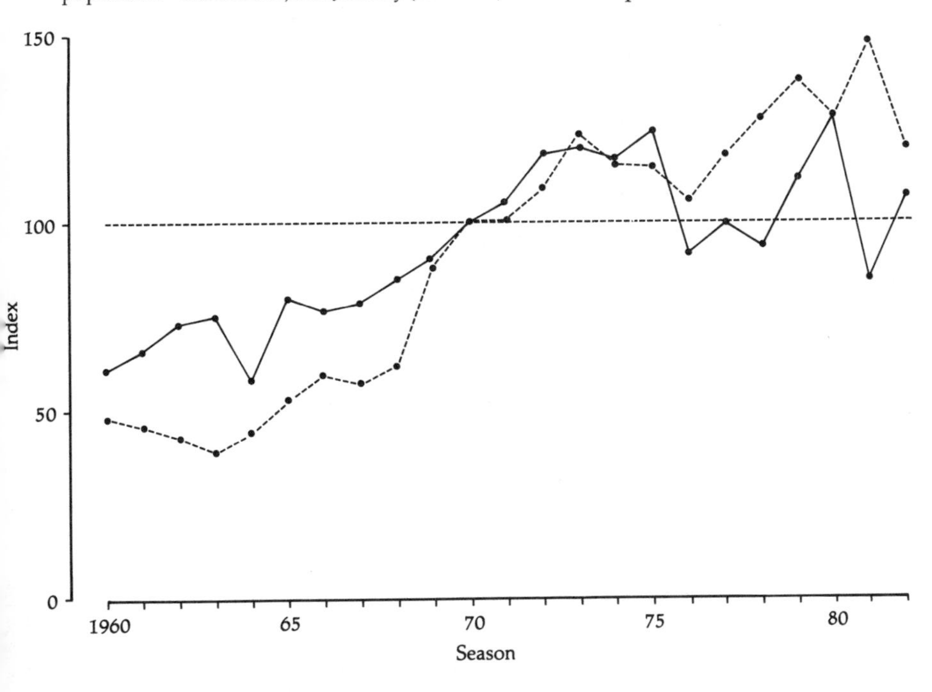

Whereas Tufted Ducks breeding in Scotland move to winter in Ireland (8) those in southern Britain are largely resident. Immigration does not start until late September so the count for that month can be regarded as an index of the native stock. Clearly the build-up to the January peak is gradual (Fig 122), while the decrease in February is more rapid. The March population is similar to that in September, suggesting that a considerable number of immigrants of this late breeding species remain. At Loch Leven the main spring arrival occurs in late March and early April.

The largest breeding concentration of Tufted Ducks in Britain is at Loch Leven, where 5-600 pairs nest mostly on St Serf's Island. With the provision of gravel pits in lowland England the species has increased its numbers rapidly. The nest is either on islands or close to the water margin, usually in tussocky grassland or low shrubs affording some cover. Tufted Ducks breed later than most other ducks, the peak of laying being in late May at Loch Leven (380) and only a week or so earlier further south. In an extensive British survey, production per breeding adult was estimated as 2.3 compared with 1.8 for Mallard (598).

The British breeding population in the late 1960s was estimated at 1,500-2,000 pairs (621), but this was revised to 4,000-5,000 in the Atlas survey (542). The September trend (Fig 123) indicates an increase of around 50% since then which suggests a current population in the region of 7,000 pairs. The average September count has been 30,000 in recent years, indicating a post-breeding population of 38,000, given that about 80% of Tufted Ducks are counted (see below). A productivity estimate of 2.3 young per adult (598) is extremely high and is likely to be an overestimate. At Loch Leven production is very poor in comparison, perhaps as low as 0.5 young per pre-breeding adult. Using an intermediate value of 1.5 young the post-breeding population yields an estimate of 7,000-8,000 breeding pairs, and this is likely to be close to the present British population.

Numerous Tufted Ducks have been ringed in Britain both in winter and summer. As well as in Britain many are recovered in France in late winter and in breeding areas stretching from Iceland east to central USSR in subsequent seasons. These represent both immigrants caught in winter and dispersal (partly through abmigration) of home-produced ducks. The distribution of summer recoveries is shown in Fig 124.

Tufted Ducks are a numerous species in north-west Europe, where some 500,000 winter. A further 350,000 are found in the Black Sea/Mediterranean region and at least half a million in western Asia (22, 529). In January most are in the Baltic, where up t 200,000 may be found in Danish waters. The Dutc IJsselmeer also holds tens of thousands, and as man as 30,000 have been counted at Lough Neagh Northern Ireland; elsewhere in the range the distribu tion is more dispersed.

In Britain the species has been recorded on som 2,500 sites, 61% of those surveyed, but only seven c these (Table 255) regularly hold more than 1,000 bird and none approach the qualifying level of 5,000 fc international importance. Two other 10km square hold concentrations of more than 1,000 as Fig 12 shows, but these birds are on several waters in th London area, where there is a very large disperse population (p.116).

The peak at Abberton Reservoir is usually in Jul and August, when large numbers gather to moul there. These are predominantly males and recoverie suggest that they originate from all parts of souther England, but some may come from overseas. Ther are also moulting concentrations in the London area totalling 6,600 in 1979, around 80% of which wer drakes (419). Loch Leven breeders also moult on th loch (8).

The trend in Tufted Duck numbers shows remarkable increase over the two decades (Fig 123) The British breeding stock has trebled since 1963. Th

Fig. 124. The origin of immigrant Tufted Ducks as indicated by ringing recoveries. Triangles – May, circles – June, diamonds – July, squares – August.

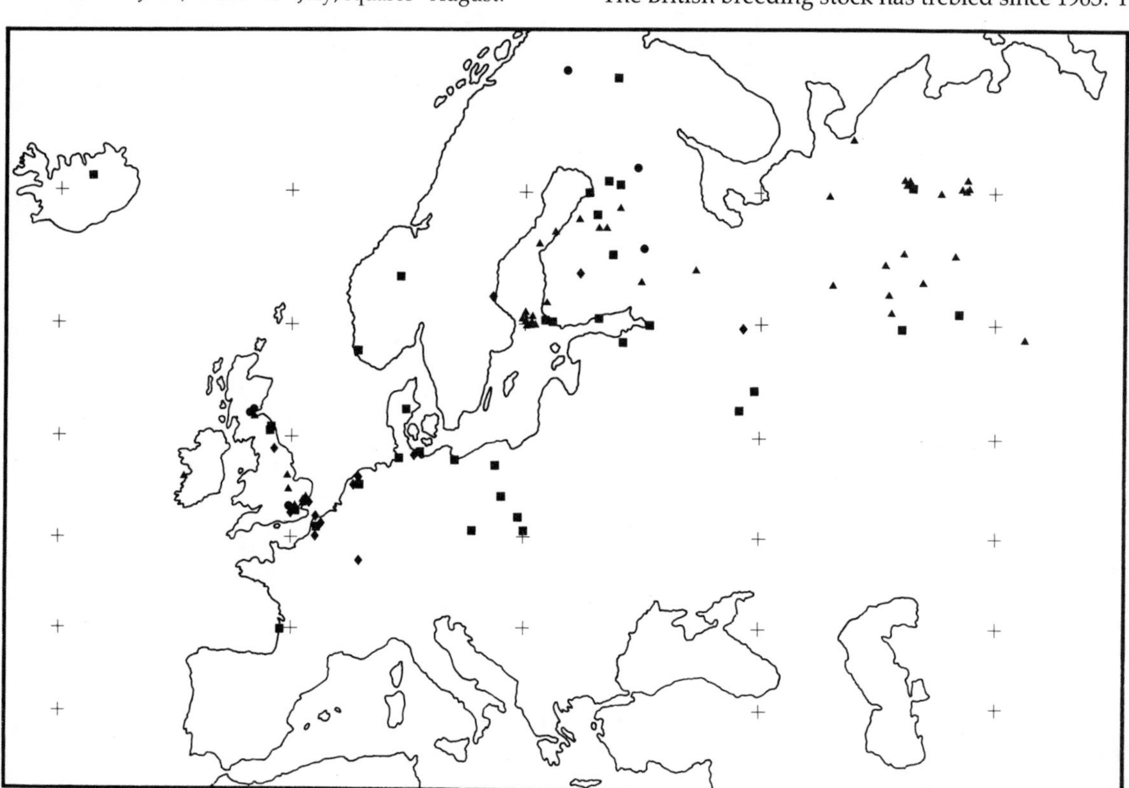

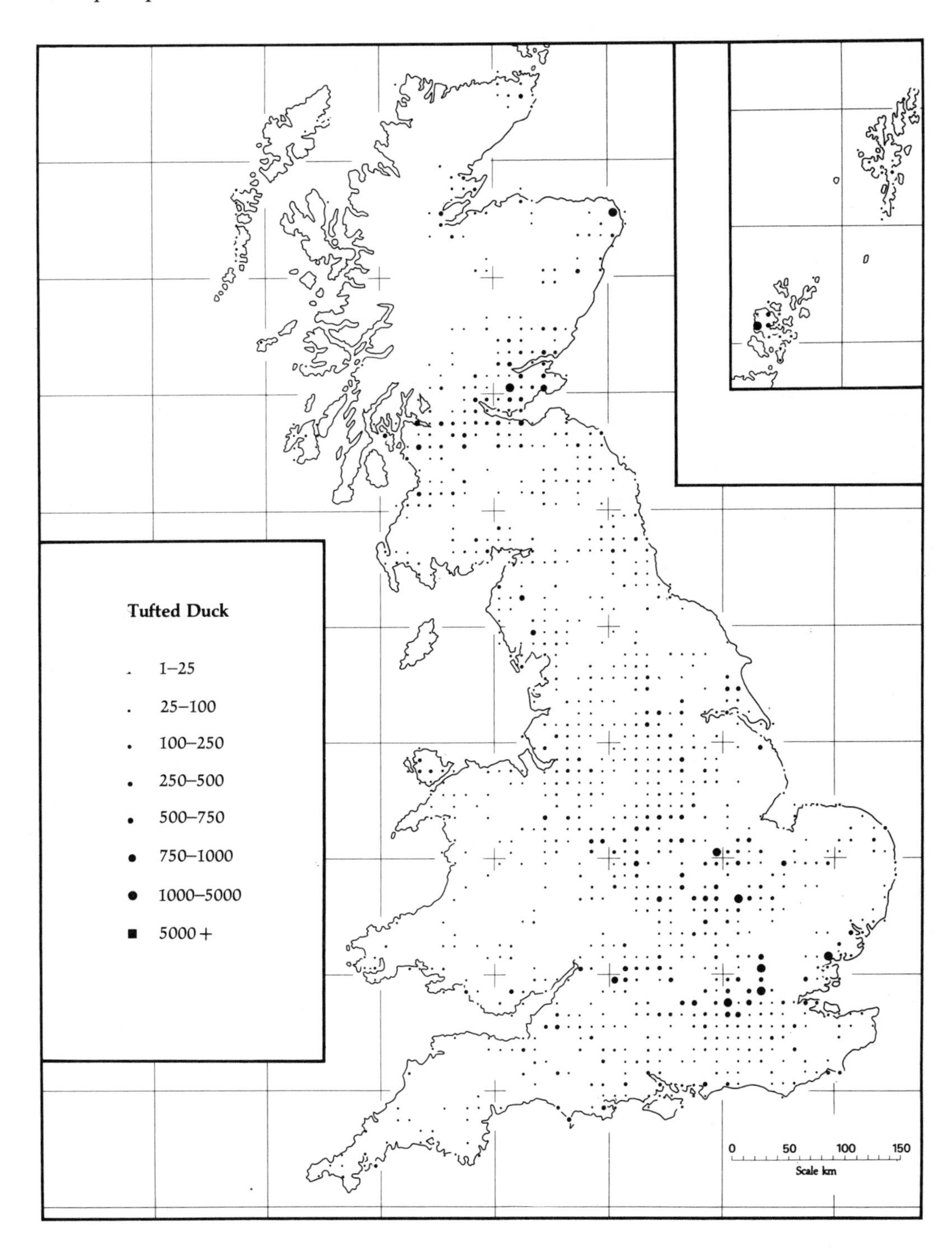

Fig. 125. Tufted Duck – British distribution.

wintering stock has increased steadily but less sharply, having almost doubled since the early 1960s, and recently seems to have levelled off. The average number counted in 1975-1979 was 30,000 in September and 40,000 in January. The 5-year mean totals indicate that 80% of Tufted Ducks on surveyed sites are counted annually. Being a dispersed species it seems likely that at least 10% may be missed altogether, so that the true numbers are, at a minimum, 45,000 in September and 62,000 in January. The increase in numbers wintering could all be accounted for by the increase in local stock while immigrants remained constant, but there is probably considerable turnover, with some of our own breeders moving south and west, and being replaced by immigrants from the Continent.

As a diving carnivore the Tufted Duck has no serious competitor for the new artificial habitats of gravel pits and reservoirs, so further expansion must be expected both in the breeding and wintering stocks. The species can live and breed in close proximity to humans, and this enables it to withstand increasing pressures from sport and recreation and even to breed in substantial numbers in parks and other leisure areas.

Table 255. *Numbers of Tufted Ducks on sites holding more than 1000 regularly. 5-year means 1960-1974 and seasonal maxima thereafter. A blank = no data.*

	Abberton Reservoir	Wraysbury Gravel Pits	Grafham Water *	Rutland Water +	Loch Leven	Loch of Strathbeg	Loch of Harray
1960-64	405#					416	829
1965-69	634	552	515		1190	546	318
1970-74	1824	551	964		2180	1068	1517
1975	2101	561	2500	252	1000	1200	1960
1976	3389	607	1300	703	1500	1500	1400
1977	2042	805	1180	949	1250	1000	1390
1978	3354	1411	1750	2287	2000	1100	1740
1979	2478	909	3050	2208	4500	1500	916
1980	1660	1528	1010	1523	4273	1160	1289
1981	2670	1343	765	1804	4560	1350	1322
1982	1560	1512	190	2380	3455	1950	2279
Mean 1975-82	2407	1085	1468	1513	2817	1346	1537

* Flooded 1964.
+ Flooded 1975.
# Winter counts, all other peaks are of moulting birds.

Scaup
Aythya marila

The most northerly of the *Aythya* ducks, the Scaup breeds largely between 60 and 70°N across Eurasia and winters in temperate maritime areas between 30 and 60°N, with only a few further north in Norway and Iceland. The Eurasian subspecies is *A.m. marila* and a North American race, *A.m. mariloides*, is also recognised. The races are indistinguishable in the field so any occurrence of birds from North America in Europe would be undetected.

The Scaup is never found in numbers on freshwater in Britain in winter, though it is occasionally in other countries. Most frequently it is found in shallow bays and estuaries, predominantly in brackish rather than fully saline water. The natural diet is made up largely of molluscs in winter, especially the Blue Mussel (*Mytilus edulis*), which accounts for more than two-thirds of the food in Denmark (324) and more than half in Sweden (386). Cockles and *Macoma* bivalves are also common foods, and fish (goby and stickleback) and crustaceans are also eaten. Plant food occurs in small quantities, chiefly seeds of shoreline plants such as *Scirpus*. In Britain major concentrations are found using rather unnatural food sources. The massive flock in the Firth of Forth in the 1960s and early 1970s concentrated near sewer outfalls bearing grain from distilleries and also providing a rich invertebrate food source (112). The flock at Loch Indaal, Islay, is frequently found near a distillery discharge and these birds probably also rely heavily on grain. There are numerous other examples of such associations (481) and we must conclude that British Scaup have largely abandoned the natural winter diet in favour of one inadvertently provided by man.

Scaup are generally found in shallow water, no more than 4m deep, though they are apparently able to dive to 10m (324). Practically all are well within the 5m contour on the Forth (112), and most dives in Sweden are below 4m. There diving times vary from 9 to 29 seconds and largely depend on depth, with means of mean 13 seconds for 0-1m and 27 seconds for 3-4m. Pauses between dives average 10-17 seconds and about two-thirds of the time is spent underwater during periods of intensive feeding (386).

The pattern of arrival and departure (Fig 126) shows the Scaup to be a mid-winter species in Britain, with substantial numbers only in December – February. The population is virtually all immigrant but the proportion arriving from different breeding areas is not exactly known. By far the largest number of recoveries are of birds ringed in Iceland, but most ringing has been done there. Of 80 Icelandic recoveries in the British Isles, 27 (34%) have been in Great Britain and the remainder in Ireland. However, of 14 British-ringed Scaup recovered, 3 have been in Iceland and 11 either in the USSR (the furthest at 72°E), Scandinavia or on the continental European wintering grounds. The Irish population is also somewhat mixed, since a bird ringed at Wexford was found in the USSR and a Finnish-ringed Scaup has been recovered in Co Tyrone. A few Icelandic ringed ducks have been found on the continent of Europe (67) but this may be the result of abmigration. There is a possibility that some Scaup from Iceland migrate to Britain via north-west Europe (67).

Although information from ringing is sparse and somewhat difficult to interpret, it seems that at least in the 1950s to early 1970s, when most ringing was carried out, the British stock originated partly from Iceland and, perhaps largely, from mainland Europe and the USSR. There is some mixing of the two stocks both in Britain and elsewhere. There has been a suggestion that since the British population has changed so radically (see below), we no longer receive migrants from the east and this seems quite likely. The Icelandic population currently numbers 15-20,000,

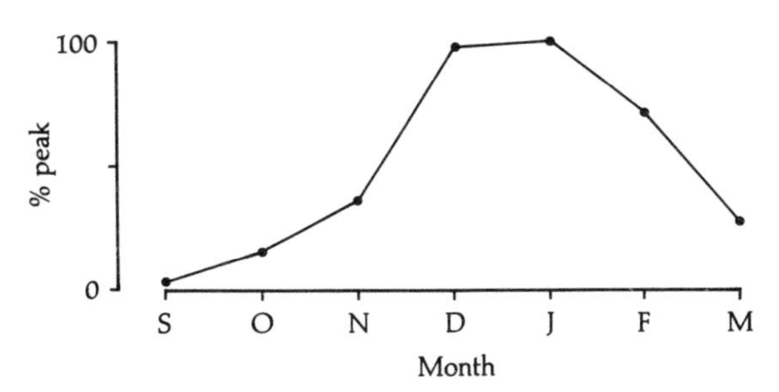

Fig. 126. Scaup – monthly pattern of numbers in Britain, expressed as a percentage of the peak count.

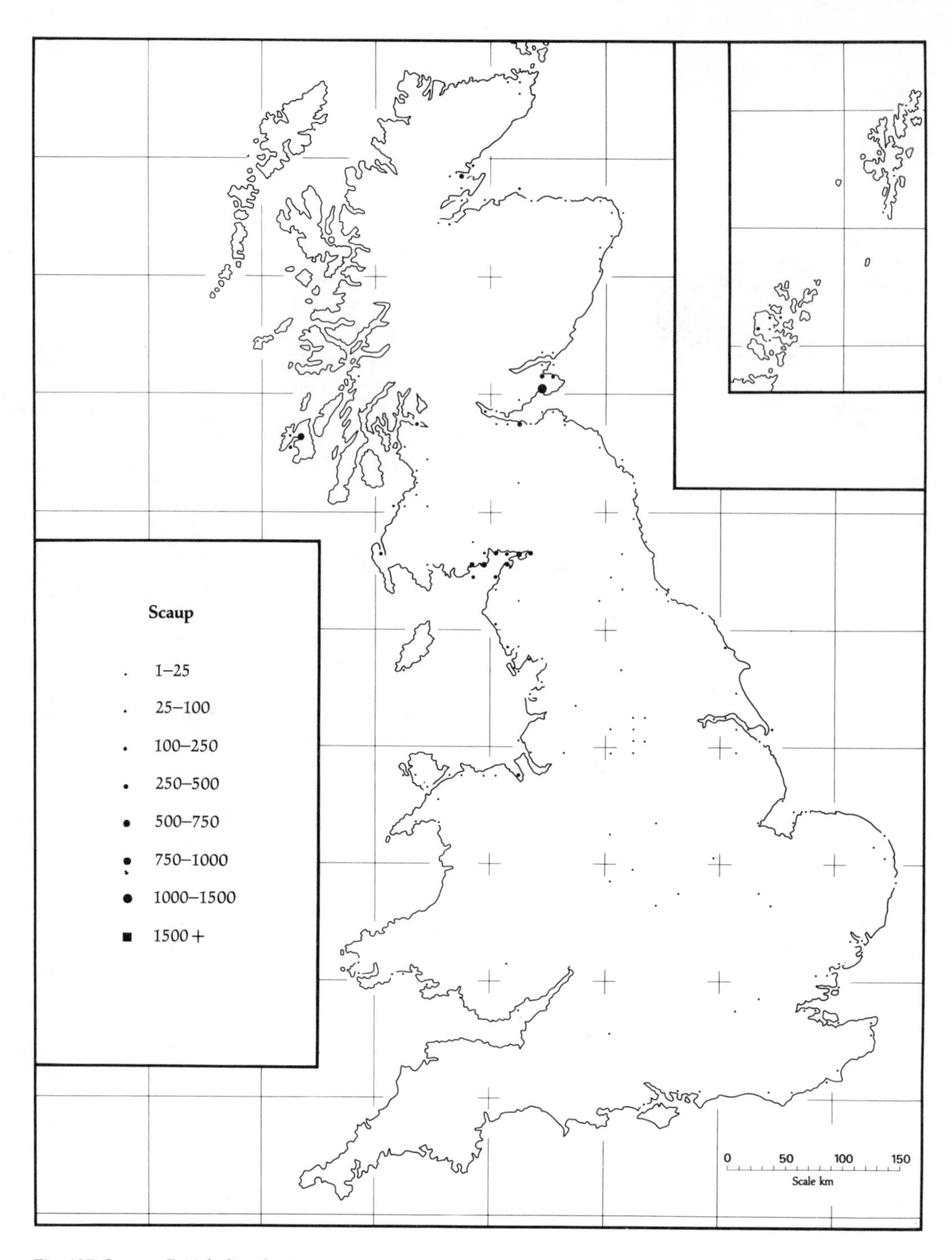

Fig. 127. Scaup – British distribution.

having shown no trend over recent decades. Numbers counted in Britain fell to below 2,000 birds in the early 1980s (but then recovered somewhat), so it seems likely that the Icelandic birds make up all our winter peak, and probably overflow to the Continent.

The Scaup breeds sporadically in northern Britain, but during the five years of the Atlas survey breeding was confirmed in only four 10km squares, one of which was in south-east England (542).

There are an estimated 150,000 Scaup in north-west Europe in mid-winter, highly concentrated in the Danish part of the Baltic and the Dutch Wadden Sea and IJsselmeer. There is a somewhat smaller group wintering in the Black Sea/Caspian region and another population in the Far East. Although Scaup are not intensively hunted in most parts of their range, because they are so concentrated they are in great danger from oil pollution (22, see also p.491).

The distribution of Scaup in Britain is shown in Fig 127. The pattern is of a few large concentrations and a scatter of isolated records. Only one site, the stretch of the Firth of Forth around Edinburgh, has qualified as internationally important. In the late 1960s and early 1970s this site held 80-90% of all the Scaup in the British Isles, with a peak of 25,000 in the late 1960s. The numbers declined markedly to around 10,000 in 1974-1976 and in the following two winters dropped again to 1,500 (112). The sewage treatment works in Edinburgh, on whose outfalls the Scaup were apparently dependent, were modernised in 1978, after the initial decline. Since then, however, the numbers of Scaup have dwindled to less than 50. With such a major fall the numbers might have been expected to increase correspondingly elsewhere, but this did not occur. Even the neighbouring site of Largo Bay, on the other side of the firth, showed only a slight increase. Nowadays this site, with 1,500-2,500, Loch Indaal, Islay (800-1,200), and Carsethorn Bay, north Solway (600-1,000), hold the bulk of British wintering Scaup.

The trend in numbers up to the early 1980s reflected the changes in fortune of the Edinburgh flock (Fig 128). The British count, however, continued to fall – to 3,200 in 1982-83 and and only 1,990 in 1983-84 (when coverage was incomplete) but increased again to 3,450 in 1984-85. Incomplete counts from Ireland, though including the main resorts, yielded 2,000 Scaup in 1980-81 (524), and counts from all sites indicate an Irish population of no more than 3,000 (279). This means that the British Isles stock is now less than 10,000 and perhaps no more than 6,000 Scaup, unless sizeable flocks remain undiscovered.

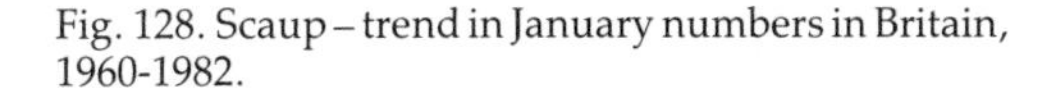

Fig. 128. Scaup – trend in January numbers in Britain, 1960-1982.

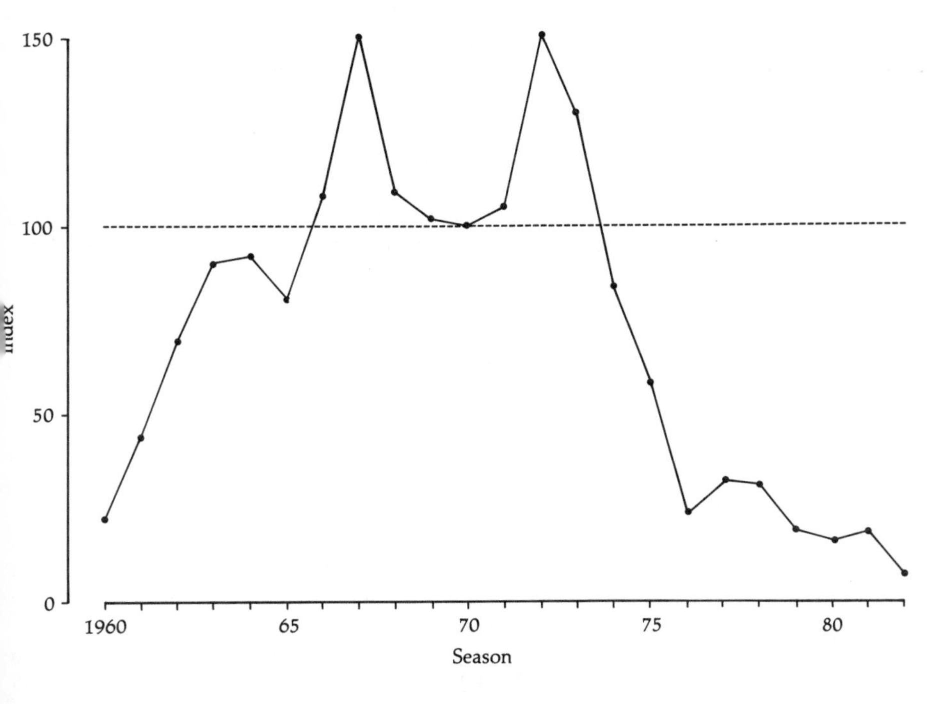

Eider
Somateria mollissima

A sea duck throughout the year, the Eider is distributed across the Arctic except for north central Siberia. There are six subspecies, three of which are exclusively North American (including the sedentary Hudson Bay subspecies *S.m. sedentaria*). Of the three eastern Atlantic races *faeroensis* is restricted to the Faeroe Islands, *borealis* occurs in the central north Atlantic, Greenland and western Canada and the nominate European Eider occupies the coastline of Europe and western Siberia. It breeds as far south as Brittany (48°N) in small numbers and as far north as 77°N in Novaya Zemlya. It winters over most of the breeding range except in western Siberia.

The Eider is rarely seen inland in Britain; it occupies shallow coastlines, bays and estuaries throughout the year, although there are regular sightings in the larger lakes of central Europe. Its distribution is largely tied to that of its main food – the Blue Mussel (*Mytilus edulis*) – which it obtains mainly by diving. Other bivalves are taken frequently as well as marine gastropods (periwinkles and whelks), crabs, sea urchins, starfish and a wide variety of other sedentary or slow moving invertebrates. Faster moving species such as *Nereis* worms and fish are eaten occasionally (324, 475). On the Firth of Forth, although occurring in the same general area as other diving ducks dependent on sewer outfalls (and which have largely disappeared as the outfalls have been closed), the Eiders concentrate on natural foods (112, 475).

In the breeding season feeding is in much shallower water and a wider variety of food is eaten. In August and September crustaceans replace molluscs as the most frequent dietary items but with the mussel still making up a major proportion. Very small ducklings concentrate on crustaceans but after the second week molluscs make an increasing contribution, until at nine weeks of age the diet resembles that of adults (470).

The feeding depth is usually no more than 4m, although it may very rarely reach 16m (470). On the Forth the preferred depth is said to be 2-3m and many birds choose to feed in shallow creeks and channels by immersing the head and neck or upending (475). In some areas ducks with broods forage among seaweed-covered rocks in very shallow water and dive little. Feeding activity is clearly influenced by the state of the tide, often ceasing completely at high water levels. Nevertheless, in the long days of mid-summer there are distinct peaks in feeding activity in the early morning and the late afternoon (470, 475).

The Eider uses its strong bill to prise molluscs and starfish from rocks and swallow them whole. It has a large, muscular gizzard which crushes the shells and grinds up the contents. The range of mussel size taken does not correspond with that available, the smaller ones being preferred. On the Forth mussels between 5 and 30mm are most commonly taken, although a few exceed 40mm (475). Exceptionally, mussels up to 80mm long have been found in Eider stomachs (324).

In recent years, as mussel cultivation has become commoner in western Scotland, Eiders have been causing problems by capitalising on this easily available food source, and licences have had to be issued to shoot them (there is no open season).

The British Eider stock is considered to be resident and to represent a reasonably discrete population moving only short distances (349). Most ringing has been carried out on the Ythan Estuary, Grampian, and in Northumberland breeding colonies. A recent analysis of more than 1,500 recoveries (33) shows only 16 abroad, all in Denmark and the Baltic. Of these 12 had been sexed and all were males, strongly suggesting abmigration. The suggestion is that Baltic Eiders sometimes winter in eastern Scotland and form pairs with native birds. The Firth of Tay flock may no longer contain immigrants since the foreign recoveries have ceased (none have been reported of birds ringed since 1971 despite the continuation of the ringing effort). Little is known of the Eiders scattered on the north and west coasts but it is assumed that they, too, make only local journeys to moult and winter, perhaps as far as Northern Ireland. Many of the wintering birds in the south of England are almost certainly part of the continental (Baltic) population, and there have been recoveries of Dutch-ringed birds on the east coast (565).

Moult migration is common in Eiders; those from the Ythan move to moulting sites at Murcar, just south of the breeding site, and at Gourdon, closer to the Tay wintering area. The Tay provides a wintering haunt for birds breeding to the south as well as to the north (33). First year birds do not join adults in the wintering flocks but disperse along the coast, not far

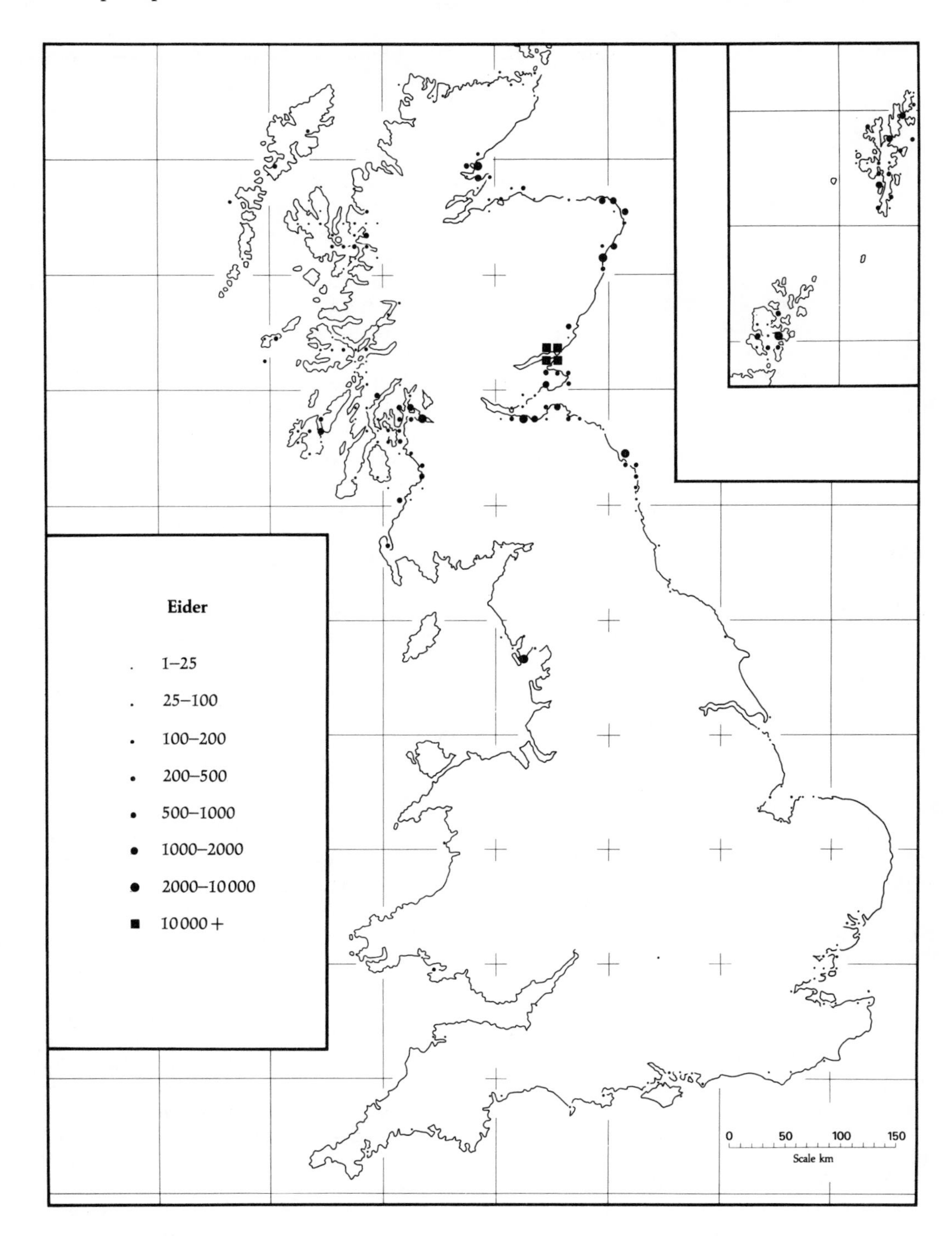

Fig. 129. Eider – British distribution.

from the breeding site. Nearly three-quarters of the Eiders on the east coast of mainland Scotland are on only 4 sites, and these concentrations are threatened by oil pollution and to some extent recreational disturbance (116).

Eiders are vulnerable to predation from land and air while nesting and prefer to breed on islands. The nest is, if possible, placed among cover of vegetation, rocks or debris, and in some areas humans capitalise on the Eider's preference for cover to concentrate large numbers of breeding pairs in shelters or huts provided (to facilitate down collection). Because of the high predation rate on unguarded nests the Eider has adopted the strategy of laying down large fat reserves to enable it to stop feeding during incubation and cover the eggs almost continuously. Eiders mature slowly; at the Sands of Forvie NNR on the Ythan Estuary about a quarter of the females breed at two years of age but most do so at three or four (34).

Soon after incubation begins, the males leave the breeding area and congregate in safe offshore moulting sites some distance away. A few nests are as much as 4km inland and the brood's first journey from the nest to the shore is a hazardous one, with predation from large gulls accounting for many of the young. Survival after reaching the water is usually also poor, with gull predation increasing markedly during wet and windy weather. The ten-year average survival between hatching and fledging at Sands of Forvie is below 10% (33, 341). The young form large creches, sometimes of several hundred ducklings, attended by adult females, with an average of about 3.5 ducklings per female (220).

The Eider population wintering in the Baltic and the Netherlands numbers in excess of a million birds, mainly found around the coasts of Denmark and Sweden. At least 250,000 are thought to occupy the coasts of Norway and the western USSR, and the *borealis* populations in Iceland and the North Atlantic islands, together with an unknown number of *faeroensis*, bring the European total to at least 2 million. The contribution made by Britain to the north Atlantic population is a minor one, estimated at about 2.5% in the early 1970s (23).

As can be seen from Fig 129, the Eider is largely concentrated on the east coast of Scotland. The main site and the only one qualifying for international importance is the mouth of the Firth of Tay. The highest numbers are present there in autumn, and have probably exceeded 20,000 on occasions, allowing for the extent to which the regular surveys (Table 163 p.000) may have underestimated the total or, in some years, missed the peak. Other large concentrations are on the Ythan Estuary, where about 1,000 winter but the breeding population numbers 5,000, Seafield on the Firth of Forth (3,000-7,000) and Lindisfarne (about 3,000, largely from the Farne Islands breeding colony). On the west coast the only notable known winter flocks are off Walney Island, Cumbria (4-5,000), and in the inner Firth of Clyde (1,500-3,000), but there may be a few major concentrations further north, where coverage is poor and there is certainly a large dispersed population.

The Eider has spread considerably in Britain since the mid 19th century from its former strongholds in the northern and western isles. Colonies were already established in East Lothian and the Farne Islands, but there was a rapid expansion in mainland Scotland, and Walney Island was colonised in 1949 (574). The numbers at most sites have been increasing in recent years, e.g. from about 1,000 pairs at the Sands of Forvie in the early 1960s to the present 2,500. The trend of Wildfowl Counts indicates an increase in the British population, but although the trend is apparently clear, the indices are not reliable because of irregularity of coverage.

The total number in Britain was estimated at 30-40,000 in the First Edition and at between 50 and 60,000 in the early 1970s (23). The five-year mean for all sites is highest in February, at 35,000, but since monthly coverage at most Eider sites is lacking this must be a considerable underestimate. There have been good aerial surveys of the east coast of Scotland, where the total was close to 30,000 in both 1970 and 1971 (353). The west coast population was thought "unlikely to exceed 20,000" in the early 1970s (23).

The Atlas survey (542) led to an estimate of 15-25,000 breeding pairs in the British Isles (very few in Ireland). Taking the production of the breeding population at Forvie as typical, there are 8.5 young per 100 adults in the post-breeding population (33). Assuming that the birds breed on average at three years of age, the non-breeding segment of a population of 20,000 pairs, allowing for the first year mortality, is in the region of 6,000. This calculation gives an estimate of 46,000 for the winter population. This is close enough to estimates made on the basis of winter counts (above) to conclude that the total number of Eiders in Britain is close to 50,000, an increase of 30-40% in the last two decades.

The steady increase in the Forvie population is continuing (33), and there is no reason to suppose that elsewhere the direction of the trend will be reversed. However, the increase in the threat of oil pollution is worrying as this could have a major influence on the future prospects for Eider stocks (p.491). Should there be a major setback the low reproductive output of the species means that recovery would be slow.

Long-tailed Duck
Clangula hyemalis

The Longtail is a numerous Holarctic sea duck, breeding throughout the Arctic regions of Eurasia and North America, south to 52°N in Hudson Bay and north to 82° in Greenland and eastern Canada. The wintering area extends southwards to 40°N in North America but in Europe its winter range is largely between 53 and 72°.

A truly marine species in winter, the Longtail is rarely found inland. The birds congregate in shallow bays or coasts predominantly over hard bottoms with extensive mussel beds. Like other sea ducks the Long-tailed Duck specialises on molluscs, and the Blue Mussel (*Mytilus edulis*) forms by far the largest proportion of the diet. In Danish coastal waters mussels predominate but a variety of other bivalves, especially cockles, clams of the genera *Spisula* and *Mya*, and gastropods such as periwinkles and whelks, are taken. Crustaceans are also eaten in quantity as well as fish such as gobies and sticklebacks. In the brackish fjords, where the species is less numerous, insect larvae, crustaceans and seeds are among the items taken (324). In Sweden the food is very similar, with about 60% of the diet consisting of mussels, another 12% of other bivalves and 21% crustaceans (385). There is no information for Britain, but since they occur in similar areas over bivalve beds their requirements are probably similar.

Long-tailed Ducks are smaller than most other sea ducks and take smaller mussels, usually less than 10mm long with the largest only 20mm (324). They are, however, expert divers, reaching depths of 30m or even more judging from depths at which fishing-net casualties have been recovered, though most commonly feeding birds are found between the 5 and 10m contours. At normal feeding depths the duration of a dive is between 30 and 60 seconds but it is even longer in deeper water (324, 385).

The Longtail breeds mainly near freshwater, in lake islands or adjacent to small pools, in North America sometimes several hundred kilometres from the sea. At Myvatn, Iceland, the species occupies the deepest parts of the lake (more than 3m deep), and feeds chiefly on chironomid larvae, also taking fish eggs and water fleas (*Cladocera*) and occasionally sticklebacks. Young ducklings feed exclusively on water fleas but when half grown change predominantly to chironomid larvae and a few molluscs, while *Cladocera* remain important (46).

Although there are a handful of breeding records in Scotland, the latest in 1926 in Orkney, the Long-tailed Duck is a mid- and late-winter species. Unfortunately, however, coverage is too sparse to indicate the pattern of arrival and departure. Very few have been ringed, so the origin of the British stock is uncertain. It is known, however, that those breeding in Iceland move west to winter in Greenland and that western Siberian migrants winter in the Baltic. There is a single recovery of a Greenland-ringed Longtail in Denmark, but it seems very likely that the British birds are part of the large population in Scandinavia and the north-western USSR (67).

There has been a great deal of confusion as to the number of Long-tailed Ducks wintering in north-west Europe, and the matter is still largely unresolved. On the one hand estimates of the number of breeding pairs in the western USSR suggest an autumn population of 1-2 million. On the other, the total counted in the whole of north-west Europe has not exceeded 113,000 ducks. This is no doubt an underestimate, but other information from known haunts suggests a total of no more than 500,000, which seems to be the best present estimate of numbers. The possibility still remains, however, that there are major, undiscovered, sites (23).

The Longtail in Britain is largely found in north-east Scotland, chiefly in the Moray Firth area (Fig 130). The main flocks are usually found in the inner firth between Cromarty and Rosemarkie, in the southern part of the outer firth at Spey Bay and Burghead Bay, and in the Dornoch Firth. Because the birds are mobile only one site, Burghead Bay, qualifies for international importance, although other parts of the firth are equally significant. 15,600 roosted in Burghead Bay in February 1982. The area as a whole held about 10,000 birds in 1977-1979 (361), 20,000 in 1981-82 and 14,000 in 1982-83 (540). Other concentrations are found in Orkney, especially at Scapa Flow, but elsewhere flocks are scattered and rather small (p.330).

The Wildfowl Count data are far too incomplete to produce estimates of the British population or trends in numbers on their own. The most recent published estimate has been a total of 20,000 in the late 1970s, with no indication of recent trends (361). Numbers were clearly considerably larger in 1981-82 (see above) and possibly also in 1982-83. Like other sea ducks and perhaps even more so in view of the concentration, Long-tailed Ducks are vulnerable to oil pollution and more regular surveys will be necessary if this threat is to be effectively monitored (p.491).

Fig. 130. Long-tailed Duck – British distribution.

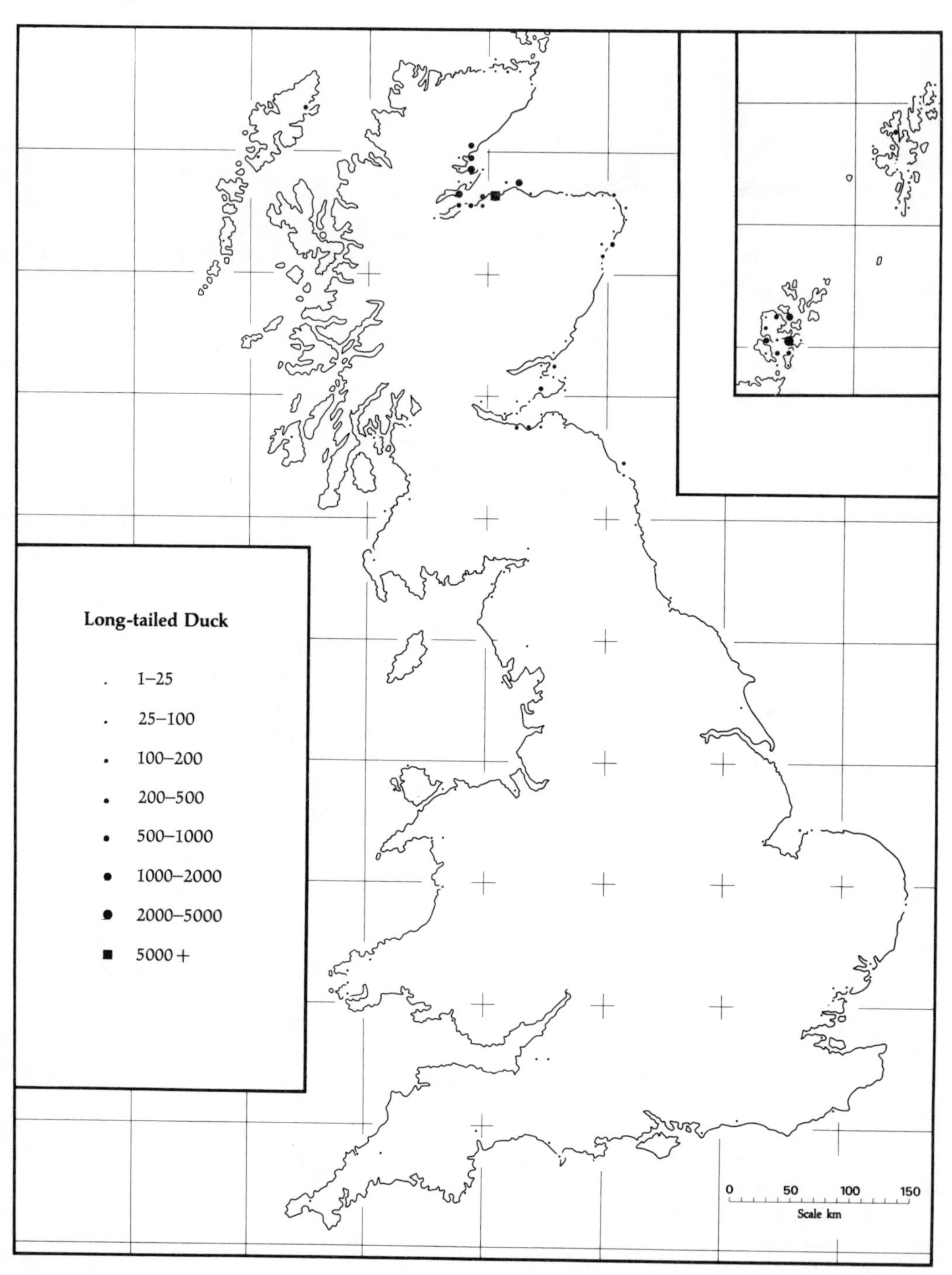

Common Scoter
Melanitta nigra

The Common or Black Scoter has two races, the ominate *nigra* occupying the eastern Atlantic and vestern Siberia, and *americana* eastern Siberia and Jorth America. The breeding range of the European ace extends from 55°N in Ireland to 74°N on north Atlantic islands. In winter it is the most southerly of ur sea ducks, with a range extending as far south as 1°N in west Africa.

Scoters are rarely found inland in winter, nhabiting shallow coastlines and bays, in both marine nd brackish conditions. Common Scoters, like other ea ducks, concentrate their feeding over mussel beds nd the Blue Mussel predominates in their diet. In Danish waters two-thirds of stomachs analysed contained only bivalves and most of these consisted of nussels. Snails such as periwinkles and whelks were lso eaten in significant quantities, but crustaceans, vorms and fish were of minor importance (324). The iet in Britain is probably similar, since the birds feed redominantly over mussel beds. There are a few reas where Common Scoters congregate near sewage r brewery discharges, but the numbers appear to be mall in comparison to those at natural sites (481).

At Myvatn, Iceland, during the breeding season Common Scoters stay in the deeper parts of the lake nd feed on chironomid larvae, fish eggs and water leas, also taking some seeds. Small ducklings concentrate on adult insects and seeds and older ones dive for hironomid larvae, water fleas and seeds (46).

Scoters stay further out at sea than most sea ducks, often in depths of 10-20m, though usually less than 10m, with dives lasting 20-40 seconds (141, 324).

There is a small British breeding stock, with onfirmed breeding in 15 10km squares during the Atlas survey, all in Scotland. Most nest in moorland ochs, either on islands or around the margin close to vater among heather or other shrub cover. Breeding is poradic in many places, but the total number of nests in the late 1960s was thought to be between 30 and 50, with about three times as many in Ireland (542). There has, however, been a marked decline since, to 10-20 pairs in Britain in the early 1980s, and the fall seems to be continuing.

The movements of Common Scoters are spectacular but not well understood. There is a massive south-westward moult migration through the Baltic in July and August, with totals estimated at several hundred thousand, though counts are difficult since many movements are at night. Some moult among the Danish islands and in north-east Jutland, but most cross the Jutland Peninsula and moult in the Danish and other sectors of the Wadden Sea, north-east Scotland and probably other areas, since the known concentrations account for only a small proportion of migration estimates (289). In June 1974 a count of Carmarthen Bay, Dyfed, yielded 6,000 and there were no fewer than 16,000 in August of the same year and 10,500 in July 1976, but many more moulting birds must be present in the southern North Sea if the Baltic counts are to be believed.

There is a movement back into the Baltic in late summer and autumn and a turnover of birds in north-east Scotland and probably elsewhere. Counts are not frequent enough to show the pattern of build-up of the wintering flocks and their origins are not well known. None winter in Iceland but two Icelandic-ringed birds have been recovered in Britain. However, Icelandic rings have also been found on the Atlantic coasts of France, Spain, Portugal and the Azores. Since the numbers in Iceland are estimated at only 500 pairs (141) most of our wintering flocks must come from the east, from breeding areas in Scandinavia and western Siberia. Where our small breeding population winters is unknown.

Despite the large numbers counted on moult migration and the estimated one and a half million on spring passage in Finland (49), no more than 200,000 Common Scoters have been counted in January. As they winter at sea underestimates would be expected, but it seems inconceivable that so many have been missed completely, and the north-west European total has been provisionally put at 400,000 to 500,000 (23). The main concentrations are in the shallows around Denmark, but there are large groups on the Atlantic seaboard of France, in western Portugal and southern Spain. Obtaining regular and reliable counts of sea ducks is very difficult because air surveys are necessary and these are often impossible in the short days and unfavourable weather of mid-winter. The locations of the flocks change from year to year, and counts from the shore yield irregular numbers, though probably the scoters are consistently in the vicinity.

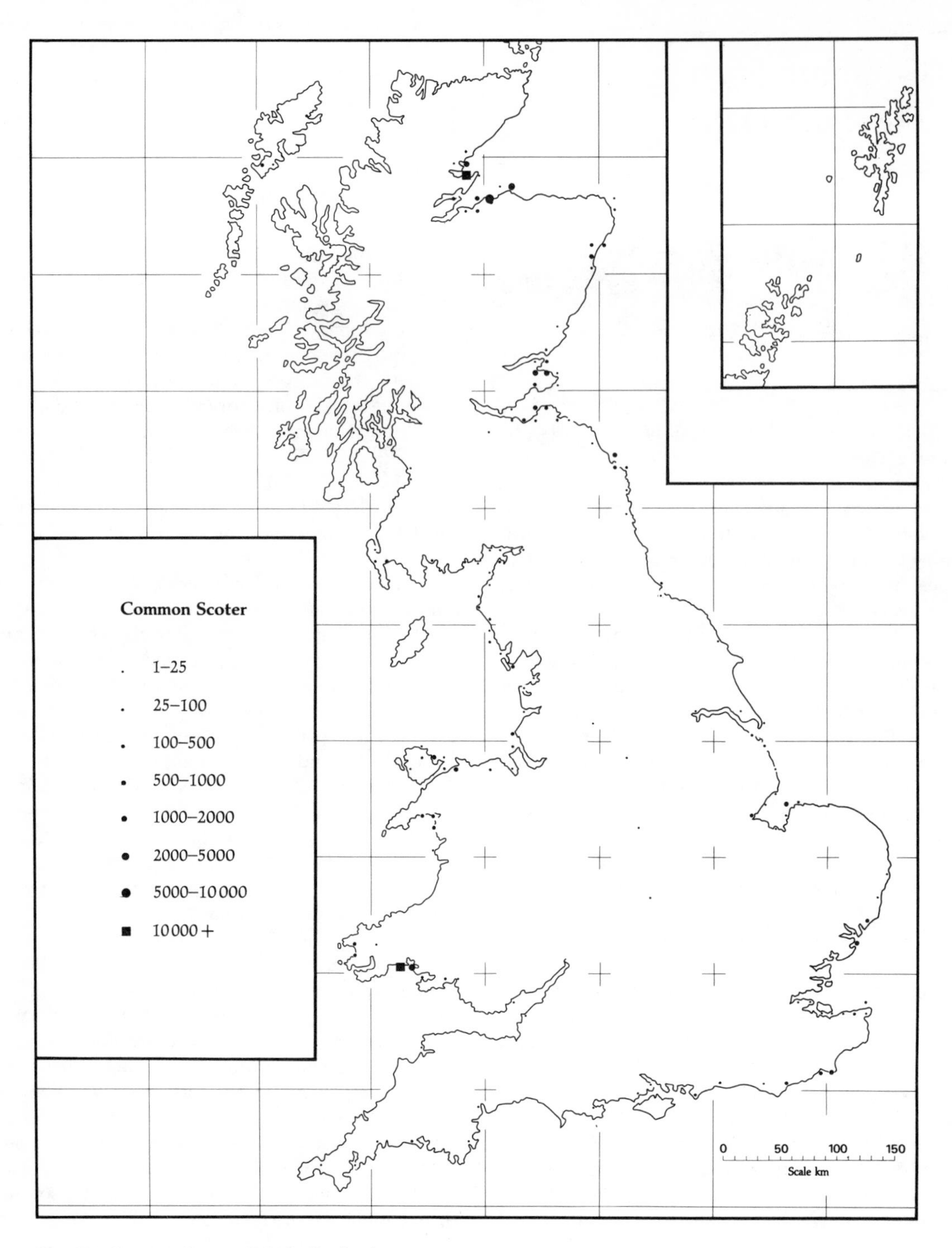

Fig. 131. Common Scoter – British distribution.

reas poorly covered are the Iberian Peninsula and e shallows off the Dutch and German coasts as well s most of southern Great Britain and Ireland (23).

The known winter distribution in Britain (Fig 31) is very localised, with the greatest concentrations eing in the Moray Firth and Carmarthen Bay. The hole of north-east Scotland was covered by a survey 1970-1972 (353) which showed an increase in the nportance of the Moray area. The whole population f north-east Scotland was estimated at 5,000 but this as not at the peak season. St Andrews Bay, Fife, used hold several thousand birds but had declined to a w hundred, a situation which has continued. The loray Firth probably holds 10,000 birds in most years, e main sites being on the south side at Spey Bay and urghead Bay (361). The flocks are mobile, their exact osition depending on weather conditions, making it ifficult to amalgamate counts made on different days. armarthen Bay is an important wintering area but adequately covered. The count of 5,000 from the hore in February 1974 led to boat and aerial surveys eing carried out, and these yielded an estimate of 5,000. How regular such concentrations are is nknown (23).

Because of the sparsity of counts it is impossible ither to make confident assessments of the British tock or to indicate trends in numbers. In 1974 it was hought the peak could have been as high as 50,000 half at Carmarthen Bay), compared with previous stimates of 20-25,000 (23). No more can be said until ifficult and expensive surveys can be made more egular. The vulnerability of this species to oil ollution and the increase in oil-related activity in the Jorth and Celtic Seas make such surveys essential lespite the problems of logistics and expense.

Velvet Scoter
Melanitta fusca

Closely related to the Common Scoter, this species is similar in winter habits and its winter range largely overlaps. There are three races: *deglandi*, the American White-winged Scoter; *stejnegeri*, in eastern Asia; and our own nominate subspecies, all very similar in appearance but showing minor variations in head and bill.

The Velvet Scoter is slightly larger than the Common Scoter and takes larger prey items. It also has a more northerly winter distribution (its larger size enables it to withstand lower temperatures), stays closer to the shore and hence has a more varied diet. Although, as with the Common Scoter, molluscs predominate, in Denmark, at least, more cockles and gastropods (e.g. periwinkles and whelks) are taken. Crustaceans, urchins, lugworms and fish are also eaten in small quantities (324). The spring diet in Finland is similar, with shellfish being taken roughly in proportion to their availability in the *Fucus* zone over which the birds feed (31).

The Velvet Scoter is an accomplished diver, feeding at depths of up to 30m though most commonly under 10m. According to depth, dive duration can vary from 20 seconds to more than a minute (141, 324).

The breeding range of the two species also overlaps in the central part of Scandinavia and western Siberia, but that of the Velvet Scoter extends further south to the brackish waters of the Gulf of Finland and the Gulf of Bothnia, and to 50°N in central USSR, compared to the Common Scoter's southerly limit in the same area of about 65°N. The Velvet Scoter consequently occupies more wooded regions but always nests close to water. Breeding has been suspected in northern Scotland but never confirmed.

The British wintering birds originate in Scandinavia and west Siberia but how far east their range extends is unknown. Birds ringed in Finland and Norway have been recovered largely in the Baltic but

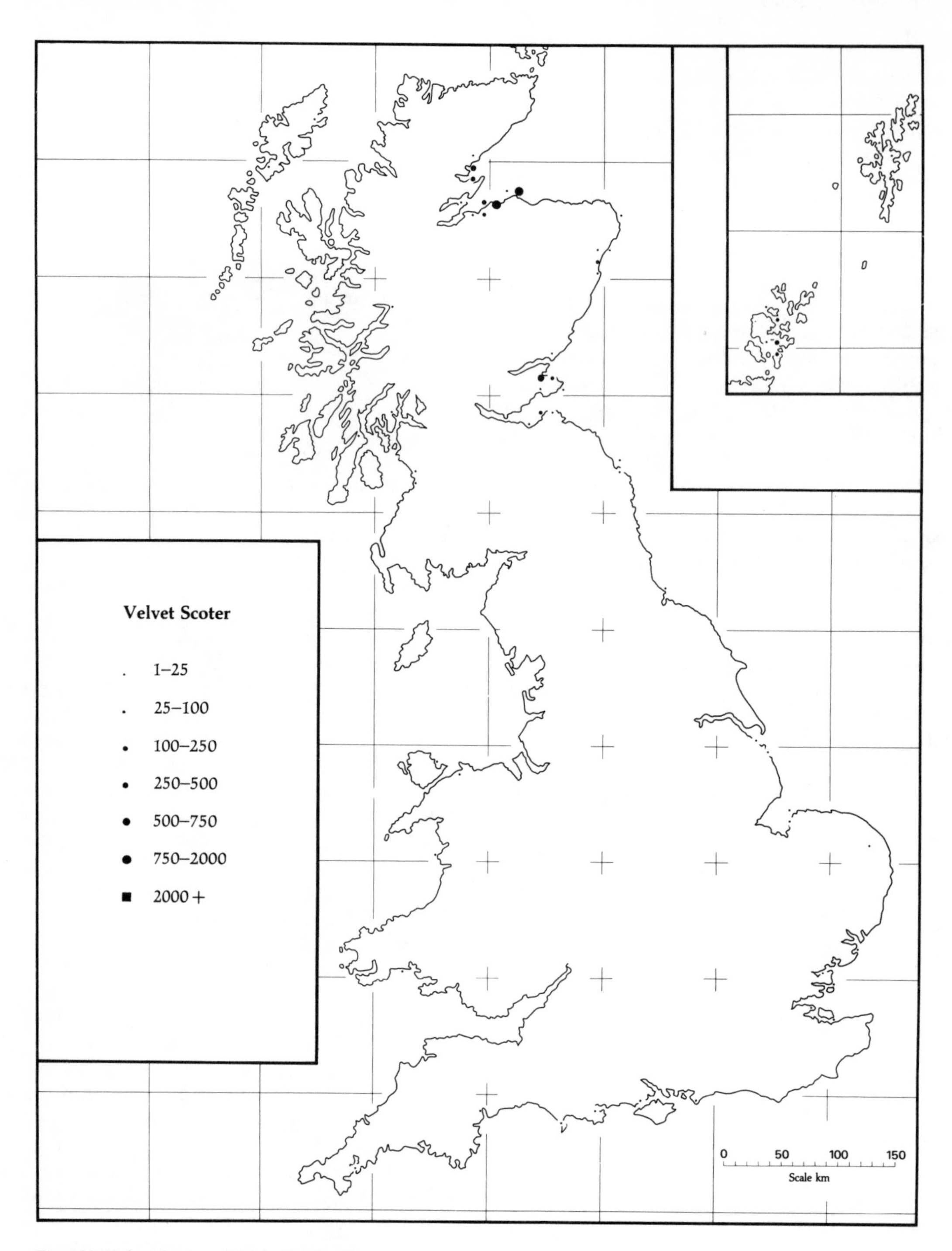

Fig. 132. Velvet Scoter – British distribution.

lso in northern France, and one each in south-east ngland and south-west Scotland. It seems likely that orth-west Europe has a single population of which ne British stock is a small part.

Only 25,000 Velvet Scoters are counted in orth-west Europe in January, but this is known to be n underestimate. About 45,000 (80% or more males) ave been estimated in Danish moulting flocks (289) nd even if these included all the males from the opulation its size in autumn is likely to exceed 00,000 birds. The mid-winter distribution is centred n the Baltic, with some numbers probably in the)utch and German parts of the North Sea (23).

In Britain the Velvet Scoters are predominantly n north-east Scotland (Fig 132), nearly always being ound intermingled with flocks of Common Scoters nd sometimes other sea ducks. Exact counts are lifficult, because the two scoters are not easy to istinguish when at rest, but an estimate based on the roportion of this species in scoter flocks in 1971-72 uggested a total of only 500 in eastern Scotland (353). here have been isolated records of much larger umbers since in the Moray Firth. In 1977-78 the vinter population there was 2-3,000, reaching a peak f 5,000 in April 1978 (361), and in March 1983 8,000 vere present. The numbers in Britain, being a minor art of the population and on the edge of the range, uctuate widely but in most winters probably reach ,000 at least for a time (361).

Because of its very localised distribution the 'elvet Scoter is probably the most vulnerable sea duck o the effects of oil pollution, and large numbers are eing lost in the Kattegat and Baltic annually (288).

Goldeneye
Bucephala clangula

A Holarctic species, the Goldeneye breeds mainly between 50°N and the Arctic Circle, across mainland Eurasia and North America. The American subspecies *americana* differs only slightly in size from the Eurasian one and would be overlooked on the British side of the Atlantic.

The Goldeneye occurs on fresh, brackish and saline water, although the largest concentrations are mostly found on the coast. It is more omnivorous than other sea ducks, especially when on inland habitats. On the coast it feeds on a wide variety of invertebrates, the most common of which are mussels, cockles and other bivalves, and snails such as periwinkles and *Hydrobia*. Crustaceans, especially crabs, are also important, as are more mobile animals such as *Nereis* worms and fish (gobies and sticklebacks). The small amount of plant food consists largely of seeds with small amounts of vegetative material and algae (324, 386). In inland Britain insect larvae are by far the most important food, particularly larvae of *Trichoptera* or chironomids. Water boatmen (*Corixa*) are regularly taken but only a small amount of plant material – the seeds of *Potamogeton* (431).

The species is very commonly associated with sewage and other discharges and the birds feed close up to the pipe outlets, unlike ducks of other species which feed some distance away (481). Many of these outfalls contain large quantities of grain from distilleries, and results from a small sample of birds from Seafield, Edinburgh, indicated that grain formed an important part of the birds' diet there (112). In 1973-74 the total number of Goldeneyes counted at outfall sites in Scotland was 5,580 (481). The average number counted in Great Britain in the same period was 8,500, and although some of those included in the outfall survey were not recorded in the Wildfowl Counts it seems likely that about half or more of our wintering

Goldeneyes congregate around effluent discharges, particularly in the Forth Estuary, near Peterhead and at Invergordon. Many schemes are under way to clean up these outfalls, and this could have a serious effect on local Goldeneye numbers and perhaps even affect the total wintering population (see also below). The numbers on the south side of the Firth of Forth have certainly declined, as have other sea ducks, following the improvement in the sewage system discharging into the firth.

Diving depths preferred are usually up to 7m but may reach 9m in some circumstances. There have been suggestions that males are better divers than females and that this separates them in winter, but the differences in winter distribution (see below) are more likely to be due to other considerations. The Goldeneye is intermediate in diving ability between the shallow divers (Tufted Ducks and Pochard) and the true sea ducks (Scoters, Eiders and Long-tailed Ducks) (386, 431).

The Goldeneye is a mid- and late-winter species in Britain, as Fig 133 shows. The main influx is not until November and 85% of peak numbers are still present in March. The species is later to leave the wintering areas and is later to lay (mid-May and June) than most ducks. Although there is a small native population, most of our Goldeneyes are immigrants from north-west Europe. An analysis of the recoveries of birds ringed in Scandinavia showed that British wintering birds come predominantly from Sweden (and probably Norway where little ringing has taken place). While 23% of recoveries outside Sweden of birds ringed there have been in the British Isles, only one (2%) of 47 foreign recoveries of birds ringed in Finland has been in Britain. The remaining Goldeneyes migrate to the Baltic and central Europe. Peak numbers are counted in inland Sweden in October (pre-migration) and especially in late April before the pairs disperse onto breeding territories. There is a difference in the winter dispersal of adult males and females, with the latter moving further south and west. In inland Sweden 45-61% of wintering Goldeneyes are males, whereas females predominate i more southerly areas. In England only 19-30% ar males and in south central Europe only 11-21% (384 This reflects the greater cold-tolerance of males, whic are larger than females and immatures, which have t migrate over longer distances, but other factors ma also be involved.

The Goldeneye is a forest breeder, nesting i natural cavities or, more commonly, in the nest hole of the larger species of woodpecker. In many areas i North America, and probably also in Europe, th shortage of suitable cavities limits nesting density, an large increases have been achieved by the introductio of nest-boxes, which the birds readily adopt. Nowa days nest-boxes are provided in much of the range c the Goldeneye and a sizeable proportion of the bird use them. There is a large amount of nest parasitism i the species, mostly of nests of other Goldeneyes. I one southern Swedish study more than one femal laid in about one-third of nests and breeding succes was lower in such nests. Shortage of sites was not th reason since only one-third of boxes were occupied The most likely explanation seemed to be that youn and inexperienced birds, which would otherwise no lay or would be too inexperienced to rear a brood were more successful by parasitising other female (176). Nest cavities may be 1-2km from water especially in areas where nest sites are sparse. Th broods are reared on small woodland ponds and lake with little vegetation but an abundance of insect larva and emerging flies, on which the downy young feed

Coniferous forests in Scotland and norther England provide very similar habitat for Goldeneye to that in Scandinavia, but breeding was not con

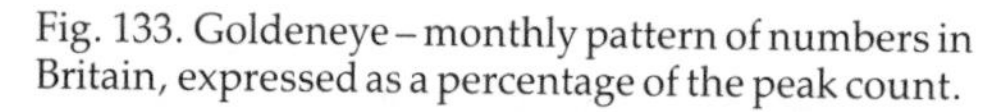
Fig. 133. Goldeneye – monthly pattern of numbers in Britain, expressed as a percentage of the peak count.

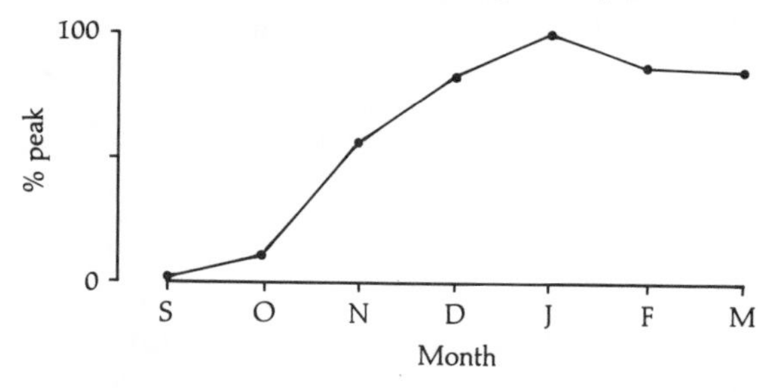

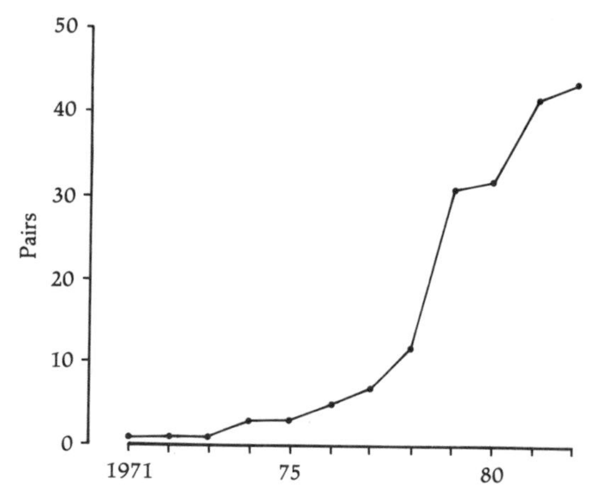

Fig. 134. The probable number of Goldeneyes breeeding in Scotland each year since 1971. From reports of Rare Breeding Birds Panel published in *British Birds*. Where a range is given the mid-point is taken, e.g. 20 to 30=25.

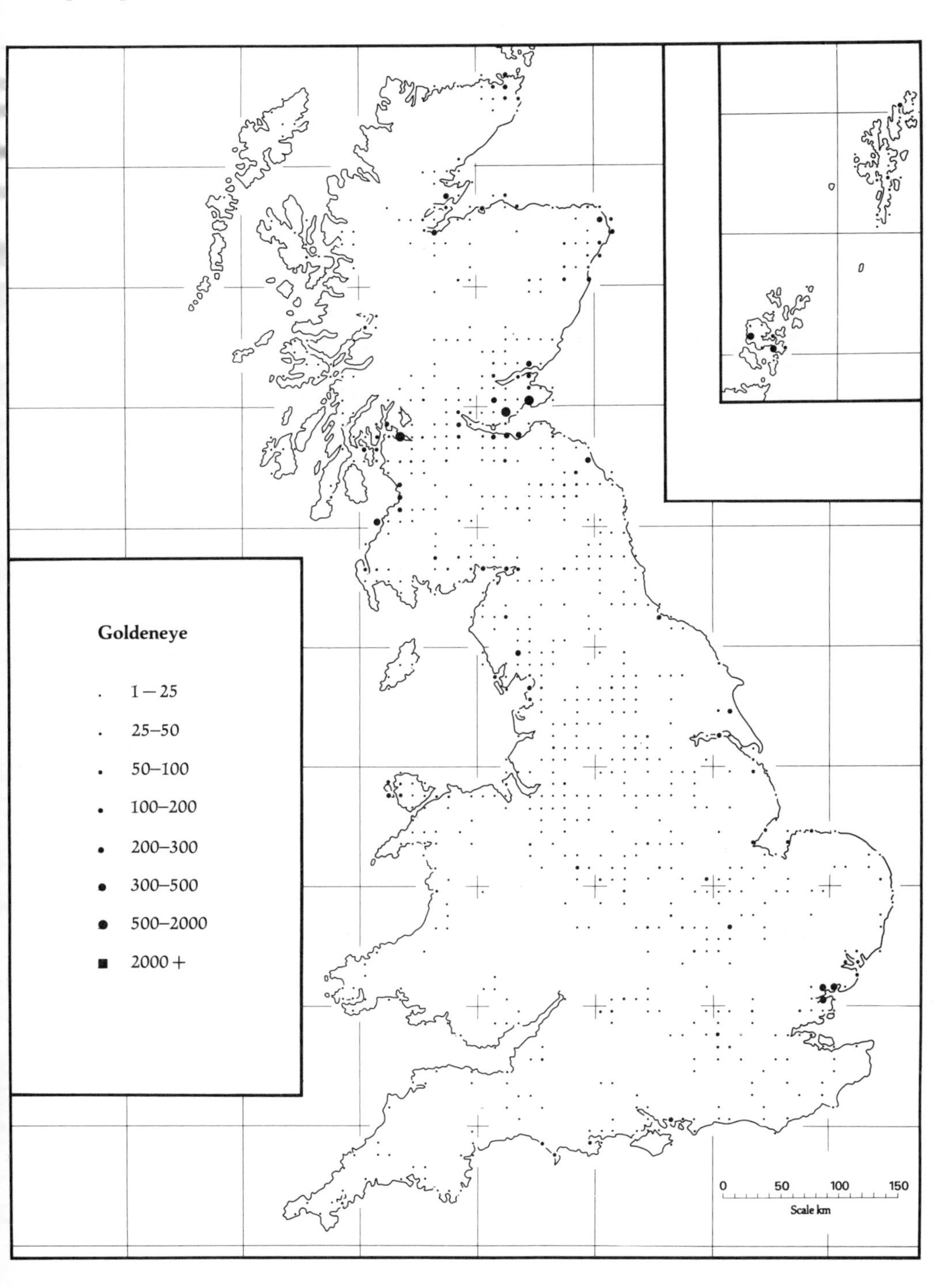

Fig. 135. Goldeneye – British distribution.

firmed here until 1970, when a brood was found in Inverness-shire following the widespread provision of nest-boxes in the 1960s. Breeding was confirmed in 2 squares in the Atlas survey (542) but there has been a remarkable increase since, as shown in Fig 134. The vast majority of these nests are around the original establishment, all in nest-boxes in most years. Breeding has also been confirmed further south in central Scotland. A few pairs summer regularly in the Lake District and breeding there will surely not be long delayed.

As with many other diving ducks, the southern and western parts of the Baltic provide the wintering habitat for the vast majority of Goldeneyes in north-west Europe. This area holds perhaps 85% of the total in January, the remainder being in the Netherlands, Scotland and Ireland, with a widely scattered, but small, proportion elsewhere. The total number of birds in this northern group is estimated at 170,000 (23). The group occupying central Europe and the western Black Sea areas numbers a further 20-30,000, but since there are recoveries of Finnish-ringed birds from all parts of this range (384) it cannot be regarded as separate from the main coastal group (23).

In Britain, despite the fact that the Goldeneye is a very widespread species, having been recorded on more than 2,000 sites, the vast majority inland, most birds are in rather few coastal areas (Fig 135). No single site qualifies for international importance (2,000 or more), although the Firth of Forth did until recently. The Goldeneye flock at Seafield, in common with those of other diving ducks, reached a peak in the early 1970s, when the numbers exceeded 2,000 in most years, and the flocks in the southern Forth altogether accounted for 3,500-4,000 Goldeneyes (112). The numbers at Seafield dropped dramatically to 1,000 in 1978-79, to 600 in 1980-81 and to 270 in 1981-82 coinciding with the cleaning up of the sewage discharges from Edinburgh which contained waste grain from distilleries, on which the birds fed. There was no concurrent large increase elsewhere but there were slight gains in other parts of the Forth. The estuary as a whole remains a key site for Goldeneye in the British context.

The consequence of the Forth decline on the population in Britain is unclear; there has been a drop in the maximum recently but this is not outside normal fluctuations in the last two decades (Fig 136). The trend shows a marked peak in the early 1970s followed by a decline, but a similar trend was also evident in the early 1960s. This pattern is difficult to interpret since the number wintering in Britain may depend largely on the weather in north-west Europe. The late 1960s and early 1970s was a period when the population as a whole was increasing (23), so the fortunes of the British stock recently may indicate a more general decline.

The sum of the January 5-year means amounts to 13,762, and the average counted annually 9,150,

Fig. 136. Goldeneye – trend in January numbers in Britain, 1960-1982.

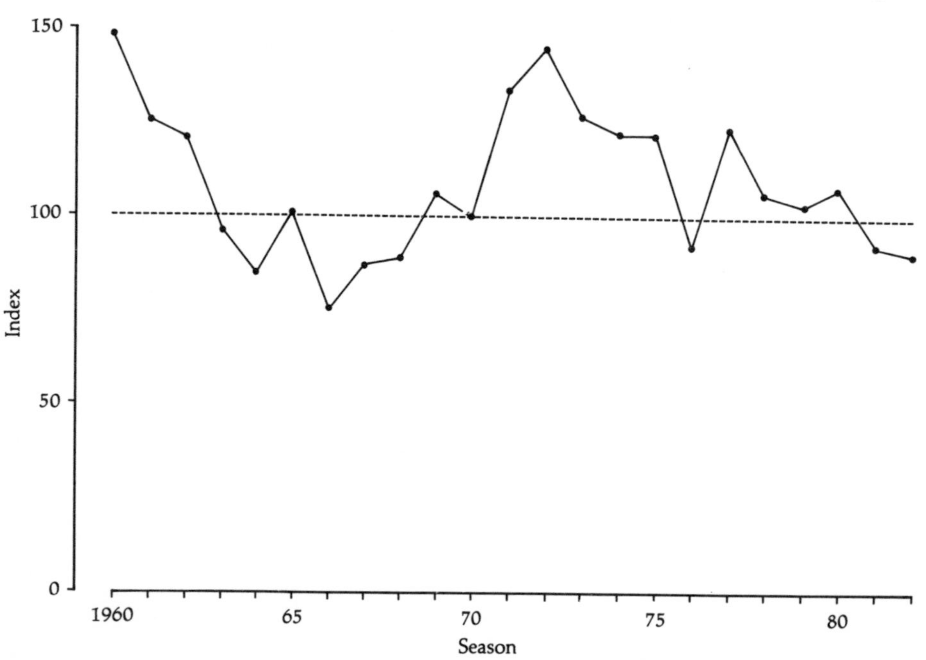

indicating that 66% of Goldeneyes on counted sites are included in the annual totals. The species occurs only in very small numbers outside the major concentrations and in most years the British stock, at peak, is likely to be in the region of 15,000. The Goldeneye prefers sheltered bays and fjords to the open sea and is consequently less vulnerable to oil pollution than the true sea ducks. With active support for breeding populations, by the provision of nest-boxes, there is no reason to suppose that the Goldeneye will not continue to spread as a breeding bird in Britain. The heavy dependence of wintering birds on sewage outfalls makes them vulnerable to changes in the sewage systems such as that which caused such a major decline around Edinburgh.

Smew
Mergus albellus

The smallest of the world's sawbills, the Smew represents a link between the Goldeneyes and the true sawbills. The Smew is restricted to Eurasia, its ecological counterpart in North America being the Hooded Merganser. It breeds in the forest zone of Eurasia, largely between 55°N and the Arctic Circle, a distribution following closely that of the Black Woodpecker (*Dryocopus martius*) in whose holes it frequently nests.

The Smew is mainly found on fresh or only slightly brackish water in winter. It is primarily a fish eater, taking a wide variety of prey. In brackish and marine areas gobies and sticklebacks form the staple diet, whereas in freshwater a wide variety of young coarse fish or trout make up the bulk of the food. The fish eaten are normally 3-6cm long, but a 14cm trout and a 29cm-long eel have been recorded. A wide variety of invertebrates, including crustaceans (prawns), molluscs (water snails), insects (caddis larvae, water boatmen, beetles) and a small amount of plant food in the form of seeds have also been found in stomachs (325, 430). Thus although largely subsisting on fish, the Smew is an opportunist and can survive on a wide variety of habitats and foods.

The Smew dives in very shallow water, usually below 2m, and surfaces on average after 18 seconds. Paired birds synchronise their diving (387). Mass feeding of a flock of at least 750 Smew in the IJsselmeer, the Netherlands, has been recorded, but this behaviour, though similar to cooperative feeding by other wintering sawbills, may be unusual for Smew (292). The species moves south in the hardest of weather; provided enough ice-free patches are available it will continue to pursue fish beneath the ice.

The Smew breeds in the forest areas of Scandinavia and the USSR, nesting in tree holes (it will also take over nest-boxes) and rearing its brood on small forest pools or flooded taiga. In spring and summer the diet changes to one chiefly of caddisfly larvae and water beetles, with dragonfly and other insect larvae and water boatmen common (153).

In winter there is a north-west European group and a larger one in the Black Sea and Caspian region. The main western concentration is in the IJsselmeer in the Netherlands and on the major rivers Rhine and Waal. There are large annual fluctuations in the various sectors but between 1967 and 1973 there was a marked increase in the western group, from 3,500-4,000 to more than 10,000 (22). Since then the numbers

Fig. 137. Smew – British distribution.

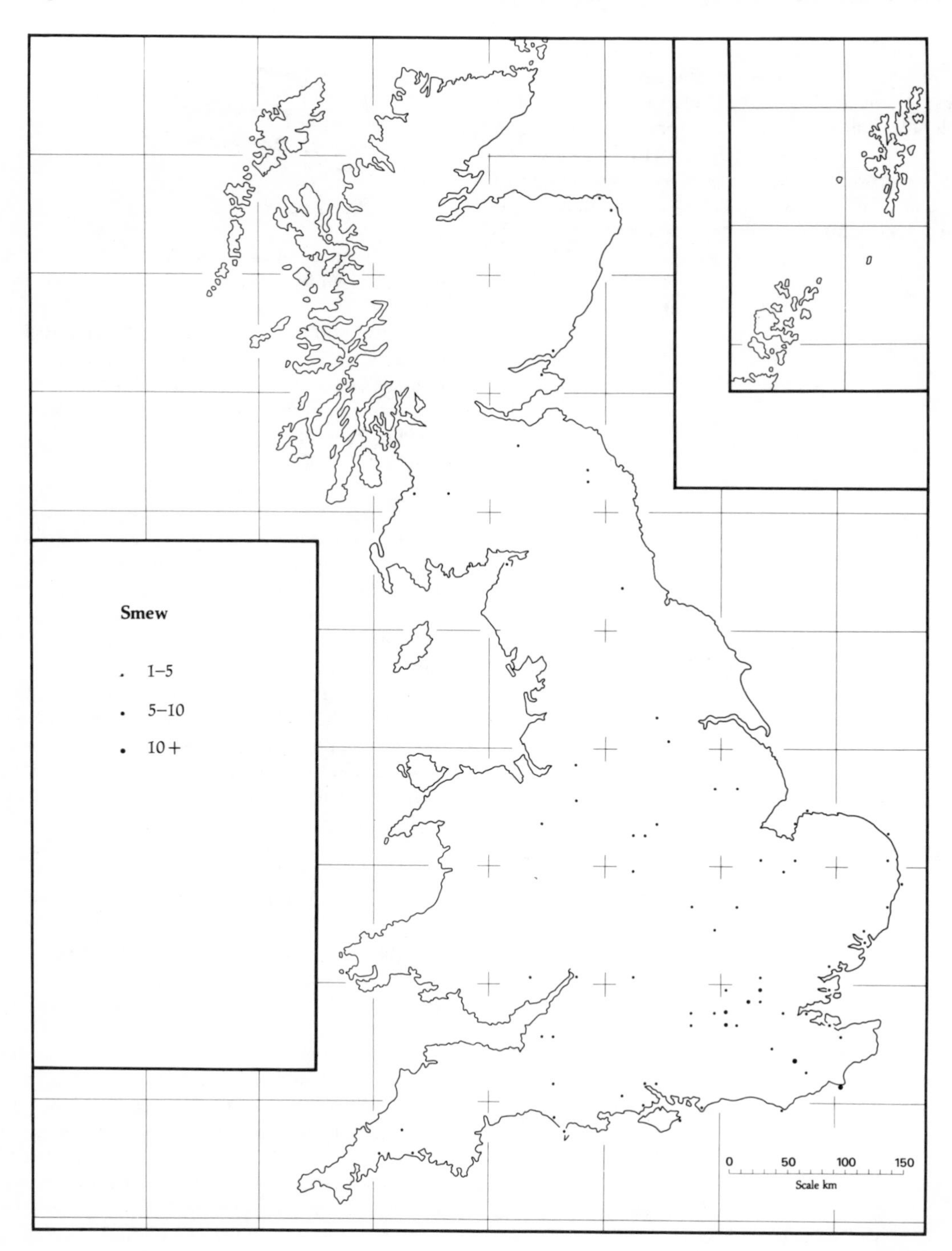

in the Netherlands have shown enormous fluctuations from only 1,100 in 1975 to more than 21,000 in 1977. The size of this group is now probably in the region of 20,000 birds.

In Britain the Smew is mainly found in south-east England (Fig 137) with only a scattering elsewhere. It occurs from November to March but with marked increases in January and February when cold weather forces some from the Netherlands. The numbers in Britain have not increased with the European stock; indeed the regular numbers have declined from the "few hundred birds" of the First Edition. The peak numbers from 1960 to 1982 are shown in Fig 138, which illustrates the overall decline, arrested only by a relatively large influx in the cold weather of 1978-79 and a smaller one in 1981-82. The 1982-83 maximum of 35 was, in fact, the lowest at least since 1947-48, when records began. Only a few sites have regularly held flocks of more than 10 Smew since the 1960s (524), but during cold weather there have been exceptional flocks of 41 at Shingle Street, Suffolk, and 36 at Wraysbury Gravel Pits, Berkshire, in January 1979, and 31 at Dungeness, Kent, in January 1982.

In the cold weather of 1984-85 there was a very large influx; 390 were counted in February, widely scattered through the country.

Internationally the main threat to Smew is seen as industrial pollution, particularly on the major rivers (22), but despite this the last 15 years have seen a quadrupling of the western population. Britain is at the edge of the range and its importance is declining except in the occasional hard winter.

Fig. 138. Smew – annual peak numbers (usually January or February) counted in Britain, 1960-1982.

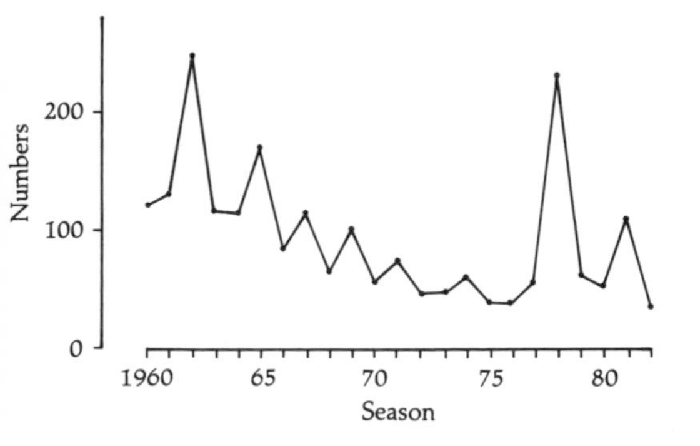

Red-breasted Merganser *Mergus serrator*

The Red-breasted Merganser is a Holarctic species, breeding throughout northern North America between 45 and 70°N, Greenland, Iceland, the British Isles and northern Eurasia to 75°N. There has been no attempt to split the species into races and the homogeneity of the stock probably means that there is regular intermixing of birds from different parts of the range.

In winter the Red-breasted Merganser is a largely coastal bird; in Britain about 85% of those counted (and probably most of those which are not), are on the coast or in estuaries. The serrated bill is well adapted for gripping fish and the Merganser is largely a fish eater at all times of the year. In estuarine and marine areas Three-spined (*Gasterosteus aculeatus*) and Ten-spined (*G. pungitius*) Sticklebacks are by far the most common item taken, with gobies also very important. A wide variety of other small marine fish are also recorded, including flatfish and herrings.

Fig. 139. Red-breasted Merganser – monthly pattern of numbers in Britain, expressed as a percentage of the peak count.

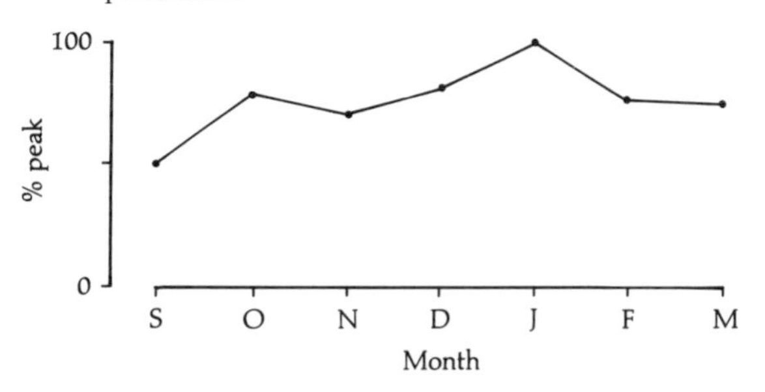

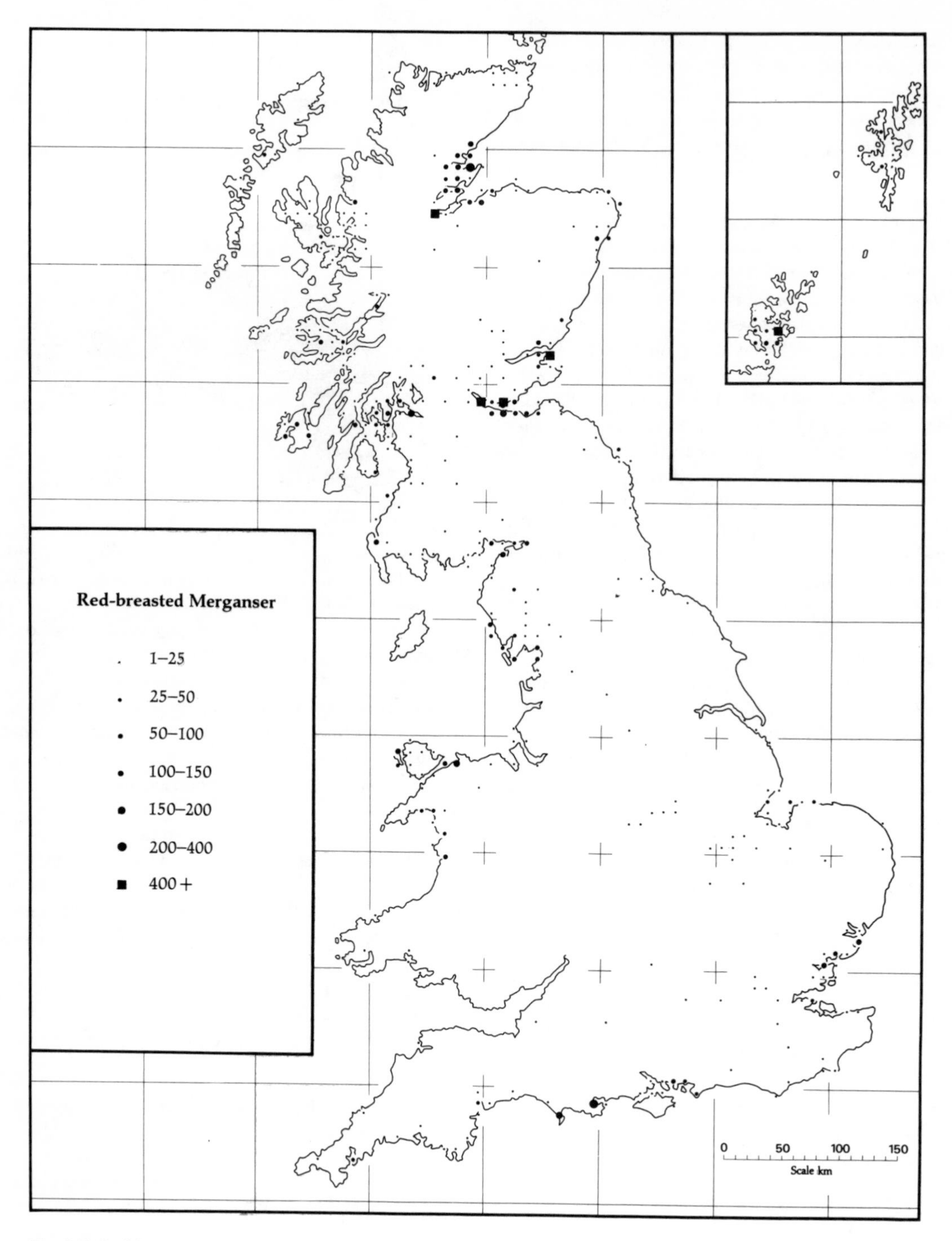

Fig. 140. Red-breasted Merganser – British distribution.

Shrimps and prawns are eaten regularly as are *Nereis* worms and a wide variety of invertebrates in small quantities (31, 141, 325).

Many Red-breasted Mergansers summer and rear their broods in freshwaters, where they come into conflict with fishing interests, since they are accused of eating large quantities of salmon and trout fry. Salmon (*Salmo salar*) is indeed the most important prey of adults in Scottish rivers, accounting for about half the fish, but Brook Lampreys (*Lampetra planeri*), minnows, gobies and eels are also important (347). Ducklings eat a wide variety of invertebrates, insect larvae and fish, and from experiments in captivity it has been calculated that a single duckling, in the 100 days it remains on the rearing water, will consume approximately 20kg of fish, which is the equivalent of 14,000 0-1 year old Perch (*Perca fluviatilis*) fry. This is probably a substantial underestimate since ducklings in captivity use much less energy in gathering their food than would those hunting fish (16). In brackish water sticklebacks, gobies and shrimps are likely to form the bulk of the food of adults and young during the breeding season.

The Red-breasted Merganser hunts in shallow water, usually below 3.5m and often much shallower, first immersing its bill and eyes and then diving after its prey. It is not uncommon for groups of Mergansers to hunt together, either encircling fish or combining to drive shoals into shallow water, presumably increasing the chances of each individual bird to make a capture. Drives by Mergansers are so effective in concentrating shoals that several species of herons sometimes associate with them to profit from this (159, 175). In summer at Myvatn, Iceland, adults and broods feed in shallow water close to shore and both feed predominantly on sticklebacks, also taking water fleas, emerging insects and some seeds (46).

The nest is usually in dense cover, among rocks or in holes or burrows. The species has long bred in Britain, with strongholds in the Scottish highlands and islands. In this century, however, the range extended rapidly and numbers increased, and this spread is continuing. Breeding was first confirmed in England in 1950 and in Wales in 1953, and it is still increasing and spreading in many areas (346, 542). The Atlas survey confirmed breeding in 365 10km squares, 44 of these in England and Wales, the most southerly being in Glamorgan.

Most of the Mergansers that winter in Britain probably also breed here, but since very few have been ringed there is rather little information on movements. Of those ringed in Iceland in summer most have been recovered in the British Isles, notably in Northern Scotland, but one moved to the Netherlands and one to East Greenland. Birds ringed in Scandinavia and Germany migrate to France, Spain, central Europe and as far east as the USSR part of the Black Sea (39, 67). One ringed in Denmark has been recovered in Britain, one caught at Loch Leven as a juvenile was recovered four years later in Denmark, and a non-flying young bird ringed in Grampian was recovered in Norway four years later. It thus appears that our wintering birds are of mixed origin; we certainly receive migrants from Iceland and recent evidence tends to confirm the suggestion (67) that some continental birds reach Britain. The two overseas recoveries of birds ringed here are suggestive of abmigration.

The monthly pattern of counts (Fig 139) shows two peaks. This is rather difficult to interpret and may purely be due to irregularity of coverage, but the sharp rise in October possibly represents the post-breeding aggregation of birds on the coast, perhaps followed by

Fig. 141. Red-breasted Merganser – trend in January numbers in Britain, 1960-1982.

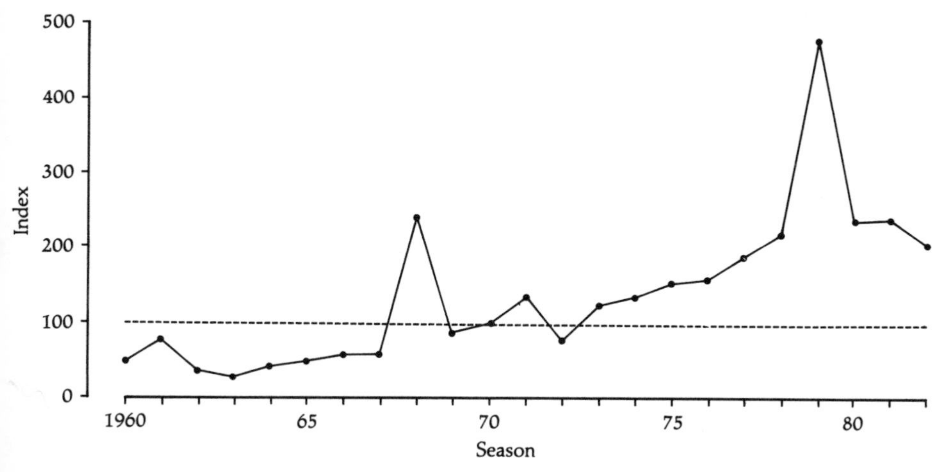

dispersal (to Ireland?). The mid-winter influx, which seems to be represented in January only, may be of continental immigrants. There is a wide variation between years, possibly because of inadequacy of coverage of the main resorts, or because of temporary shifts in distribution.

The north-west European population is estimated at about 40,000 birds, while another group of about 50,000 winters in the Black Sea and Mediterranean region (529). The centre of distribution of the north-west European stock is, as with most sea ducks, in the Baltic, largely around the coasts of Denmark (290). Only a thousand or so are counted in Sweden but this is thought to be a serious underestimate (388). Considerable numbers also winter in the Netherlands part of the Wadden Sea and the total for that country is in the region of 5-7,000, though reaching an unprecedented 19,600 in January 1977 (48).

The distribution map of Britain (Fig 140) under-represents the importance of north-west Scotland, where the species is very widespread on almost all stretches of shallow coastlines, estuaries and sea lochs (p.298). The species has, nevertheless, been counted on nearly 800 sites in Britain, including a scattering of inland records. Invariably the largest concentrations are in eastern Scotland, with 500-1,500 in the Firth of Forth and the Moray Firth. In the latter area, the Beauly Firth has been particularly important in recent years (p.277). Poole Harbour, Dorset, usually also holds 100-400, with an exceptional maximum of 540 in January 1980.

The trend of numbers shows a large and steady increase (Fig 141), reflecting the spread of the species as a breeding bird. Again annual irregularities are probably due to differences in movement between counted and uncounted areas and possibly differences in the pattern of immigration. Numbers both in Iceland and in north-west Europe as a whole appear to have changed little during the period before the 1977 influx in the Netherlands, but data on subsequent years are not available.

The total 5-year mean for all sites (5,300) may give an underestimate of the total because of the sparsity of counts in north-west Scotland. Surveys by the BTO in November 1983, however, found only 130 birds in mainland parts of this area. It may, therefore, be less important than previously suspected. At this stage we can only suggest a rough estimate for Britain in the range 6,000-10,000. Following the Atlas survey (524) the breeding population in the British Isles was estimated at 2,000-3,000 pairs. Assuming similar density in Britain and Ireland this suggests 1,500-2,200 pairs in Britain, consistent with the estimate of 1,000-2,000 in the 1960s (20). The only estimates of productivity come from the Finnish archipelago, where production varies between 0.7 and 2 young per pair (261). Taking the average of 1.4 and assuming that all birds attempt to breed at 2 years old, the post-breeding population from our native stock would be 7,000-10,000 birds. Since we also receive immigrants the wintering population is likely to be nearer the higher end of this range.

Because of problems with fisheries the Red-breasted Merganser has been regarded as a pest species in Scotland and widely shot. This has kept the increase in numbers under control while the species has been expanding its range. The 1981 Wildlife and Countryside Act protected Mergansers and Goosanders but made provision for licensing control to protect fishing interests. Since shooting is now more likely to be on a local scale this may allow the Red-breasted Merganser further to increase its numbers in Britain.

Goosander
Mergus merganser

The largest of the sawbills, the Goosander occurs throughout the Northern Hemisphere, mainly breeding between 50°N and the Arctic Circle and wintering chiefly inland in central North America and temperate Eurasia. North American (*americanus*) and southern Asian (*comatus*) races are recognised but these are virtually indistinguishable in the field.

While the breeding and wintering ranges overlap considerably with those of the Red-breasted Merganser the two species are ecologically separated in winter. The Goosander occurs largely in fresh or slightly brackish water.

A specialist fish eater, the Goosander is said to use its hooked bill to help it retrieve fish from among stones. A wide variety are taken, depending on availability; in Denmark eels and sticklebacks are most important and only a very few birds feed on anything other than fish (325). In Scottish rivers 71% of individual fish in Goosander guts are small salmon, with perch, Brown Trout (*Salmo trutta*) and Eel (*Anguilla anguilla*) also important (347). These samples were taken largely in summer; wintering birds take a larger variety of coarse fish. Nearly all kinds commonly found in freshwaters have been recorded as food for Goosanders and very few examples of items other than fish have been found. Most fish caught are less than 10cm in length, but exceptionally a 46cm long eel, 36cm salmon and 31cm Pike (*Esox lucius*) have been found (141).

Downy ducklings in the very early stages catch insect larvae underwater but soon start feeding on small fish. By the time they are 12 days old fish predominate in their diet; at first the prey is very small but by the time they are 3-4 weeks old they take salmon parr and other fish 4-10cm long. In salmon rivers the food of both adults and young is largely young salmon, and experiments in captivity have suggested that it takes about 33kg of fish or 1,600 salmon parr to rear one Goosander to full growth. Further extrapolations indicated that on some of the largest salmon-producing rivers in Canada Goosanders, by eating salmon parr, accounted for two-thirds of the potential smolts produced each year (611).

Such calculations, though very simplistic, and not taking into account compensatory mortality, have led to the wholesale shooting of Goosanders on both sides of the Atlantic. Because of the complicated life history of the salmon it is virtually impossible to determine in a controlled way whether such activity achieves the end of ensuring a plentiful supply of spawning salmon in future years. In the absence of such evidence the species will continue to be shot as a pest.

Over most of its range the Goosander is either resident or moves rather short distances between breeding and wintering areas. There has been little ringing in continental Europe but the slight evidence indicates a movement of Scandinavian birds into the Baltic, Germany and the Netherlands. Those wintering in the lakes of central Europe may have come from further east. A few Scandinavian-ringed birds have been recovered in Britain, from the north of Scotland to Suffolk – enough to suggest a regular immigration in winter, mainly to the southern half of Britain (39, 67). The monthly pattern of numbers (Fig 142) confirms that there is an influx in January, and in some years February, and that this is largely in the south.

From a small amount of ringing in Surrey in the 1930s a high proportion was recovered overseas, in Scandinavia and western Siberia (67). Since 1967 substantial numbers of Goosanders have been ringed (mainly as ducklings) in Northumberland, and so far 52 have been recovered. Of these 5 were local, 38 elsewhere in Great Britain and 9 overseas, 6 in Norway and 3 in Finland. The high proportion of overseas recoveries (17%) is surprising, especially since three of the birds were not yet of breeding age when recovered. Abmigration seems a likely explanation for the remainder; of those that were sexed two were males and one a female. Movements within Britain of birds ringed in Northumberland are largely to the north and west although three recoveries are from

Fig. 142. Goosander – monthly pattern of numbers in Britain, expressed as a percentage of the peak count.

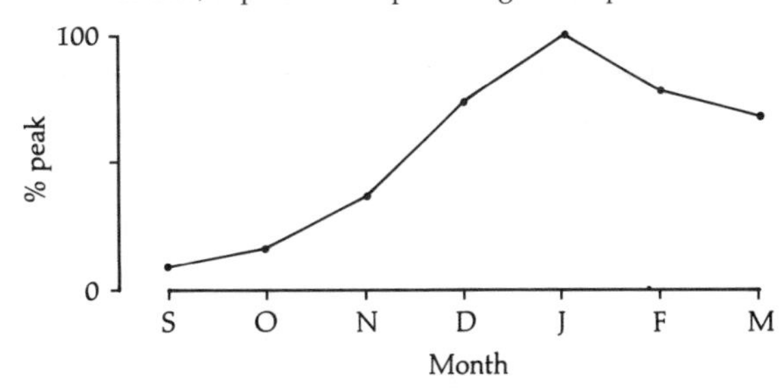

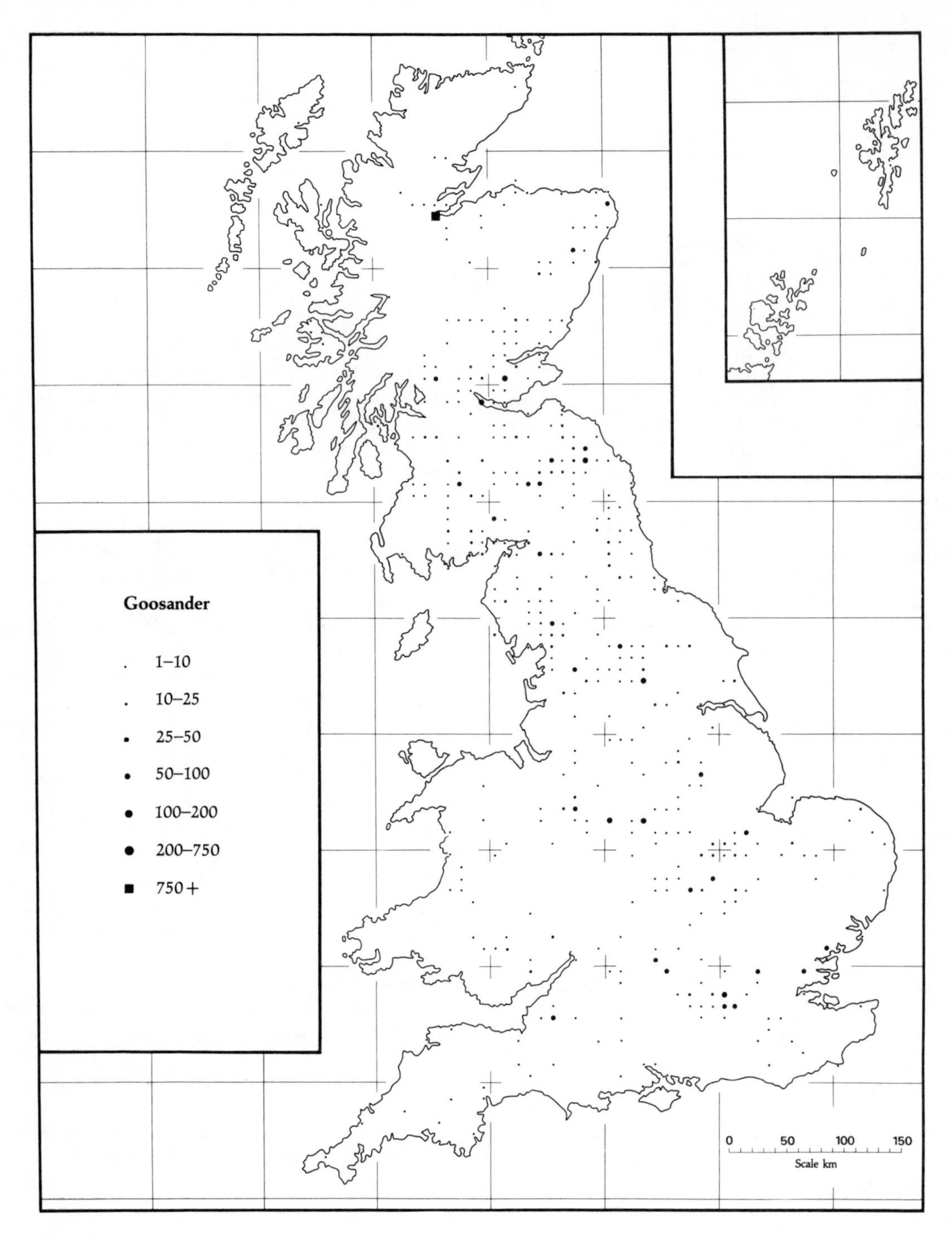

Fig. 143. Goosander – British distribution.

more southerly parts of northern England and one from Wales. Most are around the Scottish borders, on average only about 100km from the ringing place (337). Clearly, although most British Goosanders are resident there is a good deal of intermixing with the population of continental Europe, and those birds wintering in southern England, as well as some of those further north, appear to be of continental origin.

Goosanders favour clear, fast flowing rivers for breeding, exactly those also chosen by salmon. The nest is usually in a tree cavity, in burrows, or among rocks near the ground. Nest-boxes are readily used and this has been responsible for an increase in the number of birds breeding in the Finnish archipelago (227). The nest may be far from water in areas where suitable sites are sparse.

Breeding was not proved in Britain until 1871, when a pair bred in Perthshire, but thereafter the spread and increase was rapid and by the end of the century the species was nesting in many parts of north and west Scotland. Thereafter it continued its southward advance and was breeding in Northumberland in 1941. There were pairs in most parts of Cumbria and Northumberland by the late 1960s and the first record for Wales was in 1972. The British total was then assessed as 1,000-2,000 pairs (542). A detailed survey in 1975 estimated 736-952 pairs in Scotland and 176-291 (28%) in England. With two pairs in Wales the British total was put at between 914 and 1,245 pairs but probably closer to the higher figure (336).

The north-west European stock is put at 750,000, with another 10,000 in the Black Sea/Mediterranean region (529). The bulk of the northern group winters in Denmark, counts varying between 13,200 and 28,200 in 1967-1973 with the highest number in the hard winter of 1969-70. Most are on fresh or slightly brackish waters (290). Very large numbers also winter in the Netherlands, usually 3,000-15,000 but with a high of 31,000 in January 1977. Most are in the IJsselmeer and on the main rivers Rhine and Waal (48).

In Britain, as Fig 143 shows, Goosanders are almost always found inland, on a total of more than a thousand sites during our study. The only site which comes up to the internationally important level (more than 750 Goosanders) is, however, an estuary – the Beauly Firth, but in the British context flocks of 100 or more are significant. Only seven sites surpassed this between 1975 and 1982, and only the Beauly Firth has had more than 200. In February 1983 the total British count was 3,860, of which 2,400 (62%) were on the Beauly Firth, the result of a rapid increase at that site since the mid 1970s. The other places to have held seasonal maxima of over 100 between 1975 and 1982 were as follows: Eccup Reservoir, West Yorkshire (five times), Castle Loch, Lochmaben, Dumfries and Galloway (four times), Blithfield Reservoir, Staffordshire (twice), Foremark Reservoir, Derbyshire (once), Farmoor Reservoir, Oxfordshire (once) and Queen Mary Reservoir, Surrey (once). More sites may qualify but not be recognised if they are used as night-time roosts for birds feeding by day (when the counts are made) on nearby uncounted watercourses, as in at least one site in the borders of Scotland.

It is predictable that the trend (Fig 144) shows an increase, especially in the last ten years. Because of annual variations in distribution and counting problems, however, it is not possible either to attribute absolute numbers to the index or to lay any stress on one or two exceptional values, except perhaps the cold weather influx of 1962-63. The absence of Beauly Firth counts in some years might also have an influence by

Fig. 144. Goosander – trend in January numbers in Britain, 1960-1982.

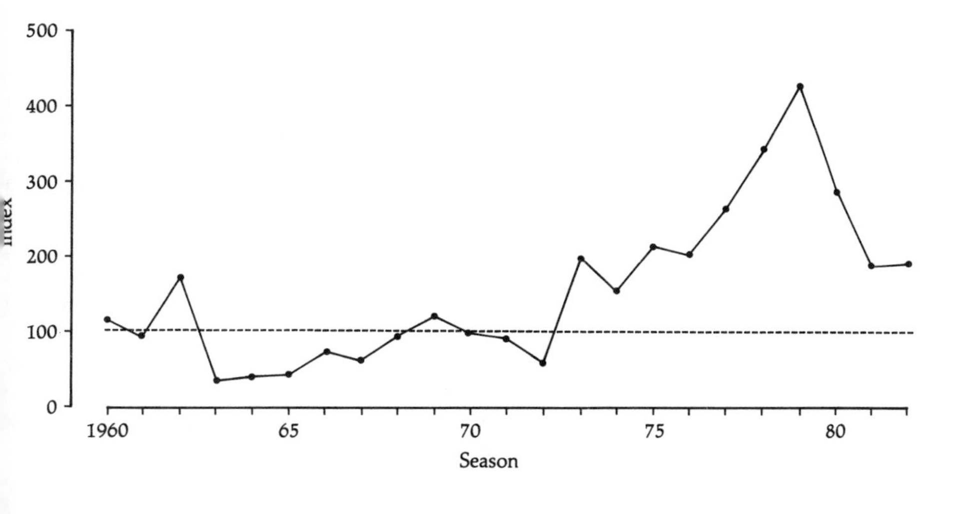

reducing the number of birds in the sample and hence making minor fluctuations more important.

The same reservations apply to the estimation of a British total. The sum of 5-year means for all sites gives 2,700, but this is likely to be a considerable underestimate. An assessment can be made from the breeding population count (336), taking 1,100 pairs as being close to the real total. Unfortunately there are no reliable figures for productivity, but if we assume the same average as for Red-breasted Merganser (1.4 young per pair) and the age at first breeding as 2 years, this indicates a native post-breeding total of about 5,000. This is supplemented by immigants in mid-winter, but probably no more than a few hundred. These figures are not inconsistent with the maximum counts in winter.

The Goosander is considered a pest in Scotland (but protected in England) and under the 1954 Protection of Birds Act it has been widely persecuted. In the north of Scotland the numbers have been kept stable or even reduced by shooting and nest destruction (336). The southward spread has brought it into areas which are not so important as commercial fisheries and the numbers have increased dramatically. The fact that England held 28% of the breeding pairs only three decades after it first bred there shows that the population's centre of gravity is shifting southwards. The 1981 Wildlife and Countryside Act protects the Goosander throughout Britain, but makes provision for shooting them under licence to protect fisheries. Whereas this may cause less intense control in some parts of Scotland, it may increase persecution further south. The spread of this species is likely to continue but the numbers may not rise in parallel.

North American Ruddy Duck *Oxyura jamaicensis*

A native of North and western South America, this diminutive diving duck (a member of the stifftail family) would probably never have reached Britain had it not been introduced by man. The species does not breed well in captivity and until recently ducklings have proved extremely difficult to hand rear. It became the practice at Slimbridge, following the first breeding there in 1949, to allow the females to rear their own young on a quarter hectare pond. Attempts were made to catch and pinion the ducklings just after hatching, but since they are very able swimmers and divers at an early age, some inevitably escaped and became free-flying. Some of these moved from Slimbridge in winter and up to 20 escaped in 1957. The core of the feral stock was established from these birds, although a few have since escaped from other collections (274).

In its native North America the Ruddy Duck breeds in freshwater and winters both inland and on the coast. In winter it is primarily a vegetarian, grazing underwater the leaves of submerged plants such as *Potamogeton*, *Ruppia* and *Zostera* and sifting out the seeds of a wide variety of aquatic plants. Whereas only a quarter of the winter diet is of animal origin the birds are carnivorous in summer. The most common items are chironomid and other insect larvae and freshwater snails, with small fish being taken occasionally (457, 545).

The stifftails are the most highly aquatic of wildfowl and hardly ever leave the water, being very ungainly on land. All their food is obtained by diving, in 1.5m of water at Chew Valley Lake, Avon, adult Ruddy Ducks have been shown to remain submerged for 14-20 seconds (mean 17.2s). Ducklings, because of their high buoyancy, are unable to dive at first, but by the age of 8 or 9 days submerge for about 10 seconds in shallow water and by 30 days are almost as proficient as their parents (306).

From the above description it seems that Ruddy

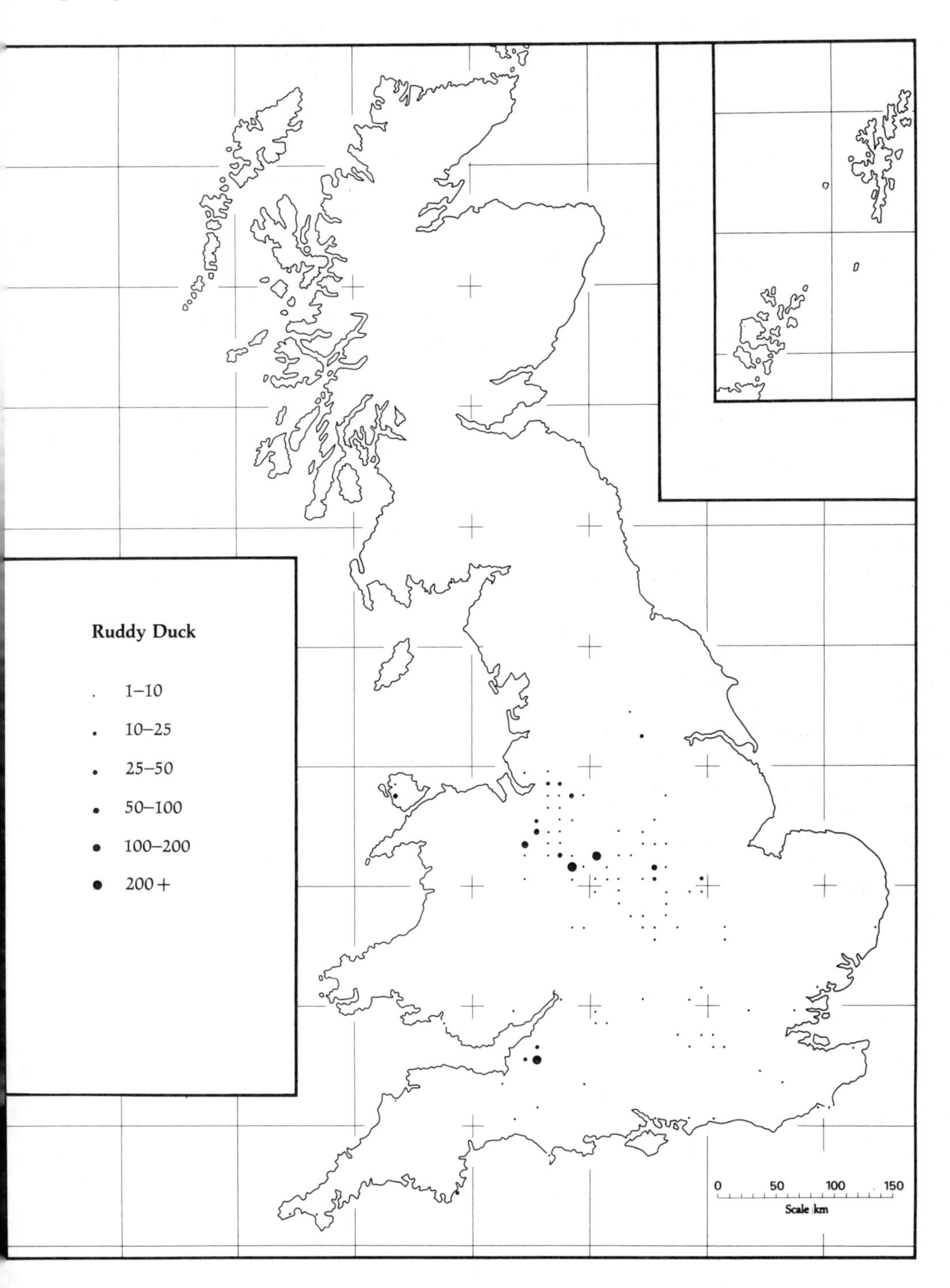

Fig. 145. North American Ruddy Duck – British distribution.

Ducks are most similar in food requirements to the Pochard in winter and to the Tufted Duck in summer. Since they have been successful here competition from either species has not been great enough to halt their progress. Differences from the Tufted Duck in summer habits, particularly in nesting sites and brood behaviour, may be sufficiently great to separate the two species ecologically and avoid competition.

The nest is a platform of reeds, usually above or very close to water. Several platforms may be built by both male and female prior to the building of the nest proper by the female (546). There are usually 6-10 relatively very large eggs and the female plucks little or no down to line the nest. In Britain predation from foxes and gulls is thought to account for many of the nests, and gulls and fish prey on young ducklings (306).

Soon after the first feral British birds were established there were separate groups in the west Midlands and in Avon. The present distribution is centred around these two areas (Fig 145) although the species has been recorded on a total of 228 sites. The main west Midland centres are Blithfield and Belvide Reservoirs, Staffordshire, while Chew Valley Lake and the nearby Blagdon Reservoir (in the same 10km square – Fig 145) hold most of the southern group. The numbers in Avon and Somerset increase in winter as birds from the west Midlands move south, especially during cold spells. Table 256 gives recent counts for the four most important sites. In the relatively hard January of 1979 a major displacement occurred, as the Staffordshire haunts were deserted and the ducks concentrated in Avon, but a few were also seen as far west as Ireland and the Isles of Scilly and east to Norfolk and Lincolnshire (602). In 1981-82 the pattern was even more striking, with 161 birds counted in Cornwall, Devon and Dorset in January, and 101 in Anglesey. The Midlands haunts were almost deserted, but by March most had returned there. Such movements are very likely to lead to further spread c the species within Britain and perhaps even overseas a few did reach France in 1981-82. The speed c colonisation of Britain is illustrated by the fact tha whereas there were records from 135 sites up t 1979-80, this increased to 228 in the next three winters

Because the numbers were so small, it i impossible to obtain a meaningful picture of the tren from the Wildfowl Counts for the early years c establishment, but the Ruddy Duck has been we studied since its introduction and its numbers ar fairly well known. Despite a setback in the cold winte of 1962-63, the numbers increased rapidly; in th period 1965-1975 the rate was about 25% annually, to total of 50-60 breeding pairs and a post-breedin population of 300-350 birds (274). In December 1978 a intensive survey yielded 770 individuals (602), a

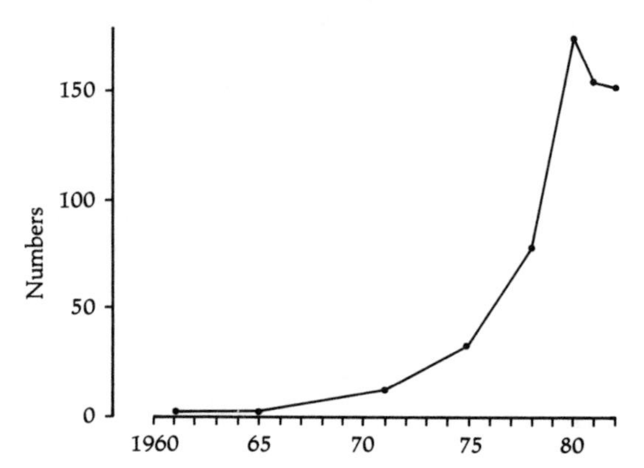

Fig. 146. The probable Ruddy Duck population in Britain in 1961-1982. Figures up to 1975 from Hudson (1976). (Those for 1965 and 1971 are extrapolated from the number of pairs estimated, assuming that the ratio of pairs to post-breeding population was similar to that in 1975, when population = pairs x 6.) 1978 from Vinicombe and Chandler (1982) and later figures from Wildfowl Counts (see text).

Table 256. *Counts of Ruddy Ducks on the four most important sites in Britain, 1976-1982. Numbers are peak counts, month in brackets.*

	Belvide Resr, Staffs	Blithfield Resr, Staffs	Blagdon Resr, Avon	Chew Valley Lake, Avon
1976	312 (O)	132 (J)	155	109 (J)
1977	186 (O)	198 (F)	125	94 (D)
1978	313 (O)	123 (D)	110	278 (M)
1979	198 (S)	297 (J)	312	167 (J)
1980	427 (F)	630 (J)	295 (J)	290 (F)
1981	340 (S)	477 (D)	330 (D)	135 (F)
1982	276 (S)	358 (J)	323 (J)	456 (F)
Mean 1976-82	293	316	236	218

ncrease over 1975 at the rate of about 33% per annum. 'he Wildfowl Count total in the same month yielded 77 Ruddy Ducks – 62% of the survey total, but in 1arch 70% were counted. In winters which are milder han this, when the birds are more concentrated, overage is better and 90% are probably accounted for. ssuming that level of coverage the 1,570 counted in ebruary 1981 represents about 1,750 ducks – an ncrease at the rate of 50% per annum since 1978. Even the 1,570 were the actual British total the annual rate f increase would have been in excess of 40%. Fig 146 lots this amazing growth, extrapolating figures for 965 and 1971 from the estimates of the number of airs (274). Clearly the growth curve is exponential up 1980-81 and the spread into new areas continues. 'he hard weather of 1981-82 caused a setback but the ncrease continued thereafter.

Being a small duck, the Ruddy Duck is suscepti-le to cold weather and severe losses might be xpected after a hard winter. A mortality rate of 13% was calculated for 1978-79 but then the period of really cold weather was rather short. Wildfowl counts for 1981-82 revealed 1,400 in December before the hard weather and 770 in March. If this represents the true effect (it is likely to be an overestimate of the mortality if the birds had remained more dispersed than in the autumn), the mortality rate of 45% is high but hardly sufficient to give the population more than a temporary setback. Indeed, it bounced back in 1982-83, when 1,380, almost up to the previous maximum, were counted in January. As in 1979, new sites have been colonised as a result of the dispersal, and in future, perhaps, some of these may even be on the Continent or in Ireland. The species seems destined to become a widespread bird in the western Palearctic. Some have voiced fears that, should it colonise the Continent, it could evict or hybridise with the White-headed Duck, *Oxyura leucocephala*, in Spain. The White-headed Duck is very rare in Europe nowadays and its existence is precarious with or without the Ruddy Duck.

Other species

Lesser White-fronted Goose *Anser erythropus*

A smaller, close relative of the Whitefront, this species breeds across northern Europe and Asia from Norway to eastern Siberia. It is an upland breeder and migrates through central Europe to winter around the Black and Caspian Seas and south-western Asia. Stragglers to Britain are almost always seen with European Whitefronts; up to 1982 there had been just over 100 records, well over half of these at the New Grounds on the upper Severn Estuary, where it was first recorded in 1945. Before then there had been only 2 sightings, the first in 1886. The largest number present in any one year was 4 at the New Grounds in 1979-80. Probable hybrids with the Whitefront are sometimes reported.

Snow Goose *Anser caerulescens*

A numerous species of North America, the Lesser Snow Goose *A. c. caerulescens* has a blue and a white phase; the Greater Snow Goose *A. c. atlanticus* has only the white form. Both races, and both phases of the Lesser Snow, occur almost annually in Scotland and occasionally in England, usually with flocks of other geese. The species is, however, commonly kept in captivity in a fully-winged state and it is impossible nowadays to establish which birds are wild. Those accompanying Whitefronts from west Greenland are more likely to be so; a group of 18 seen in the Netherlands in 1980 contained a single adult ringed in southern Hudson Bay.

Red-breasted Goose *Branta ruficollis*

The Red-breasted Goose breeds in the Taimyr Peninsula in western Siberia and winters mainly on the Caspian and Black Sea coasts. It is a very rare

vagrant to Britain, almost always occurring with flocks of White-fronted or Brent Geese, whose breeding ranges overlap. Up to 1984 there were 28 records, mainly from south and south-east England, including three, probably different individuals, in 1983-84. Although Red-breasted Geese are widely kept in captivity, few are left full-winged, so most if not all the sightings are of true wild vagrants.

Egyptian Goose *Alopochen aegyptiacus*

This species has been introduced from Africa, where it is common south of 20°N and in the Nile Valley. There are sporadic occurrences of wild birds in southern Europe but introductions have accounted for all the records from north-west Europe. Introductions were also made in France and Australasia but these did not establish viable populations (319).

Egyptian Geese were released in Britain at least as early as the 18th century and free-flying flocks were established on private estates in north Norfolk, Devon, Bedfordshire and East Lothian (542). Only those in Norfolk resulted in a viable stock, with some evidence of spread into Suffolk.

The Egyptian Goose, although closely related to the Shelducks and not a true goose, is a vegetarian and spends much of its time grazing on land. The diet consists mainly of grass leaves but seeds and flower-heads are taken when available (167). The birds also use crops such as maize and other cereals. During the moult, when they keep to the water to avoid predators, they feed on submergent and marginal plants.

The species nests in a wide variety of situations, from holes or among grass cover (the most common locations) to hollow trees or disused nests of tree-nesting birds, sometimes at a height of 25m, and cliffs (173, 474).

The least well studied of our introduced wildfowl, there is no good information on numbers, but the level of 300-400 birds estimated in the First Edition was thought to be close to the number present during the Atlas survey (542). The species is not well counted in winter, the annual maximum rarely exceeding 100 birds. Its failure to expand its numbers or its range during the long period following introduction, and its habit of nesting during the early spring when the weather is often cold and wet, make it unlikely that the Egyptian Goose will become a numerous species in Britain. Its survival will probably always depend on the well-protected strongholds in Norfolk, chiefly at Holkham, Beeston and nearby waters.

Ruddy Shelduck *Tadorna ferruginea*

This is principally an Asian species, breeding across the mid latitudes of the USSR westwards to the Black Sea and eastern Mediterranean. There is also a small, isolated population in North Africa, chiefly in the Atlas Mountains in Morocco and north-eastern Algeria. Some of the African birds used to winter in southern Spain but this movement has all but ceased in recent years.

The Ruddy Shelduck is a regular vagrant to Britain in winter. In 1886 and 1892 there were sizeable influxes and small flocks, sometimes of up to 20 birds, were reported from many places. Recently there has been an increase in the number of sightings but, since the incidence in southern Europe has declined, these are thought to be largely escapes. Of the 123 reports between 1965 and 1979, none can with certainty be regarded as other than feral (503).

American Wigeon *Anas americana*

A regular vagrant from North America, the American Wigeon is similar both physically and ecologically to the Eurasian species. Its habitat is similar both on coasts and inland, although the American species is more aquatic and spends less time grazing.

Every year there are several reports of single birds (usually males since the females are more difficult to distinguish from our own species). By 1982 there had been 116 records, 84 of these since 1957. There was an unprecedented influx in autumn 1981 with a total of about 24 in October, chiefly in south-west England. Although the American Wigeon is fairly common in collections the majority of records still refer to real wild vagrants.

North American Black Duck *Anas rubripes*

The Black Duck is a species of eastern North America similar in structure and habits to the Mallard but more able to withstand severe weather. It is not highly migratory, which accounts for the fact that there have been few European records.

The first Black Duck recorded in Britain was in Kent in 1964 and the total up to 1982 was 11. A female first seen on the Scilly Isles in October 1976 later paired with a Mallard and produced young in at least 3 subsequent seasons. Two full grown hybrids were present at Tresco in 1980 as well as the surviving female. Another, a male, stayed in Gwynedd after its arrival in 1979 and there were three hybrid young in 1980.

Blue-winged Teal *Anas discors*

This bird is a vagrant from North America which has apparently been a much more regular visitor in recent years. It is similar in size and ecology to the European Teal but more closely related to the Garganey.

The first British record was in 1858 but there were only 19 up to 1957. A further 41 were recorded to 1977 with no more than five individuals in any one year. There were no fewer than 13 birds in 1978, 12 in 1979, 7 each in 1980 and 1981, and 3 in 1982. Some of these may have been escapes but the fact that the vast majority were first seen during the autumn and spring migration season of the species suggests that most were wild – a marked increase in its incidence here.

Red-crested Pochard *Netta rufina*

A scarce winter visitor to Britain, the Red-crested Pochard breeds mainly in western Asia and southern Europe. It extended its range considerably westwards and northwards into Europe in the early part of this century, and bred in West Germany in 1920 and in the Netherlands in 1942. In the 1950s the annual maxima in the Netherlands were usually 100-300 with a peak of 650 in 1953, having increased from very few prior to 1948. There has been a considerable decline since (592). During the same period similar trends were seen in the British records, which became more frequent in the 1950s, to a peak of 44 in 1957 before declining subsequently. These birds probably originated from the Netherlands (487).

There have been several breeding records in Britain but all are thought to have originated from escapes from collections. There is a full-winged group in St James's and other London parks, and one or two pairs have nested regularly in gravel pits near the Wildfowl Trust grounds at Slimbridge and at the nearby Cotswold Water Park. Several other collections have full-winged birds and it is no longer possible to separate wild birds from these. It seems likely, however, that since the decline in north-west Europe the frequency of occurrence of vagrants has been low and most sightings nowadays are of feral or escaped ducks.

Ring-necked Duck
Aythya collaris

A native of North America, the Ring-necked Duck is similar in many respects to the Tufted Duck, being a largely inland omnivorous diving duck. It breeds in mid latitudes across the continent and winters in the southern United States and in Central America.

The Ring-necked Duck was a rare vagrant to Europe before the 1970s, there having been only one record before 1957 (in 1801) and eleven before 1970. By the winter of 1977-78 there had been another 30 records, but there were at least 22 individuals recorded in that winter, 25 in 1979 and 29 in 1980, bringing the total to 142 by the end of 1982. Ireland received a similar increase at the same time and several records were of more than one individual.

Although the first ringed in Britain (at Slimbridge) in March 1979 was later recovered on its way home in Greenland in May of the same year, many individuals stay for some time (e.g. 10 remained from 1979 to 1980). The species is well adapted to our climate and may be sufficiently different from the Tufted Duck to prevent wholesale hybridisation. There have, however, been several recent reports of possible hybrids, including a convincing case at Chew Valley Lake, Avon, in 1977. It is not known whether these originate from captivity or whether they are the progeny of breeding in the wild. The Ring-necked Duck could prove to be the first transatlantic wildfowl species naturally to colonise the British Isles.

Ferruginous Duck
Aythya nyroca

The Ferruginous Duck or Common White-eye breeds in southern Europe and western Asia, mainly between 40 and 55°N, with isolated groups to 30°N. The northerly populations are migratory and the nearest ones to Britain are in East Germany and Poland.

There are several records each year, chiefly in south-east England in autumn and spring, presumably birds from central Europe displaced while migrating to wintering areas in the Mediterranean. The species is commonly kept in captivity and a few might be escapes, but most are genuine vagrants; records to date amount to 200-300.

King Eider
Somateria spectabilis

A species of the High Arctic, the King Eider has a circumpolar distribution, breeding predominantly north of the Arctic Circle and reaching beyond 8°N in northern Greenland and north-east Canada. The European wintering range is between 63 and 71° on the Norwegian coast and in western Iceland. It is a regular vagrant to the northern parts of the British Isles, most commonly in Shetland, the area closest to the normal range. Most records are of males, but since females are not easily distinguished from those of the Common Eider, many are presumably missed. Many of those that do occur stay for some years and there are records in Iceland of hybridisation with Common Eiders. There have been 162 records in Britain up to 1982 including 5 in 1978 and 1979, with one sighting as far south as Walney Island, Cumbria. Of four seen in 1980 three had stayed from previous years; 3 and 4 respectively occurred in 1981 and 1982.

Steller's Eider
Polysticta stelleri

This is a small eider with a northerly breeding distribution in central and eastern Siberia and Alaska. The main wintering area is in the Pacific, but small numbers are regularly found in northern Norway and on the north coast of the Kola Peninsula, from where vagrants occasionally wander further south and west. There have only been 13 British records, the latest in 1979-80 in Orkney. An individual arrived in the Western Isles in 1974 and still remained in 1982. There has been an increase in the number wintering in the Baltic recently, with a flock of 77 in Finland and 19 in Poland in 1978, and a total of 150 in Sweden, mainly around the islands of Gotland and Oland, in 1979.

Harlequin Duck
Histrionicus histrionicus

The Harlequin breeds on fast flowing streams inland in Iceland, Greenland, North America and eastern Asia. It winters on the nearest coastline to the breeding area and it is a rare vagrant to north-west Europe. There have only been eight records in Britain and none since 1965, when a male and female were seen together both in Fair Isle and in Caithness.

Surf Scoter
Melanitta perspicillata

A North American species, the Surf Scoter breeds from western Alaska to Labrador, north to 70° and south to 55° in central Canada. It winters on both coasts of North America south to 30°N and it is the commonest scoter on that continent. It is a regular vagrant to western parts of northern Europe, chiefly the British Isles and Scandinavia, and some are found in summer in Arctic Eurasia. Only 75 had been recorded in Britain up to 1957 but it has become much more common since, with another 40 up to 1976 and a further 60 in the following seven years. One recent sighting was of a flock of 8 in the Moray Firth in 1979 six were seen in 1981 and the same number in 1982. Most vagrants mix with other scoter flocks and some individuals have remained in Britain for several years. Unless in flight females are very difficult to distinguish from those of Velvet Scoters, so most isolated records are of males.

Bufflehead
Bucephala albeola

A rare vagrant to Europe, the Bufflehead breeds in the wooded areas of central and northern Canada and Alaska, a distribution matching that of the medium-sized woodpeckers whose holes it uses as nest sites. It migrates to winter in coastal areas of the eastern United States, the Gulf coast and the Pacific coast. There are very few European records and only 6 in Britain. The latest, and the first for Scotland, was in South Uist in March 1980.

Hooded Merganser
Mergus cucullatus

Breeding in wooded parts of central North America, the Hooded Merganser winters, chiefly inland, on both sides of the continent. It is a very rare vagrant to Europe; there have been only four records in the British Isles, one in this century, in Co Armagh in December 1957.

Part IV

The conservation of wildfowl and their habitats

Introduction

The main purpose of the survey and analysis presented in the preceding parts of this volume was to review, twenty years after the first major assessment, the status of wildfowl in Britain and the progress that has been made in their conservation since the First Edition. This final part considers the position, examines the changes that are likely to be imposed in future, both beneficial and deleterious, and considers where best our efforts should be directed. We examine the influence of man, both as a modifier and destroyer of habitat and as a positive force in management and conservation. We consider the degree to which wildfowl stocks can be managed or controlled, and the present state of wildfowl conservation and its future needs.

As an introduction to this part we here discuss some of the considerations, both within and outside Britain, which must be borne in mind when we consider the status and conservation of our wildfowl.

The importance of Britain as a wintering area

There are five main wintering regions for wildfowl in the western Palearctic. These are well defined geographically and most are large enough and varied enough in climate to accommodate their populations in all weather conditions, so that there is normally little interchange of birds from one region or flyway to another (22). The regional boundaries, together with the main areas of concentration, are shown in Fig 147. Britain lies in the north-west European wintering area and it is in the context of this flyway population that the importance of Britain must be viewed. Britain is closely associated with the continental wintering grounds, including the Nether-

Table 257. *The north-west European and regular maximum British populations of swans and geese. Data from Atkinson-Willes 1976, Scott 1980 and this survey. Figures in brackets are percentages in Britain of the species rather than the particular population. - = percentage under 1%.*

Species		NW Europe	Britain	% Britain
Mute Swan		120000	18000	15
Bewick's Swan		16000	7500	47
Whooper Swan	(Iceland)	9-11000	4500	50(18)
	(Scandinavia/Siberia)	15000	0	0
Bean Goose		70000	200	-
Pink-footed Goose	(Iceland/Greenland)	100000	100000	100(77)
	(Spitsbergen)	30000	0	0
European White-fronted Goose		200000	6000	3
Greenland White-fronted Goose		20000	9000	45
Greylag Goose	(Iceland)	100000	100000	100(71)
	(NW Europe)	40000	0	0
Barnacle Goose	(Greenland)	25000	20000	80
	(Spitsbergen)	10000	10000	100(40)
	(Siberia)	40000	0	0
L-b Brent Goose	(Greenland/Canada)	15000	0	0
	(Spitsbergen)	3000	1500	50(8)
D-b Brent Goose		150000	60000	40
Canada Goose		83000	33000	40
Total		1047000	369700	35

lands, Germany and Denmark, but, with Ireland, is at the western end of the region, the final wintering destination of most species.

The estimated European and British populations of all the swans and geese wintering regularly in western Europe are shown in Table 257. Although we hold less than a quarter of the total number, Britain is vital for many species. We hold all the birds from three distinct goose populations and, together with Ireland, the whole wintering stock of four more. Britain supports nearly three-quarters of the world population of Pink-footed Geese, two-fifths of Greenland Whitefronts and Dark-bellied Brent and over a third of the Barnacle Geese.

A similar analysis for ducks is given in Table 258. Excluding the sea ducks, which are very largely concentrated around the Baltic and shallows around Denmark, Britain holds a fifth of north-west Europe's ducks. Britain is of key importance for dabbling ducks and Shelducks; Goosanders and Red-breasted Mergansers have populations which are largely resident. We are particularly significant as a final stopping place and can hold many more than these numbers when conditions are unfavourable in continental Europe.

For wildfowl in general, then, Britain is a very important wintering area. For some species, particularly dabbling ducks and geese, British haunts are of vital importance to the welfare of the flyway stock.

The origins of our wintering birds and their seasonal patterns

Numerically the vast majority of our wintering wildfowl originate from outside Britain, though for some species our native breeders are important. The

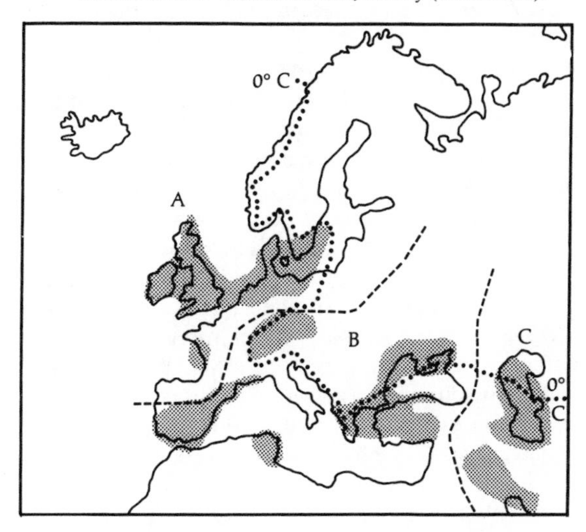

Fig. 147. A map of the main wintering areas (shaded) and groups (separated by dashed lines) of wildfowl in the western Palearctic. A = north-west Europe, B = Black Sea/Mediterranean, C = Caspian/Gulf. Also shown is the 0°C isotherm for January (dotted line).

Table 258. *The estimated north-west European and British populations of ducks. Data from Atkinson-Willes 1976, 1978, Scott 1980 and this survey.*

Species	NW Europe	Britain	% Britain
Shelduck	125000	75000	60
Wigeon	500000	200000	40
Gadwall	55000	4000	7
Teal	200000	100000	50
Mallard	3000000	500000	17
Pintail	75000	25000	33
Shoveler	100000	9000	9
Pochard	250000	50000	20
Tufted Duck	500000	60000	12
Scaup	150000	5000	3
Eider	2000000	50000	2.5
Long-tailed Duck	500000	20000	4
Common Scoter	500000	35000	7
Velvet Scoter	200000	3000	1.5
Goldeneye	200000	15000	7.5
Smew	20000	50	0.3
Red-breasted Merganser	40000	10000	25
Goosander	10000	5500	55
Total	8425000	1164050	13.8
Excl. sea ducks and sawbills	4805000	1021000	21.2

estimated post-breeding populations of those wildfowl which breed here at all extensively, and the proportions these are of the peak wintering populations are shown in Table 259. Mute Swans and Eiders are, to all intents and purposes, wholly resident and self-sustaining, with neither species showing marked changes in numbers in recent years. The Canada Goose is wholly introduced and resident, and breeding Gadwall are very largely of recently introduced stock, though some releases were made many years ago. Although post-breeding Gadwall numbers are equivalent to those at the winter peak, there is some turnover, with British juveniles dispersing southwards while continental birds move here to winter.

The two sawbills are largely resident, with rather a small immigration from the Continent in mid-winter. Both species have continued to increase in numbers and spread in recent years. The breeding population of Tufted Ducks has increased dramatically, as has the contribution of the native stock to the winter peak. There is again some turnover of birds and this may be considerable, with many British breeders wintering in Ireland and continental immigrants making up our mid-winter numbers. The British stock also makes a substantial contribution to the number of Mallard, Teal and Shoveler wintering here, though for the last two species there may be some autumn dispersal of home-bred birds southwards and westwards. The Shelduck population has grown considerably in the last two decades (p.396), though there is no indication that this has been a result of an improvement in breeding. Indeed, there is some doubt as to whether many of the dense estuarine populations are self-sustaining.

The three remaining species listed in Table 259 breed in insignificant numbers, and the additional 16 wildfowl listed in Tables 257 and 258 originate wholly from outside Britain. Nearly all our migrant Whooper

Table 259. *The estimated post-breeding numbers of British breeding wildfowl and the contribution they make to the peak winter total. Only those species breeding here in substantial numbers are included.*

Species	Post-breeding total	% of winter peak
Mute Swan	18000	100
Greylag Goose *	4000	5
Canada Goose *	33000	100
Shelduck	40000+	53
Wigeon	2000	1
Gadwall *	4000	100
Teal	20000	20
Mallard #	500000	70
Shoveler	15000	30
Pochard	2500	5
Tufted Duck	50000	83
Eider	50000	100
Red-breasted Merganser	8500	85
Goosander	5000	90

* Partly or entirely established from introduced birds.
\+ Based on 1 young fledged per pair and first year survival rate of 60% (see Patterson 1982).
\# Not including birds reared and released for shooting. Estimates of both post-breeding and total populations are very approximate.

Fig. 148. The monthly pattern of use of Britain (numbers counted expressed as a percentage of the annual maximum) by Spitsbergen Barnacle (solid line) and Siberian Brent Geese (dashed line). Figures for the Brent are based on the average counts for the three winters 1979-80 to 1981-82, those for the Barnacle are from a typical season (1979-80).

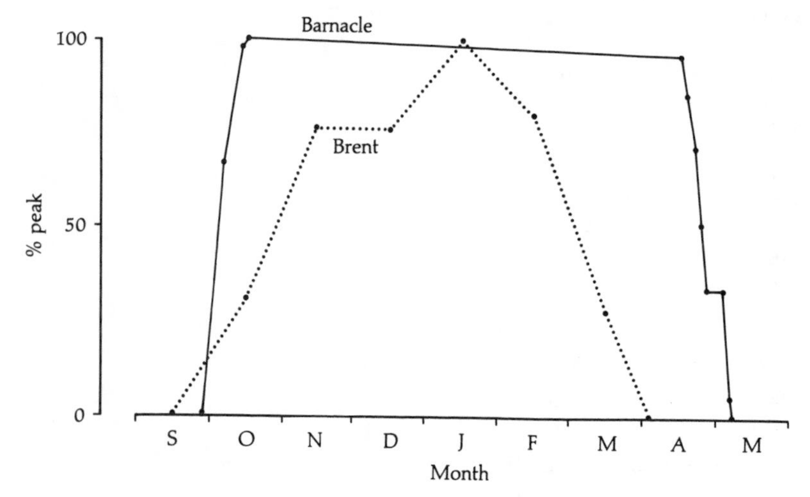

Swans come from Iceland, and their pattern of use of Britain (Fig 64) is radically different from that of the Siberian breeding Bewick's Swans, which come via continental Europe. The Bewick's Swans arrive gradually during the autumn and peak in January and February. The difference in the pattern between those birds that originate in Greenland and the north Atlantic islands and those that breed in western Siberia is well illustrated by the monthly numbers of Barnacle Geese from Spitsbergen and the Siberian Dark-bellied Brent Geese (Fig 148).

The Barnacle Goose is forced by hard weather from its autumn staging area at Bear Island in late September, and makes the 2,400km journey to its Solway wintering grounds non-stop. The whole flock arrives there within a very few days. The geese stay on the wintering grounds, suffering a 4% overwinter mortality, until late April, when the departure to the spring staging grounds 1,600km away in Norway is equally rapid (443). The Brent Goose, on the other hand, although its breeding season and departure date from the breeding grounds are similar to those of the Barnacle, makes its way through the Baltic and Netherlands in a much more leisurely fashion. The peak here is not reached until January, when Dutch birds are pushed by hard weather into England and France. No sooner are all the birds at their destination than the return migration has started, and the departure is equally gradual until nearly all the geese have left our shores by early April.

The Barnacle, and other geese of the north Atlantic, such as Greenland Whitefronts, Pinkfeet and Greylags, therefore use Britain not only as a wintering area, but also for an autumn and spring fattening place. Most of the Brent and European Whitefronts, which do not make long over-sea journeys, fatten up before arrival and after departure from Britain. This is extremely important to bear in mind when considering the conservation of our geese, especially when we are under pressure from farmers to disrupt feeding flocks or reduce numbers.

The migrations of ducks are less well known, but we receive relatively few from the north Atlantic islands and the bulk from Scandinavia, central Europe and the USSR via Germany and the Netherlands. Those from Iceland probably behave in a similar way to the north Atlantic geese; many Wigeon, for example, remain here well into April. These are probably local and Icelandic breeders – those from western Siberia have long departed for their breeding grounds some 4-5,000km away (p.400). In general ducks which do not make long over-sea migrations winter as close to their breeding places as possible, moving to their final wintering areas only when food supplies are depleted or when forced by hard weather. Dabbling ducks, especially the smaller species such as Teal and Shoveler, move here earlier than do sea and diving ducks, which remain in the open shallows of the Baltic until late in winter. Consequently the numbers and timing of arrival of sea and diving ducks tend to differ more from year to year than do those of dabbling ducks and swans.

Movements and population turnover

The Wildfowl Counts provide us with spot checks on the numbers of birds at particular sites and in Britain as a whole, but this does not necessarily tell us the number of individual birds that rely on British sites or on any one locality throughout the winter. It is well established, for example, that there is a very large turnover of wading birds in our estuaries, and the number of birds using an area may be several times the count at any one time (485). This is not so pronounced with wildfowl, since Britain tends to be a wintering rather than a staging area for most species. On a site by site basis, however, this consideration is important, although our knowledge of the subject is very limited, since intensive marking programmes are needed to discover rates of turnover to answer this question. An early study showed that Pink-footed Geese used east and central Scotland as a staging area, thereafter dispersing to their wintering grounds further south (61). More recently it was discovered that Foulness is a major gathering ground for Brent Geese arriving from the Continent. Birds are dispersing from Foulness as others are arriving, so its real value to the population is underestimated if we only consider the maximum count (519). This kind of situation is probably very widespread among wildfowl, and the dispersal patterns and seasonal requirements must be borne in mind in any conservation programme.

There is some turnover of dabbling ducks and swans, with many continuing through Britain to Ireland. Tufted Ducks from Loch Leven migrate to Ireland in autumn while others arrive here from the Continent and Iceland (p.431). British breeding Shoveler migrate to France and Spain in September and October. Autumn arrivals disperse later in the season, showing a quite different recovery pattern from those ringed in mid-winter (141, 395). There is also a significant passage of Shoveler through Britain in autumn and in spring (p.422).

With quarry species, birds which are shot early in the season are, in effect, replaced by immigrants later in the winter. This is most pronounced in Mallard, whose overall numbers change little through the season, judging by Wildfowl Count totals. The count data suggest a population of around half a million post-breeding birds in September, and only a few more at peak time in January, though this estimate includes an element of guesswork. The BASC shooting survey estimated that the national bag of ducks was about a million and that 60% of these were Mallard (238). This might suggest that the equivalent of the whole post-breeding population is shot and replaced by immigrants in mid-winter. The complicating factor is that many Mallard are hand-reared and released in autumn. That number is estimated at 400,000 (238), a figure which is regarded as realistic by the Game Conservancy. Many of these may be shot before the count in mid-September or remain on flight ponds or other uncounted waters and not register in the counts. The situation is no clearer for Teal (where there are no hand-reared releases), with an estimated bag of 175,000 compared with our estimated peak population of 100,000.

Clearly our state of knowledge is unsatisfactory and much more work is needed to establish the importance of shooting mortality and movements in the major quarry species. This can only be achieved by cooperation between the shooters and conservation organisations in analysing together annual, seasonal and local variations in the shooting bag in relation to the numbers and distribution of wildfowl.

The movements and losses of wildfowl are complex and difficult to establish in detail. A start has been made with geese and swans, and the advent of easily readable marks has facilitated this research. Ducks are rather more difficult to mark visibly and continued ringing and analysis of species ringed less commonly is needed before we can establish detailed conservation plans for them.

The effect of cold weather

The numbers and distribution of many species are likely to be radically different during extended periods of cold weather compared with normal winters. Long cold spells on the Continent result in influxes to south-east Britain, as happened during the hard weather of January 1979. Then there were major influxes of Shelducks and Wigeon onto east coast estuaries and many other species also entered the country for short periods. Many birds leave Britain if severe weather is prolonged, as has been demonstrated for Teal, which move in large numbers into France and Spain (409). There is also some redistribution within the country, with birds moving southwards and to the coasts as northerly and inland areas freeze. In January 1982 thousands of Pink-footed Geese deserted Scotland for Lancashire, where an unprecedented 36,000 (40% of that season's total population) were found late in the month. In Fig 149 the proportion of all wildfowl found on the coast in the hard winter of 1962-63 is compared with the mean for the two adjoining winters. After the cold weather arrived in January a substantially higher proportion of birds were on the coast, although a third remained inland despite widespread freezing.

The Wildfowl Counts are too infrequent to give a detailed and clear picture of cold-weather movements, knowledge necessary if we are to understand the need for and effects of restrictions on shooting or other activity during cold weather (p.530). The impact of cold weather through mortality and reduced breeding success is not well understood. In quarry species this impact is likely to be caused by making the birds more accessible to hunters or sending them to countries where shooting is more intensive and less discriminating, as well as by starvation. Despite voluntary shooting bans in 1962-63, however, many wildfowl did die of starvation and overall population levels were reduced; we can never insulate our wildfowl completely against the effects of adverse weather.

The use of different habitats

Clearly, since different species have different feeding and roosting requirements, the types of habitat used by wildfowl and their changes through the season are important factors in the overall conservation strategy. Whereas some waters are used predominantly as roosts, others provide both roost and feeding areas. Those whose main use is for roosting obviously depend for their importance on there being suitable feeding areas within easy reach. Nowadays geese feed predominantly on farmland and roost on nearby estuaries or lakes. Unless quite unforeseen dramatic changes occur in arable agriculture these habitats are in no danger of disappearing in the near future. However, changes in distribution may

Fig. 149. The proportion of counted wildfowl on the coast in each month in the cold winter of 1962-63 (solid line) and in the mean of the winters 1961-62 and 1963-64 (dotted line).

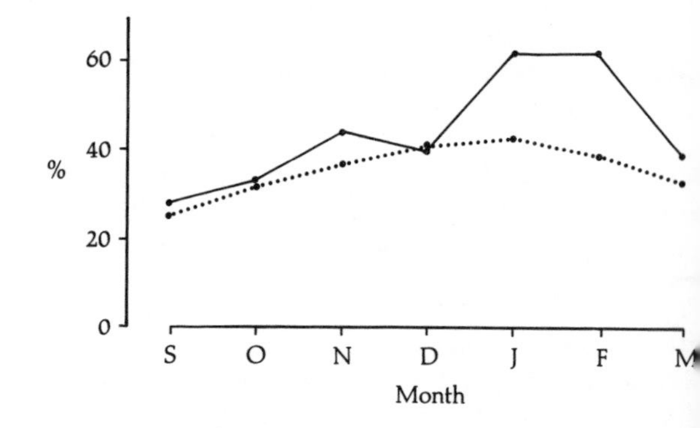

well be necessary as seasonal and agricultural factors alter food availability. Dabbling ducks and swans fly to wet pastures and small marshes from inland roosts, and these wetlands are continually being eroded by piecemeal drainage. As only the major schemes come under detailed scrutiny by conservationists, it seems likely that the importance of some roosts will decline as suitable local feeding areas disappear.

The seasonal pattern of use of the five habitats distinguished in our site classification by all ducks, geese and swans is shown in Fig 150. The three-year mean was chosen since the three seasons include a relatively hard winter and two mild ones, giving a fair representation of "average" conditions. On average 25% of wildfowl are found on natural inland waters, 17% on reservoirs, 8% on gravel pits, 9% on freshwater marshes and rivers and 40% on the coast. The proportion on the coast increases in mid-winter at the expense of natural waters and reservoirs. This is not only because birds are moving to the coast, but also because coastal species are later to arrive in numbers (p.435). The proportion on gravel pits is very similar in all months, but the use of marshes increases dramatically from December onwards as the availability of floodwater increases. By March native species, particularly Mallard, are moving into marshes to breed. About 20% of Mallard are on this habitat in March compared with less than 10% at other times.

The use of habitats by individual species is given in Table 260, which excludes the Smew (whose wintering numbers are usually very small) and those species which are more or less entirely dependent on the coast. These include the Brent Goose, Shelduck and the five sea ducks. Pintail, Teal and Wigeon rely heavily on intertidal areas for feeding, while Red-breasted Merganser and Goldeneye are found predominantly on shallow coastal waters and large estuaries. Pochard, Tufted Duck and Goosander rely on enclosed inland waters, although in the case of Goosanders these are often used mainly as day-time roosts while most of the feeding takes place on rivers. Bewick's Swans are the only species which largely concentrate on marshes and rivers. The other species occur on all habitats, but in all cases there are distinct preferences. Thus, of the 21 species of swans, ducks and Brent, 7 concentrate offshore, 5 are estuarine and 9 inland. The remaining geese largely roost on estuaries and lakes and feed inland on pastures or arable land. The maintenance of a wide range and diversity of habitats is thus vital if we are to conserve the species we have in roughly their present strength and without causing major changes in their distribution.

Fig. 150. The proportion of counted wildfowl on each of the five habitat types recognised. Each percentage is a mean of that in the three seasons 1979-80, 1980-81 and 1981-82.

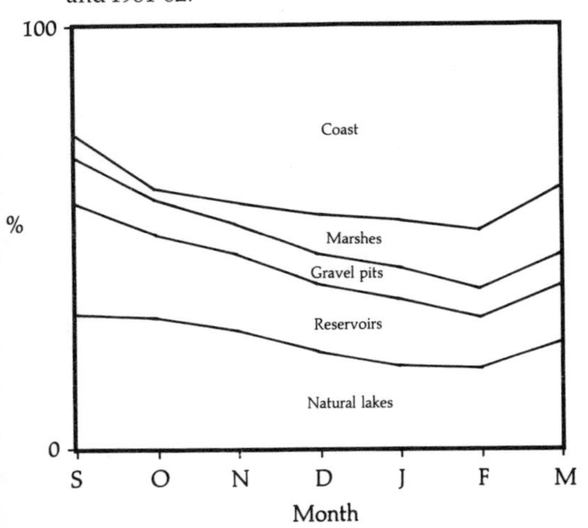

Table 260. *The average proportion of different species of wildfowl found on 5 different habitats throughout the winter. Based on the three winters 1979-80, 1980-81 and 1981-82. Those species which are exclusively coastal are not included.*

Species	Habitat 0	1	2	3	4
Mute Swan	35	7	16	16	26
Bewick's Swan	4	4	1	72	19
Whooper Swan	55	7	0	15	23
Wigeon	12	8	2	20	58
Gadwall	37	26	17	7	13
Teal	21	13	4	10	52
Mallard	33	23	11	9	24
Pintail	7	2	0	7	84
Shoveler	25	36	12	9	18
Pochard	29	28	30	7	6
Tufted Duck	29	36	27	3	5
Goldeneye	25	16	3	1	55
Red-breasted Merganser	7	1	0	0	92
Goosander	23	39	9	4	25

0 – natural or long-established artificial lakes
1 – reservoirs
2 – gravel pits and other mineral workings
3 – rivers and freshwater marshes
4 – estuaries and coast

The influence of man

Losses and threats to habitat

Wetlands have long been considered unexploited areas by man, and no energy or expense has been spared in draining them for agriculture, industry or other uses. This section briefly relates the progress of drainage and reclamation in the past, and discusses present progress and future threats.

Inland drainage for agriculture

Inland marshes exist because of impeded drainage of low-lying land and range from permanently wet areas of reed-beds, bogs and fens, to temporary floodlands along the floodplains of rivers. Progressive drainage allows marshes to be converted to wet grasslands, where summer grazing only is possible, then to relatively dry grassland, capable of reseeding and intensive grazing, and eventually to arable. The costs of lowering the water table progressively increase, but so also do the benefits; the final stage from pasture to arable land yields rich rewards for the farmer. Drained fens and marshes have a light, workable soil which is extremely productive, and this adds to the pressure for drainage.

The Fenland of East Anglia amounted to 3,400s km before drainage, consisting of 1,900sq km of si fens flooded by tides from the Wash and 1,500sq km c peat fens to the south, flooded by water from the larg catchment area of the rivers Nene, Ouse, Cam, Lark Little Ouse and Wissey. Enormous concentrations c wildfowl were found there, including a resider population of Greylag Geese, and early fenmen mad a living from wildfowling and fishing. Early attempt at drainage were ineffective, but in the mid 17t century a Dutch engineer, Vermuyden, was engage by English landowners to drain the fens. The Ol Bedford River channel was cut in 1637, giving th water from the tortuous Ouse a straight path betwee Earith and Denver. The cutting of the New Bedford i 1651 completed the construction of the Ouse Washe (p.147), which were used as a flood relief channel fo water from the catchment. The use of pumps and th cutting of new drains to link with the main channe accelerated drainage, and its progress since 1650 i shown in Fig 151. Over 70% of the fens had bee drained by the mid 18th century, and with th introduction of the steam engine around 1820 th efficiency of drainage improved markedly and most c the remaining wet areas were also drained. Nowaday only a tiny area of undrained fen remains and th

Fig. 151. The progress of drainage in the East Anglian fenland in relation to the introduction of new pumping methods. Redrawn from Thomas *et al.* (1981).

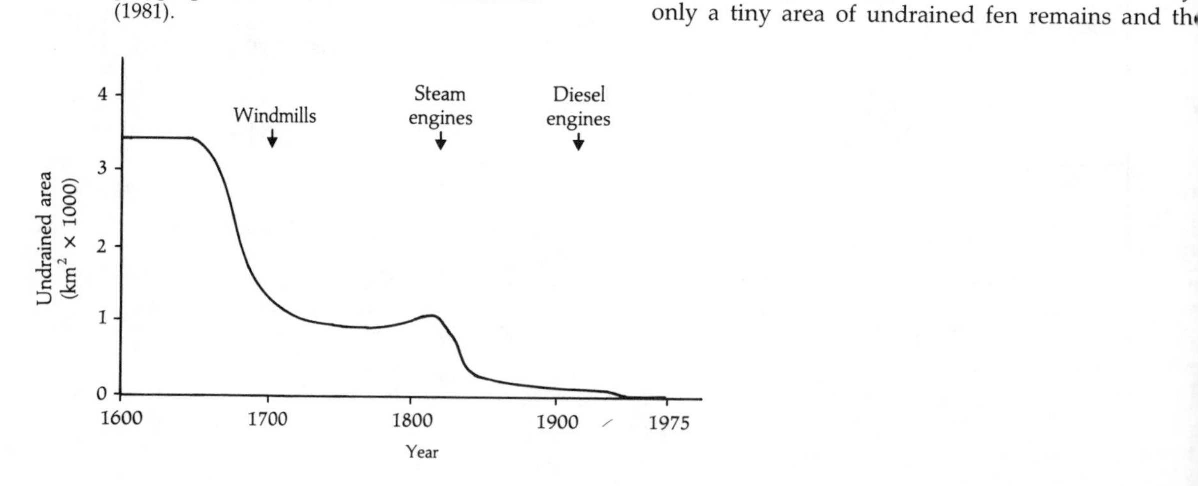

vashes of the Nene and Ouse form the only extensive vet sites. The Nene Washes have largely been eclaimed, but regular flooding there still makes arable ıgriculture unpredictable.

The Somerset Levels (p.45) consist of 680sq km of marshland along the valleys of the major rivers of west Somerset, formed in a series of saucers surounded by hill ridges and eventually draining into Bridgwater Bay. The original vegetation consisted of reed-marshes, base-rich sedge fens and acid *Sphagnum* bogs (370). Many parts were permanently waterlogged and further flooding in winter provided extensive habitat for wildfowl. Drainage continued gradually rom the 13th century, and by the early 19th century nearly all the Levels were farmed, at least seasonally. Since that time there have been continued improvements in speeding up the outflow of water and floods have been eliminated from much of the Levels. Only a few areas now flood at all regularly and the Agricultural Advisory Council consider that three-quarters of the land is capable of being converted to arable farming. The existing system of arterial drainage is capable, with increased pumping, of substantially reducing the water level to allow this (370).

Clearly much of the wetland habitat on the Levels is already lost. There was enormous opposition from farmers in the early 1980s to the scheduling of the few remaining pieces as SSSIs, and the battle to preserve what remains will be a difficult one under the pressure of continual calls for increased productivity of agricultural land. Ironically much of the land to be scheduled is not capable of being converted to arable; its improvement would only add to the large surpluses of beef and dairy products held by the European Economic Community (EEC).

The North Kent Marshes, another traditional haven for wildfowl, include large areas of grazing marsh, originally reclaimed from saline marshes centuries ago. Because they are below high tide level, runoff of water through fleets and tidal sluices was slow, tidal flooding occurred occasionally and the marshes remained wet, with summer grazing, until the early 20th century (223). In recent years, however, improved drainage has enabled much of the land to be converted to arable or developed for housing and industry. In 1935, 14,750ha of grazing marsh remained, but by 1982 this had been reduced to half, at 7,675ha. Urban development accounted for a quarter of drained marsh and 70% was converted to arable. Although the wintering wildfowl populations did not suffer dramatically there was a decrease in the number of wet-meadow breeding birds, particularly waders and Pochard, Shoveler and Garganey. The fragmentation of wet areas and reduction in the size of blocks of marshland is likely to accelerate the decline when a critical stage is reached (618).

The thousands of hectares of fenland inland of Southport, including the 1,200ha lake and marshes at Martin Mere, were progressively drained for arable use from the 18th century onwards. By the 1960s only a small area, near the centre of the former mere, remained as wet pasture. Some of the habitat was restored by the Wildfowl Trust in the next ten years, and this became an internationally important site almost overnight (p.193). Nevertheless, what remains represents a tiny fraction of the original area.

These important examples are not the whole story – drainage works are proceeding in all parts of lowland Britain, grant aided by the Ministry of Agriculture, Fisheries and Food (MAFF). In the mid 1970s grant aid was provided to under-drain some 100,000ha annually. The vast majority of natural fens and swamps have gone through the first stage of drainage, to wet meadow, and a great proportion to arable. The remaining wet grasslands are under severe threat. A report prepared by the RSPB (514) listed major drainage schemes threatening 164,000ha of marshes. The report severely condemned the present procedures for assessing the desirability of drainage schemes, which were heavily biased towards agricultural interests, took no account of conservation and did not provide organisations other than MAFF with the means of assessing their real agricultural value. The report advocated the more extensive use of public enquiries; only one enquiry had been held, at Amberley Wildbrooks in Sussex, where the inspector had found against drainage proposals because of the considerable environmental impact.

The effect of marshland drainage has been enormous on breeding and wintering dabbling ducks and swans, but most of the damage had been done before there was any means of monitoring declines. In the last two decades breeding and wintering populations have increased despite habitat loss, because new areas have been created and managed for wildfowl (p.507). Several important areas of marshland have been acquired and maintained by conservation bodies. This is, however, a tiny fraction of what could be saved if drainage cost – benefit analyses were realistic in their assessment of the economic value of grant aided schemes, and in particular if the losses of conservation value as well as agricultural considerations, were taken into account (514). The outlook at last began to look more promising in 1984 and 1985, when the UK Government announced a major reduction in grants for land drainage, and an EEC Regulation on improving the efficiency of agricultural structures was introduced which included provision for aid, on a national basis, to maintain traditional farming practices in "environmentally sensitive" areas.

Coastal reclamation

All our major estuaries had, at one time, much more extensive associated marshes than exist today. The building of sea walls has progressively reclaimed much of this area so that now only a small fraction remains in many estuaries. Reclamation around the coast of the Wash has accounted for some 47,000ha of marshes since Saxon times. This compares with 4,450ha of saltmarshes and 27,000ha of intertidal sands and mudflats remaining (118). Some of the new land has been built up as a result of earlier reclamation, but the intertidal area has been diminished overall. The area of coastal marsh in the whole of Britain in the mid 1960s was put at no more than 81,000ha (489), only a small fraction of what previously existed. In most estuaries embankment has continued to the very edges of deep-water channels and there is little room for further accretion. In the dynamic west coast systems cycles of accretion have alternated with erosion; for example, at Caerlaverock on the Solway Firth considerable accretion between 1850 and 1946 gave way to erosion and large areas have been lost subsequently (87). During the same period Morecambe Bay gained 530ha of saltmarsh, but much has been embanked and reclaimed for agriculture.

Reclamation of mudflats offers rich rewards for farmers, since the soil, unless very sandy, is productive if drained and converted to arable. A scheme to embank the 3,500ha Ribble saltmarshes was thwarted only by last-minute intervention by the Nature Conservancy Council following a public outcr (p.193). Similar schemes have been mooted for oth remaining expanses of saltmarsh on the west coas Such large schemes arouse considerable public ir terest and the outcry over the Ribble proposa enabled the NCC to succeed in persuading th government to provide money to purchase th saltmarsh. Local conservation groups have bee formed for many estuaries, and they monitor reclama tion schemes in detail and object if the conservatio case is powerful enough.

Fig. 152. A map of Teesmouth showing the areas of reclaimed land. The shaded area was reclaimed before 1971. Based on Evans (1981).

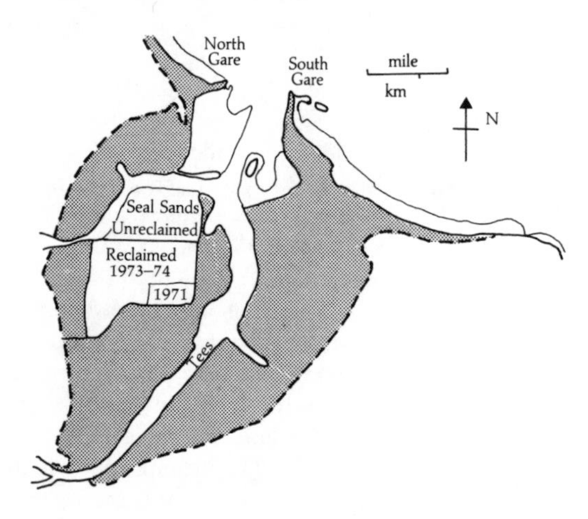

The loss of saltmarsh pasture chiefly affects th grazing birds, Wigeon and geese. With the exceptio of the Brent, geese have largely deserted the saltmarsh feeding areas in favour of agricultural lanc although saltmarsh pasture is still important, especia ly in spring, for Pinkfeet and Barnacles. Brent Gees have also taken advantage of inland habitats in recer years, but this is all on reclaimed land. The progressiv siltation of estuaries such as the Wash, accelerated b saltmarsh reclamation and structures associated wit shipping has, however, reduced the area of intertida mud, and this must have been to the detriment of th Brent. Mudflats and saltmarsh are the most importan feeding areas for Wigeon in Britain. Conditions ha deteriorated in many sites examined in the 1970s drainage and reclamation, mainly on the coast, bein the most important factors apart from shooting Deteriorating situations on the coast may partly b responsible for the move inland of this species (187)

Estuaries are attractive sites for industrial de velopment, having an outlet for disposal of wast products, access to shipping and a supply of low cos land. Developments related to the oil industry hav taken up a large area of land, especially on th Thames, at Milford Haven, at Nigg Bay on th Cromarty Firth, and on Shetland. As well as th impact on the environment, the potential danger through oil pollution are additional hazards (p.491

The best documented case of industrial reclama tion in recent years is that of Teesmouth, an enclose estuary in north-east England. Industrial and othe developments had reduced the original 2,500ha o mudflats to only about 200ha by 1969, when detailed monitoring of developments and bird numbers began The pattern of reclamation from 1969 onwards is shown in Fig 152. The recent reclamation was achieved by enclosing areas of mudflats with a slag wall and progressively building up the level of mud within it by pumping and dredging from the river. By 1974, only 140ha of intertidal land remained, 6% of the original area (187).

Teesmouth is an internationally important site for Shelducks, and a roost for dabbling ducks; other pecies occur in small numbers. The regular numbers f Shelducks and total wildfowl occurring from 1961 to 981 are shown in Fig 153. Excluding the low counts in 970-1975 when reclamation was in progress, the total vildfowl numbers have shown an increasing trend, hough in the late 1970s this was largely due to an ncrease in the number of dabbling ducks, particularly Mallard and Teal. Shelduck numbers also fell sharply n the early 1970s, but rose again to reach their highest evel ever in 1976-77. There followed a reduction in umbers to a level somewhat below the best years, but bout average for the whole period. This most recent ecline did, however, occur when the British Shel-uck population as a whole continued to increase airly rapidly (p.396); the relative importance of 'eesmouth has thus diminished.

The recovery of the Shelduck population at 'eesmouth in the mid 1970s, although the intertidal eeding area had been reduced to 35% of the 1969 evel, was extremely surprising. Feeding density was .9 times that in the late 1960s, suggesting that the use f the area was formerly well below its capacity. The ncrease in Shelducks in Britain during the same eriod may mean that the species still has scope for xpansion and that it is not as vulnerable to partial loss f habitat as was feared (22). The complete reclama-on of Seal Sands, which is now being considered, vould, of course, mean the end of its use by helducks; the species has not yet demonstrated any bility to use other wintering habitat than intertidal nudflats, though it is found some way inland during he breeding season. Other ducks which are likely to uffer from the loss of intertidal habitat are Wigeon, Teal and Pintail but, as yet, there are no examples of habitat losses causing drastic declines of any of these species on any major site. The industrial reclamation of Nigg Bay in the Cromarty Firth in the 1970s gave cause for concern, largely because of its international importance for Wigeon. Some of the reclamation was not carried out, however, and duck numbers held up remarkably well in the late 1970s and 1980s (p.281). Nigg Bay has now become part of an NNR.

A large-scale and well-publicised proposal was made in the late 1960s, namely to develop Maplin Sands at Foulness, Essex, as a site for the third London Airport and associated developments (59). Although the site is important for several species of waders and for certain breeding birds, the main concern of conservationists was for the Dark-bellied Brent Geese, whose world population numbered only 25-30,000 at the time; for a part of the autumn a quarter of these were at Foulness. The total number of Brent subsequently increased, but a very substantial proportion of the population, especially if the turnover at Foulness is taken into account, continue to use that site at one time or another between October and March.

The airport development would have resulted in the direct loss of between 30 and 60% of the *Zostera* flats at Maplin, a loss of resources sufficient to feed up to 4,500 Brent Geese, or 13% of the British population at the time. Additional effects of disturbance would have increased the area lost for both waders and Brent Geese. Subsequently, with the rise in its population, the Brent Goose has increasingly resorted to agricultural land to feed. There is good evidence that the intertidal resources were needed for the existing population and that their loss would now have increased the pressure on inland habitats and consequent problems with agricultural interests. The loss of Maplin, had it occurred, might not have had a

Fig. 153. Regular numbers of Shelducks (dotted line) and total wildfowl (solid line) at Teesmouth, 1961-1981.

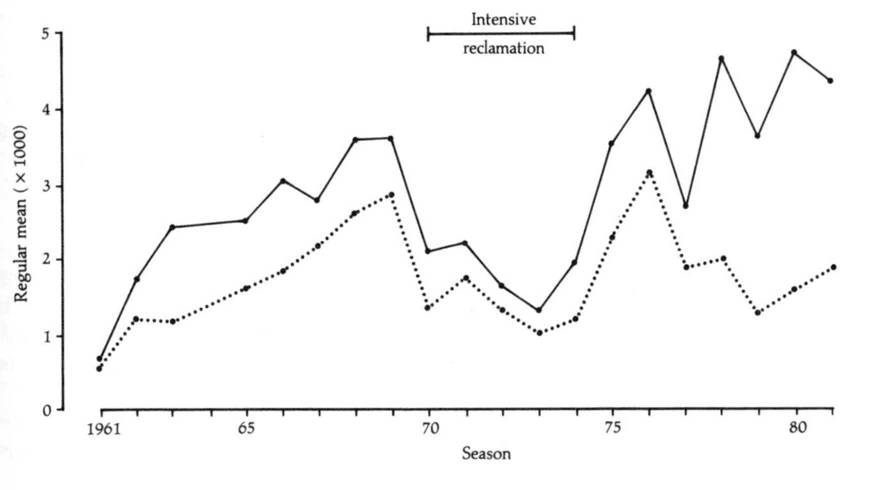

disastrous effect on the population of the subspecies, provided that refuges were created inland, or at least that feeding geese were tolerated on grass and cereal crops. Other estuarine birds might not have fared so well.

The plans for the third London Airport were shelved in 1974, but reconsidered in the early 1980s. In 1984, however, after a Public Enquiry, it was decided that Stansted, in Essex, should be the site of a third London Airport.

Estuarine barrages and reservoirs

Because it does not involve flooding of settlements or high value agricultural land, the creation of reservoirs on intertidal areas is very attractive to water authorities. There are often other advantages, such as the siting of the reservoir at the mouth of the river, where water flow is greatest, and close to major industrial developments. In some cases the creation of road and rail links and flood prevention inland add greatly to the projected cost effectiveness of a barrage. Because no major projects of this kind have yet been undertaken in this country, all evaluation of their impact on birds can only be hypothetical.

Four schemes are possibilities at present, three reservoir schemes, at Morecambe Bay, the Dee and the Wash, and one barrage, across the Severn, for the generation of tidal power. Proposals for power generation schemes have also been put forward for the Solway, but these have been shelved indefinitely. There have also been suggestions for a barrage across the Loughor Estuary near Llanelli in Glamorgan, mainly on behalf of recreational interests and to provide a road link. This proposal involved a rather small-scale construction providing a 200ha freshwater lake. With the building of the motorway further inland and the limited demand for the scheme, there has not been a feasibility study carried out and the project is unlikely to proceed (263).

The schemes investigated in detail and still possibilities involve some of our most important estuaries as far as wildfowl are concerned. Each of these schemes is briefly described before the overall implications of such projects are considered.

Morecambe Bay

The main purpose of the proposed barrage was to provide freshwater storage, and the original scheme, put forward in 1966 following a desk study, was a very ambitious one. The barrage would be 16km long across the broad part of the bay, enclosing a reservoir of 7,800ha, and polders of 6,000ha were to be reclaimed in the shallow areas, including the existing saltmarshes. A full feasibility study was initiated in 1967 and was completed in three and a half years. Thi came up with four separate schemes, the fir involving the complete barrage, the second and thir with shorter barrages across the Leven and Ker Estuaries and including pumped storage reservoirs and the fourth a more limited proposal involving sho barrages further up the estuaries and pumped storag reservoirs (134). The main features of two of th schemes are shown in Fig 154.

The effects of the various schemes obviously var considerably. The full barrage would involve the los of almost all the saltmarshes and much of the intertid area, with implications for many species of wildfow and particularly for many waders. The feedin grounds of waders would be reduced by 45% and th formation of new banks outside the barrage woul probably not fully compensate for this (483). As far a wildfowl are concerned, Dutch experience of polde creation suggests that in the short or medium term a species apart from Shelduck will adapt to feeding i the reclaimed polder areas. Indeed certain stages i the reclamation from mudflat to agricultural lan provide very suitable conditions for dabbling duck and geese. The long-term consequences depend o the intensity of development of empoldered land.

The feasibility report on Morecambe Bay wa submitted in 1972, but there was no progress for som years while the other estuarine barrages were bein considered. The North Western Water Authority hav reappraised the situation and developed furthe modifications of the water storage schemes (485). I seems unlikely that the full barrage will ever be built but one of the smaller schemes is still a possibility.

The Dee Estuary

The main impetus for the scheme, proposed in th 1960s, was the need for water storage, but th possibility of a road crossing between Merseyside an North Wales and elimination of the bottleneck a Queensferry was also a very attractive aspect of th scheme. Following the feasibility study there was considerable amount of concern and discussion, sinc early plans involved covering the vast majority of th intertidal areas and all the saltmarsh. The consultants Binnie and Partners, then produced several modifie schemes, the main features of which are shown in Fi 155.

The preliminary residual estuary shape schem involved two large reservoirs, removing 65% of th feeding grounds of Shelducks and waders as well as a the saltmarsh (102). Although continued siltatio would provide new areas, clearly the relatively smal size of the remaining estuary would limit its potentia for extensive mudflats. Two alternative road crossing

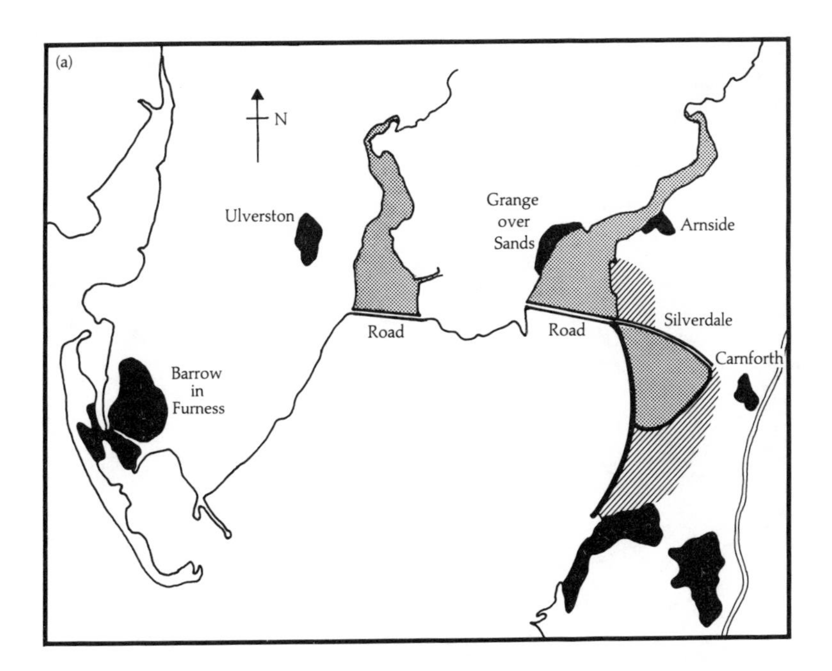

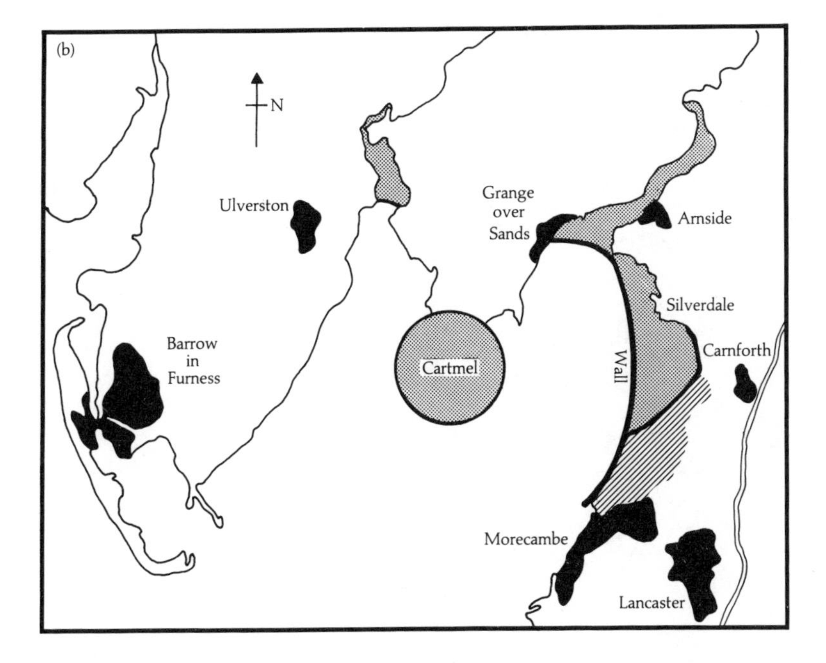

Fig. 154. Two possible modified schemes for reservoirs in Morecambe Bay:
a) twin barrages with Silverdale Reservoir, and b) river barriers with Cartmel and Silverdale Reservoirs. Stippled areas = standing freshwater; hatching = polders. Redrawn from Corlett (1972).

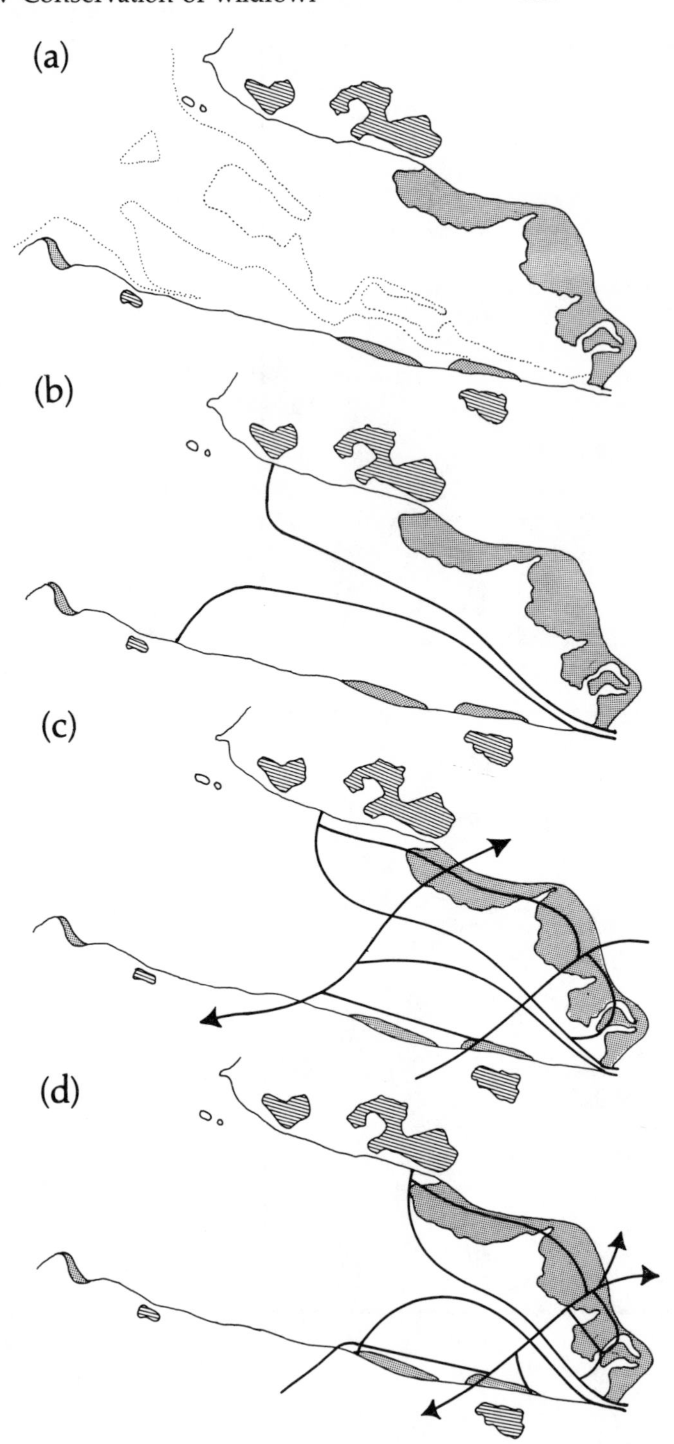

Fig. 155. The proposed barrage schemes for the Dee Estuary:
a) the intertidal and saltmarsh areas of the estuary, b) the preliminary residual estuary plan, c) the extended shape plan for Greenfield – Gayton road crossing. This scheme has an option for contracted estuary shape (see (d)), d) the Flint – Burton road crossing with contracted estuary shape and a reservoir at the head of the estuary. This scheme has an option for an extended shape (see (c)). Stippled areas are saltmarsh, heavy solid lines are bunded walls, and arrowed lines are road crossings. Redrawn from Buxton *et al*. (1977).

proposals were put forward, each in an extended estuary shape and a contracted shape. The extended reservoirs removed just under 50% of the feeding areas of waders and Shelducks, whereas in the contracted modifications the intertidal feeding grounds were reduced by a third. All schemes eliminated more or less all of the present saltmarsh.

Translating these losses of habitat to their impact on the birds is difficult, as the Teesmouth example shows. Waders and Shelducks do, however, use the whole of the intertidal area of the Dee, and their usage is related to the distribution and abundance of invertebrate foods. There is no evidence that birds suffer food shortage in normal winters on the Dee, though the effect of a reduction in intertidal area might be felt in mid-winter when feeding rates are reduced and energy demands high (102). Environmental considerations centre around the intertidal feeders. While dabbling ducks could use the new reservoirs as a roost, unless new saltmarshes quickly became established beyond the barrage, they would also undoubtedly suffer, at least at times of high population levels.

There has been little action on the development side since the feasibility study in 1977, but conservation interests have got together to form the Dee Estuary Conservation Group, which coordinates the examination of future proposals. The barrage is only one of many major threats to the Dee, but the Secretary of State for Wales is reported to have spoken of the "possibility" of the combined reservoir/road crossing scheme being implemented in the late 1980s (152).

The Wash

The original proposals made in the mid 1960s involved the embankment of the whole of the south shore of the Wash, creating four enormous reservoirs and associated polders. With the revision of the estimates for water demand these proposals were modified, and a feasibility study was set up in the early 1970s to examine the ecological and other consequences of the various schemes. The Wash is outstandingly important for wildfowl and the best British site for waders, and the ecological investigations were the most detailed for any of the studies carried out (118). The outlines of two of the modified schemes are given in Fig 156.

The original reservoir proposals involved the embankment of most of the intertidal area between the outlets of the rivers Nene and Ouse, together with a large reservoir to the east, with large polders linking the reservoirs with the mainland in each area. This scheme was modified, so that the first stage involved the creation of a bunded reservoir in one of five locations (Fig 156a). Each covered about 600-700ha, some 5% of the intertidal area of the Wash. As a second stage a larger reservoir was to be built, either separate from or contiguous with the Stage I scheme; one of the leading contenders, Westmark (Stage I) and Hull Sand (Stage II), is shown in Fig 156b.

The proposals were complex and the task of estimating their possible impact on the environment no less so (118). The area included in the proposals included some of the most productive parts of the Wash for invertebrates and consequently contained a high density of estuarine birds, including Shelducks. All the schemes included the creation of a polder from the extreme edges of the bunds to the existing sea wall, which would involve the loss of 400-700ha of saltmarsh in a two-stage scheme. The effects predicted on the intertidal feeding birds were considerable, with 25-50% of waders, or about 40,000-80,000 birds, being displaced, depending on the scheme. Shelducks would also suffer a loss of a quarter or more of their feeding grounds to a two-stage scheme, but the main Brent Goose habitat on the west side of the Wash would be largely unaffected. The saltmarsh-feeding Wigeon have declined recently in the Wash, and other widlfowl which feed outside the intertidal zone would be little affected by the proposals. Indeed, some species might benefit from the creation of a large water-body serving as a dabbling duck and goose roost and as a feeding area for diving species.

One indirect effect of the Wash reservoir scheme could be a change in the water regime on the Ouse Washes, which depends on catchment floodwater temporarily stored there and released into the Wash. This would be serious if water extraction from the Ouse eliminated flooding on the Washes. However, the Ouse Washes are designated as a Ramsar Site and half the land is owned by conservation organisations, so a workable scheme to maintain its character must be possible.

The Severn Estuary

Plans for a Severn Barrage differ from the other proposed schemes in that power generation is the main purpose of the dam. The Severn Estuary's tidal amplitude is one of the greatest in the world, and it is sited close to centres of high energy demand, making it a very favourable area for potential generation of tidal power. Although barrage schemes for various purposes (flood control, road and rail links, shipping management) had been put forward for the Severn since 1849, the main interest in the proposal was generated in the early 1970s when tidal power schemes began to be actively proposed (543). For most effective power generation the basin enclosed must be large and the dam placed in the area of greatest tidal amplitude to optimise the volume and head of water. Many schemes were proposed and three were examined in detail in a pre-feasibility study (541). The main features of these schemes are shown in Fig 157. The feasibility study selected the inner barrage without Stage II secondary barrage, generating electricity as the tide ebbed, as the preferred option, purely on the basis of cost effectiveness. Of the three, this would obviously be the preferred option from an environmental point of view as both the others influence a larger area, including Bridgwater Bay.

Unlike any other scheme, the Severn Barrage depends for its effectiveness on the maintenance by means of sluices of the tidal routine in the estuary. Nevertheless, there are several major consequences on the estuary:

a) water levels and movement patterns would change both inside and outside the basin,

b) sedimentation patterns would change.

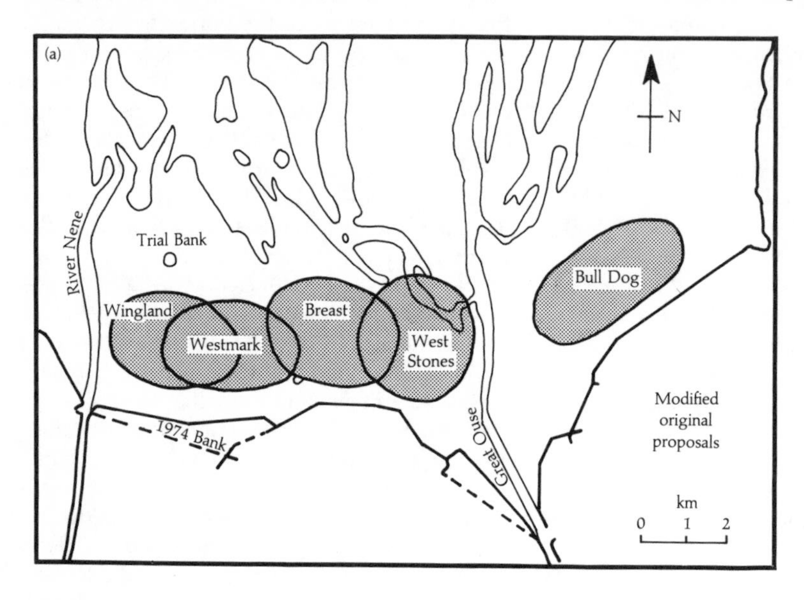

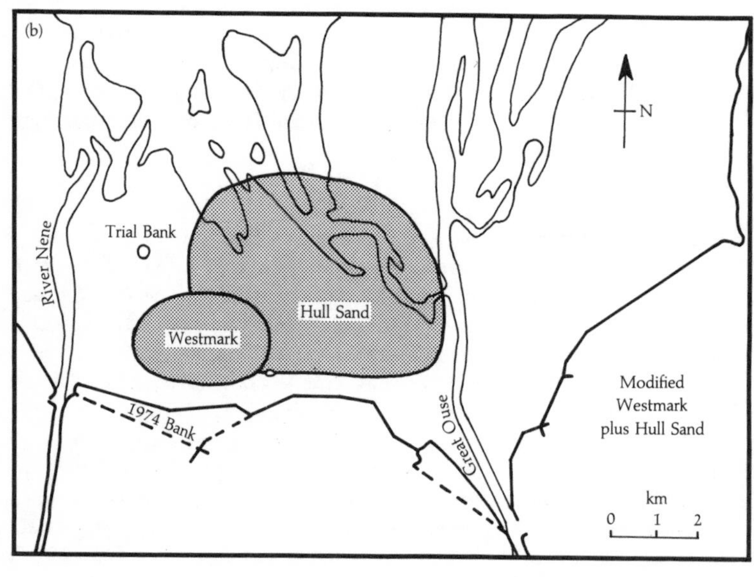

Fig. 156. Two of the possible schemes for reservoirs on the Wash:
a) modified original proposals, with 5 possible sites for reservoirs, b) modified Westmark (first stage) and Hull Sand (second stage) reservoirs.
Redrawn from Central Water Planning Unit (1976).

The tidal range outside the barrage would diminish by about 11% (about 1m) if the preferred scheme were adopted. This effect diminishes further away from the embankment, but would be felt to some extent throughout the Bristol Channel. This would clearly have a significant impact on the sedimentation patterns and on the area of mud exposed at various stages of the tides, which in turn affect invertebrates and the birds which feed on them. In the probable ebb-generation system the lowest tide within the basin would be at the present mid-tide level, and the highest tide just below that at present. The effects of this on sedimentation patterns inside the basin are difficult to predict (541).

Of the wildfowl, the Shelduck is the species most vulnerable to changes in the intertidal zone. It seems likely that the barrage would have a deleterious impact both in the basin and at Bridgwater Bay, by reducing the area of exposed mud at low tide and, within the barrage, the time for which areas are exposed. Wildfowl which roost on the estuary, such as dabbling ducks and White-fronted Geese, would be little affected. They would, however, be influenced by any changes that might occur on their feeding grounds in marshes surrounding the estuary. Higher average water levels would mean impediments in the gravity drainage of the 80,000ha of low-lying land adjacent to the estuary. The cost of installing pumped drainage schemes would be built into the costing of the barrage. These schemes would thus enable farmers, at little cost to themselves, to improve drainage and convert much of the land to arable use, to the detriment of ducks and geese, so safeguards would have to be made for important wetland sites (190, 435).

The construction of a Severn Barrage is a costly project, but the benefits in power generation are considerable and long term. The energy cost, even at today's prices, compares favourably with all other generation systems except nuclear power. However, the production of energy is linked to tidal cycles and its main function would be to save fossil fuel expenditure. If there were a large nuclear element in our energy generating policy, the barrage would not be so attractive. The Barrage Committee recommended, following the feasibility study, that the investigations be continued to the next, acceptability, stage. Although the project is enormously costly, it must still be treated as a real possibility.

The overall consequences of extensive barrages

The positive and negative effects of barrages have been examined in detail in each of the scheme feasibility studies (485). The implications will be treated only briefly here. There are four main consequences of barrage construction:

a) disturbance due to construction and operation of the barrage,
b) loss of or change in character of intertidal areas,
c) loss of or changes in salt and other marshes,
d) the presence of a large area of relatively stable water behind the dam.

The construction of large barrages takes many years (estimated as about 15 years in the case of the Severn), and the works are on a very large scale, involving a great deal of disturbance to the intertidal

Fig. 157. The three Severn Barrage schemes examined in the pre-feasibility studies:
a) outer barrage, b) the preferred inner barrage, c) inner barrage with optional second stage.
Redrawn from Severn Barrage Committee (1981).

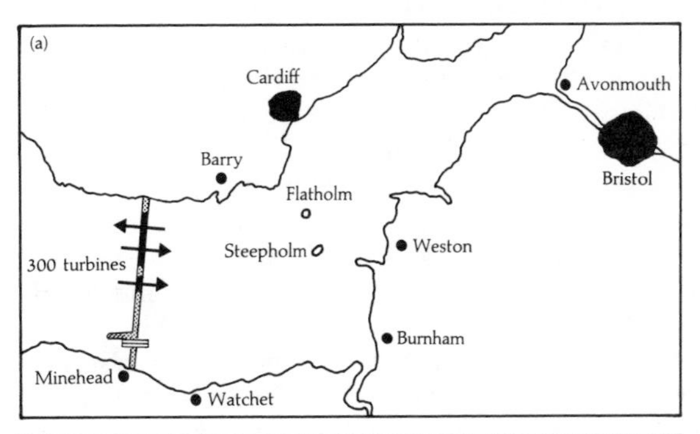

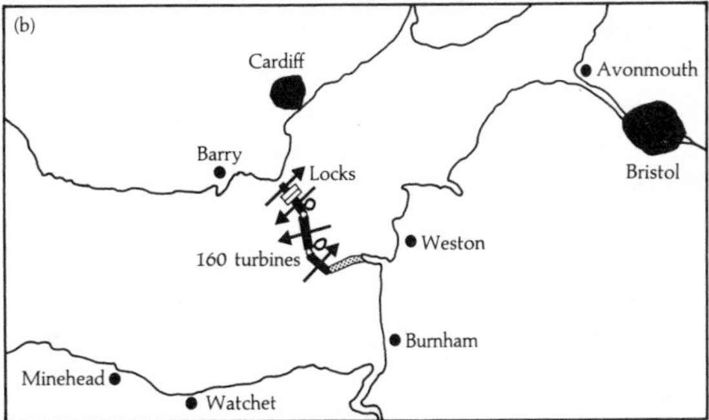

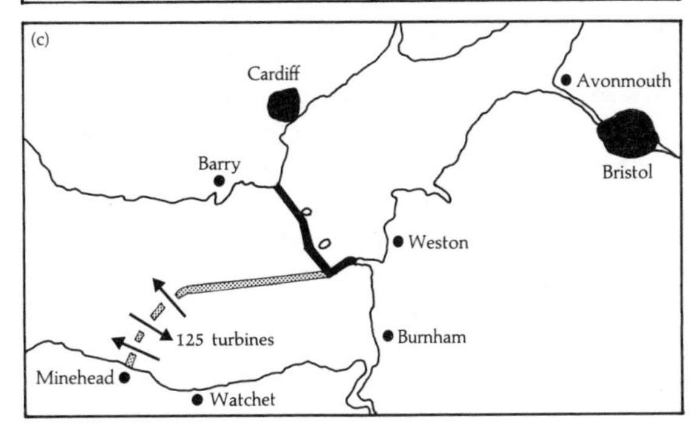

area and the approaches. These effects are difficult to quantify, but must be great on a local scale, possibly causing birds to leave certain areas of the estuary during periods of intensive works.

All the schemes involve effective loss of intertidal habitat, although the rate at which this will be replaced by subsequent accretion is difficult to predict. It would seem reasonable to suppose that a loss of habitat eventually means a reduction in the number of birds which can survive. This presupposes that it is habitat which is limiting the number of birds at present (i.e. it is effectively "full up"). This may not be true in many cases, especially for those species whose numbers may be limited by breeding success in Arctic regions, such as the Brent Goose. It was estimated in 1961 that the food resources on the English coast would support 12,000 Brent Geese throughout the winter, and only 6,000 at reasonable levels of density (96). The following two decades showed, however, that the mudflats could hold many times that number, with a little help from saltings and grass fields. Similarly, one might have predicted that the reduction of 72% of the habitat for Shelducks at Seal Sands would have reduced the wintering flock, which it did not (p.479). There is considerable circumstantial evidence that partial loss of habitat leads to losses of waders (485). There may well be subtle effects on wildfowl populations, through long-term survival or breeding success, which may not manifest themselves early enough to be attributed to habitat loss. Ultimately, of course, the fate of our estuarine birds depends on sufficient habitat being available to support them in winter and on all stages of migration; if habitat losses continue there is bound to be a stage reached when populations are seriously reduced, at least of those species which find it difficult to adapt to inland feeding.

Geese and Wigeon, the main saltmarsh feeders, have already changed successfully to inland food sources. Even if the polders associated with the reservoir schemes were to be converted to arable, this would probably not be disastrous, although if there were extensive drainage of fresh marshes, as would be likely with the Severn schemes, some accommodation for these grazing birds would have to be made.

Reservoirs provide feeding grounds and roosts for wildfowl (p.512), but the least valuable are bunded reservoirs having deep water – the kind proposed in estuarine schemes. They are likely to provide roosts for dabbling ducks and feeding areas for the common diving species, but this contribution is unlikely to provide significant compensation for the loss of estuarine species.

Other threats to habitat

The effect of the disturbance caused by recreational activities is treated in detail elsewhere (p.495), but some related activities also involve the loss of habitat. Marina development involves complete loss of some intertidal land, and in areas of dense intertidal boat-moorings the feeding grounds of some species are effectively lost.

The tipping of industrial or domestic waste material on estuaries is common, and this land is eventually reclaimed. Much land is lost to the dumping of power station ash in bunded areas on the coast, and the resultant lagoons are sterile as far as wildlife is concerned. Such small-scale loss of habitat is difficult to contest since it is seldom crucial in an individual case. However, the nibbling away of wetlands in this way may well be more dangerous than one or two grandiose schemes, since it is very widespread and continuing.

Pollution poses threats to habitats as well as to the survival of birds (see below). Offshore developments, if they are large enough to change tidal or sedimentation patterns, can also have an effect. An example is the proposal to site wave energy converters from Duncansby Head to Fraserburgh, across the mouth of the Moray Firth. This would be likely to affect the important estuarine and shallow-water habitat because the wave patterns would change, affecting sedimentation in the firth, with possible effects on intertidal zones and saltmarshes, as well as bottom living invertebrates on which sea ducks feed (486).

Pollution

A ubiquitous environmental problem, the pollution of water is extremely serious, since it is not easily localised or avoided. This section deals with the various kinds of pollution affecting wildfowl or wetlands, using examples to illustrate the nature and scale of the effects. Perhaps the most dramatic incidences are large oil spills at sea, since they involve large numbers of birds and have a direct effect on public amenity as well as on wildlife. Though less dramatic, other forms are equally important, perhaps the most underestimated being those arising from leaking of agricultural fertilisers and other chemicals into watercourses, which lead to habitat damage and occasionally to direct mortality among wildlife.

Agricultural chemicals and waste products

This aspect of environmental pollution was examined in detail by the Royal Commission on

Environmental Pollution (509) and, unless otherwise stated, the information in this section is derived from their report. Fertilisers are the most widely used chemicals in agriculture, and the use of nitrogenous fertilisers, in particular, has more than quadrupled in the last two decades, with no sign of a levelling out. Two-thirds of the nitrogen is applied to grassland, mainly as nitrates in a highly soluble form for quick response. The nitrates are washed out of the soil by rain, enriching watercourses and reservoirs. The nitrate load of drinking water has risen sufficiently for the EEC maximum permitted limit of 50mg/l to have been exceeded in a number of supplies, and the trend continues to be upwards.

The main effect of nitrates and other fertilisers on wetlands is the enrichment of the water – so-called eutrophication, and this is most marked in lakes and reservoirs. In severe cases the growth of algae can be so great as to cut off light from submerged plants, which die, setting off a decomposition process in the water. Decomposing bacteria remove oxygen from the water, eventually leading to the deaths of fish and aquatic invertebrates. Such enrichment has been the main cause of the decline in value of Llangorse Lake for aquatic plants since 1960, when 15 species of submerged plants were present (p.63). Only one remained in 1977 (147) and by 1982 even that had disappeared, although a small amount of another species was found. Algal blooms occur in the lake, accompanied by occasional fish deaths. The wildfowl population has declined partly as a result of this enrichment and partly because of the very high recreational pressure (p.498).

The Norfolk Broads have suffered greatly from eutrophication, again attributable largely to very high levels of nitrogen and phosphorus in the water. High turbidity caused by phytoplankton growth caused the death of higher aquatic plants so that "... in only 4 of more than 40 Broads do substantial stands of submerged macrophytes now remain" (357). The Broads include several SSSIs and two Ramsar Sites. Sewage and other effluents contribute to the enrichment process, but nitrates and phosphates in ground-water are the most widespread. They are also more serious, since they are much more difficult to control. The Royal Commission report indicated that removal of nitrate from drinking water would be more cost effective than curbing the use of fertilisers in agriculture, so levels of damaging nutrients in water-bodies not used for drinking water can be expected to continue to increase, with consequent detriment to wildfowl, particularly plant feeders and breeding birds.

Farm waste is another pollutant which frequently ends up in rivers and lakes. Effluents most commonly come from intensive animal units, stockyards, silage and so on, and are high in nutrients. Their effect is to enrich the water, leading to the problems described above.

Pesticides have effects on birds high up in the food chain, but there have been few examples where direct mortality of wildfowl was concerned, except for the goose and swan poisoning incidents described on p.502. These arose from organophosphorus seed dressing, whose use has now been restricted in Scotland, where most problems occurred. Chronic poisoning, no doubt, affects wildfowl as it does other birds, but neither mortality attributable to this nor any effect on breeding performance (found in the USA) have yet been demonstrated in Britain.

One pesticide incident that did have a serious effect on an important wildfowl site happened at Chew Valley Lake, Avon, in 1968 (42). The first sign that anything was wrong was a rapid bloom of the alga *Tribonema bombycinium* in mid-January which was so severe that the lake had to be temporarily taken out of service. In April an even more severe bloom of another alga *Monodus* sp took place, causing the lake again to be taken out of service. Examination of the water samples and of fish showed that the pesticides Dieldrin and Lindane had been released into the lake at some time before February 1968. This had resulted in the elimination of planktonic animals, particularly *Daphnia*, which preyed on the algae, allowing the bloom to occur. The two pesticides were commonly used in sheep dips, but the quantity released into the lake was equivalent to 50 times the content of a single dip. Following difficult investigations it was discovered that a communal sheep dip was traditionally emptied in a field some 4km from the lake. This dip found its way into a large underground reservoir and accumulated over a period of years before being released into the lake at one time, presumably in late 1967.

Following this incident water authorities gave farmers much stricter instructions as to the disposal of sheep dips and other chemicals, and it is unlikely that a similar incident will occur again. The example is given here to illustrate the way in which an accident, admittedly of some magnitude, can alter the delicate balance of a water system with what could have been very serious results.

Sewage effluent

The problem of enrichment with sewage effluent is similar to that originating from farm materials, except that in many cases the scale of the problem is greater, affecting harbours and estuaries as

well as freshwaters. The volume of effluent entering Langstone Harbour, Hampshire, increased five-fold between 1959 and 1981. At the same time the area of intertidal mud covered by green algae increased from a small proportion to more than half. Similar effects had also been seen in other parts of the Solent, notably Chichester and Portsmouth Harbours (596). The effect of this was to increase the algal food supply of Brent Geese, enabling the Solent to support larger numbers. The trend in Brent numbers at Langstone Harbour is shown, together with that of Shelduck, in Fig 158. The pattern in Brents follows the national trend, although the increase was steeper at Langstone, which (in company with other south coast haunts) now holds a much higher proportion of the population than it did in the 1960s. Shelducks, on the other hand, followed the national trend of decline and recovery from the hard winter in the early 1960s, but then declined in Langstone Harbour despite a healthy national picture (Fig 92).

The negative link between eutrophication and the numbers of Shelducks and those of some wader species was said to be due to the blanket cover of *Enteromorpha* over parts of the area, making conditions unfavourable to the mud-dwelling invertebrates on which the birds feed (595). The decline certainly coincided with the period when algal cover reached its maximum. The Southern Water Authority, conscious of the enrichment problem, began to extract phosphates and nitrates from the effluent before discharge. The 1978 and 1979 surveys showed appreciable change in algal cover (596), and Shelduck numbers did begin to recover, and continue to do so. A decline in Brent numbers might also be expected, but there is evidence that the carrying capacity of the Solent has not yet been reached, and in any case the birds have taken increasingly to feeding on inland grasslands and cereals (507, 596).

This example, although the causal link between enrichment and Shelduck numbers has not been proven, provides strong circumstantial evidence of the effect of sewage effluent on a relatively enclosed estuarine system. The pattern may be repeated in other estuaries where large quantities of sewage are discharged.

Raw sewage outfalls are often beneficial to wildfowl, either directly because they provide an easily available source of food, or indirectly by enriching the water and increasing the availability of the natural invertebrate prey which feed on the effluent. On the coast of Scotland discharge sites have had large gatherings of Eiders, Scaup, Goldeneyes

Fig. 158. The number of Brent Geese and Shelducks at Langstone Harbour, 1960-1982, expressed as 3-year running means of regular numbers. (For example, the first point is an average of the numbers for the seasons 1960 to 1962.)

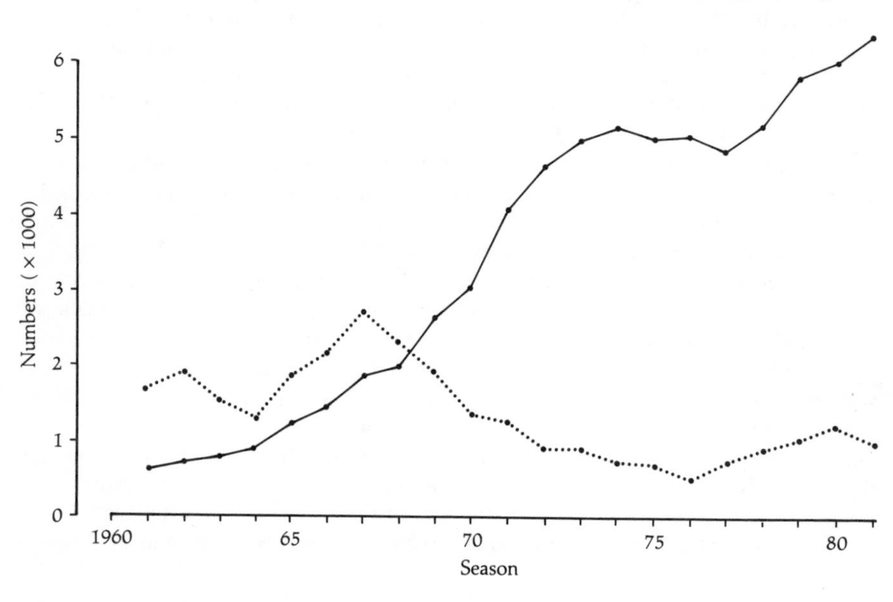

and Mute Swans around them (481). Many outfalls carry large quantities of grain from distilleries and this, at least in the Firth of Forth, has formed a great proportion of the birds' diet (112). Most of the Scaup in Britain, as well as about half of our wintering Goldeneyes, are dependent on such sites. The process of establishing treatment plants to deal with this sewage might well have an effect on these species; indeed the treatment of the effluent of Edinburgh in the late 1970s was a major cause of the drastic decline in the number of Scaup and other species wintering on the Firth of Forth.

Industrial pollution

Many industries require large amounts of water, and their discharges contain a variety of pollutants. Heavy metal pollution has caused great concern, particularly in estuaries such as the Severn, where highly elevated levels of cadmium, originating from the aluminium smelter at Avonmouth, have been found in limpets (97). Levels of copper, zinc and lead are also high in the same region and these elements are concentrated in invertebrate predators (614). The effects at the levels found are not clearly understood, but are not considered great at present, and in any case heavy metal concentrations in the Severn fauna have declined recently (541).

A major incident occurred on the Mersey Estuary in 1979-80, when more than 2,500 dead birds, including a number of ducks, were picked up. Analysis of their tissues showed that there were extremely high levels of lead and that this was largely of organic origin. Organic lead for use in petrol is manufactured on the estuary, but the exact source of the pollution has not been traced with certainty. Further deaths occurred in the following winter and lead was again implicated, though lead levels in tissues were not as high as in the previous incident. The Royal Commission on Environmental Pollution (510) has come down particularly heavily against the release of lead into the atmosphere and have recommended that lead should be outlawed as an anti-knock agent in petrol, so the threat from this form of pollution may soon be reduced.

Power stations use a large amount of water for cooling and return it to the source considerably warmer, so that there is a local increase in water temperature. The effect of this has largely been beneficial, by increasing the productivity of the water and attracting fish in particular. The feasibility study of the Severn Barrage, though considering the implications for power stations, did not examine the possibility of an ecological effect from temperature changes, so we conclude that this would be negligible.

Lead poisoning

Poisoning of wildfowl by the ingestion of spent pellets from shotgun cartridges has been recognised as a major problem in North America for many years, and the use of lead for marsh shooting is now banned in many parts of the United States. In Britain early findings indicated that levels here were considerably lower that those of North America (422), but a study at the Ouse Washes indicated that, at least locally, the problem could be more serious. There, 9% of Mallard, 10% of Pintail and smaller proportions of other duck

Table 261. Sample sizes and percentages of shot wildfowl with lead pellets in their gizzards and with elevated lead levels in liver (recent exposure) and in bone (lifetime exposure) in the 1979-1981 survey. (From Mudge 1983.)

	Gizzard pellets		Liver lead * above 10 g/g dry		Bone lead + above 20 g/g dry	
	No.	%	No.	%	No.	%
Pink-footed Goose	73	2.7	66	4.5		
Greylag Goose	42	7.1	35	8.6		
Barnacle Goose	61	0.0	21	0.0		
Wigeon	354	0.0	204	4.4	200	4.0
Teal	633	0.5	242	5.8	473	1.9
Mallard	820	4.2	572	14.7	863	21.9
Pintail	73	1.4	71	5.6	48	4.2
Pochard	64	10.9	60	16.7	67	7.5
Tufted Duck	77	11.7	63	7.9	92	2.2

* Liver lead levels are a measure of recent exposure to lead and the 10 g/g level indicates a concentration considerably higher than normal background levels.
+ Lead concentration in bone accumulates throughout life and the 20 g/g level indicates above normal lifetime exposure to lead (from all sources).

species were carrying lead (580). A substantial number of Bewick's Swans also died there from the ingestion of lead shot (445). At the same time mortality at worrying levels was demonstrated in several other European countries, particularly Denmark and the Netherlands (583).

An extensive survey was initiated in Britain in 1979, financed through the RSPB and BASC and carried out at the Wildfowl Trust. It involved the analysis of lead levels in liver and bone tissues, as well as the more traditional method of examining gizzards for traces of pellets. Soils were sampled to determine pellet density and depth, and settling rates (360). The incidence of ingested pellets and the proportions of birds with elevated lead levels in the liver and bones are given in Table 261. In general the incidence of pellets is lower than in the Ouse Washes study and, except for the two diving species, is rather low in comparison with studies from other countries. Despite having a low incidence of pellets in gizzards, 4-6% of Wigeon and Teal showed high liver levels, indicating recent exposure. Some of these cases, however, must have involved contamination from environmental lead, and were not attributable to shot ingestion.

The density of pellets in wetland soils was very variable, though adequate sampling is difficult. The highest levels were, not surprisingly, at two flight ponds, but the Ouse Washes had extremely high densities of 16.6 pellets per square metre of ground. Two-thirds of pellets were in the top 5cm of soil and most of these would be readily available to feeding birds. Settlement rates of pellets were extremely slow; in five of six plots, most pellets were less than 2cm deep after nine months, so they remain available from season to season. Settlement naturally was more rapid in softer sediments.

The three sponsoring bodies agreed that although problems from lead shot poisoning were not acute in Britain at present, deaths from this cause were unnecessary if effective substitutes could be found. BASC agreed to pursue actively the search for substitutes and to discuss possibilities with cartridge manufacturers. Intensive work in the USA indicated that soft steel shot was a practicable substitute for lead; although some of its ballistic characteristics were different its effectiveness for wildfowl shooting, if certain conditions were met, was no poorer than lead. In the meantime advice was to be given to wildfowlers on ways of minimising the deposition rates of lead in sensitive areas. The provision of grit in wetlands deficient in it can provide a useful means of reducing the likelihood of ingestion, and gritting places provided by the Wildfowl Trust at the Ouse Washes were certainly heavily used by ducks and swans.

Lead poisoning in Mute Swans arises primarily from the ingestion of split shot and weights used by fishermen. In recent years several substantial local declines in Mute Swan numbers have occurred and a high proportion of deaths have been attributable to lead poisoning. The most spectacular declines have been on the Avon at Stratford, where numbers fell from 70-80 in the early 1960s to none in the late 1970s, and on the Thames, where counts during the annual swan-upping declined from 1,212 swans in 1969 to only 153 in 1981. On the Avon, 70 swans were picked up dead between 1975 and 1978 and 38 (54%) certainly died of lead poisoning. Another 13, whose cause of death was unknown, had also been poisoned, so that a probable 73% of the Stratford flock died as a result of lead fishing weights picked up from the banks in place of grit, or together with fishing lines entangled in submerged or bankside vegetation (236).

The Stratford deaths caused great public concern and an NCC working group was set up in 1979, by request of the Government, to investigate the decline and to recommend corrective measures. The group examined the implications of lead poisoning in Britain as a whole. Post-mortems revealed that nearly 40% of swans examined had died of lead poisoning. 51% of deaths in England and Wales resulted from lead ingestion, but none was recorded from Scotland. Some areas were particularly bad, with 90% of deaths on the Trent, 77% in the Stratford area and 75% on the Thames resulting from lead (219).

The group spelt out the severe problem for the British Mute Swan population as a whole, and particularly those in the populous lowland areas, and concluded that the use of lead as split shot should be phased out within five years. In the meantime urgent efforts should be made to find an effective and acceptable substitute. Several new substitutes, based on tungsten and on stainless steel, have been developed, and tests have shown that these are not toxic to Mallard. Tungsten products approximate to the weight characteristics of lead, but are more expensive. A code of practice for anglers has been produced to minimise the input of lead onto watercourses, and widespread publicity has been given to the drive to reduce lead input. The working group recommended that grit should be provided in problem areas so that swans would be less likely to pick up split shot spilt by fishermen on banks, and, where practicable, groups should be organised to clear up visible pellets. The RSPB's Young Ornithologists' Club has organised parties to clear river banks; in

1978-79 thirty miles of bank yielded (per mile) 808 foot of line, 86 pieces of split shot (a fraction of what must have been present) and 7 hooks.

The reaction of anglers has been mixed, although responsible organisations are anxious to alleviate the problem. The Government is in any case committed to banning the use of lead at the end of 1986, if the code of practice has not proved effective. With manufacturers putting their full marketing machinery into action to promote substitutes, the outlook is promising.

To summarise, since 1979 great attention has been focused on problems of lead poisoning in British wildfowl. The ingestion rate of shotgun pellets, though a problem in some areas, is not a serious mortality factor nationwide. However, if poisoning is avoidable, these deaths are unnecessary, and very slow settlement rates of spent shot in most soils mean that the problem will become worse in future. The problem of Mute Swans is extremely serious and an urgent solution must be found. The case for replacing lead was strengthened by the report of the Royal Commission on Environmental Pollution (510), which came down very heavily on release of lead in all forms. One of its conclusions stated: "Lead shot from spent cartridges and lead fishing weights poison wildlife. We recommend that as soon as substitutes are available the Government should legislate to ensure their adoption and use." This strong recommendation will help conservationists to achieve this aim, but because of potential problems with gun barrel wear and traditional prejudices, acceptance of substitutes is much more difficult for wildfowlers than for fishermen.

Oil pollution

Heavy oil clings to the feathers of birds, making them congeal and lose their insulating properties. Many birds die as a result of ingesting oil, which they try to remove by preening; others die of exposure or starvation, being unable to feed effectively. Light oils penetrate the feathers, destroying their waterproofing properties and causing death from chilling.

Occasionally oil spills occur on inland waters and these mainly involve light fuel oils spilling from barges or waterside installations. Water authorities have a responsibility for dealing with these incidents in relation to public safety, water supply and damage to wildlife. As such cases are uncommon, small, and threaten human water supplies, they are usually quickly contained and severe impacts on wildlife have not been reported in Britain. The main problem arises from oil pollution at sea, mostly heavy crude oil. Chronic pollution comes from deliberate discharges from tankers as a result of tank cleaning operations. Acute pollution comes from tankers, either wrecks or accidental spillages, or from oil installations at sea or at coastal terminals.

Both chronic and acute pollution has increased substantially during the last twenty years because of the increased drilling activity, particularly in the North Sea. The chief rigs, pipelines and onshore installations are shown in Fig 159. The important sea duck wintering areas in the Moray Basin and around the Northern Isles are at high risk, particularly from accidents at terminals and other shore-based installations. Eiders, Longtails and scoters are the birds in most danger, though estuarine species which concentrate in large flocks, such as Scaup and Goldeneyes, could also be badly affected.

The *Torrey Canyon* disaster, which discharged 60,000 tonnes of oil around the coasts of Brittany and Cornwall in April 1967, emphasised the potential biological impacts not only of oil spillages at sea, but also of the detergents used for dispersing it, which are themselves highly toxic to wildlife (551). Largely

Fig. 159. Oil developments in the North Sea. From Scottish Development Department Map – Oil and Gas Developments – February 1981.

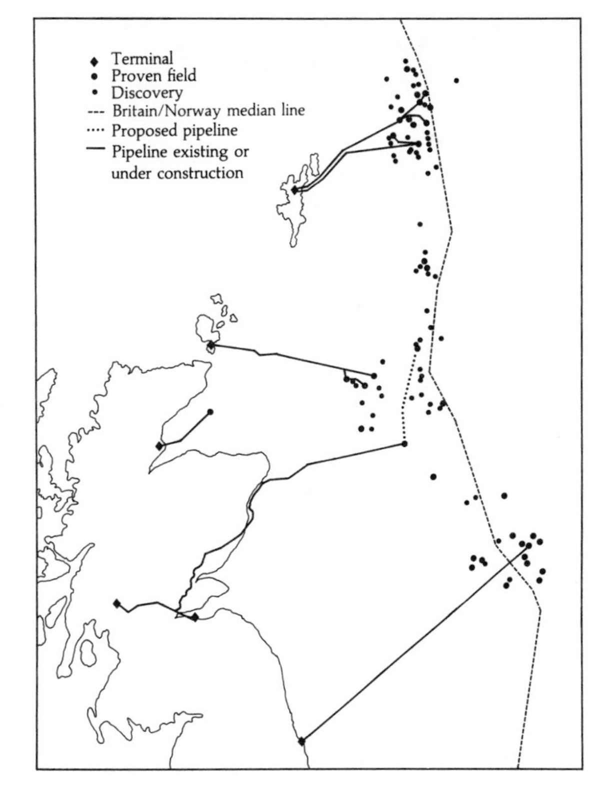

because of its timing, the *Torrey Canyon* had little effect on wildfowl, but an incident involving the *Tank Duchess* in the Firth of Tay happened on 29th February 1968, well inside the wintering period. A minimum of 1,368 birds were affected, and the large majority died as a result of oiling. These included 1,127 Eiders and 167 Common Scoters (225).

An even worse incident occurred in north-east Britain in 1970, producing one of the worst casualty totals ever in the North Sea. Most birds came ashore on the coasts of Fife, Angus and Kincardine, but oiled birds were found from Yorkshire to Aberdeen. The known toll was 12,856 birds, of which 2,799 were wildfowl, including 2,124 Eiders and 520 scoters. The exact source of the oil was not discovered, though evidence on the course of the slick suggested that it originated from some way out at sea; at least two sources were probably involved (226).

The Seabird Group and the RSPB set up a Beached Bird Survey in 1971, whereby several hundred volunteers walk selected stretches of the coastline of Britain and record the incidence and condition of birds found dead. This is aimed at monitoring the day to day effects of chronic oil pollution, but special surveys are also organised following known incidences. The results have recently been analysed in detail (554). There were 91 incidents involving mortality of 50 or more birds in Britain between November 1971 and March 1981, and their positions are shown in Fig 160. While spillages have been well spread along the coastline, apart from two major incidences at the mouth of the Bristol Channel, most of the really damaging ones have been on the North Sea coasts of England and Scotland. Two-thirds of all beached birds found in Britain have been on the east coast.

Fig. 160. The position of the 91 oil pollution incidents recorded between November 1971 and March 1981 which accounted for 50 or more dead birds recovered. Redrawn from Stowe (1982).

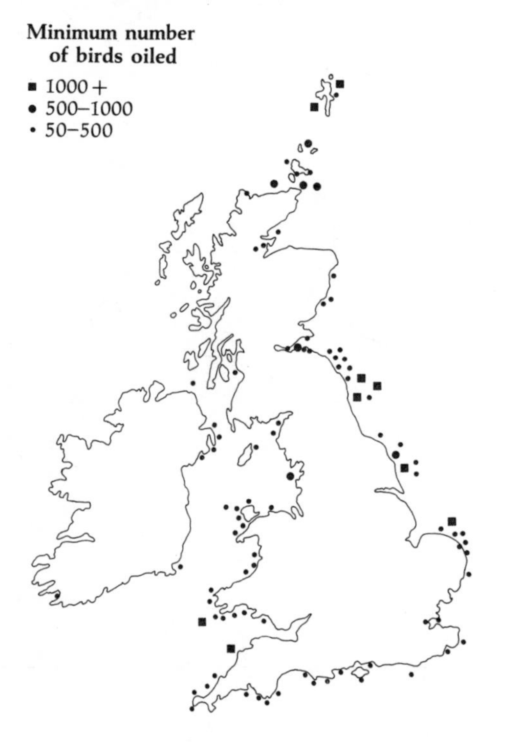

In only 23 of the 91 cases was the origin of the oil discovered. Seven of these resulted from spills at terminals, usually when unloading or loading tankers, four arose from illegal tank cleaning at sea, and four from collisions between tankers and other ships. Three were the result of defects in ships and two of grounding or wrecks. In three cases the discharge was from a land-based industrial source. The most serious single event as far as wildfowl were concerned was the *Esso Bernicia* incident at Sullom Voe, Shetland, on 30th December 1978, soon after the opening of the terminal there (258). The tanker was involved in a collision while berthing and spilled nearly 1,200 tonnes of oil. The failure of retaining booms resulted in the pollution of 65% of Yell Sound and Sullom Voe. Subsequent beached bird and boat surveys yielded 3,700 dead or dying birds, but, allowing for counting difficulties, the true number could have been twice as high. The known toll included 570 Eiders, 306 Long-tailed Ducks and 57 other wildfowl. An estimated 95% of the birds present in Sullom Voe and at least 75% of those in Yell Sound were affected. In national terms these numbers are not particularly large, but the incident does illustrate how serious the local situation can be, and the potential toll should a similar incident occur in areas with higher populations of sea ducks, such as the Moray Firth.

Overall, bird deaths from incidents rose alarmingly between 1971 and 1978, from about 2,000 deaths a year to over 10,000 (553). Casualties were lower in 1979-80 and 1980-81 (554), but the trend over the whole period is still significant, though there is no significant trend in the number of incidents recorded.

Routine beached bird surveys provide information on day to day casualties from pollution and other sources. The number of wildfowl recovered on beaches each year from 1972 to 1981 is shown in Fig 161. The length of coastline searched is relatively constant, usually between 1,800 and 2,200km each year. There is an apparent increase in the later years, though 1981 was exceptional, and there are confounding weather variables.

Other evidence on the importance of oiling as a cause of death in birds comes from ringing recoveries.

nd these confirm that auks are most at risk. Eiders are he only sea ducks ringed in substantial numbers and 7% of recoveries of this species died from oiling. ecoveries have shown that the numbers of birds ffected by the major incidents are far fewer than hose dying from chronic pollution (393).

Although the numbers of birds found dead in eached bird and other surveys appear large, the ctual toll from oil pollution is much greater, since nany dead birds will not be recovered. The proportion aries considerably depending on prevailing weather, articularly wind, conditions: 11%, 44% and 59% in hree separate experiments involving gull corpses ropped in the Irish Sea (51). An earlier experiment, nvolving auks, had yielded 20% of birds recovered in our months, and circumstantial evidence suggested hat half of the corpses dropped on the surface had unk within eleven days (268); other evidence suggested similar or smaller proportions. Clearly, experinents should be carried out in association with each najor incident to assess total casualties (554).

Many birds are killed quickly when covered by hick oil, or die soon afterwards from ingesting oil vhile trying to preen it off their feathers. Substantial umbers are, however, found alive on beaches, and onsiderable effort and money have been expended in rying to clean up oiled birds, allowing them to recover n captivity and then returning them to the sea. A esearch unit was established at Newcastle upon Tyne n 1967 to develop rehabilitation methods, but results vere disappointing. Birds could be treated only if they vere in reasonable condition and had not ingested ubstantial quantities of oil. This represents a small roportion of birds from many incidents. Cleaning is ime consuming and expensive and only about half of he auks and sea ducks arriving at the unit were released after rehabilitation. Under field conditions the success rate would be more realistically put at 20-25% (371). Survival following release is not known in detail, but many released birds are likely to die before they regain their full capabilities.

The Nature Conservancy Council examined the evidence on rehabilitation and reluctantly concluded that "... the problems in saving oiled birds are considerable. The overall conservation value of attempted rehabilitation must be regarded as minimal ..." (371). The Newcastle unit has closed and it is now considered that oiled birds found alive should be humanely destroyed.

In the remainder of north-west Europe, oil pollution is even more serious than in Britain, and much of this threatens the sea duck populations in the Baltic. Annual deaths in Denmark are counted in tens of thousands (291), and oil pollution is one of the main threats to north-west European sea duck populations. In Britain, though some incidents have been serious, substantial proportions of the British stock have not yet been involved, although if a major spill occurred in one of the principal areas for Eiders or scoters this could well happen. In summary, oil pollution has not yet affected substantially the populations of any species of wildfowl, but we must maintain our vigilance and pressure on polluters to ensure that this continues to be the case.

Shooting

Shooting is discussed elsewhere in relation to the information obtained from shot birds (p.20), to the law (p.528) and to the effect of shooting on wildfowl numbers (p.493). This section deals briefly with the extent of shooting and the bag in Britain, as far as is known.

As evidenced by this volume, a great deal of time and effort has been expended over the past two or three decades in counting wildfowl and, by ringing, monitoring their movements. Much of this interest

Fig. 161. The number of beached wildfowl found along British coasts, 1972-1981 (about 2,000km searched per year). Numbers above the columns indicate the number of wildfowl in total. From figures in Stowe (1982).

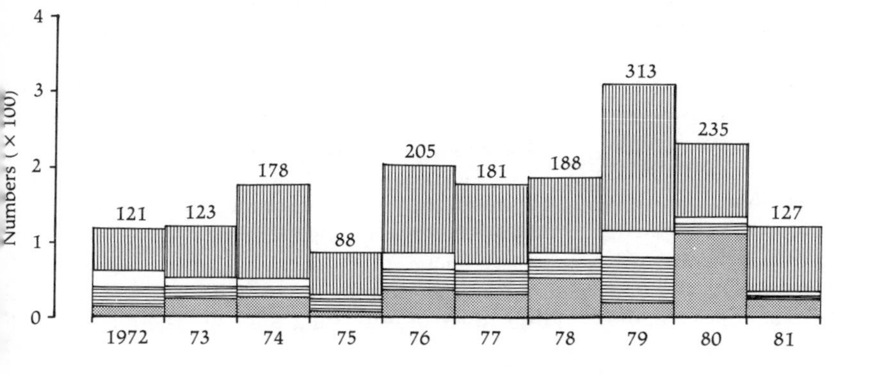

arises because wildfowl are quarry, and in some cases the concern of conservationists has centred around the effects of shooting. It is not until recently, however, that attempts have been made to estimate the size of the bag and monitor other aspects of shooting behaviour. Bag surveys, carried out for many years in other parts of Europe, are very important if we are effectively to manage the populations of quarry species and ensure that the impact of shooting is not a lasting one.

The survey organised by BASC, known as the National Shooting Survey, began in the 1979-80 season and has been carried out every year since. The first results were published in 1982 (238). There are three aspects to it – firstly a survey of BASC members, secondly a national survey to identify differences between the activity of BASC members and other shooters, and thirdly a national survey to assess the numbers and types of shooters in Britain.

Of the 850,000 people in Great Britain who own a shotgun, 160,000 shoot wildfowl and 60,000 are members of BASC. In both age and socio-economic class shooters represent a wide cross-section – not differing in either respect from the population as a whole, though, as might be expected, more males shoot and shooters are much more common in country areas than in towns. Generally those who are members of BASC shoot more often and bag more birds than those who are not, but the variance is very large (i.e. individuals within both groups differed greatly in their activity and success). These and other factors were taken into account when compiling the bag estimate, but the following results, based on averages of 1979-80 and 1980-81, are still considered provisional (238). However, a larger survey conducted in 1982-83 came to very similar conclusions (239).

Three-quarters of all ducks are bagged inland, but this is not surprising considering that Mallard comprise the majority of all kills. The survey estimated that 874,000 ducks are killed and retrieved annually, excluding those bagged by rough shooters, which would put the figure up to about a million. There is no estimate for the overall species composition of the bag, but 60% of ducks shot by BASC members were Mallard, 20% Teal and 14% Wigeon, with the other quarry species making up the rest. These figures cannot be directly applied to the total of a millio ducks shot since the composition of the non-BAS members' bag will be different. Thus the roug shooters' total will nearly all be Mallard, and th proportion of Wigeon taken inland by non-BAS members will probably be very low. Since the majc Teal concentrations are also coastal their proportion i the national bag will be less than 20%.

Although to use the available figures to estimat the number of the various species killed by hunters i highly speculative, we must have some idea of this s that we can relate the figures to the estimate population. The proportion of Mallard in the total ba is likely to be 70-75%, (700-750,000), Teal about 15% (150,000) and Wigeon 5-10% (50-100,000). We have n estimate of the crippling loss (unretrieved kill), bu figures from North America suggest this could b between 20 and 35% of retrieved birds.

With 200,000 Wigeon in the country at peak, th kill (bag plus crippling loss) of 60-120,000 seem credible, especially since most Wigeon are on the coa in shootable places. For Teal, the 180,000 or s retrieved and crippled is considerably more than th estimated peak population of 100,000. It is sti possible, of course, that both figures are correct if th turnover of birds is sufficient. Because of its habit c hiding in cover the species is likely to be unde counted and there may also be a larger number of Te in uncounted areas than we suspect, so the Britis population may be larger. Similarly, the 600,000 or s Mallard killed is equivalent to the estimated pea population. In this case, however, an estimate 400,000 ducks are said to be released by shooters, an many of these would be shot in September, or stay o small release pools where they are not counted. Thi figure, put forward by BASC, is a bit of a guess, bu Game Conservancy scientists, while admitting ther are no concrete data, agree that this total is possibility. About 60% of the bag is taken before th end of November, during the time when many bird are arriving in Britain, so this masks the shootin losses and makes it very difficult to estimate th proportion of the population that is shot.

Most geese are shot in December and January they are largely Greylags and Pinkfeet, with 18% Canada and a few Barnacle Geese. The estimates of th bag, however, were considered much too inaccurat to use to give a national total.

This account indicates a highly unsatisfactor situation; we are very far indeed from assessing th importance of shooting mortality in British ducks an geese. There are shortcomings in the assessment o bird numbers and movements as well as grea difficulties in estimating the bag. There is no doub ıat we need much better estimates of the number of ·ildfowl shot, and also an idea of crippling losses. We eed the detailed breakdown by species and by ·ason, together with confidence limits, if the bag is to e related to the number of birds in the population.

Water-based recreation

Before the early 1950s there was a small amount f water-based recreation, especially on inland wa- ·rs, but since then there has been extremely rapid rowth in most such activities. In particular, there has een a tremendous increase in sailing on inland ·aters, and the sport has become increasingly popular n the coast. Canoeing has also increased and spread om rivers, where there has long been an interest in ɔmpetition, to still waters. There has also been a ıpid growth in power-boating, especially pleasure- :uising and water-skiing. Fishing has continued to ain popularity and spread to newly created reservoirs nd gravel pits (571). The trend in the membership of ɔur of the most imporant sporting associations is ɿown in Fig 162.

Most spectacular has been the expansion of ·creation on water supply reservoirs, now providing ɔr a large proportion of water-based activities. In the 950s water engineers were against allowing access nto water because of the risk of pollution, effectively ıaking drinking water reservoirs into wildfowl re- ıges. Continuing demands and successful pilot :hemes, however, led to a general acceptance that ɔntrolled access to water was socially desirable and ot damaging to the interests of supplying clean rinking water (281). The 1973 Water Act changed the dministrative situation and the 10 Regional Water uthorities were given a statutory duty under the Act) provide for recreation, as well as conservation ıterests, on waters created for public supply.

The Water Space Amenity Commission (WSAC) ·as also set up by the 1973 Act as an advisory body on ıe use of water areas, but was disbanded in 1983. In 981 WSAC estimated that 93% of water supply ·servoirs supported some form of recreation (55); all ew projects included provision for fishing and water ports.

The greatest demand for water recreation has een in the well-populated Midlands and south-east, nd much of this has been satisfied by the new wet ravel quarries, which have shown an enormous rowth in the last two decades (p.509). Gravel ·orkings are exploited mainly in areas of high human ɔncentration such as the Thames and Trent Valleys, ecause of the associated demand for building mate- als, and to some extent this has tended to reduce the potential pressure on other waters in southern Britain.

Wildfowl and recreation are both intensive users of water space, and in some respects their requirements largely overlap. For example, waters most suitable both for breeding and wintering wildfowl are productive lowland sites, which are mainly found in southern England, where recreational pressure is greatest. On the other hand some water sports demand large expanses of open water, whereas wildfowl select shallow marginal areas for feeding and breed in secluded bays or islands.

The short-term effects

There are two main potential effects: (a) an indirect one through the alteration of the habitat, affecting the availability of food or nesting cover, and (b) a direct effect, through displacement or disturbance of the birds themselves or their nests.

Indirect effects

In some areas weed cutting to facilitate fishing activities may affect the food supply for wildfowl, but the most widespread influence on the habitat is that of

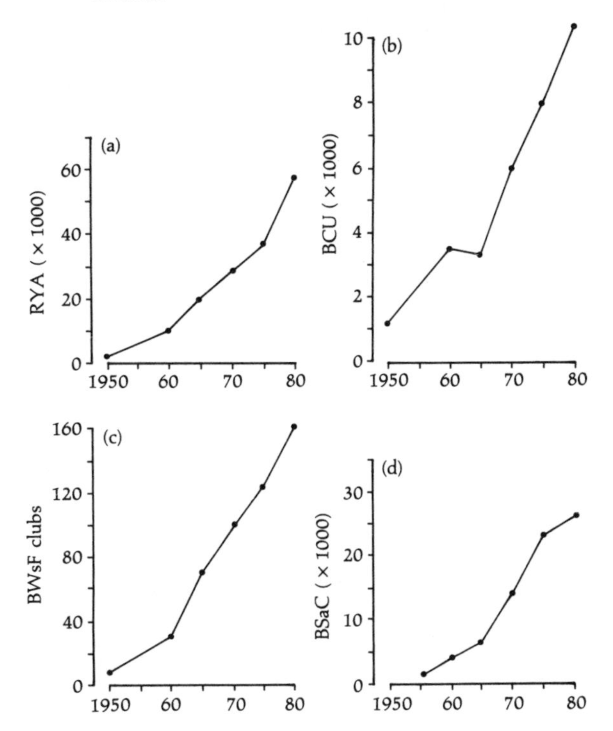

Fig. 162. The growth in participation in water sports from 1950 to 1980:
a) Members of the Royal Yachting Association. b) Members of the British Canoe Union. c) British Water Ski Federation Affiliated Clubs. d) British Sub-aqua Club members. Data from CRRAG (1978) and Sports Council.

boats, particularly by their effect on submerged and marginal plants on which the birds feed. The action of propellers and the wash from boats can cause direct damage to plants and cause banks to erode and become less suitable for marginal species. Boats also tend to increase the turbidity of the water and this may be deleterious to the habitat, though there is little direct evidence. In general the physical effects of boats on the habitat are thought to be deleterious to water plants (137, 315), and presumably to the animals which feed on them, but there is, as yet, little direct or quantitative evidence for effects on wildfowl.

Direct effects

The most important potential influence of water-based recreation arises through disturbance by those activities on the birds themselves. Some activities, such as cruising on canals or the operation of marinas (p.486), may be so intensive that they exclude wildfowl from usable habitat without destroying the habitat as such. However, it is the lower level of disturbance on a much higher proportion of wetlands that has most worried conservationists. There has been, on both sides of the argument, an abundance of speculation and circumstantial evidence, but until recently few detailed extensive or intensive studies into the impact of recreation on wildfowl. The following section

Fig. 163. The proportion of four species of waterfowl in the Sanctuary Bay area of Grafham Water in September – November 1974. The area was fished from the banks and from boats until 14th October (arrows). Numbers on each graph are the mean percentage of birds in the area before (left) and after (right) the start of the close season. Based on Cooke (1975).

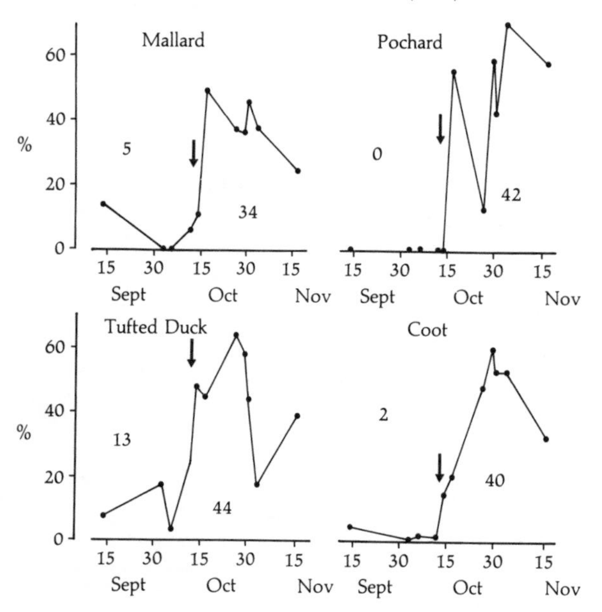

reviews recent work and discusses the implications the findings.

Pleasure-boating on rivers and canals has lo been common, but, although a few Mallard ar Tufted Ducks breed along the banks, these are n important wildfowl habitats. In saltwater, much of th recreational disturbance is concentrated on th estuaries and natural harbours of the south an south-east, many of which are key areas for wildfow especially in winter. Many estuaries are large enoug to accommodate recreation and birds, but in some the most intensively used areas a substantial amou of the water may be occupied, and in harbours whe floating or intertidal moorings are provided the effe of boats is not only evident when active recreation taking place. A study in Langstone Harbour failed demonstrate marked effects of boating on wildfov numbers (327); undue conflict was avoided because temporal separation of recreation (summer) an wildfowl (winter). The best feeding areas for wildfov are on soft mud, unsuitable for landing and moorin boats and generally avoided by boatmen. Howeve the potential effect of the presence of moored boat particularly in very shallow water or on mudflats, ha been realised (327, 485). Other kinds of recreation disturbance are common to all habitats though the effect is likely to be most marked on the small enclosed inland waters, and most of the investigation of the impact of recreation have involved such areas

The potentially damaging effect of the increa ing intensity of recreation on reservoirs became cause for concern in the 1960s. These artificial wate included some of the most important inland wildfov sites, because of the loss of much of the marshy habit and floodlands to intensive agriculture. Those r creational activities involving the entry of boats on water were considered potentially the most dama ing, and sailing was identified as the most importar threat because of the large number of participants an severe disturbance effect: "Sailing poses great problems than any other activity. The demand fc facilities is enormous ... and the degree of disturbanc is greater than most species of ducks can tolerate" (19 This analysis was based on a sound knowledge of th requirements of recreation and of birds, but there wa at that time, no empirical evidence of a deleteriou effect of recreation on wildfowl.

On Brent Reservoir, London, a marked drop i the number of Mallard coincided with the introductio of sailing in 1963 but, though birds were disturbe from the reservoir by boats, they had often returne by the following morning. The continued use of th water by ducks and grebes was thought to depend o the availability of an undisturbed part of the reservo s an effective refuge for the birds (37). The introduction of sailing did not cause a reduction in wintering wildfowl either at Scaling Dam Reservoir, Cleveland (560), or at Dinton Pastures Country Park, Berkshire (515), though in both areas an effective refuge was available to the birds.

Fishing is one of the most common activities allowed on inland waters, and the National Angling Survey in 1979 estimated that 3.7 million people participated in the sport in Great Britain in that year (217). The effect of fishing has been examined on Grafham Water, Cambridgeshire, where fishing is intensive, both from the banks and from boats (131). The proportion of birds in a section of the reservoir where fishing ceased on 14th October is shown in Fig 163. It is evident that for all four species the sanctuary area is the preferred zone of the reservoir when there is no fishing, but that fishing drastically reduces the usage of that zone by the birds. Pochard and Coot (both of which feed on submerged plant leaves) are almost comletely excluded from this area when fishing is taking place. Teal distribution (not shown) is variable when there is no fishing, but the average percentage in the sanctuary is 2.5% with, and 37% without, fishing. In at least three of the species shown there appears to be a decline in use of the sanctuary area later in the season. As well as the effect on distribution of birds, there was a suggestion that fishing also depressed the number of waterfowl on the reservoir as a whole during the day.

Of the other kinds of recreation commonly practised in and around water, power-boating and water-skiing are thought to be most disruptive, whereas activities carried out on the banks are least so. Goldeneyes are said to take flight when power-boats are as far as 700m away (278). Brent Geese and Shelducks feeding on mud take flight when humans approach on foot at about 200m in Langstone Harbour (327). In Essex humans have a similar effect on Brent Geese early in the season, but the birds become habituated to disturbance later as food becomes scarce. At times disturbance causes a 12% reduction in feeding time and a sevenfold increase in flying time (the most energy-expensive activity). In conditions of food shortage this could have a significant effect on the welfare of the birds (456).

The long-term impact

The above studies show that intensive recreation can have quite marked short-term effects on the numbers and distribution of birds on individual waters. We must, however, distinguish between the short-term and the long-term impact on species or populations, the latter being much more difficult to demonstrate but widely assumed to be serious (572). Ultimately only factors which cause an increase in mortality levels or a decrease in breeding success will have a long-term impact, and any which limit either the habitat available or the duration of its use by wildfowl in the long-term will come into this category. In 1978 the Nature Conservancy Council and the Sports Council initiated two three-year studies into the impact of recreation, one on a national scale and on a pair of sites in South Wales (597), and one in the Trent Valley (609).

The Trent study demonstrated numerous direct effects of recreation on wildfowl, including the reduction in the use of sites when recreation was intensive, and the limitation of the use of certain areas or of feeding time by the presence of boats or people. A useful method of evaluating the effect on the birds was devised, based both on the increased expenditure of energy and on limitation of feeding time. The effects on the activity of Tufted Ducks and on feeding time in Wigeon are shown in Fig 164. Tufted Ducks are forced by disturbance to increase their time swimming by 20 minutes, and this is more costly than sleeping, the predominant activity when undisturbed. Wigeon, which feed on the banks, suffer a restriction of feeding opportunity after disturbance, but return to normal after 30 minutes. Clearly frequent disturbance could be costly for both species. In general, low levels of disturbance cause short flights and the birds return within 90 minutes, resulting in an increase in energy

Fig. 164. The effect of disturbance on the activities of Tufted Ducks and Wigeon on reservoirs in the Trent Valley. Figures on the y-axis are percentages of the flock engaged in the stated activity as determined by flock scans. Based on Watmough (1983).

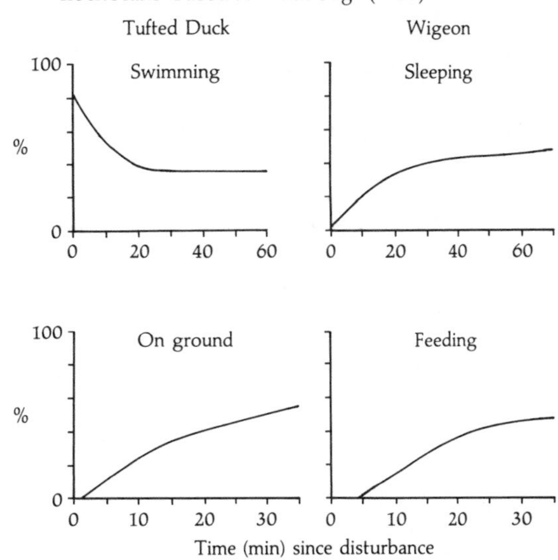

costs of about 5%. Exclusion from parts of the water by intensive recreation limits feeding opportunity and results in longer flights, causing an increase in energy costs of about 10%. Intensive recreation, causing departure from the lake, increases energy costs by 20-30%. In areas where recreation is continuous and intensive ducks can be excluded altogether; this occurs at Foremark Reservoir during trout fishing. It is probable also that the persistence of some effects throughout the season means that the population of wildfowl in the Trent Valley as a whole may be kept lower by recreation than it might otherwise be. The effect of disturbance in increasing energy costs of individual birds could cause an increase in mortality rate, especially in cold weather, if they were unable to compensate for this increased cost by more intensive feeding in mid-winter (609).

There is considerable circumstantial evidence that the very high level of recreation at Llangorse Lake, Powys, reduces its carrying capacity for wildfowl. The numbers of most species are lower there than at the nearby Talybont Reservoir, despite the fact that the lake is much more suitable for birds in most respects (147). A large amount of evidence of recreational effects, operating at various levels, has now been collected in an intensive study (599). In particular, high recreational intensity restricts the use of their preferred zones by most species, as shown in Fig 165. Six species and the total number of wildfowl show a significant negative relationship with recreational intensity (Spearman rank correlation), and in the case of Mallard, Wigeon, Mute Swan and Pochard, the birds are excluded from their preferred zones altogether when recreational intensity is high. The only species not to show a significant trend is the Tufted Duck, though the relationship is still negative. Thus recreation has a substantial deleterious effect on bird distribution, and this, together with other evidence from the pattern of use of Llangorse in relation to time, suggests that recreation is substantially reducing the carrying capacity of the lake, and consequently the local area, for wildfowl. There have been suggestions that the displacement of birds from Llangorse is not serious because they use the nearby Talybont Reservoir as a refuge (166), but the intensive studies have shown that this is not so. Although recreation is exceptionally intensive and uncontrolled at Llangorse, this example does illustrate the potential deleterious impact of water-based recreation on a site or local area scale.

The national survey of the effect of recreation on wintering wildfowl used limnological and morphological characteristics of sites to predict the number of wildfowl they should support. This expected value, which was based on countrywide distribution, wa compared with the actual value for each site on th basis of the incidence of recreation taking place (597 This is a rather coarse-grained examination, but there were a large-scale impact of recreation, such survey would be expected to show it up. An invento of site characteristics from 1,455 sites was made ar

Fig. 165. The proportion of time spent in their preferred zone (that used most when there was no disturbance) by seven species and total waterfowl at Llangorse Lake in 1980-81 under different recreation intensities. From Tuite *et al.* (1983).

Intensity classes:

1	No recreation	6	51-75 boat/people hours
2	1-10 boat/people hours	7	76-100 boat/people hour
3	11-20 boat/people hours	8	101-150 boat/people hour
4	21-30 boat/people hours	9	151-200 boat/people hour
5	31-50 boat/people hours	10	201-300 boat/people hour
		11	301 or more boat/ people hours

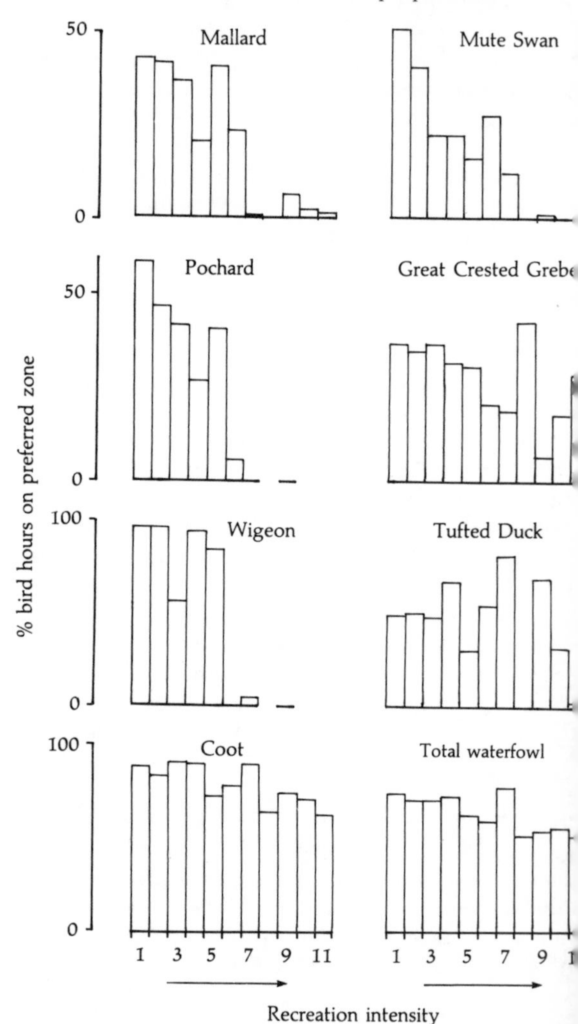

639 of these had sufficient Wildfowl Count data to make tests valid. The incidence of different types of recreational activity, in those sites where the data are available, is shown in Table 262. Both in winter and summer birdwatching/natural history is the most commonly occurring activity. This does not mean that more people are involved; in fact the person hours spent in such activities as fishing and sailing are probably higher than in birdwatching, since the former tend to be concentrated activities involving large clubs and competitions. Nevertheless, birdwatching is a highly significant active form of recreation on inland waters. The more disturbing water sports, involving powered boats, are restricted to a small proportion of sites and tend to occur together. These do, of course, interfere not only with birds but also with other forms of recreation.

Table 262. *The incidence of different kinds of recreational activities on inland waters in Great Britain according to the winter inventory and the summer survey, 1980. Data from Tuite (1982). The figures make no assumptions on the number of participants, only recording whether an activity takes place.*

	Winter inventory		Summer survey	
Activity	No. of sites	%	No. of sites	%
Birdwatching/ Natural history	359	59	273	66
Game fishing	277	45	152	37
Coarse fishing	222	36	216	52
Sailing/ Windsurfing	180	29	96	23
Informal *	168	27	230	56
Canoeing	60	10	62	15
Rowing	39	6	69	17
Sub-aqua	39	6	13	3
Shooting	32	5	0	0
Water-skiing	25	4	26	6
Power-boating	20	3	19	5
Motor cruising	11	2	5	1
Swimming	8	1	47	11
Total	613		414	

* Bankside activities – walking, picknicking, games, etc.

Large numbers of tests of the influence of recreation were carried out but few were significant, suggesting that the effects were not widespread or consistent. Two ranking lists were produced which indicated the susceptibility of various species to disturbance and the relative impact of various recreational activities, and these are reproduced in Table 263. The fact that those species high on the susceptibility ranking are those known from other studies to be most vulnerable to disturbance reassures us that the results really do illustrate meaningful relationships. Similarly, the widely practised coarse fishing and sailing, together with rowing and canoeing, would be expected to have the highest impact on birds. The low score of power-boating reflects its low incidence rather than a real lack of local effect. Among those activities showing positive scores, both birdwatching and shooting would be expected to be associated with sites with higher than average numbers of birds. It is reassuring, too, that these activities are not unduly disturbing (i.e. have few negative effects).

Table 263. *The relative susceptibility of different species of wildfowl to recreational impact (a), where the score is the number of significant tests showing lower than expected numbers of birds at sites with particular activities. The relative impact of different recreational activities (b) gives the number of species/months where tests showed significantly fewer (negative) and more (positive) wildfowl at sites where that activity took place. From Tuite (1982).*

(a) Relative susceptibility of wildfowl (higher score, more susceptible).

(b) Relative disturbance impact of activities (higher negative score, more disturbing).

			Score	
Species	Score	Activity	Negative	Positive
Goldeneye	16	Coarse fishing	12	0
Shoveler	9	Sailing	10	0
Teal	8	Rowing	8	0
Wigeon	4	Canoeing	4	0
Mallard	2	Informal	4	0
Mute Swan	0	Natural hist/Birdwatching	1	7
Goosander	0	Shooting	0	3
Tufted Duck	0	Game fishing	0	1
Pochard	0	Power-boating/skiing	0	0

Overall, these "league tables" give a valuable guide in the assessment of recreational impact and illustrate that, on the national scale, it is not necessarily the most obviously disturbing activities (powered boats) that have the greatest overall impact. In general, however, the conclusion remains that despite the enormous increase in water-based recreation in the last three decades, the effect on wintering wildfowl nationally has been small. This conclusion is supported by the species trends given in Part III. Of 12 species which occur in important numbers on enclosed inland waters, 9 have increased in number in the last 20 years, the Gadwall, Tufted Duck and Canada Goose spectacularly so. Three species have fluctuated around a stable level (though there have been some local declines in Mute Swan populations) and none have shown an overall decrease.

The summer wildfowl survey carried out in 1980 collected information on recreation as well as on bird numbers and breeding success. Again birdwatching was the most common activity, occurring on two-thirds of sites (262), with fishing and informal recreation also having a high incidence. A large number of tests were made using various measures of recreational intensity, but no consistent effects, either on bird numbers or productivity, could be demonstrated (597). This was due as much to other factors such as site area and location, which introduced variability into the data, as to the fact that there were few real effects. On a local scale disturbance to breeding birds could be highly disruptive, but there is little concrete data. On a national scale, however, the impact seems not to have been serious. This is confirmed by a comparison between the surveys in the 1960s (621) and 1980 on 124 matched sites. Of 11 species that were represented, 8 had increased in number (though the Mute Swan had been stable overall), and 2 (Teal and Pochard) had stayed the same. Only one – the Shoveler, which is, in any case, a rather uncommon breeder – has appreciably declined in this sample since the 1960s (598). The national trend in Shoveler numbers in September, however, does not confirm that there has been a decline.

Conclusions

Despite the fact that intensive and uncontrolled recreational activity can have a substantially deleterious impact on local wildfowl populations, and that the level of participation in water-based activities has dramatically increased in the last twenty years, the overall national effect appears to have been slight. How has this happy state of affairs been maintained in the face of such demand from the conflicting interests of different users of water space? Several factors have been responsible for mitigating the potentially damaging effects:

a) Most of the recreational activities involving moving boats are seasonal with very little activity during the main wildfowl wintering season. In summer the requirements for large expanses of open water for power-boating and sailing do not correspond to those of breeding birds for secluded shallows, well-vegetated bays and islands. There is, therefore, a high degree of temporal and some degree of spatial separation.

b) The last two decades have seen an expansion in the number and area of enclosed inland waters, particularly through the creation of wet gravel pits in the lowland valleys (p.509). To some extent pressure from recreational interests can be said to have been beneficial to wildfowl since it has been easier for extracting companies to justify leaving pits flooded rather than restoring them to agricultural land.

c) Many inland sites are used mainly as day-time roosts by wintering wildfowl, which feed in nearby marshes and smaller waters around dawn and dusk. This means that recreational disturbance for many species is not as serious as it would be if it limited access onto the feeding grounds or prevented the birds from feeding for a substantial proportion of time.

d) Most of our inland wildfowl are adaptable enough in their feeding habits and behaviour to compensate for temporary loss of feeding or roosting space, by flying to undisturbed parts of the site or to nearby waters which they use as a refuge.

e) Many of the more important inland waters are controlled by water authorities, which have been generally sympathetic to the needs of wildfowl, following the warnings and suggestions made by conservationists in the 1960s (19). Nearly all the important reservoirs for wildfowl either prohibit boating or limit activities by exercising a zoning arrangement which restricts recreational access to the best wildfowl habitat. Zoning has been spectacularly successful in the larger sites and all new reservoir plans now incorporate conservation as well as recreational interests.

f) The increase in birdwatching as an active form of recreation has created its own demand for facilities for birds. This has influenced conscientious gravel companies and the planning authorities which grant extraction licences for gravel. The result has been that a number of the smaller pits in many large complexes have remained undisturbed, and in many cases have been managed by agreement with individuals or Naturalists' Trusts to maintain and improve the wildlife interest.

The absence of marked recreational impact on a national scale has, therefore, partly been accidental, but has been due largely to positive efforts to accommodate the interests of birds as well as those of other users of water space. We hope that this successful compromise will be maintained, but two recent developments in the recreational field may make the situation more difficult. The first is the increasing use of wet-suits, which extends the season for all kinds of water sports well into the winter, and for the real enthusiast, all the year round. The other is the growing popularity of windsurfing with its long season. Unlike the large sailing dinghies and day boats, these boards can sail in a very few inches of water. This may be most worrying in estuaries and harbours where shallow waters and tidal hazards have hitherto restricted most waterborne activity to the deeper waters away from the mudflats and creeks where wildfowl feed.

On the local scale recreational effects can be substantial and in some cases disastrous for wildfowl, and conflict is likely to continue between the various interests when proposals are considered to introduce disturbing activies to an important and hitherto undisturbed wildfowl resort. Each case must be considered on its merits, with due consideration being given to the importance of the area for birds in a national and local context. It should also be remembered that birds themselves provide for one of the most popular forms of human recreation on inland waters.

Wildfowl and agriculture

About 80% of the land area of Great Britain is classified as agricultural land and the vast majority of the remainder is either urban or forested, with a small proportion of open water. Thus all those species which feed on land above the intertidal zone, including temporary wetlands, interact to some degree with the requirements of agriculture. In some cases, such as on wetlands farmed in summer and wet or flooded in winter, there is no conflict with present farming regimes, although there is always pressure from farmers for "improvements" deleterious to wildlife (p.476). In most other situations, however, though the effect may be local, seasonal and short term, most geese and swans and some dabbling ducks do cause conflict by feeding on farm crops.

Before man cleared much of the lowland forests of Britain and drained the marshes, each species of goose was more or less restricted to a specific wetland habitat to which it was specially adapted. The Pinkfeet roamed the larger estuaries feeding on saltmarsh pastures, Greenland Whitefronts were restricted to the raised and blanket bogs of the north-west and Ireland, while Greylags were resident on lowland fens and the migratory population visited the inner marsh zones of estuaries. Barnacle Geese grazed the short swards of west coast islands and headlands, while Brent Geese frequented the muddy estuaries of southern England and the north-east (437). Swans and dabbling ducks used the shallows and margins of lakes and estuaries. Although the estuarine habitats still exist, they are much less extensive and less diverse nowadays. Fens and marshes have all but disappeared and many of the bogs at low altitude have been drained and reclaimed for agriculture and forestry. Although in some cases birds have voluntarily deserted suitable natural or semi-natural habitat in favour of farmland, the change has mainly been enforced by pressure on their traditional areas. The creation of roosts in the form of reservoirs and gravel pits inland has accelerated the occupation of agricultural areas and caused considerable changes in the distribution of some species.

Although the change in habitat may have been forced on the birds, many species have fared well on farmland. Being birds of open land, the potential habitat was greatly increased by the clearing of the forests, once suitable roosts were available close to the feeding areas. Many foods are easier to collect on farmland monocultures and more nutritious than their wild counterparts. Pastures are heavily fertilised and provide palatable and digestible forage for grazers. Thus, whatever the reasons for the adoption of arable habitats (arable crops or cultivated grassland) and foods, these are now vitally important for some species of geese and swans. Pressure for increases in agricultural production is always great and it is essential, where there are conflicts between the interests of wildfowl and farmers, that ways of alleviating the problems are found.

Some fears have been expressed that the diet gathered from farmland by geese, while high in energy, may be deficient in certain nutrients, particularly on the approach to the breeding season. The wholesale adoption of arable foods may, therefore, have a depressing effect on reproductive performance (495). Although the breeding success in terms of percentage young has declined in many goose species (p.521), this is almost certainly due to factors operating outside Britain. Many of our populations have incresed greatly in the last few decades and these larger numbers could not have been accommodated except on arable land. The breeding success of Greenland Barnacle Geese wintering on Islay is greater than that on the semi-natural islands of Ireland; this may well be due to better feeding conditions on the improved grasslands in spring (73). The same may well be true of Greenland Whitefronts (p.370) and other farmland geese.

One acute effect of farmland feeding has been the poisoning of geese and swans by carbophenothion seed dressing aimed at protecting the crop from attacks by the wheat bulb fly. The first incident occurred in 1971, when several hundred geese (Greylags and a few Pinkfeet) died having consumed wheat grains from the surface of newly sown fields. The results of autopsies of birds sent for post-mortem examination were consistent with the organophosphorus dressing being the cause of death (32). Further incidents in south-east Scotland in November 1974 involved about 500 Greylags, while more than 200 Pinkfeet were found dead on the Humber in January 1975. Many bodies from this latter incident had been carried down the river and not recovered, and up to 1,000 geese could have been affected (233). Carbophenothion had been tested for toxicity to a number of grain feeding birds, such as pheasants and pigeons, and found to be relatively innocuous. This was an example of a few species being highly susceptible to the poison, a fact which was not detected in normal screening. Because of their very gregarious nature, geese are very vulnerable to this kind of incident and large numbers can be killed before the occurrence is detected.

The publicity given to these events and the concern of the Ministry of Agriculture, Fisheries and Food led to an agreement being reached between the Ministry and manufacturers and distributors of carbophenothion not to distribute this pesticide in central Scotland. The problem of wheat bulb fly is not acute in that region, and withdrawal has stopped the poisoning of geese in this, their most important area. About 30 Bewick's Swans were killed by the same pesticide in East Anglia in 1982 but the scaring of birds from the field after the first incident prevented more swans from being affected. There is no effective alternative to carbophenothion in this area, which is highly vulnerable to bulb fly, so vigilance will have to be maintained.

Wildfowl have long taken advantage of man's agriculture, the Bean Goose probably being the first species to visit cereal and bean stubbles in autumn (hence its name). These and waste potatoes are used extensively by a number of species without detriment to the crop. It is when the birds feed on unharvested roots in winter or on growing winter wheat and grass that problems occur. The following gives a brief species by species account of recent problems.

Ducks

There are few areas where ducks visit arable land in large numbers; regular complaints are received only from the vicinity of Loch Leven and the Ouse Washes. In the latter site most problems occur when ducks visit wheat fields drilled after potatoes have been harvested. The ducks trample the winter wheat shoots and puddle the ground while feeding on the waste tubers left from the previous crop. Fields worst affected have standing water on them. There is also some grazing of the wheat leaves but this is considered by farmers to do no harm. Mallard are the most common species field feeding but they are often accompanied by Pintail and Wigeon. The eventual yield of wheat from affected fields is within the range experienced in the remainder of each farm and it has been concluded that in investigated cases damage has been negligible. On isolated occasions Mallard may alight in unharvested cereal fields flattened by wind and rain in late seasons, but this is extremely local and in many cases much of the crop has already been lost. Ducks are generally nervous and in most cases a single disturbance by shooting has driven them away (585).

Mute Swans

Although Mutes do not gather in large flocks on agricultural land they do cause some local concern, especially on pastures close to rivers or lakes where they roost. Flocks of 30-50 birds have sometimes been involved and because of the species' large size such flocks represent substantial competition with farm stock for grazing in spring and summer. Mute Swans are very tame and are not easily discouraged, though fencing along the waterside may succeed in some instances. Attempts at displacing the birds by scaring have rarely been successful and some farmers have resorted to shooting substantial numbers of birds. Although damage is undoubtedly being caused this is regrettable in the current situation where the species is declining in several localities because of lead poisoning. Physical removal of birds and establishing them elsewhere has been proposed, but unless vacant habitat is available this is unlikely to meet with success. If areas where lead was a problem can be cleared and the use of lead by fishermen phased out, it may be possible to move swans in future, but this is unlikely to provide more than a short-term solution.

Migratory swans

Whooper Swans visit farmland in increasing numbers in Scotland, stubbles and potatoes being particularly favoured. They are accustomed to digging for their food and they will eat unharvested potatoes and swede turnips, given access to them. They move on to grass in spring and are accused of damage, but since the flocks move away in April this can be expected to be slight.

Bewick's Swans are more adapted to feeding on land than the other species, and grazing on wet meadows, and winter wheat in the vicinity of the Ouse Washes and near Slimbridge, is usual. Unless the ground is excessively wet, when the crop would be damaged to some extent anyway, their effect is slight.

Swans can be discouraged by persistent disturbance by humans or by placing scaring devices or tapes across affected fields.

Geese

The most land-based of wildfowl, geese have suffered most from the loss of traditional habitats and have most often come into conflict with farmers. The areas where geese occur in substantial numbers on agricultural land and where damage incidents are most likely are shown in Fig 166. Greenland Whitefronts are found in rather small flocks and often on marginal land and they are not justifiably accused of causing damage. European Whitefronts occur on wet grasslands mainly at a time when they are not competing with stock or affecting spring yield of pasture, silage or hay.

Pink-footed Goose

Pinkfeet have fed on farmland for decades, but it was not until their numbers increased substantially in the 1960s that they began to pose problems for farmers. On autumn stubbles and potato fields they do no harm, but they graze pastures throughout the winter and, on occasions, especially in spring, graze winter wheat and barley. The geese are present in Scotland until the end of April and it is during this month that they are accused of the most serious effects. In the latter part of April and in early May farm stock are being let out onto specially prepared "spring bite" grassland, expensively managed and fertilised. The Pinkfeet prefer this young grass to older leys and they congregate on these pastures, competing directly with stock for forage. This is expensive to the farmer since he can only replace this forage by costly bought-in foodstuffs.

Fig. 166. The distribution of geese in Britain in situations where they are likely to come into conflict with farmers.

BA = Barnacle
C = Canada
PF = Pinkfoot
BR = Brent
GL = Greylag

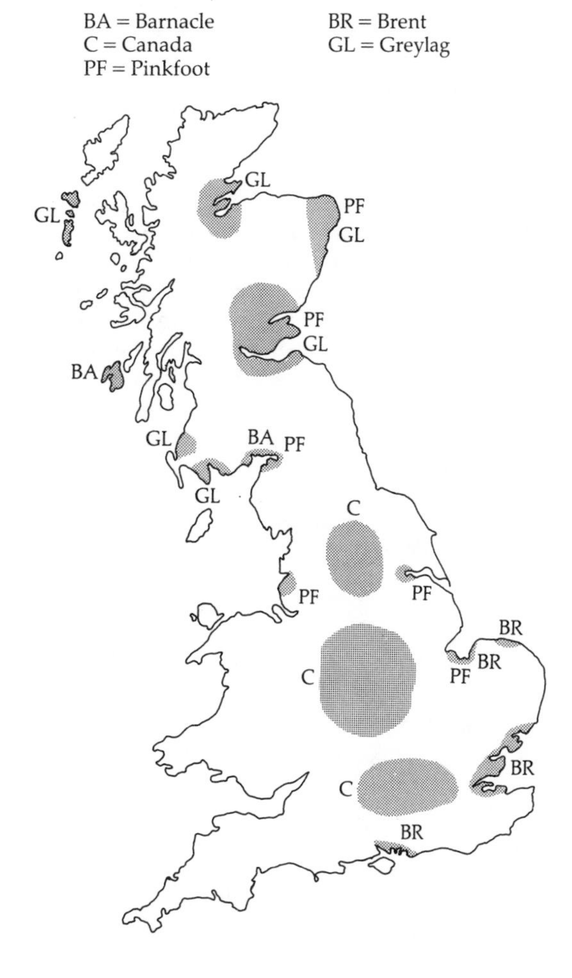

Pinkfeet are frequently accused of damaging growing winter and spring-sown cereals, but clipping trials showed that the effect was slight (297), although studies on other geese (see below) do suggest that damage from grazing and puddling of the soil may occur in waterlogged conditions on heavy soils.

The most serious allegations of damage against Pinkfeet were made in Lancashire in 1973-74, when large numbers of geese visited unharvested carrot fields and gouged out the tops of the roots to a depth of 3-4cm, making the whole of the crop unmarketable. Carrots are a high value crop grown on a very small proportion of the farm area, and since the problem was recognised most of the damage has been avoided by careful siting of carrot fields and regular patrols by the farmer. On grass Pinkfeet are rather more difficult to discourage, although small areas of high value spring bite can be protected by the intensive use of scaring devices.

Greylag Goose

Wintering Greylags occur in the same areas and have broadly similar habitats to Pinkfeet. They more commonly eat roots, however, and have caused problems when they have visited unharvested swede turnip fields or even eaten chopped turnips put out for stock. These roots are not highly nutritious, however, and these problems are local and are serious only during hard weather. There have also been isolated incidences of Greylags damaging unharvested carrots and potatoes in southern Scotland.

Breeding Greylags in the Outer Hebrides cause particular problems to growing, ripening and even harvested oats and rye. Much of the sand-dune "machair" of the Uists is sown with cereals to provide winter food for stock. Adult and young Greylags move from their breeding lochs to the machair and attack the corn, sometimes flattening considerable areas. The corn is harvested by the traditional method of cutting with a binder and leaving to finish ripening in stooks in the fields. Final drying is in small stacks in the fields. The geese attack the ripening corn at all these stages and have proved very difficult to discourage because of the highly nutritious value of the grain. This is undoubtedly a problem, and one which occurs to some extent with feral geese in England, but finding a solution, other than culling breeding geese, which is highly undesirable, is difficult (136).

Canada Goose

Canada Geese are resident wherever they occur in Britain, breeding on lake islands and feeding on adjacent pastures and arable land. Initially, when they were restricted to stately homes and large estates they caused few problems, but with the spread to gravel pits and other waters damage became more widespread. They visit grass fields throughout winter and spring, competing with outwintered stock and reducing the yield of spring bite, causing a delay in letting out inwintered cattle. The most serious damage is, however, caused to growing cereals throughout the summer.

In extreme cases damage is total for parts of fields near the breeding and moulting lake, i.e. the cereal is grazed so continuously that it is never allowed to head. Less severe grazing in April and May can cause a reduction in yield of spring barley of up to 50%, the extent depending on how heavy and how late the grazing. The ripening date is also delayed in grazed patches and this causes additional losses at harvest, because either the yield of late ripening heads is not at its maximum or there is greater harvesting wastage and bird damage on over-ripe heads (612).

Canada geese stay very close to their breeding lakes in spring, and feed on fields immediately adjacent. The most obvious way of avoiding damage to cereals is not to plant them in lakeside fields. Cereals are not so preferred by geese that they will fly in search of corn if grass is readily available at the lakeside (612). Taking some fields permanently out of cereal production, however, involves farmers in the loss of potential production, and culling the geese may be the only long-term solution. A reduction in numbers nationally may be desirable to check the continued expansion of this introduced species (p.382).

Barnacle Goose

The small Solway stock is supported for most of the winter and spring on refuges or semi-natural grasslands, and the Greenland birds on the Outer Hebrides are scattered and give little cause for concern. On the island of Islay, however, the increase in the population from 3,000 in the 1950s to 24,000 in 1976 caused considerable concern to farmers. The geese concentrate on improved pastures, which are most important for dairy farmers to provide spring bite for their stock. The gregariousness of Barnacle Geese and their ability to graze very closely means that they are potentially the most damaging of our geese to pasture. Field trials involving exclosures have indicated that very substantial losses of potential production are sustained under heavy grazing pressure up to the end of April (461). While these results have wrongly been extrapolated by farming interests to whole farms or even sections of the island, the fact remains that damage is substantial and because of the concentration of geese some farmers are very badly hit. They are, in any case, under the disadvantage over mainland farmers that the costs of transport of feeding stuffs and agricultural products are very high. Shooting of the geese has been allowed in December and January since 1955, and this was extended to the whole season in 1976. The Barnacle was protected under the 1981 Wildlife and Countryside Act and a system of licensing introduced, but this performed very unsatisfactorily in the first two years. A more effective method of alleviating the conflict must involve the creation of refuges. The establishment of an RSPB reserve at Gruinart in 1983 goes part of the way towards easing the situation, but more effective refuges are necessary, coupled with a more responsible attitude to the granting of licences to shoot by the Department of Agriculture and Fisheries for Scotland.

Brent Goose

Field feeding by Brent Geese is restricted to the Dark-bellied race and is a recent phenomenon, largely since the marked increase in numbers in the early 1970s, although they had been grazing pastures in the Netherlands in spring some years previously. The numbers occurring inland correlate with those in the total population (520), and it seems likely that Brent Geese initially only resort to farmland once estuarine foods are depleted. In the late 1970s and early 1980s 20-30,000 birds were involved at times. Winter wheat and pastures are grazed in mid-winter and spring, and are usually within 1km of the shore, although Brent have been known to fly some distance inland in Norfolk. In extreme conditions, when grazing is heavy and on waterlogged ground, very substantial losses in cereal yields can be suffered (151). Since the wintering season ends in March damage to grass is negligible, and geese can be disturbed onto grass if scaring is carried out before the birds have become accustomed to feeding in the fields. A system of small grassland refuges along the coast facilitates the disturbance of Brent from cereals, and this system is being set up in as many areas as possible (522).

As well as their effect on crop yield, geese and swans are accused of fouling grass and making it unpalatable to stock. In trials a slight but short-term effect has been demonstrated (502) and in the field the effect must be negligible. In fact cattle are known to eat goose droppings on at least one Scottish island where they are short of minerals and nutrients in winter. Although there is clearly some value in goose droppings as manure (294), as can be seen on very poor soils such as estuarine sand, this is negligible in terms of farm economics.

The prevention of damage

Some kinds of damage can be avoided by changes in farming strategy, such as planting vulnerable crops close to farm buildings or busy roads. Goose damage may be a factor in the choice of crops in vulnerable areas.

Much research has been put into the development of effective scarers, and advice on the selection and management of scarers is now readily available to farmers (157, 587). Despite this it is disappointing to see how little scarers are used in areas such as Islay where complaints of substantial losses are made by farmers. Scarers are of many different kinds, most relying on loud noises or colour and movement to deter the birds. Some imitate natural dangers such as the outline of a flying hawk. Most are effective for only a short time and a variety has to be used to deter birds for an extended period. A new approach has been to use the birds' own signalling devices to elicit flight, such as head-wagging model geese, which have been successful with Brent Geese (280). Scarers are most effective when the birds have a ready alternative, and it is important to leave them alone when they are in a situation where they do no damage.

The payment of compensation to farmers for their losses is an attractive option providing the money is available, and is a method used in other countries. In the prairie provinces of Canada the federal government buys damaged crops, and in the Netherlands compensation schemes have been in operation for some years. Apart from purely financial considerations, the difficulty of operating compensation schemes in the British situation would be enormous. The biggest difficulty would be to assess damage on such difficult crops as sprouting cereals and grass. The variability in estimates arrived at by intensive research is considerable (297), so a large number of personnel would have to be involved at enormous additional costs. Whereas both in Canada and the Netherlands there is a national source of income for compensation – derived from shooting – there is very little revenue in Britain directly attributable to wildfowl shooting.

One of the main problems with damage by wildfowl is that it is so localised, affecting a few farmers very badly indeed. On a national scale the impact is negligible. Dispersal of large winter concentrations has been seen as a way of spreading the burden on individual farmers while retaining goose populations at their present strength. Very costly and intensive dispersal programmes failed to discourage Canada Geese from a Wisconsin refuge (208), but there have been few British experiences. Dispersal from Islay was one of the supposed desirable effects of lengthening the shooting season there, and this appeared to have been achieved when autumn numbers fell from 22,000 in 1978 to 13,000 in 1982. The number shot on the island was 1,000-1,500 annually, overall mortality probably being no more than 10-15%. At the higher rate, 7,500 geese were lost in the 4 years and at the lower rate 3,600, or between 1,000 and 2,000 geese per year. The 1983 census, however, showed that the numbers elsewhere in Scotland had not increased since the 1978 spring census and those in Ireland had declined. Thus, either the numbers shot were greatly underestimated or the geese driven out of Islay had died as a result of displacement. The lesson from this is that dispersed birds must have somewhere suitable to go if they are to survive; the danger that they do not find a suitable alternative makes mass dispersal a drastic and rather risky technique.

Farmers consider that since goose numbers have increased so markedly in the last 20 years there are now "too many" and they should be culled. This goes against the objectives of conservation (p.526), besides which many of our populations are still very small in global terms. Mass slaughter, even if it could be achieved, would lead to public outcry and should not be contemplated except, perhaps, to reduce the numbers of introduced species such as the Canada Goose. The licensing system allowed under the 1981 Act is meant to enable farmers to shoot as an adjunct to scaring rather than to reduce goose numbers.

If we argue that all these ways of alleviating the problem of agricultural damage are unsatisfactory, the only alternative is to provide the birds with habitat or food supplied by conservation interests, and this has been proposed as a matter of urgency (438, 441). It is unsatisfactory to create reserves or sanctuaries on goose roosts while expecting farmers to support feeding birds on their land, but this has been the usual practice in the past, with very few exceptions. Although in Denmark food has been provided at the roost of Pink-footed Geese and successfully prevented them from visiting farmland (193), it is hardly a realistic possiblity on any considerable scale. The conservation organisations must therefore acquire farmland and manage it specifically for geese. The success of the Wildfowl Trust's refuges at Slimbridge and Caerlaverock shows that birds can be accommodated on rather small areas.

The area of land required to support the numbers of birds using farmland has been calculated, based on known usage rates of Wildfowl Trust refuges (438), and this approach has been shown to be realistic in the case of Brent Geese (507). Since numbers fluctuate from year to year the decision on a "target" figure to support must be based on several years counts. The likely areas required to accommodate the species accused of damage are given in Table 264.

The total area required to house 180,000 geese is surprisingly small in practice. Much smaller numbers of Greylags and Pinkfeet need to be accommodated since the smaller concentrations do little harm and some landowners welcome wild geese on their land. For Barnacle and Brent Geese, however, the specified area is very small considering that a fifth of the world's Barnacle and Dark-bellied Brent Geese are involved.

The smaller the perimeter of a refuge in relation to its area the better, and sites of 200-500ha would be a reasonable compromise between the requirement not to concentrate the birds in too few sites and the creation of refuges large enough to be effective (441). In the case of Brent Geese, whose intertidal zones are undisturbed, smaller areas just over the sea wall are useful, and have been seen to be successful (522). Refuges must obviously be created where the birds are at present, and it would be most sensible to protect farmland for geese contiguous with internationally or nationally important goose roosts, particularly those recognised as being eligible for designation under the Ramsar Convention.

The method of acquisition giving the most power to managers is land purchase and reletting to farmers under strict management regimes, but this is very expensive both in capital and running costs. Leasing from the landowner and reletting to tenant farmers is less expensive; this is the way in which the Wildfowl Trust's refuges are run. The cheapest method and the most easily set in motion is to make a management agreement with a landowner or tenant

Table 264. *The number of geese in the total British and in the arable feeding populations of species accused of causing agricultural damage. Also given is the estimated area of refuges (in ha) needed to support these field-feeding populations. Greylag and Pinkfeet are assumed to have suffered 15% of their 18-19% annual mortality by the damaging months. The Solway Barnacle population is excluded from the farmland total since their main arable feeding grounds are already a refuge. Based on Owen (1977, 1980).*

Species	Season	Total no.	No. on farmland	Refuge area
Pinkfoot	March – April	75000	60000	2800
Greylag	March – April	75000	75000	5000
Barnacle	January – April	28000	15000	570
Brent	February – March	60000	30000	800
Total		238000	180000	9170

whereby he receives payment for allowing geese undisturbed access to his land. The Nature Conservancy Council has powers to make such agreements under the 1981 Act, and these should be enacted as soon as possible in the most urgent cases, such as Islay.

The creation of habitat

As the foregoing sections have outlined, much of man's impact on wildfowl habitat has been negative, but there has also been a positive side, although in many cases the benefit was accidental. Many inland waters resulted from man's excavation for minerals and fuel, while others were the result of flooding and damming for ornamental purposes, water supply or industrial use. This section briefly describes the various kinds of artificially created wetlands and assesses their present value for wildfowl both on a local and national scale.

Long-established habitats

Nearly all the lakes in southern England are man-made, and the longest established of these are the Norfolk Broads, a result of deep peat digging, largely prior to the 14th century. The pits became flooded and formed a complex of waters, joined by the navigable rivers Bure, Yare and Waveney. In the hundred years from the mid-19th to the mid-20th century, ecological succession reduced the water area and produced large tracts of reed-beds and woodland carr. The rivers and broads are tidal, the water level fluctuating by between 75cm near the mouth and only 15cm inland. The Broads provide an extensive breeding area for our native ducks and support substantial winter populations (162). Introduced or re-introduced species are well established, particularly Gadwall, and Greylag and Canada Geese.

Often overlooked as man-made waters are the ornamental lakes established by damming or digging in the grounds of stately homes. These are particularly plentiful in the north Midlands of England (where there are no natural lakes), the Home Counties and East Anglia. Although few are important in their own right, groups such as the Dukeries in Nottinghamshire do support substantial populations of the commoner ducks. Ornamental lakes also provided the focus for expansion of the introduced Canada Goose in Britain and still hold a sizeable proportion. Duck decoys, often established in conjunction with estates and stately homes (p.9), provided secluded roosting waters for ducks feeding on nearby estuaries or agricultural land, but no longer do because of the proliferation of other habitats.

Drainage for agriculture has resulted in widespread deterioration of the value of wetlands, but the Ouse and Nene Washes, in Norfolk and Cambridgeshire, were created specifically as flood relief areas. These have provided wintering and breeding habitat in an area where the extensive natural marshes have almost completely disappeared. The Ouse Washes are now the most important inland site in Britain for wintering wildfowl and are a key area for breeding dabbling ducks and waders. The Nene Washes, though much more successfully drained, also support large numbers of dabbling ducks and have been used as an alternative site by the Bewick's Swans whose headquarters are on the Ouse Washes, only 15km away.

The extensive network of canals that was created in the 18th and 19th centuries greatly extended the amount of linear wetlands in inland Britain. While in commercial use and deeply dredged these provided little usable habitat, but since many are disused and partly overgrown with vegetation, they do provide wintering and breeding areas for a scattered population of Mallard, Tufted Ducks and Mute Swans. A consequence of the building of canals was the construction of a large number of canal feeder reservoirs, which have a value of their own (see below).

Mining subsidence has provided useful areas of water, the "ings" and flashes of Yorkshire, and Stodmarsh, Kent, being the best examples. There are also substantial areas in the west Midlands and in the north-east of England, where natural freshwaters are in very short supply. Fairburn Ings in North Yorkshire, now managed by the RSPB, holds a wide diversity of species and regularly supports well over 1,000 individuals, including 135 Shoveler and more than 40 Whooper Swans. Stodmarsh is a National Nature Reserve because of its mire habitat, but the flooded grasslands of the colliery subsidence are very important for wintering wildfowl, the most common being Mallard (up to 3,500) and Teal (up to 700), with a wide diversity of other species.

The old style of sewage farm, so valuable as a feeding area for waders and wildfowl, has now been replaced by other methods of sewage disposal, but treatment works still provide breeding areas, particularly for Mallard and Mute Swans, but also for less common breeders. Mallard and Teal also use modern sewage treatment areas in some numbers in winter (200).

Habitats created for conservation

With conservation organisations becoming major landowners there are several examples of habitat

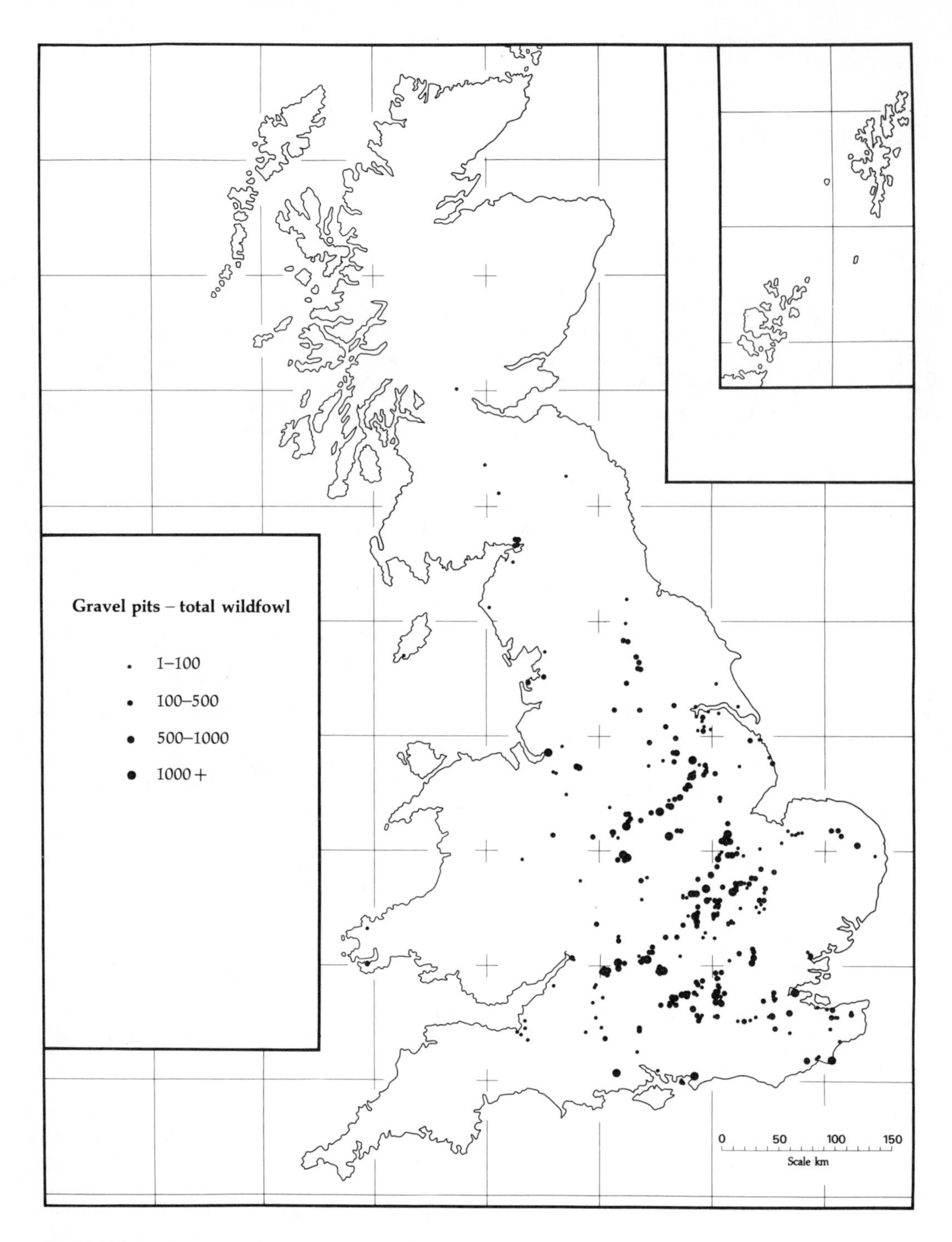

Fig. 167. The total number of wildfowl on wet mineral pits covered by Wildfowl Counts at some time between 1960 and 1982.

eing created or re-created for conservation purposes. he classic case of Minsmere, Suffolk, re-created and dapted by the RSPB is long established and well nown. A more recent example is the excavation of the nere at Martin Mere, Lancashire, by the Wildfowl rust. The area had been progressively drained for griculture and the excavation in some part restored he original habitat. In many other areas local and ational organisations are creating wetland habitat by xcavation and flooding, but this is treated in more etail under refuge management (p.537).

Mineral workings

There is a wide variety of mineral workings, ncluding brick pits, clay pits and other excavations, ut the overwhelming majority of wet pits are created y gravel excavation; in the following account the term gravel pit" is used to include all other wet mineral vorkings. Most wet pits are in the valley gravel and lluvium deposits of the Thames, the Trent, and the Ouse and its tributaries. The distribution of wet nineral sites included in Wildfowl Count data (352 in he 23 years 1960-61 to 1982-83) is shown in Fig 167. It is vident that the main complexes are concentrated in central and southern England, where rich gravel deposits coincide with areas of high human population density and consequent large demand for road and other building works. There are few pits in the sparsely populated south-west, in northern England, or in Wales and Scotland, despite the fact that substantial gravel deposits do occur there. Much of the demand from south-west England and South Wales is supplied from marine deposits in the Bristol Channel.

The demand for sand and gravel has increased dramatically in the present century, the national extraction rising from less than 5 million tonnes annually before 1925 to about 30 million in 1940. The growth in extraction rate since then is shown in Fig 168. It reached a peak of around 110 million tonnes in 1972, at the height of the era of motorway construction. Thereafter it declined slightly, but is still running at a level of around 90 million tonnes per year. The figure also shows an index of the area of wet pits, using data from six counties of southern England from where detailed information was obtained. The sample includes a large area of gravel pits and can thus be used as a reliable index of countrywide trends. The water area index follows very closely the total production during the period of growth but, as expected, it continued to increase in the late 1970s, even though production had levelled off.

In 1970 it was estimated that 1,600ha of land was worked annually and that 1,000ha (63%) of this resulted in wet pits (14). Using the same ratio of production to area, over 20,000ha of wet gravel pits have been created since 1940, and three-quarters of

Fig. 168. The tonnage of gravel and sand extracted from land deposits in Britain in the early parts of the last five decades (dotted line). The solid line is the area of wet pits in the counties of Berkshire, Cambridgeshire, Greater London, Leicestershire, Norfolk and Surrey. Data on extraction rate from Archer (1972) and updated statistics from the Institute of Geological Sciences provided by the Department of the Environment. Wet pit area details provided by the planning departments of the respective County Councils.

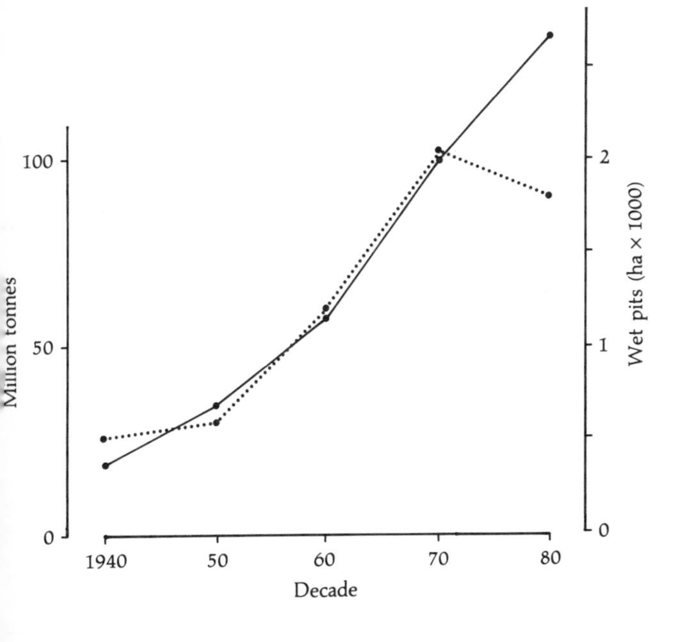

this since 1960. Despite the pressure from farmers for restoration of pits for agriculture, the difficulty of finding suitable infill and the counterbalancing demand from sporting and recreational interests are likely to ensure that at least a similar proportion of quarries will be allowed to flood in future. Even at present levels of demand, therefore, the area of wet pits could almost double again in the next two decades.

Gravel pits usually contain a large proportion of relatively shallow water and are often worked in smallish blocks. Most pit complexes therefore consist of numerous pools, some including peninsulas and islands, so that the perimeter is long relative to the water area. Valley gravels are in the fertile river plains and rich aquatic and marginal vegetation slowly develops. The process can be speeded up and the habitat improved by planting and other forms of management, discussed elsewhere (p.543). In most newly excavated pits the habitat is left to settle and mature without human manipulation, and the ecological succession in such a pit is broadly as follow During the excavation and pre-inundation stages th habitat is of little use to wintering wildfowl, although few pairs of Mallard and Coot may breed. Soon aft flooding the area becomes very productive and is use for feeding by large numbers of dabbling ducks an Coots. As the pit matures the productivity of shallo waters declines and with it the number of feedin dabbling ducks. The number of diving ducks the increases and they become the dominant waterfowl i mature pits (348).

It would be revealing to examine in detail th build-up of wildfowl populations in relation to th successional stages of a large gravel complex, bu detailed information on the area of water is n available in most cases. In others the birds on the pit are not counted in the early stages of developmen since their numbers are then insignificant. Reasonabl complete data are available for the Chichester Grav Pits (Chichester Leisure Centre, Wyke and neighbou ing pits), and these are summarised in Fig 169 Extraction at the complex began in 1930 and cease completely in 1975. Most pits remained wet, althoug 6ha were filled in in the early 1970s.

The figure shows a pattern broadly similar t that described above, with total numbers increasing i close relationship to pit area. The number of commo dabbling ducks is highest during the early stages of pi development and then falls as the area of newl inundated pits decreases. The increase in the lat 1970s and early 1980s may well be due to an increase i birds roosting on the larger waters during the day an feeding on the nearby marshes of Chichester Harbour

Fig. 169. The area of water at Chichester Gravel Pits – 1950 to 1980 (solid line) – and the number of wildfowl using the area. Wildfowl numbers are the mean of the three highest counts in each of five winters (each point is a mean of 15 counts), though lack of coverage occasionally reduces this to 4 winters. Dashed line – total wildfowl, dotted line – Mallard and Teal, dashes and dots – Pochard and Tufted Ducks, large dashes – Tufted Ducks in September. Note that the species regular means cannot be related to those of total wildfowl since they may represent different months. Data on pit area provided by West Sussex County Council.

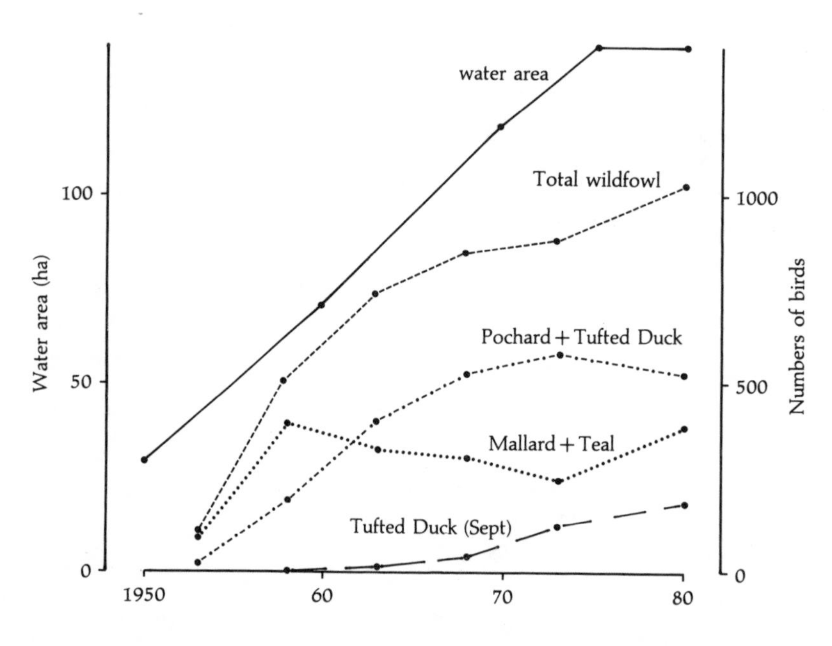

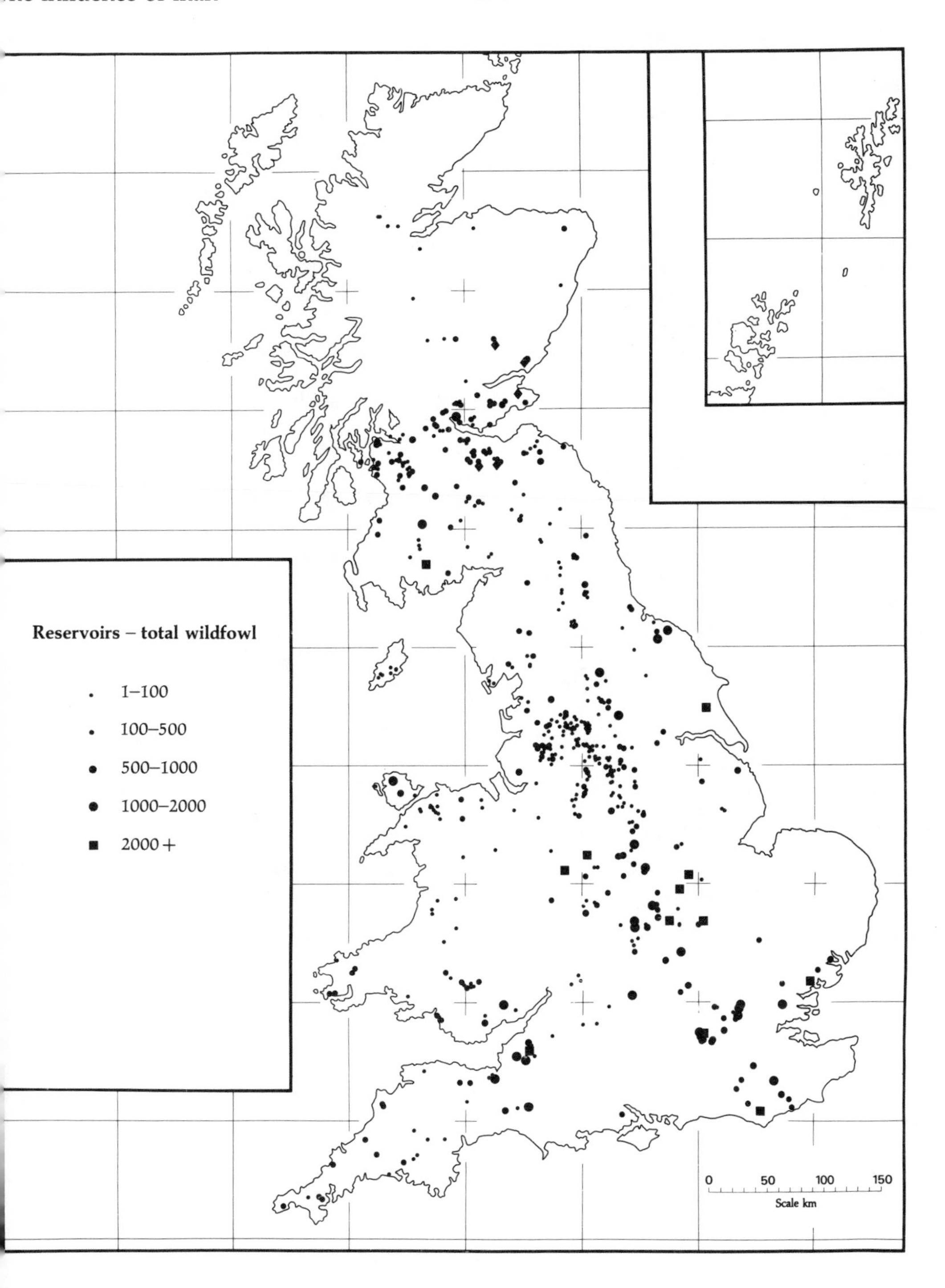

Fig. 170. The total number of wildfowl on reservoirs covered by Wildfowl Counts at some time between 1960 and 1982.

Diving duck numbers were low initially, but rose dramatically in the late 1950s, as some of the early excavations became mature. They became the most numerous birds in the later stages, but their numbers stabilised as the water area ceased to grow. The number of Tufted Ducks in September represents the production of the breeding population. Breeding Tufted Ducks require mature pits with a substantial population of invertebrates, and the Chichester Pits were not colonised until 30 years after initial excavation. Thereafter there was a gradual increase to about 200 birds, representing about 40-50 pairs and their young.

Reservoirs

Reservoirs, including those for drinking water (much the most numerous), canal feeding, industry and flood relief, vary widely, in both morphological and chemical characteristics. Most of the 600 or so reservoirs in Britain are not suitable habitats for wildfowl. Those in the uplands of north-west England and Wales are very unproductive and isolated from areas providing food and cover. Many others are too small or too close to industrial or other human activities to provide refuge for significant numbers of birds (18). The distribution of the 544 reservoirs counted in Britain at some time or other between 1960-61 and 1982-83 is shown in Fig 170, together with an indication of their importance for wildfowl.

Although the largest density of reservoirs is in the north-west and Yorkshire, the most important are concentrated in central and southern England. These are favoured by the birds not only because they are low-lying and more productive, but also because they are well placed to intercept winter immigrants from continental Europe. Many of the reservoirs in central Scotland are important because they hold large roosting concentrations of geese, which feed on the surrounding farmland.

Most reservoirs in Britain were created in the second half of the 19th century and in the first three decades of the 20th, but many of the most valuable are more recent. Since the mid 1960s, 7,300ha of land have been flooded to form new reservoirs, notably Grafham and Rutland Waters.

Embanked reservoirs, such as those in the London area, are of rather uniform depth and relatively short perimeter. They provide limited feeding areas for dabbling ducks, but the bottom is accessible to diving species over most of the water. Those reservoirs formed by flooding fertile valleys, on the other hand, have a convoluted margin providing food and cover for wintering and breeding surface feeding ducks, as well as feeding areas for diving species. The management of the margins and the availability of a variety of farmland in the vicinity is important for grazing and arable feeders.

As with gravel pits, successional changes brought about in the habitat after flooding bring about changes in the suitability of reservoirs for wildfowl, as illustrated in the case of Grafham Water, Cambridgeshire, where flooding began in December 196

Fig. 171. The build-up of wildfowl numbers at Grafham Water, Cambridgeshire, from 1965-66, the first season when water was available throughout the winter. The figures are regular means for total wildfowl (solid line), Mallard and Teal (dashed line) and Tufted Duck and Pochard (dotted line).

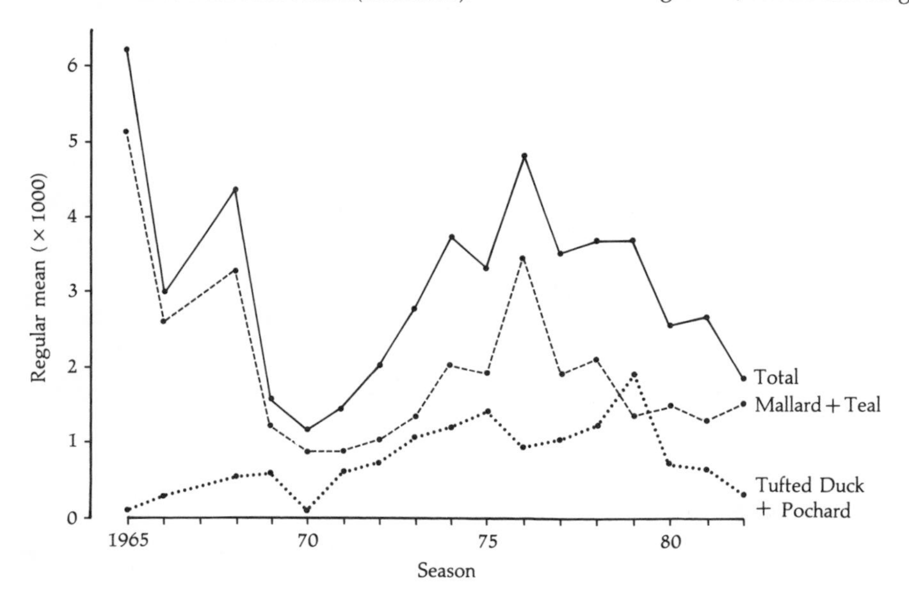

and top water level was reached in February 1966 (131). The development of the wildfowl population, from the first winter throughout which water was available, is shown in Fig 171. The pattern of total wildfowl numbers shows a sharp drop followed by a gradual increase to a level somewhat lower than at the early stages. How this was brought about is illustrated by the two predominant species, Mallard and Tufted Duck. The Mallard was very abundant during the period of flooding, when newly flooded grassland was continually available. After four years of decline the level stabilised or even increased slightly. A very similar pattern of abundance and decline is seen from 1976 onwards. The summer of 1976 was extremely dry and the water level receded steadily through the summer, until it was 6m below top water level in September. The exposed ground quickly became weed-covered and when the water rose again in mid-winter, conditions simulated those of the original flooding. These patterns had been recognised in managed wildfowl impoundments in North America, where it was recommended that the water be drawn down every five to seven years to increase productivity (613). The numbers of Tufted Ducks at Grafham showed an almost exactly opposite pattern to that of Mallard, increasing steadily during the early years and stabilising after about ten years. Their population received a setback from the drought of 1976, but subsequently recovered, although they have gone into a sudden and disastrous decline in recent years (p.512). This example provides a basis for the management of newly created reservoirs to maintain as much as possible their early value for wildfowl, and the newly formed Rutland Water illustrates what can be achieved (p.544).

The importance of man-made waters for wildfowl in Britain

Of the long-established artificial sites, ornamental lakes are individually not very significant, but because they are so numerous and widespread, they collectively make a large contribution to maintaining stocks of inland ducks and swans. Subsidence pools are not numerous, but a few are key wildfowl areas and their significance has been discussed. This section considers the contribution made by the two most important new habitats – reservoirs and gravel pits (including other mineral workings).

Wintering populations

Except for very large complexes such as those in the Thames and Trent Valleys, which cannot justifiably be regarded as single sites, no gravel pits regularly hold more than 2,000 wildfowl. Only about 20 pits hold more than 1,000 wildfowl regularly (Fig 167). Many reservoirs, on the other hand, hold very large concentrations and their importance as major resorts has long been recognised (18). The wildfowl numbers on sites regularly holding more than 2,000 birds are listed in Table 265. Only 2 of these were built before 1930 and half in the last 30 years.

As well as holding a large total and a wide diversity of wildfowl, some reservoirs support concentrations of single species of national or international importance. The contrast between the valley reservoirs, which hold large numbers of dabbling

Table 265. *The mean maximum numbers 1979-80 to 1981-82 of certain species and total wildfowl in the 12 reservoirs holding more than 2000 birds.*

Reservoir	Date flooded	Mallard	Teal	Wigeon	Gadwall	Shoveler	Pochard	Tufted Duck	Total wildfowl
Abberton, Essex	1939	2857	1963*	5066+	176*	353*	1366*	2269*	10633
Rutland Water, Leics	1975	2029	1332*	2938*	291*	340*	656*	1845*	7169
Blithfield, Staffs	1955	1489	663	987	1	39	208	121	3988
Staines, Surrey	1902	67	62	96	9	43	2198	1609*	4030
Grafham Water, Cambs	1965	1647	432	970	71*	44	113	1927*	3664
Loch Ken, Dumfries & Gall	1936	363	325	628	1	50	37	69	2789
Pitsford, Northants	1934	805	570	601	25	94	317	705*	2673
Chew Valley Lake, Avon	1956	1144	520	351	80*	341*	130	275	2557
Tophill Low, Humberside	1959	877	510	332	24	55	252	749*	2484
Belvide, Staffs	1834	1350	208	51	1	280*	217	337	2327
Arlington, Sussex	1970	560	11	1450	0	37	96	50	2196
Eyebrook, Leics	1934	742	557	947	5	58	177	210	2183

* More than 1% of the British population.
+ More than 1% of the north-west European population.

ducks, and the steep-sided London reservoirs is clearly shown by the Wildfowl Counts. The London reservoirs, as complexes, hold a considerable proportion of the British stocks of Pochard and Tufted Ducks (p.116). Large concentrations of moulting diving ducks are also found in south-east England, particularly at Abberton, Essex, where more than 3,000 flightless Pochard and Tufted Ducks have been recorded in recent years. The London area reservoirs held 6,150 moulting Tufted Ducks in August 1979 (419).

The significance of reservoirs and gravel pits on a national scale is summarised in Table 266, which gives the proportion of the British totals found on man-made habitats. Nearly a fifth of all wildfowl wintering in Britain, amounting to some 370,000 birds, are found on reservoirs and gravel pits. Even predominantly coastal species such as Goldeneye and Wigeon occur in numbers inland, and the provision of inland roosts for the latter species has opened up new habitats (454).

The potential of both reservoirs and gravel pits has long been recognised, but there have been considerable recent changes. The total numbers of wildfowl counted on reservoirs and gravel pits since 1967, when the sample of sites counted reached roughly its present level, is shown in Fig 172. There has been a 50% increase in the numbers on reservoirs, discounting the fall in January 1982, when severe weather caused the freezing of many inland sites (the continued fall in January 1983 is, however, puzzling). The increase on gravel pits has been more dramatic, the numbers increasing threefold from around 20,000 to 60,000 in 1981.

The number of wildfowl in Britain generally has increased during the last 20 years, and the relative contribution of man-made lakes is shown in Fig 173. The contribution of reservoirs has fluctuated around a stable average of about 12-13% of the counted wildfowl. Gravel pits, on the other hand, have grown in relative importance, holding only 2% of British wildfowl in the early 1960s compared with 6-8% in the late 1970s.

As Table 266 shows, Canada Goose, Gadwall and the two inland diving ducks have benefited most from gravel pits; the proportion of two of these species found on gravel pits as opposed to natural lakes – a similar but static or declining habitat – is shown in Fig 174. The number and proportion of Pochard on gravel pits has risen so that these are now equal to natural lakes, where the numbers have followed the national trends, in absolute and relative importance. Canada Geese were introduced and spread onto park lakes and other long-established waters, but following rapid increases in overall numbers the growth of the population on these waters slowed down and an increasing proportion were found on gravel pits. This growth was due both to dispersal of birds from their strongholds on park lakes and to good breeding

Table 266. *The estimated winter population (for details see species accounts in Part III) of the most important inland species, and the proportion found on man-made waters. Percentages are mean maxima for the three years 1979-80, 1980-81 and 1981-82.*

Species	Estimated population	% Reservoirs	% Gravel pits	All man-made
Mallard	500000	19.9	12.1	42.0
Teal	100000	10.0	5.8	15.8
Wigeon	200000	9.4	1.7	11.1
Gadwall	4000	19.9	26.9	46.8
Shoveler	9000	35.7	11.8	47.5
Pochard	50000	27.4	29.2	56.6
Tufted Duck	60000	37.0	25.2	62.2
Goldeneye	15000	19.4	4.2	23.6
Goosander	5500	34.5	10.7	45.2
Mute Swan	18000	5.7	16.1	21.8
Canada Goose	33000	13.2	27.5	40.7
All wildfowl	2000000	11.6	7.0	18.6

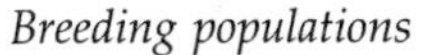

success on the new gravel pits under reduced competition.

Man-made waters, then, make a substantial contribution to the present habitat usage of our wintering wildfowl, and for some species are taking over from more traditional waters as the chief haunts. The importance of gravel pits, in particular, is likely to increase still further as the area of water available continues to increase and as the existing new pits become older and more productive.

Fig. 172. The number of wildfowl counted on man-made lakes in January from 1966-67, when the number of sites counted reached its present level, to 1982-83. Solid line – reservoirs, dotted line – gravel pits.

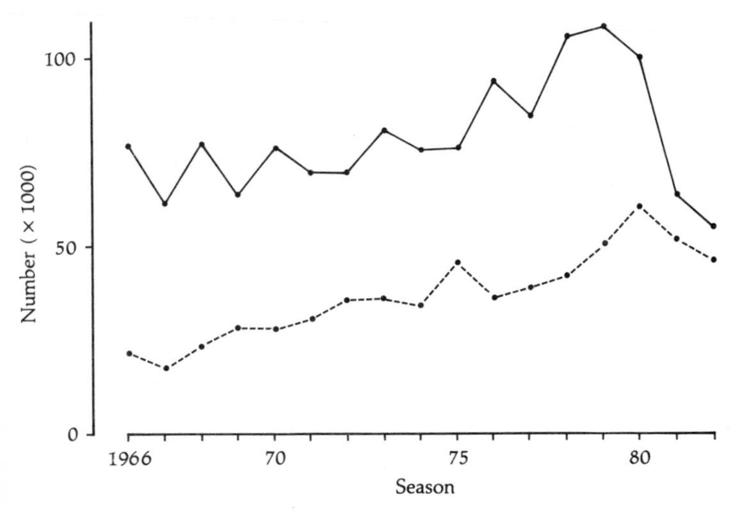

Fig. 173. The proportion of counted wildfowl in Britain in January 1961 to 1983 that were on man-made lakes. Solid line – reservoirs, dotted line – gravel pits.

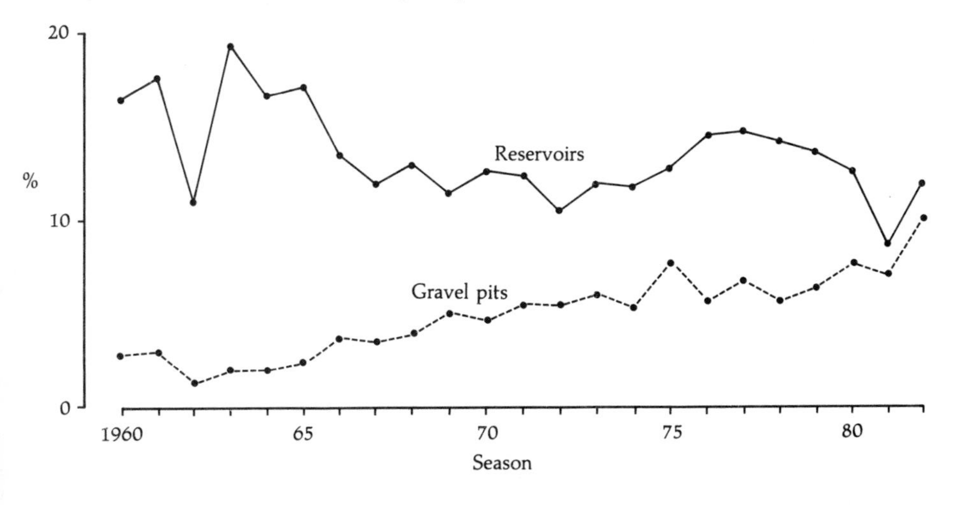

Breeding populations

In connection with the study of recreational effects on wildfowl of inland waters, a survey of breeding birds was carried out in 1980, and this enables us to examine the importance of man-made lakes in this respect. The survey was restricted to enclosed inland waters, of which 448 were counted (598). These included 60% "natural" lakes, 28% reservoirs and 12% gravel pits, proportions that are very similar to the 60%, 25% and 15% respectively included in the winter counts. The sites also included a wide geographical spread, from northern Scotland to southern England and Wales.

A summary of the numbers of adults and young found in the three types of water is given in Table 267, which includes grebes and rails as well as wildfowl.

Reservoirs and gravel pits held about 40% of the adults and young in 1980. Reservoirs had fewer birds than expected on the basis of their number or water area, and in general adult and young density decreased as site area increased. This is not surprising since the length and nature of the margin is important for nesting, and reservoirs and larger waters tend to have relatively smaller perimeters. Natural waters are smallest and hold the highest densities of adults and young.

In seven of the ten species the production per adult is higher on reservoirs than on gravel pits and in six cases higher than on natural waters. This is presumably because nesting density on reservoirs is low and there is less competition for food for the young. It is no coincidence that the three exceptions – Mute Swan, Moorhen and Coot – are vegetarians, doing better in the shallower and well-vegetated smaller natural waters and gravel pits. Gravel pits are generally less productive than long-established lakes, which is not surprising since breeding cover and invertebrate food for young is slow to develop and many pits are not mature.

Man-made habitats support a substantial proportion of inland breeding species and their value will continue to increase. In the case of Tufted Duck and Great Crested Grebe, over half the pre-breeding population is on man-made areas, along with more than 40% of Canada Geese, Coots and Little Grebes. The Tufted Duck is now a very widespread breeding species in inland Britain, owing largely to the expansion of gravel pits. In the late 1960s the total population was estimated at 1,500-2,000 pairs (621) and this increased to 4,000-5,000 pairs in the early 1970s (542). The present population is in the region of 7,000 pairs (p.431) and the trend is continuing.

In general, artificial habitats make a substantial contribution to the breeding population of those species whose native production is important to maintaining the wintering stock. This contribution continues to increase, and the Tufted Duck, in particular, has undergone a remarkable change in the last two decades, with the British birds (i.e. the September count) now amounting to double the peak wintering counts of the early 1960s.

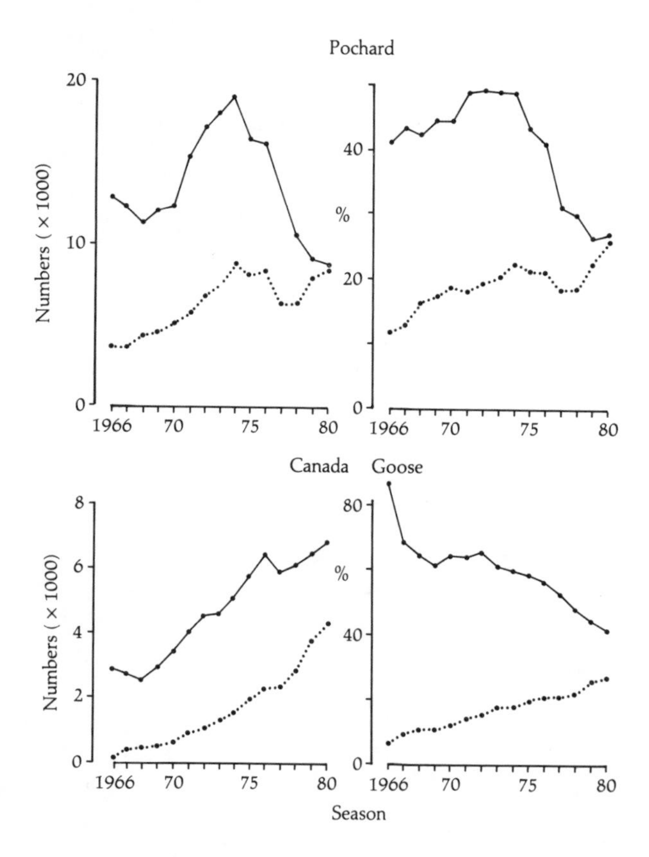

Fig. 174. The absolute and relative (% of British count) populations in January of Pochard and Canada Goose, expressed as 3-year running means, on natural lakes (solid lines) and gravel pits (dotted lines), in January from 1966-67 to 1981-82.

Table 267. *The proportion of the more important inland species of water birds found on different habitats during the 1980 summer breeding survey of enclosed inland waters. Adult numbers are based on pre-breeding populations and numbers of young on the peak brood month for each species. Y/Adult is the number of young per pre-breeding adult.*

	Reservoirs	Gravel pits	Natural lakes
Number (%) of sites	125 (27.9)	53 (11.8)	270 (60.3)
Mean site area (ha)	45.6	31.9	27.7
Percentage water area	38.8	11.3	49.9

	Total counted		% Reservoirs			% Gravel pits			Natural lakes
	Adults	Young	Adults	Young	Y/Adult	Adults	Young	Y/Adult	Y/Adult
Mallard *	6568	3902	24.6	29.1	0.703	18.7	12.9	0.410	0.608
Teal	188	100	19.1	57.0	1.58	2.1	0.0	0.0	0.290
Pochard	538	162	6.5	14.1	0.657	10.0	1.9	0.055	0.302
Tufted Duck	4793	1637	20.9	23.8	0.389	29.7	23.4	0.269	0.213
Mute Swan *	700	436	22.7	18.1	0.497	16.1	23.4	0.903	0.595
Canada Goose	2772	1385	7.8	15.6	1.005	34.2	33.8	0.494	0.434
Moorhen *	1360	394	15.6	13.2	0.244	18.0	16.5	0.265	0.305
Coot	6240	1944	25.3	21.5	0.264	20.8	19.7	0.294	0.336
Great Crested Grebe	1429	475	26.4	24.2	0.304	33.7	25.8	0.254	0.416
Little Grebe	287	120	22.9	33.3	0.61	21.2	14.2	0.28	0.391
All water birds	33289	9792	20.1	22.4	0.327	20.8	18.4	0.260	0.294

* These species have a substantial breeding population on habitats not covered by this survey (rivers and small marshes).

The management of stocks

Supplementing wild populations

Ducks lay large clutches and often re-lay if the first clutch is lost. The eggs of most species are easy to hatch in incubators, under hens or foster ducks, and the survival of ducklings is excellent under captive conditions. In the wild the weather, food shortage and competition take their toll and the brood is reduced from 8-12 at hatching to 2-4 at fledging. It would, therefore, be easy to increase the number of fledged young produced, by taking the first clutch of eggs (allowing the female to re-lay again), hatching and rearing them artificially, and releasing them onto the breeding area or similar habitat. This has been practised for centuries in Britain and intensively for decades in North America, with the aims of:

a) increasing the number of ducks available for autumn and winter shooting and

b) supplementing wild populations for future breeding seasons.
An additional aim of the WAGBI scheme in Britain (249) was

c) to encourage a responsible attitude to harvesting and an interest in conservation in every wildfowler.

Mallard originating from game-farm stock are not likely to behave like wild birds in either tameness or flying ability, so eggs taken from wild birds are preferable. If eggs are taken early the duck will re-lay and there will be a minimum loss to the wild birds. In North America young ducks are usually released at 6 weeks of age, two or three weeks before fledging, these doing better than older or younger ducklings (82). Nevertheless, losses of birds liberated in unprotected places are up to 30% between release and fledging (194). This is due largely to predators, and, although there are no data for Britain, we might expect a higher survival rate here.

The results of North American research on Mallard survival and performance following release, and early results from British work, indicated that the above aims were far from being satisfactorily achieved in the early years (63).

Mallard are very sedentary, and hand-reared ones tend to be less willing to move than wild ones. Although normal dispersal did occur from Manitoba (82) the birds there are normally forced out by cold weather, whereas early results in Britain showed that over 90% of released birds were recovered within a few kilometres of the release point. The mortality of juveniles in the first twelve months was 94% compared with 70% for wild juveniles (63), and similar to the 91% recorded in Denmark (192). Clearly the released population did not disperse significantly and contributed very little to the Mallard population at large or the following year's production.

These revelations caused a policy to be established by WAGBI that birds would be released in places where they would not be shot in their first year so that they could infiltrate into the natural population and breed in the wild (249). There followed a change in the survival pattern, with the calculated survival rate in the first year being 61% and in the subsequent years 50-54%. However, the recovery rate of WAGBI Mallard was only 7.6% compared with 20-25% for ringed wild birds. Thus only a third of released ducks get into the normal shot population and the real mortality rate in the first year could still be over 85%. Dispersal of recovered Mallard was greater from the later releases, although 89% of first year recoveries were within 10 miles (16km) of the release point. This decreased to less than 80% in subsequent years. Seven per cent of recoveries were outside Britain, mostly in France, the Netherlands and Ireland. Thus, although the policy of releasing onto non-shooting areas did result in improvements both in survival and dispersal, the proportion of the released birds that bred with the wild population was small.

In order to claim that release programmes have a lasting value for the conservation of the species one must argue that they have added to the wild stock. Such a claim has been made: "Certainly there are sound scientific grounds for believing that the WAGBI scheme has played a significant part in the marked increase which has taken place in the British Mallard population." (249). It certainly appeared that there had been an increase in the early 1960s, based on priority count indices (20), but the more extensive analysis in this volume (p.415) shows that this was not part of a general trend either in September or January. Indeed the numbers in March declined significantly in the 1960s and 1970s, when the WAGBI scheme was at its peak, releasing more than 15,000 birds annually.

This suggests that the British Mallard population is already saturating the habitat and that the release of further birds merely leads to density dependent increases in mortality and/or emigration. This has been demonstrated to be the case in an analysis of September and March figures at 35 individual sites (262). Thus overwinter loss has a stabilising effect on the breeding population at individual places and in Britain as a whole.

The releases made by WAGBI form only a small proportion of all Mallard released in Britain, most of which are put out on flight ponds by estates and individuals on a "put and take" basis, with three-quarters or more of the released birds being shot from the pond in the autumn of release. There are no good data on the total number of Mallard released, but it has been estimated at 400,000 (238). This is a massive input into the British population, which we estimate stands at above half a million birds in autumn (p.415). It is impossible to tell what the situation might be if releases were stopped, but it could well be that the pressure on wild Mallard from shooting would increase, especially in the autumn.

As long as duck shooting continues to be an active pursuit there will be demand for reared ducks and for taking eggs from the wild. Since the taking of early eggs does little harm to the wild population, there is no reason to prohibit this in the case of Mallard. The most cost-effective way of managing a release scheme is to shoot as many of the released birds as possible in the autumn and early winter, before food supplies become depleted and natural mortality begins to take its toll. It should not be argued, however, that these activities are beneficial to the long-term conservation of British Mallard. Despite this, the achievement of the educational objective of the WAGBI scheme as far as "putting something back" and conservation are concerned has served a useful practical purpose, in that BASC (WAGBI) members are now much more conscious of and concerned about the conservation of wildfowl in general than was the case in the 1950s.

The release of other duck species has been carried out on a more limited scale. Several thousand have been released, chiefly Gadwall, Pintail and Wigeon, but also Teal, Shoveler, Shelducks, Pochard and Tufted Ducks, and even a few Eiders, Common Scoters and Garganey. The results of these releases are more encouraging, with 29 (30%) of 94 Gadwall being recovered, 11 outside the 10 mile (16km) local area, including 3 in France (242). Similarly the recovery rate of Pintail, at 16%, was comparable to that of wild birds, and recoveries were very widely dispersed (566). However, WAGBI ceased their release schemes for all ducks in 1977, and in 1981 it became illegal to take the eggs of wild birds for hatching except under licence. A general licence was granted to take the eggs of Mallard before the end of March, but it is unlikely that one would be granted for large-scale rearing of other species. However, it is still legal to rear birds from captive laid eggs and release them into the wild. The rapid expansion of the British Gadwall population in recent years has been largely due to the establishment of feral populations, the pattern of recovery being similar to that of all British-ringed Gadwall (p.403). In this case it seems that a thriving British population has been established from introductions, and recent activities have greatly accelerated the process.

In the case of the geese, Greylags had long been eliminated from southern Scotland and from England and Wales, and the Canada Goose was established in only a few centres, when release and dispersal programmes began in the 1960s. The Canada Goose was disseminated widely by people who were anxious to relieve local agricultural problems as well as those wishing to promote shooting opportunities. These operations were spectacularly successful in accelerating the spread of the species and maintaining its rate of increase, to the point where it is now a widespread problem (p.382).

Greylag releases, carried out in England mainly by WAGBI in the 1970s, were also successful and the introduced population now probably numbers more than 3,000 geese. They are, however, giving rise to complaints from farmers and their continued survival in numbers depends on their depredations on crops being kept under control.

In the case of the Mallard, then, already a widespread and numerous species, attempts at increasing the population through releasing birds have not been successful. Released birds have, however, provided additional hunting for (chiefly inland) duck shooters, and the process of rearing and release has served a valuable purpose in education and awareness of conservation. Introductions and re-introductions of Gadwall, Canada Goose and Greylag, on the other hand, have succeeded in establishing thriving stocks of these species in Britain, and the Ruddy Duck has managed to expand without help. Many conservationists, however, would not agree that this is a wholly desirable result, and in the case of geese, farmers are becoming increasingly anxious about this newly created problem.

The dynamics of wildfowl populations

Populations fluctuate because of changes in recruitment – the number of individuals born – or mortality – the number dying. If we are to understand the way populations work, or attempt to control trends and fluctuations in numbers, we must know what factors influence the birth rate and death rate. In species which are shot, we must know what contribution mortality from shooting makes to overall mortality. Most important, we must find out whether shooting mortality merely replaces natural mortality (i.e. is compensatory) or adds to it (additive). In the former case the cessation of shooting would not result in an increase in numbers, whereas in the latter it would. There comes a stage, of course, as hunting mortality increases, when it overtakes natural mortality and becomes additive. Beyond this point the population begins to decline and the "maximum sustainable yield" is achieved somewhat below this level of harvesting.

Most of our wildfowl breed elsewhere and winter in Britain; usually we have little control over their breeding success, our only influence being on the mortality rate at the time they are here. Although in some cases conditions in the wintering and staging areas of our migrants can improve the recruitment, in reality our influence on their performance is very small. The following section discusses the factors which affect the number of birds in the various populations and discusses the influence of shooting on them.

Swans

Swans are long-lived and mature late, breeding for the first time at 4 or 5 years of age. Inevitably, the populations contain a large proportion of non-breeders, and observed breeding success, as measured by the percentage of young birds in the winter flocks, is low. The mortality rate is also low (all three species being protected throughout Europe), and, discounting accidents, this makes for rather stable populations or ones showing rather gradual trends.

The resident Mute Swan is the most studied wildfowl species in Britain, with numerous population studies in progress, many long term. Most of our Mute Swans are dispersed nesters, breeding some distance apart along rivers and in eutrophic lakes and reservoirs. They are rather sedentary and there is little interchange between local groups, which can be regarded almost as populations in themselves. In the Oxford area, the density of breeding pairs declined between the 1960s and the 1970s, largely as a result of a decline in the quality of the habitat and possibly the increase in adult mortality as a result of poisoning (30, 467). In this and in the colonial population at Abbotsbury, Dorset (468), survival of juveniles after fledging is 65-70% and of adults around 90%. The number of juveniles released at Abbotsbury is controlled by the artificial rearing programme, and both here and in south Staffordshire (128, 354), unlike the situation at Oxford, the production of cygnets is sufficient to maintain the population. In Staffordshire, cygnet survival post fledging is only 41% and subsequent survival only 70-80%. Collision, usually with overhead wires, is the most common cause of death, but the importance of lead poisoning is unknown. In the isolated population of the Outer Hebrides cygnet production in 1978 and 1979 was sufficient to maintain the population, but this depended heavily on the pairs breeding in eutrophic machair lakes, fledging 3 young per pair as opposed to 1.4 young in less productive waters. The number of pairs breeding differed markedly in the two years but the reason for this is unknown (552).

The breeding success of Bewicks's Swans is generally low, the proportion of juveniles varying from 6.4 to 20% at the Ouse Washes and from 7.2 to 25.6% at Slimbridge between 1969 and 1982 (104, 180). The average for the 14 years was 12.4% at the Ouse Washes and 15.8% at Slimbridge (the average sample size was 1,620 and 480 respectively). Combining the two sites, the average comes to 13.2%. This is a very low figure but is matched by a low mortality rate, estimated at less than 2% during the winter (445) and 13-18% annually (180). The north-west European population was said to be as low as 6-7,000 in 1970 (403) increasing to 12,000 by 1978 (p.356). This represented a growth rate of 8% per year, clearly inconsistent with production and mortality rates. The population may have grown slightly but improved counting is responsible for much of the apparent increase during the period. Increases since (to over 16,000 in 1984) are due to better than average breeding success in the 1980s.

Little is known of the dynamics of the Whooper Swan stock which breeds in Iceland, but it seems to perform much like the Bewick's in that production varies between 5 and 22%, averaging 15% over a 12-year period (84). There are some indications of recent population growth (p.352), so average survival must be high.

Swans, then, are characterised by low production rates arising partly from delayed maturity and partly as a result of a high number of mature non-breeders. In territorial Mutes, this may well be due to shortage of suitable habitat (30), which seems a likely explanation for the other species as well. Some pairs of Bewick's Swans which winter at Slimbridge have not arrived with cygnets in 7 years of observation, while others breed consistently well. Supplementary feeding there ensures that all birds leave in good condition so it seems likely that the limitation lies in the breeding area, with some pairs occupying good territories and others poor ones or none at all. Collisions and lead poisoning are responsible for most swan deaths but, apart from the Mute in some areas, the populations are holding their own. The removal of lead as a threat should obviously be a high priority, especially for the Mute.

Geese

Both the Greylags and Canada Geese that breed here are adapted to nest at more northerly latitudes, where conditions at laying are much more unpredictable than in Britain. They tend to have rather consistent breeding performance here. The Greylags of the Uists increased from about 500 in 1970 to 1,600-2,000 in 1982 (381, 472). Breeding success averaged well over 20% annually in the late 1960s and early 1970s. With a mortality of less than 15%, not unlikely for this sedentary group, the apparent growth rate of 11-12% seems quite possible without immigration. An increase in mortality would stem this growth, but this has to be considerable to cause numbers to decline.

There is no information on breeding performance for any whole group of British Canada Geese, but the adult mortality rate has been calculated at 16% for Yorkshire (579) and 18% for Nottinghamshire (458), and that of juveniles 17 and 35% respectively. To provide for the observed 8% increase nationally (p.382), breeding success has to be, on average, in excess of 25%. There is no sign of a slowdown in the trend; a considerable increase in mortality would be required to produce a decline. There are some indications that increased shooting is beginning to cause this, at least in Nottinghamshire (458).

Our other geese breed in the Arctic, where breeding conditions can be so severe as to reduce nesting considerably in some years. The breeding performance of each species is estimated each year soon after arrival in Britain, and the age ratio used to calculate the annual mortality. This is done by subtracting the number of goslings from the population total to give the number of survivors from the previous year. The results should be treated with care, however, since bias is likely to be introduced into the age ratio estimate because of problems in sampling from individual flocks and from the population. It should also be remembered that the mortality estimate is heavily dependent on the accuracy of both counting and age ratio assessment. Where the mortality rate is low, small percentage errors in counts and ratios may result in large errors in mortality estimates (440).

Fig. 175. The total count, numbers of young produced by the population and apparent mortality of Iceland/ Greenland Pink-footed Geese, 1950-1982. All the graphs represent 5-year running means. From Boyd and Ogilvie (1969), Ogilvie and Boyd (1976).

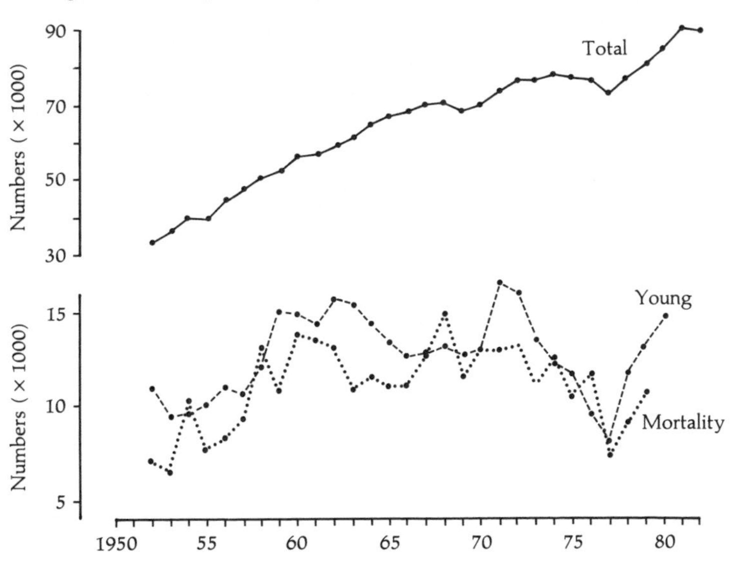

The dynamics of Icelandic Greylags and Pinkfeet have been studied in detail since the mid 1950s (77, 78, 414). In 1967 legislation was introduced to ban the sale of dead wild geese, which had an effect on the mortality rate. Both populations behaved in a similar way, showing a steady increase in numbers and a decrease in relative breeding success (see Fig 80, p.378, for Greylags). The number of juveniles, apparent mortality and population size for Pinkfeet are given in Fig 175. Although the estimates of losses are not reliable from year to year (there have been 2 "gains" in the 32 years), averaged over 5 years, as presented, they should give an accurate indication of mortality. This is assuming that counting and age ratio assessment is not becoming less accurate (unlikely since the same observer was involved throughout).

Despite the fact that the population has almost trebled in the three decades, the number of young produced has not increased, so there has been a decrease in relative performance (% young in autumn). This has been brought about in two ways: a smaller proportion of adult pairs have bred successfully and the brood size on arrival in Britain has been lower in recent years (around 2 compared with nearly 3 in the 1950s). There is evidence that the main breeding area, Thjorsarver, had reached its carrying capacity by the early 1970s (203), so the increase in the number of non-breeders could be the result of failure to find suitable habitat elsewhere. The decline in brood size is most likely due to more competition between pairs with increased nesting density, although competition in late-winter and spring staging areas is also a possibility.

The number of birds lost from the population has fluctuated, but there has been a recent downward trend, perhaps after increasing slightly in the early years. The annual death rate declined from 22% in 1960-1964, to 18% in 1965-1974 and 13% in 1975-1982. Most mortality is from shooting; the Pinkfoot has few natural hazards to overcome, so it is rather surprising that, despite the doubling of the population since the late 1950s, no more geese are being shot. This may be because of changes in shooting habits or in goose distribution. More geese have wintered in Lancashire recently and shooting pressure is probably less there than in the traditional Scottish areas, such as the Solway, where there were more Pinkfeet in autumn in the 1950s and 1960s.

There is apparently no correlation between breeding success and losses, although in many shot populations in North America the size of the bag increases markedly if the number of juveniles is high. Such a relationship, if it existed in the Pinkfoot, would undoubtedly not be apparent because counts and mortality estimates are not sufficiently accurate to indicate such differences unless these are large.

This example shows how sensitive goose populations are to changes in mortality rates. If 1,900 more Pinkfeet had been shot annually since 1950 there would have been no growth at all. An increase in the bag of 4,000 a year would have exactly reversed the trend. The population was producing the same number of young at 35% of its present size, so it could be argued that the additional birds are "surplus to requirements". They do, however, give some insurance against adversity. This is why a cessation of shooting of the much smaller Greenland Whitefront population was advocated in the late 1970s. It was calculated that a halving of the mortality rate would allow numbers to increase. Shooting at the original level could then be re-introduced without detriment.

Several of our goose population behave in a similar way. The Greylag has already been mentioned and both populations of Barnacles have shown an increase in the proportion of non-breeders with increasing numbers (410, 451). The increasing trend was quickly reversed on Islay by extending the shooting season and increasing shooting pressure, inducing a drop in the whole Greenland stock from over 30,000 to 24,000.

The European Whitefront, though its numbers have increased enormously, has not shown a decline in fecundity, and neither has the Brent, presumably because their breeding areas are much more extensive. The growth in population and annual breeding success in the Dark-bellied Brent Goose since regular counts began in 1962 are shown in Fig 176. Clearly the species either has good or bad breeding seasons, there being only one year of the 23 when the proportion of young was between 10 and 30%. The average percentage of young in good years is undiminished, despite a fivefold increase in population size. The increase was made possible by protection from shooting and the series of good breeding seasons in the early 1970s. There was a very close positive correlation between recruitment and mortality in the 1950s and 1960s, but this apparently broke down in the 1970s (418). The validity of this relationship has been questioned, however, since it could have arisen at least partly from bias in counts and age ratio estimates (440).

Clearly the Brent Goose has very erratic breeding success, but since it has been protected its numbers have declined little in poor years. Assuming that the present level of 15% mortality is maintained, this population requires 2 good years in every 5 to maintain its numbers. In the 23 years of Fig 176, 11 were good and one average; it is therefore likely that, until some density dependent mechanism begins to operate (this could be an increase in mortality through shooting as more birds are forced onto agricultural land), numbers are likely to continue to increase. The protected status of the Brent came under scrutiny in 1977 in the first Technical Meeting of the IWRB in Paris, where it was decided that stocks "... do not permit any harvesting at present" (282). Some did argue for controlled shooting above a population level of 120,000 (568), and following further population increases a level of 100,000 has been proposed (523). Shooting would aim to alleviate agricultural damage inland by reducing populations to the carrying capacity of the intertidal area. Following further discussions at an IWRB meeting in Munster, West Germany, in 1982, it was again decided that "... general shooting throughout the range is not acceptable unless it can be as closely monitored and controlled as in North America, which in Europe is not possible given lack of breeding ground surveys and the nature of current shooting legislation" (284).

We have little control over the productivity of migratory goose species, but we are in a position to change their mortality rates by altering shooting legislation or practices. There is, as yet, no evidence that shooting inflicts only compensatory mortality in any goose population, and plenty of cases where populations have increased when shooting kills have been curtailed. There is, however, evidence of density dependent effects on recruitment in several of the populations nesting in limited habitats in the North Atlantic islands and in Greenland. It seems most likely that nesting or brood rearing habitat is the limiting factor in these groups. Siberian breeders show better capabilities for expansion and their populations are likely to continue to increase as long as mortality does not rise.

Now we have protected the Whitefront in Scotland and (at least temporarily) in Ireland a major anomaly in our legislation has been removed. Stocks of Greylags, Pinkfeet and European Whitefronts are healthy enough to continue to support a harvest, in the case of the first two species at a fairly low level, not much above the present. Spitsbergen Barnacles, with their tiny population, are likely to need protection for the foreseeable future and the stage is being reached where there might have to be curtailment of shooting on the Greenland birds. There seems no biological

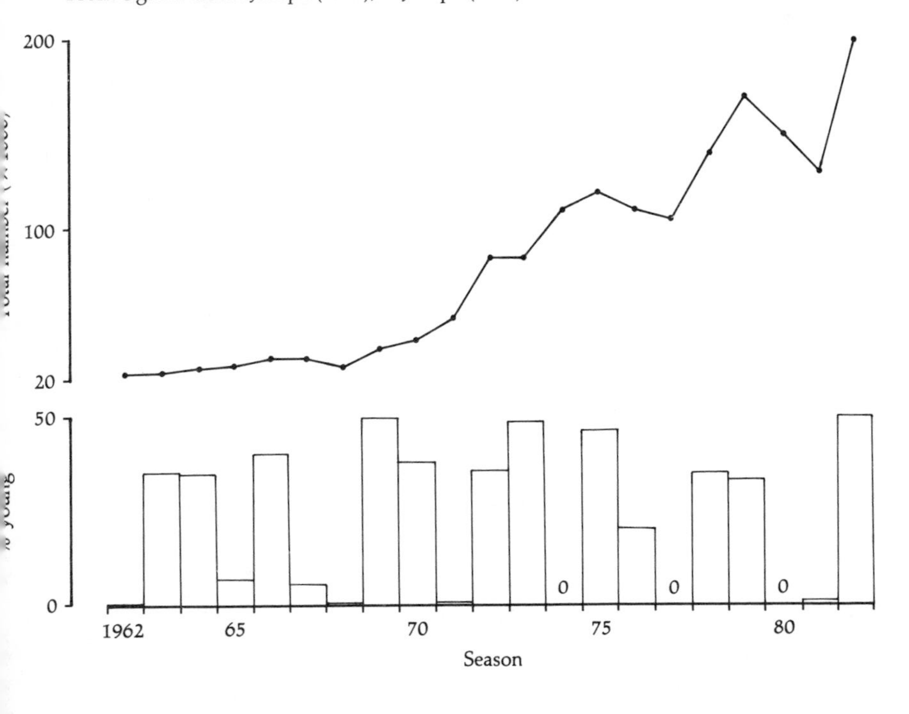

Fig. 176. The total population of Dark-bellied Brent Geese in north-west Europe and the % young in winter flocks, since regular censuses began in 1962. From Ogilvie and St Joseph (1976), St Joseph (1982).

reason for not allowing the Brent Goose to be shot on inland agricultural areas.

Ducks

Duck populations are different from those of geese and swans in that they are not as discrete, their flyways are not as well defined and there is much mixing of birds originating from different breeding areas. Ducks re-pair every winter and in groups of mixed origin males from one area follow females from another to the latter's natal area to breed (abmigration), ensuring continued genetic mixing. Thus Wigeon wintering in western Europe originate from Iceland, Scandinavia and across Siberia, and although broad groups can be delimited, these are thoroughly mixed by abmigration (160). The absence of morphological differences among Mallard from different breeding areas indicates that they are also well mixed genetically (450). In such circumstances studying population dynamics and the effect of shooting on numbers is extremely difficult, especially since seasons differ between countries.

Fig. 177. Percentage young in the wings returned by shooters for Mallard, Teal and Wigeon in Britain, 1966-1980. The dashed line indicates the overall mean. Data from Boyd *et al.* (1975), Harradine (1981) and annual reports by J.G. Harrison in *WAGBI Annual Reports*, 1973-1978.

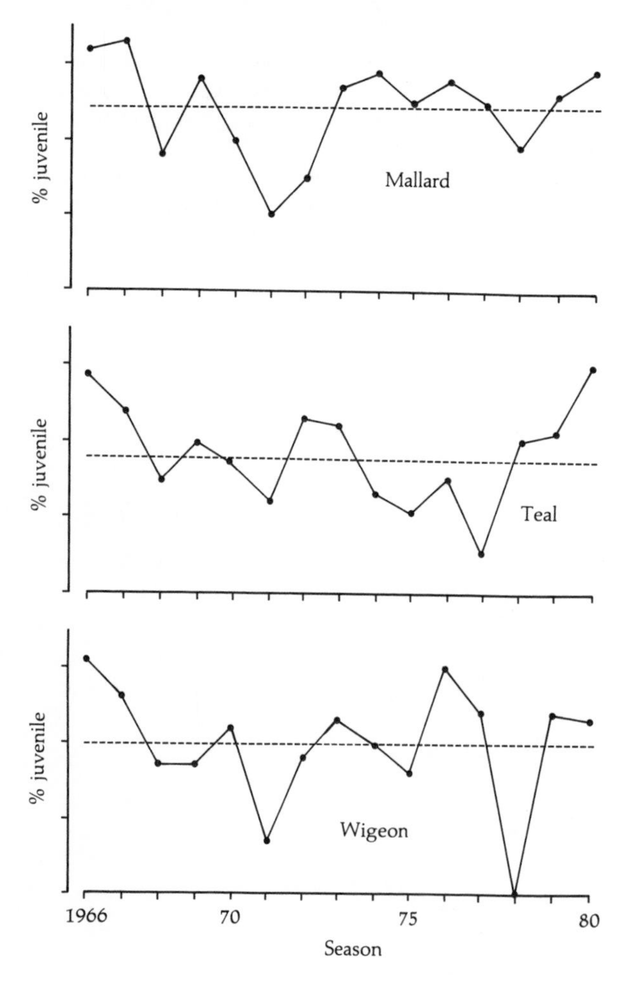

In North America an attempt is made to match the harvest of ducks closely to the characteristics of the population and to breeding success, assessed by massive surveys on the breeding grounds and by analysing the age ratio in wing returns. In Europe breeding surveys are carried out on a very limited scale and the control of shooting seasons is much less flexible than in North America, with no bag limits imposed. Duck production surveys based on an analysis of wings sent in by shooters have, however, also been carried out here for many years (p.21).

Mallard, Teal and Wigeon are the only species whose wings were returned in sufficient numbers to give reliable estimates of breeding success from year to year; the results are given in Fig 177. No species has shown a significant trend over the years, and, as might be expected from the fact that they breed at lower latitudes, Mallard success, at 64.3% (1.8 young per adult), is higher than that of Teal (58%, 1.4 y/a) and Wigeon (59.8%, 1.5 y/a), though only the difference between Mallard and Teal is significant. Success is most variable in the Wigeon (coefficient of variation 13.8% compared with Mallard 10.1% and Teal 11.7%). There is a significant positive relationship between the

Table 268. *Production and mortality rates of British ducks, assessed by wing surveys and analysis of ringing recoveries respectively. Production data from Boyd* et al. *1975, Harradine 1981 and reports by J.G.Harrison in* WAGBI Annual Reports *1973-1978. For Mallard, Teal and Wigeon, figures are averages of annual percentage for 15 years (see Fig 177). For other species the annual samples are combined and the sample size is given (n). Mortality data from Boyd 1957a, 1962, Nilsson 1971.*

Species	% Juveniles	n	% Mortality (adult)
Mallard	64.3		48
Teal	58.5		53
Wigeon	59.8		47
Gadwall	64.0	253	
Pintail	52.8	1205	48
Shoveler	55.2	609	44
Pochard	35.8	424	
Tufted Duck	48.2	794	46
Goldeneye	57.9	558	37 (Sweden)

performance of Mallard and Wigeon; the correlations between Mallard and Teal and Teal and Wigeon are positive but non-significant.

The production of the other shot species, for all years lumped together, with the calculated annual mortality rates, are given in Table 268. Because the productivity estimates are from shot birds and young ducks are more vulnerable to the gun than adults, estimates of breeding success are likely to be exaggerated. The relative performance of the various species can, however, be gauged. Most species show similar variation around the mean to the three species shown in Fig 177, but Pochard stand out as being exceptionally poor producers, rarely bringing back 1 young per adult in any year.

The mortality estimates are, unfortunately, very much out of date and the bias in age ratios makes it difficult to use the figures in any meaningful way. Those species showing the largest differential between mortality and production would be expected to have increased during the period, but this is by no means the case, Mallard and Wigeon having remained stable and Tufted Duck and Pintail having increased rather dramatically. It is obviously vital to obtain population data from the whole flyway since the position in Britain very much depends on what is happening internationally.

The most important use of duck population data is to assess the impact, if any, of shooting on quarry species. The assumption in North America, where hunting is closely regulated, has been that controlled shooting will ensure a consistently healthy breeding population in the following year. A massive analysis of available data on Mallard has shown that this assumption is not supported (9). Instead the evidence suggests that, at present levels of shooting, natural mortality completely compensates for that due to hunting, i.e. the number of ducks that survive does not depend on the number shot. This is analogous to the situation in Woodpigeons *Columba palumbus*, where in normal years a high proportion of the population dies from food shortage in mid-winter; removal of some birds by shooting in autumn is compensated for by a lower subsequent death rate (365). Obviously a point will be reached when increased shooting will have an effect, and this threshold level is estimated to be 40% for the Mallard and 10% for Canvasbacks (*Aythya valisineria*) and Redheads (*Aythya americana*) in North America (460).

The position in Europe has recently been reviewed in the light of these North American experiences (530). The harvest rate of ducks in Europe varies between 40 and 45% of the autumn population, compared with 20-25% in North America. There the harvest represents a third to a half annual mortality, whereas in Europe the shooting pressure is clearly around the threshold level for some species.

A detailed study of wildfowl counts from 35 sites in England (262), has shown that overwinter loss is density dependent in the Mallard, i.e. the more are present in September the more die or leave an area. In the Tufted Duck, however, this is not so, with high September counts leading to high numbers in the following breeding season. These results are supported by the fact that while the population size of the Mallard in Britain as a whole has remained constant over the last two decades, that of the Tufted Duck has increased. The assumption is that whereas Mallard numbers in England are now limited by habitat, Tufted Ducks have yet to reach this point.

Following consideration of the North American situation, D.A. Scott concluded: "The most logical assumption for the time being seems to be that populations of most quarry species of wildfowl are limited by the extent of utilizable habitat during mid- or late winter. If we wish to adopt a positive approach to the management of wildfowl populations, we should focus on the protection, creation and management of habitat throughout our waterfowl flyways" (530). This seems a sensible approach as far as ducks are concerned, but geese do not fit into this pattern. There is no suggestion that wintering habitat is limiting for any quarry goose species (except to the extent that farmers would like to impose a limit). It does seem that in geese, shooting mortality is additive, every species showing an increase following the curtailment of shooting.

As far as ducks are concerned, we badly need better figures on numbers, mortality and shooting kill for flyways as a whole, and with better organisation of the collection of bag statistics and continued intensive ringing, we should be better placed in the near future. In the meantime we can only continue to manage our populations by non-interference. As this discussion and the status reports in Part III show, this has been successful in maintaining or increasing populations in the last two decades. Providing that fluctuations and trends are not too violent in future and the hunting kill does not increase considerably, this should continue to be the case.

Conservation

Aims and criteria

The aims of wildfowl conservation

Wildfowl have long been appreciated by man, as a source of both food and enjoyment, either through sport or sheer appreciation of their beauty, wildness and behaviour. Given the background of pressures and possibilities described in previous chapters, the role of conservation is to ensure a future for wildfowl in a constantly changing environment. Following discussions in the late 1950s between the Nature Conservancy, the Wildfowlers' Association of Great Britain and Ireland and the Wildfowl Trust, a statement of aims was agreed in 1960. The purpose of wildfowl conservation was stated as (a) to safeguard species from the threat of extinction, and (b) to maintain existing stocks in at least their present strength and in their present distribution.

No species or subspecies was in danger of extinction in Britain in the early 1960s, and since that time changes in numbers have almost invariably been upwards or merely short-term fluctuations as a result of environmental conditions. We must reserve this responsibility as our overriding aim, however, and restate it here.

As we have seen in Part III, the aim of maintaining stocks and distribution has been more than achieved in the last 20 years, indeed several species have increased in number and spread during that time. These increases have been brought about for different reasons. In the case of Dark-bellied Brent and, to a large extent other species of geese, the increases have been a result of a lowering of shooting pressure, helped by favourable breeding seasons, and can be regarded more as a recovery than a real advance. The inland diving ducks have benefited from the increase in habitat both for breeding and wintering, whereas the introduced stocks of Canada Geese and Gadwall have continued to spread and grow, largely as a direct result of man's intervention. The sawbills have continued their increase and spread despite persecution by fishing interests, but the reason for this progress is unclear. All species undoubtedly benefited from the long run of mild winters in the 1960s and 1970s – in fact no species has apparently had severe losses due to weather since 1963.

Despite this change the maintenance of existing stocks and distribution still seems a reasonable and realistic aim, although we must see this in the context of normal fluctuations and changes in conditions. The decline in Scaup in Britain from more than 30,000 in the mid 1960s to 5,000 in the early 1980s was merely a return to the level of the early 1960s. The recent further decline, to only 2-3,000 (p.437), however, gives rise to concern, although the "missing" birds may just have been absorbed in the huge continental population. Many of our geese are in constant conflict with farmers and this has largely been the result of increases in population levels. Whereas we should not set out to reduce goose numbers, we need not be too worried if legitimate efforts to protect crops lead to a reduction of the more populous species. In the case of the alien Canada Goose, we might well argue that a reduction would be desirable to halt the dramatic progress of the species through the lowlands.

The main justification for the maintenance of stocks and distribution, as was expressed in the First Edition, is to provide human pleasure and sport without harming bird populations. Since the 1960s there has been a dramatic upsurge of interest in birdwatching, now one of the most common active leisure pursuits (p.499), and conservation, as evidenced by the growth in membership of voluntary organisations (p.547). The justification for wildfowl conservation is, therefore, greater now than ever and its momentum shows no sign of diminishing.

The conservation aims stated above make no distinction between breeding and wintering birds, although there has been an implication, at least, that except in a few cases, breeding birds are of little significance in Britain (19), and their requirements have been largely overlooked. The *Nature Conservation Review* (492) stated that "The native breeding stocks of most species are too small to contribute significantly to the winter populations, even within Britain." and "... breeding populations are widely dispersed, and relatively few sites hold enough pairs to warrant high status for this reason alone." While this is still largely true, despite the recent increase in the importance of our breeding populations, their value as a source of enjoyment merits their inclusion in an overall conservation strategy. An illustration of the public interest is the enormous concern expressed, up to ministerial level, at the large local (though only slight national) declines in the number of Mute Swans as a result of lead poisoning (219). Although consideration of breeding birds in the strategy would have little influence on such site-based surveys as the *Conservation Review*, it would certainly affect our attitude to such wide-ranging influences as river improvement works, the pollution of inland waters and water-based recreation.

We therefore believe that the aims as stated in the First Edition should be modified so that (b) reads "to maintain breeding and wintering stocks in at least ...".

The main requirement in the 1950s and 1960s was the creation of a national network of refuges to safeguard the main resorts of concentrated species, and to provide a widespread and diverse network of sites for the benefit of wildfowl as a whole. This requirement has largely been achieved, but since the major concern was to protect birds, it must be noted that most of the refuges and sanctuaries protected only the roost (particularly in the case of goose haunts) or only a small amount of feeding area. With habitat loss concentrating birds on smaller and smaller areas the need to protect feeding habitat, as well as roosts, has come to the forefront. To supplement those refuges set up under statutory arrangement or by the Nature Conservancy Council, the voluntary bodies, particularly the RSPB, have bought and managed large areas of land, and the success of this approach has been amply demonstrated (p.541).

Because most geese feed on intensively farmed land rather than on natural or semi-natural habitats, provision for their feeding needs has been more difficult, not least for financial reasons. The Wildfowl Trust and the RSPB have tackled the problem only to a limited extent and widespread conflict between farmers and geese makes the need for the establishment of protected feeding grounds urgent in some areas (438). The following account, therefore, not only discusses the need for the establishment of refuges, but also considers methods of habitat modification and management which are increasingly necessary to accommodate larger numbers of birds on smaller areas, and to allow as far as possible for the needs of agriculture and conservation on areas of interest to both.

Criteria for evaluating wildfowl sites

The protection of a network of sites remains a very important means of conserving wildfowl, which are gregarious and mobile and in many cases have special requirements fulfilled only in special habitats. Clearly the most important sites should be included in such a network, and a large amount of effort has gone into developing objective criteria for assessing sites. The main impetus for this came from the International Waterfowl Research Bureau, which promoted discussion and provided for a detailed assessment. The emphasis was on the conservation of the wetland habitat rather than wildfowl as such, although it was recognised that wildfowl were at the consuming end of the system and their state of well-being would reflect the health of plants and animals lower down the trophic scale. The number of wildfowl using a site could therefore be used as a convenient and valuable part of the objective criteria for the classification of wetlands for conservation.

A site can be considered internationally important if it makes a substantial contribution to the maintenance of a population of one or more species, and it was the determination of the size of that contribution which was the main consideration of the IWRB meeting at Heiligenhafen, West Germany, in 1974. Various numerical criteria were examined using International Wildfowl Count data from 1967 to 1973 (22). Tying a criterion to a percentage of the flyway population of a species ensures that all species receive equal emphasis, and the 1% level was considered to be the best one from the point of view of both keeping the key sites to a reasonable number and at the same time ensuring that they do safeguard a significant part of the flyway stock. This criterion has now become an integral part of our classification system for wetlands.

The Heiligenhafen conference presented a number of criteria for site classification and these were re-examined at another IWRB conference at Cagliari, Italy, in 1980 (283). The criteria, in their slightly revised form, are as follows:

1. Quantitative criteria for identifying wetlands of importance to waterfowl.
   A wetland should be considered internationally important if it:
   a regularly supports either 10,000 ducks, geese and swans; or 10,000 coots; or 20,000 waders; or
   b regularly supports 1% of the individuals in a

population of one species or subspecies of wildfowl; or

c regularly supports 1% of the breeding pairs in a population of one species or subspecies of waterfowl.

2. General criteria for identifying wetlands of importance to plants or animals.
A wetland should be considered internationally important if it:

a supports an appreciable number of rare, vulnerable or endangered species or subspecies of plant or animal; or

b is of special value for maintaining the genetic and ecological diversity of a region because of the quality and peculiarities of its flora and fauna; or

c is of special value as the habitat of plants or animals at a critical stage of their biological cycles; or

d is of special value for its endemic plant or animal species or communities.

3. Criteria for assessing the value of representative or unique wetlands.
A wetland should be considered internationally important if it is a particularly good example of a specific type of wetland characterisitic of its region.

The criteria under 2 and 3 are essentially subjective, based as they are on such terms as "appreciable number", "special value" and "good example", and the numerical criteria provide the only objective and most valuable classification guide. The criteria can also be adapted to the national or even regional scale, although clearly the use of such guides is not realistic for rare species or those at the edges of their range, such as Smew in Britain, where only a few individuals would be considered nationally important, or for such populations as the Spitsbergen Brent, which number only about 3,000 birds. In such cases a lower limit of importance has generally been set, of 100 on the international scale (529) and 50 on the national (524).

The value and objectivity of the numerical criteria depend on the quality of the data on which they are based, and this can be said to be the greatest weakness of the system. There is, then, a constant need for continued monitoring and reassessment. Following this review and with the computerisation of the British data this is much easier and quicker than has hitherto been possible for this country. Coordinating and analysing the international data is much more difficult, but, through the IWRB, increasing efforts are being made to bring the international count analyses up to date to provide a basis for national conservatic strategies. Throughout Part III the absence of rece analyses of international data, particularly of trend has clearly been a drawback in interpreting the Britis counts, but it is hoped that this information will soc be available.

Wildfowl and the law

The 1954 Protection of Birds Act provided tl first comprehensive legal treatment of the status of o wild birds in relation to shooting and other forms exploitation. It started out with the position that it an offence to kill or destroy the nests and eggs of ar wild bird and then listed exceptions to this rule. Tl Act was amended and updated by the Protection Birds Act (1967) and replaced by the Wildlife an Countryside Act (1981). This section outlines briefl the present position.

Shooting practices

Legal restrictions on the methods of shootin birds are aimed at avoiding excessive kills and minimising unretrieved losses, largely through cri pling. These not only result in unnecessary sufferin to birds but are also a waste of the resource. Most the practices that have now been made illegal hav long been voluntarily outlawed by responsible sport men in Britain, and our legislation has in many case provided an example for other countries. The interna tional position is closely monitored (308, 309).

The 1954 Act outlawed the use of nets, bird lim and baits to kill birds, and gas was added to these i 1967. The use of mechanically-propelled vehicles i immediate pursuit and of artificial light for nigl hunting were banned, and the internal diameter gun barrels was restricted to 1¾ inches (4.5cm) t outlaw the larger punt guns. The 1981 Act added t these restrictions by prohibiting the use of bows an crossbows, automatic or semi-automatic weapon night-vision devices and chemical wetting agents. Th use of sound recordings to attract birds was prohibite and a few minor loopholes in the previous Acts wer tightened up. These provisions brought Britain int line with many European countries which had alread outlawed most of these practices.

The law now protects wild birds from mo unethical hunting methods and those which ar biologically damaging, but the case of night shootin is still a contentious issue. In a discussion in the IWR Board Meeting in Hungary in 1981, most agreed tha biologically speaking, night shooting should be proh bited because most wildfowl feed at night, when the should not be unduly disturbed, and that this practic is wasteful since many ducks are lost in the darkness. Shooting from half an hour after sunset until half an hour before sunrise has been prohibited in North America since 1916 and only four countries – Bulgaria, France (some parts), Hungary (though very little practised) and the United Kingdom – still permit it. Attempts to have night shooting banned during the passage of the 1981 Act were not successful. The wildfowlers argued that night shooting was not widely practised, was carried out by responsible sportsmen who were conscious of the need for minimising crippling losses, and should be retained for reasons of tradition.

Species protection and seasons

The 1981 Act has three schedules concerned with species status. Species on Schedule 1 are afforded a high degree of protection by the imposition of special penalties for their killing or disturbance, including that of their nests, eggs and young. This schedule is aimed at protecting those birds that are scarce or threatened throughout the year, especially as breeders including:

At all times	During the close season
Bewick's Swan	Greylag Goose (north-west Scotland only)
Whooper Swan	Pintail
Garganey	Goldeneye
Scaup	
Long-tailed Duck	
Common Scoter	
Velvet Scoter	

Schedule 2 Part I lists those species which can be taken in the open season. This includes only those species for which there is a demand for shooting and whose populations are deemed to be able to withstand hunting, namely, for wildfowl:

Pink-footed Goose
White-fronted Goose (England and Wales)
Greylag Goose
Canada Goose
Wigeon
Gadwall
Teal
Mallard
Pintail
Shoveler
Pochard
Tufted Duck
Goldeneye

Schedule 3 Part III concerns those birds which may be sold when dead between lst September and 28th February, the wildfowl being:

Wigeon
Teal
Mallard
Pintail
Shoveler
Pochard
Tufted Duck

The major changes since 1954 are that the three sea ducks on Schedule 1 are now protected by special penalties at all times and that the Bean Goose is now protected. The restriction of shooting of the Whitefront to England and Wales effectively protects the Greenland Whitefront – an important conservation step and one not contested by shooting interests. The Barnacle Goose, which was protected in 1954, but which became legal quarry again (in the western islands of Scotland in December and January) in 1955, extended by order in 1976 to September – January because of agricultural damage problems, also became completely protected. The two sawbills, which used to be on the "pest" list in Scotland, have now been removed.

Exceptions and licensing

As in the previous Acts, it is not an offence to kill birds, other than those in Schedule 1, if it can be proved that the action was necessary to protect public health, prevent the spread of disease or prevent serious damage to crops or fisheries. A bird may also be killed for humane reasons if it is maimed beyond recovery, or may be taken to be treated and subsequently released. The catching of geese and ducks for ringing in nets and traps, excluding duck decoys in use before 1954, is not allowed without a licence. All other exceptions are subject to licensing by the NCC, the Department of the Environment, or the Ministry of Agriculture, Fisheries and Food (MAFF) (Department of Agriculture and Fisheries for Scotland (DAFS) in Scotland).

Licences can be general or closely controlled; for example a general licence has been granted for the collection of Mallard eggs for hatching and rearing, up to 31st March in England and Wales and 10th April in Scotland, but any collection outside this period or of eggs of any other species must be individually licensed. Licences may be granted to farmers in order to prevent damage to crops, either by protected species or by quarry species in the close season. These licences are granted by MAFF in England and Wales and by DAFS in Scotland, after they have consulted the NCC on the advisability of granting licences and their terms. Unfortunately the generality of licences and the conditions attached to them are left to the granting authorities, and in the case of licences for Brent Geese in England and Barnacle Geese on Islay, the powers were applied in quite different ways by MAFF and DAFS in 1982-83 – the first year of operation. Clearly, with such flexible legislation, it will take some years to perfect a system of licensing which both protects the farmer and conforms to the spirit of the bird protection Acts.

Other provisions relate to the birds which may be sold alive if bred in captivity and to the conditions in which captive birds are kept. For the first time the introduction of non-native birds into the wild is controlled by legislation, and this includes especially the Canada and Egyptian Geese, and Mandarin, Carolina and Ruddy Ducks, which are already established here. It is an offence to release any bird which "... is not ordinarily resident and is not a regular visitor to Great Britain ...", but the word "regular" is clearly open to interpretation. It is a defence to argue that the person made all possible attempts to prevent the release, and this is a loophole which may mean that this part of the Act is not effective against continued introductions.

The imposition of close seasons for shooting is aimed primarily at protecting the breeding stock, both from disturbance during nesting and rearing, and while on migration to the breeding area. Spring shooting is considered particularly undesirable because it attacks the population after most natural factors have taken their toll and while the potential breeders are preparing for nesting. The EEC Directive on the Conservation of Birds, in force since April 1981, prohibits shooting during spring migration, which at least in southern and western Europe begins in February. Many countries, through the IWRB, have been pressing for an end to shooting throughout north-west Europe on 31st January, and this was attempted for Britain during the discussion of the 1981 Act. The attempt failed, however, and the close season for all quarry species of wildfowl remained from 21st February to 31st August below high water mark of ordinary spring tides. Inland it does begin on 1st February.

Bans and restraints

In the absence of legislation, responsible wildfowlers do in some cases institute voluntary restraint to protect certain local populations. The local wildfowling club has, for many years, voluntarily banned shooting on the small flock of Greenland Whitefront on the Dyfi Estuary, Dyfed. Responsible wildfowlers, recognising that such restraints are in their own long-term interests, do cooperate.

The 1967 Protection of Birds Act introduced the provision for a statutory ban on shooting which could be imposed by the Secretary of State for the Environment in the whole or part of Great Britain during the open season in consequence of severe weather. This provision was included in the 1981 Act, but in neither case were the criteria for severe weather specified. A working party was established in 1979 to draw up objective criteria for the recommendation to wildfowlers for restraint and to the Secretary of State to implement a ban. It was agreed that the state of ground recorded by the Meteorological Office would be used, and thirteen weather stations, spread throughout coastal areas of Britain, were selected. If the ground was frozen or snow-covered for seven days at over half of these stations, or after ten days interspersed with one or two days of thaw, a call for restraint would be made. After fourteen days of frosty weather the statutory ban would be invoked. Three days of thaw would effectively break the period of severe weather. Since the order has to be renewed after fourteen days, the ban would then cease to be in operation unless the cold weather persisted. Since birdwatchers and ringers could be almost as disturbing as wildfowlers, their activities would also be limited by the suspension of ringing licences and by a code of conduct (38).

The system was first put to the test during the severe weather of 1981-82, and while there were shortcomings in the implementation, particularly the amount of publicity given to the ban, it was agreed that the criteria worked well. Since for a time severe conditions were not countrywide there were calls for the imposition and lifting of bans on a local scale. These were resisted, however, on the basis that birds could move into unfrozen areas if local conditions became severe, and that if only a small part of the country was ice-free, birds which had escaped to there should be protected. The system would obviously be more effective if operated on a European scale. For example, many birds leave Britain for France, Spain and Italy, where they are heavily shot, if there is a hard winter here, and they could only be protected by ban in those countries. The IWRB is working on such arrangements, but, although such protection on a European scale has improved enormously over recent years, it is likely to be some time before it is fully achieved.

The protection of habitat

At the IWRB meeting at Ramsar, Iran, in 1971, there was a significant step forward in international wetland and wildfowl conservation with the agreement of the text of a Convention on Wetlands of International Importance Especially as Waterfowl Habitat, thereafter known as the Ramsar Convention. Although not placing a binding obligation on the contracting parties the Convention placed a duty on them to preserve the ecological character of wetlands of international importance designated as Ramsar Sites. By April 1985, 38 countries were parties to the Convention and 19 million hectares of wetland had been designated under it. At the time of writing, the United Kingdom had designated 19 areas, one of which – Loughs Neagh and Beg – is in Northern Ireland, and for a further 6 (all in Britain) the process was near to completion. These, together with the other 96 British sites eligible for designation, are listed in Table 270 (p.538).

A measure of habitat and site protection is also granted to areas designated as Special Protection Areas by the EEC, and a list of all potential sites is being drawn up.

The 1981 Wildlife and Countryside Act gives some increased protection to habitat scheduled by the NCC as Sites of Special Scientific Interest. Owners of such land are required to notify the NCC of their intention of carrying out any operation which would be deleterious to the specified conservation interest. The NCC can object to such operations and can take effective action by influencing grant giving bodies if such operations qualify for grants, or by entering into a management agreement whereby an owner receives money for carrying out or not carrying out certain operations on the land. The Secretary of State was also given the power to apply an order prohibiting the owner from carrying out damaging operations on land of very high conservation value. The effectiveness of this habitat legislation has been questioned, and some unfortunate loopholes exposed, but its true worth will only become clear after some time has elapsed and a number of individual cases have been argued and resolved.

Other conservation methods

In North America it has long been the practice not only to vary the short open seasons from year to year, but also to impose stringent restrictions on hunters, which can be rapidly altered regularly and operated on a very local scale. The most commonly used method is the bag limit, which restricts the number of individuals of any species that can be shot on a daily, weekly, monthly or seasonal basis. These limits can be varied according to the size and productivity of the wildfowl population from season to season; refinements have been implemented, involving a points system which allows more of certain commoner species to be shot than others. Limits on the number of shots allowed have been imposed on some marshes; these are aimed at reducing crippling losses and the input of lead pellets onto feeding areas.

In Europe, no such refined methods of limiting harvests have yet been tried; species have been managed only by protection or changing the shooting season. A recent review (530) concluded that there is insufficient evidence from the North American experience that such restrictive regulations do produce the desired close control over population levels and harvest rates. In fact, the North Americans have been using a fixed season regime for five years without any obvious effect on duck populations. However, in general duck shooting takes a much higher toll in Europe and it may not be entirely legitimate to extrapolate from North American experience to the British scene.

A measure taken in Britain to limit the harvest rate was the restriction on the sale of dead wild geese imposed in the 1967 Act. This curbed market shooting of geese, and did result in a reduction in mortality and increase in the populations of Greylags and Pinkfeet. The restriction on the species of duck which may be sold, under the 1981 Act, may also benefit the species removed from the market place. The banning of various methods of mass destruction also has some impact on the reduction of the kill.

The importance of refuges

The Wildfowl Conservation Committee, established in the late 1950s to bring conservationists and shooters around the same table, examined the needs of conservation and the means of fulfilling them. A useful consensus was established under the chairmanship of E.M. Nicholson (then Director General of the Nature Conservancy) and in 1960 a policy was outlined which included provision for refuges for wildfowl with the following aims:

> "The Role of Refuges in Wildfowl Conservation.
> (a) In the case of a species with a limited distribution in this country Refuges may be needed to safeguard some or all of its main

resorts. Should the species be readily identifiable, statutory protection may be a more suitable measure; if this is not practicable or adequate the necessary Refuges should be given priority.

(b) More usually a Refuge will serve as a strongpoint for a number of species to ensure their status in the district. Such Refuges will form wildfowl reservoirs, increasing the stock to the advantage of sportsmen and naturalists alike, and will also provide centres from which some of the less common species may be encouraged to extend their range.

(c) Refuges may likewise be used to mitigate the local effects of overshooting or unsporting practice, while still affording controlled shooting to responsible wildfowlers."

Refuges were seen either as National, those whose existence was crucial to the national population; or Regional, being set up to maintain local numbers and distribution. The commitment to international conservation was also recognised and refuges would play an important part in our contribution. Many Wildfowl Refuges were created following the statement of aims but, since other types of reserves (NCC, voluntary body, etc.) are common nowadays, Wildfowl Refuges are no longer being created.

Wildfowl require a safe roost and feeding ground, and both of these should ideally be met in a refuge system. This has not always been the case, especially for geese, some swans and dabbling ducks, which rely on marshes and agricultural land which may be some distance from the roost. When wildfowl concentrations have increased on a roost following the creation of a refuge, this has sometimes caused problems with local farmers, as is discussed more fully on p.501.

The creation of a refuge could be the only way of protecting a scarce species which is not easily distinguishable from one that is legal quarry. For example, there were objections to the protection of the Bean Goose (now achieved under the 1981 Act) because of possible confusion with the Pinkfoot, but since the former species occurs in numbers on only one or two sites, the creation of official refuges there would avoid the necessity for legal measures. Illegal shooting is, unfortunately, still a problem in some cases, and the provision of a refuge which holds protected birds during the shooting season may be essential. For example, the Solway Barnacle Goose population was fluctuating at around 3-4,000 in the 1960s despite good breeding success. The extension and improvement of the feeding refuge in the early 1970s reduced mortality from illegal shooting and allowed numbers to increase to 7-10,000 (p.387).

Refuge creation is more effective for those species which concentrate on few sites, such as geese and the migratory swans. Thus 60% of European Whitefronts in England are on a single site, while the protection of only 400 or 500ha on the Solway and Islay would safeguard 80% of British Barnacle Geese. Pinkfeet and Greylags could be safeguarded by the protection of a few roosts, but the safeguarding of their farmland feeding grounds would be more difficult.

Ducks, which have a dispersed distribution, cannot be safeguarded by the provision of a few refuges. The distributional patterns of three dabbling ducks (Fig 178a) and three diving ducks (Fig 178b) illustrate this point. The Pintail is the most concentrated dabbling duck, with more than 50% of those counted on only seven sites and three-quarters on only 20. To include more than half the Wigeon, however, 64 sites would have to be protected, and 235 places have 300 birds or more. Mallard are even more dispersed, occurring on the vast majority of sites, and it is quite unrealistic to protect them by the establishment of refuges. Similarly Scaup are the most concentrated diving duck, whereas Tufted Ducks are the most dispersed. Only 4 sites are needed to include more than half the Scaup, but 78 for Pochard and 165 for Tufted Ducks.

Those species which have specialised requirements and a concentrated distribution are clearly in

Fig. 178. a) The proportion of counted populations of Pintail (dashed line), Wigeon (solid line) and Mallard (dotted line) which are found on a given number of sites. b) The proportion of Scaup (dashed line), Pochard (solid line) and Tufted Ducks (dotted line) found on a given number of sites.
For example, the six most important Wigeon sites hold 20% of the total British count, the top 13 sites have 28%, etc. Classification is based on the mean count for the last 5 years available, or less where coverage is limited.

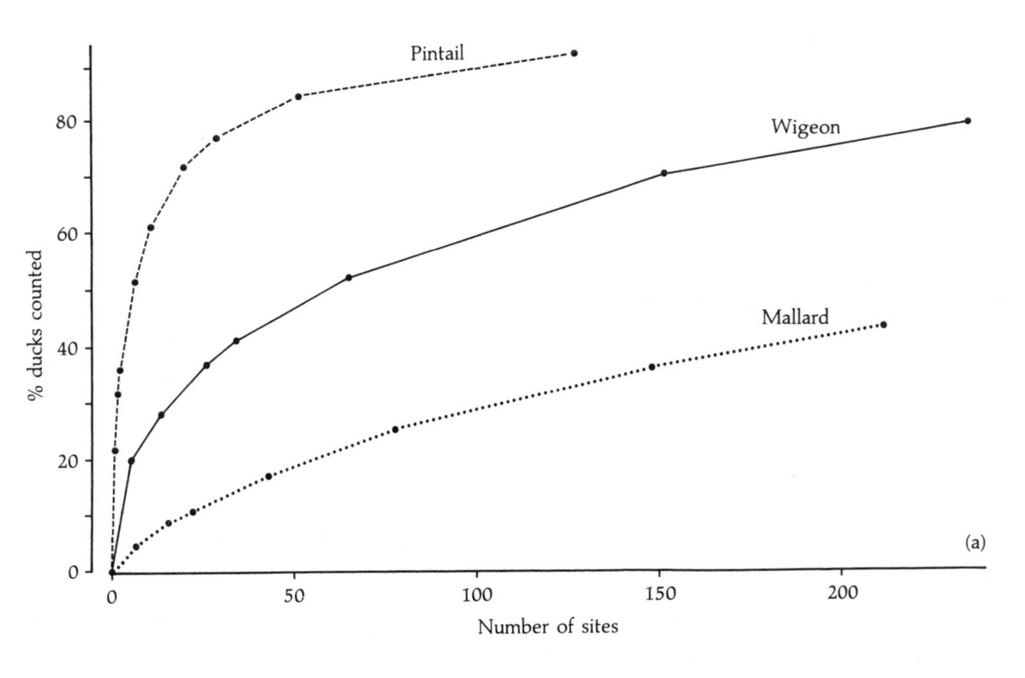

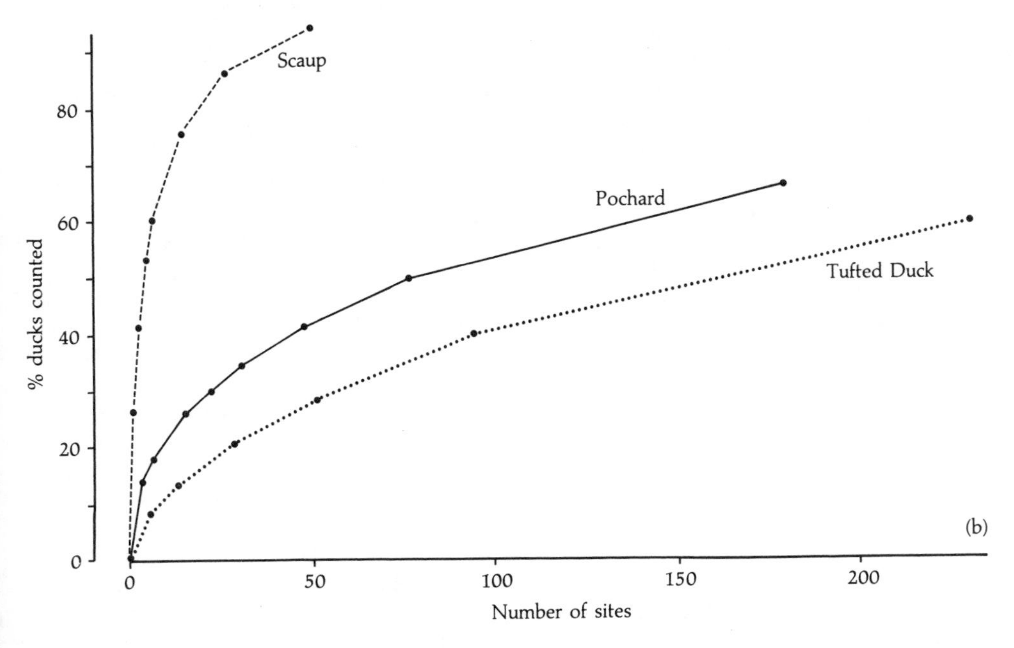

greatest need of protection, since the loss of a very few sites could be crucial to the whole population. Dispersed species are safeguarded by their adaptability and the variety of habitats which they can exploit. Thus to put the Mallard, Tufted Duck and Teal in danger a very large number of sites would have to be lost. These species occur, sometimes in large numbers, on refuges established for other species and this provides them with additional security. It is important to note that, though a refuge may be established to protect a single species, it is also of value to all the others which visit the area.

The criteria for identifying key sites for reserves rely on the number of birds that have used those places in the recent past. Site characteristics change naturally, as do the habitats of the birds, and constant review is necessary to keep the classification up to date. Some sites may have considerable potential for wildfowl which is not realised because of some limiting factor, often human induced. This potential, as well as the actual numbers of birds they hold, must be fully recognised in site assessments. The management of the Ouse Washes for conservation, for example, greatly increased its importance as a wildfowl site (p.148), and doubtless protection and management of other areas could make them into nationally or internationally important haunts.

The present position

The concept of establishing a network of refuges to protect wildfowl populations and their distribution was described in the First Edition and the objectives of that system have largely been achieved. Many of those refuges were important roosts, but, as described earlier (p.532), the safeguarding of feeding habitat is equally important, as recognised by the key role given to habitat by the Ramsar Convention. In the last two decades, as well as increases in the number of reserves established through the Nature Conservancy Council, a number of voluntary organisations, both national and local, have created wetland reserves, many of which are important for wildfowl. In this section we consider the proportion of British wildfowl housed on reserves of various kinds, as well as examining the present situation as far as the nationally and internationally important sites are concerned.

Seven categories of protection were distinguished for the analysis, listed below in decreasing order of the degree of protection afforded:

1. National Nature Reserve.
2. Owned, leased or management agreement by a national voluntary conservation body (chiefly RSPB and Wildfowl Trust).
3. Statutory wildfowl refuge, Local Nature Reserve or area covered by a Sanctuary Order (put here because protection not often afforded to feeding grounds).
4. Owned or managed by a local voluntary body (e.g. County Naturalists' Trusts).
5. Owned or managed by a body, local or national, whose main aim is not the conservation of wildlife (Forestry Commission, National Trust, Water Authority, County Council, etc.).
6. Site of Special Scientific Interest, not included in any of the above categories (many sites included in categories 2-5, and all in category 1, are also SSSIs).
7. No protection.

Of the 3,915 sites covered by counts at some time between 1960 and 1982, 662 (17%) now have some degree of protection, though they did not necessarily have this at the beginning of the period. The distribution of these is shown in Fig 179. The number of sites in each category, and the percentages of the various species counted in 1981-82 which were on protected sites, are given in Table 269. Generally a site was included in a category if a substantial part of it was protected, since, in many cases, the birds are concentrated in the conserved zones, at least during the shooting season. There are cases, however, where this does not apply and where the table is biased in favour of protected areas. For example, Langstone Harbour, which has a 554ha RSPB reserve but 1600ha of intertidal habitat, is counted as a single site and all its wildfowl are attributed to category 2. With these reservations, however, for most species the analysis gives a reasonable idea of the extent of habitat protection. Also given in the table are the results of a similar analysis of the situation in 1961-1965 (476). This shows clearly the increase in the proportion of all species on reserves, but it should be borne in mind that the protection given to a large proportion of some species is only that of SSSI status.

It is very encouraging that the proportions of wildfowl on protected areas are so high at present, although only 22% of sites counted in 1981-82 are in any protected category. Including the geese (see below), 20 out of the 24 species have more than 50% of the counted total on protected sites, and only the Canada Goose falls below 40%. As might be expected, the more dispersed species, such as Mute Swan, Canada Goose, Mallard, Pochard, Tufted Duck and Goldeneye are in smaller proportions on reserves, since these are inevitably geared to high concentrations of birds and in many cases tend to create concentrations of the gregarious species. The Red-

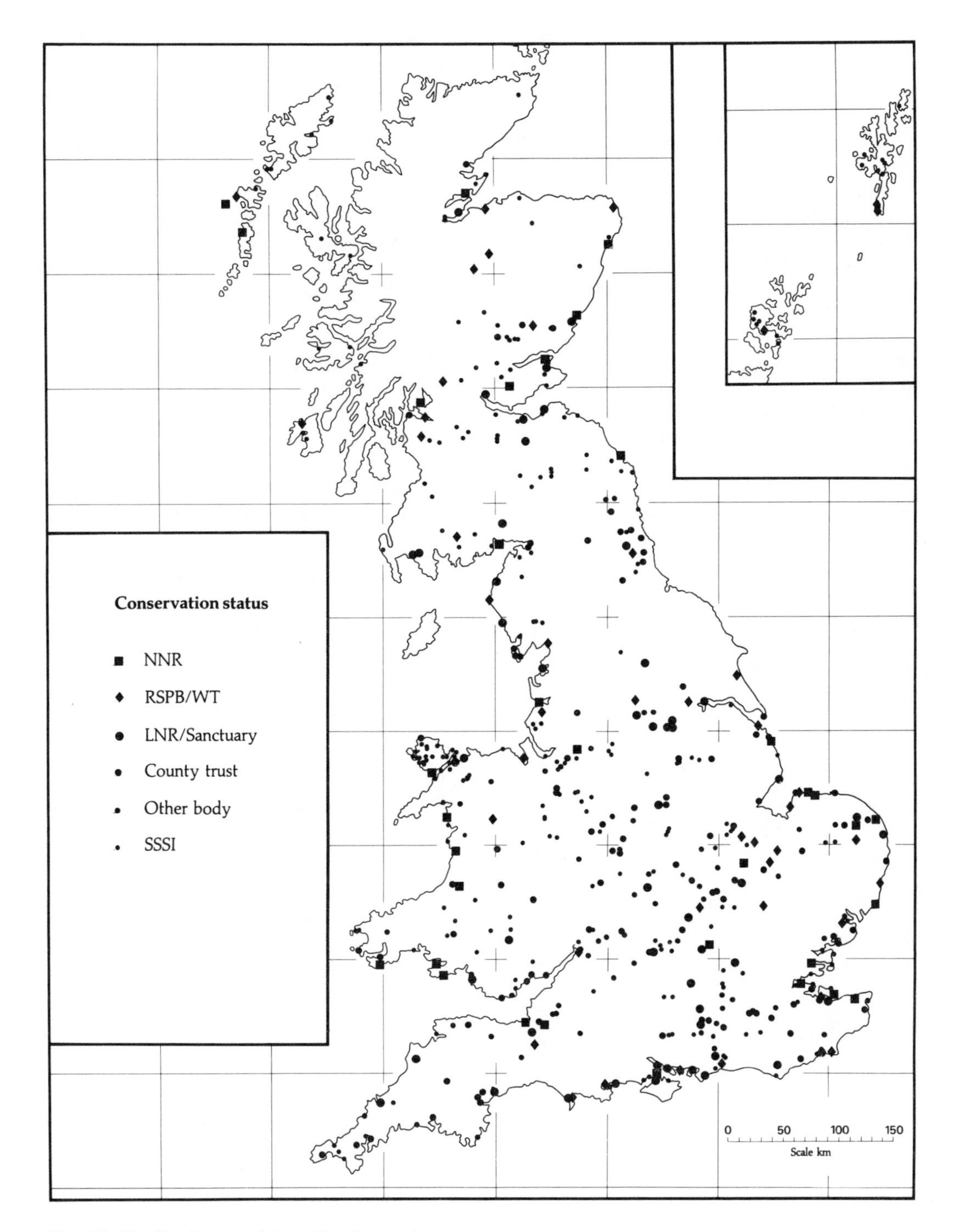

Fig. 179. The distribution of sites offered some degree of protection and covered by Wildfowl Counts, up to 1982-83.

breasted Merganser would have a much lower proportion of its number on reserves except for the fact that Poole Harbour, one of its strongholds, is classified as an RSPB site, although most of the Mergansers are outside the Arne reserve. The high values for Teal, Pintail and Goosander are slightly misleading since such a high proportion of birds are on SSSIs – the lowest degree of protection. In the case of Teal and Pintail this is because of the importance of the Mersey Estuary and, for Goosanders, the Beauly Firth. There is some urgency to have these areas designated Ramsar Sites, but no progress has yet been made (see below).

The other interesting aspect of the table is the major contribution made by the voluntary bodies, particularly the national ones, chiefly the RSPB, to wildfowl protection. Their sites hold double the numbers supported on National Nature Reserves despite the fact that some NNRs include land owned, leased or managed by the voluntary bodies and their birds are counted as in the NNR. Local organisations also play their part, especially in the case of the less numerous species such as Shoveler, Gadwall and Ruddy Duck. The contribution of the organisations not primarily concerned with wildlife conservation, chiefly the National Trust, is also considerable and is, in fact, underestimated in the table because portions of the north Norfolk coast owned by the National Trust were not counted in 1981-82. These areas are very important for Brent Geese, so the zero in the Brent column is wholly misleading. Thus, despite the fact that we must assume that the majority of uncounted wildfowl are on unprotected sites, the national picture is encouraging.

Of the species not listed, all major roosts of Greylag and Pink-footed Geese are largely protected, at least as SSSIs, and 80% of Greylags and 87% of Pinkfeet are found on internationally important sites. Only a tiny proportion of the Greylag's feeding grounds and a rather small part of the Pinkfoot's are protected. European Whitefronts are largely on reserves and their largest concentration at the New

Table 269. *The percentage of the peak count (peak months in brackets) of wildfowl species on sites with some degree of protection in 1981-82. The migratory geese, other than Brent, are excluded because not many are covered by the Wildfowl Counts; sea ducks are also omitted, but are treated separately (see text). For explanation of conservation status categories see text. Also given are the results of an analysis in the early 1960s.*

		Conservation status							
		1	2	3	4	5	6	All*	1961-65+
No. of sites		26	28	28	47	38	151	318	% reserves
Mute Swan	(O)	3.7	8.1	13.4	3.1	4.0	17.4	49.5	10
Bewick's Swan	(F)	9.9	68.0	2.2	0.6	1.3	1.5	83.4	3
Whooper Swan	(N)	9.1	30.1	0.3	1.2	0.7	16.8	58.2	18
Canada Goose	(S)	0.9	6.0	2.1	8.8	8.2	8.3	34.2	-
Brent Goose	(D)	15.4	14.0	8.1	5.0	0.0	27.9	70.5	10
Shelduck	(J)	10.8	24.8	3.7	1.2	0.8	28.9	70.1	5
Wigeon	(J)	11.6	25.8	8.7	6.4	3.3	14.1	70.0	11
Gadwall	(N)	7.7	20.2	5.5	15.0	12.5	6.4	67.3	6
Teal	(D)	9.1	9.8	4.0	3.4	2.8	45.4	74.5	6
Mallard	(J)	9.8	13.3	5.3	6.6	8.0	13.7	56.8	5
Pintail	(N)	8.0	25.6	1.2	0.1	0.2	61.5	96.7	7
Shoveler	(N)	6.0	20.0	5.3	12.7	12.2	9.1	65.2	18
Pochard	(N)	2.0	12.0	4.4	5.5	8.5	10.3	42.5	21
Tufted Duck	(S)	12.4	6.4	1.6	7.1	12.5	11.8	51.8	20
Goldeneye	(F)	3.1	10.6	3.4	5.1	9.0	17.4	48.6	4
R-b Merganser	(N)	2.0	27.1	8.5	5.4	0.0	11.5	54.6	-
Goosander	(D)	2.1	0.9	1.2	1.3	2.4	75.8	83.7	-
Ruddy Duck	(N)	0.6	0.2	0.1	14.1	21.7	43.2	79.9	-
Total wildfowl	(J)	9.8	21.4	5.7	4.9	3.2	17.0	61.9	-

Note: All the birds at a site are included under the highest category of protection afforded, e.g. if a site is owned by a voluntary body and is an NNR its wildfowl are attributed to the NNR.
* The total number of sites counted in 1981-82 was 1439, so 1121 had no protection.
+ From Pollard (1967).

Grounds is on a proposed Ramsar Site. Greenland Whitefronts have most of their roosts and major feeding grounds listed as SSSIs, and the largest flock of Bean Geese is adjacent to an RSPB reserve. On the Solway, Barnacle Geese spend most of their time on effectively protected areas, but on Islay, though the main roosts and feeding grounds are scheduled SSSIs, and the whole island is a proposed Ramsar Site, this has not prevented the Department of Agriculture and Fisheries for Scotland from granting landowners and tenants licences to shoot Barnacle Geese on those feeding grounds. An agreement to prevent this in future is being negotiated by the NCC; the RSPB reserve purchased in 1983 will also make a considerable contribution. Elsewhere in the range of the Greenland Barnacles, the internationally important islands are proposed Ramsar Sites (see below).

Those ducks which remain at sea or on estuaries outside the intertidal zone rely on habitat which is unprotected by statutory reserves. However, the Ramsar Convention includes in its definition of a wetland saltwater areas below 6m in depth, and such areas include the vast majority of sea duck feeding grounds, which can thus be protected from changes in character by Ramsar designation. This would include industrial and oil development, but accidental incidents of oil pollution would be very difficult to guard against in practice.

A preliminary list of wetlands of international importance has recently been drawn up (529) and, largely based on this, those sites in the United Kingdom which qualified were put forward by the NCC to the Department of the Environment in October 1982 as a complete list of sites eligible to be designated under the Ramsar Convention. The list for Great Britain, given in Table 270, includes 123 sites, 90 of which are included because they hold internationally important concentrations of wildfowl and/or waders, the remainder being eligible for other reasons. Of the wildfowl sites, 7 islands in north-west Scotland are not strictly wetlands, but are included because they hold internationally important numbers of Barnacle Geese (300 or more). A further 21 sites, all but one in Scotland, are important only as goose roosts, or at least have very little interest otherwise.

Of these 123 sites, only 18, together with the Lough Neagh/Beg area in Northern Ireland, have been formally designated so far; consultations between the Department of the Environment and owners have been completed on a further 6 sites, which will be designated soon. Although most of the remainder are scheduled as SSSIs (all will be by definition in the rescheduling process currently taking place), this gives them much less protection than they would have under Ramsar. The fear is that before they are designated some of their characteristics may already have been destroyed.

The management of wetlands

As is evident from the preceding section, there is continuous pressure on wetland habitats on this small and populous island. Virtually no wetlands remain which have not been influenced by man, since wetlands continue to be eroded and disappear. Those semi-natural and artificial sites that do exist must support larger and denser populations than they do now if we are to retain the numbers and diversity of our wildfowl. Wetland management is concerned with the planning and sympathetic design of new habitats as well as the modification of existing ones. In some cases this involves re-creation of habitats or types of wetland degraded or destroyed by man in the past, but in others both the techniques used and the resulting habitat are entirely artificial.

The last two decades have seen an upsurge of interest and enthusiasm for habitat manipulation, and in 1972 the IWRB established a Wetland Management Research Group to coordinate activities. The results of this came to fruition in 1982 with the publication of an extensive manual (531). The following account examines some of the techniques relevant to the British situation and illustrates their potential by describing case histories.

The potential for effective management is limited in coastal and riverine habitats. On intertidal areas the influence of the tide is overwhelming, though the control of stock grazing on saltmarshes is a useful management technique for geese (522). Non-tidal rivers are not very important for wildfowl in Britain, although they do support sizeable populations of Mallard, Mute Swans and Tufted Ducks. Conditions on rivers are largely dependent on the requirements of water authorities for water supply, waste disposal, water power, flood relief and land drainage. Management here usually means achieving a compromise on canalisation, tree removal, drainage and vegetation control, between water authority interests and those of fishing and navigation and the needs of wildlife, which are frequently conflicting. A start has been made through increasing contact between river authority engineers and conservation interests (617). Habitat management is most effective in enclosed inland waters and marshes; the techniques and examples described below have been developed and used in these situations.

Table 270. *Sites eligible to be designated under the Ramsar Convention in Great Britain, as recommended by the NCC to the Department of the Environment in October 1982. The NCR column indicates the grade of the site as given in the* Nature Conservation Review *(Ratcliffe 1977).*

(a) Sites of international importance as habitats for waterfowl (Cagliari Criterion 1).

Site	Designation	Area(ha)	NCR	Other	Comments
ENGLAND					
Abberton Reservoir	15.6.81	1228	1	SSSI	Sanctuary area 1188ha
Alt Estuary	-	500	-	SSSI	
Blackwater flats and marshes	-	10000	1	SSSI	Includes NNR (1031ha), 2 LNRs and RSPB reserve (459ha)
Bridgwater Bay (incl. Fenning and Stert I)	5.1.76	4250	1	NNR	NNR area 2559ha
Chichester and Langstone Harbours	-	5528	1	SSSI	Includes RSPB reserve (554ha) and 2 LNRs
Dee Estuary	In progress	12611	1	SSSI	Includes RSPB reserve (2040ha), LNR (49ha)
Derwent Ings (Wheldrake to Bubwith)	In progress	783	1	SSSI	
Duddon Sands and Millom	-	3817	2	SSSI	
Exe Estuary	-	1599	2	SSSI	Includes LNR (216ha)
Foulness and Maplin Sands	-	13600	1	SSSI	
Hamford Water	-	2396	1	SSSI	Includes NNR (688ha)
Holburn Moss	In progress	500	-	SSSI	Goose roost
Humber flats and marshes	-	12600	1	SSSI	Includes 186ha RSPB reserve and 1267ha Wildfowl Refuge
Leigh Marsh	-	1200	2	-	Includes 257ha NNR
Lindisfarne	5.1.76	3643	1	NNR	NNR area 3278ha
Martin Mere	-	147	-	W Trust	
Medway Estuary	-	7942	1	SSSI	
Mersey Estuary	-	6346	2	SSSI	
Morecambe Bay	-	17000	-	SSSI	Includes RSPB reserve and Wyre/Lune Sanctuary
Nene Washes	-	1200	1	SSSI	Includes RSPB reserve
New Grounds, Slimbridge	-	1800	1	SSSI	Wildfowl Trust reserve
North Norfolk Coast	5.1.76	7380	1	NNR	Includes NNRs, RSPB, National Trust and others
Ouse Washes	5.1.76	2276	1	SSSI	RSPB, Wildfowl Trust and local naturalists
Pagham Harbour	-	300	-	-	Unprotected
Poole Harbour	-	5596	1	SSSI	RSPB reserve included
Ribble Estuary	-	12500	1	NNR	NNR 2302ha
Severn Estuary	-	35000	1	SSSI	Excluding New Grounds and Bridgwater Bay
Upper Solway flats and marshes	-	29300	1	NNR	NNR 5501ha, also W Trust and Cumbria Naturalists' Trust reserve
Somerset Levels	-	68000	1	SSSI	Includes NNR (460ha), RSPB reserve (170ha)
South Thames Marshes	-	3600	2	SSSI	Includes 26ha LNR
Stour and Orwell Estuaries	-	3209	1	SSSI	Small RSPB reserve
The Swale	In progress	9102	1	NNR	Includes RSPB reserve (1350ha), NNR (165ha) and LNR (420ha)
Taw/Torridge Estuary	-	4046	2	SSSI	Includes 604ha NNR
Teesmouth flats and marshes	-	680	1	SSSI	
The Wash	-	68500	1	SSSI	Includes 400ha NNR and 1315ha RSPB reserve

Table 270 — *contd.*

Site	Designation	Area(ha)	NCR	Other	Comments
SCOTLAND					
Beauly Firth	-	1648	-	SSSI	
Caithness Lochs	-	600	-	-	Unprotected
Cameron Reservoir	-	49	-	SSSI	Goose roost
Carsebreck Lochs	-	344	1	SSSI	Goose roost
Cromarty Firth	-	12533	1	SSSI	Includes 640ha NNR
Dee Valley and Loch of Skene	-	135	1	-	Unprotected. Goose roost
Lower Dornoch Firth	-	2000	1	SSSI	
Drummond Pond	-	135	2	SSSI	Goose roost
Dupplin Lochs	-	265	1	SSSI	Goose roost
Flanders Moss	-	245	2	SSSI	Goose roost
Eden Estuary	-	700	2	-	LNR
Endrick Mouth, Loch Lomond	-	416	2	NNR	
Fala Flow	-	535	2	-	Unprotected. Goose roost
Firth of Forth	-	2750	1	-	Includes NNR (57ha), LNR (582ha) and 3 RSPB reserves
Firth of Tay	-	4260	1	SSSI	
Gladhouse Reservoir	-	302	1	-	Unprotected. Goose roost
Greenlaw Moor and Hule Moss	-	1200	2	SSSI	Goose roost
Hoselaw Loch	-	50	-	-	Unprotected. Goose roost
Inner Clyde	-	500	-	SSSI	Includes 231ha RSPB reserve
Islay	-	4500	1	SSSI	Includes 1215ha RSPB reserve. SSSI area 2880ha
Loch Eye	-	360	1	SSSI	
Lochs Forfar, Rescobie and Balgavies	-	212	2	SSSI	Balgavies Scottish Wildlife Trust. Goose roost
Lochs Hallan and Kilpheder	-	500	1	SSSI	Goose roost
Lochs Harray, Stenness and Skaill	-	1840	1	SSSI	
Loch Insh and Marshes	-	1000	1	SSSI	RSPB reserve 509ha
Loch Ken and Dee Marshes	-	1500	1	SSSI	RSPB reserve 158ha
Loch of Kinnordy	-	65	2	RSPB	Goose roost
Loch Leven	5.1.76	1597	1	NNR	
Loch of Lintrathen	15.6.81	218	2	SSSI	Goose roost
Lochyloch and Cleuch Reservoirs	-	200	-	-	Unprotected. Goose roost
Loch Mahaick	-	95	-	SSSI	Goose roost
Loch Shiel and Kentra Moss	-	1695	2	SSSI	Loch Shiel is SSSI
Loch Spynie	-	70	2	SSSI	Goose roost
Loch of Strathbeg	-	200	1	SSSI	RSPB reserve
Macrihanish and Tangy Loch	-	750	-	-	Unprotected but part SSSI
Montrose Basin	-	1125	2	SSSI	LNR, Scottish Wildlife Trust reserve
Moray Firth (south shore)	-	5050	-	SSSI	Includes 620ha RSPB reserve
Rhunahaorine	-	327	2	SSSI	Scottish Wildlife Trust reserve
Tay – Isla Valley	-	1118	1	SSSI	Goose roost
Tentsmuir Point and Abertay Sands	-	505	1	NNR	Goose roost
Tiree and Coll Islands	-	16000	1	SSSI	SSSI 1500ha
Westwater and Baddinsgill Reservoirs	-	100	-	SSSI	Goose roost
White Loch (Lochinch)	-	60	1	SSSI	Goose roost
Ythan Estuary	-	1050	1	NNR	Including Slains Lochs. NNR 974ha
WALES					
Burry Inlet	-	5000	1	SSSI	Gower coast NNR 47ha and Wildfowl Refuge
Conwy Bay	-	2610	-	SSSI	
Taf, Tywi and Gwendraeth Estuaries	-	4000	-	-	Unprotected

Table 270 — *contd.*

Additional sites which are not strictly wetlands but important for Barnacle Geese (all in SCOTLAND)

Site	Designation	Area(ha)	NCR	Other	Comments
Isay, Skye	-	80	-	-	Unprotected
Isles at mouth of Kyle of Tongue	-	80	-	-	Unprotected
Isles in Sound of Barra	-	335	-	-	Unprotected
Isles of Sound of Harris	-	150	-	-	Unprotected
Monach Isles	-	577	1	NNR	
Shiant Isles	-	225	2	SSSI	
Treshnish Isles	-	208	-	SSSI	

(b) Important wetlands on the basis of their ecological characteristics (Cagliari Criteria 2 and 3).

Site	Designation	Area(ha)	NCR	Other	Comments
ENGLAND					
Bure Marshes	5.1.76	517	1	NNR	NNR 451 ha
Chesil Beach and the Fleet	In progress	792	1	SSSI	Also nationally important for wildfowl
Esthwaite Water	-	195	1	-	
Hickling Broad and Horsey Mere	5.1.76	905	1	-	Nationally important for wildfowl. Includes 487ha NNR
Irthinghead Mires	In progress	580	1	-	
Malham Tarn	-	65	1	-	National Trust
Minsmere/Walberswick	5.1.76	1902	1	-	Walberswick NNR (514ha). Minsmere RSPB reserve (595ha)
Moor House	-	6600	1	NNR	
New Forest Valley Mires	-	440	1	SSSI	
Redgrave – South Lopham Fen	-	130	1	-	
Rostherne Mere	15.6.81	153	1	NNR	Nationally important for wildfowl
Roydon Common	-	40	1	-	
Scarning Fen	-	4	1	-	
Surlingham, Wheatfen and Rockland Broads		305	1	-	
SCOTLAND					
Blar nam Faoileag	-	4600	1	-	
Cairngorm Lochs	15.6.81	179	1	NNR	
Claish Moss	15.6.81	563	1	NNR	Important for Greenland Whitefront
Durness Lochs and Streams	-	140	1	-	Lochs Corispol, Lanlish and Borralie
Howmore Estuary	-	47	1	-	Lochs Roag and Fada
Loch an Duin	-	245	1	-	
Loch Druidbeg and A'Mhachair	5.1.76	1677	1	NNR	Breeding Greylags
Loch Fleet	-	1400	1	-	Scottish Wildlife Trust reserve. National importance
Loch Lomond	5.1.76	7100	1	-	See also Endrick Mouth in (a)
Loch Morar	-	3350	1	-	
North Roe	-	7200	1	-	
Rannoch Moor	5.1.76	10300	1	NNR	NNR 1499ha
Silver Flowe	15.6.81	608	1	NNR	NNR 191ha
Strathy Bogs	-	950	1	-	Includes 49ha NNR
WALES					
Bosherston Lake	-	80	1	-	
Cors Fochno and Dyfi	5.1.76	2497	1	NNR	Includes RSPB reserve (255ha). NNR 2095ha. National importance
Llangorse Lake	-	220	1	-	
Llyn Idwal	-	12	1	-	
Llyn Tegid	-	435	1	-	

Management for breeding

The interest as far as breeding is concerned in Britain is restricted to rather few species and management in most cases is aimed as much at increasing duck numbers for shooting as for purely conservation reasons. Breeding wildfowl also benefit from activities designed to improve conditions for wintering birds.

Water control

One of the main limitations of marshland habitat is water, essential for feeding during the whole breeding cycle, but especially for brood rearing, which takes place in the driest part of the summer. Water is usually available in the form of drainage ditches, but the flora of ditches, and hence the invertebrates which they support, varies considerably according to when they were last cleared (588). To maintain suitable conditions for ducklings a rotational pattern of ditch clearance should be practised. The provision of standing water helps considerably for many species. The creation of permanent pools at the Ouse Washes in the late 1960s and early 1970s resulted in an increase in Gadwall numbers from 1-2 to 40-50 pairs and in Shelducks from 1 to 12 pairs. Other species also benefited, though less spectacularly (583).

Artificial nesting structures

Potential nest sites are commonly limited, especially in gravel pits and reservoirs, and the creation of islands where wildfowl are safe from foxes usually increases the nesting density. When a new water area is being created it is advisable to leave islands, but in the case of gravel extraction the cost of leaving saleable material under islands can be very high (339). In shallow lakes, islands of rubble can be created, capped with soil and planted with nesting cover. In deep water, floating rafts supported by oil drums have been used as artificial islands with great success. The wooden edge to the raft top contains 15cm of soil which supports nesting cover in the form of *Juncus* and *Carex* species (251). Such rafts have provided nest sites for Canada Geese, ducks, grebes and terns and are extremely useful in lakes with a fluctuating water level, where nests on fixed structures are liable to flooding.

Nest-boxes and baskets are readily used by some species, but they will only be successful in increasing nesting density where natural sites are limited or vulnerable to predators or flooding. Nesting baskets, first designed and used in Holland more than 300 years ago, are often used to increase Mallard nesting density. The pitcher-shaped basket is mounted on stakes 1-2m above the water to keep the nest safe from predators. Many other designs have been used, mainly for Mallard, and they are successful in increasing nest success. When nest density is artificially increased using boxes, food for ducklings will often be limiting and in some situations it may be necessary to feed broods like reared birds (202).

Nest-boxes nailed to trees have proved successful in increasing the nest density of tree-nesting ducks, especially in newer plantations where natural cavities are scarce (322). Nest-boxes have been used for Goldeneyes in Scandinavia for many years and they have been conspicuously successful in increasing the British breeding population (p.134).

Providing food and cover

Ducks often nest in tussocky pastures; Shoveler prefer those which are heavily grazed, whereas Mallard nest most densely in lightly grazed meadows (583). A variety of grazing and mowing regimes produces preferred nesting habitat for a variety of species.

Newly excavated gravel pits are poor in invertebrate foods for ducklings, largely because of the shortage of organic matter in the bottom mud. This normally depends on input from streams, marginal vegetation and trees, almost non-existent in newly excavated pits. The production of invertebrates can be greatly increased by the addition of straw to the water at the rate of 10 tonnes per hectare, a technique which has proved successful at the Game Conservancy's Great Linford Reserve in Buckinghamshire (555).

Other activities, such as predator control, can add to the success of other management regimes for breeding. In areas where man has already tilted the balance in favour of the predators, e.g. by artificially concentrating nesting birds, predator control may be an essential part of management. In general, however, tampering with the ecosystem in favour of one species as opposed to another is nowadays considered by many to be unacceptable where the balance has not been artificially disturbed (564).

Management for wintering birds

Wetlands have been intensively managed for wildfowl in North America for decades, but in Britain the creation and management of habitats is a recent development, impetus being given by the pioneering work of the late J.G. Harrison at Sevenoaks. Many techniques are available which have been shown to be successful in attracting and holding birds. They can best be illustrated by means of examples from managed wetlands. The following account describes four case histories on different habitats where management has been geared towards the needs of that site and of the birds wintering there. In some cases useful techniques were developed as a result of common sense and trial and error, but most are backed by detailed research both into their application and effectiveness. Where techniques have been developed elsewhere on a similar habitat these are mentioned under the relevant case history.

The Ouse Washes, East Anglia (p.147)

The Ouse Washes had existed for 300 years and periodically supported large numbers of wildfowl, chiefly dabbling ducks, before refuges were established by the RSPB and the Wildfowl Trust in the late 1960s. Land was acquired slowly from 1964 onwards, but by the early 1970s extensive blocks had been bought to create effective refuges from disturbance. The area was previously heavily shot over most of its length and the creation of disturbance-free areas had an immediate effect in increasing the number of ducks, particularly during the shooting season. Fig 180 shows the change in wintering pattern of Wigeon – the most abundant species – in the periods before, during and after refuge establishment. This shows that before establishment numbers did not peak until March, but most were found in January in following years. The extension of the wintering season has meant that whereas the maximum count was 2.6 times higher in 1969-1977 than in 1951-1959, the amount of time spent on the Washes by Wigeon in the same period was 5.3 times higher (453).

In the past the availability of floodwater has been critical in determining the number of ducks using the Washes, and flooding has varied annually from nil to a complete cover. Shallow and gradual flooding is most favoured by ducks and in years when there was little or deep flooding wildfowl numbers were traditionally low. Permanent water was provided by the conservation bodies in the early 1970s, and areas of grassland were artificially flooded with freshwater from a tidal river. This made the Wigeon much less dependent on runoff water, though the numbers of more aquatic dabbling ducks and especially diving species still fluctuate, largely in tune with the amount of floodwater (582). Bewick's Swans have increased dramatically in the last two decades (p.358). Since their feeding is mainly outside the Washes, they are largely independent of standing water, congregating on permanent pools, particularly the lagoon in front of the Wildfowl Trust observatory, where they are attracted by supplementary feeding of grain and waste potatoes.

Much of the Washes is still shot over and the distribution of birds depends largely on the amount of shooting. Overall, the refuges support the majority of the wildfowl, although they represent less than half of the area. During the shooting season, however, the wildfowl spend more than three-quarters of the time on refuges, compared with only 30-50% after 31st January (581). It could be argued that the shooting pressure is beneficial in that it keeps some feeding areas in reserve to be used in late winter when food supplies on refuges are depleted. One of the disadvantages of this regime is that large quantities of lead shot becomes accessible to the birds in shooting areas. The incidence of lead poisoning is relatively high in some species of ducks at the Washes (580) and very high in swans (445). In the RSPB/BASC/WT survey, the density of lead pellets in the top 15cm of soil was about 166,000 per hectare, higher than any other site sampled, except for two flight ponds (360). The soil of the Washes is largely peat or fine silt, and grit is very scarce. Wildfowl probably ingest pellets as grit, and when piles of grit were provided at the Wildfowl Trust refuge these were extensively used by swans and ducks. It is not known whether this reduced the rate of ingestion or poisoning, but this is likely.

The wildfowl frequenting the Washes are largely vegetation or seed eaters. Summer management has been aimed at increasing the quality and quantity of these foods available in autumn. Swans and Wigeon prefer young, soft grasses to older stands or the coarser species (445, 453). Summer grazing of the Wildfowl Trust's Welney Refuge by cattle and sheep is

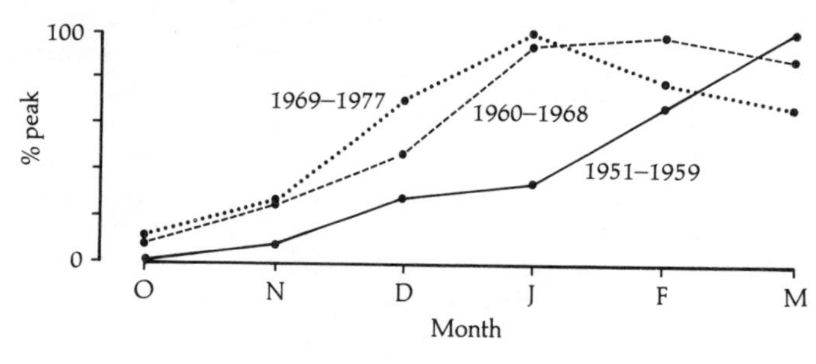

Fig. 180. The seasonal distribution of Wigeon at the Ouse Washes, expressed as a percentage of the average annual maxima in three periods: before refuge establishment, 1951-52 to 1959-60 (mean annual max = 10,300) – solid line; a period of some protection but no active management, 1960-61 to 1968-69 (mean annual max = 12,000) – dashed line; and the period of increasing protection and management, 1969-70 to 1977-78 (mean annual max = 31,500) – dotted line. From Owen and Thomas (1979).

as heavy as practicable to keep down the coarser grasses and maintain a young sward. The result has been that the refuge is by far the most heavily used area in the Washes by both Wigeon and Bewick's Swans, often holding more than half the birds on about one tenth of the area.

Heavy and early grazing is not compatible with high seed production, since cattle graze the early growth of *Eleocharis* and *Rumex* spp – the most important seed producers on the Washes. After the refuge was surveyed to determine the distribution of seed plants, the most productive washes were excluded from early grazing to allow the seed-bearing plants to grow to a stage where they were unpalatable to stock.

It is difficult to evaluate the relative contribution of the management regimes described to the spectacular improvements in the Ouse Washes since the 1960s, but control of disturbance, particularly creating a large non-shooting block, is a prerequisite to any successful refuge. For the more aquatic species water control is crucially important in attracting wildfowl, while summer management of food plants determines the amount of food available for wintering birds.

Sevenoaks, Kent (p.92)

This reserve was established in 1956, when the Kent Sand and Ballast Company agreed to allow the Wildfowlers' Association of Great Britain and Ireland (now BASC) to develop and manage a complex of newly excavated gravel pits near Sevenoaks. Largely through the efforts of the late J.G. Harrison, and in cooperation with the Wildfowl Trust, the reserve became an experimental management area, for breeding and wintering wildfowl and for waders.

The first objective was to modify the morphology of the area to provide islands and spits and to change the shore from a steep to a gradually shelving one. Sheltered sand or shingle banks devoid of vegetation were also provided as "loafing" spots for birds using the gravel pits as a roost.

Table 271. *Aquatic and marginal plants useful for introducing into wildfowl habitats, with an indication of their value. Based on the Sevenoaks (Harrison 1974) and Linford (Game Conservancy 1981) experimental projects.*

English name	Scientific name	Value
SUBMERGED AQUATIC PLANTS		
Water Crowfoot	Ranunculus spp	seed, leaves
Horned Pondweed	Zannichellia palustris	seed, leaves
Fennel Leaved Pondweed	Potamogeton pectinatus	seed, leaves
Milfoil	Myriophyllum spp	seed
Stonewort	Chara spp	whole plant
FLOATING PLANTS		
Duckweed	Lemna spp	whole plant
Amphibious Bistort	Polygonum amphibium	seed
EMERGENT PLANTS		
Spike-rush	Eleocharis spp	seed
Sea Clubrush	Scirpus maritimus	seed
Bulrush	Schoenoplectus spp	seed
Persicarias	Polygonum spp	seed
Sedges	Carex spp	seed, cover
Mare's-tail	Hippuris vulgaris	seed, cover
Bur-reed	Sparganium erectum	seed, cover
Reed-grass	Glyceria maxima	seed, cover
Arrowhead	Sagittaria sagittifolia	tubers
SCRUB PLANTS		
Bramble	Rubus fruticosus	seed, cover
Docks	Rumex spp	seed
Orache	Atriplex spp	seed
TREES		
Alder	Alnus glutinosa	seed
Oak	Quercus robur	seed
Birch	Betula spp	seed

Gravel pits are colonised naturally by plants from tubers or seeds carried by wind and birds. The process is rather slow because the pits are rarely fed by rivers, and are unconnected to other water-bodies, which precludes waterborne transmission of aquatic plants. The planting of submerged and marginal vegetation was the most important part of the management at Sevenoaks, based on extensive and detailed research carried out by the Wildfowl Trust on the food requirements of the various wildfowl species (244). A list of the most valuable plants, based on the Sevenoaks and Great Linford Projects, is given in Table 271.

In the early years of the refuge the food plants did not produce sufficient to support the wintering wildfowl population, many birds using the reserve as a roost while feeding on surrounding farmland (477). The proportion of their food gathered from the reserve did, however, increase during the early period (430) and the contribution could be expected to have grown even more as the stands matured. What was certain was that the habitat would have developed much more slowly without the intensive planting. What the project also achieved was to provide the impetus for many others to experiment with planting programmes in areas of greater potential and strategic importance for wintering wildfowl than Sevenoaks.

Rutland Water, Leicestershire (p.158)

In the early stages of planning for the construction of the large new reservoir near Empingham, Rutland, it was realised that a water-body of such a size and in such a position would be very attractive for wintering wildfowl. The NCC and the Leicestershire an Rutland Trust for Nature Conservation were con sulted at an early stage by the Welland and Nene Rive Authority (now a division of the Anglian Wate Authority) and the principle that a nature reserv would be established was agreed three years befor the reservoir was filled. A full-time warden wa appointed in April 1975, just as flooding began. Onl 160ha of water was present during the followin winter, the first year of extensive flooding bein 1976-77. During the period of filling a variety c practices, aimed at improving the eventual habitat fo wildfowl, were carried out. A shallow scrape wa excavated near to the top water level to provide wade feeding grounds, and bunds were constructed in on of the reserve sites to enclose three lagoons to hol water when the water level on the reservoir fell. Th lagoons were designed as breeding and loafing areas with islands and spits and planted marginal cover (13) The eventual reserve covered 80ha of shallow wate and 160ha of adjoining land. The reserve and re creational boundaries are shown in Fig 181.

The large size of the reservoir and its shape in many respects make it highly suitable for multiple use The shallow parts at the western end include ar interesting natural area at Burley Fishponds, but th natural history importance has been almost entirel artificially created. The most important developmen for wildfowl was the formation of lagoons, but diversity of habitat was created for other species including the planting of 3,200 trees in the first seve years (13). The pattern of wildfowl numbers in the firs seven years of flooding (starting in 1976-77 – the firs year of extensive flooding) is shown in Fig 182 together with that at Grafham Water, Cambridgeshir (see also p.512). Although the two reservoirs are no

Fig. 181. A map of Rutland Water showing the reserve and recreation areas.

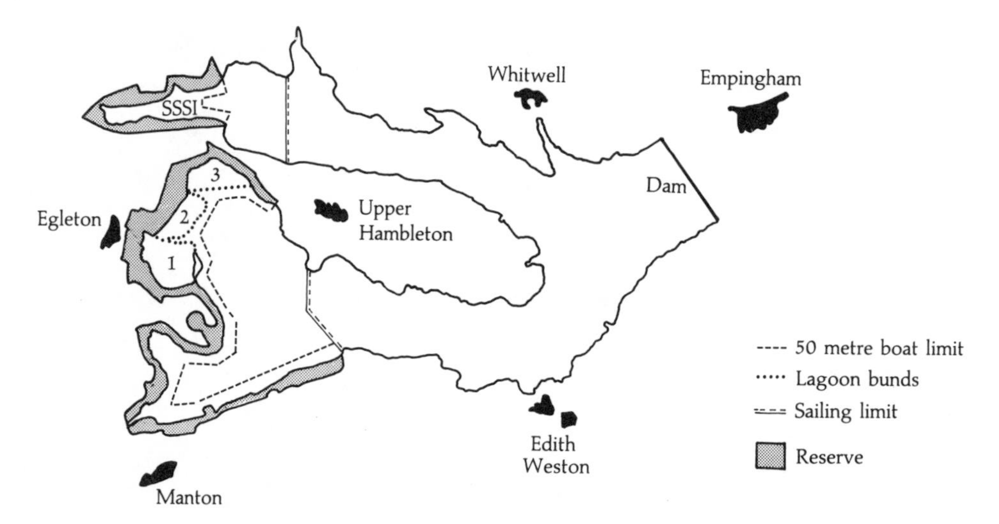

entirely comparable, they are both large, on the east side of the country and surrounded by mixed arable farmland. Grafham also has a reserve area, and although there are no strict sailing limits boats keep out of the deeper water in the reserve area (132). Fishing causes some redistribution of birds, but the season ends on 15th October (131).

The highly contrasting pattern of wildfowl numbers visiting the two reservoirs can, then, largely be attributed to the creation and management of habitat. The way in which dabbling ducks, which normally decline rapidly in new reservoirs after the initial influx, have maintained their numbers at Rutland is striking, particularly the large numbers of Wigeon that now use the water and its surrounds. The early anticipation of wildlife and other needs, the sensitive cooperation by the water authority and the expert practical organisation and activities on the ground, should provide a model for the development of any future lowland reservoirs for multi-use.

New Grounds, Gloucestershire

The New Grounds consist of 500ha of moist pasture reclaimed from the brackish marshes of the Severn Estuary several hundred years ago. Progressive drainage has dried out most of the inland pastures, but many are waterlogged in winter. There are also 66ha of saltmarsh pasture and 40ha of enclosures of the Wildfowl Trust. The area has been well keepered for

shooting purposes for at least a century and supports the largest flock of European White-fronted Geese in Britain, as well as large numbers of dabbling ducks and swans. The Wildfowl Trust took over the wardening of the refuge in 1946, but agricultural management was in private hands until 1972, when the Trust acquired the tenancy of 150ha.

All of the holding taken over by the Trust was pasture or hay meadow, and sheep were outwintered on most of the fields. Previous to its occupancy the Trust had carried out research into the movements of the geese and discovered that they moved progressively inland during the winter, using first the fields closest to the roosts on the Severn sandflats (432). The first management step was to limit disturbance by instituting a summer only grazing programme and prohibiting access to the land in winter. The area was split into three grazing zones based on the preferences of the geese. The first zone consisted of the main salting pasture, which was traditionally first used by the birds, and grazing was terminated there on 30th September. The adjacent zone was vacated by stock on 30th October and the inland zone on 30th November. The result of this regime was markedly to increase the goose usage of the managed area by between 70 and

Fig. 182. The pattern of wildfowl numbers at Rutland Water (dashed line) and Grafham Water (solid line) in the years following flooding. Values are regular means of total wildfowl (ducks + geese + swans), expressed as a percentage of numbers in the first year of flooding. For Rutland year 0 is 1976-77, when rapid filling took place (100% = 7,996 wildfowl); the equivalent for Grafham was 1965-66 (100% = 6,170).

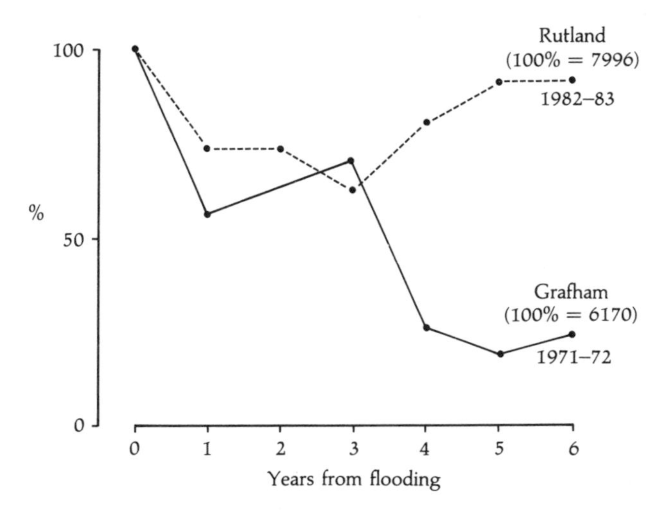

140%. The newly acquired holding supported 25-43% of the total New Grounds goose usage before management and 56-77% afterwards (436).

Wild geese prefer the more nutritious and digestible short vegetation to coarser grasses, and experimental cutting of coarse grass and autumn nitrogen fertiliser application was tried on the salting pasture. The cutting treatment, carried out in 25m-wide strips, involved removing the inflorescenses of Meadow Barley (*Hordeum secalinum*) and some of the dead grass. Nitro-chalk fertiliser (25%N) was applied at 125kg/ha in mid-October just before the arrival of the geese. The use of cut areas by geese in the following winter was 32% higher than the controls, fertilised areas 42% higher and where both treatments had been applied 87% higher. Fertilising is expensive and not justified except in special circumstances, but tipping or cutting dead grass is now regularly practised (434).

The consequences of refuge management

The very success of refuges in attracting and holding large and dense populations of wildfowl can in some cases give rise to problems which have not been anticipated and which might be difficult to counter (441).

Disease risk

In North America the enormous concentrations of wildfowl attracted to wildfowl refuges have given rise to problems of contracting and spreading disease. Botulism, a disease of warm climates, is the main killer, but this is rare in Britain except in very warm summers. There have been major outbreaks in the Netherlands, but disease is unlikely to become a problem on British refuges, which are largely managed for wintering birds. Wildfowl on agricultural diets are more susceptible to lead poisoning than those on natural foods, but this is much less of a problem in Britain, where the planting of agricultural crops for wildfowl is not practised and where the incidence of lead pellets in soil is low over most of our habitats.

Modifications of feeding habits

Foods on agricultural land are usually easier for birds to collect and in some cases to digest, and the provision of an easy source of food on refuges may cause birds to desert more natural habitats. The adoption of the field feeding habit by Bewick's Swans at the Ouse Washes coincided with the increase in numbers which followed the establishment of the refuges. Some field feeding had occurred previously, but it seems certain that the rapid spread of the habit through the wintering population was facilitated by the fact that a large refuge roost was present. Field feeding does have its risks, and 30 or more Bewick's Swans were poisoned by carbophenothion wheat dressing on land adjacent to the washes in 1982. Mallard have long used stubbles and potato fields for feeding in winter, and evidence from the Ouse Washes suggests that both Pintail and Wigeon accompany Mallard into fen fields from their common roost on the washes themselves (585).

It does seem that the artificial concentration of birds on refuges can induce or at least accelerate changes in wildfowl feeding habits which can lead to conflict with farmers. The main aim of a refuge programme should be to maintain and manage natural or semi-natural areas rather than to introduce wholly artificial feeding grounds. The provision of inland refuges for Brent Geese, a species which greatly increased in numbers in the 1970s and 1980s, would lead to "honeypots" being created on arable land, but hopefully these would only be used after the natural food supply was depleted. In future we may even have to consider discouraging the use of artificial refuges by birds at times of natural food abundance so that managed food supplies still provide insurance against future shortage.

Population changes

The ultimate aim of most refuges is to increase wildfowl populations by improving survival or breeding success, or at least stemming a decline. One frequent complaint of farmers near goose refuges is that they are self-defeating, inducing increases in goose populations which again overspill onto arable farmland. There is no doubt that the provision of unshot feeding areas during the shooting season will reduce mortality and allow the number of birds to increase. There are some spectacular examples from North America (499) and this was considered the most important factor in allowing Spitsbergen Barnacle Geese to increase in the 1970s (451).

Rather more subtle changes in numbers can also take place. A refuge in one part of the winter range can reduce the mortality of those birds, allowing them to increase, while shooting continues in other parts and an apparent redistribution occurs (493). This affects not only the wintering area, but also the breeding range if there are discrete subunits in the population.

There is growing evidence that spring feeding conditions affect the breeding performance of geese, and the provision of improved spring feeding on refuges could cause population increases by improving productivity. There is some evidence that differences in spring condition are reflected in better breeding in Barnacle Geese (441) and Brent Geese

(161). This effect would be particularly relevant for those species, such as Greenland Whitefronts and Pinkfeet, which do not have important spring staging areas outside Britain. The artificial feeding of Pinkfeet in Denmark in spring was followed by a large increase in the population, but there is, as yet, no evidence that the feeding of Bewick's Swans at Wildfowl Trust refuges causes those birds to breed better than those wintering outside refuges.

Refuges can and have produced substantial changes in goose numbers, and this may also be true of swans and ducks. If the main object of the refuge was to prevent agricultural damage on private land, the alleviation may well only be temporary, while bird numbers build up to overflow refuge boundaries again, as has been the case with the Spitsbergen Barnacle Goose (441).

Distributional shifts

Much has been made of the "short-stopping" effects of refuges in North America, whereby wildfowl are enabled to stay further north than they would normally by the provision of abundant food on refuges. Farming changes have produced this phenomenon in British Pinkfeet (294), and the provision of protected roosts and new ones on reservoirs facilitated the process (382). Every refuge, if it is successful, inevitably causes some redistribution on a local, regional, or even national scale, but in Britain such changes are unlikely to give cause for concern in the same way as they have in North America, where the deprivation of hunting opportunity for shooters in a locality is considered undesirable.

The consequences of establishing and successfully managing refuges, therefore, are not all desirable or easy to forecast, but the drawbacks are not likely to be major ones in the British context.

Public access and education

In the preceding sections we have discussed conservation for the sake of the birds, with reserves being set aside for their benefit. However, conservation costs money, and although the amount spent on it is small in relation to the national budget, we would be deluding ourselves if we thought that we could put large areas of our countryside aside without being called to question. We must convince a wide public that reserves and wildlife are worth preserving, not only to reduce problems of trespass and vandalism, but also to build up public sympathy which could be vital when a major wildlife issue is being discussed at an inquiry or in parliament. Just such a case was the threat to reclaim the Ribble Marshes, stopped at the last minute largely as a result of widespread public concern.

Over the past two decades the Wildfowl Trust and the RSPB have expended a large amount of effort on reserves. The Trust has concentrated its work on the intensive use of a few refuges, promoting education in the broad sense. The RSPB has put its money into land purchase and the protection of habitat and has increased the number of its reserves to over 100. About half its reserves are wetlands, and public access for viewing is allowed on the vast majority of them. All this has been done with a relatively small input of public money (grants from governmental organisations such as the NCC and the Countryside Commission). The growth in interest in birds and their habitats is illustrated by the increase in the membership of the two organisations (Fig 183). The RSPB figures, in particular, show a remarkable rise during the period, although the trends in both graphs now show signs of levelling off. The growth is,

Fig. 183. The number of members of two conservation organisations, the Royal Society for the Protection of Birds (upper) and the Wildfowl Trust (lower), from 1960 onwards.

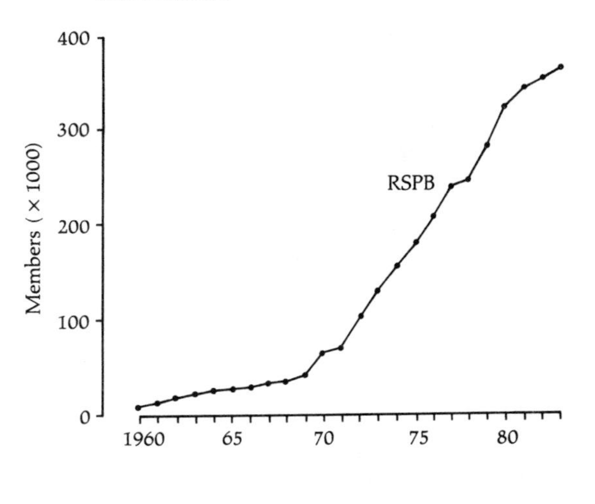

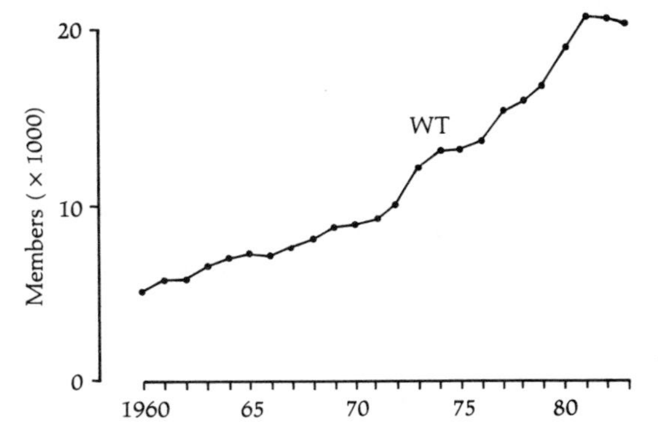

to a large extent, self-generating, i.e. it creates a demand as well as satisfying it. The fact that birdwatching is now the most widespread recreational activity on inland waters (p.499) emphasises that there has been a general growth of interest, far beyond the committed people who belong to an organisation and may be termed enthusiasts.

There is no doubt that there are certain areas that, because of the fragility of the habitat or the vulnerability of its occupants to disturbance, must be protected from public access. There are places, however, that can be opened up to the public, at least to view from the periphery, without destroying their character or disturbing their habitats. Birds have a wide appeal and are easy to see and appreciate. Wetland birds, in particular, live in open areas, often in large flocks, providing spectacular sights and exciting sounds to the casual visitor and the enthusiast alike.

The key to public access without disturbance is careful location of access points and concealment of observers at all times (536). Earth banks are the most effective form of screening in that they are permanent and easily maintained. Many reserves have embankments already, in the form of a sea wall or other flood prevention structures. The placement of obstacles on and use of these banks must, however, conform to their use for their primary purpose, usually under the control of a river authority or internal drainage board. It is usually necessary to build screening banks, which should be 2-3m high and have steep sloping sides. The other benefit of bank construction is that the excavated areas or "borrow-pits" can be developed to provide additional habitat as permanent freshwater, tidal lagoons or marshy "scrapes". Where there is too little land available, or where the soil is unsuitable for banks, reed or wooden screens can be used, though these are expensive, more obtrusive, and require more regular maintenance than banks. A belt of osiers can be used, but the area of land must be large to provide sufficient depth for effective winter screening.

Where the refuge is narrow, screening and access can be restricted to one side, so only a single bank is necessary. Where the refuge is deeper or the area of interest is too far from the perimeter for effective observation, it will be necessary to construct a double banked or screened approach to an observation point out on the refuge itself. Encroachment onto a refuge in this way must be carefully considered, since the intrusion caused by banks can destroy the viewing opportunities.

Observation points can vary from the simplest reed or wooden screen with viewing slits cut in it, to a heated observatory or tower with plate-glass windows. The choice obviously depends on the position, finance and likely use, but a hide is preferable to an open viewing point as it gives screening from birds flying overhead. The most common hides, based on a model developed by the Wildfowl Trust, have a breeze-block base, set into the bank, and a wooden superstructure with a pitched or sloping felted roof (43). The Trust has also experimented with the use of roadmen's shelters made of fibreglass modified as hides set in banks. Modifications involve painting these usually brightly coloured structures in a dull green and cutting slits to fit in observation slats or windows. Benches and elbow shelves are also fitted, and the hides come in two sizes, housing 5 and 10 people. These are now extensively used on Trust reserves, and their ease of placement and durability are great advantages.

Various refinements to hides have been developed over the years. It is important that viewing slats open inwards, to avoid disturbing birds, and that they should be kept open at various angles so that only the minimum of viewing area is provided. Screening the entrance with sacking or other material is also vital so that observers entering the hide are not framed against the light and cause disturbance. Individual seats of adjustable height are a great asset, providing comfortable viewing for children and adults. Where access allows this, hides should have ramps and low-level slits for visitors in wheelchairs.

Where there is a considerable spectacle the construction of a larger observatory may be justified. An example is the Wildfowl Trust's lagoon at Welney, where hundreds of ducks and swans gather to feed and roost. This has large plate-glass windows giving excellent views of the refuge from a comfortable, warm vantage point. People are attracted to this facility who would not normally venture onto the edge of the marsh, or could not, by reason of age or disability. Observation towers give even better and all-round views, and have the additional benefit that they allow the warden more effectively to overlook and police the refuge.

The educational aspects of viewing areas are extremely important; informative displays enhance the visitor's enjoyment and carry the ecological and conservation message. Creating an awareness in children is especially important; the Wildfowl Trust accommodates about 100,000 children in organised parties each year and the RSPB's Young Ornithologists' Club has over 100,000 members.

The management of refuges to attract wildfowl was discussed in detail in the previous section, but some techniques may be used specifically to attract birds where they can easily be seen by visitors. To a large extent this can be done by previous research and careful siting of observation points where birds are more likely to be, but if the viewing facility is to be most effective, good viewing should be predictable and large numbers of birds present whenever possible. The most common method of attracting wildfowl is to provide water, either by flooding an area in front of the observation point or by digging a scrape or lagoon. The area can be improved by providing organic matter to encourage invertebrates, by planting with food plants or by providing islands or other nesting areas.

The most intensive method, however, is directly to supply food in the form of grain, potatoes, etc. The regular and predictable food supply encourages a concentration of birds and ensures that they are frequently in view. The Rushy Pen at Slimbridge, though only 2ha in area, attracts at times 3-400 Bewick's Swans and more than a thousand ducks of various kinds. The lagoon at Welney, where food is also provided morning and evening, attracts hundreds of roosting swans and ducks, which disperse to the washes and fens during the day to obtain their main food supply.

These recent developments and techniques have provided some notable "honeypots" for bird viewers, but most reserves and refuges are not managed nearly so intensively. The provision of intensive viewing has had its critics. Many argue that the provision of food artificially, and particularly with flood-lighting, as at Slimbridge and Welney, is excessively unnatural. Some also object to the "regimentation" of birdwatching, the taking away of the "wilderness element" in natural areas.

Whether we like it or not, our wilderness areas are being lost, eroded or encroached upon by agriculture, forestry and industry; the creation of reserves by the RSPB and other voluntary organisations has played a major part, over the last 10 years, in maintaining many of the remaining portions intact. Voluntary organisations are financed by membership contributions and/or by payment for access to viewing facilities. If this is the only way operations can be financed surely it is better than letting the habitat be lost. It could be argued that the channelling of the majority of people to the "honeypots" takes the pressure off more sensitive areas, to which we can then justifiably exclude public access.

We must maintain a balanced approach to the use of our refuges by the public, on the one hand giving and encouraging access to areas designed for the purpose in sites where such viewing pressure can be accommodated, and on the other protecting sensitive sites or species. Our ability to maintain a minimum quantity and diversity of wetland and other habitats in this overcrowded island must depend on our success in building up a body of opinion to counterbalance the powerful pressures to which these areas are being continually subjected.

Conclusion

Having reviewed the considerable progress made in the conservation of wildfowl and their habitats in the 1950s, the First Edition ended on an optimistic note, concluding that science was playing an increasingly important part in the process of conservation. Reviewing the position since then, we are again justifiably pleased with the progress. Two major developments give us particular grounds for optimism. One is the emergence of the International Waterfowl Research Bureau as a major force in international conservation, leading to the establishment of the Ramsar Convention. The commitment of the Nature Conservancy Council and the government to the Convention is illustrated by the fact that the list of sites proposed for designation under Ramsar includes *all* those wetlands qualifying under the 1% criterion. The other major development in the 1970s was in the contribution made by the voluntary organisations to the conservation of our wetland habitats.

The Wildlife and Countryside Act (1981) increased the number of wildfowl given protection in law and also extended protection to habitats in the case of SSSIs. The NCC has the burden of making this legislation work in practice. There have been setbacks, but we hope that landowners and conservationists can work together to protect our most important sites. An encouraging and extremely important factor is the growth in public interest and involvement in aspects of conservation in recent years.

Our challenge now is to maintain and build on this favourable situation and safeguard our wetlands for the future. Since wildfowlers, voluntary and governmental conservation bodies and many other organisations are working together, we are hopeful that we will succeed.

References

1. Aberdein, A.F. 1982. Stanstead Abbots Gravel Pit—a review of its history, future and birdlife. Birds in the Lee Valley 1981:35-43. Lee Valley Project Group.
2. Alderton, R.E. 1976. Birds at Surrey Commercial Docks, January 1973 to December 1975. London Bird Report 40:85-90.
3. Allen, D.S. 1979. Seaducks in the Moray and Dornoch Firths, Scotland, winter of 1978/1979. Unpubl. Rep. NCC and RSPB.
4. Allen, R.H. 1974. The Mersey ducks since 1950. Cheshire Bird Report 1973:31-33.
5. Allen, R.H. and Rutter, G.E. 1956. The moult migration of the Shelduck in Cheshire in 1955. Brit. Birds 49:221-225.
6. Allen, R.H. and Rutter, G.E. 1964. The Shelduck population of the Mersey area in summer, 1957-1963. Wildfowl 15:45-46.
7. Allison, A. and Newton, I. 1974. Waterfowl at Loch Leven, Kinross. Proc. Roy. Soc. Edin. (B), 74:365-382.
8. Allison, A., Newton, I. and Campbell, C. 1974. Loch Leven National Nature Reserve. A study of waterfowl biology. Chester WAGBI.
9. Anderson, D.R. and Burnham, K.P. 1976. Population Ecology of the Mallard. V. The effect of exploitation on survival, recovery and harvest rates. US Dept. Interior Fish and Wildlife Service Resource Publ. 125.
10. Anderson, H.G. 1959. Food habits of migratory ducks in Illinois. Ill. Nat. Hist. Surv. Bull. 27:289-344.
11. Anderson, P. 1981. Attenborough Gravel Pits: an ecological basis for future management. Countryside Commission/Broxtowe Borough Council.
12. Anon. 1981. Washington Waterfowl Park 1973-1980. Circulated document, The Wilfowl Trust.
13. Appleton, T.P. 1982. Rutland Water Nature Reserve: concept, design and management. Hydrobiologia 88:211-224.
14. Archer, A.A. 1972. Sand and gravel as aggregate. Mineral Resources Division, Inst. Geol. Sci. Mineral Dossier No. 4. London, HMSO.
15. Atkinson, K.M. 1981. Changes in the wildfowl population of Windermere in recent years. Birds in Cumbria 1981:4-5.
16. Atkinson, K.M. and Hewitt, D.P. 1978. A note on the food consumption of the Red-breasted Merganser. Wildfowl 29:87-91.
17. Atkinson-Willes, G.L. 1956. National Wildfowl Counts 1954-55. Slimbridge, The Wildfowl Trust.
18. Atkinson-Willes, G.L. 1961. The importance to wildfowl of the reservoirs in England and Wales. Wildfowl Trust Ann. Rep. 12:29-33.
19. Atkinson-Willes, G.L. 1969. Wildfowl and recreation: a balance of requirements. Brit. Water Supply 11:5-15.
20. Atkinson-Willes, G.L. 1970a. Wildfowl situation in England, Scotland and Wales. Proc. Int. Reg. Mtng. Conserv. Wildfowl Resources, Leningrad 1968:101-107 Moscow.
21. Atkinson-Willes, G.L. 1970b. National Wildfowl Counts. In Sedgwick, N.M., Whitaker, P. and Harrison, J.G. (Eds). The New Wildfowler in the 1970s:237-248. London, Barrie and Jenkins.
22. Atkinson-Willes, G.L. 1976. The numerical distribution of ducks, swans and coots as a guide to assessing the importance of wetlands. Proc. Int. Conf. Conserv. Wetlands and Waterfowl, Heiligenhafen 1974:199-271. Slimbridge, IWRB.
23. Atkinson-Willes, G.L. 1978. The numbers and distribution of sea ducks in North West Europe, January 1967-73. Proc. Symp. Sea Ducks, Stockholm 1975:28-67. Nat. Swedish Environ. Prot. Board/IWRB.
24. Atkinson-Willes, G.L. 1981. The numerical distribution and the conservation requirements of *Cygnus olor, Cygnus cygnus* and *Cygnus columbianus bewickii* in north-west Europe. Proc. 2nd Int. Swan Symp., Sapporo, Japan 1980:40-48. Slimbridge, IWRB.
25. Atkinson-Willes, G.L. 1982. Trends in duck numbers in north-west Europe. Proc. 2nd Tech. Mtng. Western Palearctic Migratory Bird Management, Paris 1979:58-64. Slimbridge, IWRB.
26. Atkinson-Willes, G.L. and Matthews, G.V.T. 1960. The past status of the Brent Goose. Brit. Birds 53:352-357.
27. Atkinson-Willes, G.L. and Yarker, B. 1971. The numerical distribution of some British breeding ducks. Wildfowl 22:63-70.
28. Avon Bird Reports. In Proc. Bristol Nat. Soc.
29. Axell, H.E. 1982. Control of reeds (*Phragmites communis*) at Minsmere, England. In Scott, D.A. (Ed). Managing Wetlands and their Birds:44-50. Slimbridge, IWRB.
30. Bacon, P.J. 1980. Status and dynamics of a Mute Swan population near Oxford between 1976 and 1978. Wildfowl 31:37-50.
31. Bagge, P., Lemmetyinen, R. and Raitis, T. 1970. Saaristomeren vesiluitujen kevatravinnosta. Sommen Rusta 22:35-45.

32. Bailey, S., Bunyan, P.J., Hamilton, G.A., Jennings, D.M. and Stanley, P.I. 1972. Accidental poisoning of wild geese in Perthshire, November 1971. Wildfowl 23:88-91.

33. Baillie, S.R. 1981. Population dynamics of the Eider (*Somateria mollissima*) in north-east Scotland. PhD. Thesis, University of Aberdeen.

34. Baillie, S.R. and Milne, H. 1982. The influence of female age on breeding in the Eider *Somateria mollissima*. Bird Study 29:55-66.

35. Balfour, E. 1968. Breeding birds of Orkney. Scot. Birds 5:89-104.

36. Bannerman, D.A. 1958. The Birds of the British Isles. Vol. 6. Edinburgh and London, Oliver and Boyd.

37. Batten, L.A. 1977. Sailing on reservoirs and its effects on water birds. Biol. Cons. 11:49-58.

38. Batten, L.A. and Swift, J. 1982. British criteria for calling a ban on wildfowling in severe weather. Proc. 2nd Tech. Mtng. Western Palearctic Migratory Bird Management, Paris 1979:181-189. Slimbridge, IWRB.

39. Bauer, K.M. and Glutz von Blotzhein, U.N. 1969. Handbuch der Vogel Mitteleuropas. Vol. 3. Frankfurt am Main.

40. Baxter, E.V. and Rintoul, L.J. 1922. Some Scottish breeding duck: their arrival and dispersal. Edinburgh, Oliver and Boyd.

41. Baxter, E.V. and Rintoul, L.J. 1953. The birds of Scotland. Edinburgh, Oliver and Boyd.

42. Bays, L.R. 1969. Pesticide pollution and the effects on the biota of Chew Valley Lake. Water Treatment and Examination 18:295-326.

43. Beale, C.J. and Wright, F.S. 1968. The Slimbridge observation hides. Wildfowl 19:137-143.

44. Beer, J.V. and Boyd, H. 1962. Weights of Pink-footed Geese in autumn. Bird Study 9:91-95.

45. Bellrose, F.C. 1959. Lead poisoning as a mortality factor in waterfowl populations. Ill. Nat. Hist. Surv. Bull. 27:235-288.

46. Bengtson, S.-A. 1971. Food and feeding of diving ducks breeding at Lake Myvatn, Iceland. Ornis Fenn. 48:77-92.

47. Bennett, T. and St Joseph, A.K.M. 1974. Brent Geese feeding inland in South-east England. Unpubl. Rep. Wildfowl Trust.

48. Bergh, J.M.J. van den 1981. Verslag van de watervogeltellingen in Januari en Maart 1980. Watervogels 6:95-118.

49. Bergman, G. and Donner, K.O. 1964. An analysis of the spring migration of the Common Scoter and the Long-tailed Duck in southern Finland. Acta. Zool. Fennica 105:1-59.

50. Berry, J. 1939. The status and distribution of wild geese and wild duck in Scotland. International Wildfowl Inquiry, Vol. II. Cambridge, the University Press.

51. Bibby, C.J. and Lloyd, C.S. 1977. Experiment to determine the fate of dead birds at sea. Biol. Cons. 12:295-309.

52. Bignall, E.M. 1980. Observations on the Shelduck population of the Loch Lomond National Nature Reserve. Unpubl. Rep. NCC.

53. Birds in Cumbria. Cumbria Trust for Nature Conservation

54. Birds in Northumbria. Tyneside Bird Club.

55. Blenkarn, A. 1981. Water recreation. Roneo Draft Rep. Civils Conference.

56. Blindell, R.M. 1977. The estuarine bird populations of the region Orwell – Thames, 1972-1975. Essex Bird Report 1976:71-102.

57. Blurton Jones, N.G. 1956. Census of breeding Canada Geese 1953. Bird Study 3:153-170.

58. Bond, P. 1980. Coed Farm—Wildfowl Reserve. Nature in Wales 17:96-99.

59. Boorman, L.A. and Ranwell, D.S. 1977. Ecology of Maplin Sands. Cambridge, Institute of Terrestrial Ecology.

60. Booth, C.G. 1975. Birds in Islay. Port Charlotte, Argyll Reproductions.

61. Boyd, H. 1955. The role of tradition in determining the winter distribution of Pinkfeet in Britain. Wildfowl Trust Ann. Rep. 7:107-122.

62. Boyd, H. 1956a. Statistics of the British population of the Pink-footed Goose. J. Anim. Ecol. 25:253-273.

63. Boyd, H. 1956b. The use of hand-reared ducks for supplementing wild populations. Wildfowl Trust Ann. Rep. 8:91-94.

64. Boyd, H. 1957a. Mortality and kill amongst British-ringed Teal *Anas crecca*. Ibis 99:157-177.

65. Boyd, H. 1957b. The White-fronted Geese of England and Wales. Wildfowl Trust Ann. Rep. 8:80-84.

66. Boyd, H. 1958. The survival of White-fronted Geese (*Anser albifrons flavirostris*) ringed in Greenland. Dansk Ornith. For. Tiddsk. 52:1-8.

67. Boyd, H. 1959. Movements of marked sea and diving ducks in Europe. Wildfowl Trust Ann. Rep. 10:59-70.

68. Boyd, H. 1961a. The number of Barnacle Geese in Europe in 1959-60. Wildfowl Trust Ann. Rep. 12:116-124.

69. Boyd, H. 1961b. The flightless period of the Mallard in England. Wildfowl Trust Ann. Rep. 12:140-143.

70. Boyd, H. 1962. Population dynamics and the exploitation of ducks and geese. In le Cren, E.D. and Holdgate, N.W. (Eds). The Exploitation of Natural Animal Populations. Oxford, Blackwell.

71. Boyd, H. 1968. Barnacle Geese in the west of Scotland, 1957-67. Wildfowl 19:96-107.

72. Boyd, H. 1970. The migrations of British geese and ducks. In Sedgwick, N.M., Whitaker, P. and Harrison, J.G. (Eds). The New Wildfowler in the 1970s:75-84. London, Barrie and Jenkins.

73. Boyd, H. 1974. Comment on a proposed reduction in the number of Barnacle Geese wintering in Islay. Unpubl. Rep. NCC.

74. Boyd, H. and Eltringham, S.K. 1962. The Whooper Swan in Great Britain. Bird Study 9:217-241.

75. Boyd, H. and Ogilvie, M.A. 1961. The distribution of Mallard ringed in southern England. Wildfowl Trust Ann. Rep. 12:125-136.

76. Boyd, H. and Ogilvie, M.A. 1964. Losses of Mute Swans in England in the winter of 1962-63. Wildfowl Trust Ann. Rep. 15:37-40.

77. Boyd, H. and Ogilvie, M.A. 1969. Changes in the British-wintering population of the Pink-footed Goose from 1950 to 1975. Wildfowl 20:33-46.

78. Boyd, H. and Ogilvie, M.A. 1972. Icelandic Greylag Geese wintering in Britain, 1960-1971. Wildfowl 23:64-82.

79. Boyd, H. and Radford, J. 1958. Barnacle Geese in Western Scotland, February 1957. Wildfowl Trust Ann. Rep. 9:42-46.

80. Boyd, H., Harrison, J.G. and Allison, A. 1975. Duck Wings. Sevenoaks, WAGBI.

81. Boyd, J.M. 1958. The birds of Tiree and Coll. Brit. Birds 51:41-56, 103-118.

82. Brakhage, G.K. 1953. Migration and mortality of ducks hand-reared and wild-trapped at Delta, Manitoba. J. Wild. Mgmt. 17:465-477.

83. Brazil, M.A. 1981a. Whooper Swan migration and movements: a study using neck-bands. Unpubl. Rep. University of Stirling.

84. Brazil, M.A. 1981b. The behavioural ecology of the Whooper Swan *Cygnus cygnus*. PhD. Thesis, University of Stirling.

85. Brazil, M.A. and Kirk, J. 1981. The current status of Whooper Swans in Great Britain and Ireland. Unpubl. Rep. University of Stirling.

86. Breconshire Birds. Brecknock Naturalists' Trust.

87. Bridson, R.H. 1976. The accretion and erosion of saltmarsh at Caerlaverock National Nature Reserve, Dumfries. Unpubl. Rep. NCC.

88. Brouwer, G.A. and Tinbergen, L. 1939. De Verspreiding der Kleine Zwanen *Cygnus b. bewickii* Yarr. in de Zuiderzee, voor en na de verzoeting. Limosa 12:1-18.

89. Brown, C. and Jenkins, D. 1973. Long-tailed Ducks in the Uists. Scot. Birds 7:404-405.

90. Brownie, C., Anderson, D.R., Burnham, K.P. and Robson, D.S. 1978. Statistical inference from band recovery data—a handbook. US Dept. Interior Fish and Wildlife Service Resource Publ. 131.

91. Bryant, D.M. 1978. Moulting Shelducks on the Forth Estuary. Bird Study 25:103-108.

92. Bryant, D.M. 1981a. Moulting Shelducks on the Wash. Bird Study 28:157.

93. Bryant, D.M. 1981b. Moulting Shelducks on the Wash. Norfolk Bird and Mammal Report 1980:264-265.

94. Bryant, D.M. and Leng, J. 1975. Feeding distribution and behaviour of Shelduck in relation to food supply. Wildfowl 26:20-30.

95. Burton, J.F. 1974. Duck and waders on the Thames, 1945 to 1950. London Bird Report 37:65-66.

96. Burton, P.J.K. 1961. The Brent Goose and its food supply in Essex. Wildfowl Trust Ann. Rep. 12:104-115.

97. Butterworth, J., Lester, P. and Nickless, G. 1972. Distribution of heavy metals in the Severn estuary. Marine Pollution Bull. 3:72-74.

98. Buxton, N.E. 1978. The wildlife importance of the Stanlow and Ince Banks. Mersey Marshes Local Plan Technical Report No. 3.

99. Buxton, N.E. 1983. The wildfowl of Lewis and Harris, Outer Hebrides. Unpubl. Rep. NCC.

100. Buxton, N.E. and Cunningham, W.A.J. 1980. The status of the Whooper Swan in Lewis and Harris, Outer Hebrides. J. Western Isles NHS.

101. Buxton, N.E. and Young, C.M. 1981. The food of the Shelduck in north-east Scotland. Bird Study 28:41-48.

102. Buxton, N.E., Gillham, R. and Pugh Thomas, M. 1977. The Dee Estuary water storage scheme feasibility study. Tech. Rep. II. Report of the ornithological studies. Unpubl. Rep. to Central Water Planning, Dee and Clwyd River Division and Welsh Water Authority.

103. Cabot, D. and West, B. 1973. Population dynamics of Barnacle Geese *Branta leucopsis* in Ireland. Proc. Roy. Irish Acad. 73:415-443.

104. Cadbury, C.J. 1975. Populations of swans at the Ouse Washes, England. Wildfowl 26:148-159.

105. Cadbury, C.J. and St Joseph, A.K.M. 1976. Brent Geese in the Wash in late spring. Brit. Birds 71:268-269.

106. Cadbury, C.J. and Rooney, M.E.S. 1983. Survey of breeding wildfowl and waders of wet grasslands in Cambridgeshire, 1980 and 1982. Cambridge Bird Club Report 56:28-34.

107. Cadman, W.A. 1953. The winter food and ecological distribution of Greenland White-fronted Geese in Britain. Brit. Birds 46:374-375.

108. Cambrian Birds. Cambrian Ornithological Society.

109. Campbell, B. 1960. The Mute Swan census in England and Wales 1955-56. Bird Study 7:208-223.

110. Campbell, J.W. 1946. The food of the Wigeon and Brent Goose. Brit. Birds 39:194-200, 226-232.

111. Campbell, J.W. 1960. The Outer Hebrides. Unpubl. Rep. Nature Conservancy.

112. Campbell, L.H. 1978a. Patterns of distribution and behaviour of flocks of seaducks wintering at Leith and Musselburgh, Scotland. Biol. Cons. 14:111-123.

113. Campbell, L.H. 1978b. Report of the Forth Ornithological Working Party. Unpubl. Rep. NCC. 4 vols.

114. Campbell, L.H. 1978c. Forth and Tay Estuaries winter seaduck surveys 1977-1978. Unpubl. Rep. NCC.

115. Campbell, L.H. 1984. The impact of changes in sewage treatment on seaducks wintering in the Firth of Forth, Scotland. Biol. Cons. 28:173-180.

116. Campbell, L.H. and Milne, H. 1983. Moulting Eiders in eastern Scotland. Wildfowl 34:105-107.

117. Carney, S.M. 1964. Preliminary key to waterfowl age and sex identification by means of wing plumage. US Dept. Interior Fish and Wildlife Service Special Scientific Rep.—Wildlife, 82.

118. Central Water Planning Unit 1976. The Wash Water Storage Scheme Feasibility Study: a report on the ecological studies. Natural Environment Research Council Publ. Series C. No. 15.

119. Chandler, R.J. 1981. Influxes into Britain and Ireland of Red-necked Grebes and other waterbirds during the winter 1978/79. Brit. Birds 74:55-81.

120. Charman, K. 1975. The feeding ecology of the Brent Goose. Rep. of the Maplin Ecological Research Programme, Part II 3b:259-289. London, Dept. of the Environment (unpubl.).

121. Charman, K. 1979a. The seasonal pattern of food utilisation by *Branta bernicla* on the coast of south-east England. Proc. 1st Tech. Mtng. Western Palearctic Migratory Bird Management, Paris 1977:64-75. Slimbridge, IWRB.

122. Charman, K. 1979b. Feeding ecology and energetics of the Dark-bellied Brent Goose *Branta bernicla bernicla* in Essex and Kent. Ecological Processes in Coastal Environments. Symp. Brit. Ecol. Soc., September 1977:451-465.

123. Charman, K. and Macey, A. 1978. The winter grazing of saltmarsh vegetation by Dark-bellied Brent Geese. Wildfowl 29:153-162.

124. Clark, J.M. 1979. The Avon Valley from Ringwood to Harbridge. An ornithological appraisal. Unpubl. Rep. NCC.

125. Clark, M. 1980. Canada Geese breeding, and other geese wintering, on the Island of Colonsay. Western Naturalist 1977:103.

126. Coates, M. 1977. Nene Lake Gravel Pits. Cambridge Bird Club Report 50:55-61.

127. Cohen, E. 1963. Birds of Hampshire and the Isle of Wight. Edinburgh and London, Oliver and Boyd.

128. Coleman, A.E. and Minton, C.D.T. 1980. Mortality of Mute Swan progeny in an area of south Staffordshire. Wildfowl 31:22-28.

129. Cook, W.A. 1960. The numbers of ducks caught in Borough Fen Decoy 1776-1959. Wildfowl Trust Ann. Rep. 11:118-122.

130. Cook, W. A. and Pilcher, R.E.M. 1982. The History of Borough Fen Decoy. Ely, Providence Press.

131. Cooke, A. 1975. The effects of fishing on waterfowl on Grafham Water. Cambridge Bird Club Report 48:40-46.

132. Cooke, A. 1977. The Birds of Grafham Water. Huntingdon, Huntingdon Local Publications Group.

133. Coombes, R.A.H. 1950. The moult migration of the Shelduck. Ibis 92:405-418.

134. Corlett, J. 1972. The ecology of Morecambe Bay. 1. Introduction. J. Appl. Ecol. 9:153-159.

135. Cotswold Water Park, Aquatic Nature Reserve. Rep. of Technical Working Party. 1975.

136. Coxon, P.S. 1982. Greylag Geese and corn. Loch an Duin SSS1, North Uist. Unpubl. Rep. NCC.

137. Cragg, B.A., Fry, J.C., Bacchus, L. and Thurley, S.S. 1980. The aquatic vegetation of Llangorse Lake, Wales. Aquatic Bot. 8:187-196.

138. Cramp, S. 1957. The Census of Mute Swans, 1955 and 1956. London Bird Report 21:58-62.

139. Cramp, S. 1963. The Census of Mute Swans, 1961. London Bird Report 26:100-103.

140. Cramp, S. 1972. One hundred and fifty years of Mute Swans on the Thames. Wildfowl 23:119-124.

141. Cramp, S. and Simmons, K.E.L. 1977. Birds of the Western Palearctic Vol. I. Oxford, the University Press.

142. Cramp, S. and Teagle, W.G. 1955. A comparative study of the birds of two stretches of the Thames in Inner London, 1951-53. London Bird Report 18:42-57.

143. Cramp, S. and Tomlins, A.D. 1966. The birds of Inner London 1951-65. Brit. Birds 59:209-233.

144. Crosby, M.J. 1982. Wildfowl and wader counts in the Vyrnwy Confluence area winter 1981/82. Unpubl. Rep. RSPB.

145. Crosby, M.J. 1983. Wildfowl and waders on the Vyrnwy Confluence winter 1982/83. Unpubl. Rep. RSPB.

146. CRRAG. 1978. Digest of countryside recreation statistics. Countryside Commission.

147. Cundale, G.C. 1980. Llangorse Lake is dying: fact or fantasy? Nature in Wales 17:71-79.

148. Dare, P.J. and Schofield, P. 1976. Ecological Survey of the Lavan Sands: Ornithological Survey, 1969-74. Report of Cambrian Orn. Soc. Project No. 72-12.

149. Darling, F.F. 1940. Island Years. London, Bell.

150. Darling, F.F. and Boyd, J.M. 1969. The Highlands and Islands. London, Collins.

151. Deans, I.R. 1979. Feeding of Brent Geese on cereal fields in Essex and observations of the subsequent loss of yield. Agro-Ecosystems 5:283-288.

152. Dee Estuary Conservation Group 1979. The conservation of the Dee Estuary. Unpubl. Rep. DECG.

153. Dementiev, G.P. and Gladkov, N.A. 1952. Birds of the Soviet Union. Moscow.

154. Dennis, M.K. 1983. A review of the wetlands of metropolitan Essex. Essex Bird Report 1982:80-90.

155. Dennis, R.H. 1964. Capture of moulting Canada Geese on the Beauly Firth. Wildfowl Trust Ann. Rep. 15:71-74.

156. Dennis, R.H. 1975. Ornithological importance of Moray Firth area. Unpubl. Rep. RSPB.

157. Department of Agriculture and Fisheries for Scotland 1982. Wild Geese and Scottish Agriculture. DAFS advisory leaflet.

158. Department of the Environment 1982. Cleaning up the Mersey. Consultation paper.

159. Des Lauriers, J.R. and Brattstrom, B.H. 1965. Cooperative feeding behaviour in Red-breasted Mergansers. Auk 82:639.

160. Donker, J.K. 1959. Migration and distribution of Wigeon *Anas penelope* L., in Europe, based on ringing results. Ardea 47:1-27.

161. Drent, R.H., Ebbinge, B. and Weijand, B. 1982. Balancing the energy budgets of arctic breeding geese throughout the annual cycle: a progress report. Verh. Orn. Ges. Bayern 23:239-264.

162. Duffey, E. 1964. The Norfolk Broads: a regional study of wildlife conservation in a wetland area with high tourist attraction. Proc. MAR Conf. November 1962. IUCN Publications New Series 3:290-301.

163. Dzubin, A. 1969. Assessing breeding populations of ducks by ground counts. Canadian Wildl. Serv. Rep. Series 6.

164. Eastern Sports Council 1975. A Regional Strategy for Water Recreation. Zone 3. The Waters of Essex and Parts of Suffolk.

165. Ebbinge, B. 1979. The significance of the Dutch part of the Wadden Sea for *Branta bernicla bernicla*. Proc. 1st Tech. Mtng. Western Palearctic Migratory Bird Management, Paris 1977:77-87. Slimbridge, IWRB.

166. Edington, J.M. and Edington, M.A. 1977. Ecology and Environmental Planning. London, Chapman and Hall.

167. Edroma, E.L. and Jumbe, J. 1983. The number and daily activity of the Egyptian Goose in Rwenzori National Park, Uganda. Wildfowl 34:99-104.

168. Elder, W.H. 1955. Fluoroscopic measures of shooting pressure on Pink-footed and Greylag Geese. Wildfowl Trust Ann. Rep. 7:123-126.

169. Elkins, N. 1974. Long-tailed Ducks in the Outer Hebrides. Scot. Birds 8:201-202.

170. Ellis, P. 1979. Unpubl. Rep. NCC. (Skye Survey).

171. Ellwood, J. 1970. Goose conservation. WAGBI Ann. Rep. and Year Book 1969-70:36-39.

172. Eltringham, S.K. 1963. The British population of the Mute Swan in 1961. Bird Study 10:10-28.

173. Eltringham, S.K. 1974. The survival of broods of the Egyptian Goose in Uganda. Wildfowl 25:41-48.

174. Eltringham, S.K. and Boyd, H. 1960. The Shelduck population in the Bridgwater Bay moulting area. Wildfowl Trust Ann. Rep. 11:107-117.

175. Emlen, S.T. and Ambrose, H.W. 1970. Feeding interactions of Snowy Egrets and Red-breasted Mergansers. Auk 87:164-165.

176. Eriksson, M.O.G. and Andersson, M. 1982. Nest parasitism and hatching success in a population of Goldeneyes *Bucephala clangula*. Bird Study 29:49-54.

177. Essex Bird Reports. Essex Bird Watching and Preservation Society.

178. Evans, M.E. 1977. Recognising individual Bewick's Swans by bill pattern. Wildfowl 28:153-158.

179. Evans, M.E. 1979a. Aspects of the life cycle of the Bewick's Swan, based on recognition of individuals at a wintering site. Bird Study 26:149-162.

180. Evans, M.E. 1979b. Population composition, and return according to breeding status, of Bewick's Swans wintering at Slimbridge, 1963 to 1976. Wildfowl 30:118-128.

181. Evans, M.E. 1982. Movements of Bewick's Swans *Cygnus columbianus bewickii* marked at Slimbridge, England from 1960-1979. Ardea 70:59-75.

182. Evans, M.E. and Kear, J. 1978. Weights and measurements of Bewick's Swans during winter. Wildfowl 29:118-122.

183. Evans, M.E. and Sladen, W.J.L. 1980. A comparative analysis of the bill markings of Whistling and Bewick's Swans and out-of-range occurrences of the two taxa. Auk 97:697-703.

184. Evans, M.E., Wood, N.A. and Kear, J. 1973. Lead shot in Bewick's Swans. Wildfowl 24:56-60.

185. Evans, P.R. 1981. Reclamation of intertidal land: some effects on Shelduck and wader populations in the Tees estuary. Verh. Orn. Ges. Bayern 23:147-168.

186. Evans, P.R. and Flower, W.U. 1967. The birds of the Small Isles. Scot. Birds 4:404-445.

187. Evans, P.R., Herdson, D.M., Knight, P.J. and Pienkowski, M.W. 1979. Short term effects of reclamation on part of Seal Sands, Teesmouth, on wintering waders and Shelduck. 1. Shorebird diets and invertebrate densities. Oecologia 14:183-206.

188. Fabricius, E. 1983. Kanadagasen—en livskrafting femtioaring i Sverige. Fauna och Flora 75:205-221.

189. Fair Isle Bird Observatory Reports. FIBO Trust.

190. Ferns, P. 1980. Distribution of wading birds and wildfowl in the Severn Estuary and the coastal levels. In Shaw, T.L. (Ed). An Environmental Appraisal of Tidal Power Stations: With Particular Reference to the Severn Barrage:144-169. London, Pitman.

191. Fisher, J. 1941. Watching Birds. Harmondsworth, Penguin.

192. Fog, J. 1971. Survival and exploitation of Mallards *Anas platyrhynchos* released for shooting. Dan. Rev. Game Biol. 6, No. 4.

193. Fog, M. 1982. Baiting as a means of prevention of crop damage by Pink-footed Geese *Anser brachyrhnchus* at Vest-stadil Fjord, Denmark. In Scott, D.A. (Ed). Managing Wetlands and their Birds:233-234. Slimbridge, IWRB.

194. Foley, D. 1954. Studies on survival of three strains of Mallard ducklings in New York State. New York Fish and Game J. 1:75-83.

195. Folk, C. 1971. A study on diurnal activity rhythm and feeding habits of *A. fuligula*. Acta Sc. Nat. Brno 5:1-39.

196. Forshaw, W.D. 1983. Numbers, distribution and behaviour of Pink-footed Geese in Lancashire. Wildfowl 34:64-76.

197. Fox, A.D. and Madsen, J. 1981. The prenesting behaviour of the Greenland White-fronted Goose. Wildfowl 32:48-54.

198. Fox, A.D. and Stroud, D.A. (Eds). 1981. Report of the 1979 Greenland White-fronted Goose Study Expedition to Eqalungmiut Nunat, West Greenland. Aberystwyth, Greenland White-fronted Goose Study Group.

199. French, F.A. 1983. River Thames Swan Survey—9th January, 1983. Unpubl. Rep. Thames Fisheries Consultative Council.

200. Fuller, R.J. and Glue, D.E. 1981. Sewage works as bird habitats in Britain. Biol. Cons. 17:165-181.

201. Furness, R.W. 1981. Seabird populations of Foula. Scot. Birds 11:238-253.

202. Game Conservancy 1981. Wildfowl Management on Inland Waters. Fordingbridge, The Game Conservancy.

203. Gardarsson, A. 1976. Thjorsarver. Framleidsla Grodurs og Heidargaesar. Reykjavik, University of Iceland.

204. Gardarsson, A. and Sigurdsson, J.B. 1972. Skyrsla um rannsoknir a heidtagaes i Pjorsarvverum. Sumarid 1971. Natturufraestistifrum Islands. Reykjavik, University of Iceland.

205. Gardarsson, A. and Skarphedinsson, K.H. 1984. A census of the Icelandic Whooper Swan population. Wildfowl 35:37-47.

206. Garnett, M.G.H. 1980. Moorland breeding and moulting of Canada Geese in Yorkshire. Bird Study 27:219-226.

207. George, R.W. 1974. Birds at Surrey Commercial Docks, April 1971 to December 1972. London Bird Report 37:67-70.

208. Gilbert, B. 1977. Uncle Sam says SCRAM! Audubon 79:42-55.

209. Gillham, E.H. and Homes, R.C. 1950. Birds of the North Kent Marshes. London, Collins.

210. Gillham, M.E. 1956. Feeding habits and seasonal movements of Mute Swans in South Devon estuaries. Bird Study 3:205-212.

211. Ginn, H.B. 1969. The use of annual ringing and nest record card totals as indicators of bird population levels. Bird Study 16:210-248.

212. Glamorgan Bird Reports. Cardiff Naturalists' Society/ Gower Ornithological Society.

213. Glegg, W.E. 1943. The food of Wigeon *Mareca penelope*. Ibis 85:82-87.

214. Gloucestershire Bird Reports. Gloucestershire Naturalists' Society.

215. Glover, R. 1979. Roosting movements on Essex estuaries. Essex Bird Report 1978:70-92.

216. Glue, D.E. 1971. Saltmarsh reclamation stages and their associated bird-life. Bird Study 18:187-198.

217. Glyptis, S. 1980. National Angling Survey. Unpubl. Rep. Sports Council.

218. Goethe, F. 1961. The moult gatherings and moult migration of Shelduck in north-west Germany. Brit. Birds 54:145-160.

219. Goode, D.A. 1981. Lead poisoning in swans. Rep. of the Nature Conservancy Council Working Group. NCC.

220. Gorman, M.L. and Milne, H. 1972. Creche behaviour in the Common Eider *Somateria m. mollissima* L. Ornis Scand. 3:21-25.

221. Grant, P.J. 1971. Birds at Surrey Commercial Docks. London Bird Report 35:87-91.

222. Grant, P.J., Harrison, J.G. and Noble, K. 1974. The return of wildlife to the Inner Thames. London Bird Report 37:61-64.

223. Green, B.H. (Ed). 1971. Report of working party on wildlife conservation in the North Kent Marshes. Unpubl. Rep. NCC.

224. Greenhalgh, M.E. 1975. Wildfowl of the Ribble Estuary. Chester, WAGBI.

225. Greenwood, J.J.D. and Keddie, J.P.F. 1968. Birds killed by oil in the Tay Estuary, March and April, 1968. Scot. Birds 5:189-196.

226. Greenwood, J.J.D., Donally, R.J., Feare, C.J., Gordon, N.J. and Waterson, G. 1971. A massive wreck of oiled birds: northeast Britain, winter 1970. Scot. Birds 6:235-250.

227. Grenquist, P. 1970. Status of the species of wildfowl occurring in Finland. Proc. Int. Reg. Mtng. Conserv. Wildfowl Resources, Leningrad 1968:83-87.

228. Gwent Bird Reports. Gwent Ornithological Society.

229. Hale, W.G. 1974. Aerial counting of waders. Ibis 116:412.

230. Halliday, J.B. 1978. The feeding distribution of birds on the Clyde Estuary tidal flats. Unpubl. Rep. NCC.

231. Halliday, J.B. 1979. The feeding distribution of birds on sea lochs in the Clyde Sea Area—1979. Unpubl. Rep. NCC.

232. Halliday, J.B., Curtis, D.J., Thompson, D.B.A., Bignall, E.M. and Smythe, J.C. 1982. The abundance and feeding distribution of Clyde Estuary shorebirds. Scot. Birds 12:65-72.

233. Hamilton, G.A. and Stanley, P.I. 1975. Further cases of poisoning of wild geese by organophosphorus winter wheat seed treatment. Wildfowl 26:49-54.

234. Hammond, N. 1975. Outer Hebrides Expedition, 15-22 February 1975. Unpubl. Rep. RSPB Cumbria Group.

235. Hampshire Bird Reports. Hampshire Ornithological Society.

236. Hardman, J.A. and Cooper, D.R. 1980. Mute Swans on the Warwickshire Avon—a study of decline. Wildfowl 31:29-36.

237. Harradine, J. 1981. The Duck Production Survey 1978/79-1980/81. BASC Publ.

238. Harradine, J. 1982. Sporting shooting in the United Kingdom—some facts and figures. Proc. 2nd Mtng. IUGB Working Group on Game Statistics, Netherlands 1982.

239. Harradine, J. 1983. Facts about shooting. Shooting and Conservation, Winter 1983:28-30.

240. Harrison, G.R., Dean, A.R., Richards, A.J. and Smallshire, D. (Eds). 1982. The Birds of the West Midlands. Worcester, West Midlands Bird Club.

241. Harrison, J.G. 1960. A technique for removing wildfowl viscera for research. Wildfowl Trust Ann. Rep. 11:135-136.

242. Harrison, J.G. 1969. Some preliminary results from the release of hand-reared Gadwall. WAGBI Ann. Rep. and Year Book 1968-69:37-40.

243. Harrison, J.G. 1972. Wildfowl of the North Kent Marshes. WAGBI Conservation Publ.

244. Harrison, J.G. 1974. The Sevenoaks Gravel Pit Reserve. WAGBI Conservation Publ.

245. Harrison, J.G. 1979. A new overland migration route of *Branta bernicla bernicla* in south-east England in autumn. Proc. 1st Tech. Mtng. Western Palearctic Migratory Bird Management, Paris 1977:60-63. Slimbridge, IWRB.

246. Harrison, J.G. and Boyd, H. 1968. A duck production survey in Britain. Report of the pilot scheme 1965-68. WAGBI Ann. Rep. and Year Book 1967-68:29-33.

247. Harrison, J.G. and Grant, P.J. 1976. The Thames Transformed. London, Andre Deutsch.

248. Harrison, J.G. and Meikle, A. 1968. An assessment of the efficiency of the WAGBI—Wildfowl Trust experimental reserve. WAGBI Ann. Rep. and Year Book 1967-68:44-51.

249. Harrison, J.G. and Wardell, J. 1970. WAGBI duck to supplement wild populations. In Sedgwick, N.M., Whitaker, P. and Harrison, J.G. (Eds). The New Wildfowler in the 1970s:195-209. London, Barrie and Jenkins.

250. Harrison, J.G., Grant, P.J. and Swift, J. 1976. Thames transformed. Proc. Int. Conf. Conserv. Wetlands and Waterfowl, Heiligenhafen 1974:355-358. Slimbridge, IWRB.

251. Harrison, J.M., Harrison, J.G. and Meikle, A. 1972. The WAGBI—Wildfowl Trust Experimental Reserve—the first eleven years. Part I. Development of the wildfowl population. Wildfowl Trust Ann. Rep. 18:44-47.

252. Harrison, R. and Rogers, D.A. 1977. The Birds of Rostherne Mere. Banbury, Nature Conservancy Council.

253. Harvey, W.G. and Henderson, A.C.B. 1977-82. Surveys of breeding and wintering birds of the East Kent Lowlands. Unpubl. Reps. Kent Orn. Soc.

254. Hawker, P. 1893. The diary of Colonel Peter Hawker. Vols. I and II. London, Longmans, Green and Co.

255. Hay, G.W.J. 1982. Some aspects of the ecology and behaviour of overwintering Pochard (*Aythya ferina*) and Tufted Duck (*A. fuligula*) in the Cardiff Area. Unpublished thesis to University College, Cardiff.

256. Henny, G.J. and Holgersen, N.E. 1974. Range expansion and population increase of the Gadwall in eastern North America. Wildfowl 25:95-101.

257. Henty, C.J. 1977. The roost flights of Whooper Swans in the Devon Valley (Central Scotland). Forth Nat. and Hist. 2:31-35.

258. Heubeck, M., Richardson, M.G. and Syratt, W.J. 1983. Effects of the Esso Bernicia oil spill (December 1978) on the wintering population of diving seabirds in Sullom Voe, Shetland. Proc. Symp. Physiological Research, Clinical Application and Rehabilitation, New Jersey 1982:115-129.

259. Hewson, R. 1962. Whooper Swans at Loch Park, Banffshire, 1955-61. Bird Study 10:203-210.

260. Hewson, R. 1973. Changes in a winter herd of Whooper Swans at a Banff Loch. Bird Study 20:41-49.

261. Hilden, O. 1964. Ecology of duck populations in the island group of Valassaaret, Gulf of Bothnia. Ann. Zool. Fennici 1:153-279.

262. Hill, D.A. 1982. The comparative population ecology of Mallard and Tufted Duck. PhD. Thesis, University of Oxford.

263. Hitchings, E.C. 1977. Planning and development on the West Glamorgan shore. In Nelson-Smith, A. and Bridges, E.M. (Eds). Problems of a Small Estuary. Proc. Symp. Burry Inlet, Swansea 1976:8:2/1-8:2/14. Swansea, Institute of Marine Studies.

264. Holland, S.C. and Mardle, D.V. 1977. Birdwatching in the Cotswold Water Park. Gloucestershire County Council.

265. Hope Jones, P. 1979. Roosting behaviour of Long-tailed Ducks in relation to possible oil pollution. Wildfowl 30:155-158.

266. Hope Jones, P. and Kinnear, P.K. 1979. Moulting Eiders in Orkney and Shetland. Wildfowl 30:109-113.

267. Hope Jones, P. and Lawton Roberts, J. 1983. Birds of Denbighshire. Nature in Wales. New Series. Vol. I. Part 2:56-65.

268. Hope Jones, P., Howells, G., Rees, E.I.S. and Wilson, J. 1970. Effect of Hamilton Trader oil on birds in the Irish Sea in May 1969. Brit. Birds 63:97-110.

269. Hopkins, P.G. and Coxon, P. 1979. Birds of the Outer Hebrides: Waterfowl. Proc. Roy. Soc. Edin. (B), 77:431-444.

270. Hori, J. 1966. Observations on Pochard and Tufted Duck breeding biology with particular reference to colonisation of a home range. Bird Study 13:297-310.

271. Hori, J. 1969. Social and population studies in the Shelduck. Wildfowl 20:5-22.

272. Howie, J.E. 1980. Ornithological interest of some inland waters in Wigtownshire. Unpubl. Rep. NCC.

273. Hudgell, S. and Smith, J.T. 1974. The birds of Hanningfield Reservoir. Essex Bird Report 1973:63-78.

274. Hudson, R. 1976. Ruddy Ducks in Britain. Brit. Birds 69:132-143.

275. Hughes, S.W.M. 1973. The Canada Goose in Sussex. Sussex Bird Report 25:51-66.

276. Hughes, S.W.M. and Codd, D. 1982. A further assessment of the status of the Mandarin in Sussex. Sussex Bird Report 34:84.

277. Humber Advisory Group/University of Hull 1979. The Humber Estuary. Natural Environment Research Council Publ. Series C. No.20.

278. Hume, R.A. 1976. Reactions of Goldeneyes to boating. Brit. Birds 69:178-179.

279. Hutchinson, C. 1979. Ireland's Wetlands and their Birds. Dublin, Irish Wildbird Conservancy.

280. Inglis, I.R. 1980. Visual birds scarers: an ethological approach. In Wright, E.N., Inglis, I.R. and Feare, C.J. (Eds). Bird Problems in Agriculture: 121-143. Croydon, British Crop Protection Council Publ.

281. Institution of Water Engineers 1963. Final Report to the Council on the Recreational Use of Waterworks. London, Inst. Water Engineers.

282. International Waterfowl Research Bureau 1979. Recommendations. In Smart, M. (Ed). Proc. 1st Tech. Mtng. Western Palearctic Migratory Bird Management, Paris 1977:208-215. Slimbridge, IWRB.

283. International Waterfowl Research Bureau 1981. Conference on the Conservation of Wetlands of International Importance Especially as Waterfowl Habitat, Cagliari 1980. Slimbridge, IWRB.

284. International Waterfowl Research Bureau 1982. Summary record of the third technical meeting on western Palearctic migratory bird management. In Scott, D.A. (Ed). Managing Wetlands and their Birds:333-343. Slimbridge, IWRB.

285. James, H. 1970. Famous wildfowlers of the past. In Sedgwick, N.M., Whitaker, P. and Harrison, J.G. (Eds). The New Wildfowler in the 1970s:319-331. London, Barrie and Jenkins.

286. Jenkins, D. 1971. Eiders nesting in East Lothian. Scot. Birds 6:251-255.

287. Jenkins, D., Newton, I. and Brown, C. 1976. Structure and dynamics of a Mute Swan population. Wildfowl 27:77-82.

288. Joensen, A.H. 1972. Studies on oil pollution and seabirds in Denmark, 1968-1971. Dan. Rev. Game Biol. 6, No. 9.

289. Joensen, A.H. 1973. Moult migration and wing feather moult of seaducks in Denmark. Dan. Rev. Game Biol. 8, No. 4.

290. Joensen, A.H. 1974. Waterfowl populations in Denmark, 1965-1973. Dan. Rev. Game. Biol. 9, No. 1.

291. Joensen, A.H. 1977. Oil pollution and seabirds in Denmark, 1971-1976. Dan. Rev. Game Biol. 10:1-31.

292. Kallander, H., Mawdsley, T., Nilsson, L. and Waden, K. 1970. Mass-feeding by Smews. Brit. Birds 63:32-33.

293. Kear, J. 1962. Feeding habits of the Greylag Goose *Anser anser* on the island of Bute. Scot. Birds 2:233-239.

294. Kear, J. 1963a. Wildfowl and agriculture. In Atkinson-Willes, G.L. (Ed). Wildfowl in Great Britain:315-328. London, HMSO.

295. Kear, J. 1963b. The history of potato eating by

wildfowl in Britain. Wildfowl Trust Ann. Rep. 14:54-65.

296. Kear, J. 1963c. The agricultural importance of wild goose droppings. Wildfowl Trust Ann. Rep. 14:72-77.

297. Kear, J. 1970. The experimental assessment of goose damage to agricultural crops. Biol. Cons. 2:206-212.

298. Kear, J. 1972. Reproduction and family life. In Scott, P. and the Wildfowl Trust. The Swans:80-124. London, Michael Joseph.

299. Kent Bird Reports. Kent Ornithological Society.

300. Kerbes, R.H., Ogilvie, M.A. and Boyd. H. 1971. Pink-footed Geese of Iceland and Greenland: a population review based on an aerial survey of Thjorsarver in June 1970. Wildfowl 22:5-17.

301. King, B. 1961. Feral North American Ruddy Ducks in Somerset. Wildfowl 12:167-188.

302. Kinnear, P.K. 1974. Sea duck in Shetland in the winter of 1973/74. Unpubl. Rep. NCC.

303. Kinnear, P.K. and Richardson, M.G. 1977. Waterfowl populations in Sullom Voe/Yell Sound and the Hascosay/Fetlar sea areas. Unpubl. Rep. NCC.

304. Klima, M. 1966. A study on diurnal activity rhythm in the European Pochard *Aythya ferina* (L) in nature. Zool. Listy 15:317-332.

305. Lack, D. 1942. The breeding birds of Orkney. Ibis 84:461-484 and 85:1-27.

306. Ladhams, D.E. 1977. Behaviour of Ruddy Ducks in Avon. Brit. Birds 70:137-146.

307. Lambert, J.E.A. and Cook, W.A. 1967. Dersingham Decoy. Wildfowl 18:22-23.

308. Lampio, T. 1974. Protection of waterfowl in Europe. Finnish Game Res. 34:16-33.

309. Lampio, T., Valentincic, S. and Michaelis, H.K. 1974. Methods and practices of waterfowl hunting rationalization. Finnish Game Res. 34:36-46.

310. Laughlin, K.F. 1974. Bioenergetics of Tufted Duck (*Aythya fuligula*) at Loch Leven, Kinross. Proc. Roy. Soc. Edin. (B), 74:383-390.

311. Laughlin, K.F. 1975. The bioenergetics of the Tufted Duck *Aythya fuligula* (L). PhD. Thesis, University of Stirling.

312. Lea, D. 1974. The birds of Scapa Flow in winter. Unpubl. Rep. RSPB.

313. Lea, D. 1977. Food of seaducks wintering in Scapa Flow. Unpubl. Rep. NCC.

314. Lea, D. 1980. Seafowl in Scapa Flow, Orkney 1974-1978. Unpubl. Rep. by RSPB to NCC. CST Report No. 292.

315. Liddle, M.J. and Scorgie, H.R.A. 1980. The effects of recreation on freshwater plants and animals: a review Biol. Cons. 17: 183-206.

316. Lincolnshire County Council 1981. Development on the Lincolnshire Coast. Draft Subject Plan LCC.

317. London Bird Report. London Natural History Society.

318. London Natural History Society 1957. The Birds of the London Area. London, Collins.

319. Long, J.L. 1981. Introduced Birds of the World. Newton Abbot, David and Charles.

320. Lovegrove, R. 1977. Birds as a resource: their future conservation on the Burry Inlet. In Nelson-Smith, A. and Bridges, E.M. (Eds). Problems of a Small Estuary Proc. Symp. Burry Inlet, Swansea 1976: 5:2/1-11. Swansea, Institute of Marine Studies.

321. Lovegrove, R., Hume, R.A. and McLean, I. 1980. The status of breeding wildfowl in Wales. Nature in Wales 17:4-10.

322. Lumsden, H. 1982. Artificial nesting structures for water birds. In Scott, D.A. (Ed). Managing Wetlands and their Birds: 179-199. Slimbridge, IWRB.

323. MacMillan, A.T. 1970. Goldeneye breeding in East Inverness-shire. Scot. Birds 6:197-198.

324. Madsen, F.J. 1954. On the food habits of diving ducks in Denmark. Dan. Rev. Game Biol. 2:157-266.

325. Madsen, F.J. 1957. On the food habits of some fish-eating birds in Denmark. Dan. Rev. Game Biol. 3:19-82.

326. Madsen, J. 1982. Observations on the Svalbard population of *Anser brachyrhynchus* in Denmark: population trends 1931-1980. Aquila 89:133-140.

327. Martin, G.H. 1973. Ecology and conservation in Langstone Harbour, Hampshire. PhD. Thesis, University of Southampton.

328. Massey M.E. 1975. Waterfowl populations in the Brecon Beacons National Park.Unpubl. Rep. NCC.

329. Massey, M.E. 1976. Winter populations of waterfowl in the Brecon Beacons National Park. Nature in Wales 15:15-21.

330. Mathiasson, S. 1973. A moulting population of Mute Swans with special reference to flight feather moult, feeding ecology and habitat selection. Wildfowl 24:43-53.

331. Matthews, G.V.T. 1958. Feeding grounds for wildfowl. Wildfowl Trust Ann. Rep. 9:51-57.

332. Matthews, G.V.T. 1960. An examination of basic data from wildfowl counts. Proc. XII Int. Orn. Cong., Helsinki 1958:483-491.

333. Matthews, G.V.T. 1963. "Nonsense" orientation as a population variant. Ibis 105:185-197.

334. Matthews, G.V.T. 1969. Nacton Decoy and its catches. Wildfowl 20:131-137.

335. Matthews, G.V.T. and Campbell, C.R.G. 1969. Weights and measurements of Greylag Geese in Scotland. Wildfowl 20:86-93.

336. Meek, E.R. and Little, B. 1977a. The spread of the Goosander in Britain and Ireland. Brit. Birds 70:229-237.

337. Meek, E.R. and Little, B. 1977b. Ringing studies of Goosander in Northumberland. Brit. Birds 70:273-283.

338. Meek, E.R. and Little, B. 1978. Past status of the Brent Goose in Northumberland. Brit. Birds 71:44-46.

339. Meikle, A. 1972. The WAGBI—Wildfowl Trust Experimental Reserve—the first eleven years. Part IV. Expenditure for habitat improvement. Wildfowl Trust Ann. Rep. 18:62-63.

340. Meiklejohn, M.F.M. and Stanford, J.K. 1954. June notes on the birds of Islay. Scot. Nat. 66:129-145.

341. Mendenhall, V.M. 1975. Growth and mortality factors of Eider ducklings (*Somateria m. mollissima*) in north-east Scotland. PhD. Thesis, University of Aberdeen.

342. Merne, O.J. 1977. The changing distribution of the Bewick's Swan in Ireland. Irish Birds 1:3-15.

343. Middle Thames Bird Report. Middle Thames Natural History Society.

344. Millais, J.G. 1901. The Wildfowler in Scotland. London, Longmans.

345. Millais, J.G. 1913. British Diving Ducks. Vol. I. London, Longmans.

346. Mills, D.H. 1962a. The Goosander and Red-breasted Merganser in Scotland. Wildfowl Trust Ann. Rep 13:79-92.

347. Mills, D.H. 1962b. The Goosander and Red-breasted Merganser as predators of salmon in Scottish water. Freshwater and Salmon Fisheries Research 29.

348. Milne, B.S. 1974. Ecological succession and bird life at a newly excavated gravel-pit. Bird Study 21:263-278.

349. Milne, H. 1965. Seasonal movements and distributions of Eiders in north-east Scotland. Bird Study 12:170-180.

350. Milne, H. 1974. Breeding numbers and reproductive rate of Eiders at the Sands of Forvie National Nature Reserve, Scotland. Ibis 116:135-154.

351. Milne, H. 1977. Air survey of seaducks in Outer Hebrides. Unpubl. Rep. to NCC and Inst. Terrestrial Ecology.

352. Milne, H. 1977-1981. Air surveys of seaducks on east coast of Scotland. Unpubl. Reps. NCC.

353. Milne, H. and Campbell, L.H. 1973. Wintering sea-ducks off the east coast of Scotland. Bird Study 20:153-172.

354. Minton, C.D.T. 1968. Pairing and breeding of Mute Swans. Wildfowl 19:41-60.

355. Montier, D.J. (Ed). 1977. Atlas of Breeding Birds of the London Area. London, Batsford.

356. Morgan, N.C. 1974. Historical background to the International Biological Programme Project at Loch Leven, Kinross. Proc. Roy. Soc. Edin. (B), 74:45-56.

357. Moss, B. 1977. Conservation problems in the Norfolk Broads and rivers of East Anglia, England—phytoplankton, boats and the causes of turbidity. Biol. Cons. 12:95-114.

358. Moyse, J. and Thomas, D.K. 1977. Sea-duck of Carmarthen Bay. In Nelson-Smith, A. and Bridges, E.M. (Eds). Problems of a Small Estuary. Proc. Symp. Burry Inlet, Swansea 1976: 5:3/1-13. Swansea, Institute of Marine Studies.

359. Mudge, G.P. 1978. Seaducks in the Moray and Dornoch Firths, Scotland, winter 1977/1978. Unpubl. Rep. to NCC and RSPB.

360. Mudge, G.P. 1983. The incidence and significance of ingested lead pellet poisoning in British wildfowl. Biol. Cons. 27:333-372.

361. Mudge, G.P. and Allen, D.S. 1980. Wintering seaducks in the Moray and Dornoch Firths, Scotland. Wildfowl 31:123-130.

362. Mullie, W.C. and Poorter, E.P.R. 1977. Aantallen, verspreiding en terreinkeus van de Kleine Zwaan bij vijf landelijke tellingen in 1976 en 1977. Watervogels 2:97-101.

363. Murfitt, R.C. and Chown, D.J. 1983. Somerset Moors and Levels Winter Bird Surveys, 1981-82 and 1982-83. Unpubl. Rep. RSPB.

364. Murray, J. and Pullar, L. 1908. Bathymetrical Survey of the Freshwater Lochs of Scotland. London, Roy. Geog. Soc.

365. Murton, R.K., Westwood, N.J. and Isaacson, A.J. 1974. A study of Woodpigeon shooting: The exploitation of a natural animal population. J. Appl. Ecol. 11:61-81.

366. National Trust 1978. Properties of the National Trust. National Trust.

367. Natural Environment Research Council 1980. The Solent estuarine system: an assessment of present knowledge. Natural Environment Research Council Publ. Series C. No. 22.

368. Nature Conservancy 1965. Report on Broadland.

369. Nature Conservancy 1972. A Prospectus for Nature Conservation within the Moray Firth. NERC.

370. Nature Conservancy Council 1977a. The Somerset Wetlands Project. A consultation paper. Unpubl. Rep. NCC South West Region.

371. Nature Conservancy Council 1977b. Oil Pollution Manual. London, NCC.

372. Nature Conservancy Council 1977c. Dee Estuary Research Review.

373. Nature Conservancy Council 1978a. Bosherston and Stackpole research project. 4th Ann. Rep.:70. London, HMSO.

374. Nature Conservancy Council 1978b. Nature Conservation within the Moray Firth Area. A Revised Prospectus. NCC.

375. Nature Conservancy Council 1979. The Ribble Estuary—a test case for nature conservation. 5th Ann. Rep.:41-43. London, HMSO.

376. Nature Conservancy Council 1980. Fifth Annual Report. April 1978 – March 1979. London, HMSO.

377. Nature Conservancy Council 1981. Llandegfedd Reservoir Site of Special Scientific Interest. A nature conservation appraisal. Consultation Report. NCC South Wales Region.

378. Nature Conservancy Council 1983. Eighth Report. April 1981—March 1982. London, HMSO.

379. Newton, I. and Campbell, C.R.G. 1973. Feeding of geese on farmland in east-central Scotland. J. Appl. Ecol. 10:781-801.

380. Newton, I. and Campbell, C.R.G. 1975. Breeding of ducks at Loch Leven, Kinross. Wildfowl 26:83-103.

381. Newton, I. and Kerbes, R.H. 1974. Breeding of Greylag Geese (*Anser anser*) on the Outer Hebrides, Scotland. J. Anim. Ecol. 43:771-783.

382. Newton, I., Thom, V.M. and Brotherston, W. 1973. Behaviour and distribution of wild geese in south-east Scotland. Wildfowl 24:111-121.

383. Nicholson, E.M. 1957. Britain's Nature Reserves. London, Country Life.

384. Nilsson, L. 1969. The migration of the Goldeneye in north-west Europe. Wildfowl 20:112-118.

385. Nilsson, L. 1971. Flyttning, hemorstrohet saunt livslangd hos svenska knipor *Bucephala clangula*. Var Fagelvarld 30:180-184.

386. Nilsson, L. 1971. Habitat selection, food choice and feeding habits of diving ducks in coastal waters of

South Sweden during the non-breeding season. Ornis Scand. 3:55-78.

387. Nilsson, L. 1974. The behaviour of wintering Smew in southern Sweden. Wildfowl 25:84-88.

388. Nilsson, L. 1975. Midwinter distribution and numbers of Swedish Anatidae, Ornis Scand. 6:83-107.

389. Nisbet, I.C.T. 1955. Bewick's Swans in the fenlands: the past and present status. Brit. Birds 48:533-537.

390. Nisbet, I.C.T. 1959. Bewick's Swans in the British Isles in the winters 1954-55 and 1955-56. Brit. Birds 52:393-416.

391. Norderhaug, M. 1970. The present status of the Brent Goose *Branta bernicla hrota* in Svalbard. Norsk Polarinstitutt Arbok 1968:7-23.

392. Norris, D. and Wilson, J. 1983. Greenland White-fronted Goose Research Project progress report for 1982-83. Unpubl. Rep. Irish Forest and Wildlife Service.

393. O'Connor, R.J. and Mead, C. 1980. Oiled seabirds. BTO News October 1980:4.

394. O'Dell, A.D. and Walton, K. 1962. The Highlands and Islands of Scotland. London, Nelson.

395. Ogilvie, M.A. 1962. The movements of Shoveler ringed in Britain. Wildfowl Trust Ann. Rep. 13:65-69.

396. Ogilvie, M.A. 1964. A nesting study of the Mallard in Berkeley New Decoy, Slimbridge. Wildfowl Trust Ann. Rep. 15:84-88.

397. Ogilvie, M.A. 1967. Population changes and mortality of the Mute Swan in Britain. Wildfowl Trust Ann. Rep. 18:64-73.

398. Ogilvie, M.A. 1968. The numbers and distribution of European White-fronted Geese in Great Britain. Bird Study 15:2-15.

399. Ogilvie, M.A. 1969a. The status of the Canada Goose in Britain 1967-69. Wildfowl 20:79-85.

400. Ogilvie, M.A. 1969b. Bewick's Swans in Britain and Ireland during 1956-1969. Brit. Birds 62:505-522.

401. Ogilvie, M.A. 1970. The status of wild geese in Britain and Ireland. In Sedgwick, N.M., Whitaker, P. and Harrison, J.G. (Eds). The New Wildfowler in the 1970s:249-259. London, Barrie and Jenkins.

402. Ogilvie, M.A. 1972a. Large numbered leg bands for individual identification of swans. J. Wildl. Mgmt. 36:1261-1265.

403. Ogilvie, M.A. 1972b. Distribution, numbers and migration. In Scott, P. and the Wildfowl Trust. The Swans:30-55. London, Michael Joseph.

404. Ogilvie, M.A. 1976. Report on aerial surveys of moulting Shelduck in Bridgwater Bay, Somerset, summer 1976. Unpubl. Rep. Wildfowl Trust.

405. Ogilvie, M.A. 1977. The numbers of Canada Geese in Britain, 1976. Wildfowl 28:27-34.

406. Ogilvie, M.A. 1978. Wild Geese. Berkhamsted, Poyser.

407. Ogilvie, M.A. 1979. General introduction to *Branta bernicla bernicla* with particular reference to the wintering area. Proc. 1st. Tech. Mtng. Western Palearctic Migratory Bird Management, Paris 1977:10-18. Slimbridge, IWRB.

408. Ogilvie, M.A. 1981a. The Mute Swan in Britain, 1978. Bird Study 28:87-106.

409. Ogilvie, M.A. 1981b. The hard weather movements [of] *Anas crecca* ringed in western Europe—a preliminary computer analysis. Proc. Symp. Mapping of Waterfowl Distribution, Migration and Habitat, Alushta 1976:119-135. Moscow.

410. Ogilvie, M.A. 1983a. The numbers of Greenland Barnacle Geese in Britain and Ireland. Wildfowl 34:77-88.

411. Ogilvie, M.A. 1983b. Wildfowl of Islay. In Morton Boyd, J. and Bowes, D.R. (Eds). Proc. Roy. Soc. Edin./NCC Symposium on the Inner Hebrides:473-489. Edinburgh, RSE.

412. Ogilvie, M.A. and Booth, C.G. 1970. An oil spillage o[n] Islay in October 1969. Scot. Birds 6:149-153.

413. Ogilvie, M.A. and Boyd, H. 1975. Greenland Barnacl[e] Geese in the British Isles. Wildfowl 26:139-146.

414. Ogilvie, M.A. and Boyd, H. 1976. The number of Pink-footed and Greylag Geese wintering in Britain: observations 1969-75 and predictions 1976-80. Wildfowl 27:63-75.

415. Ogilvie, M.A. and Cook, W.A. 1971. Differential migration of the sexes and other aspects of the recovery overseas of Mallard ringed at Borough Fen Decoy, Northamptonshire. Wildfowl 22:89-97.

416. Ogilvie, M.A. and Cook, W.A. 1972. British recoveri[es] of Mallard ringed at Borough Fen Decoy, Northam[p]tonshire. Wildfowl 23:103-110.

417. Ogilvie, M.A. and Matthews, G.V.T. 1969. Brent Geese, mudflats and Man. Wildfowl 20:119-125.

418. Ogilvie, M.A and St Joseph A.K.M. 1976. Dark-bellie[d] Brent Geese in Britain and Europe, 1955-76. Brit. Bird[s] 69:422-439.

419. Oliver, P.J. 1980. Moulting Tufted Ducks and Pochards in the London area. London Bird Report 44:80-84.

420. Oliver, P.J. 1982. The decline of the Mute Swan in the London area. London Bird Report 46:87-91.

421. Olney, P.J.S. 1957. Food and feeding habits of wildfowl. Wildfowl Trust Ann. Rep. 9:47-51.

422. Olney, P.J.S. 1960. Lead poisoning in wildfowl. Wildfowl Trust Ann. Rep. 11:123-134.

423. Olney, P.J.S. 1962. The food habits of a hand-reared Mallard population. Wildfowl Trust Ann. Rep. 13:11[?]-125.

424. Olney, P.J.S. 1963a. The food and feeding habits of th[e] Tufted Duck *Aythya fuligula*. Ibis 105:55-62.

425. Olney, P.J.S. 1963b. The food and feeding habits of Teal *Anas crecca crecca* L. Proc. Zool. Soc. Lond. 140:169-210.

426. Olney, P.J.S. 1964. The food of Mallard collected fro[m] coastal and estuarine areas. Proc. Zool. Soc. Lond. 142:397-418.

427. Olney, P.J.S. 1965. The autumn and winter feeding biology of certain sympatric ducks. Trans. VI Cong. Int. Union Game Biol., Bournemouth 1963:309-322.

428. Olney, P.J.S. 1965. The food and feeding habits of the Shelduck *Tadorna tadorna*. Ibis 107:527-532.

429. Olney, P.J.S. 1968. The food and feeding habits of Pochard. Biol. Cons. 1:71-76.

430. Olney, P.J.S. 1972. The WAGBI—Wildfowl Trust Experimental Reserve - the first eleven years. Part II.

The feeding ecology of local Mallard and other wildfowl. Wildfowl Trust Ann. Rep. 18:47-55.

431. Olney, P.J.S. and Mills, D.H. 1963. The food and feeding habits of the Goldeneye *Bucephala clangula*. Ibis 105:293-300.

432. Owen, M. 1972. Movements and feeding ecology of White-fronted Geese at the New Grounds, Slimbridge. J. Appl. Ecol. 9:385-398.

433. Owen, M. 1973. The winter feeding ecology of Wigeon at Bridgwater Bay, Somerset. Ibis 115:227-242.

434. Owen, M. 1975a. Cutting and fertilizing grassland for winter goose management. J. Wildl. Mgmt. 39:163- 167.

435. Owen, M. 1975b. Implications for Wildfowl. In Shaw, T.L. 1975. An Environmental Appraisal of the Severn Barrage:73-77. Bristol, T.L. Shaw.

436. Owen, M. 1976a. The selection of winter food by White-fronted Geese. J.Appl. Ecol. 13:715-729.

437. Owen, M. 1976b. Factors affecting the distribution of geese in the British Isles. Wildfowl 27:143-147.

438. Owen, M. 1977. The role of wildfowl refuges on agricultural land in lessening the conflict between farmers and geese in Britain. Biol. Cons. 11:209-222.

439. Owen, M. 1978. The Greenland White-fronted Goose—the case for protection. Unpubl. Rep. Wildfowl Trust.

440. Owen, M. 1980a. Wild Geese of the World. Batsford, London.

441. Owen, M. 1980b. The role of refuges in wildfowl management. In Wright, E.N., Inglis, I.R. and Feare, S.J. (Eds). Bird Problems in Agriculture:144-156. Croydon, British Crop Protection Council.

442. Owen, M. 1982a. Management of summer grazing and winter disturbance on goose pasture at Slimbridge, England. In Scott, D.A. (Ed). Managing Wetlands and their Birds:67-72. Slimbridge, IWRB.

443. Owen, M. 1982b. Population dynamics of Svalbard Barnacles 1970-1980. The rate, pattern and causes of mortality as determined by individual marking studies. Aquila 89:229-247.

444. Owen, M. 1984. Dynamics and age structure of an increasing goose population, the Svalbard Barnacle Goose. Norsk Polarinstitutt Skrifter 181:37-47.

445. Owen, M. and Cadbury, C.J. 1975. The ecology and mortality of swans at the Ouse Washes, England. Wildfowl 26:31-42.

446. Owen, M. and Cook W.A. 1977. Variations in body weight, wing length and condition of Mallard *Anas platyrhynchos platyrhynchos* and their relationships to enviromental changes. J. Zool. Lond. 183:377-395.

447. Owen, M. and Kear, J. 1972. Food and feeding habits. In Scott, P. and the Wildfowl Trust. The Swans:58-77. London, Michael Joseph.

448. Owen, M. and Kerbes, R.H. 1971. On the autumn food of Barnacle Geese at Caerlaverock National Nature Reserve. Wildfowl 22:114-119.

449. Owen, M. and King, R. 1979. The duration of the flightless period in freeliving Mallard. Bird Study 27:267-269.

450. Owen, M. and Montgomery, S. 1978. Body measurements of Mallard caught in Britain. Wildfowl 29:123-134.

451. Owen, M. and Norderhaug, M. 1977. Population dynamics of Barnacle Geese *Branta leucopsis* breeding in Svalbard 1948-76. Ornis Scand. 8:161-174.

452. Owen, M. and Ogilvie, M.A. 1979. Wing molt and weights of Barnacle Geese in Spitsbergen. Condor 81:42-52.

453. Owen, M. and Thomas, G.J. 1979. The feeding ecology and conservation of Wigeon wintering at the Ouse Washes, England. J. Appl. Ecol. 16:795-809.

454. Owen, M. and Williams, G.M. 1976. Winter distribution and habitat requirements of Wigeon in Britain. Wildfowl 27:83-90.

455. Owen, M., Drent, R.H., Ogilvie, M.A. and van Spanje, T.M. 1978. Numbers, distribution and catching of Barnacle Geese on the Nordenskiold kysten, Svalbard in 1972. Norsk Polarinstitutt Arbok 1977:247-258.

456. Owens, N.W. 1977. Responses of wintering Brent Geese to human disturbance. Wildfowl 28:5-14.

457. Palmer, R.S. 1976. Handbook of North American Birds. Vol. 3. New Haven, Yale University Press.

458. Parkin, D.T. and White-Robinson, R. in press. The effect of age on survival in the Canada Goose *Branta canadensis*. In Morgan, B.J.T. and North, P.M. (Eds). Statistics in Ornithology.

459. Patterson, I.J. 1982. The Shelduck. Cambridge, the University Press.

460. Patterson, J.H. 1979. Experiences in Canada. Trans. 44th N. Amer. Wildl. Conf. 1979: 130-139.

461. Patton, D.L.H. and Frame, J. 1981. The effect of grazing in winter by wild geese on improved grassland in West Scotland. J. Appl. Ecol. 18:311-325.

462. Payne-Gallwey, R.W.F. 1886. The Book of Duck Decoys, their Construction, Management and History. London.

463. Pennie, I.D. 1959. Spread of Eider in East Sutherland. Scot. Birds 1:66-67.

464. Perdeck, A.C. 1977. The analysis of ringing data: pitfalls and prospects. Die Vogelwarte 29:33-44.

465. Perdeck, A.C. and Clason, C. 1980. Some results of waterfowl ringing in Europe. IWRB Special Publ. No. 1. Slimbridge, IWRB.

466. Perrett, D.H. 1953. Shelduck observations, 1952. Mid-Somerset Nat. Soc. Rep. 2:16-47.

467. Perrins, C.M. and Reynolds, C.M. 1967. A preliminary study of the Mute Swan, *Cygnus olor*. Wildfowl Trust Ann. Rep. 18:74-84.

468. Perrins, C.M. and Ogilvie, M.A. 1981. A study of the Abbotsbury Mute Swans. Wildfowl 32:35-47.

469. Petersen, A. 1970. Fugalif i Skogum a osholmasvoedi Heradsvatna i Skagafirdi. Nakturufraedingnum 40:26-46.

470. Pethon, P. 1967. Food and feeding habits of the Common Eider (*Somateria mollissima*). Nytt. Mag. Zool. 15:97-111.

471. Philippona, J. 1972. Die Blessgans. A. Ziemsen. Wittenberg Lutherstadt, Die Neue Brehn-Bucherei.

472. Pickup, C.H. 1983. A survey of Greylag Geese (*Anser anser*) in the Uists, 18 July—28 August 1982. Unpubl. Rep. NCC.

473. Pienkowski, M. and Evans, P. 1979. The origins of Shelducks moulting on the Forth. Bird Study 26:195-196.

474. Pitman, C.R.S. 1965. The nesting and some other habits of *Alopochen, Nettapus Plecopterus* and *Sarkidiornis*. Wildfowl Trust Ann. Rep. 16:115-121.

475. Player, P.V. 1971. Food and feeding habits of the Common Eider at Seafield, Edinburgh, in winter. Wildfowl 22:100-106.

476. Pollard, D.F.W. 1967a. Wildfowl conservation. Forestry Suppl. 1967:78-84.

477. Pollard, D.F.W. 1967b. The WAGBI—Wildfowl Trust Experimental Reserve—the first eleven years. Part III. An appraisal of the planting programme 1959-66. Wildfowl Trust Ann. Rep. 18:55-62.

478. Pollard, D.F.W. and Walters-Davies, P.W. 1968. A preliminary study of the feeding of the Greenland White-fronted Goose, *Anser albifrons flavirostris* in Cardiganshire. Wildfowl 19:108-116.

479. Portsmouth Polytechnic 1976. Langstone Harbour Study: the Effects of Sewage Effluent on the Ecology of the Harbour. Portsmouth, Southern Water Authority.

480. Pounder, B. 1971. Wintering Eiders in the Tay Estuary. Scot. Birds 6:407-419.

481. Pounder, B. 1976a. Waterfowl at effluent discharges in Scottish coastal waters. Scot. Birds 9:5-36.

482. Pounder, B. 1976b. Wintering flocks of Goldeneyes at sewage outfalls in the Tay Estuary. Bird Study 23:121-131.

483. Prater, A.J. 1972. The ecology of Morecambe Bay. III The food and feeding habits of Knot (*Calidris canutus*) in Morecambe Bay. J. Appl. Ecol. 9:179-194.

484. Prater, A.J. 1979. Trends in accuracy of counting birds. Bird Study 26:198-200.

485. Prater, A.J. 1981. Estuary Birds of Britain and Ireland. Poyser, Calton.

486. Probert, P.K. and Mitchell, R. 1980. Nature conservation implications of siting wave energy converters off the Moray Firth. Unpubl. Rep. NCC.

487. Pyman, G.A. 1959. The status of the Red-crested Pochard in the British Isles. Brit. Birds 52:42-56.

488. Radley, J. and Simms, C. 1970. Yorkshire Flooding—Some Effects on Man and Nature. York, Ebor Press.

489. Ranwell, D.S. 1968. Coastal marshes in perspective. Regional Studies Group. Bull. Strathclyde No. 9:1-26.

490. Ranwell, D.S and Downing, B.M. 1959. Brent Goose (*Branta bernicla* (L.)) winter feeding pattern and *Zostera* resources at Scolt Head Island, Norfolk. Anim. Behav. 7:42-56.

491. Rasmussen, C.D. 1982. Eiders round Bute. Unpubl. Rep. Wildfowl Trust.

492. Ratcliffe, D.A. (Ed). 1977. A Nature Conservation Review. Cambridge, the University Press. 2 vols.

493. Raveling, D.G. 1978. Dynamics of distribution of Canada Geese in winter. Trans. 43rd N. Amer. Wildl. Conf. 1978:206-225.

494. Rawcliffe, C.P. 1958. The Scottish Mute Swan census 1955-56. Bird Study 5:45-55.

495. Reed, A. 1976. Geese, nutrition and farmland. Wildfowl 27:143-146.

496. Rees, E.C. 1981. The recording and retrieval of bill pattern variations in *Cygnus columbianus bewickii*. Proc 2nd Int. Swan Symp., Sapporo, Japan 1980:105-119. Slimbridge, IWRB.

497. Rees, E.I.S. 1981. Wildfowl Counts in Anglesey and North Caernarvon 10-25 January 1981. Circulated document.

498. Rees, E.I.S. 1983. Decline in wildfowl usage of a eutrophic Anglesey lake. Circulated document.

499. Reeves, H.M., Dill, R.H. and Hawkins, A.S. 1968. A case study in Canada Goose management: the Mississippi Valley Population. In Hine, R.L. and Schoenfeld C. (Eds). Canada Goose Management. Madison, Dunbar Educational Research Services.

500. Reynolds, P. 1981. Whooper Swan Survey, Orkney, 7th/8th November 1981. Unpubl. Rep. NCC.

501. Richardson, M.G., Dunnet, G.M. and Kinnear, P.K. 1981. Monitoring seabirds in Shetland. Proc. Roy. Soc Edin. (B) 80:157-179.

502. Rochard, J.B.A. and Kear, J. 1970. Field trials of the reactions of sheep to goose droppings. Wildfowl 21:108-109.

503. Rogers, M.L. 1982. Ruddy Shelducks in Britain in 1965-79. Brit. Birds 75:446-455.

504. Rooth, J., Ebbinge, B., van Haperen, A., Lok, M., Timmerman, A., Philippona, J., and van den Bergh, L 1981. Numbers and distribution of wild geese in the Netherlands, 1974-1979. Wildfowl 32:146-155.

505. Round, P.D. 1978. An Ornithological Survey of the Somerset Levels 1976-77. Wessex Water Authority/RSPB Report.

506. Round, P.D. 1980. Survey of the Yare Basin. Norfolk Bird and Mammal Report 1979:98-102.

507. Round, P.D. 1982. Inland feeding by Brent Geese *Branta bernicla* in Sussex, England. Biol. Cons. 23:15-32.

508. Roux, F., Jarry, G., Maheo, R. and Tamisier, A. 1976. Importance, structure et origine des populations d'Anatides hivernant dans la delta du Senegal. L'Oiseau et RFO. 46:299-336.

509. Royal Commission on Environmental Pollution 1979. Seventh Report— Agriculture and Pollution. London, HMSO.

510. Royal Commission on Environmental Pollution 1983. Ninth Report—Lead in the Environment. London, HMSO.

511. RSPB 1977. An ornithological assessment of standing freshwater sites in Wales. Unpubl. Rep. RSPB Wales Office.

512. RSPB 1978. Scoter in Carmarthen Bay—Second Report to NCC. Unpublished.

513. RSPB 1983a. RSPB Nature Reserves. Sandy, RSPB.

514. RSPB 1983b. Land Drainage in England and Wales: An Interim Report. Sandy, RSPB.

515. Russell, M.G. 1982. The effect of sailing on the wintering wildfowl on Black Swan Lake, Dinton Pastures Country Park. Rec. Mgmt. Diploma Project, Loughborough University.

516. Ruttledge, R.F. and Ogilvie, M.A. 1979. The past and current status of the Greenland White-fronted Goose in Ireland and Britain. Irish Birds 1:293-364.

517. Ruxton, J. 1973. Wildfowl of Morecambe Bay. Chester, WAGBI.

518. St John, C.W.G. 1863. Natural History and Sport in Moray. Edinburgh.

519. St Joseph, A.K.M. 1979a. The seasonal distribution and movements of *Branta bernicla* in Western Europe. Proc. 1st Tech. Mtng. Western Palearctic Migratory Bird Management, Paris 1977:45-57. Slimbridge, IWRB.

520. St Joseph, A.K.M. 1979b. The development of inland feeding by *Branta bernicla bernicla* in south-eastern England. Proc. 1st Tech. Mtng. Western Palearctic Migratory Bird Management, Paris 1977: 132-145. Slimbridge, IWRB.

521. St Joseph, A.K.M. 1981. Research on Brent Geese, September 1980 to March 1981. Unpubl. Rep. to the Min. of Agriculture.

522. St Joseph, A.K.M. 1982a. The management of a protected species *Branta b. bernicla* in relation to population size, habitat loss and field feeding habit. Aquila 89:271-276.

523. St Joseph, A.K.M. 1982b. The status of *Branta bernicla*. Aquila 89:163-165.

524. Salmon, D.G. (Ed). 1980, 1981, 1982, 1983. Wildfowl and wader counts 1979-80, 1980-81, 1981-82, 1982-83. Slimbridge, Wildfowl Trust.

525. Salomonsen, F. 1958. The present status of the Brent Goose in western Europe. Vidensk. Medd. Dansk Naturh. Foren. 120:43-80.

526. Salomonsen, F. 1968. The moult migration. Wildfowl 19:5-24.

527. Savage, C. 1952. The Mandarin Duck. London, Black.

528. Schroder, H. 1969. Beobachtungen an Wild gansen der Gattung Anser im Giebiet der Mecklenburger Grossenplatte. Beitr. Vogelk. 17:349-359.

529. Scott, D.A. 1980. A preliminary inventory of wetlands of international importance for waterfowl in Western Europe and North-west Africa. IWRB. Special Publ. No. 2. Slimbridge, IWRB.

530. Scott, D.A. 1982a. Problems in the management of waterfowl populations. Proc. 2nd Tech. Mtng. Western Palearctic Migratory Bird Management, Paris 1979:89-106. Slimbridge, IWRB.

531. Scott, D.A. (Ed). 1982b. Managing Wetlands and their Birds. Slimbridge, IWRB.

532. Scott, D.K. 1978. Social behaviour of wintering Bewick's Swans. PhD. Thesis, University of Cambridge.

533. Scott, D.K. 1980. The behaviour of Bewick's Swans at the Welney Wildfowl Refuge, Norfolk, and on the surrounding fens: a comparison. Wildfowl 31:5-18.

534. Scott, P. 1948. The decoy. Wildfowl Trust Ann. Rep. 1:52-55.

535. Scott, P. 1966. The Bewick's Swans at Slimbridge. Wildfowl 17:20-26.

536. Scott, P. and Matthews, G.V.T. 1976. Public access to wetlands: control and education. Proc. Int. Conf. Conserv. Wetlands and Waterfowl, Heiligenhafen 1974:370-375 Slimbridge, IWRB.

537. Scott, P., Boyd, H. and Sladen, W.I.L. 1955. The Wildfowl Trust's second expedition to central Iceland, 1953. Wildfowl Trust Ann. Rep. 7:63-98.

538. Scott, R.E. 1966. Lydd Wildfowl Reserve 1965. WAGBI Ann. Rep. and Year Book 1965-66:82-84.

539. Scott, R.E. 1978. Breeding birds of Cliffe Marsh 1976. Kent Bird Report 25:91-2.

540. Scottish Bird Reports. Scottish Ornithologists' Club.

541. Severn Barrage Committee 1981. Tidal power from the Severn Estuary. Vol. I. Report to the Secretary of State for Energy. London, HMSO.

542. Sharrock, J.T.R. 1976. The Atlas of Breeding Birds in Britain and Ireland. Tring, BTO.

543. Shaw, T.L. 1974. Tidal energy from the Severn Estuary. Nature 249:730-733.

544. Shrubb, M. 1979. The Birds of Sussex. Chichester: Phillimore.

545. Siegfried, W.R. 1973a. Summer foods and feeding of the Ruddy Duck in Manitoba. Can. J. Zool. 51:1293-1297.

546. Siegfried, W.R. 1973b. Platform building by male and female Ruddy Ducks. Wildfowl 24:150-153.

547. Smallshire, D. and Richards, A. 1976. The Birds of Belvide Reservoir. West Midland Bird Club.

548. Smith, K.W. 1983. The status and distribution of waders breeding on wet lowland grasslands in England and Wales. Bird Study 30:177-192.

549. Snow, D.W. (Ed). 1971. The Status of Birds in Britain and Ireland. Oxford, Blackwell Scientific Publ.

550. Somerset Birds. Somerset Ornithological Society.

551. Spooner, M.F. 1967. Biological effects of the Torrey Canyon disaster. J. Devon Trust for Nature Conservation Supplement:12-19.

552. Spray, C.J. 1981. An isolated population of *Cygnus olor* in Scotland. Proc. 2nd Int. Swan Symp., Sapporo, Japan 1980:191-208. Slimbridge, IWRB.

553. Stowe, T.J. 1979. Oil pollution—the increasing toll. Birds 7 No. 8:46-47.

554. Stowe, T.J. 1982. Beached Bird Surveys. Report to NCC. Sandy, RSPB.

555. Street, M. 1982. The use of waste straw to promote the production of invertebrate foods for waterfowl in man-made wetlands. In Scott, D.A. (Ed). Managing Wetlands and their Birds:93-103. Slimbridge, IWRB.

556. Street, M. 1983. The Great Linford Wildfowl Research Project—a case history. Proc. Symp. Wildlife on Man-made Waters:21-42. Game Conservancy, Fordingbridge.

557. Stroud, D.A. 1981. Nests and nest site selection. In Fox, A.D. and Stroud, D.A. (Eds). Report of the 1979 Greenland White-fronted Goose Study Expedition to Eqalungmuit Nunat, West Greenland:74-77. Aberystwyth, Greenland White-fronted Goose Study Group.

558. Stroud, D.A. 1984. Status of Greenland White-fronted Geese in Britain, 1982/83. Bird Study 31:111-116.

559. Suffolk Birds. Suffolk Naturalists' Society.

560. Summers-Smith, D. 1977. The effect of sailing on wintering wildfowl at Scaling Dam Reservoir. Unpubl. Rep.

561. Sussex Bird Reports. Sussex Ornithological Society.

562. Sutcliffe, S.J. 1975. Common Scoter in Carmarthen—an oiling incident. Nature in Wales 14:243-249.

563. Swaine, C.M. 1982. The birds of Gloucestershire. Gloucester, A. Sutton.

564. Swanson, G.A. and Ryder, R.A. 1982. Predation—discussion. In Scott, D.A. (Ed). Managing Wetlands and their Birds:239-240. Slimbridge, IWRB.

565. Swennen, C. 1978. Geographical distribution of recoveries of Eiders (*Somateria mollissima*) ringed in the Dutch Wadden Sea area. Proc. Symp. Sea Ducks, Stockholm 1975:112-120. Nat. Swedish Environ. Prot. Board/IWRB.

566. Swift, J. 1974. Pintail: A project assessing the release of hand-reared birds. WAGBI Ann. Rep. and Year Book 1973-74:55-58.

567. Swift, J. 1976. Appendix: Waterfowl feeding on the Thames. In Harrison, J.G. and Grant, P.J. 1976. The Thames Transformed:226-227. London, Andre Deutsch.

568. Swift, J. and Harrison, J.G. 1979. Shooting of *Branta bernicla* in Europe. Proc. 1st Tech. Mtng. on Western Palearctic Migratory Bird Management, Paris 1977:152-165. Slimbridge, IWRB.

569. Szijj, J. 1965. Okologische untersuchungen an entenvogeln (Anatidae) des Ermatinger Beckens (Bodensee). Die Vogelwarte 23:24-71.

570. Tamisier, A. 1974. Etho-ecological studies of Teal wintering in the Camargue (Rhone Delta, France). Wildfowl 25:123-133.

571. Tanner, M.F. 1973. Water Resources and Recreation. Sports Council Study 3. London, The Sports Council.

572. Tanner, M.F. 1979. Wildfowl, Reservoirs and Recreation. Water Space Amenity Commission Report 5. London, WSAC.

573. Tasker, M.L. 1982. Moulting Shelducks on the Humber. Bird Study 29:164-166.

574. Taverner, J.H. 1959. The spread of the Eider in Great Britain. Brit. Birds 52:245-258.

575. Taverner, J.H. 1967. Wintering Eiders in England during 1960-65. Brit. Birds 60:509-515.

576. Taylor, D.W., Davenport, D.L. and Flegg, J.J.M. 1981. The Birds of Kent: A Review of their Status and Distribution. Kent Ornithological Society.

577. Taylor, J. 1953. A possible moult migration of Pink-footed Geese. Ibis 95:638-642.

578. Temperley, G.W. 1938. Notes on the bird life of the Island of Raasay. Scot Nat., No. 229:11-27.

579. Thomas, C.B. 1977. The mortality of Canada Geese. Wildfowl 28:35-47.

580. Thomas, G.J. 1976a. Ingested lead pellets in waterfowl at the Ouse Washes, England 1968-73. Wildfowl 26:31-42.

581. Thomas, G.J. 1976b. Habitat usage of wintering ducks at the Ouse Washes, England. Wildfowl 27:148-152.

582. Thomas, G.J. 1978. Breeding and feeding ecology of waterfowl at the Ouse Washes, England. PhD. Thesis, Council for National Academic Awards.

583. Thomas, G.J. 1980a. The ecology of breeding waterfowl at the Ouse Washes, England. Wildfowl 31:73-88.

584. Thomas, G.J. 1980b. Review of lead poisoning in waterfowl. IWRB Bull. 46:43-60.

585. Thomas, G.J. 1981. Field feeding by dabbling ducks around the Ouse Washes, England. Wildfowl 32:69-78.

586. Thomas, G.J. 1983. Management of vegetation at wetlands. In Scott, D.A. (Ed). Managing Wetlands and their Birds:21-37. Slimbridge, IWRB.

587. Thomas, G.J. and Owen, M. 1982. Wildfowl and Agriculture. Advisory leaflet. Sandy, RSPB.

588. Thomas, G.J., Allen, D.A. and Gross, M.P.B. 1981. The demography and flora of the Ouse Washes, England. Biol. Cons. 21:197-229.

589. Thompson, D.B.A. 1981. Feeding behaviour of Shelduck in the Clyde Estuary. Wildfowl 32:88-98.

590. Thompson, D.B.A. 1982. The abundance and distribution of intertidal invertebrates, and an estimation of their selection by Shelduck. Wildfowl 33:151-158.

591. Thomson, A.L. 1941. Results of ringing duck: general survey of data from all sources. In Internation Wildfowl Inquiry. Factors Affecting the General Status of Wild Geese and Wild Duck. Cambridge, the University Press.

592. Timmerman, A. 1962. De Krooneend (*Netta rufina*) in Nederland. Limosa 35:28-39.

593. Timmerman, A., Morzer Brujns, M.F. and Philippona, J. 1976. Survey of the winter distribution of Palearctic geese in Europe, Western Asia and North Africa. Limosa 49:230-292.

594. Treca, B. 1981. Regime alimentaire de la Sarcelle d'ete (*Anas querquedula* L.) dans le delta du Senegal. L'Oiseau et RFO 51:33-58.

595. Tubbs, C.R. 1977. Wildfowl and waders in Langstone Harbour. Brit. Birds 70:177-199.

596. Tubbs, C.R. and Tubbs, J.M. 1982. Brent Geese *Branta bernicla bernicla* and their food in the Solent, Southern England. Biol. Cons. 23:33-54.

597. Tuite, C.H. 1982. The Impact of Water-based Recreation on the Waterfowl of Enclosed Inland Waters in Britain. Report to the Sports Council and the Nature Conservancy Council.

598. Tuite, C.H. and Owen, M. 1984. Breeding waterfowl on British inland waters in 1980. Wildfowl 35:157-172.

599. Tuite, C.H., Owen, M. and Paynter, D. 1983. Interaction between wildfowl and recreation at Llangorse Lake and Talybont Reservoir, South Wales. Wildfowl 34:48-63.

600. Venables, L.S.V. and Venables, U.M. 1955. Birds and Mammals of Shetland. Edinburgh: Oliver and Boyd.

601. Venables, L.S.V. and Venables, U.M. 1972. Our vanishing swans. Nature in Wales 13:128-131.

602. Vinicombe, K.E. and Chandler, R.J. 1982. Movements of Ruddy Ducks during the hard winter of 1978/79. Brit. Birds 75:1-11.

603. WAGBI and the Wildfowl Trust. Undated. Know your Wildfowl Food Plants. Liverpool, WAGBI.

604. Wainwright, C.B. 1957. How to make and use a duck trap. Wildfowl Trust Ann. Rep. 8:44-47.

605. Wainwright, C.B. 1967. Results of wildfowl ringing at Abberton Reservoir, Essex, 1949-1966. Wildfowl Trust Ann. Rep. 18:28-35.

606. Walker, A.F.G. 1970. The moult migration of Yorkshire Canada Geese. Wildfowl 21:99-104.

607. Wall, T. 1983. A review of winter wildfowl counts at Rostherne Mere, Cheshire, 1947/48 to 1979/80. Unpubl. Rep. NCC.

608. Walmsley, J.G. 1981. Interpopulations-Bewegungen von Brandgansen *Tadorna tadorna* (L). Beitr. Naturk. Niedersachsens 34:140-147.

609. Watmough, B. 1983. The effects of recreation on waterfowl. Unpubl. Rep. to the Sports Council/Nature Conservancy Council.

610. West Wales Naturalists' Trust Bulletin No 9. Spring 1974. WWNT.

611. White, H.C. 1957. Food and natural history of Mergansers on salmon waters in the maritime provinces of Canada. Fish. Res. Board Canada Bull. 116.

612. White-Robinson, R. 1984. The ecology of Canada Geese *Branta canadensis* in Nottinghamshire and their importance in relation to agriculture. PhD. Thesis, University of Nottingham.

613. Whitman, W.R. 1976. Impoundments for waterfowl. Canadian Wildl. Service. Occ. Paper 22. Ottawa, CWS.

614. Wigham, G.D. 1977. Heavy metal loads of Bristol Channel biota. In Nelson-Smith, A. and Bridges, E.M. (Eds). Problems of a Small Estuary. Proc. Symp. Burry Inlet, Swansea 1976:3:3/4-14. Swansea, Institute of Marine Studies.

615. Willi, P. 1970. Zugverhalten, Aktivitat, Nahrung und Nahrungschwerb auf dem Klingrauer Strausse haufig auftretender Anatiden, insbesondere von Krickente, Tafelente und Reiherente. Orn. Beob. 67:141-217.

616. Williams, G. 1979. Bird studies on the North Kent Marshes. Unpubl. Rep. NCC.

617. Williams, G. 1982. Rivers and Wildlife. Sandy, RSPB.

618. Williams, G., Henderson, A., Goldsmith, L. and Spreadborough, A. 1983. The effects on birds of land drainage improvements in the North Kent Marshes. Wildfowl 34:33-47.

619. Williams, G.A. 1977. The status and distribution of wildfowl in the Dee Estuary. Nature in Wales 15:166-179.

620. Wood, A.R. 1973. The Hythe, Colchester. A short history of its birdlife. Essex Bird Report 1972:61-70.

621. Yarker, B. and Atkinson-Willes, G.L. 1972. The numerical distribution of some British breeding ducks. Wildfowl 22:63-70.

622. Young, J.G. 1972. Breeding biology of feral Greylag Geese in south-west Scotland. Wildfowl 23:83-87.

623. Young, J.G. 1973. Distribution, status and movements of feral Greylag Geese in south-west Scotland. Scot. Birds 7:170-182.

624. Zwarts, L. 1972. De grauwe ganzen *Anser anser* ven het brakke get ijdegebied de ventjagersplaten. Limosa 45:119-134.

Site index

This index lists all the sites, with their 10km national grid reference, from which National Wildfowl Count data are available and which have been used in the preparation of Part II. If they have been mentioned by name in Part II their page reference is given.

General Index

This index includes wildfowl sites mentioned in Parts I, III and IV as well as places mentioned in Part II which are not covered by the National Wildfowl Counts. The SITE INDEX (p567) gives a list of all the places covered by the counts mentioned, if at all, in Part II, together with their page references.